BASIC OP AMP MODULES

CIRCUIT	BLOCK DIAGRAM	GAINS
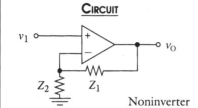 Noninverter	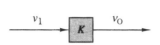	$K = \dfrac{Z_1 + Z_2}{Z_2}$
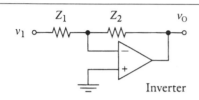 Inverter		$K = -\dfrac{Z_2}{Z_1}$
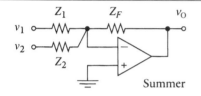 Summer	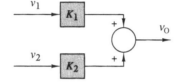	$K_1 = -\dfrac{Z_F}{Z_1}$ $\quad$ $K_2 = -\dfrac{Z_F}{Z_2}$
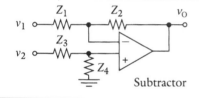 Subtractor	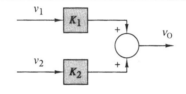	$K_1 = -\dfrac{Z_2}{Z_1}$ $\quad$ $K_2 = \left(\dfrac{Z_1 + Z_2}{Z_1}\right)\left(\dfrac{Z_4}{Z_3 + Z_4}\right)$
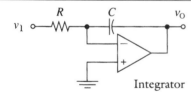 Integrator	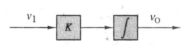	$K = -\dfrac{1}{RC}$
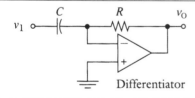 Differentiator		$K = -RC$

Survey Question/Answer

Application or Analogy

Summary

HW Problem Concept

Reading Question/Answer

THE ANALYSIS AND DESIGN OF LINEAR CIRCUITS

ROLAND E. THOMAS
Professor Emeritus
United States Air Force Academy

ALBERT J. ROSA
Professor Emeritus
University of Denver

GREGORY J. TOUSSAINT
Civilian Employee
United States Air Force

JOHN WILEY & SONS, INC.

VP & PUBLISHER	Don Fowley
ASSOCIATE PUBLISHER	Daniel Sayre
EDITORIAL ASSISTANT	Charlotte Cerf
MARKETING MANAGER	Christopher Ruel
MARKETING ASSISTANT	Ashley Tomeck
EXECUTIVE MEDIA EDITOR	Tom Kulesa
MEDIA EDITOR	Wendy Ashenberg
SENIOR PRODUCTION MANAGER	Janis Soo
SENIOR PRODUCTION EDITOR	Joyce Poh
DESIGNER	Seng Ping Ngieng

OrCAD and PSPICE are registered trademarks of Cadence Design Systems, Inc.

MATLAB® is a registered trademark of The MathWorks, Inc.
Excel is a registered trademark of Microsoft Corporation in the United States and/or other countries.

This book was set in 10/12 Times Ten Roman by Thomson Digital and printed and bound by RR Donnelley. The cover was printed by RR Donnelley.

This book is printed on acid free paper.

Founded in 1807, John Wiley & Sons, Inc. has been a valued source of knowledge and understanding for more than 200 years, helping people around the world meet their needs and fulfill their aspirations. Our company is built on a foundation of principles that include responsibility to the communities we serve and where we live and work. In 2008, we launched a Corporate Citizenship Initiative, a global effort to address the environmental, social, economic, and ethical challenges we face in our business. Among the issues we are addressing are carbon impact, paper specifications and procurement, ethical conduct within our business and among our vendors, and community and charitable support. For more information, please visit our website: www.wiley.com/go/citizenship.

Evaluation copies are provided to qualified academics and professionals for review purposes only, for use in their courses during the next academic year. These copies are licensed and may not be sold or transferred to a third party. Upon completion of the review period, please return the evaluation copy to Wiley. Return instructions and a free of charge return shipping label are available at www.wiley.com/go/ returnlabel. Outside of the United States, please contact your local representative.

Library of Congress Cataloging-in-Publication Data

Thomas, Roland E., 1930-
   The analysis and design of linear circuits / Roland E. Thomas, Albert J. Rosa, Gregory J. Toussaint.—7th ed.
       p. cm.
   Includes index.
   ISBN 978-1-118-06558-7 (hardback)
   1. Electric circuits, Linear—Design and construction. 2. Electric circuit analysis. I. Rosa, Albert J., 1942- II. Toussaint, Gregory J.   III. Title.
   TK454.T466 2012
   621.39'5—dc23

                                                                          2011042676

Printed in the United States of America
10 9 8 7 6 5 4 3 2 1

To our wives
Juanita, Kathleen, and Tricia

PREFACE

WHY THIS TEXT?

We approach the art of teaching circuits in our textbook differently than most others. We recognize that studying circuits can be rewarding and useful, even for students who are not majoring in electrical or computer engineering. Electric circuits dominate so much of our modem technology that many students are interested in learning about them. Most students who pursue engineering studies want to be creative and design things. Most circuits texts do not focus on this basic desire, rather spend their pages teaching why electric circuits work without affording the student an opportunity to put this learning into practice. The longer it takes students to try their hand in designing stuff, the more likely it is that they will become disillusioned and perhaps change to a different major.

The authors have long believed that an early introduction to design and design evaluation raises the excitement level and greatly increases student interest in their chosen discipline. Over 50 years of combined teaching experience at the USAF Academy, the University of Denver, the University of Colorado at Denver, and the Air Force Institute of Technology, have only served to strengthen our belief. This new edition furthers this strategy by adding more design and evaluation examples, exercises, homework problems, and real-world applications. In addition, students today solve problems using computers, as well as, by hand and with a calculator. Access to personal computers is nearly ubiquitous, and key software used in circuit analysis and design has become available for free or at very deep discounts for students. This edition of our text includes many software examples, exercises, and discussions geared to making the study of circuits more in line with today's student. Our text has always included software, but generally as an extension for solving circuits by hand. This edition continues our effort begun with the sixth edition by integrating software intimately into the solution of circuit problems whenever and wherever it really helps to solve the problems. It still recognizes that using computers does not replace the intuition that engineers must develop to analyze, design, and make smart judgments about different working solutions or designs.

The seventh edition of *The Analysis and Design of Linear Circuits* improves on the sixth edition and remains friendly to users who prefer a *Laplace Early* approach or those favoring the more traditional *Phasor First* approach to ac circuits. A later section discusses how to use this text to pursue either a traditional Phaser First approach or the Laplace Early approach using three different focuses. In this edition we have added many skills-level examples, exercises, and problems that can help develop the student's confidence in mastering the different objectives. The seventh edition assumes the same student prerequisites as past editions and continues to rely on students having access to personal computers—although computer access is not essential for using this textbook, we believe it improves learning. This edition targets students of all engineering disciplines who need an introductory circuit analysis course of one or two terms. The seventh edition continues the authors' combined commitment to providing a modem and innovative approach to teaching circuit analysis and design, as well as circuit evaluation.

CONTINUING FEATURES

OBJECTIVES

This text remains structured around a sequence of carefully defined learning objectives and related evaluation tools based on Bloom's Taxonomy of Educational Objectives. The initial learning objectives focus on enabling skills at the *knowledge*, *comprehension*, and *application* levels of the taxonomy that we call Chapter Learning Objectives or CLOs. As students demonstrate mastery of these lower levels, they are introduced to higher-level objectives involving *analysis, synthesis* (design), and *evaluation*. Each learning objective is explicitly stated in terms of expected student proficiency in the homework sections, and each is followed by at least 10 homework problems specifically designed to evaluate student mastery of the objective. This framework has been a standard feature of all seven editions of this book and has allowed us to maintain a consistent level of expected student performance over the years. We also list our objectives in the chapter openers to orient the student to the expected outcomes. These objectives make it easier to assess student learning and prepare for accreditation reviews. To fulfill ABET Criterion 3 *"the program must have documented student outcomes that prepare graduates to attain the program educational objectives."* And to fulfill Criterion 4 *"the program must regularly use appropriate, documented processes for assessing and evaluating the extent to which both the program educational objectives and the student outcomes are being attained. There must be measurable outcomes and an assessment and evaluation process that periodically documents and demonstrates the degree to which the program outcomes are attained."*

CIRCUIT ANALYSIS AND DESIGN

Our experience convinces us that an interweaving of analysis and design topics reinforces a student's grasp of circuit analysis fundamentals. Early involvement in design provides motivation as students apply their newly acquired analysis tools to practical situations. Using computer simulation software to verify their designs gives students an early degree of confidence that they have actually created a design that meets stated specifications. Ideally, a supporting laboratory program where students actually build and test their designs provides the final confirmation that they can create useful products. Design efforts as described in this text are very useful in helping to meet ABET's design Criterion 3(c)—*specifically, an ability to design a system, component, or process to meet desired needs....* We identify design examples, exercises, and homework problems with an icon **D**.

DESIGN EVALUATION

Realistic design problems do not have unique solutions, so it is natural for students to wonder how their designs compare with those of others. An often asked question is "is my design a good one?" Using judgment to compare alternative solutions is a fundamental trait of good engineering. The evaluation of alternative designs introduces students to real-world engineering practice. We also include judgment problems that ask students to evaluate an "off-the-shelf" design and ask if it could meet a specific need. In such problems, we ask the student "Would you buy it?"

Including design and the evaluation of design in an introductory course helps to convince students that circuits perform useful functions and that circuit design courses are not simply vehicles for teaching routine skills such as node-voltage and mesh-current analyses. Although this philosophy is implicit in all previous editions, this edition offers expanded coverage of design and evaluation among the worked examples, exercises, and homework problems. We use software extensively to help students visualize specifications and results. This, in turn, helps them to create better designs and make smart choices between competing designs. Evaluation generally involves the practical side of design and can support ABET Criterion 3(c)— specifically, ... *to meet desired needs within realistic constraints, such as economic, environmental, social, political, ethical, health and safety, manufacturability, and sustainability.* We identify evaluation examples, exercises, and homework problems with an icon ◀**E**.

THE OP AMP

An early introduction and integrated treatment of the OP AMP continues to be a central feature of this text. The modular form of OP AMP circuits simplifies analog circuit analysis and design by minimizing the effects of loading. This feature allows the interconnection of simple building blocks to produce complex signal processing functions that are especially useful to instrumentation and signal shaping applications. The close agreement between theory, simulation, and hardware allows students to analyze, design, and successfully build useful OP AMP circuits in the laboratory. The text covers numerous OP AMP applications, such as digital-to-analog conversion, transducer interface circuits, comparator circuits, block diagram realization, first-order filters, and multiple-pole active filters. These applications, which have been again expanded in this new edition, are especially useful to students from other engineering disciplines that require knowledge of instrumentation, filtering, or signal processing.

LAPLACE TRANSFORMS

In electrical engineering, Laplace transforms are used to treat important concepts such as zero-state and zero-input responses, impulse and step responses, convolution, frequency response, and filter design. An important pedagogical question is where Laplace transforms should be taught—in the Circuits course, the Signals and Systems course, a Differential Equations course, or elsewhere? The traditional approach has been to first teach phasors and use them to study ac circuit analysis, steady-state ac power, polyphase circuit analysis, magnetically coupled circuits, and frequency response. This extended treatment of phasor analysis means that Laplace transforms are often delayed to the last weeks of the second semester and treated as an advanced topic along with Fourier methods and two-port networks. Typically, then, Laplace transforms are taught in earnest in a Signals and Systems course, where their linkage to phasors is often overlooked. The authors have long advocated an early Laplace approach, one in which Laplace transforms are introduced and applied to circuit analysis before phasors are introduced. The advantage of treating Laplace-based circuit analysis first is that once mastered, it makes learning phasor-based analysis easier and more intuitive. Students quickly make the connection between phasor analysis and the concepts of network functions, transient response, and sinusoidal steady-state response developed through *s*-domain circuit analysis. We do not claim that Laplace analysis is more fundamental or even more important than phasor analysis. We do claim that the learning effort needed to master both phasor analysis and Laplace analysis is not a zero-sum game. Our experience is that less classroom time is needed to ensure mastery of both methods of analysis when

Laplace transform analysis is treated before phasor analysis. Emphasizing transform methods in the circuit course also better prepares students to handle the profusion of transforms they will encounter in subsequent Signals and Systems courses.

SIGNAL PROCESSING

We begin our treatment of dynamic circuits with a separate chapter on waveforms and signal characteristics. This chapter gives students early familiarity with important input and output signals encountered in the study of linear circuits. Introducing signals at the beginning of dynamic circuit analysis lets students become comfortable with time-varying signals without having to simultaneously deal with new concepts. A further emphasis on signal processing and systems is achieved through the use of block diagrams, input-output relationships, and transforms methods. The ultimate goal is for students to understand that time domain waveforms and frequency-domain transforms are simply alternative ways to characterize signals and signal processing. Viewing signals in both domains naturally leads to discussions of important concepts like signal bandwidth, signal sampling, and reciprocal spreading. It is also useful knowledge in choosing alternative design approaches to filters.

COMPUTER TOOLS

Our philosophy recognizes that students come to the Circuits course already knowing how to use several computer tools. Although many know how to use a spreadsheet and may be familiar with math solvers and possibly simulation software, our goal is to help them learn how to effectively use these tools. Knowing when to use these tools and how to interpret the results is essential to understanding circuits. Three types of computer programs are used in this text to illustrate computer-aided circuit analysis, namely spreadsheets (Excel®), math solvers (MATLAB®), and circuit simulators (OrCAD®). Examples, exercises, and homework problems related to computer-aided circuit analysis are integrated into all chapters, beginning with Chapter 1. The purpose of the examples is to help students develop a problem-solving style that includes the intelligent use of the productivity tools routinely used by practicing engineers. Exercises following the examples help students immediately practice the software skill demonstrated in the example. Homework problems that require the use of a particular computer tool are identified by a computer icon (🖱). In this edition we moved much of the discussion on how to use software tools to a new web appendix and included additional examples. This approach streamlines the text for more accomplished students and permits those needing more guidance to have access to additional exposure. This approach directly supports ABET's Criterion 3(k)—*an ability to use the techniques, skills, and modern engineering tools necessary for engineering practice.*

APPLICATION EXAMPLES

The text has many examples that link directly to practical uses. The purpose of these examples is to show the student that the topics being covered are more than a pedagogical exercise. These real-world examples find use in common applications and products. We have increased to thirty-one the number of Application Examples. They include topics, such as batteries, source-load interfacing, bipolar junction transistor (BJT) operation, digital multimeters, common-mode rejection ratio (CMRR) in instrumentation amplifiers, electrocardiograph (ECG) and clock timing waveforms, how to obtain a waveform equation from an oscilloscope, sample-hold circuits, clock waveform timing circuits, resonance, impedance bridge, gain-bandwidth product, digital filtering, frequency content of a full-wave rectifier, isolation and auto-transformers, and others. These examples can be used to support ABET Criterion 3(j)—*a knowledge of contemporary issues.*

TEXT AND WEB APPENDICES

We have included an appendix on complex numbers since many students need to review this essential mathematical knowledge. There are four web appendices: One on the solution of linear equations (A), one on Butterworth and Chebyshev Poles (B), a new appendix on software tools (C), and a new one with all the Exercises worked out (D). These appendices are available at www.wiley.com/college/thomas.

NEW FEATURES OF THE SEVENTH EDITION
SKILLS-BUILDING EXAMPLES, EXERCISES, AND PROBLEMS

Users have asked that we include easier, skills-building examples, exercises, and problems as a means of helping students build confidence. Throughout the text, but especially in the early chapters, we have added several one-concept examples and exercises to key sections. In addition, we added numerous such problems in support of each learning objective. These skill-building items are at the Bloom's Taxonomy "Comprehension" level, rather than the more advanced "Application" and "Analysis" levels. Answers to Exercises have always been provided with the Exercise, but with this edition we have added solutions to each Exercise in a special Web Appendix D. The authors would appreciate feedback as to the usefulness of these added skills-building efforts.

CIRCUIT DESIGN AND EVALUATION

There are dozens of new design examples and exercises. In addition, there are 17 evaluation examples (listed below) and as many evaluation exercises, many of which are new. Our emphasis on creating solutions and choosing the better or best one has been strengthened with the inclusion of 212 design and 54 evaluation homework problems.

Page	Example	Evaluation Example Topic
130	3-26	Power loss comparison
132	3-28	Interface circuit comparison
201	4-21	Block diagram realization comparison
204	4-22	Comparison of two instrumentation systems
205	4-23	Instrumentation I/O comparison
215	4-24	Temperature transducer interface design comparison
337	7-14	First-order transient circuit comparison
424	8-27	High-pass filter evaluation
559	11-7	Cascade connection loading mitigation
571	11-14	Hum elimination circuit evaluation
592	11-24	Active step-response circuit design comparison
623	12-8	High-pass filter comparison
629	12-10	Band-stop filter comparison
661	12-23	Piezoelectric pressure transducer filtering
755	14-5	Notch filter requirement comparison
778	14-9	Multipole low-pass filter time-frequency responses tradeoff
837	16-7	Residential 3-wire system evaluation

OP AMPs

The treatment of OP AMPs in Chapter 4 has been revised to further emphasize design and evaluation. Compared to the sixth edition, several OP AMP sections have been expanded to include even more building blocks and practical examples for interfacing transducers. We have also added more design examples, including one on instrumentation amplifiers.

Frequency Response and Active Filters

We have improved Chapter 12 on frequency response and Chapter 14 on active filters. These chapters are excellent means of understanding the frequency behavior of circuits. We have continued our integration of software to assist the student in understanding frequency behavior through Bode diagrams and pole-zero diagrams in both chapters. Users have told us that Chapter 14 often proves useful to students in subsequent design courses where knowledge of active filters may be needed. As a result, we have expanded our coverage of active multipole notch and tuned filters. And we added a brief overview of Butterworth, Chebyshev, and Elliptic filters. Both chapters have more design and evaluation examples as well as more homework problems.

Mutual Inductance and Transformers

We have replaced the transformer equivalent circuit discussion with examples of the application of the transformer's electrical isolation properties to include a careful definition of isolation.

AC Power Systems

We have improved Chapter 16, on three-phase ac power circuits, to make it more in line with what today's students should know. We continued our focus away from phasor circuit analysis toward power flow and systems, both single-phase and three-phase, with new examples and exercises. We simplified the power flow discussion while placing greater emphasis on the fact that power flow does not require phasors. Power flow in a balanced three-phase circuit can be determined using only the magnitudes of the line voltage and line current.

Two-Ports

In response to several users, we have updated and moved the chapter on two-ports (Chapter 17) from the web to the main text. This chapter is fully integrated into the text, with examples, exercises, and problems. It has index references and answers to selected homework problems. We have added discussion, examples, and exercises to illustrate that two-port parameters are not just another way to find voltage and current responses. Rather, their primary utility is to determine global circuit properties such as voltage gain, current gain, feedback, and Thévenin equivalence. We would appreciate feedback on the content of this chapter for consideration in future editions.

Using This Edition for Laplace Early

The seventh edition is designed so that it can be used as a Laplace Early version as well as a traditional Phasor First version. The phasor analysis chapter (Chapter 8) comes before the study of Laplace transform techniques (Chapters 9 to 11). Those wishing to follow the traditional approach can follow the seventh edition chapter organization through Chapter 8, on phasor analysis, with a possible delaying of Chapter 7 until the second semester. Those choosing a Laplace Early approach can follow the present chapter organization through Chapter 7, skip Chapter 8, and proceed directly to the Laplace chapters. The current edition includes an introduction to phasor analysis in Sect. 11–5, dealing with the sinusoidal steady state. As a result, Laplace Early users can study phasor analysis in Chapter 8 at any point after Chapter 11. The following table shows suggested chapter sequencing for the traditional and Laplace Early approaches for three different subject matter emphases. The second author used the Traditional–Electronics sequence at the USAF Academy and has used the Laplace Early–Systems sequence at the University of Denver. Enough material is available in the printed text and in the Web appendix to allow the construction of other topic sequences. Other organizational options are available in the Instructor Manual.

EMPHASIS	SEMESTER 1							SEMESTER 2						
Traditional (Phasor First)														
Power	1	2	3	4	5	6	8/7	7/8	15	16	9	10	11	12	13
Systems								7/8	9	10	11	12	13	14	15
Electronics								7/8	9	10	11	12	14	15	17
Laplace Early														
Power	1	2	3	4	5	6	7	9	10	11	12	13	8	15	16
Systems								9	10	11	12	13	14	8	15
Electronics								9	10	11	12	14	8	15	17

USE OF SOFTWARE IN THIS EDITION

Software use throughout the text has been increased and strengthened to include many new MATLAB, OrCAD, and Excel examples to help practice using the software. We have added a new expanded web appendix to simplify students' use of software. There are 257 homework problems that suggest solutions using MATLAB, OrCAD, or both.

ACKNOWLEDGMENTS

Individuals at John Wiley & Sons, Inc. involved with this edition include Dan Sayre, Associate Publisher; Charlotte Cerf, Editorial Assistant; Chris Ruel, Marketing Manager; Joyce Poh, Senior Production Editor; and Don Fowley, VP and Executive Publisher. Over the years numerous faculty, engineers, staff, students, and others have helped shape this work in too many ways to list. In particular we would like to acknowledge the contributions of the following individuals to whom we are indebted: Robert M. Anderson, Iowa State University; Doran J. Baker, Utah State University; James A. Barby, University of Waterloo; William E. Bennett, United States Naval Academy; Maqsood A. Chaudhry, California State University at Fullerton; Michael Chier, Milwaukee School of Engineering; Don E. Cottrell, University of Denver; Robert Curtis, Ohio University; Michael L. Daley, University of Memphis; Ronald R. Delyser, University of Denver; Prasad Enjeti, Texas A&M University; John C. Getty, University of Denver; James G. Gottling, Ohio State University; Frank Gross, Florida State University; Robert Kotiuga, Boston University; Hans H. Kuehl, University of Southern California; K.S.P. Kumar, University of Minnesota; Nicholas Kyriakopoulos, George Washington University; Michael Lightner, University of Colorado at Boulder; Jerry I. Lubell, Jaycor; Reinhold Ludwig, Worcester Polytechnic Institute; Lloyd W. Massengill, Vanderbilt University; Frank L. Merat, Case Western Reserve University; Richard L. Moat, Motorola; Gene Moriarty, San Jose State University; Dudley Outcalt, Milwaukee School of

Engineering; Anil Pahwa, Kansas State University; Michael Polis, Oakland University; Pradeep Ramuhalli, Michigan State University; William Rison, New Mexico Institute of Mining and Technology; Martin S. Roden, California State University at Los Angeles; Pat Sannuti, the State University of New Jersey; Alan Schneider, University of California at San Diego; Ali O. Shaban, California Polytechnic State University; Jacob Shekel, Northeastern University; Kadagattur Srinidhi, Northeastern University; Peter J. Tabolt, University of Massachusetts at Boston; Len Trombetta, University of Houston; David Voltmer, Rose-Hulman Institute of Technology; Bruce F. Wollenberg, University of Minnesota; Robert Whitman, University of Denver; Albert Batten, USAF Academy; Daniel Pack, USAF Academy; Michael Drew, USAF Academy; Alan R. Klayton, USAF Academy; and Robert A. Weller, Vanderbilt University. Special recognition goes to Suzanne Ingrao Production Manager for editions three through six of this text, and to James K. Kang of California State University at Pomona for his meticulous editing of the homework solutions for the fourth, fifth, sixth, and seventh editions.

CONTENTS

CHAPTER 1 INTRODUCTION

The electromotive action manifests itself in the form of two effects which I believe must be distinguished from the beginning by a precise definition. I will call the first of these ''electric tension,'' the second ''electric current.''

André-Marie Ampère, 1820,
French Mathematician/Physicist

Some History Behind This Chapter

André Ampère (1775–1836) was the first to recognize the importance of distinguishing between the electrical effects we now call voltage and current. He also invented the galvanometer, the forerunner of today's voltmeter and ammeter. A natural genius, he had mastered all the then-known mathematics by age 12. He is best known for defining the mathematical relationship between electric current and magnetism, now known as Ampère's law.

Why This Chapter Is Important Today

Welcome to the study of Linear Circuits. In this chapter you are introduced to the lexicon of electrical engineering. You will learn both the terminology and the variables that will be used throughout the book. Important concepts introduced here are voltage and current, the reference marks used to define them, and a voltage benchmark called ground. In addition, you will gain an initial appreciation of the value of the computational software that is common in the electrical engineering profession.

Chapter Sections

Chapter Learning Objectives

1-1 Electrical Symbols and Units (Sect. 1–2)

Given an electrical quantity described in terms of words, scientific notation, or decimal prefix notation, convert the quantity to an alternative description.

1-2 Circuit Variables (Sect. 1–3)

Given any two of the three signal variables (i, v, p) or the two basic variables (q, w), find the magnitude and direction (sign) of the unspecified variables.

1-3 Software Introductions (Sect. 1-4, Web Appendix C)

Given a simple computational problem, use MATLAB as an appropriate engineering tool to solve the problem. (We will introduce the use of OrCAD to solve simulation problems starting in Chapter 2.)

1–1 ABOUT THIS BOOK

The basic purpose of this book is to introduce the analysis and design of linear circuits. Circuits are important in electrical engineering because they process electrical signals that carry energy and information. For the present we can define a **circuit** as an interconnection of electrical devices and a **signal** as a time-varying electrical entity. For example, the information stored on a Digital Versatile Disc (DVD) is recovered in the DVD-ROM player as electronic signals, initially stored as discrete (digital) data that are processed by circuits to generate continuous (analog) audio and video outputs. In an electrical power system some form of stored energy (coal, nuclear, hydro, chemical, etc.) is converted to electrical form and transferred to loads, where the energy is converted into the form (mechanical, light, heat, etc.) required by the customer. The DVD-ROM player and the electrical power system both involve circuits that process and transfer electrical signals carrying energy and information.

In this text we are primarily interested in **linear circuits**. An important feature of a linear circuit is that the amplitude of the output signal is proportional to the input signal amplitude. The proportionality property of linear circuits greatly simplifies the process of circuit analysis and design. Most circuits are linear only within a restricted range of signal levels. When driven outside this range they become nonlinear, and proportionality no longer applies. Although we will treat a few examples of nonlinear circuits, our attention is focused on circuits operating within their linear range.

Our study also deals with interface circuits. For the purposes of this book, we define an **interface** as a pair of accessible terminals at which signals may be observed or specified. The interface idea is particularly important with integrated circuit (IC) technology. Integrated circuits involve many thousands, indeed millions, of interconnections, but only a small number are accessible to the user. Designing systems using integrated circuits involves interconnecting complex circuits that have only a few accessible terminals. This often includes relatively simple circuits whose purpose is to change signal levels or formats. Such interface circuits are intentionally introduced to ensure that the appropriate signal conditions exist at the connections between complex integrated circuits.

Today's engineers analyze and design circuits using software tools. Using mathematical analysis tools such as MATLAB, MathCad, and Mathematica and PSpice simulation tools such as Electronic Workbench and Cadence (OrCAD), engineers can improve their understanding and results. As you proceed through this text, we help you develop the software skills necessary to become practiced in linear circuit design. Although there are many different software programs that you can use effectively to develop these skills, we will concentrate on MATLAB and OrCAD.

COURSE OBJECTIVES

This book is designed to help you develop the knowledge and application skills needed to solve three types of circuit problems: analysis, design, and evaluation. An **analysis** problem involves finding the output signals of a given circuit with known input signals. Circuit analysis is the foundation for understanding the interaction of signals and circuits. A **design** problem involves devising one or more circuits that perform a given signal-processing function. There usually are several possible solutions to a design problem. This leads to an **evaluation** problem, which involves picking the best solution from among several candidates using factors such as cost, power consumption, and part counts. In real life the engineer's role is a blend of analysis, design, and evaluation, and in practice the boundaries between these categories are often blurred.

This text contains many worked examples to help you develop your problem-solving skills. The **examples** include a problem statement and provide the intermediate steps needed to obtain the final answer. The examples often treat analysis problems, although design and evaluation examples are included. This text also contains a number of **exercises** that include only the problem statement and the final answer. You should use the exercises to test your understanding of the circuit concepts discussed in the preceding section. Solutions to all Exercises are available on Web Appendix D.

Throughout we will show you where it is useful to turn to software to help solve problems, be they analysis, design, or evaluation. The mouse icon ⌕ identifies examples, exercises, and problems that are best solved using software tools.

CHAPTER OBJECTIVES

At the start of each chapter we provide three motivational aspects for what you are about to learn. First, we present a brief perspective of a key historical figure important to the content of the chapter. Second, we give an overview of why this chapter is important to your study. Third, we introduce you to the learning objectives for the chapter.

The **chapter learning objectives** are a carefully structured set of enabling skills. They are introduced in the chapter opener and repeated in more detail at the end of each chapter. Collectively, these objectives represent the basic knowledge and understanding needed to master the topics covered in each chapter. In the problems section the objectives explicitly state the expected behavior, followed by a graduated set of homework problems designed to help you assess your level of achievement. Each objective also lists worked examples and exercises in the text that help you work the related homework problems. Once you understand the chapter learning objectives, you can move on to the integrating problems at the very end of the problems section. These problems require mastery of several chapter learning objectives from the present and prior chapters and provide an opportunity to test your ability to deal with comprehensive, integrative problems. Throughout the text, when appropriate, we label the primary purpose of the example, exercise, course-learning problem, or chapter-integrating problem with the symbol **A** for analysis, **D** for design, or **E** for evaluation.

ASSESSMENT AND ACCREDITATION

Material in this text can be used effectively in a properly designed course to support accreditation criteria associated with comprehension, use of modern tools, design, evaluation, and real-world constraints. Additional accreditation and assessment guidance is provided in the Instructors Manual.

1–2 SYMBOLS AND UNITS

Throughout this text we will use the international system (SI) of units. The SI system includes six fundamental units: meter (m), kilogram (kg), second (s), ampere (A), kelvin (K), and candela (cd). All the other units of measure can be derived from these six.

Like all disciplines, electrical engineering has its own terminology and symbology. The symbols used to represent some of the more important physical quantities and their units are listed in Table 1–1. It is not our purpose to define these quantities here or to offer this list as an item for memorization. Rather, the purpose of this table is merely to list in one place all the electrical quantities used in this book.

Numerical values in engineering range over many orders of magnitude. Consequently, the system of standard decimal prefixes in Table 1–2 is used. These prefixes on a unit abbreviation symbol indicate the power of 10 that is applied to the numerical value of the quantity.

T A B L E 1–1 **SOME IMPORTANT QUANTITIES, THEIR SYMBOLS, AND UNIT ABBREVIATIONS**

Quantity	Symbol	Unit	Unit Abbreviation
Time	t	second	s
Frequency	f	hertz	Hz
Radian frequency	ω	radian/second	rad/s
Phase angle	θ, ϕ	degree or radian	° or rad
Energy	w	joule	J
Power	p	watt	W
Charge	q	coulomb	C
Current	i	ampere	A
Electric field	$\mathscr{E}$	volt/meter	V/m
Voltage	v	volt	V
Impedance	Z	ohm	Ω
Admittance	Y	siemens	S
Resistance	R	ohm	Ω
Conductance	G	siemens	S
Reactance	X	ohm	Ω
Susceptance	B	siemens	S
Inductance, self	L	henry	H
Inductance, mutual	M	henry	H
Capacitance	C	farad	F
Magnetic flux	ϕ	weber	wb
Flux linkages	λ	weber-turns	wb-t
Power ratio	P	bel	B

T A B L E 1–2 **STANDARD DECIMAL PREFIXES**

Multiplier	Prefix	Abbreviation
10^{18}	exa	E
10^{15}	peta	P
10^{12}	tera	T
10^{9}	giga	G
10^{6}	mega	M
10^{3}	kilo	k
10^{-1}	deci	d
10^{-2}	centi	c
10^{-3}	milli	m
10^{-6}	micro	μ
10^{-9}	nano	n
10^{-12}	pico	p
10^{-15}	femto	f
10^{-18}	atto	a

Exercise 1–1 _____

Given the pattern in the statement $1\,\text{k}\Omega = 1\,\text{kilohm} = 1 \times 10^3$ ohms, fill in the blanks in the following statements using the standard decimal prefixes.

(a) ____ = ____ = 5×10^{-3} watts
(b) $10.0\,\text{dB} = $ ____ = ____
(c) $3.6\,\text{ps} = $ ____ = ____
(d) ____ = 0.03 microfarads = ____
(e) ____ = ____ = 6.6×10^9 hertz

Answers:

(a) $5.0\,\text{mW} = 5\,\text{milliwatts}$
(b) $10.0\,\text{decibels} = 1.0\,\text{bel}$
(c) $3.6\,\text{picoseconds} = 3.6 \times 10^{-12}\,\text{seconds}$
(d) $30\,\text{nF or}\,0.03\,\mu\text{F} = 30.0 \times 10^{-9}\,\text{farads}$
(e) $6.6\,\text{GHz} = 6.6\,\text{gigahertz}$

1–3 CIRCUIT VARIABLES

The underlying physical variables in the study of electronic systems are **charge** and **energy**. The idea of electrical charge explains the very strong electrical forces that occur in nature. To explain both attraction and repulsion, we say that there are two kinds of charge—positive and negative. Like charges repel, whereas unlike charges attract one another. The symbol q is used to represent charge. If the amount of charge is varying with time, we emphasize the fact by writing $q(t)$. In the international system (SI), charge is measured in **coulombs** (abbreviated C). The smallest quantity of charge in nature is an electron's charge ($q_E = -1.6 \times 10^{-19}\,\text{C}$). Thus, there are $1/|q_E| = 6.25 \times 10^{18}$ electrons in 1 coulomb of charge.

Electrical charge is a rather cumbersome variable to measure in practice. Moreover, in most situations the charges are moving, so we find it more convenient to measure the amount of charge passing a given point per unit time. If $q(t)$ is the cumulative charge passing through a point, we define a signal variable i called **current** as follows:

$$i = \frac{dq}{dt} \tag{1–1}$$

Current is a measure of the flow of electrical charge. It is the time rate of change of charge passing a given point in a circuit. The physical dimensions of current are coulombs per second. In the SI system, the unit of current is the **ampere** (abbreviated A). That is,

$$1\,\text{coulomb/second} = 1\,\text{ampere} = 1\,\text{A}$$

Since there are two types of electrical charge (positive and negative), there is a bookkeeping problem associated with the direction assigned to the current. In engineering it is customary to define the direction of current as the direction of the net flow of positive charge. Since electrons have negative charge, they move in the opposite direction of the current.

A second signal variable called **voltage** is related to the change in energy that would be experienced by a charge as it passes through a circuit. The symbol w is commonly used to represent energy. In the SI system of units, energy carries the units of **joules** (abbreviated J). If a small charge dq were to experience a change in energy dw in passing from point A to point B in a circuit, then the voltage v between A and B is defined as the change in energy per unit charge. We can express this definition in differential form as

$$v = \frac{dw}{dq} \tag{1–2}$$

Voltage does not depend on the path followed by the charge dq in moving from point A to point B. Furthermore, there can be a voltage between two points even if there is no charge motion, since voltage is a measure of how much energy dw would be involved if a charge dq was moved. The dimensions of voltage are joules per coulomb. The unit of voltage in the SI system is the **volt**[1] (abbreviated V). That is,

$$1 \, \text{joule/coulomb} = 1 \, \text{volt} = 1 \, \text{V}$$

The general definition of the physical variable called **power** is the time rate of change of energy:

$$p = \frac{dw}{dt} \tag{1–3}$$

The dimensions of power are joules per second, which in the SI system is called a **watt**[2] (abbreviated W). In electrical circuits it is useful to relate the power associated with a device or element to the signal variables current and voltage. Using the chain rule, Eq. (1–3) can be written as

$$p = \left(\frac{dw}{dq}\right)\left(\frac{dq}{dt}\right) \tag{1–4}$$

Now using Eqs. (1–1) and (1–2), we obtain

$$p = vi \tag{1–5}$$

The electrical power associated with a situation is determined by the product of voltage and current. The total energy transferred during the period from t_1 to t_2 is found by solving for dw in Eq. (1–3) and then integrating

$$w_{\text{T}} = \int_{w_1}^{w_2} dw = \int_{t_1}^{t_2} p \, dt \tag{1–6}$$

In sum, the three key circuit variables—current, voltage, and power—are measured as follows: current at individual points, voltage always between two points, and power at an element or device.

EXAMPLE 1–1

The electron beam in the cathode-ray tube shown in Figure 1–1 carries 10^{14} electrons per second and is accelerated by a voltage of 50 kV. Find the power in the electron beam.

FIGURE 1–1

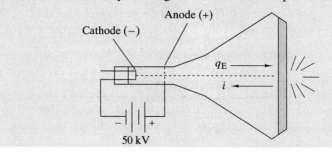

Cathode (−)   Anode (+) q_{E} i 50 kV

SOLUTION:
Since current is the rate of positive charge flow, its direction is opposite that of the electron beam, as shown in Figure 1–1. The electrons are flowing to the right from

---

[1]The volt is named after Italian physicist, Alessandro Volta (1745–1827), for the discovery of a practical source of current - the battery.

[2]The watt is named after the Scottish inventor and mechanical engineer, James Watt (1736–1819), who is credited for inventing the steam engine and enabling the Industrial Revolution.

the cathode toward the anode, but the current i is flowing to the left toward the cathode. We can find the magnitude of the current by multiplying the magnitude of the charge of an electron q_E by the rate of electron flow dn_E/dt.

$$i = |q_E|\frac{dn_E}{dt} = (1.6 \times 10^{-19})(10^{14}) = 1.6 \times 10^{-5}\text{A} = 16\mu\text{A}$$

Therefore, the beam power is

$$p = vi = (50 \times 10^3)(1.6 \times 10^{-5}) = 0.8\,\text{W} = 800\,\text{mW} \qquad ■$$

EXAMPLE 1–2

The current through a circuit element is 50 mA. Find the total charge and the number of electrons transferred during a period of 100 ns.

SOLUTION:
The relationship between current and charge is given in Eq. (1–1) as

$$i = \frac{dq}{dt}$$

Since the current i is given, we calculate the charge transferred by solving this equation for dq and then integrating

$$
\begin{aligned}
q_T &= \int_{q_1}^{q_2} dq = \int_0^{10^{-7}} i\,dt \\
&= \int_0^{10^{-7}} 50 \times 10^{-3}\,dt = 50 \times 10^{-10}\,\text{C} = 5\,\text{nC}
\end{aligned}
$$

There are $1/|q_E| = 6.25 \times 10^{18}$ electrons/coulomb, so the number of electrons transferred is

$$n_E = (5 \times 10^{-9}\,\text{C})(6.25 \times 10^{18}\,\text{electrons/C}) = 31.25 \times 10^9\,\text{electrons} \qquad ■$$

Exercise 1–2

A device dissipates 100 W of power. How much energy is delivered to it in 10 seconds?

Answer: 1 kJ

Note: 1 W-s = 1 J

Exercise 1–3

The graph in Figure 1–2(a) shows the charge $q(t)$ flowing past a point in a wire as a function of time.

(a) Find the current $i(t)$ at $t = 1, 2.5, 3.5, 4.5,$ and 5.5 ms.
(b) Sketch the variation of $i(t)$ versus time.

Answers:

(a) $-10\,\text{nA}, +40\,\text{nA}, 0\,\text{nA}, -20\,\text{nA}, 0\,\text{nA}$.
(b) The variations in $i(t)$ are shown in Figure 1–2(b).

THE PASSIVE SIGN CONVENTION

We have defined three circuit variables (current, voltage, and power) using two basic variables (charge and energy). Charge and energy, like mass, length, and time, are basic concepts of physics that provide the scientific foundation for electrical

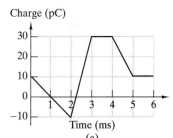
(a)

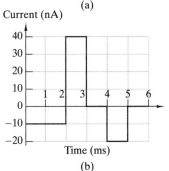
(b)

FIGURE 1–2

engineering. However, engineering problems rarely involve charge and energy directly, but are usually stated in terms of voltage, current, and power. The reason for this is simple: The circuit variables are much easier to measure and therefore are the most useful working variables in engineering practice.

At this point, it is important to stress the physical differences between current and voltage variables. Current is a measure of the time rate of charge passing a point in a circuit. We think of current as a *through variable*, since it describes the flow of electrical charge through a point in a circuit. On the other hand, voltage is not measured at a single point, but rather between two points or across an electrical device. Consequently, we think of voltage as an *across variable* that inherently involves two points.

The arrow below the $i(t)$ and the plus and minus symbols across the $v(t)$ in Figure 1–3 are *reference marks* that define the positive directions for the current and voltage associated with an electrical device. These reference marks do not represent an assertion about what is happening physically in the circuit. The response of an electrical circuit is determined by physical laws, not by the reference marks assigned to the circuit variables.

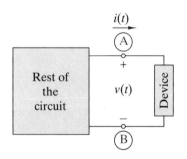

FIGURE 1–3 *Voltage and current reference marks for a two-terminal device.*

The reference marks are benchmarks assigned at the beginning of the analysis. When the actual direction and reference direction agree, the answers found by circuit analysis will have positive algebraic signs. When they disagree, the algebraic signs of the answers will be negative. For example, if circuit analysis reveals that the current variable in Figure 1–3 is positive [i.e., $i(t) > 0$], then the sign of this answer, together with the assigned reference direction, indicates that the current passes through point A in Figure 1–3 from left to right. Conversely, when analysis reveals that the current variable is negative, then this result, combined with the assigned reference direction, tells us that the current passes through point A from right to left. In summary, the algebraic sign of the answer together with arbitrarily assigned reference marks tell us the actual directions of a voltage or current variable.

In Figure 1–3, the current reference arrow enters the device at the terminal marked with the plus voltage reference mark. This orientation is called the **passive sign convention**. Under this convention, the power $p(t)$ is positive when the device absorbs power and is negative when it delivers power to the rest of the circuit. Since $p(t) = v(t) \times i(t)$, a device absorbs power when the voltage and current variables have the same algebraic sign and delivers power when they have opposite signs. Certain devices, such as heaters (a toaster, for example), can only absorb power. The voltage and current variables associated with these devices must always have the same algebraic sign. On the other hand, a battery absorbs power [$p(t) > 0$] when it is being charged and delivers power [$p(t) < 0$] when it is discharging. Thus, the voltage and current variables for a battery can have either the same or opposite algebraic signs.

In a circuit some devices absorb power and others deliver power, but the sum of the power in all of the devices in the circuit is zero. This is more than just a conservation-of-energy concept. When electrical devices are interconnected to form a circuit, the only way that power can enter or leave the circuit is via the currents and voltages at device terminals. The existence of a power balance in a circuit is one method of checking calculations.

The passive sign convention is used throughout this book. It is also the convention used by circuit simulation computer programs.[3] To interpret correctly the results of circuit analysis, it is important to remember that the reference marks (arrows and plus/minus signs) are reference directions, not indications of the circuit response. The actual direction of a response is determined by comparing its reference direction with the algebraic signs of the result predicted by circuit analysis based on physical laws.

---

[3]We discuss computer-aided circuit analysis in Section 1–4.

GROUND

Since voltage is defined between two points, it is often useful to define a common voltage reference point called **ground**. The voltages at all other points in a circuit are then defined with respect to this common reference point. We indicate the voltage reference point using the ground symbol shown in Figure 1–4. Under this convention we sometimes refer to the variables $v_A(t)$, $v_B(t)$, and $v_C(t)$ as the voltages at points A, B, and C, respectively. This terminology appears to contradict the fact that voltage is an across variable that involves two points. However, the terminology means that the variables $v_A(t)$, $v_B(t)$, and $v_C(t)$ are the voltages defined between points A, B, and C and the common voltage reference point at point G.

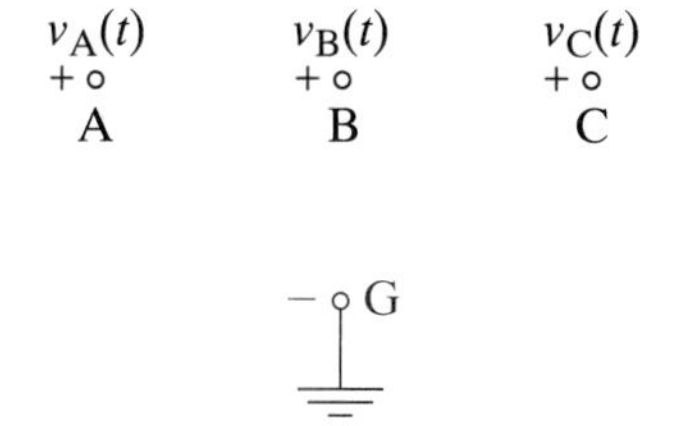

Using a common reference point for across variables is not an idea unique to electrical circuits. For example, the elevation of a mountain is the number of feet or meters between the top of the mountain and a common reference point at sea level. If a geographic point lies below sea level, then its elevation is assigned a negative algebraic sign. So it is with voltages. If circuit analysis reveals that the voltage variable at point A is negative [i.e., $v_A(t) < 0$], then this fact together with the reference marks in Figure 1–4 indicate that the potential at point A is less than the ground potential.

$$v_A(t) \quad\quad v_B(t) \quad\quad v_C(t)$$
$$+\circ \quad\quad\quad +\circ \quad\quad\quad +\circ$$
$$A \quad\quad\quad B \quad\quad\quad C$$

$$-\circ\, G$$

FIGURE 1–4 *Ground symbol indicates a common voltage reference point.*

EXAMPLE 1–3 CP HW1E2

Figure 1–5 shows a circuit formed by interconnecting five devices, each of which has two terminals. A voltage and current variable has been assigned to each device using the passive sign convention. The working variables for each device are observed to be as follows:

	DEVICE 1	DEVICE 2	DEVICE 3	DEVICE 4	DEVICE 5
v	+100 V	?	+25 V	+75 V	−75 V
i	?	+5 mA	+5 mA	?	+5 mA
p	−1 W	+0.5 W	?	0.75 W	?

(a) Find the missing variable for each device and state whether the device is absorbing or delivering power.
(b) Check your work by showing that the sum of the device powers is zero.

FIGURE 1–5

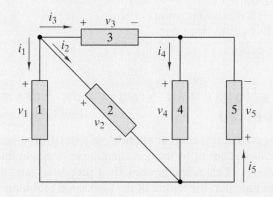

SOLUTION:

(a) We use $p = vi$ to solve for the missing variable since two of the three circuit variables are given for each device.

Device 1: $i_1 = p_1/v_1 = -1/100 = -10\,\text{mA}$ $[p(t) < 0, \text{delivering power}]$

Device 2: $v_2 = p_2/i_2 = 0.5/0.005 = 100\,\text{V}$ $[p(t) > 0, \text{absorbing power}]$

Device 3: $p_3 = v_3 i_3 = 25 \times 0.005 = 0.125\,\text{W}$ $[p(t) > 0, \text{absorbing power}]$

Device 4: $i_4 = p_4/v_4 = 0.75/75 = 10\,\text{mA}$ $[p(t) > 0, \text{absorbing power}]$

Device 5: $p_5 = v_5 i_5 = -75 \times 0.005 = -0.375\,\text{W}$ $[p(t) < 0, \text{delivering power}]$

(b) Summing the device powers yields

$$p_1 + p_2 + p_3 + p_4 + p_5 = -1 + 0.5 + 0.125 + 0.75 - 0.375$$
$$= +1.375 - 1.375 = 0$$

This example shows that the sum of the power absorbed by devices is equal in magnitude to the sum of the power supplied by devices. A power balance always exists in the types of circuits treated in this book and can be used as an overall check of circuit analysis calculations. ■

Exercise 1–4

The working variables of a set of two-terminal electrical devices are observed to be as follows:

	DEVICE 1	DEVICE 2	DEVICE 3	DEVICE 4	DEVICE 5
v	+10 V	?	−15 V	+5 V	?
i	−3 A	−3 A	+10 mA	?	−12 mA
p	?	+40 W	?	+10 mW	−120 mW

Using the passive sign convention, find the magnitude and sign of each unknown variable and state whether the device is absorbing or delivering power.

Answers:

Device 1: $p = -30\,\text{W}$ (delivering power)
Device 2: $v = -13.3\,\text{V}$ (absorbing power)
Device 3: $p = -150\,\text{mW}$ (delivering power)
Device 4: $i = +2\,\text{mA}$ (absorbing power)
Device 5: $v = +10\,\text{V}$ (delivering power)

1–4 COMPUTATIONAL AND SIMULATION SOFTWARE INTRODUCTION

The problems presented in this chapter are relatively straightforward and are solvable with simple hand calculations or by using a scientific or engineering calculator. As we increase our knowledge of electrical devices, their properties, and the circuits created by connecting them together, the nature of the associated problems will become more difficult. To solve these problems, tools that are more sophisticated are required. Extremely valuable tools for electrical and computer engineers are computational and simulation software that can perform all of the mathematical calculations associated with a problem and displays the results in several ways that assist in the analysis and design of circuits. To have a complete education, an engineer must be familiar with the contemporary software tools common to the profession. Although there are many competitive choices available, a common tool in engineering for mathematical computations is MATLAB, created by The Math Works, Inc., and one for simulation is OrCAD by Cadence Design Systems, Inc.

We introduce MATLAB as an efficient, programmable tool to perform calculations associated with engineering problems. This software is well established in the industry and has complete documentation to support users that range from novices to experts. As with any new tool, it requires an investment of time and effort to learn how to apply it correctly and efficiently. This investment will pay off in the end by offering the following advantages that come from using and mastering the MATLAB software:

- Calculating results faster with higher accuracy
- Performing simulations efficiently
- Solving more complex problems
- Examining many design options quickly
- Visualizing results professionally

Similarly, we introduce OrCAD as a method to graphically draw circuits and have the software analyze the circuit in several ways. Like MATLAB, learning OrCAD requires a bit of effort but the end results are a quick way to study the behavior of the circuits in question. Simulation software offers the following advantages:

- Rapid analysis of circuit variables, such as voltage, current, and power
- Validation of circuit designs
- Ability to modify designs quickly and accurately to explore and evaluate many possible solutions
- Producing visible graphical results of desired analyses

Web Appendix C offers a brief introduction to using the MATLAB and OrCAD software for engineering problems and provides worked examples for key topics in each chapter. Combined with the help files and tutorials available with each software package, the appendix provides a starting point of where to find applications of these powerful computational or simulation tools. Mastering each tool will require significant practice and experience. To provide opportunities to practice and additional exposure to these software products, every chapter in this book will demonstrate specific features of the appropriate software and how they can be applied to efficiently solve electrical engineering problems. In addition, selected problems at the end of each chapter specifically request solutions using MATLAB or OrCAD to reinforce the important features of the software in a deliberate, developmental manner.

SUMMARY

- Circuits are important in electrical engineering because they process signals that carry energy and information. A **circuit** is an interconnection of electrical devices. A **signal** is an electrical current or voltage that carries energy or information. An **interface** is a pair of accessible terminals at which signals may be observed or specified.

- This book defines overall course objectives at the analysis, design, and evaluation levels. In **circuit analysis** the circuit and input signals are given and the object is to find the output signals. The object of **circuit design** is to devise one or more circuits that produce prescribed output signals for given input signals. The **evaluation** problem involves appraising alternative circuit designs using criteria such as cost, power consumption, and parts count.

- Charge (q) and energy (w) are the basic physical variables involved in electrical phenomena. Current (i), voltage (v), and power (p) are the derived variables used in circuit analysis and design. In the SI system, charge is measured in coulombs (C), energy in joules (J), current in amperes (A), voltage in volts (V), and power in watts (W).

- **Current** is defined as dq/dt and is a measure of the flow of electrical charge. **Voltage** is defined as dw/dq and is a measure of the energy required to move a small charge from one point to another. **Power** is defined as dw/dt and is a measure of the rate at which energy is being transferred. Power is related to current and voltage as $p = vi$.

- The **reference marks** (arrows and plus/minus signs) assigned to a device are reference directions, not indications of the way a circuit responds. The actual direction of the response is determined by comparing the reference direction and the algebraic sign of the answer found by circuit analysis using physical laws.

- Under the **passive sign convention**, the current reference arrow is directed toward the terminal with the positive voltage reference mark. Under this convention, the device power is positive when it absorbs power and is negative when it delivers power. When current and voltage have the same (opposite) algebraic signs, the device is absorbing (delivering) power.

- Engineers use computational software, such as MATLAB, to increase the speed and accuracy of calculations. Engineers use simulation software, such as OrCAD, to model the behavior of circuits. Software is useful for performing circuit simulations and for expanding the complexity of problems that can be solved in a reasonable amount of time. Learning and exploiting the advantages of computational and simulation software are critical skills for engineers, and MATLAB and OrCAD are common tools for electrical and computer engineers.

PROBLEMS

OBJECTIVE 1–1 ELECTRICAL SYMBOLS AND UNITS (SECT. 1–2)

Given an electrical quantity described in terms of words, scientific notation, or decimal prefix notation, convert the quantity to an alternate description.
See Exercise 1–1

1–1 Express the following quantities to the nearest standard prefix using no more than three digits.
  (a) 1,000,000 Hz
  (b) 102.5×10^9 W
  (c) 0.333×10^{-7} s
  (d) 10×10^{-12} F

1–2 Express the following quantities to the nearest standard prefix using no more than three digits.
  (a) 0.000222 H
  (b) 20.5×10^5 J
  (c) 72.25×10^3 C
  (d) 3,264 Ω

1–3 An ampere-hour (Ah) meter measures the time-integral of the current in a conductor. During an 8-hour period, a certain meter records 3300 Ah. Find the number of coulombs that flowed through the meter during the recording period.

1–4 Electric power companies measure energy consumption in kilowatt-hours, denoted kWh. One kilowatt-hour is the amount of energy transferred by 1 kW of power in a period of 1 hour. A power company billing statement reports a user's total energy usage to be 2500 kWh. Find the number of joules used during the billing period.

1–5 Fill in the blanks in the following statements.
  (a) To convert capacitance from picofarads to microfarads, multiply by ___.
  (b) To convert resistance from megohms to kilohms, multiply by ___.
  (c) To convert voltage from millivolts to volts, multiply by ___.
  (d) To convert energy from megajoules to joules, multiply by ___.

OBJECTIVE 1–2 CIRCUIT VARIABLES (SECT. 1–3)

Given any two of the three signal variables (i, , p) or the two basic variables (q, w), find the magnitude and direction (sign) of the unspecified variables.
See Examples 1–1, 1–2, 1–3 and Exercises 1–2, 1–3, 1–4

1–6 A wire carries a constant current of 30 mA. How many coulombs flow past a given point in the wire in 5 s?

1–7 The net positive charge flowing through a device is $q(t) = 20 + 4t$ mC. Find the current through the device.

1–8 Figure P1–8 shows a plot of the net positive charge flowing in a wire versus time. Sketch the corresponding current during the same period of time.

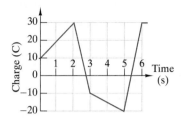

FIGURE P1–8

1–9 The net negative charge flowing through a device varies as $q(t) = 3t^2$ C. Find the current through the device at $t = 0$ s, $t = 0.5$ s, and $t = 1$ s.

1–10 A cell-phone charger outputs 9.6 V and is protected by a 50 mA fuse. A 1.5 W cell phone is connected to it to be charged. Will the fuse blow?

1–11 For $0 \leq t \leq 5$ s, the current through a device is $i(t) = 4t$ A. For $5 < t \leq 10$ s, the current is $i(t) = 40 - 4t$ A, and $i(t) = 0$ A for $t > 10$ s. Sketch $i(t)$ versus time and find the total charge flowing through the device between $t = 0$ s and $t = 10$ s.

1–12 The charge flowing through a device is $q(t) = 1 - e^{-1000t}$ μC. How long will it take the current to reach 200 μA?

1–13 The 12-V automobile battery in Figure P1–13 has an output capacity of 100 ampere-hours (Ah) when connected to a head lamp that absorbs 200 watts of power. The car engine is not running and therefore not charging the battery. Assume the battery voltage remains constant.
 (a) Find the current supplied by the battery and determine how long can the battery power the headlight.
 (b) A 100 W device is connected through the utility port. How long can the battery power both the headlight and the device?

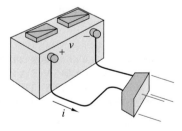

FIGURE P1–13

1–14 An incandescent lamp absorbs 100 W when connected to a 120-V source. A energy-efficient compact fluorescent lamp (CFL) producing the same amount of light absorbs 16 W when connected to the same source.

How much cheaper is it to operate the CFL versus the incandescent bulb over 1,000 hours when electricity costs 7.8 cents/kWh?

1–15 The current through a device is zero for $t < 0$ and is $i(t) = 3e^{-2t}$ A for $t \geq 0$. Find the charge $q(t)$ flowing through the device for $t \geq 0$.

1–16 A string of holiday lights is protected by a 5-A fuse and has 25 bulbs, each of which is rated at 7 W. How many strings can be connected end-to-end across a 120 V circuit without blowing a fuse?

1–17 When illuminated the i-v relationship for a photocell is $i = e^v - 10$ A. For $v = -2, 2$ and 3 V find the device power and state whether it is absorbing or delivering power.

1–18 A new 6 V Alkaline lantern battery delivers 237.5 kJ of energy during its lifetime. How long will the battery last in an application that draws 15 mA continuously. Assume the battery voltage is constant.

1–19 The maximum power a device can dissipate is 0.25 W. Determine the maximum current allowed by the device power rating when the voltage is 9 V.

1–20 Traffic lights are being converted from incandescent bulbs to LED arrays to save operating and maintenance costs. Typically each incandescent light uses three 100-W bulbs, one for each color R, Y, G. A competing LED array consists of 61 LEDs with each LED requiring 9 V and drawing 20 mA of current. There are three arrays per light – R, Y, G. A small city has 1560 traffic signals. Since one light is always on 24/7, how much can a city save in one year if the city buys their electricity at 7.2¢ per kWh?

1–21 Two electrical devices are connected as shown in Figure P1–21. Using the reference marks shown in the figure, find the power transferred and state whether the power is transferred from A to B or B to A when
 (a) $v = +11$ V and $i = -1.1$ A
 (b) $v = +80$ V and $i = +20$ mA
 (c) $v = -120$ V and $i = -12$ mA
 (d) $v = -1.5$ V and $i = -600$ μA

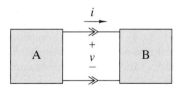

FIGURE P1–21

1–22 Figure P1–22 shows an electric circuit with a voltage and a current variable assigned to each of the six devices. The device voltages and currents are observed to be

	v(V)	i(A)		v(V)	i(A)
Device 1	15	−1	Device 4	−10	−1
Device 2	5	1	Device 5	20	−3
Device 3	10	2	Device 6	20	2

Find the power associated with each device and state whether the device is absorbing or delivering power. Use the power balance to check your work.

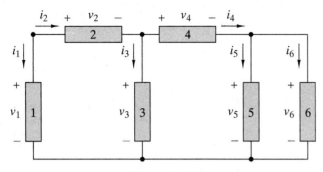

FIGURE P1–22

1–23 Figure P1–22 shows an electric circuit with a voltage and a current variable assigned to each of the six devices. Use power balance to find v_4 when $v_1 = 20$ V, $i_1 = -2$ A, $p_2 = 20$ W, $p_3 = 10$ W, $i_4 = 1$ A, and $p_5 = p_6 = 2.5$ W. Is device 4 absorbing or delivering power?

1–24 Suppose in Figure P1–22 a ground is connected to the minus $(-)$ side of element 6 and another to the junction of elements 2, 3 and 4. Further, assume that the voltage v_4 is 5 V and v_1 is 10 V. What are the voltages v_2, v_3, v_5 and v_6?

1–25 For $t \geq 0$ the voltage across and power absorbed by a two-terminal device are $v(t) = 2e^{-t}$ V and $p(t) = 40e^{-2t}$ mW. Find the total charge delivered to the device for $t \geq 0$.

OBJECTIVE **1–3** S**OFTWARE** I**NTRODUCTION** (S**ECT.** **1–4,** W**EB** A**PPENDIX** C)

Given a simple computational problem, use MATLAB as an appropriate engineering tool to solve the problem. (We will introduce OrCAD problems starting in Chapter 2.) Examples and Exercises throughout the text. See Web Appendix C.

1–26 Repeat Problem 1–22 using MATLAB to perform the calculations. Create a vector for the voltage values, `v = [15 5 10 –10 20 20 ]`, and a vector for the current values, `i=[–1 1 2 –1 –3 2]`. Compute the corresponding vector for the power values, p, using element-by-element multiplication (`.*`) and then use the `sum` command to verify the power balance.

1–27 Using the passive sign convention, the voltage across a device is $v(t) = 170\cos(377t)$ V and the current through the device is $i(t) - 2\sin(377t)$ A. Using MATLAB, create a short script (m-file) to assign a value to the time variable, t, and then calculate the voltage, current, and power at that time. Run the script for $t = 5$ ms and $t = 10$ ms and for each result state whether the device is absorbing or delivering power.

I**NTEGRATING** P**ROBLEMS**

1–28 Power Ratio (PR) in dB (**A**)
A stereo amplifier takes the output of a CD player, for example, and increases the power to an audible level. Suppose the output of the CD player is 50 mW and the desired audible output is 100 W per stereo channel, find the power ratio of the amplifier per channel in decibels (dB), where the power ratio in dB is

$$PR_{\text{dB}} = 10\,\text{Log}_{10}(p_2/p_1).$$

1–29 AC to DC Converter (**A**)
A manufacturer's data sheet for the converter in Figure P1–29 states that the output voltage is $v_{\text{dc}} = 5$ V when the input voltage $v_{\text{ac}} = 120$ V. When the load draws a current $i_{\text{dc}} = 40$ A the input power is $p_{\text{ae}} = 300$ W. Find the efficiency of the converter.

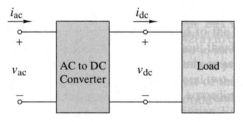

FIGURE P1–29

1–30 Charge-Storage Device (**A**)
A capacitor is a two-terminal device that can store electric charge. In a linear capacitor the amount of charge stored is proportional to the voltage across the device. For a particular device the proportionality is $q(t) = 10^{-7} v(t)$. If $v(t) = 0$ for $t < 0$ and $v(t) = 10(1 - e^{-5000t})$ for $t \geq 0$, find the energy stored in the device at $t = 200$ μs.

1–31 Computer Data Sheet (**A**)
A manufacturer's data sheet for a notebook computer lists the power supply requirements as 7.5 A @ 5 V, 2 A @ 15 V, 2.5 A @ -15 V, 2.25 A @ -5 V and 0.5 A @ 12 V. The data sheet also states that the overall power consumption is 115 W. Are these data consistent? Explain.

CHAPTER 2 BASIC CIRCUIT ANALYSIS

The equation S = A/L shows that the current of a voltaic circuit is subject to a change, by each variation originating either in the magnitude of a tension or in the reduced length of a part, which latter is itself again determined, both by the actual length of the part as well as its conductivity and its section.

Georg Simon Ohm, 1827,
German Mathematician/Physicist

Some History Behind This Chapter

Georg Simon Ohm (1789–1854) discovered the law that now bears his name in 1827. His results drew heavy criticism and were not generally accepted for many years. Fortunately, the importance of his contribution was eventually recognized during his lifetime. He was honored by the Royal Society of England in 1841 and appointed a Professor of Physics at the University of Munich in 1849. The unit of resistance was named after Professor Ohm in 1862 by the British Association Committee.

Why This Chapter Is Important Today

A circuit is an interconnection of electric devices that performs a useful function. This chapter introduces some basic tools you will need to analyze and design electric circuits. You will also be introduced to several important electric devices that control currents and voltages in a circuit. These devices range from everyday things like batteries to special integrated circuits that meter out predetermined voltages or currents.

To analyze these circuits efficiently, we can use computer-based tools, such as MATLAB for mathematical computations and OrCAD for circuit simulations. These programs are powerful tools that are common in both academic and commercial applications.

Chapter Sections

Chapter Learning Objectives

2-1 Element Constraints (Sect. 2–1)
Given a two-terminal element with one or more electrical variables specified, use the element $i - v$ constraint to find the magnitude and direction of the unknown variables.

2-2 Connection Constraints (Sect. 2–2)
Given a circuit composed of two-terminal elements:
(a) Identify nodes and loops in the circuit.
(b) Identify elements connected in series and in parallel.
(c) Use Kirchhoff's laws (KCL and KVL) to find selected signal variables.

2-3 Combined Constraints (Sect. 2–3)
Given a linear resistance circuit, use the element constraints and connection constraints to find selected signal variables.

2-4 Equivalent Circuits (Sect. 2–4)
(a) Given a circuit consisting of linear resistors, find the equivalent resistance between a specified pair of terminals.
(b) Given a circuit consisting of a source-resistor combination, find an equivalent source-resistor circuit.

2-5 Voltage and Current Division (Sect. 2–5)
(a) Given a linear resistance circuit with elements connected in series or in parallel, use voltage or current division to find specified voltages or currents.
(b) Design a voltage or current divider that delivers specified output signals.

2-6 Circuit Reduction (Sect. 2–6)
Given a linear resistance circuit, find selected signal variables using successive application of series and parallel equivalence, source transformations, and voltage and current division.

2-7 Computer-Aided Circuit Analysis (Sect. 2–7)
Given an appropriate linear circuit, use circuit simulation and/or computational software to solve for the desired response.

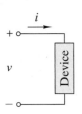

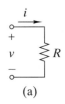

FIGURE 2–1 *Voltage and current reference marks for a two-terminal device.*

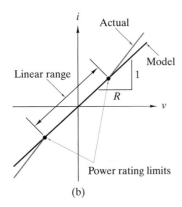

(a)

(b)

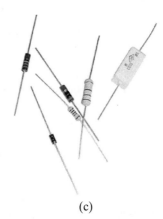

(c)

FIGURE 2–2 *The resistor: (a) Circuit symbol. (b) i–v characteristics. (c) Carbon or film resistors. (d) Wire-wound resistors.*

2–1 ELEMENT CONSTRAINTS

A **circuit** is a collection of interconnected electrical devices. An electrical **device** is a component that is treated as a separate entity. The rectangular box in Figure 2–1 is used to represent any one of the two-terminal devices used to form circuits. A two-terminal device is described by its *i–v* **characteristic**, that is, by the relationship between the voltage across and current through the device. In most cases the relationship is complicated and nonlinear, so we use a linear model that approximates the dominant features of a device.

To distinguish between a device (the real thing) and its model (an approximate stand-in), we call the model a circuit **element**. Thus, a device is an article of hardware described in manufacturers' catalogs and parts specifications. An element is a model described in textbooks on circuit analysis. This book is no exception, and a catalog of circuit elements will be introduced as we go on.

What would a non-linear resistor be like?

THE LINEAR RESISTOR

The first element in our catalog is a linear model of the device described in Figure 2–2. The actual *i–v* characteristic of this device is shown in Figure 2–2(b). To model this curve accurately across the full operating range shown in the figure would require at least a cubic equation. However, the graph in Figure 2–2(b) shows that a straight line is a good approximation to the *i–v* characteristic if we operate the device within its linear range. The power rating of the device limits the range over which the *i–v* characteristics can be represented by a straight line through the origin.

For the passive sign convention used in Figure 2–2(a), the equations describing the *linear resistor* element are

$$v = Ri \quad \text{or} \quad i = Gv \tag{2–1}$$

where R and G are positive constants that are reciprocally related:

$$G = \frac{1}{R} \tag{2–2}$$

The relationships in Eq. (2–1) are collectively known as **Ohm's law**. The parameter R is called **resistance** and has the unit **ohms**, Ω. The parameter G is called **conductance**, with the unit **siemens**, S. In earlier times the unit of conductance was cleverly called the mho, with the unit abbreviation symbol $\mho$ ("ohm" spelled backward and the ohm symbol upside down). Note that Ohm's law presumes that the passive sign convention is used to assign the reference marks to voltage and current.

The Ohm's law model is represented graphically by the black straight line in Figure 2–2(b). The *i–v* characteristic for the Ohm's law model defines a circuit element that is said to be linear and bilateral. **Linear** means that the defining characteristic is a straight line through the origin. Elements whose characteristics do not pass through the origin or are not a straight line are said to be **nonlinear**. **Bilateral** means that the *i–v* characteristic curve has odd symmetry about the origin.[1] With a bilateral resistor, reversing the polarity of the applied voltage reverses the direction but not the magnitude of the current, and vice versa. The net result is that we can connect a bilateral resistor into a circuit without regard to which terminal is which. This is important because devices such as diodes and batteries are not bilateral, and we must carefully identify each terminal.

Figures 2–2(c) and (d) show photos of discrete resistor devices.

---

[1]A curve $i = f(v)$ has odd symmetry if $f(-v) = -f(v)$.

The power associated with the resistor can be found from $p = vi$. Using Eq. (2–1) to eliminate v from this relationship yields

$$p = i^2 R \qquad (2\text{–}3)$$

or using the same equations to eliminate i yields

$$p = v^2 G = \frac{v^2}{R} \qquad (2\text{–}4)$$

Since the parameter R is positive, these equations tell us that the power is always nonnegative. Under the passive sign convention, this means that the resistor **always absorbs or consumes power**.

(d)

FIGURE 2–2 *(Continued)*

EXAMPLE 2 – 1

A 2.2-kΩ resistor has 12 V impressed across its terminals. Find the current through the resistor and the power it dissipates.

SOLUTION:
By Ohm's law $i = v/R$. Therefore, $i = 12/2200 = 0.00545 = 5.45$ mA. The power dissipated is found using Eq. (2–3) $p = i^2 R = (0.00545)^2 \, 2200 = 65.4$ mW which is the same answer we get if we were to use Eq. (2–4) $p = v^2/R = 12^2/2200 = 65.4$ mW or Eq. (1–5) $p = iv = 0.00545 \times 12 = 65.4$ mW. ∎

Exercise 2–1

A 6-V lantern battery powers a light bulb that draws 3 mA of current. What is the resistance of the lamp? How much power does the lantern use?

Answers: $R = 2$ kΩ; $p = 18$ mW

EXAMPLE 2 – 2

A resistor operates as a linear element as long as the voltage and current are within the limits defined by its power rating. Suppose we have a 47 kΩ resistor with a power rating of 0.25 W. Determine the maximum current and voltage that can be applied to the resistor and have it remain within its linear operating range.

SOLUTION:
Using Eq. (2–3) to relate power and current, we obtain

$$I_{\text{MAX}} = \sqrt{\frac{P_{\text{MAX}}}{R}} = \sqrt{\frac{0.25}{47 \times 10^3}} = 2.31 \text{ mA}$$

Similarly, using Eq. (2–4) to relate power and voltage, we obtain

$$V_{\text{MAX}} = \sqrt{R P_{\text{MAX}}} = \sqrt{47 \times 10^3 \times 0.25} = 108 \text{ V} \qquad ∎$$

Exercise 2–2

What is the maximum current that can flow through a $\frac{1}{8}$-W, 6.8-kΩ resistor? What is the maximum voltage that can be across it?

Answers: $I_{\text{MAX}} = 4.287$ mA; $V_{\text{MAX}} = 29.15$ V

OPEN AND SHORT CIRCUITS *what is an open circuit?*

The next two circuit elements can be thought of as limiting cases of the linear resistor. Consider a resistor R with a voltage v applied across it. Let's calculate the

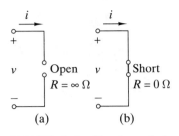

FIGURE 2–3 *Circuit symbols:*
(a) Open-circuit symbol.
(b) Short-circuit symbol.

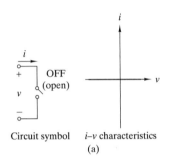

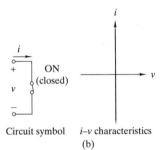

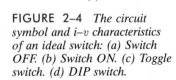

FIGURE 2–4 *The circuit
symbol and i–v characteristics
of an ideal switch: (a) Switch
OFF. (b) Switch ON. (c) Toggle
switch. (d) DIP switch.*

current i through the resistor for different values of resistance. If $v = 10$ V and $R = 1\,\Omega$, using Ohm's law we readily find that $i = 10$ A. If we increase the resistance to $100\,\Omega$, we find i has decreased to 0.1 A or 100 mA. If we continue to increase R to $1\,M\Omega$, i becomes a very small $10\,\mu$A. Continuing this process, we arrive at a condition where R is very nearly infinite and i just about zero. When the current $i = 0$, we call the special value of resistance (i.e., $R = \infty\Omega$) an **open circuit**. Similarly, if we reduce R until it approaches zero, we find that the voltage is very nearly zero. When $v = 0$, we call the special value of resistance (i.e., $R = 0\,\Omega$) a **short circuit**. The circuit symbols for these two elements are shown in Figure 2–3. In circuit analysis the elements in a circuit model are assumed to be interconnected by zero-resistance wire (that is, by short circuits).

THE IDEAL SWITCH how is it different from a real switch?

A switch is a familiar device with many applications in electrical engineering. The **ideal switch** can be modeled as a combination open- and short-circuit element. Figure 2–4(a) and (b) show the circuit symbol and the $i–v$ characteristic of an ideal switch. When the switch is closed,

$$v = 0 \quad \text{and} \quad i = \text{any value} \qquad (2\text{--}5a)$$

and when it is open,

$$i = 0 \quad \text{and} \quad v = \text{any value} \qquad (2\text{--}5b)$$

When the switch is closed, the voltage across the element is zero and the element will pass any current that may result. When open, the current is zero and the element will withstand any voltage across its terminals. The power is always zero for the ideal switch, since the product $vi = 0$ when the switch is either open ($i = 0$) or closed ($v = 0$). Actual switch devices have limitations, such as the maximum current they can safely carry when closed and the maximum voltage they can withstand when open. The switch is operated (opened or closed) by some external influence, such as a mechanical motion, temperature, pressure, or an electrical signal.

Figure 2–4(c) shows an actual toggle switch, which is simply used to turn things ON and OFF, and Figure 2–4(d) shows a DIP (dual in-line package) switch designed for use on a printed circuit board.

The ideal switch is also a basic concept in digital circuits, where OFF usually represents a logic 0 state and ON represents a logic 1 state.

APPLICATION   EXAMPLE 2–3

The **analog switch** is an important device found in analog-to-digital interfaces. Figures 2–5(a) and (b) show the two basic versions of the device. In either type the switch is actuated by applying a voltage to the terminal labeled *gate*. The switch in Figure 2–5(a) is said to be *normally open* because it is open when no voltage is applied to the gate terminal and closes when voltage is applied. The switch in Figure 2–5(b) is said to be *normally closed* because it is closed when no voltage is applied to the controlling gate and opens when voltage is applied.

Figure 2–5(c) shows an application in which complementary analog switches are controlled by the same gate. When gate voltage is applied, the upper switch closes and the lower opens so that point A is connected to point C. Conversely, when no gate voltage is applied, the upper switch opens and the lower switch closes to connect point B to point C. In the analog world this arrangement is called a double throw switch since point C can be connected to two other points. In the digital world it is called a two-to-one multiplexer (or MUX) because it allows you to select the analog input at point A or point B under control of the digital signal applied to the gate.

In many applications an analog switch can be treated as an ideal switch. In other cases, it may be necessary to account for its nonideal characteristics. When the switch is open an analog switch acts like a very large resistance (R_{OFF}), as suggested in Figure 2–5(d). This resistance is negligible because it ranges from perhaps 10^9 to 10^{11} Ω. When the switch is closed it acts like a small resistor (R_{ON}), as suggested in Figure 2–5(e). Depending on other circuit resistances, it may be necessary to account for R_{ON}, because it ranges from perhaps a few mΩ to as high as 100 Ω.

This example illustrates how ideal switches and resistors can be combined to model another electrical device. It also suggests that no single model can serve in all applications. It is up to the engineer to select a model that adequately represents the actual device in each application.

IDEAL SOURCES *what makes them ideal?*

The signal and power sources required for the operation of electronic circuits are modeled using two elements: **voltage sources** and **current sources**. These sources can produce either constant or time-varying signals. The circuit symbols and the i–v characteristic of an ideal voltage source are shown in Figure 2–6, while the circuit symbol and i–v characteristic of an ideal current source are shown in Figure 2–7. The symbol in Figure 2–6(a) represents either a time-varying or constant voltage source. The battery symbol in Figure 2–6(b) is used exclusively for a constant voltage source. There is no separate symbol for a constant current source.

The i–v characteristic of an **ideal voltage source** in Figure 2–6(c) is described by the following element equations:

$$v = v_S \quad \text{and} \quad i = \text{any value} \tag{2–6}$$

The element equations mean that the ideal voltage source produces v_S volts across its terminals and will supply whatever current may be required by the circuit to which it is connected.

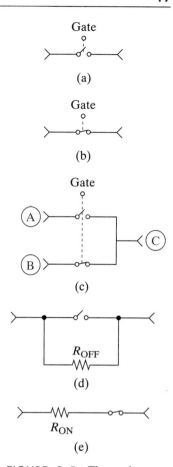

FIGURE 2–5 *The analog switch: (a) Normally open model. (b) Normally closed model. (c) Double throw model. (d) Model with finite OFF resistance. (e) Model with finite ON resistance.*

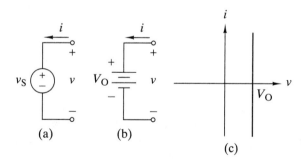

FIGURE 2–6 *Circuit symbols and i–v characteristic of an ideal independent voltage source: (a) Time-varying. (b) Constant (Battery). (c) Constant source i–v characteristics.*

The i–v characteristic of an **ideal current source** in Figure 2–7(b) is described by the following element equations:

$$i = i_S \quad \text{and} \quad v = \text{any value} \tag{2–7}$$

The ideal current source produces i_S amperes in the direction of its arrow symbol and will furnish whatever voltage is required by the circuit to which it is connected.

(a)

(b)

FIGURE 2–7 *Circuit symbol and i–v characteristic of an ideal independent current source: (a) Time-varying or constant source. (b) Constant source i–v characteristics.*

The voltage or current produced by these ideal sources is called a **forcing function** or a **driving function** because it represents an input that causes a circuit response. When the voltage or current varies with time, it is customary to write $v(t)$ or $i(t)$.

EXAMPLE 2 – 4

Given an ideal voltage source with the time-varying voltage shown in Figure 2–8(a), sketch its $i–v$ characteristic at the times $t = 0$, 1, and 2 ms.

FIGURE 2–8

w+h is this figure?

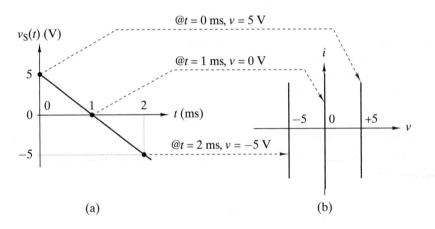

(a)   (b)

SOLUTION:

At any instant in time, the time-varying source voltage has only one value. We can treat the voltage and current at each instant of time as constants representing a snapshot of the source $i–v$ characteristic. For example, at $t = 0$, the equations defining the $i–v$ characteristic are $v_S = 5$ V and $i =$ any value. Figure 2–8(b) shows the $i–v$ relationship at the other instants of time. Curiously, the voltage source $i–v$ characteristic at $t = 1$ ms ($v_S = 0$ and $i =$ any value) is the same as that of a short circuit [see Eq. (2–5a) or Figure 2–3(b)]. ■

Exercise 2–3 _____

A digital clock is a voltage that switches between two values at a constant rate that is used to time digital circuits. A particular clock switches between 0 V and 5 V every 10 μs. Sketch the clock's i-v characteristics for the times when the clock is at 0 V and at 5 V.

Answers: On a standard i-v graph a vertical line through the origin for the times the clock is at 0 V and a vertical line crossing at 5 V when the clock is at 5 V.

PRACTICAL SOURCES

The practical models for voltage and current sources in Figure 2–9 may be more appropriate in some situations than the ideal models used up to this point. These circuits are called practical models because they more accurately represent the characteristics of real-world sources than do the ideal models. It is important to remember that models are interconnections of elements, not devices. For example, the resistance in a model does not always represent an actual resistor. As a case in point, the resistances R_S in the practical source models in Figure 2–9 do not represent physical resistors but are circuit elements used to account for resistive effects within the devices being modeled.

The linear resistor, open circuit, short circuit, ideal switch, ideal voltage source, and ideal current source are the initial entries in our catalog of circuit elements. In Chapter 4 we will develop models for active devices such as the transistor and OP AMP. Models for dynamic elements such as capacitors and inductors are introduced in Chapter 6.

2–2 CONNECTION CONSTRAINTS

In the previous section, we considered individual devices and models. In this section, we turn our attention to the constraints introduced by interconnections of devices to form circuits. The laws governing circuit behavior are based on the meticulous work of the German scientist Gustav Kirchhoff (1824–1887). **Kirchhoff's laws** are derived from conservation laws as applied to circuits. They tell us that element voltages and currents are forced to behave in certain ways when the devices are interconnected to form a circuit. These conditions are called **connection constraints** because they are based only on the circuit connections, not on the specific devices in the circuit.

In this book, we will indicate that crossing wires are connected (electrically tied together) using the dot symbol, as in Figure 2–10(a). Sometimes crossing wires are not connected (electrically insulated) but pass over or under each other. Since we are restricted to drawing wires on a planar surface, we will indicate unconnected crossovers by *not* placing a dot at their intersection, as indicated in the left of Figure 2–10(b). Other books sometimes show unconnected crossovers using the semicircular "hopover" shown on the right of

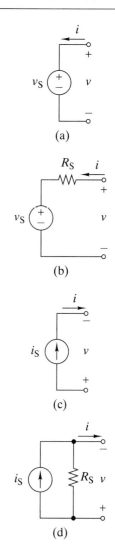

FIGURE 2–9 *Circuit symbols for ideal and practical independent sources: (a) Ideal voltage source. (b) Practical voltage source. (c) Ideal current source. (d) Practical current source.*

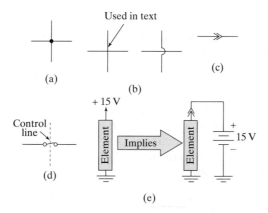

FIGURE 2–10 *Symbols used in circuit diagrams: (a) Electrical connection. (b) Crossover with no connection. (c) Jack connection. (d) Control line. (e) Power supply connection.*

Figure 2–10(b). In engineering systems two or more separate circuits are often tied together to form a larger circuit (for example, the interconnection of two integrated circuit packages). Interconnecting different circuits forms an interface between the circuits. The special jack or interface symbol in Figure 2–10(c) is used in this book because interface connections represent important points at which the interaction between two circuits can be observed or specified. On certain occasions a control line is required to show a mechanical or other nonelectrical dependency. Figure 2–10(d) shows how this dependency is indicated in this book. Figure 2–10(e) shows how power supply connections are often shown in electronic circuit diagrams. The implied power supply connection is indicated by an arrow pointing to the supply voltage, which may be given in numerical $(+15\,\text{V})$ or symbolic form $(+V_{\text{CC}})$.

The treatment of Kirchhoff's laws uses the following definitions:

- A **circuit** is an interconnection of electrical devices.
- A **node** is an electrical juncture of two or more devices.
- A **loop** is a closed path formed by tracing through an ordered sequence of nodes without passing through any node more than once.

While it is customary to designate the juncture of two or more elements as a node, it is important to realize that a node is not confined to a point but includes all the zero-resistance wire from the point to each element. In the circuit of Figure 2–11, there are only three different nodes: A, B, and C. The points 2, 3, and 4, for example, are part of node B, while the points 5, 6, 7, and 8 are all part of node C.

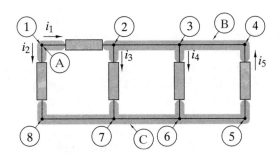

FIGURE 2–11 *Circuit for demonstrating Kirchhoff's current law.*

KIRCHHOFF'S CURRENT LAW

Kirchhoff's first law is based on the principle of conservation of charge. **Kirchhoff's current law (KCL)** states that

the algebraic sum of the currents entering a node is zero at every instant.

In forming the algebraic sum of currents, we must take into account the current reference direction associated with each device. If the current reference direction is into the node, then we assign a positive sign to the corresponding current in the algebraic sum. If the reference direction is away from the node, we assign a negative sign. Applying this convention to the nodes in Figure 2–11, we obtain the following set of KCL connection equations:

$$\text{Node A: } -i_1 - i_2 \qquad\qquad = 0$$
$$\text{Node B: } i_1 - i_3 - i_4 + i_5 = 0 \qquad\qquad (2\text{--}8)$$
$$\text{Node C: } i_2 + i_3 + i_4 - i_5 = 0$$

How could using this law have helped when 1st learning electronics?

The KCL equation at node A does not mean that the currents i_1 and i_2 are both negative. The minus signs in this equation simply mean that the reference direction for each current is directed away from node A. Likewise, the equation at node B could be written as

$$i_3 + i_4 = i_1 + i_5 \qquad (2-9)$$

This form illustrates an alternative statement of KCL:

> *The sum of the currents entering a node equals the sum of the currents leaving the node.*

There are two algebraic signs associated with each current in the application of KCL. First is the sign given to a current in writing a KCL connection equation. This sign is determined by the orientation of the current reference direction relative to a node. The second sign is determined by the actual direction of the current relative to the reference direction. The actual direction is found by solving the set of KCL equations, as illustrated in the following example.

EXAMPLE 2–5

Given $i_1 = +4\,\text{A}$, $i_3 = +1\,\text{A}$, $i_4 = +2\,\text{A}$ in the circuit shown in Figure 2–11, find i_2 and i_5.

SOLUTION:
Using the node A constraint in Eq. (2–8) yields

$$-i_1 - i_2 = -(+4) - i_2 = 0$$

The sign outside the parentheses comes from the node A KCL connection constraint in Eq. (2–8). The sign inside the parentheses comes from the actual direction of the current. Solving this equation for the unknown current, we find that $i_2 = -4\,\text{A}$. In this case, the minus sign indicates that the actual direction of the current i_2 is directed upward in Figure 2–11, which is opposite to the reference direction assigned. Using the second KCL equation in Eq. (2–8), we can write

$$i_1 - i_3 - i_4 + i_5 = (+4) - (+1) - (+2) + i_5 = 0$$

which yields the result $i_5 = -1\,\text{A}$.

Again, the signs inside the parentheses are associated with the actual direction of the current, and the signs outside come from the node B KCL connection constraint in Eq. (2–8). The minus sign in the final answer means that the current i_5 is directed in the opposite direction from its assigned reference direction. We can check our work by substituting the values found into the node C constraint in Eq. (2–8). These substitutions yield

$$+i_2 + i_3 + i_4 - i_5 = (-4) + (+1) + (+2) - (-1) = 0$$

as required by KCL. Given three currents, we determined all the remaining currents in the circuit using only KCL without knowing the element constraints. ∎

In Example 2–5, the unknown currents were found using only the KCL constraints at nodes A and B. The node C equation was shown to be valid, but it did not add any new information. If we look back at Eq. (2–8), we see that the node C equation is the negative of the sum of the node A and B equations. In other words, the KCL connection constraint at node C is not independent of the two previous equations. This example illustrates the following general principle:

> *In a circuit containing a total of N nodes there are only $N-1$ independent KCL connection equations.*

Current equations written at $N - 1$ nodes contain all the independent connection constraints that can be derived from KCL. To write these equations, we select one node as the reference or ground node and then write KCL equations at the remaining $N - 1$ nonreference nodes.

Exercise 2–4

Refer to Figure 2–12.
(a) Write KCL equations at nodes A, B, C, and D.
(b) Given $i_1 = -1\,\text{mA}$, $i_3 = 0.5\,\text{mA}$, $i_6 = 0.2\,\text{mA}$, find i_2, i_4, and i_5.

FIGURE 2–12

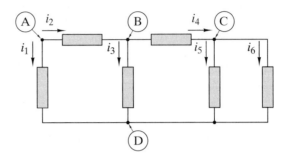

Answers:
(a) Node A: $-i_1 - i_2 = 0$; node B: $i_2 - i_3 - i_4 = 0$; node C: $i_4 - i_5 - i_6 = 0$; node D: $i_1 + i_3 + i_5 + i_6 = 0$
(b) $i_2 = 1\,\text{mA}$; $i_4 = 0.5\,\text{mA}$; $i_5 = 0.3\,\text{mA}$

KIRCHHOFF'S VOLTAGE LAW

The second of Kirchhoff's circuit laws is based on the principle of conservation of energy. **Kirchhoff's voltage law (KVL)** states that

the algebraic sum of all the voltages around a loop is zero at every instant.

For example, three loops are shown in the circuit of Figure 2–13. In writing the algebraic sum of voltages, we must account for the assigned reference marks. As a loop is traversed, a positive sign is assigned to a voltage when we go from a " + " to " – " reference mark. When we go from " – " to " + ", we use a minus sign. Traversing the three loops in Figure 2–13 in the indicated clockwise direction yields the following set of KVL connection equations:

$$\begin{aligned}
\text{Loop 1:} \quad & -v_1 + v_2 + v_3 && = 0 \\
\text{Loop 2:} \quad & -v_3 + v_4 + v_5 && = 0 \\
\text{Loop 3:} \quad & -v_1 + v_2 + v_4 + v_5 && = 0
\end{aligned}$$

(2–10)

There are two signs associated with each voltage. The first is the sign given the voltage when writing the KVL connection equation. The second is the sign

where have you seen systems with these kinds of laws before?

FIGURE 2–13 *Circuit for demonstrating Kirchhoff's voltage law.*

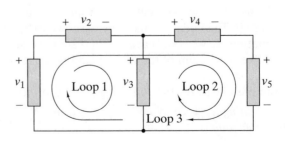

determined by the actual polarity of a voltage relative to its assigned reference polarity. The actual polarities are found by solving the set of KVL equations, as illustrated in the following example.

EXAMPLE 2–6

Given $v_1 = 5\,\text{V}$, $v_2 = -3\,\text{V}$, and $v_4 = 10\,\text{V}$ in the circuit shown in Figure 2–13, find v_3 and v_5.

SOLUTION:
Inserting the given numerical values into Eq. (2–10) yields the following KVL equation for loop 1:

$$-v_1 + v_2 + v_3 = -(+5) + (-3) + (v_3) = 0$$

The sign outside the parentheses comes from the loop 1 KVL constraint in Eq. (2–10). The sign inside comes from the actual polarity of the voltage. This equation yields $v_3 = +8\,\text{V}$. Using this value in the loop 2 KVL constraint in Eq. (2–10) produces

$$-v_3 + v_4 + v_5 = -(+8) + (+10) + v_5 = 0$$

The result is $v_5 = -2\,\text{V}$. The minus sign here means that the actual polarity of v_5 is the opposite of the assigned reference polarity indicated in Figure 2–13. The results can be checked by substituting all the aforementioned values into the loop 3 KVL constraint in Eq. (2–10). These substitutions yield

$$-(+5) + (-3) + (+10) + (-2) = 0$$

as required by KVL.   ■

In Example 2–6, the unknown voltages were found using only the KVL constraints for loops 1 and 2. The loop 3 equation was shown to be valid, but it did not add any new information. If we look back at Eq. (2–10), we see that the loop 3 equation is equal to the sum of the loop 1 and 2 equations. In other words, the KVL connection constraint around loop 3 is not independent of the previous two equations. This example illustrates the following general principle:

> *In a circuit containing a total of E two-terminal elements and N nodes, there are only $E - N + 1$ independent KVL connection equations.*

Writing voltage summations around a total of $E - N + 1$ *different* loops produces all the independent connection constraints that can be derived from KVL. A *sufficient condition* for loops to be different is that each contains at least one element that is not contained in any other loop. In simple circuits with no crossovers, the open space between elements or "window panes" produces $E - N + 1$ independent loops. However, finding all the loops in a more complicated circuit can be a nontrivial problem.

DISCUSSION: *For planar circuits, the kind encountered in this text, the number of independent loops is simply the number of "window panes." Hence, in Figure 2–13 there are three loops but only two window panes. The same is true for Figures 2–14 and 2–15. Figure 2–18(c) has three window panes but seven loops. The number of window panes simply identifies the number of independent KVL equations one can write. One can use any of the different loops as long as one does not exceed the number of independent equations. Choosing the right loops can at times simplify the solution to the problem.*

Exercise 2–5

Find the voltages v_x and v_y in Figure 2–14.

FIGURE 2–14

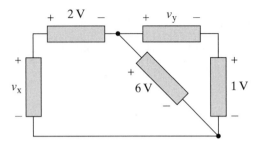

Answers: $v_x = +8\,\text{V}$; $v_y = +5\,\text{V}$

Exercise 2–6

Find the voltages v_x, v_y, and v_z in Figure 2–15.

FIGURE 2–15

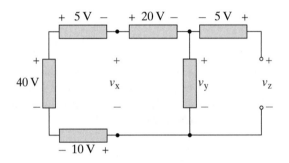

Answers: $v_x = +25\,\text{V}$; $v_y = +5\,\text{V}$; $v_z = +10\,\text{V}$. *Note:* KVL yields the voltage v_z even though it appears across an open circuit.

Is there any new info here?

PARALLEL AND SERIES CONNECTIONS

Two types of connections occur so frequently in circuit analysis that they deserve special attention. Elements are said to be connected in **parallel** when they form a loop containing no other elements. For example, loop A in Figure 2–16 contains only elements 1 and 2. As a result, the KVL connection constraint around loop A is

$$-v_1 + v_2 = 0 \tag{2–11}$$

which yields $v_1 = v_2$. In other words, in a parallel connection KVL requires equal voltages across the elements. Loop B in Figure 2–16 contains only two elements and although visibly they do not look as if they are in parallel, a KVL analysis quickly shows that $v_2 = v_3$ and, therefore, elements 2 and 3 are in parallel. The parallel connection is not restricted to two elements. As a result, in this circuit we have $v_1 = v_2 = v_3$, and we say that elements 1, 2, and 3 are connected in parallel. In general, then, any number of elements connected between two common nodes are in parallel, and as a result, the same voltage appears across each of them. Existence of a

FIGURE 2–16 *A parallel connection.*

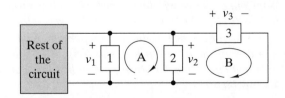

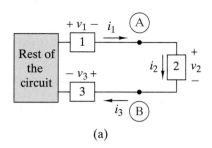

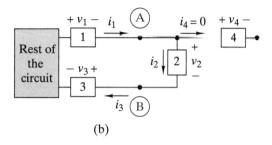

FIGURE 2–17 *A series connection.*

parallel connection does not depend on the graphical position of the elements. For example, the position of elements 1 and 3 could be switched, and the three elements are still connected in parallel.

Two elements are said to be connected in **series** when they have one common node to which no other element with current flowing through it is connected. In Figure 2–17(a) elements 1 and 2 are connected in series, since only these two elements are connected at node A. Applying KCL at node A yields

$$i_1 - i_2 = 0 \quad \text{or} \quad i_1 = i_2 \qquad (2\text{–}12)$$

In a series connection, KCL requires equal current through each element. Any number of elements can be connected in series. For example, element 3 in Figure 2–17(a) is connected in series with element 2 at node B, and KCL requires $i_2 = i_3$. Therefore, in this circuit $i_1 = i_2 = i_3$, we say that elements 1, 2, and 3 are connected in series, and the same current exists in each of the elements. In general, elements are connected in series when they form a single path between two nodes such that only elements in the path are connected to the intermediate nodes along the path.

Figure 2–17(b) shows a common connection variation. Element 4 is connected at node A between elements 1 and 2. In general, elements 1 and 2 would not be in series since more than two elements connect at the same node. However, in this case we see that the current through element 4 is zero. KCL tells us that all of the current that flows through element 1 must flow through element 2. In this case, elements 1 and 2 behave as if they were in series.

EXAMPLE 2–7

Identify the elements connected in parallel and in series in each of the circuits in Figure 2–18.

SOLUTION:

In Figure 2–18(a) elements 1 and 2 are connected in series at node A and elements 3 and 4 are connected in parallel between nodes B and C. In Figure 2–18(b) elements 1 and 2 are connected in series at node A, as are elements 4 and 5 at node D. There are no single elements connected in parallel in this circuit. In Figure 2–18(c) there are no elements connected in either series or parallel. It is important to realize that in some circuits there are elements that are not connected in either series or in parallel. ■

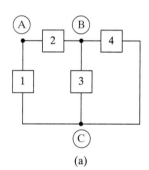

(a)

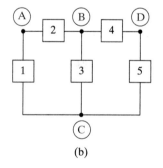

(b)

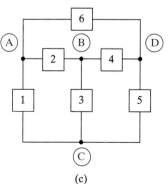

(c)

FIGURE 2–18

Exercise 2–7

Identify the elements connected in series or parallel when a short circuit is connected between nodes A and B in each of the circuits of Figure 2–18.

Answers:

Circuit in Figure 2–18(a): Elements 1, 3, and 4 are all in parallel.
Circuit in Figure 2–18(b): Elements 1 and 3 are in parallel; elements 4 and 5 are in series.
Circuit in Figure 2–18(c): Elements 1 and 3 are in parallel; elements 4 and 6 are in parallel.

Note that in all three circuits the short circuit is connected across element 2 causing the voltages at nodes A and B to be the same. Since the voltage across an element is zero there can be no power produced or absorbed by that element. One can say that element 2 has been "shorted out" and effectively removes element 2 from the circuit.

Exercise 2–8

Identify the elements in Figure 2–19 that are connected in (a) parallel, (b) series, or (c) neither.

FIGURE 2–19

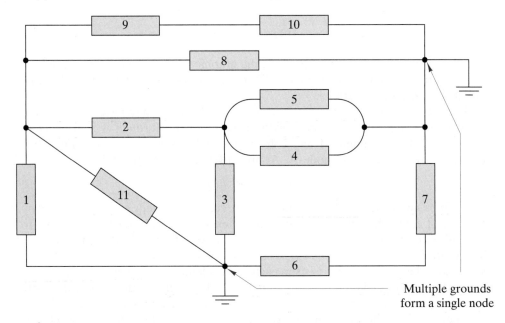

Multiple grounds form a single node

Answers:

(a) The following elements are in parallel: 1, 8, and 11; 3, 4, and 5.
(b) The following elements are in series: 9 and 10; 6 and 7.
(c) Only element 2 is not in series or parallel with any other element.

DISCUSSION: *The ground symbol indicates the reference node. When ground symbols are shown at several nodes, the nodes are effectively connected together by a short circuit to form a single node.*

2–3 COMBINED CONSTRAINTS

The usual goal of circuit analysis is to determine the currents or voltages at various places in a circuit. This analysis is based on constraints of two distinctly different types. The element constraints are based on the models of the specific devices connected in the circuit. The connection constraints are based on Kirchhoff's laws and the circuit connections. The element equations are independent of the circuit connections. Likewise, the connection equations are independent of the devices in the circuit. Taken together, however, the combination of the element and connection constraints supply the equations needed to describe a circuit.

What types of new problems can be solved w/ combined constraints?

Our study of the combined constraints begins by considering the simple but important example in Figure 2–20(a). This circuit is driven by a current source i_S and the resulting responses are current/voltage pairs (i_x, v_x) and (i_O, v_O). The reference marks for the response pairs have been assigned using the passive sign convention.

To solve for all four responses, we must write four equations. The first two are the element equations

$$i_x = i_S$$
$$v_O = Ri_O \tag{2–13}$$

The first element equation states that the response current i_x and the input driving force i_S are equal in magnitude and direction. The second element equation is Ohm's law relating v_O and i_O under the passive sign convention.

The connection equations are obtained by applying Kirchhoff's laws. The circuit in Figure 2–20 has two elements $(E = 2)$ and two nodes $(N = 2)$, so we need $E - N + 1 = 1$ KVL equation and $N - 1 = 1$ KCL equation. Selecting node B as the reference node, we apply KCL at node A and apply KVL around the loop to write

$$\text{KCL:} \quad -i_x - i_O = 0$$
$$\text{KVL:} \quad -v_x + v_O = 0 \tag{2–14}$$

We now have two element constraints in Eq. (2–13) and two connection constraints in Eq. (2–14), so we can solve for all four responses in terms of the input driving force i_S. Combining the KCL connection equation and the first element equation yields $i_O = -i_x = -i_S$. Substituting this result into the second element equation (Ohm's law) produces

$$v_O = -Ri_S \tag{2–15}$$

The minus sign in this equation does not mean that v_O is always negative. Nor does it mean the resistance is negative; it cannot be. It means that when the input driving force i_S is positive, then the response v_O is negative, and vice versa. This sign reversal is a result of the way we assigned reference marks at the beginning of our analysis. The reference marks defined the circuit input and outputs in such a way that i_S and v_O always have opposite algebraic signs. Put differently, Eq. (2–15) is an input-output relationship, not an element i–v relationship.

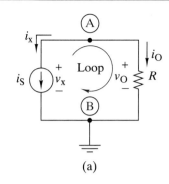

FIGURE 2–20 (a) Circuit used to demonstrate combined constraints.

(a)

EXAMPLE 2 – 8

(a) Find the responses i_x, v_x, i_O, and v_O in the circuit in Figure 2–20(a) when $i_S = +2\,\text{mA}$ and $R = 2\,\text{k}\Omega$.
(b) Repeat for $i_S = -2\,\text{mA}$.

SOLUTION:

(a) From Eq. (2–13) we have $i_x = i_S = +2\,\text{mA}$ and $v_O = 2000i_O$. From Eq. (2–14) we have $i_O = -i_x = -2\,\text{mA}$ and $v_x = v_O$. Combining these results, we obtain

$$v_x = v_O = 2000i_O = 2000(-0.002) = -4\,\text{V}$$

(b) In this case $i_x = i_S = -2\,\text{mA}$, $i_O = -i_x = -(-0.002) = +2\,\text{mA}$, and

$$v_x = v_O = 2000i_O = 2000(+0.002) = +4\,\text{V}$$

This example confirms that the algebraic signs of the outputs v_x, v_O, and i_O are always the opposite from that of the input driving force i_S. ∎

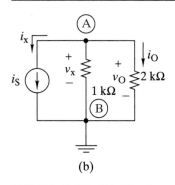

FIGURE 2–20 *(b) Circuit used to demonstrate combined constraints.*

Exercise 2–9

A 1-kΩ resistor is connected between nodes A and B in Figure 2–20(a) as shown in Figure 2–20(b). Find i_x, v_x, i_O and v_O.

Answers: $i_x = 1\,\text{mA}$; $i_O = -333\,\mu\text{A}$; $v_O = v_x = -667\,\text{mV}$

Analyzing the circuit in Figure 2–21 illustrates the formulation of combined constraints. We first assign reference marks for all the voltages and currents using the passive sign convention. Then, using these definitions we can write the element constraints as

$$v_A = V_O$$
$$v_1 = R_1 i_1 \qquad (2\text{–}16)$$
$$v_2 = R_2 i_2$$

These equations describe the three devices and do not depend on how the devices are connected in the circuit.

The connection equations are obtained from Kirchhoff's laws. To apply these laws, we must first label the different loops and nodes. The circuit contains $E = 3$ elements and $N = 3$ nodes, so there are $E - N + 1 = 1$ independent KVL constraints and $N - 1 = 2$ independent KCL constraints. There is only one loop, but there are three nodes in this circuit. We will select one node as the reference point and write KCL equations at the other two nodes. Any node can be chosen as the reference, so we select node C as the reference node and indicate this choice by drawing the ground symbol there. The connection constraints are

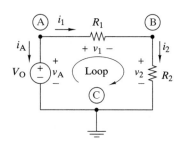

FIGURE 2–21 *Circuit used to demonstrate combined constraints.*

$$
\begin{aligned}
\text{KCL: Node A} \qquad & -i_A - i_1 = 0 \\
\text{KCL: Node B} \qquad & i_1 - i_2 = 0 \qquad (2\text{–}17) \\
\text{KVL: Loop} \; -v_A &+ v_1 + v_2 = 0
\end{aligned}
$$

These equations are independent of the specific devices in the circuit. They depend only on Kirchhoff's laws and the circuit connections.

This circuit has six unknowns: three element currents and three element voltages. Taken together, the element and connection equations give us six independent equations. For a network with (N) nodes and (E) two-terminal elements, we can write $(N - 1)$ independent KCL connection equations, $(E - N + 1)$ independent KVL connection equations, and (E) element equations. The total number of equations generated is

Element equations	E
KCL equations	$N - 1$
KVL equations	$E - N + 1$
Total	$2E$

The grand total is then $(2E)$ combined connection and element equations, which is exactly the number of equations needed to solve for the voltage across and current through every element—a total of $(2E)$ unknowns.

EXAMPLE 2–9

Find all of the element currents and voltages in Figure 2–21 for $V_O = 10\,\text{V}$, $R_1 = 2000\,\Omega$, and $R_2 = 3000\,\Omega$.

SOLUTION:
Substituting the element constraints from Eq. (2–16) into the KVL connection constraint in Eq. (2–17) produces

$$-V_O + R_1 i_1 + R_2 i_2 = 0$$

This equation can be used to solve for i_1 since the second KCL connection equation requires that $i_2 = i_1$.

$$i_1 = \frac{V_O}{R_1 + R_2} = \frac{10}{2000 + 3000} = 2\,\text{mA}$$

In effect, we have found all of the element currents since the elements are connected in series. Hence, collectively the KCL connection equations require that

$$-i_A = i_1 = i_2$$

Substituting all of the known values into the element equations gives

$$v_A = 10\,\text{V} \quad v_1 = R_1 i_1 = 4\,\text{V} \quad v_2 = R_2 i_2 = 6\,\text{V}$$

Every element voltage and current has been found. Note the analysis strategy used. We first found all the element currents and then used these values to find the element voltages.   ■

Exercise 2–10

The wire connecting R_1 to node B in Figure 2–21 is broken. What would you measure for i_A, v_1, i_2, and v_2? Is KVL violated? Where does the source voltage appear across?

Answers: $i_A = i_2 = 0\,\text{A}$, $v_1 = v_2 = 0\,\text{V}$. KVL is not violated. The voltage V_O appears across he open (broken) circuit.

EXAMPLE 2–10

Use element and connection equations to find the voltages across the resistors in Figure 2–22.

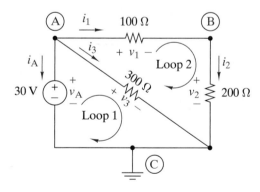

FIGURE 2–22

SOLUTION:
A complete description of this circuit involves four element equations and four connection equations. The element equations are

$$v_1 = 100 i_1$$
$$v_2 = 200 i_2$$
$$v_3 = 300 i_3$$
$$v_A = 30\,\text{V}$$

The four connection equations are

$$\begin{aligned}
\text{KCL: Node A} && -i_A - i_1 - i_3 &= 0 \\
\text{KCL: Node B} && i_1 - i_2 &= 0 \\
\text{KVL: Loop 1} && -v_A + v_3 &= 0 \\
\text{KVL: Loop 2} && -v_3 + v_1 + v_2 &= 0
\end{aligned}$$

Combining the last element equation and the KVL equation around loop 1 shows that

$$v_3 = v_A = 30\,\text{V}$$

which is nothing more than a statement that the voltage source and R_3 are connected in parallel. Using this result in the loop 2 equation yields $v_1 + v_2 = v_3 = 30\,\text{V}$. Substituting the first two element equations into this equation produces

$$100i_1 + 200i_2 = 30$$

But the KCL equation at node B points out that $i_1 = i_2$, and this result reduces to $300i_2 = 30$ or $i_1 = i_2 = 0.1\,\text{A}$. Finally, the first two element equations yield

$$v_1 = 100i_1 = 10\,\text{V and } v_2 = 200i_2 = 20\,\text{V}$$

In summary, the voltages across the three resistors are $v_1 = 10\,\text{V}$, $v_2 = 20\,\text{V}$, and $v_3 = 30\,\text{V}$. ∎

Exercise 2–11 _____

Repeat the problem of Example 2–10 if the 30–V voltage source is replaced with a 2-mA current source with the arrow pointing up toward node A.

Answers: $v_1 = 100$ mV, $v_2 = 200$ mV, $v_3 = 300$ mV

EXAMPLE 2−11

Use element and connection equations to find the voltages across and the currents through each of the elements in Figure 2–23.

FIGURE 2–23

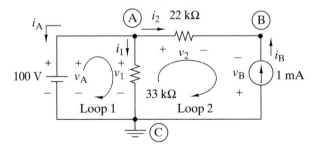

SOLUTION:
A complete description of this circuit involves $2E$ equations where $E = 4$ (two Rs and one each voltage and current sources). There are four element equations and four connection equations. The four element equations are

$$\begin{aligned}
v_A &= 100 \text{ V} \\
v_1 &= 33 \text{ k}\Omega\, i_1 \\
v_2 &= 22 \text{ k}\Omega\, i_2 \\
i_B &= 1 \text{ mA}
\end{aligned}$$

The four connection equations are

KCL: Node A	$-i_A - i_1 - i_2 = 0$
KCL: Node B	$i_2 + i_B = 0$
KVL: Loop 1	$-v_A + v_1 = 0$
KVL: Loop 2	$-v_1 + v_2 - v_B = 0$

Combining the equation for loop 1 and the v_A element constraint yields

$$v_A = v_1 = 100 \text{ V}$$

From the element equation for the 33-kΩ resistor we find

$$i_1 = \frac{100}{33\text{k}} = 3.03 \text{ mA}$$

Using the KCL equation at Node B we find

$$i_2 = -i_B = -1 \text{ mA}$$

Using the element constraint for the 22-kΩ resistor we find

$$v_2 = (-1 \text{ m})(22 \text{ k}) = -22 \text{ V}$$

Now from the loop 2 equation we can find the voltage across the current source

$$-100 + (-22) = v_B = -122 \text{ V}$$

And from the KCL equation for Node A we find the current through the voltage source

$$i_A = -3.03 \text{ m} - (-1 \text{ m}) = -2.03 \text{ mA}$$

In sum, the following values have been found

$$
\begin{aligned}
v_A &= 100 \text{ V} \\
i_A &= -2.03 \text{ mA} \\
v_1 &= 100 \text{ V} \\
i_1 &= 3.03 \text{ mA} \\
v_2 &= -22 \text{ V} \\
i_2 &= -1 \text{ mA} \\
v_B &= -122 \text{ V} \\
i_B &= 1 \text{ mA}
\end{aligned}
$$

∎

Exercise 2–12

In Figure 2–24 on the next page $i_1 = 200$ mA and $i_3 = -100$ mA. Find the voltage v_x.

Answer: $v_x = 35$ V

ASSIGNING REFERENCE MARKS

In all of our previous examples and exercises, the reference marks for the element currents (arrows) and voltages (+ and −) were given. When reference marks are not shown on a circuit diagram, they must be assigned by the person solving the problem.

How much will assigning these help me?

FIGURE 2–24

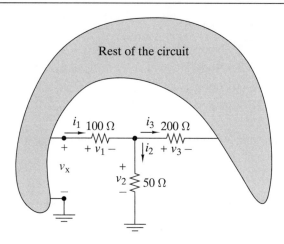

Beginners sometimes wonder how to assign reference marks when the actual voltage polarities and current directions are unknown. It is important to remember that the reference marks do not indicate what is actually happening in the circuit. They are benchmarks assigned at the beginning of the analysis. If it turns out that the actual direction and reference direction agree, then the algebraic sign of the response will be positive. If they disagree, the algebraic sign will be negative. In other words, the sign of the answer together with assigned reference marks tell us the actual voltage polarity or current direction.

In this book the reference marks always follow the passive sign convention. This means that for any given two-terminal element we can arbitrarily assign either the + voltage reference mark or the current reference arrow, but not both. For example, we can arbitrarily assign the voltage reference marks to the terminals of a two-terminal device. Once the voltage reference is assigned, however, the passive sign convention requires that the current reference arrow be directed into the element at the terminal with the + mark. On the other hand, we could start by arbitrarily selecting the terminal at which the current reference arrow is directed into the device. Once the current reference is assigned, however, the passive sign convention requires that the + voltage reference be assigned to the selected terminal.

Following the passive sign convention avoids confusion about the direction of power flow in a device. In addition, the element constraints, such as Ohm's law, assume that the passive sign convention is used to assign the voltage and current reference marks to a device.

The next example illustrates the assignment of reference marks.

EXAMPLE 2–12

Find the voltages across the resistors and current sources in Figure 2–25(a).

SOLUTION:

No voltage reference marks are given in Figure 2–25(a), so we assign those shown in Figure 2–25(b). Other choices are possible, of course, but once the voltage marks for v_1, v_2, and v_3 are assigned, the passive sign convention requires that the current reference directions for i_1, i_2, and i_3 be assigned as shown in Figure 2–25(c). KCL can be used to find the resistor currents directly. Using KCL at node A gives $2 - i_1 - 3 = 0$; hence $i_1 = -1$ A. KCL applied at node C yields $3 + i_3 - 5 = 0$; hence $i_3 = 2$ A. Finally, at node B KCL requires $i_1 - i_2 - i_3 = 0$; hence $i_2 = i_1 - i_3 = -1 - 2 = -3$ A. Given the three resistor

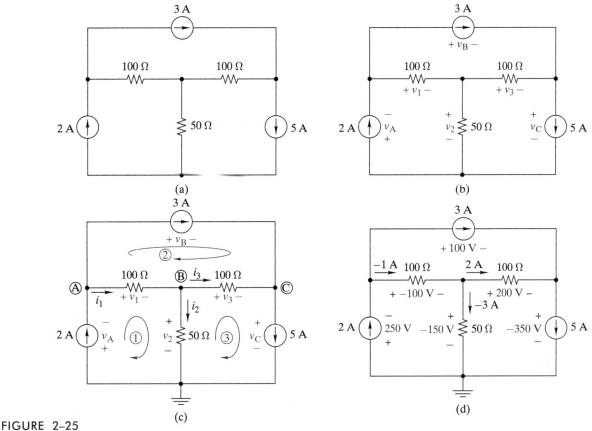

FIGURE 2–25

currents, we use Ohm's law to find the three resistor voltages.

$$v_1 = 100i_1 = -100\,\text{V}$$
$$v_2 = 50i_2 = -150\,\text{V}$$
$$v_3 = 100i_3 = +200\,\text{V}$$

The plus on the numerical value of v_3 means that the assigned reference marks agree with the actual voltage polarity. The minus sign on the numerical values of v_1 and v_2 mean that the assigned marks and physical reality disagree. This disagreement does not mean that the assigned marks for v_1 and v_2 are wrong. Reference marks are not predictions. They are definitions that allow us to correctly formulate circuit equations and interpret the numerical results of circuit analysis.

The voltages across the current sources can now be found by applying KVL around the three loops shown in Figure 2–25(c).

Loop 1 $v_A + v_1 + v_2 = 0$ or $v_A = -v_1 - v_2$
Loop 2 $v_B - v_1 - v_3 = 0$ or $v_B = v_1 + v_3$
Loop 3 $v_C - v_2 + v_3 = 0$ or $v_C = v_2 - v_3$

Using the resistor voltages found above we have

$$v_A = -(-100) - (-150) = 250\,\text{V}$$
$$v_B = -100 + 200 = 100\,\text{V}$$
$$v_C = -150 - 200 = -350\,\text{V}$$

Figure 2–25(d) shows the numerical values of all the voltages and currents, some of which are negative. Again, the negative values do not mean that the voltage reference marks originally assigned in Figure 2–25(b) are incorrect.   ∎

Exercise 2–13

In Figure 2–25(a), the 2-A source is replaced by a 100-V source with the + terminal at the top, and the 3-A source is removed. Find the current and its direction through the voltage source.

A n s w e r : $i_{\text{Voltage Source}} = 2.33$ A up.

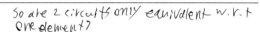

So are 2 circuits only equivalent w.r.t one element?

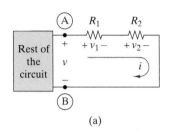

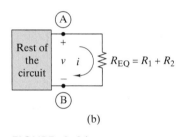

FIGURE 2–26 *A series resistance circuit: (a) Original circuit. (b) Equivalent circuit.*

2–4 EQUIVALENT CIRCUITS

The analysis of a circuit can often be made easier by replacing part of the circuit with one that is equivalent but simpler. The underlying basis for two circuits to be equivalent is contained in their i–v relationships.

> *Two circuits are said to be equivalent if they have identical i–v characteristics at a specified pair of terminals.*

In other words, when two circuits are equivalent the voltage and current at an interface do not depend on which circuit is connected to the interface.

EQUIVALENT RESISTANCE

The two resistors in Figure 2–26(a) are connected in series between a pair of terminals A and B. The objective is to simplify the circuit without altering the electrical behavior of the rest of the circuit.

The KVL equation around the loop from A to B is

$$v = v_1 + v_2 \tag{2–18}$$

Since the two resistors are connected in series, the same current i exists in both. Applying Ohm's law, we get $v_1 = R_1 i$ and $v_2 = R_2 i$. Substituting these relationships into Eq. (2–18) and then simplifying yields

$$v = R_1 i + R_2 i = i(R_1 + R_2)$$

We can write this equation in terms of an equivalent resistance R_{EQ} as

$$v = i R_{\text{EQ}} \quad \text{where} \quad R_{\text{EQ}} = R_1 + R_2 \tag{2–19}$$

This result means the circuits in Figures 2–26(a) and (b) have the same i–v characteristic at terminals A and B. As a result, the response of the rest of the circuit is unchanged when the series connection of R_1 and R_2 is replaced by a resistance R_{EQ}.

The parallel connection of two conductances in Figure 2–27(a) is the dual[2] of the series circuit in Figure 2–26(a). Again, the objective is to replace the parallel connection by a simpler equivalent circuit without altering the response of the rest of the circuit.

A KCL equation at node A produces

$$i = i_1 + i_2 \tag{2–20}$$

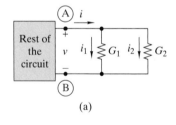

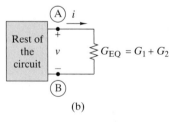

FIGURE 2–27 *A parallel resistance circuit: (a) Original circuit. (b) Equivalent circuit.*

Since the conductances are connected in parallel, the voltage v appears across both. Applying Ohm's law, we obtain $i_1 = G_1 v$ and $i_2 = G_2 v$. Substituting these relationships into Eq. (2–20) and then simplifying yields

$$i = v G_1 + v G_2 = v(G_1 + G_2)$$

---

[2]Dual circuits have identical behavior patterns when we interchange the roles of the following parameters: (1) voltage and current, (2) series and parallel, and (3) resistance and conductance. In later chapters we will see duality exhibited by other circuit parameters as well.

This result can be written in terms of an equivalent conductance G_{EQ} as follows:

$$i = vG_{EQ}, \quad \text{where} \quad G_{EQ} = G_1 + G_2 \tag{2-21}$$

This result means the circuits in Figures 2–27(a) and (b) have the same i–v characteristic at terminals A and B. As a result, the response of the rest of the circuit is unchanged when the parallel connection of G_1 and G_2 is replaced by a conductance G_{EQ}.

Since conductance is not normally used to describe a resistor, it is sometimes useful to rewrite Eq. (2–21) as an equivalent resistance $R_{EQ} = 1/G_{EQ}$. That is,

$$R_1 \| R_2 = R_{EQ} = \frac{1}{G_{EQ}} = \frac{1}{G_1 + G_2} = \frac{1}{\frac{1}{R_1} + \frac{1}{R_2}} = \frac{R_1 R_2}{R_1 + R_2} \tag{2-22}$$

where the symbol "$\|$" is shorthand for "in parallel." The expression on the far right in Eq. (2–22) is called the product over the sum rule for two resistors in parallel. This rule is useful in derivations in which the resistances are left in symbolic form (see Example 2–14), but it is not an efficient algorithm for calculating numerical values of equivalent resistance.

Caution: The product over sum rule applies only to two resistors connected in parallel. When more than two resistors are in parallel, we must use the following general result to obtain the equivalent resistance:

$$R_{EQ} = \frac{1}{G_{EQ}} = \frac{1}{G_1 + G_2 + G_3 + \cdots} = \frac{1}{\frac{1}{R_1} + \frac{1}{R_2} + \frac{1}{R_3} + \cdots} \tag{2-23}$$

EXAMPLE 2-13

Find the equivalent resistance for the circuits in Figure 2–28 (a) and (b).

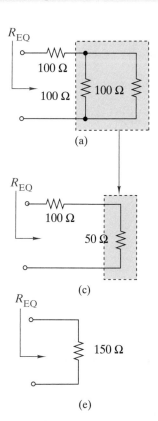

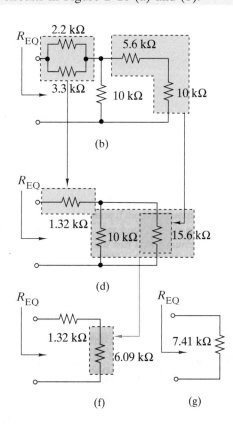

FIGURE 2-28

SOLUTION:

In the circuit of Figure 2–28(a) the two 100-Ω resistors on the right of the circuit are in parallel and combine using Eq. (2–22) to 50 Ω as shown in Fig. 2–28(c). This new resistor is connected to the remaining 100-Ω resistor. These two resistor are in series and add as per Eq. (2–19) as shown in Fig. 2–28(e). The equivalent resistance R_{EQ} of the circuit of Figure 2–28(a) is 150 Ω.

The circuit of Figure 2–28(b) requires a few extra steps. Starting at the farthest right we see that the 5.6-kΩ resistor and the 10-kΩ resistor are in series. They add to equal 15.6 kΩ as shown in Fig. 2–28(d). This new resistor is in parallel with the 10-kΩ resistor in the center of the circuit. The new 15.6-kΩ resistor and the 10-kΩ resistor combine to yield 6.09 kΩ as shown in Fig. 2–28(f). The two resistors at the leftmost part of the circuit, that is, the 2.2-kΩ and the 3.3-kΩ resistors are in parallel. These can combine as a 1.32-kΩ resistor as shown in Fig. 2–28(d). Finally, the 1.32-kΩ resistor and the 6.09-kΩ resistor are in series and can be combined resulting in a $R_{EQ} = 7.41$ kΩ as shown in Fig. 2–28(g). ∎

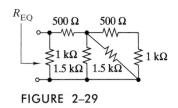

FIGURE 2–29

Exercise 2–14

Find the equivalent resistance for the circuit in Figure 2–29.

Answer: $R_{EQ} = 500$ Ω

EXAMPLE 2–14

Given the circuit in Figure 2–30(a),

(a) Find the equivalent resistance R_{EQ1} connected between terminals A and B.
(b) Find the equivalent resistance R_{EQ2} connected between terminals C and D.

FIGURE 2–30

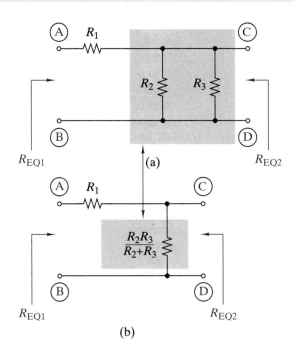

(b)

SOLUTION:

First we note that resistors R_2 and R_3 are connected in parallel. Applying the product over sum rule [Eq. (2–22)], we obtain

$$R_2 \| R_3 = \frac{R_2 R_3}{R_2 + R_3}$$

As an interim step, we redraw the circuit, as shown in Figure 2–30(b).

(a) To find the equivalent resistance between terminals A and B, we note that R_1 and the equivalent resistance $R_2\|R_3$ are connected in series. The total equivalent resistance R_{EQ1} between terminals A and B is

$$R_{\mathrm{EQ1}} = R_1 + (R_2\|R_3)$$

$$R_{\mathrm{EQ1}} = R_1 + \frac{R_2 R_3}{R_2 + R_3}$$

$$R_{\mathrm{EQ1}} = \frac{R_1 R_2 + R_1 R_3 + R_2 R_3}{R_2 + R_3}$$

(b) Looking between terminals C and D yields a different result. In this case R_1 is not involved, since there is an open circuit (an infinite resistance) between terminals A and B. Therefore, only $R_2\|R_3$ affect the resistance between terminals C and D. As a result, R_{EQ2} is simply

$$R_{\mathrm{EQ2}} = R_2\|R_3 = \frac{R_2 R_3}{R_2 + R_3}$$

This example shows that equivalent resistance depends on the pair of terminals involved. ■

Exercise 2–15

Find the equivalent resistance between terminals A–C, B–D, A–D, and B–C in the circuit in Figure 2–30.

Answers: $R_{\mathrm{A-C}} = R_1$; $R_{\mathrm{B-D}} = 0\,\Omega$ (a short circuit); $R_{\mathrm{A-D}} = R_1 + R_2\|R_3$; $R_{\mathrm{B-C}} = R_2\|R_3$

Exercise 2–16

Find the equivalent resistance between terminals A–B, A–C, A–D, B–C, B–D, and C–D in the circuit of Figure 2–31. For example: $R_{\mathrm{A-B}} = (80\|80) + 60 = 100\,\Omega$.

Answers: $R_{\mathrm{A-C}} = 70\,\Omega$; $R_{\mathrm{A-D}} = 65\,\Omega$; $R_{\mathrm{B-C}} = 90\,\Omega$; $R_{\mathrm{B-D}} = 85\,\Omega$; $R_{\mathrm{C-D}} = 55\,\Omega$

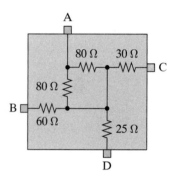

FIGURE 2–31

One final note on checking numerical calculations of equivalent resistance. When several resistances are connected in parallel, the equivalent resistance must be smaller than the smallest resistance in the connection. Conversely, when several resistances are connected in series, the equivalent resistance must be larger than the largest resistance in the connection.

EQUIVALENT SOURCES *what is the connection to Liu alg here?*

The practical source models introduced previously are shown in Figure 2–32. These models consist of an ideal voltage source in series with a resistance and an ideal current source in parallel with a resistance. We now determine the conditions under which the practical voltage source and the practical current sources are equivalent.

Figure 2–32 shows the two practical sources connected between terminals labeled A and B. A parallel analysis of these circuits yields the conditions for equivalency at terminals A and B. First, Kirchhoff's laws are applied as

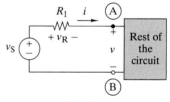

Circuit A

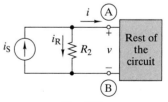

Circuit B

FIGURE 2–32 *Practical source models that are equivalent when Eq. (2–24) is satisfied.*

Circuit A	Circuit B
KVL	KCL
$v_{\mathrm{S}} = v_{\mathrm{R}} + v$	$i_{\mathrm{S}} = i_{\mathrm{R}} + i$

FIGURE 2–33 *The i–v characteristics of practical sources in Figure 2–32.*

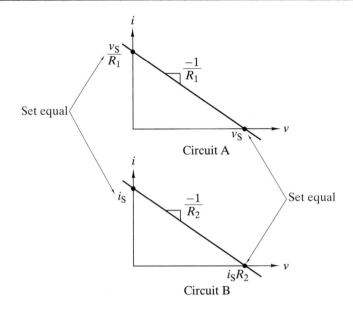

Circuit A

Circuit B

Next, Ohm's law is used to obtain

$$\begin{array}{cc} \text{Circuit A} & \text{Circuit B} \\ v_R = R_1 i & i_R = \dfrac{v}{R_2} \end{array}$$

Combining these results, we find that the *i–v* relationships of each of the circuits at terminals A and B are

$$\begin{array}{cc} \text{Circuit A} & \text{Circuit B} \\ i = -\dfrac{v}{R_1} + \dfrac{v_S}{R_1} & i = -\dfrac{v}{R_2} + i_S \end{array}$$

These *i–v* characteristics take the form of the straight lines shown in Figure 2–33. The two lines are identical when the intercepts are equal. This requires that $v_S/R_1 = i_S$ and $v_S = i_S R_2$, which, in turn, requires that

$$R_1 = R_2 = R \quad \text{and} \quad v_S = i_S R \tag{2–24}$$

When conditions in Eq. (2–24) are met, the response of the rest of the circuit is unaffected when we replace a practical voltage source by an equivalent practical current source, or vice versa. Exchanging one practical source model for an equivalent model is called *source transformation.*

Caution: Source transformation means that either model will deliver the same voltage and current to the rest of the circuit. Hence, a circuit connected to either cannot tell which practical circuit it is connected to. It does not mean the two models are identical in every way. For example, when the rest of the circuit is an open circuit, there is no current in the resistance of the practical voltage source, and hence no $i^2 R$ power loss. However, the current in the practical current source is not zero when the load is an open circuit. Thus, equivalent sources do not usually have the same internal power loss even though they deliver the same current and voltage to the rest of the circuit. Suppose a problem requires the determination of the power supplied by a practical voltage source connected to a resistive load. It would be incorrect to do a source transformation and use the transformed current times the original voltage to find the power. It would also be incorrect to find the power supplied by the practical current source, since it usually is not the same as that delivered by the equivalent voltage source. (See Example 2–15.)

EXAMPLE 2 – 15

(a) Convert the practical voltage source to the left of nodes A and B in Figure 2–34(a) to an equivalent current source.
(b) Suppose the practical voltage source is connected to a 5-Ω load across nodes A and B. How much power is provided by the voltage source?

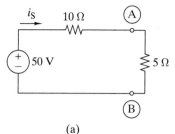

(a)

SOLUTION:
(a) Using Eq. (2–24), we have

$$R_1 = R_2 = R = 10\,\Omega$$

$$i_S = \frac{v_S}{R} = \frac{50}{10} = 5\,\text{A}$$

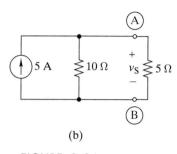

The equivalent practical current source is shown in Figure 2–34(b) to the left of nodes A and B.

(b) The power provided by the source is equal to the 50-V source times the current flowing through it. The current is found by a loop equation as follows:

(b)

FIGURE 2–34

$$50 = 10 \times i_S + 5 \times i_S$$

$$i_S = \frac{50}{15} = 3.33\,\text{A}$$

The power provided by the voltage source then is

$$p_S = v_S \times i_S = 50 \times 3.33 = 166\,\text{W} \qquad \blacksquare$$

DISCUSSION: *As noted earlier, multiplying the voltage of the voltage source by the current of the transformed current source does not produce the correct answer: $50 \times 5 = 250\,W$, not 166 W. However, one may be fooled into thinking that the power delivered by the transformed current source is equal to that delivered by the original voltage source. To demonstrate that this is not correct, combine the 5-Ω load connected in parallel with the 10-Ω source resistance to get an equivalent resistance of 3.33 Ω. Find the voltage across the equivalent 3.33-Ω resistance using Ohm's law:*

$$v_S = i_S \times R_{EQ} = 5 \times 3.3\overline{3} = 16.\overline{6}\,\text{V}$$

The power provided by the equivalent current source then is

$$p_S = v_S \times i_S = 16.\overline{6} \times 5 = 83.33\,\text{W}$$

This is not equivalent to the correct answer of 166 W.

Transformed sources are valuable tools, as we will see, but one must remember that the transformation creates an equivalent circuit only from the perspective of the rest of the circuit and not between the transformed sources.

Exercise 2–17

A practical current source consists of a 2-mA ideal current source in parallel with a 500-Ω resistance. Find the equivalent practical voltage source.

Answer: The equivalent is a 1-V ideal voltage source in series with a 500-Ω resistance.

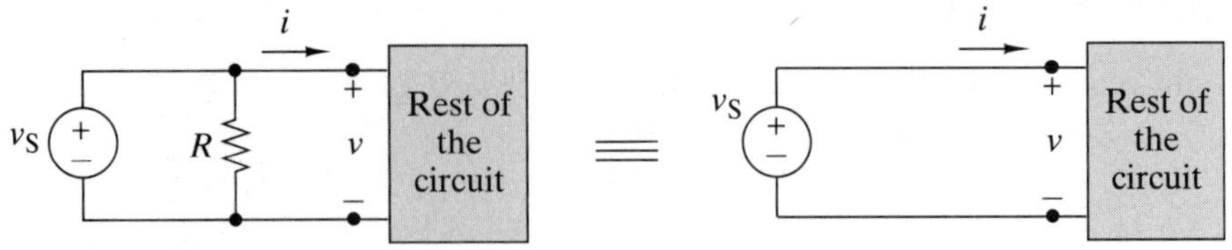

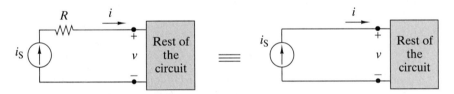

FIGURE 2–35 *Equivalent circuit of a voltage source and a resistor in parallel.*

Figure 2–35 shows another source transformation in which a voltage source and resistor in parallel is replaced by a voltage source acting alone. The two circuits are equivalent because the i–v constraint at the input to the rest of the circuit is $v = v_S$ in both circuits. In other words, the response of the rest of the circuit is unchanged if a resistor in parallel with a voltage source is removed from the circuit. However, removing the resistor does reduce the total current supplied by the voltage source by v_S/R. While the resistor does not affect the current and voltage delivered to the rest of the circuit, it does dissipate power that must be supplied by the source.

The dual situation is shown in Figure 2–36. In this case a current source connected in series with a resistor can be replaced by a current source acting alone because the i–v constraint at the input to the rest of the circuit is $i = i_S$ for both circuits. In other words, the response of the rest of the circuit is unchanged if a resistor in series with a current source is removed from the circuit.

FIGURE 2–36 *Equivalent circuit of a current source and a resistor in series.*

SUMMARY OF EQUIVALENT CIRCUITS

Figure 2–37 is a summary of two-terminal equivalent circuits involving resistors and sources connected in series or parallel. The series and parallel equivalences in the first row and the source transformations in the second row are used regularly in subsequent discussions. The last row in Figure 2–37 presents additional source transformations that reduce series or parallel connections to a single ideal current or voltage source. Proof of these equivalences involves showing that the final single-source circuits have the same i–v characteristics as the original connections. The details of such a derivation are left as an exercise for the reader.

There are several other circuit combinations that involve equivalent circuits that we should mention. The first is what happens if we connect two or more voltage sources in parallel? Practical sources, such as real, same-value batteries, are often connected in parallel to achieve more current. However, ideal sources are capable, by their definition, of providing whatever current the load requires, hence, multiple, same-value, voltage sources can be replaced by a single voltage source of that value. However, one cannot connect ideal voltage sources of different values, including a short circuit, in parallel since this would violate KVL. The dual is also true for ideal current sources connected in series. Same-value current sources can be replaced by a single current source of that value, while connecting ideal current sources of different values would violate KCL.

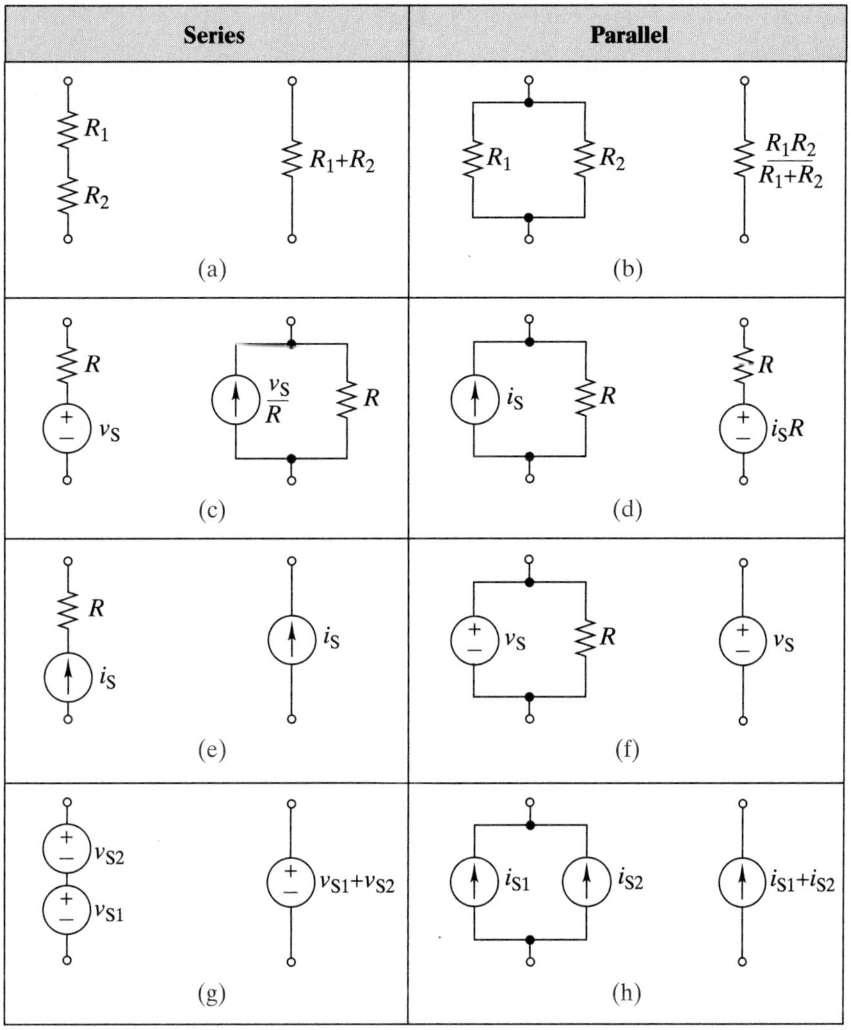

Series		Parallel	

FIGURE 2–37 *Summary of two-terminal equivalent circuits.*

Exercise 2–18

Find the equivalent circuit for each of the following

(a) Three ideal 1.5-V batteries connected in series.
(b) A 5-mA current source in series with a 100-kΩ resistor.
(c) A 40-A ideal current source in parallel with an ideal 10-A current source.
(d) A 100-V source in parallel with two 10-kΩ resistors.
(e) An ideal 15-V source in series with an ideal 10-mA source.
(f) A 15-V ideal source and a 5-V ideal source connected in parallel.

Answers:

(a) One 4.5-V voltage source.
(b) A single 5-mA current source.
(c) One 50-A current source.
(d) A single 100-V voltage source.
(e) A 15-V source in series with a 10-mA source—ideal voltage and current sources cannot be combined.
(f) This is not a possible combination since KVL would be violated.

Is this another "trick" for avoiding tedious solving of equations?

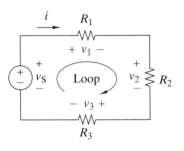

FIGURE 2–38 *A voltage divider circuit.*

2–5 VOLTAGE AND CURRENT DIVISION

We complete our treatment of series and parallel circuits with a discussion of voltage and current division. These two analysis tools find wide application in circuit analysis and design.

VOLTAGE DIVISION

Voltage division provides a simple way to find the voltage across each element in a series circuit. Figure 2–38 shows a circuit that lends itself to solution by voltage division. Applying KVL around the loop in Figure 2–38 yields

$$v_S = v_1 + v_2 + v_3 \tag{2–25}$$

The elements in Figure 2–38 are connected in series, so the same current i exists in each of the resistors. Using Ohm's law, we find that

$$v_S = R_1 i + R_2 i + R_3 i \tag{2–26}$$

Solving for i yields

$$i = \frac{v_S}{R_1 + R_2 + R_3} \tag{2–27}$$

Once the current in the series circuit is found, the voltage across each resistor is computed using Ohm's law:

$$v_1 = R_1 i = \left(\frac{R_1}{R_1 + R_2 + R_3}\right) v_S \tag{2–28}$$

$$v_2 = R_2 i = \left(\frac{R_2}{R_1 + R_2 + R_3}\right) v_S \tag{2–29}$$

$$v_3 = R_3 i = \left(\frac{R_3}{R_1 + R_2 + R_3}\right) v_S \tag{2–30}$$

Looking over these results, we see an interesting pattern. In a series connection, the voltage across each resistor is equal to its resistance divided by the equivalent series resistance of the connection times the voltage across the series circuit. Thus, the general expression of the *voltage division rule* is

$$v_k = \left(\frac{R_k}{R_{EQ}}\right) v_{TOTAL} \tag{2–31}$$

In other words, the total voltage divides among the series resistors in proportion to their resistance over the equivalent resistance of the series connection. The following examples show several applications of this rule.

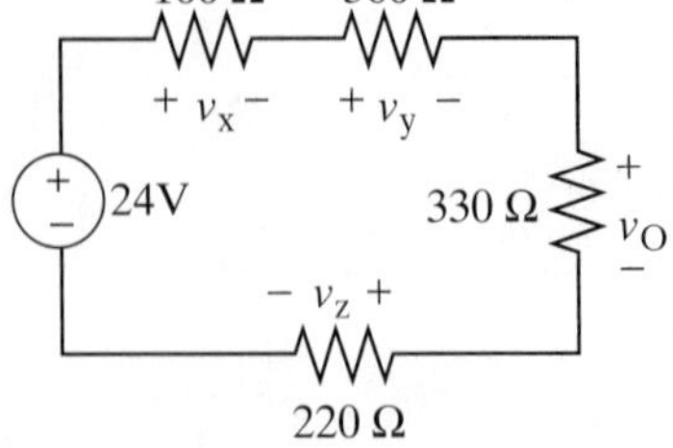

EXAMPLE 2–16

Find the voltage across the 330-Ω resistor in the circuit of Figure 2–39.

SOLUTION:
Applying the voltage division rule, we find that

$$v_O = \left(\frac{330}{100 + 560 + 330 + 220}\right) 24 = 6.55 \text{ V}$$

FIGURE 2–39

Exercise 2–19

Find the voltages v_x, v_y, and v_z in the circuit of Figure 2–39. Show that the sum of all the voltages across each of the individual resistors equals the source voltage.

Answers: $v_x = 1.98$ V, $v_y = 11.11$ V, and $v_z = 4.36$ V. $1.98 + 11.11 + 4.36 + 6.55 = 24$ V, Q.E.D.

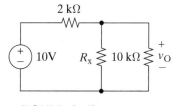

EXAMPLE 2–17

Select a value for the resistor R_x in Figure 2–40 so $v_O = 8$ V.

FIGURE 2–40

SOLUTION:

The unknown resistor is in parallel with the 10-kΩ resistor. Since voltages across parallel elements are equal, the voltage $v_O = 8$ V appears across both. We first define an equivalent resistance $R_{EQ} = R_x \| 10\,\text{k}\Omega$ as

$$R_{EQ} = \frac{R_x \times 10000}{R_x + 10000}$$

We write the voltage division rule in terms of R_{EQ} as

$$v_O = 8 = \left(\frac{R_{EQ}}{R_{EQ} + 2000} \right) 10$$

which yields $R_{EQ} = 8$ kΩ. Finally, we substitute this value into the equation defining R_{EQ} and solve for R_x to obtain $R_x = 40$ kΩ. ■

Exercise 2–20

In Figure 2–40 $R_x = 10$ kΩ. The output voltage $v_O = 20$ V. Find the voltage source that would produce that output. (*Hint:* It is not 10 V.)

Answer: Voltage source = 28 V.

EXAMPLE 2–18

Use the voltage division rule to find the output voltage v_O of the circuit in Figure 2–41.

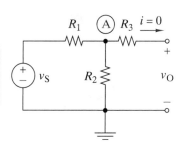

FIGURE 2–41

SOLUTION:

At first glance it appears that the voltage division rule does not apply, since the resistors are not connected in series. However, the current through R_3 is zero since the output of the circuit is an open circuit. Therefore, Ohm's law shows that $v_3 = R_3 i_3 = 0$. Applying KCL at node A shows that the same current exists in R_1 and R_2, since the current through R_3 is zero. Applying KVL around the output loop shows that the voltage across R_2 must be equal to v_O since the voltage across R_3 is zero. In essence, it is as if R_1 and R_2 were connected in series. Therefore, voltage division can be used and yields the output voltage as

$$v_O = \left(\frac{R_2}{R_1 + R_2} \right) v_S$$

The reader should carefully review the logic leading to this result because voltage division applications of this type occur frequently. ■

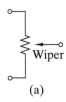

Wiper

(a)

Single-turn potentiometer.

Multiple-turn potentiometer.

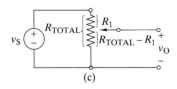

Trim pots.

(b)

(c)

FIGURE 2–42 *The potentiometer: (a) Circuit symbol. (b) Actual devices. (c) An application.*

Exercise 2–21

In Figure 2–41, suppose that a resistor R_4 is connected across the output. What value should R_4 be if we want $\frac{1}{2}v_S$ to appear between node A and ground?

Answer: $R_4 = \dfrac{R_1R_3 + R_1R_2 - R_3R_2}{R_2 - R_1}$

APPLICATION   **EXAMPLE 2–19**

The operation of a potentiometer is based on the voltage division rule. The device is a three-terminal element that uses voltage (potential) division to meter out a fraction of the applied voltage. Simply stated, a **potentiometer** is an adjustable voltage divider. Figure 2–42 shows the circuit symbol of a potentiometer, photos of three different types of actual potentiometers, and a typical application.

The voltage v_O in Figure 2–42(c) can be adjusted by turning the shaft on the potentiometer to move the wiper arm contact. Using the voltage division rule, the voltage v_O is found as

$$v_O = \left(\frac{R_{\text{TOTAL}} - R_1}{R_{\text{TOTAL}}}\right)v_S \qquad (2\text{–}32)$$

Adjusting the movable wiper arm all the way to the top makes R_1 zero, and voltage division yields

$$v_O = \left(\frac{R_{\text{TOTAL}} - 0}{R_{\text{TOTAL}}}\right)v_S = v_S \qquad (2\text{–}33)$$

In other words, 100% of the applied voltage is delivered to the rest of the circuit. Moving the wiper all the way to the bottom makes R_1 equal to R_{TOTAL}, and voltage division yields

$$v_O = \left(\frac{R_{\text{TOTAL}} - R_{\text{TOTAL}}}{R_{\text{TOTAL}}}\right)v_S = 0 \qquad (2\text{–}34)$$

This opposite extreme delivers zero voltage. By adjusting the wiper arm position, we can obtain an output voltage anywhere between zero and the applied voltage v_S. When the wiper is positioned halfway between the top and bottom, we naturally expect to obtain half of the applied voltage. Setting $R_1 = \frac{1}{2}R_{\text{TOTAL}}$ yields

$$v_O = \left(\frac{R_{\text{TOTAL}} - \frac{1}{2}R_{\text{TOTAL}}}{R_{\text{TOTAL}}}\right)v_S = \frac{v_S}{2} \qquad (2\text{–}35)$$

as expected. The many applications of the potentiometer include volume controls, voltage balancing, and fine-tuning adjustment.

Exercise 2–22

Ten volts (v_S) are connected across the 10-kΩ potentiometer (R_{TOTAL}) shown in Figure 2–42(c). A load resistor of 10 kΩ is connected across its output. At what resistance should the wiper ($R_{\text{TOTAL}} - R_1$) be set so that 2 V appears at the output, v_O?

Answer: $R_{\text{TOTAL}} - R_1 = 2.36\,\text{k}\Omega$.

D **DESIGN EXAMPLE 2–20**

Design a voltage divider that will provide $5.5\,V \pm 5\%$ from a 9-V battery using only the $\pm 10\%$ standard-value resistors (see inside back cover). The current from the source should be at or below 0.5 mA to avoid draining the source too quickly.

SOLUTION:
The percentage of the 9-V source desired is

$$\frac{5.5}{9} = 0.6111$$

Using the voltage division rule, we want

$$\frac{R_2}{R_1 + R_2} = 0.6111$$

$$R_2 = 0.6111R_1 + 0.6111\,R_2$$

$$R_2 = \frac{0.6111R_1}{0.3888} = 1.57R_1$$

Selecting $R_1 = 1\,k\Omega$ and $R_2 = 1.5\,k\Omega$, both of which are standard values, provides an output voltage of

$$v_O = \frac{R_2 \times 9}{R_1 + R_2} = 5.4\,V$$

This result is well within the range of acceptable values of $5.225\,V \le v_O \le 5.775\,V$.
Checking to see what the current drain on the source is, we use Ohm's law.

$$i = \frac{v_S}{R_1 + R_2} = \frac{9}{1000 + 1500} = 3.6\,mA$$

This fails to meet the second requirement, a current less than or equal to 0.5 mA. Our choice of resistors is too small. Selecting $R_1 = 10\,k\Omega$ and $R_2 = 15\,k\Omega$, both also standard values, provides the same ratio but a current of only 0.36 mA, thus meeting both requirements. Some other $R_1 - R_2$ pairs that would work are $22\,k\Omega$ and $33\,k\Omega$, $47\,k\Omega$ and $68\,k\Omega$, and $100\,k\Omega$ and $150\,k\Omega$. Figure 2–43 shows the design task and our chosen result.

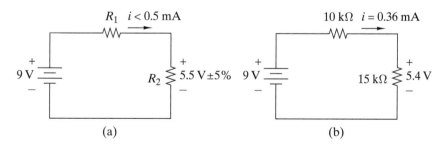

FIGURE 2–43

(a)   (b)

In design, one must start somewhere and make an assumption. Testing the assumption may result in the realization that the assumption was wrong and a new assumption must be made. Design by its nature is often an iterative process. ∎

CURRENT DIVISION

Current division is a simple way to find the current through each element of a parallel circuit. It is the dual of voltage division, so we will observe some similarities in the form

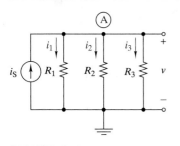

FIGURE 2–44 *A current divider circuit.*

of the equations for the two approaches. Figure 2–44 shows a parallel circuit that lends itself to solution by current division. Applying KCL at node A yields

$$i_S = i_1 + i_2 + i_3$$

The voltage v appears across all three resistances since they are connected in parallel. Using Ohm's law, we can write

$$i_S = v\frac{1}{R_1} + v\frac{1}{R_2} + v\frac{1}{R_3} = v\left(\frac{1}{R_1} + \frac{1}{R_2} + \frac{1}{R_3}\right)$$

and solve for v as

$$v = i_S \frac{1}{\dfrac{1}{R_1} + \dfrac{1}{R_2} + \dfrac{1}{R_3}}$$

Given the voltage v, the current through any element is found using Ohm's law as

$$i_1 = v\frac{1}{R_1} = \frac{\dfrac{1}{R_1} i_S}{\dfrac{1}{R_1} + \dfrac{1}{R_2} + \dfrac{1}{R_3}} = \frac{G_1 i_S}{G_1 + G_2 + G_3} \tag{2–36}$$

$$i_2 = v\frac{1}{R_2} = \frac{\dfrac{1}{R_2} i_S}{\dfrac{1}{R_1} + \dfrac{1}{R_2} + \dfrac{1}{R_3}} = \frac{G_2 i_S}{G_1 + G_2 + G_3} \tag{2–37}$$

$$i_3 = v\frac{1}{R_3} = \frac{\dfrac{1}{R_3} i_S}{\dfrac{1}{R_1} + \dfrac{1}{R_2} + \dfrac{1}{R_3}} = \frac{G_3 i_S}{G_1 + G_2 + G_3} \tag{2–38}$$

These results show that the source current is divided among the parallel resistors in proportion to their *conductances* divided by the equivalent conductances in the parallel connection. Thus, the general expression for the current through the kth resistor is given by the *current division rule* as

$$i_k = \left(\frac{\dfrac{1}{R_k}}{\dfrac{1}{R_1} + \dfrac{1}{R_2} + \cdots + \dfrac{1}{R_n}}\right) i_{TOTAL} = \left(\frac{G_k}{G_{EQ}}\right) i_{TOTAL} \tag{2–39}$$

Comparing this equation with Eq. (2–31) for voltage division, we see how similar they look. In essence, if you know one rule you know the other: Replace v with i and R with G in the voltage division rule to obtain the current division rule, and vice versa. This is one of the strengths of the concept of duality.

For the two-resistor case in Figure 2–45, the current i_1 is found using current division as

$$i_1 = \left(\frac{G_1}{G_1 + G_2}\right) i_S = \frac{\dfrac{1}{R_1}}{\dfrac{1}{R_1} + \dfrac{1}{R_2}} i_S = \left(\frac{R_2}{R_1 + R_2}\right) i_S \tag{2–40}$$

Similarly, the current i_2 in Figure 2–45 is found to be

$$i_2 = \left(\frac{G_2}{G_1 + G_2}\right) i_S = \frac{\dfrac{1}{R_2}}{\dfrac{1}{R_1} + \dfrac{1}{R_2}} i_S = \left(\frac{R_1}{R_1 + R_2}\right) i_S \tag{2–41}$$

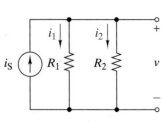

FIGURE 2–45 *Two-path current divider circuit.*

These two results lead to the following *two-path current division rule:* When a circuit can be reduced to two equivalent resistances in parallel, the current through one resistance is equal to the other resistance divided by the sum of the two resistances times the total current entering the parallel combination.

Caution: Equations (2–40) and (2–41) apply only when the circuit is reduced to two parallel paths in which one path contains the desired current and the other path is the equivalent resistance of all other paths.

EXAMPLE 2 – 21

Find the current i_x in Figure 2–46(a).

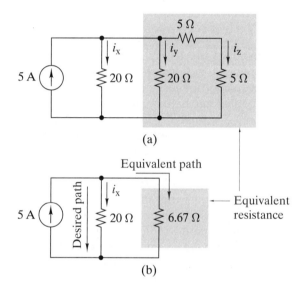

FIGURE 2–46

(a)

(b)

SOLUTION:

To find i_x, we reduce the circuit to two paths, a path containing i_x and a path equivalent to all other paths, as shown in Figure 2–46(b). Now we can use the two-path current divider rule as

$$i_x = \frac{6.67}{20 + 6.67} \times 5 = 1.25 \text{ A}$$

Exercise 2–23

(a) Find i_y and i_z in the circuit of Figure 2–46(a).
(b) Show that the sum of i_x, i_y, and i_z equals the source current.

Answers:

(a) $i_y = 1.25$ A; $i_z = 2.5$ A
(b) $i_x + i_y + i_z = 5$ A

Exercise 2–24

The circuit in Figure 2–47 shows a delicate device that is modeled by a 90-Ω equivalent resistance. The device requires a current of 1 mA to operate properly. A 1.5-mA fuse is inserted in series with the device to protect it from overheating. The resistance of the fuse is 10 Ω. Without the shunt resistance R_x, the source would deliver 5 mA to the device, causing the fuse to blow. Inserting a shunt resistor R_x diverts a portion of the available source current around the fuse and device. Select a value of R_x so only 1 mA is delivered to the device.

FIGURE 2–47

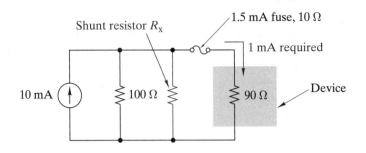

Answer: $R_x = 12.5\,\Omega$

APPLICATION   EXAMPLE 2–22

Batteries chemically produce electricity that is used to power many portable devices including cell phones, laptops, iPods, flashlights, hearing aids, back-up power, and automobiles to mention just a few. Batteries are rated in both voltage and ampere-hours. A typical car battery delivers a nominal 12 V for 70 A-hrs while a "D-cell" delivers 1.5 V for 4.5 A-hrs. Batteries are not ideal sources; they are all real or practical sources of energy and must be modeled as an ideal voltage source in series with a resistor. The series resistor models the batteries *internal resistance*, which can vary from a few milliohms for a car battery to as high as 100 ohms for a hearing-aid battery. As a battery ages, especially in use, its internal resistance increases decreasing its usefulness. Therefore, it is good practice to measure a battery's voltage while under load since measuring the battery's voltage with an open circuit could indicate a good voltage regardless of its internal resistance. The following example looks at the effect of a battery's internal resistance on the current available to a load.

A car battery of 12.6 V and an internal resistance of 25 mΩ delivers 100 mA to accessories and 210 A to the starter motor of a 6-cylinder car. Find the resistance of the starter motor.

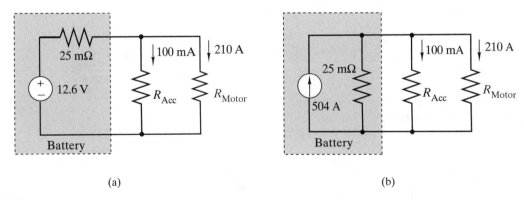

(a)   (b)

FIGURE 2–48

SOLUTION:
Figure 2–48(a) shows the circuit in question. There are several ways to tackle this problem. Let's solve it two ways. For our first approach we will do a source transformation as shown in Figure 2–48(b). We know that the total current delivered

to the loads is 210.1 A. we can use a current divider, Eq. (2–41), to find the total resistance of the loads.

$$I_{\text{Loads}} = \frac{R_s \times I_{\text{Total}}}{R_s + R_{\text{Loads}}} = \frac{0.025 \times 504}{0.025 + R_{\text{Loads}}} = 210.1 \text{ A}$$

Solving for R_{Loads} yields 34.97 mΩ

Ohm's law can be used to find the voltage across the total load

$$V_{\text{Loads}} = 210.1 \times .03497 = 7.3475 \text{ V}$$

Ohm's law can again be used to find the resistance of the starter motor

$$R_{\text{Motor}} = \frac{7.3475}{210} = 34.98 \text{ m}\Omega$$

A second, somewhat simpler, approach is to realize that 210.1 A are flowing through the source resistance. This results in a voltage drop of

$$V_{\text{Source}} = 210.1 \times 0.25 = 5.2525 \text{ V}$$

By KVL this leaves 12.6 − 5.2525 = 7.3475 V across the loads. We can then solve for R_{Motor} as before and find it to be 34.98 mΩ. ∎

Exercise 2–25

Repeat the problem of Example 2–22 if the battery's internal resistance increases to 70 mΩ. Will there be sufficient current available to start the car?

Answer: A source transformation can quickly show that there will be a maximum of only 180 A available for the motor and accessories. The battery is insufficient to power the starting motor and accessories and driver will hear that stomach-wrenching sound. Rrrrrr-rrr-rr-ugh!

APPLICATION EXAMPLE 2–23

The *R*-2*R* ladder circuit in Figure 2–49 is a *binary current divider* that finds applications in digital-to-analog signal conversion. The operation of this circuit can be explained using current division together with series and parallel equivalent resistance. The equivalent resistance connected to ground at node 3 is $2R\|2R = R$, which means that the equivalent resistance seen to the right of node 2 of $R + R = 2R$. This in turn means that the total equivalent resistance connected to ground at node 2 is $2R\|2R = R$ and hence the equivalent resistance seen to the right of node 1 of $R + R = 2R$. The net result is that the equivalent resistance seen to the right of each numbered node is 2*R*.

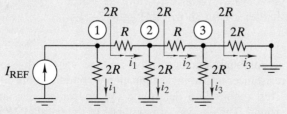

FIGURE 2–49 *R-2R binary circuit divider.*

The reference current I_{REF} entering node 1 divides equally between the two available 2*R* paths with the result that $i_1 = I_{\text{REF}}/2$, and the current into node 2 is also $i_1 = I_{\text{REF}}/2$. At node 2 this current again divides equally between the two 2*R* paths with the result that $i_2 = i_1/2 = I_{\text{REF}}/4$ and the current into node 3 is $i_2 = I_{\text{REF}}/4$.

Finally, at node 3 this current divides equally once more so that $i_3 = i_2/2 = I_{REF}/8$. In sum, the currents in the $2R$ resistors connected to ground are all of the form $i_k = I_{REF}/2^k$, where k is the node number to which the resistor is connected. Thus, the R-$2R$ ladder circuit produces signals (currents in this case) that decrease in a binary fashion as we proceed down the ladder.

Clearly the R-$2R$ ladder can be extended to a larger number of nodes. Commercially available integrated circuit ladders have as many as 8 numbered nodes producing binary currents ranging from $I_{REF}/2$ to $I_{REF}/256$. The advantage of this circuit is that it produces this wide range of precisely related signals using only two values of resistance, namely R and $2R$. This greatly simplifies the fabrication of the R-$2R$ ladder in integrated circuit form.

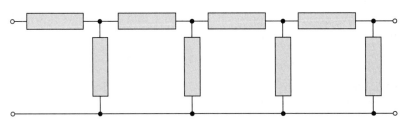

FIGURE 2–50 *A ladder circuit.*

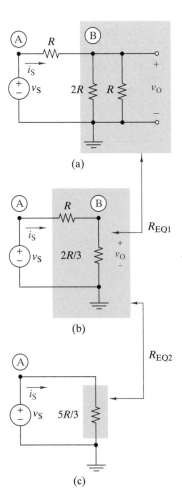

(a)

(b)

(c)

FIGURE 2–51

how would this be used on an R-2R ladder?

2–6 CIRCUIT REDUCTION

The concepts of series/parallel equivalence, voltage/current division, and source transformations can be used to analyze **ladder circuits** of the type shown in Figure 2–50. The basic analysis strategy is to reduce the circuit to a simpler equivalent in which the output is easily found by voltage or current division or Ohm's law. There is no fixed pattern to the reduction process, and much depends on the insight of the analyst. In any case, with circuit reduction we work directly with the circuit model, and so the process gives us insight into circuit behavior.

With circuit reduction the desired unknowns are found by simplifying the circuit and, in the process, eliminating certain nodes and elements. However, we must be careful not to eliminate a node or element that includes the desired unknown voltage or current. The next three examples illustrate circuit reduction. The final example shows that rearranging the circuit can simplify the analysis.

EXAMPLE 2–24

Use series and parallel equivalence to find the output voltage v_O and the input current i_S in the ladder circuit shown in Figure 2–51(a).

SOLUTION:

One approach is to combine parallel resistors and use voltage division to find v_O, and then combine all resistances into a single equivalent to find the input current i_S. Figure 2–51(b) shows the step required to determine the equivalent resistance between the terminals B and ground. The equivalent resistance of the parallel $2R$ and R resistors is

$$R_{EQ1} = \frac{R \times 2R}{R + 2R} = \frac{2}{3}R$$

The reduced circuit in Figure 2–51(b) is a voltage divider. Notice that the two nodes needed to find the voltage v_O, nodes B and ground, have been retained. The unknown voltage is found in terms of the source voltage as

$$v_O = \frac{\frac{2}{3}R}{\frac{2}{3}R + R}\, v_S = \frac{2}{5}\, v_S$$

The input current is found by combining the equivalent resistance found previously with the remaining resistor R to obtain

$$R_{EQ2} = R + R_{EQ1}$$
$$= R + \frac{2}{3}R = \frac{5}{3}R$$

Application of series/parallel equivalence has reduced the ladder circuit to the single equivalent resistance shown in Figure 2–51(c). Using Ohm's law, the input current is

$$i_S = \frac{v_S}{R_{EQ2}} = \frac{3}{5}\frac{v_S}{R}$$

Notice that the reduction step between Figures 2–51(b) and 2–51(c) eliminates node B, so the output voltage v_O must be calculated before this reduction step is taken. ∎

Exercise 2–26

In Figure 2–51, $R = 15\,\text{k}\Omega$. The voltage source $v_S = 5\,\text{V}$. Find the power delivered to the circuit by the source.

Answer: $p_S = 1\,\text{mW}$

EXAMPLE 2–25

Use source transformations to find the output voltage v_O and the input current i_S in the ladder circuit shown in Figure 2–52(a).

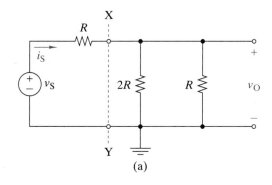

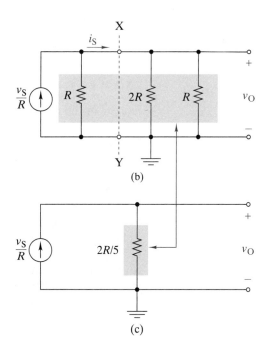

FIGURE 2–52

SOLUTION:

Figure 2–52 shows another way to reduce the circuit analyzed in Example 2–24. Breaking the circuit at points X and Y in Figure 2–52(a) produces a voltage source v_S in series with a resistor R. Using source transformation, this combination can be replaced by an equivalent current source in parallel with the same resistor, as shown in Figure 2–52(b).

Caution: The current source v_S/R is *not* the input current i_S, as is indicated in Figure 2–52(b). Applying the two-path current division rule to the circuit in Figure 2–52(b) yields the input current i_S as

$$i_S = \frac{R}{\frac{2}{3}R + R} \times \frac{v_S}{R} = \frac{v_S}{\frac{5}{3}R} = \frac{3}{5}\frac{v_S}{R}$$

The three parallel resistances in Figure 2–52(b) can be combined into a single equivalent conductance without eliminating the node pair used to define the output voltage v_O. Using parallel equivalence, we obtain

$$G_{EQ} = G_1 + G_2 + G_3$$
$$= \frac{1}{R} + \frac{1}{2R} + \frac{1}{R} = \frac{5}{2R}$$

which yields the equivalent circuit in Figure 2–52(c). The current source v_S/R determines the current through the equivalent resistance in Figure 2–52(c). The output voltage is found using Ohm's law.

$$v_O = \left(\frac{v_S}{R}\right) \times \left(\frac{2R}{5}\right) = \frac{2}{5}v_S$$

Of course, these results are the same as the result obtained in Example 2–24, except that here they were obtained using a different sequence of circuit reduction steps. ■

Exercise 2–27

In Figure 2–52(a), find the current through the 2R resistor.

Answer: $i_{2R} = \frac{v_S}{5R}$ A

EXAMPLE 2–26

Find v_x in the circuit shown in Figure 2–53(a).

SOLUTION:

In the two previous examples the unknown responses were defined at the circuit input and output. In this example the unknown voltage appears across a 10-Ω resistor in the center of the network. The approach is to reduce the circuit at both ends while retaining the 10-Ω resistor defining v_x. Applying a source transformation to the left of terminals X–Y and a series reduction to the two 10-Ω resistors on the far right yields the reduced circuit shown in Figure 2–53(b). The two pairs of 20-Ω resistors connected in parallel can be combined to produce the circuit in Figure 2–53(c). At this point there are several ways to proceed. For example, a source transformation at the points W–Z in Figure 2–53(c) produces the circuit in Figure 2–53(d). Using voltage division in Figure 2–53(d) yields v_x.

$$v_x = \frac{10}{10 + 10 + 10} \times 7.5 = 2.5\,\text{V}$$

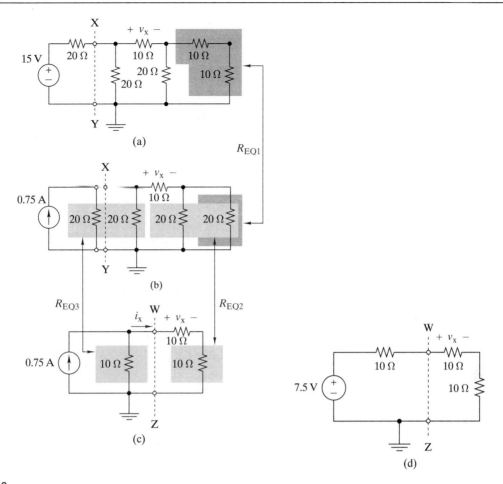

FIGURE 2–53

Yet another approach is to use the two-path current division rule in Figure 2–53(c) to find the current i_x.

$$i_x = \frac{10}{10 + 10 + 10} \times \frac{3}{4} = \frac{1}{4} \text{ A}$$

Then, applying Ohm's law to obtain v_x,

$$v_x = 10 \times i_x = 2.5 \text{ V}$$ ∎

Exercise 2–28

Find v_x and i_x using circuit reduction on the circuit in Figure 2–54.

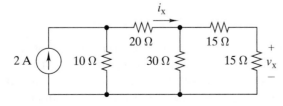

FIGURE 2–54

Answers: $v_x = 3.33 \text{ V}; \; i_x = 0.444 \text{ A}$

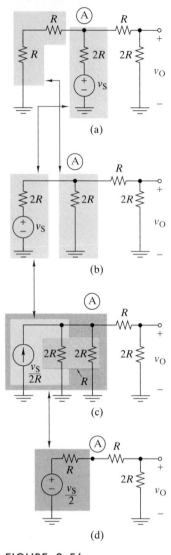

(a)

(b)

(c)

(d)

FIGURE 2–56

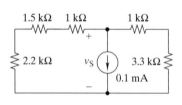

FIGURE 2–57

Exercise 2–29

Find v_x and v_y using circuit reduction on the circuit in Figure 2–55.

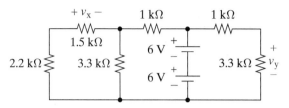

FIGURE 2–55

Answers: $v_x = -3.09\text{ V}$; $v_y = 9.21\text{ V}$

EXAMPLE 2–27

Using circuit reduction, find v_O in Figure 2–56(a).

SOLUTION:

One way to solve this problem is to notice that the source branch and the leftmost two-resistor branch are connected in parallel between node A and ground. Switching the order of these branches and replacing the two resistors by their series equivalent yields the circuit of Figure 2–56(b). A source transformation yields the circuit in Figure 2–56(c). This circuit contains a current source $v_S/2R$ in parallel with two $2R$ resistances whose equivalent resistance is

$$R_{EQ} = 2R\|2R = \frac{2R \times 2R}{2R + 2R} = R$$

Applying a source transformation to the current source $v_S/2R$ in parallel with R_{EQ} results in the circuit of Figure 2–56(d), where

$$V_{EQ} = \left(\frac{v_S}{2R}\right) \times R_{EQ} = \left(\frac{v_S}{2R}\right)R = \frac{v_S}{2}$$

Finally, applying voltage division in Figure 2–52(d) yields

$$v_O = \left(\frac{2R}{R + R + 2R}\right)\frac{v_S}{2} = \frac{v_S}{4}$$

Exercise 2–30

Find the voltage across the current source in Figure 2–57.

Answer: $v_S = -0.225\text{ V}$

2–7   COMPUTER-AIDED CIRCUIT ANALYSIS

In this text, we use three types of computer programs to illustrate computer-aided circuit analysis, namely spreadsheets, math solvers, and circuit simulators. Practicing engineers routinely use these tools to analyze and design circuits, so it is important to learn how to use them effectively. The purpose of including computer examples in this book is to help you develop an analysis style that includes the intelligent use of computer tools. As you develop your style, always keep in mind that computer tools are not problem solvers. *You* are the problem solver. Computer tools can be very useful, even essential, once you have defined the problem. However, they do not substitute for an understanding of the fundamentals needed to formulate the problem, identify a practical approach, and interpret analysis results.

There are about 100 worked examples and exercises in the text that use computer tools. The spreadsheet examples use Microsoft Excel. The math solver examples use MATLAB Release 2010b by The MathWorks, Inc. The circuit simulation examples use OrCAD Capture CIS Release 16.3 by Cadence Design Systems, Inc.

Our objective is to illustrate the effective use of computer tools rather than develop your ability to operate these specific software programs. Although this book provides examples as helpful starting points, it does not emphasize the details of how to operate any of these software tools. We assume that you learned how to operate computer tools in previous courses or have enough familiarity with your computer's operating system to learn how to do so using online tutorials or any of a number of commercially available manuals.

The following discussion gives a brief overview of circuit simulation and provides one example of applying a math solver to a circuit analysis problem. Many more examples will follow in subsequent chapters. Web Appendix C provides additional information on computer programs that support circuit analysis.

CIRCUIT SIMULATION USING OrCAD

Most circuit simulation programs are based on a circuit analysis package called SPICE, which is an acronym for **S**imulation **P**rogram with **I**ntegrated **C**ircuit **E**mphasis. Figure 2–58 is a block diagram summarizing the major features of a SPICE-based circuit simulation program. The inputs are a circuit diagram and the type of analysis required. In contemporary programs the circuit diagram is drawn on the monitor screen using a graphical schematic editor. When the circuit diagram is complete, the input processor performs a *schematic capture,* a process that documents the circuit in what is called a *netlist.* To initiate circuit simulation, the input processor sends the netlist and analysis commands to the simulation processor. If the netlist file is not properly prepared, the simulation will not run or (worse) will return erroneous results. Hence, it is important to check the netlist to be sure that the circuit it defines is the one you want to analyze.

The simulation processor uses the netlist together with data from the device library to formulate a set of equations that describes the circuit. The simulation

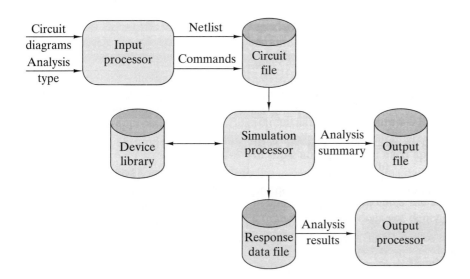

FIGURE 2–58 *Flow diagram for circuit simulation programs.*

processor then solves the equations, writes a dc analysis summary to a standard SPICE output file, and writes the other analysis results to a response data file. For simple dc analysis, the desired response data are accessible by examining the SPICE output file. For other types of analysis, the output processor can be used to generate graphical plots of the data in the response data file.

In the OrCAD Capture CIS Demo edition, the input processor is called OrCAD *Capture*, the simulation processor is a version of SPICE called *PSPICE*, and the output processor is called *Probe*. These three programs allow the circuit designer to draw a circuit graphically and perform numerous analyses on the circuit. The analysis results are then viewed on a computer monitor. Web Appendix C provides examples that use OrCAD to simulate circuit performance. In general, OrCAD is used to view currents, voltages, and powers at various nodes or devices in a circuit. The following example illustrates the main steps in solving a circuit analysis problem via simulation.

EXAMPLE 2–28

Use OrCAD to find all the voltages and currents for the circuit in Figure 2–22 (Example 2–10).

SOLUTION:

The main steps required to solve this problem using OrCAD are as folows:

1. Graphically draw the circuit schematic using OrCAD Capture.
2. Tell the program what simulation is desired by setting the profile parameters.
3. Simulate the circuit and record the results.

Create the circuit schematic by using the "Place Parts" menu option and placing the appropriate parts (three resistors, a dc voltage source, and a ground — the "0/ source") on the workspace. Click on the resistor values and adjust each to match the problem. Do the same for the voltage source. Connect each part using the wiring tool. Figure 2–59 shows the resulting OrCAD Capture schematic.

FIGURE 2–59

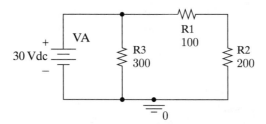

Note that the physical layout of the OrCAD schematic looks different than that of Figure 2–22. OrCAD does not allow for diagonal part placement. Although they look different, the two circuits have the same connections and are electrically and functionally equivalent. You can verify this fact by writing device and connection equations for both of the circuits.

The next step requires that you tell OrCAD what type of simulation you wish to run. Under the "PSpice" menu option, select "New Simulation Profile." After you provide a name for the simulation, the window shown in Figure 2–60 will open. Select "Bias Point" under *Analysis type* and "General Settings" under *Options*. Then select "Run" under the PSpice menu, or simply click on the triangular run button to run the simulation.

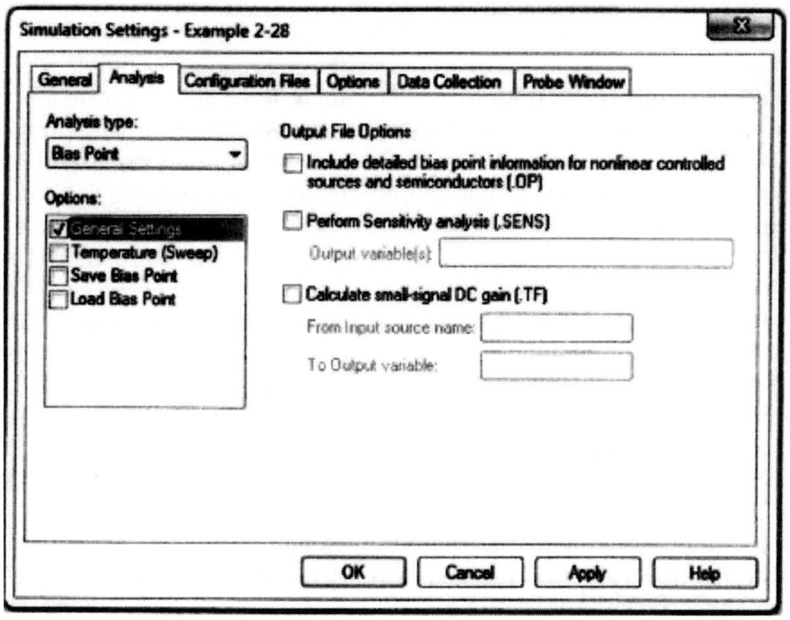

FIGURE 2–60

Copyright © Cadence Design Systems, Inc. Used with permission.

OrCAD allows you to see the simulation results two ways. You can click on the Probe icon that appears as you run the simulation. A black screen appears. Click on the top menu "View" and select "Output File." The voltages and currents appear along with other information about the circuit. Or, simpler, on the screen with the circuit diagram, click the "V" and "I" buttons and the results appear right on the schematic as shown on Figure 2–61.

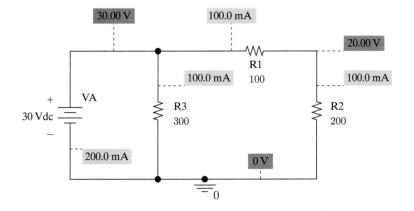

FIGURE 2–61

Interpreting the results from the simulation, we have

$$v_1 = 10\,\text{V} \qquad v_2 = 20\,\text{V} \qquad v_3 = 30\,\text{V} \qquad v_A = 30\,\text{V}$$
$$i_1 = 100\,\text{mA} \quad i_2 = 100\,\text{mA} \quad i_3 = 100\,\text{mA} \quad i_A = -200\,\text{mA}$$

which are consistent with the answers found in Example 2–10.

The problem in Example 2–28 could also be solved using MATLAB. You can find its solution in Web Appendix C. The next section provides a discussion of which approach one should consider when using software to solve circuit problems. ∎

COMPARISON OF COMPUTER-AIDED TECHNIQUES

We applied two computer-based tools to solve the circuit analysis problem originally presented as Example 2–10. Both tools efficiently arrived at a complete and correct solution and may offer some advantages over the manual approach presented in Example 2–10. As you explore these tools, you will develop the experience and judgment to know when each tool is appropriate to assist in solving a problem. The circuit simulation tools allow for visual representations of the circuits and efficiently compute numerical solutions when all of the parameter values are known. A math solver quickly manipulates a set of equations to find a solution, but you must relate the numerical results back to the original schematic to get a full representation of the answer. Math solver software can offer distinct advantages over circuit simulation software when the circuit contains parameter values expressed as variables. The circuit simulation software cannot directly handle this type of the problem, but the math solver software accommodates it with only minor changes, as we will explore in Chapter 3. As we develop circuit analysis and design techniques throughout the text, we will continue to emphasize the advantages and disadvantages of the various computer-based tools so that you can expand your judgment and efficiently apply the appropriate tools for each problem. No matter what technique one uses, good "engineering sense" must predominate. Results that just do not seem right should be challenged and verified. As you grow in knowledge and experience, you will develop and enhance this important engineering quality.

SUMMARY

- An **electrical device** is a real physical entity, while a **circuit element** is a mathematical or graphical model that approximates major features of the device.

- Two-terminal circuit elements are represented by a circuit symbol and are characterized by a single constraint imposed on the associated current and voltage variables.

- An **electrical circuit** is an interconnection of electrical devices. The interconnections form nodes and loops.

- A **node** is an electrical juncture of the terminals of two or more devices. A **loop** is a closed path formed by tracing through a sequence of devices without passing through any node more than once.

- Device interconnections in a circuit lead to two connection constraints: **Kirchhoff's current law (KCL)** states that the algebraic sum of currents at a node is zero at every instant; and **Kirchhoff's voltage law (KVL)** states that the algebraic sum of voltages around any loop is zero at every instant.

- A pair of two-terminal elements are connected in **parallel** if they form a loop containing no other elements. The same voltage appears across any two elements connected in parallel.

- A pair of two-terminal elements are connected in **series** if they are connected at a node to which no other elements are connected. The same current exists in any two elements connected in series.

- Two circuits are said to be **equivalent** if they each have the same i–v constraints at a specified pair of terminals.

- Series and parallel equivalence and **voltage** and **current division** are important tools in circuit analysis and design.

- **Source transformation** changes a voltage source in series with a resistor into an equivalent current source in parallel with a resistor, or vice versa.

- **Circuit reduction** is a method of solving for selected signal variables in ladder circuits. The method involves sequential application of the series/parallel equivalence rules, source transformations, and the voltage/current division rules. The reduction sequence used depends on the variables to be determined and the structure of the circuit and is not unique.

- Computer-aided circuit analysis applies spreadsheets, circuit simulation, or math solver software to analyze circuit problems efficiently. The tools allow for visual solutions and can eliminate the need to perform tedious or lengthy manual calculations.

PROBLEMS

OBJECTIVE 2-1 ELEMENT CONSTRAINTS (SECT. 2-1)

Given a two-terminal element with one or more electrical variables specified, use the element i-v constraint to find the magnitude and direction of the unknown variables.
See Examples 2-1, 2-2, 2-3, and 2-4 and Exercises 2-1, 2-2, and 2-3.

2-1 The current through a 56-kΩ resistor is 2.2 mA. Find the voltage across the resistor.

2-2 The voltage across a particular resistor is 6.23 V and the current is 2.75 mA. What is the actual resistance of the resistor? Using the inside back cover, what is the likely standard value of the resistor?

2-3 A 100-kΩ resistor dissipates 100 mW. Find the current through the resistor.

2-4 The conductance of a particular resistor is 0.5 mS. Find the current through the resistor when connected across a 5-V source.

2-5 In Figure P2-5 the resistor dissipates 25 mW. Find R_x.

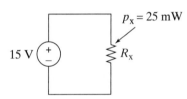

FIGURE P2-5

2-6 In Figure P2-6 find R_x and the power delivered to the resistor.

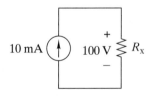

FIGURE P2-6

2-7 A resistor found in the lab has three orange stripes followed by a gold stripe. An ohmmeter measures its resistance as 34.9 kΩ. Is the resistor properly color coded? (See inside back cover for color code.)

2-8 The i-v characteristic of a nonlinear resistor is $v = 82i + 0.18i^3$.
(a) Calculate v and p for $i = \pm 0.5, \pm 1, \pm 2, \pm 5,$ and ± 10 A.
(b) Find the maximum error in v when the device is treated as an 82-Ω linear resistance on the range $|i| < 0.5$ A.

2-9 A 100-kΩ resistor has a power rating of 0.125 W. Find the maximum voltage that can be applied to the resistor.

2-10 A certain type of film resistor is available with resistance values between 10 Ω and 100 MΩ. The maximum ratings for all resistors of this type are 500 V and 0.25 W. Show that the voltage rating is the controlling limit for $R > 1$ MΩ, and that the power rating is the controlling limit when $R < 1$ MΩ.

2-11 Figure P2-11 shows the circuit symbol for a class of two-terminal devices called diodes. The i-v relationship for a specific pn junction diode is $i = 2 \times 10^{-16} (e^{40v} - 1)$ A

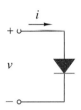

FIGURE P2-11

(a) Use this equation to find i and p for $v = 0, \pm 0.1, \pm 0.2, \pm 0.4,$ and ± 0.8 V. Use these data to plot the i-v characteristic of the element.
(b) Is the diode linear or nonlinear, bilateral or nonbilateral, and active or passive?
(c) Use the diode model to predict i and p for $v = 5$ V. Do you think the model applies to voltages in this range? Explain.
(d) Repeat (c) for $v = -5$ V.

2-12 A thermistor is a temperature-sensing element composed of a semiconductor material that exhibits a large change in resistance proportional to a small change in temperature. A particular thermistor has a resistance of 5 kΩ at 25 °C. Its resistance is 340 Ω at 100 °C. Assuming a straight-line relationship between these two values, at what temperature will the thermistor's resistance equal 1 kΩ?

OBJECTIVE 2-2 CONNECTION CONSTRAINTS (SECT. 2-2)

Given a circuit composed of two-terminal elements:
(a) Identify nodes and loops in the circuit
(b) Identify elements connected in series and in parallel.
(c) Use Kirchhoff's laws (KCL and KVL) to find selected signal variables.
See Examples 2-5, 2-6, and 2-7 and Exercises 2-4, 2-5, 2-6, 2-7, and 2-8

2-13 In Figure P2-13 $i_2 = -5$ A and $i_3 = 2$ A. Find i_1 and i_4.

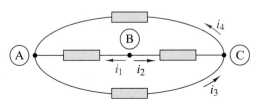

FIGURE P2-13

2–14 In Figure P2–14 $v_1 = 3$ V and $v_3 = 5$ V. Find v_2, v_4 and v_5.

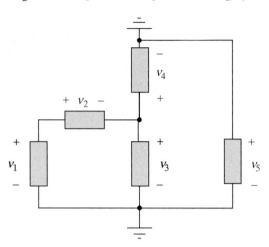

FIGURE P2–14

2–15 For the circuit in Figure P2–15:

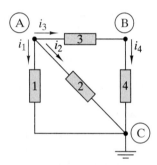

FIGURE P2–15

(a) Identify the nodes and at least two loops.
(b) Identify any elements connected in series or in parallel.
(c) Write KCL and KVL connection equations for the circuit.

2–16 In Figure P2–15 $i_2 = -20$ mA and $i_4 = 10$ mA. Find i_1, and i_3.

2–17 For the circuit in Figure P2–17:

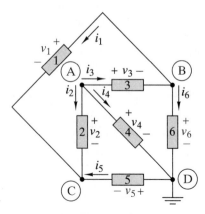

FIGURE P2–17

(a) Identify the nodes and at least three loops in the circuit.
(b) Identify any elements connected in series or in parallel.
(c) Write KCL and KVL connection equations for the circuit.

2–18 In Figure P2-17 $v_2 = 10$ V, $v_3 = -10$ V, and $v_4 = 3$ V. Find v_1, v_5, and v_6.

2–19 In many circuits the ground is often the metal case that houses the circuit. Occasionally a failure occurs whereby a wire connected to a particular node touches the case causing that node to become connected to ground. Suppose that in Figure P2-17 Node B accidently touches ground. How would that affect the voltages found in problem 2–18?

2–20 The circuit in Figure P2–20 is organized around the three signal lines A, B, and C.

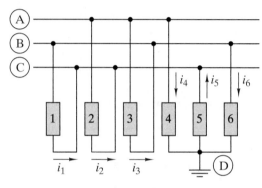

FIGURE P2–20

(a) Identify the nodes and at least three loops in the circuit.
(b) Write KCL connection equations for the circuit.
(c) If $i_1 = -20$ mA, $i_2 = -12$ mA, and $i_3 = 50$ mA, find i_4, i_5, and i_6.
(d) Show that the circuit in Figure P2–20 is identical to that in Figure P2–17.

2–21 In Figure P2–21 $v_2 = 10$ V, $v_4 = 5$ V, and $v_5 = 15$ V. Find v_1, v_3, and v_6.

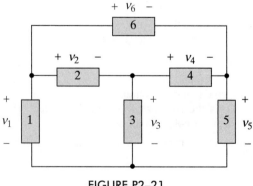

FIGURE P2–21

2–22 In Figure P2-22 $i_1 = 25$ mA, $i_2 = 10$ mA, and $i_3 = -15$ mA. Find i_4 and i_5.

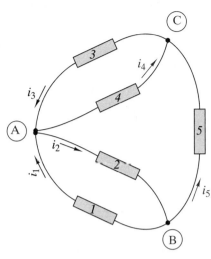

FIGURE P2–22

2–23 (a) Use the passive sign convention to assign voltage variables consistent with the currents in Figure P2-22. Write three KVL connection equations using these voltage variables.
(b) If $v_4 = 0$ V, what can be said about the voltages across all the other elements?

2–24 The KCL equations for a three-node circuit are:

Node A $\quad -i_1 + i_2 - i_4 = 0$
Node B $\quad -i_1 - i_3 + i_5 = 0$
Node C $\quad i_1 + i_3 + i_4 - i_5 = 0$

Draw the circuit diagram and indicate the reference directions for the element currents.

OBJECTIVE 2–3 COMBINED CONSTRAINTS (SECT. 2–3)

Given a linear resistance circuit, use the element constraints and connection constraints to find selected signal variables. See Examples 2–8, 2–9, 2–10, 2–11, and 2–12 and Exercise 2–9, 2–10, 2–11, 2–12, and 2–13.

2–25 Find v_x and i_x in Figure P2–25.

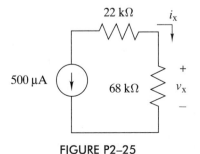

FIGURE P2–25

2–26 Find v_x and i_x in Figure P2–26.

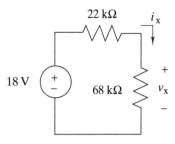

FIGURE P2–26

2–27 Find v_x and i_x in Figure P2–27. Compare the results of your answers with those in problem 2–26. What effect did adding the 33-kΩ resistor have on the overall circuit? Why isn't i_y zero?

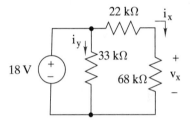

FIGURE P2–27

2–28 A modeler wants to light his model building using miniature grain-of-wheat light bulbs connected in parallel as shown in Figure P2–28. He uses two 1.5-V "C-cells" to power his lights. He wants to use as many lights as possible, but wants to limit his current drain to 500 μA to preserve the batteries. If each light has a resistance of 36 kΩ, how many lights can he install and still be under his current limit?

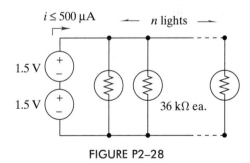

FIGURE P2–28

2–29 Find v_x and i_x in Figure P2–29.

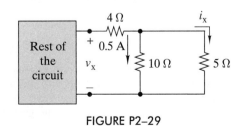

FIGURE P2–29

2–30 In Figure P2–30:

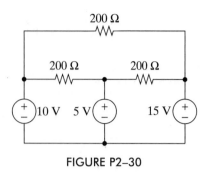

FIGURE P2–30

(a) Assign a voltage and current variable to every element.
(b) Use KVL to find the voltage across each resistor.
(c) Use Ohm's law to find the current through each resistor.
(d) Use KCL to find the current through each voltage source.

2–31 Find the power provided by the source in Figure P2–31.

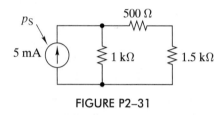

FIGURE P2–31

2–32 Figure P2–32 shows a subcircuit connected to the rest of the circuit at four points.

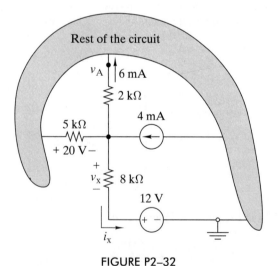

FIGURE P2–32

(a) Use element and connection constraints to find v_x and i_x.
(b) Show that the sum of the currents into the rest of the circuit is zero.
(c) Find the voltage v_A with respect to the ground in the circuit.

2–33 In Figure P2–33 $i_x = 0.5$ mA. Find the value of R.

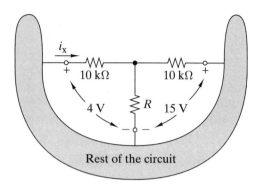

FIGURE P2–33

2–34 Figure P2–34 shows a resistor with one terminal connected to ground and the other connected to an arrow. The arrow symbol is used to indicate a connection to one terminal of a voltage source whose other terminal is connected to ground. The label next to the arrow indicates the source voltage at the ungrounded terminal. Find the voltage across, current through, and power dissipated in the resistor.

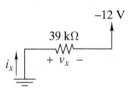

FIGURE P2–34

OBJECTIVE 2–4 EQUIVALENT CIRCUITS (SECT. 2–4)

(a) Given a circuit consisting of linear resistors, find the equivalent resistance between a specified pair of terminals.
(b) Given a circuit consisting of a source-resistor combination, find an equivalent source-resistor circuit.
See Example 2–13, 2–14, and 2–15 and Exercises 2–14, 2–15, 2–16, 2–17, and 2–18.

2–35 Find the equivalent resistance R_{EQ} in Figure P2–35.

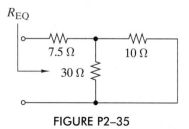

FIGURE P2–35

2–36 Find the equivalent resistance R_{EQ} in Figure P2–36.

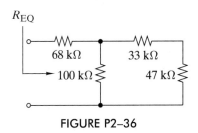

FIGURE P2–36

2–37 Find the equivalent resistance if R_{EQ} in Figure P2–37.

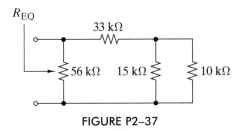

FIGURE P2–37

2–38 Equivalent resistance is defined at a particular pair of terminals. In the following figure the same circuit is looked at from two different terminal pairs. Find the equivalent resistances R_{EQ1} and R_{EQ2} in Figure P2–38. Note that in calculating R_{EQ2} the 33-kΩ resistor is connected to an open circuit and therefore doesn't affect the calculation.

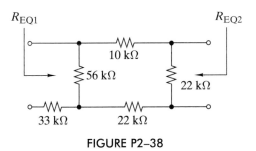

FIGURE P2–38

2–39 Find R_{EQ} in Figure P2–39 when the switch is open. Repeat when the switch is closed.

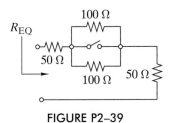

FIGURE P2–39

2–40 Find R_{EQ} between nodes A and B for each of the circuits in Figure P2–40. What conclusion can you draw about resistors of the same value connected in parallel?

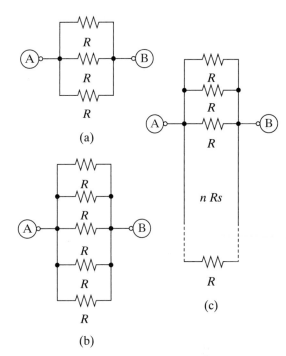

(a)

(c)

(b)

FIGURE P2–40

2–41 Show how the circuit in Figure P2–41 could be connected to achieve a resistance of 100 Ω, 200 Ω, 150 Ω, 50 Ω, 25 Ω, 33.3 Ω, and 133.3 Ω.

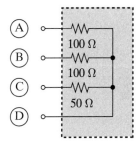

FIGURE P2–41

2–42 In Figure P2–42 find the equivalent resistance between terminals A-B, A-C, A-D, B-C, B-D, and C-D.

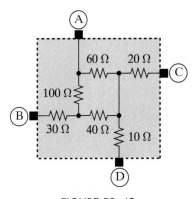

FIGURE P2–42

2–43 In Figure P2–43 find the equivalent resistance between terminals A-B, A-C, A-D, B-C, B-D, and C-D.

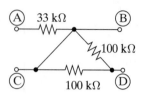

FIGURE P2–43

2–44 Select a value of R_L in Figure P2–44 so that $R_{EQ} = 6$ kΩ. Repeat for $R_{EQ} = 5$ kΩ.

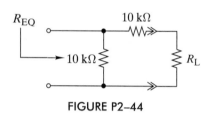

FIGURE P2–44

2–45 Using no more than four 1-kΩ resistors, show how the following equivalent resistors can be constructed: 2 kΩ, 500 Ω, 1.5 kΩ, 333 Ω, 200 Ω, and 400 Ω.

2–46 Do a source transformation at terminals A and B for each practical source in Figure P2–46.

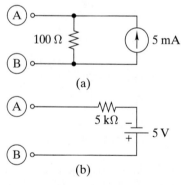

FIGURE P2–46

2–47 Find the equivalent practical voltage source at terminals A and B in Figure P2–47.

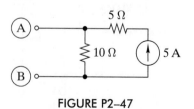

FIGURE P2–47

2–48 In Figure P2–48 the *i-v* characteristic of network N is $v + 50i = 5$ V. Find the equivalent practical current source for the network.

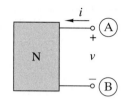

FIGURE P2–48

2–49 Select the value of R_x in Figure P2–49 so that $R_{EQ} = 75$ kΩ.

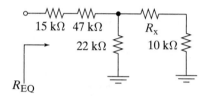

FIGURE P2–49

2–50 Two 10-kΩ potentiometers (a variable resistor whose value between the two ends is 10 kΩ and between one end and the wiper — the third terminal — can range from 0 Ω to 10 k Ω) are connected as shown in Figure P2–50. What is the range of R_{EQ}?

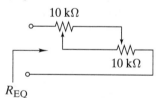

FIGURE P2–50

2–51 Select the value of R in Figure P2–51 so that $R_{AB} = R_L$.

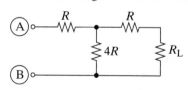

FIGURE P2–51

2–52 What is the range of R_{EQ} in Figure P2–52?

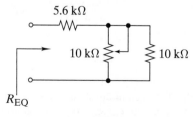

FIGURE P2–52

2–53 Find the equivalent resistance between terminals A and B in Figure P2–53.

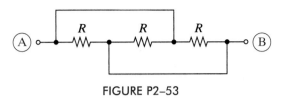

FIGURE P2–53

OBJECTIVE 2–5 VOLTAGE AND CURRENT DIVISION (SECT. 2–5)

(a) Given a linear resistive circuit with elements connected in series or parallel, use voltage or current division to find specified voltages or currents.

(b) Design a voltage or current divider that delivers specified outputs signals.

See Examples 2–16, 2–17, 2–18, 2–19, 2–20, 2–21, 2–22, and 2–23 and Exercises 2–19, 2–20, 2–21, 2–22, 2–23, 2–24, and 2–25.

2–54 Use voltage division in Figure P2–54 to find v_x.

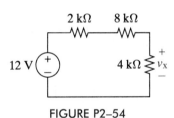

FIGURE P2–54

2–55 Use voltage division in Figure P2–55 to obtain an expression for v_L in terms of R, R_L, and v_S.

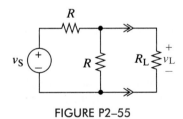

FIGURE P2–55

2–56 Use current division in Figure P2–56 to find i_x and v_x.

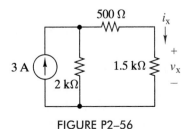

FIGURE P2–56

2–57 Use current division in Figure P2–57 to find an expression for v_L in terms of R, R_L and i_S.

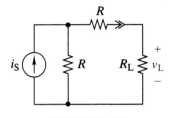

FIGURE P2–57

2–58 Find i_x, i_y, and i_z in Figure P2–58.

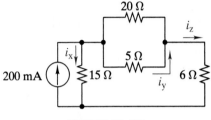

FIGURE P2–58

2–59 Find v_O in the circuit of Figure P2–59.

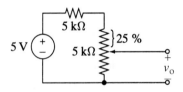

FIGURE P2–59

2–60 ▲ The 1-kΩ load in Figure P2–60 needs 5 V across it to operate correctly. Where should the wiper on the potentiometer be set (R_x) to obtain the desired output voltage?

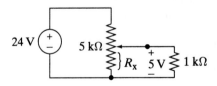

FIGURE P2–60

2–61 Find the range of values of v_O in Figure P2–61.

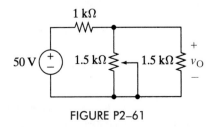

FIGURE P2–61

2–62 Use current division in the circuit of Figure P2–62 to find R_x so that the voltage out is 3 V.

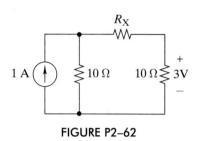

FIGURE P2–62

2–63 **A** Figure P2–63 shows a voltage bridge circuit, that is, two voltage dividers in parallel with a source v_S. One resistor R_X is variable. The goal is often to "balance" the bridge by making $v_X = 0$ V. Derive an expression for R_X in terms of the other resistors for when the bridge is balanced.

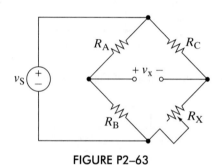

FIGURE P2–63

2–64 **A** Ideally, a voltmeter has infinite internal resistance and can be placed across any device to read the voltage without affecting the result. A particular digital multimeter (DMM), a common laboratory tool, is connected across the circuit shown in Figure P2–64. The expected voltage was 10.2 V. However, the DMM reads 7.73 V. The large, but finite, internal resistance of the DMM was "loading" the circuit and causing a wrong measurement to be made. Find the value of the internal resistance R_M of this DMM.

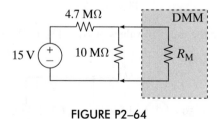

FIGURE P2–64

2–65 **D** Select values for R_1, R_2, and R_3 in Figure P2–65 so the voltage divider produces the two output voltages shown.

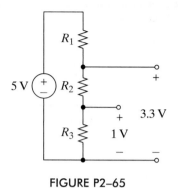

FIGURE P2–65

2–66 **D** Select a value of R_x in Figure P2–66 so that $v_L = 2$V.

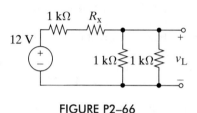

FIGURE P2–66

2–67 **D** Select a value of R_X in Figure P2–67 so that $v_L = 2$ V. Repeat for 4 V and 6 V. Caution: R_x must be positive.

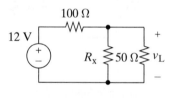

FIGURE P2–67

OBJECTIVE 2–6 CIRCUIT REDUCTION (SECT. 2–6)

Given a linear resistive circuit, find selected signal variables using successive application of series and parallel equivalence, source transformations, and voltage and current division. See Examples 2–24, 2–25, 2–26, and 2–27 and Exercises 2–26, 2–27, 2–28, 2–29, and 2–30.

2–68 Use circuit reduction to find v_x and i_x in Figure P2–68.

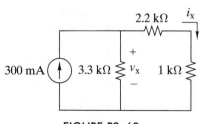

FIGURE P2–68

2–69 Use circuit reduction to find v_x, i_x, and p_x in Figure P2–69. Repeat using OrCAD.

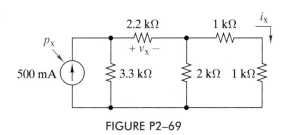

FIGURE P2–69

2–70 Use circuit reduction to find v_x and i_x in Figure P2–70.

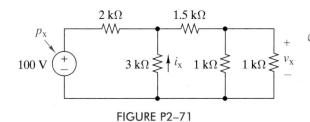

FIGURE P2–70

2–71 Use circuit reduction to find v_x, i_x, and p_x in Figure P2–71.

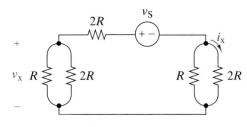

FIGURE P2–71

2–72 Use circuit reduction to find v_x and i_x in Figure P2–72.

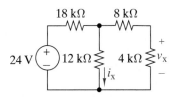

FIGURE P2–72

2–73 Use source transformation to find i_x in Figure P2–73.

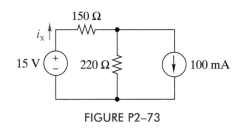

FIGURE P2–73

2–74 Select a value for R_x so that $i_x = 0$ A in Figure P2–74.

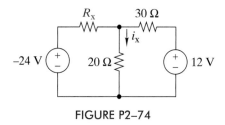

FIGURE P2–74

2–75 Use source transformations in Figure P2–75 to relate v_O to v_1, v_2, and v_3.

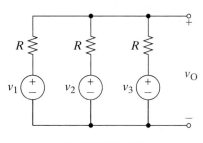

FIGURE P2–75

2–76 The current through R_L in Figure P2–76 is 100 mA. Use source transformations to find R_L. Validate your answer using OrCAD.

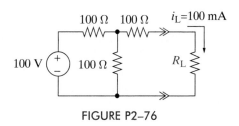

FIGURE P2–76

2–77 Select R_x so that 50 V are across it in Figure P2–77.

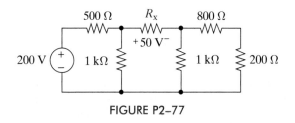

FIGURE P2–77

2–78 The box in the circuit in Figure P2–78 is a resistor whose value can be anywhere between 8 kΩ and 80 kΩ. Use circuit reduction to find the range of values of v_x.

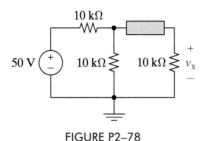

FIGURE P2–78

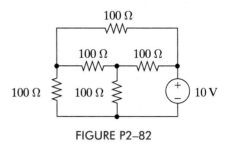

FIGURE P2–82

OBJECTIVE 2–7 COMPUTER-AIDED CIRCUIT ANALYSIS (SECT. 2–7)

Given an appropriate linear circuit, use circuit simulation and/or computational software to solve for the desired response. See Example 2–28 and Web Appendix C.

2–79 A circuit is found to have the following element and connection equations:

$$v_1 = 24 \text{ V}$$
$$v_2 = 8k\, i_2$$
$$v_3 = 5k\, i_3$$
$$v_4 = 4k\, i_4$$
$$v_5 = 16k\, i_5$$
$$-v_1 + v_2 + v_3 = 0$$
$$-v_3 + v_4 + v_5 = 0$$
$$i_1 + i_2 = 0$$
$$-i_2 + i_3 + i_4 = 0$$
$$-i_4 + i_5 = 0$$

Use MATLAB to solve for all of the unknown voltages and currents associated with this circuit. Sketch one possible schematic that matches the given equations.

2–80 Consider the circuit of Figure P2–80. Use MATLAB to find all of the voltages and currents in the circuit and find the power provided by the source.

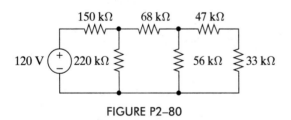

FIGURE P2–80

2–81 Consider the circuit of Figure P2–80 again. Use OrCAD to find all of the voltages, currents and power provided used or provided. Verify that the sum of all power in the circuit is zero.

2–82 The circuit of Figure P2–82 is called a *"bridge-T"* circuit. Use OrCAD to find all of the voltages and currents in the circuit.

INTEGRATING PROBLEMS

2–83 Nonlinear Device Characteristics **A**
The circuit in Figure P2–83 is a parallel combination of a 50-Ω linear resistor and a varistor whose *i-v* characteristic is $i_y = 2.6 \times 10^{-5}v^3$. For a small voltage, the varistor current is quite small compared to the resistor current. For large voltages, the varistor dominates because its current increases more rapidly with voltage.

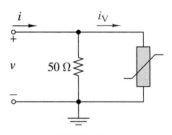

FIGURE P2–83

(a) Plot the *i-v* characteristic of the parallel combination.

(b) State whether the parallel combination is linear or non-linear, active or passive, and bilateral or nonbilateral.

(c) Find the range of voltages over which the resistor current is at least 10 times as large as the varistor current.

(d) Find the range of voltages over which the varistor current is at least 10 times as large as the resistor current.

2–84 Transistor Biasing **D**
The circuit shown in Figure P2–84 is a typical biasing arrangement for a BJT-type transistor. The actual transistor for this problem can be modeled as 0.7-V battery in series with a 200-kΩ resistor. Biasing allows signals that have both a positive and negative variation to be properly amplified by the transistor. Select the two biasing resistors R_A and R_B so that 3 V ± 0.1 V appears across R_B.

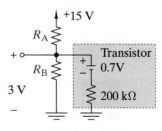

FIGURE P2–84

2–85 Center Tapped Voltage Divider **A**
Figure P2–85 shows a voltage divider with the center tap connected to ground. Derive equations relating v_A and v_B to v_S, R_1, and R_2.

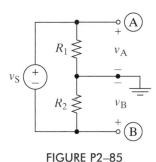

FIGURE P2–85

2–86 Active Transducer **A**
Figure P2–86 shows an active transducer whose resistance R (V_T) varies with the transducer voltage V_T as $R(V_T) = 0.5V_T^2 + 1$. The transducer supplies a current to a 12–Ω load. At what voltage will the load current equal 100 mA?

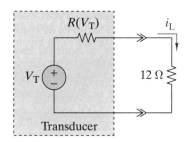

FIGURE P2–86

2–87 Interface Circuit Choice **E**
You have a practical voltage source that can be modeled as a 5-V ideal source in series with a 1-kΩ source resistor. You need to use your source to drive a 1-kΩ load that requires exactly 2 V across it. Two solutions are provided to you as shown in Figure P2–87. Validate that both meet the requirement then select the best solution and give the reason for your choice. Consider part count, standard parts, accuracy of meeting the spec, power consumed by the source, etc.

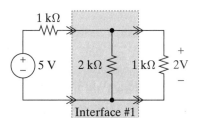
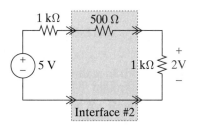

FIGURE P2–87

2–88 Programmable Voltage Divider **A**
Figure P2–88 shows a programmable voltage divider in which digital inputs b_0 and b_1 control complementary analog switches connecting a multitap voltage divider to the analog output v_O. The switch positions in the figure apply when digital inputs are low. When inputs go high the switch positions reverse. Find the analog output voltage for $(b_1, b_0) = (0, 0)$, $(0, 1)$, $(1, 0)$, and $(1, 1)$ when $V_{REF} = 12$ V.

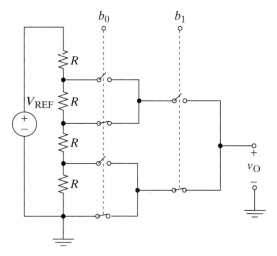

FIGURE P2–88

2–89 Analog Voltmeter Design **A D E**
Figure P2–89(a) shows a voltmeter circuit consisting of a D'Arsonval meter, two series resistors, and a two-position selector switch. A current of $I_{FS} = 400$ μA produces full-scale deflection of the D'Arsonval meter, whose internal resistance is R_M-25 Ω.

(a) **D** Select the series resistance R_1 and R_2 so a voltage v_x $= 100$ V produces full-scale deflection when the switch is in position A, and voltage $v_x = 10$ V produces full-scale deflection when the switch is in position B.

(b) **A** What is the voltage across the 20-kΩ resistor in Figure P2–89(b)? What is the voltage when the voltmeter in part (a) is set to position A and connected across the 20-kΩ resistor? What is the percentage error introduced connecting the voltmeter?

(c) **E** A different D'Arsonval meter is available with an internal resistance of 100 Ω and a full-scale deflection current of 100 μA. If the voltmeter in part (a) is redesigned using this D'Arsonval meter, would the error found in part (b) be smaller or larger? Explain.

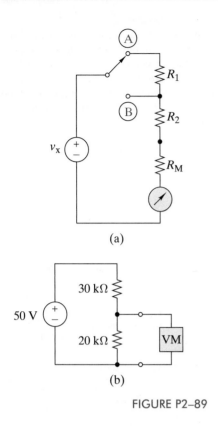

(a)

(b)

FIGURE P2–89

2–90 MATLAB Function for Parallel Equivalent Resistors **A**
Create a MATLAB function to compute the equivalent resistance of a set of resistors connected in parallel. The function has a single input, which is a vector containing the values of all of the resistors in parallel, and it has a single output, which is the equivalent resistance. Name the function "EQparallel" and test it with at least three different resistor combinations. At least one test should have three or more resistor values.

2–91 Finding an Equivalent Resistance using OrCAD **A**
Use OrCAD to find the equivalent resistance at terminals A and B of the resistor mesh shown in Figure P2–91. (*Hint:* use a 1-V dc source and measure the current provided by the source.)

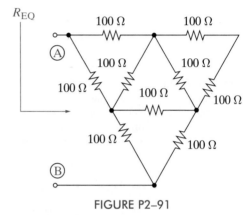

FIGURE P2–91

CHAPTER 3 CIRCUIT ANALYSIS TECHNIQUES

Assuming any system of linear conductors connected in such a manner that to the extremities of each one of them there is connected at least one other, a system having electromotive forces E_1, E_2. . . . E_3, no matter how distributed, we consider two points A and A' belonging to the system and having potentials V and V'. If the points A and A' are connected by a wire ABA', which has a resistance r, with no electromotive forces, the potentials of points A and A' assume different values from V and V', but the current i flowing through this wire is given by $i = (V - V')/(r + R)$ in which R represents the resistance of the original wire, this resistance being measured between the points A and A', which are considered to be electrodes.

Leon Charles Thévenin, 1883,
French Telegraph Engineer

Some History Behind This Chapter

Leon Charles Thévenin (1857–1926), a distinguished French telegraph engineer and teacher, was led to his theorem in 1883 following an extensive study of Kirchhoff's laws. Norton's theorem, the dual of Thévenin's theorem, was not proposed until 1926 by Edward L. Norton, an American electrical engineer working on long-distance telephony. Curiously, it turns out that the basic concept had been discovered earlier by Herman von Helmholtz, while studying electricity in animal tissue. Electrical engineering tradition credits Thévenin and Norton, perhaps because they worked in areas that offered practical applications for their results.

Why This Chapter Is Important Today

In this chapter you advance to studying general methods of analyzing circuits and to the major theorems that describe linear circuits. These theorems are conceptual tools that give new insight into circuit behavior. Most importantly, you will be introduced to the design of interface circuits; your first exposure to devising a circuit to perform a predetermined function.

Chapter Sections

Chapter Learning Objectives

3-1 General Circuit Analysis (Sects. 3–1 to 3–2)

Given a linear resistance circuit:
(a) (Formulation) Write node-voltage or mesh-current equations for the circuit.
(b) (Solution) Solve the equations from (a) for selected signal variables or input-output relationships using classical or software computational techniques.

3-2 Linearity Properties (Sect. 3–3)

Given a linear resistance circuit:
(a) Use the proportionality principle to find selected signal variables.
(b) Use the superposition principle to find selected signal variables.

3-3 Thévenin and Norton Equivalent Circuits (Sect. 3–4)

Given a linear resistance circuit:
(a) Find the Thévenin or Norton equivalent at a specified pair of terminals.
(b) Use the Thévenin or Norton equivalent to find the signals delivered to linear or nonlinear loads.

3-4 Maximum Signal Transfer (Sect. 3–5)

Given a linear resistance circuit:
(a) Find the maximum voltage, current, and power available at a specified pair of terminals.
(b) Find the resistive loads required to obtain the maximum available signal levels.

3-5 Interface Circuit Design (Sect. 3–6)

Given the signal transfer goals at a source-load interface, design one or more two-port interface circuits to achieve the goals and evaluate the alternative design solutions.

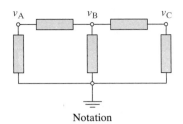

Notation

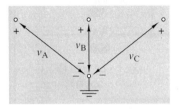

Interpretation

FIGURE 3–1 *Node-voltage definition and notation.*

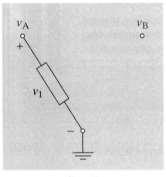

Case A

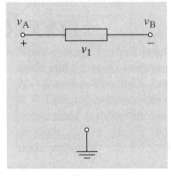

Case B

FIGURE 3–2 *Two possible connections of a two-terminal element.*

3–1 NODE-VOLTAGE ANALYSIS

Before describing node-voltage analysis, we first review the foundation for every method of circuit analysis. As noted in Sect. 2–3, circuit behavior is based on constraints of two types: (1) connection constraints (Kirchhoff's laws) and (2) device constraints (element i–v relationships). As a practical matter, however, using element voltages and currents to express the circuit constraints produces a large number of equations that must be solved simultaneously to find the circuit responses. For example, a circuit with only six devices requires us to treat 12 equations with 12 unknowns. Although this is not an impossible task using software tools like MATLAB, it is highly desirable to reduce the number of equations that must be solved simultaneously.

You should not abandon the concept of element and connection constraints. This method is vital because it provides the foundation for all methods of circuit analysis. In subsequent chapters, we use element and connection constraints many times to develop important ideas in circuit analysis.

Using node voltages instead of element voltages as circuit variables can reduce the number of equations that must be treated simultaneously. To define a set of node voltages, we first select a reference node. The **node voltages** are then defined as the voltages between the remaining nodes and the selected reference node. Figure 3–1 shows the notation used to define node-voltage variables. In this figure the reference node is indicated by the ground symbol and the node voltages are identified by a voltage symbol next to all the other nodes. This notation means that the positive reference mark for the node voltage is located at the node in question while the negative mark is at the reference node. Obviously, any circuit with N nodes involves N–1 node voltages since one node, the reference node, is known.

A fundamental property of node voltages needs to be covered at the outset. Suppose we are given a two-terminal element whose element voltage is labeled v_1. Suppose further that the terminal with the plus reference mark is connected to a node, say node A. The two cases shown in Figure 3–2 are the only two possible ways the other element terminal can be connected. In case A, the other terminal is connected to the reference node, in which case KVL requires $v_1 = v_A$. In case B, the other terminal is connected to a non-reference node, say node B, in which case KVL requires $v_1 = v_A - v_B$. This example illustrates the following fundamental property of node voltages:

> *If the Kth two-terminal element is connected between nodes X and Y, then the element voltage can be expressed in terms of the two node voltages as*

$$v_K = v_X - v_Y \tag{3–1}$$

> *where X is the node connected to the positive reference for element voltage v_K.*

Equation (3–1) is a KVL constraint at the element level. If node Y is the reference node, then by definition $v_Y = 0$ and Eq. (3–1) reduces to $v_K = v_X$. On the other hand, if node X is the reference node, then $v_X = 0$ and therefore $v_K = -v_Y$. The minus sign occurs here because the positive reference for the element is connected to the reference node. In any case, the important fact is that the voltage across any two-terminal element can be expressed as the difference of two node voltages, one of which may be zero.

Exercise 3–1 _____

The reference node and node voltages in the bridge circuit of Figure 3–3 are $v_A = 5\,\text{V}$, $v_B = 10\,\text{V}$, and $v_C = -3\,\text{V}$. Find the element voltages.

FIGURE 3–3

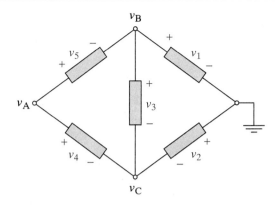

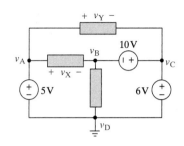

Answers: $v_1 = 10\,\text{V}$; $v_2 = 3\,\text{V}$; $v_3 = 13\,\text{V}$; $v_4 = 8\,\text{V}$; $v_5 = -5\,\text{V}$

FIGURE 3–4

Exercise 3–2

For the circuit in Figure 3–4, first find the node voltages and then find v_x and v_y.

Answers: $v_A = 5$ V, $v_B = -4$ V, $v_C = 6$ V, $v_D = 0$ V (ground). $v_X = v_A - v_B = 9$ V and $v_Y = v_A - v_C = -1$ V.

FORMULATING NODE-VOLTAGE EQUATIONS

To formulate a circuit description using node voltages we use element and connection analysis, except that the KVL connection equations are not explicitly written. Instead, we use the fundamental property of node analysis to express the element voltages in terms of the node voltages.

The circuit in Figure 3–5 demonstrates the formulation of node-voltage equations. In Figure 3–5 we have identified a reference node (indicated by the ground symbol), four element currents (i_0, i_1, i_2, and i_3), and two node voltages (v_A and v_B).

The KCL constraints at the two nonreference nodes are

$$\text{Node A: } -i_0 - i_1 - i_2 = 0$$
$$\text{Node B: } \qquad i_2 - i_3 = 0 \tag{3-2}$$

Using the fundamental property of node analysis, we use the element equations to relate the element currents to the node voltages.

$$\text{Resistor } R_1: i_1 = \frac{1}{R_1} v_A$$

$$\text{Resistor } R_2: i_2 = \frac{1}{R_2}(v_A - v_B)$$

$$\text{Resistor } R_3: i_3 = \frac{1}{R_3} v_B \tag{3-3}$$

$$\text{Current source: } i_0 = -i_S$$

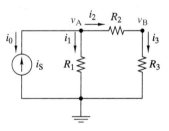

FIGURE 3–5 *Circuit for demonstrating node-voltage analysis.*

We have written six equations in six unknowns—four element currents and two node voltages. The element equations on the right side of Eqs. (3–3) involve unknown node voltages and the input signal i_S. Substituting the device constraints in Eqs. (3–3) into the KCL connection constraints in Eqs. (3–2) yields

$$\text{Node A: } i_S - \frac{1}{R_1} v_A - \frac{1}{R_2}(v_A - v_B) = 0$$

$$\text{Node B: } \qquad \frac{1}{R_2}(v_A - v_B) - \frac{1}{R_3} v_B = 0$$

which can be arranged into the following standard form:

$$\text{Node A:} \quad \left(\frac{1}{R_1}+\frac{1}{R_2}\right)v_A - \frac{1}{R_2}v_B = i_S$$

$$\text{Node B:} \quad -\frac{1}{R_2}v_A + \left(\frac{1}{R_2}+\frac{1}{R_3}\right)v_B = 0$$

(3–4)

In this standard form all of the unknown node voltages are grouped on one side and the independent sources on the other. This will facilitate the analysis.

By systematically eliminating the element currents, we have reduced the circuit description to two linear equations in the two unknown node voltages. The coefficients in the equations on the left side $(1/R_1 + 1/R_2, 1/R_2, 1/R_2 + 1/R_3)$ depend only on circuit parameters, while the right side contains the known input driving force i_S.

As noted previously, every method of circuit analysis must satisfy KVL, KCL, and the device i–v relationships. In developing the node-voltage equations in Eqs. (3–4), it may appear that we have not used KVL. However, KVL is satisfied because the equations $v_1 = v_A, v_2 = v_A - v_B,$ and $v_3 = v_B$ were used to write the right side of the element equations in Eqs. (3–3). The KVL constraints do not appear explicitly in the formulation of node equations, but they are implicitly included when the fundamental property of node analysis is used to write the element voltages in terms of the node voltages.

In summary, four steps are needed to develop node-voltage equations.

STEP 1   Select a reference node. Identify a node voltage at each of the remaining $N - 1$ nodes and a current with every element in the circuit.

STEP 2   Write KCL connection constraints in terms of the element currents at the $N - 1$ nonreference nodes.

STEP 3   Use the i–v relationships of the elements and the fundamental property of node analysis to express the element currents in terms of the node voltages.

STEP 4   Substitute the element constraints from step 3 into the KCL connection constraints from step 2 and arrange the resulting $N - 1$ equations in a standard form.

Writing node-voltage equations leads to $N - 1$ equations that must be solved simultaneously. If we write the element and connection constraints in terms of element voltages and currents, we must solve $2E$ simultaneous equations. The node-voltage method reduces the number of linear equations that must be solved simultaneously to $N - 1$. The reduction from $2E$ to $N - 1$ is particularly impressive in circuits with a large number of elements (large E) connected in parallel (small N).

EXAMPLE 3–1

Formulate node-voltage equations for the bridge circuit in Figure 3–6.

SOLUTION:

STEP 1   The reference node, node voltages, and element currents are shown in Figure 3–6.

STEP 2   The KCL constraints at the three nonreference nodes are:

$$\text{Node A: } i_0 - i_1 - i_2 = 0$$

$$\text{Node B: } i_1 - i_3 + i_5 = 0$$

$$\text{Node C: } i_2 - i_4 - i_5 = 0$$

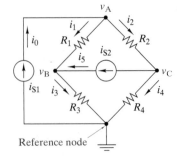

FIGURE 3–6

STEP 3   We write the element equations in terms of the node voltages and input signal sources.

$$i_0 = i_{S1} \qquad\qquad i_3 = \frac{1}{R_3} v_B$$

$$i_1 = \frac{1}{R_1}(v_A - v_B) \quad i_4 = \frac{1}{R_4} v_C$$

$$i_2 = \frac{1}{R_2}(v_A - v_C) \quad i_5 = i_{S2}$$

STEP 4   Substituting the element equations into the KCL constraints and arranging the result in standard form yields three equations in the three unknown node voltages.

$$\text{Node A: } \left(\frac{1}{R_1} + \frac{1}{R_2}\right) v_A - \frac{1}{R_1} v_B - \frac{1}{R_2} v_C = i_{S1}$$

$$\text{Node B: } -\frac{1}{R_1} v_A + \left(\frac{1}{R_1} + \frac{1}{R_3}\right) v_B + 0 v_C = i_{S2}$$

$$\text{Node C: } -\frac{1}{R_2} v_A + 0 v_B + \left(\frac{1}{R_2} + \frac{1}{R_4}\right) v_C = -i_{S2}$$

The three equations in three unknowns can be written in matrix form. Missing terms are zeros in the matrix.

$$\begin{bmatrix} \left(\frac{1}{R_1} + \frac{1}{R_2}\right) & -\frac{1}{R_1} & -\frac{1}{R_2} \\ -\frac{1}{R_1} & \left(\frac{1}{R_1} + \frac{1}{R_3}\right) & 0 \\ -\frac{1}{R_2} & 0 & \left(\frac{1}{R_2} + \frac{1}{R_4}\right) \end{bmatrix} \begin{bmatrix} v_A \\ v_B \\ v_C \end{bmatrix} = \begin{bmatrix} i_{S1} \\ i_{S2} \\ -i_{S2} \end{bmatrix}$$

This matrix equation is of the form $\mathbf{Ax} = \mathbf{b}$, where $\mathbf{A}$ is a 3×3 square matrix describing the circuit, $\mathbf{x}$ is a 3×1 column matrix of unknown node voltages, and $\mathbf{b}$ is a 3×1 column matrix of known inputs. Note that matrix $\mathbf{A}$ is symmetrical, that is, the terms on either side of the major diagonal are the same.   ∎

Exercise 3–3

For the circuit in Figure 3–6 replace the current source i_{S2} with a resistor R_5.

(a) Using the same node designations and reference node, formulate node-voltage equations for the modified circuit. Place the result in matrix form $\mathbf{Ax} = \mathbf{b}$.
(b) Is the resulting $\mathbf{A}$ matrix symmetrical?

Answers:

(a) $$\begin{bmatrix} \left(\frac{1}{R_1} + \frac{1}{R_2}\right) & -\frac{1}{R_1} & -\frac{1}{R_2} \\ -\frac{1}{R_1} & \left(\frac{1}{R_1} + \frac{1}{R_3} + \frac{1}{R_5}\right) & -\frac{1}{R_5} \\ -\frac{1}{R_2} & -\frac{1}{R_5} & \left(\frac{1}{R_2} + \frac{1}{R_4} + \frac{1}{R_5}\right) \end{bmatrix} \begin{bmatrix} v_A \\ v_B \\ v_C \end{bmatrix} = \begin{bmatrix} i_{S1} \\ 0 \\ 0 \end{bmatrix}$$

(b) Yes; see terms on either side of the diagonal above.

Exercise 3–4

Formulate node-voltage equations for the circuit of Figure 3–7 and place the results in matrix form $\mathbf{Ax}=\mathbf{b}$. Is the resulting matrix $\mathbf{A}$ symmetrical?

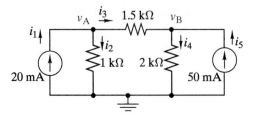

FIGURE 3–7

Answers:

$$\mathbf{A}\mathbf{x} = \mathbf{b} = \begin{bmatrix} 1.66 & -0.666 \\ -0.666 & 1.16 \end{bmatrix} \begin{bmatrix} v_A \\ v_B \end{bmatrix} = \begin{bmatrix} 20 \\ 50 \end{bmatrix}. \text{ Yes, the matrix is symmetrical.}$$

WRITING NODE-VOLTAGE EQUATIONS BY INSPECTION

The node-voltage equations derived in Example 3–1 have a symmetrical pattern, as shown by the dashed line drawn along the major diagonal. The coefficient of v_B in the node A equation and the coefficient of v_A in the node B equation are each the negative of the conductance connected between the nodes $(-1/R_1)$. Likewise, the coefficients of v_C in the node A equation and v_A in the node C equation are $-1/R_2$. The coefficients of v_A in the node A equation, v_B in the node B equation, and v_C in the node C equation are the sum of the conductances connected to the node in question. The two missing terms or zeros also fall symmetrically about the diagonal.

This symmetrical pattern always occurs in circuits containing only resistors and independent current sources. To understand why, consider any general two-terminal resistance R with one terminal connected to, say, node A. Then according to the fundamental property of node analysis there are only two possibilities. Either the other terminal of R is connected to the reference node, in which case the current *leaving* node A via resistance R is

$$i = \frac{1}{R}(v_A - 0) = \frac{1}{R}v_A$$

or else it is connected to another nonreference node, say, node B, in which case the current *leaving* node A via R is

$$i = \frac{1}{R}(v_A - v_B)$$

The pattern for node equations follows from these observations. The sum of the currents *leaving* any node A via resistances is

1. v_A times the sum of conductances connected to node A.
2. Minus v_B times the sum of conductances connected between nodes A and B.
3. Minus similar terms for all other nodes connected to node A by conductances.

Because of KCL, the sum of currents leaving node A via resistances plus the sum of currents directed away from node A by independent current sources must equal zero.

Understanding the aforementioned process allows us to write node-voltage equations by inspection without going through the intermediate steps involving the KCL constraints and the element equations or even labeling currents. For example, the circuit in Figure 3–8 contains two independent current sources and four resistors. Starting with node A, the sum of conductances connected to node A is $1/R_1 + 1/R_2$. The conductance between nodes A and B is $1/R_2$. The reference direction for the source current i_{S1} is into node A and for i_{S2} is directed away from node A. Pulling all of the observations together, we write the sum of currents directed out of node A as

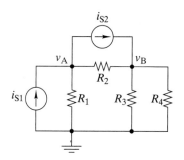

FIGURE 3–8 *Circuit for demonstrating how to write node-voltage equations by inspection.*

$$\text{Node A}: \quad \left(\frac{1}{R_1}+\frac{1}{R_2}\right)v_A - \frac{1}{R_2}v_B - i_{S1} + i_{S2} = 0 \qquad (3\text{–}5)$$

Similarly, the sum of conductances connected to node B is $1/R_2 + 1/R_3 + 1/R_4$, the conductance connected between nodes B and A is again $1/R_2$, and the source current i_{S2} is directed toward node B. These observations yield the following node-voltage equation.

$$\text{Node B}: \quad \left(\frac{1}{R_2}+\frac{1}{R_3}+\frac{1}{R_4}\right)v_B - \frac{1}{R_2}v_A - i_{S2} = 0 \qquad (3\text{–}6)$$

Rearranging Eqs. (3–5) and (3–6) in standard form yields

$$\text{Node A}: \quad \left(\frac{1}{R_1}+\frac{1}{R_2}\right)v_A - \frac{1}{R_2}v_B = i_{S1} - i_{S2}$$

$$\text{Node B}: \quad -\frac{1}{R_2}v_A + \left(\frac{1}{R_2}+\frac{1}{R_3}+\frac{1}{R_4}\right)v_B = i_{S2}$$

$$(3\text{–}7)$$

In the matrix form $\mathbf{Ax} = \mathbf{b}$, where $\mathbf{A}$ is a 2×2 square matrix describing the circuit, $\mathbf{x}$ is a 2×1 column matrix of unknown node voltages, and $\mathbf{b}$ is a 2×1 column matrix of known inputs, we can express the equations as

$$\begin{bmatrix} \left(\frac{1}{R_1}+\frac{1}{R_2}\right) & -\frac{1}{R_2} \\ -\frac{1}{R_2} & \left(\frac{1}{R_2}+\frac{1}{R_3}+\frac{1}{R_4}\right) \end{bmatrix} \begin{bmatrix} v_A \\ v_B \end{bmatrix} = \begin{bmatrix} i_{S1} - i_{S2} \\ i_{S2} \end{bmatrix}$$

We have two symmetrical equations in the two unknown voltages. The equations are symmetrical as seen by the dashed line because the resistance R_2 connected between nodes A and B appears as the cross-coupling term in each equation.

EXAMPLE 3–2

Formulate node-voltage equations for the circuit in Figure 3–9

SOLUTION:
The total conductance connected to node A is $1/(2R) + 2/R = 2.5/R$, to node B is $1/(2R) + 2/R + 1/(2R) = 3/R$, and to node C is $2/R + 1/R + 1/(2R) = 3.5/R$. The conductance connected between nodes A and B is $1/(2R)$, between nodes A and C is $2/R$, and between nodes B and C is $1/(2R)$. The independent current source is directed into node A. By inspection, the node-voltage equations are

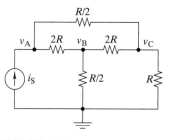

FIGURE 3–9

$$\text{Node A}: \quad \frac{2.5}{R}v_A - \frac{0.5}{R}v_B - \frac{2}{R}v_C = i_S$$

$$\text{Node B}: \quad -\frac{0.5}{R}v_A + \frac{3}{R}v_B - \frac{0.5}{R}v_C = 0$$

$$\text{Node C}: \quad -\frac{2}{R}v_A - \frac{0.5}{R}v_B + \frac{3.5}{R}v_C = 0$$

Written in matrix form $\mathbf{Ax} = \mathbf{b}$,

$$
\begin{bmatrix}
\dfrac{2.5}{R} & -\dfrac{0.5}{R} & -\dfrac{2}{R} \\[2mm]
-\dfrac{0.5}{R} & \dfrac{3}{R} & -\dfrac{0.5}{R} \\[2mm]
-\dfrac{2}{R} & -\dfrac{0.5}{R} & \dfrac{3.5}{R}
\end{bmatrix}
\begin{bmatrix}
v_A \\[2mm] v_B \\[2mm] v_C
\end{bmatrix}
=
\begin{bmatrix}
i_S \\[2mm] 0 \\[2mm] 0
\end{bmatrix}
$$

Note that the $\mathbf{A}$ matrix is symmetrical.   ■

Exercise 3–5

Formulate node-voltage equations for the circuit in Figure 3–10.

FIGURE 3–10

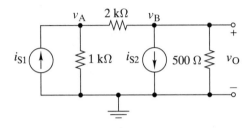

Answers:
$$
(1.5 \times 10^{-3})v_A - (0.5 \times 10^{-3})v_B = i_{S1}
$$
$$
-(0.5 \times 10^{-3})v_A + (2.5 \times 10^{-3})v_B = -i_{S2}
$$

SOLVING LINEAR ALGEBRAIC EQUATIONS

So far we have dealt only with the problem of formulating node-voltage equations. To complete a circuit analysis problem, we must solve these linear equations for selected responses. As introduced in Chapter 2 (see Example 2–28), a system of linear algebraic equations can be solved through manual calculations or with the aid of a computer. Skilled engineers should be familiar with both techniques and should develop the judgment to know when each type of approach is appropriate. In choosing between manual and computer-aided methods, you may consider the following four factors:

- The number of unknown responses (the order of the problem)
- The number of parameters in symbolic versus numeric form
- Your skill in performing manual calculations
- The availability of and your proficiency with computer tools

In general, you can efficiently use manual techniques to solve lower-order problems with numeric parameters. Cramer's rule and Gaussian elimination are standard mathematical tools commonly used for manual solutions. Web Appendix A provides a brief review of these techniques.

As the order of the problem increases or if symbolic parameters are present, it usually becomes more efficient to use an advanced calculator or computer tools, such as MATLAB, to find the desired responses. Many scientific handheld calculators have a built-in capability to solve linear equations when all of the parameter values are numeric. MATLAB can efficiently solve these types of problems, and it requires only minor adjustments to accommodate symbolic parameters also. To include symbolic parameters in MATLAB code, use either the sym or syms command to define the parameters and then proceed with the normal matrix-based solution to the problem. See Web Appendix C for additional guidance using MATLAB.

Earlier in this section we formulated node-voltage equations for the circuit in Figure 3–5 [see Eqs. (3–4)].

$$\text{Node A:} \quad \left(\frac{1}{R_1} + \frac{1}{R_2}\right) v_A - \frac{1}{R_2} v_B = i_S$$

$$\text{Node B:} \quad -\frac{1}{R_2} v_A + \left(\frac{1}{R_2} + \frac{1}{R_3}\right) v_B = 0$$

This problem is formulated with two unknown responses (v_A and v_B) and has three symbolic parameters ($R_1, R_2,$ and R_3). In this case, Cramer's rule would be a suitable manual technique for finding the responses, but we will illustrate the solution using MATLAB to handle the symbolic parameters. First, define the symbolic parameters:

```
syms R1 R2 R3 iS real
```

Now formulate the problem in matrix notation, using the **Ax** = **b** structure as follows:

```
A = [(1/R1 + 1/R2) -1/R2; -1/R2 (1/R2 + 1/R3)];
B = [iS; 0];
x = A\B;
vA = x(1)
vB = x(2)
```

The resulting solutions are

```
vA = (R1*iS*(R2 + R3))/(R1 + R2 + R3)
vB = (R1*R3*iS)/(R1 + R2 + R3)
```

We can write these solutions compactly as

$$v_A = \frac{R_1 i_S (R_2 + R_3)}{R_1 + R_2 + R_3}$$

$$v_B = \frac{R_1 R_3 i_S}{R_1 + R_2 + R_3}$$

The results express the two node voltages in terms of the circuit parameters and the input signals. Given the two node voltages v_A and v_B, we can now determine every element voltage and every current using Ohm's law and the fundamental property of node voltages.

$$v_1 = v_A \qquad v_2 = v_A - v_B \qquad v_3 = v_B$$

$$i_1 = \frac{v_A}{R_1} \qquad i_2 = \frac{v_A - v_B}{R_2} \qquad i_3 = \frac{v_B}{R_3}$$

In solving the node equations, we left everything in symbolic form to emphasize that the responses depend on the values of the circuit parameters ($R_1, R_2,$ and R_3) and the input signal (i_S). Even when numerical values are given, it is sometimes useful to leave some parameters in symbolic form to obtain input-output relationships or to reveal the effect of specific parameters on the circuit response.

E X A M P L E   3 – 3

Given the circuit in Figure 3–11, find the input resistance R_{IN} seen by the current source and the output voltage v_O.

FIGURE 3–11

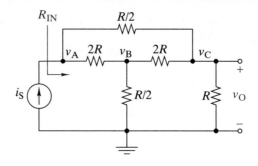

SOLUTION:

In Example 3–2 we formulated node-voltage equations for this circuit as follows:

$$\text{Node A:} \quad \frac{2.5}{R}v_A - \frac{0.5}{R}v_B - \frac{2}{R}v_C = i_S$$

$$\text{Node B:} \quad -\frac{0.5}{R}v_A + \frac{3}{R}v_B - \frac{0.5}{R}v_C = 0$$

$$\text{Node C:} \quad -\frac{2}{R}v_A - \frac{0.5}{R}v_B + \frac{3.5}{R}v_C = 0$$

First we will solve the problem using Cramer's rule and then we will solve it again using MATLAB.

The input resistance R_{IN} is the ratio $\frac{v_A}{i_S}$, while the output voltage is simply v_C. We solve for both v_A and v_C using Cramer's rule as follows:

Rewrite the three node equations after multiplying both sides of each equation by R

$$2.5v_A - 0.5v_B - 2v_C = Ri_S$$
$$-0.5v_A + 3v_B - 0.5v_C = 0$$
$$-2v_A - 0.5v_B + 3.5v_C = 0$$

Then write as a matrix $\mathbf{Ax = b}$

$$\begin{bmatrix} 2.5 & -0.5 & -2 \\ -0.5 & 3 & -0.5 \\ -2 & -0.5 & 3.5 \end{bmatrix} \begin{bmatrix} v_A \\ v_B \\ v_C \end{bmatrix} = \begin{bmatrix} Ri_S \\ 0 \\ 0 \end{bmatrix}$$

Then

$$v_A = \frac{\begin{vmatrix} Ri_S & -0.5 & -2 \\ 0 & 3 & -0.5 \\ 0 & -0.5 & 3.5 \end{vmatrix}}{\begin{vmatrix} 2.5 & -0.5 & -2 \\ -0.5 & 3 & -0.5 \\ -2 & -0.5 & 3.5 \end{vmatrix}} = \frac{Ri_S(10.25)}{11.75} = 0.872\,Ri_S\ \text{V}$$

$$R_{IN} = \frac{v_A}{i_S} = 0.872\frac{Ri_S}{i_S} = 0.872\,R\ \Omega$$

And

$$v_C = v_O = \frac{\begin{vmatrix} 2.5 & -0.5 & Ri_S \\ -0.5 & 3 & 0 \\ -2 & -0.5 & 0 \end{vmatrix}}{\begin{vmatrix} 2.5 & -0.5 & -2 \\ -0.5 & 3 & -0.5 \\ -2 & -0.5 & 3.5 \end{vmatrix}} = \frac{Ri_S(6.25)}{11.75} = 0.532\,Ri_S\ \text{V}$$

Now using MATLAB, we solve for he voltages v_A and v_C beginning with v_A as follows

```
syms vA vB vC R iS Rin real
A = [2.5/R -1/2/R -2/R; ...
     -1/2/R 3/R -1/2/R; ...
     -2/R -1/2/R 3.5/R];
B = [iS; 0; 0];
x = A\B;
vA = x(1)
```

The result is given by

```
vA = (41*R*iS)/47
```

which can be expressed compactly as:

$$v_A = \frac{41 R i_S}{47} = 0.872\, i_S R$$

The input resistance can be found using

```
Rin = vA/iS
```

which yields

$$R_{IN} = \frac{41\,R}{7} = 0.872\,R$$

The output voltage is v_C, which we found at the same time as v_A above.

```
vC = x(3)
```

This result is given by

```
vC = (25*R*iS)/47
```

which can be expressed compactly as

$$v_C = \frac{25\,R i_S}{47} = 0.532\,R i_S$$

■

Exercise 3–6

Solve the node-voltage equations in Exercise 3–5 for v_O in Figure 3–10.

Answer: $$v_O = 1000(i_{S1} - 3i_{S2})/7$$

Exercise 3–7

Use node-voltage equations to solve for v_1, v_2, and i_3 in Figure 3–12(a).

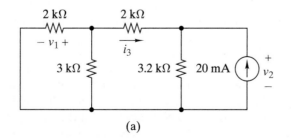

FIGURE 3–12

(a)

Answers: $v_1 = 12\,\text{V}$; $v_2 = 32\,\text{V}$; $i_3 = -10\,\text{mA}$

Exercise 3–8

Solve Exercise 3–7 using OrCAD.

FIGURE 3–12 *(Continued)*

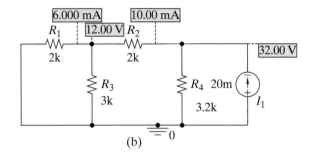

(b)

A n s w e r :   The results are shown in Figure 3–12(b) directly on the OrCAD circuit.

NODE ANALYSIS WITH VOLTAGE SOURCES

Up to this point we have analyzed circuits containing only resistors and independent current sources. Applying KCL in such circuits is simplified because the sum of currents at a node only involves the output of current sources or resistor currents expressed in terms of the node voltages. Adding voltage sources to circuits modifies node analysis procedures because the current through a voltage source is not directly related to the voltage across it. While initially it may appear that voltage sources complicate the situation, they actually simplify node analysis by reducing the number of equations required.

Figure 3–13 shows three ways to deal with voltage sources in node analysis. Method 1 uses a source transformation to replace the voltage source and series resistance with an equivalent current source and parallel resistance. We can then formulate node equations at the remaining nonreference nodes in the usual way. The source transformation eliminates node C, so there are only $N - 2$ nonreference nodes left in the circuit. Obviously, method 1 only applies when there is a resistance in series with the voltage source.

Method 2 in Figure 3–13 can be used whether or not there is a resistance in series with the voltage source. When node B is selected as the reference node, then by definition $v_B = 0$ and the fundamental property of node voltages says that $v_A = v_S$. We do not need a node-voltage equation at node A because its voltage is known to be equal to the source voltage. We write the node equations at the remaining $N - 2$ nonreference nodes in the usual way. In the final step, we move all terms involving v_A to the right side, since it is a known input and not an unknown response. Method 2 reduces the number of node equations by 1 since no equation is needed at node A.

The third method in Figure 3–13 is needed when neither node A nor node B can be selected as the reference and the source is not connected in series with a resistance. In this case we combine nodes A and B into a **supernode**, indicated by the boundary in Figure 3–13. We use the fact that KCL applies to the currents penetrating this boundary to write a node equation at the supernode. We then write node equations at the remaining $N - 3$ nonreference nodes in the usual way. We now have $N - 3$ node equations plus one supernode equation, leaving us one equation

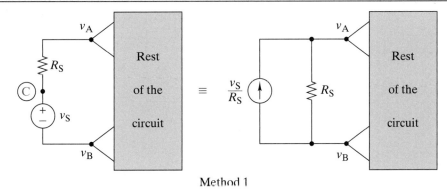

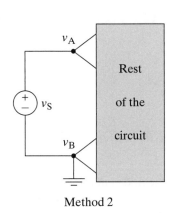

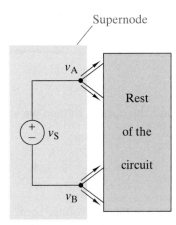

FIGURE 3–13 *Three methods of treating voltage sources in node analysis.*

Method 1

Method 2   Method 3

short of the $N - 1$ required. Using the fundamental property of node voltages, we can write

$$v_A - v_B = v_S \qquad (3\text{–}8)$$

The voltage source inside the supernode constrains the difference between the node voltages at nodes A and B. The voltage source constraint provides the additional relationship needed to write $N - 1$ independent equations in $N - 1$ node voltages.

For reference purposes we will call these modified node equations, since we either modify the circuit (method 1), use voltage source constraints to define node voltage at some nodes (method 2), or combine nodes to produce a supernode (method 3). The three methods are not mutually exclusive. We frequently use a combination of methods, as illustrated in the following examples.

EXAMPLE 3 – 4

Use node-voltage analysis to find v_O in the circuit in Figure 3–14(a).

SOLUTION:

The circuit in Figure 3–14(a) has four nodes, so we appear to need $N - 1 = 3$ node-voltage equations. However, applying source transformations to the two voltage sources (method 1) produces the two-node circuit in Figure 3–14(b). For the modified circuit we need only one node equation,

$$\left(\frac{1}{R_1} + \frac{1}{R_2} + \frac{1}{R_3} \right) v_B = \frac{v_{S1}}{R_1} + \frac{v_{S2}}{R_2}$$

FIGURE 3–14

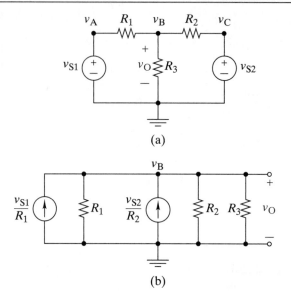

(a)

(b)

To find the output voltage we solve for v_B:

$$v_O = v_B = \frac{\dfrac{v_{S1}}{R_1} + \dfrac{v_{S2}}{R_2}}{\dfrac{1}{R_1} + \dfrac{1}{R_2} + \dfrac{1}{R_3}} = \frac{\dfrac{v_{S1}}{R_1} + \dfrac{v_{S2}}{R_2}}{\dfrac{R_2R_3 + R_1R_3 + R_1R_2}{R_1R_2R_3}} = \frac{R_2R_3v_{S1} + R_1R_3v_{S2}}{R_2R_3 + R_1R_3 + R_1R_2}$$

Because of the two voltage sources, we need only one node equation in what appears to be a four-node circuit. The two voltage sources have a common node, so the number of unknown node voltages is reduced from three to one. The general principle illustrated is that the number of independent KCL constraints in a circuit containing N nodes and N_V voltage sources is $N - 1 - N_V$. ∎

As an alternative to source transformations, for this circuit we can also apply the second method for treating voltages sources in node analysis. Since both voltage sources are connected to ground nodes, v_A and v_C are defined by the two voltage sources, i.e., $v_A = v_{S1}$ and $v_C = v_{S2}$. This then leaves v_B as the only unknown. We can write the equation for v_B as follows

$$\frac{v_B - v_{S1}}{R_1} + \frac{v_B}{R_3} + \frac{v_B - v_{S2}}{R_2} = 0$$

Solving for v_B

$$v_B\left(\frac{1}{R_1} + \frac{1}{R_2} + \frac{1}{R_3}\right) = \frac{v_{S1}}{R_1} + \frac{v_{S2}}{R_2}$$

Which, as before, simplifies to

$$v_B = v_O = \frac{R_2R_3v_{S1} + R_1R_3v_{S2}}{R_2R_3 + R_1R_3 + R_1R_2}$$

Exercise 3–9 _____

In Figure 3–14(a) $v_{S1} = 24\,\text{V}$, $v_{S2} = -12\,\text{V}$, $R_1 = 3.3\,\text{k}\Omega$, $R_2 = 5.6\,\text{k}\Omega$, and $R_3 = 10\,\text{k}\Omega$. Find v_O using OrCAD.

Answer: 8.82 V

EXAMPLE 3-5 *CP P₂ Hw2*

Use node-voltage analysis to find v_X in the circuit of Figure 3–15.

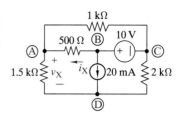

FIGURE 3–15

SOLUTION:

Since the circuit does not have a ground designated, we must choose a reference node to start the node-voltage analysis. In theory, any node can be chosen as a ground, but some ground choices are better than others. Choosing a ground wisely can simplify the calculations required to find a solution. Recognize that the ground node is our zero voltage reference. Hence, selecting a node at the minus terminal of a voltage source immediately determines the voltage at the positive node to be the value of the source. In our circuit, selecting node C as our ground automatically tells us that node B is equal to 10 V. If we chose node A or node D as our reference, we would only know that the voltage between nodes B and C is equal to 10 volts, which does not simplify our analysis. We could choose node B as our ground and we would automatically know that node C is equal to −10 V. This choice is acceptable, but we prefer to work with positive voltages, so node C is a better option.

Having wisely chosen node C as our ground, we can write the following two node-voltage equations

$$\text{Node A:} \quad \frac{V_A - 10}{0.5k} + \frac{V_A - V_D}{1.5k} + \frac{V_A}{1k} = 0$$

$$\text{Node D:} \quad \frac{V_D - V_A}{1.5k} - 0.020 + \frac{V_D}{2k} = 0$$

Collecting like terms, moving the known sources to the right side, and multiplying both equations by 1000

$$V_A\left[\frac{1}{0.5} + \frac{1}{1.5} + \frac{1}{1}\right] - V_D\left[\frac{1}{1.5}\right] = \frac{10}{0.5} = 3.66V_A - 0.666V_D = 20$$

$$-V_A\left[\frac{1}{1.5}\right] + V_D\left[\frac{1}{1.5} + \frac{1}{2}\right] = 20 = -0.666V_A + 1.16V_D = 20$$

We can write these two equations as a matrix and solve it using Cramer's rule

$$V_A = \frac{\begin{vmatrix} 20 & -0.666 \\ 20 & 1.16 \end{vmatrix}}{\begin{vmatrix} 3.66 & -0.666 \\ -0.666 & 1.16 \end{vmatrix}} = 9.565 \text{ V}$$

$$V_D = \frac{\begin{vmatrix} 3.66 & 20 \\ -0.666 & 20 \end{vmatrix}}{\begin{vmatrix} 3.66 & -0.666 \\ -0.666 & 1.16 \end{vmatrix}} = 22.609 \text{ V}$$

Therefore, $v_X = V_A - V_D = -13.044$ V. ■

Exercise 3-10 _____

For the circuit of Figure 3–15, find i_X.

Answer: $i_X = 869.6 \ \mu A$

EXAMPLE 3–6

Find the input resistance of the circuit in Figure 3–16(a).

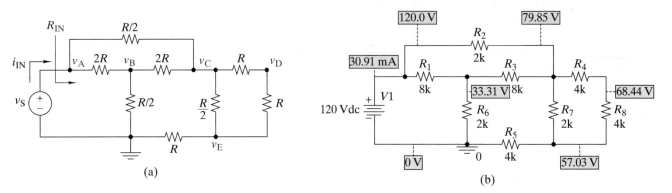

(a)   (b)

FIGURE 3–16

SOLUTION:

Method 1 for handling voltage sources will not work here because the source in Figure 3–16 is not connected in series with a resistor. Method 2 will work in this case because the voltage source is connected to the reference node. As a result, we can eliminate one node equation since the node A voltage is $v_A = v_S$. By inspection, the four remaining node equations are

$$\text{Node B: } -v_A\left(\frac{1}{2R}\right) + v_B\left(\frac{1}{2R}+\frac{2}{R}+\frac{1}{2R}\right) - v_C\left(\frac{1}{2R}\right) = 0$$

$$\text{Node C: } -v_A\left(\frac{2}{R}\right) - v_B\left(\frac{1}{2R}\right) + v_C\left(\frac{2}{R}+\frac{1}{2R}+\frac{2}{R}+\frac{1}{R}\right) - v_D\left(\frac{1}{R}\right) - v_E\left(\frac{2}{R}\right) = 0$$

$$\text{Node D: } -v_C\left(\frac{1}{R}\right) + v_D\left(\frac{1}{R}+\frac{1}{R}\right) - v_E\left(\frac{1}{R}\right) = 0$$

$$\text{Node E: } -v_C\left(\frac{2}{R}\right) - v_D\left(\frac{1}{R}\right) + v_E\left(\frac{1}{R}+\frac{1}{R}+\frac{2}{R}\right) = 0$$

We apply the equation $v_A = v_S$ to incorporate the source voltage and eliminate v_A from the equations. With this change, we can simplify the equations and write them in matrix form (note the symmetry) as follows:

$$
\begin{bmatrix}
\dfrac{3}{R} & -\dfrac{1}{2R} & 0 & 0 \\[2mm]
\dfrac{1}{2R} & \dfrac{11}{2R} & -\dfrac{1}{R} & -\dfrac{2}{R} \\[2mm]
0 & -\dfrac{1}{R} & \dfrac{2}{R} & -\dfrac{1}{R} \\[2mm]
0 & -\dfrac{2}{R} & -\dfrac{1}{R} & \dfrac{4}{R}
\end{bmatrix}
\begin{bmatrix}
v_B \\[2mm] v_C \\[2mm] v_D \\[2mm] v_E
\end{bmatrix}
=
\begin{bmatrix}
\dfrac{v_S}{2R} \\[2mm]
\dfrac{2v_S}{R} \\[2mm]
0 \\[2mm]
0
\end{bmatrix}
$$

In this standard form of $\mathbf{Ax} = \mathbf{b}$, we can now use MATLAB to efficiently solve for all of the node voltages. The MATLAB code required to do this is shown below.

```
syms vS vB vC vD vE R
A = [3/R -1/2/R 0 0; ...
    -1/2/R 11/2/R -1/R -2/R; ...
    0 -1/R 2/R -1/R; ...
    0 -2/R -1/R 4/R];
B = [vS/2/R; 2*vS/R; 0; 0];
x = A\B;
vB = x(1)
vC = x(2)
vD = x(3)
vE = x(4)
```

The resulting solutions are

```
vB = (73*vS)/263
vC = (175*vS)/263
vD = (150*vS)/263
vE = (125*vS)/263
```

We can write these results compactly as

$$v_B = \frac{73v_S}{263} \quad v_C = \frac{175v_S}{263} \quad v_D = \frac{150v_S}{263} \quad v_E = \frac{125v_S}{263}$$

To solve for the input resistance, we need to calculate the input current:

$$i_{IN} = \frac{v_S - v_C}{R/2} + \frac{v_S - v_B}{2R}$$

Using MATLAB, we get

$$i_{IN} = \frac{271v_S}{263R}$$

such that

$$R_{IN} = \frac{v_{IN}}{i_{IN}} = \frac{263R}{271} \qquad \blacksquare$$

Exercise 3–11

For the circuit in Figure 3–16(a) let $v_S = 120$ V and $R = 4$ kΩ.

(a) Use OrCAD to simulate the circuit, and find all of the node voltages and the input current.
(b) Verify that the results for the node voltages agree with the symbolic expressions determined in the solution of Example 3–6.
(c) Use the input current to calculate R_{IN} and compare it with that found in Example 3–6.

Answers:

(a) See OrCAD results shown in Figure 3–16(b).
(b) From the calculations in Example 3–5 using $v_S = 120$ V, we get

$$v_B = \frac{73v_S}{263} = 33.31 \text{ V} \quad v_C = \frac{175v_S}{263} = 79.85 \text{ V} \quad v_D = \frac{150v_S}{263} = 68.44 \text{ V}$$

$$v_E = \frac{125v_S}{263} = 57.03 \text{ V}$$

The OrCAD voltages are the same.
(c) Using the input voltage and input current found in the OrCAD simulation, R_{IN} for $R = 4$ kΩ results in $R_{IN} = v_{IN}/i_{IN} = \frac{120}{30.91\,m} = 3.882$ kΩ, compared to $R_{IN} = \frac{263}{271}R = 3.882$ kΩ. Again, they are the same.

EXAMPLE 3–7

For the circuit in Figure 3–17,

(a) Formulate node-voltage equations.
(b) Solve for the output voltage v_O using $R_1 = R_4 = 2\,\text{k}\Omega$ and $R_2 = R_3 = 4\,\text{k}\Omega$.

FIGURE 3–17

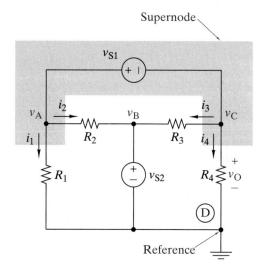

SOLUTION:

(a) The voltage sources in Figure 3–17 do not have a common node, and we cannot select a reference node that includes both sources. Selecting node D as the reference forces the condition $v_B = v_{S2}$ (method 2) but leaves the other source v_{S1} ungrounded. We surround the ungrounded source, and all wires leading to it, by the supernode boundary shown in Figure 3–17 (method 3). KCL applies to the four element currents that penetrate the supernode boundary, and we can write

$$i_1 + i_2 + i_3 + i_4 = 0$$

These currents can easily be expressed in terms of the node voltages:

$$\frac{v_A}{R_1} + \frac{(v_A - v_B)}{R_2} + \frac{(v_C - v_B)}{R_3} + \frac{v_C}{R_4} = 0$$

But since $v_B = v_{S2}$, the standard form of this equation is

$$\left(\frac{1}{R_1} + \frac{1}{R_2}\right)v_A + \left(\frac{1}{R_3} + \frac{1}{R_4}\right)v_C = \left(\frac{1}{R_2} + \frac{1}{R_3}\right)v_{S2}$$

We have one equation in the two unknown node voltages v_A and v_C. Applying the fundamental property of node voltages inside the supernode, we can write

$$v_A - v_C = v_{S1}$$

That is, the ungrounded voltage source constrains the difference between the two unknown node voltages inside the supernode. It thereby supplies the relationship needed to obtain two equations in two unknowns.

(b) Inserting the given numerical values yields

$$(7.5 \times 10^{-4})v_A + (7.5 \times 10^{-4})v_C = (5 \times 10^{-4})v_{S2}$$

$$v_A - v_C = v_{S1}$$

To find the output v_O, we need to solve these equations for v_C. The second equation yields $v_A = v_C + v_{S1}$, which, when substituted into the first equation, gives the required output:

$$v_O = v_C = \frac{v_{S2}}{3} - \frac{v_{S1}}{2} \qquad \blacksquare$$

Exercise 3–12

Find v_O in Figure 3–18 when the element E is

(a) A 10-kΩ resistance
(b) A 4-mA independent current source with reference arrow pointing left

Answers:

(a) 2.53 V
(b) −17.3 V

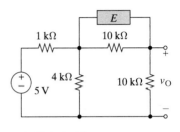

FIGURE 3–18

Exercise 3–13

Find v_O in Figure 3–18 when the element E is

(a) An open circuit
(b) A 10-V independent voltage source with the plus reference on the right

Answers:

(a) 1.92 V
(b) 12.96 V

SUMMARY OF NODE-VOLTAGE ANALYSIS

We have seen that node-voltage equations are very useful in the analysis of a variety of circuits. These equations can always be formulated using KCL, the element constraints, and the fundamental property of node voltages. When in doubt, always fall back on these principles to formulate node equations in new situations. With practice and experience, however, we eventually develop an analysis approach that allows us to recognize shortcuts in the formulation process. The following guidelines summarize our approach and may help you develop your own analysis style:

1. Simplify the circuit by combining elements in series and parallel wherever possible.

2. If not specified, select a reference node so that as many voltage sources as possible are directly connected to the reference.

3. Node equations are required at supernodes and all other nonreference nodes except those that are directly connected to the reference by voltage sources.

4. Use KCL to write node equations at the nodes identified in step 3. Express element currents in terms of node voltages or the currents produced by independent current sources.

5. Write expressions relating the node voltages to the voltages produced by independent voltage sources.

6. Substitute the expressions from step 5 into the node equations from step 4 and arrange the resulting equations in standard form.

7. Solve the equations from step 6 for the node voltages of interest. Manual techniques may be efficient for lower-order problems. Computer tools, such as MATLAB, are usually more practical and faster for higher-order problems.

3–2 MESH-CURRENT ANALYSIS

Mesh currents are analysis variables that are useful in circuits containing many elements connected in series. To review terminology, a loop is a closed path formed by passing through an ordered sequence of nodes without passing through any node more than once. A mesh is a special type of loop that does not enclose any elements. For example, loops A and B in Figure 3–19 are meshes, while the loop X is not a mesh because it encloses an element.

Mesh-current analysis is restricted to planar circuits. A **planar circuit** can be drawn on a flat surface without crossovers in the "window pane" fashion shown in Figure 3–19. To define a set of variables, we associate a **mesh current** (i_A, i_B, i_C, etc.) with each window pane and assign a reference direction. The reference directions for all mesh currents are customarily taken in a clockwise sense. There is no momentous reason for this, except perhaps tradition.

We think of these mesh currents as circulating through the elements in their respective meshes, as suggested by the reference directions shown in Figure 3–19. We should emphasize that this viewpoint is not based on the physics of circuit behavior. There are not red and blue electrons running around that somehow get assigned to mesh currents i_A or i_B. Mesh currents are variables used in circuit analysis. They are only somewhat abstractly related to the physical operation of a circuit and may be impossible to measure directly. For example, there is no way to cut the circuit in Figure 3–19 to insert an ammeter that measures only i_E.

Mesh currents have a unique feature that is the dual of the fundamental property of node voltages. If we examine Figure 3–19, we see the elements around the perimeter are contained in only one mesh, while those in the interior are in two meshes. In a planar circuit any given element is contained in at most two meshes. When an element is in two meshes, the two mesh currents circulate through the element in opposite directions. In such cases KCL declares that the net current through the element is the difference of the two mesh currents.

These observations lead us to the fundamental property of mesh currents:

> *If the Kth two-terminal element is contained in meshes X and Y, then the element current can be expressed in terms of the two mesh currents as*
> $$i_K = i_X - i_Y \qquad (3–9)$$
> *where X is the mesh whose reference direction agrees with the reference direction of i_K.*

Equation (3–9) is a KCL constraint at the element level. If the element is contained in only one mesh, then $i_K = i_X$ or $i_K = -i_Y$, depending on whether the reference

FIGURE 3–19 *Meshes in a planar circuit.*

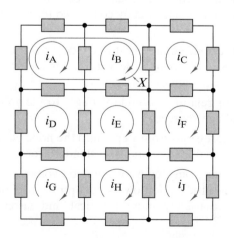

direction for the element current agrees or disagrees with the reference direction of the mesh current. The key idea is that the current through every two-terminal element in a planar circuit can be expressed in terms of no more than two mesh currents.

Exercise 3–14

In Figure 3–20 the mesh currents are $i_A = 10\,A$, $i_B = 5\,A$, and $i_C = -3\,A$. Find the element currents i_1 through i_6 and show that KCL is satisfied at nodes A, B, and C.

Answers: $i_1 = -10\,A$; $i_2 = 13\,A$; $i_3 = 5\,A$; $i_4 = 8\,A$; $i_5 = 5\,A$; $i_6 = -3\,A$.
at A $i_1 + i_2 + i_6 = 0$; at B $-i_2 + i_3 + i_4 = 0$; at C $-i_4 + i_5 - i_6 = 0$
$\quad 10 + 13 - 3 = 0$;$\qquad -13 + 5 + 8 = 0 \qquad -8 + 5 - (-3) = 0$

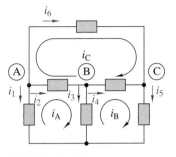

FIGURE 3–20

To use mesh currents to formulate circuit equations, we use elements and connection constraints, except that the KCL constraints are not explicitly written. Instead, we use the fundamental property of mesh currents to express the element voltages in terms of the mesh currents. By doing so we avoid using the element currents and work only with the element voltages and mesh currents.

For example, the planar circuit in Figure 3–21 can be analyzed using the mesh-current method. In the figure we have defined two mesh currents and five element voltages. We write KVL constraints around each mesh using the element voltages.

$$\text{Mesh A: } -v_0 + v_1 + v_3 = 0$$
$$\text{Mesh B: } -v_3 + v_2 + v_4 = 0 \tag{3–10}$$

Using the fundamental property of mesh currents, we write the element voltages in terms of the mesh currents and input voltages:

$$
\begin{aligned}
v_1 &= R_1 i_A & v_0 &= v_{S1} \\
v_2 &= R_2 i_B & v_4 &= v_{S2} \\
v_3 &= R_3(i_A - i_B)
\end{aligned}
\tag{3–11}
$$

We substitute these element equations into the KVL connection equations and arrange the result in standard form.

$$
\begin{aligned}
(R_1 + R_3)i_A - R_3 i_B &= v_{S1} \\
-R_3 i_A + (R_2 + R_3)i_B &= -v_{S2}
\end{aligned}
\tag{3–12}
$$

We have completed the formulation process with two equations in two unknown mesh currents.

As we have previously noted, every method of circuit analysis must satisfy KCL, KVL, and the element i–v relationships. When formulating mesh equations, it may appear that we have not used KCL. However, writing the element constraints in the form in Eq. (3–11) requires the KCL equations $i_1 = i_A$, $i_2 = i_B$, and $i_3 = i_A - i_B$. Mesh-current analysis implicitly satisfies KCL when the element constraints are expressed in terms of the mesh currents. In effect, the fundamental property of mesh currents ensures that the KCL constraints are satisfied.

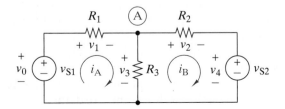

FIGURE 3–21 *Circuit for demonstrating mesh-current analysis.*

We use MATLAB to solve for the mesh currents in Eq. (3–12):

```
syms R1 R2 R3 vS1 vS2 iA iB v3 real
A = [R1+R3 -R3; -R3 R2+R3];
B = [vS1; -vS2];
x = A\B;
iA = x(1)
iB = x(2)
```

The resulting mesh currents are:

$$i_A = \frac{(R_2 + R_3)v_{S1} - R_3 v_{S2}}{R_1 R_2 + R_1 R_3 + R_2 R_3} \tag{3–13}$$

$$i_B = \frac{R_3 v_{S1} - (R_1 + R_3)v_{S2}}{R_1 R_2 + R_1 R_3 + R_2 R_3} \tag{3–14}$$

Equations (3–13) and (3–14) can now be substituted into the element constraints in Eq. (3–11) to solve for every voltage in the circuit. For instance, the voltage across R_3 is

$$v_3 = R_3(i_A - i_B) = \frac{R_2 R_3 v_{S1} + R_1 R_3 v_{S2}}{R_1 R_2 + R_1 R_3 + R_2 R_3} \tag{3–15}$$

You are invited to show that the result in Eq. (3–15) agrees with the node analysis result obtained in Example 3–4 for the same circuit.

The mesh-current analysis approach just illustrated can be summarized in four steps:

STEP 1   Identify a mesh current with every mesh and a voltage across every circuit element.

STEP 2   Write KVL connection constraints in terms of the element voltages around every mesh.

STEP 3   Use KCL and the i–v relationships of the elements to express the element voltages in terms of the mesh currents.

STEP 4   Substitute the element constraints from step 3 into the connection constraints from step 2 and arrange the resulting equations in standard form.

The number of mesh-current equations derived in this way equals the number of KVL connection constraints in step 2. When discussing combined constraints in Chapter 2, we noted that there are $E - N + 1$ independent KVL constraints in any circuit. Using the window panes in a planar circuit generates $E - N + 1$ independent mesh currents. Mesh analysis works best when the circuit has many elements (E large) connected in series (N also large).

Since OrCAD uses node-voltage analysis to solve for node voltages, the currents that it solves for are branch currents that may be composed of one or two mesh currents. One can calculate the mesh currents from the information provided by OrCAD if desired.

WRITING MESH-CURRENT EQUATIONS BY INSPECTION

The mesh equations in Eq. (3–12) have a symmetrical pattern that is similar to the coefficient symmetry observed in node equations. The coefficients of i_B in the first equation and i_A in the second equation are the negative of the resistance common to meshes A and B. The coefficients of i_A in the first equation and i_B in the second equation are the sum of the resistances in meshes A and B, respectively.

This pattern will always occur in planar circuits containing resistors and independent voltage sources when the mesh currents are defined in the window panes of a planar circuit, as shown in Figure 3–19. To see why, consider a general resistance R that is contained in, say, mesh A. There are only two possibilities. Either R is not contained in any other mesh, in which case the voltage across it is

$$v = R(i_A - 0) = Ri_A$$

or else it is also contained in only one adjacent mesh, say mesh B, in which case the voltage across it is

$$v = R(i_A - i_B)$$

These observations lead to the following conclusions. The voltage across resistance in mesh A involves the following terms:

1. i_A times the sum of the resistances in mesh A.
2. Minus i_B times the sum of resistances common to mesh A and mesh B.
3. Minus similar terms for any other mesh adjacent to mesh A with a common resistance.

The sum of the voltages across resistors plus the sum of the independent voltage sources around mesh A must equal zero.

The aforementioned process makes it possible for us to write mesh-current equations by inspection without going through the intermediate steps involving the KVL connection constraints and the element constraints.

EXAMPLE 3–8

For the circuit of Figure 3–22,

(a) Formulate mesh-current equations.
(b) Find the output v_O using $R_1 = R_4 = 2\,k\Omega$ and $R_2 = R_3 = 4\,k\Omega$.

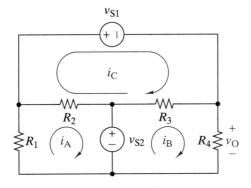

FIGURE 3–22

SOLUTION:
(a) To write mesh-current equations by inspection, we note that the total resistances in meshes A, B, and C are $R_1 + R_2$, $R_3 + R_4$, and $R_2 + R_3$, respectively. The resistance common to meshes A and C is R_2. The resistance common to meshes B and C is R_3. There is no resistance common to meshes A and B. Using these observations, we write the mesh equations as

$$\text{Mesh A:}\quad (R_1 + R_2)i_A - 0\,i_B - R_2 i_C + v_{S2} = 0$$
$$\text{Mesh B:}\quad (R_3 + R_4)i_B - 0\,i_A - R_3 i_C - v_{S2} = 0$$
$$\text{Mesh C:}\quad (R_2 + R_3)i_C - R_2 i_A - R_3 i_B + v_{S1} = 0$$

The algebraic signs assigned to voltage source terms follow the passive convention for the mesh current in question. Arranged in standard form, these equations become

$$(R_1 + R_2)i_A - R_2i_C = -v_{S2}$$
$$(R_3 + R_4)i_B - R_3i_C = +v_{S2}$$
$$-R_2i_A - R_3i_B + (R_2 + R_3)i_C = -v_{S1}$$

Coefficient symmetry greatly simplifies the formulation of these equations compared with the more fundamental, but time-consuming, process of writing element and connection constraints. Inserting the numerical values into these equations yields

$$6000\,i_A \qquad\qquad - 4000\,i_C = -v_{S2}$$
$$6000\,i_B - 4000\,i_C = v_{S2}$$
$$-4000\,i_A - 4000\,i_B + 8000\,i_C = -v_{S1}$$

Putting these three mesh equations in matrix form produces a symmetrical matrix

$$\begin{bmatrix} 6000 & 0 & -4000 \\ 0 & 6000 & -4000 \\ -4000 & -4000 & +8000 \end{bmatrix} \begin{bmatrix} i_A \\ i_B \\ i_C \end{bmatrix} = \begin{bmatrix} -v_{S2} \\ v_{S2} \\ -v_{S1} \end{bmatrix}$$

This is a matrix equation of the form $\mathbf{Ax} = \mathbf{B}$, where

$$\mathbf{A} = \begin{bmatrix} 6000 & 0 & -4000 \\ 0 & 6000 & -4000 \\ -4000 & -4000 & +8000 \end{bmatrix} \quad \mathbf{x} = \begin{bmatrix} i_A \\ i_B \\ i_C \end{bmatrix} \quad \mathbf{B} = \begin{bmatrix} -v_{S2} \\ v_{S2} \\ -v_{S1} \end{bmatrix}$$

Using MATLAB to solve for the mesh currents, we first enter the $\mathbf{A}$ matrix with the statement

```
A=[6000 0 -4000; 0 6000 -4000; -4000 -4000 8000];
```

The elements in the $\mathbf{B}$ matrix are the symbolic variables v_{S1} and v_{S2}. These quantities are not unknowns, but symbols that represent all possible values of the input voltages. Define the symbolic variables with the statement

```
syms VS1 VS2
```

and then create the $\mathbf{B}$ matrix:

```
B=[-VS2; VS2; -VS1];
```

Solve for the unknown mesh currents using matrix division,

$$\mathbf{x} = \mathbf{A}\backslash\mathbf{B}$$

which yields

```
x =
  -VS1/4000 - VS2/6000
   VS2/4000 - VS1/4000
     -(3*VS1)/8000
```

The elements of the column vector $\mathbf{x}$ are the three unknown mesh currents expressed in terms of the input voltages. The output voltage in Figure 3–22 is written in terms of the mesh currents as $v_O = R_4i_B$. We know $R_4 = 2\,\text{k}\Omega$, so in MATLAB we can compute the output voltage as

```
Vo = 2000*x(2)
```

which yields

```
Vo =
VS2/3 - VS1/2
```

We can write this answer compactly as

$$v_O = \frac{v_{S2}}{3} - \frac{v_{S1}}{2}$$

The result from the mesh-current analysis obtained here is the same as the node-voltage result obtained in Example 3–7. Either approach produces the same answer, but which method do you think is easier? ∎

Exercise 3–15

Using Figure 3–18 (see Exercises 3–12 and 3–13), find the current through the 4-kΩ resistor when the element E is

(a) A 10-kΩ resistance
(b) A 4-mA independent current source with reference arrow pointing left

Answers:

(a) 949.4 μA
(b) 1.346 mA

MESH EQUATIONS WITH CURRENT SOURCES

In developing mesh analysis, we assumed that circuits contain only voltage sources and resistors. This assumption simplifies the formulation process because the sum of voltages around a mesh is determined by voltage sources and the mesh currents through resistors. A current source complicates the picture because the voltage across it is not directly related to its current. We need to adapt mesh analysis to accommodate current sources just as we revised node analysis to deal with voltage sources.

There are three ways to handle current sources in mesh analysis:

1. If the current source is connected in parallel with a resistor, then it can be converted to an equivalent voltage source by source transformation. Each source conversion eliminates a mesh and reduces the number of equations required by 1. This method is the dual of method 1 for node analysis.

2. If a current source is contained in only one mesh, then that mesh current is determined by the source current and is no longer an unknown. We write mesh equations around the remaining meshes in the usual way and move the known mesh current to the source side of the equations in the final step. The number of equations obtained is one less than the number of meshes. This method is the dual of method 2 for node analysis.

3. Neither of the first two methods will work when a current source is contained in two meshes or is not connected in parallel with a resistance. In this case we create a **supermesh** by excluding the current source and any elements connected in series with it, as shown in Figure 3–23. We write one mesh equation around the supermesh using the currents i_A and i_B. We then write mesh equations of the remaining meshes in the usual way. This leaves us one equation short because parts of meshes A and B are included in the supermesh. However, the fundamental property of mesh currents relates the currents $i_S, i_A,$ and i_B as

$$i_A - i_B = i_S$$

This equation supplies the one additional relationship needed to get the requisite number of equations in the unknown mesh currents.

The aforementioned three methods are not mutually exclusive. We can use more than one method in a circuit, as the following examples illustrate.

FIGURE 3–23 *Example of a supermesh.*

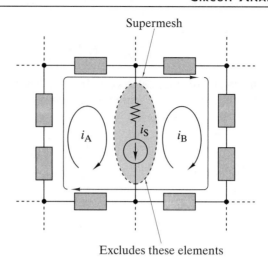

EXAMPLE 3 – 9

Use mesh-current equations to find i_O in the circuit in Figure 3–24(a).

FIGURE 3–24

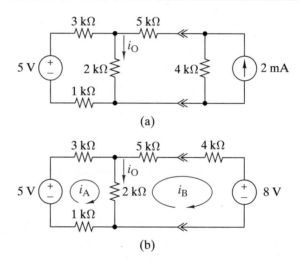

(a)

(b)

SOLUTION:

The current source in this circuit can be handled by a source transformation (method 1). The 2-mA source in parallel with the 4-kΩ resistor in Figure 3–24(a) can be replaced by an equivalent 8-V voltage source in series with the same resistor, as shown in Figure 3–24(b). In this circuit the total resistance in mesh A is 6kΩ, the total resistance in mesh B is 11 kΩ, and the resistance contained in both meshes is 2 kΩ. By inspection, the mesh equations for this circuit are

$$(6000)i_A - (2000)i_B = 5$$
$$-(2000)i_A + (11000)i_B = -8$$

Solving for the two mesh currents yields $i_A = 0.6290$ mA and $i_B = -0.6129$ mA. By KCL the desired current is $i_O = i_A - i_B = 1.2419$ mA. The given circuit in Figure 3–24 (a) has three meshes and one current source. The source transformation leading to Figure 3–24(b) produces a circuit with only two meshes. The general principle illustrated is that the number of independent mesh equations in a circuit containing E elements, N nodes, and N_I current sources is $E - N + 1 - N_I$. ∎

Exercise 3–16

In Figure 3–24 replace the 5-V source with a 1-mA dc current source with the arrow pointing up. Use source transformations to reduce the circuit to a single mesh and then solve for i_O.

Answer: 1.545 mA. If you got a different answer, check it with OrCAD.

EXAMPLE 3 – 1 0

Use mesh-current equations to find the v_O in Figure 3–25.

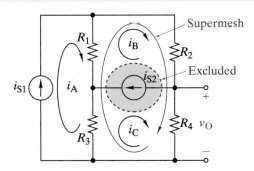

FIGURE 3–25

SOLUTION:
Source transformation (method 1) is not possible here since neither current source is connected in parallel with a resistor. The current source i_{S2} is in both mesh B and mesh C, so we exclude this element and create the supermesh (method 3) shown in the figure. The sum of voltages around the supermesh is

$$R_1(i_B - i_A) + R_2(i_B) + R_4(i_C) + R_3(i_C - i_A) = 0$$

The supermesh voltage constraint yields one equation in the three unknown mesh currents. Applying KCL to each of the current sources yields

$$i_A = i_{S1}$$
$$i_B - i_C = i_{S2}$$

Because of KCL, the two current sources force constraints that supply two more equations. Using these two KCL constraints to eliminate i_A and i_B from the super-mesh KVL constraint yields

$$(R_1 + R_2 + R_3 + R_4)i_C = (R_1 + R_3)i_{S1} - (R_1 + R_2)i_{S2}$$

Hence, the required output voltage is

$$v_O = R_4i_C = R_4 \times \left[\frac{(R_1 + R_3)i_{S1} - (R_1 + R_2)i_{S2}}{R_1 + R_2 + R_3 + R_4} \right]$$
■

Exercise 3–17

Use mesh analysis to find the current i_O in Figure 3–26 when the element E is

(a) A 5-V voltage source with the positive reference at the top
(b) A 10-kΩ resistor

Answers:

(a) −0.136 mA
(b) −0.538 mA

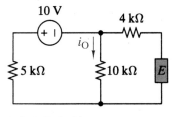

FIGURE 3–26

Exercise 3–18

Use mesh analysis to find the current i_O in Figure 3–26 when the element E is

(a) A 1-mA current source with the reference arrow directed down
(b) Two 20-kΩ resistors in parallel

Answers:

(a) $-1\,\text{mA}$
(b) $-0.538\,\text{mA}$

Exercise 3–19

Write a set of mesh-current equations for the circuit in Figure 3–27. Do not solve the equations.

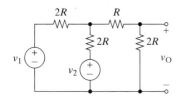

FIGURE 3–27

Answers:

$$-v_1 + 2Ri_A + 2R(i_A - i_B) + v_2 = 0; \qquad 4Ri_A - 2Ri_B = v_1 - v_2$$
$$-v_2 + 2R(i_B - i_A) + Ri_B + 2Ri_B = 0; \qquad -2Ri_A + 5Ri_B = v_2$$

Exercise 3–20

Use mesh-current equations to find v_O in Figure 3–27.

Answer:

$$v_O = (v_1 + v_2)/4$$

SUMMARY OF MESH-CURRENT ANALYSIS

Mesh-current equations can always be formulated from KVL, the element constraints, and the fundamental property of mesh currents. When in doubt, always fall back on these principles to formulate mesh equations in new situations. The following guidelines summarize an approach to formulating mesh equations for resistance circuits:

1. Simplify the circuit by combining elements in series or parallel wherever possible.

2. Mesh equations are required for supermeshes and all other meshes except those where current sources are contained in only one mesh.

3. Use KVL to write mesh equations for the meshes identified in step 2. Express element voltages in terms of mesh currents or the voltage produced by independent voltage sources.

4. Write expressions relating the mesh currents to the currents produced by independent current sources.

5. Substitute the expressions from step 4 into the mesh equations from step 3 and place the result in standard form.

6. Solve the equations from step 5 for the mesh currents of interest. Manual techniques may be efficient for lower-order problems. Computer tools, such as MATLAB, are usually more practical and faster for higher-order problems.

3–3 LINEARITY PROPERTIES

This book treats the analysis and design of **linear circuits**. A circuit is said to be linear if it can be adequately modeled using only linear elements and independent sources. The hallmark feature of a linear circuit is that outputs are linear functions of the inputs. Circuit **inputs** are the signals produced by external sources, and **outputs** are any other designated signals. Mathematically, a function is said to be linear if it possesses two properties—homogeneity and additivity. In linear circuits, **homogeneity** means that the output is proportional to the input. **Additivity** means that the output due to two or more inputs can be found by adding the outputs obtained when each input is applied separately. Mathematically, these properties are written as follows:

$$f(Kx) = Kf(x) \quad \text{(homogeneity)} \qquad (3\text{--}16)$$

and

$$f(x_1 + x_2) = f(x_1) + f(x_2) \quad \text{(additivity)} \qquad (3\text{--}17)$$

where K is a scalar constant. In circuit analysis the homogeneity property is called **proportionality**, while the additivity property is called **superposition**.

THE PROPORTIONALITY PROPERTY

The **proportionality property** applies to linear circuits with one input. For linear resistive circuits, proportionality states that every input-output relationship can be written as

$$y = Kx \qquad (3\text{--}18)$$

where x is the input current or voltage, y is an output current or voltage, and K is a constant. The block diagram in Figure 3–28 describes this linear input–output relationship. In a block diagram the lines headed by arrows indicate the direction of signal flow. The arrow directed into the block indicates the input, while the output is indicated by the arrow directed out of the block. The variable names written next to these lines identify the input and output signals. The scalar constant K written inside the block indicates that the input signal x is multiplied by K to produce the output signal as $y = Kx$.

The concept of proportionality is a central and recurring theme in linear circuit design. The ratio of the output to the input is a concept that dominates much of circuit analysis and design. We will analyze and design circuits that achieve desired K values. In this chapter, we will be restricted to $K \le 1$, but beginning in the next chapter we will treat circuits with K values that can be greater than one. In later chapters we will learn to design and analyze circuits using complex ratios that vary with time or frequency.

Caution: Proportionality only applies when the input and output are current or voltage. It does not apply to output power since power is equal to the product of current and voltage. In other words, output power is not linearly related to the input current or voltage.

We have already seen several examples of proportionality. For instance, using voltage division in Figure 3–29(a) produces

$$v_O = \left(\frac{R_2}{R_1 + R_2} \right) v_S$$

which means

$$x = v_S \qquad y = v_O$$
$$K = \frac{R_2}{R_1 + R_2}$$

as shown in Figure 3–29(b). Similarly, applying current division to the circuit of Figure 3–29(c)

$$i_O = \left(\frac{R_1}{R_1 + R_2} \right) i_S$$

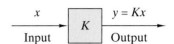

FIGURE 3–28 *Block diagram representation of the proportionality property.*

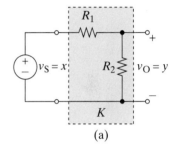

(a)

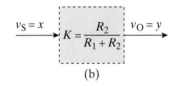

(b)

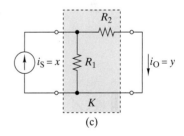

(c)

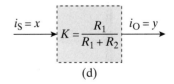

(d)

FIGURE 3–29 *Examples of circuit exhibiting proportionality: (a) Voltage divider. (b) Block diagram for (a). (c) Current divider. (d) Block diagram for (c).*

so that

$$x = i_S \qquad y = i_O$$

$$K = \frac{R_1}{R_1 + R_2}$$

as shown in Figure 3-29(d). In these two examples the proportionality constant K is dimensionless because the input and output have the same units. In other situations K could have the units of ohms or siemens when the input and output have different units.

▶ DESIGN EXAMPLE 3–11

Design a circuit that has a $K = v_O/v_S = 0.67$ using standard value resistors. (See inside back cover.)

SOLUTION:
Since K is less than one we can use the voltage divider shown in Figure 3–29(a), i.e.,

$$K = 0.67 = \frac{R_2}{R_1 + R_2}$$

There are many possible combinations. If we have access to 5% tolerance resistors, one choice is to select $R_2 = 20$ kΩ and solve for R_1. In this case, $R_1 = 10$ kΩ. If we must use 20% tolerance resistors, choose $R_2 = 68$ kΩ and $R_1 = 33$ kΩ. ∎

▶ Design Exercise 3–21

Design a circuit that has $K = i_O/i_A = 0.9$ using 5% tolerance standard value resistors. (See inside back cover.)

Answer: There are many solutions using the circuit of Figure 3–29(c). One is to select $R_1 = 91$ kΩ and $R_2 = 10$ kΩ.

The next example illustrates that the proportionality constant K can be positive, negative, or even zero.

EXAMPLE 3–12

You are given the bridge circuit of Figure 3–30(a).

(a) Find the proportionality constant K in the input-output relationship $v_O = K v_S$.
(b) Find the sign of K when $R_2 R_3 > R_1 R_4$, $R_2 R_3 = R_1 R_4$, and $R_2 R_3 < R_1 R_4$.
(c) Draw a block diagram of this relationship.

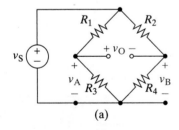

SOLUTION:
(a) We observe that the circuit consists of two voltage dividers. Applying the voltage division rule to each side of the bridge circuit yields

$$v_A = \frac{R_3}{R_1 + R_3} v_S \quad \text{and} \quad v_B = \frac{R_4}{R_2 + R_4} v_S$$

The fundamental property of node voltages allows us to write

$$v_O = v_A - v_B$$

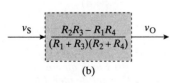

(b)

FIGURE 3–30

Substituting the equations for v_A and v_B into this KVL equation yields

$$v_O = \left(\frac{R_3}{R_1 + R_3} - \frac{R_4}{R_2 + R_4} \right) v_S$$
$$= \left(\frac{R_2 R_3 - R_1 R_4}{(R_1 + R_3)(R_2 + R_4)} \right) v_S$$
$$= (K) v_S$$

(b) The proportionality constant K can be positive, negative, or zero. Specifically,

If $R_2 R_3 > R_1 R_4$ then $K > 0$
If $R_2 R_3 = R_1 R_4$ then $K = 0$
If $R_2 R_3 < R_1 R_4$ then $K < 0$

When the products of the resistances in opposite legs of the bridge are equal, then $K = 0$ and the bridge is said to be balanced.

(c) See Figure 3–30(b). ■

Exercise 3–22

In Figure 3–30(a) select values of R so that $K = -0.333$.

Answer: $R_1 = R_2 = R_3 = 1\,k\Omega$ and $R_4 = 5\,k\Omega$. Many other solutions are possible.

UNIT OUTPUT METHOD

The **unit output method** is an analysis technique based on the proportionality property of linear circuits. The method involves finding the input-output proportionality constant K by assuming an output of one unit and determining the input required to produce that unit output. This technique is most useful when applied to ladder circuits, and it involves the following steps:

1. A unit output is assumed; that is, $v_O = 1\,V$ or $i_O = 1\,A$.
2. The input required to produce the unit output is then found by successive application of KCL, KVL, and Ohm's law.
3. Because the circuit is linear, the proportionality constant relating input and output is

$$K = \frac{\text{Output}}{\text{Input}} = \frac{1}{\text{Input for unit output}}$$

Given the proportionality constant K, we can find the output for any input using Eq. (3–18).

In a way, the unit output method solves the circuit response problem backwards—that is, from output to input—as illustrated by the next example.

EXAMPLE 3–13

Use the unit output method to find v_O in the circuit shown in Figure 3–31(a).

SOLUTION:
We start by assuming $v_O = 1$, as shown in Figure 3–31(b). Then, using Ohm's law, we find i_O.

$$i_O = \frac{v_O}{20} = 0.05\,A$$

FIGURE 3–31

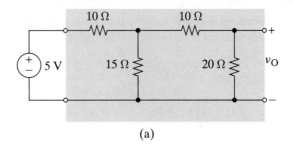

(a)

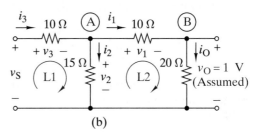

(b)

Next, using KCL at node B, we find i_1.

$$i_1 = i_O = 0.05 \text{ A}$$

Again, using Ohm's law, we find v_1.

$$v_1 = 10i_1 = 0.5 \text{ V}$$

Then, writing a KVL equation around loop L2, we find v_2 as

$$v_2 = v_1 + v_O = 0.5 + 1.0 = 1.5 \text{ V}$$

Using Ohm's law once more produces

$$i_2 = \frac{v_2}{15} = \frac{1.5}{15} = 0.1 \text{ A}$$

Next, writing a KCL equation at node A yields

$$i_3 = i_1 + i_2 = 0.05 + 0.1 = 0.15 \text{ A}$$

Using Ohm's law one last time,

$$v_3 = 10i_3 = 1.5 \text{ V}$$

We can now find the source voltage v_S by applying KVL around loop L1:

$$v_S|_{\text{for } v_O=1\text{V}} = v_3 + v_2 = 1.5 + 1.5 = 3 \text{ V}$$

A 3–V source voltage is required to produce a 1-V output. From this result, we calculate the proportionality constant K to be

$$K = \frac{v_O}{v_S} = \frac{1}{3}$$

Once K is known, the output for the specified 5-V input is $v_O = (1/3)\, 5 = 1.667$ V. ■

Exercise 3–23 _____

Find v_O in the circuit of Figure 3–31(a) when v_S is −5 V, 10 mV, and 3 kV.

Answers: $v_O = -1.667$ V; 3.333 mV; 1 kV

Exercise 3–24

Use the unit output method to find $K = i_O/i_{IN}$ for the circuit in Figure 3–32. Then use the proportionality constant K to find i_O for the input current shown in the figure.

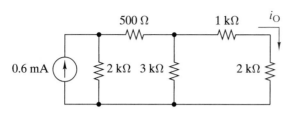

FIGURE 3–32

Answers: $K = \frac{1}{4}$; $i_O = 0.15\,\text{mA}$

ADDITIVITY PROPERTY

The **additivity property** states that any output current or voltage of a linear resistive circuit with multiple inputs can be expressed as a linear combination of the several inputs:

$$y = K_1x_1 + K_2x_2 + K_3x_3 + \cdots \qquad (3\text{–}19)$$

where $x_1, x_2, x_3, \ldots$ are current or voltage inputs, and $K_1, K_2, K_3, \ldots$ are constants that depend on the circuit parameters. Figure 3–33 shows how we represent this relationship in block diagram form. Again the arrows indicate the direction of signal flow and the K's within the blocks are scalar multipliers. The circle in Figure 3–33 is a new block diagram element called a summing point that implements the operation $y = \Sigma K_i x_i$. Although the block diagram in Figure 3–33 is nothing more than a pictorial representation of Eq. (3–19), the diagram often helps us gain a clearer picture of how signals interact in different parts of a circuit.

To illustrate this property, we analyze the two-input circuit in Figure 3–34(a) using node-voltage analysis. Applying KCL at node A, we obtain

$$\frac{v_A - v_S}{R_1} - i_S + \frac{v_A}{R_2} = 0$$

Moving the inputs to the right side of this equation yields

$$\left[\frac{1}{R_1} + \frac{1}{R_2}\right]v_A = \frac{v_S}{R_1} + i_S$$

Since $v_O = v_A$, we obtain the input-output relationship in the form

$$v_O = \left[\frac{R_2}{R_1 + R_2}\right]v_S + \left[\frac{R_1R_2}{R_1 + R_2}\right]i_S \qquad (3\text{–}20)$$

$$y = [K_1]x_1 + [K_2]x_2$$

This result shows that the output is a linear combination of the two inputs. Note that K_1 is dimensionless since its input and output are voltages, and that K_2 has the units of ohms since its input is a current and its output is a voltage. A representative block diagram is shown in Figure 3–34(b).

SUPERPOSITION PRINCIPLE

Since the output in Eq. (3–19) is a linear combination, the contribution of each input source is independent of all other inputs. This means that the output can be found by finding the contribution from each source acting alone and then adding the

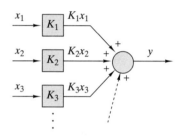

FIGURE 3–33 *Block diagram representation of the additivity property.*

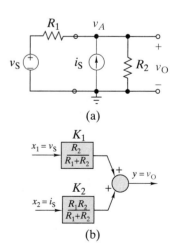

(a)

(b)

FIGURE 3–34 *(a) Circuit used to demonstrate superposition. (b) Block diagram.*

FIGURE 3–35 *Turning off an independent source: (a) Voltage source. (b) Current source.*

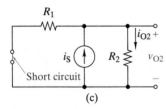

(a)

(b)

individual responses to obtain the total response. This suggests that the output of a multiple-input linear circuit can be found by the following steps:

STEP 1 "Turn off" all independent sources except one and find the output of the circuit due to that source acting alone.

STEP 2 Repeat the process in step 1 until each independent source has been turned on and the output due to that source found.

STEP 3 The total output with all independent sources turned on is the algebraic sum of the outputs caused by each source acting alone.

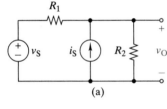

(a)

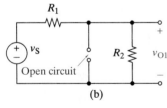

(b)

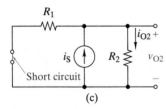

(c)

FIGURE 3–36 *Circuit analysis using superposition: (a) Original circuit. (b) Current source off. (c) Voltage source off.*

These steps describe a circuit analysis technique called the **superposition principle**. Before applying this method, we must discuss what happens when a voltage or current source is "turned off."

The i–v characteristics of voltage and current sources are shown in Figure 3–35. A voltage source is "turned off" by setting its voltage to zero ($v_S = 0$). This step translates the voltage source i–v characteristic to the i-axis, as shown in Figure 3–35(a). In Chapter 2 we found that a vertical line on the i-axis is the i–v characteristic of a short circuit. Similarly, "turning off" a current source ($i_S = 0$) in Figure 3–35(b) translates its i–v characteristic to the v-axis, which is the i–v characteristic of an open circuit. Therefore, when a voltage source is "turned off" we replace it by a short circuit, and when a current source is "turned off" we replace it by an open circuit.

The superposition principle is now applied to the circuit in Figure 3–34 to duplicate the response in Eq. (3–20), which was found by node analysis. Figure 3–36 shows the steps involved in applying superposition to the circuit in Figure 3–34. Figure 3–36(a) shows that the circuit has two input sources. We will first "turn off" i_S and replace it with an open circuit, as shown in Figure 3–36(b). The output of the circuit in Figure 3–36(b) is called v_{O1} and represents that part of the total output caused by the voltage source. Using voltage division in Figure 3–36(b) yields v_{O1} as

$$v_{O1} = \frac{R_2}{R_1 + R_2} v_S$$

Next we "turn off" the voltage source and "turn on" the current source, as shown in Figure 3–36(c). Using Ohm's law, we get $v_{O2} = i_{O2}R_2$. We use current division to express i_{O2} in terms of i_S to obtain v_{O2}

$$v_{O2} = i_{O2}R_2 = \left[\frac{R_1}{R_1 + R_2}\, i_S\right] R_2 = \frac{R_1 R_2}{R_1 + R_2}\, i_S$$

Applying the superposition principle, we find the response with both sources "turned on" by adding the two responses v_{O1} and v_{O2}.

$$v_O = v_{O1} + v_{O2}$$

$$v_O = \left[\frac{R_2}{R_1 + R_2}\right] v_S + \left[\frac{R_1 R_2}{R_1 + R_2}\right] i_S$$

This superposition result is the same as the circuit reduction result given in Eq. (3–20) and can be represented by the same block diagram shown in Figure 3.34(b).

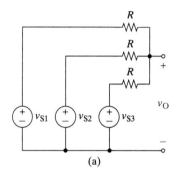

EXAMPLE 3–14

Figure 3–37(a) shows a resistance circuit used to implement a signal-summing function. Use superposition to show that the output v_O is a weighted sum of the inputs v_{S1}, v_{S2}, and v_{S3}.

SOLUTION:
To determine v_O using superposition, we first turn off sources 1 and 2 ($v_{S1} = 0$ and $v_{S2} = 0$) to obtain the circuit in Figure 3–37(b). This circuit is a voltage divider in which the output leg consists of two equal resistors in parallel. The equivalent resistance of the output leg is $R/2$, so the voltage division rule yields

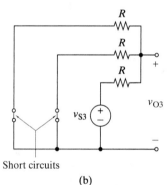

$$v_{O3} = \frac{R/2}{R + R/2}\, v_{S3} = \frac{v_{S3}}{3}$$

Because of the symmetry of the circuit, it can be seen that the same technique applies to all three inputs; therefore,

$$v_{O2} = \frac{v_{S2}}{3} \quad \text{and} \quad v_{O1} = \frac{v_{S1}}{3}$$

Applying the superposition principle, the output with all sources "turned on" is

$$v_O = v_{O1} + v_{O2} + v_{O3}$$
$$= \frac{1}{3}[v_{S1} + v_{S2} + v_{S3}]$$

FIGURE 3–37

That is, the output is proportional to the sum of the three input signals with $K_1 = K_2 = K_3 = \frac{1}{3}$. ∎

Exercise 3–25

The circuit of Figure 3–38 contains two R-$2R$ modules. Use superposition to find v_O.

FIGURE 3–38

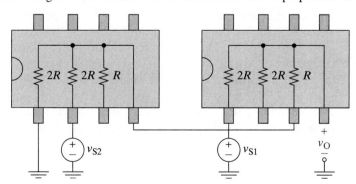

Answer: $v_O = \dfrac{1}{2} v_{S1} + \dfrac{1}{4} v_{S2}$

Exercise 3–26 _____

Repeat Exercise 3–25 with the voltage source v_{S2} replaced by a current source i_{S2} with the current reference arrow directed toward ground.

Answer: $v_O = 3v_{S1}/5 - 4i_{S2}R/5$

_____

EXAMPLE 3 – 1 5

Use the principle of superposition to find the voltage v_X in Figure 3–39(a).

FIGURE 3–39

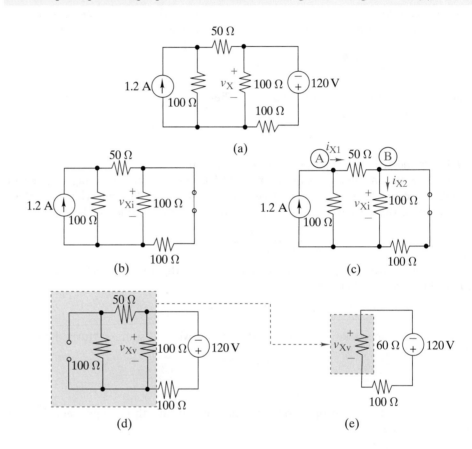

(a)

(b)   (c)

(d)   (e)

SOLUTION:

To find v_X we will turn off each source one at a time starting with the voltage source as shown in Figure 3-39(b). We chose to find the voltage v_{Xi} by first determining the current i_{X2} and then using ohm's law as shown in Figure 3-39(c). The circuit to the right of the 50-Ω resistor consists of two 100-Ω resistors in parallel. This result, then, is in series with the 50-Ω resistor. This combination equals 100 Ω. At node A, the current i_{X1} is exactly half of the source current or 0.6 A, because the source current divides equally between two paths with the same equivalent resistances. At node B, the current divides equally again, since both path resistances equal 100 Ω, resulting in $i_{X2} = 0.3$ A. Then by Ohm's law

$$v_{Xi} = 0.3 \times 100 = 30 \text{ V}$$

To find the contribution due to the voltage source, we set the current source to zero as shown in Figure 3–39(d). The circuit to the left of the source's 100-Ω resistor can

9/15/14

be combined resulting in an equivalent resistance of 60 Ω. The problem of finding v_{Xv} then reduces to a simple voltage divider as shown in Figure 3-39(e).

$$v_{Xv} = \frac{60}{100 + 60}(-120) = -45 \text{ V}$$

The desired voltage v_x can be found by algebraically adding the two contributions

$$v_X = v_{Xi} + v_{Xv} = 30 + (-45) = -15 \text{ V}$$

■

Exercise 3–27

Use the principle of superposition to find the current i_X in Figure 3-40.

Answers:
$$i_{Xi} = 0.8 \text{ A}, \; i_{Xv} = 0.4 \text{ A}, \; i_X = 1.2 \text{ A}$$

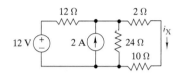

FIGURE 3–40

J. Equiv. Networks

l. load networks

The preceding examples and exercises illustrate use of the superposition theorem to analyze multiple-input linear circuits. You should not conclude that this is the primary application of this concept. In fact, superposition is not a particularly attractive method, since a circuit with N signal sources requires N different circuit analyses to obtain the final result. Unless the circuit is relatively simple, super-position may not reduce the analysis effort compared with, say, node-voltage analysis or a software tool. Rather, superposition is an important property of linear circuits used primarily as a conceptual tool to develop other circuit analysis and design techniques and as an aid in understanding the effect that different sources have on an output.

3–4 THÉVENIN AND NORTON EQUIVALENT CIRCUITS

An *interface* is a connection between circuits. Circuit interfaces occur frequently in electrical and electronic systems, so special analysis methods are used to handle them. For the two-terminal interface shown in Figure 3–41, we normally think of one circuit as the source S and the other as the load L. We think of signals as being produced by the source circuit and delivered to the load circuit. The source-load interaction at an interface is one of the central problems of circuit analysis and design.

assume has only no indep. sources

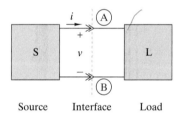

Source Interface Load

FIGURE 3–41 *A two-terminal interface.*

The Thévenin and Norton equivalent circuits shown in Figure 3–42 are valuable tools for dealing with circuit interfaces. The conditions under which these equivalent circuits exist can be stated as a theorem:

> *If the source circuit in a two-terminal interface is linear, then the interface signals v and i do not change when the source circuit is replaced by its Thévenin or Norton equivalent circuit.*

The equivalence requires the source circuit to be linear but places no restriction on the linearity of the load circuit. Later in this section we consider cases in which the load is nonlinear. In subsequent chapters we will study circuits in which the loads are linear energy storage elements called capacitors and inductors.

The Thévenin equivalent circuit consists of a voltage source (v_T) in series with a resistance (R_T). The Norton equivalent circuit is a current source (i_N) in parallel with a resistance (R_N). Note that the Thévenin and Norton equivalent circuits are practical sources in the sense discussed in Chapter 2.

The two circuits have the same i–v characteristics, since replacing one by the other leaves the interface signals unchanged. To derive the equivalency conditions, we

FIGURE 3–42 *Equivalent circuits for the source:* (a) Thévenin equivalent. (b) Norton equivalent.

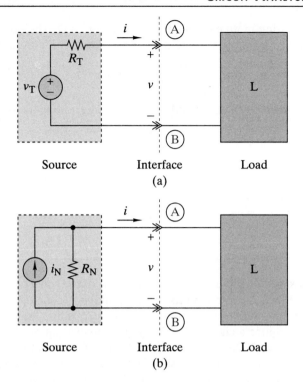

apply KVL and Ohm's law to the Thévenin equivalent in Figure 3–42(a) to obtain its i–v relationship at the terminals A and B:

$$v = v_T - iR_T \tag{3–21}$$

Next, applying KCL and Ohm's law to the Norton equivalent in Figure 3–42(b) yields its i–v relationship at terminals A and B:

$$i = i_N - \frac{v}{R_N} \tag{3–22}$$

Solving Eq. (3–22) for v yields

$$v = i_N R_N - iR_N \tag{3–23}$$

The Thévenin and Norton circuits have identical i–v relationships. Comparing Eqs. (3–21) and (3–23), we conclude that

$$R_N = R_T$$
$$i_N R_N = v_T \tag{3–24}$$

In essence, the Thévenin and Norton equivalent circuits are related by the source transformation studied in Chapter 2. We do not need to find both equivalent circuits. Once one of them is found, the other can be determined by a source transformation. The Thévenin and Norton circuits involve four parameters (v_T, R_T, i_N, R_N), and Eq. (3–24) provides two relations between the four parameters. Therefore, only two parameters are needed to specify either equivalent circuit. Since Thévenin and Norton equivalent circuits are, well, equivalent, one uses whichever one is more useful for the task.

In circuit analysis problems it is convenient to use the short-circuit current and open-circuit voltage to specify Thévenin and Norton circuits. The circuits in Figure 3–43(a) show that when the load is an open circuit the interface voltage equals the Thévenin voltage; that is, $v_{OC} = v_T$, since there is no voltage across R_T when $i = 0$. Similarly, the circuits in Figure 3–43(b) show that when the load is a short circuit

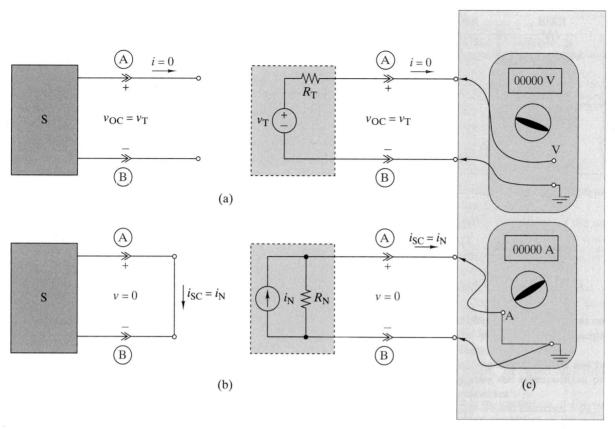

FIGURE 3–43 *Loads used to find Thévenin and Norton equivalent circuits: (a) Open circuit yields the Thévenin voltage. (b) Short circuit yields the Norton current. (c) Measuring v_{OC} and i_{SC} using a DMM.*

the interface current equals the Norton current; that is, $i_{SC} = i_N$, since all the source current i_N is diverted through the short-circuit load.

In summary, the parameters of the Thévenin and Norton equivalent circuits at a given interface can be found by determining the open-circuit voltage and the short-circuit current.

$$v_T = v_{OC}$$
$$i_N = i_{SC} \qquad \text{(3–25)}$$
$$R_N = R_T = v_{OC}/i_{SC}$$

APPLICATION   EXAMPLE 3–16

Thévenin measurements can be made in the laboratory or in the field with a simple digital multimeter (DMM). Look at Figure 3–43(c). If one sets the DMM to read voltage, then the multimeter acts like an open circuit[1] and the meter reading will be the circuit's open-circuit voltage v_{OC}. Setting the DMM to read current makes the DMM act like a short circuit[2] and the meter reading will be the circuit's short-circuit current. The circuit's Thévenin resistance can then be found using Eq. (3–25).

---

[1]The open-circuit resistance of a DMM is not infinite, but varies with the quality of the DMM from 10^7 to 10^{11} ohms.
[2]Similarly, the short-circuit resistance of a DMM is not zero, but varies with the quality of the DMM from 10^{-1} to 10 ohms.

APPLICATIONS OF THÉVENIN AND NORTON EQUIVALENT CIRCUITS

Replacing a complex circuit by its Thévenin or Norton equivalent can greatly simplify the analysis and design of interface circuits. For example, suppose we need to select a load resistance in Figure 3–44(a) so the source circuit to the left of the interface A–B delivers 4 volts to the load. This task is easily handled once we have the Thévenin or Norton equivalent for the source circuit.

To obtain the Thévenin and Norton equivalents, we need v_{OC} and i_{SC}. The open-circuit voltage v_{OC} is found by disconnecting the load at the terminals A and B, as shown in Figure 3–44(b). The voltage across the 15-Ω resistor is zero because the open circuit causes the current through the resistor to be zero. The open-circuit voltage at the interface is the same as the voltage across the 10-Ω resistor. Using voltage division, this voltage is

$$v_T = v_{OC} = \frac{10}{10+5} \times 15 = 10 \text{ V}$$

Next we find the short-circuit current i_{SC} using the circuit in Figure 3–44(c). The total current i_X delivered by the 15-V voltage source is

$$i_X = 15/R_{EQ}$$

FIGURE 3–44 *Example of finding the Thévenin and Norton equivalent circuits: (a) The given circuit. (b) Open circuit yields the Thévenin voltage. (c) Short circuit yields the Norton current.(d) Thévenin equivalent circuit. (e) Norton equivalent circuit.*

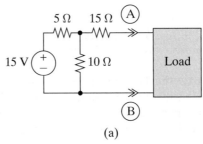

(a)

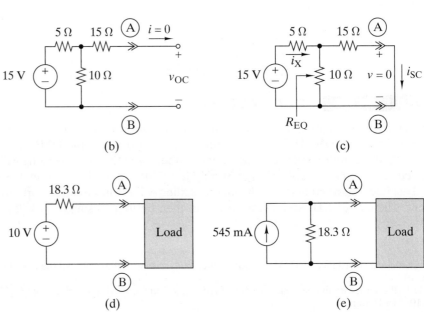

(b)   (c)

(d)   (e)

where R_{EQ} is the equivalent resistance seen by the voltage source with a short circuit at the interface.

$$R_{EQ} = 5 + \cfrac{1}{\cfrac{1}{10} + \cfrac{1}{15}} = 11\,\Omega$$

We find $i_X = 15/11 = 1.36\,A$. Given i_X, we now use current division to obtain the short-circuit current

$$i_N = i_{SC} = \frac{10}{10 + 15} \times i_X = 0.545\,A$$

Finally, we compute the Thévenin and Norton resistances

$$R_T = R_N = \frac{v_{OC}}{i_{SC}} = 18.3\,\Omega$$

The resulting Thévenin and Norton equivalent circuits are shown in Figures 3–44(d) and 3–44(e).

It now is an easy matter to select a load R_L so that 4 V is supplied to the load. Using the Thévenin equivalent circuit, the problem reduces to a voltage divider

$$\frac{R_L}{R_L + R_T} \times v_T = \frac{R_L}{R_L + 18.3} \times 10 = 4\,V$$

Solving for R_L yields $R_L = 12.2\,\Omega$.

The Thévenin or Norton equivalent can always be found from the open-circuit voltage and short-circuit current at the interface. The following examples illustrate other methods of determining these equivalent circuits.

So use this as fallback method

EXAMPLE 3 – 17

Find the Thévenin equivalent at nodes A and B for the circuit in Figure 3–45(a).

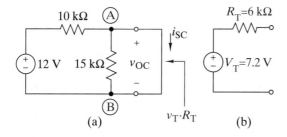

(a)   (b)

FIGURE 3–45

SOLUTION:
The open-circuit voltage can be found using a voltage divider

$$v_{OC} = v_T = \frac{15\,k \times 12}{15\,k + 10\,k} = 7.2\,V$$

The short-circuit current is found by placing a short circuit connecting nodes A and B and finding the current flowing through that short circuit. The short circuit is in parallel with the 15-kΩ resistor thereby effectively removing it from the circuit. The short-circuit current is thus found using Ohm's law

$$i_{SC} = \frac{12}{10\,k} = 1.2\,mA$$

The Thévenin resistance is then found as

$$R_T = \frac{v_{OC}}{i_{SC}} = \frac{7.2}{1.2\,m} = 6\,k\Omega$$

The Thévenin equivalent circuit is shown in Figure 3–45(b).   ■

Exercise 3–28

Find the Thévenin equivalent at nodes A and B for the circuit in Figure 3–46.

FIGURE 3–46

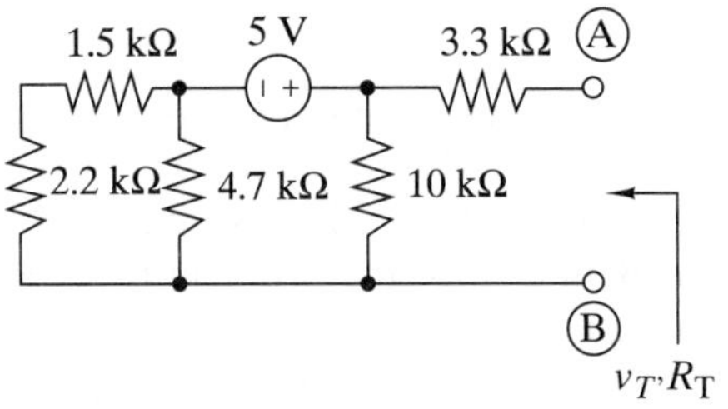

Answers: $v_T = 4.14$ V, $R_T = 5.01$ kΩ

EXAMPLE 3 – 1 8

(a) Find the Thévenin equivalent circuit of the source circuit to the left of the interface in Figure 3–47(a).
(b) Use the Thévenin equivalent to find the power delivered to two different loads. The first load is a 10-kΩ resistor and the second is a 5-V voltage source whose positive terminal is connected to the upper interface terminal.

SOLUTION:
(a) To find the Thévenin equivalent, we use the sequence of circuit reductions shown in Figure 3–47. In Figure 3–47(a) the 15-V voltage source in series with the 3-kΩ to the left of terminals A and B is replaced by a 3-kΩ resistor in parallel with an equivalent current source with $i_S = 15/3000 = 5\,mA$. In Figure 3–47(b), looking to the left of terminals C and D, we see two resistors in parallel whose equivalent resistance is $(3\,k\Omega)\|(6\,k\Omega) = 2\,k\Omega$. We also see two current sources in parallel whose equivalent is $i_S = 5\,mA - 2\,mA = 3\,mA$. This equivalent current source is shown in Figure 3–47(c) to the left of terminals C and D. Figure 3–47(d) shows this current source converted to an equivalent voltage source $v_S = 3\,mA \times 2\,k\Omega = 6\,V$ in series with $2\,k\Omega$. In Figure 3–47(d) the three resistors are connected in series and can be replaced by an equivalent resistance $R_{EQ} = 2\,k\Omega + 3\,k\Omega + 4\,k\Omega = 9\,k\Omega$. This step produces the Thévenin equivalent shown in Figure 3–47(e).

Note: The steps leading from Figure 3–47(a) to 3–47(e) involve circuit reduction techniques studied in Chapter 2, so we know that this approach works best on ladder circuits like the one in Figure 3–47(a).

FIGURE 3–47

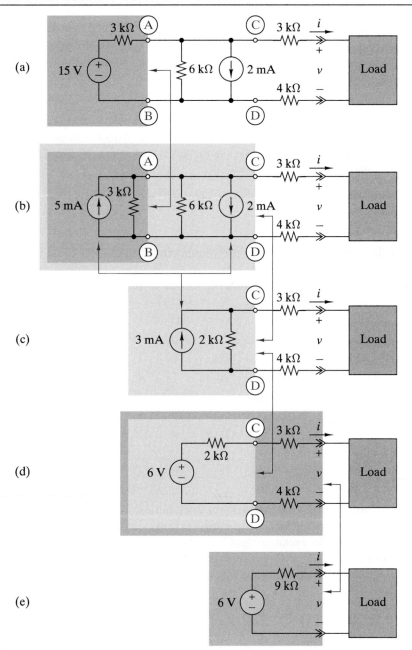

(a)

(b)

(c)

(d)

(e)

(b) Figure 3–48 shows the Thévenin equivalent found in part (a) (Figure 3–47(e)) and the two loads. When the load is a 10-kΩ resistor, the interface current is $i = (6)/(9000 + 10,000) = 0.3158 \, \text{mA}$, and the power delivered to the load is $i^2 R_L = 0.9973 \, \text{mW}$. When the load is a 5-V source, the interface voltage and current are $v = 5 \, \text{V}$ and $i = (6 - 5)/9000 = 0.1111 \, \text{mA}$, and the power to the load is $v \times i = 0.5555 \, \text{mW}$. Since $p > 0$ in the latter case, we see that the voltage source load is absorbing rather than delivering power. A practical example of this situation is a battery charger.

Caution: The Thévenin equivalent allows us to calculate the power delivered to a load, but it does not tell us what power is dissipated in the source circuit. For instance, if the load in Figure 3–47(e) is an open circuit, then no power is dissipated in the

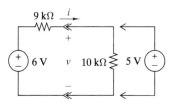

FIGURE 3–48

Thévenin equivalent since $i = 0$. This does not mean that the power dissipated in the source circuit is zero, as we can easily see by looking back at Figure 3–47(a). The Thévenin equivalent circuit has the same i–v characteristic at the interface, but it does not duplicate the internal characteristics of the source circuit.   ■

Exercise 3–29

For the Thévenin circuit of Figure 3–48, select a load R_L so that 2.5 V are delivered across it. Choose R_L from the 10% values given in the inside back cover. What will be the actual value of voltage delivered to the load?

Answers: $R_L = 6.8\,\mathrm{k\Omega}$, $v_L = 2.58 \pm 10\%\,\mathrm{V}$

EXAMPLE 3–19

(a) Find the Norton equivalent of the source circuit to the left of the interface in Figure 3–49.
(b) Find the interface current i when the power delivered to the load is 5 W.

FIGURE 3–49

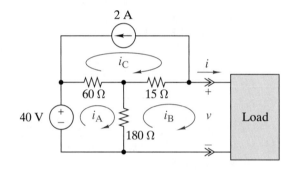

SOLUTION:

(a) The circuit reduction method will not work here since the source circuit is not a ladder. In this example we write mesh-current equations and solve directly for the source circuit i–v relationship. We only need to write equations for meshes A and B since the 2-A current source determines the mesh C current. The voltage sums around these meshes are

$$\text{Mesh A:} \quad -40 + 60(i_A - i_C) + 180(i_A - i_B) = 0$$
$$\text{Mesh B:} \quad -180(i_A - i_B) + 15(i_B - i_C) + v = 0$$

But since $i_B = i$ and the current source forces the condition $i_C = -2$, these equations have the form

$$240i_A - 180i = -80$$
$$-180i_A + 195i = -30 - v$$

Solving for the currents in terms of v yields

$$
\begin{bmatrix} i_A \\ i \end{bmatrix} =
\begin{bmatrix} 240 & -180 \\ -180 & 195 \end{bmatrix}^{-1}
\begin{bmatrix} -80 \\ -30 - v \end{bmatrix} =
\begin{bmatrix} -\dfrac{35}{24} - \dfrac{v}{80} \\[2mm] -\dfrac{3}{2} - \dfrac{v}{60} \end{bmatrix}
$$

$$i = -\frac{3}{2} - \frac{v}{60}$$

At the interface the i–v relationship of the source circuit is $i = -1.5 - v/60$. Equation (3–22) gives the i–v relationship of the Norton circuit as $i = i_N - v/R_N$. By direct comparison, we conclude that $i_N = -1.5\,\text{A}$ and $R_N = 60\,\Omega$. This equivalent circuit is shown in Figure 3–50.

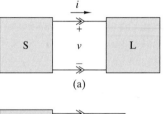

FIGURE 3–50

(b) When 5 W is delivered to the load, we have $vi = 5$ or $v = 5/i$. Substituting $v = 5/i$ into the source i–v relationship $i = -1.5 - v/60$ yields the quadratic equation

$$12i^2 + 18i + 1 = 0$$

whose roots are $i = -0.05778\,\text{A}$ and $-1.442\,\text{A}$. Thus, there are two values of interface current that deliver 5 W to the load.   ∎

DERIVATION OF THÉVENIN'S THEOREM

The derivation of Thévenin's theorem is based on the superposition principle. We begin with the circuit in Figure 3–51(a), where the source circuit S is linear. Our approach is to use superposition to show that the source circuit and the Thévenin circuit have the same i–v relationship at the interface. To find the source circuit i–v relationship, we first disconnect the load and apply a current source i_{TEST}, as shown in Figure 3–51(b). Using superposition to find v_{TEST}, we first turn i_{TEST} off and leave all the sources inside S on, as shown in Figure 3–51(c). Turning a current source off leaves an open circuit, so

$$v_{\text{TEST1}} = v_{\text{OC}}$$

Next we turn i_{TEST} back on and turn off all of the sources inside S. Since the source circuit S is linear, it reduces to the equivalent resistance shown in Figure 3–51(d) when all internal sources are turned off. Using Ohm's law, we write

$$v_{\text{TEST2}} = (R_{\text{EQ}})(-i_{\text{TEST}})$$

The minus sign in this equation results from the reference directions originally assigned to i_{TEST} and v_{TEST} in Figure 3–51(b). Using the superposition principle, we find the i–v relationship of the source circuit at the interface to be

$$v_{\text{TEST}} = v_{\text{TEST1}} + v_{\text{TEST2}}$$
$$= v_{\text{OC}} - R_{\text{EQ}}i_{\text{TEST}}$$

This equation has the same form as the i–v relationship of the Thévenin equivalent circuit in Eq. (3–21) when $v_{\text{TEST}} = v$, $i_{\text{TEST}} = i$, $v_{\text{OC}} = v_T$, and $R_T = R_{\text{EQ}}$.

The derivation points out another method of finding the Thévenin resistance. As indicated in Figure 3–51(d), when all the sources are turned off, the i–v relationship of the source circuit reduces to $v = -iR_{\text{EQ}}$. Similarly, the i–v relationship of a Thévenin equivalent circuit reduces to $v = -iR_T$ when $v_T = 0$. We conclude that

$$R_T = R_{\text{EQ}} \qquad (3\text{–}26)$$

We can find the value of R_T by determining the resistance seen looking back into the source circuit with all sources turned off. For this reason the Thévenin resistance R_T is sometimes called the **lookback resistance**.

The next example shows how lookback resistance contributes to finding a Thévenin equivalent circuit.

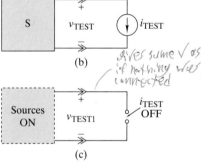

(a)

(b)

gives same V as if nothing was connected

(c)

makes source like a load

also R_{TH} (d)

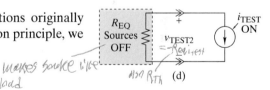

FIGURE 3–51 *Using superposition to prove Thévenin's theorem.*

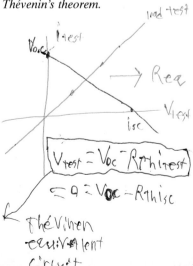

$V_{test} = V_{oc} - R_{Th}(i_{test})$
$\quad \subset \, 0 = V_{oc} - R_{Th}i_{sc}$

Thévenin equivalent circuit

EXAMPLE 3–20

(a) Find the Thévenin equivalent of the source circuit to the left of the interface in Figure 3–52.
(b) Use the Thévenin equivalent to find the voltage delivered to the load.

FIGURE 3–52

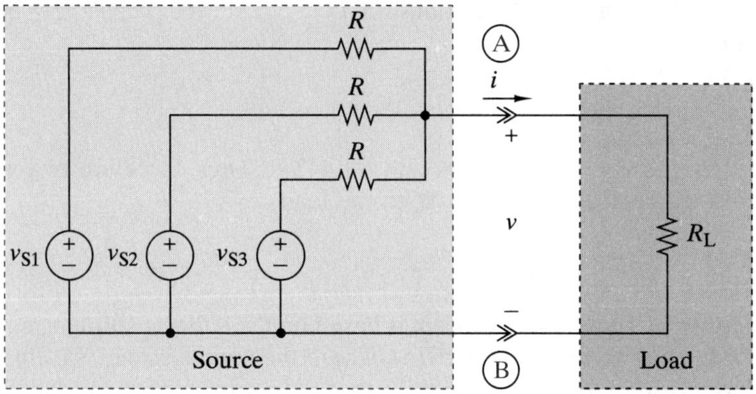

also could do

$V_{test} - V_{o.c} = -R_{Th} i_{test}$

$i_{test} = \frac{V_{o.c}}{R_{Th}} - \frac{V_{test}}{R_{Th}}$

$i_{test} = i_{sc} - \frac{V_{test}}{R_{Th}}$

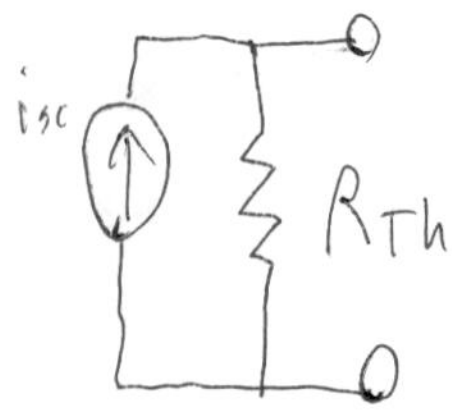

Norton equiv circuit

SOLUTION:

(a) The source circuit in Figure 3–52 is treated in Example 3–14 by using superposition to calculate the open-circuit voltage between terminals A and B. Using the results from Example 3–14, we have

$$v_T = v_{OC} = \frac{1}{3}(v_{S1} + v_{S2} + v_{S3})$$

Turning all sources off in Figure 3–52 leads to the resistance circuit in Figure 3–53. Looking back into the source circuit in Figure 3–53, we see three equal resistances connected in parallel whose equivalent resistance is $R/3$. Hence, the Thévenin resistance is

$$R_T = R_{EQ} = \frac{R}{3}$$

(b) Given the Thévenin circuit parameters v_T and R_T, we apply voltage division in Figure 3–52 to find the interface voltage.

$$v = \frac{R_L}{R_L + R_T} v_T = \left(\frac{R_L}{R_L + R/3}\right)\left(\frac{v_{S1} + v_{S2} + v_{S3}}{3}\right)$$
$$= \left(\frac{R_L}{3R_L + R}\right)(v_{S1} + v_{S2} + v_{S3})$$

FIGURE 3–53

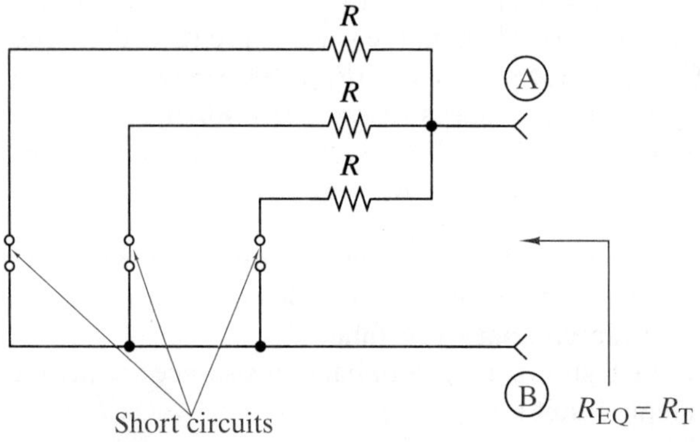

NOTE:
guaranteed to be a Thévinen or Norton Problem on 1st quiz

The interface voltage is proportional to the sum of the three source voltages. The proportionality constant $K = R_L/(3R_L + R)$ depends on both the source and the load since these two circuits are connected at the interface. ■

Exercise 3–30

(a) Find the Thévenin and Norton equivalent circuits seen by the load in Figure 3–54.

(b) Find the voltage, current, and power delivered to a 50-Ω load resistor.

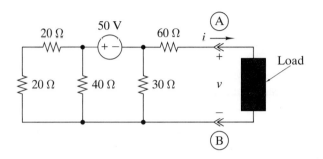

FIGURE 3–54

Answers:

(a) $v_T = -30\,\text{V}$; $i_N = -417\,\text{mA}$; $R_N = R_T = 72\,\Omega$

(b) $v = -12.3\,\text{V}$; $i = -246\,\text{mA}$; $p = 3.03\,\text{W}$

Exercise 3–31

Find the current and power delivered to an unknown load in Figure 3–54 when $v = +6\,\text{V}$.

Answers: $i = -\frac{1}{2}\,\text{A}$; $p = -3\,\text{W}$

APPLICATION TO NONLINEAR LOADS

Thévenin and Norton equivalent circuits can be used to find the response of a two-terminal nonlinear element (NLE). The method of analysis is a straight forward application of device and interface $i–v$ characteristics. An interface is defined at the terminals of the nonlinear element, and the linear part of the circuit is reduced to the Thévenin equivalent in Figure 3–55(a). The $i–v$ relationship of the Thévenin equivalent can be written with interface current as the dependent variable:

$$i = \left(-\frac{1}{R_T}\right)v + \left(\frac{v_T}{R_T}\right) \tag{3–27}$$

This is the equation of a straight line in the $i–v$ plane shown in Figure 3–55(b). The line intersects the i-axis $(v = 0)$ at $i = v_T/R_T = i_{SC}$ and intersects the v-axis $(i = 0)$ at $v = v_T = v_{OC}$. This line could logically be called the source line since it is determined by the Thévenin parameters of the source circuit. Logic notwithstanding, electrical engineers call this the **load line** for reasons that have blurred with the passage of time.

The nonlinear element has the $i–v$ characteristic shown in Figure 3–55(c). Mathematically, this nonlinear characteristic has the form

$$i = f(v) \tag{3–28}$$

To find the circuit response, we must solve Eqs. (3–27) and (3–28) simultaneously. Computer software tools like MATLAB can easily solve this problem when a numerical expression for the function $f(v)$ is known explicitly. However, in practice

FIGURE 3–55 *Graphical analysis of a nonlinear circuit: (a) Given circuit. (b) Load line. (c) Nonlinear device's i–v characteristics. (d) Q-point.*

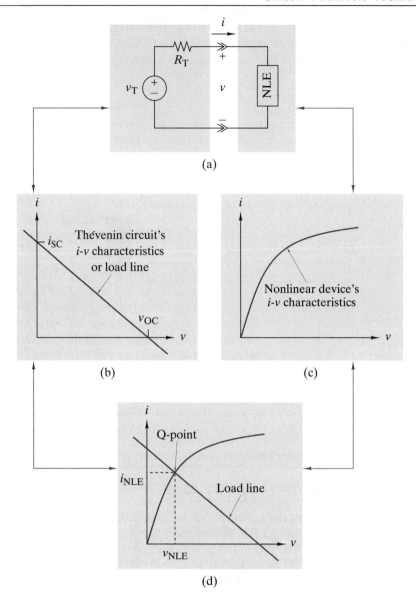

(a)

(b)

(c)

(d)

an approximate graphical solution is often adequate, particularly when $f(v)$ is given only in graphical form.

The developers of SPICE software packages like OrCAD work hard to create software models of the myriad of electronic devices used in circuit design—yet, even with their best efforts, devices are often very non-linear and difficult to model accurately. No matter what software approach one uses, there is always a possibility of error. That is why circuits are first modelled using software and then built and tested in the laboratory to validate the design.

In Figure 3–55(d) we superimpose the load line on the i–v characteristic curve of the nonlinear element. The two curves intersect at the point $i = i_{\text{NLE}}$ and $v = v_{\text{NLE}}$, which yields the values of interface variables that satisfy both the source constraints in Eq. (3–27) and the nonlinear element constraints in Eq. (3–28). In the terminology of electronics, the point of intersection is called the operating point or **Q-point**, where "Q" stands for "quiescent."

EXAMPLE 3–21

Find the voltage, current, and power delivered to the diode in Figure 3–56(a). The diode's i–v characteristics are given in Figure 3–56(b).

FIGURE 3–56

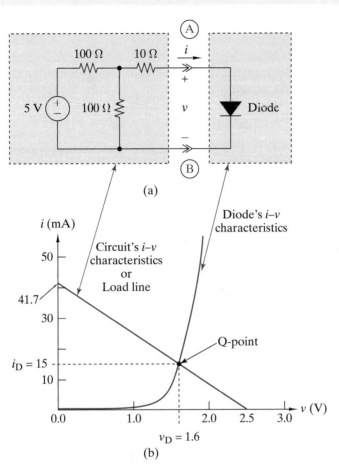

(a)

(b)

SOLUTION:
We first find the Thévenin equivalent of the circuit to the left of terminals A and B. By voltage division, the open-circuit voltage is

$$v_T = v_{OC} = \frac{100}{100 + 100} \times 5 = 2.5\,\text{V}$$

When the voltage source is turned off, the lookback equivalent resistance seen between terminals A and B is

$$R_T = 10 + 100 \| 100 = 60\,\Omega$$

The source circuit load line is given by

$$i = -\frac{1}{60}v + \frac{1}{60} \times 2.5$$

This line intersects the i-axis ($v = 0$) at $i = i_{SC} = 2.5/60 = 41.7\,\text{mA}$ and intersects the v-axis ($i = 0$) at $v = v_{OC} = 2.5\,\text{V}$. Figure 3–56(b) superimposes the source circuit load line on the diode's i–v curve. The intersection (Q-point) is at $i = i_D = 15\,\text{mA}$ and

$v = v_D = 1.6\,\text{V}$. This is the point (i_D, v_D) at which both the source and diode device constraints are satisfied. Finally, the power delivered to the diode is

$$p_D = i_D v_D = (15 \times 10^{-3})(1.6) = 24\,\text{mW}$$

Because of the nonlinear element, the proportionality and superposition properties do not apply to this circuit. For instance, if the source voltage in Figure 3–56 is decreased from 5 V to 2.5 V, the diode current and voltage do not decrease by one-half. Try it. ■

Exercise 3–32 _____

Suppose for the circuit shown in Figure 3–56(a) that the diode's i–v characteristic can be modeled by the following equation:

$$i = \frac{e^{4v} - 1}{100,000}$$

where v is given in volts and i is given in amperes. Note that this i–v characteristic is for a different diode from that shown in Figure 3–56(b). Use manual calculations or the MATLAB function `solve` to find the exact operating point for this circuit.

Answers: $v_D = 1.77\,\text{V}$, $i_D = 12.1\,\text{mA}$

_____

In summary, any two of the following parameters determine the Thévenin or Norton equivalent circuit at a specified interface:

- The open-circuit voltage at the interface
- The short-circuit current at the interface
- The source circuit lookback resistance

Alternatively, for ladder circuits the Thévenin or Norton equivalent circuit can be found by a sequence of circuit reductions (see Example 3–18). For general circuits they can always be found by directly solving for the i–v relationship of the source circuit using node-voltage or mesh-current equations that include the interface current and voltage as unknowns (see Example 3–19).

3–5   MAXIMUM SIGNAL TRANSFER

An interface is a connection between two circuits at which the signal levels may be observed or specified. In this regard an important consideration is the maximum signal levels that can be transferred across a given interface. This section defines the maximum voltage, current, and power available at an interface between a *fixed source* and an *adjustable load.*

For simplicity we will treat the case in which both the source and load are linear resistance circuits. The source can be represented by a Thévenin equivalent and the load by an equivalent resistance R_L, as shown in Figure 3–57. For a fixed source,

FIGURE 3–57 *Two-terminal interface for deriving the maximum signal transfer conditions.*

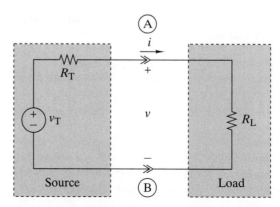

the parameters v_T and R_T are given and the interface signal levels are functions of the load resistance R_L.

By voltage division, the interface voltage is

$$v = \frac{R_L}{R_L + R_T} v_T \tag{3–29}$$

For a fixed source and a variable load, the voltage will be a maximum if R_L is made very large compared with R_T. Ideally, R_L should be made infinite (an open circuit), in which case

$$v_{MAX} = v_T = v_{OC} \tag{3–30}$$

Therefore, the maximum voltage available at the interface is the source open-circuit voltage v_{OC}.

The current delivered at the interface is

$$i = \frac{v_T}{R_L + R_T} \tag{3–31}$$

For a fixed source and a variable load, the current will be a maximum if R_L is made very small compared with R_T. Ideally, R_L should be zero (a short circuit), in which case

$$i_{MAX} = \frac{v_T}{R_T} = i_N = i_{SC} \tag{3–32}$$

Therefore, the maximum current available at the interface is the source short-circuit current i_{SC}.

The power delivered at the interface is equal to the product $v \times i$. Using Eqs. (3–29) and (3–31), the power is

$$p = v \times i$$
$$= \frac{R_L v_T^2}{(R_L + R_T)^2} \tag{3–33}$$

For a given source, the parameters v_T and R_T are fixed and the delivered power is a function of a single variable R_L. The condition for maximum voltage $(R_L \to \infty)$ and the condition for maximum current $(R_L = 0)$ both produce zero power. The value of R_L that maximizes the power lies somewhere between these two extremes. To find this value, we differentiate Eq. (3–33) with respect to R_L and solve for the value of R_L for which $dp/dR_L = 0$.

$$\frac{dp}{dR_L} = \frac{[(R_L + R_T)^2 - 2R_L(R_L + R_T)]v_T^2}{(R_L + R_T)^4} = \frac{R_T - R_L}{(R_L + R_T)^3} v_T^2 = 0 \tag{3–34}$$

Clearly, the derivative is zero when $R_L = R_T$. Therefore, **maximum power transfer** occurs when the load resistance equals the Thévenin resistance of the source. When the condition $R_L = R_T$ exists, the source and load are said to be **matched**.

Substituting the condition $R_L = R_T$ back into Eq. (3–33) shows the maximum power to be

$$p_{MAX} = \frac{v_T^2}{4R_T} \tag{3–35}$$

Since $v_T = i_N R_T$, this result can also be written as

$$p_{MAX} = \frac{i_N^2 R_T}{4} \tag{3–36}$$

FIGURE 3–58 *Normalized plots of current, voltage, and power versus R_L/R_T.*

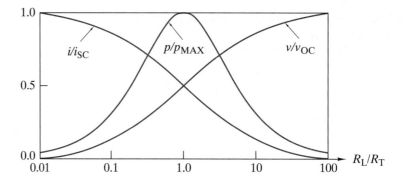

or

$$p_{MAX} = \frac{v_T i_N}{4} = \left[\frac{v_{OC}}{2}\right]\left[\frac{i_{SC}}{2}\right] \qquad (3\text{–}37)$$

These equations are consequences of what is known as the **maximum power transfer theorem**:

> *A source with a fixed Thévenin resistance R_T delivers maximum power to an adjustable load R_L when $R_L = R_T$.*[1]

To summarize, at an interface with a fixed source,

1. The maximum available voltage is the open-circuit voltage.
2. The maximum available current is the short-circuit current.
3. The maximum available power is the product of one-half the open-circuit voltage times one-half the short-circuit current.

Figure 3–58 shows plots of the interface voltage, current, and power as functions of R_L/R_T. The plots of v/v_{OC}, i/i_{SC}, and p/p_{MAX} are normalized to the maximum available signal levels, so the ordinates in Figure 3–58 range from 0 to 1. The plot of the normalized power p/p_{MAX} in the neighborhood of the maximum is not a particularly strong function of R_L/R_T. Changing the ratio R_L/R_T by a factor of 2 in either direction from the maximum reduces p/p_{MAX} by less than 20%. The normalized voltage v/v_{OC} is within 20% of its maximum when $R_L/R_T = 4$. Similarly, the normalized current is within 20% of its maximum when $R_L/R_T = 1/4$. In other words, for engineering purposes we can get close to the maximum signal levels with load resistances that only approximate the theoretical requirements.

EXAMPLE 3–22

A source circuit with $v_T = 2.5\,\text{V}$ and $R_T = 60\,\Omega$ drives a load with $R_L = 30\,\Omega$.

(a) Determine the maximum signal levels available from the source circuit.
(b) Determine the actual signal levels delivered to the load.

SOLUTION:
(a) The maximum available voltage and current are

$$v_{MAX} = v_{OC} = v_T = 2.5\,\text{V}\ (R_L \to \infty)$$

$$i_{MAX} = i_{SC} = \frac{v_T}{R_T} = 41.7\,\text{mA}\ (R_L = 0)$$

---

[1]An ideal voltage source has zero internal resistance, hence $R_T = 0$. Equation (3–35) points out that $R_T = 0$ implies an infinite p_{MAX}. Infinite power is a physical impossibility, which reminds us that all ideal circuit models have some physical limitations.

The maximum available power is found using Eq. (3–37).

$$p_{\text{MAX}} = \left[\frac{v_{\text{OC}}}{2}\right]\left[\frac{i_{\text{SC}}}{2}\right] = 26.0\,\text{mW} \quad (R_{\text{L}} = R_{\text{T}} = 60\,\Omega)$$

(b) The actual signal levels delivered to the 30-Ω load are

$$v_{\text{L}} = \frac{30}{30 + 60}\,2.5 = 0.833\,\text{V}$$

$$i_{\text{L}} = \frac{2.5}{30 + 60} = 27.8\,\text{mA}$$

$$p_{\text{L}} = v_L i_L = 23.1\,\text{mW}$$

Although these levels are less than the maximum available values, the power delivered to the 30-Ω load is nearly 90% of the maximum. ∎

Exercise 3–33

A source circuit delivers 4 V when a 50-Ω resistor is connected across its output and 5 V when a 75-Ω resistor is connected. Find the maximum voltage, current, and power available from the source.

Answers: 10 V, 133 mA; 333 mW

Remember that the maximum signal levels just derived are for a fixed source resistance and an adjustable load resistance. This situation often occurs in communication systems where devices such as antennas, transmitters, and signal generators have fixed source resistances such as 50, 75, 300, or 600 ohms. In such cases the load resistance is selected to achieve the desired interface conditions, which often involves matching.

Matching source and load applies when the load resistance R_{L} in Figure 3–57 is adjustable and the Thévenin source resistance R_{T} is fixed. When R_{L} is fixed and R_{T} is adjustable, then Eqs. (3–29), (3–31), and (3–33) point out that the maximum voltage, current, and power are delivered when the Thévenin source resistance is zero. If the source circuit at an interface is adjustable, then ideally the Thévenin source resistance should be zero. In the next chapter we will see that OP AMP circuits approach this ideal.

3–6 INTERFACE CIRCUIT DESIGN

The maximum signal levels discussed in the previous section place bounds on what is achievable at an interface. However, those bounds are based on a fixed source and an adjustable load. In practice, there are circumstances in which the source or the load, or both, can be adjusted to produce prescribed interface signal levels or neither can be adjusted. Sometimes it is necessary to insert a circuit between the source and the load to achieve the desired results. Figure 3–59 shows the general situations and some examples of resistive interface circuits. By its nature, the inserted circuit has two terminal pairs, or interfaces, at which voltage and current can be observed or specified. These terminal pairs are also called *ports*, and the interface circuit is referred to as a **two-port network**. The port connected to the source is called the input, and the port connected to the load is called the output. The purpose of this two-port network is to make certain that the source and load interact in a prescribed way.

There is near-infinite number of interface circuits that one can use or devise to meet the many interfacing challenges an analog design engineer faces. The five shown in Figure 3-59 represent the simplest ones. The pass-through is used when the load can be varied by the designer—in practice it may actually not even be shown

FIGURE 3–59 *A general interface circuit and a few examples. (a) Simple pass-through (often omitted), (b) series resistor, (c) parallel resistor, (d) L-pad left, (e) L-pad right.*

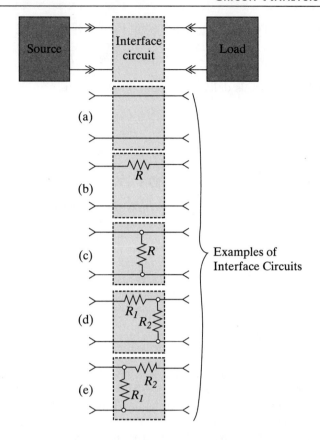

FIGURE 3–60 *L-pad for matching amplifier output to speakers*

since it is a simple connection between the source and the load (Examples 23 and 24). The series and parallel interfaces are used to help deliver a particular current, voltage, or power to a fixed load from a fixed source. In general, the series interface works best when the load is powered by a voltage source and a parallel interface works best when the load is powered by a current source (Examples 25 and 26). The L-pads are used in helping to match the fixed source and fixed load resistances and deliver specific voltages, currents, or power (Examples 27, 28, and 29). A common example is matching a stereo amplifier to a speaker—see Figure 3-60. A more complex interface circuit is the bridge-T considered in Example 3–31. That interface serves as an attenuation pad to reduce the signal by a fixed amount while maintaining the input and output resistances. In the following section we will see how to design these interface circuits for various applications.

BASIC CIRCUIT DESIGN CONCEPTS

Before we treat examples of different interface situations, you should recognize that we are now discussing a limited form of circuit design, as contrasted with circuit analysis. Although we use circuit analysis tools in design, there are important differences. A linear circuit analysis problem generally has a unique solution. A circuit design problem may have many solutions or even no solution. The maximum available signal levels found in the preceding section provide bounds that help us test for the existence of a solution. Generally there will be several ways to meet the interface constraints, and it then becomes necessary to evaluate the alternatives using other factors, such as cost, power consumption, or reliability.

At this point in our study, resistors are the only elements we can use to design interface circuits. In subsequent chapters we will introduce other devices, such as

OP AMPs (Chapter 4), capacitors and inductors (Chapter 6) and transformers (Chapter 15). In a design situation the engineer must choose the resistance values in a proposed circuit. This decision is influenced by a host of practical considerations, such as standard values and tolerances, power ratings, temperature sensitivity, cost, and fabrication methods. We will occasionally introduce some of these considerations into our design examples. Gaining a full understanding of these practical matters is not one of our objectives. Rather, our goal is simply to illustrate how different constraints can influence the design process.

D DESIGN EXAMPLE 3–23

Select the load resistance in Figure 3–61 so that the interface signals are in the range defined by $v \geq 4\,V$ and $i \geq 30\,mA$.

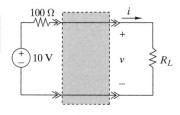

FIGURE 3–61

SOLUTION:

In this design problem, the source circuit is given and we are free to select the load. For a fixed source, the maximum signal levels available at the interface are

$$v_{MAX} = v_T = 10\,V$$
$$i_{MAX} = \frac{v_T}{R_T} = 100\,mA$$

The bounds given as design requirements are below the maximum available signal levels, so we should be able to find a suitable resistor. Using voltage division, the interface voltage constraint requires

$$\frac{R_L}{100 + R_L} \times 10 \geq 4$$

or

$$10R_L \geq 4R_L + 400$$

This condition yields $R_L \geq 400/6 = 66.7\,\Omega$. The interface current constraint can be written as

$$\frac{10}{100 + R_L} \geq 0.03$$

or

$$10 \geq 3 + 0.03\,R_L$$

which requires $R_L \leq 7/0.03 = 233\,\Omega$. In theory, any value of R_L between 66.7 Ω and 233 Ω will work. However, to allow for parameter variations we select $R_L = 150\,\Omega$ because it lies at the arithmetic midpoint of the allowable range and is a standard value of resistance (see inside back cover). ■

D Design Exercise 3–34 _____

Select R_L in Figure 3–61 so that $190 \pm 10\%$ mW are delivered to the load. Select a 10% resistor from inside the back cover that will provide the desired voltage.

Answer: $R_L = 33\,\Omega\,(187\,mW)$ or $270\,\Omega\,(197\,mW)$. The 270-Ω load requires the source to deliver less power to the total circuit, $P_{S270} = 270\,mW$ versus $P_{S33} = 752\,mW$, and generally is the better solution.

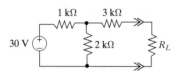

FIGURE 3–62

Select a 5% standard-value load resistor R_L for the circuit in Figure 3-62 that will result in 10 V ±5% delivered across it.

SOLUTION:
Start by finding the Thévenin equivalent circuit that R_L sees. The Thévenin voltage is found using a voltage divider

$$v_T = \left(\frac{2\,\text{k}}{1\,\text{k} + 2\,\text{k}}\right)30 = 20\,\text{V}$$

We can find R_T by the look-back method
$$(2\,\text{k}||1\,\text{k}) + 3\,\text{k} = 666 + 3000 = 3.66\,\text{k}\Omega$$

Using the Thévenin circuit we just found, find the maximum voltage possible by choosing R_L to be an open circuit. By inspection we determine that $v_{\text{MAX}} = 20$ V. Hence, it is possible to find a suitable resistor to deliver 10 V.

Using voltage division with the Thévenin circuit, we find R_L

$$10 = \frac{R_L}{R_L + 3.66\,\text{k}}\,20$$

$$R_L = \frac{36.6\,\text{k}}{10} = 3.66\,\text{k}\Omega$$

From the table in the inside rear cover we find that a 5% 3.6 kΩ resistor is available. Using it as our choice, we find that the resistance can vary as $3.42\,\text{k}\Omega \leq R_L \leq 3.78\,\text{k}\Omega$. Therefore, the output voltage can vary as $9.66\,\text{V} \leq v_L \leq 10.16\,\text{V}$. This is within the 10 V ±5% required.

Our interface is a simple pass-through as shown in Figure 3–59(a). ■

D Design Exercise 3–35 _____

For the circuit of Figure 3–62, select a load resistor, if possible, so that 6 mA flows through it.

A n s w e r: The maximum current available is 5.46 mA, hence there is no resistor available that can result in 6mA flowing through it.

D Design Exercise 3–36 _____

For the circuit of Figure 3–62, determine the maximum power available, and if sufficient, select a resistive load that will dissipate 20 mW.

A n s w e r s: $P_{\text{MAX}} = 27.3$ mW, $R_L = 1.169$ kΩ or 11.50 kΩ

A light-emitting diode (LED) converts electric current into an optical signal. LEDs operate at low signal levels with voltages from about 1 V to perhaps 3 V and at currents between about 10 mA and 40 mA. Voltages or currents above these levels may damage or destroy the device.

Figure 3–63 shows an LED operating at $v = 1.5$ V and connected to a 5-V source by an interface circuit. Design the interface circuit so that the LED current is $i = 15$ mA ± 10% using one or more of the following standard resistors: 110 Ω, 160 Ω, 240 Ω, 360 Ω, and 510 Ω. These resistors all have a tolerance of ±5%, which you must account for in your design.

FIGURE 3–63

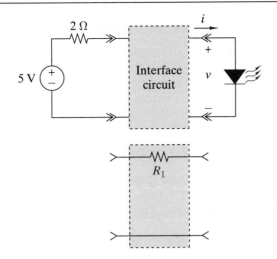

SOLUTION:

If the source is directly connected to the LED, the delivered current would be

$$i = \frac{5 - 1.5}{2} = 1.75 \text{ A}$$

This much current would destroy (vaporize?) the device. The series resistor R_1 interface Figure 3.59(b) is needed to limit the current to the prescribed level. Applying KVL around the series loop yields

$$-5 + (2 + R_1)i + 1.5 = 0$$

Setting $i = 15 \text{ mA}$ and solving for R_1 yields

$$R_1 = \frac{3.5}{0.015} - 2 = 231 \text{ } \Omega$$

The nearest standard value listed is $240 \text{ } \Omega \pm 5\%$, which means that R_1 would fall in the range $228 \leq R_1 \leq 252 \text{ } \Omega$. At the end points of this range, the LED current is

$$i = \frac{3.5}{252 + 2} = 13.8 \text{ mA} \quad \text{and} \quad i = \frac{3.5}{228 + 2} = 15.2 \text{ mA}$$

Both of these values are within the $15 \text{ mA} \pm 10\%$ tolerance on the LED current. ∎

D Design Exercise 3–37

Suppose the source circuit in Figure 3–63 is now 12 V in series with a 5-Ω source resistor. The same LED is used. How does the solution change?

A n s w e r : $R_1 = 695 \text{ } \Omega$. If we are restricted to using the resistors specified, then there are at least two reasonable solutions. First, we could use a 510-Ω resistor in series with a 160-Ω resistor to give a total resistance of 670 Ω. Another option would be to combine a 510-Ω resistor in series with a parallel combination of two 360-Ω resistors to give a total resistance of 690 Ω. These options satisfy the constraints, even if we account for the resistor tolerances.

◀ ▶   D E S I G N   A N D   E V A L U A T I O N   E X A M P L E   3 – 2 6

Design two versions of the interface circuit in Figure 3–64 that deliver $v_2 = 5\,\text{V}$ to the 200-Ω load. Evaluate the two designs in terms of power loss in the interface circuit.

FIGURE 3–64

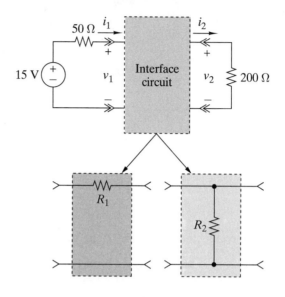

SOLUTION:

If the 15-V source is directly connected to the load, the delivered voltage would be

$$v_2 = \left(\frac{200}{50 + 200}\right) 15 = 12\,\text{V}$$

An interface circuit is required to reduce this voltage to the prescribed 5-V level.

The figure shows two possible interface circuits Figure 3–59(b) and (c). In either case

$$i_2 = \frac{5}{200} = 0.025\,\text{A}$$

In the series case $i_1 = i_2 = 0.025$, and the same current flows through all elements in the loop. Applying KVL to the series loop

$$-15 + 0.025 \times (50 + R_1) + 5 = 0$$

and solving for R_1,

$$R_1 = \frac{10}{0.025} - 50 = 350\,\Omega$$

In the parallel case $v_1 = v_2 = 5\,\text{V}$. Applying KCL to the parallel resistor R_2

$$i_1 - \frac{v_2}{R_2} - i_2 = \frac{15 - 5}{50} - \frac{5}{R_2} - 0.025 = 0$$

and solving for R_2,

$$R_2 = \frac{5}{0.2 - 0.025} = 28.57\,\Omega$$

We have two alternative designs, both of which deliver $v_2 = 5\,\text{V}$ to the 200-Ω load.

In practice, engineers use additional factors to evaluate alternatives that meet the same design goal. The power dissipation in the interface circuit is an important factor for two reasons. First, less interface dissipation means less power demand on the source. Second, less dissipation in the interface resistors means they can have lower power ratings, which are generally less expensive.

In the series case the power dissipated in R_1 is $i_2^2 R_1 = 0.219$ W. In the parallel case the power dissipated in R_2 is $v_2^2/R_2 = 0.875$ W. Clearly the power dissipation factor strongly favors the series design in this case.   ■

D E Design and Evaluation Exercise 3–38

A Norton source of 300 mA in parallel with a 50-Ω source resistor provides current to a load $R_L = 200\,\Omega$. Your task is to design an interface so that $5\,V \pm 10\%$ are delivered to the load.

(a) Using the series resistor R_S interface shown in Figure 3–59(b), select a 10% resistor from the inside back cover that will provide the desired voltage.
(b) Using the parallel resistor R_P interface shown in Figure 3–59(c), select a 10% resistor from the inside back cover that will provide the desired voltage.
(c) Select the solution, series or parallel, that causes the source to provide the desired voltage while delivering the least power. Calculate the power in each case to defend your choice.

Answers:

(a) $R_S = 350\,\Omega$; hence choose 330-Ω resistor ($v_L = 5.17$ V).
(b) $R_P = 28.57\,\Omega$; hence choose 27-Ω resistor ($v_L = 4.84$ V).
(c) $p_S = 4.11$ W, $p_p = 1.45$ W; hence choose the parallel solution.

DISCUSSION: *This is the same problem with the same numbers as in Example 3–26 except that a Norton source provides the energy rather than the voltage source (do a source transformation on the Thévenin source in Example 3–26). As expected, the interfaces are the same. Yet in this case the parallel solution requires the source to deliver less power than the series solution. What these two exercises point out is that equivalent sources are equivalent only to the rest of the circuit and not internally. One cannot do a source transformation and then use the transformed source to determine the power the original source provides.*

D DESIGN EXAMPLE 3 – 27

Design the interface circuit in Figure 3–65 so that the 40-V source delivers $v_2 = 2$ V to the output load and the resistance seen at the input port is $R_{IN} = 300\,\Omega$. Note that this means that input resistance of the two ports matches the source resistance.

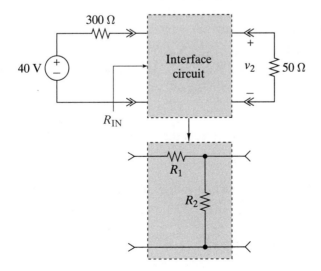

FIGURE 3–65

SOLUTION:

This example places constraints at both the output port and the input port of the interface circuit. In most cases two independent constraints cannot be satisfied using only one resistor in the interface circuit. To see why, suppose we use a single 650-Ω series resistor in the interface circuit. By voltage division, the output voltage would be

$$v_2 = \left(\frac{50}{300 + 650 + 50}\right)40 = 2\,\text{V}$$

as required. However, the input resistance would be $R_{\text{IN}} = 650 + 50 = 700\,\Omega$, which does not meet the input port requirement of $300\,\Omega$.

To meet both requirements, we need a two-resistor L-circuit such as the one shown Figure 3.59(d). To design this circuit, we first define $R_{\text{EQ}} = R_2\|50$. Using this notation, the input port constraint is $R_{\text{IN}} = R_1 + R_{\text{EQ}} = 300\,\Omega$ and the output port constraint becomes

$$v_2 = \left(\frac{R_{\text{EQ}}}{300 + R_1 + R_{\text{EQ}}}\right)40 = 2\,\text{V}$$

But $R_1 + R_{\text{EQ}} = 300$; hence, the output constraint reduces to $40R_{\text{EQ}} = 2 \times 600$, which means that $R_{\text{EQ}} = 30\,\Omega$. By definition,

$$R_{\text{EQ}} = \frac{50R_2}{50 + R_2} = 30$$

which leads to $50R_2 = 1500 + 30R_2$, or $R_2 = 75\,\Omega$. Finally, since $R_{\text{EQ}} = 30\,\Omega$, the input port constraint then tells us that $R_1 = 300 - R_{\text{EQ}} = 270\,\Omega$. In sum, the L-circuit in the figure with $R_1 = 270\,\Omega$ and $R_2 = 75\,\Omega$ will meet both the input port and the output port constraints. ∎

◑ Design Exercise 3–39

Repeat Example 3–27 with the desired $v_2 = 10\,\text{V}$ instead of 2 V.

Answer: $R_2 = -75\,\Omega$; therefore, it is not possible.

◐ EVALUATION EXAMPLE 3 – 28

In Example 3–27 we designed the interface circuit in Figure 3–65 to meet the requirements $v_2 = 2\,\text{V}$ and $R_{\text{IN}} = 300\,\Omega$. It is claimed that the interface circuit in Figure 3–66 meets the same requirements.

(a) Verify that the circuit in Figure 3–66 produces $v_2 = 2\,\text{V}$ and $R_{\text{IN}} = 300\,\Omega$.
(b) It is desired that the 50-Ω load "see" a low output resistance. Which of these two circuits best meets this requirement?

SOLUTION:

(a) The circuit in Figure 3–66 meets the input port constraint since

$$R_{\text{IN}} = 750\|(450 + 50) = 750\|500 = 300\,\Omega$$

as required. Using this fact and voltage division, we find the voltage at the input port of the interface circuit to be

$$v_1 = \left(\frac{R_{\text{IN}}}{300 + R_{\text{IN}}}\right)40 = \left(\frac{300}{600}\right)40 = 20\,\text{V}$$

FIGURE 3–66

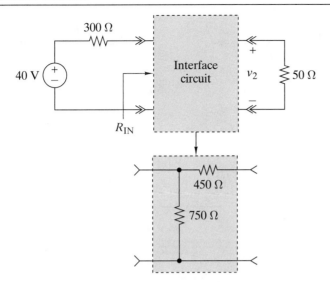

Using this voltage as the input to the voltage divider made up of the 450-Ω series resistor and the 50-Ω load gives us

$$v_2 = \left(\frac{50}{450 + 50}\right) v_1 = \left(\frac{50}{500}\right) 20 = 2\text{ V}$$

This verifies that the circuit in Figure 3–66 produces $v_2 = 2$ V and $R_{IN} = 300\,\Omega$.

(b) To compare the output resistances, we turn the 40-V source off (replace it by a short) and find the lookback resistance seen at the output port. For the circuit in Figure 3–65

$$R_{OUT} = R_2 \| (R_1 + 300) = 75 \| (270 + 300) = 66.3\,\Omega$$

For the circuit in Figure 3–66

$$R_{OUT} = 450 + 750\|300 = 664\,\Omega$$

Clearly the circuit in Figure 3–65 has a much lower output resistance. ∎

E Evaluation Exercise 3–40 _____

Use OrCAD to determine which solution, Figure 3–65 or Figure 3–66, requires less power from the source.

Answer: Both require the same power, 2.667 W.

DISCUSSION: *In fact, one need not do any analysis or simulation if the only question to answer is which is lower. In the preceding example, it was shown that both circuits met the same requirements, in particular that $R_{IN} = 300\,\Omega$. Since both circuits met that requirement exactly, both would have the same source current, namely $40/(300 + 300) = 66.67$ mA, and therefore, the same power. Of course, the simulation would verify that fact.*

D DESIGN EXAMPLE 3–29

Design the interface circuit in Figure 3–67 so the 50-Ω load "sees" a Thévenin resistance of 50 Ω between terminals C and D, while simultaneously the input voltage source "sees" an input resistance of 300 Ω between terminals A and B. Meeting these two constraints produces matched conditions at the input and output ports of the interface circuit.

FIGURE 3–67

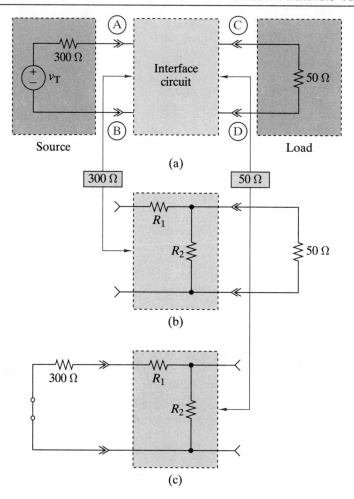

(a)

(b)

(c)

SOLUTION:

To meet the two constraints in this example, the interface circuit should be a two-resistors L-circuit. We have chosen the L-circuit configuration shown in Figure 3–59(d) for the following reasons. The source must see a larger resistance (300 Ω) at the input port than the load sees at the output port (50 Ω). This indicates that the source should "look" into a large series resistor R_1 while the load "looks" into a smaller parallel resistor R_2.

The design constraints in this example can be expressed in equation form. At the input port (terminals A and B) the equation is

$$R_1 + \frac{50R_2}{R_2 + 50} = 300\,\Omega$$

At the output port (terminals C and D) the equation is

$$\frac{(R_1 + 300)R_2}{R_1 + 300 + R_2} = 50\,\Omega$$

The design requirements reduce to two equations in two unknowns. What could be simpler?

These equations can easily be solved using a program like MATLAB. Solving them using pencil and paper is a bit of a chore. At this point we encourage you to think about the problem in physical terms. For instance, if we simply set $R_2 = 50\,\Omega$, then the conditions at terminals C and D will be met, at least approximately. With

$R_2 = 50\,\Omega$ the requirement at terminals A and B reduces to $R_{AB} = R_1 + 50\|50 = R_1 + 25 = 300\,\Omega$. In other words, by physical reasoning we conclude that $R_1 = 275\,\Omega$ and $R_2 = 50\,\Omega$ is an approximate solution. How good is the approximation?

These values yield input and output resistances of $R_{AB} = 300\,\Omega$ as required, and

$$R_{CD} = 50\|(275 + 300) = 50\|575 = 46\,\Omega$$

This value is not exactly $50\,\Omega$, but it is within 10% of the desired value. Since electrical components may have tolerances in the 10% range, a design based on our first-guess approximation might be adequate.

Our first-guess solution can also serve as the starting place for improving the design. The fact that $R_{CD} = 46\,\Omega$ tells us that $R_2 = 50\,\Omega$ is just a bit too low. Suppose we increase R_2 slightly to, say, $R_2 = 56\,\Omega$ (a standard value—see inside back cover). Then at the input port we have

$$R_{AB} = R_1 + 50\|56 = R_1 + 26.42 = 300\,\Omega$$

which would require $R_1 = 273.58\,\Omega$. The nearest standard value to this is $R_1 = 270\,\Omega$. Using the standard values $R_1 = 270\,\Omega$ and $R_2 = 56\,\Omega$ as our second guess, the input and output resistances are

$$R_{AB} = 270 + 50\|56 = 270 + 26.42 = 296.42\,\Omega$$
$$R_{CD} = 56\|(270 + 300) = 56\|570 = 50.99\,\Omega$$

both of which are within 2% of the desired values. Thus, finding an approximate solution can serve as the first step in the design process. Finding that first step is often the most creative and challenging part of circuit design.   ∎

D Design Exercise 3–41 _____

A common problem is interfacing a TV antenna's 300-Ω line to a 75-Ω cable input on an HDTV set. Repeat Example 3–29 for this particular interface. See Figure 3–68 for a photo of such a device.

Answers: $R_1 = 259.8\,\Omega$, $R_2 = 86.6\,\Omega$

APPLICATION   **EXAMPLE 3–30**

The source-load interface in Figure 3–69 serves to introduce an important concept that we will encounter many times in subsequent chapters. By simple voltage division, the interface voltage is

$$v = \frac{R_L}{R_L + R_T}\, v_T$$

www.ShowMeCables.com

FIGURE 3–68 *300 Ω to 75 Ω adapter*

FIGURE 3–69

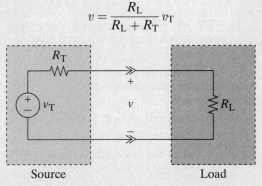

Source   Load

If the source is ideal ($R_T = 0$), then the interface voltage is $v = v_T$ regardless of the value of the load R_L. Conversely, if the load is an open circuit ($R_L = \infty$), then the interface voltage is $v = v_T$ regardless of the value of the source resistance R_T. Real-world applications typically fall between these two extremes with the result that

$v < v_T$. Since $v_T = v_{OC} = v_{MAX}$, interface voltage is generally less than the maximum available voltage. The reduction in interface voltage is an example of an effect called *loading*. In general

> *Loading is the reduction in load voltage due to the effect of load resistance on the signal source driving it.*

A loading problem is fundamentally different than the fixed-source, maximum power transfer problem. With loading, the source and load are both adjustable and the question is how they should be chosen to *minimize loading*. The undesirable effects of loading can be mitigated by making $R_T \ll R_L$, either by reducing the output resistance of the source or increasing the load resistance, or both. As a rule of thumb, the loading effect is less than 10% when $R_L \geq 10R_T$ and less than 1% when $R_L \geq 100R_T$.

Exercise 3–42 _____

Suppose $R_T = 200\,\Omega$ and the loading effect should be less than 1%. What should be the smallest value for R_L?

Answer: $R_L \geq 20\,\mathrm{k\Omega}$

_____

APPLICATION   **EXAMPLE 3 – 31**

An **attenuation pad** is a two-port resistance circuit that provides a nonadjustable reduction in signal level while also providing resistance matching at the input and output ports. Figure 3–70(a) shows an example of an attenuation pad. A typical switchable in-line attenuator is shown in Figure 3–70(b). The manufacturer's data sheet for this pad specifies the following characteristics at the input and output ports.

Port	Characteristics	Condition	Value	Units
Output	Thévenin voltage	600-Ω source connected at the input port	$v_S/4$	V
Output	Thévenin resistance	600-Ω source connected at the input port	600	Ω
Input	Input resistance	600-Ω load connected at the output port	600	Ω

Use OrCAD to verify these characteristics.

SOLUTION:

OrCAD Capture can calculate the two-port characteristics of the pad in Figure 3–70. A circuit diagram created using OrCAD is shown at the upper left of Figure 3–71. In this schematic a 600-Ω source is connected at the pad's input port. A 600-Ω load is *not* connected at the output port. The reason is that we need to find the open-circuit (Thévenin) voltage at the output port. We have labeled the pad's output port node D. OrCAD has labeled our node as N00323. This OrCAD assignment is not under our control, and the labeling will change every time we draw a new OrCAD circuit.[3]

Selecting "PSpice" and "New Simulation Profile" opens the "Simulation Settings" window shown at the upper right of Figure 3–71. Select "Bias Point" for the Analysis type and then check the "Calculate small-signal DC gain (.TF)" box as shown in the

_____

[3]Every time a schematic is drawn (even the same schematic) OrCAD generates a new random number for each node name. To display the OrCAD assigned names, double-click on each node to open a new window that gives information about the node. Click on the node's name, then select **Display**. This opens a second window. Select **Value Only**. Then click **OK**. Finally select **Apply** to place it on the drawing. Alternatively, the OrCAD assigned names can be viewed in the netlist file.

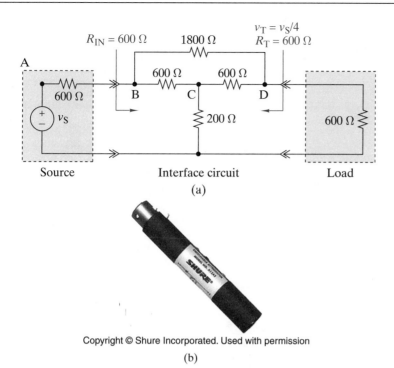

FIGURE 3–70 *Switchable In-Line Attenuator.*

Copyright © Shure Incorporated. Used with permission

(b)

figure. To calculate a gain, OrCAD must know the input and output variables. We inform OrCAD by entering the variables' names in "From Input source name:" (enter V1) and "To Output variable:" [enter V(N00323)]. Note that V(N00323) is the voltage between node D and ground—the output port voltage. OrCAD will calculate three things: (1) the small-signal gain V(N00323)/V_V1, (2) the input resistance seen by V_V1, and (3) the output or Thévenin resistance.

From the main menu, we run the simulation by pressing the blue triangle. OrCAD performs the requested dc analysis and writes the results to the output file. Scrolling down past the netfile, we find the desired results listed. These are shown at the bottom left of Figure 3–71. PSpice reports that the small-signal gain is 2.500E–01 or one-fourth of the source voltage as stated on the data sheet. The output resistance is 6.000E+02 or 600 Ω also as stated in the data sheet. The input resistance seen by the source is 1.280E + 03, or 1.28 kΩ. The data sheet states that the input resistance should be 600 Ω with a 600-Ω load connected. We ran the simulation with an open circuit to match the specifications for the other two parameters we were seeking to validate. If we add a 600-Ω load between node D and ground and rerun the simulation, OrCAD will return 1.200E + 03, or 1.2 kΩ. This result represents what the source V1 sees, including the 600-Ω source resistor. Subtracting 600 Ω from the total shows that all the specifications are as stated on the data sheet. ∎

Exercise 3–43

Modify the OrCAD simulation in Example 3–31 to include the 600-Ω load resistor. Create an OrCAD simulation profile to compute the gain between the input voltage and the load voltage. Verify that the results agree with the analysis and discussion presented in Example 3–31.

Answer: $v_L/v_S = 0.125$. This result is correct because we can model the entire circuit to the left of the load as a Thévenin source, with $v_T = v_S/4$, in series with the Thévenin resistance $R_T = 600 \, \Omega$. We then have a voltage division problem and the output voltage will be half of the Thévenin voltage, or $v_S/8$, because the load resistance matches the Thévenin resistance.

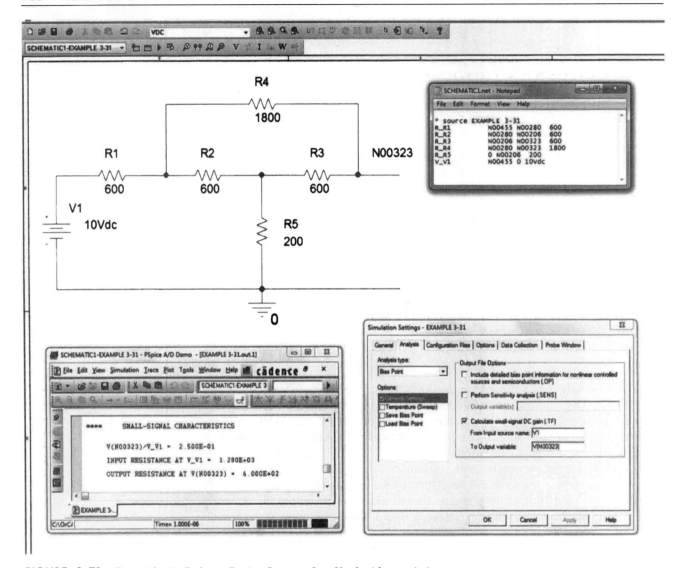

FIGURE 3–71   *Copyright © Cadence Design Systems, Inc. Used with permission.*

SUMMARY

- Node-voltage analysis involves identifying a reference node and the node to datum voltages at the remaining $N-1$ nodes. The KCL connection constraints at the $N-1$ nonreference nodes combined with the element constraints written in terms of the node voltages produce $N-1$ linear equations in the unknown node voltages.

- Mesh-current analysis involves identifying mesh currents that circulate around the perimeter of each mesh in a planar circuit. The KVL connection constraints around $E-N+1$ meshes combined with the element constraints written in terms of the mesh currents

produce $E-N+1$ linear equations in the unknown mesh currents.

- Node and mesh analysis can be modified to handle both types of independent sources using a combination of three methods: (1) source transformations, (2) selecting circuit variables so independent sources specify the values of some of the unknowns, and (3) using supernodes or supermeshes.

- A circuit is linear if it contains only linear elements and independent sources. For single-input linear circuits, the proportionality property states that any output is proportional to the input. For multiple-input linear circuits, the superposition principle states that any output can be found by summing the output produced when each input acts alone.

- A Thévenin equivalent circuit consists of a voltage source in series with a resistance. A Norton equivalent circuit consists of a current source in parallel with a resistance. The Thévenin and Norton equivalent circuits are related by a source transformation.

- The parameters of the Thévenin and Norton equivalent circuits can be determined using any two of the following: (1) the open-circuit voltage at the interface, (2) the short-circuit current at the interface, and (3) the equivalent resistance of the source circuit with all sources turned off.

- The parameters of the Thévenin and Norton equivalent circuits can also be determined using circuit reduction methods or by directly solving for the source i-v relationship using node-voltage or mesh-current analysis.

- For a fixed source and an adjustable load, the maximum interface signal levels are $v_{MAX}=v_{OC}$ ($R_L = \infty$), $i_{MAX} = i_{SC}(R_L = 0)$, and $p_{MAX} = v_{OC}i_{SC}/4(R_L = R_T)$. When $R_L = R_T$, the source and load are said to be matched.

- Interface signal transfer conditions are specified in terms of the voltage, current, or power delivered to the load. The design constraints depend on the signal conditions specified and the circuit parameters that are adjustable. Some design requirements may require a two-port interface circuit. An interface design problem may have one, many, or no solutions.

PROBLEMS

OBJECTIVE 3–1 GENERAL CIRCUIT ANALYSIS (SECT. 3–1 TO 3–2)

Given a circuit consisting of linear resistors and independent sources,

(a) (Formulation) Write node-voltage or mesh-current equations for the circuit.

(b) (Solution) Solve the equations from (a) for selected signal variables or input-output relationships using classical or software computational techniques.

Node-voltage method: See Examples 3–1, 3–2, 3–3, 3–4, 3–5, 3–6, 3–7 and Exercises 3–2, 3–3, 3–4, 3–5, 3–6, 3–7, 3–8, 3–9, 3–10, 3–11, 3–12, 3–13.

Mesh-current method: See Examples 3–8, 3–9, 3–10 and Exercises 3–14, 3–15, 3–16, 3–17, 3–18, 3–19, 3–20.

3–1 (a) Formulate node-voltage equations for the circuit in Figure P3–1. Arrange the results in matrix from $\mathbf{Ax} = \mathbf{b}$.

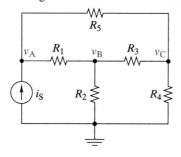

FIGURE P3–1

3–2 (a) Formulate node-voltage equations for the circuit in Figure P3–2. Arrange the results in matrix form $\mathbf{Ax} = \mathbf{b}$.

(b) Solve these equations for v_A and v_B.

(c) Use these results to find v_x and i_x.

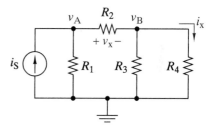

FIGURE P3–2

3–3 (a) Formulate node-voltage equations for the circuit in Figure P3–3. Arrange the results in matrix form $\mathbf{Ax} = \mathbf{b}$.

(b) Solve these equations for v_A and v_B.

(c) Use these results to find v_x and i_x.

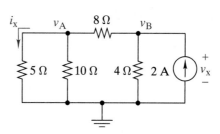

FIGURE P3–3

3–4 (a) Formulate node-voltage equations for the circuit in Figure P3–4.
(b) Solve these equations for v_A and v_B.
(c) Use these results to find v_x and i_x.

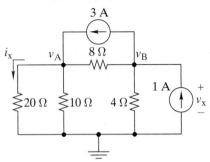

FIGURE P3–4

3–5 (a) Formulate node-voltage equations for the circuit in Figure P3–5. Arrange the results in matrix form $\mathbf{Ax} = \mathbf{b}$.
(b) Solve these equations for v_A and v_C.
(c) Use these results to find v_x and i_x.

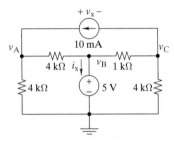

FIGURE P3–5

3–6 (a) Choose a ground wisely and formulate node-voltage equations for the circuit in Figure P3–6.
(b) Solve for v_x and i_x when $R_1 = R_2 = R_3 = R_4 = 10\,\text{k}\Omega$, $v_s = 25$ V, and $i_s = 1$ mA.

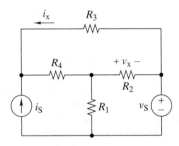

FIGURE P3–6

3–7 The following are a set of node-voltage equations; draw the circuit they represent.

$$v_A = v_S$$
$$\frac{v_B - v_A}{R_1} + \frac{v_B - v_C}{R_2} - i_S = 0$$
$$\frac{v_C - v_A}{R_3} + \frac{v_C - v_B}{R_2} + \frac{v_C}{R_4} = 0$$
$$v_D = 0$$

3–8 (a) Choose a ground wisely and formulate node-voltage equations for the circuit in Figure P3–8.
(b) Solve for v_x and i_x.

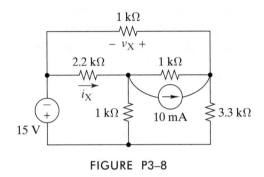

FIGURE P3–8

3–9 (a) Formulate node-voltage equations for the circuit in Figure P3–9.
(b) Use MATLAB to find symbolic expressions for the node voltages in terms of the parameters in the circuit.
(c) Find numeric values for v_A, v_B, and v_C when $R_1 = 1.5\,\text{k}\Omega$, $R_2 = 2.2\,\text{k}\Omega$, $R_3 = 3.3\,\text{k}\Omega$, $R_4 = 4.7\,\text{k}\Omega$, and $i_{S1} = i_{S2} = 2$ mA.
(d) Use OrCAD to verify your solution to part (c) is correct.

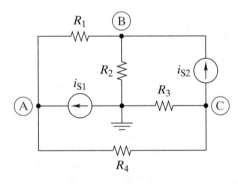

FIGURE P3–9

3–10 (a) Formulate node-voltage equations for the circuit in Figure P3–10.
(b) Solve for v_x and i_x when $R_1 = R_4 = 2\,\text{k}\Omega$, $R_2 = R_3 = 500\,\Omega$, $R_x = 750\,\Omega$, and $v_s = 15$ V.
(c) Repeat (b) when R_4 is a variable resistor that varies from 10 Ω to 10,000 Ω. At what value of R_4 is the voltage across $R_x = 0$ V? *Hint:* Use OrCAD to vary R_4 by trial and error to approach the answer.

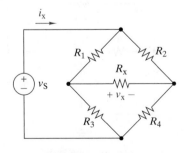

FIGURE P3–10

3–11 (a) Formulate node-voltage equations for the circuit in Figure P3–11.

(b) Solve for v_x and i_x when $R_1 = R_4 = 2$ kΩ, $R_2 = R_3 = 500$ Ω, $R_x = 750$ Ω, and $v_S = 15$ V.

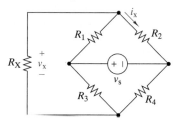

FIGURE P3–11

3–12 (a) Formulate node-voltage equations for the circuit in Figure P3–12. (*Hint:* use a supernode.)

(b) Solve for v_x and i_x.

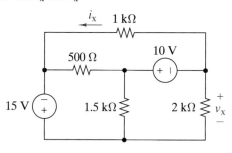

FIGURE P3–12

3–13 (a) Formulate mesh-current equations for the circuit in Figure P3–13. Arrange the results in matrix form $\mathbf{Ax = b}$.

(b) Solve for i_A and i_B.

(c) Use these results to find v_x and i_x.

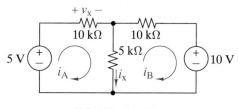

FIGURE P3–13

3–14 (a) Formulate mesh-current equations for the circuit in Figure P3–14. Arrange the results in matrix form $\mathbf{Ax = b}$.

(b) Solve for i_A, i_B and i_C.

(c) Use these results to find v_x and i_x.

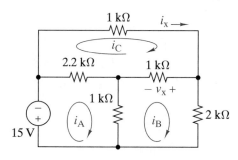

FIGURE P3–14

3–15 (a) Formulate mesh-current equations for the circuit in Figure P3–15. Arrange the results in matrix form $\mathbf{Ax = b}$.

(b) Solve for i_A and i_B.

(c) Use these results to find v_x and i_x.

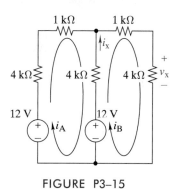

FIGURE P3–15

3–16 (a) Formulate mesh-current equations for the circuit in Figure P3–16. Arrange the results in matrix form $\mathbf{Ax = b}$.

(b) Solve for i_A and i_B.

(c) Use these results to find v_x and i_x.

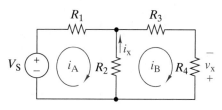

FIGURE P3–16

3–17 (a) Formulate mesh-current equations for the circuit in Figure P3–17.

(b) Formulate node-voltage equations for the circuit in Figure P3–17.

(c) Which set of equations would be easier to solve? Why?

(d) Using MATLAB, find v_x and i_x in terms of the mesh-current variables.

(e) Using MATLAB, find v_x and i_x in terms of the node-voltage variables.

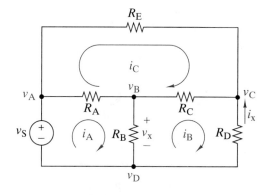

FIGURE P3–17

3–18 **(a)** Formulate mesh-current equations for the circuit in Figure P3–18.
(b) Solve for v_x and i_x when $R_1 = 2$ kΩ, $R_2 = 3$ kΩ, $R_3 = 500$ Ω, $R_4 = 2.5$ kΩ, $R_5 = 2$ kΩ, $i_S = 10$ mA, and $v_S = 24$ V.
(c) Find the total power dissipated in the circuit.

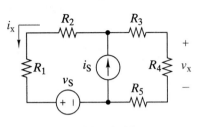

FIGURE P3–18

3–19 **(a)** For the circuit of Figure P3–19 solve for i_A, i_B and i_C using two supermesh equations.
(b) Use these results to find v_x.

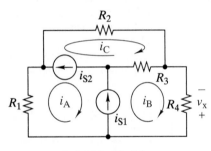

FIGURE P3–19

3–20 **(a)** Formulate mesh-current equations for the circuit in Figure P3–20.
(b) Use MATLAB to find symbolic expressions for v_x and i_x in terms of the parameters in the circuit.
(c) Find numeric values for v_x and i_x when $R_1 = R_2 = 5.6$ kΩ, $R_3 = 2.2$ kΩ, $R_4 = 1.5$ kΩ, $i_S = 3$ mA, $v_{S1} = 15$ V, and $v_{S2} = 5$ V.
(d) Find the power supplied by v_{S1}.
(e) Use OrCAD to verify your solutions to parts (c) and (d) are correct.

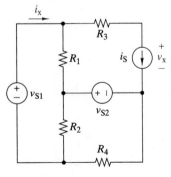

FIGURE P3–20

3–21 The circuit in Figure P3–21 seems to require two supermeshes since both current sources appear in two meshes. However, sometimes rearranging the circuit diagram will eliminate the need for a supermesh.
(a) Show that supermeshes can be avoided in Figure P3–21 by rearranging the connection of resistor R_6.
(b) Formulate mesh-current equations for the modified circuit as redrawn in (a).
(c) Solve for v_x when $R_1 = R_2 = R_3 = R_4 = 4$ kΩ, $R_5 = R_6 = 2$ kΩ, $i_{S1} = 80$ mA, and $i_{S2} = 40$ mA.

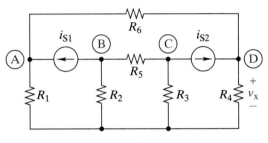

FIGURE P3–21

3–22 **(a)** Formulate mesh-current equations for the circuit in Figure P3–22.
(b) Formulate node-voltage equations for the circuit in Figure P3–22.
(c) Which set of equations would be easier to solve? Why?
(d) Find v_x and i_x using whichever method you prefer.

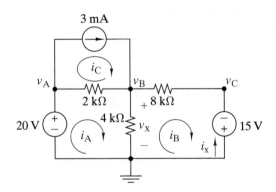

FIGURE P3–22

3–23 Use mesh-current, node-voltage or simple engineering intuition to find the input resistance of the circuit in Figure P3–23.

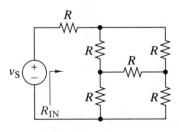

FIGURE P3–23

3–24 In Figure P3–24 all of the resistors are 1 kΩ and $v_S = 10$ V. The voltage at node C is found to be $v_C = -2$ V when node B is connected to ground. Find he node voltages v_A and v_D, and the mesh currents i_A and i_B.

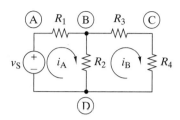

FIGURE P3–24

3–25 Use Figure P3–24 and MATLAB to solve the following problems:
(a) Using mesh-current analysis, find a symbolic expression for i_A in terms of the circuit parameters.
(b) Computer the ratio i_A/v_S.
(c) Find a symbolic expression for the equivalent resistance of the circuit by combining resistors in series and parallel. Compare your answer to the results from part (b).

3–26(a) Formulate mesh-current equations for the circuit in Figure P3–26.
(b) Formulate node-voltage equations for the circuit in Figure P3–26.
(c) Which set of equations would be easier to solve? Why?
(d) Use OrCAD to find the node voltages v_A and v_B and the mesh-currents i_A, i_B and i_C in Figure P3–26.

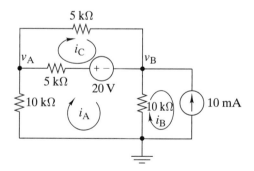

FIGURE P3–26

3–27(a) Formulate mesh-current equations for the circuit in Figure P3–27. Arrange the results in matrix form **Ax = b**.
(b) Use MATLAB and mesh-current analysis to solve for the mesh currents i_A, i_B, i_C, and i_D in Figure P3–27.
(c) Formulate node-voltage equations for the circuit in Figure P3–27. Arrange the results in matrix form **Ax = b**.
(d) Use MATLAB and node-voltage analysis to solve for the mesh currents i_A, i_B, i_C, and i_D in Figure P3–27 and compare the effort required with each technique, i.e. mesh-current versus node-voltage.
(e) Use OrCAD to verify your results in the previous two parts.

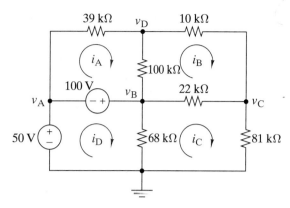

FIGURE P3–27

3–28 Find the proportionality constant $K = v_O/v_S$ for the circuit in Figure P3–28.

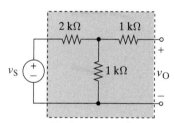

FIGURE P3–28

3–29 Find the proportionality constant $K = i_O/v_S$ for the circuit in Figure P3–29.

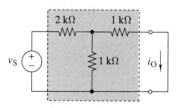

FIGURE P3–29

3–30 Find the proportionality constant $K = v_O/i_S$ for the circuit in Figure P3–30.

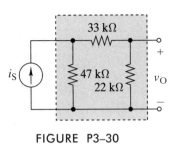

FIGURE P3–30

3–31 Find the proportionality constant $K = i_O/i_S$ for the circuit in Figure P3–31.

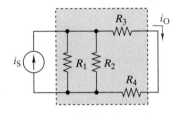

FIGURE P3–31

3–32 Find the proportionality constant $K = v_O/v_S$ for the circuit in Figure P3–32.

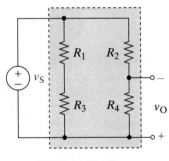

FIGURE P3–32

3–33 Use the unit output method to find K and v_O in Figure P3–33.

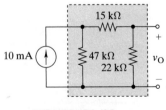

FIGURE P3–33

3–34 Use the unit output method to find K and v_O in Figure P3–34.

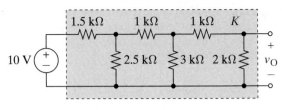

FIGURE P3–34

3–35 Use the unit output method to find K and i_O in Figure P3–35.

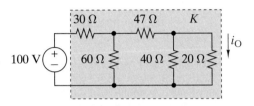

FIGURE P3–35

3–36 Use the superposition principle to find v_O in Figure P3–36.

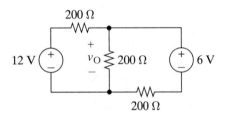

FIGURE P3–36

3–37 Use the superposition principle to find i_O in Figure P3–37. Verify your answer using OrCAD.

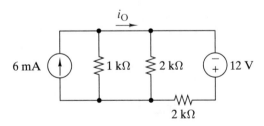

FIGURE P3–37

3–38 Use the superposition principle to find v_O in Figure P3–38.

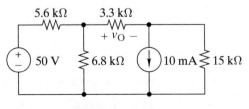

FIGURE P3–38

3–39 Use the superposition principle to find v_O in Figure P3–39.

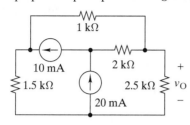

FIGURE P3–39

3–40 Use the superposition principle to find i_O in Figure P3–40. Verify your answer using OrCAD.

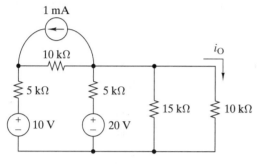

FIGURE P3–40

3–41 (a) Use the superposition principle to find v_O in terms of v_1, v_2, and R in Figure P3–41 (This circuit is a 2-bit R-2R network)
(b) Use MATLAB and node-voltage analysis to verify your answer symbolically.

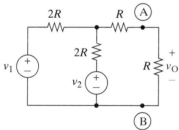

FIGURE P3–41

3–42 (a) Use the superposition principle to find v_O in terms of v_S, i_S, and R in Figure P3–42.
(b) Use MATLAB and node-voltage analysis to verify your answer symbolically.

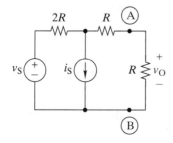

FIGURE P3–42

3–43 A linear circuit containing two sources drives a 100-Ω load resistor. Source number 1 delivers 1 W to the load when source

number 2 is off. Source number 2 delivers 4 W to the load when source number 1 is off. Find the power delivered to the load when both sources are on. *Hint:* The answer is not 5 W. Why?

3–44 A linear circuit is driven by an independent voltage source $v_S = 10$ V and an independent current source $i_S = 10$ mA. The output voltage is $v_O = 2$ V when the voltage source is on and the current source off. The output is $v_O = 1$ V when both sources are on. Find the output voltage when $v_S = 20$ V and $i_S = -20$ mA.

3–45 A certain linear circuit has four input voltages and one output voltage v_O. The following table lists the output for different values of the four inputs. Find the input-output relationship for the circuit. Specifically, find an expression for v_O in terms of the four input voltages.

$v_{S1}(V)$	$v_{S2}(V)$	$v_{S3}(V)$	$v_{S4}(V)$	$v_O(V)$
2	4	−4	1	20
1	2	2	1.5	−4
1	4	2	2	−1
0	5	3	−1	3

3–46 When the current source is turned off in the circuit of Figure P3–46 the voltage source delivers 25 W to the load. How much power does it deliver to the load when both sources are on? Explain your answer.

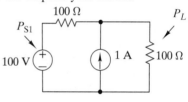

FIGURE P3–46

OBJECTIVE 3–3 THÉVENIN AND NORTON EQUIVALENT CIRCUITS (SECT. 3–4)

Given a circuit containing linear resistors and independent sources,
(a) Find the Thévenin or Norton equivalent at a specified pair of terminals.
(b) Use the Thévenin or Norton equivalent to find the signals delivered to linear or nonlinear loads.
See Examples 3–16, 3–17, 3–18, 3–19, 3–20, 3–21 and Exercises 3–28, 3–29, 3–30, 3–31, 3–32.

3–47 For the circuit in Figure P3–47 find its Thévenin and Norton equivalent circuits.

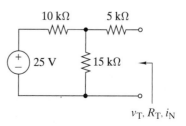

FIGURE P3–47

3–48 For the circuit in Figure P3–48 find its Thévenin and Norton equivalent circuits.

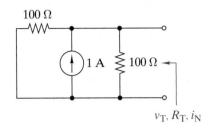

FIGURE P3–48

3–49 (a) Find the Thévenin or Norton equivalent circuit seen by R_L in Figure P3–49.
(b) Use the equivalent circuit found in (a) to find i_L if $R_L = 22$ kΩ.

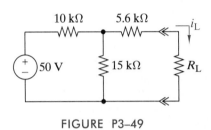

FIGURE P3–49

3–50 (a) Find the Thévenin or Norton equivalent circuit seen by R_L in Figure P3–50.
(b) Use the equivalent circuit found in (a) to find i_L in terms of i_S, R_1, R_2, and R_L.
(c) Check your answer in (b) using current division.

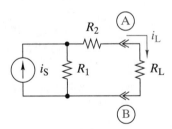

FIGURE P3–50

3–51 Find the Thévenin equivalent circuit seen by R_L in Figure P3–51. Find the voltage across the load when $R_L = 5$ Ω, 10 Ω, and 20 Ω.

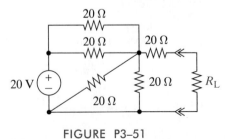

FIGURE P3–51

3–52 Find the Norton equivalent seen by R_L in Figure P3–52. Find the current through the load for when $R_L = 4.7$ kΩ, 15 kΩ, and 68 kΩ.

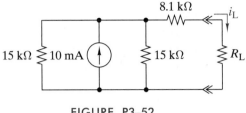

FIGURE P3–52

3–53 You need to determine the Thévenin Equivalent circuit of a more complex linear circuit. A technician tells you she made two measurements using her DMM. The first was with a 10-kΩ load and the load current was 91 μA, the second was with a 10-kΩ load and the load voltage was 5 V. Calculate the Thévenin equivalent circuit as shown in Figure P3–53.

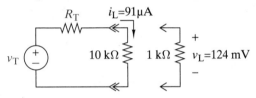

FIGURE P3–53

3–54 Find the Thévenin equivalent seen by R_L in Figure P3–54. Find the power delivered to the load when $R_L = 50$ kΩ and 200 kΩ.

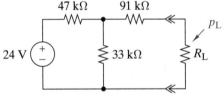

FIGURE P3–54

3–55 (a) Use OrCAD to find the Norton equivalent at terminals A and B in Figure P3–55. *Hint:* Find the open-circuit voltage and short-circuit current at the requisite terminals.
(b) Use the Norton equivalent circuit found in (a) to determine the power dissipated in R_L when it is equal to 37 kΩ.
(c) Use OrCAD to simulate both the original and the Norton equivalent circuits with $R_L = 37$ kΩ. Verify that the power dissipated by the load is the same in both situations.

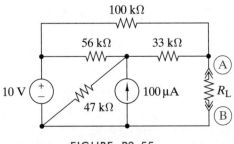

FIGURE P3–55

3–56 The purpose of this problem is to use Thévenin equivalent circuits to find the voltage v_X in Figure P3–56. Find the Thévenin equivalent circuit seen looking to the left of terminals A and B. Find the Thévenin equivalent circuit seen looking to the right of terminals A and B. Connect these equivalent circuits together and find the voltage v_X.

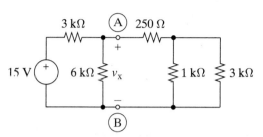

FIGURE P3–56

3–57 The circuit in Figure P3–57 was solved earlier using supermeshes (Prob.(3–19). In this problem solve for the voltage across R_L by first finding the Thévenin equivalent that the load resistor sees, the for $R_L = 2.5$ kΩ find v_L.

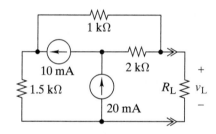

FIGURE P3–57

3E2

3–58 Assume that Figure P3–58 represents a model of the auxiliary output port of a car. The output current is $i = 1$ A when $v = 0$ V. The output voltage is $v = 12$ V when a 20-Ω resistor is connected between the terminals. Suppose you wanted to charge a 12-V battery by connecting the battery at the port. How much current would the port deliver to the battery?

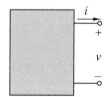

FIGURE P3–58

3–59 The i-v characteristic of the active circuit in Figure P3–58 is $5v + 500i = 100$. Find the output voltage when a 500-Ω resistive load is connected.

3P1

3–60 **E** You successfully completed Circuits I and as an undergraduate work-study, your past professor asked you to help her grade a Circuits I quiz. On the quiz, students were asked to find the power supplied by the source to both the 10 kΩ load (R_L)

and to the entire circuit as shown in Figure P3–60. Your professor asks to help her by creating a grading sheet.

(a) Solve he quiz and establish reasonable A, B . . . etc. cuts for incorrect solutions.

(b) A particular student correctly finds the Thévenin equivalent circuit that the resistive load sees and calculates the power to the load using V_T^2/R_L. He then does a source transformation correctly finding the Norton equivalent of the circuit. He calculates the source power by $V_T \times I_N$. What grade would you give him?

(c) Another student finds $P_L = 5.625$ mW and $P_S = -22.5$ mW, but provides no work to justify her answers. What grade would you give her?

(d) A third student first finds the Norton equivalent, finds the current through the load using a current divider and calculates the power in the load using $I_L^2 R_L$. He figures correctly what the parallel voltage would be across the Norton circuit and the load v_L, and then calculates $P_s = I_N \times v_L$. What grade would you give him?

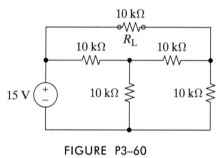

FIGURE P3–60

3–61 The Thévenin equivalent parameters of a practical voltage source are $v_T = 30$ V $R_T = 300$ Ω. Find the smallest load resistance for which the load voltage exceeds 10 V.

3–62 Use a sequence of source transformations to find the Thévenin equivalent at terminals A and B in Figure P3–62. Then select a resistor to connect across A and B so that 5 V appears across it.

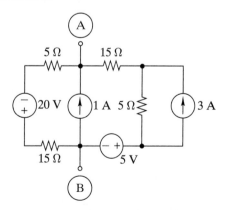

FIGURE P3–62

3–63 The circuit in Figure P3–63 provides power to a number of loads connected in parallel. The circuit is protected by a $^3/_4$ mA fuse with a nominal 100 Ω resistance. Each load is 10 kΩ.

What is the maximum number of loads the circuit can drive without blowing the fuse?

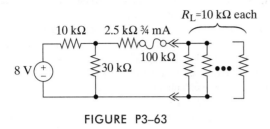

FIGURE P3–63

3–64 Find the Thévenin equivalent at terminals A and B in Figure P3–64.

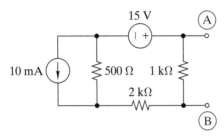

FIGURE P3–64

3–65 A nonlinear resistor is connected across a two-terminal source whose Thévenin equivalent is $v_T = 5$ V and $R_T = 500\ \Omega$. The $i\text{-}v$ characteristic of he resistor is $i = 10^{-4}\ (v + 2\ v^{3.3})$. Use the MATLAB function `solve` to find the operating point for this circuit and determine the voltage across, the current through and the power dissipated in the non-linear resistor.

3–66 A blue Light-emitting diode (LED) is connected across a two-terminal source whose Thévenin equivalent is $v_T = 4$ V and $R_T = 20\ \Omega$. The $i\text{-}v$ characteristic of the LED is $i = 10^{-12}\ (e^{10v} - 1)$. Figure P3–66 shows the LED's $i\text{-}v$ characteristic. Either using MATLAB or a graphical approach determine the voltage across and current through the LED.

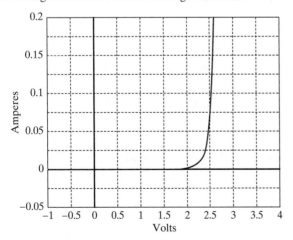

FIGURE P3–66

3–67 Find the Norton equivalent seen by R_L in Figure P3–67. Select the value of R_L so that:

(a) 150 mA is delivered to the load.
(b) 6 V is delivered to the load.
(c) 100 mW is delivered to the load.

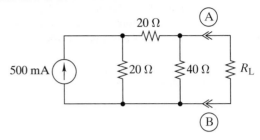

FIGURE P3–67

3–68 Find the Thévenin equivalent seen by R_L in Figure P3–68.

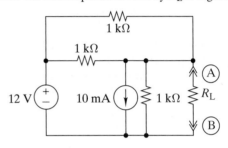

FIGURE P3–68

3–69 Find the Thévenin equivalent seen by R_L in Figure P3–69.

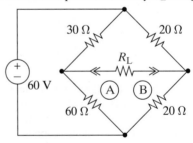

FIGURE P3–69

OBJECTIVE 3–4 MAXIMUM SIGNAL TRANSFER (SECT. 3–5)

Given a circuit containing linear resistors and independent sources,

(a) Find the maximum voltage, current, and power available at a specified pair of terminals.
(b) Find the resistive loads required to obtain the maximum available signal levels.
(c) See Example 3–22 and Exercise 3–33.

3–70 For the circuit of Figure P3–70 find the value of R_L that will result in:

(a) Maximum voltage. What is that voltage?
(b) Maximum current. What is that current?
(c) Maximum power. What is that power?

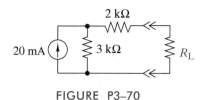

FIGURE P3–70

3–71 For the circuit of Figure P3–71 find the value of R_L that will result in:
(d) Maximum voltage. What is that voltage?
(e) Maximum current. What is that current?
(f) Maximum power. What is that power?

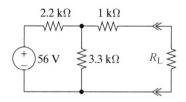

FIGURE P3–71

3–72 The resistance R in Figure P3–72 is adjusted until maximum power is delivered to the load consisting of R and the 15-kΩ resistor in parallel.
(a) Find the required value of R.
(b) How much power is delivered to the load?

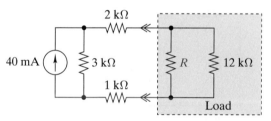

FIGURE P3–72

3–73 When a 5-kΩ resistor is connected across a two-terminal source a current of 15 mA is delivered to the load. When a second 5-kΩ resistor is connected in parallel with the first, a total current of 20 mA is delivered. Find the maximum power available from the source.

3–74 Find the value of R in the circuit of Figure P3–74 so that maximum power is delivered to the load. What is the value of the maximum power?

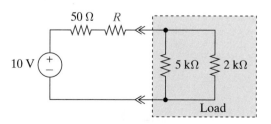

FIGURE P3–74

3–75 For the circuit of Figure P3–75 find the value of R_L that will result in:

(g) Maximum voltage. What is that voltage?
(h) Maximum current. What is that current?
(i) Maximum power. What is that power?

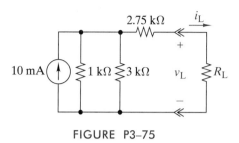

FIGURE P3–75

3–76 A 100 Ω-load needs 10 mA to operate correctly. Design a practical power source to provide the needed current. The smallest source resistance you can practically design for is 50 Ω, but you can add any other series resistance if you need to.

3–77 A practical source delivers 50 mA to a 300-Ω load. The source delivers 12 V to a 120-Ω load. Find the maximum power available from the source.

OBJECTIVE 3–5 INTERFACE CIRCUIT DESIGN AND EVALUATION (SECT. 3–6)

(a) **D** Given the signal transfer objective at a source-load interface, adjust the circuit parameters or design one or more two-port interface circuits to achieve the specified objectives within stated constraints.
(b) **E** Given two or more circuits that perform the same interface function, rank-order the circuits using stated criteria.
See Examples 3–23, 3–24, 3–25, 3–26, 3–27, 3–28, 3–29, 3–30, 3–31 and Exercises 3–34, 3–35, 3–36, 3–37, 3–38, 3–39, 3–40, 3–41, 3–42, 3–43.

3–78 **D** Select R_L and design an interface circuit for the circuit shown in Figure P3–78 so that the load voltage is 2 V.

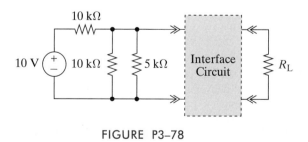

FIGURE P3–78

3–79 **D** The source in Figure P3–79 has a 100 mA output current limit. Design an interface circuit so that the load voltage is $v_2 = 20$ V and the source current is $i_1 < 50$ mA.

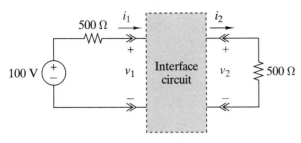

FIGURE P3–79

3–80 Ⓓ Figure P3–80 shows an interface circuit connecting a 15-V source to a diode load. The *i-v* characteristic of the diode is $i = 10^{-14} \left(e^{40v} - 1\right)$.
(a) Design an interface circuit so that $v = 0.7$ V.
(b) Validate your answer using MATLAB.

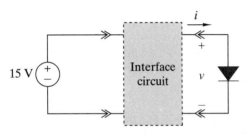

FIGURE P3–80

3–81 Ⓓ Design the interface circuit in Figure P3–81 so that the voltage delivered to the load is $v = 10$ V $\pm10\%$. Use only one or more of the following standard resistors: 1.3 kΩ, 2 kΩ, 3 kΩ, 4.3 kΩ, 6.2 kΩ, and 9.1 kΩ. These resistors all have a tolerance of $\pm5\%$ which you must account for in your design.

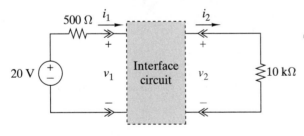

FIGURE P3–81

3–82 Ⓓ, Ⓔ In this problem you will design two interface circuits that deliver 150 V to the 5-kΩ load.
(a) Ⓓ Design a parallel resistor interface to meet the requirements.
(b) Ⓓ Convert the source circuit to its Thévenin equivalent and then design a series interface to meet the requirement.
(c) Ⓔ If minimizing the power that the current source delivers is the primary consideration, which of your two designs best meets the requirement?

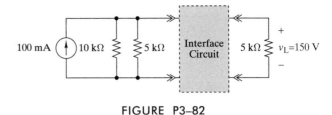

FIGURE P3–82

3–83 Ⓔ Two teams are competing to design the interface circuit in Figure P3–83 so that the 25 mW $\pm10\%$ is delivered to the 1-kΩ load resistor. Their designs are shown in Figure P3–83. Which solution is better considering the use of standard values, number of parts, and power required by the source? Would your choice be different if the power had to be within $\pm5\%$?

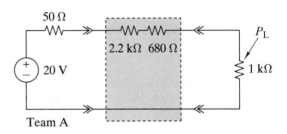

Team A

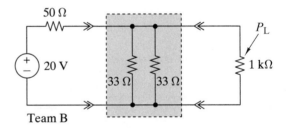

Team B

FIGURE P3–83

3–84 Ⓔ The bridge-T attenuation pad shown in Figure P3–84 was found in a drawer. You need an attenuation pad that would match to a 75-Ω source and a 75-Ω load and provide for a -12 dB drop of signal (reduction of four times). Use OrCAD to determine if the device will work.

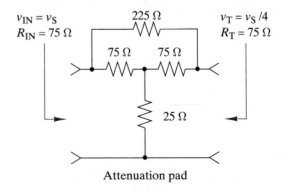

Attenuation pad

FIGURE P3–84

3–85 Design the interface circuit in Figure P3–85 so that the power delivered to the load is 100 mW.

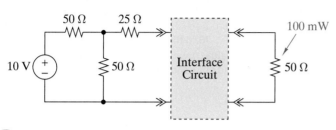

FIGURE P3–85

3–86 Design the interface circuit in Figure P3–85 so that the voltage delivered to the load is 2.5 V.

3–87 Design the interface circuit in Figure P3–87 so that $R_{IN} = 100\ \Omega$ and the current delivered to the 50-Ω load is $i = 50$ mA. Hint: Use an L-pad.

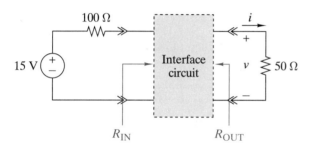

FIGURE P3–87

3–88 Design the interface circuit in Figure P3–87 so that $R_{OUT} = 50\ \Omega$ and the voltage delivered to the 50-Ω load is $v = 2.5$ V. Hint: Use an L-pad.

3–89 The circuit in Figure P3–89 has a source resistance of 75 Ω and a load resistance of 300 Ω. Design the interface circuit so that the input resistance is $R_{IN} = 75\ \Omega \pm 10\%$ and the output resistance is $R_{OUT} = 300\ \Omega \pm 10\%$.

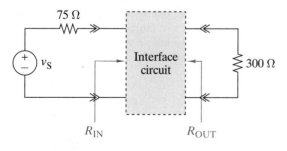

FIGURE P3–89

3–90 It is claimed that both interface circuits in Figure P3–90 will deliver $v = 4$ V to the 75-Ω load. Verify this claim. Which interface circuit consumes the least power? Which has an output resistance that best matches the 75-Ω load?

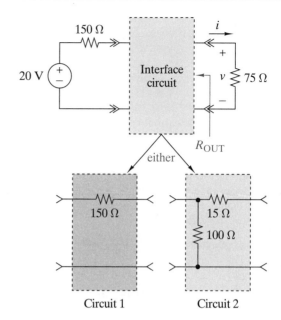

FIGURE P3–90

INTEGRATING PROBLEMS

3–91 Audio Speaker Resistance-Matchign Network

A company is producing an interface network that they claim would result in an R_{IN} of 600 $\Omega \pm 2$ % and R_{OUT} of 16, 8, or 4 $\Omega \pm 2\%$ – depending on whether the connected speakers are 16, 8, or 4 Ω – selectable via a built-in switch. Their design is shown in Figure P3–91. Prove or disprove their claim.

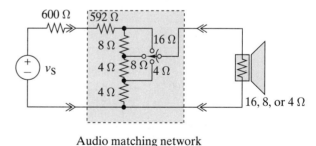

Audio matching network

FIGURE P3–91

3–92 Attenuation Analysis

In Figure P3–92 a two-port attenuator connects a 600-Ω source to a 600-Ω load. Find the power delivered to the load in terms of v_S. Remove the attenuator and find the power delivered to the load when the source is directly connected to the load. By what fraction does the attenuator reduce the power delivered to the 600-Ω load? Express the fraction in dB.

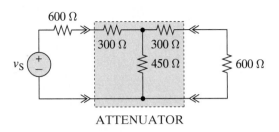

ATTENUATOR

FIGURE P3–92

3–93 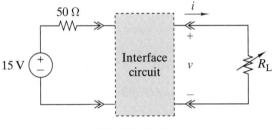 Interface Circuit Design

Using no more than three 50-Ω resistors, design the interface circuit in Figure P3–93 so that $v \leq 5$ V and $i \leq 100$ mA regardless of the value of R_L.

FIGURE P3–93

3–94 **D** Battery Design

A satellite requires a battery with an open-circuit voltage $v_\mathrm{OC} = 36$ V and a Thévenin resistance $R_\mathrm{T} \leq 10$ Ω. The battery is to be constructed using series and parallel combinations of one of two types of cells. The first type has $v_\mathrm{OC} = 9$ V, $R_\mathrm{T} = 4$ Ω, and a weight of 30 grams. The second type has $v_\mathrm{OC} = 4$ V, $R_\mathrm{T} = 2$ Ω, and a weight of 18 grams. Design a minimum weight battery that meets the open-circuit voltage and Thévenin resistance requirements.

3–95 **E** Design Evaluation

A requirement exists for a circuit to deliver 0 to 5 V to a 100-Ω load from a 20-V source rated at 2.5 W. Two proposed circuits are shown in Figure P3–95. Which one would you choose and why?

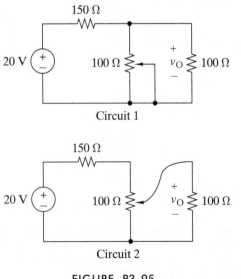

FIGURE P3–95

3–96 **E** Design Interface Competition.

The output of a transistorized power supply is modeled by the Norton equivalent circuit shown Figure P3–96. Two teams are competing to design the interface circuit so that the 25 mW $\pm 10\%$ is delivered to the 1-kΩ load resistor. Their designs are shown in Figure P3–96. Which solution is better considering the use of standard values, number of parts, and power required by the source? Would your choice be different if the power had to be within $\pm 5\%$?

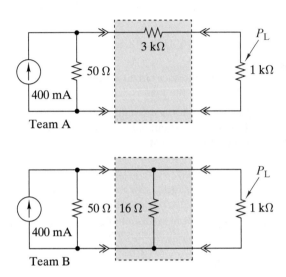

FIGURE P3–96

3–97 **A, E** Analysis of Competing Interface Circuits Using MATLAB.

Figure P3–97 displays two generalized interface circuit designs. In both circuits, resistors R_1 and R_2 connect a Thévenin equivalent circuit to a load resistor. Using MATLAB, develop symbolic expressions for the load resistance, i_L, and the input resistance, R_IN, for each circuit in terms of the given parameters. Using these two expressions, now use the MATLAB command solve to solve for R_1 and R_2 in terms of i_L and R_IN. Let $v_\mathrm{T} = 15$ V, $R_\mathrm{T} = 100$ Ω, $R_\mathrm{L} = 50$ Ω, $i_\mathrm{L} = 50$ mA, and $R_\mathrm{IN} = 100$ Ω. Can you use both types of interface circuits to find suitable values for R_1 and R_2 to meet these specifications? Compare your interface design(s) with the solution to Problem 3–87.

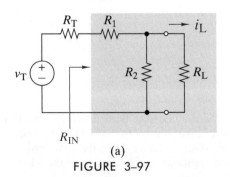

(a)

FIGURE 3–97

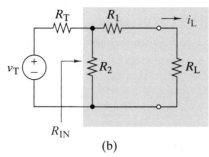

(b)

FIGURE P3–97 *(Continued)*

3–98 A Maximum Power Transfer Using OrCAD

Figure P3–98 shows a circuit with two sources, a fixed load and a resistor R. Select R for maximum power transfer to the load. The result is not an obvious one. Hint: Simulate in OrCAD using the PARAMETER function and sweep R from 0.1 Ω to 100 MΩ.

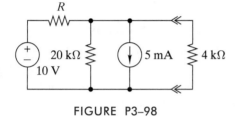

FIGURE P3–98

3–99 A Non-inverting Summer

A *non-inverting summer* interface device is shown in Figure P3–99. Of importance is that the input to the device has infinite resistance – i.e. no current flows into the device. The output voltage for this configuration is two times whatever the input voltage is. Develop a relationship for the voltage v_L across R_L with respect to the two input voltages and the two input resistances.

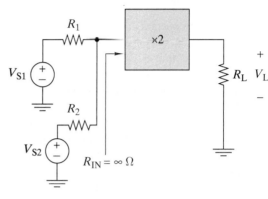

FIGURE P3–99

CHAPTER 4 ACTIVE CIRCUITS

Then came the morning of Tuesday, August 2, 1927, when the concept of the negative feedback amplifier came to me in a flash while I was crossing the Hudson River on the Lacakawana Ferry, on my way to work.

Harold S. Black, 1927,
American Electrical Engineer

Some History Behind This Chapter

The integrated circuit operational amplifier (OP AMP) is the workhorse of present-day linear electronic circuits. However, to operate as a linear amplifier, the OP AMP must be provided with "negative feedback." The negative feedback amplifier is one of the key inventions of all time. During the 1920s Harold S. Black (1898–1983) had been working for several years without much success on the problem of improving the performance of vacuum tube amplifiers in telephone systems. The feedback amplifier solution came to him suddenly on the way to work. He realized that by utilizing negative feedback, he could obtain a desired reduction in distortion at the expense of a sacrifice in amplification. He documented his invention by writing the key concepts of negative feedback on his morning copy of the *New York Times*. His invention paved the way for the development of worldwide communication systems and spawned completely new areas of technology, such as feedback control systems and robotics. Two other achievements were necessary before low-noise amplifiers could become practical and commonplace: the invention of the transistor by John Bardeen, Walter Brattain, and William Shockley in 1947, and the invention of the integrated circuit by Jack S. Kilby in 1958.

Why This Chapter Is Important Today

This is an important chapter for all engineering disciplines. You will be introduced to modern electronic devices and how they can be modeled. The utility of these devices will be apparent when you design OP AMP circuits that provide signal conditioning in instrumentation systems. You will also be introduced to criteria used to evaluate alternative designs. That is, you will begin to function as an engineer making judgments about the best solution to a problem.

Chapter Sections

Chapter Learning Objectives

4-1 Linear Active Circuits (Sects. 4–1, 4–2)

Given a linear resistance circuit containing dependent sources, find selected output signals, input-output relationships, or input-output resistances.

4-2 OP AMP Circuit Analysis (Sects. 4–3, 4–4)

Given a linear resistance circuit containing OP AMPs, find selected output signals or input-output relationships.

4-3 OP AMP Circuit Design (Sect. 4–5)

Given an input-output relationship, design resistive OP AMP circuits that implement the relationship. Evaluate the alternative designs using stated criteria.

4–1 LINEAR DEPENDENT SOURCES

This chapter treats the analysis and design of circuits containing active devices, such as transistors or operational amplifiers (OP AMPs). An **active device** is a component that requires an external power supply to operate correctly. An **active circuit** is one that contains one or more active devices. An important property of active circuits is that they are capable of providing signal amplification, one of the most important signal-processing functions in electrical engineering. Linear active circuits are governed by the proportionality property, so their input-output relationships are of the form $y = Kx$. The term **signal amplification** means the proportionality factor K is greater than 1 when the input x and output y have the same dimensions. Thus, active circuits can deliver more signal voltage, current, and power at their output than they receive from the input signal. The passive resistance circuits studied thus far cannot produce voltage, current, or power gains greater than unity.

Active devices operating in a linear mode are modeled using resistors and one or more of the dependent sources shown in Figure 4–1. A **dependent source** is a voltage or current source whose output is controlled by a voltage or current in a different part of the circuit. As a result, there are four possible types of dependent sources: a current-controlled voltage source (CCVS), a voltage-controlled voltage source (VCVS), a current-controlled current source (CCCS), and a voltage-controlled current source (VCCS). The properties of these dependent sources are very different from those of the independent sources described in Chapter 2. The output voltage (current) of an independent voltage (current) source is a specified value that does not depend on the circuit to which it is connected. To distinguish between the two types of sources, the dependent sources are represented by the diamond symbols in Figure 4–1, in contrast to the circle symbols used for independent sources.

Caution: This book uses the diamond symbol to indicate a dependent source and the circle to show an independent source. However, some books and circuit analysis programs use the circle symbol for both dependent and independent sources. Some authors use a rectangle for dependent sources.

A **linear dependent source** is one whose output is proportional to the controlling voltage or current. The defining relationships for dependent sources in Figure 4–1 are all of the form $y = Kx$, where x is the controlling variable, y is the source output variable, and K is the proportionality factor. Each type of dependent source is characterized by a proportionality factor, either μ, β, r, or g. These parameters are often called simply the **gain** of the controlled source. Strictly speaking, the parameters μ and β are dimensionless quantities called the **voltage gain** and **current gain**, respectively. The parameter r has the dimensions of ohms and is called the **transresistance**, a contraction of transfer resistance. The parameter g is then called **transconductance** and has the dimensions of siemens.

Although dependent sources are elements used in circuit analysis, they are conceptually different from the other circuit elements we have studied. The linear resistor and ideal switch are models of actual devices called resistors and switches. However, you will not find dependent sources listed in electronic part catalogs. For this reason, dependent sources are more abstract, since they are not models of identifiable physical devices. Dependent sources are used in combination with other circuit elements to create models of active devices.

In Chapter 3 we found that a voltage source acts as a short circuit when it is turned off. Likewise, a current source behaves as an open circuit when it is turned off. The same results apply to dependent sources, with one important difference. Dependent sources cannot be turned on and off individually because they depend on excitation supplied by independent sources.

Some consequences of this dependency are illustrated in Figure 4–2. When the independent current source is turned on, KCL requires that $i_1 = i_S$. Through controlled

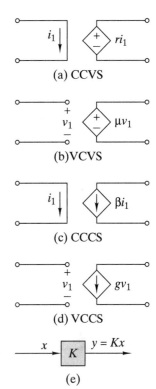

FIGURE 4–1 *Dependent source circuit symbols: (a) Current-controlled voltage source. (b) Voltage-controlled voltage source. (c) Current-controlled current source. (d) Voltage-controlled current source. (e) Block diagram of a gain stage.*

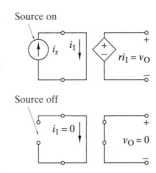

FIGURE 4–2 *Turning off the independent source affects the dependent source.*

source action, the current controlled voltage source is on and its output is

$$v_O = ri_1 = ri_S$$

When the independent current source is off ($i_S = 0$), it acts as an open circuit and KCL requires that $i_1 = 0$. The dependent source is now off and its output is

$$v_O = ri_1 = 0$$

When the independent current source is off, the dependent voltage source acts as a short circuit.

In other words, turning the independent source on and off turns the dependent source on and off as well. We must be careful when applying the superposition principle and Thévenin's theorem to active circuits, since the state of a dependent source depends on the excitation supplied by independent sources. To account for this possibility, we modify the superposition principle to state that the response due to all *independent* sources acting simultaneously is equal to the sum of the responses due to each *independent* source acting one at a time.

4–2 ANALYSIS OF CIRCUITS WITH DEPENDENT SOURCES

With certain modifications, the analysis tools developed for passive circuits apply to active circuits as well. Circuit reduction applies to active circuits, but in so doing we must not eliminate the control variable for a dependent source. As noted previously, when applying the superposition principle or Thévenin's theorem, we must remember that dependent sources cannot be turned on and off independently since their states depend on excitation supplied by one or more independent sources. Applying a source transformation to a dependent source is sometimes helpful, but again we must not lose the identity of a controlling signal for a dependent source. Methods like node and mesh analysis can be adapted to include dependent sources as well.

However, the main difference is that the properties of active circuits can be significantly different from those of the passive circuits treated in Chapters 2 and 3. Our analysis examples are chosen to highlight these differences.

Consider the circuit of Figure 4–3. In this dependent-source circuit the dependent source, a voltage-controlled voltage source, is shown highlighted by the shaded box. To the left, usually, there is a *source circuit* that provides the input to the dependent source. To the right is the *load circuit* that receives the result of the dependent source. Let us analyze this circuit and find the voltage gain $K = v_O/v_S$.

We recognize the load circuit as a voltage divider. That is,

$$v_O = \frac{R_L}{R_L + R_C}(-\mu v_x)$$

FIGURE 4–3 *A circuit with a dependent source.*

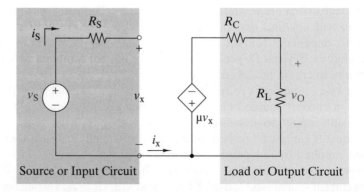

Source or Input Circuit   Load or Output Circuit

Note that the dependent source's output is negative, with the plus sign on the bottom and the negative sign on the top. This was done intentionally to point out that in many active devices there is a sign inversion that is modeled by inverting the sign of the dependent source. To continue our analysis we need to find a relationship between the dependent source's control variable v_x and the input v_S. AKVL around the input loop is

$$-v_S + R_S i_S + v_x = 0$$

Now, since the circuit is an open at v_x, $i_S = 0$. This makes $v_x = v_S$ and therefore,

$$v_O = \frac{R_L}{R_L + R_C}(-\mu v_S)$$

And therefore,

$$K = \frac{v_O}{v_S} = \frac{-\mu R_L}{R_L + R_C}$$

Let us look at what happens when we give values to the various parameters. Let all of the resistors equal 1 kΩ and $\mu = 10^5$. K in this example would be $-50,000$. If in our example we let $v_S = 100$ μV, our output v_O would be -5 V. The circuit has amplified the input by 50,000 times! What makes this even more remarkable is that due to the open circuit i_S, the source current, is zero. The source is not providing any power to the circuit. Nor is there any current i_x flowing between the input and output circuits. Yet it should be clear that there is power provided to the load. This example serves to point out that there is something else at play. The dependent source does not rely on the input for its output power. Yes, the input provides the signal that will be amplified or attenuated by the dependent source, but the output power is obtained from a secondary source driving the dependent source that makes all this possible.[1]

EXAMPLE 4-1

Determine the current, voltage, and power delivered to the 500-Ω output load in Figure 4–4. Then find the power gain defined as p_O/p_S.

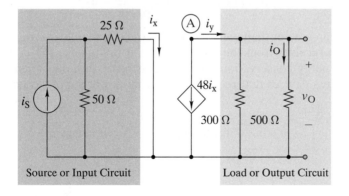

FIGURE 4–4

Source or Input Circuit   Load or Output Circuit

---

[1]Usually the external power supply is not shown in circuit diagrams. When using a dependent source to model an active circuit, we assume that the external supply and the active device itself can handle whatever power is required by the circuit. When designing the actual circuit, the engineer must make certain that the active device and its power supply operate within their power ratings.

SOLUTION:

The control current i_x is found using current division in the input circuit:

$$i_x = \left(\frac{50}{50 + 25}\right) i_S = \frac{2}{3} i_S \tag{4–1}$$

Similarly, the output current i_O is found using current division in the output circuit:

$$i_O = \left(\frac{300}{300 + 500}\right) i_y = \frac{3}{8} i_y \tag{4–2}$$

At node A, KCL requires that $i_y = -48i_x$. Combining this result with Eqs. (4–1) and (4–2) yields the output current:

$$i_O = \left(\frac{3}{8}\right)(-48)i_x = (-18)\left(\frac{2}{3} i_S\right) \tag{4–3}$$
$$= -12i_S$$

The output voltage v_O is found using Ohm's law:

$$v_O = i_O 500 = -6000i_S \tag{4–4}$$

The input-output relationships in Eqs. (4–3) and (4–4) are of the form $y = Kx$ with $K < 0$. The proportionality constants are negative because the reference direction for i_O in Figure 4–4 is the opposite of the orientation of the dependent source reference arrow. As noted earlier and worth repeating active circuits often produce negative values of K, which means that the input and output signals have opposite algebraic signs. Circuits for which $K < 0$ are said to provide **signal inversion**. In the analysis and design of active circuits, it is important to keep track of signal inversions.

Using Eqs. (4–3) and (4–4), the power delivered to the 500-Ω load in Figure 4–4 is

$$p_O = v_O i_O = (-6000i_S)(-12i_S) = 72,000i_S^2 \tag{4–5}$$

The independent source at the input delivers its power to the parallel combination of 50 Ω and 25 Ω. Hence, the input power supplied by the independent source is

$$p_S = (50\|25)i_S^2 = \left(\frac{50}{3}\right)i_S^2$$

Given the input power and output power, we find the power gain in the circuit:

$$\text{Power gain} = \frac{p_O}{p_S} = \frac{72,000i_S^2}{(50/3)i_S^2} = 4320$$

A power gain greater than unity means that the circuit delivers more power at its output than it receives from the input source. At first glance this appears to be a violation of energy conservation, until we remember that dependent sources are models of active devices that require an external power supply to operate. ∎

Exercise 4–1

Find the output v_O in terms of the input v_S in the circuit in Figure 4–5.

FIGURE 4–5

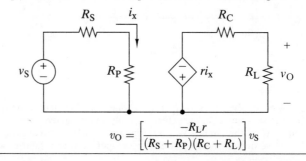

Answer:
$$v_O = \left[\frac{-R_L r}{(R_S + R_P)(R_C + R_L)}\right] v_S$$

NODE-VOLTAGE ANALYSIS WITH DEPENDENT SOURCES

Node analysis of active circuits is much the same as for passive circuits except that we must account for the additional constraints caused by the dependent sources.

For example, let us look at using node analysis to continue our study of the circuit first discussed in Figure 4–3. As shown in Figure 4–6, we have inserted a resistor R_F between the source circuit and the load circuit. This simple insertion is an example of the major reason for the great value of active circuits. Let us analyze this circuit to find $K = v_O/v_S$ and see why.

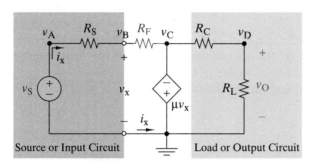

FIGURE 4–6 *Circuit used to demonstrate effects of feedback*

There are five nodes in this circuit. We have selected a ground and labeled the remaining four nodes. We notice that all four nodes can be related to other parameters in the circuit as follows

$$v_A = v_S$$
$$v_B = v_x$$
$$v_C = -\mu v_x$$
$$v_D = v_O$$

There are only two unknown nodes, B and D. At node B we can write the following node equation

$$\frac{v_x - v_S}{R_S} + \frac{v_x - (-\mu v_x)}{R_F} = 0$$

At node D we write

$$\frac{v_O}{R_L} + \frac{v_O - (-\mu v_x)}{R_C} = 0$$

Solving for v_x in the node B equation, we get

$$v_x = \frac{v_S/R_S}{\dfrac{1}{R_S} + \left(\dfrac{1+\mu}{R_F}\right)} = \frac{v_S}{1 + \dfrac{(1+\mu)R_S}{R_F}}$$

Solving the node D equation for v_o

$$v_O\left[\frac{1}{R_L} + \frac{1}{R_C}\right] = \frac{-\mu v_x}{R_C}$$

$$v_O = \frac{-\mu v_x R_L}{R_L + R_C}$$

Substituting our results for v_x and solving for v_O we get

$$v_O = \left[\frac{-\mu R_L}{R_L + R_C}\right]\left[\frac{v_S}{1 + \frac{(1+\mu)R_S}{R_F}}\right]$$

And K equals

$$K = \frac{v_O}{v_S} = \left[\frac{R_L}{R_L + R_C}\right]\left[\frac{-\mu}{1 + \dfrac{(1+\mu)R_S}{R_F}}\right]$$

This equation looks a bit intimidating at first, but with some analysis it makes a lot of sense. If we let R_F become an open circuit, that is, look exactly like our circuit in Figure 4–3, our gain $K = v_O/v_S$ becomes

$$K = \frac{-\mu R_L}{R_L + R_C}$$

which is exactly the same response we found earlier. More importantly, by selecting different values of R_F we can control the gain up to this maximum value. For example, using the same values as before, i.e., all resistors except for R_F set to 1 kΩ, and $\mu = 10^5$, and varying R_F, we can construct the following table

R_F	K
∞	−50,000
1 MΩ	−495.05
100 kΩ	−49.9
10 kΩ	−4.99
5 kΩ	−2.49
2 kΩ	−0.99
1 kΩ	−0.499
0	0

From this example one can see that the circuit designer can control the amount of gain the circuit produces up to the maximum the circuit can provide, −50,000 in this example, by selecting the proper R_F. The key element R_F is called the feedback resistor. Feedback is the reason for the success of many analog circuit designs. It is important to realize that feedback often causes the current i_x to not be zero. With feedback the output circuit directs back some of its voltage or current to the input circuit and helps the circuit designer achieve the output desired. We will study many more examples of feedback later in this and subsequent chapters.

Exercise 4–2

With all other resistors set to 1 kΩ and $\mu = 10^5$, select an appropriate value for R_F in Figure 4–6 so that the gain $|K|$ can never be larger than 10,000, or smaller than 50.

Answer: Use a 100.1-kΩ fixed resistor in series with a 24-MΩ variable resistor for R_F.

EXAMPLE 4–2

For the circuit of Figure 4–7, use node-voltage analysis to find expressions for the unknown node voltages. Write your results as a matrix in $\mathbf{Ax} = \mathbf{b}$ form.

FIGURE 4–7

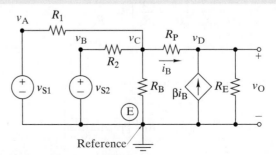

SOLUTION:

The circuit in Figure 4–7 has five nodes. Selecting node E as the reference, each independent voltage source has one terminal connected to ground.

These connections force the conditions $v_A = v_{S1}$ and $v_B = v_{S2}$. Therefore, we only need to write node equations at nodes C and D because voltages at nodes A and B are already known.

Node analysis involves expressing element currents in terms of the node voltages and applying KCL at each unknown node. The sum of the currents *leaving* node C is

$$\frac{1}{R_1}(v_C - v_{S1}) + \frac{1}{R_2}(v_C - v_{S2}) + \frac{1}{R_B}v_C + \frac{1}{R_P}(v_C - v_D) = 0$$

Similarly, the sum of currents leaving node D is

$$\frac{1}{R_P}(v_D - v_C) + \frac{1}{R_E}v_D - \beta i_B = 0$$

These two node equations can be rearranged into the form

Node C: $\quad \left(\dfrac{1}{R_1} + \dfrac{1}{R_2} + \dfrac{1}{R_B} + \dfrac{1}{R_P}\right) v_C - \dfrac{1}{R_P} v_D = \dfrac{1}{R_1} v_{S1} + \dfrac{1}{R_2} v_{S2}$

Node D: $\quad -\dfrac{1}{R_P} v_C + \left(\dfrac{1}{R_P} + \dfrac{1}{R_E}\right) v_D = \beta i_B$

(4–6)

Note that we could write these two symmetrical node equations by inspection if the dependent current source βi_B had been an independent source. But it is not independent, so we must express its constraint in terms of the unknown node voltages. Applying the fundamental property of node voltages and Ohm's law, the current i_B can be written in terms of the node voltages as follows:

$$i_B = \frac{1}{R_P}(v_C - v_D)$$

Substituting this expression for i_B into Eqs. (4–6) and putting the results in standard form yields

Node C: $\quad \left(\dfrac{1}{R_1} + \dfrac{1}{R_2} + \dfrac{1}{R_B} + \dfrac{1}{R_P}\right) v_C - \dfrac{1}{R_P} v_D = \dfrac{1}{R_1} v_{S1} + \dfrac{1}{R_2} v_{S2}$

Node D: $\quad -(\beta + 1)\dfrac{1}{R_P} v_C + \left((\beta + 1)\dfrac{1}{R_P} + \dfrac{1}{R_E}\right) v_D = 0$

(4–7)

And in matrix form

$$\begin{pmatrix} \dfrac{1}{R_1} + \dfrac{1}{R_2} + \dfrac{1}{R_B} + \dfrac{1}{R_P} & -\dfrac{1}{R_P} \\ -\dfrac{(\beta + 1)}{R_P} & \dfrac{(\beta + 1)}{R_P} + \dfrac{1}{R_E} \end{pmatrix} \begin{pmatrix} v_C \\ v_D \end{pmatrix} = \begin{pmatrix} \dfrac{1}{R_1} v_{S1} + \dfrac{1}{R_2} v_{S2} \\ 0 \end{pmatrix}$$

The result in Eqs. (4–7) involves two equations in two unknowns—the node voltages—and includes the effect of the dependent source. However, notice that the matrix is *not* symmetrical. The dependent source constraint destroys the coefficient symmetry. The resultant equations can be readily solved using simple substitution or MATLAB.

This example illustrates a general approach to writing node-voltage equations for circuits with dependent sources. We start out treating the dependent sources as if they are independent sources and write node equations for the resulting passive circuit using the inspection method developed in Chapter 3. This step produces a set of symmetrical node-voltage equations with the independent and dependent source terms on the right-hand side. Then we express the dependent source terms in terms of the unknown node voltages and move them to the left-hand side of the equations with the other terms involving the unknown node voltages. This step destroys the coefficient symmetry but leads to a set of node-voltage equations that describe the active circuit. ∎

Exercise 4-3

For the circuit in Figure 4–7, use the node-voltage equations in Eqs. (4–7) to find the output voltage v_O when $R_1 = 1\,\text{k}\Omega, R_2 = 3\,\text{k}\Omega, R_B = 100\,\text{k}\Omega, R_P = 1.3\,\text{k}\Omega, R_E = 3.3\,\text{k}\Omega$, and $\beta = 50$.

Answer:
$$v_O = v_D = 0.736 v_{S1} + 0.245 v_{S2}$$

This circuit is a signal summer that does not involve a signal inversion. The fact that the output is a linear combination of the two inputs reminds us that the circuit is linear.

EXAMPLE 4 – 3

For the circuit in Figure 4–8 find the voltage gain $K_v = v_O/v_S$, and the current gain $K_i = i_O/i_S$ using node-voltage analysis.

FIGURE 4–8

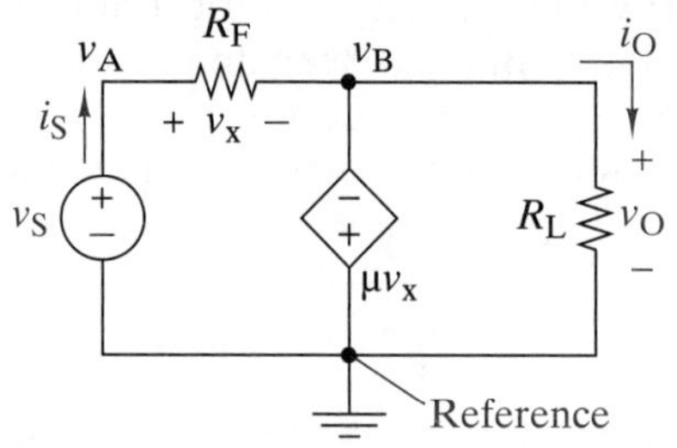

SOLUTION:
There are three nodes in this circuit. If we select the bottom node as our reference, the remaining two nodes are defined by the voltage sources. That is, $v_A = v_S$ and $v_B = -\mu v_x$.

By node analysis, the voltage v_x is

$$v_x = v_A - v_B = v_S - (-\mu v_x)$$

Solving for v_x yields

$$v_x = \frac{v_S}{1 - \mu}$$

The output voltage is directly across the dependent source, hence,

$$v_O = -(\mu v_x)$$

Substituting our expression for v_x, we get

$$v_O = -\mu \left(\frac{v_S}{1 - \mu} \right)$$

The voltage gain is

$$K_v = \left(\frac{\mu}{\mu - 1} \right)$$

This result tells us that for positive values of μ greater than one, the voltage gain is always more than one regardless of μ.

To find the current gain we note that by Ohm's law

$$i_O = \frac{v_O}{R_L} = \frac{-(\mu v_x)}{R_L}$$

And

$$i_S = \frac{v_S - (-\mu v_x)}{R_F} = \frac{v_S + \mu v_x}{R_F}$$

Substituting for v_x, we get

$$i_O = \frac{-\mu\left(\frac{v_S}{1-\mu}\right)}{R_L}$$

And

$$i_S = \frac{v_S + \mu\left(\dfrac{v_S}{1-\mu}\right)}{R_F} = \frac{v_S\left(1 + \left(\dfrac{\mu}{1-\mu}\right)\right)}{R_F} = \frac{v_S\left(\dfrac{1-\mu+\mu}{1-\mu}\right)}{R_F} = \frac{\dfrac{v_S}{1-\mu}}{R_F}$$

The current gain is

$$K_i = \left(\frac{\dfrac{-\mu\left(\frac{v_S}{1-\mu}\right)}{R_L}}{\dfrac{\left(\frac{v_S}{1-\mu}\right)}{R_F}}\right) = \frac{-\mu R_F}{R_L}$$

The magnitude of the current gain could, in theory, range from 0 to ∞, depending on μ, R_F, and R_L. ∎

Exercise 4–4

(a) Formulate node-voltage equations for the circuit in Figure 4–9.

(b) Solve the node-voltage equations for v_O and i_O in terms of i_S.

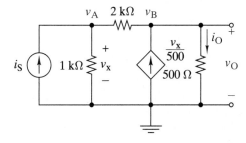

v_A 2 kΩ v_B

FIGURE 4–9

Answers:

(a) $\begin{cases} (1.5 \times 10^{-3})v_A - (0.5 \times 10^{-3})v_B = i_S \\ -(2.5 \times 10^{-3})v_A + (2.5 \times 10^{-3})v_B = 0 \end{cases}$

(b) $v_O = 1000i_S$, $i_O = 2i_S$

EXAMPLE 4–4

Determine the output voltage of the circuit shown in Figure 4–10(a) using OrCAD.

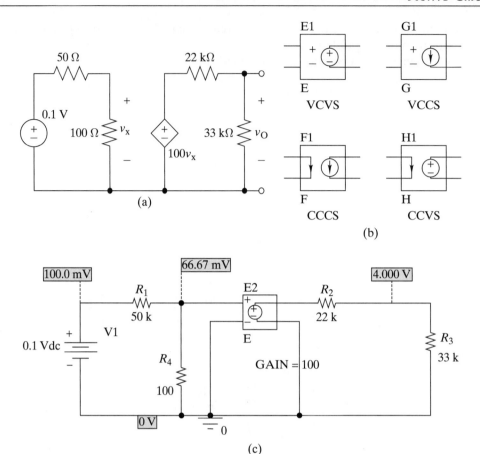

(a)

(b)

(c)

FIGURE 4–10

SOLUTION:
Figure 4–10(b) shows the symbols used in OrCAD to represent the four dependent sources. These elements are found in the **Analog** symbol library as part names starting with the letters:

E for a voltage-controlled voltage source

F for a current-controlled current source

G for a voltage-controlled current source

H for a current-controlled voltage source

The input ports are open circuits for the voltage-controlled elements and short circuits for the current-controlled devices. The output ports are voltage sources or current sources depending on the controlled variable. Note that the controlled sources are indicated by circles rather than diamonds. All four dependent sources are characterized by a single parameter called GAIN. The gain is set by calling up the Property Editor (double-click on the dependent source)—it is nominally set at 1— and entering the value for μ, β, g, or r depending on the type of dependent source. The dimensions of the gain depend on the dimensions of the signals at the input and output ports. As we will see in subsequent examples, these elements are combined with circuit elements to model active devices such as transistors and OP AMPs.

For the problem at hand, we used E, the voltage-controlled voltage source dependent model, and set the gain at 100. The resulting circuit is shown in Figure 4–10(c) with the desired output response of $v_O = 4$ V. ∎

Exercise 4–5 _____

Find i_O using OrCAD for the circuit of Figure 4–11(a).

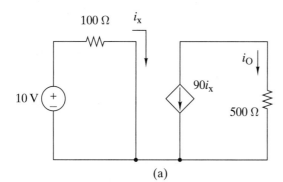

FIGURE 4–11

(a)

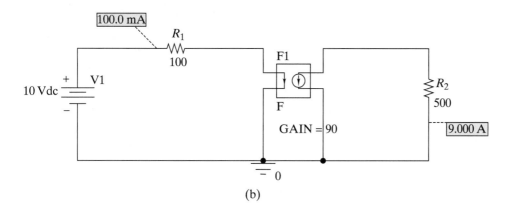

(b)

Answer: $i_O = -9$ A. See Figure 4–11(b).

EXAMPLE 4–5

The circuit in Figure 4–12(a) is a model of an inverting OP AMP circuit.

(a) Use node-voltage analysis to find the output v_O in terms of the input v_S.

(b) Evaluate the input-output relationship found in part (a) as the gain μ becomes very large.

(c) Assume $R_1 = R_2 = 100\,\Omega$, $\mu = 1000$, and $R_4 = 1\,k\Omega$. Use OrCAD to show the effect of the feedback resistor R_3 by plotting the output gain $K = v_O/v_S$ for R_3 varying from 10 Ω to 10 MΩ.

SOLUTION:

(a) Applying a source transformation to the independent source leads to the modified three-node circuit shown in Figure 4–12(b). With the indicated reference node the dependent voltage source constrains the voltage at node B. The control voltage is $v_x = v_A$, and the controlled source forces the node B voltage to be

$$v_B = -\mu v_x = -\mu v_A$$

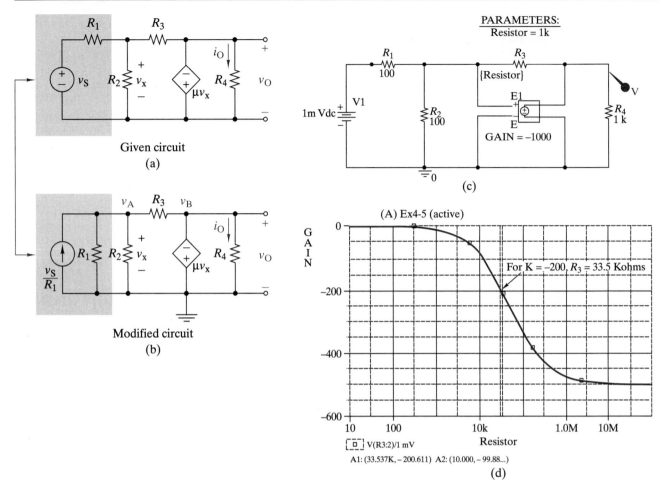

Given circuit
(a)

Modified circuit
(b)

PARAMETERS:
Resistor = 1k

(c)

(A) Ex4-5 (active)

For K = −200, R_3 = 33.5 Kohms

V(R3:2)/1 mV   Resistor

A1: (33.537K, − 200.611) A2: (10.000, − 99.88...)

(d)

FIGURE 4–12

Thus, node A is the only independent node in the circuit. We can write the node A equation by inspection as

$$\left(\frac{1}{R_1}+\frac{1}{R_2}+\frac{1}{R_3}\right)v_A - \frac{1}{R_3}\,v_B = \frac{1}{R_1}\,v_S$$

Substituting in the control source constraint yields the standard form for this equation:

$$\left(\frac{1}{R_1}+\frac{1}{R_2}+\frac{1}{R_3}+\mu\,\frac{1}{R_3}\right)v_A = \frac{1}{R_1}\,v_S$$

We end up with only one node equation even though at first glance the given circuit appeared to need three node equations. The reason is that there are two voltage sources in the original circuit in Figure 4–12(a). Since the two sources share the reference node, the number of unknown node voltages is reduced from three to one. The general principle illustrated is that the number of independent KCL constraints in a circuit containing N nodes and N_V voltage sources (dependent or independent) is $N - 1 - N_V$. The one-node equation can easily be solved for the output voltage $v_O = v_B$:

$$v_O = v_B = -\mu v_A = \left(\frac{-\mu\dfrac{1}{R_1}}{\dfrac{1}{R_1}+\dfrac{1}{R_2}+(1+\mu)\dfrac{1}{R_3}}\right)v_S$$

The minus signs means the circuit provides signal inversion, which is caused by the reference polarity of the controlled source. The output voltage does not depend on the load resistor R_4, since the load is connected across an ideal (though dependent) voltage source.

(b) For large gains μ, we have $(1 + \mu)(1/R_3) \gg [(1/R_1) + (1/R_2)]$ and the input-output relationship reduces to

$$v_O = \left[\frac{-\mu \dfrac{1}{R_1}}{(1 + \mu) \dfrac{1}{R_3}} \right] v_S \approx - \left[\frac{R_3}{R_1} \right] v_S$$

That is, when the active device gain is large, the voltage gain of the active circuit depends on the ratio of two resistances. We will encounter this situation again with OP AMP circuits.

(c) Figure 4–12(c) shows the circuit drawn in OrCAD. The "Special" library call "Param" is used to allow us to sweep the feedback resistor using the "DC Sweep" simulation from 10 Ω to 10 MΩ using a log scale. We selected the input to be 1 mV to keep the output at realistic levels. A voltage probe on the output allows us to display the output as a function of the feedback resistor R_3. The result of the sweep is shown in Figure 4–12(d). Note that the output, a voltage, is divided by 1 mV, the input. The ratio then is the gain K of the circuit as R_3 varies. If, for example, we wanted the circuit to have a gain of –200, we find from the graph that we would need a feedback resistor of about 33.5 kΩ. What would you need for a gain of – 400? (200 kΩ) ∎

Exercise 4–6

Use node-voltage analysis to find v_O for the circuit in Figure 4–13.

Answer:

$$v_O = \frac{\dfrac{1}{R_x} + \mu \dfrac{1}{R_2}}{\dfrac{1}{R_x} + \dfrac{1}{R_L} + (\mu + 1)\dfrac{1}{R_2}} v_S$$

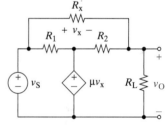

FIGURE 4–13

MESH-CURRENT ANALYSIS WITH DEPENDENT SOURCES

Mesh-current analysis of active circuits follows the same pattern noted for node-voltage analysis. We initially treat the dependent sources as independent sources and write the mesh equations of the resulting passive circuit using the inspection method from Chapter 3. We then account for the dependent sources by expressing their constraints in terms of unknown mesh currents. The following example illustrates the method.

EXAMPLE 4 – 6

(a) Formulate mesh-current equations for the circuit in Figure 4–14.

(b) Use the mesh equations to find v_O and R_{IN} when $R_1 = 50\,\Omega, R_2 = 1\,k\Omega$, $R_3 = 100\,\Omega, R_4 = 5\,k\,\Omega$, and $g = 100\,mS$.

SOLUTION:

(a) Applying source transformation to the parallel combination of R_3 and gv_x in Figure 4–14(a) produces the dependent voltage source $R_3 g v_x = \mu v_x$ in Figure 4–14(b).

FIGURE 4–14

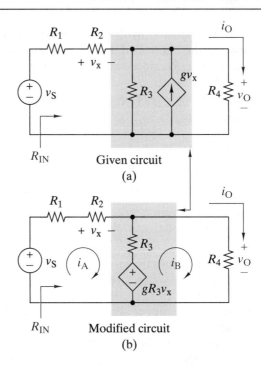

Given circuit
(a)

Modified circuit
(b)

In the modified circuit we have identified two mesh currents. Initially treating the dependent source $(gR_3)v_x$ as an independent source leads to two symmetrical mesh equations.

Mesh A: $(R_1 + R_2 + R_3)i_A - R_3 i_B = v_S - (gR_3)v_x$
Mesh B: $-R_3 i_A + (R_3 + R_4)i_B = (gR_3)v_x$

The control voltage v_x can be written in terms of mesh currents.

$$v_x = R_2 i_A$$

Substituting this equation for v_x into the mesh equations and putting the equations in standard form yields

$$(R_1 + R_2 + R_3 + gR_2R_3)i_A - R_3 i_B = v_S$$
$$-(R_3 + gR_2R_3)i_A + (R_3 + R_4)i_B = 0$$

The resulting mesh equations are not symmetrical because of the controlled source. (b) Substituting the numerical values into the mesh equations gives

$$(1.115 \times 10^4)i_A - (10^2)i_B = v_S$$
$$-(1.01 \times 10^4)i_A + (5.1 \times 10^3)i_B = 0$$

Solving for the two mesh currents using Cramer's rule yields

$$i_A = \frac{\Delta_A}{\Delta} = \frac{\begin{vmatrix} v_S & -10^2 \\ 0 & 5.1 \times 10^3 \end{vmatrix}}{\begin{vmatrix} 1.115 \times 10^4 & -10^2 \\ -1.01 \times 10^4 & 5.1 \times 10^3 \end{vmatrix}} = \frac{5.1 \times 10^3 v_S}{5.5855 \times 10^7}$$

$$= 0.9131 \times 10^{-4} v_S$$

$$i_B = \frac{\Delta_B}{\Delta} = \frac{\begin{vmatrix} 1.115 \times 10^4 & v_S \\ -1.01 \times 10^4 & 0 \end{vmatrix}}{5.5885 \times 10^7} = 1.808 \times 10^{-4} v_S$$

The output voltage and input resistance are found using Ohm's law.

$$v_O = R_4 i_B = 0.904 v_S$$

$$R_{IN} = \frac{v_S}{i_A} = 10.95 \, k\Omega$$ ∎

Exercise 4-7

Write a set of node-voltage equations and use them to find v_O and R_{IN} for the circuit in Figure 4-14. Use the parameter values given in Example 4-6.

Answers:

(a) Labeling node v_A between the 50-Ω and 1-kΩ resistors, we write the node-voltage equations and the relationship of v_x and the node voltages.

$$\frac{v_A - v_S}{50} + \frac{v_A - v_O}{1000} = 0$$

$$\frac{v_O - v_A}{1000} + \frac{v_O}{100} + \frac{v_O}{5000} - 0.1 v_x = 0$$

$$v_x = v_A - v_O$$

(b) Solving these equations yields $v_O = 0.904 v_S$ and $R_{IN} = 10.95 \, k\Omega$.

EXAMPLE 4-7

The circuit in Figure 4-15 is a model of a bipolar junction transistor operating in the active mode. Use mesh analysis to find the transistor base current i_B.

Excluded

FIGURE 4-15

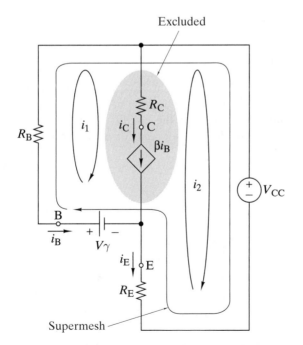

SOLUTION:
The two mesh currents in Figure 4-15 are labeled i_1 and i_2 to avoid possible confusion with the transistor base current i_B. As drawn, the circuit requires a supermesh since the dependent current source βi_B is included in both meshes and is not connected in parallel with a resistance. A supermesh is created by combining meshes 1 and 2 after

excluding the series subcircuit consisting of βi_B and R_C. Beginning at the bottom of the circuit, we write a KVL mesh equation around the supermesh using unknowns i_1 and i_2:

$$i_2 R_E - V_\gamma + i_1 R_B + V_{CC} = 0$$

This KVL equation provides one equation in the two unknown mesh currents. Since the two mesh currents have opposite directions through the dependent current source βi_B, the currents i_1, i_2, and βi_B are related by KCL as

$$i_1 - i_2 = \beta i_B$$

This constraint supplies the additional relationship needed to obtain two equations in the two unknown mesh-current variables. Since $i_B = -i_1$, the preceding KCL constraint means that $i_2 = (\beta + 1)i_1$. Substituting $i_2 = (\beta + 1)i_1$ into the supermesh KVL equation and solving for i_B yields

$$i_B = -i_1 = \frac{V_{CC} - V_\gamma}{R_B + (\beta + 1)R_E}$$ ∎

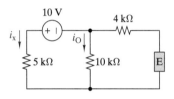

FIGURE 4–16

Exercise 4–8

Use mesh analysis to find the current i_O in Figure 4–16 when the element E is a dependent current source $2i_x$ with the reference arrow directed down.

Answer: −0.857 mA

Exercise 4–9

Use mesh analysis to find the current i_O in Figure 4–16 when the element E is a dependent voltage source $2000i_x$ with the plus reference at the top.

Answer: −0.222 mA

EXAMPLE 4–8

The circuit in Figure 4–17(a) represents a small-signal model of a field effect transistor (FET) amplifier with two inputs, v_{S1} and v_{S2}. Use OrCAD to solve for the input-output relationship of the circuit. (*Hint:* Use superposition to find the respective gains due to the two sources, for example, $v_{S1} = 1\,V$ and $v_{S2} = 0\,V$, and then vice versa.)

FIGURE 4–17

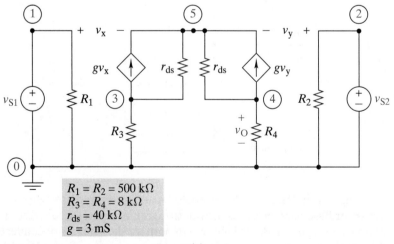

(a)

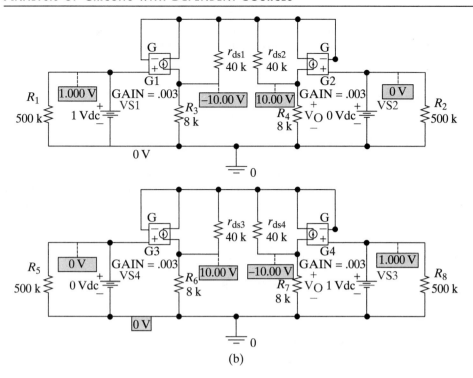

(b)

SOLUTION:

Since the circuit is linear, the input-output relationship is of the form

$$v_O = K_1 v_{S1} + K_2 v_{S2}$$

Using the superposition principle as suggested, let us find the gain K_1 by setting $v_{S1} = 1\,V$ and $v_{S2} = 0\,V$ and solving for v_O (node 4). The gain K_2 is then found by setting $v_{S1} = 0\,V$ and $v_{S2} = 1\,V$ and again solving for v_O (node 4). Figure 4–17(b) shows an OrCAD circuit diagram for both cases and the resulting outputs. From these simulations you can determine that $K_1 = 10$ and $K_2 = -10$.

The input-output relationship for the circuit is

$$v_O = 10(v_{S1} - v_{S2})$$

The circuit is a model of a *differential* amplifier of a type often used as the input stage of an OP AMP. ■

Exercise 4–10

It is very important in designing difference amplifiers that the two transistors be matched in every way so that the outputs are balanced. Use OrCAD to determine the relationship in the circuit of Figure 4–17(a) if the transconductance of FET G2 is 2.9 mS rather than 3 mS.

Answer: $v_O = 9.832 v_{S1} - 9.829 v_{S2}$. Equal inputs will not receive equal gains.

THÉVENIN EQUIVALENT CIRCUITS WITH DEPENDENT SOURCES

To find the Thévenin equivalent of an active circuit, we must leave the independent sources on or else supply excitation from an external test source. This means that the Thévenin resistance cannot be found by the lookback method because that method requires that all independent sources be turned off. Turning off the independent sources deactivates the dependent sources as well and can result in a profound change in the input and output characteristics of an active circuit. Thus, there are two ways of finding active circuit Thévenin equivalents. We can either find the open-circuit voltage and short-circuit current at the interface or directly solve for the interface i–v relationship.

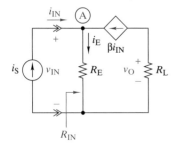

FIGURE 4–18

EXAMPLE 4–9

Find the input resistance of the circuit in Figure 4–18.

SOLUTION:
With the independent source turned off ($i_{IN} = i_s = 0$), the resistance seen at the input port is R_E since the dependent current source βi_{IN} is inactive and acts like an open circuit. Applying KCL at node A with the input source turned on yields

$$i_E = i_{IN} + \beta i_{IN} = (\beta + 1)i_{IN}$$

By Ohm's law, the input voltage is

$$v_{IN} = i_E R_E = (\beta + 1)i_{IN}R_E$$

Hence, the active input resistance is

$$R_{IN} = \frac{v_{IN}}{i_{IN}} = (\beta + 1)R_E$$

The circuit in Figure 4–18 is a model of a transistor circuit in which the gain parameter β typically lies between 50 and 250. The input resistance with external excitation is $(\beta + 1)R_E$, which is significantly higher from the value of R_E without external excitation. A higher R_{IN} helps the transistor reduce the effects of loading on the input source.   ■

Exercise 4–11

For the circuit of Figure 4–18 find the voltage gain $K = v_O/v_{IN}$.

Answer:
$$K = -\frac{\beta R_L}{(\beta + 1)R_E}$$

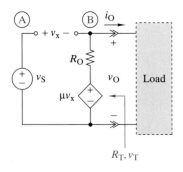

FIGURE 4–19

EXAMPLE 4–10

Find the Thévenin Equivalent at the output interface of the circuit in Figure 4–19.

SOLUTION:
In this circuit the controlled voltage v_x appears across an open circuit between nodes A and B. By the fundamental property of node voltages, $v_x = v_s - v_O$. With the load disconnected and the input source turned off ($v_x = 0$), the dependent voltage source μv_x acts like a short circuit, and the Thévenin resistance looking back into the output port is R_O. With the load connected and the input source turned on, the sum of currents leaving node B is

$$\frac{v_O - \mu v_x}{R_O} + i_O = 0$$

Using the relationship $v_x = v_s - v_O$ to eliminate v_x and then solving for v_O produces the i–v characteristic at the output interface as

$$v_O = \frac{\mu v_s}{\mu + 1} - i_O \left[\frac{R_O}{\mu + 1}\right]$$

The i–v relationship of a Thévenin circuit is $v = v_T - iR_T$. By direct comparison, we find the Thévenin parameters of the active circuit to be

$$v_T = \frac{\mu v_S}{\mu + 1} \quad \text{and} \quad R_T = \frac{R_O}{\mu + 1}$$

The circuit in Figure 4–19 is a model of an OP AMP circuit called a voltage follower. The resistance R_O for a general-purpose OP AMP is around 100 Ω, while the gain μ is about 10^5. Thus, the active Thévenin resistance of the voltage follower is not 100 Ω, as the lookback method suggests, but is only a milliohm. A low output resistance reduces the loading effect caused by correcting a load to the output. ∎

Exercise 4–12

Find the input resistance and output Thévenin equivalent circuit of the circuit in Figure 4–20.

Answers:

$$R_{IN} = (1 + \mu)R_F$$

$$v_T = \frac{\mu}{\mu + 1}\, v_S$$

$$R_T = R_O$$

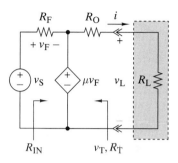

FIGURE 4–20

EXAMPLE 4–11

We have defined four linear dependent sources and shown how to analyze circuits containing these active elements. In this and the next section we show how dependent sources are used to model semiconductor devices like transistors and OP AMPs. The transistor model used here describes the voltages and currents at its external terminals. The model does not describe the transistor's physical structure or internal charge flow. Those subjects are left to subsequent courses in semiconductor materials and devices.

The two basic transistor types are the *bipolar junction transistor* (BJT) and the *field effect transistor* (FET). Both types have several possible operating modes, each with a different set of $i - v$ characteristics. This is something new in our study. Up to this point the characteristics of circuit elements have been fixed. With the transistor we encounter a device whose $i - v$ characteristics can change. We concentrate on the BJT because its $i - v$ characteristics are much easier to understand than the FET. Because it is easier to understand, the simpler BJT best serves as a prelude to our study of the OP AMP—an important semiconductor device that also has several possible operating modes.

The circuit symbol of the BJT is shown in Figure 4–21(a). The device has three terminals called the **emitter (E)**, the **base (B)**, and the **collector (C)**. The voltages v_{BE} and v_{CE} are called the *base-emitter* and *collector-emitter* voltages, respectively. The three currents i_E, i_B, and i_C are called the emitter, base, and collector currents. Photos of real devices are shown in Figure 4–21(b) through 4–21(e).

Applying KCL to the BJT as a whole yields

$$i_E = i_B + i_C$$

which means that only two of the three currents can be independently specified. We normally work with i_B and i_C, and use KCL to find i_E when it is needed.

The BJT's large-signal model is defined in terms of input signals i_B and v_{BE}, and output signals i_C and v_{CE}. For the BJT shown in Figure 4–21(a), the model applies to a region in which these signals are never negative. Within this region there are three

Collector (C)

i_C

Base
(B)

i_B

v_{CE}

v_{BE}

i_E

Emitter (E)

(a)

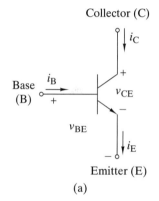

(b)

(c)

FIGURE 4–21 *(a) Circuit symbol for the BJT. (b) BD243C Power transistor. (c) 2N3442 High voltage power transistor. (d) 2N3904 Small signal transistor. (e) 2N2907 Switching transistor.*

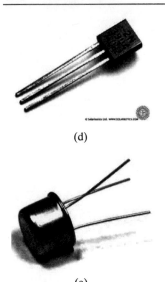

(d)

(e)

FIGURE 4–21   *(Continued)*

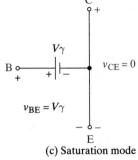

(a) Active mode

(b) Cutoff mode

(c) Saturation mode

FIGURE 4–22 *Circuit models for BJT operating modes. (a) Active mode. (b) Cutoff mode. (c) Saturation mode.*

possible operating modes. The **active mode** is the dominant feature of a BJT. In this mode the collector current i_C is controlled by the base current i_B and v_{BE} is constant.

$$\text{Active mode}: \quad i_C = \beta i_B \quad \text{and} \quad v_{BE} = V_\gamma \qquad (4\text{–}8)$$

The proportionality factor β is called the *forward current gain* and typically ranges from about 50 to several hundred. The constant V_γ is called the *threshold voltage*, which is normally less than a volt. Figure 4–22(a) shows the circuit elements that model the active mode i–v characteristics as defined in Eq. (4–8). In the active mode i_B and v_{CE} are determined by the interaction of these i–v characteristics with the rest of the circuit.

Two additional operating modes exist at the boundary of the BJT's operating region. When $i_B = 0$ and $i_C = 0$, the transistor is in the **cutoff mode** and the device acts like an open circuit between the collector and emitter. When $v_{CE} = 0$ and $v_{BE} = V_\gamma$ the transistor is in the **saturation mode**, and the device acts like a short circuit between the collector and emitter. These two modes are summarized by writing

$$\begin{aligned} \text{Cutoff mode}: \quad & i_B = 0 \quad \text{and} \quad i_C = 0 \\ \text{Saturation mode}: \quad & v_{CE} = 0 \quad \text{and} \quad v_{BE} = V_\gamma \end{aligned} \qquad (4\text{–}9)$$

Figures 4–22(b) and 4–22(c) show the circuit elements that model the i–v characteristics defined in Eq. (4–9).

The circuit in Figure 4–22(b) points out that in the cutoff mode, v_{CE} must equal the open-circuit voltage available from the external circuit. The circuit in Figure 4–22(c) points out that in the saturation mode, i_C must equal the short-circuit current available from the external circuit. The net result is that the BJT's output variables must fall within the following bounds:

	Cutoff bounds		Saturation bounds		
0	$\leq$	i_C	$\leq$	i_{SC}	(4–10)
v_{OC}	$\geq$	v_{CE}	$\geq$	0	

where v_{OC} and i_{SC} are the open-circuit voltage and short-circuit current available between the collector and emitter terminals. In the cutoff mode the transistor outputs i_C and v_{CE} are equal to their respective cutoff bounds. In saturation mode the outputs equal their saturation bounds. In the active mode the outputs fall between the cutoff and saturation bounds.

With this background we are prepared to analyze the transistor circuit in Figure 4–23.[2] Our analysis objective is to find the outputs i_C and v_{CE}. To do this we must know the transistor's operating mode. To find the operating mode we make use of two facts:

1. The lower bounds in Eq. (4–10) mean that i_C and v_{CE} cannot be negative.

2. The upper bounds in Eq. (4–10) depend on the rest of the circuit.

For the circuit in Figure 4–23 these upper bounds are $v_{OC} = V_{CC}$ and $i_{SC} = V_{CC}/R_C$.

Our analysis strategy *assumes* the device is in the active mode and uses the active mode device equations to find i_C. According to Eq. (4–8) the active mode element equations are $v_{BE} = V_\gamma$ and $i_C = \beta i_B$. Using these element constraints and applying KVL around the input loop in Figure 4–23 yields the collector current as

$$i_C = \beta i_B = \beta \left(\frac{v_S - V_\gamma}{R_B} \right) \qquad (4\text{–}11)$$

---

[2]This circuit is called the common-emitter configuration because the emitter terminal is common to the input and output loops.

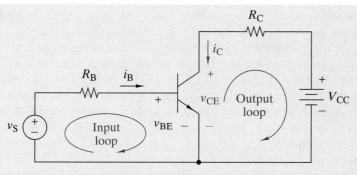

FIGURE 4–23 *BJT common-emitter circuit.*

This equation indicates that if $v_S > V_\gamma$ then, $i_C > 0$. However, if $v_S < V_\gamma$, then $i_C < 0$, which would violate its cutoff bound. Thus, if the input voltage v_S is greater than the threshold voltage V_γ, then the BJT can be in the active mode. But if $v_S < V_\gamma$, the BJT is in the cutoff mode and the outputs equal their cutoff bounds in Eq. (4–10), namely $i_C = 0$ and $v_{CE} = v_{OC} = V_{CC}$.

When $v_S > V_\gamma$, Eq. (4–11) predicts a positive collector current that increases linearly with v_S. To find the collector-emitter voltage we apply KVL around the output loop in Figure 4–23 to obtain

$$v_{CE} = V_{CC} - i_C R_C \tag{4–12}$$

This equation predicts that $v_{CE} > 0$ as long as $i_C < V_{CC}/R_C$. But V_{CC}/R_C is the short-circuit current available from the external circuit. Thus, as long as $v_S > V_\gamma$ and $i_C < i_{SC}$, the BJT is in the active mode and Eqs. (4–11) and (4–12) correctly predict the outputs i_C and v_{CE}. However, if Eq. (4–11) predicts that $i_C > i_{SC}$, then Eq. (4–12) says that $v_{CE} < 0$. Both of these results violate the saturation bounds in Eq. (4–10). When this happens, the BJT is actually in the saturation mode and the outputs equal their saturation bounds in Eq. (4–10), namely $i_C = i_{SC} = V_{CC}/R_C$ and $v_{CE} = 0$.

Figure 4–24 summarizes this discussion using graphs of the outputs v_{CE} and i_C versus the input voltage v_S. When $v_S < V_\gamma$, the BJT is in the cutoff mode and the outputs are $i_C = 0$ and $v_{CE} = v_{OC} = V_{CC}$. When $v_S > V_\gamma$, the BJT enters the active mode and the outputs i_C and v_{CE} are governed by Eqs. (4–11) and (4–12). Under these equations, i_C increases linearly as v_S increases, with the result that v_{CE} decreases linearly. The collector current continues to increase as v_S increases until it reaches its saturation bound at $i_C = i_{SC}$. At that point the transistor switches into the saturation mode and thereafter the outputs remain constant at $i_C = i_{SC} = V_{CC}/R_C$ and $v_{CE} = 0$.

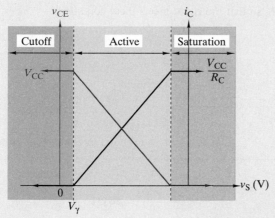

FIGURE 4–24 *Output responses of the BJT circuit in Figure 4–23. This is called the circuit's transfer characteristics.*

In digital applications the input voltage drives the transistor between the cutoff and saturation modes passing through the active mode as quickly as possible. In analog circuit applications the transistor remains in the active mode where the slope of the transfer characteristic provides voltage amplification. In the next section we find that the OP AMP has similar transfer characteristics.

Suppose that the circuit parameters in Figure 4–23 are $\beta = 100$, $V_\gamma = 0.7$ V, $R_B = 100$ kΩ, $R_C = 1$ kΩ, and $V_{CC} = 5$ V. Find i_C and v_{CE} when $v_S = 2$ V. Repeat when $v_S = 6$ V.

SOLUTION:
Since $v_S = 2$ is greater than $V_\gamma = 0.7$, the transistor is *not* in the cutoff mode. We assume that it is in the active mode and use Eq. (4–11) to calculate i_C.

$$i_C = \beta \left(\frac{v_S - V_\gamma}{R_B} \right) = 100 \left(\frac{2 - 0.7}{100 \times 10^3} \right) = 1.3 \text{ mA}$$

The available short-circuit current is $i_{SC} = V_{CC}/R_C = 5$ mA. Since the calculated i_C is less than i_{SC}, the transistor is in fact in the active mode and we use Eq. (4–12) to find v_{CE}.

$$v_{CE} = V_{CC} - i_C R_C = 5 - 1.3 \times 10^{-3} \times 1000 = 3.7 \text{ V}$$

For $v_S = 2$ V, the transistor is in the active mode and the outputs are $i_C = 1.3$ mA and $v_{CE} = 3.7$ V.

For $v_S = 6$ V, we again assume that the transistor is in the active mode and calculate the collector current from Eq. (4–11).

$$i_C = \beta \left(\frac{v_S - V_\gamma}{R_B} \right) = 100 \left(\frac{6 - 0.7}{100 \times 10^3} \right) = 5.3 \text{ mA}$$

The calculated i_C is greater than the available i_{SC}. For this input the transistor is in the saturation mode and the outputs equal their saturation bounds, namely $i_C = i_{SC} = 5$ mA and $v_{CE} = 0$. ■

◀ Design Exercise 4–13 _____

The known parameters in Figure 4–25 are $\beta = 100$, $V_\gamma = 0.7$ V, $R_C = 1$ kΩ, and $V_{CC} = 5$ V. The circuit is to function as a digital inverter that meets two conditions:

1. An input of $v_S = 0$ V must produce an output $v_{CE} = 5$ V.

2. An input of $v_S = 5$ V must produce an output $v_{CE} = 0$ V

Select a value of R_B so that the circut meets these conditions.

FIGURE 4–25

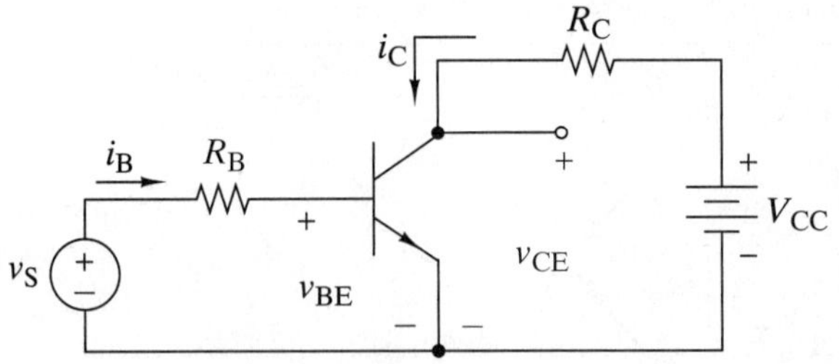

Answer: $$R_B < \frac{100 \times 4.3}{5 \times 10^{-3}} = 86 \text{ k}\Omega$$

Any reasonable value less than 86 kΩ (say 56 kΩ, a standard value) will work.

4-3 THE OPERATIONAL AMPLIFIER

The integrated circuit operational amplifier is the premier linear active device in present-day analog circuit applications. The term *operational amplifier* was apparently first used in a 1947 paper by John R. Ragazzini and his colleagues, who reported on work carried out for the National Defense Research Council during World War II. The paper described high-gain dc amplifier circuits that perform mathematical operations (addition, subtraction, multiplication, division, integration, etc.); hence the name *operational amplifier*. For more than a decade the most important applications were general- and special-purpose analog computers using vacuum tube amplifiers. In the early 1960s general-purpose, discrete-transistor operational amplifiers became readily available, and by the mid-1960s the first commercial integrated circuit OP AMPs entered the market. The transition from vacuum tubes to integrated circuits decreased the size, power consumption, and cost of OP AMPs by nearly three orders of magnitude. By the early 1970s the integrated circuit version became the dominant active device in analog circuits.

The device itself is a complex array of transistors, resistors, diodes, and capacitors, all fabricated and interconnected on a tiny silicon chip. Figure 4–26 shows examples of ways OP AMPs are packaged for use in circuits. In spite of its complexity, the device can be modeled by rather simple *i–v* characteristics. We do not need to concern ourselves with what is going on inside the package; rather, we treat the OP AMP using a behavioral model that constrains the voltages and currents at the external terminals of the device.

OP AMP NOTATION

Certain matters of notation and nomenclature must be discussed before developing a circuit model for the OP AMP. The OP AMP is a five-terminal device, as shown in Figure 4–27(a). The "+" and "−" symbols identify the input terminals and are a shorthand notation for the noninverting and inverting input terminals, respectively. These "+" and "−" symbols identify the two input terminals and have nothing to do with the polarity of the voltages applied. The other terminals are the output and the positive and negative supply voltages, usually labeled $+V_{CC}$ and $-V_{CC}$. While some OP AMPs have more than five terminals, these five are always present and are the only ones we will use in this text. Figure 4–27(b) shows how these terminals are arranged in a common eight-pin integrated circuit package.

The two power supply terminals in Figure 4–27 are not usually shown in circuit diagrams. Be assured that they are always there because the external power supplies are required for the OP AMP to operate as an active device. The power required for signal amplification comes through these terminals from an external power source. The $+V_{CC}$ and $-V_{CC}$ voltages applied to these terminals also determine the upper and lower limits on the OP AMP output voltage.

Figure 4–28(a) shows a complete set of voltage and current variables for the OP AMP, while Figure 4–28(b) shows the abbreviated set of signal variables we will use. All voltages are defined with respect to a common reference node, usually ground. Voltage variables v_P, v_N, and v_O are defined by writing a voltage symbol beside the corresponding terminals. This notation means the "+" reference mark is at the terminal in question and the "−" reference mark is at the reference or ground terminal. In this book the reference directions for the currents are directed in at input terminals and out at the output. At times the abbreviated set of current variables may appear to violate KCL. For example, a global KCL equation for the complete set of variables in Figure 4–28(a) is

$$i_O = I_{C+} + I_{C-} + i_P + i_N \quad \text{(correct equation)} \tag{4–13}$$

©Rapid Electronics

(a)

©TI

(b)

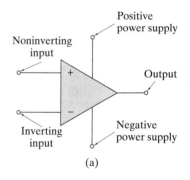

(c)

FIGURE 4–26 *Examples of OP AMP packages. (a) Dual in-line 14-pin and 8-pin packages (DIP). (b) A discrete-component, high-performance audio package. (c) Low-power surface-mount package.*

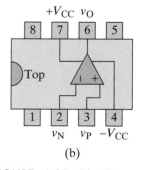

(a)

(b)

FIGURE 4–27 *The OP AMP: (a) Circuit symbol. (b) Pin-out diagram for an eight-pin DIP package.*

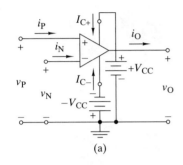

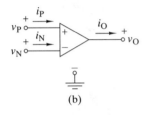

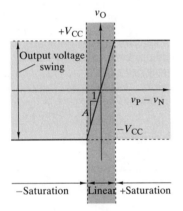

FIGURE 4–28 *OP AMP voltage and current definitions:*
(a) Complete set.
(b) Shorthand set.

FIGURE 4–29 *OP AMP transfer characteristics.*

A similar equation using the shorthand set of current variables in Figure 4–28(b) reads

$$i_O = i_N + i_P \quad \text{(incorrect equation)} \tag{4–14}$$

This equation is *not* correct, since it does not include all the currents. What is more important, it implies that the output current comes from the inputs. In fact, this is wrong. The input currents are very small, ideally zero. The output current comes from the supply voltages, as Eq. (4–13) points out, even though these terminals are not shown on the abbreviated circuit diagram.

TRANSFER CHARACTERISTICS

The dominant feature of the OP AMP is the transfer characteristic shown in Figure 4–29. This characteristic provides the relationships between the **noninverting input** v_P the **inverting input** v_N, and the **output voltage** v_O. The transfer characteristic is divided into three regions or modes called +**saturation**, −**saturation**, and **linear**. In the linear region the OP AMP is a **differential amplifier** because the output is proportional to the difference between the two inputs. The slope of the line in the linear range is called the voltage gain. In this linear region the input-output relation is

$$v_O = A(v_P - v_N) \tag{4–15}$$

The voltage gain of an OP AMP is very large, usually greater than 10^5. As long as the net input $(v_P - v_N)$ is very small, the output will be proportional to the input. However, when $A|v_P - v_N| > V_{CC}$, the OP AMP is saturated and the output voltage is limited by the supply voltages (less some small internal losses).

In the previous section, we found that the transistor has three operating modes. The input-output characteristic in Figure 4–29 points out that the OP AMP also has three operating modes:

1. +Saturation mode when $A(v_P - v_N) > V_{CC}$ and $v_O = +V_{CC}$.
2. −Saturation mode when $A(v_P - v_N) < -V_{CC}$ and $v_O = -V_{CC}$.
3. Linear mode when $A|v_P - v_N| < V_{CC}$ and $v_O = A(v_P - v_N)$.

Usually we analyze and design OP AMP circuits using the model for the linear mode. When the operating mode is not given, we use a self-consistent approach similar to the one used for the transistor. That is, we assume that the OP AMP is in the linear mode and then calculate the output voltage v_O. If it turns out that $-V_{CC} < v_O < +V_{CC}$, then the assumption is correct and the OP AMP is indeed in the linear mode. If $v_O < -V_{CC}$, then the assumption is wrong and the OP AMP is in the −saturation mode with $v_O = -V_{CC}$. If $v_O > +V_{CC}$, then the assumption is wrong and the OP AMP is in the +saturation mode with $v_O = +V_{CC}$.

IDEAL OP AMP MODEL

A dependent-source model of an OP AMP operating in its linear range is shown in Figure 4–30. This model includes an input resistance (R_I), an output resistance (R_O), and a voltage-controlled voltage source whose gain is A.[3] Numerical values of these

---

[3] The parameter A is used in OP AMP notation to define the device's "open-loop, gain, which is equal to μ for VCVS models. In real OP AMPs, the open-loop gain listed in reference manuals is generally a minimum rather than an exact value. Hence, a listing for A equal to, say, 10^5 may mean that the device can be expected to have at least that amount of open-loop gain, but could exceed it by a factor of 10 or more. In general, we like to have A as large as possible. As we will see, it is the "closed-loop" gain that we want to be exact.

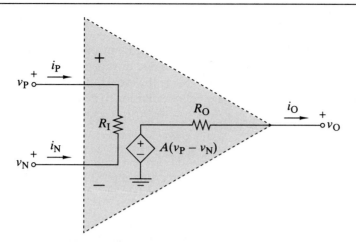

FIGURE 4–30 *Dependent-source model of an OP AMP operating in the linear mode.*

OP AMP parameters typically fall in the following ranges:

$$10^6 < R_I < 10^{12}\,\Omega$$
$$10 < R_O < 100\,\Omega$$
$$10^5 < A < 10^8$$

Clearly, high input resistance, low output resistances, and high voltage gain are the key attributes of an OP AMP.

The dependent-source model can be used to develop the i–v relationships of the ideal model. For the OP AMP to operate in its linear mode, the output voltage is bounded by

$$-V_{CC} \le v_O \le +V_{CC}$$

Using Eq. (4–15), we can write this bound as

$$-\frac{V_{CC}}{A} \le (v_P - v_N) \le +\frac{V_{CC}}{A}$$

The supply voltage V_{CC} is typically about 15 V, while A is a very large number, usually 10^5 or greater. Consequently, linear operation requires that $v_P \approx v_N$. In the ideal OP AMP model, the voltage gain is assumed to be infinite ($A \to \infty$), in which case linear operation requires $v_P = v_N$. The input resistance R_I of the ideal OP AMP is assumed to be infinite, so the currents entering both input terminals are both zero. In summary, the i–v relationships of the **ideal model** of the OP AMP are

$$v_P = v_N$$
$$i_P = i_N = 0 \tag{4–16}$$

The implications of these element equations are illustrated on the OP AMP circuit symbol in Figure 4–31.

At first glance the element constraints of the ideal OP AMP appear to be fairly useless. They look more like connection constraints and are totally silent about the output quantities (v_O and i_O), which are usually the signals of greatest interest. They seem to say that the OP AMP input terminals are simultaneously a short circuit ($v_P = v_N$) and an open circuit ($i_P = i_N = 0$). In practice, however, the ideal model of the OP AMP is very useful because in linear applications feedback is always present. That is, for the OP AMP to operate in a linear mode, it is necessary for there to be feedback paths from the output to one or both of the inputs. These feedback paths ensure that $v_P \approx v_N$ and make it possible for us to analyze OP AMP circuits using the ideal OP AMP element constraints in Eq. (4–16).

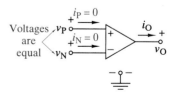

FIGURE 4–31 *Ideal OP AMP characteristics.*

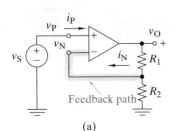

(a)

v_S $\boxed{K}$ v_O

(b)

FIGURE 4–32 *The noninverting amplifier circuit.*

NONINVERTING OP AMP

To illustrate the effects of feedback, let us find the input-output characteristics of the circuit in Figure 4–32. In this circuit the voltage divider provided a feedback path from the output to the inverting input.[4] Since the ideal OP AMP draws no current at either input ($i_P = i_N = 0$), we can use voltage division to determine the voltage at the inverting input:

$$v_N = \frac{R_2}{R_1 + R_2} v_O \tag{4–17}$$

The input source connection at the noninverting input requires the condition

$$v_P = v_S \tag{4–18}$$

The ideal OP AMP element constraints demand that $v_P = v_N$; therefore, we can equate the right sides of Eqs. (4–17) and (4–18) to obtain the input-output relationship of the overall circuit.

$$v_O = \frac{R_1 + R_2}{R_2} v_S \tag{4–19}$$

The preceding analysis illustrates a general strategy for analyzing OP AMP circuits. We use normal circuit analysis methods to express the OP AMP input voltages v_P and v_N in terms of circuit parameters. We then use the ideal OP AMP constraint $v_P = v_N$ to solve for the overall circuit input-output relationship.

The circuit in Figure 4–32(a) is called a **noninverting amplifier**. The input-output relationship is of the form $v_O = Kv_S$, which reminds us that the circuit is linear. Figure 4–32(b) shows the functional building block for this circuit, where the proportionality constant K is

$$K = \frac{R_1 + R_2}{R_2} \tag{4–20}$$

In an OP AMP circuit the proportionality constant K is sometimes called the **closed-loop** gain, because it defines the input-output relationship when the feedback loop is connected (closed).

When discussing OP AMP circuits, it is necessary to distinguish between two types of gains. The first is the large open-loop voltage gain provided by the OP AMP device itself. The second is the closed-loop voltage gain of the OP AMP circuit with a negative feedback path. Note that Eq. (4–20) indicates that the circuit gain is determined by the resistors in the feedback path, not by the value of the OP AMP gain. The gain in Eq. (4–20) is really the voltage division rule upside down. Variation of the value of K depends on the tolerance on the resistors in the feedback path, not the variation in the value of the OP AMP's gain. In effect, feedback converts the OP AMP's very large but variable gain into a much smaller but well-defined gain.

Let us look at a first example of a noninverting OP AMP. Consider the circuit shown in Figure 4–33(a). Let us find the output voltage, the output current, the voltage gain, the output power, and the power gain.

---

[4]The feedback must always be to the inverting terminal. Otherwise the circuit will be unstable for reasons that we cannot explain based on what we have learned thus far. Further complicating this understanding is that when using the ideal OP AMP model in OrCAD, the software does not distinguish between feedback to the positive or negative terminals. This, of course, is not true for a real OP AMP like a uA741 as used in the laboratory and in circuit applications.

FIGURE 4–33

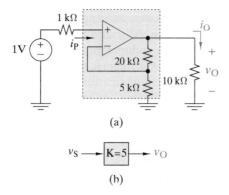

(a)

$$v_S \longrightarrow \boxed{K=5} \longrightarrow v_O$$

(b)

The circuit we just studied is contained within the shaded area. There is an input source and an output load. The gain of the OP AMP circuit is found using Eq. (4–20) as

$$K = \frac{R_1 + R_2}{R_2} = \frac{20\,k + 5\,k}{5\,k} = 5$$

The output voltage is found by substituting into Eq. (4–19), but there is a possible complication. In the derivation of Eq. (4–19) the input was connected directly to the noninverting terminal. In this circuit there is a 1-kΩ resistor between the source and the OP AMP. Doing a KVL at the input yields

$$-v_S + 1\,k\,i_P + v_P = 0$$

But we recall that for an ideal OP AMP $i_P = i_N = 0$. Therefore, there is no voltage drop across the 1-kΩ resistor so that the input voltage is impressed directly across the OP AMP. Eq. (4–19) holds. The output voltage is found as

$$v_O = 5 \times 1 = 5\,V$$

We calculate the output current by using Ohm's Law

$$i_O = \frac{v_O}{R_L} = \frac{5}{10\,k} = 500\,\mu A$$

The output power is simply $p_O = v_O \times i_O = 5 \times 500\,\mu = 2.5$ mW. It does not make sense to talk about power gain with these types of circuits. Since the input current i_P is zero, the source does not provide any power to the circuit. This would appear to say that this circuit has infinite power gain. Of course that is not true. There is a power source driving the OP AMP that is indeed providing the necessary power to the load.

DESIGN EXAMPLE 4–12

Design an amplifier with a gain of $K = 10$.

SOLUTION:
Using a noninverting OP AMP circuit, the design problem is to select the values of the resistors in the feedback path. From Eq. (4–20) the design constraint is

$$10 = \frac{(R_1 + R_2)}{R_2}$$

We have one constraint with two unknowns. Arbitrarily selecting $R_2 = 10\,k\Omega$, we find $R_1 = 90\,k\Omega$. These resistors would normally have low tolerances ($\pm 1\%$ or less) to produce a precisely controlled closed-loop gain.

Comment: The problem of choosing resistance values in OP AMP circuit design problems deserves some discussion. Although values of resistance from a few ohms to several hundred megohms are commercially available, we generally limit our-selves to the range from about 1 kΩ to perhaps 1 MΩ. The lower limit of 1 kΩ is imposed in part because of power dissipation in the resistors. Typically, we use resistors with $\frac{1}{4}$ -W power ratings or less. The maximum voltage in OP AMP circuits is often around 15 V. The smallest $\frac{1}{4}$ -W resistance we can use is $R_{MIN} \geq (15)^2/0.25 = 900\ \Omega$, or about 1 kΩ. The upper bound of 1 MΩ comes about because surface leakage makes it difficult to maintain the tolerance in a high-value resistance. High-value resistors are also noisy, which leads to problems when they are connected in the feedback path. The 1-kΩ to 1-MΩ range should be used as a guideline, not an inviolate design rule. Actual design choices are influenced by system-specific factors and changes in technology.   ■

◐ Design Exercise 4–14 _____

Design a noninverting amplifier circuit with a gain of $7 \pm 10\%$ using standard 10% resistors. (See inside back cover for standard values)

A n s w e r :   Referring to Figure 4–32(a), $R_1 = 22\,k\Omega$ and $R_2 = 3.3\,k\Omega$, or $R_1 = 27\,k\Omega$ and $R_2 = 4.7\,k\Omega$. Other combinations are possible.

Exercise 4–15 _____

The noninverting amplifier circuit in Figure 4–32(a) is operating with $R_1 = 2R_2$ and $V_{CC} = \pm12\,V$. Over what range of input voltages v_S is the OP AMP in the linear mode?

A n s w e r : $-4\,V < v_S < +4\,V$

Effects of Finite OP AMP Gain

The ideal OP AMP model has an infinite gain. Actual OP AMP devices have very large, but finite voltage gains. We now address the effect of large but finite gain on the input-output relationships of OP AMP circuits.

The circuit in Figure 4–34 shows a finite gain OP AMP circuit model in which the input resistance R_I is infinite. The actual values of OP AMP input resistance range from 10^6 to 10^{12} Ω, so no important effect is left out by ignoring this resistance. Examining the circuit, we see that the noninverting input voltage is determined by the independent voltage source. The inverting input can be found by voltage division, since the current i_N is zero. In other words, Eqs. (4–17) and (4–18) apply to this circuit as well.

FIGURE 4–34 *The noninverting amplifier circuit with the dependent-source model.*

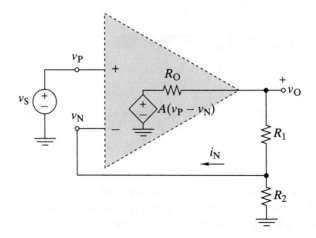

We next determine the output voltage in terms of the controlled-source voltage using voltage division on the series connection of the three resistors R_O, R_1, and R_2:

$$v_O = \frac{R_1 + R_2}{R_O + R_1 + R_2} A(v_P - v_N)$$

Substituting v_P and v_N from Eqs. (4–17) and (4–18) yields

$$v_O = \left[\frac{R_1 + R_2}{R_O + R_1 + R_2}\right] A \left[v_S - \frac{R_2}{R_1 + R_2} v_O\right] \tag{4–21}$$

The intermediate result in Eq. (4–21) shows that feedback is present since v_O appears on both sides of the equation. Solving for v_O yields

$$v_O = \frac{A(R_1 + R_2)}{R_O + R_1 + R_2(1 + A)} v_S \tag{4–22}$$

In the limit, as $A \to \infty$, Eq. (4–22) reduces to

$$v_O = \frac{R_1 + R_2}{R_2} v_S = K v_S$$

where K is the closed-loop gain we previously found using the ideal OP AMP model.

To see the effect of a finite A, we ignore R_O in Eq. (4–22) since it is generally quite small compared with $R_1 + R_2$. With this approximation Eq. (4–22) can be written in the following form:

$$v_O = \frac{K}{1 + (K/A)} v_S \tag{4–23}$$

When written in this form, we see that the closed-loop gain reduces to K as $A \to \infty$. Moreover, we see that the finite-gain model yields a good approximation to the ideal model results as long as $K \ll A$. In other words, the ideal model yields good results as long as the closed-loop gain is much less than the open-loop gain of the OP AMP device. One practical rule of thumb is to limit the closed-loop gain to less than 1% of the OP AMP gain (i.e., $K < A/100$).

The feedback path also affects the active output resistance. To see this, we construct a Thévenin equivalent circuit using the open-circuit voltage and the short-circuit current. Equation (4–23) is the open-circuit voltage, and we need only find the short-circuit current. Connecting a short-circuit at the output in Figure 4–34 forces $v_N = 0$ but leaves $v_P = v_S$. Therefore, the short-circuit current is

$$i_{SC} = A(v_S/R_O)$$

As a result, the Thévenin resistance is

$$R_T = \frac{v_{OC}}{i_{SC}} = \frac{K/A}{1 + K/A} R_O$$

When $K \ll A$, this expression reduces to

$$R_T = \frac{K}{A} R_O \approx 0\,\Omega$$

The OP AMP circuit with feedback has an output Thévenin resistance that is much smaller than the output Thévenin resistance of the OP AMP device itself. In fact, the Thévenin resistance is very small since R_O is typically less than $100\,\Omega$ and A is greater than 10^5.

At this point we can summarize our discussion. We introduced the OP AMP as an active five-terminal device including two supply terminals not normally shown on the circuit diagram. We then developed an ideal model of this device that is used to analyze and design circuits that have feedback. Feedback must be present for the device to operate in the linear mode. The most dramatic feature of the ideal model is the assumption of infinite gain. Using a finite-gain model, we found that the ideal model predicts the circuit input-output relationship quite closely as long as the circuit gain K is much smaller than the OP AMP gain A. We also discovered that the Thévenin output resistance of an OP AMP with feedback is essentially zero.

In the rest of this book we use the ideal i–v constraints in Eq. (4–16) to analyze OP AMP circuits. The OP AMP circuits have essentially zero output resistance, which means that the output voltage does not change with different loads. Unless otherwise stated, from now on the term *OP AMP* refers to the ideal model.

4–4 OP AMP CIRCUIT ANALYSIS

This section introduces OP AMP circuit analysis using examples that are building blocks for analog signal-processing systems. We have already introduced one of these circuits—the noninverting amplifier discussed in the preceding section. The other basic circuits are the voltage follower, the inverting amplifier, the summer, and the subtractor. The key to using the building block approach is to recognize the feedback pattern and to isolate the basic circuit as a building block. The first example illustrates this process.

EXAMPLE 4–13

Find the input-output relationship of the circuit in Figure 4–35(a).

FIGURE 4–35

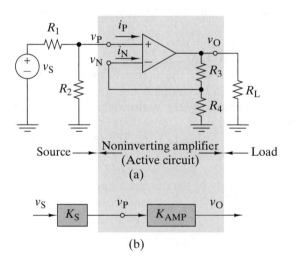

(a)

(b)

SOLUTION:
When the circuit is partitioned as shown in Figure 4–35(a), we recognize two building block gains: (1) K_S, the proportionality constant of the source circuit, and (2) K_{AMP}, the gain of the noninverting amplifier as shown in Figure 4–35(b). The OP AMP circuit input current $i_P = 0$; hence, we can use voltage division to find K_S as

$$K_S = \frac{v_P}{v_S} = \frac{R_2}{R_1 + R_2}$$

Since the noninverting amplifier has zero output resistance, the load R_L has no effect on the output voltage v_O. Using Eq. (4–19), the gain of the noninverting amplifier circuit is

$$K_{\text{AMP}} = \frac{v_O}{v_P} = \frac{R_3 + R_4}{R_4}$$

The overall circuit gain is found as

$$K_{\text{CIRCUIT}} = \frac{v_O}{v_S} = \left[\frac{v_P}{v_S} \right] \left[\frac{v_O}{v_P} \right]$$

$$= K_S \times K_{\text{AMP}}$$

$$= \left[\frac{R_2}{R_1 + R_2} \right] \left[\frac{R_3 + R_4}{R_4} \right]$$

The gain K_{CIRCUIT} is the product of K_S times K_{AMP} because the amplifier circuit does not load the source circuit since $i_P = 0$. ∎

Exercise 4–16

(a) Find v_O in Figure 4–35(a) when $R_1 = R_2 = 1\,\text{k}\Omega$, $v_S = 1\,\text{V}$, $R_3 = R_4 = 1\,\text{k}\Omega$ and $R_L = 100\,\Omega$. (b) Repeat for when R_3 is a short circuit and the other values are the same.

Answers:

(a) $v_O = 1\,\text{V}$

(b) $v_O = 0.5\,\text{V}$

VOLTAGE FOLLOWER

The OP AMP in Figure 4–36(a) is connected as a **voltage follower** or **buffer**. In this case, the feedback path is a direct connection from the output to the inverting input. The feedback connection forces the condition $v_N = v_O$. The input current $i_P = 0$, so there is no voltage across the source resistance R_S. Applying KVL, we have the input condition $v_P = v_S$. The ideal OP AMP model requires $v_P = v_N$, so we conclude that $v_O = v_S$. By inspection, the closed-loop gain is $K = 1$. Since the output exactly equals the input, we say that the output follows the input (hence the name *voltage follower*).

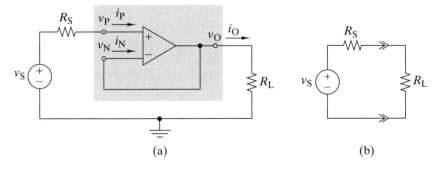

(a)   (b)

FIGURE 4–36 (a) Source-load interface with a voltage follower. (b) Interface without the voltage follower.

The voltage follower is used in interface circuits because it isolates the source from the load. Note that the input-output relationship $v_O = v_S$ does not depend on the source or load resistance. When the source is connected directly to the load, as in Figure 4–36(b), the voltage delivered to the load depends on R_S and R_L. The source and load interaction limits the signals that can be transferred across the interface, as discussed in Chapter 3. When the voltage follower is inserted between the source and load, the signal levels are limited by the capability of the OP AMP.

By Ohm's law, the current delivered to the load is $i_O = v_O/R_L$. But since $v_O = v_S$, the output current can be written in the form

$$i_O = v_S/R_L$$

Applying KCL at the reference node, we discover an apparent dilemma:

$$i_P = i_O$$

For the ideal model $i_P = 0$, but the preceding equations say that i_O cannot be zero unless v_S is zero. It appears that KCL is violated.

The dilemma is resolved by noting that the circuit diagram does not include the supply terminals. The output current comes from the power supply, not from the input. This dilemma arises only at the reference node (the ground terminal). In OP AMP circuits, as in all circuits, KCL must be satisfied. However, we must be alert to the fact that a KCL equation at the reference node could yield misleading results because the power supply terminals are not usually included in circuit diagrams.

APPLICATION   **E X A M P L E   4 – 1 4**

Digital (or even analog) Multimeters (DMMs) are ubiquitous engineering tools. They can be purchased for as little as $5 to as much as several thousand dollars. Most popular models are in the range of $100 to $200 and are hand-held portables. Clearly there must be differences to warrant such a wide range of prices. These differences are durability, accuracy, functions, input resistance, and battery or plug-in powered, to mention a few. For this discussion let us focus on input resistance.

Consider the circuit of Figure 4–37(a). A particular DMM is used to measure the voltage across the 10-MΩ resistor. The anticipated voltage was 8 V. However, the meter reads 6 V. The meter has an internal resistance that is in parallel with the resistor that is being measured and is significantly altering the reading. One

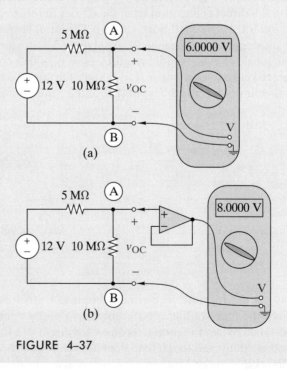

FIGURE 4–37

can calculate the internal resistance of the meter R_M using voltage division as follows

$$6 = \frac{(R_M \| 10\,M)}{(R_M \| 10\,M) + 5\,M} 12$$

Where

$$R_M \| 10\,M = \frac{10\,M \times R_M}{10\,M + R_M}$$

Solving for R_M, we find that $R_M = 10\,M\Omega$.

The reason for the problem is that the current splits at node A sending some through the DMM. In theory, with a perfect DMM all of the current would flow through the load.

The problem can be solved by inserting a follower in front of the DMM as shown in Figure 4–37(b). Better DMMs have a follower in the front end to permit them to measure voltages across high resistances. The down side is that a follower requires power to operate. It should be noted that most voltage measurements are made across resistors significantly smaller than 10 MΩ, negating the need for a follower.

Exercise 4–17

A DMM with a known internal resistance of 12.5 MΩ is used to measure voltages across several resistors in the circuit shown in Figure 4–38. What voltage will be measured on the DMM across each resistor?

FIGURE 4–38

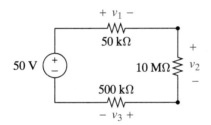

Answers:

RESISTOR	EXPECTED VOLTAGE	MEASURED VOLTAGE	PERCENT ERROR
50 kΩ	0.237 V	0.236V	0.41%
10 MΩ	47.4 V	45.5 V	4.01%
500 kΩ	2.37 V	2.28 V	3.68%

Exercise 4–18

The circuits in Figure 4–36 have $v_S = 1.5\,V$, $R_S = 2\,k\Omega$, and $R_L = 1\,k\Omega$. Compute the maximum power available from the source. Compute the power absorbed by the load resistor in the direct connection in Figure 4–36(b) and in the voltage follower circuit in Figure 4–36(a). Discuss any differences.

Answers: 281 μW; 250 μW; 2250 μW

DISCUSSION: *With the direct connection, the power delivered to the load is less than the maximum power available. With the voltage follower circuit, the power delivered to the load is greater than the maximum value specified by the maximum power transfer theorem. However, the maximum power transfer theorem does not apply to the voltage follower circuit since the load power comes from the OP AMP power supply rather than the signal source.*

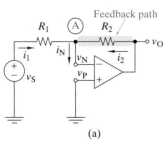

(a)

(b)

FIGURE 4–39 *The inverting amplifier circuit.*

THE INVERTING AMPLIFIER

The circuit in Figure 4–39 is called an **inverting amplifier**. The key feature of this circuit is that the input signal and the feedback are both applied at the inverting input. Since the noninverting input is grounded, we have $v_P = 0$, an observation we will use shortly. The sum of currents entering node A can be written as

$$\frac{v_S - v_N}{R_1} + \frac{v_O - v_N}{R_2} - i_N = 0 \qquad (4\text{–}24)$$

The element constraints for the OP AMP are $v_P = v_N$ and $i_P = i_N = 0$. Since $v_P = 0$, it follows that $v_N = 0$. Substituting the OP AMP constraints into Eq. (4–24) and solving for the input-output relationship yields

$$v_O = -\left(\frac{R_2}{R_1}\right) v_S \qquad (4\text{–}25)$$

This result is of the form $v_O = K v_S$, where K is the closed-loop gain. However, in this case the voltage gain $K = -R_2/R_1$ is negative, indicating a signal inversion (hence the name *inverting amplifier*). We use the block diagram symbol in Figure 4–39(b) to indicate either the inverting or the noninverting OP AMP configuration.

Exercise 4–19

The switch in Figure 4–40 moves from A to B. What is the output voltage v_O when the switch is in position A and in position B?

FIGURE 4–40

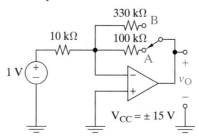

Answers: Switch in position A, $v_O = -10$ V.
Switch in position B, $v_O = -15$ V, because the OP AMP is saturated.

● Design Exercise 4–20

A 2-mV signal v_S needs to be amplified by a gain of $-450 \pm 10\%$ using standard 10% resistors from the inside back cover. Design an appropriate circuit to amplify the signal.

Answer: Because the gain needs to be negative, an inverting amplifier like that shown in Figure 4–39(a) is used with $R_2 = 470\,\text{k}\Omega$ and $R_1 = 1\,\text{k}\Omega$. This produces a gain of -470, well within the 10% tolerance.

The OP AMP constraints mean that the input current i_1 in Figure 4–39(a) is

$$i_1 = \frac{v_S - v_N}{R_1} = \frac{v_S}{R_1}$$

This, in turn, shows that the input resistance seen by the source v_S is

$$R_{IN} = \frac{v_S}{i_1} = R_1 \qquad (4\text{–}26)$$

In other words, the inverting amplifier has a finite input resistance determined by the external resistor R_1.

The next example shows that the finite input resistance must be taken into account when analyzing circuits with OP AMPs in the inverting amplifier configuration.

EXAMPLE 4–15

Find the input-output relationship of the circuit in Figure 4–41(a).

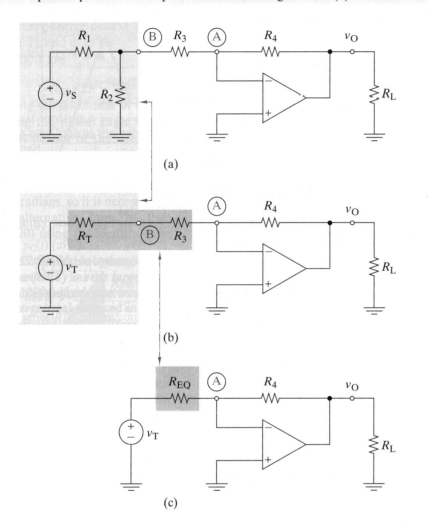

FIGURE 4–41

SOLUTION:
The circuit to the right of node B is an inverting amplifier. The load resistance R_L has no effect on the circuit transfer characteristics since the OP AMP has zero output resistance. However, the source circuit to the left of node B is influenced by the input resistance of the inverting amplifier circuit. The effect can be seen by constructing a Thévenin equivalent of the circuit to the left of node B, as shown in Figure 4–41(b). By inspection of Figure 4–41(a),

$$v_T = \frac{R_2}{R_1 + R_2} v_S$$

$$R_T = \frac{R_1 R_2}{R_1 + R_2}$$

In Figure 4–41(b) this Thévenin resistance is connected in series with the input resistor R_3, yielding the equivalence resistance $R_{EQ} = R_T + R_3$ shown in Figure 4–41(c). This reduced circuit is in the form of an inverting amplifier, so we can write the input-output relationship relating v_O and v_T as

$$K_1 = \frac{v_O}{v_T} = -\frac{R_4}{R_{EQ}} = -\frac{R_4(R_1 + R_2)}{R_1 R_2 + R_1 R_3 + R_2 R_3}$$

The overall input-output relationship from the input source v_S to the OP AMP output v_O is obtained by writing

$$K_{CIRCUIT} = \frac{v_O}{v_S} = \left[\frac{v_O}{v_T}\right]\left[\frac{v_T}{v_S}\right]$$

$$= -\left[\frac{R_4(R_1 + R_2)}{R_1 R_2 + R_1 R_3 + R_2 R_3}\right]\left[\frac{R_2}{R_1 + R_2}\right]$$

$$= -\left[\frac{R_2 R_4}{R_1 R_2 + R_1 R_3 + R_2 R_3}\right]$$

It is important to note that the overall gain is *not* the product of the source circuit voltage gain $R_2/(R_1 + R_2)$ and the inverting amplifier gain $-R_4/R_3$. In this circuit the two building blocks interact because the input resistance of the inverting amplifier circuit loads the source circuit. ∎

Exercise 4–21

Find v_O in Figure 4–42 when $v_s = 2\,V$. Repeat for $v_s = -4\,V$ and $v_s = 6\,V$.

FIGURE 4–42

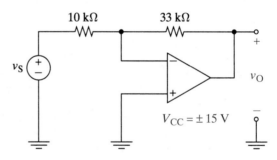

Answers: $-6.6\,V$; $13.2\,V$; $-15\,V$—the OP AMP is saturated.

THE SUMMING AMPLIFIER

The **summing amplifier** or **adder** circuit is shown in Figure 4–43(a). This circuit has two inputs connected at node A, which is called the **summing point**. Since the

FIGURE 4–43 *The inverting summer.*

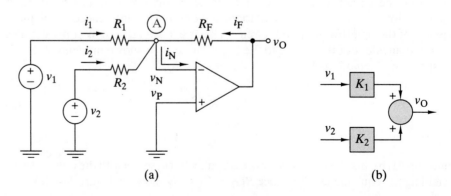

(a)   (b)

noninverting input is grounded, we have the condition $v_P = 0$. This configuration is similar to the inverting amplifier, so we start by applying KCL to write the sum of currents entering the node A summing point.

$$\frac{v_1 - v_N}{R_1} + \frac{v_2 - v_N}{R_2} + \frac{v_O - v_N}{R_F} - i_N = 0 \tag{4–27}$$

With the noninverting input grounded, the OP AMP element constraints are $v_N = v_P = 0$ and $i_N = 0$. Substituting these OP AMP constraints into Eq. (4–27), we can solve for the circuit input-output relationship.

$$v_O = \left(-\frac{R_F}{R_1}\right)v_1 + \left(-\frac{R_F}{R_2}\right)v_2$$

$$= (K_1)v_1 + (K_2)v_2 \tag{4–28}$$

The output is a weighted sum of the two inputs. The scale factors (or gains, as they are called) are determined by the ratio of the feedback resistor R_F to the input resistor for each input: that is, $K_1 = -R_F/R_1$ and $K_2 = -R_F/R_2$. In the special case $R_1 = R_2 = R$, Eq. (4–28) reduces to

$$v_O = -\frac{R_F}{R}(v_1 + v_2)$$

In this special case the output is proportional to the sum of the two inputs (hence the name *summing amplifier* or, more precisely, *inverting summer*). A block diagram representation of this circuit is shown in Figure 4–43(b).

The summing amplifier in Figure 4–43 has two inputs, so there are two gains to contend with, one for each input. The input-output relationship in Eq. (4–28) is easily generalized to the case of n inputs as

$$v_O = \left(-\frac{R_F}{R_1}\right)v_1 + \left(-\frac{R_F}{R_2}\right)v_2 + \cdots + \left(-\frac{R_F}{R_n}\right)v_n$$

$$= K_1v_1 + K_2v_2 + \cdots + K_nv_n \tag{4–29}$$

where R_F is the feedback resistor and $R_1, R_2, \ldots, R_n$ are the input resistors for the n input voltages $v_1, v_2, \ldots, v_n$. You can easily verify this result by expanding the KCL sum in Eq. (4–27) to include n inputs, invoking the OP AMP constraints, and then solving for v_O.

Exercise 4–22

In Figure 4–43, $v_1 = 0.6\,\text{V}$, $v_2 = 0.4\,\text{V}$, $R_1 = 3.3\,\text{k}\Omega$, $R_2 = 4.7\,\text{k}\Omega$, and $R_F = 15\,\text{k}\Omega$. Find v_O.

Answer: $v_O = -4.0\,\text{V}$

◖ DESIGN EXAMPLE 4–16

Design an inverting summer that implements the input-output relationship

$$v_O = -(5v_1 + 13v_2)$$

SOLUTION:
The design problem involves selecting the input and feedback resistors so that

$$\frac{R_F}{R_1} = 5 \quad \text{and} \quad \frac{R_F}{R_2} = 13$$

One solution is arbitrarily to select $R_F = 65\,\text{k}\Omega$, which yields $R_1 = 13\,\text{k}\Omega$ and $R_2 = 5\,\text{k}\Omega$. The resulting circuit is shown in Figure 4–44(a). The design can be modified to use standard resistance values for resistors with $\pm 5\%$ tolerance (see inside back cover). Selecting the standard value $R_F = 56\,\text{k}\Omega$ requires $R_1 = 11.2\,\text{k}\Omega$ and $R_2 = 4.31\,\text{k}\Omega$. The nearest standard values are $11\,\text{k}\Omega$ and $4.3\,\text{k}\Omega$. The resulting circuit shown in Figure 4–44(b) incorporates standards value resistors and produces gains of $K_1 = -56/11 = -5.09$ and $K_2 = -56/4.3 = -13.02$. These nominal gains are within 2% of the values in the specified input-output relationship.   ■

FIGURE 4–44

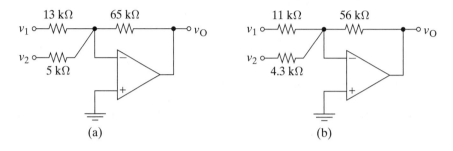

(a)   (b)

Exercise 4–23

(a) Find v_O in Figure 4–44(a) when $v_1 = 2\,\text{V}$ and $v_2 = -0.5\,\text{V}$.

(b) If $v_1 = 400\,\text{mV}$ and $V_{CC} = \pm 15\,\text{V}$, what is the maximum value of v_2 for linear mode operation?

(c) If $v_1 = 500\,\text{mV}$ and $V_{CC} = \pm 15\,\text{V}$, what is the minimum value of v_2 for linear mode operation?

Answers:
(a) $-3.5\,\text{V}$; (b) 1 V; (c) $-1.346\,\text{V}$

APPLICATION   E X A M P L E   4 – 1 7

Inverting amplifiers of the type discussed above have the ability to be easily designed with outputs being the weighted sum of the various inputs. While this type of OP AMP Summer is the most common by far, there is occasionally the need for a non-inverting summer. In this example we analyze a non-inverting summer and discuss the advantages and disadvantages of this summer versus the inverting summer.

Consider the circuit of Figure 4–45 and find the input-output relationship.

One should recognize the shaded area of the figure as being that of a non-inverting amplifier with gain $K = \left[\dfrac{R_B + R_A}{R_A}\right]$, and $v_O = K \times v_P$, Where v_P acts as the summing terminal of the circuit as well as the non-inverting terminal of the OP AMP.

FIGURE 4–45

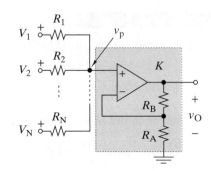

We can find the voltage at v_p using superposition. For the contribution of v_1 we set all sources except v_1 to zero and solve for the voltage v_{p1}

$$v_{P1} = v_1\left(\frac{R_2\|R_3\|\cdots R_N}{R_1 + R_2\|R_3\|\cdots R_N}\right) = v_1 K_1$$

Similarly for the contribution of v_2

$$v_{P2} = v_2\left(\frac{R_1\|R_3\|\cdots R_N}{R_2 + R_1\|R_3\|\cdots R_N}\right) = v_2 K_2$$

And so forth, until all input sources have been accounted for. Then using the additive property of linear circuits, we obtain the final result

$$v_P = v_{P1} + v_{P2} + \ldots v_{PN}$$

$$v_P = v_1 K_1 + v_2 K_2 + \ldots v_N K_N$$

$$v_o = K v_P = K(v_1 K_1 + v_2 K_2 + \ldots v_N K_N)$$

While this appears to be a simple result, calculating weighted gains for each input is not so simple, especially if there are more than two or three inputs. If all the inputs are to have the same gain then selecting all source resistors to be equal, makes the circuit useful and easy to use. For example, consider a three-input non-inverting summer with all source resistors equal. Using our previous result for N = 3, we get

$$K_1 = \frac{R_2\|R_3}{R_1 + R_2\|R_3} = \frac{R/2}{R + R/2} = \frac{1}{3}$$

$$K_2 = \frac{R_1\|R_3}{R_2 + R_1\|R_3} = \frac{R/2}{R + R/2} = \frac{1}{3}$$

$$K_3 = \frac{R_1\|R_2}{R_3 + R_1\|R_2} = \frac{R/2}{R + R/2} = \frac{1}{3}$$

So that our final result is

$$v_O = K\left(\frac{v_1}{3} + \frac{v_2}{3} + \frac{v_3}{3}\right) = \frac{K}{3}(v_1 + v_2 + v_3)$$

The designer can choose the OP AMP gain K to produce the overall gain desired. If, for example, one wanted an overall gain of 10, one would choose K = 30, resulting in

$$v_O = \frac{30}{3}(v_1 + v_2 + v_3) = 10(v_1 + v_2 + v_3)$$

In summary, the non-inverting summer has the advantage of being able to sum multiple inputs with all positive gains using a single OP AMP. Its main disadvantage is that the design process of selecting resistors to achieve specific gain values can be significantly more complicated.

Exercise 4–24

Design a non-inverting summer for four inputs with equal gains of 50.

Answers: Select all source resistors equal and then make the gain of the OP AMP 200.

THE DIFFERENTIAL AMPLIFIER

The circuit in Figure 4–46(a) is called a **differential amplifier** or **subtractor**. Like the summer, this circuit has two inputs, one applied at the inverting input and one at the noninverting input of the OP AMP. The input-output relationship can be obtained using the superposition principle.

FIGURE 4–46 The differential amplifier.

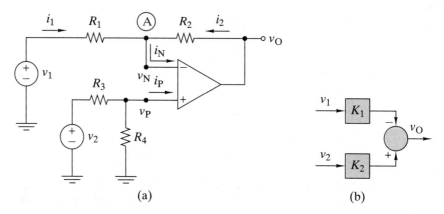

(a)   (b)

First, we turn off source v_2, in which case there is no excitation at the noninverting input and $v_P = 0$. In effect, the noninverting input is grounded and the circuit acts like an inverting amplifier with the result that

$$v_{O1} = -\frac{R_2}{R_1} v_1 \tag{4–30}$$

Next, turning v_2 back on and turning v_1 off, we see that the circuit looks like a noninverting amplifier with a voltage divider connected at its input. This case was treated in Example 4–13, so we can write

$$v_{O2} = \left[\frac{R_4}{R_3 + R_4}\right]\left[\frac{R_1 + R_2}{R_1}\right] v_2 \tag{4–31}$$

Using superposition, we add outputs in Eqs. (4–30) and (4–31) to obtain the output with both sources on:

$$v_O = v_{O1} + v_{O2}$$

$$= -\left[\frac{R_2}{R_1}\right] v_1 + \left[\frac{R_4}{R_3 + R_4}\right]\left[\frac{R_1 + R_2}{R_1}\right] v_2 \tag{4–32}$$

$$= [-K_1]v_1 + [K_2]v_2$$

where $-K_1$ and K_2 are the inverting and noninverting gains. Figure 4–46(b) shows how the differential amplifier is represented in a block diagram.

For the special case of $R_3/R_1 = R_4/R_2$, Eq. (4–32) reduces to

$$v_O = \frac{R_2}{R_1}(v_2 - v_1) \tag{4–33}$$

In this case the output is proportional to the difference between the two inputs (hence the name *differential amplifier* or *subtractor*).

Exercise 4–25

(a) Find the input-output relationship of the subtractor circuit in Figure 4–47.

(b) If $v_{CC} = \pm 15V$ and $v_1 = 3\,V$, what is the allowable range of v_2 for linear operation of the OP AMP?

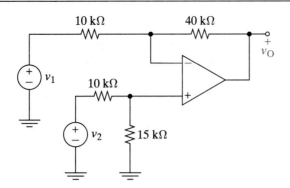

FIGURE 4–47

Answers:
(a) $v_O = -4v_1 + 3v_2$; (b) $-1\,\text{V} \le v_2 \le 9\,\text{V}$

Circuit	Block diagram	Gains
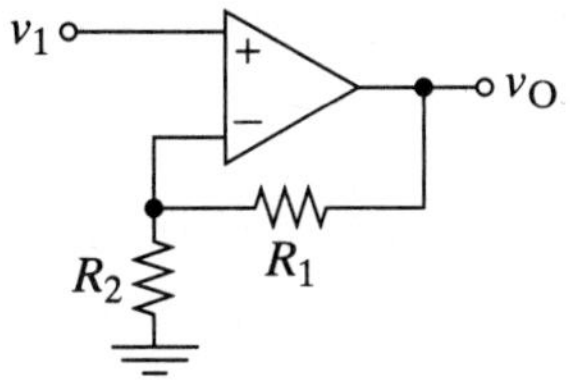 v_1 — op amp with R_2, R_1 → v_O	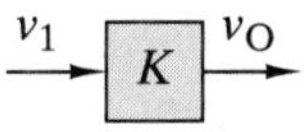 $v_1 \to \boxed{K} \to v_O$ NONINVERTING	$K = \dfrac{R_1 + R_2}{R_2}$
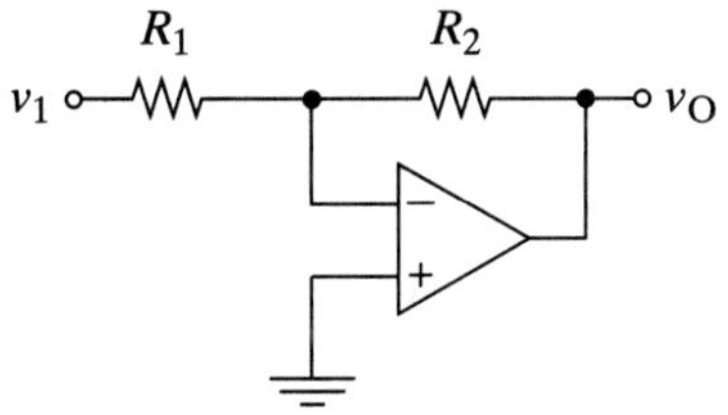 v_1 — R_1, R_2 → v_O (inverting)	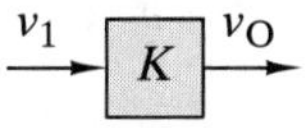 $v_1 \to \boxed{K} \to v_O$ INVERTING	$K = -\dfrac{R_2}{R_1}$
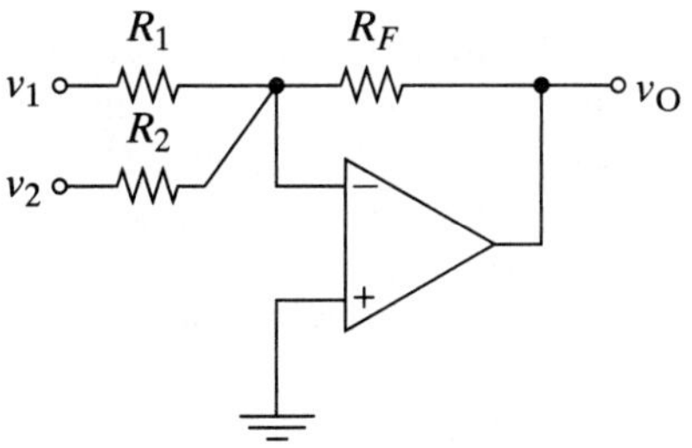 v_1 — R_1; v_2 — R_2; R_F → v_O	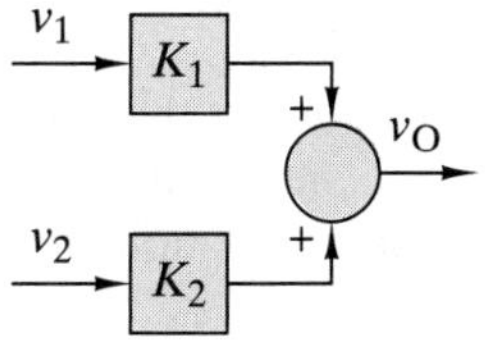 $v_1 \to \boxed{K_1} \xrightarrow{+}$; $v_2 \to \boxed{K_2} \xrightarrow{+} \to v_O$ SUMMER	$K_1 = -\dfrac{R_F}{R_1}$ $K_2 = -\dfrac{R_F}{R_2}$
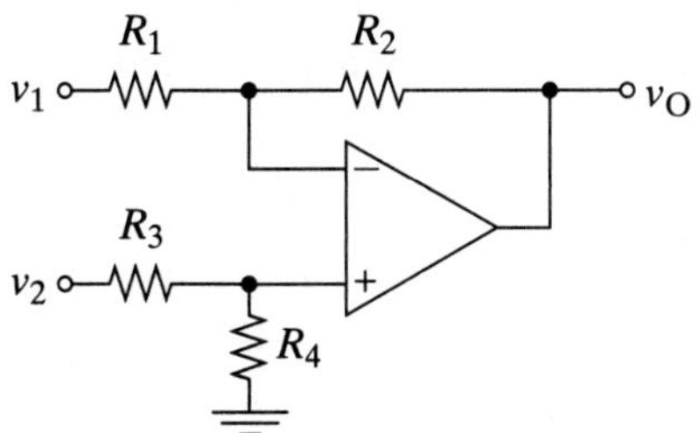 v_1 — R_1, R_2; v_2 — R_3, R_4 → v_O	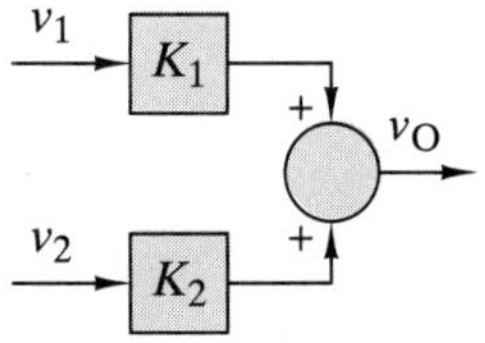 $v_1 \to \boxed{K_1} \xrightarrow{+}$; $v_2 \to \boxed{K_2} \xrightarrow{+} \to v_O$ SUBTRACTOR	$K_1 = -\dfrac{R_2}{R_1}$ $K_2 = \left(\dfrac{R_1 + R_2}{R_1}\right)\left(\dfrac{R_4}{R_3 + R_4}\right)$

FIGURE 4–48   *Summary of basic OP AMP signal-processing circuits.*

BASIC OP AMP BUILDING BLOCKS

The block diagram representations of the basic OP AMP circuit configurations are shown in Figure 4–48. The noninverting and inverting amplifiers are represented as gain blocks. The summing amplifier and differential amplifier require both gain blocks and the summing point symbol. Considerable care must be used when translating from a block diagram to a circuit, or vice versa, since some gain blocks involve negative gains. For example, the gains of the inverting summer are negative. The required minus sign is sometimes moved to the summing point and the value of K within the gain block changed to a positive number. Since there is no standard convention for doing this, it is important to keep track of the signs associated with gain blocks and summing points.

The OP AMP building blocks in Figure 4–48 can be interconnected to obtain complex signal-processing functions. These interconnects do not change the input-output relationships of each block, provided the connections are all between outputs and inputs. Each of the building blocks in Figure 4–48 is a feedback circuit with an OP AMP output. These feedback circuits have insignificant output resistances and can drive any load within the OP AMP's output current capacity.[5] In other words, the building block outputs act like ideal voltage sources just like the voltage sources connected to the block inputs. This observation leads to the following conclusion:

> *Connecting the output of one building block circuit to an input of another does not change the signal-processing function performed by either circuit.*

This property allows us to find the function performed by an interconnection using the functions performed by the individual building blocks. Conversely, this property allows us to design an interconnection by breaking a required function down into separate building block functions.

EXAMPLE 4–18

Derive an expression for v_O in Figure 4–49(a) in terms of the two inputs. Draw a block diagram representative of the circuit.

SOLUTION:
The circuit is an interconnection of two basic building blocks: a three-input summer and a noninverting amplifier. The circuit meets the connection requirement since building block outputs are connected to other building block inputs. The node voltage v_A in the figure is the output of the summer. The summer inputs are a fixed 5-V source, a signal source v_S and the noninverting amplifier output v_O Using the inverting summer input-output relationship in Eq. (4–29), we have

$$v_A = -5 - 2v_S - \frac{1}{2} v_O$$

The summer output v_A is the input to the noninverting amplifier whose voltage gain is $K = 3$. Using the input-output relationship of the noninverting amplifier, we have

$$v_O = 3v_A$$

$$= -15 - 6v_S - \frac{3}{2} v_O$$

---

[5]Maximum OP AMP output currents are typically around 20 mA and generally range from 1 to 100 mA.

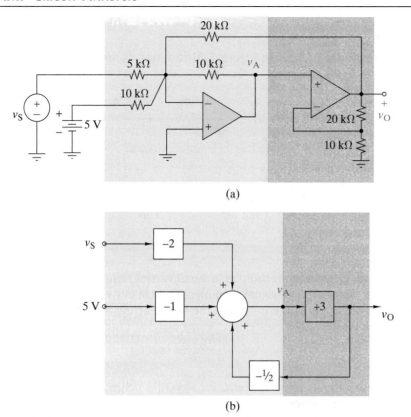

FIGURE 4–49 *(a) Circuit.*
(b) Block diagram.

(a)

(b)

Solving for v_O yields

$$v_O = -6 - 2.4\,v_S$$

The signal-processing function on the interconnection was found using the input-output relationships of the individual building blocks. The method even works with a feedback path because the 20-kΩ resistor is connected from the noninverting amplifier output to an input of the inverting summer. The block diagram is shown in Figure 4–49(b). ■

Exercise 4–26

Derive an expression for v_O in Figure 4–50 in terms of the inputs v_1 and v_2.

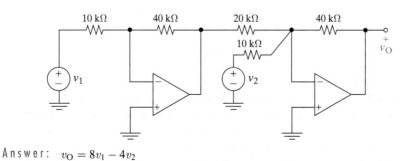

FIGURE 4–50

Answer: $v_O = 8v_1 - 4v_2$

EXAMPLE 4–19

Derive an expression for v_O in Figure 4–51 in terms of the two inputs v_1 and v_2.

FIGURE 4–51

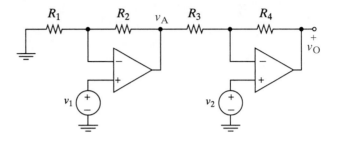

SOLUTION:

The circuit is an interconnection of a noninverting amplifier and an inverting amplifier with an additional signal v_2 applied at its noninverting input. The voltage v_A shown in the figure is the output of the noninverting amplifier:

$$v_A = \left[\frac{R_1 + R_2}{R_1}\right] v_1$$

We use superposition to find v_O. First, set $v_2 = 0$, which connects the noninverting input of the second stage to ground. In this case the second stage acts like an inverting amplifier whose input is v_A and whose output is the response due to v_1 acting alone:

$$v_{O1} = \left[-\frac{R_4}{R_3}\right] v_A$$
$$= \left[-\frac{R_4}{R_3}\right]\left[\frac{R_1 + R_2}{R_1}\right] v_1$$

Next, set $v_1 = 0$ which sets $v_A = 0$ in turn. In effect this connects resistor R_3 to ground. In this case the second stage acts like a noninverting amplifier whose input is v_2 and whose output is the response due to v_2 acting alone:

$$v_{O2} = \left[\frac{R_3 + R_4}{R_3}\right] v_2$$

Applying superposition, the total output is

$$v_O = v_{O1} + v_{O2}$$
$$= -\left[\frac{R_4}{R_3}\right]\left[\frac{R_1 + R_2}{R_1}\right] v_1 + \left[\frac{R_3 + R_4}{R_3}\right] v_2$$
$$= [-K_1]v_1 + [K_2]v_2$$

where $-K_1$ and K_2 are the inverting and noninverting gains, respectively.

The dual OP AMP circuit in Figure 4–51 performs the same signal-processing function as the single OP AMP subtractor circuit in Fig. 4–48. Why use two OP AMPs to obtain a function that can be achieved using only one? The answer is that both input signals in Figure 4–51 are applied to noninverting OP AMP inputs that have very high input resistances. This means that the two–OP AMP subtractor does not load the input signal sources. The basic subtractor in Figure 4–48 has finite input resistances that may load the input signal sources. This difference could be important when the input signal sources have high Thévenin resistances. ■

◗ Design Exercise 4–27

Using the circuit and analysis shown in Example 4–19, design a circuit using standard 5% resistors from the inside back cover that realizes the expression $v_O = -20v_1 + 10v_2$.

Answer: $R_1 = 9.1\,\text{k}\Omega$, $R_2 = 11\,\text{k}\Omega$, $R_3 = 1\,\text{k}\Omega$, and $R_4 = 9.1\,\text{k}\Omega$. Other solutions are possible.

NODE-VOLTAGE ANALYSIS WITH OP AMPS

There are many useful OP AMP circuits that are not simply interconnections of basic building blocks. In such cases we use a modified form of node-voltage analysis that is based on the OP AMP connections in Figure 4–52. The overall circuit contains N nodes, including the three associated with the OP AMP. Normally the objective is to find the OP AMP output voltage v_O relative to the reference node (ground). We assign node voltage variables to the $N - 1$ nonreference nodes, including a variable at the OP AMP output. However, an ideal OP AMP acts like a dependent voltage source connected between the output terminal and ground. As a result, the OP AMP output voltage is determined by the other node voltages, so we do not need to write a node equation at the OP AMP output node.

FIGURE 4–52 *General OP AMP circuit analysis.*

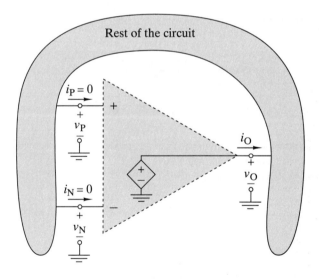

We formulate node equations at the other $N - 2$ nonreference nodes in the usual way. Since there are $N - 1$ node voltages, we seem to have more unknowns than equations. However, the OP AMP forces the condition $v_P = v_N$ in Figure 4–52. This eliminates one unknown node voltage since these two nodes are forced to have identical voltages. Finally, remember that the ideal OP AMP draws no current at its inputs ($i_P = i_N = 0$) in Figure 4–52, so these currents can be ignored when formulating node equations.

The following steps outline an approach to the formulation of node equations for OP AMP circuits:

STEP 1   Identify a node voltage at all nonreference nodes, including OP AMP outputs, but do *not* formulate node equations at the OP AMP output nodes.

STEP 2 Formulate node equations at the remaining nonreference nodes and then use the ideal OP AMP voltage constraint $v_P = v_N$ to reduce the number of unknowns.

EXAMPLE 4–20

Derive an expression for v_O in Figure 4–53 in terms of the inputs v_1 and v_2.

SOLUTION:

This circuit is not an interconnection of OP AMP building blocks because the resistor R_2 is connected between two inputs. We use the node-voltage method outlined above to find the required input-output relationship.

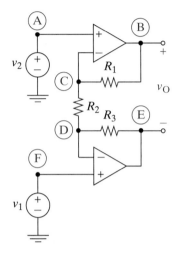

FIGURE 4–53

STEP 1 The circuit has a total of six nonreference nodes as shown in the figure. Nodes B and E are OP AMP outputs while nodes A and F are connected to ground by the input voltage sources v_2 and v_1, respectively. As a result, we only need node equations at C and D. Writing the sum of currents leaving these nodes

$$\text{Node C}: \quad \frac{v_C - v_B}{R_1} + \frac{v_C - v_D}{R_2} = 0$$

$$\text{Node D}: \quad \frac{v_D - v_C}{R_2} + \frac{v_D - v_E}{R_3} = 0$$

yields two equations in the four node voltages v_B, v_C, v_D, and v_E.

STEP 2 The noninverting OP AMP inputs are connected to independent voltages sources v_2 and v_1. The OP AMP voltage constraints $(v_P = v_N)$ mean that $v_D = v_1$ and $v_C = v_2$. These constraints eliminate v_C and v_D as unknowns and reduce the two node equations to

$$\text{Node C}: \quad \frac{v_2 - v_B}{R_1} + \frac{v_2 - v_1}{R_2} = 0$$

$$\text{Node D}: \quad \frac{v_1 - v_2}{R_2} + \frac{v_1 - v_E}{R_3} = 0$$

The node-voltage formulation method outlined above leads to two equations in the two unknown node voltages v_B and v_E.

We solve the node C equation for v_B

$$v_B = v_2 + \frac{R_1}{R_2}(v_2 - v_1)$$

and the node D equation for v_E

$$v_E = v_1 + \frac{R_3}{R_2}(v_1 - v_2)$$

Using the fundamental property of node voltages yields the required output as

$$v_O = v_B - v_E = \left[\frac{R_1 + R_2 + R_3}{R_2}\right](v_2 - v_1)$$

The circuit in Figure 4–53 is a differential amplifier of the form $v_O = K(v_2 - v_1)$ in which the input voltages v_1 and v_2 are applied at noninverting OP AMP inputs that have very high input resistances. The circuit is the first stage of a commercially available integrated circuit called an instrumentation amplifier. We will use this circuit later. ■

◀ D Design Exercise 4–28 _____

Select values of R_1, R_2, and R_3 in Figure 4–53 so that $v_O = 50(v_2 - v_1)$.

Answer: Selecting $R_1 = R_3 = 24.5$ kΩ and $R_2 = 1$kΩ is one of many possible solutions.

_____

4–5 OP AMP Circuit Design

This section is dedicated to OP AMP circuit *design*. Unlike circuit *analysis*, where we are asked to find the single correct input-output relationship for a given circuit configuration, circuit design asks us to create a circuit that will realize a desired input-output relationship. Successful OP AMP circuit designs are accomplished by interconnecting the various building blocks studied (inverter, noninverter, follower, summer, subtractor, and more to come). The design process is greatly simplified by the nearly one-to-one correspondence between the actual OP AMP circuits and their building-block equivalents. However, unlike analysis problems, design problems rarely have a unique solution since there are usually several OP AMP circuits that can meet the design objective. This prompts the question, which is the better or best design solution? Sometimes the choice is obvious; at other times it is not. Making the right choice is the highest level of learning—a key objective of every nascent engineer—and is called *evaluation* (the art of making smart engineering decisions). We have already seen some application of this skill in Chapter 3 with respect to choosing the better interface. Engineers are often faced with selecting the best solution from several possibilities. In making their recommendations, engineers consider many factors, including economic considerations (purchase, maintenance, disposal, and replacement costs), environmental issues (power used, recyclability, pollution produced), ergonomics and aesthetics, reliability (durability, maintainability, and accuracy), number of parts, uniqueness of parts, physical properties (size, shape, and weight), dependability of the vendor, and availability over time. In making a decision, it is essential that one knows the appropriate factors to consider and their respective importance to the application. In this section and the next, we will look at some of these issues.

E ⬤ **EVALUATION EXAMPLE 4–21**

In a particular application, it is necessary to implement the block diagram shown in Figure 4–54. The maximum individual OP AMP gain cannot exceed ±2000. The input resistance of first signal stage must be at least 10 kΩ. The nominal input signal v_1 is 1 μV.

FIGURE 4–54

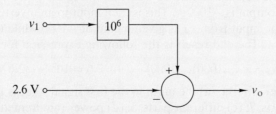

Two vendors have provided competitive solutions, shown in Figure 4–55. Choose the best solution based on achieving the desired output, cost, parts count, variety of parts needed, and power usage.

FIGURE 4–55

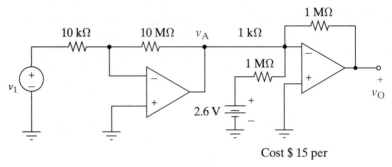

Cost $ 15 per

(a) Vendor A

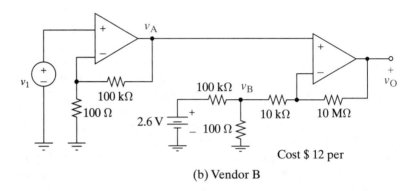

Cost $ 12 per

(b) Vendor B

SOLUTION:
Referring to the block diagram, we determine what output the circuit is supposed to deliver. We see that the input is multiplied by 10^6 and then added to -2.6 V. Hence, we must have a circuit that has the following input-output relation:

$$v_O = 10^6 v_1 - 2.6\,\text{V}$$

Also, the input signal that must be amplified is small, only around $1\,\mu\text{V}$, and therefore it needs a large gain to make it usable ($K = 10^6$). The input resistance must be at least $10\,\text{k}\Omega$, probably to avoid loading from v_1. The individual OP AMPs cannot have a gain magnitude exceeding 2000. At some point in the design, the 2.6-V source will need to be subtracted from the amplified v_1.

Let us analyze each vendor's submission to see first if the circuits meet the required specifications.

Vendor A

This vendor chose a straightforward approach, splitting the gain across two OP AMP stages. At v_A the output is $-1000v_1$. This is then input to an inverting summer along with the 2.6 V. The input from v_A is given a gain of -1000 while the 2.6-V input is given a gain of -1. The end result is the following expression for v_O:

$$v_O = v_1(-1000)(-1000) + (-1)2.6 = 10^6 v_1 - 2.6\,\text{V}$$

Checking the other criteria; (a) the input to the first signal stage is $10\,\text{k}\Omega$, right at the limit; (b) total parts, 7; (c) different parts, 5; (d) power requirement, two OP AMPs; and (e) cost, $15.

Vendor B

This vendor took a nontraditional approach. The vendor split the gain across two OP AMPs but used a noninverter for the first stage followed by a subtractor. The input to the inverting terminal seems uncommon and needs to be analyzed to see if it works. The first stage provides an output gain at v_A of 1001. We need to see what the inputs to the subtractor are before we can determine the output. We start by doing a Thévenin equivalence at v_B. We get $v_T = 2.597\,\text{mV}$ and $R_T = 100\,\Omega\|100\,\text{k}\Omega = 99.9\,\Omega$.

Next, we look at the subtractor relationship:

$$v_O = -\frac{R_2}{R_1}v_1 + \left(\frac{R_1 + R_2}{R_1}\right)\left(\frac{R_4}{R_3 + R_4}\right)v_2$$

We see that $R_2 = 10\,\text{M}\Omega$. The other resistance will take some work to calculate. R_1 is the input resistance at the inverting terminal. We see that it is 10 kΩ plus R_T, or 10,099.9 Ω. Now R_3 is the series resistance connected to the noninverting terminal, or 0 Ω, while R_4 is the resistance from the noninverting terminal to ground—an open circuit, or ∞ Ω. Substituting these values into our subtractor equation, we get

$$v_O = -\frac{10^7}{10,099.9}0.002597 + \left(\frac{10,099.9 + 10^7}{10,099.9}\right)\left(\frac{\infty}{0 + \infty}\right)(1001)v_1$$

$$v_O = -2.57 + 0.992 \times 10^6 v_1$$

This vendor's solution is close but not exact. The gain is off by 0.8% and the constant is off by 1.2%. Checking the other criteria: (a) the input to the first signal stage is into a noninverting stage or about ∞ Ω, so there are no concerns with loading; (b) total parts, 8, (c) different parts, 5, (d) power requirement two OP AMPs; and (e) cost, $12.

DISCUSSION: *Vendor A's circuit meets all of the specifications exactly, but it costs more than Vendor B's circuit. Vendor B's solution, on the other hand, is not exact, but the errors are small. Vendor B uses one extra part, but three of the parts are the same, while Vendor A's circuit has only two parts that are the same. Both use two OP AMPs and, since the power supplies to the OP AMPs usually are the largest sources of power usage, they are equal on this criterion. The bottom line: if the number of circuits being purchased is small, then Vendor A provides the simpler and most accurate design. If the number is large so that cost becomes a significant factor and the small errors can be tolerated, then choose Vendor B.* ∎

Exercise 4–29 _____

Verify the solutions found in Evaluation Example 4–21 using OrCAD.

Answer: Figure 4–56 displays the results.

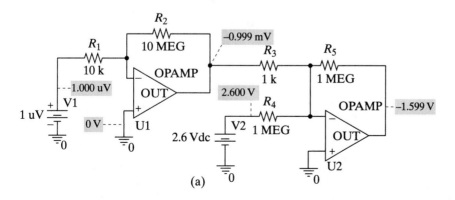

FIGURE 4–56

(a)

FIGURE 4–56 *(Continued)*

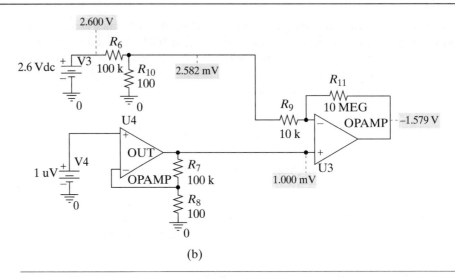

(b)

D E DESIGN AND EVALUATION EXAMPLE 4 – 22

Design the interface circuit in Figure 4–57 so that 200 mW is delivered to the 500-Ω load.

SOLUTION:
The maximum power available from the 5-V source is

$$p_{\text{MAX}} = \frac{v_T^2}{4R_T} = \frac{5^2}{4 \times 50} = 125 \, \text{mW}$$

which is less than the required 200-mW output. This means that the interface circuit must contain an active device that provides voltage gain. To determine the required voltage gain, we express the output power as

$$p_2 = \frac{v_2^2}{500} = 0.2 \, \text{W}$$

which yields the required output voltage as $v_2 = \pm\sqrt{0.2 \times 500} = \pm 10 \, \text{V}$. Since the input is a 5-V source, we need overall voltage gains of $K = \pm 2$. We can get these gains using either the noninverting or the inverting amplifier shown in the figure.

FIGURE 4–57

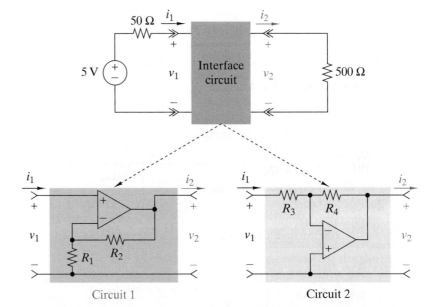

Circuit 1   Circuit 2

1st Design

The noninverting amplifier shown in the figure has very high input resistance; hence, $i_1 = 0$ and $v_1 = 5\,\text{V}$. To get a voltage gain of $K = +2$ we need $(1 + R_2/R_1) = 2$. Selecting $R_1 = R_2 = 10\,\text{k}\Omega$ completes the design. Many other choices of resistors are possible.

2nd Design

The inverting amplifier shown in the figure has an input resistance of R_3; hence, the input current is not zero, and $v_1 = 5 - 50i_1 < 5\,\text{V}$. In other words, the interface circuit loads the input source. However, if we specify that $R_3 \gg 50\,\Omega$, the effect of loading can be ignored since $i_1 \approx 0$ and the input voltage is essentially 5 V. Neglecting the effect of loading, a voltage gain of $K = -2$ requires $R_4/R_3 = 2$. Selecting $R_3 = 25\,\text{k}\Omega$ and $R_4 = 50\,\text{k}\Omega$ completes the design. Many other choices of resistors are possible as long as $R_3 \gg 50\,\Omega$. ∎

Exercise 4–30 _____

Using the circuits and analyses shown in Design Example 4–22, how much power is being provided by the signal source in each design? What is providing the power to the load?

Answers: In circuit 1 the signal source is providing zero power. In circuit 2 the signal source is providing $p_S = v_1^2/R_3\,\text{W}$. In each case, the power for the load is being supplied by the OP AMP's $\pm V_{CC}$ sources.

D E ◀ **DESIGN AND EVALUATION EXAMPLE 4 – 23** ▶

Design a circuit using OP AMPs that implement the input-output relationship

$$v_O = 5v_1 + 4v_2 - 2v_3$$

SOLUTION:
We will show two ways to solve this design problem and then discuss each design.

First Design

Rewriting the required input-output relationship as

$$v_O = -5(-v_1) - 4(-v_2) - 2v_3$$

suggests an inverting summer with three inputs $-v_1$, $-v_2$, and v_3 with summing gains of -5, -4, and -2, respectively. Figure 4–58(a) is a block diagram of this approach and Figure 4–58(b) is an OP AMP circuit that implements the block diagram. This design requires three OP AMPs and eight resistors.

Second Design

Rewriting the required input-output relationship as

$$v_O = -2[-2.5v_1 - 2v_2] - 2v_3$$

suggests an inverting summer with two inputs $[-2.5v_1 - 2v_2]$ and v_3, both with a summing gain of -2. The input $[-2.5v_1 - 2v_2]$ can be obtained using a second inverting summer whose inputs are v_1 and v_2. Figure 4–58(c) is a block diagram of this approach and Figure 4–58(d) is an OP AMP circuit that implements the block diagram. This design requires two OP AMPs and six resistors.

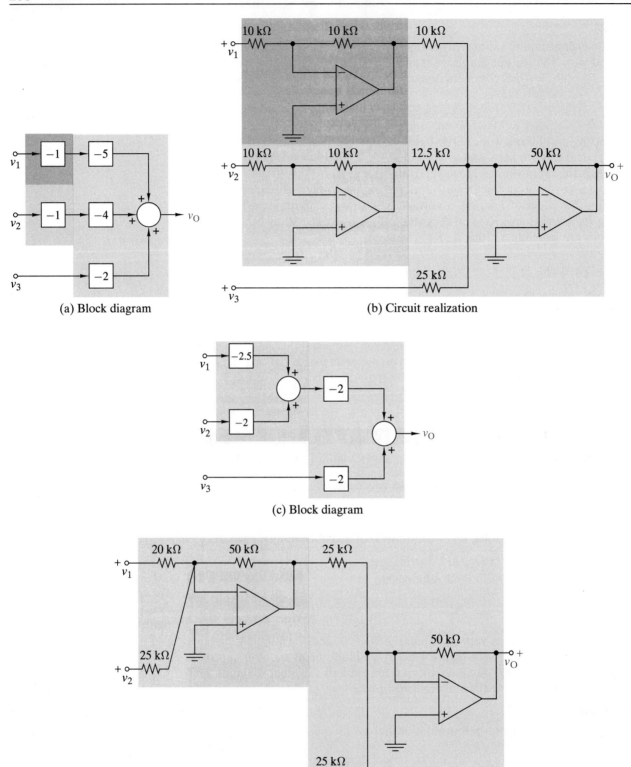

(a) Block diagram

(b) Circuit realization

(c) Block diagram

(d) Circuit realization

FIGURE 4–58

Evaluation Discussion

This example illustrates again that there are often several ways to solve a given design problem. This leads to the question of how to choose between different design solutions. Candidate designs are evaluated using criteria beyond the fact that they implement a given signal-processing function. One common measure is to choose the design that uses the fewest components. Applying this "smallest part count" criteria in this example leads us to choose the second design.

Using this criterion, it is important to understand the impact of part count on the cost of fabricating a circuit. Reducing part count does much more than simply reduce component costs. It also (and often more importantly) reduces printed circuit board space, assembly costs, packaging costs, testing, and logistics costs. ■

E Evaluation Exercise 4–31

A requirement exists for a circuit that implements the block diagram in Figure 4–59(a). The circuit in Figure 4–59(b) is a proposed solution. A breadboard prototype of this circuit failed to pass preliminary testing. Why?

FIGURE 4–59

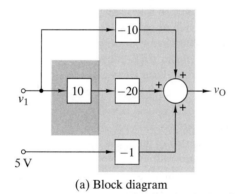

(a) Block diagram

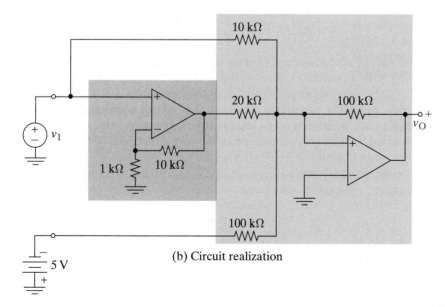

(b) Circuit realization

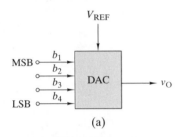

(a)

(b)

(c)

www.mikroe.com

(d)

FIGURE 4–60 *A digital-to-analog converter (DAC). (b) Ceramic thin-film/IC 8- to 20-bit devices. (c) 16-bit DAC-8734. (d) 12-bit DAC development board.*

Answers: (1) The noninverter produces a gain of 11, not the desired 10. (2) The gain from the noninverter into the inverting summer is -5, not -20. (3) The 5-V source is inverted, resulting in a final value of $+5$ V instead of the desired -5 V. (4) A common error for new students, the inverting summer has the "+" and "−" terminals reversed.

4–6 OP AMP CIRCUIT APPLICATIONS

OP AMP circuits are fundamental building blocks in a wide range of signal-processing applications, especially instrumentation, status monitoring, process control, filtering, digital-to-analog conversion, and analog-to-digital conversion. This section provides a brief introduction to three of these applications.

DIGITAL-TO-ANALOG CONVERSION

A digital-to-analog converter (DAC) has become a very common device. Its purpose is to convert digital signals—binary 1s and 0s—to an analog or continuous format. DACs are mixed-signal devices: the input to a DAC is a digital signal and the output is an analog signal. DACs are used extensively in robotics, audio applications, and communications, especially in high-definition digital television and cell phone applications. Figure 4–60 shows a simplified block diagram of this device along with photos of commercially available DACs. For the purpose of the following discussion we will focus on four-bit devices, although typical devices are generally eight-bit, 10-bit, 12-bit, or even 24-bit. The block diagram in Figure 4–60(a) has a four-bit digital input (b_1, b_2, b_3, b_4) and analog output v_O, and a fixed reference voltage V_{REF}. The input bits can each have only one of two values: a high (1) or a low (0). The input-output relationship of four-bit converter can be written as follows:

$$v_O = KV_{REF}\left(b_1 + \frac{b_2}{2} + \frac{b_3}{4} + \frac{b_4}{8}\right) \tag{4–34}$$

Bit b_1 is called the most significant bit (MSB) because it carries the largest weight in this sum. Conversely, bit b_4 is called the least significant bit (LSB) because it carries the smallest weight.

For example, a four-bit DAC with $K = 0.5$ and $V_{REF} = 10$ V and a digital input (0, 1, 0, 1) produces an analog output of

$$v_O = 0.5 \times 10 \left(0 + \frac{1}{2} + \frac{0}{4} + \frac{1}{8}\right) = 3.125 \text{ V}$$

Similarly, inputs of (0, 1, 1, 1) and (1, 1, 0, 0) produce outputs of 4.375 V and 7.5 V, respectively. With a four-bit converter, there are $2^4 = 16$ possible input codes and hence 16 possible analog output levels. The *full-scale output* of a DAC is defined as the output when the input bits are all 1 ($v_O = 9.375$ V in this example). The *resolution* of a DAC is defined as the change in the analog output caused by an input change of one LSB (0.625 V in this example). We can think of resolution as the voltage increment between adjacent output levels.

The N-bit converter generalization of the four-bit input-output relationship in Eq. (4–34) is

$$v_O = KV_{REF} \sum_{k=1}^{N} \frac{b_k}{2^{k-1}} \tag{4–35}$$

Integrated circuit DACs are available with inputs of $N = 8$ to $N = 24$ bits. Increasing the number of bits improves the conversion precision since the resolution is inversely proportional to 2^{N-1}. For this reason, DAC resolution is often somewhat loosely quoted in terms of bits.

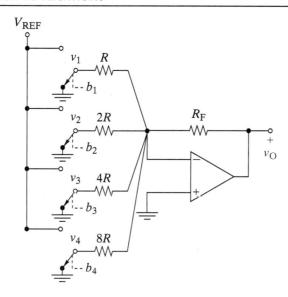

FIGURE 4–61 *A binary-weighted summer DAC.*

We now discuss two OP AMP circuits that implement Eq. (4–35) for $N = 4$. The first method uses the four-input OP AMP summer in Figure 4–61. The summer inputs v_1, v_2, v_3, and v_4 are applied to the binary-weighted input resistors R, $2R$, $4R$, and $8R$, respectively. The output of the summer is related to these inputs by the input-output relationship in Eq. (4–29).

$$v_O = -\frac{R_F}{R} v_1 - \frac{R_F}{2R} v_2 - \frac{R_F}{4R} v_3 - \frac{R_F}{8R} v_4$$

$$= -\frac{R_F}{R}\left(v_1 + \frac{v_2}{2} + \frac{v_3}{4} + \frac{v_4}{8}\right)$$

The input voltages are determined by switches controlled by the input bits b_1, b_2, b_3, and b_4. When a bit is low (0), the switch is in the lower position, connecting the related input to ground. When a bit is high (1), the switch is in the upper position, connecting the related input to the reference voltage V_{REF}. In other words, an input voltage is zero when its control bit is 0 and equal to V_{REF} when its control bit is 1. In effect, the input voltages are related to the input bits as $v_k = b_k V_{REF}$. As a result, we can write the summer output as

$$v_O = -\frac{R_F}{R}\left(b_1 V_{REF} + \frac{b_2 V_{REF}}{2} + \frac{b_3 V_{REF}}{4} + \frac{b_4 V_{REF}}{8}\right)$$

$$= -\frac{R_F}{R} V_{REF}\left(b_1 + \frac{b_2}{2} + \frac{b_3}{4} + \frac{b_4}{8}\right)$$

This result is of the form of Eq. (4–35) with $N = 4$ and $K = -R_F/R$.

A four-bit DAC can also be realized using the *R-2R* ladder circuit in Figure 4–62. In this case the voltage inputs v_1, v_2, v_3, and v_4 are applied to the $2R$ legs of the ladder. The OP AMP's noninverting input is connected to ground ($v_P = 0$). Since $v_N = v_P = 0$, node A in Figure 4–62 is a "virtual" ground. In other words, the R-$2R$ ladder is effectively shorted to ground at node A. The short-circuit current the ladder delivers to this virtual short can be shown to be

$$i_{SC} = \frac{v_1}{2R} + \frac{v_2}{4R} + \frac{v_3}{8R} + \frac{v_4}{16R}$$

FIGURE 4–62 *An R-2R ladder DAC.*

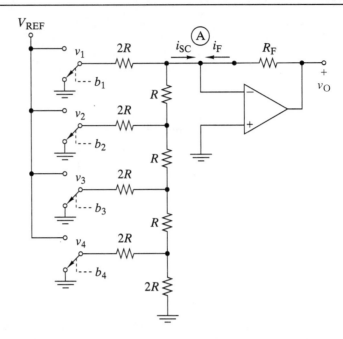

Since node A is a virtual ground, the current through the feedback resistor is $i_F = v_O/R_F$. For an ideal OP AMP, $i_N = 0$, so the KCL constraint at node A requires $i_F + i_{SC} = 0$, or

$$\frac{v_O}{R_F} + \frac{v_1}{2R} + \frac{v_2}{4R} + \frac{v_3}{8R} + \frac{v_4}{16R} = 0$$

Solving for v_O leads to the following results

$$v_O = -\frac{R_F}{2R}\,v_1 - \frac{R_F}{4R}\,v_2 - \frac{R_F}{8R}\,v_3 - \frac{R_F}{16R}\,v_4$$

$$= -\frac{R_F}{2R}\left(v_1 + \frac{v_2}{2} + \frac{v_3}{4} + \frac{v_4}{8}\right)$$

The input voltages v_1, v_2, v_3, and v_4 are defined by the same switching arrangement as the inverting summer DAC in Figure 4–61. Hence, they are related to the input bits as $v_k = b_k V_{REF}$, and the output voltage becomes

$$v_O = -\frac{R_F}{2R}\left(b_1 V_{REF} + \frac{b_2 V_{REF}}{2} + \frac{b_3 V_{REF}}{4} + \frac{b_4 V_{REF}}{8}\right)$$

$$= -\frac{R_F}{2R} V_{REF}\left(b_1 + \frac{b_2}{2} + \frac{b_3}{4} + \frac{b_4}{8}\right)$$

This result is of the form of Eq. (4–35) with $N = 4$ and $K = R_F/2R$.

In theory, the four-bit inverting summer DAC in Figure 4–61 could be extended to a larger number of bits. However, the range of input resistances rapidly gets out of hand. For example, a 10-bit DAC would require input resistances ranging from R to $1024R$. With integrated circuit technology it is virtually impossible to maintain tight resistance tolerances over a wide range of resistance values.

The four-bit R-$2R$ ladder DAC in Figure 4–62 can be extended to a larger number of bits using only two values of resistance, namely R and $2R$. The absolute value of R does not matter. What matters is that the added resistances stand in a two-to-one ratio. Controlling the ratio of two resistances is much easier to accomplish with integrated circuit technology than controlling the absolute value of a resistance. This fact accounts for the widespread use of R-$2R$ ladder architecture in integrated circuit DACs.

Exercise 4–32 _____

The $R-2R$ ladder DAC in Figure 4–62 has $R_F = 40\,\text{k}\Omega$, $R = 10\,\text{k}\Omega$, and $V_{REF} = -3\,\text{V}$. Find the full-scale output and resolution of the converter.

Answers: 11.25 V; 0.75 V

Exercise 4–33 _____

A student chooses to design a 6-bit weighted sum DAC for a project. The requirement is to have errors of 1% or less. What tolerance (accuracy) must the resistors used in the design have to meet the requirement?

Answers: Assume we can control the value for R in the design. Whatever value the designer chooses for R, the $32R$ resistor must meet the tolerance specification. Therefore, the tolerance for R is at most 0.03125%. An R-$2R$ design is probably a better option.

INSTRUMENTATION SYSTEMS

One of the most interesting and useful applications of OP AMP circuits is for instrumentation systems that collect and process data about physical phenomena. The instrumentation system typically connects an input transducer to an output transducer in a meaningful way. An *input transducer* is a device that converts some physical quantity, such as temperature, light intensity, stress, strain, pressure, rotation, acceleration, or velocity, to an electrical signal. An *output transducer* performs the opposite conversion, using electrical signals to generate a physical quantity. In an instrumentation system, an input transducer generates an electrical signal that describes some ongoing physical phenomenon. The transducer signal is processed by OP AMP circuits and provided to an output transducer for observation, recording, or, frequently, processing by an analog-to-digital converter (ADC). In the latter case, the output of an ADC is usually sent to a computer or microprocessor. In addition, the output signal can be used in a feedback control system to monitor and regulate the physical process itself.

The starting point in instrumentation system design is the transducer. In general, there are two basic types of transducers: *active transducers*, which produce a voltage or current proportional to the physical parameter being sensed, and *passive transducers*, which change a parameter, such as resistance or conductance, in proportion to the physical parameter being sensed but do not directly produce an output voltage or current. Active transducers usually produce outputs that are quite small—in the millivolt or microvolt range or in the microampere or nano-ampere range. Active transducers often need considerable amplification to make their signals useful. Passive transducers, on the other hand, often have effects that are more dramatic. A photoresistor, for example, can change the resistance over four, five, or even six orders of magnitude, but it does not produce an output on its own. Passive transducers need to have an external source to produce a voltage or current that is proportional to the physical parameter.

The block diagram in Figure 4–63 shows an instrumentation system in its simplest form. The objective of the system is to deliver an output signal, v_O, that is related to

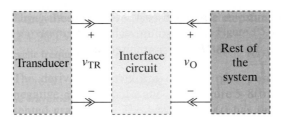

FIGURE 4–63   *Transducer interface circuit.*

the physical quantity measured by the input transducer, v_{TR}. In many instrumentation systems, the transducer is known and is assumed to be linear over the desired range of measurements. The desired output is also assumed to be linear. The task is to design the interface circuit between the input and output. Suppose the input transducer converts a physical variable x into an electrical voltage v_{TR}. For many transducers this voltage is of the form

$$v_{TR} = mx + b$$

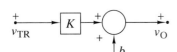

FIGURE 4–64

where m is a calibration constant and b is a constant offset or bias. With active transducers, the voltage is usually quite small and must be amplified by some gain K to meet the needs of the rest of the system. The amplified signal includes a component proportional to the physical variable, $K(mx)$, and an amplified bias component, $K(b)$. We can introduce another bias voltage, V_b, to compensate for the $K(b)$ term and yield an output voltage, v_O, which is aligned with the needs of the rest of the system. Figure 4–64 is a block diagram of the typical signal processing steps that need to be designed. To summarize, the input transducer voltage, v_{TR}, is amplified by K and then a bias voltage, V_b, is applied to make the signal the proper level for the output v_O:

$$v_O = K v_{TR} + V_b \qquad (4\text{–}36)$$

The gain K is the ratio of the desired output voltage range to the available voltage range at the input transducer:

$$K = \frac{\text{Desired range}}{\text{Available range}} \qquad (4\text{–}37)$$

A concern in calculating the desired K is getting the sign correct. In this regard, it is important to keep track of the limits of the range. To do so, it may be helpful to create a table similar to Table 4–1, where p_1 and p_2 are the physical values measured by the input transducer. The gain K is then calculated as

$$K = \frac{\text{Desired range}}{\text{Available range}} = \frac{v_O(p_1) - v_O(p_2)}{v_{TR}(p_1) - v_{TR}(p_2)}$$

T A B L E **4–1**

Physical Range	Lower Physical Value, P_1	Upper Physical Value, P_2
Desired range for v_O	$v_O(p_1)$	$v_O(p_2)$
Available range at v_{TR}	$v_{TR}(p_1)$	$v_{TR}(p_2)$

For example, if you are designing a system that will read the temperature from 0°C to 100°C and you want 0°C to produce 5 V and 100°C to produce 0 V, create a table similar to Table 4–2 where $v_{TR}(0°)$ and $v_{TR}(100°)$ are the voltage signals the input transducer produces when the temperature is 0° and 100°, respectively. The gain is then computed as

$$K = \frac{\text{Desired range}}{\text{Available range}} = \frac{\begin{matrix}\text{Physical range} & 0° & 100° \\ 5 & - & 0\end{matrix}}{v_{TR}(0°) - v_{TR}(100°)}$$

T A B L E **4–2**

PHYSICAL RANGE	0°	100°
Desired Range for v_O	$v_O(0°) = 5\,V$	$v_O(100°) = 0\,V$
Available Range at v_{TR}	$v_{TR}(0°)$	$v_{TR}(100°)$

The physical range is shown above the gain equation for reference and to help keep the voltage values aligned correctly. If you wanted 100°C to produce 5 V and 0°C to produce 0 V, you would reverse the terms on the top of the ratio:

$$K = \frac{\text{Desired range}}{\text{Available range}} = \frac{\begin{matrix}\text{Physical range} & 0° & 100° \\ 0 & - & 5\end{matrix}}{v_{TR}(0°) - v_{TR}(100°)}$$

ACTIVE TRANSDUCERS

To put this procedure into practice, we will consider the specific example of an active transducer. Figure 4–65 shows the characteristics of a light intensity transducer that

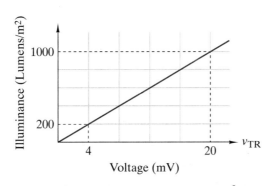

FIGURE 4–65 *Photocell transducer characteristics.*

converts luminance in the range from 200 to 1000 lumens/m² to an electrical signal in the range from 4 to 20 mV. The output of the interface circuit is required to drive an ADC whose full-scale input range is 0 to 5 V. The interface circuit must convert a range of 4 mV to 20 mV to a output range of 0 V to 5 V, respectively. Inserting these values into Eq. (4–36) and Table 4–3 gives

$$K \times 0.004 + V_b = 0\,V \text{ @200@lumens/m}^2$$
$$K \times 0.02 + V_b = 5\,V \text{ @1000@lumens/m}^2$$

T A B L E **4–3**

PHYSICAL RANGE	200 LUMENS/M²	1000 LUMENS/M²
Desired range for v_O	0 V	5 V
Available range at v_{TR}	4 mV	20 mV

Applying Eq. (4–37) and carefully aligning our range end points, we get

$$K = \frac{\text{Luminance range}}{\text{Available range}} = \frac{\begin{array}{ccc} 200\,\text{lumens/m}^2 & & 1000\,\text{lumens/m}^2 \\ 0 & - & 5 \end{array}}{\begin{array}{ccc} 0.004 & - & 0.02 \end{array}} = 312.5$$

Inserting this value of K into the first equation and solving for the bias gives $V_b = -0.004K = -1.25\,\text{V}$ Thus, the interface circuit must implement an input-output relationship defined by

$$v_O = (312.5)v_{TR} - 1.25 \qquad (4\text{–}38a)$$

To design the interface circuit, we partition the required gain as $K = (-25) \times (-12.5) = 312.5$. This allows us to get the overall gain of $+312.5$ using two inverting amplifier stages with gains of -25 and -12.5. We also rewrite the bias voltage as $V_b = (-0.25)5 = -1.25\,\text{V}$. This allows us to get the required bias using an inverting gain of -0.25 and a standard 5-V power supply as the reference source. Inserting these results into Eq. (4–38a) yields

$$v_{TR} = (-25)(-12.5)v_{TR} - (0.25)5 \qquad (4\text{–}38b)$$

Figure 4–66(a) shows a block diagram of Eq. (4–38b) while Figure 4–66(b) shows an OP AMP circuit that implements the block diagram. Obviously

FIGURE 4–66 *Photocell interface circuit realization.*

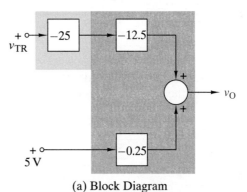

(a) Block Diagram

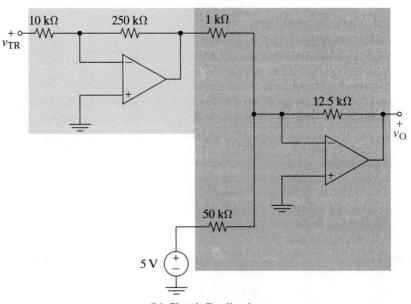

(b) Circuit Realization

this design is not unique since we can rearrange Eq. (4–38a) in many different ways.

◀▶ ◀▶ DESIGN AND EVALUATION EXAMPLE 4 – 24

A commercially available temperature transducer has the characteristics shown in Figure 4–67. Design an OP AMP circuit to interface the transducer output for temperatures in the range from $-20°C$ to $120°C$ with a panel meter whose full-scale input range is 0 to 3 V. Use a standard 1.5-V battery as the reference source for the required bias.

FIGURE 4–67

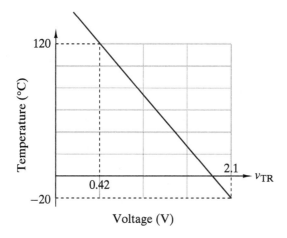

SOLUTION:
Figure 4–67 shows that the transducer output voltage ranges from 2.1 V to 0.42 V for the temperature range from $-20°C$ to $120°C$. The interface circuit must convert a range of 2.1 V to 0.42 V to a range of 0 V to 3 V, respectively. Inserting these values into Eq. (4–36) gives

$$0 = K \times 2.1 + V_b \quad \text{at } -20°C$$
$$3 = K \times 0.42 + V_b \quad \text{at } 120°C$$

Calculating the gain K,

$$\text{Temp. range} \quad \overset{-20°C}{} \quad \overset{120°C}{}$$
$$K = \frac{0 - 3}{2.1 - 0.42} = -1.786$$

Inserting this value of K into the first equation and solving for the bias gives $V_b = -2.1K = 3.75\,\text{V}$. The bias voltage can be rewritten as $V_b = 3.75 = (2.5)1.5\,\text{V}$, which means we need a gain of 2.5 to get the required bias from the specified 1.5-V battery reference source. Thus, the interface circuit must realize the following input-output relationship.

$$v_O = -1.786 v_{TR} + (2.5)1.5$$

We now use each of the subtractor circuits in Figure 4–68 to realize this relationship.

First Design

The required function is of the form $v_O = -K_1 v_1 + K_2 v_2$, where $K_1 = 1.786, v_1 = v_{TR}, K_2 = 2.5$, and $v_2 = 1.5\,\text{V}$. This function can be realized using the basic subtractor building block in Figure 4–68(a). Equation (4–32)

FIGURE 4–68

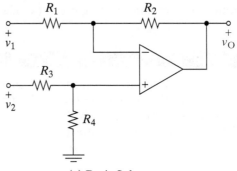

(a) Basic Subtractor

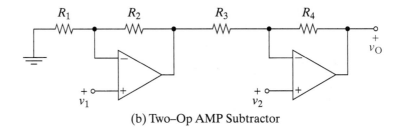

(b) Two–Op AMP Subtractor

relates the two gains to the subtractor circuit parameters as

$$K_1 = \frac{R_2}{R_1} \quad \text{and} \quad K_2 = \left(\frac{R_1 + R_2}{R_1}\right)\left(\frac{R_4}{R_3 + R_4}\right)$$

The gain $K_1 = 1.786$ can be realized by selecting $R_2 = 10\,\text{k}\Omega$ and $R_1 = 0.56R_2 = 5.6\,\text{k}\Omega$. Using these values in the K_2 equation produces

$$K_2 = \left(1 + \frac{10}{5.6}\right)\left(\frac{R_4}{R_3 + R_4}\right) = 2.5$$

Solving for R_4 yields $R_4 = 8.75R_3$. Selecting $R_3 = 1\,\text{k}\Omega$ requires that $R_4 = 8.75\,\text{k}\Omega$. This completes the first design. Obviously, many other choices are possible

Second Design

The required function $v_O = -K_1 v_1 + K_2 v_2$ can also be realized using the two–OP AMP subtractor shown in Figure 4–68(b). This circuit is analyzed in Example 4–19, where the two gains are related to circuit parameters as

$$K_1 = \frac{R_4}{R_3}\left(1 + \frac{R_2}{R_1}\right) \quad \text{and} \quad K_2 = 1 + \frac{R_4}{R_3}$$

The gain $K_2 = 2.5$ can be realized by selecting $R_3 = 10\,\text{k}\Omega$ which requires $R_4 = 1.5R_3 = 15\,\text{k}\Omega$. Using these values in the K_1 equation produces

$$K_1 = \frac{15}{10}\left(1 + \frac{R_2}{R_1}\right) = 1.786$$

Solving for R_1 yields $R_1 = 5.25R_2$. Selecting $R_2 = 10\,\text{k}\Omega$, requires that $R_1 = 52.5\,\text{k}\Omega$. This completes the second design. Many other choices are possible.

Evaluation Discussion

In terms of parts count, both designs use four resistors. The first design uses a one–OP AMP subtractor while the second design uses the two–OP AMP version. This difference makes the first design the best choice under a "smallest parts count" criteria. However, with the two–OP AMP subtractor, both input signals are applied directly to noninverting OP AMP inputs. These inputs have very high input resistances, which means that the second design does not load the transducer or drain energy from the 1.5-V battery. This difference could be an important advantage of the second design if the transducer has a high Thévenin resistance or if the interface circuit must operate for extended periods of time with no servicing of the battery. ■

◑ Design Exercise 4–34 _____

A pressure transducer must be connected to a boiler. The selected transducer is linear between 100 and 1000 psi. Specifically, it has the following characteristics: At 100 psi it produces 10 μV, and at 1000 psi it produces 100 μV. The output needs to be connected to a 0–10 V meter so that 100 psi will give a reading of 0 V and 1000 psi a reading of 10 V. Design a suitable interface using OP AMPs that have a maximum closed-loop gain of 2000.

A n s w e r : $K = 111{,}111$ and $V_b = -1.11$ V. One of many possible solutions is shown in Figure 4–69.

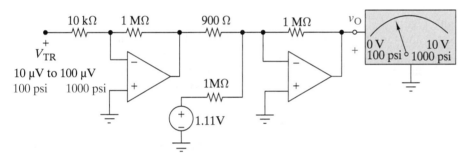

FIGURE 4–69

(a)

PASSIVE TRANSDUCERS

As mentioned previously, passive transducers require an external source to produce a voltage or current that is proportional to the physical parameter being measured. Photoresistors, thermistors, strain gauges, and rotation sensors are but a few of the passive transducers that use a variance in resistance to sense a physical parameter. Figure 4–70 shows photos of some passive transducers. In all cases we will look at how the resistance varies with the physical parameter.

There are two ways to apply an external source to a passive transducer. The first is using a simple voltage divider, shown in Figure 4–71(a). This technique is useful when the accuracy of the sensor is not critical and the change in parameter value is significant, as with photoresistors, thermistors, or rotational devices such as potentiometers. The second approach is to use a bridge circuit as shown in Figure 4–71(b). This technique allows for very precise measurement of small changes in parameter values, such as those created by strain gauges. We will show an example of each approach.

(b)

(c)

◑ ⬭ DESIGN EXAMPLE 4–25 ⬭

A particular photoresistor varies from 10 MΩ in total darkness to 10 kΩ in bright light. Design an interface system that will produce 5 V when the photoresistor senses bright light and 0 V when it is in total darkness.

FIGURE 4–70 *Passive transducers: (a) Photoresistors. (b) Strain gauge. (c) Thermistors.*

FIGURE 4–71

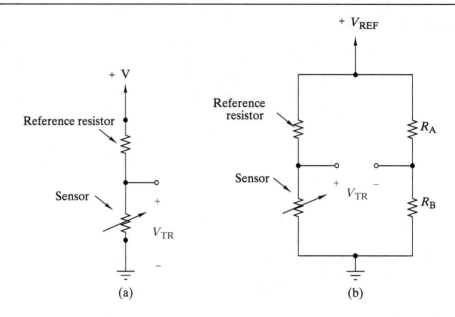

(a)   (b)

SOLUTION:

Using Figure 4–71(a), we select a 10-V source in series with a 10-kΩ reference resistor and the photoresistor as shown in Figure 4–72(a). Figure 4–72(b) shows the Thévenin equivalent of the circuit. The v_T varies from 5 V in bright light to about 10 V in darkness and R_T varies from 5 kΩ to about 10 kΩ, respectively. The interface circuit must convert a range of 5 V to 10 V to an output range of 5 V to 0 V, respectively.

Inserting these values into Eq. (4–36) gives

$$K \times 10 + V_b = 0\,\text{V in darkness}$$
$$K \times 5 + V_b = 5\,\text{V in brightness}$$

FIGURE 4–72

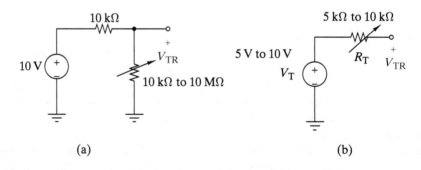

(a)   (b)

Applying Eq. (4–37) and aligning our range end points, we get

$$K = \frac{\text{Desired range}}{\text{Available range}} = \frac{\overset{\text{Sensor range}}{\overset{\text{Bright}}{5} - \overset{\text{Dark}}{0}}}{5 \; - \; 10} = -1$$

The bias is readily calculated as + 10 V.

Figure 4–73 shows a possible interface circuit. Note that the 10-V bias was inverted so that it will add as a positive value. If this is not possible, use an inverter to change the sign of the bias before applying it to the summer. The large 1-MΩ resistors avoid loading the input to the inverter.

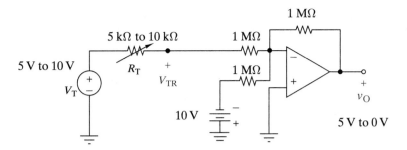

FIGURE 4–73

D Design Exercise 4–35 _____

Use a subtractor to design the interface for the example above.

Answer: Use all 1-MΩ resistors in a standard subtractor circuit. Apply the transducer input to the wire leading to the negative terminal of the OP AMP and apply the 10-V bias to the input leading to the positive terminal of the OP AMP.

D Design Exercise 4–36 _____

A 2-kΩ potentiometer is connected to the flaps of an unmanned aerial vehicle, or UAV, to detect their position. When the flaps are at their maximum upward extension of +45°, the potentiometer is at its maximum resistance of $2\,k\Omega$; when the flaps are flat or 0°, the potentiometer is set at $1\,k\Omega$; and when the flaps are at their minimum downward extension of −45°, the potentiometer is at its minimum resistance of $0\,k\Omega$. Design an interface to a 0–5 V ADC that gives +45° at 5 V and −45° at 0 V.

Answer: One possible design is shown in Figure 4–74. If there is a concern about loading the input to the ADC, insert a buffer after the potentiometer. Sometimes simplicity is the best policy.

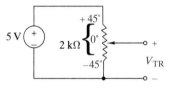

FIGURE 4–74

The next example demonstrates the use of a bridge circuit to detect small changes in a passive transducer.

D DESIGN EXAMPLE 4 – 26

A strain gauge is a resistive device that measures the elongation (strain) of a solid material caused by applied forces (stress). A typical strain gauge consists of a thin film of conducting material deposited on an insulating substrate. When the gauge is bonded to a member under stress, its resistance changes by

$$\Delta R = 4R_{\mathrm{G}} \frac{\Delta L}{L}$$

where R_{G} is the resistance of the gauge with no applied stress and $\Delta L/L$ is the elongation of the material expressed as a fraction of the unstressed length L. The change in resistance ΔR is only a tenth of an ohm or so, which is far too little to be measured with an ohmmeter. To detect such a small change, the strain gauge is placed in a Wheatstone bridge circuit like the one shown in Figure 4–71(b). The bridge contains fixed resistors R_{A} and R_{B}, two matched strain gauges R_{G1} [the reference resistor in Figure 4–71(b)] and R_{G2} [the passive transducer or sensor in Figure 4–71(b)], and a precisely controlled reference voltage V_{REF}. The values of R_{A} and R_{B} are chosen so that the bridge is balanced ($v_{\mathrm{TR}} = 0$) when no stress is applied. When stress is applied, the resistance of the stressed gauge changes to $R_{\mathrm{G2}} + \Delta R$ and the bridge is unbalanced ($v_{\mathrm{TR}} \neq 0$). The differential signal (v_{TR}) indicates the strain resulting from the applied stress. See Figure 4–75.

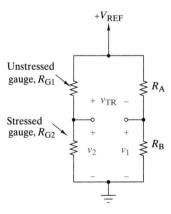

FIGURE 4–75

Design an OP AMP circuit to translate strains in the range $0 < \Delta L/L < 0.02\%$ into an output voltage in the range $0\,V < v_O < 5\,V$, for $R_G = 120\,\Omega$ and $V_{REF} = 25\,V$.

SOLUTION:
With external stress applied, the resistance R_{G2} changes to $R_{G2} + \Delta R$. Applying voltage division to each leg of the bridge yields

$$v_2 = \frac{R_{G2} + \Delta R}{R_{G1} + R_{G2} + \Delta R} V_{REF}$$

$$v_1 = \frac{R_B}{R_A + R_B} V_{REF}$$

The differential voltage $(v_2 - v_1) = v_{TR}$ can be written as

$$v_2 - v_1 = v_{TR} = V_{REF} \left[\frac{R_{G2} + \Delta R}{R_{G1} + R_{G2} + \Delta R} - \frac{R_B}{R_A + R_B} \right]$$

To achieve bridge balance in the unstressed state, we select $R_{G1} = R_{G2} = R_A = R_B = R_G$, in which case the differential voltage reduces to

$$v_2 - v_1 = v_{TR} = V_{REF} \left[\frac{\Delta R}{4R_G + 2\Delta R} \right] \cong V_{REF} \left[\frac{\Delta R}{4R_G} \right] = V_{REF} \left[\frac{\Delta L}{L} \right]$$

Thus, the differential voltage is directly proportional to the strain $\Delta L/L$. However, for $V_{REF} = 25\,V$ and $\Delta L/L = 0.02\%$, the differential voltage is only $(V_{REF})(\Delta L/L) = 25 \times 0.0002 = 5\,mV$. To obtain the required 5-V output we need a voltage gain of 1000:

$$\begin{array}{cc} \text{Strain} & 0\% \qquad 0.02\% \\ K = & \dfrac{0 \;-\; 5}{0 \;-\; 0.005} = 1000 \end{array}$$

The bias voltage, V_b, is zero in this case.

The OP AMP subtractor is specifically designed to amplify differential signals. Selecting $R_1 = R_3 = 10\,k\Omega$ and $R_2 = R_4 = 10\,M\Omega$ produces an input-output relationship for the subtractor circuit of

$$v_O = 1000 v_{TR}$$

Figure 4–76 shows the basic design.

The input resistance of the subtractor circuit must be large to avoid loading the bridge circuit. The Thévenin resistance looking back into the bridge circuit is

$$R_T = R_{G1}\|R_{G2} + R_A\|R_B$$
$$R_T = R_G\|R_G + R_G\|R_G$$
$$R_T = R_G = 120\,\Omega$$

This value is small compared with the 10-kΩ input resistance of the subtractor's inverting input.

DISCUSSION: *The transducer in this example is the resistor R_{G2}. In the unstressed state, the voltage across this resistor is $v_2 = 12.5\,V$. In the stressed state the voltage*

FIGURE 4–76

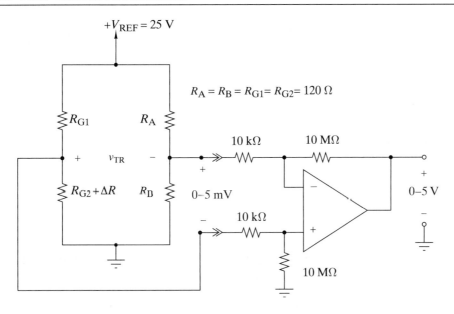

is $v_2 = 12.505$ V. *In other words, the transducer's 5–mV signal component is accompanied by a very large bias. We cannot amplify the 12.5-V bias component by $K = 1000$ before subtracting it out since it will saturate the OP AMP. By using a bridge circuit in which $v_1 = 12.5$ V, the bias is eliminated at the input and the OP AMP processes the differential signal $v_2 - v_1 = v_{TR}$. The situation illustrated in this example is actually quite common. Consequently, the first amplifier stage in most instrumentation systems is a differential amplifier that removes the transducer bias.* ∎

Exercise 4–37

Using the circuit shown in Figure 4–76, simulate the effect in OrCAD of varying the strain gauge R_{G2} from $120\,\Omega$ to $120.1\,\Omega$ in increments of $0.001\,\Omega$. Plot the resulting output versus percent of stress ($\Delta L/L$).

FIGURE 4–77

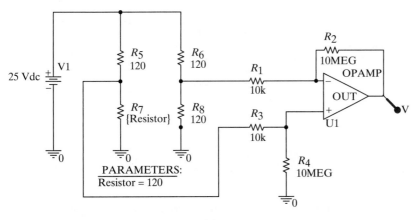

(a)

FIGURE 4–77 *(Continued)*

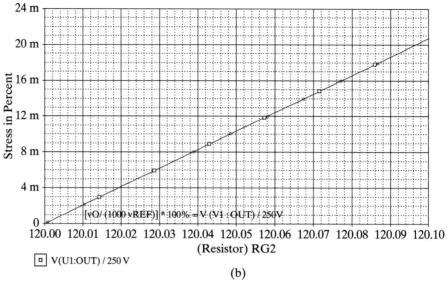

(b)

A n s w e r : See Figure 4–77.

APPLICATION   **E X A M P L E   4 – 2 7**

Instrumentation amplifiers are used as we have seen to provide signal processing of gain and bias to transducer outputs in preparation for use in ADC and other output devices. A well-designed instrumentation amplifier provides very high gain, very high common-mode rejection ratio, and very high input resistance, among other more subtle features best left for future study. Using active transducers can be difficult if they have a high source resistance or a source resistance that varies with the transducer's output voltage. A traditional difference amplifier shown in Figure 4–46 has the potential problem of loading the transducer because of its finite input resistance. A better solution is a two-stage configuration using the OP AMP circuit analyzed in Example 4-20 followed by a difference amplifier, as shown in Figure 4–78.

FIGURE 4–78

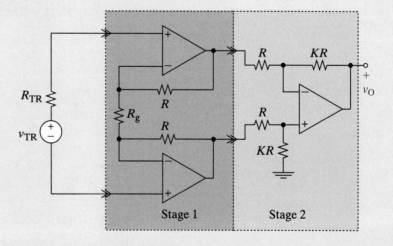

The instrumentation amplifier consists of two differential amplifiers each with a potential gain. The first stage acts like a pair of buffers (remove R_g and that can be easily seen) to provide a very high input resistance thereby minimizing loading problems. The second stage is a common difference amplifier that enables easy resistance connection to the output. Both stages can have gain. The first stage's gain was found in Example 4-20 as $K_1 = \frac{2R}{R_g} + 1$, while the second stage has gain $K_2 = K$. A necessary condition for a high quality instrumentation amplifier is a high common mode rejection ratio (CMRR). The CMRR of a differential amplifier measures the ability of the amplifier to reject input signals common to both inputs. A high CMRR is important in applications where the signal of interest, say, the transducer's small voltage change, is superimposed on a larger voltage offset, or when the key information is contained in the voltage difference between two signals. To ensure this quality, the resistors R must be carefully matched. If not, the gain of one input side over the other could mask the actual information change desired. Since most instrumentation amplifiers are manufactured on a single semiconductor chip, resistors can be made nearly identical and, if necessary, laser-trimmed. An interesting and useful feature is that the resistor R_g can be external of the chip and a potentiometer used to vary the gain. Since this resistance occurs only once in the gain relationship, it does not need to be carefully matched. The advantages of this two-stage configuration are that the input has very high input resistance and the two stages can help distribute the gain if a very high gain is needed. Many commercial manufacturers employ this architecture for their instrumentation amplifiers, such as Analog Devices, Linear Technology, Maxim Integrated Products, National Semiconductor, and Texas Instruments.

◑ Design Exercise 4–38

Design an instrumentation amplifier using the configuration of Figure 4-78 that has a gain of 10^5. Note that no single stage can have a gain greater than 10^4.

A n s w e r : Choose $R = 10 \text{ k}\Omega$, $KR = 5 \text{ M}\Omega$, $R_g = 100 \,\Omega$ as one possible solution.

COMPARATOR CIRCUITS

A **comparator** is a mixed-signal device whose digital output is either high or low depending on the relative amplitudes of two analog inputs. For example, the open-loop OP AMP circuit in Figure 4–79 functions as a comparator. In the absence of feedback, the OP AMP is driven into one of its two saturation modes by the analog inputs v_P and v_N. Specifically, if $v_P > v_N$, the OP AMP is in +saturation with $v_O = +V_{CC}$. Conversely, if $v_P < v_N$, the OP AMP is in –saturation with $v_O = -V_{CC}$.

In digital logic terminology, the comparator output is said to be high (1) when $v_P > v_N$ and low (0) when $v_P < v_N$. The output voltage levels associated with the high and low states are usually denoted V_{OH} and V_{OL}, respectively. These levels are determined by the positive and negative power supply voltages, which can be different from the $\pm 15 \text{ V}$ commonly used in linear applications of OP AMPs.

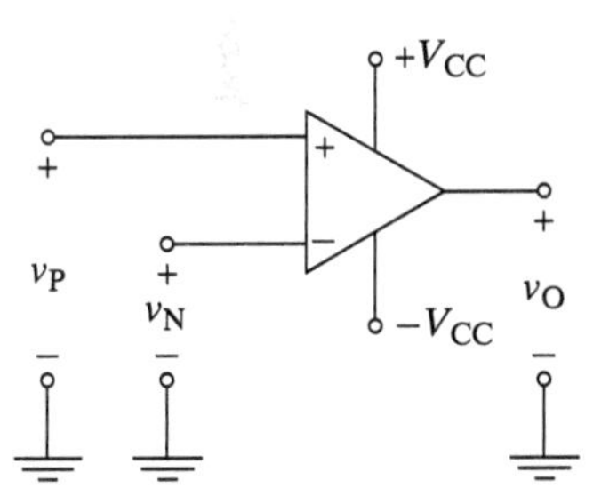

FIGURE 4–79 *An OP AMP comparator.*

Exercise 4–39

Find the comparator output voltage in Figure 4–80 for the following:
(a) $v_1 = 2 \text{ V}$, $v_2 = 3 \text{ V}$, $V_{CC} = 5 \text{ V}$
(b) $v_1 = 0 \text{ V}$, $v_2 = -3 \text{ V}$, $V_{CC} = 10 \text{ V}$
(c) $v_1 = -2 \text{ V}$, $v_2 = -3 \text{ V}$, $V_{CC} = 3 \text{ V}$

A n s w e r s : (a) $v_O = 0 \text{ V}$; (b) $v_O = 10 \text{ V}$; (c) $v_O = 3 \text{ V}$

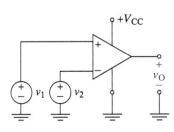

FIGURE 4–80

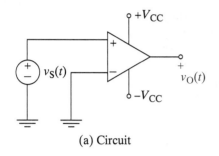

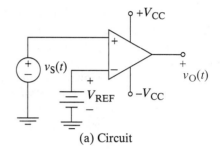

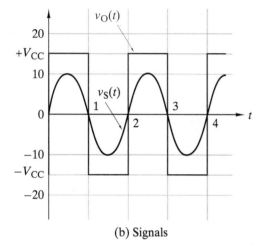

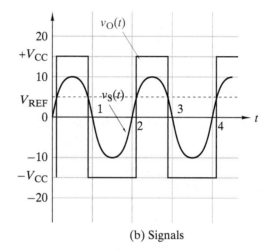

FIGURE 4–81   *Zero-crossing detector.*   FIGURE 4–82   *Modified zero-crossing detector.*

Figure 4–81(a) shows a comparator circuit often called a *zero-crossing detector*. A time-varying analog signal is applied to the noninverting input $[v_P = v_S(t)]$, and the inverting input is connected to ground $(v_N = 0)$. In this configuration the comparator output is $v_O = +V_{CC}$ when $v_S(t) > 0$ and $v_O = -V_{CC}$ when $v_S(t) < 0$. Figure 4–81(b) shows plots of the input and output signals versus time for $v_S(t) = 10 \sin(\pi t)$ V and $V_{CC} = 15$ V. In digital logic terms, the comparator output changes state (toggles) whenever the analog input passes through zero; hence the name *zero-crossing detector*. This circuit is also called a *polarity detector* because the digital output is high when the polarity of the analog input is positive and low when the polarity is negative.

A modified version of the zero-crossing detector is shown in Figure 4–82(a). A time-varying analog signal is still applied to the noninverting input $[v_P = v_S(t)]$ as before, and the modification connects a fixed reference voltage to the inverting input $(v_N = V_{REF})$. In this configuration the comparator output is $v_O = +V_{CC}$ when $v_S(t) > V_{REF}$ and $v_O = -V_{CC}$ when $v_S(t) < V_{REF}$. In effect the applied reference voltage shifts the switching threshold from zero to V_{REF}. Figure 4–82(b) shows how the threshold shift changes the output signal for $v_S(t) = 10 \sin(\pi t)$ V, $V_{CC} = 15$ V, and $V_{REF} = 5$ V.

An analog-to-digital converter (ADC) is a mixed-signal device that converts an analog input into a multibit digital output. Figure 4–83 is a block diagram of an ADC with an analog input $v_S(t)$, a four-bit digital output (b_1, b_2, b_3, b_4), and a fixed reference voltage V_{REF}. The output bits can have only one of two values: a high (1) or a low (0). The code used by these bits to represent the analog input depends on the architecture of the ADC.

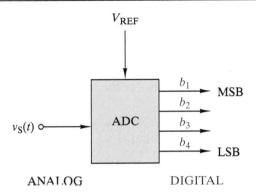

FIGURE 4–83 *An analog-to-digital converter (ADC).*

The circuit in Figure 4–84(a) is a four-comparator ADC that converts the analog input $v_S(t)$ into a four-bit digital output. The analog input is applied to the noninverting input of each of the comparators. A fixed reference voltage is applied to a voltage divider string. Successive taps on the voltage divider supply a reference voltage to the inverting input of the comparators. These reference voltages are all different, and each is larger than the reference voltage of the comparator immediately below it.

The output of any one comparator is high (1) when the analog input exceeds its applied reference voltage; otherwise it is low (0). When $v_S(t) < 0.2V_{REF}$, the input is smaller than all of the reference voltages, so the digital output is (0, 0, 0, 0). When $0.2V_{REF} < v_S(t) < 0.4V_{REF}$, the digital output is (0, 0, 0, 1). In this range the input exceeds the reference voltage of the lower comparator, so its output switches to $b_4 = 1$. The outputs of the other comparators remain at zero since the input is smaller than their reference voltages. When $0.4V_{REF} < v_S(t) < 0.6V_{REF}$, the output is (0, 0, 1, 1) because the input exceeds the reference voltages of the bottom two comparators but not the top two. These observations reveal a pattern that is summarized in Table 4–4.

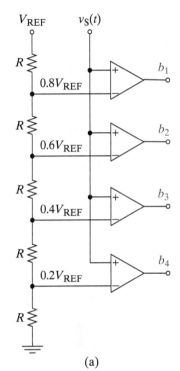

(a)

TABLE **4–4**

ANALOG INPUT RANGE	DIGITAL OUTPUTS			
	b_1	b_2	b_3	b_4
$0 < v_S < 0.2V_{REF}$	0	0	0	0
$0.2V_{REF} < v_S < 0.4V_{REF}$	0	0	0	1
$0.4V_{REF} < v_S < 0.6V_{REF}$	0	0	1	1
$0.6V_{REF} < v_S < 0.8V_{REF}$	0	1	1	1
$0.8V_{REF} < v_S < V_{REF}$	1	1	1	1

The circuit in Figure 4–84(a) is called a *flash converter* because the comparators operate in parallel and the conversion takes place almost instantaneously. The circuit divides the input amplitude range into five bins and converts each bin into a unique four-bit code. The *full-scale input range* is the voltage range over which the input amplitude falls within one of the bins (0 to V_{REF} in this example). This range can obviously be increased by increasing the reference voltage. The price of doing so is reduced *resolution*, defined as the largest input voltage change that falls entirely within one bin ($0.2V_{REF}$ in this example). Resolution can be improved by expanding the voltage divider string and adding more comparators. Some integrated circuit flash converters have as many as 256 comparators. An example of an 8-bit flash converter is shown in Figure 4–84(b).

(b)

FIGURE 4–84 *(a) Flash ADC diagram (b) 8-bit flash ADC IC.*

The pattern of 1s and 0s in Table 4–4 is called a *thermometer code* because the number of 1s increases monotonically as v_s increases, like the way the mercury column in a thermometer increases with temperature. Integrated circuit flash converters have built-in decoders that convert the thermometer code into a standard binary code. The flash converter is by far the fastest ADC architecture.

Exercise 4–40 _____

The reference voltage in Figure 4–84(a) is $V_{REF} = 15$ V. What are the output codes corresponding to $v_S = 1, 2, 5, 10,$ and 14 V?

Answers: $(0, 0, 0, 0); (0, 0, 0, 0); (0, 0, 0, 1); (0, 1, 1, 1); (1, 1, 1, 1)$

SUMMARY

- A linear dependent source generates a voltage or a current whose value is proportional to a voltage or current at another point in a circuit. There are four such sources: the current-controlled voltage source, the voltage-controlled voltage source, the current-controlled current source, and the voltage-controlled current source.

- Circuits containing dependent sources can be analyzed by node-voltage or mesh-current methods. Such circuits can have input-output relationships that produce voltage, current, or power gain. The presence of feedback can dramatically influence the input and output resistances of active circuits.

- The OP AMP is an active device with five terminals called the inverting input, the noninverting input, the output, and two power supply terminals. The device is a high-gain differential amplifier with three possible operating modes: +saturation, −saturation, and linear. The output predicted by the linear mode circuit model is compared with known bounds to determine the actual operating mode.

- The ideal OP AMP model has an infinite voltage gain, an infinite input resistance, and zero output resistance. The i–v characteristics of an ideal OP AMP are $i_P = i_N = 0$ and $v_P = v_N$. The ideal model is a good working approximation in linear applications.

- The four basic OP AMP circuit building blocks are the inverting amplifier, the noninverting amplifier, the inverting summer, and the subtractor. The analysis or design of complex OP AMP circuits can be based on these four building blocks provided the interconnections are made between the output of one to the input of another.

- Important applications of OP AMPs include digital-to-analog converters, instrumentation systems, and comparator circuits.

PROBLEMS

OBJECTIVE 4–1 LINEAR ACTIVE CIRCUITS (SECTS. 4–1, 4–2)

Given a linear resistance circuit containing dependent sources, find selected output signals, input-output relationships, or input-output resistances.
See Examples 4–1, 4–2, 4–3, 4–4, 4–5, 4–6, 4–7, 4–8, 4–9, 4–10, 4–11 and Exercises 4–1, 4–2, 4–3, 4–5, 4–6, 4–7, 4–8, 4–9, 4–10, 4–11, 4–12, 4–13.

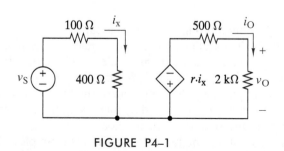

FIGURE P4–1

4–1 Find the voltage gain v_O/v_S and current gain i_O/i_x in Figure P4–1 for $r = 5$ kΩ.

4–2 Find the voltage gain v_O/v_1 and the current gain i_O/i_S in Figure P4–2. For $i_S = 5$ mA, find the power supplied by the input current source and the power delivered to the 2-kΩ load resistor.

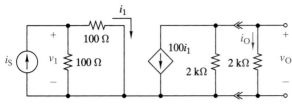

FIGURE P4–2

4–3 Find the voltage gain v_O/v_S and current gain i_O/i_x in Figure P4–3 for $g = 2 \times 10^{-3}$ S. For $v_S = 5$ V, find the power supplied by the input voltage source and the power delivered to the 2-kΩ load resistor.

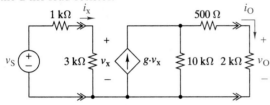

FIGURE P4–3

4–4 (a) Find the voltage gain v_O/v_S and current gain i_O/i_x in Figure P4–4.

(b) Validate your answers by simulating the circuit in OrCAD.

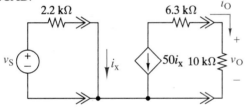

FIGURE P4–4

4–5 Find the voltage gain v_O/v_S in Figure P4–5.

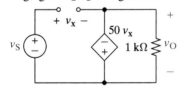

FIGURE P4–5

4–6 Find an expression for the current gain i_O/i_S in Figure P4–6. *Hint:* Apply KCL at node A.

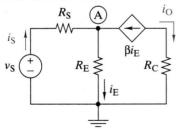

FIGURE P4–6

4–7 (a) Find the voltage v_O in Figure P4–7.

(b) Validate your answer by simulating the circuit in OrCAD.

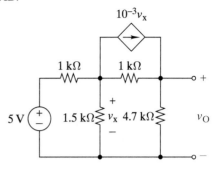

FIGURE P4–7

4–8 (a) Find an expression for the gain i_O/v_S in Figure P4–8 in terms of R_X.

(b) Select a value for R_X so that the gain is -0.227.

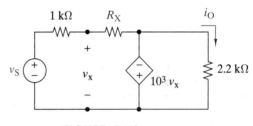

FIGURE P4–8

4–9 Find an expression for the voltage gain v_O/v_S in Figure P4–9.

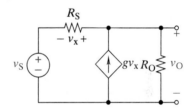

FIGURE P4–9

4–10 (a) Find an expression for the voltage gain v_O/v_S in Figure P4–10.

(b) Let $R_S = 10$ kΩ, $R_L = 10$ kΩ and $\mu = 100$. Find the voltage gain v_O/v_S as a function of R_F. What is the voltage gain when R_F is an open circuit, a short circuit, and for $R_F = 100$ Ω?

(c) Simulate the circuit in OrCAD by varying R_F from 1 Ω to 10 MΩ. Read your output $R_F = 100$ Ω. How does your answer compare with part (b)?

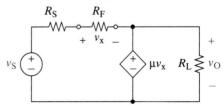

FIGURE P4–10

4–11 Select g in the circuit of Figure P4–11 so that the output voltage is 10 V.

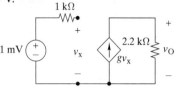

FIGURE P4–11

4–12 D Design a dependent source circuit that has a closed-loop voltage gain of -10 using a VCVS with a μ of 100. The source circuit is a voltage source v_S in series with a 1-kΩ resistor, and the load is a 3.3-kΩ resistor. (*Hint:* See Figure P4–10.)

4–13 Find the Thévenin equivalent circuit that the load R_L sees in Figure P4–13. Repeat the problem with R_F replaced by an open circuit.

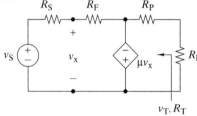

FIGURE P4–13

4–14 Find the Thévenin equivalent circuit that the load R_L sees in Figure P4–14.

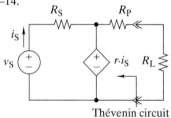

Thévenin circuit

FIGURE P4–14

4–15 Find R_{IN} in Figure P4–15

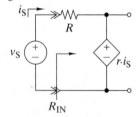

FIGURE P4–15

4–16 Find R_{IN} in Figure P4–16.

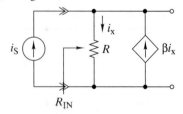

FIGURE P4–16

4–17 Find the Norton Equivalent circuit seen by the load in Figure P4–17.

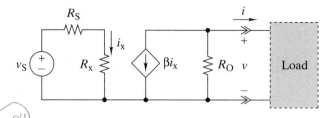

FIGURE P4–17

4–18 Find the Thévenin Equivalent circuit seen by the load in Figure P4–18.

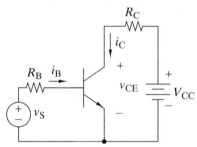

FIGURE P4–18

4–19 The circuit parameters in Figure P4–19 are $R_B = 100$ kΩ, $R_C = 3.3$ kΩ, $\beta = 100$, $V_\gamma = 0.7$ V, and $V_{CC} = 15$ V. Find i_C and v_{CE} for $v_S = 1$ V. Repeat for $v_S = 5$ V.

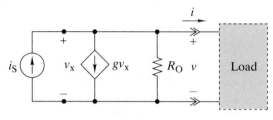

FIGURE P4–19

4–20 D The circuit parameters in Figure P4–19 are $R_C = 3$ kΩ, $\beta = 120$, $V_\gamma = 0.7$ V, and $V_{CC} = 5$ V. Select a value of R_B such that the transistor is in the saturation mode when $v_S \geq 2$ V.

4–21 The parameters of the transistor in Figure P4–21 are $\beta = 60$ and $V_\gamma = 0.7$ V. Find i_C and v_{CE} for $v_S = 0.8$ V. Repeat for $v_S = 2$ V.

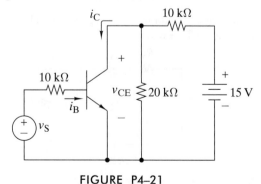

FIGURE P4–21

4–22 **D** An emergency indicator light uses a 10 V, 2-W incandescent lamp. It is to be ON when a digital output is high (5 V). The digital circuit does not have sufficient power to turn on the lamp directly. However, as is common practice, a transistor driver is used as a digital switch. Select R_B in the circuit of Figure P4–22 so to drive the transistor into saturation causing it to act as a short circuit between the lamp and ground when the digital output is high. The Thévenin equivalent for the digital circuit is also shown in the figure.

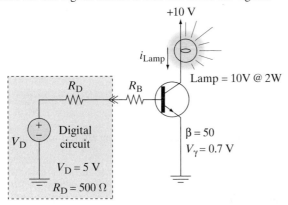

FIGURE P4–22

OBJECTIVE 4–2 OP AMP CIRCUIT ANALYSIS (SECTS. 4–3, 4–4)

Given a linear resistance circuit containing OP AMPs, find selected output signals or input-output relationships.
See Examples 4–12, 4–13, 4–14, 4–15, 4–16, 4–17, 4–18, 4–19, 4–20 and Exercises 4–14, 4–15, 4–16, 4–17, 4–18, 4–19, 4–20, 4–21, 4–22, 4–23, 4–24, 4–25, 4–26, 4–27, 4–28.

4–23 Find the voltage gain of each OP AMP circuit shown in Figure P4–23.

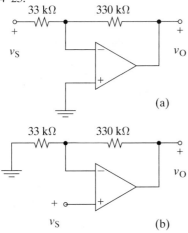

FIGURE P4–23

4–24 **D** Considering simplicity and standard 10% tolerance resistors as major constraints, design OP AMP circuits that produce the following voltage gains ±10%: − 100, + 200, + 1, −0.5, +0.5.

4–25 **E** Two OP AMP circuits are shown in Figure P4–25. Both claim to produce a gain of either ± 100.
   (a) Show that the claim is true.
   (b) A practical source with a series resistor of 1 kΩ is connected to the input of each circuit. Does the original claim still hold? If it does not, explain why?

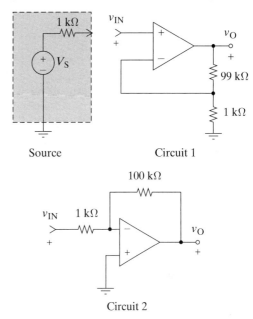

FIGURE P4–25

4–26 **E** Suppose the output of the practical source shown in Figure P4–25 needs to be amplified by -10^4 and you can use only the two circuits shown. How would you connect the circuits to achieve this? Explain why.

4–27 (a) Find the voltage gain v_O/v_S in Figure P4–27.
   (b) Validate your answer by simulating the circuit in OrCAD.

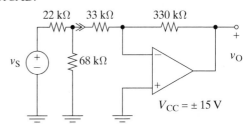

FIGURE P4–27

4–28 What is the range of the gain v_O/v_S in Figure P4–28?

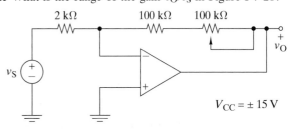

FIGURE P4–28

4–29 **D** Design a simple OP AMP circuit that has a variable gain from +10 to +100.

4–30 For the circuit in Figure P4–30:
(a) Find v_O in terms v_S.
(b) Find i_O for $v_S = 1$ V. Repeat for $v_S = 3$ V.

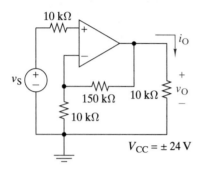

FIGURE P4–30

4–31 For the circuit in Figure P4–31:
(a) Find v_O in terms v_S.
(b) Find i_O for $v_S = 0.5$ V. Repeat for $v_S = 2$ V.

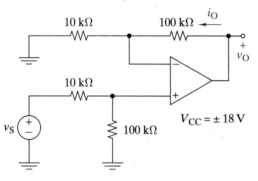

FIGURE P4–31

4–32 **E** A young designer needed to amplify a 2-V signal by the factors of 1, 5, and 10. Find the problem with the design shown in Figure P4–32. Recommend a fix.

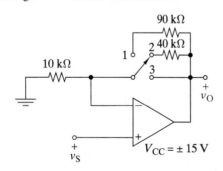

FIGURE P4–32

4–33 **D** Design two circuits to produce the following output: $v_O = 2v_1 - 3v_2$.
(a) In your first design use a standard subtractor.
(b) In your second design both inputs must be into high input resistance amplifiers to avoid loading.

4–34 For the circuit in Figure P4–34:
(a) Find v_O in terms of the inputs v_1 and v_2.
(b) If $v_1 = 1$ V, what is the range of values v_2 can have without saturating the OP AMP?

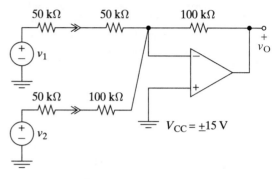

FIGURE P4–34

4–35 The input-output relationship for a three-input inverting summer is

$$v_O = -[v_1 + 10v_2 + 100v_3]$$

The resistance of the feedback resistor is 100 kΩ. Find the values of the input resistors R_1, R_2, and R_3.

4–36 Find v_O in terms of the inputs v_1 and v_2 in Figure P4–36.

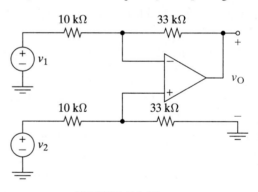

FIGURE P4–36

4–37 The switch in Figure P4–37 is open, find v_O in terms of the inputs v_{S1} and v_{S2}. Repeat with the switch closed.

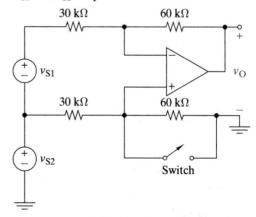

FIGURE P4–37

4–38 Find v_O in terms of v_{S1} and v_{S2} in Figure P4–38.

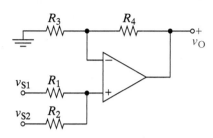

FIGURE P4–38

4–39 It is claimed that $v_O = v_S$ when the switch is closed in Figure P4–39 and that $v_O = -v_S$ when the switch is open. Prove or disprove this claim.

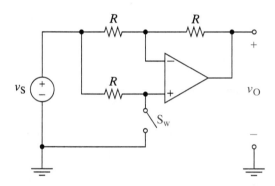

FIGURE P4–39

4–40 The circuit in Figure P4–40 has a diode in its feedback path and is called a "log-amp" because its output is proportional to the natural log of the input.
  (a) Show that $v_O = -V_T \ell n(1 + v_S/(R_S I_O))$ if the i-v characteristics of the diode is $i_D = I_O(e^{v_D/v_T} - 1)$.
  (b) Using MATLAB plot v_O versus v_S for $R_S = 10$ kΩ, $I_O = 2 \times 10^{-14}$ A and $V_T - 0.026$ V. Plot your results on a semilog plot for 10^{-6} V $\leq v_S \leq 100$ V.

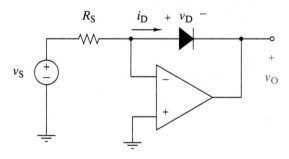

FIGURE P4–40

4–41 (a) Use node voltage analysis to find the input-output relationship or K of the circuit in Figure P4–41.

(b) Select values for the resistors so that $K = -6$.

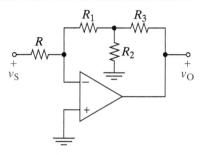

FIGURE P4–41

4–42 Use node voltage analysis in Figure P4–42 to show that $i_O = -v_S/2R$ regardless of the load. That is, show that the circuit is a voltage-controlled current source.

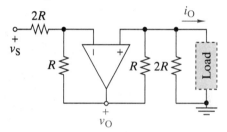

FIGURE P4–42

4–43 For the circuit of Figure P4–43:
  (a) Find the output in terms of v_1 and v_2.
  (b) Draw a block diagram for the circuit.

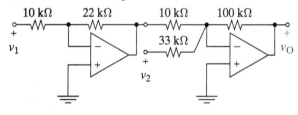

FIGURE P4–43

4–44 Find v_O in terms of the inputs v_{s1} and v_{s2} in Figure P4–44.

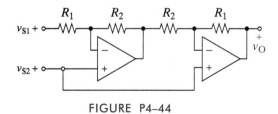

FIGURE P4–44

4–45 For the block diagram of Figure P4–45:
  (a) Find an expression for v_O in terms of v_1 and v_2.
  (b) Design a suitable circuit that realizes the block diagram using only one OP AMP.

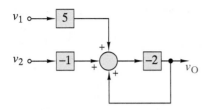

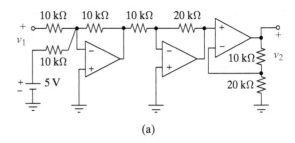

(a)

FIGURE P4–45

4–46 For the block diagram of Figure P4–46:
   (a) Find an expression for v_O in terms of v_S and the input voltage source.
   (b) ◗ Design a suitable circuit that realizes the block diagram using only one OP AMP and the 0.5-V source.

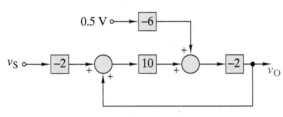

FIGURE P4–46

4–47 For the circuit in Figure P4–47
   (a) Find v_O in terms of v_S and the 1-V source.
   (b) Prove that the block diagram provides the same output.
   (c) ◗ Redesign the circuit using only one OP AMP.
   (d) Validate your design using OrCAD.

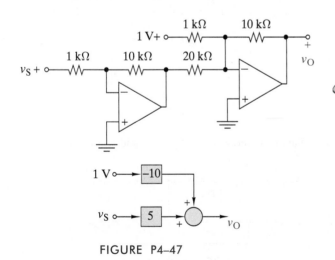

FIGURE P4–47

4–48 ⓔ On an exam, students were asked to design an efficient solution for the following relationship: $v_2 = 3v_1 + 15$. Two of the designs are shown in Figure P4–48. Which, if any, of the designs are correct and what grade would you award each student?

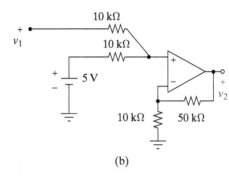

(b)

FIGURE P4–48

4–49 For the circuit of Figure P4–49:
   (a) Find the output v_2 in terms of the input v_1 in Figure P4–49.
   (b) Draw a representative block diagram for the circuit.

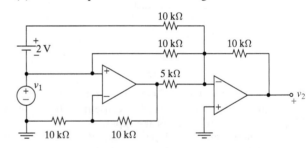

FIGURE P4–49

4–50 For the circuit of Figure P4–50:
   (a) Use node-voltage analysis to find the output v_O in terms of the input v_S.
   (b) Draw a representative block diagram for the circuit.
   (c) Verify your answer using OrCAD.

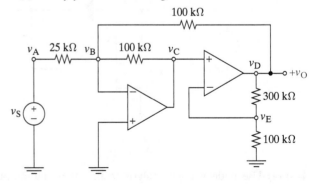

FIGURE P4–50

4–51 🖹 Faced with having to construct the circuit in Figure P4–51(a), a student offers to build the circuit in Figure P4–51(b) claiming it performs the same task. As the teaching assistant in the course, do you agree with the students claim?

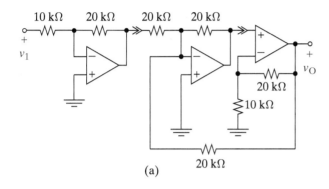

(a)

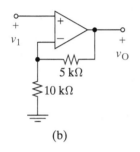

(b)

FIGURE P4–51

4–52 Design a single OP AMP amplifier with a voltage gain of −100 and an input resistance greater than 5 kΩ using standard 5% resistance values less than 1 MΩ.

4–53 Design an OP AMP amplifier with a voltage gain of 5 using only 10-kΩ resistors and one OP AMP.

4–54 Using a single OP AMP, design a circuit with inputs v_1 and v_2 and an output $v_O = v_2 - 3v_1$. The input resistance seen by each input should be greater than 5 kΩ.

4–55 Design a difference amplifier with inputs v_1 and v_2 and an output $v_O = 30(v_2 - v_1)$ using only one OP AMP. All resistances must be between 5 kΩ and 200 kΩ.

4–56 Using no more than two OP AMPs, design an OP AMP circuit with inputs v_1, v_2, and 50 mV and an output $v_O = -3v_1 + 2v_2 - 250$ mV.

4–57 Design a two-input non-inverting summer that will produce an output $v_O = 500 (v_1 + v_2)$.

4–58 Design a four- input non-inverting summer that will produce an output $v_O = 6 (v_1 + v_2 + v_3 + 1V)$.

4–59 Design a cascaded OP AMP circuit that will produce the following output $v_O = 2 \times 10^7 v_S + 3.5$ V. The maximum gain for an OP AMP is 10,000. The input stage must have an input resistance of 1 kΩ or greater.

4–60 Design a cascaded OP AMP circuit that will produce the following output $v_O = -5 \times 10^6 v_S + 1.8$ V. The maximum gain for an OP AMP is 10,000. The input stage must have an input resistance of 1 kΩ or greater. The only voltage source available is the ±15 V used to power the OP AMPs.

4–61 Using the difference amplifier shown in Figure 4–53 (Example 4–20), design a circuit that will produce the following output $v_O = 500 (v_1 - v_2)$.

4–62 Design the interface circuit in Figure P4–62 so that 10 mW is delivered to the 100-Ω. load. Repeat for a 100-kΩ load. Verify your designs using OrCAD.

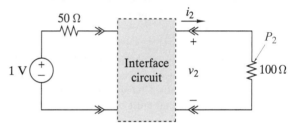

FIGURE P4–62

4–63 Design the interface circuit in Figure P4–63 so that the output is $v_2 = 150v_1 + 1.5$ V.

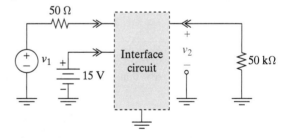

FIGURE P4–63

4–64 **(a)** Design a circuit that can produce $v_O = 2000v_{TR} - 2.6$V using two OP AMPs. The input resistance must be greater than 5 kΩ for v_{TR}.
(b) Repeat using only one OP AMP.

4–65 🖹 A requirement exists for an OP AMP circuit with the input-output relationship

$$v_O = 5v_{S1} - 2v_{S2}$$

Two proposed designs are shown in Figure P4–65. As the project engineer, you must recommend one of these circuits for production. Which of these circuits would you recommend for production and why? *Hint:* First, verify that the circuits perform the required function.

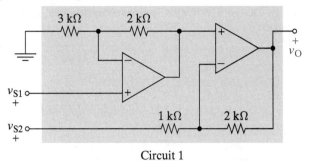

Circuit 1

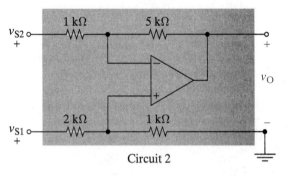

Circuit 2

FIGURE P4–65

4–66 🌐 A requirement exists for an OP AMP circuit to deliver 12 V to a 1-kΩ load using a 4–V source as an input voltage. Two proposed designs are shown in Figure P4–66. Some characteristics of the OP AMP that must be used in the design are

CHARACTERISTIC	MIN	TYPICAL	MAX	UNITS
Open-loop Gain	10^5	2×10^5	-	V/mV
Input Resistance	10^{10}	10^{11}	-	Ω
Output Voltage	−12	-	+02	V
Output Current	-	-	25	MA

Which of these circuits would you recommend for production and why? *Hint:* First, verify that the circuits perform the required function.

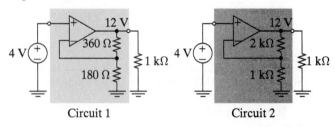

Circuit 1   Circuit 2

FIGURE P4–66

4–67 🌐 A particular application requires that an instrumentation interface deliver $v_O = 200 v_{TR} - 5\ V \pm 2\ \%$ to a DAC. The solution currently in use requires two OP AMPs and is constantly draining the supply batteries. A young engineer

designed another tentative solution using just one OP AMP shown in Figure P4–67. As her supervisor, you must determine if it works. Does her design meet the specifications?

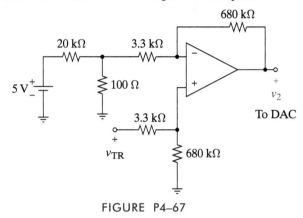

FIGURE P4–67

APPLICATION PROBLEMS

4–68 The analog output of a five-bit DAC is 3.59 V when the input code is (1, 0, 1, 1, 1). What is the full-scale output of the DAC? How much does the analog output change when the input LSB changes?

4–69 The full-scale output of a six-bit DAC is 5.0 V. What is the analog output when the input code is (0, 1, 0, 1, 0, 0)? What is the resolution of this DAC?

4–70 An *R-2R* DAC is shown in Figure P4–70. The digital voltages v_1, v_2, etc. can be either 5 V for a logic 1 or 0 V for a logic 0. What is the DAC's output when the logic input is (1, 1, 0 1)?

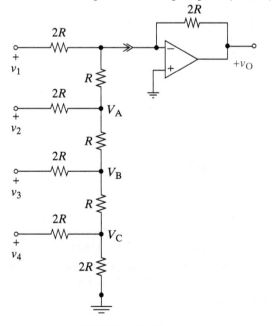

FIGURE P4–70

4–71 A 5th bit is added to the *R-2R* DAC shown in Figure P4–70. What is the maximum possible magnitude of the output voltage? What is the resolution of the revised DAC?

4–72 A Chromel-Constantan thermocouple (curve E) has the characteristics shown in Figure P4–72. Design an interface that will produce a 0- to 5-V output where 0 V refers to 0° C and 5 V refers to 1000° C. The transducer can be modeled as a voltage source in series with a 15-Ω resistor.

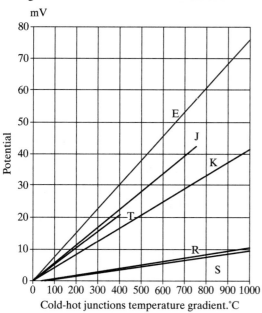

FIGURE P4–72

4–73 A Chromel-Alumel thermocouple (curve K in Figure P4–72) is used to measure the temperature of an electric oven used in the semiconductor industry. Design an interface that will produce a 0- to 5-V output where 0 V refers to 200° C and 5 V refers to 1200° C (assume a straight line out to 1200° C). The transducer can be modeled as a voltage source in series with a 100-Ω resistor.

4–74 An analog accelerometer produces a continuous voltage that is proportional to acceleration in gravitational units or g. Figure P4–74 shows the characteristics of the accelerometer in question. The black curve is the actual characteristics; the colored curve is an acceptable linearized model. Design an instrumentation system that will output 0 V for −1 g and 5 V for 1.5 g. Note that this accelerometer has an output resistance of 32 kΩ

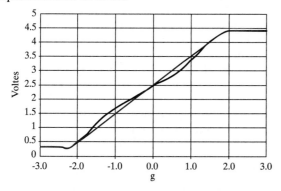

FIGURE P4–74

4–75 A small pressure transducer has the characteristics shown in Figure P4–75. Design an interface that will operate between 7- and 32-psi. An input of 7 psi should produce −5 V and 32 psi should produce 5 V. The transducer is modeled as a voltage source in series with a 500 Ω resistor that can vary ±75 Ω depending on the pressure. The OP AMPs you must use have a maximum closed-loop gain of 2000. Your available bias source is 5 V.

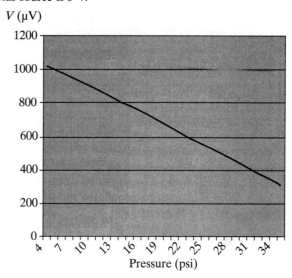

FIGURE P4–75

4–76 A medical grade pressure transducer has been developed for use in invasive blood pressure monitoring. The output voltage of the transducer is $v_{TR} = (0.06P - 0.75)$ mV, where P is pressure in mmHg. The output resistance of the transducer is 1 kΩ. The blood pressure measurement is to be an input to an existing multisensor monitoring system. This system treats a 1-V input as a blood pressure of 20 mmHg and a 10-V input as 200 mmHg. Design an OP AMP circuit to interface the new pressure transducer with the existing monitoring system.

4–77 The acid/alkaline balance of a fluid is measured by the pH scale. The scale runs from 0 (extremely acid) to 14 (extremely alkaline,) with pH 7 being neutral. A pH electrode is a sensor that produces a small voltage that is directly proportional to the pH of the fluid in a test chamber. For a certain pH electrode, the proportionality factor is 60 mV/pH. A preamplifier is needed to interface this sensor with a variety of laboratory instruments. The output of the preamp must be 1 V when the sensor is immersed in a test solution with pH = 4 and 1.75 V when it is immersed in a solution with pH = 7. Design an amplifier to meet these requirements.

4–78 A photoresistor varies from 50 Ω, in bright sunlight to 500 kΩ in total darkness. Design a suitable circuit using the photoresistor so that total darkness produces 5 V, while bright sunlight produces −5 V, regardless of the load. You have a 5-V source and a ± 15-V source to power any OP AMP you may need.

4–79 Your engineering firm needs an instrumentation amplifier that provides the following input-output relationship: $v_O = 10^6 v_{TR} - 3.5$ V. The transducer is modeled as a voltage source in series with a resistor that varies with the transducer voltage from 40 Ω to 750 Ω. A vendor is offering the amplifier shown in Figure P4–79 and the vendor agrees to make a single change to the amplifier, if needed, for no cost. Would you recommend buying it? Explain your rationale.

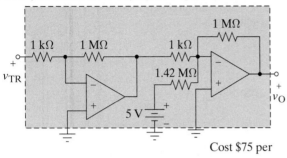

Cost $75 per

Instrumentation Amplifier Model 54

FIGURE P4–79

4–80 **D** Your supervisor drew the Figure of P4–80 on the back of an envelope to show you what he expects as an output to a signal that varies between ±5 V. Design a suitable comparator circuit to achieve his expectation.

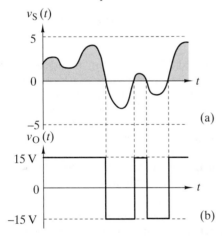

FIGURE P4–80

4–81 The OP AMP in Figure P4–81 operates as a comparator. Find the output voltage when $v_S = 2$ V. Repeat for $v_S = -2$ V and $v_S = 6$ V.

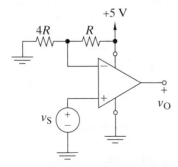

FIGURE P4–81

4–82 The circuit in Figure P4–82 has $V_{CC} = 15$ V and $v_N = -3$ V. Sketch the output voltage v_O on the range $0 \leq t \leq 2$ s for $v_S(t) = 10 \sin(2\pi t)$ V.

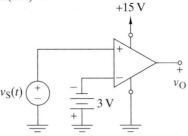

FIGURE P4–82

4–83 **D** A signal varies as $v_S(t) = 10(1 - e^{-100t})$ V. Design a comparator circuit that will switch from 0 to 5 V at 2 ms. Use a single OP AMP that has $+ V_{CC} = + 5$V and $- V_{CC} = 0$ V.

4–84 A 5-bit flash ADC in Figure P4–84 uses a reference voltage of 6 V. Find the output code for the analog inputs $v_S = 3.5$ V, 2.3 V, and 5.3 V. If the reference voltage is changed to 9 V, which of these codes would change?

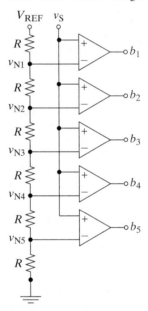

FIGURE P4–84

INTEGRATING PROBLEMS

4–85 **A** Bipolar Power Supply Voltages
The circuit in Figure P4–85 produces bipolar power supply voltages $V_{POS} > 0$ and $V_{NEG} < 0$ from a floating unipolar voltage source $V_{REF} > 0$. Note that the OP AMP output is grounded and that its $+ V_{CC}$ and $-V_{CC}$ terminals are connected to V_{POS} and V_{NEG}, respectively.
   (a) Show that $V_{POS} = +V_{REF}/2$ and $V_{NEG} = -V_{REF}/2$ even if the load resistors R_{POS} and R_{NEG} are not equal.
   (b) If R_{POS} and R_{NEG} are not equal, a current i_G must flow into or out of ground. How does the ungrounded voltage source V_{REF} supply this ground current?

(c) In effect, the OP AMP creates a "virtual ground" at point A ($V_A = 0$) but draws no current in doing so. Why not just connect point A to a "real ground" and do away with the OP AMP?

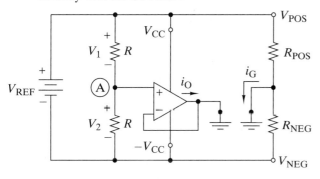

FIGURE P4–85

4–86 ⓓ Thermometer Design Problem

There is a need to design a thermometer that can read from 250° C to 1000° C to monitor the temperature of a rocket's nose cone. The output will feed a 0- to 5-V ADC prior to transmission of the temperature data to the ground. A reading of 250° C will deliver 0 V to the ADC, while 1000° C will deliver 5 V. Select from Figure P4–72 an appropriate thermocouple with sufficient voltage spread and design the instrumentation amplifier. Other criteria are: the transducers have a 15-Ω series resistance, the fewer OP AMPs the better, since power is limited, the maximum OP AMP gain is 1000, and the bias voltage, if needed, must be from the available 15-V supply.

4–87 ⓓ High Bias Design Problem

A particular pressure sensor is designed to operate under constant pressure. The task is to detect a pressure increase and sound an alarm. The sensor produces 1 mV at 100 psi, its usual operating pressure, and increases by 1 μV per psi. The design must sound an alarm if the pressure reaches 150 psi. The transducer is modeled as a voltage source with a series resistor that varies with pressure from 50 Ω at 100 psi to 150 Ω at 150 psi. Two types of OP AMPs are available. Type 1 are single-sided, meaning they have a $+V_{CC}$ of 5 V and a $-V_{CC}$ of 0 V. Type 2 OP AMPs are double-sided with a V_{CC} of ±15 V and have a maximum closed-loop gain of 5000. The alarm can be driven with the +5 V available from a saturated type #1 OP AMP. Only ±15-V and +5-V sources are available. Design the alarm circuit.

4–88 ⓐ Current Switching DAC

The circuit in Figure P4–88 is a 4–bit digital-to-analog converter (DAC). The DAC output is the voltage v_O and the input is the binary code represented by bits b_1, b_2, b_3, and b_4. The input bits are either 0 (low) or 1 (high), and each controls one of the four switches in the figure. When bits are low, their switches are in the left position, directing the $2R$ leg currents to ground. When bits are high, their switches move to the right position, directing the $2R$ leg currents to the OP AMP's inverting input. The $2R$ leg currents do not change when switching from left to right because the inverting input is a

virtual ground ($v_N = v_P = 0$). The purpose of this problem is to show that this constant-current switching produces the following input-output relationship.

$$v_O = -\frac{R_F}{2R} V_{REF} \left(b_1 + \frac{b_2}{2} + \frac{b_3}{4} + \frac{b_4}{8} \right)$$

(a) Since the inverting input is a virtual ground, show that the currents in the $2R$ legs are $i_1 = V_{REF}/2R$, $i_2 = V_{REF}/4R$, $i_3 = V_{REF}/8R$, and $i_4 = V_{REF}/16R$, regardless of switch positions.

(b) Show that the sum of currents at the inverting input is

$$b_1 i_1 + b_2 i_2 + b_3 i_3 + b_4 i_4 \, | \, i_F - 0$$

where bits b_k ($k = 1, 2, 3, 4$) are either 0 or 1.

(c) Use the results in parts (a) and (b) to show that the OP AMP output voltage is

$$v_O = -\frac{R_F}{2R} V_{REF} \left(b_1 + \frac{b_2}{2} + \frac{b_3}{4} + \frac{b_4}{8} \right)$$

as stated.

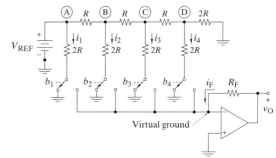

FIGURE P4–88

4–89 (ⓐ, ⓓ) OP AMP Circuit Analysis and Design

(a) ⓐ Find the input-output relationship of the circuit in Figure P4–89.

(b) ⓓ Design a circuit that realizes the relationship found in part (a) using only 10-kΩ resistors and one OP AMP.

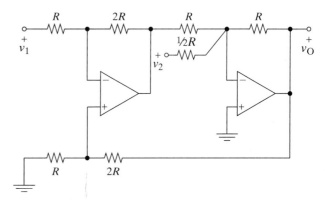

FIGURE P4–89

4–90 Instrumentation Amplifier with Alarm

Strain gauges measuring the deflection of a sintered metal column are connected to a Wheatstone bridge. The output of the bridge is balanced when there is no strain producing 0 V output. As the column is deflected, the bridge produces 200 μV per ohm change caused by the strain gauges. The maximum possible defection would result in a 5-Ω change. Design an instrumentation amplifier that can take the voltage created in the bridge and send it to a 0-5V ADC, where no defection produces 0 V at the ADC and maximum deflection produces 5 V. To avoid loading, send the output of the bridge to a very high input resistance difference amplifier. The columns being tested are brittle and can shatter violently if the conditions cause the strain gauges to change by more than 4.5 Ω. For safety, connect the output of your instrumentation amplifier to a comparator circuit that will trigger an alarm when the strain causes the resistance to change by 4.3 Ω. The alarm needs to be triggered by 15 V.

4–91 Resistance Temperature Transducer

A resistive transducer uses a sensing element whose resistance varies with temperature. For a particular transducer, the resistance varies as $R_{TR} = 0.375T + 100\ \Omega$, where T is temperature in °C. This transducer is to be included in a circuit to measure temperatures in the range from $-200°$ C to $800°$ C. The circuit must convert the transducer resistance variation over this temperature range into an output voltage in the range from 0 to 5 V. Two proposed circuit designs are shown in Figure P4–91.

Which of these circuits would you recommend for production and why? *Hint:* First verify that the circuits perform the required function. Use OrCAD to verify your results.

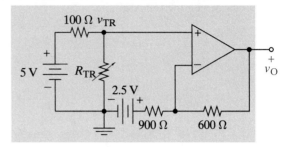

Circuit 1

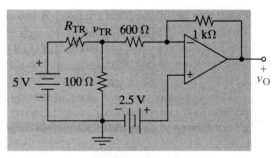

Circuit 2

FIGURE P4–91

CHAPTER 5 SIGNAL WAVEFORMS

Under the sea, under the sea mark how the telegraph motions to me. Under the sea, under the sea signals are coming along.

James Clerk Maxwell, 1873,
Scottish Physicist and
Occasional Humorous Poet.

Some History Behind This Chapter

James Clerk Maxwell (1831–1879) is considered the unifying founder of the mathematical theory of electromagnetics. This genial Scotsman often communicated his thoughts to friends and colleagues via whimsical poetry. In the short excerpt given above, Maxwell reminds us that the purpose of a communication system (the submarine cable telegraph in this case) is to transmit signals and that those signals must be changing, or *in motion* as he put it.

Why This Chapter Is Important Today

Up to this point we have only treated dc signals that are time invariant. These constant signals are a logical place to begin the study of circuit analysis and design. However, to carry information, signals must change, otherwise they keep telling us the same thing over and over again. This chapter introduces the three basic, time-varying signals used in the analysis and design of linear circuits.

The chapter also demonstrates how signals can be combined to create more complex signals. Finally, we look at key, though partial, properties of signals that allow us to determine how they might perform without describing the complete signal.

Chapter Sections

Chapter Learning Objectives

5-1 Basic Waveforms (Sects. 5–2, 5–3, 5–4)

Given an equation, graph, or word description of step, ramp, exponential, or sinusoid waveforms:
(a) Construct an alternative description of the waveform.
(b) Find the parameters or properties of the waveform.
(c) Construct new waveforms by integrating or differentiating the given waveform.
(d) Generate the basic waveforms in MATLAB or OrCAD and use them appropriately to solve problems or simulate circuits. (See Web Appendix C)

5-2 Composite Waveforms (Sect. 5–5)

Given an equation, graph, or word description of a composite waveform:
(a) Construct an alternative description of the waveform.
(b) Find the parameters or properties of the waveform.
(c) Generate the composite waveforms in MATLAB or OrCAD and use them appropriately to solve problems or simulate circuits. (See Web Appendix C)

5-3 Waveform Partial Descriptors (Sect. 5–6)

Given a complete description of a basic or composite waveform:
(a) Classify the waveform as periodic or aperiodic and causal or noncausal.
(b) Find the applicable partial waveform descriptors.
(c) Use appropriate software tools to calculate applicable partial waveform descriptors (See Web Appendix C).

5-1   INTRODUCTION

We normally think of a signal as an electrical current $i(t)$ or voltage $v(t)$. The time variation of the signal is called a waveform. More formally,

> **A waveform** *is an equation or graph that defines the signal as a function of time.*

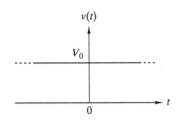

Up to this point our study has been limited to the type of waveform shown in Figure 5–1. The three dots at the start of the waveform indicate that the waveform is unchanged from the beginning, while the three dots at the end imply that it will remain the same forever. Waveforms that are constant for all time are called **dc signals**. The abbreviation *dc* stands for direct current, but it applies to either voltage or current. Mathematical expressions for a dc voltage $v(t)$ or current $i(t)$ take the form

FIGURE 5–1 *A constant or dc waveform.*

$$\left.\begin{array}{l} v(t) = V_0 \\ i(t) = I_0 \end{array}\right\} \quad \text{for } -\infty < t < \infty \qquad (5\text{–}1)$$

This equation is only a model. No physical signal can remain constant forever. It is a useful model, however, because it approximates the signals produced by physical devices such as batteries.

There are two matters of notation and convention that must be discussed before continuing. First, quantities that are constant (non-time-varying) are usually represented by uppercase letters (V_A, I, T_O) or lowercase letters in the early part of the alphabet (a, b_7, f_0). Time-varying electrical quantities are represented by the lowercase letters i, v, p, q, and w. The time variation is expressly indicated when we write these quantities as $v_1(t)$, $i_A(t)$, or $w_C(t)$. Time variation is implicit when they are written as v_1, i_A, or w_C.

Second, in a circuit diagram, signal variables are normally accompanied by the reference marks (+, −) for voltage and (→) for current. It is important to remember that these reference marks *do not* indicate the polarity of a voltage or the direction of current. The marks provide a baseline for determining the sign of the numerical value of the actual waveform. When the actual voltage polarity or current direction coincides with the reference directions, the signal has a positive value. When the opposite occurs, the value is negative. Figure 5–2 shows examples of voltage waveforms, including some that assume both positive and negative values. The bipolar waveforms indicate that the actual voltage polarity is changing as a function of time.

The waveforms in Figure 5–2 are examples of signals used in electrical engineering. Since there are many such signals, it may seem that the study of signals involves the uninviting task of compiling a lengthy catalog of waveforms. However, it turns out that a long list is not needed. In fact, we can derive most of the waveforms of interest using just three basic signal models: the step, exponential, and sinusoidal functions. The small number of basic signals illustrates why models are so useful to engineers. In reality, waveforms are very complex, but their time variation can be approximated adequately using only a few basic building blocks.

Finally, in this chapter we will focus on the use of voltage $v(t)$ to represent a signal waveform. Remember, however, that a signal can be either a voltage $v(t)$ or current $i(t)$.

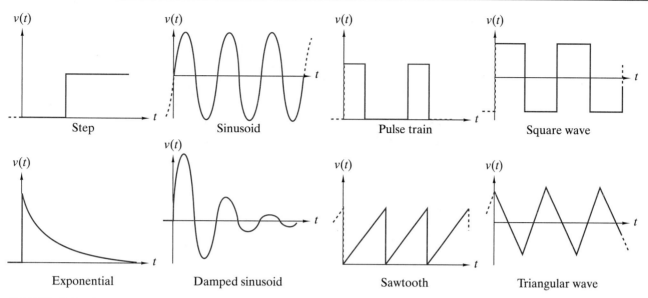

FIGURE 5–2 *Some example waveforms.*

5–2 THE STEP WAVEFORM

The first basic signal in our catalog is the step waveform. The general step function is based on the **unit step function**[1] defined as

$$u(t) = \begin{cases} 0 & \text{for } t < 0 \\ 1 & \text{for } t \geq 0 \end{cases} \qquad (5\text{–}2)$$

The step function waveform is equal to zero when its argument t is negative, and is equal to unity when its argument is positive. Mathematically, the function $u(t)$ has a jump discontinuity at $t = 0$.

Strictly speaking, it is impossible to generate a true step function since signal variables like current and voltage cannot jump from one value to another in zero time. Practically speaking, we can generate very good approximations to the step function. What is required is that the transition time be short compared with other response times in the circuit. Actually, the generation of approximate step functions is an everyday occurrence since people frequently turn things like TVs, stereos, and lights on and off.

Figure 5–3(a) shows how a step might be physically constructed on a circuit diagram. It is assumed that the transition from off to on occurs instantaneously at the time the switch is thrown, that is, at $t = 0$ in the figure. The same process is shown more compactly in Figure 5–3(b).

On the surface, it may appear that the step function is not a very exciting waveform or, at best, only a source of temporary excitement. However, the step waveform is a versatile signal used to construct a wide range of useful waveforms. Multiplying $u(t)$ by a constant V_A produces the waveform

$$V_A u(t) = \begin{cases} 0 & \text{for } t < 0 \\ V_A & \text{for } t \geq 0 \end{cases} \qquad (5\text{–}3)$$

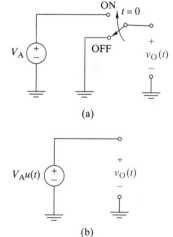

FIGURE 5–3 *(a) Approximation to a step function. (b) Typical representation.*

---

[1]The step function is also referred to as a heaviside function, after Oliver Heaviside, English electrical engineer (1850–1925).

Replacing t by $(t - T_S)$ produces a waveform $V_A u(t - T_S)$, which takes on the values

$$V_A u(t - T_S) = \begin{cases} 0 & \text{for } t < T_S \\ V_A & \text{for } t \geq T_S \end{cases} \qquad (5-4)$$

The **amplitude** V_A scales the size of the step discontinuity, and the **time-shift** parameter T_S advances or delays the time at which the step occurs, as shown in Figure 5–4. The step function transitions when the value of t makes the argument of the function equal to zero. For example, the function $v(t) = 5\,u(t + 6)$ V shifts $+5$ volts at $t = -6$ s, and $i(t) = 2\,u(t - 1)$ mA shifts $+2$ mA at $t = 1$ s.

FIGURE 5–4 *Effect of time shifting on the step function waveform.*

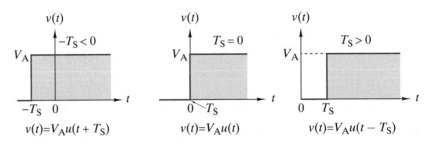

Amplitude and time-shift parameters are required to define the general step function. The amplitude V_A carries the units of volts. The amplitude of the step function in electric current is I_A and carries the units of amperes. The constant T_S carries the units of time, usually seconds. The parameters V_A (or I_A) and T_S can be positive, negative, or zero. By combining several step functions, we can represent a number of important waveforms. One possibility is illustrated in the following example:

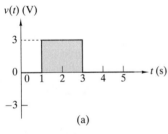

(a)

(b)

FIGURE 5–5

EXAMPLE 5–1

Express the waveform in Figure 5–5(a) in terms of step functions.

SOLUTION:

The amplitude of the pulse jumps to a value of 3 V at $t = 1$ s; therefore, $3u(t - 1)$ is part of the equation for the waveform. The pulse returns to zero at $t = 3$ s, so an equal and opposite step must occur at $t = 3$ s. Putting these observations together, we express the rectangular pulse as

$$v(t) = 3u(t - 1) - 3u(t - 3) \text{ V}$$

Figure 5–5(b) shows how the two step functions combine to produce the given rectangular pulse. ∎

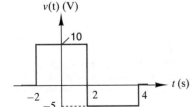

FIGURE 5–6

Exercise 5–1

Write an expression using unit step functions for the waveform in Figure 5–6.

Answer: $v(t) = 10u(t + 2) - 15u(t - 2) + 5u(t - 4)V$

THE IMPULSE FUNCTION

The generalization of Example 5–1 is the waveform

$$v(t) = V_A[u(t - T_1) - u(t - T_2)] \text{ V}$$

This waveform is a rectangular pulse of amplitude V_A that turns on at $t = T_1$ and off at $t = T_2$. The pulse train and square wave signals in Figure 5–2 can be generated by a series of these pulses. Pulses that turn on at some time T_1 and off at some later time T_2 are sometimes called **gating functions** because they are used in conjunction with electronic switches to enable or inhibit the passage of another signal.

A unit-area pulse centered on $t = 0$ is written in terms of step functions as

$$v(t) = \frac{1}{T}\left[u\left(t + \frac{T}{2}\right) - u\left(t - \frac{T}{2}\right)\right] \text{ V} \tag{5–5}$$

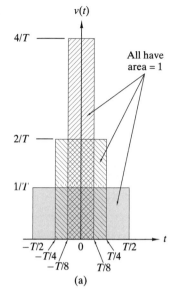

The pulse in Eq. (5–5) is zero everywhere except in the range $-T/2 \le t \le T/2$, where its value is $1/T$. The area under the pulse is 1 because its scale factor is inversely proportional to its duration. As shown in Figure 5–7(a), the pulse becomes narrower and higher as T decreases but maintains its unit area. In the limit as $T \to 0$ the scale factor approaches infinity but the area remains 1. The function obtained in the limit is called a **unit impulse**[2], symbolized as $\delta(t)$. The graphical representation of $\delta(t)$ is shown in Figure 5–7(b). The impulse is an idealized model of a large-amplitude, short-duration pulse.

A formal definition of the unit impulse is

$$\delta(t) = 0 \text{ for } t \ne 0 \quad \text{and} \quad \int_{-\infty}^{t} \delta(x)dx = u(t) \tag{5–6}$$

The first condition says the impulse is zero everywhere except at $t = 0$. The second condition suggests that the unit impulse is the derivative of a unit step function:

$$\delta(t) = \frac{du(t)}{dt} \tag{5–7}$$

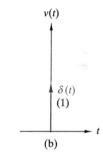
(b)

FIGURE 5–7 *Rectangular pulse waveforms and the impulse.*

The conclusion in Eq. (5–7) cannot be justified using elementary mathematics since the function $u(t)$ has a discontinuity at $t = 0$ and its derivative at that point does not exist in the usual sense. However, the concept can be justified using limiting conditions on continuous functions, as discussed in texts on signals and systems.[3] Accordingly, we defer the question of mathematical rigor to later courses and think of the unit impulse as the derivative of a unit step function. Note that this means that the unit impulse $\delta(t)$ has units of reciprocal time, or s^{-1}.

An impulse of strength K is denoted $v(t) = K\delta(t)$. Consequently, the scale factor K has the units of V-s and is the area under the impulse $K\delta(t)$. In the graphical representation of the impulse the value of K is written in parentheses beside the arrow, as shown in Figure 5–7(b).

EXAMPLE 5–2

Calculate and sketch the derivative of the pulse in Figure 5–8(a).

SOLUTION:
In Example 5–1 the pulse waveform was written as

$$v(t) = 3u(t - 1) - 3u(t - 3) \text{ V}$$

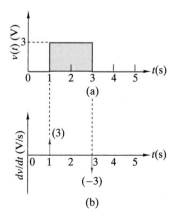

(b)

FIGURE 5–8

---

[2]The unit impulse is also referred to as the Dirac delta, after Paul Dirac, the British physicist who introduced it.

[3]For example, see Alan V. Oppenheim and Allan S. Willsky, *Signals and Systems Analysis* (Englewood Cliffs, N.J.: Prentice Hall, 1983), pp. 22–23.

Using the derivative property of the step function, we write

$$\frac{dv(t)}{dt} = 3\delta(t-1) - 3\delta(t-3)$$

The derivative waveform consists of a positive-going impulse at $t=1$ s and a negative-going impulse at $t=3$ s. Figure 5–8(b) shows how the impulse train is represented graphically. The waveform $v(t)$ has the units of volts (V), so its derivative $dv(t)/dt$ has the units of V/s. ∎

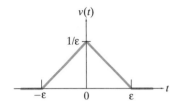

FIGURE 5–9

Exercise 5–2

Figure 5–9 purports to be an alternative description of an impulse function as $\varepsilon \to 0$. Prove or disprove the claim.

Answer: From the figure, the area of the triangle is $\frac{1}{2}(2\varepsilon)(1/\varepsilon) = 1$. As $\varepsilon \to 0$, the base of the triangle shrinks to zero but the amplitude grows to infinity. Yet the area always remains at 1. Hence, this is equivalent to the definition of an impulse and proves the claim.

THE RAMP FUNCTION

The **unit ramp** is defined as the integral of a step function:

$$r(t) = \int_{-\infty}^{t} u(x)dx = tu(t) \qquad (5\text{–}8)$$

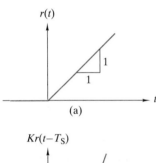

(a)

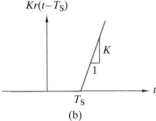

(b)

FIGURE 5–10 *(a) Unit ramp waveform. (b) General ramp waveform.*

The unit ramp waveform $r(t)$ in Figure 5–10(a) is zero for $t<0$ and is equal to t for $t>0$. Notice that the slope of $r(t)$ is 1 and has the units of time, or s. A ramp of strength K is denoted $v(t) = Kr(t)$, where the scale factor K has the units of V/s and is the slope of the ramp. The general ramp waveform shown in Figure 5–10(b), written as $v(t) = Kr(t - T_S)$, is zero for $t < T_S$ and equal to $K(t - T_S)$ for $t \geq T_S$. By adding a sequence of ramps, we can create the triangular and sawtooth waveforms shown in Figure 5–2.

SINGULARITY FUNCTIONS

The unit impulse, unit step, and unit ramp form a triad of related signals that are referred to as **singularity functions**. They are related by integration as

$$u(t) = \int_{-\infty}^{t} \delta(x)dx$$

$$r(t) = \int_{-\infty}^{t} u(x)dx \qquad (5\text{–}9)$$

or by differentiation as

$$\delta(t) = \frac{du(t)}{dt}$$

$$u(t) = \frac{dr(t)}{dt} \qquad (5\text{–}10)$$

These signals are used to generate other waveforms and as test inputs to linear systems to characterize their responses. When applying the singularity functions in circuit analysis, it is important to remember that $u(t)$ is a dimensionless function. Eqs. (5–9) and (5–10) point out that $\delta(t)$ carries the units of s^{-1} and $r(t)$ carries units of seconds.

EXAMPLE 5-3

Derive an expression for the waveform for the integral of the pulse in Figure 5–11(a).

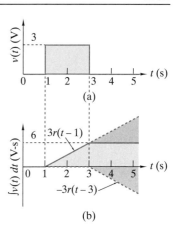

(a)

SOLUTION:

In Example 5–1 the pulse waveform was written as

$$v(t) = 3u(t-1) - 3u(t-3) \text{ V}$$

Using the integration property of the step function, we write

$$\int_{-\infty}^{t} v(x)dx = 3r(t-1) - 3r(t-3)$$

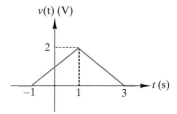

(b)

FIGURE 5–11

The integral is zero for $t < 1$ s. For $1 < t < 3$ the waveform is $3(t-1)$. For $t > 3$ it is $3(t-1) - 3(t-3) = 6$. These two ramps produce the pulse integral shown in Figure 5–11(b). The waveform $v(t)$ has the units of volts (V), so the units of its integral are V-s. ∎

Exercise 5–3

Write an expression using ramp functions to describe the waveform shown in Figure 5–12

Answer: $v(t) = r(t+1) - 2r(t-1) + r(t-3)V$

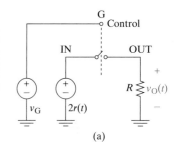

FIGURE 5–12

EXAMPLE 5-4

Figure 5–13(a) shows an ideal electronic switch whose input is a ramp $2r(t)$, where the scale factor $K = 2$ carries the units of V/s. Find the switch output $v_O(t)$ when the gate function in Example 5–1 is applied to the control terminal (G) of the switch.

SOLUTION:

In Example 5–1 the gate function was written as

$$v_G(t) = 3u(t-1) - 3u(t-3) \text{ V}$$

The gate function turns the switch on at $t = 1$ s and off at $t = 3$ s. The output voltage of the switch is

$$v_O(t) = \begin{cases} 0 & t < 1 \\ 2t & 1 < t < 3 \\ 0 & 3 < t \end{cases}$$

Only the portion of the input waveform within the gate interval appears at the output. Figures 5–13(b), 5–13(c), and 5–13(d) show how the gate function $v_G(t)$ controls the passage of the input signal through the electronic switch.

This waveform can be written as a sum of singularity functions as follows. First we write $v_O(t)$ in terms of a gate function:

$$v_O(t) = 2t \underbrace{[u(t-1) - u(t-3)]}_{\text{Gate function}}$$

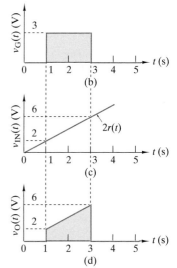

FIGURE 5–13

We then manipulate this equation as follows:

$$v_O(t) = 2tu(t-1) - 2tu(t-3)$$

$$= 2(t-1+1)u(t-1) - 2(t-3+3)u(t-3)$$

$$= 2\underbrace{(t-1)u(t-1)}_{r(t-1)} + 2u(t-1) - 2\underbrace{(t-3)u(t-3)}_{r(t-3)} - 6u(t-3)$$

So finally,

$$v_O(t) = 2r(t-1) + 2u(t-1) - 2r(t-3) - 6u(t-3)$$

which describes the gated ramp in terms of step and ramp waveforms.   ■

Exercise 5–4

Express the following signals in terms of singularity functions:

(a) $v_1(t) = \begin{cases} 0 & t < 2 \\ 4 & 2 < t < 4 \\ -4 & 4 < t \end{cases}$ (b) $v_2(t) = \begin{cases} 0 & t < 2 \\ 4 & 2 < t < 4 \\ -2t + 12 & 4 < t \end{cases}$

(c) $v_3(t) = \int_{-\infty}^{t} v_1(x)dx$ (d) $v_4(t) = \dfrac{dv_2(t)}{dt}$

Answers:

(a) $v_1(t) = 4u(t-2) - 8u(t-4)$ (b) $v_2(t) = 4u(t-2) - 2r(t-4)$
(c) $v_3(t) = 4r(t-2) - 8r(t-4)$ (d) $v_4(t) = 4\delta(t-2) - 2u(t-4)$

5–3 THE EXPONENTIAL WAVEFORM

The **exponential waveform** is a step function whose amplitude factor gradually decays to zero. The equation for this waveform is

$$v(t) = \left[V_A e^{-t/T_C} \right] u(t) \qquad (5\text{–}11)$$

A graph of $v(t)$ versus t/T_C is shown in Figure 5–14. The exponential starts out like a step function. It is zero for $t < 0$ and jumps to a maximum amplitude of V_A at $t = 0$. Thereafter it monotonically decays toward zero as time marches on. The two parameters that define the waveform are the **amplitude** V_A (in volts) and the **time constant** T_C (in seconds). The amplitude of a current exponential would be written I_A and carry the units of amperes.

FIGURE 5–14 *The exponential waveform.*

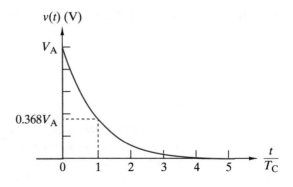

The time constant is of special interest, since it determines the rate at which the waveform decays to zero. An exponential decays to about 36.8% of its initial amplitude $v(0) = V_A$ in one time constant, because at $t = T_C, v(T_C) = V_A e^{-1}$, or approximately $0.368 \times V_A$. At $t = 5T_C$, the value of the waveform is $V_A e^{-5}$, or approximately $0.00674\, V_A$. An exponential signal decays to less than 1% of its initial amplitude in a time span of five time constants. In theory, an exponential endures forever, but practically speaking after about $5T_C$ the waveform amplitude becomes negligibly small. We define the **duration** of a waveform to be the interval of time outside of which the waveform is everywhere less than a stated value. Using this concept, we say the duration of an exponential waveform is $5T_C$.

Figure 5–15 shows how an exponential waveform can be constructed. Note that the unit step multiplies the exponential function and turns it on at $t = 0$.

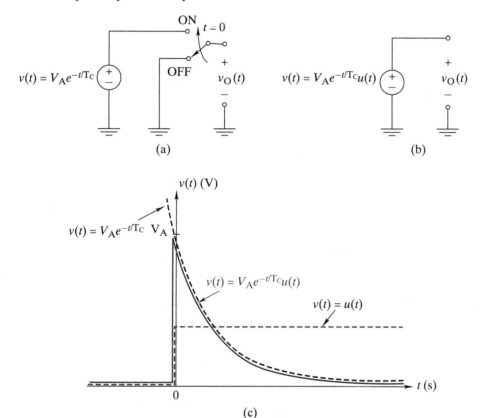

FIGURE 5–15

Note: The waveforms actually fall atop each other but are shown slightly shifted so they can be distinguished.

EXAMPLE 5 – 5

Plot the waveform $v(t) = [-17e^{-100t}]u(t)$ V.

SOLUTION:
From the form of $v(t)$, we recognize that $V_A = -17$ V and $T_C = 1/100$ s or 10 ms. The minimum value of $v(t)$ is $v(0) = -17$ V, and the maximum value is approximately 0 V as t approaches $5T_C = 50$ ms. These observations define appropriate scales for plotting the waveform. Spreadsheet programs are especially useful for the repetitive calculations and graphical functions involved in waveform plotting. Figure 5–16 shows how this example can be handled using Excel. We begin by filling in the A column with time values ranging from $t = 0$ to $t = 50$ ms. The data in the B column are obtained by calculating the value of $-17e^{-100t}$ for each of the values of t (ms) in the A column.

FIGURE 5–16

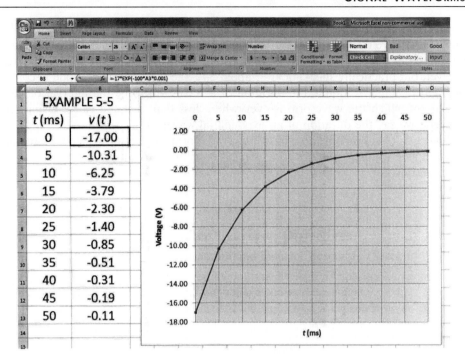

Then, using the graphing tool, we create a plot showing that the waveform starts out at $v(t) = -17\,\text{V}$ at $t = 0$ and then increases toward $v(t) = 0$ as $t \to 50$ ms.   ■

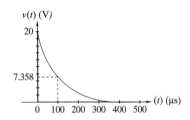

FIGURE 5–17

Exercise 5–5

Sketch the waveform described by

$$v(t) = 20e^{-10,000t}u(t)\,\text{V}$$

A n s w e r :   See Figure 5–17.

PROPERTIES OF EXPONENTIAL WAVEFORMS

The **decrement property** describes the decay rate of an exponential signal. For $t > 0$ the exponential waveform is given by

$$v(t) = V_{\text{A}}e^{-t/T_{\text{C}}}\,\text{V} \tag{5–12}$$

The step function can be omitted since it is unity for $t > 0$. At time $t + \Delta t$ the amplitude is

$$v(t + \Delta t) = V_{\text{A}}e^{-(t+\Delta t)/T_{\text{C}}} = V_{\text{A}}e^{-t/T_{\text{C}}}e^{-\Delta t/T_{\text{C}}}\,\text{V} \tag{5–13}$$

The ratio of these two amplitudes is

$$\frac{v(t + \Delta t)}{v(t)} = \frac{V_{\text{A}}e^{-t/T_{\text{C}}}\,e^{-\Delta t/T_{\text{C}}}}{V_{\text{A}}e^{-t/T_{\text{C}}}} = e^{-\Delta t/T_{\text{C}}} \tag{5–14}$$

The decrement ratio is independent of amplitude and time. In any fixed time period Δt, the fractional decrease depends only on the time constant. The decrement property states that the same percentage decay occurs in equal time intervals.

The slope of the exponential waveform (for $t > 0$) is found by differentiating Eq. (5–12) with respect to time:

$$\frac{dv(t)}{dt} = -\frac{V_A}{T_C}e^{-t/T_C} = -\frac{v(t)}{T_C} \tag{5–15}$$

The **slope property** states that the time rate of change of the exponential waveform is inversely proportional to the time constant. Small time constants lead to large slopes or rapid decays, while large time constants produce shallow slopes and long decay times.

Equation (5–15) can be rearranged as

$$\frac{dv(t)}{dt} + \frac{v(t)}{T_C} = 0 \tag{5–16}$$

When $v(t)$ is an exponential of the form in Eq. (5–12), then $dv/dt + v/T_C = 0$. That is, the exponential waveform is a solution of the first-order linear differential equation in Eq. (5–16). We will make use of this fact in Chapter 7.

The time-shifted exponential waveform is obtained by replacing t on the right side of in Eq. (5–11) by $t - T_S$. The general exponential waveform is written as

$$v(t) = [V_A e^{-(t-T_S)/T_C}]\, u(t - T_S)\ \text{V} \tag{5–17}$$

where T_S is the time-shift parameter for the waveform. Figure 5–18 shows exponential waveforms with the same amplitude and time constant but different values of T_S. Time shifting translates the waveform to the left or right depending on whether T_S is negative or positive. *Caution*: The factor $t - T_S$ must appear in both the argument of the step function and the exponential, as shown in Eq. (5–17).

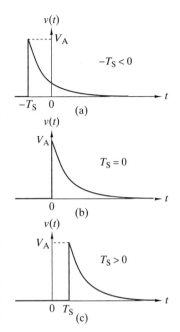

FIGURE 5–18 *Effect of time shifting on the exponential waveform.*

APPLICATION   **E X A M P L E   5 – 6**

An oscilloscope is a laboratory instrument that displays the instantaneous value of a waveform versus time. Figure 5–19 shows an oscilloscope display of a portion of an exponential waveform. In the figure, the vertical (amplitude) axis is calibrated at 2 V per division, and the horizontal (time) axis is calibrated at 1 ms per division. Find the time constant of the exponential.

FIGURE 5–19

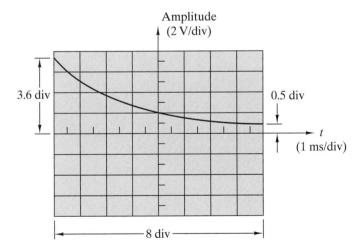

SOLUTION:
For $t > T_S$ the general expression for an exponential in Eq. (5–17) becomes

$$v(t) = V_A e^{-(t-T_S)/T_C}\ \text{V}$$

We have only a portion of the waveform, so we do not know the location of the $t = 0$ time origin; hence, we cannot find the amplitude V_A or the time shift T_S from the display. But, according to the decrement property, we should be able to determine the time constant since the decrement ratio is independent of amplitude and time. Specifically, Eq. (5–14) points out that

$$\frac{v(t + \Delta t)}{v(t)} = e^{-\Delta t/T_C}$$

Solving for the time constant T_C yields

$$T_C = \frac{\Delta t}{\ln\left[\dfrac{v(t)}{v(t + \Delta t)}\right]}$$

Taking the starting point at the left edge of the oscilloscope display yields

$$v(t) = (3.6\,\text{div})(2\,\text{V/div}) = 7.2\,\text{V}$$

Next, defining Δt to be the full width of the display produces

$$\Delta t = (8\,\text{div})(1\,\text{ms/div}) = 8\,\text{ms}$$

and

$$v(t + \Delta t) = (0.5\,\text{div})(2\,\text{V/div}) = 1\,\text{V}$$

As a result, the time constant of the waveform is found to be

$$T_C = \frac{\Delta t}{\ln\left[\dfrac{v(t)}{v(t + \Delta t)}\right]} = \frac{8 \times 10^{-3}}{\ln(7.2/1)} = 4.05\,\text{ms}$$

∎

Exercise 5–6

Figure 5–20 contains three exponential waveforms. Match each curve with the appropriate expression.

1. $v_1(t) = 100\,e^{-(t/100\mu)}\,u(t - 100\mu)\,\text{V}$

2. $v_2(t) = 100\,e^{-(t/100\mu)}\,u(t)\,\text{V}$

3. $v_3(t) = 100\,e^{-(t-100\mu/100\mu)}\,u(t - 100\mu)\,\text{V}$

FIGURE 5–20

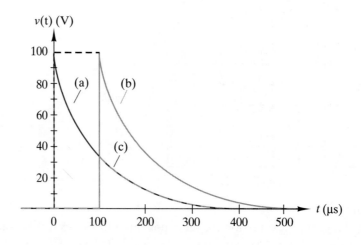

Answers: Waveform (a) $= 2$, (b) $= 3$, and (c) $= 1$.

Exercise 5–7

(a) An exponential waveform has $v(0) = 1.2$ V and $v(3) = 0.5$ V. What are V_A and T_C for this waveform?

(b) An exponential waveform has $v(0) = 5$ V and $v(2) = 1.25$ V. What are the values of $v(t)$ at $t = 1$ s and $t = 4$ s?

(c) An exponential waveform has $v(0) = 5$ V and an initial $(t = 0)$ slope of -25 V/s. What are V_A and T_C for this waveform?

(d) An exponential waveform decays to 10% of its initial value in 3 ms. What is T_C for this waveform?

(e) A waveform has $v(2) = 4$ V, $v(6) = 1$ V, and $v(10) = 0.5$ V. Is it an exponential waveform?

Answers:

(a) $V_A = 1.2$ V, $T_C = 3.43$ s

(b) $v(1) = 2.5$ V, $v(4) = 0.3125$ V

(c) $V_A = 5$ V, $T_C = 200$ ms

(d) $T_C = 1.303$ ms

(e) No, it violates the decrement property.

Exercise 5–8

Find the amplitude and time constant for each of the following exponential signals:

(a) $v_1(t) = [-15e^{-1000t}]u(t)$ V

(b) $v_2(t) = [+12e^{-t/10}]u(t)$ mV

(c) $i_3(t) = [15e^{-500t}]u(-t)$ mA

(d) $i_4(t) = [4e^{-200(t-100)}]u(t - 100)$ A

Answers:

(a) $V_A = -15$ V, $T_C = 1$ ms

(b) $V_A = 12$ mV, $T_C = 10$ s

(c) $I_A = 15$ mA, $T_C = 2$ ms

(d) $I_A = 4$ A, $T_C = 5$ ms

5–4 THE SINUSOIDAL WAVEFORM

The cosine and sine functions are important in all branches of science and engineering. The corresponding time-varying waveform in Figure 5–21 plays an especially prominent role in electrical engineering.

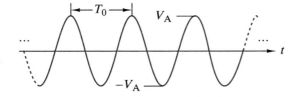

FIGURE 5–21 *The eternal sinusoid.*

In contrast with the step and exponential waveforms studied earlier, the sinusoid, like the dc waveform in Figure 5–1, extends indefinitely in time in both the positive and negative directions. The sinusoid has neither a beginning nor an end. Of course, real signals have finite durations. They were turned on at some finite time in the past and will be turned off at some time in the future. While it may seem unrealistic to have a signal model that lasts forever, it turns out that the eternal sinewave is a very good approximation in many practical applications.

The sinusoid in Figure 5–21 is an endless repetition of identical oscillations between positive and negative peaks. The **amplitude** V_A (in volts) defines the maximum and minimum values of the oscillations. The **period** T_0 (usually seconds) is the time required to complete one cycle of the oscillation. The sinusoid can be expressed mathematically using either the sine or the cosine function. The choice between the two depends on where we choose to define $t = 0$. If we choose $t = 0$ at a point where the sinusoid is zero, then it can be written as

$$v(t) = V_A \sin(2\pi t/T_0) \text{ V} \tag{5–18a}$$

On the other hand, if we choose $t = 0$ at a point where the sinusoid is at a positive peak, we can write an equation for it in terms of a cosine function:

$$v(t) = V_A \cos(2\pi t/T_0) \text{ V} \tag{5–18b}$$

Although either choice will work, it is common practice to choose $t = 0$ at a positive peak; hence Eq. (5–18b) applies. Thus, we will continue to call the waveform a sinusoid even though we use a cosine function to describe it.

As in the case of the step and exponential functions, the general sinusoid is obtained by replacing t by $(t - T_S)$. Inserting this change in Eq. (5–18) yields a general expression for the sinusoid as

$$v(t) = V_A \cos[2\pi(t - T_S)/T_0] \text{ V} \tag{5–19}$$

where the constant T_S is the time-shift parameter. Figure 5–22 shows that the sinusoid shifts to the right when $T_S > 0$ and to the left when $T_S < 0$. In effect, time shifting causes the positive peak nearest the origin to occur at $t = T_S$.

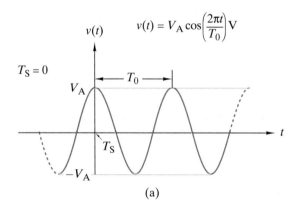

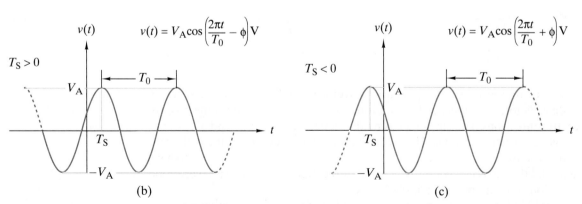

FIGURE 5–22   *Effect of time shifting on the sinusoidal waveform.*

The time-shifting parameter can also be represented by an angle:

$$v(t) = V_A \cos[2\pi t/T_0 + \phi] \text{ V} \tag{5–20}$$

The parameter ϕ is called the **phase angle**. The term *phase angle* is based on the circular interpretation of the cosine function. We think of the period as being divided into 2π radians, or $360°$. In this sense the phase angle is the angle between $t = 0$ and the nearest positive peak. Comparing Eqs. (5–19) and (5–20), we find the relation between T_S and ϕ to be

$$\phi = -2\pi \frac{T_S}{T_0} = -360° \frac{T_S}{T_0} \tag{5–21}$$

Changing the phase angle moves the waveform to the left or right, revealing different phases of the oscillating waveform (hence the name *phase angle*).

The phase angle should be expressed in radians, but is more often reported in degrees. Care must be used when numerically evaluating the argument of the cosine $(2\pi t/T_0 + \phi)$ to ensure that both terms have the same units. The term $2\pi t/T_0$ has the units of radians, so it is necessary to convert ϕ to radians when it is given in degrees.

An alternative form of the general sinusoid is obtained by expanding Eq. (5–20) using the identity $\cos(x + y) = \cos(x)\cos(y) - \sin(x)\sin(y)$,

$$v(t) = [V_A \cos\phi] \cos(2\pi t/T_0) + [-V_A \sin\phi] \sin(2\pi t/T_0) \text{ V}$$

The quantities inside the brackets in this equation are constants; therefore, we can write the general sinusoid in the form

$$v(t) = a\cos(2\pi t/T_0) + b\sin(2\pi t/T_0) \text{ V} \tag{5–22}$$

The two amplitudelike parameters a and b have the same units as the waveform (volts in this case) and are called Fourier coefficients. By definition, the Fourier coefficients are related to the amplitude and phase parameters by the equations

$$a = V_A \cos\phi$$
$$b = -V_A \sin\phi \tag{5–23}$$

The inverse relationships are obtained by squaring and adding the expressions in Eq. (5–23):

$$V_A = \sqrt{a^2 + b^2} \tag{5–24}$$

and by dividing the second expression in Eq. (5–23) by the first:

$$\phi = \tan^{-1}\frac{-b}{a} \tag{5–25}$$

Caution: The inverse tangent function on a calculator has a $\pm180°$ ambiguity that can be resolved by considering the signs of the Fourier coefficients a and b.

It is customary to describe the time variation of the sinusoid in terms of a frequency parameter. **Cyclic frequency** f_0 is defined as the number of periods per unit time. By definition, the period T_0 is the number of seconds per cycle; consequently, the number of cycles per second is

$$f_0 = \frac{1}{T_0} \tag{5–26}$$

where f_0 is the cyclic frequency or simply the frequency. The unit of frequency (cycles per second) is the **hertz** (Hz). The **angular frequency** ω_0 in radians per second is related to the cyclic frequency by the relationship

$$\omega_0 = 2\pi f_0 = \frac{2\pi}{T_0} \tag{5–27}$$

because there are 2π radians per cycle.

There are two ways to express the concept of sinusoidal frequency: cyclic frequency (f_0, hertz) and angular frequency (ω_0, radians per second). When working with signals, we tend to use the former. For example, radio stations transmit carrier signals at frequencies specified as 690 kHz (AM band) or 101 MHz (FM band). Radian frequency is more convenient when describing the characteristics of circuits driven by sinusoidal inputs.

In summary, there are several equivalent ways to describe the general sinusoid:

$$v(t) = V_A \cos\left[\frac{2\pi(t - T_S)}{T_0}\right] = V_A \cos\left(\frac{2\pi t}{T_0} + \phi\right) = a \cos\left(\frac{2\pi t}{T_0}\right) + b \sin\left(\frac{2\pi t}{T_0}\right) \text{ V}$$

$$= V_A \cos[2\pi f_0(t - T_S)] = V_A \cos(2\pi f_0 t + \phi) = a \cos(2\pi f_0 t) + b \sin(2\pi f_0 t) \text{ V}$$

$$= V_A \cos[\omega_0(t - T_S)] = V_A \cos(\omega_0 t + \phi) = a \cos(\omega_0 t) + b \sin(\omega_0 t) \text{ V}$$

To use any one of these expressions, we need three types of parameters:

1. *Amplitude:* either V_A or the Fourier coefficients a and b
2. *Time shift:* either T_S or the phase angle ϕ
3. *Time/frequency:* either T_0, f_0, or ω_0

In different parts of this book we use different forms to represent a sinusoid. Therefore, it is important for you to understand thoroughly the relationships among the various parameters in Eqs. (5–21) through (5–27).

EXAMPLE 5–7

Figure 5–23 shows an oscilloscope display of a sinusoid. The vertical axis (amplitude) is calibrated at 5 V per division, and the horizontal axis (time) is calibrated at 0.1 ms per division. Derive an expression for the sinusoid displayed in Figure 5–23.

FIGURE 5–23

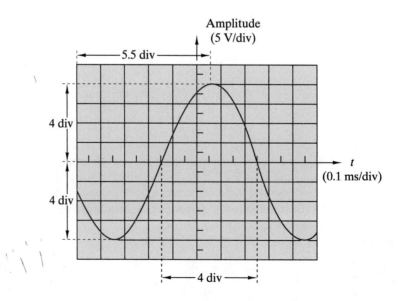

SOLUTION:

The maximum amplitude of the waveform is seen to be four vertical divisions; therefore,

$$V_A = (4\,\text{div})(5\,\text{V/div}) = 20\,\text{V}$$

There are four horizontal divisions between successive zero crossings, which means there are a total of eight divisions in one cycle. The period of the waveform is

$$T_0 = (8\,\text{div})(0.1\,\text{ms/div}) = 0.8\,\text{ms}$$

The two frequency parameters are $f_0 = 1/T_0 = 1.25\,\text{kHz}$ and $\omega_0 = 2\pi f_0 = 7854\,\text{rad/s}$. The parameters V_A, T_0, f_0, and ω_0 do not depend on the location of the $t = 0$ axis.

To determine the time shift T_S, we need to define a time origin. The $t = 0$ axis is arbitrarily taken at the left edge of the display in Figure 5–23. The positive peak shown in the display is 5.5 divisions to the right of $t = 0$, which is more than half a cycle (four divisions). The positive peak closest to $t = 0$ is not shown in Figure 5–23 because it must lie beyond the left edge of the display. However, the positive peak shown in the display is located at $t = T_S + T_0$ since it is one cycle after $t = T_S$. We can write

$$T_S + T_0 = (5.5\,\text{div})(0.1\,\text{ms/div}) = 0.55\,\text{ms}$$

which yields $T_S = 0.55 - T_0 = -0.25$ ms. As expected, T_S is negative because the nearest positive peak is to the left of $t = 0$.

Given T_S, we can calculate the remaining parameters of the sinusoid as follows:

$$\phi = -\frac{2\pi T_S}{T_0} = 1.96\,\text{rad or }112.5°$$

$$a = V_A \cos\phi = -7.65\,\text{V}$$

$$b = -V_A \sin\phi = -18.5\,\text{V}$$

Finally, the three alternative expressions for the displayed sinusoid are

$$v(t) = 20\cos[(7854\,t) + 0.25 \times 10^{-3}]\,\text{V}$$

$$= 20\cos(7854\,t + 112.5°)\,\text{V}$$

$$= -7.65\cos 7854t - 18.5\sin 7854t\,\text{V}$$

∎

Exercise 5–9

Derive an expression for the sinusoid displayed in Figure 5–23 when $t = 0$ is placed in the middle of the display.

Answer: $v(t) = 20\cos(7854t - 22.5°)\,\text{V}$

Exercise 5–10

Sketch the waveform described by

$$v(t) = 10\cos(2000\pi t - 60°)\,\text{V}$$

Answer: See Figure 5–24.

FIGURE 5–24

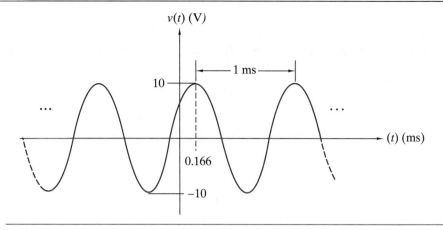

PROPERTIES OF SINUSOIDS

In general, a waveform is said to be **periodic** if

$$v(t + T_0) = v(t)$$

for all values of t. The constant T_0 is called the period of the waveform if it is the smallest nonzero interval for which $v(t + T_0) = v(t)$. Since this equality must be valid for all values of t, it follows that periodic signals must have eternal waveforms that extend indefinitely in time in both directions. Signals that are not periodic are called **aperiodic**.

The sinusoid is a periodic signal since

$$v(t + T_0) = V_A \cos[2\pi(t + T_0)/T_0 + \phi]$$
$$= V_A \cos[2\pi(t)/T_0 + \phi + 2\pi]$$

But $\cos(x + 2\pi) = \cos(x)$. Consequently,

$$v(t + T_0) = V_A \cos(2\pi t/T_0 + \phi) = v(t)$$

for all t.

The **additive property** of sinusoids states that summing two or more sinusoids with the same frequency yields a sinusoid with different amplitude and phase parameters but the same frequency. To illustrate, consider two sinusoids

$$v_1(t) = a_1 \cos(2\pi f_0 t) + b_1 \sin(2\pi f_0 t) \text{ V}$$
$$v_2(t) = a_2 \cos(2\pi f_0 t) + b_2 \sin(2\pi f_0 t) \text{ V}$$

The waveform $v_3(t) = v_1(t) + v_2(t)$ can be written as

$$v_3(t) = (a_1 + a_2) \cos(2\pi f_0 t) + (b_1 + b_2) \sin(2\pi f_0 t) \text{ V}$$

because cosine and sine are linearly independent functions. We obtain the Fourier coefficients of the sum of two sinusoids by adding their Fourier coefficients, provided the two have the same frequency. *Caution:* The summation must take place with the sinusoids in Fourier coefficient form. Sums of sinusoids *cannot* be found by adding amplitudes and phase angles.

The **derivative** and **integral** properties state that when we differentiate or integrate a sinusoid, the result is another sinusoid with the same frequency:

$$\frac{d(V_A \cos \omega t)}{dt} = -\omega V_A \sin \omega t = \omega V_A \cos(\omega t + \pi/2)$$

$$\int V_A \cos(\omega t)dt = \frac{V_A}{\omega} \sin \omega t = \frac{V_A}{\omega} \cos(\omega t - \pi/2)$$

These operations change the amplitude and phase angle but do not change the frequency. The fact that differentiation and integration preserve the underlying waveform is a key property of the sinusoid. No other periodic waveform has this shape-preserving property.

EXAMPLE 5−8

(a) Find the period and the cyclic and radian frequencies for each of the following sinusoids:

$$v_1(t) = 17 \cos(2000t - 30°) \, \text{V}$$
$$v_2(t) = 12 \cos(2000t + 30°) \, \text{V}$$

(b) Find the waveform of $v_3(t) = v_1(t) + v_2(t) \, \text{V}$.

SOLUTION:

(a) The two sinusoids have the same frequency $\omega_0 = 2000$ rad/s since a term $2000t$ appears in the arguments of $v_1(t)$ and $v_2(t)$. Therefore, $f_0 = \omega_0/2\pi = 318.3$ Hz and $T_0 = 1/f_0 = 3.14$ ms.

(b) We use the additive property, since the two sinusoids have the same frequency. Beyond this checkpoint, the frequency plays no further role in the calculation. The two sinusoids must be converted to the Fourier coefficient form using Eq. (5–23).

$$a_1 = 17 \cos(-30°) = +14.7 \, \text{V}$$
$$b_1 = -17 \sin(-30°) = +8.50 \, \text{V}$$
$$a_2 = 12 \cos(30°) = +10.4 \, \text{V}$$
$$b_2 = -12 \sin(30°) = -6.00 \, \text{V}$$

The Fourier coefficients of the signal $v_3 = v_1 + v_2$ are found as

$$a_3 = a_1 + a_2 = 25.1 \, \text{V}$$
$$b_3 = b_1 + b_2 = 2.50 \, \text{V}$$

The amplitude and phase angle of $v_3(t)$ are found using Eqs. (5–24) and (5–25):

$$V_A = \sqrt{a_3^2 + b_3^2} = 25.2 \, \text{V}$$
$$\phi = \tan^{-1}(-2.5/25.1) = -5.69°$$

Two equivalent representations of $v_3(t)$ are

$$v_3(t) = 25.1 \cos(2000t) + 2.5 \sin(2000t) \, \text{V}$$

and

$$v_3(t) = 25.2 \cos(2000t - 5.69°) \, \text{V}$$

■

Exercise 5–11

Write an equation for the waveform obtained by integrating and differentiating the following signals:

(a) $v_1(t) = 30\cos(10t - 60°)$ V
(b) $v_2(t) = 3\cos(4000\pi t) - 4\sin(4000\pi t)$ V

Answers:

(a) $\dfrac{dv_1}{dt} = 300\cos(10t + 30°)$ V/s

$\displaystyle\int v_1(t)dt = 3\cos(10t - 150°)$ V-s

(b) $\dfrac{dv_2}{dt} = 2\pi \times 10^4\cos(4000\pi t + 143.1°)$ V/s

$\displaystyle\int v_2 dt = \dfrac{1}{800\pi}\cos(4000\pi t - 36.87°)$ V-s

Exercise 5–12

A sinusoid has a period of 5 μs. At $t = 0$ the amplitude is 12 V. The waveform reaches its first positive peak after $t = 0$ at $t = 4$ μs. Find its amplitude, frequency, and phase angle.

Answers: $V_A = 38.8$ V; $f_0 = 200$ kHz; $\phi = +72°$

5–5 COMPOSITE WAVEFORMS

In the previous sections we introduced the step, exponential, and sinusoidal waveforms. These waveforms are basic signals because they can be combined to synthesize all other signals used in this book. Signals generated by combining the three basic waveforms are called **composite signals**. This section provides examples of composite waveforms.

EXAMPLE 5−9

Characterize the composite waveform generated by

$$v(t) = V_A u(t) - V_A u(-t) \text{ V}$$

SOLUTION:
The first term in this waveform is simply a step function of amplitude V_A that occurs at $t = 0$. The second term involves the function $u(-t)$, whose waveform requires some discussion. Strictly speaking, the general step function $u(x)$ is unity when $x > 0$ and zero when $x < 0$. That is, $u(x)$ is unity when its argument is positive and zero when it is negative. Under this rule the function $u(-t)$ is unity when $-t > 0$ and zero when $-t < 0$, that is,

$$u(-t) = \begin{cases} 1 & \text{for } t < 0 \\ 0 & \text{for } t > 0 \end{cases}$$

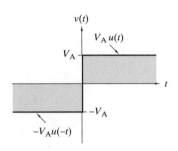

FIGURE 5–25 *The signum waveform.*

which is the reverse of the step function $u(t)$. Figure 5–25 shows how the two components combine to produce a composite waveform that extends indefinitely in both directions and has a jump discontinuity of $2V_A$ at $t = 0$. This composite waveform is called a **signum** function. ∎

Exercise 5–13 _____

Describe the following waveform:

$$v(t) = [V_A u(t) - V_A u(-t)][\delta(t+1) + \delta(t) + \delta(t-1)] \text{ V}$$

Answer: There are only two terms present: $v(t) = -V_A \delta(t+1) + V_A \delta(t-1)$ V

DISCUSSION: *Each impulse function exists only at one point in time. Since all three impulse functions have a weight of 1 they simply multiply the value of the signum function at the time the impulses exist, namely at $t = -1, 0,$ and $+1$. The values are $-V_A, 0,$ and $+V_A$, respectively. The selective nature of multiplying a function by an impulse is called the* **sifting property**.

_____

EXAMPLE 5–10

Characterize the composite waveform generated by subtracting an exponential from a step function with the same amplitude.

SOLUTION:
The equation for this composite waveform is

$$v(t) = V_A u(t) - [V_A e^{-t/T_C}]u(t) \text{ V}$$

$$= V_A[1 - e^{-t/T_C}]u(t) \text{ V}$$

For $t < 0$ the waveform is zero because of the step function. At $t = 0$ the waveform is still zero since the step and exponential cancel:

$$v(0) = V_A[1 - e^0](1) = 0$$

For $t \gg T_C$ the waveform approaches a constant value V_A because the exponential term decays to zero. For practical purposes $v(t)$ is within less than 1% of its final value V_A when $t = 5T_C$. At $t = T_C$, $v(T_C) = V_A(1 - e^{-1}) = 0.632V_A$. The waveform rises to about 63% of its final value in one time constant. All of the observations are summarized in the plot shown in Figure 5–26. This waveform is called an **exponential**

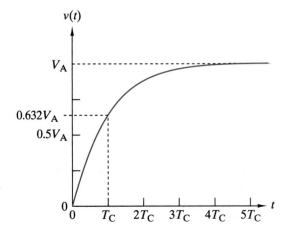

FIGURE 5–26 *The exponential rise waveform.*

rise. It is also sometimes referred to as a "charging exponential," since it represents the behavior of signals that occur during the buildup of voltage in resistor-capacitor circuits studied in Chapter 7.   ■

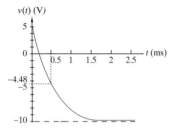

FIGURE 5–27

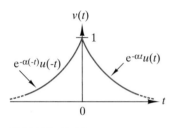

FIGURE 5–28

Exercise 5–14

(a) Sketch the waveform described by the following:

$$v(t) = [15e^{-2000t} - 10]\,u(t)\ \text{V}$$

(b) What is the value of the voltage at $t = T_C$?

Answers:

(a) See Figure 5–27.
(b) -4.48 V

Exercise 5–15

Figure 5–28 contains a waveform that is called a double-sided exponential, which is defined as the sum of a normal exponential and a reversed exponential. This waveform is 1 at $t = 0$ and decays exponentially to zero in both directions along the time axis. Write an expression for this waveform.

Answer: $v(t) = e^{-a|t|}$ V.

EXAMPLE 5–11

Characterize the composite waveform obtained by multiplying the ramp $r(t)/T_C$ times an exponential.

SOLUTION:
The equation for this composite waveform is

$$v(t) = \frac{r(t)}{T_C}\,[V_A e^{-t/T_C}]u(t)\ \text{V}$$

$$= V_A[(t/T_C)e^{-t/T_C}]u(t)\ \text{V}$$

For $t < 0$ the waveform is zero because of the step function. At $t = 0$ the waveform is zero because $r(0) = 0$. For $t > 0$ there is a competition between two effects—the ramp increases linearly with time while the exponential decays to zero. Since the composite waveform is the product of these terms, it is important to determine which effect dominates. In the limit, as $t \to \infty$, the product of the ramp and exponential takes on the indeterminate form of infinity times zero. A single application of *l'Hôpital's rule*, then, shows that the exponential dominates, forcing the $v(t)$ to zero as t becomes large. That is, the exponential decay overpowers the linearly increasing ramp, as shown by the graph in Figure 5–29. The waveform obtained by multiplying a ramp by a decaying exponential is called a **damped ramp**.

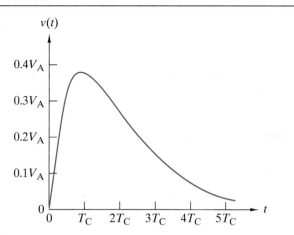

FIGURE 5–29 *The damped ramp waveform.*

Exercise 5–16

(a) Determine the time that the damped ramp reaches its maximum value.
(b) What is the value of $v(t)$ at the maximum?

$$v(t) = V_A \left[\left(\frac{t}{T_C} \right) e^{-t/T_C} \right] u(t) \text{ V}$$

Answers:

(a) The maximum occurs at $t = T_C$.
(b) $v(T_C) = 0.368 V_A$.

EXAMPLE 5–12

Characterize the composite waveform obtained by multiplying $\sin \omega_0 t$ by an exponential.

SOLUTION:
In this case the composite waveform is expressed as

$$v(t) = \sin \omega_0 t [V_A e^{-t/T_C}] u(t) \text{ V}$$
$$= V_A [e^{-t/T_C} \sin \omega_0 t] u(t) \text{ V}$$

Figure 5–30 shows a graph of this waveform for $T_0 = 2T_C$. For $t < 0$ the step function forces the waveform to be zero. At $t = 0$, and periodically thereafter, the waveform

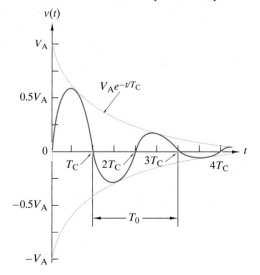

FIGURE 5–30 *The damped sine waveform.*

passes through zero because sin $(n\pi) = 0$. The waveform is not periodic, however, because the decaying exponential gradually reduces the amplitude of the oscillation. for all practical purposes the oscillations become negligibly small for $t > 5T_C$. The waveform obtained by multiplying a sinusoid by a decaying exponential is called a **damped sine**.   ■

APPLICATION   **EXAMPLE 5 – 1 3**

Underdamped second-order systems produce the damped sinusoidal waveform shown in Figure 5–30. When presented with such a display, it may be necessary to determine an expression for the resulting waveform. Consider the damped sinusoid shown in Figure 5–31. We will find an approximate expression.

FIGURE 5–31

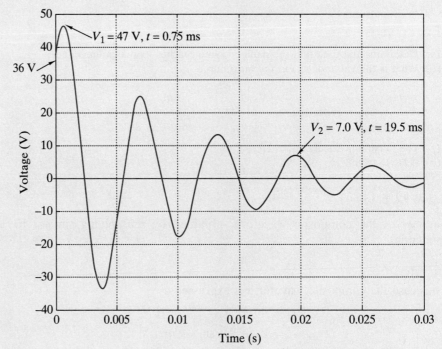

We need to determine the waveform's amplitude V_A, its time constant T_C, and its oscillatory frequency, ω_0. We start by estimating the coordinates of the first peak, (0.00075 s, 47 V). To calculate the time constant we need another peak. We could select the second peak, but if we choose a later peak we can get a more accurate result. Hence, we choose the fourth peak, (0.0195 s, 7.0 V). We can find the time constant from the decrement property noted earlier in Example 5–6.

$$T_C = \frac{\Delta t}{\ln\left(\dfrac{v(t)}{v(t + \Delta t)}\right)} = \frac{0.0195 - 0.00075}{\ln\left(\dfrac{47}{7.0}\right)} = \frac{0.01875}{1.90} = 9.87\,\text{ms}$$

$$\frac{1}{T_C} = 101.3\,\text{s}^{-1}$$

We can use the same two points for finding T_0. However, we must divide the result by 3 since there are three cycles involved.

$$T_0 = \frac{\Delta t}{3} = \frac{0.01875}{3} = 6.25\,\text{ms}$$

Then we find the radian frequency ω_0 from the period:

$$\omega_0 = \frac{2\pi}{T_0} = 1005 \, \text{rad/s}$$

Since the waveform has a phase shift, we find the time shift T_S by measuring the time from when the function is zero to the first peak of the cosine. This was our first peak, 0.00075 s. We can then calculate the phase shift from Eq. (5–21):

$$\phi = \frac{-0.00075}{0.00625} \times 360° = -43.2°$$

We can then write what we have found thus far in our waveform equation:

$$v(t) = V_A \, e^{-101.3t} \cos(1005t - 43.2°) \, u(t) \, \text{V}$$

We can find V_A by substituting a value for $v(t)$ at a time we know. The easiest is at $t = 0$, where $v(0) \approx 36 \, \text{V}$.

$$v(0) = 36 = V_A e^0 \cos(0 - 43.2°) = V_A 0.729$$
$$V_A = 49.4$$

Finally, our desired waveform is

$$v(t) = 49.4 \, e^{-101.3t} \cos(1005t - 43.2°) \, u(t) \, \text{V}$$

DISCUSSION: *The function used to generate the waveform in Figure 5–31 was*

$$v(t) = 50 \, e^{-100t} \cos(1000t - 45°) \, u(t) \, \text{V}$$

The errors are all small, with the phase angle having the largest error of 4%. The size of the errors is, of course, dependent on how accurately one can read the display.

Exercise 5–17

For the damped sinusoid waveform shown in Figure 5–32, determine an approximate expression for $v(t)$.

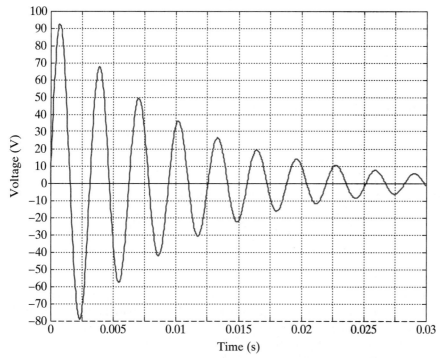

FIGURE 5–32

Answer: The generating waveform is $v(t) = 100 \, e^{-100t} \sin(2000t) \, u(t) \, \text{V}$.

EXAMPLE 5–14

Characterize the composite waveform obtained as the difference of two exponentials with the same amplitude.

SOLUTION:
The equation for this composite waveform is

$$v(t) = [V_A e^{-t/T_1}]u(t) - [V_A e^{-t/T_2}]u(t)$$
$$= V_A(e^{-t/T_1} - e^{-t/T_2})u(t)$$

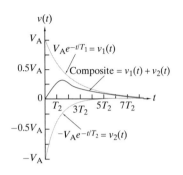

For $T_1 > T_2$ the resulting waveform is illustrated in Figure 5–33 (plotted for $T_1 = 2T_2$). For $t < 0$ the waveform is zero. At $t = 0$ the waveform is still zero, since

$$v(0) = V_A(e^{-0} - e^{-0})$$
$$= V_A(1 - 1) = 0$$

FIGURE 5–33 *The double exponential waveform.*

For $t \gg T_1$ the waveform returns to zero because both exponentials decay to zero. For $5T_1 > t > 5T_2$ the second exponential is negligible and the waveform essentially reduces to the first exponential. Conversely, for $t \ll T_1$ the first exponential is essentially constant, so the second exponential determines the early time variation of the waveform. The waveform is called a **double exponential**, since both exponential components make important contributions to the waveform.   ■

Exercise 5–18

A double exponential waveform is given as

$$v(t) = 10[e^{-1000t} - e^{-2500t}]u(t) \text{ V}$$

(a) What is the value of $v(t)$ at the maximum, and at what time does it occur?
(b) What is the time constant of the dominant (longer-lasting) exponential?

Answers:

(a) The maximum value is 3.257 V, and it occurs at $t = 611$ μs.
(b) The dominant exponential is the first one, since it has the longer time constant of 1 ms versus 400 μs for the second exponential.

EXAMPLE 5–15

Characterize the composite waveform defined by

$$v(t) = 5 - \frac{10}{\pi}\sin(2\pi 500t) - \frac{10}{2\pi}\sin(2\pi 1000t) - \frac{10}{3\pi}\sin(2\pi 1500t) \text{ V}$$

SOLUTION:
The waveform is the sum of a constant (dc) term and three sinusoids at different frequencies. The first sinusoidal component is called the **fundamental** because it has the lowest frequency. As a result the frequency $f_0 = 500$ Hz is called the **fundamental frequency**. The other sinusoidal terms are said to be harmonics because their frequencies are integer multiples of f_0. Specifically, the second sinusoidal term is called the **second harmonic** ($2f_0 = 1000$ Hz) while the third term is the **third**

harmonic ($3f_0 = 1500$ Hz). Figure 5–34 shows a plot of this waveform. Note that the waveform is periodic with a period equal to that of the fundamental component, namely, $T_0 = 1/f_0 = 2$ ms. The decomposition of a periodic waveform into a sum of harmonic sinusoids is called a **Fourier series**, a topic we will study in detail in Chapter 13. In fact, the waveform in this example is the first four terms in the Fourier series for a 10-V sawtooth wave of the type shown in Figure 5–2. ■

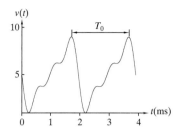

FIGURE 5–34

Exercise 5–19

Find the maximum amplitude and the approximate duration of the following composite waveforms:

(a) $v_1(t) = [25 \sin 1000t][u(t) - u(t - 10)]$ V
(b) $v_2(t) = [50 \cos 1000t][e^{-200t}]u(t)$ V
(c) $i_3(t) = [3000te^{-1000t}]u(t)$ mA
(d) $i_4(t) = 10e^{|5000t|}$ A

Answers:

(a) 25 V, 10 s
(b) 50 V, 25 ms
(c) 1.10 mA, 5 ms
(d) 10 A, 2 ms

Exercise 5–20

Characterize the following waveform defined by

$$v(t) = \sum_{n=1}^{\infty} [b_n \sin(2\pi n f_O t)] \text{V}$$

where $b_n = 4V_A/\pi n$, $V_A = 10$ V, $f_O = 1000$ Hz, and $n = 1, 3, 5, 7, 9, 11 \ldots$ by plotting the first six n terms for a half period of the fundamental frequency f_o. (Hint: Vary t from 0 to 500 μs in 50 μs steps. Excel is useful here.) What waveform in Figure 5–2 does this function best resemble?

Answers: A plot using Excel of the odd terms 1 through 11 is shown in Figure 5–35. The plot resembles a half-cycle of a 1000 Hz square wave.

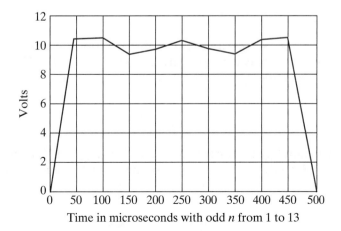

FIGURE 5–35

5–6 WAVEFORM PARTIAL DESCRIPTORS

An equation or graph defines a waveform for all time. The value of a waveform $v(t)$ or $i(t)$ at time t is called the **instantaneous value** of the waveform. We often use parameters called **partial descriptors** that characterize important features of a waveform but do not give a complete description. These partial descriptors fall into two categories: (1) those that describe temporal features and (2) those that describe amplitude features.

TEMPORAL DESCRIPTORS

Temporal descriptors identify waveform attributes relative to the time axis. For example, waveforms that repeat themselves at fixed time intervals are said to be **periodic**. Stated formally,

> *A signal $v(t)$ is periodic if $v(t + T_0) = v(t)$ for all t, where the period T_0 is the smallest value that meets this condition. Signals that are not periodic are called* **aperiodic**.

The fact that a waveform is periodic provides important information about the signal but does not specify all of its characteristics. Thus, the fact that a signal is periodic is itself a partial description, as is the value of the period. The eternal sinewave is the premier example of a periodic signal. The square wave and triangular wave in Figure 5–2 are also periodic. Examples of aperiodic waveforms are the step function, exponential, and damped sine.

Waveforms that are identically zero prior to some specified time are said to be **causal**. Stated formally,

> *A signal $v(t)$ is causal if there exists a value of T such that $v(t) = 0$ for all $t < T$; otherwise it is noncausal.*

It is usually assumed that a causal signal is zero for $t < 0$, since we can always use time shifting to make the starting point of a waveform at $t = 0$. Examples of causal waveforms are the step function, exponential, and damped sine. The eternal sinewave is, of course, noncausal.

Causal waveforms play a central role in circuit analysis. When the input driving force $x(t)$ is causal, the circuit response $y(t)$ must also be causal. That is, a physically realizable circuit cannot anticipate and respond to an input before it is applied. Causality is an important temporal feature, but only a partial description of the waveform.

AMPLITUDE DESCRIPTORS

Amplitude descriptors are positive scalars that describe signal strength. Generally, a waveform varies between two extreme values denoted as V_{MAX} and V_{MIN}. The **peak-to-peak value** (V_{pp}) describes the total excursion of $v(t)$ and is defined as

$$V_{\text{pp}} = V_{\text{MAX}} - V_{\text{MIN}} \tag{5–28}$$

Under this definition V_{pp} is always positive even if V_{MAX} and V_{MIN} are both negative. The **peak value** (V_{p}) is the maximum of the absolute value of the waveform. That is,

$$V_{\text{p}} = \text{MAX}\{|V_{\text{MAX}}|, |V_{\text{MIN}}|\} \tag{5–29}$$

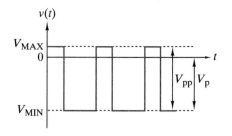 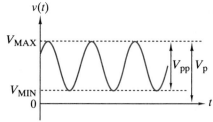

FIGURE 5–36 *Maximum value* (V_{MAX}), *Minimum value* (V_{MIN}), *Peak value* (V_{p}) *and peak-to-peak value* (V_{pp}).

The peak value is a positive number that indicates the maximum absolute excursion of the waveform from zero. Figure 5–36 shows examples of these two amplitude descriptors.

The peak and peak-to-peak values describe waveform variation using the extreme values. The average value smooths things out to reveal the underlying waveform baseline. Average value is the area under the waveform over some period of time T, divided by that time period. Mathematically, we define **average value** (V_{avg}) over the time interval T as

$$V_{\text{avg}} = \frac{1}{T} \int_{t}^{t+T} v(x)\,dx \qquad (5\text{–}30)$$

For periodic signals the period T_0 is used as the averaging interval T.

For some periodic waveforms the integral in Eq. (5–30) can be estimated graphically. The net area under the waveform is the area above the time axis minus the area below the time axis. For example, the two waveforms in Figure 5–36 obviously have nonzero average values. The top waveform has a negative average value because the negative area below the time axis more than cancels the area above the axis. Similarly, the bottom waveform clearly has a positive average value.

The average value indicates whether the waveform contains a constant, non-time-varying component. The average value is also called the **dc component** because dc signals are constant for all t. On the other hand, the **ac components** have zero average value and are periodic. For example, the waveform in Example 5–15

$$V(t) = 5 - \frac{10}{\pi} \sin(2\pi 500 t) - \frac{10}{2\pi} \sin(2\pi 1000 t) - \frac{10}{3\pi} \sin(2\pi 1500 t) \text{ V}$$

has a 5-V average value due to its dc component. The three sinusoids are ac components because they are periodic and have zero average value. Sinusoids have zero average value because over any given cycle the positive area above the time axis is exactly canceled by the negative area below.

EXAMPLE 5–16

Find the peak, peak-to-peak, and average values of the periodic input and output waveforms in Figure 5–37 of a half-wave rectifier.

SOLUTION:
The input waveform is a sinusoid whose amplitude descriptors are

$$V_{\text{pp}} = 2V_{\text{A}} \qquad V_{\text{p}} = V_{\text{A}} \qquad V_{\text{avg}} = 0$$

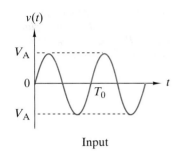

Input

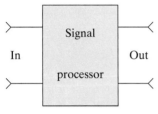

In ———— Out

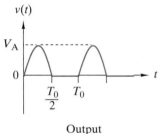

Output

FIGURE 5–37

The output waveform is obtained by clipping off the negative half-cycle of the input sinusoid. The amplitude descriptors of the output waveform are

$$V_{pp} = V_p = V_A$$

The output has a nonzero average value, since there is a net positive area under the waveform. The upper limit in Eq. (5–30) can be taken as $T_0/2$, since the waveform is zero from $T_0/2$ to T_0.

$$V_{avg} = \frac{1}{T_0} \int_0^{T_0/2} V_A \sin(2\pi t/T_0)\,dt = -\frac{V_A}{2\pi} \cos(2\pi t/T_0)\Big|_0^{T_0/2}$$

$$= \frac{V_A}{\pi}$$

The signal processor produces an output with a dc value from an input with no dc component. Rectifying circuits described in electronics courses produce waveforms like the output in Figure 5–37. ■

Exercise 5–21

For the pulse waveform in Figure 5–38 find V_{MAX}, V_{MIN}, V_p, V_{pp}, and V_{avg}.

FIGURE 5–38 *http://www
.salemstate.edu/academics/
schools/3698.php*

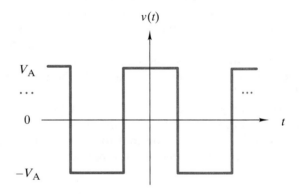

Answers: $V_{MAX} = V_A, V_{MIN} = -V_A, V_p = V_A, V_{pp} = 2\,V_A, V_{avg} = 0\,\text{V}$

ROOT-MEAN-SQUARE VALUE

The **root-mean-square value** (V_{rms}) is a measure of the average power carried by the signal. The instantaneous power delivered to a resistor R by a voltage $v(t)$ is

$$p(t) = \frac{1}{R}[v(t)]^2 \qquad (5\text{–}31)$$

The average power delivered to the resistor in time span T is defined as

$$P_{avg} = \frac{1}{T} \int_t^{t+T} p(t)\,dt \qquad (5\text{–}32)$$

Combining Eqs. (5–31) and (5–32) yields

$$P_{avg} = \frac{1}{R}\left[\frac{1}{T} \int_t^{t+T} [v(t)]^2\,dt\right] \qquad (5\text{–}33)$$

The quantity inside the large brackets in Eq. (5–33) is the average value of the square of the waveform. The units of the bracketed term are volts squared. The square root of this term defines the amplitude partial descriptor V_{rms}:

$$V_{\text{rms}} = \sqrt{\frac{1}{T} \int_{t}^{t+T} [v(t)]^2 \, dt} \qquad (5\text{–}34)$$

The amplitude descriptor V_{rms} is called the root-mean-square (rms) value because it is obtained by taking the square root of the average (mean) of the square of the waveform. For periodic signals the averaging interval is one cycle since such a waveform repeats itself every T_0 seconds.

We can express the average power delivered to a resistor in terms of V_{rms} as

$$P_{\text{avg}} = \frac{1}{R} V_{\text{rms}}^2 \qquad (5\text{–}35)$$

The equation for average power in terms of V_{rms} has the same form as the power delivered by a dc signal. For this reason the rms value was originally called the **effective value**, although this term is no longer common. If the waveform amplitude is doubled, its rms value is doubled, and the average power is quadrupled. Commercial electrical power systems use transmission voltages in the range of several hundred kilovolts (rms) to efficiently transmit power over long distances.

EXAMPLE 5–17

Find the average and rms values of the sinusoid and sawtooth in Figure 5–39.

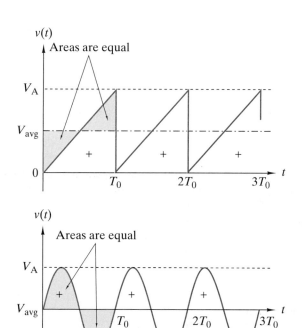

FIGURE 5–39

SOLUTION:

As noted previously, the sinusoid has an average value of zero. The sawtooth clearly has a positive average value. By geometry, the net area under one cycle of the sawtooth waveform is $V_A T_0/2$, so its average value is $(1/T_0)(V_A T_0/2) = V_A/2$. To obtain the rms value of the sinusoid we apply Eq. (5–34) as

$$V_{rms} = \sqrt{\frac{(V_A)^2}{T_0} \int_0^{T_0} \sin^2(2\pi t/T_0)dt}$$

$$= \sqrt{\frac{(V_A)^2}{T_0}\left[\frac{t}{2} - \frac{\sin(4\pi t/T_0)}{8\pi/T_0}\right]_0^{T_0}} = \frac{V_A}{\sqrt{2}}$$

For the sawtooth the rms value is found as:

$$V_{rms} = \sqrt{\frac{1}{T_0}\int_0^{T_0}(V_A t/T_0)^2 dt} = \sqrt{\frac{(V_A)^2}{T_0^3}\left[\frac{t^3}{3}\right]_0^{T_0}} = \frac{V_A}{\sqrt{3}}$$ ∎

Exercise 5–22

Find the peak, peak-to-peak, average, and rms values of the periodic waveform in Figure 5–40.

FIGURE 5–40

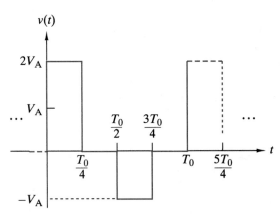

Answers: $V_p = 2V_A$; $V_{pp} = 3V_A$; $V_{avg} = \dfrac{V_A}{4}$; $V_{rms} = \dfrac{\sqrt{5}}{2}V_A$

Exercise 5–23

Classify each of the following signals as periodic or aperiodic and causal or noncausal. Then calculate the average and rms values of the periodic waveforms, and the peak and peak-to-peak values of the other waveforms.

(a) $v_1(t) = 99\cos 3000t - 132\sin 3000t$ V
(b) $v_2(t) = 34[\sin 800\pi t][u(t) - u(t - 0.03)]$ V
(c) $i_3(t) = 120[u(t + 5) - u(t - 5)]$ mA
(d) $i_4(t) = 50$ A

Answers:

(a) Periodic, noncausal, $V_{avg} = 0$ and $V_{rms} = 117$ V
(b) Aperiodic, causal, $V_p = 34$ V and $V_{pp} = 68$ V
(c) Aperiodic, causal, $V_p = V_{pp} = 120$ mA
(d) Aperiodic, noncausal, $V_p = 50$ A and $V_{pp} = 0$

EXAMPLE 5–18

The operation of a digital system is coordinated and controlled by a periodic waveform called a clock. The *clock waveform* provides a standard timing reference to maintain synchronization between signal-processing results that become valid at different times during the clock cycle. Because of differences in digital circuit delays, there must be agreed-upon instants of time when circuit outputs can be treated as valid. The clock defers further signal processing until slower and faster outputs settle down when the clock signals the start of the next signal-processing cycle.

Figure 5–41 shows an idealized clock waveform as a periodic sequence of rectangular pulses. While we could easily write an exact expression for the clock waveform, we are interested here in discussing its partial descriptors. The first descriptor is the period T_0 or equivalently the **clock frequency** $f_0 = 1/T_0$. Clock frequency is a common measure of signal-processing speed and can take values into the GHz range. The pulse duration T is the time interval in each cycle when the pulse amplitude is high (not zero). In waveform terminology the ratio of the time in the high state to the period, that is, T/T_0, is called the **duty cycle**, usually expressed as a percentage. The **pulse edges** are the transition points at which the pulse changes states. There is a **rising edge** at the low-to-high transition and a **falling edge** at the high-to-low transition.

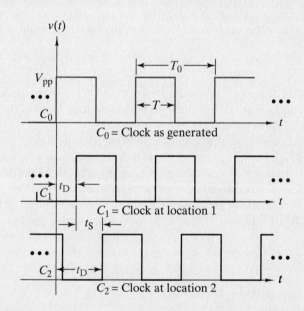

FIGURE 5–41

The pulse edges define the agreed-upon time instants at which the circuit outputs can be treated as valid inputs to other circuits. This means that circuit outputs must settle down during the time period between successive edges. Some synchronous operations are triggered by the rising edge and others by the falling edge. To provide equal settling times for both cases requires equal time between edges. In other words, it is desirable for the clock duty cycle to be 50%. As a result, the clock waveform is essentially a raised square wave whose dc offset equals one half of the peak-to-peak value.

The system clock C_0 in Figure 5–41 is generated at some point in a circuit and then distributed to other locations. The clock distribution network almost invariably introduces delays, as illustrated by C_1 and C_2 in Figure 5–41. **Clock delay (t_D)** is defined as the time difference between a clock edge at a given location and the corresponding edge in the system clock at the point where it was generated. Delay is not necessarily a bad thing unless unequal delays cause the edges to be skewed, as indicated by the offset between C_1 and C_2 in Figure 5–41. When delays are significantly different, there is uncertainty as to instants of time at which further signal processing can safely proceed. This delay dispersion is called **clock skew (t_S)**, defined as the time difference between a clock edge at a given location and the corresponding edge at another location. Controlling clock skew is an important consideration in the design of the clock distribution network in high-speed very large scale integrated (VLSI) circuits.

Thus, partial descriptors of clock waveforms include *frequency, duty cycle, edges, delay,* and *skew.* The coming chapters treat dynamic circuits that modify input waveforms to produce outputs with different partial descriptors. In particular, dynamic circuit elements cause changes in a clock waveform, especially the partial descriptors of edges, delay, and skew.

APPLICATION   EXAMPLE 5–19

An electrocardiogram (ECG) is a valuable diagnostic tool used in cardiovascular medicine. The ECG is based on the fact that the heart emits measurable bioelectric signals that can be recorded to evaluate the functioning of the heart as a mechanical pump. These signals were first observed in the late 19th century and subsequent signal processing developments have led to the advanced technology of present day ECG equipment.

The bioelectric signals of the heart muscle are measured and recorded through the placement of skin electrodes at various sites on the surface of the body. The site selection as well as discussion of the functions of the cardiac muscle are beyond the scope of this example. Rather, our purpose is to introduce some of the useful partial descriptors of ECG waveforms.

In bioelectric terminology the normal ECG waveform in Figure 5–42 is composed of a P wave, a QRS complex, and a T wave. This sequence of pulses depicts the electrical activity that stimulates the correct functioning of the cardiac muscle. The flat baseline between successive events is called isoelectric, which means there is no bioelectric activity and the heart muscle returns to a resting state. The body's natural pacemaker produces a nominally periodic waveform under the resting conditions used with ECG tests.

FIGURE 5–42

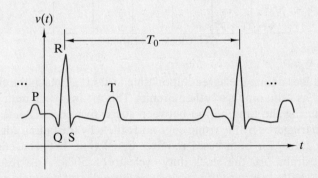

Partial waveform descriptors used to analyze ECG waveforms include:

1. The **heart rate**$(1/T_0)$, which is normally between 60 and 100 beats per minute.

2. The **PR interval** (normally 0.12–0.20 seconds), which is the time between the start of the P wave and the start of the QRS complex.

3. The **QRS interval** (normally 0.06–0.10 seconds), which is the time between onset and end of the QRS complex.

4. The **ST segment** is the signal level between the end of the QRS complex and the start of the T wave. This level should be the same as the isoelectric baseline between successive pulses.

Departures from these normal conditions serve as diagnostic tools in cardiovascular medicine. Some of the abnormal waveform features of concern include an irregular heart rate, a missing P wave, a prolonged QRS interval, or an elevated ST segment. Departures from nominal conditions allow the trained clinician to diagnose the situation, especially when abnormal features occur in certain combinations. However, it is not our purpose to discuss the medical interpretation of ECG waveform abnormalities. Rather, this example illustrates that bioelectric signals carry information and that the information is decoded by analyzing the signal's partial waveform descriptors.

SIGNALS AND SOFTWARE TOOLS

The value of software tools is readily apparent when circuits are tested using a variety of signals. The OrCAD software will allow us to rapidly simulate circuit behavior under a variety of excitations such as steps, exponentials, sinusoids, and composite signals. Web Appendix C discusses how to use the various signal sources available in OrCAD. In addition, the web appendix examines the calculating, plotting, and integration functions available in MATLAB to create professional graphs and to efficiently calculate partial waveform descriptors. In the web appendix there are examples, exercises, sample problems, and MATLAB software routines that you can use to help solve problems found in the text.

SUMMARY

- A waveform is an equation or graph that describes a voltage or current as a function of time. Most signals of interest in electrical engineering can be derived using three basic waveforms: the step, exponential, and sinusoid.

- The step function is defined by its amplitude and time-shift parameters. The impulse, step, and ramp are called singularity functions and are often used as test inputs for circuit analysis purposes.

- The exponential waveform is defined by its amplitude, time constant, and time-shift parameter. For practical purposes, the duration of the exponential waveform is five time constants.

- A sinusoid can be defined in terms of three types of parameters: *amplitude* (either V_A or the Fourier coefficients a and b), *time shift* (either T_S or the phase angle ϕ), and *time/frequency* (either T_0 or f_0 or ω_0).

- Many composite waveforms can be derived using the three basic waveforms. Some examples are the impulse, ramp, damped ramp, damped sinusoid, exponential rise, and double exponential.

- Partial descriptors are used to classify or describe important signal attributes. Two important temporal attributes are periodicity and causality. Periodic waveforms repeat themselves every T_0 seconds. Causal signals are zero for $t < 0$. Some important amplitude descriptors are peak value V_p, peak-to-peak value V_{pp}, average value V_{avg}, and root-mean-square value V_{rms}.

- Software programs like OrCAD and MATLAB can generate appropriate signals for use in simulation, analysis, and computation of circuit responses. MATLAB can perform numeric and symbolic integration to assist with signal analysis.

PROBLEMS

OBJECTIVE 5–1 BASIC WAVEFORMS (SECTS. 5–2, 5–3, 5–4)

Given an equation, graph, or word description of a linear combination of step, ramp, exponential, or sinusoid waveforms,

(a) Construct an alternative description of the waveform.
(b) Find the parameters or properties of the waveform.
(c) Construct new waveforms by integrating, or differentiating the given waveform.
(d) Generate the basic waveforms in MATLAB or OrCAD and use them appropriately to solve problems or simulate circuits. (See Web Appendix C.)

See Examples 5–1, 5–2, 5–3, 5–4, 5–5, 5–6, 5–7, 5–8 and Exercises 5–1, 5–2, 5–3, 5–4, 5–5, 5–6, 5–7, 5–8, 5–9, 5–10, 5–11, 5–12.

5–1 Sketch the following waveforms:

(a) $v_1(t) = 5u(t) - 5u(t-1)$ V

(b) $v_2(t) = 3u(t+2) - 2u(t-2)$ V

(c) $v_3(t) = \int_{-\infty}^{t} v_1(x)\,dx$

(d) $v_4(t) = \dfrac{dv_2(t)}{dt}$

5–2 Using appropriate step functions, write an expression for each waveform in Figure P5–2.

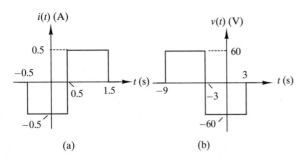

FIGURE P5–2

5–3 Sketch the following waveforms
(a) $v_1(t) = 2 - u(t)$ V
(b) $v_2(t) = -2u(t + 200) + 3u(t + 100) - u(t)$ V
(c) Construct the waveforms in parts (a) and (b) using OrCAD.

5–4 Sketch the following waveforms:
(a) $v_1(t) = 4r(t + 1) - 4r(t - 1)$ V

(b) $v_2(t) = 2 + r(t + 1) - 2r(t - 1) + r(t - 3)$ V

(c) $v_3(t) = \dfrac{d\,v_1(t)}{dt}$

(d) $v_4(t) = \dfrac{d^2\,v_2(t)}{dt^2}$

5–5 Express each of the following signals as a sum of singularity functions.

(a) $v_1(t) = \begin{cases} 3 & t < 1 \\ -2 & 1 \le t < 2 \\ 0 & 2 \le t \end{cases}$

(b) $v_2(t) = \begin{cases} 0 & t < 0 \\ -6t & 0 \le t < 2 \\ -18 + 3t & 2 \le t < 6 \\ 0 & 6 \le t \end{cases}$

5–6 Express the waveform in Figure P5–6 as a sum of step functions.

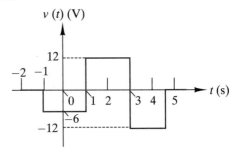

FIGURE P5–6

5–7 Express each of the waveforms in Figure P5–7 as a sum of singularity functions.

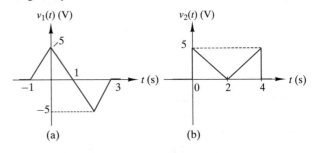

FIGURE P5–7

5–8 Sketch the waveform described by the following:

$$v(t) = \left[\frac{1}{\varepsilon^2}t + \frac{1}{\varepsilon}\right][u(t+\varepsilon) - u(t)]$$
$$+ \left[-\frac{1}{\varepsilon^2}t + \frac{1}{\varepsilon}\right][u(t) - u(t-\varepsilon)]\,\text{V}$$

5–9 Sketch the waveform described by the following:

$$v(t) = 10\delta(t) - 5\delta(t-5) + 2\delta(t-10)\,\text{V}$$

5–10 Generate on OrCAD a waveform $v(t)$ that starts at $t = 1$ ms and consists of a pulse train of 1 V pulses with a 1 ms pulse width that repeat every 3 ms.

5–11 Sketch the following exponential waveforms. Find the amplitude and time constant of each waveform.

(a) $v_1(t) = [100\,e^{-100t}]\,u(t)$ V
(b) $v_2(t) = [20\,e^{-t/100}]\,u(t-2)$ V

(c) $v_3(t) = [5 e^{-2(t-5)}] u(t-5)$ V

(d) $v_4(t) = [-50 e^{-10,000t}] u(t)$ V

5–12 Write expressions for the derivative $(t > 0)$ and integral (from 0 to t) of the exponential waveform $i(t) = [500 e^{-5000t}] u(t)$ μA.

5–13 An exponential waveform decays to 50% of its initial $(t=0)$ amplitude in 25 ms. Find the time constant of the waveform.

5–14 Write an expression for the waveform in Figure P5–14.

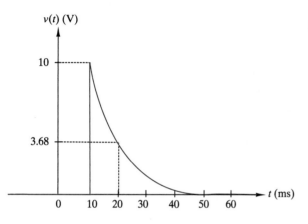

FIGURE P5–14

5–15 The amplitude of an exponential waveform is 6 V at $t = 0$ and 3.5 V at $t = 3$ ms. What is its time constant?

5–16 Construct an exponential waveform that fits entirely within the nonshaded region in Figure P5–16.

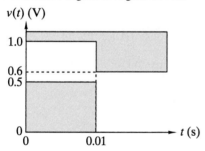

FIGURE P5–16

5–17 Construct an exponential waveform that fits entirely within the nonshaded region in Figure P5–17.

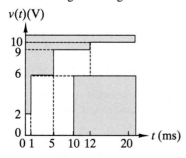

FIGURE P5–17

5–18 By direct substitution show that the exponential function $v(t) = V_A e^{-\alpha t}$ satisfies the following first-order differential equation.

$$\frac{dv(t)}{dt} + \alpha v(t) = 0$$

5–19 Find the period, frequency, amplitude, time shift, and phase angle of the following sinusoids.

(a) $v_1(t) = 240 \cos(120\pi t) + 240 \sin(120\pi t)$ V

(b) $v_2(t) = -40 \cos(100\ k\pi t) + 30 \sin(100\ k\pi t)$ V

5–20 (a) Plot the waveform of each sinusoid in Problem 5–19

(b) Use OrCAD to produce the waveform in Problem 5–19(b).

5–21 Write an expression for the sinusoid in Figure P5–21. What are the phase angle and time shift of the waveform?

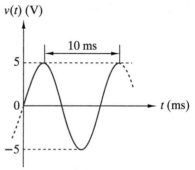

FIGURE P5–21

5–22 Write an expression for the sinusoid in Figure P5–22. What are the phase angle and time shift of the waveform?

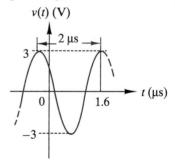

FIGURE P5–22

5–23 Find the Fourier coefficients, cyclic frequency, and radian frequency of the following sinusoids:

(a) $v(t) = 120 \cos(120\ \pi t + 36.9°)$ V

(b) $i(t) = 12 \cos(120\pi t - 270°)$ A

5–24 Use MATLAB or Excel to display two cycles of the following waveform:

$$v_S = 19.1 \sin 1000\ \pi t + 6.37 \sin 3000\ \pi t$$
$$+ 3.82 \sin 5000\ \pi t + 2.73 \sin 7000\ \pi t$$
$$+ 2.1 \sin 9000\ \pi t \text{ V}$$

What are the period and amplitude of the resulting waveform? What common waveform is this waveform approximating?

5–25 For the following sinusoid:

$$v(t) = 100 \cos(2\pi 400 t + 30°) \text{ V}$$

(a) Find the Fourier coefficients, cyclic frequency, and radian frequency

(b) Plot the waveform.

(c) Use MATLAB to produce the waveform.

5–26 Find the time shift of each sinusoid

(a) in Problem 5–23.

(b) in Problem 5–25.

OBJECTIVE 5–2 COMPOSITE WAVEFORMS (SECT. 5–5)

Given an equation, graph, or word description of a composite waveform,

(a) Construct an alternative description of the waveform.

(b) Find the parameters or properties of the waveform.

(c) Generate the composite waveforms in MATLAB or OrCAD and use them appropriately to solve problems or simulate circuits. (See Web Appendix C.)

See Examples 5–9, 5–10, 5–11, 5–12, 5–13, 5–14, 5–15 and Exercises 5–13, 5–14, 5–15, 5–16, 5–17, 5–18, 5–19, 5–20.

5–27 Consider the following composite waveforms.

(a) $v_1(t) = 3\left[1 - e^{-10,000t}\right] u(t) \text{ V}$

(b) $v_2(t) = 15\left[e^{-10t} - e^{-5t}\right] u(t) \text{ V}$

Sketch each on paper and then generate each using OrCAD and MATLAB and plot the results.

5–28 Consider the following composite waveforms.

(a) $i_1(t) = 5 - 5 \sin(1000\pi t) u(t) \text{ A}$

(b) $i_2(t) = 350\left[e^{-1000t} + \cos(2000\pi t)\right] u(t) \text{ mA}$

Sketch each on paper and then generate each using OrCAD and MATLAB and plot the results.

5–29 Sketch the damped ramp $v(t) = te^{-30t}u(t)$. Find the maximum value of the waveform and the time at which it occurs.

5–30 The value of the waveform $v(t) = (V_A - V_B e^{\alpha t}) u(t)$ is 5 V at $t = 0$, 8 V at $t = 5$ ms, and approaches 12 V as $t \to \infty$. (a) Find V_A, V_B, and α, and then sketch the waveform, (b) Validate your answers by plotting your result in MATLAB.

5–31 Write an expression for the composite sinusoidal waveform in Figure P5–31.

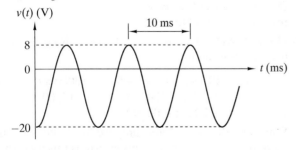

FIGURE P5–31

5–32 Write an expression for the composite sinusoidal waveform in Figure P5–32.

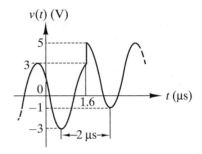

FIGURE P5–32

5–33 A waveform of the form $v(t) = 5-10 \cos(\beta t - 45°)$ periodically reaches a minimum every 2 ms. (a) Find the values of V_{MAX}, V_{MIN}, and β, and then sketch the waveform, (b) Generate the waveform in OrCAD.

5–34 Write an expression for the composite exponential waveform in Figure P5–34.

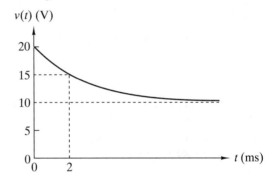

FIGURE P5–34

5–35 Write an expression for the composite exponential waveform in Figure P5–35.

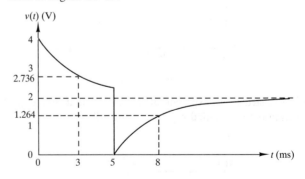

FIGURE P5–35

5–36 For the double exponential $v(t) = 20\,(e^{-100t}-e^{-1000t})u(t) \text{ V}$

(a) Find the maximum value of the waveform and the time at which it occurs.

(b) Determine the dominant exponential.

(c) Generate the waveform in MATLAB and validate the result.

5–37 Write an expression for the damped sine waveform in Figure P5–37.

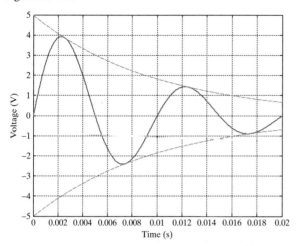

FIGURE P5–37

5–38 A circuit response is shown in Figure P5–38. Determine an approximate expression for the waveform.

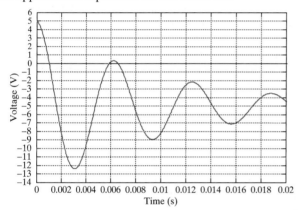

FIGURE P5–38

OBJECTIVE 5–3 WAVEFORM PARTIAL DESCRIPTORS (SECT. 5–6)

Given a complete description of a basic or composite waveform,
(a) Classify the waveform as periodic or aperiodic and causal or noncausal.
(b) Find the applicable partial waveform descriptors.
(c) Use appropriate software tools to calculate applicable partial waveform descriptors.
See Examples 5–16, 5–17, 5–18, 5–19 and Exercises 5–21, 5–22, 5–23.

5–39 Find V_p, V_{pp}, V_{avg}, and V_{rms} for each of the following sinusoids.
  (a) $v_1(t) = 120 \cos(377t) + 120 \sin(377t)$ V

  (b) $v_2(t) = -40 \cos(2000\pi t) - 30 \sin(2000\pi t)$ V
  (c) $v_3(t) = 10 + 24 \cos(10{,}000\pi t - 45°)$ V

5–40 An exponential waveform given by $v(t) = 15e^{-1000t}u(t)$ V repeats every five time constants.
  (a) Find V_p, V_{pp}, V_{MAX}, and V_{MIN}.
  (b) Find V_{avg} and V_{rms}.
  (c) Find the period T_O of the waveform.

5–41 Find V_p, V_{pp}, V_{avg}, V_{rms}, and T_O for the periodic waveform in Figure P5–41.

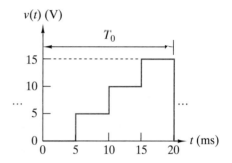

FIGURE P5–41

5–42 Find V_p, V_{pp}, V_{avg}, V_{rms}, and T_O for the periodic waveform in Figure P5–42.

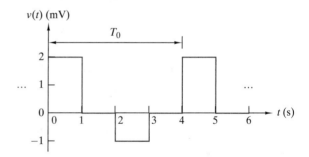

FIGURE P5–42

5–43 Figure P5–43 is the result of the sum of a fundamental and one of its harmonics. Find V_p, V_{pp}, V_{avg}, V_{rms}, and T_O for the waveform.

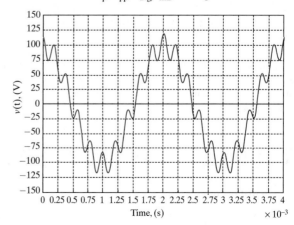

FIGURE P5–43

5–44 Figure P5–44 displays the response of a circuit to a square wave signal. The response is a periodic sequence of exponential waveforms. Each exponential has a time constant of 1.6 ms.
(a) Find V_p, V_{pp}, and T_O for the waveform.
(b) Use MATLAB to find V_{avg}.

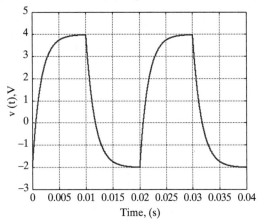

FIGURE P5–44

5–45 Find V_{avg} and V_{rms} of the offset sine wave $v(t) = V_0 + V_A \cos(2\pi t/T_O)$ in terms of V_0 and V_A.

5–46 Find V_{avg} and V_{rms} of the full-wave rectified sine wave $v(t) = V_A |\sin(2\pi t/T_0)|$ in terms of V_A.

5–47 The first cycle ($t > 0$) of a periodic waveform with $T_0 = 50$ ms can be expressed as

$$v(t) = 3u(t) - 6u(t - .02) + 3u(t - .04)\text{V}$$

Sketch the waveform and find V_{max}, V_{min}, V_p, V_{pp}, and V_{avg}.

5–48 A periodic waveform can be expressed as

$$v(t) = 10 + 8\cos 1000\pi t - 4\sin 2000\pi t + 2\cos 4000\pi t \text{ V}$$

What is the period of the waveform? What is the average value of the waveform? What is the amplitude of the fundamental (lowest frequency) component? What is the highest frequency in the waveform?

5–49 Using OrCAD, create a triangular wave that has an amplitude of 75 V and a period of 10 ms. (Hint: The pulse width can be very small but not zero. See also Web Appendix C for use of OrCAD.)

5–50 Consider the waveforms described in the cited problems and determine for each if they are causal or non-causal. Problem 5–24, Problem 5–28, Problem 5–39, Problem 5–44, and Problem 5–48.

INTEGRATING PROBLEMS

5–51 Gated Function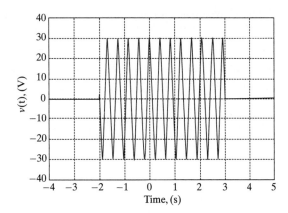

Some radars use a modulated pulse to determine range and target information. A gated modulated pulse is shown in Figure P5–51. Determine an expression for the waveform.

FIGURE P5–51

5–52 Exponential Signal Descriptors
Several of the time descriptors used in digital data communication systems are based on exponential signals. In this problem we explore three of these descriptors.

(a) The *time constant of fall* is defined as the time required for a pulse to fall from 70.7% to 26.0% of its maximum value. Assuming that the pulse decreases as $e^{-\frac{t}{T_C}}$, find the relationship between the time constant of fall and the time constant of the exponential decay.

(b) The *rise time* of a pulse is the time required for a pulse to rise from 10% to 90% of its maximum value. Assuming the pulse increases as $1 - e^{-\frac{t}{T_C}}$, find the relationship between rise time and the time constant of the exponential rise.

(c) The *leading-edge pulse time* is defined as the time at which a pulse rises to 50% of its maximum value. Assuming the pulse increases as $1 - e^{-\frac{t}{T_C}}$, find the relationship between leading-edge pulse time and the time constant of the exponential rise.

5–53 Defibrillation Waveforms
Ventricular fibrillation is a life threatening loss of synchronous activity in the heart. To restore normal activity, a defibrillator delivers a brief but intense pulse of electrical current through the patient's chest. The pulse waveform is of interest because different waveforms may lead to different outcomes. Figure P5-53 show a waveform known as a *biphasic truncated exponential* used in implantable defibrillators. The waveform is an exponential current whose direction of flow reverses after 4 ms and terminates after 8 ms. Write an expression for this waveform using the basic signals discussed in this chapter.

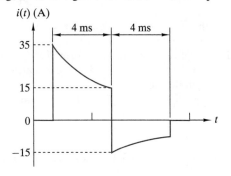

FIGURE P5–53

5–54 Digital Clock Generator **D**

Timing digital circuits is vital to the operation of any digital device. Using an ideal OP AMP (running open-loop, i.e., without feedback) and appropriate resistors, design a way to convert a sinusoid into a square wave that varies from 0 to 5V with a period of 1 μs. The only non-sinusoidal supply available is $V_{CC} = \pm 15$ V.

5–55 Undesired Oscillations **A** **E**

A test is being run in a wind tunnel when a sensor on the trailing edge of a wing produces the response shown in Figure P5–55. When the sensor output reached 1 V, the test was terminated. You are asked to analyze the results. The oscillation could be tolerated if it never reaches 20 V because the on-board computer can mitigate it. However, the response time for the computer to actuate the compensating aileron is 80 ms. Will the compensation occur in time?

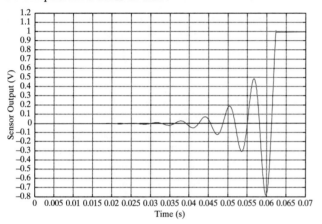

FIGURE P5–55

5–56 Voltmeter Calibration **A**

Most dc voltmeters measure the average value of the applied signal. A dc meter that measures the average value can be adapted to indicate the rms value of an ac signal. The input is passed through a rectifier circuit. The rectifier output is the absolute value of the input and is applied to a dc meter whose deflection is proportional to the average value of the rectified signal. The meter scale is calibrated to indicate the rms value of the input signal. A calibration factor is needed to convert the average absolute value into the rms value of the ac signal. What is the required calibration factor for a sinusoid? Would the same calibration factor apply to a square wave?

5–57 MATLAB Signal Analyzer **A**

Create a MATLAB function to analyze signals represented numerically. The function should have the following two inputs: (1) a vector containing equally spaced samples of the signal of interest and (2) the time step used to sample the signal contained in the vector. The function should display the following descriptors of the signal: V_{MAX}, V_{MIN}, V_p, V_{pp}, V_{avg}, and V_{rms}. The function should also plot the waveform assuming the signal starts at $t = 0$.

CHAPTER 6 CAPACITANCE AND INDUCTANCE

From the foregoing facts, it appears that a current of electricity is produced, for an instant, in a helix of copper wire surrounding a piece of soft iron whenever magnetism is induced in the iron; also that an instantaneous current in one or the other direction accompanies every change in the magnetic intensity of the iron.

Joseph Henry, 1831,
American Physicist

Some History Behind This Chapter

Joseph Henry (1797–1878) and the British physicist Michael Faraday (1791–1867) independently discovered magnetic induction almost simultaneously. The quotation above is Henry's summary of the experiments leading to his discovery of magnetic induction. Although Henry and Faraday used similar apparatus and observed almost the same results, Henry was the first to fully recognize the importance of the discovery. The units of circuit inductance (henrys) honors Henry, while the mathematical generalization of magnetic induction is called Faraday's law.

Why This Chapter Is Important Today

Electric circuits owe much of their utility to devices that store energy, even if only for a short period of time. In this chapter you will be introduced to two new circuit elements: the capacitor and the inductor. These energy storing elements lead to circuits that perform various mathematical operations like integration and differentiation. The energy storing capability makes possible the signal processing operations in modern communications systems and audio equipment.

Chapter Sections

6–1 The Capacitor

6–2 The Inductor

6–3 Dynamic OP AMP Circuits

6–4 Equivalent Capacitance and Inductance

Chapter Learning Objectives

6-1 Capacitor and Inductor Responses (Sects. 6–1, 6–2)

(a) Given the current through a capacitor or an inductor, find the voltage across the element.

(b) Given the voltage across a capacitor or an inductor, find the current through the element.

(c) Find the power and energy associated with a capacitor or inductor.

6-2 Dynamic OP AMP Circuits (Sect. 6–3)

(a) Given an OP AMP integrator or differentiator, determine the output for specified inputs.

(b) Given an RC circuit containing OP AMPs, find the input-output relationship and construct a block diagram.

(c) Design an RC circuit containing OP AMPs that implements a given input-output relationship.

6-3 Equivalent Inductance and Capacitance (Sect. 6–4)

(a) Derive equivalence properties of inductors and capacitors or use equivalence properties to simplify LC circuits.

(b) Given an RLC circuit with dc inputs, find the dc currents and voltage responses.

6-1 THE CAPACITOR

A capacitor is a dynamic element involving the time variation of an electric field produced by a voltage. Figure 6–1(a) shows the parallel plate capacitor, which is the simplest physical form of a capacitive device. Figure 6–1 also shows two alternative circuit symbols. Photos of actual devices are shown in Figure 6–2.

Note: Capacitor manufacturers typically do not use nanofarads and prefer to rate their capacitors as fractions of microfarads or multiples of picofarads. Some standard values for commercially available capacitors are found on the inside rear cover.

Electrostatics shows that a uniform electric field $\mathscr{E}(t)$ exists between the metal plates in Figure 6–1(a) when a voltage exists across the capacitor.[1] The electric field produces charge separation with equal and opposite charges appearing on the capacitor plates. When the separation d is small compared with the dimension of the plates, the electric field between the plates is

$$\mathscr{E}(t) = \frac{q(t)}{\varepsilon A} \qquad (6\text{--}1)$$

where ε is the permittivity of the dielectric, A is the area of the plates, and $q(t)$ is the magnitude of the electric charge on each plate. The relationship between the electric field and the voltage across the capacitor $v_C(t)$ is given by

$$\mathscr{E}(t) = \frac{v_C(t)}{d} \qquad (6\text{--}2)$$

Substituting Eq. (6–2) into Eq. (6–1) and solving for the charge $q(t)$ yields

$$q(t) = \left[\frac{\varepsilon A}{d}\right] v_C(t) \qquad (6\text{--}3)$$

The proportionality constant inside the brackets in this equation is the **capacitance** C of the capacitor. That is, by definition,

$$C = \frac{\varepsilon A}{d} \qquad (6\text{--}4)$$

The unit of capacitance is the **farad** (F), a term that honors the British physicist Michael Faraday. Values of capacitance range from a few pF (10^{-12} F) in semiconductor devices to tens of mF (10^{-3} F) in industrial capacitor banks. Using Eq. (6–4), the defining relationship for the capacitor becomes

$$q(t) = C v_C(t) \qquad (6\text{--}5)$$

Figure 6–3(a) graphically displays the element constraint in Eq. (6–5). The graph points out that the capacitor is a linear element since the defining relationship between voltage and charge is a straight line through the origin.

I–V RELATIONSHIP

To express the element constraint in terms of voltage and current, we differentiate Eq. (6–5) with respect to time t:

$$\frac{dq(t)}{dt} = \frac{d[C v_C(t)]}{dt}$$

---

[1]An electric field is a vector quantity. In Figure 6–1(a) the field is confined to the space between the two plates and is perpendicular to the plates.

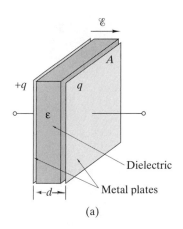

(a)

(b)

FIGURE 6–1 *The capacitor: (a) Parallel plate de ice. (b) Circuit symbols.*

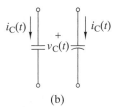

(a)

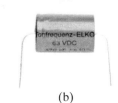

(b)

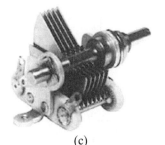

(c)

FIGURE 6–2 *Photos of real capacitors: (a) Ceramic. (b) Electrolytic. (c) Air-tunable. (d) Film capacitors. (e) High- oltage. (f) Trim capacitors.*

(d)

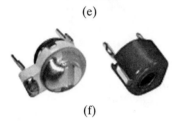

(e)

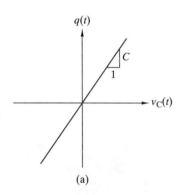

(f)

FIGURE 6-2 *(Continued)*

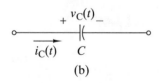

(a)

(b)

FIGURE 6-3 *(a) Graph of the defining relationship of a linear capacitor. (b) Circuit symbol showing capacitor voltage and current.*

Since C is constant and $i_C(t)$ is the time derivative of $q(t)$, we obtain a capacitor $i–v$ relationship in the form

$$i_C(t) = C\frac{dv_C(t)}{dt} \tag{6–6}$$

The relationship assumes that the reference marks for the current and voltage follow the passive sign convention shown in Figure 6–3(b).

The time derivative in Eq. (6–6) means the current is zero when the voltage across the capacitor is constant, and vice versa. In other words, the capacitor acts like an open circuit ($i_C = 0$) when dc excitations are applied. The capacitor is a dynamic element because the current is zero unless the voltage is changing. However, a discontinuous change in voltage would require an infinite current, which is physically impossible. Therefore, the capacitor voltage must be a continuous function of time.

Equation (6–6) relates the capacitor current to the rate of change of the capacitor voltage. To express the voltage in terms of the current, we multiply both sides of Eq. (6–6) by dt, solve for the differential dv_C, and integrate:

$$\int dv_C = \frac{1}{C}\int i_C(t)dt$$

Selecting the integration limits requires some discussion. We assume that at some time t_0 the voltage across the capacitor $v_C(t_0)$ is known and we want to determine the voltage at some later time $t > t_0$. Therefore, the integration limits are

$$\int_{v_C(t_0)}^{v_C(t)} dv_C = \frac{1}{C}\int_{t_0}^{t} i_C(x)dx$$

where x is a dummy integration variable. Integrating the left side of this equation yields

$$v_C(t) = v_C(t_0) + \frac{1}{C}\int_{t_0}^{t} i_C(x)dx \tag{6–7}$$

In practice, the time t_0 is established by a physical event such as closing a switch or the start of a particular clock pulse. Nothing is lost in the integration in Eq. (6–7) if we arbitrarily define t_0 to be zero. Using $t_0 = 0$ in Eq. (6–7) yields

$$v_C(t) = v_C(0) + \frac{1}{C}\int_{0}^{t} i_C(x)dx \tag{6–8}$$

Equation (6–8) is the integral form of the capacitor $i–v$ constraint. Both the integral form and the derivative form in Eq. (6–6) assume that the reference marks for current and voltage follow the passive sign convention in Figure 6–3(b).

POWER AND ENERGY

With the passive sign convention the capacitor power is

$$p_C(t) = i_C(t)v_C(t) \tag{6–9}$$

Using Eq. (6–6) to eliminate $i_C(t)$ from Eq. (6–9) yields the capacitor power in the form

$$p_C(t) = Cv_C(t)\frac{dv_C(t)}{dt} = \frac{d}{dt}\left[\frac{1}{2}Cv_C^2(t)\right] \tag{6–10}$$

This equation shows that the power can be either positive or negative because the capacitor voltage and its time rate of change can have opposite signs. With the passive sign convention, a positive sign means the element absorbs power, while a negative sign means the element delivers power. The ability to deliver power implies that the capacitor can store energy.

To determine the stored energy, we note that the expression for power in Eq. (6–10) is a perfect derivative. Since power is the time rate of change of energy, the quantity inside the brackets must be the energy stored in the capacitor. Mathematically, we can infer from Eq. (6–10) that the energy at time t is

$$w_C(t) = \frac{1}{2} C v_C^2(t) + \text{constant}$$

The constant in this equation is the value of stored energy at some instant t when $v_C(t) = 0$. At such an instant the electric field is zero; hence the stored energy is also zero. As a result, the constant is zero and we write the capacitor energy as

$$w_C(t) = \frac{1}{2} C v_C^2(t) \qquad (6\text{--}11)$$

The stored energy is never negative, since it is proportional to the square of the voltage. The capacitor absorbs power from the circuit when storing energy and returns previously stored energy when delivering power to the circuit.

The relationship in Eq. (6–11) also implies that voltage is a continuous function of time, since an abrupt change in the voltage implies a discontinuous change in energy. Since power is the time derivative of energy, a discontinuous change in energy implies infinite power, which is physically impossible. The capacitor voltage is called a **state variable** because it determines the energy state of the element.

To summarize, the capacitor is a dynamic circuit element with the following properties:

1. *The current through the capacitor is zero unless the voltage is changing. The capacitor acts like an open circuit to dc excitations.*

2. *The voltage across the capacitor is a continuous function of time. A discontinuous change in capacitor voltage would require infinite current and power, which is physically impossible.*

3. *The capacitor absorbs power from the circuit when storing energy and returns previously stored energy when delivering power. The net energy transfer is nonnegative, indicating that the capacitor is a passive element.*

The following examples illustrate these properties.

EXAMPLE 6 – 1

The voltage in Figure 6–4(a) appears across a $\frac{1}{2}$-μF capacitor. Find the current through the capacitor.

SOLUTION:
The capacitor current is proportional to the time rate of change of the voltage. For $0 < t < 2\,\text{ms}$ the slope of the voltage waveform has a constant value

$$\frac{dv_C}{dt} = \frac{10}{2 \times 10^{-3}} = 5000\,\text{V/s}$$

The capacitor current during this interval is

$$i_C(t) = C\frac{dv_C}{dt} = (0.5 \times 10^{-6}) \times (5 \times 10^3) = 2.5\,\text{mA}$$

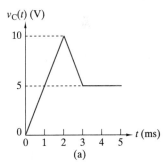

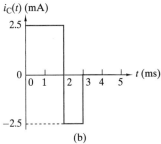

FIGURE 6–4

For $2 < t < 3$ ms the rate of change of the voltage is -5000 V/s. Since the rate of change of voltage is negative, the current changes direction and takes on the value $i_C(t) = -2.5$ mA. For $t > 3$ ms, the voltage is constant, so its slope is zero; hence the current is zero. The resulting current waveform is shown in Figure 6–4(b). Note that the voltage across the capacitor (the state variable) is continuous, but the capacitor current can be, and in this case is, discontinuous. ∎

Exercise 6–1

A 1-μF capacitor has no voltage across it at $t = 0$. A current flowing through the capacitor is given as $i_C(t) = 2u(t) - 3u(t - 2) + u(t - 4)$ μA. Find the voltage across the capacitor at $t = 4$ s.

Answer: $v_C(t) = 2$V.

EXAMPLE 6–2

The $i_C(t)$ in Figure 6–5(a) is given by

$$i_C(t) = I_0(e^{-t/T_C})u(t) \text{ A}$$

Find the voltage across the capacitor if $v_C(0) = 0$ V.

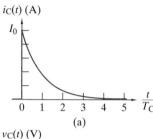

(a)

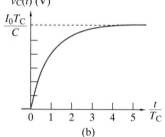

(b)

FIGURE 6–5

SOLUTION:
Using the capacitor i–v relationship in integral form,

$$v_C(t) = v_C(0) + \frac{1}{C} \int_0^t i_C(x)dx$$

$$= 0 + \frac{1}{C} \int_0^t I_0 e^{-x/T_C} dx = \frac{I_0 T_C}{C}(-e^{-x/T_C})\Big|_0^t$$

$$= \frac{I_0 T_C}{C}(1 - e^{-t/T_C}) \text{ V}$$

The graph in Figure 6–5(b) shows that the voltage is continuous while the current is discontinuous. ∎

Exercise 6–2

(a) The voltage across a 10-μF capacitor is $25[\sin 2000t]u(t)$V. Derive an expression for the current through the capacitor.
(b) At $t = 0$ the voltage across a 100-pF capacitor is -5 V. The current through the capacitor is $10[u(t) - u(t - 10^{-4})]$ μA. What is the voltage across the capacitor for $t > 0$?

Answers:

(a) $i_C(t) = 0.5[\cos 2000t]u(t)$ A
(b) $v_C(t) = -5 + 10^5 t$ V for $0 < t < 0.1$ ms and $v_C(t) = 5$ V for $t > 0.1$ ms

Exercise 6–3

For $t \geq 0$ the voltage across a 200-pF capacitor is $5e^{-4000t}$ V.

(a) What is the charge on the capacitor at $t = 0$ and $t = +\infty$?
(b) Derive an expression for the current through the capacitor for $t \geq 0$.
(c) For $t > 0$ is the device absorbing or delivering power?

Answers:

(a) 1 nC and 0 C
(b) $i_C(t) = -4e^{-4000t}$ μA
(c) Delivering

EXAMPLE 6–3

Figure 6–6(a) shows the voltage across a 0.5-μF capacitor. Find the capacitor's energy and power.

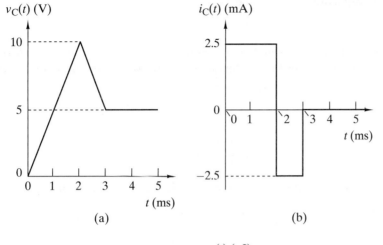

FIGURE 6–6

(a)

(b)

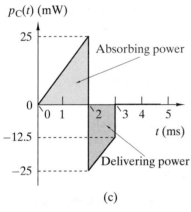

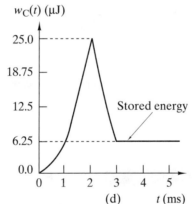

(c)

(d)

SOLUTION:

The current through the capacitor was found in Example 6–1. The power waveform is the point-by-point product of the voltage and current waveforms. The energy is found either by integrating the power waveform or by calculating $\frac{1}{2}C[v_C(t)]^2$ point by point. The current, power, and energy are shown in Figures 6–6(b), 6–6(c), and 6–6(d). Note that the capacitor energy increases when it is absorbing power $[p_C(t) > 0]$ and decreases when delivering power $[p_C(t) < 0]$. ∎

Exercise 6–4

Find the power and energy for the capacitors in Exercise 6–2.

Answers:

(a) $p_C(t) = 6.25[\sin 4000t]u(t)$ W
$w_C(t) = 3.125 \sin^2 2000t$ mJ

(b) $p_C(t) = -0.05 + 10^3 t$ mW for $0 < t < 0.1$ ms
$p_C(t) = 0$ for $t > 0.1$ ms
$w_C(t) = 1.25 - 5 \times 10^4 t + 5 \times 10^8 t^2$ nJ for $0 < t < 0.1$ ms
$w_C(t) = 1.25$ nJ for $t > 0.1$ ms

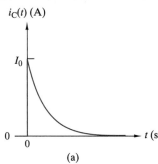

(a)

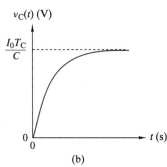

(b)

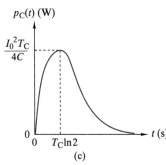

(c)

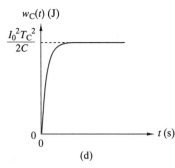

(d)

FIGURE 6–7

EXAMPLE 6–4

The current through a capacitor is given by

$$i_C(t) = I_0[e^{-t/T_C}]u(t) \text{ A}$$

Find the capacitor's energy and power.

SOLUTION:

The current and voltage were found in Example 6–2 and are shown in Figures 6–7(a) and 6–7(b). The power waveform is found as the product of current and voltage:

$$p_C(t) = i_C(t)v_C(t)$$
$$= [I_0e^{-t/T_C}]\left[\frac{I_0T_C}{C}(1 - e^{-t/T_C})\right]$$
$$= \frac{I_0^2T_C}{C}(e^{-t/T_C} - e^{-2t/T_C})$$

The waveform of the power is shown in Figure 6–7(c). The energy is

$$w_C(t) = \frac{1}{2}Cv_C^2(t) = \frac{(I_0T_C)^2}{2C}(1 - e^{-t/T_C})^2$$

The time history of the energy is shown in Figure 6–7(d). In this example, both power and energy are always positive. ∎

Exercise 6–5

Find the power and energy for the capacitor in Exercise 6–3.

Answers:
$$p_C(t) = -20e^{-8000t} \text{ μW}$$
$$w_C(t) = 2.5e^{-8000t} \text{ nJ}$$

APPLICATION   EXAMPLE 6–5

A **sample-and-hold circuit** is usually found at the input to an analog-to-digital converter (ADC). The purpose of the circuit is to sample a time-varying input waveform at a specified instant and then hold that value constant until conversion to digital form is complete. This example discusses the role of a capacitor in such a circuit.

The basic sample-and-hold circuit in Figure 6–8(a) includes an input buffer, a digitally controlled electronic switch, a holding capacitor, and an output buffer. The input buffer is a voltage follower whose output replicates the analog input $v_S(t)$ and supplies charging current to the capacitor. The output buffer is also a voltage follower whose output replicates the capacitor voltage.

To see how the circuit operates, we describe one cycle of the sample-and-hold process. At time t_1 shown in Figure 6–8(b), the digital control $v_G(t)$ goes high, which causes the switch to close. Thereafter, the input buffer supplies a charging current $i_C(t)$ to drive the capacitor voltage to the level of the analog input. At time t_2 shown in Figure 6–8(b) the digital control goes low, the switch opens, and thereafter the capacitor current $i_C(t) = 0$. Zero current means that capacitor voltage is constant since dv_C/dt is zero. In sum, closing the switch causes the capacitor voltage to track the input and opening the switch causes the capacitor voltage to hold a sample of the input.

Figure 6–8(b) shows several more cycles of the sample-and-hold process. Samples of the input waveform are acquired during the time intervals labeled t_C. During these intervals the control signal is high, the switch is closed, and the capacitor charges or discharges in order to track the analog input voltage. Analog-to-digital conversion of the circuit output voltage takes place during the time intervals t_{ADC}. During these intervals the control signal is low, the switch is open, and the capacitor holds the output voltage constant.

Sample-and-hold circuits are available as monolithic integrated circuits (see Figure 6–8(c)) that include the two buffers, the electronic switch, but not the holding capacitor. The capacitor is supplied externally, and its selection involves a trade-off. In an ideal sample-and-hold circuit, the capacitor voltage tracks the input when the switch is closed (sample mode) and holds the value indefinitely when the switch is open (hold mode). In real circuits the input buffer has a maximum output current, which means that some time is needed to charge the capacitor in the sample mode. Minimizing this sample acquisition time argues for a small capacitor. On the other hand, in the hold mode the output buffer draws a small current which gradually discharges the capacitor, causing the output voltage to slowly decrease. Minimizing this output droop calls for a large capacitor. Thus, selecting the capacitance of the holding capacitor involves a compromise between the sample acquisition time and the output voltage droop in the hold mode.

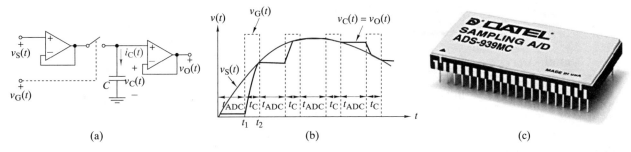

(a) (b) (c)

FIGURE 6–8

6–2 THE INDUCTOR

The inductor is a dynamic circuit element involving the time variation of the magnetic field produced by a current. Magnetostatics shows that a magnetic flux ϕ surrounds a wire carrying an electric current. When the wire is wound into a coil, the lines of flux concentrate along the axis of the coil, as shown in Figure 6–9(a). In a linear magnetic medium, the flux is proportional to both the current and the number of turns in the coil. Therefore, the total flux is

$$\phi(t) = k_1 N i_L(t) \qquad (6\text{–}12)$$

where k_1 is a constant of proportionality.

The magnetic flux intercepts or links the turns of the coil. The flux linkage in a coil is represented by the symbol λ, with units of webers (Wb), named after the German scientist Wilhelm Weber (1804–1891). The flux linkage is proportional to the number of turns in the coil and to the total magnetic flux, so $\lambda(t)$ is

$$\lambda(t) = N\phi(t) \qquad (6\text{–}13)$$

Substituting Eq. (6–12) into Eq. (6–13) gives

$$\lambda(t) = [k_1 N^2] i_L(t) \qquad (6\text{–}14)$$

The proportionality constant inside the brackets in this equation is the **inductance** L of the coil. That is, by definition

$$L = k_1 N^2 \qquad (6\text{–}15)$$

The unit of inductance is the henry (H) (plural: henrys), a name that honors American scientist Joseph Henry. Figure 6–9(b) shows the circuit symbol for an inductor. Photos of some examples of actual devices are shown in Figure 6–10. Some standard values for commercially available inductors are found in the inside rear cover.

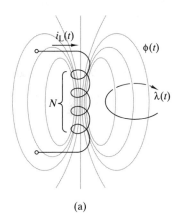

(a)

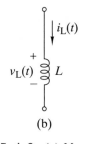

(b)

FIGURE 6–9 *(a) Magnetic flux surrounding a current-carrying coil. (b) Circuit symbol showing inductor current and voltage.*

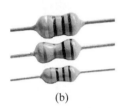

www.butlerwinding.com
(c)

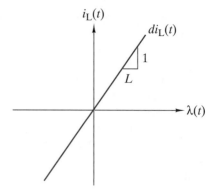

FIGURE 6–11 *Graph of the defining relationship of a linear inductor.*

Using Eq. (6–15), the defining relationship for the inductor becomes

$$\lambda(t) = Li_L(t) \tag{6–16}$$

Figure 6–11 graphically displays the inductor's element constraint in Eq. (6–16). The graph points out that the inductor is a linear element since the defining relationship is a straight line through the origin.

I-V RELATIONSHIP

Equation (6–16) is the inductor element constraint in terms of current and flux linkage. To obtain the element characteristic in terms of voltage and current, we differentiate Eq. (6–16) with respect to time:

$$\frac{d[\lambda(t)]}{dt} = \frac{d[Li_L(t)]}{dt} \tag{6–17}$$

The inductance L is a constant. According to Faraday's law, the voltage across the inductor is equal to the time rate of change of flux linkage. Therefore, we obtain an inductor i–v relationship in the form

$$v_L(t) = L\frac{di_L(t)}{dt} \tag{6–18}$$

The time derivative in Eq. (6–18) means that the voltage across the inductor is zero unless the current is time varying. Under dc excitation the current is constant and $v_L = 0$, so the inductor acts like a short circuit. The inductor is a dynamic element because only a changing current produces a nonzero voltage. However, a discontinuous change in current would produce an infinite voltage, which is physically impossible. Therefore, the current $i_L(t)$ must be a continuous function of time t.

Equation (6–18) relates the inductor voltage to the rate of change of the inductor current. To express the inductor current in terms of the voltage, we multiply both sides of Eq. (6–18) by dt, solve for the differential di_L, and integrate:

$$\int di_L(t) = \frac{1}{L}\int v_L(t)dt \tag{6–19}$$

To set the limits of integration, we assume that the inductor current $i_L(t_0)$ is known at some time t_0. Under this assumption the integration limits are

$$\int_{i_L(t_0)}^{i_L(t)} di_L(t) = \frac{1}{L}\int_{t_0}^{t} v_L(x)dx \tag{6–20}$$

where x is a dummy integration variable. The left side of Eq. (6–20) integrates to produce

$$i_L(t) = i_L(t_0) + \frac{1}{L}\int_{t_0}^{t} v_L(x)dx \tag{6–21}$$

www.butlerwinding.com
(e)

www.butlerwinding.com
(f)

FIGURE 6–10 *Examples of real inductors: (a) Toroidal. (b) Axial. (c) Choke. (d) Surface mount. (e) Air Coil. (f) Bobbin Wound.*

The reference time t_0 is established by some physical event, such as closing or opening a switch. Without losing any generality, we can assume $t_0 = 0$ and write Eq. (6–21) in the form

$$i_L(t) = i_L(0) + \frac{1}{L}\int_0^t v_L(x)\,dx \tag{6–22}$$

Equation (6–22) is the integral form of the inductor i–v characteristic. Both the integral form and the derivative form in Eq. (6–18) assume that the reference marks for the inductor voltage and current follow the passive sign convention shown in Figure 6–9(b).

POWER AND ENERGY

With the passive sign convention the inductor power is

$$p_L(t) = i_L(t)v_L(t) \tag{6–23}$$

Using Eq. (6–18) to eliminate $v_L(t)$ from this equation puts the inductor power in the form

$$p_L(t) = [i_L(t)]\left[L\frac{di_L(t)}{dt}\right] = \frac{d}{dt}\left[\frac{1}{2}Li_L^2(t)\right] \tag{6–24}$$

This expression shows that power can be positive or negative because the inductor current and its time derivative can have opposite signs. Therefore, like a capacitor, an inductor can both absorb and deliver power. The ability to deliver power indicates that the inductor can store energy.

To find the stored energy, we note that the power relation in Eq. (6–24) is a perfect derivative. Since power is the time rate of change of energy, the quantity inside the brackets must represent the energy stored in the magnetic field of the inductor. From Eq. (6–24), we infer that the energy at time t is

$$w_L(t) = \frac{1}{2}Li_L^2(t) + \text{constant}$$

As is the case with capacitor energy, the constant in this expression is zero since it is the energy stored at an instant t at which $i_L(t) = 0$. As a result, the energy stored in the inductor is

$$w_L(t) = \frac{1}{2}Li_L^2(t) \tag{6–25}$$

The energy stored in an inductor is never negative because it is proportional to the square of the current. The inductor stores energy when absorbing power and returns previously stored energy when delivering power, so that the net energy transfer is never negative.

Equation (6–25) implies that inductor current is a continuous function of time because an abrupt change in current causes a discontinuity in the energy. Since power is the time derivative of energy, an energy discontinuity implies infinite power, which is physically impossible. Current is called the **state variable** of the inductor because it determines the energy state of the element.

In summary, the inductor is a dynamic circuit element with the following properties:

1. *The voltage across the inductor is zero unless the current through the inductor is changing. The inductor acts like a short circuit for dc excitations.*

2. *The current through the inductor is a continuous function of time. A discontinuous change in inductor current would require infinite voltage and power, which is physically impossible.*

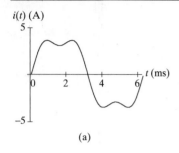

i(t) (A)

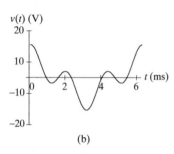

(a)

v(t) (V)

FIGURE 6–12 *(a) Current (b) Voltage.*

3. *The inductor absorbs power from the circuit when storing energy and delivers power to the circuit when returning previously stored energy. The net energy is nonnegative, indicating that the inductor is a passive element.*

EXAMPLE 6–6

The current through a 2-mH inductor is $i_L(t) = 4\sin 1000t + 1\sin 3000t$ A. Find the resulting inductor voltage.

SOLUTION:
The voltage is found from the derivative form of the i–v relationship:

$$v_L(t) = L\frac{di_L(t)}{dt} = 0.002[4 \times 1000\cos 1000t + 1 \times 3000\cos 3000t]$$

$$= 8\cos 1000t + 6\cos 3000t \text{ V}$$

The current and voltage waveforms are shown in Figure 6–12. Note that the current and voltage each contain two sinusoids at different frequencies. However, the relative amplitudes of the two sinusoids are different. In $i_L(t)$ the ratio of the amplitude of the ac component at $\omega = 3$ krad/s to the ac component at $\omega = 1$ krad/s is 1-to-4, whereas in $v_L(t)$ this ratio is 3-to-4. The fact that the ac responses of energy storage elements depend on frequency allows us to create frequency-selective signal processors called filters. ∎

Exercise 6–6

For $t > 0$, the voltage across a 4-mH inductor is $v_L(t) = 20e^{-2000t}$ V. The initial current is $i_L(0) = 0$.

(a) What is the current through the inductor for $t > 0$?
(b) What is the power for $t > 0$?
(c) What is the energy for $t > 0$?

Answers:

(a) $i_L(t) = 2.5(1 - e^{-2000t})$ A
(b) $p_L(t) = 50(e^{-2000t} - e^{-4000t})$ W
(c) $w_L(t) = 12.5(1 - 2e^{-2000t} + e^{-4000t})$ mJ

EXAMPLE 6–7

Figure 6–13 shows the current through and voltage across an unknown energy storage element.

(a) What is the element and what is its numerical value?

(b) If the energy stored in the element at $t = 0$ is zero, how much energy is stored in the element at $t = 1$ s?

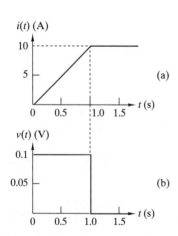

i(t) (A)

(a)

v(t) (V)

(b)

FIGURE 6–13

SOLUTION:
(a) By inspection, the voltage across the device is proportional to the derivative of the current, so the element is a linear inductor. During the interval $0 < t < 1$ s, the slope of the current waveform is 10 A/s. During the same interval the voltage is a constant 100 mV. Therefore, the inductance is

$$L = \frac{v}{di/dt} = \frac{0.1 \text{ V}}{10 \text{ A/s}} = 10 \text{ mH}$$

(b) The energy stored at $t = 1$ s is

$$w_L(1) = \frac{1}{2}Li_L^2(1) = 0.5(0.01)(10)^2 = 0.5 \text{ J}$$ ∎

Exercise 6-7

For $t < 0$, the current through a 100-mH inductor is zero. For $t \geq 0$, the current is $i_L(t) = 20e^{-2000t} - 20e^{-4000t}$ mA.

(a) Derive an expression for the voltage across the inductor for $t > 0$.
(b) Find the time $t > 0$ at which the inductor voltage passes through zero.
(c) Derive an expression for the inductor power for $t > 0$.
(d) Find the time interval over which the inductor absorbs power and the interval over which it delivers power.

Answers:

(a) $v_L(t) = -4e^{-2000t} + 8e^{-4000t}$ V
(b) $t = 0.347$ ms
(c) $p_L(t) = -80e^{-4000t} + 240e^{-6000t} - 160e^{-8000t}$ mW
(d) Absorbing for $0 < t < 0.347$ ms, delivering for $t > 0.347$ ms

EXAMPLE 6-8

The current through a 2.5-mH inductor is a damped sine $i(t) = 10e^{-500t}\sin 2000t$. Plot the waveforms of the element current, voltage, power, and energy.

SOLUTION:

Following are the MATLAB code and the resulting plots for the current, voltage, power, and energy. The code uses symbolic variables, differentiation, and multiplication to calculate the expressions for the four signals. An appropriate time vector is then substituted into the symbolic expressions to create numerical vectors that can be plotted. The subplot function allows multiple plots to be placed in a single figure window. In the plots shown in Figure 6–14, note that the current, voltage, and power alternate signs, whereas the energy signal is always positive.

The MATLAB code uses a numerical time vector defined by the expression [tt = 0:0.00001:0.004;]. The sinusoid has a period of $T_0 = 2\pi/2000 = 0.00314$, so plotting from $t = 0$ ms to $t = 4$ ms is sufficient to capture a full period of the sinusoid. The time step size of 10 μs is a convenient choice and generated 401 points to plot, which produced a smooth curve for each signal. ■

MATLAB Code
```
% Create symbolic variables
syms t iL vL pL wL real

% Create the inductor current signal
iL = 10*exp(-500*t)*sin(2000*t);

% Define the inductance
L = 0.0025;

% Calculate the voltage, power, and energy
vL = L*diff(iL, 't');
pL = iL*vL;
wL = L*iL^2/2;

% Create a time vector
tt = 0:0.00001:0.004;

% Substitute the time vector into the symbolic expressions
iLtt = subs(iL, t, tt);
vLtt = subs(vL, t, tt);
pLtt = subs(pL, t, tt);
```

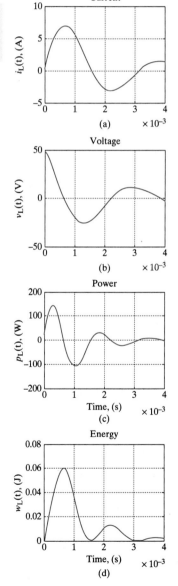

FIGURE 6-14

```
wLtt = subs(wL, t, tt);

% Plot the results using the subplot function
subplot(2,2,1)
plot(tt,iLtt, 'b','LineWidth',2);
grid on
ylabel('i_L(t), (A)')
title('Current')

subplot(2,2,2)
plot(tt,vLtt,'b','LineWidth',2);
grid on
ylabel('v_L(t), (V)')
title('Voltage')

subplot(2,2,3)
plot(tt,pLtt,'b','LineWidth',2);
grid on
xlabel('Time, (s)');
ylabel('p_L(t), (W)')
title('Power')
subplot(2,2,4)
plot(tt,wLtt,'b','LineWidth',2);
grid on
xlabel('Time, (s)');
ylabel('w_L(t), (J)')
title('Energy')
```

Exercise 6–8

A 50-mH inductor has an initial current of $i_L(0)=0$ A. The following voltage is applied across the inductor starting at $t=0$:

$$v_L(t) = -\frac{5}{\pi}\sin(400\pi t) - \frac{5}{2\pi}\sin(800\pi t) - \frac{5}{3\pi}\sin(1200\pi t)\,\text{V}$$

For $t \geq 0$, use MATLAB to determine the inductor current, power, and energy. Plot those three signals and the inductor voltage for $0 \leq t \leq 10$ ms.

Answer: The plots are shown in Figure 6–15.

MORE ABOUT DUALITY

The capacitor and inductor characteristics are quite similar. Interchanging C and L, and i and v converts the capacitor equations into the inductor equations, and vice versa. This interchangeability illustrates the principle of duality. The dual concepts seen so far are as follows:

KVL	$\leftrightarrow$	KCL
Loop	$\leftrightarrow$	Node
Resistance	$\leftrightarrow$	Conductance
Voltage source	$\leftrightarrow$	Current source
Thevenin	$\leftrightarrow$	Norton
Short circuit	$\leftrightarrow$	Open circuit
Series	$\leftrightarrow$	Parallel
Capacitance	$\leftrightarrow$	Inductance
Flux linkage	$\leftrightarrow$	Charge

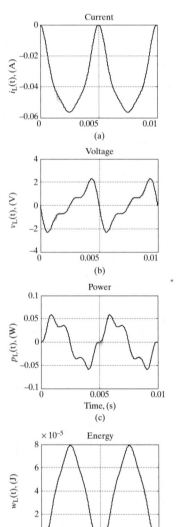

FIGURE 6–15

The term in one column is the dual of the term in the other column. The **principle of duality** states that

> *If every electrical term in a correct statement about circuit behavior is replaced by its dual, then the result is another correct statement.*

This principle may help beginners gain confidence in their understanding of circuit analysis. When the concept in one column is understood, the dual concept in the other column becomes easier to remember and apply.

6–3 DYNAMIC OP AMP CIRCUITS

The dynamic characteristics of capacitors and inductors produce signal processing functions that cannot be obtained using resistors. The OP AMP circuit in Figure 6–16 (a) is similar to the inverting amplifier circuit except for the capacitor in the feedback path. To determine the signal-processing function of the circuit, we need to find its input-output relationship.

We begin by writing a KCL equation at node A.

$$i_R(t) + i_C(t) = i_N(t)$$

The resistor and capacitor device equations are written using their i–v relationships and the fundamental property of node voltages:

$$i_C(t) = C \frac{d[v_O(t) - v_A(t)]}{dt}$$

$$i_R(t) = \frac{1}{R}[v_S(t) - v_A(t)]$$

The ideal OP AMP device equations are $i_N(t)=0$ and $v_A(t)=0$. Substituting all of the element constraints into the KCL connection constraint produces

$$\frac{v_S(t)}{R} + C \frac{dv_O(t)}{dt} = 0$$

To solve for the output $v_O(t)$, we multiply this equation by dt, solve for the differential dv_O, and integrate:

$$\int dv_O(t) = -\frac{1}{RC} \int v_S(t)dt$$

Assuming the output voltage is known at time $t_0=0$, the integration limits are

$$\int_{v_C(0)}^{v_C(t)} dv_O(t) = -\frac{1}{RC} \int_0^t v_S(x)dx$$

which yields

$$v_O(t) = v_O(0) - \frac{1}{RC} \int_0^t v_S(x)dx$$

The initial condition $v_O(0)$ is actually the voltage on the capacitor at $t = 0$, since by KVL, we have $v_C(t) = v_O(t) - v_A(t)$. But $v_A = 0$ for the OP AMP, so in general $v_O(t) = v_C(t)$. When the voltage on the capacitor is zero at $t = 0$, the circuit input-output relationship reduces to

$$v_O(t) = -\frac{1}{RC} \int_0^t v_S(x)dx \qquad (6–26)$$

The output voltage is proportional to the integral of the input voltage when the initial capacitor voltage is zero. The circuit in Figure 6–16(a) is an **inverting integrator** since

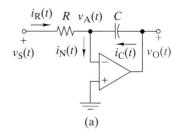

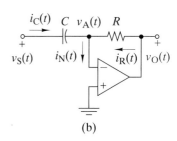

FIGURE 6–16 *(a) The inverting OP AMP integrator. (b) The inverting OP AMP differentiator.*

the proportionality constant is negative. The constant $1/RC$ has the units of reciprocal seconds (s^{-1}) so that both sides of Eq. (6–26) have the units of volts.

Interchanging the resistor and capacitor in Figure 6–16(a) produces the OP AMP differentiator in Figure 6–16(b). To find the input-output relationship of this circuit, we start by writing the element and connection equations. The KCL connection constraint at node A is

$$i_R(t) + i_C(t) = i_N(t)$$

The device equations for the input capacitor and feedback resistor are

$$i_C(t) = C\frac{d[v_S(t) - v_A(t)]}{dt}$$

$$i_R(t) = \frac{1}{R}[v_O(t) - v_A(t)]$$

The device equations for the OP AMP are $i_N(t) = 0$ and $v_A(t) = 0$. Substituting all of these element constraints into the KCL connection constraint produces

$$\frac{v_O(t)}{R} + C\frac{dv_S(t)}{dt} = 0$$

Solving this equation for $v_O(t)$ produces the circuit input-output relationship:

$$v_O(t) = -RC\frac{dv_S(t)}{dt} \qquad (6\text{–}27)$$

The output voltage is proportional to the derivative of the input voltage. The circuit in Figure 6–16(b) is an **inverting differentiator** since the proportionality constant $(-RC)$ is negative. The units of the constant RC are seconds so that both sides of Eq. (6–27) have the units of volts.

There are OP AMP inductor circuits that produce the inverting integrator and differentiator functions; however, they are of little practical interest because of the physical size and resistive losses in real inductor devices.

Figure 6–17 shows OP AMP circuits and block diagrams for the inverting integrator and differentiator, together with signal-processing functions studied in Chapter 4. The term *operational amplifier* results from the various mathematical operations implemented by these circuits. The following examples illustrate using the collection of circuits in Figure 6–17 on the next page in the analysis and design of signal-processing functions.

EXAMPLE 6–9

The input to the circuit in Figure 6-18(a) is $v_s(t) = 10\,u(t)$ V. Derive an expression for the output voltage. The OP AMP saturates when $v_O(t) = \pm15$ V.

SOLUTION:
The circuit is an inverting integrator with an initial voltage of 0 V across the capacitor. Assuming the OP AMP is operating in the linear mode, the output voltage is

$$v_O(t) = v_O(0) - \frac{1}{RC}\int_0^t v_s(x)dx$$

$$v_O(t) = 0 - \frac{1}{10^6 \times 10^{-6}}\int_0^t 10\,dx$$

$$v_O(t) = -\int_0^t 10\,dx = -10t \text{ V} \quad t > 0$$

The output contains a ramp with a slope of -10 V/s. The negative slope is due to the fact that the OP AMP is an inverting integrator. The output will continue to decrease until

Circuit	Block diagram	Gains
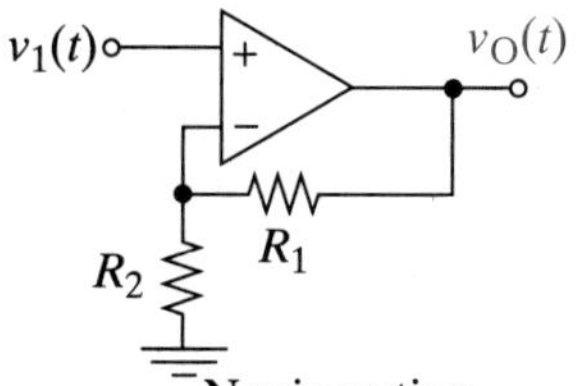 Noninverting	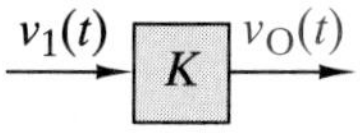	$K = \dfrac{R_1 + R_2}{R_2}$
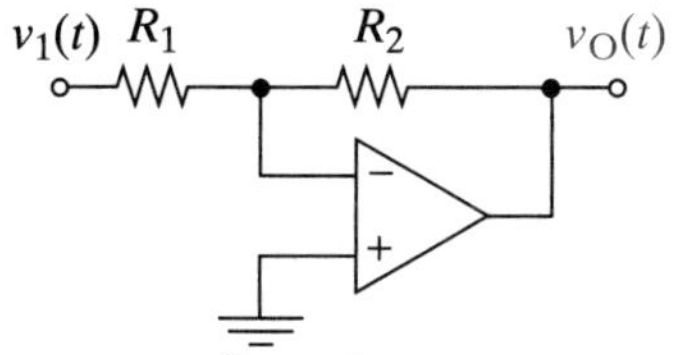 Inverting	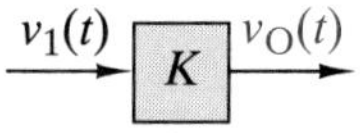	$K = -\dfrac{R_2}{R_1}$
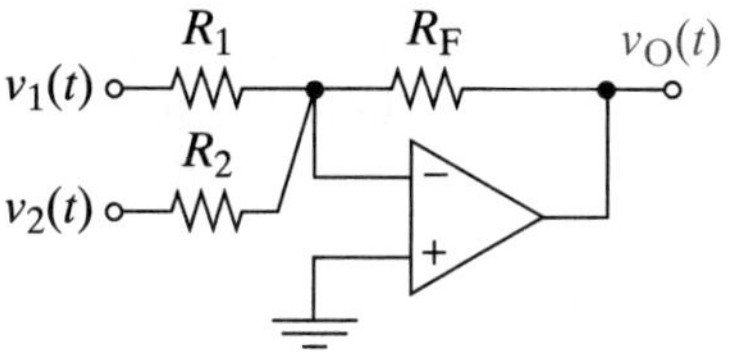 Summer	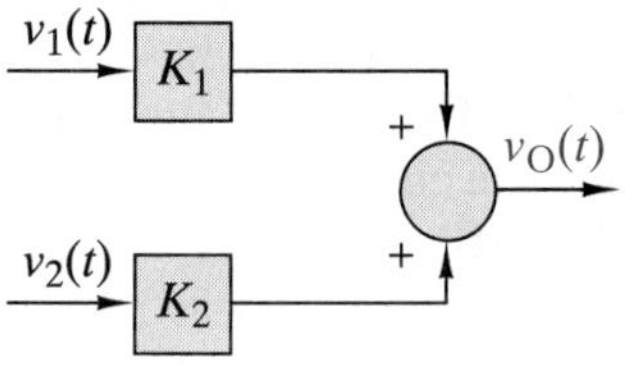	$K_1 = -\dfrac{R_F}{R_1}$ $K_2 = \dfrac{-R_F}{R_2}$
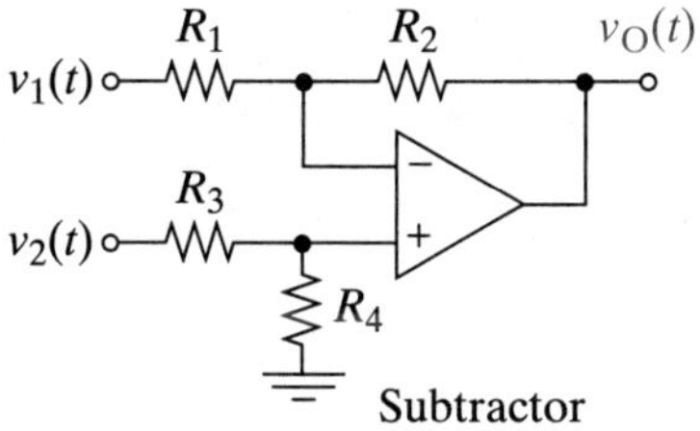 Subtractor	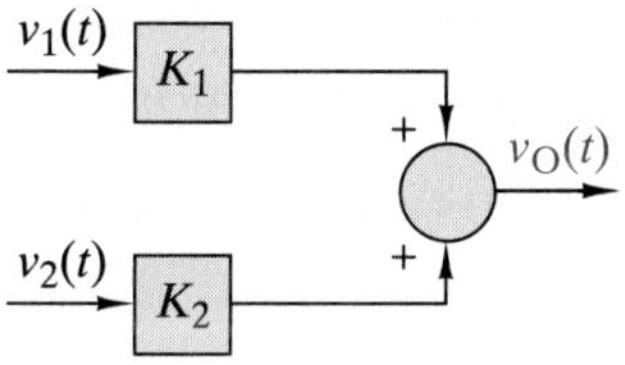	$K_1 = -\dfrac{R_2}{R_1}$ $K_2 = \left(\dfrac{R_1 + R_2}{R_1}\right)\left(\dfrac{R_4}{R_3 + R_4}\right)$
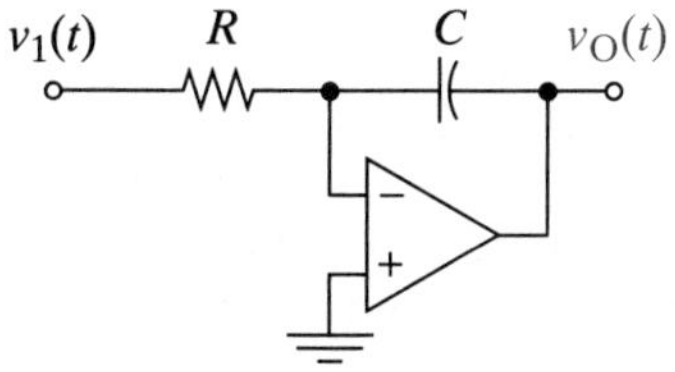 Integrator	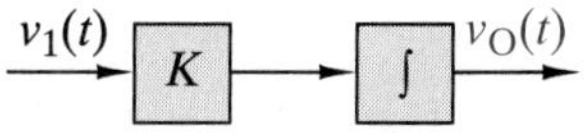	$K = -\dfrac{1}{RC}$
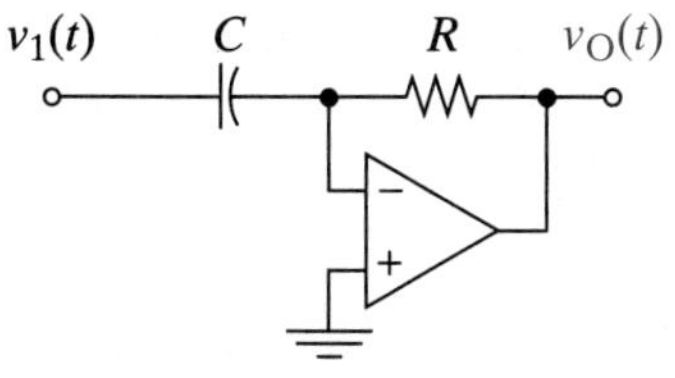 Differentiator	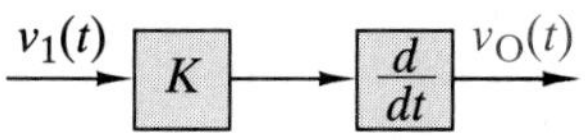	$K = -RC$

FIGURE 6–17 *Summary of basic OP AMP signal-processing circuits.*

FIGURE 6-18

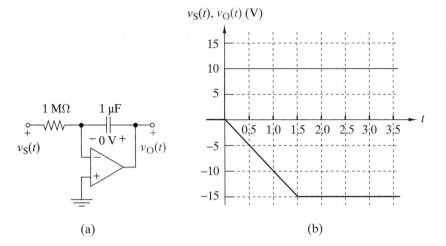

(a)   (b)

the OP AMP saturates at -15 at $t = 1.5$ s. Beyond that time the output is a constant -15 V. Figure 6-18(b) shows both the input and the output of the integrator.

This example illustrates that dynamic circuits with bounded inputs may have unbounded responses. The circuit input here is a 10-V step function that has a bounded amplitude. The circuit output is a ramp whose output would be unbounded except that the OP AMP saturates.   ∎

Exercise 6–9

The input to the circuit in Figure 6–18 is $v_S(t) = 10\,[e^{-5t}]u(t)$ V.

(a) For $v_C(0)=0$, derive an expression for the output voltage, assuming the OP AMP is in its linear range.

(b) Does the OP AMP saturate with the given input?

Answers:

(a) $v_O(t) = 2(e^{-5t} - 1)u(t)$ V
(b) Does not saturate

EXAMPLE 6–10

The input to the circuit in Figure 6–19(a) is the trapezoidal waveform shown in Figure 6–19(b). Find the output waveform. The OP AMP saturates when $v_O(t) = \pm 15$ V.

SOLUTION:
The circuit is the inverting differentiator with the following input-output relationship:

$$v_O(t) = -RC\frac{dv_S(t)}{dt} = -\frac{1}{1000}\frac{dv_S(t)}{dt}$$

The output voltage is constant over each of the following three time intervals:

1. For $0 < t < 1$ ms, the input slope is 5000 V/s and the output is $v_O = -5$ V.

2. For $1 < t < 3$ ms, the input slope is zero, so the output is zero as well.

3. For $3 < t < 5$ ms, the input slope is -2500 V/s and the output is $+2.5$ V.

The resulting output waveform is shown in Figure 6–19(c).

FIGURE 6–19

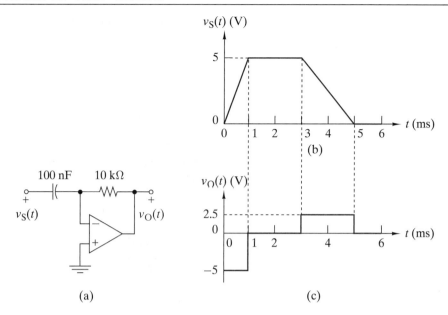

The output voltage remains within ±15-V limits, so the OP AMP operates in the linear mode. ∎

Exercise 6–10

The input to the circuit in Figure 6–19(a) is $v_S(t) = V_A \cos 2000t$ V. The OP AMP saturates when $v_O(t) = \pm 15$ V.

(a) Derive an expression for the output, assuming that the OP AMP is in the linear mode.
(b) What is the maximum value of V_A for linear operation?

Answers:

(a) $v_O(t) = 2V_A \sin 2000t$ V
(b) $|V_A| \leq 7.5$ V

D DESIGN EXAMPLE 6–11

Use the functional blocks in Figure 6–17 to design an OP AMP circuit to implement the following input-output relationship:

$$v_O(t) = 10v_S(t) + \int_0^t v_S(t)dt$$

SOLUTION:
There is no unique solution to this design problem. We begin by drawing the block diagram in Figure 6–20(a), which shows that we need a gain block, an integrator, and a summer. However, the integrator and summer in Figure 6–17 are inverting circuits. Figure 6–20(b) shows how to overcome this problem by including an even number of signal inversions in each path. The inverting building blocks are realizable using OP AMP circuits in Figure 6–17, and the overall transfer characteristic is noninverting as required. One (of many) possible circuit realization of these processors is shown in Figure 6–20(c). The element parameter constraint on this circuit is $RC = 1$. Selecting the OP AMPs and the values of R and C depends on many additional factors, such as accuracy, internal resistance of the input source, and output load. ∎

FIGURE 6–20

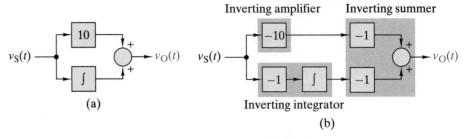

(a)

(b)

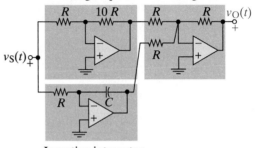

(c)

Exercise 6–11

Find the input-output relationship of the circuit in Figure 6–21.

Answer:
$$v_O(t) = v_O(0) + \frac{1}{RC} \int_0^t (v_{S1}(x) - v_{S2}(x))dx$$

FIGURE 6–21

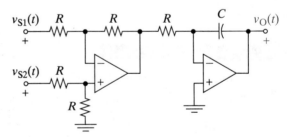

D **DESIGN EXAMPLE 6 – 12**

Differential, integral, and integrodifferential equations can be solved using dynamic circuits. Suppose a second-order system can be described by the following differential equation:

$$5\frac{d^2 v_O(t)}{dt^2} + 100\frac{dv_O(t)}{dt} + 5000v_O(t) = 250 \text{ V}$$

Develop a block-diagram representation of the equation and then design a circuit to solve the equation. That is, design a circuit where the output is $v_O(t)$.

FIGURE 6–22

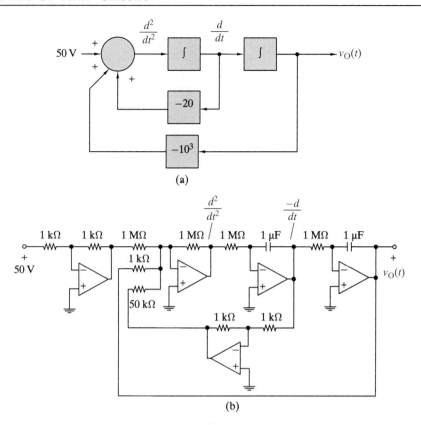

(a)

(b)

SOLUTION:

The first step is to solve the equation for the highest-order derivative with a coefficient of 1. Hence,

$$\frac{d^2 v_O(t)}{dt^2} = -20\frac{d v_O(t)}{dt} - 1000 v_O(t) + 50 \text{ V}$$

Next, we realize that the highest-order derivative is simply the sum of three inputs. If we integrate the second derivative we obtain the first derivative, and if we integrate the first derivative we obtain the desired output, $v_O(t)$. We can use the outputs of the integrators, scale them, and feed them back into the summer along with the 50-V driving function. This can be seen in the block diagram of Figure 6–22(a).

We can design the circuit by implementing the block diagram. For each integrator we select the gain to be -1. Because all of our building blocks are inverting amplifiers, we need to include two inverters to make the signs come out correctly. The summer is used to achieve the appropriate gains. Figure 6–22(b) shows one possible circuit design.

DISCUSSION: *Not too long ago, before digital computers became ubiquitous, these types of problems were sol ed using analog computers. Analog computers sol ed complex, coupled, differential equations using integrators, differentiators, summers, in erters, and nonin erters. Our early space efforts, for example, relied hea ily on analog computers. Today, of course, fast, redundant, and relati ely cheap digital computers are used and analog computers ha e been relegated to the annals of history.* ∎

○ Design Exercise 6–12

Design a circuit to solve the following differential equation:

$$10\frac{dv_\text{x}(t)}{dt} + 50v_\text{x}(t) = 1\text{ V}$$

Answer: There are several possible solutions. Figure 6–23 shows one.

FIGURE 6–23

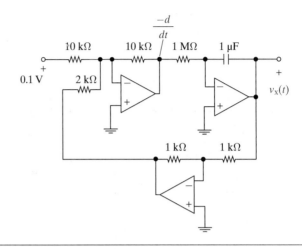

APPLICATION EXAMPLE 6 – 13

Dynamic OP AMP circuits can function as integrators or differentiators. Given an equation or graph of an input waveform, we can predict output waveforms using the mathematical operations of integration or differentiation. Integrators and differentiators work flawlessly on ideal signal models. In practice there are some limitations caused by relatively small imperfections in real-world signals.

First consider an inverting integrator with an input $v_\text{S}(t) + V_0$, where $v_\text{S}(t)$ is the desired signal and V_0 is a relatively small dc offset. For this input an ideal inverting integrator has an output of

$$v_\text{O}(t) = -\frac{1}{RC}\left[\underbrace{\int_0^t v_s(x)dx}_{\text{Desired output}} + \underbrace{V_0t}_{\text{Ramp}}\right]$$

The desired output is accompanied by a ramp waveform caused by a small dc offset. Even if V_0 is very small, the ramp V_0t will eventually overwhelm the desired signal and saturate the OP AMP.

The dc offset problem is dealt with using the reset switch in Figure 6–24(a) to limit the time interval during which the circuit performs integration. When the reset signal is high, the switch is closed, any voltage on the capacitor is removed, and the integrator output voltage is forced to zero. When the reset signal goes low, the switch opens and the circuit operates as an integrator. At the end of a fixed time interval, the reset signal goes high again and the switch closes, forcing the output to zero. Practical integrator applications use *time-limited integration* in which the integrator output is periodically reset to zero before any offset driven ramp becomes important.

Next, consider an inverting differentiator with an input $v_\text{S}(t) + V_\text{A} \times \sin(\omega t)$, where $v_\text{S}(t)$ is the desired signal and $V_\text{A}\sin(\omega t)$ represents relatively small

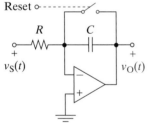

(a) Time-limited integrator

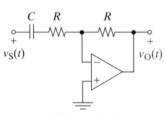

(b) Band-limited differentiator

FIGURE 6–24

high-frequency noise. For this input an ideal inverting differentiator has an output of

$$v_O(t) = -RC\left[\underbrace{\frac{dv_S(t)}{dt}}_{\text{Desired output}} + \underbrace{\omega V_A \cos(\omega t)}_{\text{Noise}}\right]$$

The desired output is accompanied by an amplified noise component. Even if the input noise amplitude V_A is very small, the term ωV_A can be large for high-frequency noise. The basic problem is that differentiation amplifies high-frequency noise to a degree that can overwhelm the desired signal.

The high-frequency noise problem is dealt with by adding the series resistor shown in Figure 6–24(b). This addition limits the frequency range over which the circuit actually performs differentiation. At low frequencies the capacitor is the dominant element, the series resistor plays no role, and the circuit performs differentiation. At high frequencies the roles reverse: The added resistor dominates, the capacitor can be ignored, and the circuit functions as an inverter. As a result the modified circuit only differentiates signals in a low-frequency band.[2] Practical applications use *band-limited differentiation* to avoid high-frequency noise problems.

6–4 EQUIVALENT CAPACITANCE AND INDUCTANCE

In Chapter 2 we found that resistors connected in series or parallel can be replaced by equivalent resistances. The same principle applies to connections of capacitors and inductors—for example, to the parallel connection of capacitors in Figure 6–25(a). Applying KCL at node A yields

$$i(t) = i_1(t) + i_2(t) + \cdots + i_N(t)$$

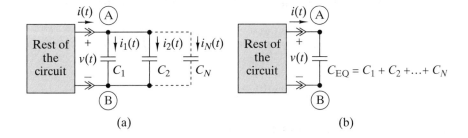

FIGURE 6–25 *Capacitors connected in parallel. (a) Given circuit. (b) Equivalent circuit.*

Since the elements are connected in parallel, KVL requires

$$v_1(t) = v_2(t) = \cdots = v_N(t) = v(t)$$

Because the capacitors all have the same voltage, their i–v relationships are all of the form $i_k(t) = C_k\, dv(t)/dt$. Substituting the i–v relationships into the KCL equation yields

$$i(t) = C_1\frac{dv(t)}{dt} + C_2\frac{dv(t)}{dt} + \cdots + C_N\frac{dv(t)}{dt}$$

Factoring the derivative out of each term produces

$$i(t) = (C_1 + C_2 + \cdots + C_N)\frac{dv(t)}{dt}$$

---

[2]In Chapter 12 we find that this low-frequency band falls below $\omega_C = 1/RC$.

This equation states that the responses $v(t)$ and $i(t)$ in Figure 6–25(a) do not change when the N parallel capacitors are replaced by an equivalent capacitance:

$$C_{EQ} = C_1 + C_2 + \cdots + C_N \qquad \text{(parallel connection)} \qquad (6\text{--}28)$$

The equivalent capacitance simplification is shown in Figure 6–25(b). The initial voltage, if any, on the equivalent capacitance is $v(0)$, the common voltage across all of the original N capacitors at $t=0$.

Next consider the series connection of N capacitors in Figure 6–26(a). Applying KVL around loop 1 in Figure 6–26(a) yields the equation

$$v(t) = v_1(t) + v_2(t) + \cdots + v_N(t)$$

FIGURE 6–26 *Capacitors connected in series. (a) Gi en circuit. (b) Equi alent circuit.*

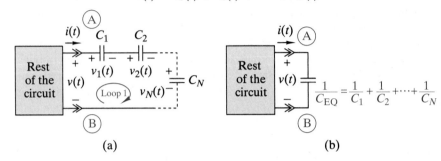

(a)   (b)

Since the elements are connected in series, KCL requires

$$i_1(t) = i_2(t) = \cdots = i_N(t) = i(t)$$

Since the same current exists in all capacitors, their i–v relationships are all of the form

$$v_k(t) = v_k(0) + \frac{1}{C_k} \int_0^t i(x)dx$$

Substituting these i–v relationships into the loop 1 KVL equation yields

$$v(t) = v_1(0) + \frac{1}{C_1} \int_0^t i(x)dx + v_2(0) + \frac{1}{C_2} \int_0^t i(x)dx$$

$$+ \cdots + v_N(0) + \frac{1}{C_N} \int_0^t i(x)dx$$

We can factor the integral out of each term to obtain

$$v(t) = [v_1(0) + v_2(0) + \cdots + v_N(0)] + \left(\frac{1}{C_1} + \frac{1}{C_2} + \cdots + \frac{1}{C_N} \right) \int_0^t i(x)dx$$

This equation indicates that the responses $v(t)$ and $i(t)$ in Figure 6–26(a) do not change when the N series capacitors are replaced by an equivalent capacitance:

$$\frac{1}{C_{EQ}} = \frac{1}{C_1} + \frac{1}{C_2} + \cdots + \frac{1}{C_N} \qquad \text{(series connection)} \qquad (6\text{--}29)$$

The equivalent capacitance is shown in Figure 6–26(b). The initial voltage on the equivalent capacitance is the sum of the initial voltages on each of the original N capacitors.

The equivalent capacitance of a parallel connection is the sum of the individual capacitances. The reciprocal of the equivalent capacitance of a series connection is the sum of the reciprocals of the individual capacitances. Since the capacitor and inductor are dual elements, the corresponding results for inductors are found by interchanging the series and parallel equivalence rules for the capacitor. That is, in a series connection the equivalent inductance is the sum of the individual inductances:

$$L_{EQ} = L_1 + L_2 + \cdots + L_N \qquad \text{(series connection)} \qquad (6\text{--}30)$$

For the parallel connection, the reciprocals add to produce the reciprocal of the equivalent inductance:

$$\frac{1}{L_{EQ}} = \frac{1}{L_1} + \frac{1}{L_2} + \cdots + \frac{1}{L_N} \qquad \text{(parallel connection)} \qquad (6\text{--}31)$$

Derivation of Eqs. (6–30) and (6–31) uses the approach given previously for the capacitor except that the roles of voltage and current are interchanged. Completion of the derivation is left as a problem for the reader.

EXAMPLE 6–14

Find the equivalent capacitance and inductance of the circuits in Figure 6–27.

SOLUTION:

(a) For the circuit in Figure 6–27(a), the two 0.5-μF capacitors in parallel combine to yield an equivalent $0.5 + 0.5 = 1$-μF capacitance. This 1-μF equivalent capacitance is in series with a 1-μF capacitor, yielding an overall equivalent of $C_{EQ} = 1/(1 + 1/1) = 0.5$ μF.

(b) For the circuit of Figure 6–27(b), the 10-mH and the 30-mH inductors are in series and add to produce an equivalent inductance of 40 mH. This 40-mH equivalent inductance is in parallel with the 80-mH inductor. The equivalent inductance of the parallel combination is $L_{EQ} = 1/(1/40 + 1/80) = 26.67$ mH.

(c) The circuit of Figure 6–27(c) contains both inductors and capacitors. In later chapters, we will learn how to combine all of these into a single equivalent element. For now, we combine the inductors and the capacitors separately. The 5-pF capacitor in parallel with the 0.1-μF capacitor yields an equivalent capacitance of 0.100005 μF. For all practical purposes, the 5-pF capacitor can be ignored, leaving two 0.1-μF capacitors in series with equivalent capacitance of 0.05 μF. Combining this equivalent capacitance in parallel with the remaining 0.05-μF capacitor yields an overall equivalent capacitance of 0.1 μF. The parallel 700-μH and 300-μH inductors yield an equivalent inductance of $1/(1/700 + 1/300) = 210$ μH. This equivalent inductance is effectively in series with the 1-mH inductor at the bottom, yielding $1000 + 210 = 1210$ μH as the overall equivalent inductance.

Figure 6–28 shows the simplified equivalent circuits for each of the circuits of Figure 6–27. ∎

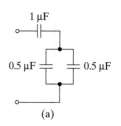

(a)

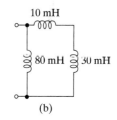

(b)

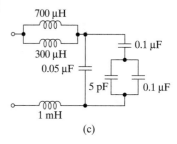

(c)

FIGURE 6–27

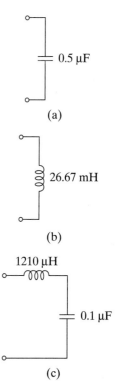

(a)

(b)

(c)

FIGURE 6–28

Exercise 6–13

(a) A 150-μH inductor is in parallel with two other, identical inductors. That combination is in series with a 100-μH inductor. What is the equivalent inductance of the connection?

(b) A capacitor bank consists of five hundred 400-VDC, 10-mF capacitors in parallel. How much energy can the bank store when the capacitors are fully charged?

Answers:

(a) $L_{EQ} = 150$ μH
(b) $w = 400$ kJ

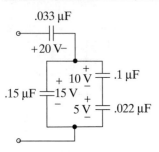

FIGURE 6–29

Exercise 6–14

Find the equivalent capacitance and the initial stored voltage for the circuit in Figure 6–29.

Answer: $C_{Eq} = 0.0276 \ \mu F$ with $v_C(0) = 35$ V.

Exercise 6–15

The current through a series connection of two 1-μF capacitors is a rectangular pulse with an amplitude of 2 mA and a duration of 10 ms. At $t = 0$ the voltage across the first capacitor is +10 V and across the second is zero.

(a) What is the voltage across the series combination at $t = 10$ ms?
(b) What is the maximum instantaneous power delivered to the series combination?
(c) What is the energy stored on the first capacitor at $t = 0$ and $t = 10$ ms?

Answers:

(a) 50 V
(b) 100 mW at $t = 10$ ms
(c) 50 μJ and 450 μJ

DC EQUIVALENT CIRCUITS

Sometimes we need to find the dc response of circuits containing capacitors and inductors. In the first two sections of this chapter, we found that under dc conditions a capacitor acts like an open circuit and an inductor acts like a short circuit. In other words, under dc conditions, an equivalent circuit for a capacitor is an open circuit and an equivalent circuit of an inductor is a short circuit.

To determine dc responses, we replace capacitors by open circuits and inductors by short circuits and analyze the resulting resistance circuit using any of the methods in Chapters 2 through 4. The circuit analysis involves only resistance circuits and yields capacitor voltages and inductor currents along with any other variables of interest. Computer programs like OrCAD use this type of dc analysis to find the initial operating point of a circuit to be analyzed. The dc capacitor voltages and inductor currents become initial conditions for a transient response that begins at $t = 0$ when something in the circuit changes, such as the position of a switch.

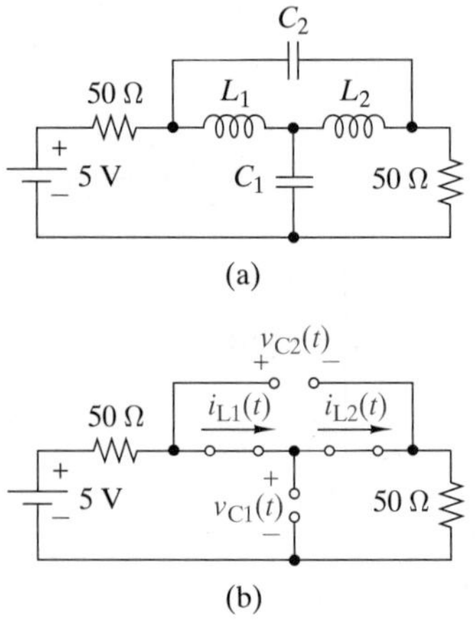

EXAMPLE 6–15

Determine the voltage across the capacitors and current through the inductors in Figure 6–30(a).

FIGURE 6–30

(a)

(b)

SOLUTION:
The circuit is driven by a 5-V dc source. Figure 6–30(b) shows the equivalent circuit under dc conditions. The current in the resulting series circuit is $5/(50 + 50) = 50$ mA. This dc current exists in both inductors, so $i_{L1} = i_{L2} = 50$ mA. By voltage division the voltage across the 50-Ω output resistor is $v = 5 \times 50/(50 + 50) = 2.5$ V; therefore, $v_{C1}(0) = 2.5$ V. The voltage across C_2 is zero because of the short circuits produced by the two inductors. ∎

Exercise 6–16

Find the OP AMP output voltage in Figure 6–31.

Answer:

$$v_O(0) = \frac{R_2 + R_1}{R_1} v_{dc}$$

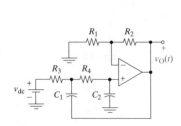

FIGURE 6–31

SUMMARY

- The linear capacitor and inductor are dynamic circuit elements that can store energy. The instantaneous element power is positive when they are storing energy and negative when they are delivering previously stored energy. The net energy transfer is never negative because inductors and capacitors are passive elements.

- The current through a capacitor is zero unless the voltage is changing. A capacitor acts like an open circuit to dc excitations.

- The voltage across an inductor is zero unless the current is changing. An inductor acts like a short circuit to dc excitations.

- Capacitor voltage and inductor current are called state variables because they define the energy state of a circuit. Circuit state variables are continuous functions of time as long as the circuit driving forces are finite.

- OP AMP capacitor circuits perform signal integration or differentiation. These operations, together with the summer and gain functions, provide the building blocks for designing dynamic input-output characteristics.

- Capacitors or inductors in series or parallel can be replaced with an equivalent element found by adding the individual capacitances or inductances or their reciprocals. The dc response of a dynamic circuit can be found by replacing all capacitors with open circuits and all inductors with short circuits.

PROBLEMS

OBJECTIVE 6–1 CAPACITOR AND INDUCTOR RESPONSES (SECTS. 6–1, 6–2)

(a) Given the current through a capacitor or an inductor, find the voltage across the element.
(b) Given the voltage across a capacitor or an inductor, find the current through the element.
(c) Find the power and energy associated with a capacitor or inductor.
(d) See Examples 6–1, 6–2, 6–3, 6–4, 6–6, 6–7, 6–8 and Exercises 6–1, 6–2, 6–3, 6–4, 6–5, 6–6, 6–7, 6–8.

6–1 For $t \geq 0$ the voltage across a 3.3-μF capacitor is $v_C(t) = 5\,u(t)$ V. Derive expressions for $i_C(t)$ and $p_C(t)$. Is the capacitor absorbing power, delivering power, or both?

6–2 For $t \geq 0$ the voltage across a 0.022-μF capacitor is $v_C(t) = 10\,e^{-1000t}u(t)$ V. Derive expressions for $i_C(t)$ and $p_C(t)$. Is the capacitor absorbing power, delivering power, or both?

6–3 The voltage across a 1000-pF capacitor is $v_C(t) = 50 \cos(2\pi 10^4 t)$ V. Derive expressions for $i_C(t)$ and $p_C(t)$. Is the capacitor absorbing power, delivering power, or both?

6–4 The current through a 0.2-μF capacitor is a rectangular pulse with an amplitude of 3 mA and a duration of 5 ms. Find the capacitor voltage at the end of the pulse when the capacitor voltage at the beginning of the pulse is -1 V.

6–5 For $t \geq 0$ the current through a capacitor is $i_C(t) = 10\,r(t)$ mA. At $t = 0$ the capacitor voltage is 3 V. At $t = 1$ ms the voltage is 8 V. Find the capacitance of the device.

6–6 The voltage across a 0.1 μF capacitor is shown in Figure P6–6. Prepare sketches of $i_C(t)$, $p_C(t)$, and $w_C(t)$. Is the capacitor absorbing power, delivering power, or both?

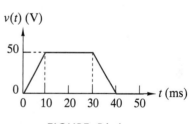

FIGURE P6–6

6–7 The voltage across a 0.01 μF capacitor is shown in Figure P6–7. Prepare sketches of $i_C(t)$, $p_C(t)$, and $w_C(t)$. Is the capacitor absorbing power, delivering power, or both?

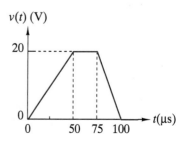

FIGURE P6–7

6–8 The current through a 1500-pF capacitor is shown in Figure P6–8. Given that $v_C(0) = -5$ V, at what time will the voltage $v_C(t)$ first reach 50 V.

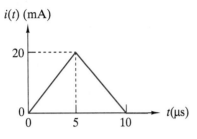

FIGURE P6–8

6–9 For $t \geq 0$ the current through a 0.22- μF capacitor is $i_C(t) = 6 \sin(1000\pi t)$ mA. Plot $v_C(t)$ versus time when $v_C(0) = -10$ V. [*Hint:* In OrCAD use the ISIN source — the frequency is in Hz—and set the other parameters appropriately. Use the Property Editor to set the capacitor's initial condition to −10 V. Then plot the output using the transient function. Place a huge, 1000MEG resistor in parallel with the capacitor to avoid "floating nodes". *Caution:* In OrCAD rotating the capacitor to make it vertical places the capacitor's number 2 terminal at the top. In the identification of an initial voltage on the capacitor, the applied polarity of the voltage is between the number 1 terminal (considered positive) and the number 2 terminal (considered negative). Hence, the capacitor needs to be "flipped vertically" to place the desired −10 V across the capacitor.]

6–10 The current through a 10-mH inductor is shown in Figure P6–10. Prepare sketches of $v_L(t)$, $p_L(t)$, and $w_L(t)$.

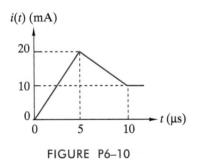

FIGURE P6–10

6–11 For $t \geq 0$ the current through a 560- μH inductor is $i_L(t) = 1\ e^{-10000t}$ A. Find $v_L(t)$, $p_L(t)$ and $w_L(t)$ for $t \geq 0$. Is the inductor absorbing power, delivering power, or both?

6–12 For $t \geq 0$ the voltage across a 10-mH inductor is $v_L(t) = 20\ e^{-400t}$ V. Plot $i_L(t)$ versus time when $i_L(0) = -5$ A. [*Hint:* In OrCAD use the VEXP source, set TD1 to zero and TD2 to 10 s, and set the other parameters appropriately. Use the Property Editor to set the inductor's initial condition to −5 A. Then plot the output using the transient function. Place a tiny, 1-nΩ resistor in series with the inductor to avoid an "inductor loop".]

6–13 A voltage $v_L(t) = 5 \cos(1000\ n\ t)$ V appears across a 50-mH inductor, where n is a positive integer that controls the frequency of the input signal. The amplitude of the input signal is constant. Assume $i_L(0) = 0$ A. Use MATLAB and symbolic variables to compute an expression for $i_L(t)$. On the same axes, plot $i_L(t)$ versus time for $n = 1, 2, 3, 4$, and 5, over an appropriate time scale. On another set of axes, plot the amplitude of $i_L(t)$ versus the coefficient n. As n approaches infinity, what happens to the amplitude of the current? What type of circuit element does the inductor behave like as n approaches infinity?

6–14 For $t \geq 0$ the voltage across a 50-mH inductor is $v_L(t) = 1000r(t)$ V. At $t = 5$ ms the inductor current is observed to be zero. Find the value of $i_L(0)$.

6–15 For $t \geq 0$ the current through a 200-mH inductor is $i_L(t) = 200te^{-2000t}$ A. Derive an expression for $v_L(t)$. Is the inductor absorbing power or delivering power or both?

6–16 The capacitor in Figure P6–16 carries an initial voltage $v_C(0) = 25$ V. At $t = 0$, the switch is closed, and thereafter the voltage across the capacitor is $v_C(t) = 100-75\ e^{-2000t}$ V. Derive expressions for $i_C(t)$ and $p_C(t)$ for $t > 0$. Is the capacitor absorbing power, delivering power, or both?

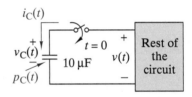

FIGURE P6–16

6–17 A 50-μF capacitor and a 100-mH inductor are connected in parallel with a closed switch as shown in Figure P6–17. The inductor has 25 mA flowing through it at $t = 0-$. The switch opens at $t = 0$.
(a) Find the initial voltage across the capacitor at $t = 0$.
(b) Write an equation for the voltage across the elements for $t > 0$. Do not solve it.
(c) Simulate the circuit using OrCAD. Connect an inductor in parallel with a capacitor and assign the appropriate initial conditions and run a transient analysis. Plot the voltage across the elements for at least 20 ms.
(d) Characterize the response signal.

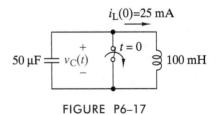

FIGURE P6–17

6–18 The inductor in Figure P6–18 carries an initial current of $i_L(0) = 0.5$ A. At $t = 0$, the switch opens, and thereafter the current into the rest of the circuit is $i(t) = -0.5\ e^{-2000t}$ A.

Derive expressions for $v_L(t)$ and $p_L(t)$ for $t > 0$. Is the inductor absorbing or delivering power?

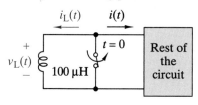

FIGURE P6–18

6–19 The inductor in Figure P6–18 carries an initial current of $i_L(0) = 30$ mA. At $t = 0$, the switch opens, and thereafter the voltage across the inductor is $v_L(t) = -8 e^{-1000t}$ mV. Derive expressions for $i_L(t)$ and $p_L(t)$ for $t > 0$. Is the inductor absorbing or delivering power?

6–20 A 0.056-μF capacitor is connected in series with a 1-kΩ resistor. The voltage across the capacitor is $v_C(t) = 10 \cos(5000t)$ V. What is the voltage across the resistor?

6–21 A 50-mH inductor is connected in parallel with a 33-kΩ resistor. The current through the inductor is $i_L(t) = 20 e^{-1000t}$ mA. What is the current through the resistor?

6–22 For $t > 0$ the voltage across an energy storage element is $v(t) = 5 e^{-100t}$ V and the current through the element is $i(t) = 10 - 5 e^{-100t}$ A. What is the element, the element value, and its initial condition?

6–23 For $t > 0$ the voltage across an energy storage element is $v(t) = 5 - 20 e^{-500t}$ V and the current through the element is $i(t) = 2000t + 16 e^{-500t}$ mA. What is the element, the element value, and its initial condition?

OBJECTIVE 6–2 DYNAMIC OP AMP CIRCUITS (SECT. 6–3)

(a) Given an RC OP AMP integrator or differentiator, determine the output for specified inputs.
(b) Given a general RC OP AMP circuit, determine its input-output relationship and construct a block diagram.
(c) Design an RC OP AMP circuit to implement a given input-output relationship or a block diagram.
(d) See Examples 6–9, 6–10, 6–11, 6–12, 6–13 and Exercises 6–9, 6–10, 6–11, 6–12.

6–24 The OP AMP integrator in Figure P6–24 has $R = 33$ kΩ, $C = 0.056$ μF, and $v_O(0) = 10$ V. The input is $v_S(t) = 5 e^{-500t} u(t)$ V. Find $v_O(t)$ for $t > 0$.

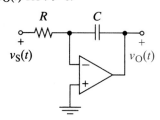

FIGURE P6–24

6–25 An OP AMP integrator with $R = 1$ MΩ, $C = 1$ μF, and $v_O(0) = 0$ V has the input waveform shown in Figure P6–25. Sketch $v_O(t)$ for $t > 0$.

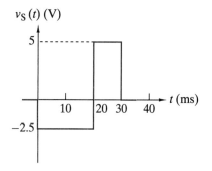

FIGURE P6–25

6–26 An OP AMP circuit from Figure 6–17 is in the box shown in Figure P6–26. The input and outputs are given. What is the function of the circuit in the box if:
(a) $v_S(t) = \cos 500t$ mV and $v_O(t) = 5 \sin 500t$ V.
(b) $v_S(t) = \cos 500t$ mV and $v_O(t) = -20 \sin 500t$ μV.
(c) $v_S(t) = \cos 500t$ mV and $v_O(t) = -2 \cos 500t$ V.

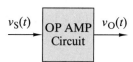

FIGURE P6–26

6–27 ⬦ Design appropriate OP AMP circuits that will realize each of the functions in problem 6–26.

6–28 The OP AMP integrator in Figure P6–24 has $R = 47$ kΩ, $C = 150$ μF, and $v_O(0) = -2$ V. The input is $v_S(t) = 10 u(t)$ V. Use OrCAD to determine how long it takes for the OP AMP to saturate when $V_{CC} = \pm 15$ V.

6–29 The OP AMP integrator in Figure P6–24 has $R = 22$ kΩ, $C = 0.001$ μF, and $v_O(0) = 0$ V. The input is $v_S(t) = 2 \sin(\omega t) u(t)$ V. Derive an expression for $v_O(t)$ and find the smallest allowable value of ω for linear operation of the OP AMP. Assume $V_{CC} = \pm 15$ V.

6–30 The OP AMP differentiator in Figure P6–30 with $R = 15$ kΩ and $C = 0.56$ μF has the input $v_S(t) = 6(1 - e^{-50t})u(t)$ V. Find $v_O(t)$ for $t > 0$.

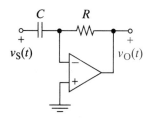

FIGURE P6–30

6–31 The OP AMP differentiator in Figure P6–30 with $R = 47$ kΩ and $C = 0.022$ μF has the input waveform shown in Figure P6–31. Sketch $v_O(t)$ for $t > 0$.

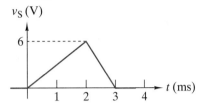

v_S (V)

FIGURE P6–31

6–32 The OP AMP differentiator shown in Figure P6–30 has $R = 100$ kΩ and $C = 0.1$ μF and an output $v_O(t) = 2[\sin{(100t)}] u(t)$ V. What is its input $v_S(t)$?

6–33 🄳 The input to the OP AMP differentiator in Figure P6–30 is $v_S(t) = 10 [\sin{(2\pi \times 10^6 t)}] u(t)$ mV. Select R and C so that the output sinusoid has a maximum of at least ± 14 V but does not saturate the OP AMP at ± 15 V.

6–34 The OP AMP differentiator in Figure P6–30 with $R = 5$ kΩ and $C = 220$ pF has the input $v_S(t) = 2[\sin{(\omega t)}]u(t)$ V. Determine the frequency at which the OP AMP saturates at ± 15 V. Validate your answer using OrCAD. [*Hint:* Use VAC source and an AC sweep.]

6–35 Find the input-output relationship of the RC OP AMP circuit in Figure P6–35.

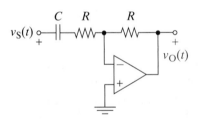

FIGURE P6–35

6–36 Find the input-output relationship of the RC OP AMP circuit in Figure P6–36.

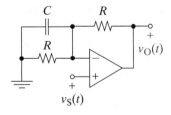

FIGURE P6–36

6–37 Show that the RC OP AMP circuit in Figure P6–37 is a noninverting integrator whose input-output relationship is

$$v_O(t) = \frac{1}{RC} \int_0^t v_S(x)dx + v_O(0)$$

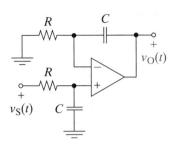

FIGURE P6–37

6–38 🄳 Design an RC OP AMP circuit to implement the block diagram in Figure P6–38.

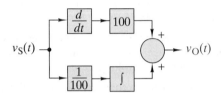

FIGURE P6–38

6–39 🄳 Repeat problem 6–38 except use an RL OP AMP circuit.

6–40 🄳 In this problem you will design an oscillator. The equation for your oscillator is

$$\frac{d^2 v_O(t)}{dt^2} + v_O(t) = 0 \text{ V}$$

(a) Draw a block diagram to solve your equation ($v_O(t)$ should be your output) using differentiators.
(b) Draw a block diagram to solve your equation ($v_O(t)$ should be your output) using integrators.
(c) Design a circuit using OP AMP integrators to realize your block diagram.
(d) Using OrCAD, simulate your circuit and show that your output is an oscillator. Give both capacitors in the circuit the same non-zero initial conditions. What is its oscillating frequency? What could you do to alter the oscillating frequency?

6–41 🄳 Design an RC OP AMP circuit to implement the input-output relationship

$$v_O(t) = -15 v_S(t) + 100 \int_0^t v_S(x)dx$$

6–42 🄳 Design an OP AMP circuit to solve the following differential equation

$$v_O(t) = 10 v_S(t) + \frac{1}{10} \frac{dv_S(t)}{dt}$$

6–43 🄳 Design an RC circuit using only one OP AMP and only one capacitor that implements the input-output relationship

$$v_O(t) = -15 \int_0^t v_{S1}(x)dx - 20 \int_0^t v_{S2}(x)dx$$

OBJECTIVE 6–3 EQUIVALENT INDUCTANCE AND CAPACITANCE (SECT. 6–4)

(a) Derive equivalence properties of inductors and capacitors or use equivalence properties to simplify *LC* circuits.
(b) Given an *RLC* with dc input signals, find the dc current and voltage responses.
See Examples 6–14, 6–15 and Exercise 6–13, 6–14, 6–15, 6–16.

6–44 Find a single equivalent element for each circuit in Figure P6–44.

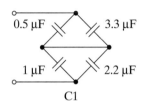

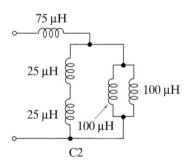

FIGURE P6–44

6–45 ⏺ A 50-μH inductor is connected in series with a 2-mH inductor and the combination connected in parallel with a 1-mH inductor. All are ± 5%. Find the equivalent inductance of the connection. Which inductor played no effective role in this combination and could have been ignored?

6–46 Use the look-back method to find the equivalent capacitance of the circuit shown in Figure P6–46.

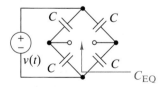

FIGURE P6–46

6–47 Verify Eqs. (6–30) and (6–31).

6–48 What is the equivalent capacitance and initial voltage of a series connection of a 10-μF capacitor with 100 V stored and a 33-μF capacitor with 200 V stored?

6–49 What is the equivalent capacitance and initial voltage for the capacitor bank shown in Figure P6–49?

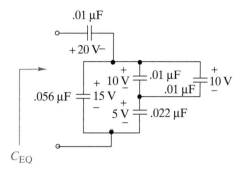

FIGURE P6–49

6–50 What is the equivalent inductance and initial current for the inductors shown in Figure P6–50?

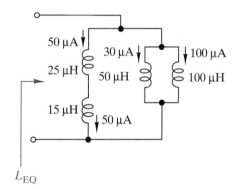

FIGURE P6–50

6–51 For the circuit in Figure P6–51, find an equivalent circuit consisting of one inductor and one capacitor.

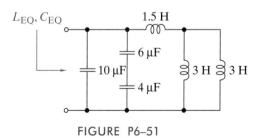

FIGURE P6–51

6–52 Figure P6–52 is the equivalent circuit of a two-wire feed-through capacitor.
(a) What is the capacitance between terminal 1 and ground when terminal 2 is open?
(b) What is the capacitance between terminal 1 and ground when terminal 2 is grounded?

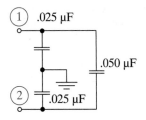

FIGURE P6–52

6–53 🅓 A capacitor bank is required that can be charged to 5 kV and store at least 250 J of energy. Design a series/parallel combination that meets the voltage and energy requirements using 22-μF capacitors each rated at 2 kV max.

6–54 🅔 A switching power supply requires an inductor that can store at least 1 mJ of energy. A list of available inductors is shown below. Select the inductor that best meets the requirement. Consider both meeting the specifications and cost. Explain your choice.

$L(\mu H)$	$I_{MAX}(A)$	COST EACH
10	9.2	$4.75
20	7.0	$5.00
50	5.5	$4.50
100	4.3	$4.75
150	3.8	$4.50
250	2.5	$4.75
500	2.1	$5.00

6–55 The circuits in Figure P6–55 are driven by dc sources. Find the current through the 330-Ω resistor under dc conditions.

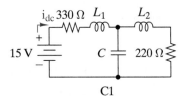

C1

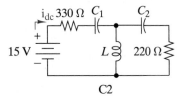

C2

FIGURE P6–55

6–56 The circuit in Figure P6–56 is driven by 15-V dc source. Find the energy stored in the capacitor and inductor under dc conditions.

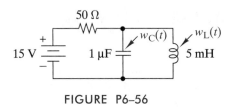

FIGURE P6–56

6–57 The OP AMP circuit in Figure P6–57 has a capacitor in its feedback loop. Determine the circuit gain at dc and as the frequency approaches ∞ rad/s.

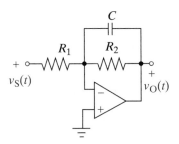

FIGURE P6–57

6–58 The OP AMP circuit in Figure P6–58 has a capacitor in its feedback loop. Determine the circuit gain at dc and as the frequency approaches ∞ rad/s.

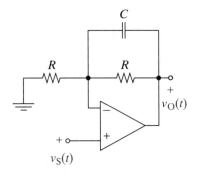

FIGURE P6–58

INTEGRATING PROBLEMS

6–59 Piezoelectric Transducer 🅐

Piezoelectric transducers (sensors) measure dynamic phenomena such as pressure and force. These phenomena cause stresses that "squeeze" a quantity of electric charge from piezoelectric material in the transducer (the term piezo means "squeeze" in Greek). The amount of charge $q(t)$ is directly proportional to the measured variable $x(t)$, that is $q(t) - ax(t)$. Signal amplification is needed because the amount of charge produced is on the order of pC. Figure P6–59 shows an OP AMP charge amplifier that provides the necessary gain. First show that the OP AMP output is $v_O(t) = Kq(t)$. Then select a value of C so that the charge amplifier gain is $K = 6$ mV/pC.

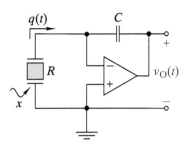

FIGURE P6–59

6–60 LC Circuit Response

At $t = 0$ the switch in Figure P6–60 is closed and thereafter the voltage across the capacitor is

$$v_C(t) = (20 + 10000t)\, e^{-8000t}\ \text{V}$$

Use MATLAB to solve all of the following problems.

(a) Use the capacitor's i- characteristic to find the current $i(t)$ for $t \geq 0$.

(b) Use the inductor's i- characteristic and $i(t)$ to find $v_L(t)$ for $t \geq 0$.

(c) Use $v_C(t)$, $v_L(t)$, and KVL to find the voltage $v(t)$ delivered to the rest of the circuit.

(d) The $v(t)$ found in (c) should be proportional to the $i(t)$ found in (a). If so, what is the equivalent resistance looking into the rest of the circuit?

(e) On the same axes, plot $v_C(t)$, $v_L(t)$, and $v(t)$. Use a different color for each waveform. Use the plots to verify KVL for the circuit.

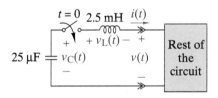

FIGURE P6–60

6–61 Supercapacitor

Supercapacitors have very large capacitances (typically from 0.1 to 50 F,) very long charge holding times, and small sizes, making them useful in non-battery backup power applications. To measure its capacitance, a supercapacitor is charged to an initial voltage $V_0 = 5.5$ V. At $t = 0$ the device undergoes a constant current discharge of $i_D = 2$ mA. At $t = 2500$ s the voltage remaining on the capacitor is 3 V. Find the device capacitance.

6–62 Analog Computer Solution

Design an OP AMP circuit that solves the following second-order differential equation for $v_O(t)$. Solve for the response for $v_O(t)$ using OrCAD, *Caution:* Avoid saturating the OP AMPs by distributing the gain across several OP AMPs.

$$10^{-6}\frac{d^2 v_O(t)}{dt^2} + \tfrac{1}{2} \times 10^{-3}\frac{d v_O(t)}{dt} + v_O(t) = 1.5u(t)$$

6–63 *RC* OP AMP Circuit Design

An upgrade to one of your company's robotics products requires a proportional plus integral compensator that implements the input-output relationship

$$v_O(t) = v_S(t) + 50 \int_o^t v_S(x)dx$$

The input voltage $v_S(t)$ comes from an OP AMP, and the output voltage $v_O(t)$ drives a 10-kΩ resistive load. Two competing designs are shown in Figure P6–63. As the project engineer, you are responsible for recommending one of these designs for production. Which design would you recommend and why? (Your mentor, a wise senior engineer, suggests that you first check that both designs implement the required signal processing function.)

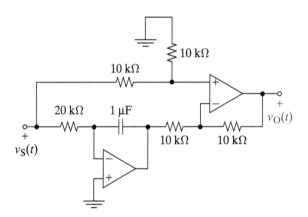

Design #1

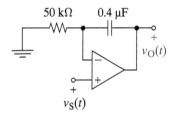

Design #2

FIGURE P6–63

6–64 Tunable Capacitor

In Section 6–1, Figure 6–2(c) shows an air-tunable capacitor as one example of the capacitor types. This type of device can vary its capacitance similarly to how a potentiometer can vary its resistance. Changing the capacitance in a circuit can change the frequency at which it operates. Suppose we are given the circuit in Figure P6–64 with a capacitor connected in parallel with an inductor. There are no other devices in the circuit. The capacitor has an initial voltage $v_C(0) = V_0$ V and the inductor's initial current is $i_L(0) = 0$ A.

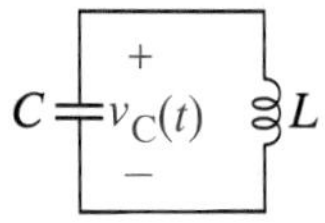

FIGURE P6–64

The differential equation for the voltage across the capacitor in this circuit is given by

$$\frac{d^2 v_C(t)}{dt^2} + \frac{1}{LC} v_C(t) = 0$$

We will learn more about solving this type of differential equation in the next chapter and beyond. The solution to this differential equation is

$$v_C(t) = V_0 \cos\left(\frac{t}{\sqrt{LC}}\right), \; t \geq 0$$

Using MATLAB, plot on a semi-log scale (logarithmic on the horizontal and linear on the vertical,) the radian frequency of $v_C(t)$ versus the capacitance of the circuit. Use capacitances scaled logarithmically from 0.001 μF, to 1 μF. In a separate MATLAB figure, use the `subplot` command to plot $v_C(t)$ versus time for $C = 0.001$ μF, 0.01 μF, 0.1 μF, and 1 μF. Using the plots you created and your knowledge of how a capacitor stores charge, explain why changing the capacitance of a parallel LC circuit changes the frequency at which is oscillates.

6–65 Equivalent Capacitance Bridge

Find the equivalent capacitance of the capacitance bridge shown in Figure P6–65. (*Hint:* Use Node Analysis.)

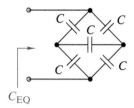

C_{EQ}

FIGURE P6–65

CHAPTER 7 FIRST- AND SECOND-ORDER CIRCUITS

When a mathematician engaged in investigating physical actions and results has arrived at his own conclusions, may they not be expressed in common language as fully, clearly and definitely as in mathematical formula? If so, would it not be a great boon to such as we to express them so—translating them out of their hieroglyphics that we also might work upon them by experiment.

Michael Faraday, 1857,
British Physicist

Some History Behind This Chapter

Michael Faraday (1791–1867) was appointed a Fellow in the Royal Society at age 32 and was a lecturer at the Royal Institution in London for more than 50 years. During this time he published over 150 papers on chemistry and electricity. The most important of these papers was the series *Experimental Researches in Electricity*, which included a description of his discovery of magnetic induction. A gifted experimentalist, Faraday apparently felt that mathematics obscured the physical truths he discovered through experimentation.

Why This Chapter Is Important Today

OK, this is a tough chapter. It concentrates on the classical methods of finding the transient response of circuits containing resistors, capacitors, and inductors. Mathematically this requires us to solve first- and second-order differential equations. These solutions help us understand applications such as timing circuits and digital gate delays. It is important to understand first- and second-order transients because we will revisit these concepts frequently in subsequent chapters.

Chapter Sections

Chapter Learning Objectives

7-1 First-order Circuit Analysis (Sects. 7–1, 7–2, 7–3, 7–4)

Given a first-order *RC* or *RL* circuit:
(a) Find the circuit differential equation, the circuit time constant, and the initial conditions (if not given).
(b) Find the zero-input response.
(c) Find the complete response for step function, exponential, and sinusoidal inputs.

7-2 First-order Circuit Design (Sects. 7–1, 7–2, 7–3, 7–4)

Given responses in a first-order *RC* or *RL* circuit:
(a) Find the circuit parameters or other responses.
(b) Design a circuit to produce the given responses.

7-3 Second-order Circuit Analysis (Sects. 7–5, 7–6, 7–7)

Given a second-order circuit:
(a) Find the circuit differential equation.
(b) Find the circuit natural frequencies and the initial conditions (if not given).
(c) Find the zero-input response.
(d) Find the complete response for a step function input.

7-4 Second-order Circuit Design (Sects. 7–5, 7–6, 7–7)

Given responses in a second-order *RLC* circuit:
(a) Find the circuit parameters or other responses.
(b) Design a circuit to produce the given responses.

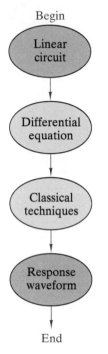

Begin

Linear
circuit

Differential
equation

Classical
techniques

Response
waveform

End

FIGURE 7–1 *Flow diagram for dynamic circuit analysis.*

7–1 *RC* AND *RL* CIRCUITS

The flow diagram in Figure 7–1 shows the two major steps in the analysis of a dynamic circuit. In the first step we use device and connection equations to formulate a differential equation describing the circuit. In the second step we solve the differential equation to find the circuit response. In this chapter we examine basic methods of formulating circuit differential equations and the time-honored, classical methods of solving for responses. Solving for the responses of simple dynamic circuits gives us insight into the physical behavior of the basic circuit modules of the more complex networks in subsequent chapters. This insight will help us correlate circuit behavior with the results obtained by other methods of dynamic circuit analysis. The treatment of other methods includes phasor circuit analysis (Chapter 8) and the Laplace transform methods (beginning in Chapter 9).

FORMULATING *RC* AND *RL* CIRCUIT EQUATIONS

RC and *RL* circuits contain linear resistors and a single capacitor or a single inductor. Figure 7–2 shows how we can divide *RC* and *RL* circuits into two parts: (1) the dynamic element and (2) the rest of the circuit, containing only linear resistors and sources. To formulate the equation governing either of these circuits, we replace the resistors and sources by their Thévenin or Norton equivalents shown in Figure 7–2.

Dealing first with the series *RC* circuit in Figure 7–2(a), we note that the Thévenin equivalent source is governed by the KVL constraint

$$R_{\mathrm{T}}i(t) + v(t) = v_{\mathrm{T}}(t) \qquad (7\text{–}1)$$

The capacitor *i–v* constraint is

$$i(t) = C\frac{dv(t)}{dt} \qquad (7\text{–}2)$$

Substituting the *i–v* constraint into the source constraint yields

$$R_{\mathrm{T}}C\frac{dv(t)}{dt} + v(t) = v_{\mathrm{T}}(t) \qquad (7\text{–}3)$$

The unknown in Eq. (7–3) is the capacitor voltage $v(t)$ that determines the amount of energy stored in the *RC* circuit and is referred to as the **state variable**.

FIGURE 7–2 *First-order circuits: (a) RC series circuit. (b) RL parallel circuit.*

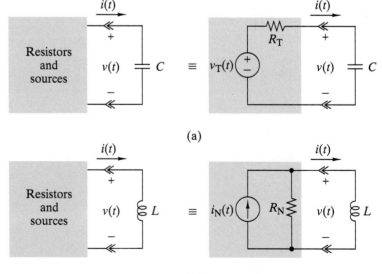

(a)

(b)

Mathematically, Eq. (7–3) is a first-order linear differential equation with constant coefficients. The equation is first order because the first derivative of the dependent variable is the highest-order derivative in the equation. The product $R_\mathrm{T}C$ is a constant coefficient because it depends on fixed circuit parameters. The signal $v_\mathrm{T}(t)$ is the Thévenin equivalent of the independent sources driving the circuit. The voltage $v_\mathrm{T}(t)$ is the input, and the capacitor voltage $v(t)$ is the circuit response.

The Norton equivalent source in the *RL* parallel circuit in Figure 7–2(b) is governed by the KCL constraint

$$\frac{v(t)}{R_\mathrm{N}} + i(t) = i_\mathrm{N}(t) \tag{7–4}$$

The element constraint for the inductor can be written

$$v(t) = L\frac{di(t)}{dt} \tag{7–5}$$

Combining the element and source constraints produces the differential equation for the *RL* circuit:

$$\frac{L}{R_\mathrm{N}}\frac{di(t)}{dt} + i(t) = i_\mathrm{N}(t) \tag{7–6}$$

The response of the *RL* circuit is also governed by a first-order linear differential equation with constant coefficients. The dependent variable in Eq. (7–6) is the inductor current. The circuit parameters enter as the constant product L/R_N, and the driving forces are represented by a Norton equivalent current $i_\mathrm{N}(t)$. The unknown in Eq. (7–6) is the inductor current $i(t)$. This current determines the amount of energy stored in the *RL* circuit and is referred to as the **state variable**.

The state variables in first-order circuits are the capacitor voltage in the *RC* circuit and the inductor current in the *RL* circuit. As we will see, these state variables contain sufficient information about the past to determine future circuit responses.

We observe that Eqs. (7–3) and (7–6) have the same form. In fact, interchanging the quantities

$$R_\mathrm{T} \leftrightarrow G_\mathrm{N} = \frac{1}{R_\mathrm{N}} \qquad C \leftrightarrow L \qquad v \leftrightarrow i \qquad v_\mathrm{T} \leftrightarrow i_\mathrm{N} \qquad \text{series} \leftrightarrow \text{parallel}$$

converts one equation into the other. This interchange is another example of the principle of duality. Because of duality we do not need to study the *RC* and *RL* circuits as independent problems. Everything we learn by solving the *RC* circuit, for example can be applied to the *RL* circuit as well.

We refer to the *RC* and *RL* circuits as **first-order circuits** because they are described by a first-order differential equation. The first-order differential equations in Eqs. (7–3) and (7–6) describe general *RC* and *RL* circuits shown in Figure 7–2. Any circuit containing a single uncombinable capacitor or inductor and linear resistors and sources is a first-order circuit.

ZERO-INPUT RESPONSE OF FIRST-ORDER CIRCUITS

The response of a first-order circuit is found by solving the circuit differential equation. For the *RC* circuit the response $v(t)$ must satisfy the differential equation in Eq. (7–3) and the initial condition $v(0)$. By examining Eq. (7–3) we see that the response depends on three factors:

1. The inputs driving the circuit $v_\mathrm{T}(t)$
2. The values of the circuit parameters R_T and C
3. The value of $v(t)$ at $t = 0$ (i.e., the initial condition)

The first two factors apply to any linear circuit, including resistance circuits. The third factor relates to the initial energy stored in the circuit. The initial energy can cause the circuit to have a nonzero response even when the input $v_T(t) = 0$ for $t \geq 0$. The existence of a response with no input is something new in our study of linear circuits.

To explore this discovery we find the **zero-input response**. Setting all independent sources in Figure 7–2 to zero makes $v_T(t) = 0$ in Eq. (7–3):

$$R_T C \frac{dv(t)}{dt} + v(t) = 0 \tag{7–7}$$

Mathematically, Eq. (7–7) is a **homogeneous equation** because the right side is zero. The classical approach to solving a linear homogeneous differential equation is to try a solution in the form of an exponential

$$v(t) = Ke^{st} \tag{7–8}$$

where K and s are constants to be determined.

The form of the homogeneous equation suggests an exponential solution for the following reasons. Equation (7–7) requires that $v(t)$ plus $R_T C$ times its derivative must add to zero for all time $t \geq 0$. This can only occur if $v(t)$ and its derivative have the same form. In Chapter 5 we saw that an exponential signal and its derivative are both of the form e^{-t/T_C}. Therefore, the exponential is a logical starting place.

If Eq. (7–8) is indeed a solution, then it must satisfy the differential equation in Eq. (7–7). Substituting the trial solution into Eq. (7–7) yields

$$R_T CKse^{st} + Ke^{st} = 0$$

or

$$Ke^{st}(R_T Cs + 1) = 0$$

The exponential function e^{st} cannot be zero for all t. The condition $K = 0$ is a trivial solution because it implies that $v(t)$ is zero for all time t. The only nontrivial way to satisfy the equation involves the condition

$$R_T Cs + 1 = 0 \tag{7–9}$$

Equation (7–9) is the circuit **characteristic equation** because its root determines the attributes of $v(t)$. The characteristic equation has a single root at $s = -1/R_T C$, so the zero-input response of the RC circuit has the form

$$v(t) = Ke^{-t/R_T C} \qquad t \geq 0$$

The constant K can be evaluated using the value of $v(t)$ at $t = 0$. Using the notation $v(0) = V_0$ yields

$$v(0) = Ke^0 = K = V_0$$

The final form of the zero-input response is

$$v(t) = V_0 e^{-t/R_T C} \quad t \geq 0 \tag{7–10}$$

The zero-input response of the RC circuit is the familiar exponential waveform shown in Figure 7–3. At $t = 0$ the exponential response starts out at $v(0) = V_0$ and then decays to zero as $t \to \infty$. The time constant $T_C = R_T C$ depends only on fixed circuit parameters. From our study of the exponential signals in Chapter 5, we know that the $v(t)$ decays to about 36.8% of its initial amplitude in one time constant and to essentially zero after about five time constants. The zero-input response of the RC circuit is determined by two quantities: (1) the circuit time constant and (2) the value of the capacitor voltage at $t = 0$.

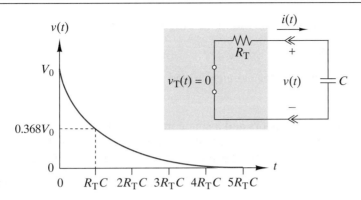

FIGURE 7–3 *First-order RC circuit zero-input response.*

Let us now look at the behavior of an RL circuit and compare it to the RC circuit just studied.

The zero-input response of the RL circuit in Figure 7–2(b) is found by setting the Norton current $i_N(t) = 0$ in Eq. (7–6).

$$\frac{L}{R_N}\frac{di(t)}{dt} + i(t) = 0 \tag{7–11}$$

The unknown in this homogeneous differential equation is the inductor current $i(t)$. Equation (7–11) has the same form as the homogeneous equation for the RC circuit, which suggests a trial solution of the form

$$i(t) = Ke^{st}$$

where K and s are constants to be determined. Substituting the trial solution into Eq. (7–11) yields the RL circuit characteristic equation

$$\frac{L}{R_N}s + 1 = 0 \tag{7–12}$$

The root of this equation is $s = -R_N/L$. Denoting the initial value of the inductor current by I_0, we evaluate the constant K:

$$i(0) = I_0 = Ke^0 = K$$

The final form of the zero-input response of the RL circuit is

$$i(t) = I_0e^{-R_Nt/L} \quad t \geq 0 \tag{7–13}$$

For the RL circuit the zero-input response of the state variable $i(t)$ is an exponential function with a time constant of $T_C = G_NL = L/R_N = L/R_T$. This response connects the initial state $i(0) = I_0$ with the final state $i(\infty) = 0$.

The zero-input responses in Eqs. (7–10) and (7–13) show the duality between first-order RC and RL circuits. These results point out that the zero-input response in a first-order circuit depends on two quantities: (1) the circuit time constant and (2) the value of the state variable at $t = 0$. Capacitor voltage and inductor current are called state variables because they determine the amount of energy stored in the circuit at any time t. The following examples show that the zero-input response of the state variable provides enough information to determine the zero-input response of every other voltage and current in the circuit.

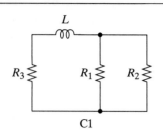

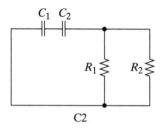

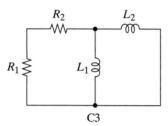

FIGURE 7–4

But first, let us look at finding the time constants of RC or RL circuits.

EXAMPLE 7–1

Find the time constants for circuits C1 and C2 in Figure 7–4.

SOLUTION:

Circuit C1 has three resistors and a single inductor. The question one should ask is "What is the equivalent resistance that the inductor sees?" The time constant then will be L/R_{EQ}. In this example, the inductor "sees" two resistors R_1 and R_2 in parallel and that combination in series with R_3. Therefore, R_{EQ} is $\dfrac{R_1 R_2}{R_1 + R_2} + R_3$. Hence, the time constant is

$$T_C = \frac{L}{R_{EQ}} = \frac{L}{\dfrac{R_1 R_2}{R_1 + R_2} + R_3} = \frac{L(R_1 + R_2)}{R_1 R_2 + R_3 R_1 + R_3 R_2} \text{ s}$$

Circuit C2 has two capacitors that can be combined into one equivalent capacitor and two resistors in parallel that can also be combined. The time constant for an RC circuit is $R_{EQ} C_{EQ}$. For this circuit we have $C_{EQ} = \dfrac{C_1 C_2}{C_1 + C_2}$, and the equivalent resistance that C_{EQ} sees is $R_{EQ} = \dfrac{R_1 R_2}{R_1 + R_2}$. Therefore, the time constant is

$$T_C = R_{EQ} C_{EQ} = \left(\frac{R_1 R_2}{R_1 + R_2}\right)\left(\frac{C_1 C_2}{C_1 + C_2}\right) \text{ s}$$ ∎

Exercise 7–1

Find the time constant T_C for circuit C3 in Figure 7–4.

Answer:

$$T_C = \frac{L_1 L_2}{(L_1 + L_2)(R_1 + R_2)} \text{ s}$$

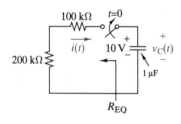

FIGURE 7–5

EXAMPLE 7–2

The switch in Figure 7–5 closes at $t = 0$, connecting a 1-μF capacitor with 10 V initially stored across it to two resistors. Find the responses $v_C(t)$ and $i(t)$ for $t \geq 0$. Write an equation for the power absorbed by the equivalent resistance.

SOLUTION:

The solution involves the zero-input response of an RC circuit since the only energy in the circuit is the voltage stored on the capacitor at $t = 0$. To start this solution we must first determine the circuit time constant *after* the switch closes. The equivalent resistance seen by the capacitor is

$$R_{EQ} = 100 \text{ k} + 200 \text{ k} = 300 \text{ k}\Omega$$

For $t \geq 0$ the time constant is

$$T_C = R_T C = R_{EQ} C = 300 \text{ k} \times 1 \text{ μ} = 0.3 \text{ s}$$

The initial capacitor voltage is $V_0 = 10$ V. Using Eq. (7–10), the zero-input response of the capacitor voltage is

$$v_C(t) = 10e^{-t/0.3} = 10e^{-3.33t} \text{ V} \quad t \geq 0$$

This result allows us to readily find the current $i(t)$, since once the switch closes we have a series circuit. The current, then, is found by calculating the current through the capacitor using the capacitor's i-v relationship

$$i(t) = i_C(t) = C\frac{dv_C(t)}{dt} = 1 \times 10^{-6}\frac{d(10e^{-3.33t})}{dt} = -33.3e^{-3.33t}\,\mu\text{A}$$

The minus sign tells us that the current is opposite of the referenced direction and is actually flowing out of the capacitor.

The power absorbed by the equivalent resistance can be found as follows

$$p_R(t) = i_R^2(t)R_{EQ} - (-33.3 \times 10^{-6}e^{-3.33t})^2(300\,\text{k}) = 333e^{-6.66t}\,\mu\text{W}$$

Notice the analysis pattern. We first determine the zero-input response of the capacitor voltage. The state variable response together with resistance circuit analysis techniques were then used to find the current. The circuit time constant and the value of the state variable at $t = 0$ provide enough information to determine the zero-input response of every voltage or current in the circuit. ∎

Exercise 7–2

The switch in Figure 7–6 closes at $t = 0$. For $t \geq 0$ the current through the resistor is $i_R(t) = e^{-100t}$ mA.

(a) What is the capacitor voltage at $t = 0$?
(b) Write an equation for $v(t)$ for $t \geq 0$.
(c) Write an equation for the power absorbed by the resistor for $t \geq 0$.
(d) How much energy does the resistor dissipate for $t \geq 0$?
(e) How much energy is stored in the capacitor at $t = 0$?

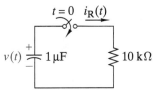

FIGURE 7–6

Answers:

(a) 10 V
(b) $v(t) = 10e^{-100t}$ V
(c) $p_R(t) = 10e^{-200t}$ mW
(d) 50 μJ
(e) 50 μJ

EXAMPLE 7–3

Find the response of the state variable of the RL circuit in Figure 7–7 using $L_1 = 10\,\text{mH}$, $L_2 = 30\,\text{mH}$, $R_1 = 2\,\text{k}\Omega$, $R_2 = 6\,\text{k}\Omega$, and $i_L(0) = 100\,\text{mA}$.

SOLUTION:
The inductors are connected in series and can be replaced by an equivalent inductor

$$L_{EQ} = L_1 + L_2 = 10 + 30 = 40 \text{ mH}$$

Likewise, the resistors are connected in parallel and the resistance seen by L_{EQ} is

$$R_{EQ} = \frac{R_1R_2}{R_1 + R_2} = 1.5\,\text{k}\Omega$$

Figure 7–7 shows the resulting equivalent circuit. The interface signals $v(t)$ and $i(t)$ are the voltage across and current through $L_{EQ} = L_1 + L_2$. The time constant of the equivalent RL circuit is

$$T_C = \frac{L_{EQ}}{R_{EQ}} = \frac{0.04}{1500} = 26.6 \text{ μs} = \frac{1}{37,500} \text{ s}$$

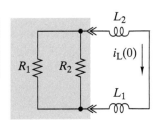

Given circuit

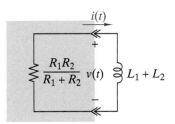

Equivalent circuit

FIGURE 7–7

The initial current through L_{EQ} is $i_L(0) = 0.1$ A. Using Eq. (7–13) with $I_0 = 0.1$ yields the zero-state response of the inductor current.

$$i(t) = 0.1e^{-37,500t} \text{ A} \qquad t \geq 0$$

Given the state variable response, we can find every other response in the original circuit. For example, by KCL and current division the current through R_1 is

$$i_{R_1}(t) = \frac{R_2}{R_1 + R_2}i(t) = 0.075\ e^{-37,500t} \text{ A} \qquad t \geq 0 \qquad \blacksquare$$

Exercise 7–3

Find the current through R_2 in Example 7–3.

Answer: $i_{R_2}(t) = \dfrac{R_1}{R_1 + R_2}i(t) = 0.025e^{-37,500t} \text{ A} \qquad t \geq 0$

Example 7–3 illustrates an important point. The RL circuit in Figure 7–7 is a first-order circuit even though it contains two inductors. The two inductors are connected in series and can be replaced by a single equivalent inductor. In general, capacitors or inductors in series and parallel can be replaced by a single equivalent element. Thus, any circuit containing the *equivalent* of a single inductor or a single capacitor is a first-order circuit.

EXAMPLE 7–4

The switch in Figure 7–8 is closed at $t = 0$, connecting a capacitor with an initial voltage of 30 V to the resistances shown. Find the responses $v_C(t)$, $i(t)$, $i_1(t)$, and $i_2(t)$ for $t \geq 0$.

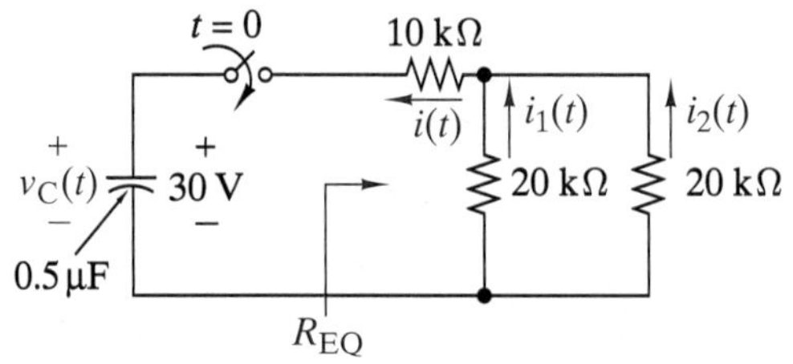

FIGURE 7–8

SOLUTION:
This problem involves the zero-input response of an RC circuit since there is no independent source in the circuit. To find the required responses, we first determine the circuit time constant with the switch closed ($t \geq 0$). The equivalent resistance seen by the capacitor is

$$R_{EQ} = 10 + (20\|20) = 20 \text{ k}\Omega$$

For $t \geq 0$ the circuit time constant is

$$T_C = R_T C = 20 \times 10^3 \times 0.5 \times 10^{-6} = 10 \text{ ms}$$

The initial capacitor voltage is given by $V_0 = 30$ V. Using Eq. (7–10), the zero-input response of the capacitor voltage is

$$v_C(t) = 30e^{-100t} \text{ V} \qquad t \geq 0$$

The capacitor voltage provides the information needed to solve for all other zero-input responses. The current $i(t)$ through the capacitor is

$$i(t) = C\frac{dv_C(t)}{dt} = (0.5 \times 10^{-6})(30)(-100)e^{-100t}$$
$$= -1.5\,e^{-100t} \text{ mA} \qquad t \geq 0$$

The minus sign means that the actual current, shown in Figure 7–8, is opposite the referenced direction. The capacitor is delivering power to the resistors in a manner similar to an exponentially decaying source. We can use current division to find the other currents.

$$i_1(t) = i_2(t) = \frac{20}{20+20}i(t) = -0.75\,e^{-100t} \text{ mA} \qquad t \geq 0$$ ■

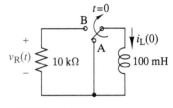

Exercise 7–4

The switch in the RL circuit of Figure 7–9 moves from position A to position B at $t = 0$. If the current flowing through the inductor at $t = 0$ is 1 mA, how long after the switch moves to position B does it take for the voltage across the resistor to reach -5 V?

Answer: $v_R(t) = -5$ V at $t = 6.93$ μs

FIGURE 7–9

Sometimes it may be difficult to determine the Thévenin or Norton equivalent seen by the dynamic element in a first-order circuit. In such cases we use other circuit analysis techniques to derive the differential equation in terms of a more convenient signal variable. For example, the OP AMP RC circuit in Figure 7–10 is a first-order circuit because it contains a single capacitor.

From previous experience we know that the key to analyzing an inverting OP AMP circuit is to write a KCL equation at the inverting input. The sum of currents entering the inverting input is

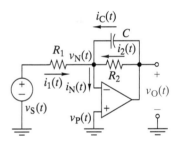

FIGURE 7–10 *First-order OP AMP RC circuit.*

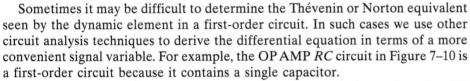

$$\underbrace{\frac{1}{R_1}(v_S(t) - v_N(t))}_{i_1(t)} + \underbrace{\frac{1}{R_2}(v_O(t) - v_N(t))}_{i_2(t)} + \underbrace{C\frac{d(v_O(t) - v_N(t))}{dt}}_{i_C(t)} - i_N(t) = 0$$

The element equations for the OP AMP are $i_N(t) = 0$ and $v_N(t) = v_P(t)$. However, the noninverting input is grounded; hence $v_N(t) = v_P(t) = 0$. Substituting the OP AMP element constraints into the KCL constraint yields

$$\frac{v_S(t)}{R_1} + \frac{v_O(t)}{R_2} + C\frac{dv_O(t)}{dt} = 0$$

which can be rearranged in standard form as

$$R_2C\frac{dv_O(t)}{dt} + v_O(t) = -\frac{R_2}{R_1}v_S(t) \qquad (7\text{–}14)$$

The unknown in Eq. (7–14) is the OP AMP output voltage rather than the capacitor voltage. The form of the differential equation indicates that the circuit time constant is $T_C = R_2C$.

EXAMPLE 7–5

Use OrCAD to display the responses $v_C(t)$, $i_C(t)$, $i_1(t)$, and $i_2(t)$ in the circuit of Figure 7–8 for $t \geq 0$.

FIGURE 7–11
Copyright © Cadence Design Systems, Inc. Used with permission.

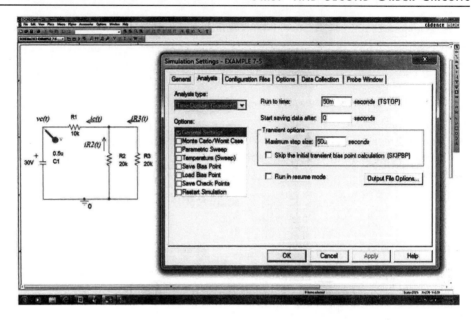

SOLUTION:
An analytical solution for this problem was given in Example 7–4. OrCAD can be used to display the results of a transient behavior of a circuit. The simulation option used is called the *Time Domain (Transient)* analysis. This analysis displays in Probe the time variation of selected currents and voltages. Figure 7–11 shows the circuit as drawn in OrCAD. Note that there is no switch included, since OrCAD assumes that you want to display the results starting at $t = 0$. The labels on the diagram $v_C(t)$, $i_C(t)$, $i_{R2}(t)$, and $i_{R3}(t)$ have been added using the Place menu for reference to the problem and are not necessary for circuit simulation. Since the only source in the circuit is the energy stored in the capacitor, the initial condition of $+30$ V needs to be placed on the capacitor model. This is done by double-clicking on the capacitor symbol on the circuit diagram and updating the Property Editor by placing 30 in the IC box ensuring the "+" or "1" terminal of the capacitor is up. Once this is done the Apply button is pressed to update the model. After the circuit diagram is complete, we set up a transient analysis. Select PSpice to open the simulation profile and choose the *Time Domain (Transient)* simulation. An OrCAD simulation defaults to start at $t = 0$ and will continue until "Run to time" which you must enter. You need to know something about the circuit to make a proper entry. We know that the time constant is 10 ms, so let us display five time constants or 50 ms. The "Start saving data after" setting defaults to 0 s unless you want to see the transient after some time has elapsed. "Maximum step size" refers to how small a time period you want OrCAD to increment to calculate each response. The smaller the step size, the more accurate and smooth the response, but the longer the simulation will take. A rough rule of thumb is to have OrCAD make about 1000 calculations per display. We selected 50 μs. Figure 7–11 shows the simulation settings used for this example.

Pressing the triangle on the menu bar starts the simulation. Probe will display a blank black screen. Since we want to display a voltage and currents versus time, we will need two separate y-axes. Start by selecting Trace, then Add Trace. From the menu of available traces, select V(C1:1). This will display the variation of capacitor voltage versus time—a decaying exponential. We now select Plot and Add Y Axis. This will allow us to display the currents on a separate axis although all the curves will appear on the same graph. We again select Trace, then Add Trace. This time we

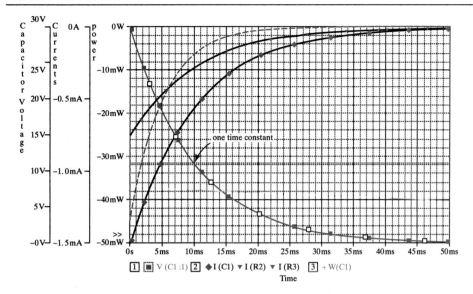

FIGURE 7–12

choose the currents associated with the circuit elements we want to display, I(C1), I(R2), and I(R3). The results are shown in Figure 7–12. The voltage starts at 30 V and decays to zero. Note that the currents are initially zero, instantaneously jump to a value, and then decay to zero. The currents are negative, meaning that the capacitor is providing power to the rest of the circuit. Using the cursor, found under the Trace menu, we find that at one time constant, or 10 ms, the capacitor voltage is exactly 30 V times e^{-1}, or 11.036 V.

We have also found the power delivered by adding yet a third y-axis and selecting the trace W(C1), which is simply V(C1:1) times I(C1) (dashed line). ■

Exercise 7–5

A 100-mH inductor and two resistors are all connected in parallel. One resistor is 100 Ω and the second is 470 Ω. At time $t = 0$, the inductor has 100 mA flowing through it. Use OrCAD to find a plot of the current through the inductor for $t \geq 0$. Then on a separate y-axis, plot the current through each resistor.

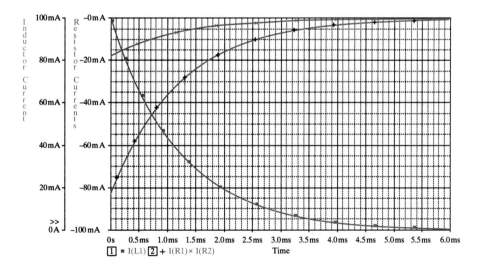

FIGURE 7–13

Answer: The Probe plot in Figure 7–13 shows the desired transient responses.

7–2 FIRST-ORDER CIRCUIT STEP RESPONSE

Linear circuits often are characterized by their response to certain classic signals. The step, for example, helps characterize the circuit's transient behavior, while the sinusoid helps characterize the circuit's frequency response. This chapter deals with the classical approach to the transient behavior of circuits. First, in this section, we will look at the step response of first-order circuits. Next we will look at the transient behavior from applying a sinusoid or an exponential to a first-order circuit. We will conclude our study of transient responses by studying the zero-input and step responses of second-order circuits.

The step response analysis we are about to undertake introduces the concepts of forced, natural, zero-state, and zero-input responses, which appear extensively in later chapters. Ironically, designing a circuit to meet transient response specifications requires making compromises with respect to the circuit's steady-state performance. Understanding why circuits behave as they do under both transient and steady-state excitations is essential to finding the optimum design and poses an interesting challenge for circuit designers.

Our development of the first-order step response treats the RC circuit in detail and then summarizes the corresponding results for its dual, the RL circuit. When the input to the RC circuit in Figure 7–2 is a step function, we can write the Thévenin source as $v_T(t) = V_A u(t)$. The circuit differential equation in Eq. (7–3) becomes

$$R_T C \frac{dv(t)}{dt} + v(t) = V_A u(t) \qquad (7\text{–}15)$$

The step response is a function $v(t)$ that satisfies this differential equation for $t \geq 0$ and meets the initial condition $v(0)$. Since $u(t) = 1$ for $t \geq 0$ we can write Eq. (7–15) as

$$R_T C \frac{dv(t)}{dt} + v(t) = V_A \qquad \text{for } t \geq 0 \qquad (7\text{–}16)$$

Mathematics provides a number of approaches to solving this equation, including separation of variables and integrating factors. However, because the circuit is linear, we chose a method that uses superposition to divide the solution for $v(t)$ into two components:

$$v(t) = v_N(t) + v_F(t) \qquad (7\text{–}17)$$

The first component, $v_N(t)$, is the **natural response** and is the general solution of Eq. (7–16) when the input is set to zero. The natural response has its origin in the physical characteristic of the circuit and does not depend on the form of the input. The component $v_F(t)$ is the **forced response** and is a particular solution of Eq. (7–16) when the input is the step function. We call this the forced response because it represents what the circuit is compelled to do by the form of the input.

Finding the natural response requires the general solution of Eq. (7–16) with the input set to zero:

$$R_T C \frac{dv_N(t)}{dt} + v_N(t) = 0 \qquad t \geq 0$$

But this is the homogeneous equation that produces the zero-input response in Eq. (7–8). Therefore, we know that the natural response takes the form

$$v_N(t) = K e^{-t/R_T C} \qquad t \geq 0 \qquad (7\text{–}18)$$

This is a general solution of the homogeneous equation because it contains an arbitrary constant K. At this point we cannot evaluate K from the initial condition, as we did for the zero-input response. The initial condition applies to the total response (natural plus forced), and we have yet to find the forced response.

Turning now to the forced response, we seek a particular solution of the equation

$$R_\mathrm{T}C\frac{dv_\mathrm{F}(t)}{dt}+v_\mathrm{F}(t)=V_\mathrm{A} \qquad t\geq0 \tag{7–19}$$

The equation requires that a linear combination of $v_\mathrm{F}(t)$ and its derivative equal a constant V_A for $t\geq0$. Setting $v_\mathrm{F}(t)=V_\mathrm{A}$ meets this condition since $dv_\mathrm{F}/dt=dV_\mathrm{A}/dt=0$. Substituting $v_\mathrm{F}(t)=V_\mathrm{A}$ into Eq. (7–19) reduces it to the identity $V_\mathrm{A}=V_\mathrm{A}$.

Now combining the forced and natural responses, we obtain

$$v(t)=v_\mathrm{N}(t)+v_\mathrm{F}(t)$$
$$=Ke^{-t/R_\mathrm{T}C}+V_\mathrm{A} \qquad t\geq0$$

This equation is the general solution for the step response because it satisfies Eq. (7–16) and contains an arbitrary constant K. This constant can now be evaluated using the initial condition:

$$v(0)=V_0=Ke^0+V_\mathrm{A}=K+V_\mathrm{A}$$

The initial condition requires that $K=(V_0-V_\mathrm{A})$. Substituting this conclusion into the general solution yields the step response of the RC circuit.

$$v(t)=(V_0-V_\mathrm{A})e^{-t/R_\mathrm{T}C}+V_\mathrm{A} \qquad t\geq0 \tag{7–20}$$

A typical plot of $v(t)$ is shown in Figure 7–14.

The RC circuit step response in Eq. (7–20) starts out at the initial condition V_0 and is driven to a final condition V_A, which is determined by the amplitude of the step function input. That is, the initial and final values of the response are

$$\lim_{t\to0+}v(t)=(V_0-V_\mathrm{A})e^{-0}+V_\mathrm{A}=V_0$$
$$\lim_{t\to\infty}v(t)=(V_0-V_\mathrm{A})e^{-\infty}+V_\mathrm{A}=V_\mathrm{A}$$

The path between the two end points is an exponential waveform whose time constant is the circuit time constant. We know from our study of exponential signals that the step response will reach its final value after about five time constants. In other words, after about five time constants the natural response decays to zero and we are left with a constant forced response caused by the step function input.

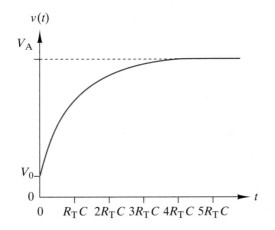

FIGURE 7–14 *Step response of a first-order RC circuit.*

The RL circuit in Figure 7–2 is the dual of the RC circuit, so the development of its step responses follows the same pattern discussed previously. Briefly sketching the main steps, the Norton equivalent input is a step function $I_A u(t)$, and for $t \geq 0$ the RL circuit differential equation Eq. (7–6) becomes

$$\frac{L}{R_N} \frac{di(t)}{dt} + i(t) = I_A \qquad t \geq 0 \tag{7–21}$$

The solution of this equation is found by superimposing the natural and forced components. The natural response is the solution of the homogeneous equation [right side of Eq. (7–21) set to zero] and takes the same form as the zero-input response found in the previous section.

$$i_N(t) = Ke^{-R_N t/L} \qquad t \geq 0$$

where K is a constant to be evaluated from the initial condition once the complete response is known. The forced response is a particular solution of the equation

$$\frac{L}{R_N} \frac{di_F(t)}{dt} + i_F(t) = I_A \qquad t \geq 0$$

Setting $i_F(t) = I_A$ satisfies this equation since $dI_A/dt = 0$.

Combining the forced and natural responses, we obtain the general solution of Eq. (7–21) in the form

$$i(t) = i_N(t) + i_F(t)$$
$$= Ke^{-R_N t/L} + I_A \qquad t \geq 0$$

The constant K is now evaluated from the initial condition:

$$i(0) = I_0 = Ke^{-0} + I_A = K + I_A$$

The initial condition requires that $K = I_0 - I_A$, so the step response of the RL circuit is

$$i(t) = (I_0 - I_A)e^{-R_N t/L} + I_A \qquad t \geq 0 \tag{7–22}$$

The RL circuit step response has the same form as the RC circuit step response in Eq. (7–20). At $t = 0$ the starting value of the response is $i(0) = I_0$, as required by the initial condition. The final value is the forced response $i(\infty) = i_F(t) = I_A$, since the natural response decays to zero as time increases.

A step function input to the RC or RL circuit drives the state variable from an initial value determined by what happened prior to $t = 0$ to a final value determined by the amplitude of the step function applied at $t = 0$. The time needed to transition from the initial to the final value is about $5T_C$, where T_C is the circuit time constant. We conclude that the step response of a first-order circuit depends on three quantities:

1. The amplitude of the step input (V_A or I_A)
2. The circuit time constant ($R_T C$ or L/R_N)
3. The value of the state variable at $t = 0$ (V_0 or I_0)

EXAMPLE 7–6

Find the response of the RC circuit in Figure 7–15.

SOLUTION:
The circuit is first order, since the two capacitors in series can be replaced by a single equivalent capacitor

$$C_{EQ} = \frac{1}{\dfrac{1}{C_1} + \dfrac{1}{C_2}} = 0.0833 \ \mu F$$

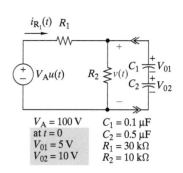

$V_A = 100$ V
at $t = 0$
$V_{01} = 5$ V
$V_{02} = 10$ V

$C_1 = 0.1 \ \mu F$
$C_2 = 0.5 \ \mu F$
$R_1 = 30 \ k\Omega$
$R_2 = 10 \ k\Omega$

FIGURE 7-15

The initial voltage on C_{EQ} is the sum of the initial voltages on the original capacitors.

$$V_0 = V_{01} + V_{02} = 5 + 10 = 15 \ V$$

To find the Thévenin equivalent seen by C_{EQ}, we first find the open-circuit voltage. Disconnecting the capacitors in Figure 7-15 and using voltage division at the interface yields

$$v_T(t) = v_{OC} = \frac{R_2}{R_1 + R_2} V_A u(t) = \frac{10}{40} 100u(t) = 25u(t) \ V$$

Replacing the voltage source by a short circuit and looking to the left at the interface, we see R_1 in parallel with R_2. The Thévenin resistance of this combination is

$$R_T = \frac{1}{\dfrac{1}{R_1} + \dfrac{1}{R_2}} = 7.5 \ k\Omega$$

The circuit time constant is

$$T_C = R_T C_{EQ} = (7.5 \times 10^3)(8.33 \times 10^{-8}) = \frac{1}{1600} \ s$$

For the Thévenin equivalent circuit, the initial capacitor voltage is $V_0 = 15$ V, the step input is $25u(t)$, and the time constant is $1/1600$ s. Using the RC circuit step response in Eq. (7–20) yields

$$v(t) = (15 - 25)e^{-1600t} + 25$$
$$= 25 - 10e^{-1600t} \ V \qquad t \geq 0$$

The initial ($t = 0$) value of $v(t)$ is $25 - 10 = 15$ V, as required. The equivalent capacitor voltage is driven to a final value of 25 V by the step input in the Thévenin equivalent circuit. For practical purposes, $v(t)$ reaches 25 V after about $5T_C = 3.125$ ms. ∎

Exercise 7–6

Use the results from Example 7–6 and find the current through R_1 in Figure 7–15 for $t \geq 0$.

Answer: $$i_{R_1}(t) = 2.5 + 0.33e^{-1600t} \ mA \qquad t \geq 0$$

EXAMPLE 7–7

Find the step response of the RL circuit in Figure 7–16(a). The initial condition is $i(0) = I_0$.

SOLUTION:
We first find the Norton equivalent to the left of the interface. By current division, the short-circuit current at the interface is

$$i_{SC}(t) = \frac{R_1}{R_1 + R_2} I_A u(t)$$

FIGURE 7–16

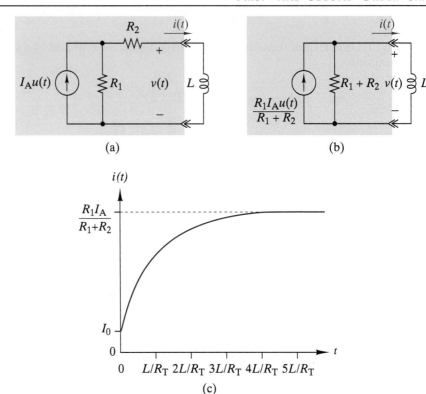

(a)

(b)

(c)

Looking to the left at the interface with the current source off (replaced by an open circuit), we see R_1 and R_2 in series producing a Thévenin resistance

$$R_T = R_1 + R_2$$

The time constant of the Norton equivalent circuit in Figure 7–16(b) is

$$T_C = \frac{L}{(R_1 + R_2)}$$

The natural response of the Norton equivalent circuit is

$$i_N(t) = Ke^{-(R_1+R_2)t/L} \qquad t \geq 0$$

The short-circuit current $i_{SC}(t)$ is the step function input in the Norton circuit. Therefore, the forced response is

$$i_F(t) = i_{SC}(t) = \frac{R_1}{(R_1 + R_2)} I_A u(t)$$

Superimposing the natural and forced responses yields

$$i(t) = Ke^{-(R_1+R_2)t/L} + \frac{R_1 I_A}{R_1 + R_2} \qquad t \geq 0$$

The constant K can be evaluated from the initial condition:

$$i(0) = I_0 = K + \frac{R_1 I_A}{R_1 + R_2}$$

which requires that

$$K = I_0 - \frac{R_1 I_A}{R_1 + R_2}$$

So the circuit step response is

$$i(t) = \left[I_0 - \frac{R_1 I_A}{R_1 + R_2} \right] e^{-(R_1 + R_2)t/L} + \frac{R_1 I_A}{R_1 + R_2} \qquad t \geq 0$$

An example of this response is shown in Figure 7–16(c). ∎

Exercise 7–7 _____

Use the results from Example 7–7 and find the voltage across the current source in Figure 7–16(a) for $t \geq 0$.

Answer: $v_{source}(t) - R_1 I_A \left[1 - \dfrac{R_1}{R_1 + R_2} \right] + \left[\dfrac{R_1 I_A}{R_1 + R_2} - I_0 \right] R_1 e^{-(R_1+R_2)t/L}$ V $\quad t \geq 0$

_____

EXAMPLE 7–8

The state variable response of a first-order RC circuit for a step function input is

$$v_C(t) = 20e^{-200t} - 10 \text{ V} \qquad t \geq 0$$

(a) What is the circuit time constant?
(b) What is the initial voltage across the capacitor?
(c) What is the amplitude of the forced response?
(d) At what time is $v_C(t) = 0$?

SOLUTION:
(a) The natural response of a first-order circuit is of the form Ke^{-t/T_C}. Therefore, the time constant of the given responses is $T_C = 1/200 = 5$ ms.
(b) The initial $(t = 0)$ voltage across the capacitor is

$$v_C(0) = 20e^{-0} - 10 = 20 - 10 = 10 \text{ V}$$

(c) The natural response decays to zero, so the forced response is the final value of $v_C(t)$.

$$v_C(\infty) = 20e^{-\infty} - 10 = 0 - 10 = -10 \text{ V}$$

(d) The capacitor voltage must pass through zero at some intermediate time, since the initial value is positive and the final value negative. This time is found by setting the step response equal to zero:

$$20e^{-200t} - 10 = 0 \quad \text{or} \quad e^{-200t} = 1/2$$

which yields the condition $e^{200t} = 2$ or $t = \ln 2/200 = 3.47$ ms. ∎

Exercise 7–8 _____

Given the first-order circuit step response

$$v_C(t) = 20 - 20e^{-1000t} \text{ V} \qquad t \geq 0$$

(a) What is the amplitude of the step input?
(b) What is the circuit time constant?
(c) What is the initial value of the state variable?
(d) What is the circuit differential equation?

Answers:

(a) 20 V
(b) 1 ms
(c) 0 V
(d) $10^{-3} \, dv_C(t)/dt + v_C(t) = 20u(t)$

_____

Exercise 7–9 _____

Find the solutions of the following first-order differential equations:

(a) $10^{-4}\dfrac{dv_C(t)}{dt} + v_C(t) = -5u(t)$ $v_C(0) = 5$ V

(b) $5 \times 10^{-2}\dfrac{di_L(t)}{dt} + 2000i_L(t) = 10u(t)$ $i_L(0) = -5$ mA

Answers:

(a) $v_C(t) = -5 + 10e^{-10,000t}$ V $t \geq 0$

(b) $i_L(t) = 5 - 10e^{-40,000t}$ mA $t \geq 0$

ZERO-STATE RESPONSE

Additional properties of dynamic circuit responses are revealed by rearranging the RC and RL circuit step responses in Eqs. (7–20) and (7–22) in the following way:

$$RC \text{ circuit: } v(t) = \underbrace{V_0 e^{-t/R_T C}}_{\text{zero-input response}} + \underbrace{V_A(1 - e^{-t/R_T C})}_{\text{zero-state response}} \qquad t \geq 0$$

$$RL \text{ circuit: } i(t) = \overbrace{I_0 e^{-R_N t/L}} + \overbrace{I_A(1 - e^{-R_N t/L})} \qquad t \geq 0$$

We recognize the first term on the right in each equation as the zero-input response discussed in Sect. 7–1. By definition, the **zero-input response** occurs when the input is zero ($V_A = 0$ or $I_A = 0$). The second term on the right in each equation is called the **zero-state response** because this part occurs when the initial state of the circuit is zero ($V_0 = 0$ or $I_0 = 0$).

The zero-state response is proportional to the amplitude of the input step function (V_A or I_A). However, the total response (zero input plus zero state) is not directly proportional to the input amplitude. When the initial state is not zero, the circuit appears to violate the proportionality property of linear circuits. However, bear in mind that the proportionality property applies to linear circuits with only one input.

The RC and RL circuits can store energy and have memory. In effect, they have two inputs: (1) the input that occurred before $t = 0$, and (2) the step function applied at $t = 0$. The first input produces the initial energy state of the circuit at $t = 0$, and the second causes the zero-state response for $t \geq 0$. In general, for $t \geq 0$, the total response of a dynamic circuit is the sum of two responses: (1) the zero-input response caused by the initial conditions produced by inputs applied before $t = 0$, and (2) the zero-state response caused by inputs applied after $t = 0$.

APPLICATION   **EXAMPLE 7–9**

The operation of a digital system is controlled by a clock waveform that provides a standard timing reference. At its source a clock waveform can be described by a rectangular pulse of the form

$$v_S(t) = V_A[u(t) - u(t - T)]$$

In this example the pulse amplitude is $V_A = 5$ V and the pulse duration is $T = 10$ ns. This clock pulse drives a digital device that can be modeled by the circuit in Figure 7–17(a). In this model $v_S(t)$ is the rectangular clock pulse defined above and $v(t)$ is the clock waveform as received at the input to the digital device. The presence of a clock pulse at the device input will be detected only if $v(t)$ exceeds a specified logic "1" threshold level.

Find the zero-state response of the voltage $v(t)$ when $RC = 10$ ns. Will the clock pulse be detected if the logic "1" threshold level is 3.7 V?

FIGURE 7–17

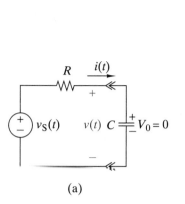

(a)

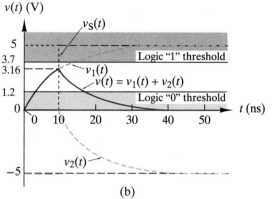

(b)

SOLUTION:
The rectangular pulse input $v_S(t)$ is indicated by dashed lines in Figure 7–17(b). The initial capacitor voltage is zero because we seek the zero-state response. The total response can be found as the sum of the zero-state responses caused by two inputs:

1. A positive 5-V step function applied at $t = 0$
2. A negative 5-V step function applied at $t = 10$ ns

The first input causes a zero-state response of

$$v_1(t) = V_A(1 - e^{-t/RC})u(t)$$

$$= 5(1 - e^{-10^8 t})u(t)$$

The second input causes a zero-state response of

$$v_2(t) = -V_A(1 - e^{-(t-T)/RC})u(t - T)$$

$$= -5[1 - e^{-10^8(t-10^{-8})}]u(t - 10^{-8})$$

Notice that $v_2(t) = -v_1(t - T)$, that is, $v_2(t)$ is obtained by inverting and delaying $v_1(t)$ by $T = 10$ ns. The total response is the superposition of these two responses.

$$v(t) = v_1(t) + v_2(t)$$

Figure 7–17(b) shows how the two responses combine to produce the overall pulse response of the circuit. The response $v_1(t)$ begins at zero and eventually reaches a final value of +5 V. At $t = T = 10$ ns the first response reaches $v_1(T) = 5(1 - e^{-1}) = 3.16$ V. The second response $v_2(t)$ begins at $t = T = 10$ ns, and thereafter is equal and opposite to $v_1(t)$ except that it is delayed by $T = 10$ ns. The net result is that the total response reaches a maximum of $v(T) = 3.16$ V. In this example the clock pulse will not be detected because the logic "1" threshold level is 3.7 V. Clock pulse detection would be made possible by increasing the pulse duration so that $v(T) > 3.7$ V. This requires that

$$5(1 - e^{-10^8 T}) > 3.7 \text{ V}$$

or $1.3 > 5 e^{-10^8 T}$, which yields $T > 1.347 \times 10^{-8}$. For the digital device in this example, the minimum detectable clock pulse duration is about 13.5 ns. ∎

Exercise 7–10

The switch in Figure 7–18 closes at $t = 0$. Find the zero-state response of the capacitor voltage for $t \geq 0$.

Answer:
$$v_C(t) = 2.5(1 - e^{-200t}) \text{ V}$$

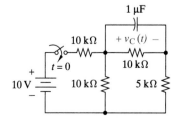

FIGURE 7–18

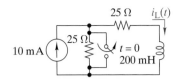

FIGURE 7–19

Exercise 7–11

The switch in Figure 7–19 opens at $t = 0$. Find the zero-state response of the inductor current for $t \geq 0$.

Answer:
$$i_L(t) = 5(1 - e^{-250t}) \text{ mA}$$

7–3 INITIAL AND FINAL CONDITIONS

Reviewing the first-order step responses of the last section shows that for $t \geq 0$ the state variable responses can be written in the form

$$RC \text{ circuit:} \quad v_C(t) = [v_C(0) - v_C(\infty)]e^{-t/T_C} + v_C(\infty) \qquad t \geq 0$$

$$RL \text{ circuit:} \quad i_L(t) = [i_L(0) - i_L(\infty)]e^{-t/T_C} + i_L(\infty) \qquad t \geq 0$$

(7–23)

In both circuits the step response is of the general form

$$
\begin{array}{c}
\text{State} \\
\text{variable} \\
\text{response}
\end{array}
=
\left[
\begin{array}{cc}
\text{Initial} & \text{Final} \\
\text{value of the} & - \quad \text{value of the} \\
\text{state variable} & \text{state variable}
\end{array}
\right]
\times e^{-t/T_C} +
\begin{array}{c}
\text{Final} \\
\text{value of the} \\
\text{state variable}
\end{array}
$$

To determine the step response of a first-order circuit, we need three quantities: the initial value of the state variable, the final value of the state variable, and the time constant. Since we know how to get the time constant directly from the circuit, it would be useful to have a direct way to determine the initial and final values by inspecting the circuit itself.

The final value can be calculated directly from the circuit by observing that for $t > 5T_C$ the step responses approach a constant value or dc value. Under dc conditions a capacitor acts like an open circuit and an inductor acts like a short circuit. As a result, the final value of the state variable is found by applying dc analysis methods to the circuit configuration for $t > 0$, with capacitors replaced by open circuits (OC) and inductors replaced by short circuits (SC).

We can also use dc analysis to determine the initial value in many practical situations. A common situation is a circuit containing dc sources and a switch that is in one position for a period of time much greater than the circuit time constant, and then is moved to a new position at $t = 0$. For example, if the switch is closed for a long period of time, then the dc sources drive the state variable to a final value. If the switch is now opened at $t = 0$, then the dc sources drive the state variable to a new final condition appropriate to the new circuit configuration for $t > 0$.

Note: The initial condition at $t = 0$ is the dc value of the state variable for the circuit configuration that existed *before* the switch changed positions at $t = 0$. The switching action cannot cause an instantaneous change in the initial condition because capacitor voltage and inductor current are continuous functions of time. In other words, opening a switch at $t = 0$ marks the boundary between two eras. The final condition of the state variable for the $t < 0$ era is the initial condition for the $t > 0$ era that follows.

The usual way to state a switched circuit problem is to say that a switch has been closed (open) for a long time and then is opened (closed) at $t = 0$. In this context, a long time means at least five time constants. Time constants rarely exceed a few hundred milliseconds in electrical circuits, so a long time passes rather quickly.

The state variable response in switched dynamic circuits is found using the following steps:

STEP 1 Find the initial value by applying dc analysis to the circuit configuration for $t < 0$ with the capacitor (inductor) replaced with an open (short) circuit.

STEP 2 Find the final value by applying dc analysis to the circuit configuration for $t > 0$ with the capacitor (inductor) replaced with an open (short) circuit.

STEP 3 Find the time constant T_C of the circuit in the configuration for $t > 0$.

STEP 4 Write the step response directly using Eq. (7–23) without formulating and solving the circuit differential equation.

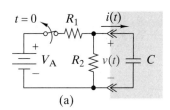

For example, the switch in Figure 7–20(a) has been closed for a long time and is opened at $t = 0$. We want to find the capacitor voltage $v(t)$ for $t \geq 0$.

STEP 1 The initial condition is found by dc analysis of the circuit configuration in Figure 7–20(b), where the switch is closed. Using voltage division, the initial capacitor voltage is found to be

$$v(0) = \frac{R_2 V_A}{R_1 + R_2}$$

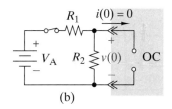

STEP 2 The final condition is found by dc analysis of the circuit configuration in Figure 7–20(c), where the switch is open. When the switch is open the circuit has no dc excitation, so the final value of the capacitor voltage is zero.

STEP 3 The circuit in Figure 7–20(c) also gives us the time constant. Looking back at the interface, we see an equivalent resistance of R_2, since R_1 is connected in series with an open switch. For $t \geq 0$ the time constant is R_2C. Using Eq. (7–23), the capacitor voltage for $t \geq 0$ is

$$v(t) = [v(0) - v(\infty)]e^{-t/T_C} + v(\infty)$$
$$= \frac{R_2 V_A}{R_1 + R_2} e^{-t/R_2 C} \qquad t \geq 0$$

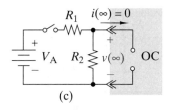

FIGURE 7–20 *Solving a switched dynamic circuit using the initial and final conditions.*

The result is a zero-input response, since there is no excitation for $t \geq 0$. But now we see how the initial condition for the zero-input response could be produced physically by opening a switch that has been closed for a long time.

To continue the analysis, we find the capacitor current using its element constraint:

$$i(t) = C\frac{dv(t)}{dt} = -\frac{V_A}{R_1 + R_2}e^{-t/R_2 C} \qquad t \geq 0$$

This is the capacitor current for $t \geq 0$. For $t < 0$ the circuit in Figure 7–20(b) points out that the capacitor current is zero since the capacitor acts like an open circuit.

The capacitor voltage and current responses are plotted in Figure 7–21. The capacitor voltage is continuous at $t = 0$, but the capacitor current has a jump discontinuity at $t = 0$. In other words, state variables are continuous, but nonstate variables can have discontinuities at $t = 0$. Since the state variable is continuous, we first find the circuit state variable and then solve for other circuit variables using the element and connection constraints.

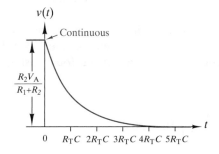

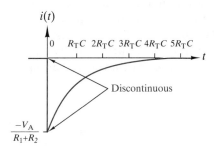

FIGURE 7–21 *Two responses in the RC circuit of Figure 7–22.*

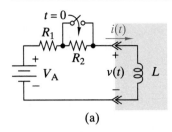

(a)

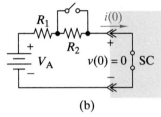

(b)

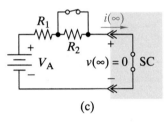

(c)

FIGURE 7–22

EXAMPLE 7–10

The switch in Figure 7–22(a) has been open for a long time and is closed at $t = 0$. Find the inductor current for $t \geq 0$.

SOLUTION:
We first find the initial condition using the circuit in Figure 7–22(b). By series equivalence the initial current is

$$i(0) = \frac{V_A}{R_1 + R_2}$$

The final condition and the time constant are determined from the circuit in Figure 7–22(c). Closing the switch shorts out R_2, and the final condition and time constant for $t > 0$ are

$$i(\infty) = \frac{V_A}{R_1} \quad \text{and} \quad T_C = \frac{L}{R_N} = \frac{L}{R_1}$$

Using Eq. (7–23), the inductor current for $t \geq 0$ is

$$i(t) = [i(0) - i(\infty)]e^{-t/T_C} + i(\infty)$$
$$= \left[\frac{V_A}{R_1 + R_2} - \frac{V_A}{R_1}\right]e^{-R_1 t/L} + \frac{V_A}{R_1} \qquad t \geq 0$$ ∎

Exercise 7–12

The switch in Figure 7–22(a) has been closed for a long time. The switch opens at $t = 0$. Find the inductor current for $t \geq 0$.

Answer: $i(t) = \left[\frac{V_A}{R_1} - \frac{V_A}{R_1 + R_2}\right]e^{-(R_1+R_2)t/L} + \frac{V_A}{R_1 + R_2}$ A $\qquad t \geq 0$

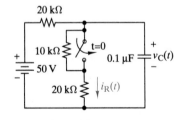

FIGURE 7–23

EXAMPLE 7–11

The switch in the circuit of Figure 7–23 has been open for a long time. It closes at $t = 0$. Find the current $i_R(t)$ for $t \geq 0$.

SOLUTION:
Although the example asks for the current through the 20-kΩ resistor, it is important that we solve for the voltage across the capacitor first, since it is the state variable. The voltage across the capacitor is continuous whereas the current may not be. Once we have found the state variable, we can apply the usual connection and element constraints to find the desired variable.

We start by finding the initial voltage across the capacitor at $t = 0-$, a whisker of time before the switch is thrown. At $t = 0-$, the capacitor is being excited by a dc source, hence it behaves like an open circuit. The voltage can be found using a voltage divider

$$v_C(0-) = \frac{(20\,k + 10\,k)}{20\,k + 20\,k + 10\,k}50 = 30V$$

The final condition can be found similarly. At $t = \infty$, the closed switch shorts out the 10-kΩ resistor. Hence, our voltage divider becomes

$$v_C(\infty) = \frac{20\,\mathrm{k}}{20\,\mathrm{k} + 20\,\mathrm{k}} 50 = 25\mathrm{V}$$

Next we need to find the time constant. For an RC circuit the time constant is given as $R_T C$, where R_T is the Thévenin resistance that the capacitor sees *after* the switch has been thrown. Using the look-back method with the switch shorting out the 10-kΩ resistor and setting the source to zero by replacing it with a short circuit, we see that R_T is

$$R_T = \frac{20\,\mathrm{k} \times 20\,\mathrm{k}}{20\,\mathrm{k} + 20\,\mathrm{k}} = 10\,\mathrm{k}\Omega$$

And the time constant is

$$T_C = R_T C = 10\,\mathrm{k} \times 0.1\,\mu = 1\,\mathrm{ms}$$

Appling these calculations to Eq. (7–23), we find the equation for the state variable

$$v_C(t) = (30 - 25)e^{-\frac{t}{.001}} + 25$$

$$v_C(t) = 5e^{-1000t} + 25\,\mathrm{V}, \quad t \geq 0$$

To find the current through the 20-kΩ resistor, we recognize that the resistor is in parallel with the capacitor, thereby sharing the same voltage. Applying Ohm's law yields the desired result

$$i_R(t) = \frac{v_C(t)}{R_{20\,\mathrm{k}}} = \frac{5e^{-1000t} + 25}{20\,\mathrm{k}} = 0.25e^{-1000t} + 1.25\,\mathrm{mA}, \ t \geq 0 \qquad \blacksquare$$

Exercise 7–13

The switch in the circuit of Figure 7–23 has been closed for a long time. It opens at $t = 0$. Find the voltage $v_C(t)$ and the current $i_R(t)$ for $t \geq 0$.

Answers:

$$v_C(t) = -5e^{-833t} + 30\,\mathrm{V}, \quad t \geq 0$$
$$i_R(t) = 0.167e^{-833t} + 1\,\mathrm{mA}, \quad t \geq 0$$

▶ DESIGN EXAMPLE 7–12

Design a first-order RC circuit using standard parts (see inside rear cover) that will produce the following voltage across the capacitor: $v_C(t) = 50 - 100e^{-2000t}$ V.

SOLUTION:
We know that the circuit will look like Figure 7–24(a). We need to select a suitable source $v_S(t)$, a resistor R, and a capacitor C.

FIGURE 7–24

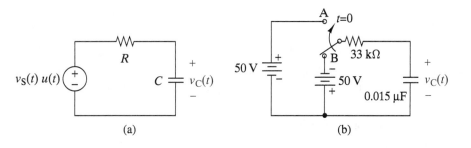

(a)

(b)

We also know that the general form of a state-variable response is

$$v_C(t) = [v_C(0) - v_C(\infty)]e^{\frac{-t}{T_C}} + v_C(\infty) \text{ V}, \quad t \geq 0$$

Comparing the general form to our given desired response $v_C(t) = 50 - 100e^{-2000t}$ V, we calculate the time constant as $1/2000 = 500$ μs. Hence, we want $T_C = RC = 500$ μs. Selecting $C = 0.015$ μF, a standard value, yields $R = 33.3$ kΩ. There is an $R = 33$ kΩ, a standard value that should be close enough considering the tolerances of the components. We now must determine the initial and final voltages. Substituting $t = 0$ into our desired output equation, we find our initial voltage is −50 V. Similarly, substituting $t = \infty$, we find our final voltage is +50 V. This suggests two voltages, switching a $t = 0$ from −50 V to +50V. We can now design our circuit as shown in Figure 7–24(b).   ∎

◀ Design Exercise 7–14 _____

Design a first-order RL circuit that will produce the following current through the inductor: $i_L(t) = 5 - 5e^{-500t}$ mA for $t \geq 0$. Use standard values for the components.

Answer: A parallel circuit consisting of a current source $i_S(t) = 5\,u(t)$ mA, an inductor of 2 mH, and a resistor of 1 Ω is one possible solution.

EXAMPLE 7–13

For $t \geq 0$ the state variable response of the RL circuit in Figure 7–25(a) is observed to be

$$i_L(t) = 50 + 100e^{-5000t} \text{ mA}$$

(a) Identified the forced and natural components of the response.
(b) Find the circuit time constant.
(c) Find the Thévenin equivalent circuit seen by the inductor.

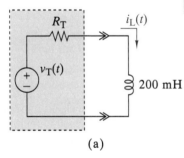

(a)

FIGURE 7–25

SOLUTION:
(a) The natural component is the exponential term $100e^{-5000t}$ mA. The forced component is what remains after the natural component dies out as $t \to \infty$, namely $i_L(\infty) = 50$ mA. The forced response is a constant 50 mA, which means that the Thévenin equivalent is a dc source.
(b) The time constant is the reciprocal of the coefficient of t in e^{-5000t}, i.e., $T_C = 5000^{-1} = 0.2$ ms.
(c) Expressed in terms of circuit parameters the time constant is $T_C = L/R_T$, which yields the Thévenin resistance as $R_T = L/T_C = 1$ kΩ. For dc excitation the inductor acts like a short circuit at $t = \infty$. Hence, $i_L(\infty) = v_T/R_T$ and the Thévenin voltage is

$$v_T(t) = R_T i_L(\infty) = 1 \text{ kΩ} \times 0.05 = 50 \text{ V}$$ ∎

Exercise 7–15 _____

Use OrCAD to simulate a transient response for the circuit in Figure 7–25 and find $i_L(t)$ and $v_L(t)$. (Hint: Be certain to add the initial condition to the inductor and to orient it correctly.)

Answer: See Figure 7–25(b).

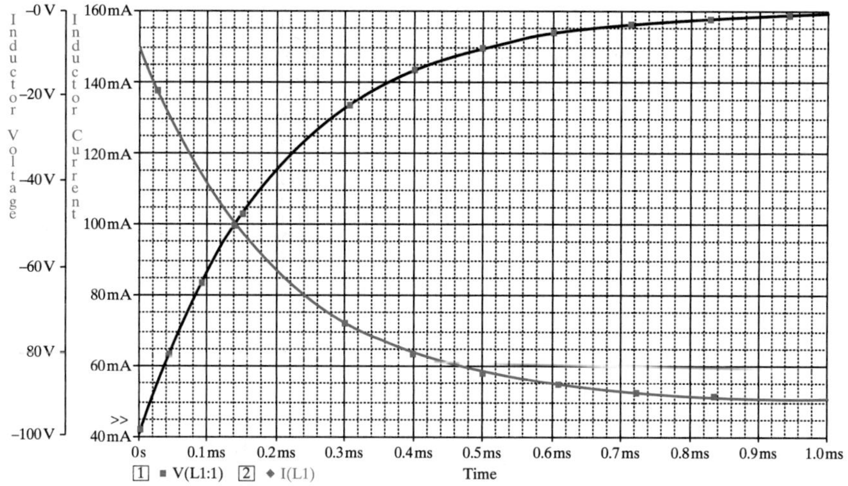

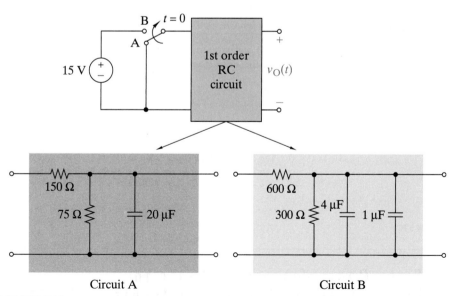

FIGURE 7–25 *(Continued)*

(b)

EVALUATION EXAMPLE 7–14

The switch in Figure 7–26 moves from position A to position B at $t = 0$. The first-order RC circuit in the figure must be designed to produce an output of

$$v_O(t) = 5(1 - e^{-1000t})\text{V} \qquad t \geq 0$$

Evaluate the two proposed circuit designs shown in the figure using the following criteria.

(a) A design must produce the required output.
(b) If both produce the desired output, then compare part counts and use of standard values to identify the best design.

FIGURE 7–26

Circuit A

Circuit B

SOLUTION:

(a) The desired output is a first-order step response with $v_O(0) = 0$, $v_O(\infty) = 5$ V, and $T_C = 1$ ms. For $t < 0$ the switch is in position B and there is zero input; hence $v_O(0) = 0$ for both circuits. For $t \geq 0$ the switch is in position A. The final condition $v_O(\infty)$ is found using voltage division on the circuit with the capacitors replaced by open circuits. The circuit time constant is found using the Thévenin resistance seen by the capacitors. The final value and time constant of circuit A are

$$\text{Circuit A: } v_O(\infty) = \frac{75}{150 + 75} 15 = 5 \text{ V}$$

$$T_C = \frac{150 \times 75}{150 + 75} 20 \times 10^{-6} = 10^{-3} \text{ s}$$

The equivalent capacitance in circuit B is $C_{EQ} = 4 + 1 = 5 \mu\text{F}$. The final value and time constant of circuit B are

$$\text{Circuit B: } v_O = \frac{300}{600 + 300} 15 = 5 \text{ V}$$

$$T_C = \frac{600 \times 300}{600 + 300} 5 \times 10^{-6} = 10^{-3} \text{ s}$$

Both circuits produce the desired output.

(b) Circuit A uses three components: a standard 75-Ω resistor, a standard 150-Ω resistor, and a standard 20-μF capacitor (see inside back cover for standard values). Circuit B uses four components: a standard 300-Ω resistor, a nonstandard 600-Ω resistor, a standard 1-μF capacitor, and a nonstandard 4-μF capacitor. Circuit A is a better design than circuit B in terms of both the number of parts and the use of standard values. ∎

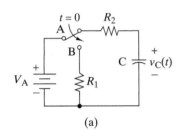

(a)

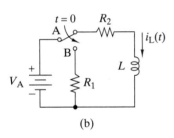

(b)

FIGURE 7–27

Exercise 7–16

In each circuit shown in Figure 7–27 the switch has been in position A for a long time and is moved to position B at $t = 0$. Find the circuit state variable for each circuit for $t \geq 0$.

Answers:

(a) $v_C(t) = V_A e^{-t/(R_1 + R_2)C}$ V

(b) $i_L(t) = \dfrac{V_A}{R_2} e^{-(R_1 + R_2)t/L}$ A

Exercise 7–17

In the circuit in Figure 7–28 the switch has been in position A for a long time and is moved to position B at $t = 0$. For $t \geq 0$ find the output voltage $v_O(t)$.

FIGURE 7–28

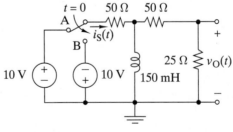

Answer: $v_O(t) = -4e^{-200t}$ V

7–4 FIRST-ORDER CIRCUIT RESPONSE TO EXPONENTIAL AND SINUSOIDAL INPUTS

A myriad of different signals can excite linear circuits. As we develop our understanding of circuit behavior, note that several signals provide important insight into circuit behavior. Thus far, we have looked at the response of first-order circuits to step functions and developed analysis techniques that are quick and methodical. We will now look at the transient response of first-order circuits to two different, common, and useful signals: the exponential and the sinusoid. In later chapters, we will learn other techniques to analyze these and more complex circuits; however, knowing the classical time-domain approach shown in Figure 7–1 will make understanding these subsequent techniques easier.

Let us consider an RC circuit excited by a signal other than a step. If the input signal to the RC circuit in Figure 7–2 starts at $t = 0$, then we can write the circuit differential equation in Eq. (7–3) as

$$R_T C \frac{dv(t)}{dt} + v(t) = v_T(t)u(t) \qquad (7\text{–}24)$$

The $u(t)$ implies that the driving signal $v_T(t)$ has a finite start time arbitrarily selected as $t = 0$. This implies that there is an initial condition, $v(0) = V_0$, that will have to satisfy Eq. (7–24).

As with the step response, we find the solution in two parts: natural response and forced response. The natural response is of the form

$$v_N(t) = Ke^{-t/R_T C} \qquad t \geq 0$$

The natural response of a first-order circuit always has this form because it is a general solution of the homogeneous equation with the input set to zero. The form of the natural response depends on the physical characteristics of the circuit and is independent of the input.

The forced response $v_F(t)$ depends on both the circuit and the nature of the forcing function (the input). The forced response is a particular solution of the equation

$$R_T C \frac{dv_F(t)}{dt} + v_F(t) = v_T(t) \qquad t \geq 0$$

This equation states that whatever function we pick as $v_F(t)$ plus $R_T C$ times the first derivative of that same function must equal $v_T(t)$. This requires that our choice of $v_F(t)$ have the same form as that of the forcing function $v_T(t)$. Table 7–1 shows the form of the forced response that should be used based on the form of the forcing function.

T A B L E 7–1

FORM OF THE FORCING FUNCTION, $v_T(t)$ $(t \geq 0)$	FORM OF THE FORCED RESPONSE, $v_F(t)$ $(t \geq 0)$
V_A	K_F
$V_A e^{-\alpha t}$	$K_F e^{-\alpha t}$
$V_A(\cos \omega t), V_A(\sin \omega t)$ or $V_A(\cos \omega t) + V_B(\sin \omega t)$	$a \cos \omega t + b \sin \omega t$

EXAMPLE 7–15

Find the response of the RC circuit in Figure 7–2 to an exponential forcing function. The initial capacitor voltage is $v(0) = V_0$.

SOLUTION:
As before, the natural response is

$$v_N(t) = Ke^{-t/R_T C} \qquad t \geq 0$$

The forced response to an exponential input is found using Eq. (7–24) as

$$R_{T}C\frac{dv_{F}(t)}{dt}+v_{F}(t)=V_{A}e^{-\alpha t} \qquad t\geq 0$$

where we select $v_{F}(t)=K_{F}e^{-\alpha t}$ from Table 7–1. Substituting for $v_{F}(t)$,

$$R_{T}C\frac{dK_{F}e^{-\alpha t}}{dt}+K_{F}e^{-\alpha t}=V_{A}e^{-\alpha t} \qquad t\geq 0$$

Performing the differentiation gives

$$R_{T}CK_{F}(-\alpha)e^{-\alpha t}+K_{F}e^{-\alpha t}=V_{A}e^{-\alpha t} \qquad t\geq 0$$

Canceling out the exponentials leaves

$$-\alpha K_{F}R_{T}C+K_{F}=V_{A}$$

Solving for K_{F} yields

$$K_{F}=\frac{V_{A}}{1-\alpha R_{T}C}$$

Substituting back into our solution for the forced response and combining it with the natural response, we get

$$v(t)=v_{N}(t)+v_{F}(t)=Ke^{-t/R_{T}C}+\frac{V_{A}}{1-\alpha R_{T}C}e^{-\alpha t} \qquad t\geq 0$$

This leaves only the constant K from the natural response to be determined. We find K by using the initial condition $v(0)=V_{0}$.

$$v(0)=V_{0}=Ke^{-0/R_{T}C}+\frac{V_{A}}{1-\alpha R_{T}C}e^{-\alpha 0}$$

$$K=V_{0}-\frac{V_{A}}{1-\alpha R_{T}C}$$

Putting it all together, we find the total solution as

$$v(t)=\left[V_{0}-\frac{V_{A}}{1-\alpha R_{T}C}\right]e^{-t/R_{T}C}+\frac{V_{A}}{1-\alpha R_{T}C}e^{-\alpha t}\ \mathrm{V} \qquad t\geq 0$$

The resulting waveform is the sum of two decaying exponentials. The exponential with the longest time constant will outlast the other and is called the *dominant exponential*. ∎

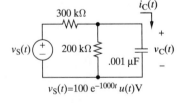

$v_{S}(t)=100\ e^{-1000t}\ u(t)\mathrm{V}$

FIGURE 7–29

Exercise 7–18

The capacitor in the circuit of Figure 7–29 is in the zero state. Find the voltage across and the current through the capacitor for $t\geq 0$.

Answers:

$$v_{C}(t)=45.5e^{-1000t}-45.5e^{-8333t}\ \mathrm{V}, \quad t\geq 0$$

$$i_{C}(t)=-45.5e^{-1000t}+379e^{-8333t}\ \mu\mathrm{A}, \quad t\geq 0$$

Exercise 7–19

The circuit in Figure 7–2 has $R_{T}=100\ \mathrm{k\Omega}$, and $C=0.1\ \mu\mathrm{F}$, and it is driven by $v_{T}(t)=10e^{-50t}\ \mathrm{V}$. The capacitor has an initial voltage of -5 V.

(a) Use OrCAD to find the transient response of the capacitor voltage.
(b) Determine which is the dominant exponential.

FIGURE 7–30

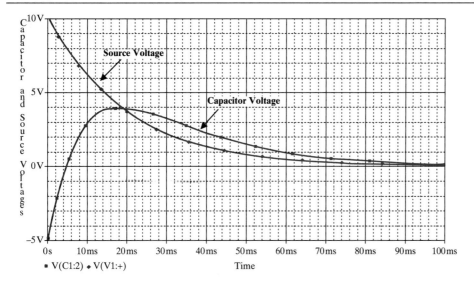

Answers:

(a) See Figure 7–30 for the OrCAD response of $v_C(t)$.
(b) The forcing function has the dominant exponential with a 20-ms time constant. The circuit's time constant is only 10 ms and it quickly decays away, leaving only that of the forcing function.

In the following discussion, we solve for the capacitor voltage in the RC circuit of Figure 7–2 when the input source is a sinusoid. The solution follows the same process as for the exponential input just studied. It is important to realize that the input on the right side of Eq. (7–24) is a sinusoid that starts at $t = 0$ through some action such as closing a switch. This means that there will be an initial condition $v(0) = V_0$.

As before, the natural response is

$$v_N(t) = Ke^{-t/R_T C} \qquad t \geq 0$$

As with the exponential input, the forced response depends on both the circuit and the nature of the forcing function. The forced response is a particular solution of the following differential equation:

$$R_T C \frac{dv_F(t)}{dt} + v_F(t) = V_A \cos \omega t \qquad t \geq 0$$

This equation requires that $v_F(t)$ plus $R_T C$ times its first derivative add to produce a cosine function for $t \geq 0$. The only way this can happen is for $v_F(t)$ and its derivative to be sinusoids of the same frequency. This requirement brings to mind the derivative property of the sinusoid. So we try a solution in the form of a general sinusoid. As noted in Chapter 5, a general sinusoid can be written in amplitude and phase angle form as

$$v_F(t) = V_F \cos(\omega t + \phi) \qquad (7\text{–}25a)$$

or in terms of Fourier coefficients as

$$v_F(t) = a \cos \omega t + b \sin \omega t \qquad (7\text{–}25b)$$

While either form will work, it is somewhat easier to work with the Fourier coefficient format. (See Table 7–1.)

The approach we are using is called the method of undetermined coefficients, where the unknown coefficients are the Fourier coefficients a and b in Eq. (7–25b). To find these unknowns we insert the proposed forced response in Eq. (7–25b) into the differential equation to obtain

$$R_T C \frac{d}{dt}(a\cos\omega t + b\sin\omega t) + (a\cos\omega t + b\sin\omega t) = V_A \cos\omega t \qquad t \geq 0$$

Performing the differentiation gives

$$R_T C(-\omega a\sin\omega t + \omega b\cos\omega t) + (a\cos\omega t + b\sin\omega t) = V_A \cos\omega t$$

We next gather all sine and cosine terms on one side of the equation.

$$[R_T C\omega b + a - V_A]\cos\omega t + [-R_T C\omega a + b]\sin\omega t = 0$$

The left side of this equation is zero for all $t \geq 0$ only when the coefficients of the cosine and sine terms are identically zero. This requirement yields two linear equations in the unknown coefficients a and b:

$$a + (R_T C\omega)b = V_A$$
$$-(R_T C\omega)a + b = 0$$

The solutions of these linear equations are

$$a = \frac{V_A}{1 + (\omega R_T C)^2} \qquad b = \frac{\omega R_T C V_A}{1 + (\omega R_T C)^2}$$

These equations express the unknowns a and b in terms of known circuit parameters $(R_T C)$ and known input signal parameters (ω and V_A).

We combine the forced and natural responses as

$$v(t) = Ke^{-t/R_T C} + \frac{V_A}{1 + (\omega R_T C)^2}(\cos\omega t + \omega R_T C\sin\omega t) \qquad t \geq 0 \qquad (7\text{–}26)$$

The initial condition requires

$$v(0) = V_0 = K + \frac{V_A}{1 + (\omega R_T C)^2}$$

which means K is

$$K = V_0 - \frac{V_A}{1 + (\omega R_T C)^2}$$

We substitute this value of K into Eq. (7–26) to obtain the function $v(t)$ that satisfies the differential equation and the initial conditions.

$$v(t) = \underbrace{\left[V_0 - \frac{V_A}{1 + (\omega R_T C)^2}\right]e^{-t/R_T C}}_{\text{natural response}} + \underbrace{\frac{V_A}{1 + (\omega R_T C)^2}(\cos\omega t + \omega R_T C\sin\omega t)}_{\text{forced response}} \text{ V} \quad t \geq 0$$

This expression seems somewhat less formidable when we convert the forced response to an amplitude and phase angle format

$$v(t) = \underbrace{\left[V_0 - \frac{V_A}{1 + (\omega R_T C)^2}\right]e^{-t/R_T C}}_{\text{natural response}} + \underbrace{\frac{V_A}{\sqrt{1 + (\omega R_T C)^2}}\cos(\omega t + \theta)}_{\text{forced response}} \text{ V} \quad t \geq 0 \qquad (7\text{–}27)$$

where

$$\theta = \tan^{-1}(-b/a) = \tan^{-1}(-\omega R_T C)$$

Equation (7–27) is the complete response of the RC circuit for an initial condition V_0 and a sinusoidal input $[V_A \cos \omega t] u(t)$. Several aspects of the response deserve comment:

1. After roughly five time constants the natural response decays to zero but the sinusoidal forced response persists.

2. The forced response is a sinusoid with the same frequency (ω) as the input but with a different amplitude and phase angle.

3. The forced response is proportional to V_A. This means that the amplitude of the forced component has the proportionality property because the circuit is linear.

In the terminology of electrical engineering, the forced component is called the **sinusoidal steady-state response**. The words *steady state* may be misleading since together they seem to imply a constant or "steady" value, whereas the forced response is a sustained oscillation. To electrical engineers *steady state* means the conditions reached after the natural response has died out. The sinusoidal steady-state response is also called the **ac steady-state response**. Often the words *steady state* are dropped and it is called simply the **ac response**. Hereafter, ac response, sinusoidal steady-state response, and the forced response for a sinusoidal input will be used interchangeably.

Finally, the forced response due to a step function input is called the **zero-frequency** or **dc steady-state response**. The zero-frequency terminology means that we think of a step function as a cosine $V_A [\cos \omega t] u(t)$ with $\omega = 0$. The reader can easily show that inserting $\omega = 0$ reduces Eq. (7–27) to the RC circuit step response in Eq. (7–20).

EXAMPLE 7–16

The switch in Figure 7–31(a) has been open for a long time and is closed at $t = 0$. Find the voltage $v(t)$ for $t \geq 0$ when $v_S(t) = [20 \sin 1000t] u(t)$ V.

SOLUTION:
We first derive the circuit differential equation. By voltage division, the Thévenin voltage seen by the capacitor is

$$v_T(t) = \frac{4}{4+4} v_S(t) = 10 \sin 1000t \ \text{V}$$

The Thévenin resistance (switch closed and source off) looking back into the interface is two 4-kΩ resistors in parallel, so $R_T = 2 \ \text{k}\Omega$. The circuit time constant is

$$T_C = R_T C = (2 \times 10^3)(1 \times 10^{-6}) = 2 \times 10^{-3} = 1/500 \ \text{s}$$

Given the Thévenin equivalent seen by the capacitor and the circuit time constant, the circuit differential equation is

$$2 \times 10^{-3} \frac{dv(t)}{dt} + v(t) = 10 \sin 1000t \qquad t \geq 0$$

Note that the right side of the circuit differential equation is the Thévenin voltage $v_T(t)$, not the original source input $v_S(t)$. The natural response is of the form

$$v_N(t) = K e^{-500t} \qquad t \geq 0$$

The forced response with undetermined Fourier coefficients is

$$v_F(t) = a \cos 1000t + b \sin 1000t$$

Substituting the forced response into the differential equation produces

$$2 \times 10^{-3}(-1000 \, a \sin 1000t + 1000b \cos 1000t) + a \cos 1000t$$
$$+ b \sin 1000t = 10 \sin 1000t$$

Collecting all sine and cosine terms on one side of this equation yields

$$(a + 2b) \cos 1000t + (-2a + b - 10) \sin 1000t = 0$$

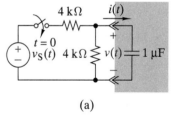

(a)

FIGURE 7–31

The left side of this equation is zero for all $t \geq 0$ only when the coefficients of the sine and cosine terms vanish:

$$a + 2b = 0$$
$$-2a + b = 10$$

The solutions of these two linear equations are $a = -4$ and $b = 2$. We combine the forced and natural responses

$$v(t) = Ke^{-500t} - 4\cos 1000t + 2\sin 1000t \qquad t \geq 0$$

The constant K is found from the initial conditions

$$v(0) = V_0 = K - 4$$

The initial condition is $V_0 = 0$ because with the switch open the capacitor had no input for a long time prior to $t = 0$. The initial condition $v(0) = 0$ requires $K = 4$, so we can now write the complete response in the form

$$v(t) = 4e^{-500t} - 4\cos 1000t + 2\sin 1000t \text{ V} \qquad t \geq 0$$

or, in an amplitude, phase angle format as

$$v(t) = 4e^{-500t} + 4.47\cos(1000t + 153°) \text{ V} \quad t \geq 0$$

Figure 7–31(b) shows an Excel worksheet that generates plots of the natural response, forced response, and total response. Column A is the time at 0.25-ms intervals. Columns B and C calculate the natural response ($4e^{-500t}$) and the forced response ($-4\cos 1000t + 2\sin 1000t$) at each of the times given in column A. The total response in column D is the sum of the entries in columns B and C. The plots show that the total response merges into the sinusoidal forced response since the

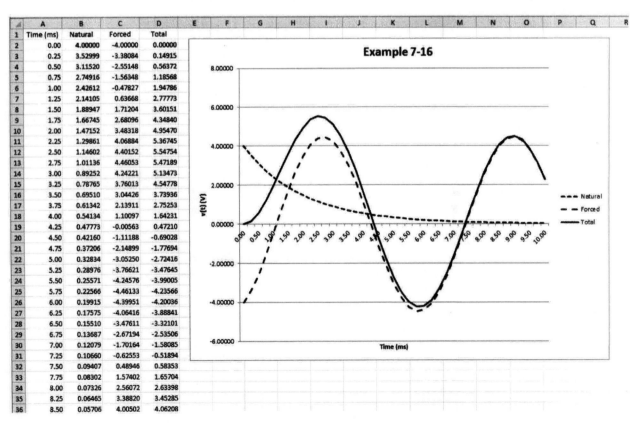

(b)

FIGURE 7–31 *(Continued)*

natural response decays to zero after about $5T_C = 10$ ms. That is, after about 10 ms or so the circuit settles down to an ac steady-state condition. ∎

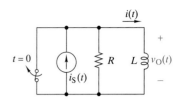

FIGURE 7–32

Exercise 7–20

Find the sinusoidal steady-state response of the output voltage $v_O(t)$ in Figure 7–32 when the input current is $i_S(t) = [I_A \cos \omega t]u(t)$ A.

Answer:

$$v_O(t) = \frac{I_A \omega L}{\sqrt{1 + \left(\dfrac{\omega L}{R}\right)^2}} \cos(\omega t + \theta) \text{ V} \qquad t \geq 0$$

where $\theta = \tan^{-1}(R/\omega L)$. The output voltage is a sinusoid with the same frequency as the input signal, but with a different amplitude and phase angle. In fact, in the sinusoidal steady state every voltage and current in a linear circuit is sinusoidal with the same frequency. We will make use of this fact in Chapter 8.

Exercise 7–21

Find the forced component solution of the differential equation

$$10^{-3} \frac{dv(t)}{dt} + v(t) = 10 \cos \omega t \text{ V}$$

for the following frequencies:

(a) $\omega = 500$ rad/s
(b) $\omega = 1000$ rad/s
(c) $\omega = 2000$ rad/s

Answers:

(a) $v_F(t) = 8 \cos 500t + 4 \sin 500t$ V $\qquad t \geq 0$
(b) $v_F(t) = 5 \cos 1000t + 5 \sin 1000t$ V $\qquad t \geq 0$
(c) $v_F(t) = 2 \cos 2000t + 4 \sin 2000t$ V $\qquad t \geq 0$

DISCUSSION: *Converting these answers to an amplitude and phase angle as*

(a) $v_F(t) = 8.94 \cos(500t - 26.6°)$ V $\qquad t \geq 0$
(b) $v_F(t) = 7.07 \cos(1000t - 45°)$ V $\qquad t \geq 0$
(c) $v_F(t) = 4.47 \cos(2000t - 63.4°)$ V $\qquad t \geq 0$

we see that increasing the frequency of the input sinusoid decreases the amplitude and phase angle of the sinusoidal steady-state output of the circuit. We will see in later chapters that this is a low pass filter.

Exercise 7–22

The circuit in Figure 7–33 is operating in the sinusoidal steady state with

$$v_O(t) = 10 \cos (100t - 45°) \text{ V}$$

Find the source voltage $v_S(t)$.

Answer: $\qquad v_S(t) = 10\sqrt{2} \cos 100t$ V

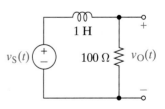

FIGURE 7–33

7–5 THE SERIES *RLC* CIRCUIT

Second-order circuits contain two energy storage elements that cannot be replaced by a single equivalent element. They are called **second-order circuits** because the circuit differential equation involves the second derivative of the dependent variable. Although there is an endless number of such circuits, in this chapter we will concentrate on two classical forms: (1) the series *RLC* circuit and (2) the parallel *RLC* circuit. These two circuits illustrate almost all of the basic concepts of second-order circuits and serve as vehicles for studying the solution of second-order

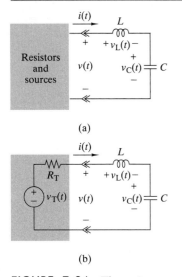

(a)

(b)

FIGURE 7–34   *The series RLC circuit.*

differential equations. In subsequent chapters we use Laplace transform techniques to analyze any second-order circuit.

FORMULATING SERIES *RLC* CIRCUIT EQUATIONS

We begin with the circuit in Figure 7–34(a), where the inductor and capacitor are connected in series. The source-resistor circuit can be reduced to the Thévenin equivalent shown in Figure 7–34(b). The result is a circuit in which a voltage source, resistor, inductor, and capacitor are connected in series (hence the name **series *RLC* circuit**).

The first task is to develop the equations that describe the series *RLC* circuit. The Thévenin equivalent to the left of the interface in Figure 7–34(b) produces the KVL constraint

$$v(t) + R_{\mathrm{T}}i(t) = v_{\mathrm{T}}(t) \tag{7–28}$$

Applying KVL around the loop on the right side of the interface yields

$$v(t) = v_{\mathrm{L}}(t) + v_{\mathrm{C}}(t) \tag{7–29}$$

Finally, the *i–v* characteristics of the inductor and capacitor are

$$v_{\mathrm{L}}(t) = L\frac{di(t)}{dt} \tag{7–30}$$

$$i(t) = C\frac{dv_{\mathrm{C}}(t)}{dt} \tag{7–31}$$

Equations (7–28) through (7–31) are four independent equations in four unknowns $(i(t), v(t), v_{\mathrm{L}}(t), v_{\mathrm{C}}(t))$. Collectively, this set of equations provides a complete description of the dynamics of the series *RLC* circuit. To find the circuit response using classical methods, we must derive a circuit equation containing only one of these unknowns.

We use circuit state variables as solution variables because they are continuous functions of time. In the series *RLC* circuit in Figure 7–34(b), there are two state variables: (1) the capacitor voltage $v_{\mathrm{C}}(t)$ and (2) the inductor current $i(t)$. We first show how to describe the circuit using the capacitor voltage as the solution variable:

To derive a single equation in $v_{\mathrm{C}}(t)$, we substitute Eqs. (7–29) and (7–31) into Eq. (7–28).

$$v_{\mathrm{L}}(t) + v_{\mathrm{C}}(t) + R_{\mathrm{T}}C\frac{dv_{\mathrm{C}}(t)}{dt} = v_{\mathrm{T}}(t) \tag{7–32}$$

These substitutions eliminate the unknowns except v_{C} and v_{L}. To eliminate the inductor voltage, we substitute Eq. (7–31) into Eq. (7–30) to obtain

$$v_{\mathrm{L}}(t) = LC\frac{d^2v_{\mathrm{C}}(t)}{dt^2}$$

Substituting this result into Eq. (7–32) produces

$$LC\frac{d^2v_{\mathrm{C}}(t)}{dt^2} + R_{\mathrm{T}}C\frac{dv_{\mathrm{C}}(t)}{dt} + v_{\mathrm{C}}(t) = v_{\mathrm{T}}(t) \tag{7–33}$$

$$v_{\mathrm{L}}(t) \quad + \quad v_{\mathrm{R}}(t) \quad + v_{\mathrm{C}}(t) = v_{\mathrm{T}}(t)$$

In effect, this is a KVL equation around the loop in Figure 7–34(b), where the inductor and resistor voltages have been expressed in terms of the capacitor voltage.

Equation (7–33) is a second-order linear differential equation with constant coefficients. It is a second-order equation because the highest-order derivative is the second derivative of the dependent variable $v_{\mathrm{C}}(t)$. The coefficients are constant because the circuit parameters L, C, and R_{T} do not change. The Thévenin voltage $v_{\mathrm{T}}(t)$ is a known driving force. The initial conditions

$$v_{\mathrm{C}}(0) = V_0 \quad \text{and} \quad \frac{dv_{\mathrm{C}}}{dt}(0) = \frac{1}{C}i(0) = \frac{I_0}{C} \tag{7–34}$$

are determined by the values of the capacitor voltage and inductor current at $t = 0$, V_0 and I_0.

In summary, the second-order differential equation in Eq. (7–33) characterizes the response of the series RLC circuit in terms of the capacitor voltage $v_C(t)$. Once the solution $v_C(t)$ is found, we can solve for every other voltage or current, including the inductor current, using the element and connection constraints in Eqs. (7–28) to (7–31).

Alternatively, we can characterize the series RLC circuit using the inductor current. We first write the capacitor i–v characteristics in integral form:

$$v_C(t) = \frac{1}{C} \int_0^t i(x)dx + v_C(0) \qquad (7\text{–}35)$$

Equations (7–35), (7–30), and (7–29) are inserted into the interface constraint of Eq. (7–28) to obtain a single equation in the inductor current $i(t)$:

$$L\frac{di(t)}{dt} + \frac{1}{C} \int_0^t i(x)dx + v_C(0) + R_T i(t) = v_T(t) \qquad (7\text{–}36)$$
$$v_L(t) + \qquad\qquad v_C(t) \qquad\quad + v_R(t) = v_T(t)$$

In effect, this is a KVL equation around the loop in Figure 7–30(b), where the capacitor and resistor voltages have been expressed in terms of the inductor current.

Equation (7–36) is a second-order linear integrodifferential equation with constant coefficients. It is second order because it involves the first derivative and the first integral of the dependent variable $i(t)$. The coefficients are constant because the circuit parameters L, C, and R_T do not change. The Thévenin equivalent voltage $v_T(t)$ is a known driving force, and the initial conditions are $v_C(0) = V_0$ and $i(0) = I_0$.

Equations (7–33) and (7–36) involve the same basic ingredients: (1) an unknown state variable, (2) three circuit parameters (R_T, L, C), (3) a known input $v_T(t)$, and (4) two initial conditions (V_0 and I_0). The only difference is that one expresses the sum of voltages around the loop in terms of the capacitor voltage, while the other uses the inductor current. Either equation characterizes the dynamics of the series RLC circuit because once a state variable is found, every other voltage or current can be found using the element and connection constraints.

ZERO-INPUT RESPONSE OF THE SERIES *RLC* CIRCUIT

The circuit dynamic response for $t \geq 0$ can be divided into two components: (1) the zero-input response caused by the initial conditions and (2) the zero-state response caused by driving forces applied after $t = 0$. Because the circuit is linear, we can solve for these responses separately and superimpose them to get the total response. We first deal with the zero-input response.

With $v_T(t) = 0$ (zero-input) Eq. (7–33) becomes

$$LC\frac{d^2v_C(t)}{dt^2} + R_T C\frac{dv_C(t)}{dt} + v_C(t) = 0 \qquad (7\text{–}37)$$

This result is a second-order homogeneous differential equation in the capacitor voltage. Alternatively, we set $v_T = 0$ in Eq. (7–36) and differentiate once to obtain the following homogeneous differential equation in the inductor current:

$$LC\frac{d^2i(t)}{dt^2} + R_T C\frac{di(t)}{dt} + i(t) = 0 \qquad (7\text{–}38)$$

We observe that Eqs. (7–37) and (7–38) have exactly the same form except that the dependent variables are different. The zero-input response of the capacitor voltage and inductor current have the same general form. We do not need to study both to understand the dynamics of the series RLC circuit. In other words, in the series RLC circuit we can use either state variable to describe the zero-input response.

In the following discussion we will concentrate on the capacitor voltage response. Equation (7–37) requires the capacitor voltage, plus RC times its first derivative, plus LC times its second derivative to add to zero for all $t \geq 0$. The only way this can happen is for $v_C(t)$, its first derivative, and its second derivative to have the same waveform. No matter how many times we differentiate an exponential of the form e^{st}, we are left with a signal with the same waveform. This observation, plus our experience with first-order circuits, suggests that we try a solution of the form

$$v_C(t) = Ke^{st}$$

where the parameters K and s are to be evaluated. When the trial solution is inserted in Eq. (7–37), we obtain the condition

$$Ke^{st}(LCs^2 + R_TCs + 1) = 0$$

The function e^{st} cannot be zero for all $t \geq 0$. The condition $K = 0$ is not allowed because it is a trivial solution declaring that $v_C(t)$ is zero for all t. The only useful way to meet the condition is to require

$$LCs^2 + R_TCs + 1 = 0 \qquad \text{(7–39)}$$

Equation (7–39) is the **characteristic equation** of the series RLC circuit. The characteristic equation is a quadratic because the circuit contains two energy storage elements. Inserting Ke^{st} into the homogeneous equation of the inductor current in Eq. (7–38) produces the same characteristic equation. Thus, Eq. (7–39) relates the zero-input response to circuit parameters for both state variables (hence the name *characteristic equation*).

In general, a quadratic characteristic equation has two roots:

$$s_1, \ s_2 = \frac{-R_TC \pm \sqrt{(R_TC)^2 - 4LC}}{2LC} \qquad \text{(7–40)}$$

From the form of the expression under the radical in Eq. (7–40), we see that there are three distinct possibilities:

Case A: If $(R_TC)^2 - 4LC > 0$, there are two real, unequal roots ($s_1 = -\alpha_1 \neq s_2 = -\alpha_2$).

Case B: If $(R_TC)^2 - 4LC = 0$, there are two real, equal roots ($s_1 = s_2 = -\alpha$).

Case C: If $(R_TC)^2 - 4LC < 0$, there are two complex conjugate roots ($s_1 = -\alpha - j\beta$ and $s_2 = -\alpha + j\beta$).

The symbol j represents the imaginary number $\sqrt{-1}$.[1] Before dealing with the form of the zero-input response for each case, we consider an example.

EXAMPLE 7–17

A series RLC circuit has a $C = 0.25\,\mu\text{F}$ and $L = 1\,\text{H}$. Find the roots of the characteristic equation for $R_T = 8.5\,\text{k}\Omega$, $4\,\text{k}\Omega$, and $1\,\text{k}\Omega$.

SOLUTION:
For $R_T = 8.5\,\text{k}\Omega$, the characteristic equation is

$$0.25 \times 10^{-6}s^2 + 2.125 \times 10^{-3}s + 1 = 0$$

---

[1]Mathematicians use the letter i to represent $\sqrt{-1}$. Electrical engineers use j, since the letter i represents electric current.

whose roots are

$$s_1, s_2 = -4250 \pm \sqrt{(3750)^2} = -500, \quad -8000$$

These roots illustrate case A. The quantity under the radical is positive, and there are two real, unequal roots at $s_1 = -500$ and $s_2 = -8000$.

For $R_T = 4\,k\Omega$, the characteristic equation is

$$0.25 \times 10^{-6}s^2 + 10^{-3}s + 1 = 0$$

whose roots are

$$s_1, s_2 = -2000 \pm \sqrt{4 \times 10^6 - 4 \times 10^6} = -2000$$

This is an example of case B. The quantity under the radical is zero, and there are two real, equal roots at $s_1 = s_2 = -2000$.

For $R_T = 1\,k\Omega$ the characteristic equation is

$$0.25 \times 10^{-6}s^2 + 0.25 \times 10^{-3}s + 1 = 0$$

whose roots are

$$s_1, s_2 = -500 \pm 500\sqrt{-15}$$

The quantity under the radical is negative, illustrating case C.

$$s_1, s_2 = -500 \pm j500\sqrt{15}$$

In case C the two roots are complex conjugates.   ∎

Exercise 7-23

For a series *RLC* circuit:

(a) Find the roots of the characteristic equation when $R_T = 2\,k\Omega$, $L = 100$ mH, and $C = 0.4\,\mu$F.

(b) For $L = 100$ mH, select the values of R_T and C so the roots of the characteristic equation are $s_1, s_2 = -1000 \pm j2000$.

(c) Select the values of R_T, L, and C so $s_1 = s_2 = -10^4$.

Answers:

(a) $s_1 = -1340, s_2 = -18,660$

(b) $R_T = 200\,\Omega, C = 2\,\mu$F

(c) There is no unique answer to part (c) since the requirement

$$(2 \times 10^{-4}s + 1)^2 = LCs^2 + R_TCs + 1$$
$$= 10^{-8}s^2 + 2 \times 10^{-4}s + 1$$

gives two equations, $R_TC = 10^{-4}$ and $LC = 10^{-8}$, in three unknowns. One solution is to select $C = 1\,\mu$F, which yields $L = 10$ mH and $R_T = 200\,\Omega$.

We have not introduced complex numbers simply to make things complex. Complex numbers arise quite naturally in practical physical situations involving nothing more than factoring a quadratic equation. The ability to deal with complex numbers is essential to our study. For those who need a review of such matters, there is a concise discussion in Appendix A.

FORM OF THE ZERO-INPUT RESPONSE

Since the characteristic equation has two roots, there are two solutions to the homogeneous differential equation:

$$v_{C1}(t) = K_1 e^{s_1 t}$$
$$v_{C2}(t) = K_2 e^{s_2 t}$$

That is,

$$LC \frac{d^2}{dt^2}(K_1 e^{s_1 t}) + R_T C \frac{d}{dt}(K_1 e^{s_1 t}) + K_1 e^{s_1 t} = 0$$

and

$$LC \frac{d^2}{dt^2}(K_2 e^{s_2 t}) + R_T C \frac{d}{dt}(K_2 e^{s_2 t}) + K_2 e^{s_2 t} = 0$$

The sum of these two solutions is also a solution since

$$LC \frac{d^2}{dt^2}(K_1 e^{s_1 t} + K_2 e^{s_2 t}) + R_T C \frac{d}{dt}(K_1 e^{s_1 t} + K_2 e^{s_2 t}) + K_1 e^{s_1 t} + K_2 e^{s_2 t} = 0$$

Therefore, the general solution for the zero-input response is of the form

$$v_C(t) = K_1 e^{s_1 t} + K_2 e^{s_2 t} \qquad (7\text{--}41)$$

The constants K_1 and K_2 can be found using the initial conditions given in Eq. (7–34). At $t = 0$ the condition on the capacitor voltage yields

$$v_C(0) = V_0 = K_1 + K_2 \qquad (7\text{--}42)$$

To use the initial condition on the inductor current, we differentiate Eq. (7–41).

$$\frac{dv_C(t)}{dt} = K_1 s_1 e^{s_1 t} + K_2 s_2 e^{s_2 t}$$

Using Eq. (7–34) to relate the initial value of the derivative of the capacitor voltage to the initial inductor current $i(0)$ yields

$$\frac{dv_C(0)}{dt} = \frac{I_0}{C} = K_1 s_1 + K_2 s_2 \qquad (7\text{--}43)$$

Equations (7–42) and (7–43) provide two equations in the two unknown constants K_1 and K_2:

$$K_1 + K_2 = V_0$$
$$s_1 K_1 + s_2 K_2 = I_0/C$$

The solutions of these equations are

$$K_1 = \frac{s_2 V_0 - I_0/C}{s_2 - s_1} \quad \text{and} \quad K_2 = \frac{-s_1 V_0 + I_0/C}{s_2 - s_1}$$

Inserting these solutions back into Eq. (7–41) yields

$$v_C(t) = \frac{s_2 V_0 - I_0/C}{s_2 - s_1} e^{s_1 t} + \frac{-s_1 V_0 + I_0/C}{s_2 - s_1} e^{s_2 t} \qquad t \geq 0 \qquad (7\text{--}44)$$

Equation (7–44) is the general zero-input response of the series RLC circuit. The response depends on two initial conditions V_0 and I_0, and the circuit parameters R_T, L, and C since s_1 and s_2 are the roots of the characteristic equation $LCs^2 + R_T Cs + 1 = 0$. The response takes on different forms depending on whether the roots s_1 and s_2 fall under case A, B, or C.

For case A the two roots are real and distinct. Using the notation $s_1 = -\alpha_1$ and $s_2 = -\alpha_2$, the form of the zero-input response for $t \geq 0$ is

$$v_C(t) = \left[\frac{\alpha_2 V_0 + I_0/C}{\alpha_2 - \alpha_1}\right] e^{-\alpha_1 t} + \left[\frac{-\alpha_1 V_0 - I_0/C}{\alpha_2 - \alpha_1}\right] e^{-\alpha_2 t} \qquad (7\text{--}45)$$

For case A the response is the sum of two exponential functions similar to the double exponential signal treated in Example 5–14. The function has two time constants $1/\alpha_1$ and $1/\alpha_2$. The time constants can be greatly different, or nearly equal, but they cannot be equal because then we would have case B.

With case B the roots are real and equal. Using the notation $s_1 = s_2 = -\alpha$, the general form in Eq. (7–44) becomes

$$v_C(t) = \frac{(\alpha V_0 + I_0/C)e^{-\alpha t} + (-\alpha V_0 - I_0/C)e^{-\alpha t}}{\alpha - \alpha}$$

We immediately see a problem here because the denominator vanishes. However, a closer examination reveals that the numerator vanishes as well, so the solution reduces to the indeterminate form 0/0. To investigate the indeterminacy, we let $s_1 = -\alpha$ and $s_2 = -\alpha + x$, and we explore the situation as x approaches zero. Inserting s_1 and s_2 in this notation in Eq. (7–44) produces

$$v_C(t) = V_0 e^{-\alpha t} + \left[\frac{-\alpha V_0 - I_0/C}{x}\right]e^{-\alpha t} + \left[\frac{\alpha V_0 + I_0/C}{x}\right]e^{-\alpha t}e^{xt}$$

which can be arranged in the form

$$v_C(t) = e^{-\alpha t}\left[V_0 - (\alpha V_0 + I_0/C)\frac{1 - e^{xt}}{x}\right]$$

We see that the indeterminacy comes from the term $(1 - e^{xt})/x$, which reduces to 0/0 as x approaches zero. Application of l'Hôpital's rule reveals

$$\lim_{x \to 0}\frac{1 - e^{xt}}{x} = \lim_{x \to 0}\frac{-te^{xt}}{1} = -t$$

This result removes the indeterminacy, and as x approaches zero the zero-input response reduces to

$$v_C(t) = V_0 e^{-\alpha t} + (\alpha V_0 + I_0/C)\, te^{-\alpha t} \qquad t \geq 0 \qquad\qquad (7\text{–}46)$$

For case B the response includes an exponential and the damped ramp studied in Example 5–11. The damped ramp is required, rather than two exponentials, because in case B the two equal roots produce the same exponential function.

Case C produces complex conjugate roots of the form

$$s_1 = -\alpha - j\beta \quad \text{and} \quad s_2 = -\alpha + j\beta$$

Inserting these roots into Eq. (7–44) yields

$$v_C(t) = \left[\frac{(-\alpha + j\beta)V_0 - I_0/C}{j2\beta}\right]e^{-\alpha t}e^{-j\beta t} + \left[\frac{(\alpha + j\beta)V_0 + I_0/C}{j2\beta}\right]e^{-\alpha t}e^{j\beta t}$$

which can be arranged in the form

$$v_C(t) = V_0 e^{-\alpha t}\left[\frac{e^{j\beta t} + e^{-j\beta t}}{2}\right] + \frac{\alpha V_0 + I_0/C}{\beta}e^{-\alpha t}\left[\frac{e^{j\beta t} - e^{-j\beta t}}{j2}\right] \qquad (7\text{–}47)$$

The expressions within the brackets have been arranged in a special way for the following reasons. Euler's relationships for an imaginary exponential are written as

$$e^{j\theta} = \cos\theta + j\sin\theta$$

and

$$e^{-j\theta} = \cos\theta - j\sin\theta$$

When we add and subtract these equations, we obtain

$$\cos\theta = \frac{e^{j\theta} + e^{-j\theta}}{2} \quad \text{and} \quad \sin\theta = \frac{e^{j\theta} - e^{-j\theta}}{2j}$$

Comparing these expressions for $\sin\theta$ and $\cos\theta$ with the complex terms in Eq. (7–47) reveals that we can write $v_C(t)$ in the form

$$v_C(t) = V_0 e^{-\alpha t}\cos\beta t + \frac{\alpha V_0 + I_0/C}{\beta}e^{-\alpha t}\sin\beta t \qquad t \geq 0$$

For case C the response contains the damped sinusoid studied in Example 5–12. The real part of the roots (α) provides the exponent coefficient in the exponential function, while the imaginary part (β) defines the frequency of the sinusoidal oscillation.

In summary, the roots of the characteristic equation affect the form of the zero-input response in the following ways. In case A the two roots are real and unequal ($s_1 = -\alpha_1 \neq s_2 = -\alpha_2$) and the zero-input response is the sum of two exponentials of the form

$$v_C(t) = K_1 e^{-\alpha_1 t} + K_2 e^{-\alpha_2 t} \tag{7–48a}$$

In case B the two roots are real and equal ($s_1 = s_2 = -\alpha$) and the zero-input response is the sum of an exponential and a damped ramp.

$$v_C(t) = K_1 e^{-\alpha t} + K_2 t\, e^{-\alpha t} \tag{7–48b}$$

In case C the two roots are complex conjugates ($s_1 = -\alpha - j\beta, s_2 = -\alpha + j\beta$) and the zero-input response is the sum of a damped cosine and a damped sine.

$$v_C(t) = K_1 e^{-\alpha t} \cos \beta t + K_2 e^{-\alpha t} \sin \beta t \tag{7–48c}$$

In determining the zero-input response we use the parameters s, α, and β. At various points in the development, these parameters appear in expressions such as e^{st}, $e^{-\alpha t}$, and $e^{j\beta t}$. Since the exponent of e must be dimensionless, the parameters s, α, and β all have the dimensions of the reciprocal of time, or equivalently, frequency. Collectively, we say that s, α, and β define the **natural frequencies** of the circuit. When it is necessary to distinguish between these three parameters we say that s is the **complex frequency**, α is the **neper frequency**, and β is the **radian frequency**. The importance of this notation will become clear as we proceed through subsequent chapters of this book. To be consistent with expressions such as $s = -\alpha + j\beta$, we specify numerical values of s, α, and β in units of radians per second (rad/s).[2]

The constants K_1 and K_2 in Eqs. (7–48a), (7–48b), and (7–48c) are determined by the initial conditions on two state variables, as illustrated in the following example.

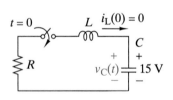

FIGURE 7–35

EXAMPLE 7–18

The circuit of Figure 7–35 has $C = 0.25\ \mu\text{F}$ and $L = 1$ H. The switch has been open for a long time and is closed at $t = 0$. Find the capacitor voltage for $t \geq 0$ for (a) $R = 8.5\ \text{k}\Omega$, (b) $R = 4\ \text{k}\Omega$, and (c) $R = 1\ \text{k}\Omega$. The initial conditions are $I_0 = 0$ and $V_0 = 15$ V.

SOLUTION:
The roots of the characteristic equation for these three values of resistance are found in Example 7–17. We are now in a position to use those results to find the corresponding zero-input responses.
(a) In Example 7–17 the value $R = 8.5\ \text{k}\Omega$ yields case A with roots $s_1 = -500$ and $s_2 = -8000$. The corresponding zero-input solution takes the form in Eq. (7–48a).
$$v_C(t) = K_1 e^{-500t} + K_2 e^{-8000t}$$
The initial conditions yield two equations in the constants K_1 and K_2:

$$v_C(0) = V_0 = 15 = K_1 + K_2$$

$$\frac{dv_C(0)}{dt} = \frac{I_0}{C} = 0 = -500K_1 - 8000K_2$$

---

[2]The term *neper frequency* honors the sixteenth-century mathematician John Napier, who invented the base e or natural logarithms. The term *complex frequency* was apparently first used about 1900 by the British engineer Oliver Heaviside.

Solving these equations yields $K_1 = 16$ and $K_2 = -1$, so that the zero-input response is

$$v_C(t) = 16e^{-500t} - e^{-8000t} \text{ V} \qquad t \geq 0$$

(b) In Example 7–17 the value $R = 4\,\text{k}\Omega$ yields case B with roots $s_1 = s_2 = -2000$. The zero-input response takes the form in Eq. (7–48b):

$$v_C(t) = K_1 e^{-2000t} + K_2 t e^{-2000t}$$

The initial conditions yield two equations in the constants K_1 and K_2:

$$v_C(0) = V_0 = 15 = K_1$$
$$\frac{dv_C(0)}{dt} = \frac{I_0}{C} = 0 = -2000K_1 + K_2$$

Solving these equations yields $K_1 = 15$ and $K_2 = 2000 \times 15$, so the zero-input response is

$$v_C(t) = 15e^{-2000t} + 15(2000t)e^{-2000t} \text{ V} \qquad t \geq 0$$

(c) In Example 7–17 the value $R = 1\,\text{k}\Omega$ yields case C with roots $s_1, s_2 = -500 \pm j500\sqrt{15}$. The zero-input response takes the form in Eq. (7–48c):

$$v_C(t) = K_1 e^{-500t} \cos\left(500\sqrt{15}\right)t + K_2 e^{-500t} \sin\left(500\sqrt{15}\right)t$$

The initial conditions yield two equations in the constants K_1 and K_2:

$$v_C(0) = V_0 = 15 = K_1$$
$$\frac{dv_C(0)}{dt} = \frac{I_0}{C} = 0 = -500K_1 + 500\sqrt{15}K_2$$

which yield $K_1 = 15$ and $K_2 = \sqrt{15}$, so the zero-input response is

$$v_C(t) = 15e^{-500t} \cos(500\sqrt{15})t + \sqrt{15}e^{-500t} \sin(500\sqrt{15})t \text{ V} \qquad t \geq 0$$

Figure 7–36 shows plots of these responses. All three responses start out at 15 V (the initial condition) and all eventually decay to zero. The temporal decay of the responses is caused by energy loss in the circuit and is called **damping**. The case A response does not change sign and is called the **overdamped** response. The case C response undershoots and then oscillates about the final value. This response is said to be **underdamped** because there is not enough damping to prevent these oscillations. The case B response is said to be **critically damped** since it is a special case at the boundary between overdamping and underdamping.

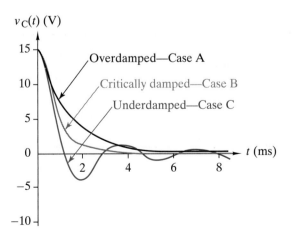

FIGURE 7–36

D Design Exercise 7–24 _____

The circuit in Figure 7–35 has $C = 0.02\,\mu F$ and $L = 100\,mH$. Select a value for R that will produce the critically damped case.

Answer: $R = 4.472\,k\Omega$

EXAMPLE 7–19

In a series RLC circuit the zero-input voltage across the 1-μF capacitor is

$$v_C(t) = 10e^{-1000t}\sin 2000t\;V \qquad t \geq 0$$

(a) Find the circuit characteristic equation.
(b) Find R and L.
(c) Find $i_L(t)$ for $t \geq 0$.
(d) Find the initial values of the state variables.

SOLUTION:
(a) The circuit is underdamped because the zero-input response is a damped sine with $\alpha = 1000$ and $\beta = 2000$ rad/s. The characteristic equation is

$$(s + 1000 - j2000)(s + 1000 + j2000) = s^2 + 2000s + 5 \times 10^6 = 0$$

(b) The characteristic equation of a series RLC circuit [Eq. (7–39)] can be written as

$$s^2 + \frac{R}{L}s + \frac{1}{LC} = 0$$

Comparing this term by term to the result in (a) yields the constraints

$$\frac{R}{L} = 2000 \quad \text{and} \quad \frac{1}{LC} = 5 \times 10^6$$

Since $C = 1\,\mu F$, we find that $L = 0.2\,H$ and $R = 400\,\Omega$.
(c) In a series circuit KCL requires $i_L(t) = i_C(t)$. Hence, the inductor current is

$$i_L(t) = C\frac{dv_C(t)}{dt} = 10^{-6}\frac{d}{dt}\left[10e^{-1000t}\sin 2000t\right]$$
$$= -10e^{-1000t}\sin 2000t + 20e^{-1000t}\cos 2000t\;mA \qquad t \geq 0$$

(d) By inspection, the initial values of the state variables are $v_C(0) = 0$ and $i_L(0) = 20$ mA. ∎

Exercise 7–25 _____

In a series RLC circuit, $R = 250\,\Omega$, $L = 10\,mH$, $C = 1\,\mu F$, $V_0 = 0$, and $I_0 = 30\,mA$. Find the capacitor voltage and inductor current for $t \geq 0$.

Answers:
$$v_C(t) = 2e^{-5000t} - 2e^{-20,000t}\;V$$
$$i_L(t) = -10e^{-5000t} + 40e^{-20,000t}\;mA$$

Exercise 7–26 _____

In a series RLC circuit the zero-input responses are

$$v_C(t) = 2000te^{-500t}\;V$$
$$i_L(t) = 3.2e^{-500t} - 1600te^{-500t}\;mA$$

(a) Find the circuit characteristic equation.
(b) Find the initial values of the state variables.
(c) Find R, L, and C.

Answers:

(a) $s^2 + 1000s + 25 \times 10^4 = 0$
(b) $V_0 = 0, I_0 = 3.2\,\text{mA}$
(c) $R = 2.5\,\text{k}\Omega, L = 2.5\,\text{H}, C = 1.6\,\mu\text{F}$

7–6 THE PARALLEL *RLC* CIRCUIT

The inductor and capacitor in Figure 7–37(a) are connected in parallel. The source-resistor circuit can be reduced to the Norton equivalent shown in Figure 7–37(b). The result is a **parallel *RLC* circuit** consisting of a current source, resistor, inductor, and capacitor. Our first task is to develop a differential equation for this circuit. We expect to find a second-order differential equation because there are two energy storage elements.

The Norton equivalent to the left of the interface introduces the constraint

$$i(t) + \frac{v(t)}{R_N} = i_N(t) \tag{7–49}$$

Writing a KCL equation at the interface yields

$$i(t) = i_L(t) + i_C(t) \tag{7–50}$$

The $i-v$ characteristics of the inductor and capacitor are

$$i_C(t) = C\frac{dv(t)}{dt} \tag{7–51}$$

$$v(t) = L\frac{di_L(t)}{dt} \tag{7–52}$$

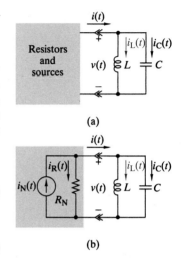

FIGURE 7–37 *The parallel RLC circuit.*

Equations (7–49) through (7–52) provide four independent equations in four unknowns ($i(t)$, $v(t)$, $i_L(t)$, $i_C(t)$). Collectively these equations describe the dynamics of the parallel *RLC* circuit. To solve for the circuit response using classical methods, we must derive a circuit equation containing only one of these four variables.

We prefer using state variables because they are continuous. To obtain a single equation in the inductor current, we substitute Eqs. (7–50) and (7–52) into Eq. (7–49):

$$i_L(t) + i_C(t) + \frac{L}{R_N}\frac{di_L(t)}{dt} = i_N(t) \tag{7–53}$$

The capacitor current can be eliminated from this result by substituting Eq. (7–52) into Eq. (7–51) to obtain

$$i_C(t) = LC\frac{d^2 i_L(t)}{dt^2} \tag{7–54}$$

Inserting this equation into Eq. (7–53) produces

$$LC\frac{d^2 i_L(t)}{dt^2} + \frac{L}{R_N}\frac{di_L(t)}{dt} + i_L(t) = i_N(t) \tag{7–55}$$

$$i_C(t) \quad + \quad i_R(t) \; + i_L(t) = i_N(t)$$

This result is a KCL equation in which the resistor and capacitor currents are expressed in terms of the inductor current.

Equation (7–55) is a second-order linear differential equation of the same form as the series RLC circuit equation in Eq. (7–33). In fact, if we interchange the following quantities:

$$v_C \leftrightarrow i_L \qquad L \leftrightarrow C \qquad R_T \leftrightarrow G_N = \frac{1}{R_N} \qquad v_T \leftrightarrow i_N \qquad \text{series} \leftrightarrow \text{parallel}$$

we change one equation into the other. The two circuits are duals, which means that the results developed for the series case apply to the parallel circuit with the preceding duality interchanges.

However, it is still helpful to outline the major features of the analysis of the parallel RLC circuit. The initial conditions in the parallel circuit are the initial inductor current I_0 and capacitor voltage V_0. The initial inductor current provides the condition $i_L(0) = I_0$ for the differential equation in Eq. (7–55). By using Eq. (7–52), the initial capacitor voltage specifies the initial rate of change of the inductor current as

$$\frac{di_L(0)}{dt} = \frac{1}{L} v_C(0) = \frac{1}{L} V_0$$

These initial conditions are the dual of those obtained for the series RLC circuit in Eq. (7–34).

To solve for the zero-input response, we set $i_N = 0$ in Eq. (7–55) and obtain a homogeneous equation in the inductor current:

$$LC\frac{d^2 i_L(t)}{dt^2} + \frac{L}{R_N}\frac{di_L(t)}{dt} + i_L(t) = 0$$

A trial solution of the form $i_L = Ke^{st}$ leads to the characteristic equation

$$LCs^2 + \frac{L}{R_N}s + 1 = 0 \tag{7–56}$$

The characteristic equation is quadratic because there are two energy storage elements in the parallel RLC circuit. The characteristic equation has two roots:

$$s_1, s_2 = \frac{-\dfrac{L}{R_N} \pm \sqrt{\left(\dfrac{L}{R_N}\right)^2 - 4LC}}{2LC}$$

and, as in the series case, there are three distinct cases:

Case A: If $\left(\dfrac{L}{R_N}\right)^2 - 4LC > 0$, there are two unequal real roots $s_1 = -\alpha_1$ and $s_2 = -\alpha_2$ and the zero-input response is the overdamped form

$$i_L(t) = K_1 e^{-\alpha_1 t} + K_2 e^{-\alpha_2 t} \qquad t \geq 0 \tag{7–57}$$

Case B: If $\left(\dfrac{L}{R_N}\right)^2 - 4LC = 0$, there are two real equal roots $s_1 = s_2 = -\alpha$ and the zero-input response is the critically damped form

$$i_L(t) = K_1 e^{-\alpha t} + K_2 t e^{-\alpha t} \qquad t \geq 0 \tag{7–58}$$

Case C: If $\left(\dfrac{L}{R_N}\right)^2 - 4LC < 0$, there are two complex, conjugate roots $s_1, s_2 = -\alpha \pm j\beta$ and the zero-input response is the underdamped form

$$i_L(t) = K_1 e^{-\alpha t} \cos \beta t + K_2 e^{-\alpha t} \sin \beta t \qquad t \geq 0 \tag{7–59}$$

The analysis results for the series RLC circuit apply to the parallel RLC case with the appropriate duality replacements. In particular, the form of overdamped, critically damped, and underdamped response applies to both circuits. The forms of the responses in Eqs. (7–57), (7–58), and (7–59) have been written with two arbitrary constants K_1 and K_2. The following example shows how to evaluate these constants using the initial conditions for the two state variables.

EXAMPLE 7 – 20

In a parallel RLC circuit $R_T = 500\,\Omega, C = 1\,\mu F, L = 0.2\,H$. The initial conditions are $I_0 = 50$ mA and $V_0 = 0$. Find the zero-input response of inductor current, resistor current, and capacitor voltage.

SOLUTION:

From Eq. (7–56) the circuit characteristic equation is

$$LCs^2 + \frac{L}{R_N}s + 1 = 2 \times 10^{-7} s^2 + 4 \times 10^{-4} s + 1 = 0$$

The roots of the characteristic equation are

$$s_1, s_2 = \frac{-4 \times 10^{-4} \pm \sqrt{16 \times 10^{-8} - 8 \times 10^{-7}}}{4 \times 10^{-7}} = -1000 \pm j2000$$

Since the roots are complex conjugates, we have the underdamped case. The zero-input response of the inductor current takes the form of Eq. (7–59).

$$i_L(t) = K_1 e^{-1000t} \cos 2000t + K_2 e^{-1000t} \sin 2000t \qquad t \geq 0$$

The constants K_1 and K_2 are evaluated from the initial conditions. At $t = 0$ the inductor current reduces to

$$i_L(0) = I_0 = K_1 e^0 \cos 0 + K_2 e^0 \sin 0 = K_1$$

We conclude that $K_1 = I_0 = 50$ mA. To find K_2 we use the initial capacitor voltage. In a parallel RLC circuit the capacitor and inductor voltages are equal, so we can write

$$v_L(t) = L \frac{di_L(t)}{dt} = v_C(t)$$

In this example the initial capacitor voltage is zero, so the initial rate of change of inductor current is zero at $t = 0$. Differentiating the zero-input response produces

$$\frac{di_L(t)}{dt} = -2000K_1 e^{-1000t} \sin 2000t - 1000K_1 e^{-1000t} \cos 2000t$$
$$-1000K_2 e^{-1000t} \sin 2000t + 2000K_2 e^{-1000t} \cos 2000t$$

Evaluating this expression at $t = 0$ yields

$$\frac{di_{L(0)}}{dt} = -2000K_1 e^0 \sin 0 - 1000K_1 e^0 \cos 0$$
$$-1000K_2 e^0 \sin 0 + 2000K_2 e^0 \cos 0$$
$$= -1000K_1 + 2000K_2 = 0$$

The derivative initial condition gives condition $K_2 = K_1/2 = 25$ mA. Given the values of K_1 and K_2, the zero-input response of the inductor current is

$$i_L(t) = 50e^{-1000t} \cos 2000t + 25e^{-1000t} \sin 2000t \text{ mA} \qquad t \geq 0$$

The zero-input response of the inductor current allows us to solve for every voltage and current in the parallel RLC circuit. For example, using the $i-v$ characteristic of the inductor, we obtain the inductor voltage:

$$v_L(t) = L\frac{di_L(t)}{dt} = -25e^{-1000t}\sin 2000t \text{ V} \qquad t \geq 0$$

Since the elements are connected in parallel, we obtain the capacitor voltage and resistor current as

$$v_C(t) = v_L(t) = L\frac{di_L(t)}{dt} = -25e^{-1000t}\sin 2000t \text{ V} \qquad t \geq 0$$

$$i_R(t) = \frac{v_L(t)}{R} = -50e^{-1000t}\sin 2000t \text{ mA} \qquad t \geq 0 \qquad \blacksquare$$

Exercise 7–27

A parallel RLC circuit has $R = 1\,\text{k}\Omega$, $C = 1\,\mu\text{F}$, and $L = 100\,\text{mH}$. The initial conditions are $I_0 = 100$ mA and $V_0 = 0$ V.

(a) Use OrCAD to plot the zero-input response of the inductor, resistor, and capacitor currents on one y-axis.

(b) From your current plots, show that Kirchhoff's current law (KCL) holds for every instant.

FIGURE 7–38

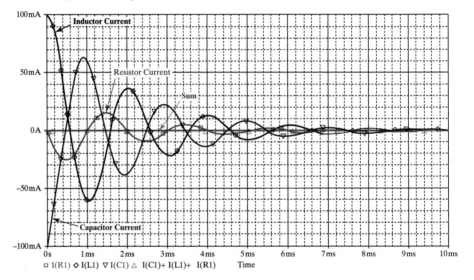

Answers:

(a) Figure 7–38 shows the plots requested.

(b) At any point in time, the three currents always sum to zero, thereby validating Kirchhoff's current law. See the plot of the sum of all three currents in Figure 7–38.

EXAMPLE 7–21

The switch in Figure 7–39 has been open for a long time and is closed at $t = 0$.

(a) Find the initial conditions at $t = 0$.

(b) Find the inductor current for $t \geq 0$.

(c) Find the capacitor voltage and current through the switch for $t \geq 0$.

FIGURE 7–39

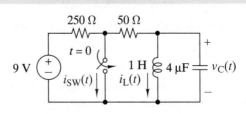

SOLUTION:

(a) For $t < 0$ the circuit is in the dc steady state, so the inductor acts like a short circuit and the capacitor like an open circuit. Since the inductor shorts out the capacitor, the initial conditions just prior to closing the switch at $t = 0$ are

$$v_C(0) = 0 \qquad i_L(0) = \frac{9}{250 + 50} = 30 \text{ mA}$$

(b) For $t \geq 0$ the circuit is a zero-input parallel *RLC* circuit with initial conditions found in (a). The circuit characteristic equation is

$$LCs^2 + \frac{L}{R}s + 1 = 4 \times 10^{-6}s^2 + 2 \times 10^{-2}s + 1 = 0$$

The roots of this equation are

$$s_1 = -50.51 \quad \text{and} \quad s_2 = -4950$$

The circuit is overdamped (case A), since the roots are real and unequal. The general form of the inductor current zero-input response is

$$i_L(t) = K_1 e^{-50.51t} + K_2 e^{-4950t} \qquad t \geq 0$$

The constants K_1 and K_2 are found using the initial conditions. At $t = 0$ the zero-input response is

$$i_L(0) = K_1 e^0 + K_2 e^0 = K_1 + K_2 = 30 \times 10^{-3}$$

The initial capacitor voltage establishes an initial condition on the derivative of the inductor current since

$$L\frac{di_L}{dt}(0) = v_C(0) = 0$$

The derivative of the inductor response at $t = 0$ is

$$\frac{di_L}{dt}(0) = (-50.51K_1 e^{-50.51t} - 4950K_2 e^{-4950t})|_{t=0}$$

$$= -50.51K_1 - 4950K_2 = 0$$

The initial conditions on inductor current and capacitor voltage produce two equations in the unknown constants K_1 and K_2:

$$K_1 + K_2 = 30 \times 10^{-3}$$
$$-50.51K_1 - 4950K_2 = 0$$

Solving these equations yields $K_1 = 30.3$ mA and $K_2 = -0.309$ mA. The zero-input response of the inductor current is

$$i_L(t) = 30.3e^{-50.51t} - 0.309e^{-4950t} \text{ mA} \qquad t \geq 0$$

(c) Given the inductor current in (b), the capacitor voltage is

$$v_C(t) = L\frac{di_L(t)}{dt} = -1.53e^{-50.51t} + 1.53e^{-4950t} \text{ V} \qquad t \geq 0$$

For $t \geq 0$ the current $i_{SW}(t)$ is the current through the 50-Ω resistor plus the current through the 250-Ω resistor.

$$i_{SW}(t) = i_{250}(t) + i_{50}(t) = \frac{9}{250} + \frac{v_C(t)}{50}$$

$$= 36 - 30.6e^{-50.51t} + 30.6e^{-4950t} \text{ mA} \qquad t \geq 0 \qquad \blacksquare$$

Exercise 7–28

The zero-input responses of a parallel RLC circuit are observed to be

$$i_L(t) = 10te^{-2000t} \text{ A}$$
$$v_C(t) = 10e^{-2000t} - 20000te^{-2000t} \text{ V} \qquad t \geq 0$$

(a) What is the circuit characteristic equation?
(b) What are the initial values of the state variables?
(c) What are the values of R, L, and C?
(d) Write an expression for the current through the resistor.

Answers:

(a) $s^2 + 4000s + 4 \times 10^6 = 0$
(b) $i_L(0) = 0$, $v_C(0) = 10 \text{ V}$
(c) $L = 1 \text{ H}$, $C = 0.25 \, \mu\text{F}$, $R = 1 \text{ k}\Omega$
(d) $i_R(t) = 10e^{-2000t} - 20000te^{-2000t} \text{ mA} \qquad t \geq 0$

7–7   SECOND-ORDER CIRCUIT STEP RESPONSE

The step response provides important insights into the response of dynamic circuits in general. So it is natural that we investigate the step response of second-order circuits. In Chapter 11 we will develop general techniques for determining the step response of any linear circuit. However, in this introduction we use classical methods, as in Figure 7–1, of solving differential equations to find the step response of second-order circuits.

The general second-order linear differential equation with a step function input has the form

$$a_2 \frac{d^2y(t)}{dt^2} + a_1 \frac{dy(t)}{dt} + a_0y(t) = Au(t) \tag{7–60}$$

where $y(t)$ is a voltage or current response, $Au(t)$ is the step function input, and a_2, a_1, and a_0 are constant coefficients. The step response is the general solution of this differential equation for $t \geq 0$. The step response can be found by partitioning $y(t)$ into forced and natural components:

$$y(t) = y_N(t) + y_F(t) \tag{7–61}$$

The natural response $y_N(t)$ is the general solution of the homogeneous equation (input set to zero), while the forced response $y_F(t)$ is a particular solution of the equation

$$a_2 \frac{d^2y_F(t)}{dt^2} + a_1 \frac{dy_F(t)}{dt} + a_0y_F(t) = A \qquad t \geq 0$$

Since A is a constant, it follows that dA/dt and d^2A/dt^2 are both zero, so it is readily apparent that $y_F(t) = A/a_0$ is a particular solution of this differential equation. So much for the forced response.

Turning now to the natural response, we seek a general solution of the homogeneous equation. The natural response has the same form as the zero-state response studied in the previous section. In a second-order circuit the zero-state and natural responses take one of the three possible forms: overdamped, critically damped, or underdamped. To describe the three possible forms, we introduce two new parameters: ω_0 (omega zero) and ζ (zeta). These parameters are defined in terms of the coefficients of the general second-order equation in Eq. (7–60):

$$\omega_0^2 = \frac{a_0}{a_2} \quad \text{and} \quad 2\zeta\omega_0 = \frac{a_1}{a_2} \tag{7–62}$$

The parameter ω_0 is called the **undamped natural frequency** and ζ is called the **damping ratio**. Using these two parameters, the general homogeneous equation is written in the form

$$\frac{d^2 y_N(t)}{dt^2} + 2\zeta\omega_0 \frac{dy_N(t)}{dt} + \omega_0^2 y_N(t) = 0 \tag{7–63}$$

The left side of Eq. (7–63) is called the **standard form** of the second-order linear differential equation. When a second-order equation is arranged in this format, we can determine its damping ratio and undamped natural frequency by equating its coefficients with those in the standard form. For example, in standard form the homogeneous equation for the series RLC circuit in Eq. (7–37) is

$$\frac{d^2 v_C(t)}{dt^2} + \frac{R_T}{L}\frac{dv_C(t)}{dt} + \frac{1}{LC}\, v_C(t) = 0$$

Equating like terms yields

$$\omega_0^2 = \frac{1}{LC} \quad \text{and} \quad 2\zeta\omega_0 = \frac{R_T}{L}$$

for the series RLC circuit. In an analysis situation the circuit element values determine the value of the parameters ω_0 and ζ. In a design situation we select L and C to obtain a specified ω_0 and then select R_T to obtain a specified ζ or vice versa.

To determine the form of the natural response using ω_0 and ζ, we insert a trial solution $y_N(t) = Ke^{st}$ into the standard form in Eq. (7–63). The trial function Ke^{st} is a solution provided that

$$Ke^{st}[s^2 + 2\zeta\omega_0 s + \omega_0^2] = 0$$

Since $K = 0$ is the trivial solution and $e^{st} \neq 0$ for all $t \geq 0$, the only useful way for the left side of this equation to be zero for all t is for the quadratic expression within the brackets to vanish. The quadratic expression is the characteristic equation for the general second-order differential equation:

$$s^2 + 2\zeta\omega_0 s + \omega_0^2 = 0$$

The roots of the characteristic equation are

$$s_1, s_2 = \omega_0\left(-\zeta \pm \sqrt{\zeta^2 - 1}\right)$$

We begin to see the advantage of using the parameters ω_0 and ζ. The constant ω_0 is a scale factor that designates the size of the roots. The expression under the radical defines the form of the roots and depends only on the damping ratio ζ. As a result, we can express the three possible forms of the natural response in terms of the damping ratio.

Case A: For $\zeta > 1$ the discriminant is positive, there are two unequal, real roots

$$s_1, s_2 = -\alpha_1, \ -\alpha_2 = \omega_0\left(-\zeta \pm \sqrt{\zeta^2 - 1}\right) \tag{7–64a}$$

and the natural response has the overdamped form

$$y_N(t) = K_1 e^{-\alpha_1 t} + K_2 e^{-\alpha_2 t} \qquad t \geq 0 \tag{7–64b}$$

Case B: For $\zeta = 1$ the discriminant vanishes, there are two real, equal roots

$$s_1 = s_2 = -\alpha = -\zeta\omega_0 \tag{7–65a}$$

and the natural response has the critically damped form

$$y_N(t) = K_1 e^{-\alpha t} + K_2 t e^{-\alpha t} \qquad t \geq 0 \tag{7–65b}$$

Case C: For $\zeta < 1$ the discriminant is negative, leading to two complex, conjugate roots $s_1, s_2 = -\alpha \pm j\beta$, where

$$\alpha = \zeta\omega_0 \quad \text{and} \quad \beta = \omega_0\sqrt{1 - \zeta^2} \tag{7–66a}$$

and the natural response has the underdamped form

$$y_N(t) = K_1 e^{-\alpha t} \cos \beta t + K_2 e^{-\alpha t} \sin \beta t \qquad t \geq 0 \qquad \text{(7–66b)}$$

Equations (7–64a), (7–65a), and (7–66a) provide relationships between the natural frequency parameters α and β and the new parameters ζ and ω_0. The reasons for using two equivalent sets of parameters to describe the natural frequencies of a second-order circuit will become clear as we continue our study of dynamic circuits. Since the units of complex frequency s are radians per second, the standard form of the characteristic equation $s^2 + 2\zeta\omega_0 s + \omega_0^2$ shows that ω_0 is specified in radians per second and ζ is dimensionless.

Combining the forced and natural responses yields the step response of the general second-order differential equation in the form

$$y(t) = y_N(t) + A/a_0 \qquad t \geq 0 \qquad \text{(7–67)}$$

The factor A/a_0 is the forced response. The natural response $y_N(t)$ takes the form in Eqs. (7–64b), (7–65b), or (7–66b), depending on the value of the damping ratio. The constants K_1 and K_2 in the natural response can be evaluated from the initial conditions.

In summary, the step response of a second-order circuit is determined by

1. The amplitude of the step function input $Au(t)$
2. The damping ratio ζ and natural frequency ω_0
3. The initial conditions $y(0)$ and $dy/dt\,(0)$

In this regard the damping ratio and natural frequency play the same role for second-order circuits that the time constant plays for first-order circuits. That is, these circuit parameters determine the basic form of the natural response, just as the time constant defines the form of the natural response in a first-order circuit. It is not surprising that a second-order circuit takes two parameters, since it contains two energy storage elements.

EXAMPLE 7–22

The series RLC circuit in Figure 7–40(a) is driven by a step function and is in the zero state at $t = 0$. Find the capacitor voltage for $t \geq 0$.

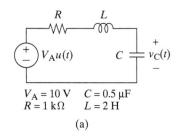

$V_A = 10\text{ V} \quad C = 0.5\ \mu\text{F}$
$R = 1\text{ k}\Omega \quad L = 2\text{ H}$

(a)

FIGURE 7–40

SOLUTION:
This is a series RLC circuit, so the differential equation for the capacitor voltage is

$$10^{-6} \frac{d^2 v_C(t)}{dt^2} + 0.5 \times 10^{-3} \frac{dv_C(t)}{dt} + v_C(t) = 10 \qquad t \geq 0$$

By inspection, the forced response is $v_{CF}(t) = 10\text{ V}$. In standard format the homogeneous equation is

$$\frac{d^2 v_{CN}(t)}{dt^2} + 500 \frac{dv_{CN}(t)}{dt} + 10^6\, v_{CN}(t) = 0 \qquad t \geq 0$$

Comparing this format to the standard form in Eq. (7–63) yields

$$\omega_0^2 = 10^6 \quad \text{and} \quad 2\zeta\omega_0 = 500$$

so $\omega_0 = 1000$ and $\zeta = 0.25$. Since $\zeta < 1$, the natural response is underdamped (case C). Using Eqs. (7–66a) and (7–66b), we have

$$\alpha = \zeta\omega_0 = 250$$

$$\beta = \omega_0 \sqrt{1 - \zeta^2} = 968$$

$$v_{CN}(t) = K_1 e^{-250t} \cos 968t + K_2 e^{-250t} \sin 968t$$

The general solution of the circuit differential equation is the sum of the forced and natural responses:

$$v_C(t) = 10 + K_1 e^{-250t} \cos 968t + K_2 e^{-250t} \sin 968t \qquad t \geq 0$$

The constants K_1 and K_2 are determined by the initial conditions. The circuit is in the zero state at $t = 0$, so the initial conditions are $v_C(0) = 0$ and $i_L(0) = 0$. Applying the initial condition constraints to the general solution yields two equations in the constants K_1 and K_2:

$$v_C(0) = 10 + K_1 = 0$$

$$\frac{dv_C}{dt}(0) = -250 K_1 + 968 K_2 = 0$$

These equations yield $K_1 = -10$ and $K_2 = -2.58$. The capacitor voltage step response is

$$v_C(t) = 10 - 10 e^{-250t} \cos 968t - 2.58 e^{-250t} \sin 968t \text{ V} \qquad t \geq 0$$

A plot of $v_C(t)$ versus time is shown in Figure 7–40(b). The response and its first derivative at $t = 0$ satisfy the initial conditions. The natural response decays to zero, so the forced response determines the final value of $v_C(\infty) = 10$ V. Beginning at $t = 0$ the response climbs rapidly but overshoots and undershoots the mark before eventually settling down to the final value. The damped sinusoidal behavior results from the fact that $\zeta < 1$, producing an underdamped natural response. ■

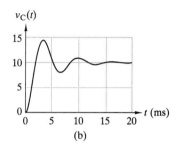

(b)

FIGURE 7–40 (Continued)

Exercise 7–29

The series *RLC* circuit of Figure 7–40(a) is excited by a 10-V step and has the initial conditions $i_L(0) = 0$ A and $v_C(0) = -5$ V. Use OrCAD to plot the response of the capacitor voltage for $t \geq 0$.

FIGURE 7–41

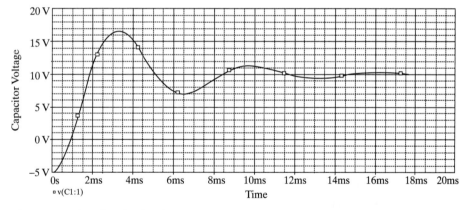

Answer: See the plot in Figure 7–41.

EXAMPLE 7 – 23

The circuit in Figure 7–42 is in the zero state. Find the current through the resistor for $t \geq 0$.

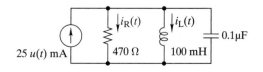

FIGURE 7–42

SOLUTION:

This is a parallel *RLC* circuit. The approach we will take is to first find an expression for the current through the inductor. Then we will use the inductor's *i–v* relationship to find the voltage across the inductor. Since the inductor is in parallel with the capacitor and has the same voltage, we can use the capacitor's *i–v* relationship to find the current through the capacitor. Finally, using KCL at the top node we can find the current through the resistor.

The characteristic equation for a parallel *RLC* circuit is

$$s^2 + \frac{1}{RC}s + \frac{1}{LC} = 0$$

Substituting the values for *R*, *L*, and *C*, we get

$$s^2 + 21{,}276s + 10^8 = 0$$

The roots of this equation are $s_1 = -14{,}300$, $s_2 = -7000$. Thus the natural response is overdamped (Case A).

We can write our basic equation for $i_L(t)$ as

$$i_L(t) = \underbrace{K_1 e^{-14300t} + K_2 e^{-7000t}}_{\text{Natural Response}} + \overbrace{25}^{\substack{\text{Forced} \\ \text{Response}}} \text{ mA}$$

We make use of the initial conditions to solve for the constants K_1 and K_2.

$$i_L(0) = 0 = k_1 e^0 + k_2 e^0 + 25 \text{ mA}$$

$$\frac{di_L(0)}{dt} = \frac{v_L(0)}{L} = \frac{v_C(0)}{L} = \frac{0}{L} = K_1(-14{,}300)e^0 + K_2(-7000)e^0$$

Solving these equations for K_1 and K_2 yields $K_1 = 24$ mA and $K_2 = -49$ mA. Therefore, the total response for the current through the inductor is

$$i_L(t) = 24e^{-14300t} - 49e^{-7000t} + 25 \text{ mA} \quad t \geq 0$$

The voltage across the inductor is found using its *i–v* relationship

$$v_L(t) = L\frac{di_L(t)}{dt} = -34.3e^{-14300t} + 34.3e^{-7000t} \text{ V} \quad t \geq 0$$

And the current through the capacitor is found using its *i–v* relationship

$$i_C(t) = C\frac{dv_C(t)}{dt} = C\frac{dv_L(t)}{dt} = 49e^{-14300t} - 24e^{-7000t} \text{ mA} \quad t \geq 0$$

We can now apply KCL at the top node and find an expression for the current through the resistor

$$i_R(t) = i_S(t) - i_L(t) - i_C(t)$$
$$i_R(t) = 25 - 24e^{-14300t} + 49e^{-7000t} - 25 - 49e^{-14300t} + 24e^{-7000t} \text{ mA} \quad t \geq 0$$
$$i_R(t) = -73e^{-14300t} + 73e^{-7000t} \text{ mA} \quad t \geq 0$$

Checking the result shows that there is no current flowing through the resistor at $t = 0$, or a $t = \infty$. Considering the behavior of inductors under dc stimulation, we realize that at $t = 0-$ and again at $t = \infty$ the inductor acts like a short circuit with zero volts across it. Since the inductor is in parallel with the resistor, the voltage across the resistor is zero and, by Ohm's law, the current through it is zero. ∎

Exercise 7–30

Use OrCAD to plot the currents through the three elements in circuit of Figure 7–42. Show that sum of all three element currents equals the source current for all time.

Answer: See Figure 7–43.

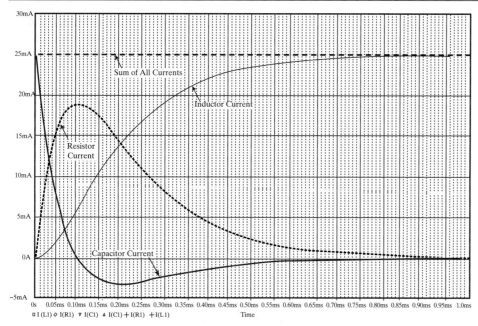

FIGURE 7–43

Exercise 7–31

Find the zero-state response of $v_O(t)$ in Figure 7–44 for $v_S(t) = 60u(t)$ V.

Answer:
$$v_O(t) = 100\left(e^{-2000t} - e^{-8000t}\right) \text{ V}$$

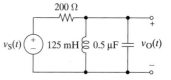

FIGURE 7–44

D Design Exercise 7–32

(a) Select a value for R that will cause the RLC circuit of Figure 7–44 to produce a critically damped response. All parameters except R remain the same.
(b) Use OrCAD to determine the maximum value of $v_O(t)$ and the time at which it reaches that value.
(c) What is the value of the maximum power delivered by the source?

Answers:

(a) $R = 250\,\Omega$ for a critically damped response.
(b) From an OrCAD simulation, the voltage maximum occurs at 250 μs and the peak value is 44.15 V.
(c) The maximum current of 240 mA occurs at $t = 0$ and again as $t \to \infty$. Hence, the maximum power delivered by the source is 14.4 W.

D DESIGN EXAMPLE 7–24

Design a series RLC circuit whose zero-state step response is

$$v_C(t) = V_A - \frac{5}{4}V_A e^{-400t} + \frac{1}{4}V_A e^{-2000t} \quad t > 0$$

where V_A is the amplitude of the step function input.

SOLUTION:
To obtain the required response, the numerical characteristic equation must be
$$(s + 400)(s + 2000) = s^2 + 2400s + 8 \times 10^5 = 0$$
Using circuit parameters, the symbolic characteristic equation of a series RLC circuit is
$$s^2 + \frac{R}{L}s + \frac{1}{LC} = 0$$

Equating coefficients in these two equations leads to two design constraints, namely $R/L=2400$ and $1/LC = 8 \times 10^5$. Let $R = 3\,\text{k}\Omega$; then $L = R/2400 = 1.25\,\text{H}$ and $C = 1/(L \times 8 \times 10^5) = 1 \times 10^{-6}\,\text{F}$. Many other choices are possible. ∎

Exercise 7–33

The step response of a series RLC circuit is observed to be

$$v_C(t) = 15 - 15e^{-1000t} \cos 1000t \text{ V} \quad t \geq 0$$
$$i_L(t) = 45e^{-1000t} \cos 1000t + 45e^{-1000t} \sin 1000t \text{ mA} \quad t \geq 0$$

(a) What is the circuit characteristic equation?
(b) What are the initial values of the state variables?
(c) What is the amplitude of the step input?
(d) What are the values of R, L, and C?
(e) What is the voltage across the resistor?

Answers:

(a) $s^2 + 2000s + 2 \times 10^6 = 0$
(b) $v_C(0) = 0, i_L(0) = 45\,\text{mA}$
(c) $V_A = 15\,\text{V}$
(d) $R = 333\,\Omega, L = 167\,\text{mH}, C = 3\,\mu\text{F}$
(e) $v_R(t) = 15e^{-1000t} \cos 1000t + 15e^{-1000t} \sin 1000t$ V

EXAMPLE 7–25

What range of source resistance will produce an underdamped natural response in a parallel RLC circuit with $L = 200$ mH and $C = 0.032\,\mu\text{F}$?

SOLUTION:
According to Eq. (7–56) the characteristic equation of a parallel RLC circuit is

$$LCs^2 + \frac{L}{R_N}s + 1 = 0$$

Dividing the LC puts this equation in the form

$$s^2 + \frac{1}{R_N C}s + \frac{1}{LC} = 0$$

Comparing this with the standard form $s^2 + 2\zeta\omega_0 s + \omega_0^2$ leads to

$$\omega_0 = \frac{1}{\sqrt{LC}} = \frac{1}{\sqrt{200 \times 10^{-3} \times 32 \times 10^{-9}}} = 12.5 \times 10^3 \text{ rad/s}$$

and $2\zeta\omega_0 = 1/R_N C$. Underdamped response requires that $\zeta < 1$; hence

$$R_N > \frac{1}{2\omega_0 C} = \frac{1}{25 \times 10^3 \times 32 \times 10^{-9}} = 1250\,\Omega$$ ∎

Exercise 7–34

(a) What range of source resistance will produce an underdamped natural response in a series RLC circuit with $L = 200$ mH and $C = 0.032\,\mu\text{F}$?
(b) Compare your answer with the parallel circuit solution in Example 7–25.

Answers:

(a) For an underdamped response, R should be less than 5 kΩ.
(b) In a parallel circuit, as R_N increases, it has less influence on the circuit response—an open circuit would ideally let the circuit oscillate forever. For a series circuit, the dual is true: as R_T decreases, it has less influence on the energy oscillating between the inductor and the capacitor, and a short circuit would ideally let the circuit oscillate forever.

◀D 〔 **DESIGN EXAMPLE 7 – 26** 〕

Design a parallel *RLC* circuit with $\zeta = 0.5$ and $\omega_0 = 25$ krad/s.

SOLUTION:
The characteristic equation of a parallel *RLC* circuit can be written as

$$s^2 + \frac{1}{R_N C}s + \frac{1}{LC} = 0$$

Comparing this with the standard form $s^2 + 2\zeta\omega_0 s + \omega_0^2$ leads to two design constraints:

$$\frac{1}{R_N C} = 2\zeta\omega_0 = 25 \times 10^3 \quad \text{and} \quad \frac{1}{LC} = \omega_0^2 = 6.25 \times 10^8$$

Let the resistance $R_N = 10\,\text{k}\Omega$; then $C = 1/25 \times 10^7 = 4 \times 10^{-9} = 4000\,\text{pF}$ and $L = 25 \times 10^7/6.25 \times 10^8 = 0.4\,\text{H}$. Many other choices are possible since there are three circuit parameters and only two constraints. ■

◀D **Design Exercise 7–35** _____

Design a series *RLC* circuit with $\zeta = 1.5$ and $\omega_0 = 50$ krad/s. You must use a 0.1-µF capacitor.

Answer: With *C* defined, there is only one solution: $R = 600\ \Omega$ and $L = 4$ mH.

SUMMARY

- Circuits containing linear resistors and the equivalent of one capacitor or one inductor are described by first-order differential equations in which the unknown is the circuit state variable.

- The zero-input response in a first-order circuit is an exponential whose time constant depends on circuit parameters. The amplitude of the exponential is equal to the initial value of the state variable.

- For linear circuits the total response is the sum of the forced and natural responses. The natural response is the general solution of the homogeneous differential equation obtained by setting the input to zero. The forced response is a particular solution of the differential equation for the given input.

- For linear circuits the total response is the sum of the zero-input and zero-state responses. The zero-input response is caused by the initial energy stored in capacitors or inductors. The zero-state response results from the input driving forces.

- The initial and final values of the step response of a first-or second-order circuit can be found by replacing capacitors by open circuits and inductors by short circuits and then using resistance circuit analysis methods.

- The transient response to a first-order circuit when the input is other than a step requires that the forced

response solution be of the same form as the input. Hence, an exponential input suggests an exponential forced response; a sinusoidal input suggests a sinusoidal forced response, and so forth.

- For a sinusoidal input the forced response is called the sinusoidal steady-state response, or the ac response. The ac response is a sinusoid with the same frequency as the input but with a different amplitude and phase angle. The ac response can be found from the circuit differential equation using the method of undetermined coefficients.

- Circuits containing linear resistors and the equivalent of two energy storage elements are described by second-order differential equations in which the dependent variable is one of the state variables. The initial conditions are the values of the two state variables at $t = 0$.

- The zero-input response of a second-order circuit takes different forms depending on the roots of the characteristic equation. Unequal real roots produce the overdamped response, equal real roots produce the critically damped response, and complex conjugate roots produce underdamped responses.

- The circuit damping ratio ζ and undamped natural frequency ω_0 determine the form of the zero-input and natural responses of any second-order circuit. The response is overdamped if $\zeta > 1$, critically damped if $\zeta = 1$, and underdamped if $\zeta < 1$. Active circuits can produce undamped ($\zeta = 0$) and unstable ($\zeta < 0$) responses.

• Software tools like MATLAB and OrCAD can help produce numerical and graphical solutions for circuit

transient behavior. In applying these tools one must have some knowledge of analytical methods and an estimate of the general form of the expected response in order to effectively use these tools and interpret their results. Web Appendix C discusses the use of these tools and contains additional examples and exercises.

PROBLEMS

OBJECTIVE 7–1 FIRST-ORDER CIRCUIT ANALYSIS (SECTS. 7–1, 7–2, 7–3, 7–4)

Given a first-order RC or RL circuit
Find the circuit differential equation, the circuit time constant, and the initial conditions (if not given).
(a) Find the zero-input response.
(b) Find the complete response for step function, exponential, and sinusoidal inputs.
(c) See Examples 7–1, 7–2, 7–3, 7–4, 7–5, 7–6, 7–7, 7–8, 7–9, 7–10, 7–11, 7–13, 7–15, 7–16 and Exercises 7–1, 7–2, 7–3, 7–4, 7–5, 7–6, 7–7, 7–8, 7–9, 7–10, 7–11, 7–12, 7–13, 7–15, 7–16, 7–17, 7–18, 7–19, 7–20; 7–21; 7–22.

7–1 Find the function $i(t)$ that satisfies the following differential equation and the initial condition:

$$500\frac{di(t)}{dt} + 25\,\text{k}\,i(t) = 0, \; i(0) = 15\,\text{mA}$$

7–2 Find the function $v(t)$ that satisfies the following differential equation and initial condition:

$$10^{-3}\frac{dv(t)}{dt} + v(t) = 0, \; v(0) = 50\,\text{V}$$

7–3 Find the time constants of the circuits in Figure P7–3.

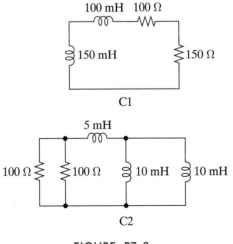

FIGURE P7–3

7–4 Find the time constants of the circuits in Figure P7–4.

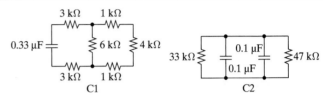

FIGURE P7–4

7–5 Each of the two circuits in Figure P7–5 have a switch that affects their time constants. For circuit C1 find the time constant when the switch is in position A and repeat for position B. For circuit C2 find the time constant when the switch is closed and repeat when it is open.

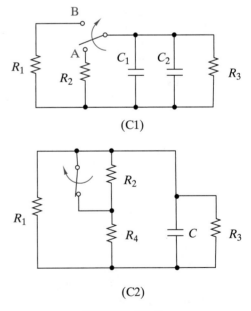

FIGURE P7–5

7–6 The switch in Figure P7–6 is closed at $t = 0$. The initial voltage on the capacitor is $v_C(0) = 250$ V.
(a) Find $v_C(t)$ and $i_O(t)$ for $t \geq 0$.
(b) Use MATLAB to plot the waveforms for $v_C(t)$ and $i_O(t)$. Simulate the problem using OrCAD and compare the results to the plots in part (b).

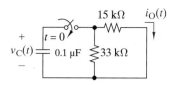

FIGURE P7–6

7–7 In Figure P7–7 the initial current through the inductor is $i_L(0) = 3$ mA. Find $i_L(t)$ and $v_O(t)$ for $t \geq 0$.

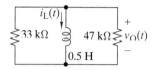

FIGURE P7–7

7–8 The switch in Figure P7–8 has been in position A for a long time and is moved to position B at $t = 0$. Find $i_L(t)$ for $t \geq 0$.

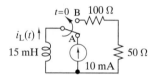

FIGURE P7–8

7–9 The circuit was in the zero state when the source in Figure P7–9 is applied. Find $v_C(t)$ for $t \geq 0$.

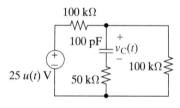

FIGURE P7–9

7–10 The switch in Figure P7–10 has been in position A for a long time and is moved to position B at $t = 0$. Find $v_C(t)$ for $t \geq 0$.

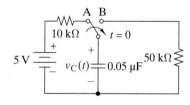

FIGURE P7–10

7–11 The switch in Figure P7–11 has been open for a long time and is closed at $t = 0$. Find $i_L(t)$ for $t \geq 0$.

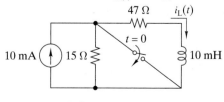

FIGURE P7–11

7–12 The circuit in Figure P7–12 is in the zero state. Find the voltage $v_O(t)$ for $t \geq 0$ when an input of $i_S(t) = I_A u(t)$ applied. Identify the forced and natural components in the output.

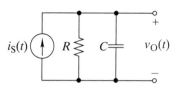

FIGURE P7–12

7–13 The circuit in Figure P7–13 is in the zero state when the input $v_S(t) = v_A u(t)$ is applied. Find $v_O(t)$ for $t \geq 0$. Identify the forced and natural components in the output.

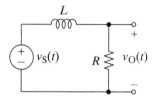

FIGURE P7–13

7–14 The circuit in Figure P7–14 is in the zero state when the input $v_S(t) = 100 \, u(t)$ is applied. If $C = 0.01 \, \mu F$ and $R = 100 \, k\Omega$, find $v_O(t)$ for $t \geq 0$. Identify the forced and natural components in the output.

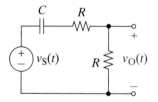

FIGURE P7–14

7–15 The circuit in Figure P7–15 is in the zero state when the input $v_S(t) = 20 \, u(t)$ is applied. If $L = 100 \, mH$ and $R = 1 \, k\Omega$, find $v_O(t)$ for $t \geq 0$. Identify the forced and natural components in the output. On a single set of axes, use MATLAB to plot the forced response, the natural response, and the complete response.

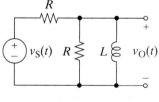

FIGURE P7–15

7–16 The switch in Figure P7–16 has been in position A for a long time and is moved to position B at $t = 0$. Find $v_C(t)$ for $t \geq 0$. Identify the forced and natural components in the response.

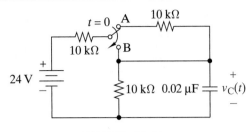

FIGURE P7–16

7–17 Repeat Problem 7–16. However, after the switch is moved to position B at $t = 0$, the switch is moved back to position A at $t = 100$ µs. Find $v_C(t)$ for $t \geq 0$.

7–18 Find the function that satisfies the following differential equation and the initial condition for an input $v_S(t) = 50 \cos (200t)$ V:

$$\frac{dv(t)}{dt} + 50v(t) = v_S(t), \quad v(0) = 0 \text{ V}$$

7–19 Repeat Problem 7–18 for $v_S(t) = 100e^{-0.02t}u(t)$ V.

7–20 The switch in Figure P7–20 has been open long enough for $i_L(0)$ to reach 0 A and is closed at $t = 0$.
 (a) If $v_S(t) = 10 \, u(t)$ V, find $v_L(t)$ for $t \geq 0$.
 (b) If $v_S(t) = 10 \cos(100t)$ V, find $v_L(t)$ for $t \geq 0$.
 (c) If $v_S(t) = 10 \, e^{-100t}$ V, find $v_L(t)$ for $t \geq 0$.

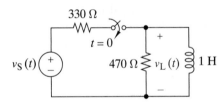

FIGURE P7–20

7–21 Repeat Problem 7–20 using OrCAD.

7–22 Repeat Problem 7–20 using MATLAB to plot the waveforms

7–23 The switch in Figure P7–23 has been in position A for a long time and is moved to position B at $t = 0$. Find $i_L(t)$ for $t \geq 0$.

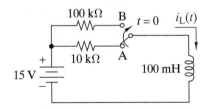

FIGURE P7–23

7–24 Repeat Problem 7–23 using MATLAB to plot the waveform.

7–25 The switch in Figure P7–25 has been in position A for a long time and is moved to position B at $t = 0$. The switch suddenly returns to position A after 10 ms. Find $v_C(t)$ for $t \geq 0$ and sketch its waveform.

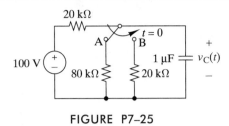

FIGURE P7–25

7–26 Use OrCAD to solve Problem 7–25. (*Hint:* OrCAD switches are located in the EVAL library and are named *Sw_tClose* and *Sw_tOpen*, depending on if you want the switch to close or open at a certain time.)

7–27 Switches 1 and 2 in Figure P7–27 have both been in position A for a long time. Switch 1 is moved to position B at $t = 0$ and Switch 2 is moved to position B at $t = 20$ ms. Find the voltage across the 0.1-µF capacitor for $t \geq 0$ and sketch its waveform.

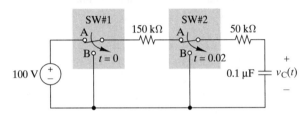

FIGURE P7–27

7–28 The switch in Figure P7–28 has been open for a long time and is closed at $t = 0$. The switch is reopened at $t = 3$ ms. Find $v_C(t)$ for $t \geq 0$.

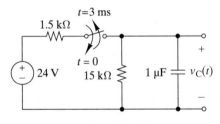

FIGURE P7–28

7–29 Find the sinusoidal steady-state response of $v_C(t)$ in Figure P7–29 when $R = 100 \, \text{k}\Omega$, $C = 0.01$ µF, and the input voltage is $v_S(t) = 10 \cos(100t) \, u(t)$ V. Repeat for an input voltage of $v_S(t) = 10 \cos(1000t) \, u(t)$ V, and one more time for an input voltage of $v_S(t) = 10 \cos(10kt) \, u(t)$ V. Describe how changing the frequency affects the output's amplitude and phase. You may choose to use MATLAB and plot the steady-state responses of each input on a single set of axes.

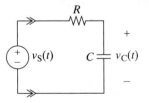

FIGURE P7–29

7–30 On the circuit of Figure P7–29 the input is $v_S(t) = 10\,e^{-1000t}$ $u(t)$ V. Find the output $v_C(t)$ when $R = 100\,\text{k}\Omega$, $C = 0.01\,\mu\text{F}$, and $v_C(0) = 0$ V.

OBJECTIVE 7–2 FIRST-ORDER CIRCUIT DESIGN (SECTS. 7–1, 7–2, 7–3, 7–4)

Given responses in a first-order RC or RL circuit:
(a) Find the circuit parameters or other responses.
(b) Design a circuit to produce the given response.
See Examples 7–8, 7–12, 7–13, 7–14 and Exercises 7–2, 7 8, 7–14, 7–22

7–31 For $t \geq 0$ the zero-input response of the circuit in Figure P7–31 is $v_C(t) = 20e^{-10kt}$ V.
(a) Find C and $i_C(t)$ when $R = 10\,\text{k}\Omega$.
(b) Find the energy stored in the capacitor at $t = 2$ ms.

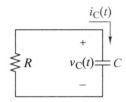

FIGURE P7–31

7–32 For $t \geq 0$ the zero-input response of the circuit in Figure P7–32 is $i_L(t) = 150e^{-500t}$ mA.
(a) Find L and $v_L(t)$ when $R = 500\,\Omega$.
(b) Find the energy stored in the inductor at $t = 0.5$ ms.

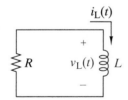

FIGURE P7–32

7–33 ◗ Design a series RC circuit using a dc voltage source that delivers the following voltage across the capacitor for $t > 0$.

$$v_C(t) = 100e^{-2000t}\ \text{V} \quad t \geq 0$$

7–34 ◗ Design a parallel RL circuit using a dc current source that delivers the following voltage across the resistor for $t > 0$.

$$v_R(t) = 100e^{-2000t}\ \text{V} \quad t \geq 0$$

7–35 ◗ Design a series RC circuit using a dc voltage source that delivers a voltage across the capacitor for $t > 0$ that fits entirely within the non-shaded region of Figure P7–35.

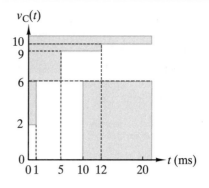

FIGURE P7–35

7–36 ◗ Design a series RC circuit using dc voltage sources that delivers a voltage across the capacitor for $t > 0$ that fits entirely within the non-shaded region of Figure P7–36.

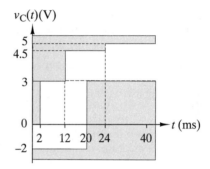

FIGURE P7–36

7–37 For $t \geq 0$ the step response of the voltage across the capacitor in Figure P7–37 is $v_C(t) = 5 - 10e^{-1000t}$ V. Find the IV, FV, T_C, R, and $i_C(t)$ when $C = 1.5\,\mu\text{F}$.

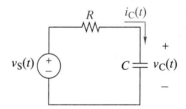

FIGURE P7–37

7–38 ◗ Design a first-order RC circuit using standard parts (see inside rear cover) that will produce the following voltage across the capacitor: $v_C(t) = 5 - 10e^{-10000t}$ V.

7–39 For $t \geq 0$ the step responses of the current through and voltage across the inductor in Figure P7–39 are $i_L(t) = 5 - 10e^{-2000t}$ mA and $v_L(t) = e^{-2000t}$ V. Find IV, FV, T_C, R, and L.

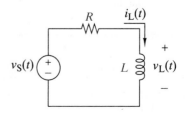

FIGURE P7–39

7–40 Design a first-order RL circuit that will produce the following current through the inductor: $i_L(t) = 50 + 100e^{-20000t}$ mA for $t \geq 0$.

7–41 The switch in Figure P7–41 has been in position B for a long time and is moved to position A at $t = 0$. Design the first-order RC interface circuit such that $v_O(t) = 15 - 15e^{-5000t}$ V.

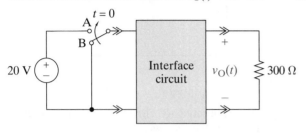

FIGURE P7–41

7–42 The switch in Figure P7–41 has been in position A for a long time and is moved to position B at $t = 0$. Design the first-order RC interface circuit such that $v_O(t) = 15e^{-5000t}$ V.

7–43 A timing circuit is required that feeds into an OP AMP's non-inverting terminal (i.e. draws no current.) The circuit's output response $v_O(t)$ must be

$$v_O(t) = 5(1 - e^{-1000t})u(t) \text{ V}$$

Figure P7–43 shows two commercial products and the vendors claim each will meet the requirement. Which will you select and why?

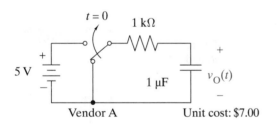

Vendor A   Unit cost: $7.00

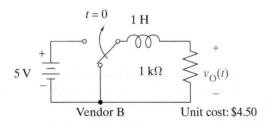

Vendor B   Unit cost: $4.50

FIGURE P7–43

7–44 A product line needs an RC circuit that will meet the following response specifications $\pm 5\%$:

IV	FV	T_C	R	C
−10 V	+5V	100 μs	≥1 kΩ	≤0.1 μF

Design a circuit to meet the specifications and validate your results using OrCAD.

OBJECTIVE 7–3 SECOND-ORDER CIRCUIT ANALYSIS (SECTS. 7–5, 7–6, 7–7)

Given a second-order circuit:
(a) Find the circuit differential equation.
(b) Find the characteristic equation and the initial conditions (if not given).
(c) Find the zero-input response.
(d) Find the complete response for a step function input.
See Examples 7–17, 7–18, 7–20, 7–21, 7–22, 7–23, 7–25 and Exercises 7–23, 7–24, 7–25, 7–27, 7–29, 7–30, 7–31, 7–33, 7–34.

7–45 Find the $v(t)$ that satisfies the following differential equation and initial conditions:

$$\frac{d^2 v(t)}{dt^2} + 5\frac{dv(t)}{dt} + 36v(t) = 0, \ v(0) = 0 \text{ V}, \ \frac{dv(0)}{dt} = 20 \text{ V/s}$$

7–46 Find the $v(t)$ that satisfies the following differential equation and initial conditions:

$$\frac{d^2 v(t)}{dt^2} + 10\frac{dv(t)}{dt} + 100v(t) = 0, \ v(0) = 5 \text{ V}, \ \frac{dv(0)}{dt} = 0 \text{ V/s}$$

7–47 Find the $v(t)$ that satisfies the following differential equation and initial conditions:

$$\frac{d^2 v(t)}{dt_2} + 10\frac{dv(t)}{dt} + 125v(t) = 250\,u(t), \ v(0) = 10\text{V}, \ \frac{dv(0)}{dt} = 25 \text{ V/s}$$

7–48 Find the $i(t)$ that satisfies the following differential equation and initial conditions:

$$\frac{d^2 i(t)}{dt^2} + 8\frac{di(t)}{dt} + 16i(t) = 48u(t), \ i(0) = 0, \ \frac{di(0)}{dt} = 0$$

7–49 The switch in Figure P7–49 has been open for a long time and is closed at $t = 0$. The circuit parameters are $L = 1$ H, $C = 0.5$ μF, $R = 1$ kΩ, and $v_C(0) = 10$ V.
(a) Find $v_C(t)$ and $i_L(t)$ for $t \geq 0$.
(b) Is the circuit overdamped, critically damped, or underdamped?
(c) Use OrCAD to simulate your results.

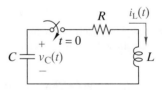

FIGURE P7–49

7–50 The switch in Figure P7–50 has been open for a long time and is closed at $t = 0$. The circuit parameters are $L = 1$ H, $C = 1$ μF, $R = 500$ Ω, and $v_C(0) = 60$ V.
(a) Find $v_C(t)$ and $i_L(t)$ for $t \geq 0$.
(b) Is the circuit overdamped, critically damped, or underdamped?
(c) Use OrCAD to simulate your results.

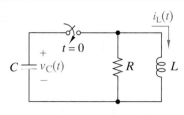

FIGURE P7–50

7–51 Use OrCAD to study how the voltage across the circuit in Figure P7–50 changes as the value of the resistor is varied. Use the Parameter function (PARAM) found in the Special Library and do a Transient Analysis for 10 ms. Select Parametric Sweep and then select Global Parameter. Use the same name as for the Parametric function and vary the resistor from 10 Ω to 10 kΩ using the logarithmic choice and 3 points per decade. This will generate a family of curves corresponding to different values of resistance. By clicking on the various curves, determine if the response becomes more or less damped as R varies from 10 Ω to 10 kΩ.

7–52 The switch in Figure P7–52 has been open for a long time and is closed at $t = 0$. The circuit parameters are $L = 0.8$ H, $C = 1.25$ μF, $R_1 = 2$ kΩ, $R_2 = 1$ kΩ, and $V_A = 12$ V.
(a) Find $v_C(t)$ and $i_L(t)$ for $t \geq 0$.
(b) Is the circuit overdamped, critically damped, or underdamped?
(c) Use OrCAD to simulate your results.

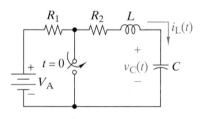

FIGURE P7–52

7–53 The switch in Figure P7–53 has been open for a long time and is closed at $t = 0$. The circuit parameters are $L = 1.25$ H, $C = 0.05$ μF, $R_1 = 20$ kΩ, $R_2 = 20$ kΩ, and $V_A = 20$ V.
(a) Find $v_C(t)$ and $i_L(t)$ for $t \geq 0$.
(b) Is the circuit overdamped, critically damped, or underdamped?
(c) Use OrCAD to simulate your results.

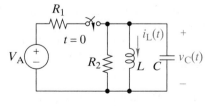

FIGURE P7–53

7–54 Repeat Problem 7–53 with $R_1 = 4$ kΩ, $R_2 = 4$ kΩ.

7–55 The switch in Figure P7–55 has been in position A for a long time. At $t = 0$ it is moved to position B. The circuit

parameters are $R_1 = 20$ kΩ, $R_2 = 4$ kΩ, $L = 1.6$ H, $C = 1.25$ μF, and $V_A = 24$ V.
(a) Find $v_C(t)$ and $i_L(t)$ for $t \geq 0$.
(b) Is the circuit overdamped, critically damped, or underdamped?
(c) Use OrCAD to simulate your results.

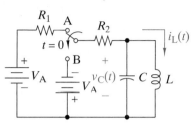

FIGURE P7–55

7–56 You have a need for an interface circuit that will connect your source to a load with a very high input as shown in Figure P7–56(a). Your interface must have a response that fits within the boundaries shown in Figure P7–56(b). A vendor offers a suitable circuit shown in Figure P7–56(a) and says that they are willing to change one component without additional cost.

Would you purchase the circuit and, if you will buy it, what change, if any, would you require? Use MATLAB or OrCAD to verify your result.

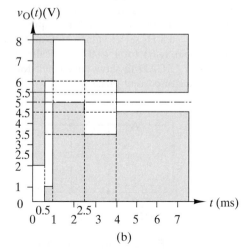

(a)

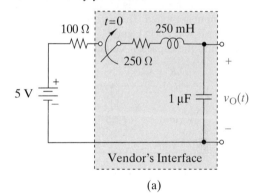

(b)

FIGURE P7–56

7–57 The switch in Figure P7–57 has been in position A for a long time and is moved to position B at $t=0$. The circuit parameters are $R_1 = 500\,\Omega$, $R_2 = 220\,\Omega$, $L = 250\,\text{mH}$, $C = 3.3\,\mu\text{F}$, and $V_A = 15\text{V}$.
(a) Find $v_C(t)$ and $i_L(t)$ for $t \geq 0$.
(b) Is the circuit overdamped, critically damped, or underdamped?
(c) Use OrCAD to simulate your results.

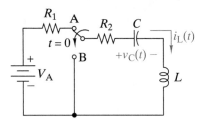

FIGURE P7–57

7–58 The switch in Figure P7–57 has been in position B for a long time and is moved to position A at $t = 0$. The circuit parameters are $R_1 = 500\,\Omega$, $R_2 = 500\,\Omega$, $L = 250\,\text{mH}$, $C = 1\,\mu\text{F}$, and $V_A = 5\text{V}$.
(a) Find $v_C(t)$ and $i_L(t)$ for $t \geq 0$.
(b) Is the circuit overdamped, critically damped, or underdamped?
(c) Use OrCAD to simulate your results.

7–59 The circuit in Figure P7–59 is in the zero state when the step function input is applied. The circuit parameters are $L = 250\,\text{mH}$, $C = 1\,\mu\text{F}$, $R = 3.3\,\text{k}\Omega$, and $V_A = 10\text{V}$. Find $v_O(t)$ for $t \geq 0$. (*Hint.* Find the capacitor voltage first.)

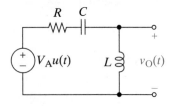

FIGURE P7–59

7–60 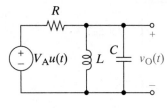 The circuit in Figure P7–60 is in the zero state when the step function input is applied. If the input source is $V_A = 70\text{V}$ and $L = 0.5\text{H}$, select values of R and C so that the circuit's output $v_O(t)$ for $t \geq 0$ is critically damped. Use MATLAB or OrCAD to show your result for $v_O(t)$ (*Hint:* Find the inductor current first.)

FIGURE P7–60

7–61 The circuit in Figure P7–61 is in the zero state when the step function input is applied.

(a) If $V_A = 24\text{ V}$, $R = 1.5\,\text{k}\Omega$, $L = 250\,\text{mH}$, and $C = 0.25\,\mu\text{F}$, derive an expression for the voltage $v_O(t)$ for $t \geq 0$.
(b) Validate your solution by plotting it using MATLAB and comparing it to an OrCAD simulation of the same circuit.

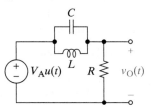

FIGURE P7–61

7–62 Derive expressions for the damping ratio and undamped natural frequency of the circuit in Figure P7–62 in terms of the circuit parameters R, L, and C. Which parameter(s) affect the damping ratio? Can you change the damping ratio without affecting the undamped natural frequency?

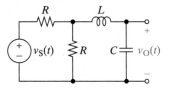

FIGURE P7–62

7–63 Derive expressions for the damping ratio and undamped natural frequency of the circuit in Figure P7–63 in terms of the circuit parameters R, L, and C. Which parameter(s) affect the damping ratio? Can you change the damping ratio without affecting the undamped natural frequency?

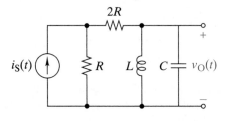

FIGURE P7–63

OBJECTIVE 7–4 SECOND-ORDER CIRCUIT DESIGN (SECTS. 7–5, 7–6, 7–7)

Given one or more responses of a second-order RLC circuit,
(a) Find the circuit parameters or other responses.
(b) Design a circuit to produce the given responses.
See Examples 7–19, 7–24, 7–26 and Exercises 7–23, 7–24, 7–26, 7–28, 7–32, 7–35.

7–64 In a series RLC circuit the step response across the 1-μF capacitor is

$$v_C(t) = 15 - e^{-200t}[15\cos(1000t) + 3\sin(1000t)]\text{ V}\quad t \geq 0$$

(a) Find R and L.
(b) Find $i_L(t)$ for $t \geq 0$.

7–65 In a parallel RLC circuit the zero-input response in the 220-mH inductor is

$$i_L(t) = 50e^{-6000t} - 40e^{-3000t} \text{ mA} \quad t \geq 0$$

(a) Find R and C.
(b) Find $v_C(t)$ for $t \geq 0$.

7–66 In a parallel RLC circuit the state variable responses are
$v_C(t) = e^{-100t} [5 \cos(300t) + 15 \sin(300t)] \text{ V} \quad t \geq 0$
$i_L(t) = 20 - 25e^{-100t} \cos(300t) \text{ mA} \, t \geq 0$
Find R, L, and C.

7–67 The zero-input response of a series RLC circuit with $R = 80 \, \Omega$ is

$$v_C(t) = 2e^{-2000t}\cos(1000t) - 4e^{-2000t}\sin(1000t) \text{ V} \quad t \geq 0$$

If the initial conditions remain the same, what is the zero-input response when $R = 40 \, \Omega$?

7–68 In a parallel RLC circuit the inductor current is observed to be

$$i_L(t) = 20e^{-20\,t}\sin(20t) \text{ mA} \quad t \geq 0$$

Find $v_C(t)$ when $v_C(0) = 0.5$ V.

7–69 Design a parallel RLC circuit whose natural response has the form

$$v_L(t) = K_1 e^{-10000t} + K_2 t\, e^{-10000t} \text{ V} \quad t \geq 0$$

7–70 Design a series RLC circuit with $\zeta = 0.1$ and $\omega_0 = 50$ krad/s.
(a) What is the form of the natural response of $v_C(t)$ for your design?
(b) Simulate your circuit in OrCAD.

7–71 Design a series RLC circuit with $\zeta = 1$ and $\omega_0 = 100$ krad/s.
(a) What is the form of the natural response of $v_C(t)$ for your design?
(b) Simulate your circuit in OrCAD.

7–72 Design a series RLC circuit whose output voltage resides entirely within the non-shaded region of Figure P7–72. Validate your design using MATLAB or OrCAD.

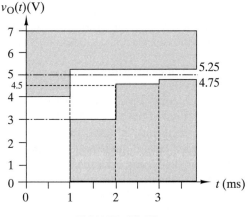

FIGURE P7–72

7–73 A circuit is needed to produce the following step response

$$v_C(t) = 10 - 13.3e^{-200t} + 3.3e^{-800t} \text{ V} \quad t > 0$$

A vendor has proposed using the circuit shown in Figure P7–73 to produce the desired response. The vendor realizes that the proposed circuit does not exactly meet the desired response and is willing to make a single change for no extra charge. What change should the vendor make? (*Hint:* Use MATLAB to generate the desired response and then simulate the vendor's corrected circuit using OrCAD and verify that the responses match.)

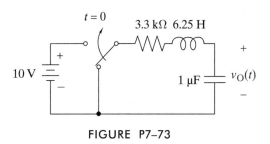

FIGURE P7–73

7–74 What range of damping ratios is available in the circuit in Figure P7–74?

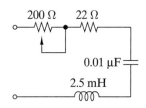

FIGURE P7–74

7–75 A variable capacitor is used in the circuit of P7–75 to vary the damping ratio. What range of damping ratios is available in the circuit?

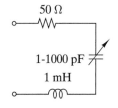

FIGURE P7–75

7–76 A particular parallel RLC circuit has the step response observed on an oscilloscope and shown in Figure P7–76. Four points on the waveform were measured and are shown. Determine the circuit's initial value, final value, the dominant exponential's time constant, and the likely case (A, B, or C) of the circuit response.

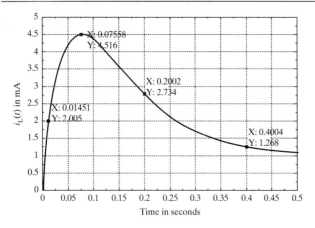

FIGURE P7–76

INTEGRATING PROBLEMS

7–77 First-order OP AMP Circuit Step Response A
Find the zero-state response of the OP AMP output voltage in Figure P7–77 when the input is $v_S(t) = V_A u(t)$ V.

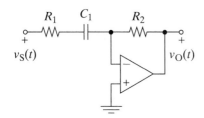

FIGURE P7–77

7–78 Intermittent Timing Circuit for Windshield Wipers D
A car maker needs an RC timing circuit to trigger the windshield wiper relay. The circuit should be driver selectable to trigger at 1, 2, 5, and 10 s ±5%. The source circuit is the car voltage of 12 V with a series resistance of 10 Ω. You must use standard parts (see inside rear cover). You have a single-sided OP AMP available with a V_{CC} of 0V and +12 V that you can use to trigger the windshield wiper relay. The relay triggers with an input of +12 V±1 V. You may assume that the circuit resets after the trigger. You should validate your design using OrCAD.

7–79 RC Circuit Design D
Design the first-order RC circuit in Figure P7–79 so an input $v_S(t) = 15\ u(t)$ V produces a zero-state response $v_O(t) = 15 - 5e^{-1000t}$ V. Validate your design using MATLAB or OrCAD.

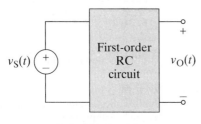

FIGURE P7–79

7–80 Sample Hold Circuit D
Figure P7–80 is a simplified diagram of a sample/hold circuit. When the switch is in position A, the circuit is in the sample mode and the capacitor voltage must charge to at least 99% of the source voltage V_A in less than 1 μs. When the switch is moved to position B, the circuit is in the hold mode and the capacitor must retain at least 99% of V_A for at least 1 ms. Select a capacitance that meets these constraints.

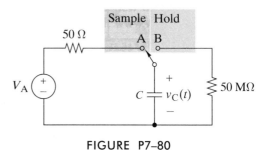

FIGURE P7–80

7–81 Super Capacitor A
Super capacitors have very large capacitance (typically from 0.1 to 50 F), small sizes, and very long charge holding times, making them useful in non-battery backup power applications. The charge holding quality of a super capacitor is measured using the circuit in Figure P7–81. The switch is closed for a long time (say, 24 hours) and the capacitor charged to 5 V. The switch is then opened and the capacitor allowed to self-discharge through any leakage resistance for 24 hours. Suppose that after 24 hours the voltage across a 0.47 F super capacitor is 4.5 V. What is the equivalent leakage resistance in parallel with the capacitor?

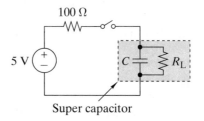

FIGURE P7–81

7–82 Cost-Conscious RLC Circuit Design D
You are assigned a task to design a series, passive RLC circuit with a characteristic equation of $s^2 + 2000s + 5 \times 10^6 = 0$. To save money, your supervisor wants you to use a previously purchased 150 mH inductor with a 10-Ω parasitic resistance. The RLC circuit will be used to interface to a Thévenin source with a 75-Ω series output resistance. Your circuit must demonstrate the desired response with the source circuit connected.

7–83 Combined First- and Second-Order Response A
The switch in Figure P7–83 has been in position A for a long time and is moved to position B at $t = 0$ and then to position C when $t = 10$ ms. For $0 < t < 10$ ms the capacitor voltage is a charging exponential $v_C(t) = 10(1 - e^{-100t})$ V. For $t > 10$ ms the capacitor voltage is a sinusoid $v_C(t) = 6.321 \cos [1000(t - 0.01)]$ V.

(a) Suppose the resistance is reduced to 1 kΩ and the switching sequence repeated.

Will the amplitude of the sinusoid increase, decrease, or stay the same?

Will the frequency of the sinusoid increase, decrease or stay the same?

(b) Suppose the inductance is reduced to 100 mH and the switching sequence repeated.

Will the amplitude of the sinusoid increase, decrease, or stay the same?

Will the frequency of the sinusoid increase, decrease or stay the same?

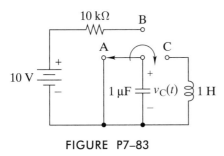

FIGURE P7–83

7–84 Undesired Ringing **A**, **D**

A digital clock has become corrupted by a ringing (undesired oscillations) as shown in Figure P7–84(a). The unwanted oscillations can cause false triggers and must be reduced. The clock can be modeled as an *RLC* series circuit as shown in Figure P7–84(b) with the voltage taken at node A. The parasitic capacitance is estimated at 0.01 μF and the Thévenin resistance at 330 Ω. From the graph determine the inductance *L*. Design an interface circuit that significantly reduces the ringing without significantly reducing the rise time (the time it takes the pulse to go from low to high or vice-versa.) The transition must occur in less than 80 us and the "overshoot" (deviation from 0 or 5 V) must be less than ±1 V. From the graph determine *L*. Then design a suitable interface to meet the output specifications. Use standard value components. Use OrCAD to validate your design.

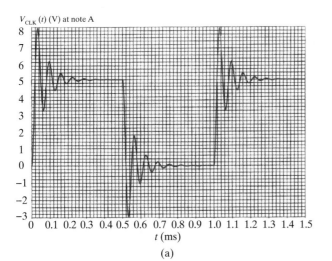

(a)

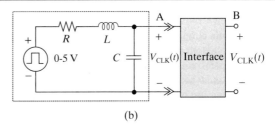

(b)

FIGURE P7–84

7–85 Optimum Fusing **E**

A sensitive instrument that can be modeled by the series *RLC* circuit shown in Figure P7–85 is to be protected by a fuse. The voltage across the capacitor is

$$v_C(t) = e^{-10t}(-\cos 103t + 0.0971\sin 103t)\, u(t) \text{ V}$$

The peak current was found to occur at about 13 ms after $t = 0$. Engineer A suggests a 1-mA fuse, Engineer B suggests a 5-mA fuse while Engineer C suggests a 10-mA fuse. The criterion is to select a fuse that is closest to the peak value of current expected without blowing. Which engineer is correct? Provide evidence for your decision.

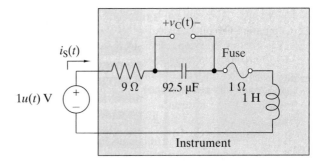

FIGURE P7–85

7–86 Lightning Pulser Design **D**

The circuit in Figure P7–86 is a simplified diagram of a pulser that delivers simulated lightning transients to the test article at the output interface. Closing the switch must produce a short-circuit current of the form $i_{SC}(t) = I_A e^{-\alpha t}\cos(\beta t)$, with $\alpha = 100$ krad/s, $\beta = 200$ krad/s, and $I_A = 2$ kA. Select the values of *L*, *C*, and V_0.

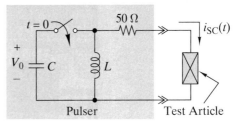

FIGURE P7–86

7–87 *RLC* Circuit Design **E**

Losses in real inductors can be modeled by a series resistor as shown in Figure P7–87. In this problem, we include the effect of this resistor on the design of the series *RLC* circuit shown in the

figure. The design requirements include a source resistance of 50 Ω, an undamped natural frequency of 50 kHz, and a damping ratio less than 0.1.
The characteristics of the available inductors are listed below.

L(mH)	$R_1(\Omega)$	L(mH)	$R_1(\Omega.)$
10.0	651	4.7	240
7.5	471	3.9	190
6.8	356	3.3	161

Which inductor would you use in your design and why?

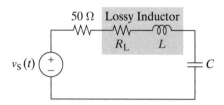

FIGURE P7–87

7–88 Competing Circuit Designs
Figure P7–88 shows the step responses $v_C(t)$ of two competing series *RLC* circuits from two different vendors. The circuits are designed to switch from 0 to 10 V and to meet a specification for a desired circuit with the following characteristic equation:

$$s^2 + 10s + 100 = 0$$

Not all of the details of the actual circuits have been provided because of proprietary reasons, but each vendor has given some information about their respective circuit. Both claim they meet the required specifications.

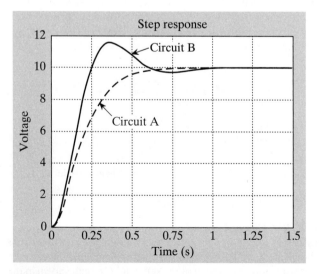

FIGURE 7–88

VENDOR A	VENDOR B
$R_T = 100$ Ω	$R/L = 10$ rad/s
$1/LC = 100$ (rad/s)^2	$\omega_0 = 10$ rad/s
$C = 2$mF	

Which vendor would you choose and why?

7–89 Solving Differential Equations with MATLAB
MATLAB has a built-in function for solving ordinary differential equations called `dsolve`. We can use this function to quickly explore the solution to a second-order differential equation when the forcing function is a sinusoidal or exponential signal. Suppose we have a series *RLC* circuit in the zero state connected to a voltage source $v_T(t)$. The parameter values are $R = 4$ kΩ, L = 1 H, and C = 1 µF. The differential equation for the voltage across the capacitor is given by Eq. (7–33). If $v_T(t) = 10\ u(t)$ V, we can use the following MATLAB code to solve for the capacitor voltage and plot the results

```
% Define the symbolic time variable
syms t real
% Define the parameter values
R = 4000;
L = 1;
C = 1e-6;
% Solve the differential equation for the
  series RLC circuit with zero
% initial conditions
vCt = dsolve ('L*C*D2v + R*C*Dv + v = 10',
   'v(0) =0 ',' Dv(0) =0 ', ' t');
% Create a time vector for plotting and sub-
  stitute in numerical values
tt = 0:0.0001:0.04;
vCt = subs(vCt);
vCtt = subs(vCt,t,tt);
% Plot the results
figure
plot(tt,vCtt,'b','LineWidth',2)
grid on
xlabel('Time, (s)')
ylabel('v_C(t), (V)')
title('Problem 7-89')
```

Run the given MATLAB code and examine the results. Modify the code to solve the same problem when the input voltage is $v_T(t) = 10 \cos(200 \pi t)$ V. Solve the problem a third time for $v_T(t) = 10\ e^{-2000t}$ V. Compare and comment on the responses for the three different types of inputs signals.

CHAPTER 8 SINUSOIDAL STEADY-STATE RESPONSE

The vector diagram of sine waves gives the best insight into the mutual relationships of alternating currents and emf's.

Charles P. Steinmetz, 1893,
American Engineer

Some History Behind This Chapter

The vector description of sinusoids was first discussed in detail by Charles Steinmetz (1865–1923) at the International Electric Congress of 1893. Although Oliver Heaviside may have used the vector concept earlier, Steinmetz is credited with popularizing the approach by demonstrating its many applications. By the turn of the century the vector method was well established in engineering practice and education. In the 1950s Steinmetz's vectors came to be called phasors to avoid possible confusion with the space vectors used to describe electromagnetic fields.

Why This Chapter Is Important Today

In this chapter we learn how to analyze ac circuits driven by a single-frequency sinusoid. We do this using a neat technique called phasor analysis that allows us to deal with ac circuits using the same tools we used on dc circuits. Phasor analysis is the key to understanding the electrical power systems that supply our homes and businesses. Using complex numbers is the price we pay for the simplicity of phasor analysis. Yet complex numbers are easy to master; after all, you were first introduced to them in high school. Fortunately, very complex (pun intended) problems can be solved using software tools.

Chapter Sections

Chapter Learning Objectives

8-1 Sinusoids and Phasors (Sect. 8–1)

Use the additive and derivative properties of phasors to convert sinusoids into phasors and vice versa.

8-2 Impedance (Sects. 8–2, 8–3)

Given a linear circuit in the sinusoidal steady state:
(a) Convert R, L, and C elements into impedances in the phasor domain.
(b) Use series and parallel equivalence to find the equivalent impedance at a specified pair of terminals.

8-3 Basic Phasor Circuit Analysis and Design (Sects. 8–3, 8–4)

(a) Given a linear circuit in the sinusoidal steady state, find phasor responses using equivalent circuits, circuit reduction, Thévenin or Norton equivalent circuits, proportionality, or superposition.
(b) Given a desired phasor response and a sinusoidal input, design a circuit in the sinusoidal steady state that produces the desired response.

8-4 General Circuit Analysis (Sect. 8–5)

Given a linear circuit in the sinusoidal steady state, find equivalent impedances and phasor responses using node-voltage or mesh-current analysis.

8-5 Average Power and Maximum Power Transfer (Sect. 8–6)

Given a linear circuit in the sinusoidal steady state:
(a) Find the average power delivered at a specified interface.
(b) Find the maximum average power available at a specified interface.
(c) Find the load impedance required to draw the maximum available power.

8-1 SINUSOIDS AND PHASORS

The phasor concept is the foundation for the analysis of linear circuits in the sinusoidal steady state. Simply put, a **phasor** is a complex number representing the amplitude and phase angle of a sinusoidal voltage or current. The connection between sinewaves and complex numbers is provided by Euler's relationship:

$$e^{j\theta} = \cos\theta + j\sin\theta \qquad (8\text{–}1)$$

Equation (8–1) relates the sine and cosine functions to the complex exponential $e^{j\theta}$. To develop the phasor concept, it is necessary to adopt the point of view that the cosine and sine functions can be written in the form

$$\cos\theta = \text{Re}\{e^{j\theta}\} \qquad (8\text{–}2)$$

and

$$\sin\theta = \text{Im}\{e^{j\theta}\} \qquad (8\text{–}3)$$

where Re stands for the "real part of" and Im for the "imaginary part of." Development of the phasor concept can begin with either Eq. (8–2) or (8–3). The choice between the two involves deciding whether to describe the sinewave using a sine or cosine function. In Chapter 5 we chose the cosine, so we will reference phasors to the cosine function.

When Eq. (8–2) is applied to the general sinusoid defined in Chapter 5, we obtain

$$v(t) = V_A \cos(\omega t + \phi)$$
$$= V_A \text{Re}\{e^{j(\omega t+\phi)}\} = V_A \text{Re}\{e^{j\omega t}e^{j\phi}\} \qquad (8\text{–}4)$$
$$= \text{Re}\{(V_A e^{j\phi})e^{j\omega t}\} = \text{Re}\{\mathbf{V}e^{j\omega t}\}$$

In the last line of Eq. (8–4), moving the amplitude V_A inside the real part operation does not change the final result because it is a real constant.

By definition, the quantity $V_A e^{j\phi}$ in the last line of Eq. (8–4) is the **phasor representation** of the sinusoid $v(t)$. The phasor **V** is written as

$$\mathbf{V} = \underbrace{V_A e^{j\phi}}_{\text{Polar}} = \underbrace{V_A \cos\phi + jV_A \sin\phi}_{\text{Rectangular}} \qquad (8\text{–}5)$$

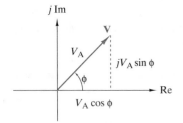

FIGURE 8–1 *Phasor diagram.*

Note that **V** is a complex number determined by the amplitude and phase angle of the sinusoid. Figure 8–1 shows a graphical representation commonly called a phasor diagram.

As shown in Eq. (8–5), a phasor is a complex number that can be written in either polar or rectangular form. An alternative way to write the polar form is to replace the exponential $e^{j\phi}$ by the shorthand notation $\angle\phi$. In subsequent discussions, we will often express phasors as $\mathbf{V} = V_A \angle\phi$, which is equivalent to the polar form in Eq. (8–5). When written this way, one states that the phasor has an amplitude of V_A at an angle of ϕ.

Two features of the phasor concept need emphasis:

1. Phasors are written in boldface type like **V** or $\mathbf{I}_1$ to distinguish them from signal waveforms such as $v(t)$ and $i_1(t)$.

2. A phasor is determined by amplitude and phase angle and does not contain any information about the frequency of the sinusoid.

The first feature points out that signals can be described in different ways. Although the phasor **V** and waveform $v(t)$ are related concepts, they have different physical interpretations and our notation must clearly distinguish between them. The absence of frequency information in the phasors results from the fact that in the sinusoidal steady state, all currents and voltages are sinusoids with the same frequency. Carrying frequency information in the phasor would be redundant, since it is the same for all phasors in any given steady-state circuit problem.

In summary, given a sinusoidal signal $v(t) = V_A \cos(\omega t + \phi)$, the corresponding phasor representation is $\mathbf{V} = V_A e^{j\phi} = V_A \angle \phi$. Conversely, given the phasor $\mathbf{V} = V_A e^{j\phi}$, the corresponding sinusoid is found by multiplying the phasor by $e^{j\omega t}$ and reversing the steps in Eq. (8–4) as follows:

$$
\begin{aligned}
v(t) = \mathrm{Re}\{\mathbf{V}e^{j\omega t}\} &= \mathrm{Re}\{(V_A e^{j\phi})e^{j\omega t}\} \\
&= V_A \mathrm{Re}\{e^{j(\omega t+\phi)}\} = V_A \mathrm{Re}\{\cos(\omega t + \phi) + j\sin(\omega t + \phi)\} \qquad (8\text{–}6) \\
&= V_A \cos(\omega t + \phi)
\end{aligned}
$$

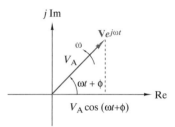

FIGURE 8–2 *Complex exponential* $\mathbf{V}e^{j\omega t}$.

The frequency ω in the complex exponential $\mathbf{V}e^{j\omega t}$ in Eq. (8–6) must be expressed or implied in a problem statement, since by definition it is not contained in the phasor. Figure 8–2 shows a geometric interpretation of the complex exponential $\mathbf{V}e^{j\omega t}$ as a vector in the complex plane of length V_A, which rotates counterclockwise with a constant angular velocity of ω, the radian frequency of the sinusoid. The real part operation projects the rotating vector onto the horizontal (real) axis and thereby generates $v(t) = V_A \cos(\omega t + \phi)$. The complex exponential is sometimes called a **rotating phasor**, and the phasor $\mathbf{V}$ is viewed as a snapshot of the situation at $t = 0$.

Exercise 8–1 _____

Convert the following sinusoids to phasors in polar and rectangular form:

(a) $v(t) = 20 \cos(150t - 60°)$ V
(b) $v(t) = 10 \cos(1000t + 180°)$ V
(c) $i(t) = -4 \cos 3t + 3 \cos(3t - 90°)$ A

Answers:

(a) $\mathbf{V} = 20 \angle -60° = 10 - j17.3$ V
(b) $\mathbf{V} = 10 \angle 180° = -10 + j0$ V
(c) $\mathbf{I} = 5 \angle -143° = -4 - j3$ A

Exercise 8–2 _____

Convert the following phasors to sinusoids:

(a) $\mathbf{V} = 169 \angle -45°$ V at $f = 60$ Hz
(b) $\mathbf{V} = 10 \angle 90° + 66 - j10$ V at $\omega = 10$ krad/s
(c) $\mathbf{I} = 15 + j5 + 10 \angle 180°$ mA at $\omega = 1000$ rad/s

Answers:

(a) $v(t) = 169 \cos(377t - 45°)$ V
(b) $v(t) = 66 \cos 10^4 t$ V
(c) $i(t) = 7.07 \cos(1000t + 45°)$ mA

PROPERTIES OF PHASORS

Two important properties of phasors play key roles in circuit analysis. First, the **additive property** states that the phasor representing a sum of sinusoids of the same frequency is obtained by adding the phasor representations of the component sinusoids. To establish this property we write the expression

$$
\begin{aligned}
v(t) &= v_1(t) + v_2(t) + \cdots + v_N(t) \\
&= \mathrm{Re}\{\mathbf{V}_1 e^{j\omega t}\} + \mathrm{Re}\{\mathbf{V}_2 e^{j\omega t}\} + \cdots + \mathrm{Re}\{\mathbf{V}_N e^{j\omega t}\}
\end{aligned}
\tag{8-7}
$$

where $v_1(t)$, $v_2(t)$, ..., $v_N(t)$ are sinusoids of the same frequency whose phasor representations are $\mathbf{V}_1, \mathbf{V}_2, \ldots, \mathbf{V}_N$. The real part operation is additive, so the sum of real parts equals the real part of the sum. Consequently, Eq. (8–7) can be written in the form

$$
\begin{aligned}
v(t) &= \mathrm{Re}\{\mathbf{V}_1 e^{j\omega t} + \mathbf{V}_2 e^{j\omega t} + \cdots + \mathbf{V}_N e^{j\omega t}\} \\
&= \mathrm{Re}\{(\mathbf{V}_1 + \mathbf{V}_2 + \cdots + \mathbf{V}_N)e^{j\omega t}\}
\end{aligned}
\tag{8-8}
$$

Comparing the last line in Eq. (8–8) with the definition of a phasor, we conclude that the phasor $\mathbf{V}$ representing $v(t)$ is

$$
\mathbf{V} = \mathbf{V}_1 + \mathbf{V}_2 + \cdots + \mathbf{V}_N
\tag{8-9}
$$

The result in Eq. (8–9) applies only if the component sinusoids all have the same frequency so that $e^{j\omega t}$ can be factored out as shown in the last line in Eq. (8–8).

In Chapter 5 we found that the time derivative of a sinusoid is another sinusoid with the same frequency. Since they have the same frequency, the signal and its derivative can be represented by phasors. The **derivative property** of phasors allows us easily to relate the phasor representing a sinusoid to the phasor representing its derivative.

Equation (8–6) relates a sinusoid function and its phasor representation as

$$
v(t) = \mathrm{Re}\{\mathbf{V}e^{j\omega t}\}
$$

Differentiating this equation with respect to time t yields

$$
\begin{aligned}
\frac{dv(t)}{dt} &= \frac{d}{dt}\,\mathrm{Re}\{\mathbf{V}e^{j\omega t}\} = \mathrm{Re}\left\{\mathbf{V}\frac{d}{dt}e^{j\omega t}\right\} \\
&= \mathrm{Re}\{(j\omega\mathbf{V})e^{j\omega t}\}
\end{aligned}
\tag{8-10}
$$

From the definition of a phasor, we see that the quantity $(j\omega\mathbf{V})$ on the right side of this equation is the phasor representation of the time derivative of the sinusoidal waveform. This phasor can be written in the form

$$
\begin{aligned}
j\omega\mathbf{V} &= (\omega e^{j90°})(V_A e^{j\theta}) \\
&= \omega V_A e^{j(\theta + 90°)}
\end{aligned}
\tag{8-11}
$$

which points out that differentiating a sinusoid changes its amplitude by a multiplicative factor ω and shifts the phase angle by 90°.

In summary, the additive property states that adding phasors is equivalent to adding sinusoids of the same frequency. The derivative property states that multiplying a phasor by $j\omega$ is equivalent to differentiating the corresponding sinusoid. The following examples show applications of these two properties of phasors.

EXAMPLE 8–1

(a) Construct the phasors for the following signals:

$$
\begin{aligned}
v_1(t) &= 10\cos(1000t - 45°)\,\mathrm{V} \\
v_2(t) &= 5\cos(1000t + 30°)\,\mathrm{V}
\end{aligned}
$$

(b) Use the additive property of phasors and the phasors found in (a) to find $v(t) = v_1(t) + v_2(t)$.

SOLUTION:
(a) The phasor representations of $v_1(t)$ and $v_2(t)$ are

$$\mathbf{V}_1 = 10e^{-j45°} = 10\cos(-45°) + j10\sin(-45°)$$
$$= 7.07 - j7.07$$
$$\mathbf{V}_2 = 5\,e^{+j30°} = 5\cos(30°) + j5\sin(30°)$$
$$= 4.33 + j2.5$$

(b) The two sinusoids have the same frequency, so the additive property of phasors can be used to obtain their sum:

$$\mathbf{V} = \mathbf{V}_1 + \mathbf{V}_2 = 11.4 - j4.57 = 12.3e^{-j21.8°}$$

The waveform corresponding to this phasor sum is

$$v(t) = \mathrm{Re}\{(12.3e^{-j21.8°})e^{j1000t}\}$$
$$= 12.3\cos(1000t - 21.8°)\,\mathrm{V}$$

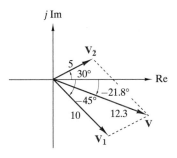

FIGURE 8–3

The phasor diagram in Figure 8–3 shows that summing sinusoids can be viewed geometrically in terms of phasors.

MATLAB can be a valuable tool in working with phasors. The following code is useful for this example:

```
% Create the phasor values
V1 = 10*exp(-j*45*pi/180)
V2 = 5*exp(j*30*pi/180)
% Form the sum
V = V1 + V2
% Compute the polar form
MagV = abs(V);
PhaseV = angle(V);
PhaseVDeg = 180*PhaseV/pi;
% Display the results
disp(['V =' ,num2str(MagV,'% 3.4g'),. . .
    exp(',num2str(PhaseVDeg,'% 3.4g'),'j)'])
```

Recall that the MATLAB complex exponential function operates on values in radians, not degrees. The code produces the following results, which agree with the solution shown above:

```
V1 =
    7.0711 - 7.0711i
V2 =
    4.3301 + 2.5000i
V =
    11.4012 - 4.5711i
V = 12.28 exp(-21.85j)
```

∎

Exercise 8–3

(a) Construct the phasors for the following signals:

$$i_1(t) = 100\cos(2000t)\,\mathrm{mA}$$
$$i_2(t) = 50\cos(2000t - 60°)\,\mathrm{mA}$$

(b) Use the additive property to find $i(t) = i_1(t) + i_2(t)$ and check the results using MATLAB.

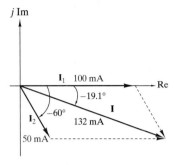

FIGURE 8–4

Answers:

(a) See Figure 8–4. $\mathbf{I}_1 = 100$, $\mathbf{I}_2 = 25 - j43.30$.

(b) $\mathbf{I} = 125 - j43.3 = 132\angle -19.1°$ mA, so $i(t) = 132\cos(2000t - 19.1°)$ mA

EXAMPLE 8–2

(a) Construct the phasors representing the following signals:

$$i_A(t) = 5\cos(377t + 50°)\,\text{A}$$
$$i_B(t) = 5\cos(377t + 170°)\,\text{A}$$
$$i_C(t) = 5\cos(377t - 70°)\,\text{A}$$

(b) Use the additive property of phasors and the phasors found in (a) to find the sum of these waveforms.

SOLUTION:

(a) The phasor representations of the three sinusoidal currents are

$$\mathbf{I}_A = 5e^{j50°} = 5\cos(50°) + j5\sin(50°) = 3.21 + j3.83\,\text{A}$$
$$\mathbf{I}_B = 5e^{j170°} = 5\cos(170°) + j5\sin(170°) = -4.92 + j0.87\,\text{A}$$
$$\mathbf{I}_C = 5e^{-j70} = 5\cos(-70°) + j5\sin(-70°) = 1.71 - j4.70\,\text{A}$$

(b) The currents have the same frequency, so the additive property of phasors applies. The phasor representing the sum of these currents is

$$\mathbf{I}_A + \mathbf{I}_B + \mathbf{I}_C = (3.21 - 4.92 + 1.71) + j(3.83 + 0.87 - 4.70)$$
$$= 0 + j0\,\text{A}$$

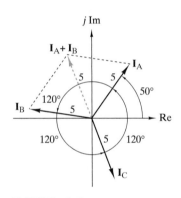

FIGURE 8–5

It is not obvious by examining the waveforms that these three currents add to zero. However, the phasor diagram in Figure 8–5 makes this fact clear, since the sum of any two phasors is equal and opposite to the third. Phasors of this type occur in balanced three-phase power systems. The balanced condition occurs when three equal-amplitude phasors are displaced in phase by exactly 120°. ■

Exercise 8–4

Show that the phasors $\mathbf{I}_A, \mathbf{I}_B$, and $\mathbf{I}_C$ would still sum to zero if they were all rotated 90° counterclockwise.

Answer:

$$\mathbf{I}_A + \mathbf{I}_B + \mathbf{I}_C = (-3.83 + j3.21) + (-0.87 - j4.92) + (4.70 + j1.71)$$
$$= 0 + j0\,\text{A}$$

EXAMPLE 8–3

Use the derivative property of phasors to find the time derivative of $v(t) = 15\cos(200t - 30°)$ V.

SOLUTION:

The phasor for the sinusoid is $\mathbf{V} = 15\angle-30°$. According to the derivative property, the phasor representing dv/dt is found by multiplying $\mathbf{V}$ by $j\omega$.

$$(j200) \times (15°\angle30°) = (200e^{j90°}) \times (15e^{-j30°}) = 3000e^{j60°}$$

The sinusoid corresponding to the phasor $j\omega\mathbf{V}$ is

$$\frac{dv(t)}{dt} = \text{Re}\{(3000e^{j60°})e^{j200t}\} = 3000\,\text{Re}\{e^{j(200t+60°)}\}$$

$$= 3000\cos(200t + 60°) \text{ V/s}$$

Finding the derivative of a sinusoid is easily carried out in phasor form, since it only involves manipulating complex numbers. ∎

Exercise 8–5

Find the phasor corresponding to the time derivative of the waveform:

$$v(t) = 100\cos(1000t) \text{ V}$$

Answer: $\qquad\qquad \mathbf{V} = 10^5\angle90°$ V/s

EXAMPLE 8–4

(a) Convert the following phasors into sinusoidal waveforms:

$$\mathbf{V}_1 = 20 + j20\,\text{V}, \omega = 500\,\text{rad/s}$$

$$\mathbf{V}_2 = 10\sqrt{2}\,e^{-j45°}\,\text{V}, \omega = 500\,\text{rad/s}$$

(b) Use phasor addition to find the sinusoidal waveform $v_3(t) = v_1(t) + v_2(t)$.

SOLUTION:

(a) Since $\mathbf{V}_1 = 20 + j20 = 20\sqrt{2}e^{j45°}$, the waveforms corresponding to the phasors $\mathbf{V}_1$ and $\mathbf{V}_2$ are

$$v_1(t) = \text{Re}\{20\sqrt{2}\,e^{j45°}\,e^{j500t}\} = 20\sqrt{2}\cos(500t + 45°)\text{ V}$$

$$v_2(t) = \text{Re}\{10\sqrt{2}\,e^{-j45°}\,e^{j500t}\} = 10\sqrt{2}\cos(500t - 45°)\text{ V}$$

(b) Since $\mathbf{V}_2 = 10\sqrt{2}e^{-j45°} = 10 - j10$, the additive property of phasors yields

$$\mathbf{V}_3 = \mathbf{V}_1 + \mathbf{V}_2 = 20 + j20 + 10 - j10$$

$$= 30 + j10 = 31.6\,e^{j18.4°}$$

Hence,

$$v_3(t) = \text{Re}\{31.6\,e^{j18.4°}\,e^{j500t}\} = 31.6\cos(500t + 18.4°)\text{ V}$$

The graphical solution is shown in Figure 8–6. ∎

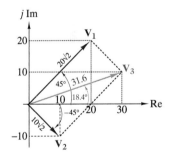

FIGURE 8–6

Exercise 8–6

Find the phasor corresponding to the waveform $v(t) = V_A\cos(\omega t) + 2V_A\sin(\omega t)$.

Answer: $\qquad\qquad \mathbf{V} = V_A - j2V_A = \sqrt{5}V_A\angle-63.4°\text{V}$

8–2 PHASOR CIRCUIT ANALYSIS

Phasor circuit analysis is a method of finding sinusoidal steady-state responses directly from the circuit without using differential equations. How do we perform phasor circuit analysis? At several points in our study we have seen that circuit analysis is based on two kinds of constraints: (1) connection constraints (Kirchhoff's laws) and (2) device constraints (element equations). To analyze phasor circuits, we must see how these constraints are expressed in phasor form.

CONNECTION CONSTRAINTS IN PHASOR FORM

The sinusoidal steady-state condition is reached after the circuit's natural response decays to zero. In the steady state all of the voltages and currents are sinusoids with the same frequency as the driving force. Under these conditions, the application of KVL around a loop could take the form

$$V_1 \cos(\omega t + \phi_1) + V_2 \cos(\omega t + \phi_2) + \cdots + V_N \cos(\omega t + \phi_N) = 0$$

These sinusoids have the same frequency but have different amplitudes and phase angles. The additive property of phasors discussed in the preceding section shows that there is a one-to-one correspondence between waveform sums and phasor sums. Therefore, if the sum of the waveforms is zero, then the corresponding phasors must also sum to zero.

$$\mathbf{V}_1 + \mathbf{V}_2 + \cdots + \mathbf{V}_N = 0$$

Clearly the same result applies to phasor currents and KCL. In other words, we can state Kirchhoff's laws in phasor form as follows:

KVL: The algebraic sum of phasor voltages around a loop is zero.

KCL: The algebraic sum of phasor currents at a node is zero.

DEVICE CONSTRAINTS IN PHASOR FORM

Turning now to the device constraints, we note that the i–v characteristics of the three passive elements are

$$\text{Resistor:} \quad v_\text{R}(t) = Ri_\text{R}(t)$$
$$\text{Inductor:} \quad v_\text{L}(t) = L\frac{di_\text{L}(t)}{dt} \tag{8–12}$$
$$\text{Capacitor:} \quad i_\text{C}(t) = C\frac{dv_\text{C}(t)}{dt}$$

In the sinusoidal steady state, all of these currents and voltages are sinusoids. Given that the signals are sinusoid, how do these i–v relationships constrain the corresponding phasors?

In the sinusoidal steady state, the voltage and current of the resistor can be written in terms of phasors as $v_\text{R}(t) = \text{Re}\{\mathbf{V}_\text{R}e^{j\omega t}\}$ and $i_\text{R}(t) = \text{Re}\{\mathbf{I}_\text{R}e^{j\omega t}\}$. Consequently, the resistor i–v relationship in Eq. (8–12) can be expressed in terms of phasors as follows:

$$\text{Re}\{\mathbf{V}_\text{R}e^{j\omega t}\} = R \times \text{Re}\{\mathbf{I}_\text{R}e^{j\omega t}\}$$

Since R is a real constant, moving it inside the real part operation on the right side of this equation does not change things:

$$\text{Re}\{\mathbf{V}_\text{R}e^{j\omega t}\} = \text{Re}\{R\mathbf{I}_\text{R}e^{j\omega t}\}$$

This relationship holds only if the phasor voltage and current for a resistor are related as

$$\mathbf{V}_\text{R} = R\mathbf{I}_\text{R} \tag{8–13}$$

To explore this relationship, we assume that the current through a resistor is $i_R(t) = I_A \cos(\omega t + \phi)$. Then the phasor current is $\mathbf{I}_R = I_A e^{j\phi}$, and according to Eq. (8–13), the phasor voltage across the resistor is

$$\mathbf{V}_R = RI_A e^{j\phi}$$

This result shows that the voltage has the same phase angle (ϕ) as the current. Phasors with the same phase angle are said to be **in phase**; otherwise they are said to be **out of phase**. Figure 8–7 shows the phasor diagram of the resistor current and voltage. Two scale factors are needed to construct a phasor diagram showing both voltage and current, since the two phasors do not have the same dimensions.

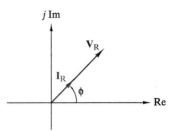

FIGURE 8–7 *Phasor i–v characteristics of the resistor.*

In the sinusoidal steady state, the voltage and current for the inductor can be written in terms of phasors as $v_L(t) = \mathrm{Re}\{\mathbf{V}_L e^{j\omega t}\}$ and $i_L(t) = \mathrm{Re}\{\mathbf{I}_L e^{j\omega t}\}$. Using the derivative property of phasors, the inductor $i–v$ relationship in Eq. (8–12) can be expressed as follows:

$$\mathrm{Re}\{\mathbf{V}_L e^{j\omega t}\} = L \times \mathrm{Re}\{j\omega \mathbf{I}_L e^{j\omega t}\}$$
$$= \mathrm{Re}\{j\omega L \mathbf{I}_L e^{j\omega t}\}$$

Since L is a real constant, moving it inside the real part operation does not change things. Written this way, we see that the phasor voltage and current for an inductor are related as

$$\mathbf{V}_L = j\omega L \mathbf{I}_L \qquad (8–14)$$

When the current is $i_L(t) = I_A \cos(\omega t + \phi)$, the corresponding phasor is $\mathbf{I}_L = I_A e^{j\phi}$ and the $i–v$ constraint in Eq. (8–14) yields

$$\mathbf{V}_L = j\omega L \mathbf{I}_L = (\omega L e^{j90°})(I_A e^{j\phi})$$
$$= \omega L I_A e^{j(\phi + 90°)}$$

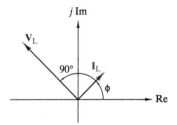

FIGURE 8–8 *Phasor i–v characteristics of the inductor.*

The resulting phasor diagram in Figure 8–8 shows that the inductor voltage and current are 90° out of phase. The voltage phasor is advanced by 90° counterclockwise, which is in the direction of rotation of the complex exponential $e^{j\omega t}$. When the voltage phasor is advanced counterclockwise (that is, ahead of the rotating current phasor), we say that the voltage phasor *leads* the current phasor by 90° or, equivalently, the current *lags* the voltage by 90°.

Finally, the capacitor voltage and current in the sinusoidal steady state can be written in terms of phasors as $v_C(t) = \mathrm{Re}\{\mathbf{V}_C e^{j\omega t}\}$ and $i_C(t) = \mathrm{Re}\{\mathbf{I}_C e^{j\omega t}\}$. Using the derivative property of phasors, the $i–v$ relationship of the capacitor in Eq. (8–12) becomes

$$\mathrm{Re}\{\mathbf{I}_C e^{j\omega t}\} = C \times \mathrm{Re}\{j\omega \mathbf{V}_C e^{j\omega t}\}$$
$$= \mathrm{Re}\{j\omega C \mathbf{V}_C e^{j\omega t}\}$$

Moving the real constant C inside the real part operation does not change the final result, so we conclude that the phasor voltage and current for a capacitor are related as

$$\mathbf{I}_C = j\omega C \mathbf{V}_C$$

Solving for $\mathbf{V}_C$ yields

$$\mathbf{V}_C = \frac{1}{j\omega C} \mathbf{I}_C \qquad (8–15)$$

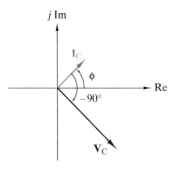

FIGURE 8–9 *Phasor i–v characteristics of the capacitor.*

When $i_C(t) = I_A \cos(\omega t + \phi)$, then according to Eq. (8–15), the phasor voltage across the capacitor is

$$\mathbf{V}_C = \frac{1}{j\omega C}\mathbf{I}_C = \left(\frac{1}{\omega C}e^{-j90°}\right)(I_A e^{j\phi})$$

$$= \frac{I_A}{\omega C}e^{j(\phi - 90°)}$$

The resulting phasor diagram in Figure 8–9 shows that voltage and current are 90° out of phase. In this case, the voltage phasor is retarded by 90° clockwise, which is in a direction opposite to rotation of the complex exponential $e^{j\omega t}$. When the voltage is retarded clockwise (that is, behind the rotating current phasor), we say that the voltage phasor *lags* the current phasor by 90° or, equivalently, the current *leads* the voltage by 90°.

THE IMPEDANCE CONCEPT

The phasor **I–V** constraints in Eqs. (8–13), (8–14), and (8–15) are all of the form

$$\mathbf{V} = Z\mathbf{I} \tag{8–16}$$

where Z is called the impedance of the element. Equation (8–16) is analogous to Ohm's law in resistive circuits. **Impedance** is the proportionality constant relating phasor voltage and phasor current in linear, two-terminal elements. The impedances of the three passive elements are

$$\text{Resistor:} \quad Z_R = R$$
$$\text{Inductor:} \quad Z_L = j\omega L \tag{8–17}$$
$$\text{Capacitor:} \quad Z_C = \frac{1}{j\omega C} = -\frac{j}{\omega C}$$

Since impedance relates phasor voltage to phasor current, it is a complex quantity whose units are ohms. Although impedance can be a complex number, it is not a phasor. Phasors represent sinusoidal signals, while impedances characterize circuit elements in the sinusoidal steady state. Finally, it is important to remember that the generalized two-terminal device constraint in Eq. (8–16) assumes that the passive sign convention is used to assign the reference marks to the voltage and current.

EFFECTS OF FREQUENCY ON IMPEDANCE

It is important to stop and reflect on what the concept of impedance means to us. The behavior of these three elements versus frequency is a fundamental concept of electrical engineering. Figure 8–10 shows a plot of the magnitude of the impedance $|Z|$ of each of the three circuit elements versus frequency ω.

FIGURE 8–10 *Magnitude of the impedance of passive circuit elements versus frequency.*

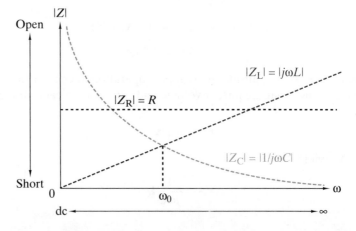

The magnitude of the impedance of the resistor $|Z_R|$ is unaffected by frequency. It is simply R, the resistor's resistance. It plots as a horizontal line on the figure. The magnitude of the impedance of an inductor is a function of frequency, $|Z_L| = \omega L$. At dc, when nothing is changing, an inductor behaves like a short circuit (i.e., zero ohms); we saw this in Chapter 6, where we discussed dc equivalent circuits and modeled an inductor when nothing was changing—that is, $\omega = 0$—as a short circuit. We made use of this substitution in Chapter 7 when we analyzed first-order RL circuits excited by a step function. However, as the frequency exciting the inductor begins to increase, the impedance of the inductor increases linearly with the frequency. The magnitude, then, appears as an increasing straight line on the graph. As the frequency approaches infinity, the nature of the inductor morphs into an open circuit of infinite ohms.

The impedance of a capacitor, however, exhibits behavior quite different from that of either a resistor or an inductor. The magnitude of the impedance of a capacitor is an inverse function of frequency, $|Z_C| = 1/\omega C$. At dc, when nothing is changing or $\omega = 0$, the capacitor behaves like an open circuit (i.e., $\infty\,\Omega$); again, we saw this in Chapter 6, where we discussed dc equivalent circuits and used it in Chapter 7 when analyzing RC circuits excited by a step function. However, as the frequency begins to increase, the impedance of the capacitor decreases inversely with the frequency. The magnitude, then, appears as a hyperbola on the graph. As the frequency approaches infinity, the nature of the capacitor completely changes into a short circuit of zero ohms. This opposing behavior of inductors and capacitors is what permits the realization of so many different circuits.

From the graph in Figure 8–10 we can see that there is a frequency, ω_0, where the magnitude of the impedance of the inductor equals that of the capacitor. This frequency is called the *resonant frequency* and it is an important property of circuits in the sinusoidal steady state that we can use to our advantage. As we progress in our study of circuits, we will learn how to use these divergent properties to analyze, design, and evaluate ac circuits.

EXAMPLE 8–5

The circuit in Figure 8–11(a) is operating in the sinusoidal steady state with $i(t) = 4\cos(5000t)$ A. Find the steady-state voltage $v(t)$.

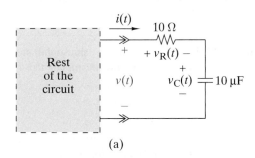

(a)

FIGURE 8–11

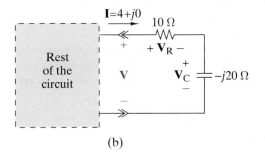

(b)

SOLUTION:

Our first step is to convert our circuit form the time-domain shown in Figure 8.11(a) into phasors. We proceed as follows. The resistor and the capacitor have impedances

$$ Z_R = R = 10 + j0 \quad \text{and} \quad Z_C = \frac{1}{j\omega C} = \frac{1}{j5000 \times 10^{-5}} = 0 - j20 $$

Applying KCL at node A shows that $i(t) = 4\cos(5000t)$ is the current through both the resistor and the capacitor. The corresponding phasor current is $\mathbf{I} = 4 + j0$. Using the element impedances, the phasor voltages across the resistor and the capacitor are

$$ \mathbf{V}_R = Z_R\mathbf{I} = 10 \times (4 + j0) = 40 + j0 $$
$$ \mathbf{V}_C = Z_C\mathbf{I} = (-j20) \times (4 + j0) = 0 - j80 $$

We can redraw our circuit using phasors as shown in Figure 8–11(b).

Applying KVL around the loop yields $\mathbf{V} = \mathbf{V}_R + \mathbf{V}_C$; hence,

$$ \mathbf{V} = 40 - j80 = 89.4\angle -63.4° \text{ V} $$

and the steady-state voltage waveform is

$$ v(t) = \text{Re}\{89.4\,e^{-j63.4°}\,e^{j5000t}\} $$
$$ = 89.4\cos(5000t - 63.4°)\text{V} $$ ■

Exercise 8–7

A series circuit is composed of a 1-kΩ resistor, a 1-μF capacitor, and a 100-mH inductor.

(a) At what frequency will the magnitude of the impedance of the inductor equal that of the resistor?

(b) At what frequency will the magnitude of the impedance of the capacitor equal that of the resistor?

(c) At what frequency will the magnitude of the impedance of the inductor equal the magnitude of the impedance of the capacitor? What is the frequency called?

Answers:

(a) $|Z_L| = |Z_R|$ when $\omega = 10\,\text{krad/s}$.

(b) $|Z_C| = |Z_R|$ when $\omega = 1\,\text{krad/s}$.

(c) $|Z_L| = |Z_C|$ when $\omega = 3.162\,\text{krad/s}$. The resonant frequency.

Exercise 8–8

The circuit in Figure 8–12 is operating in the sinusoidal steady state with $v(t) = 50\cos(500t)$ V and $i(t) = 4\cos(500t - 60°)$ A. Find the impedance of the elements in the box.

Answer:

$Z = 6.25 + j10.8\,\Omega$

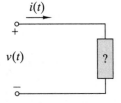

FIGURE 8–12

Exercise 8–9

A series connection consists of a 12-mH inductor and a 20-pF capacitor. The current flowing through the circuit is $i_L(t) = 20\cos(10^6 t)$ mA.

(a) Find the impedance of each element.

(b) Find the phasor voltage across each element.

(c) Find an equation for the waveform of the voltage across each element.

(d) Does the current in the inductor lead or lag the voltage across it?

(e) Does the current in the capacitor lead or lag the voltage across it?

Answers:

(a) $Z_L = j12\,\text{k}\Omega$, $Z_C = -j50\,\text{k}\Omega$

(b) $\mathbf{V}_L = 240\angle 90°$ V, $\mathbf{V}_C = 1000\angle -90°$V

(c) $v_L(t) = 240 \cos(10^6 t + 90°) \, V$, $v_C(t) = 1 \cos(10^6 t - 90°) \, kV$
(d) The current lags the voltage in the inductor by 90°
(e) The current leads the voltage in the capacitor by 90°

8–3 BASIC PHASOR CIRCUIT ANALYSIS AND DESIGN

Functions of time like $v(t) = V_A \cos(\omega t + \phi_V)$ and $i(t) = I_A \cos(\omega t + \phi_I)$ are time-domain representations of sinusoidal signals. Producing the corresponding phasors can be thought of as a transformation that carries $v(t)$ and $i(t)$ into a complex-number domain where signals are represented as phasors **V** and **I**. We call this complex number domain the **phasor domain**. When we analyze circuits in this phasor domain, we obtain sinusoidal steady-state responses in terms of phasors like **V** and **I**. Performing the inverse phasor transformation as $v(t) = \text{Re}\{Ve^{j\omega t}\}$ and $i(t) = \text{Re}\{Ie^{j\omega t}\}$ carries the responses back into the time domain. To perform ac circuit analysis in this way, we obviously need to develop methods of analyzing circuits in the phasor domain.

In the preceding section, we showed that KVL and KCL apply in the phasor domain and that the phasor element constraints all have the form **V** = **ZI**. These element and connection constraints have the same format as the underlying constraints for resistance circuit analysis as developed in Chapters 2, 3, and 4. Therefore, familiar algebraic circuit analysis tools, such as series and parallel equivalence, voltage and current division, proportionality and superposition, and Thévenin and Norton equivalent circuits, are applicable in the phasor domain. In other words, we do not need new analysis techniques to handle circuits in the phasor domain. The only difference is that circuit responses are phasors (complex numbers) rather than dc signals (real numbers).

We can think of phasor-domain circuit analysis in terms of the flow diagram in Figure 8–13. The analysis begins in the time domain with a linear circuit operating in the sinusoidal steady state and involves three major steps:

STEP 1 The circuit is transformed into the phasor domain by representing the input and response sinusoids as phasors and the passive circuit elements by their impedances.

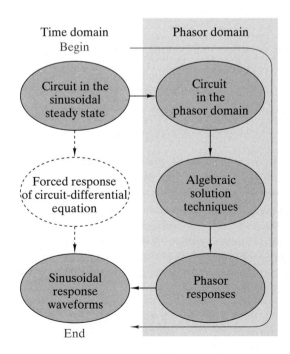

FIGURE 8–13 *Flow diagram for phasor circuit analysis.*

STEP 2   Standard algebraic circuit analysis techniques are applied to solve the phasor-domain circuit for the desired unknown phasor responses.

STEP 3   The phasor responses are inverse transformed back into time-domain sinusoids to obtain the response waveforms.

The third step assumes that the required end product is a time-domain waveform. However, a phasor is just another representation of a sinusoid. With some experience, we learn to think of the response as a phasor without converting it back into a time-domain waveform.

Figure 8–13 points out that there is another route to time-domain response using the classical differential equation method from Chapter 7. However, the phasor-domain method works directly with the circuit model and is far simpler. More important, phasor-domain analysis provides insights into ac circuit analysis that are essential to understanding much of the terminology and viewpoint of electrical engineering.

SERIES EQUIVALENCE AND VOLTAGE DIVISION

We begin the study of phasor-domain analysis with two basic analysis tools—series equivalence and voltage division. In Figure 8–14 the two-terminal elements are connected in series, so by KCL, the same phasor current $\mathbf{I}$ exists in impedances $Z_1, Z_2, \ldots, Z_N$. Using KVL and the element constraints, the voltage across the series connection can be written as

$$\begin{aligned}
\mathbf{V} &= \mathbf{V}_1 + \mathbf{V}_2 + \cdots + \mathbf{V}_N \\
&= Z_1\mathbf{I} + Z_2\mathbf{I} + \cdots + Z_N\mathbf{I} \\
&= (Z_1 + Z_2 + \cdots + Z_N)\mathbf{I}
\end{aligned} \tag{8-18}$$

FIGURE 8–14 *A series connection of impedances.*

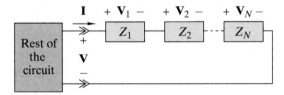

The last line in this equation points out that the phasor responses $\mathbf{V}$ and $\mathbf{I}$ do not change when the series connected elements are replaced by an equivalent impedance:

$$Z_{\text{EQ}} = Z_1 + Z_2 + \cdots + Z_N \tag{8-19}$$

In general, the equivalent impedance Z_{EQ} is a complex quantity of the form

$$Z_{\text{EQ}} = R + jX$$

where R is the real part and X is the imaginary part. The real part of Z is called **resistance** and the imaginary part (X, not jX) is called **reactance**. Both resistance and reactance are expressed in ohms (Ω), and both can be functions of frequency (ω). For passive circuits, resistance is always positive, while reactance X can be either positive or negative. A positive X is called an **inductive** reactance because the reactance of an inductor is ωL, which is always positive. A negative X is called a **capacitive** reactance because the reactance of a capacitor is $-1/\omega C$, which is always negative.

Combining Eqs. (8–18) and (8–19), we can write the phasor voltage across the kth element in the series connection as

$$\mathbf{V}_k = Z_k\mathbf{I} = \frac{Z_k}{Z_{\text{EQ}}}\mathbf{V} \tag{8-20}$$

Equation (8–20) is the phasor version of the voltage division principle. The phasor voltage across any element in a series connection is equal to the ratio of its impedance to the equivalent impedance of the connection times the total phasor voltage across the connection.

EXAMPLE 8 – 6

The circuit in Figure 8–15(a) is operating in the sinusoidal steady state with $v_S(t) = 35 \cos 1000t$ V.

(a) Transform the circuit into the phasor domain.
(b) Solve for the phasor current **I**.
(c) Solve for the phasor voltage across each element.
(d) Find the waveforms corresponding to the phasors found in (b) and (c).

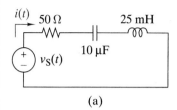

(a)

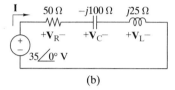

(b)

FIGURE 8–15

SOLUTION:

(a) The phasor representing the input source voltage is $\mathbf{V}_S = 35\angle 0°$ V. The impedances of the three passive elements are

$$Z_R = R = 50\,\Omega$$

$$Z_L = j\omega L = j1000 \times 25 \times 10^{-3} = j25\,\Omega$$

$$Z_C = \frac{1}{j\omega C} = \frac{1}{j1000 \times 10^{-5}} = -j100\,\Omega$$

Using these results, we obtain the phasor-domain circuit in Figure 8–15(b).

(b) The equivalent impedance of the series connection is

$$Z_{EQ} = 50 + j25 - j100 = 50 - j75 = 90.1\angle{-56.3°}\,\Omega$$

The current in the series circuit is

$$\mathbf{I} = \frac{\mathbf{V}_S}{Z_{EQ}} = \frac{35\angle 0°}{90.1\angle{-56.3°}} = 0.388\angle 56.3°\,\text{A}$$

(c) The current **I** exists in all three series elements, so the voltage across each passive element is

$$\mathbf{V}_R = Z_R\mathbf{I} = 50 \times 0.388\angle 56.3° = 19.4\angle 56.3°\,\text{V}$$

$$\mathbf{V}_L = Z_L\mathbf{I} = j25 \times 0.388\angle 56.3° = 9.70\angle 146.3°\,\text{V}$$

$$\mathbf{V}_C = Z_C\mathbf{I} = -j100 \times 0.388\angle 56.3° = 38.8\angle{-33.7°}\,\text{V}$$

Note that the voltage across the resistor is in phase with the current, the voltage across the inductor leads the current by 90°, and the voltage across the capacitor lags the current by 90°.

(d) The sinusoidal steady-state waveforms corresponding to the phasors in (b) and (c) are

$$i(t) = \text{Re}\{0.388e^{j56.3°}e^{j1000t}\} = 0.388\cos(1000t + 56.3°)\,\text{A}$$

$$v_R(t) = \text{Re}\{19.4e^{j56.3°}e^{j1000t}\} = 19.4\cos(1000t + 56.3°)\,\text{V}$$

$$v_L(t) = \text{Re}\{9.70e^{j146.3°}e^{j1000t}\} = 9.70\cos(1000t + 146.3°)\,\text{V}$$

$$v_C(t) = \text{Re}\{38.8e^{-j33.7°}e^{j1000t}\} = 38.8\cos(1000t - 33.7°)\,\text{V}$$

Exercise 8–10

The circuit in Figure 8–15(a) is operating in the sinusoidal steady state with $v_S(t) = 100 \cos(2000t - 45°)$ V.

(a) Transform the circuit into the phasor domain.
(b) Solve for the phasor current **I**.
(c) Solve for the phasor voltage across each element.
(d) Find the waveforms corresponding to the phasors found in (b) and (c).
(e) Draw a phasor diagram of all three voltages and the current.

Answers:

(a) See Figure 8–15(c).
(b) $\mathbf{I} = 2\angle -45°$ A
(c) $\mathbf{V}_R = 100\angle -45°$ V
    $\mathbf{V}_L = 100\angle +45°$ V
    $\mathbf{V}_C = 100\angle -135°$ V
(d) $i(t) = 2 \cos(2000t - 45°)$ A
    $v_R(t) = 100 \cos(2000t - 45°)$ V
    $v_L(t) = 100 \cos(2000t + 45°)$ V
    $v_C(t) = 100 \cos(2000t - 135°)$ V
(e) See Figure 8–15(d)

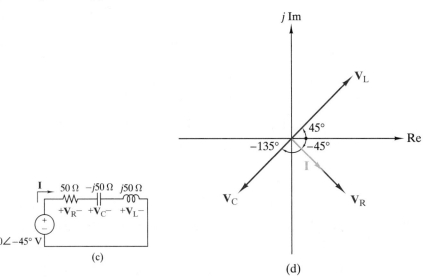

(c)

(d)

FIGURE 8–15 *(Continued)*

▶ DESIGN EXAMPLE 8–7

Design the voltage divider in Figure 8–16(a) so that an input $v_s = 15 \cos 2000t$ V produces a steady-state output $v_O(t) = 2 \sin 2000t$ V.

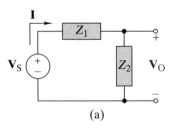

(a)

FIGURE 8–16

SOLUTION:
Using voltage division, we can relate the input and output phasors as follows:

$$\mathbf{V}_O = \frac{Z_2}{Z_1 + Z_2} \mathbf{V}_S$$

The phasor representation of the input voltage is $\mathbf{V}_S = 15\angle 0 = 15 + j0$. Using the identity $\cos(x - 90°) = \sin x$, we write the required output phasor as $\mathbf{V}_O =$

$2\angle-90° = 0 - j2$. The design problem is to select the impedances Z_1 and Z_2 so that

$$0 - j2 = \frac{Z_2}{Z_1 + Z_2}(15 + j0)$$

Solving this design constraint for Z_1 yields

$$Z_1 = \frac{15 + j2}{-j2}Z_2$$

Evidently, we can choose Z_2 and then solve for Z_1. In making this choice, we must keep some physical realizability conditions in mind. In general, an impedance has the form $Z = R + jX$. The reactance X can be either positive (an inductor) or negative (a capacitor), but the resistance R must be positive. With these constraints in mind, we select $Z_2 = -j1000$ (a capacitor) and solve for $Z_1 = 7500 + j1000$ (a resistor in series with an inductor). Figure 8–16(b) shows the resulting phasor circuit. To find the values of L and C, we note that the input is $v_S = 15\cos 2000t$; hence, the frequency is $\omega = 2000$. The inductive reactance $\omega L = 1000$ requires $L = 0.5\,\text{H}$, while the capacitive reactance requires $-(\omega C)^{-1} = -1000$ or $C = 0.5\,\mu\text{F}$. Other possible designs are obtained by selecting different values of Z_2. To be physically realizable, the selected value of Z_2 must produce $R \geq 0$ for Z_1 and Z_2. ■

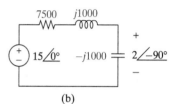

(b)

● Design Exercise 8–11

Design the voltage divider in Figure 8–16(a) so that an input $v_S(t) = 50\cos(2000t)\,\text{V}$ produces an output $v_O(t) = 25\cos(2000t - 30°)\,\text{V}$.

A n s w e r : We assumed $Z_2 = 1000 - j1000\,\Omega$ and solved for $Z_1 = 1732 + j268\,\Omega$. The corresponding elements are $R_1 = 1.73\,\text{k}\Omega$, $L = 134\,\text{mH}$, $R_2 = 1\,\text{k}\Omega$, and $C = 0.5\,\mu\text{F}$. Figure 8–16 (c) shows our phasor design. Another solution is to assume $Z_2 = 2165 - j1250\,\Omega$ and solved for $Z_1 = 2835 + j1250\,\Omega$. The corresponding elements are $R_1 = 2835\,\Omega$, $L = 625\,\text{mH}$, $R_2 = 2165\,\Omega$, and $C = 0.4\,\mu\text{F}$. Many other correct solutions are possible.

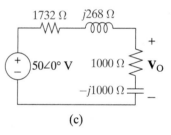

(c)

FIGURE 8–16 (*Continued*)

● Design Exercise 8–12

The circuit shown in Figure 8–17(a) is operating in the sinusoidal steady state at a frequency of 100 krad/s. It requires a load of $Z_L = 1500\angle-57.5°\,\Omega$ to operate properly. Design the load using standard parts to within $\pm 5\%$ of the desired values.

FIGURE 8–17

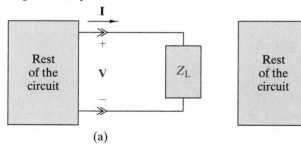

(a)　　　　　　　　(b)

A n s w e r : The load requires an 806-Ω resistor and a 7910-pF capacitor. See Figure 8–17(b) for closest standard values.

APPLICATION   **EXAMPLE 8 – 8**

The purpose of the impedance bridge in Figure 8–18 is to measure the unknown impedance Z_X by adjusting known impedances Z_1, Z_2, and Z_3 until the detector

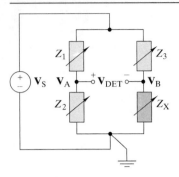

FIGURE 8–18 *Impedance bridge.*

voltage $\mathbf{V}_{DET}$ is zero. The circuit consists of a sinusoidal source $\mathbf{V}_S$ driving two voltage dividers connected in parallel. Using the voltage division principle, we find that the detector voltage is

$$\mathbf{V}_{DET} = \mathbf{V}_A - \mathbf{V}_B = \frac{Z_2}{Z_1 + Z_2}\mathbf{V}_S - \frac{Z_X}{Z_3 + Z_X}\mathbf{V}_S$$

$$= \left[\frac{Z_2 Z_3 - Z_1 Z_X}{(Z_1 + Z_2)(Z_3 + Z_X)}\right]\mathbf{V}_S$$

This equation shows that the detector voltage will be zero when $Z_2 Z_3 = Z_1 Z_X$. When the branch impedances are adjusted so that the detector voltage is zero, the unknown impedance can be written in terms of the known impedances as follows:

$$Z_X = R_X + jX_X = \frac{Z_2 Z_3}{Z_1}$$

This equation is called the bridge balance condition. Since the equality involves complex quantities, at least two of the known impedances must be adjustable to balance both the resistance R_X and the reactance X_X of the unknown impedance. In practice, bridges are designed assuming that the sign of the unknown reactance is known. Bridges that measure only positive reactance are called inductance bridges, while those that measure only negative reactance are called capacitance bridges.

The Maxwell inductance bridge in Figure 8–19 is used to measure the resistance R_X and inductance L_X of an inductive device by alternately adjusting resistances R_1 and R_2 to balance the bridge circuit. The impedances of the legs of this bridge are

$$Z_1 = \frac{1}{j\omega C_1 + \dfrac{1}{R_1}}$$

$$Z_2 = R_2 \quad Z_3 = R_3$$

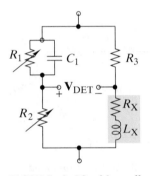

FIGURE 8–19 *Maxwell bridge.*

For the Maxwell bridge, the balance condition $Z_X = Z_2 Z_3/Z_1$ yields

$$R_X + j\omega L_X = \frac{R_2 R_3}{R_1} + j\omega C_1 R_2 R_3$$

Equating the real and imaginary parts on each side of this equation yields the parameters of the unknown impedance in terms of the known impedances:

$$R_X = \frac{R_2 R_3}{R_1} \quad \text{and} \quad L_X = R_2 R_3 C_1$$

Note that adjusting R_1 affects only R_X. The Maxwell bridge measures inductance by balancing the positive reactance of an unknown inductive device with a calibrated fraction of negative reactance of the known capacitor C_1. If the reactance of the unknown device is actually capacitive (negative), then the Maxwell bridge cannot be balanced.

Exercise 8–13

Consider the Maxwell bridge shown in Figure 8–18. Suppose we know that the unknown impedance is an unknown capacitor C_X in parallel with an unknown resistance R_X. Let Z_1 be a resistance R_1 in series with an inductance L_1. Let Z_2 be a resistance R_2 and Z_3 be a resistance R_3. Find the relationships that will allow the bridge to be balanced.

Answer:
$$R_X = \frac{R_2 R_3}{R_1}, \quad C_X = \frac{L_1}{R_2 R_3}$$

PARALLEL EQUIVALENCE AND CURRENT DIVISION

In Figure 8–20 the two-terminal elements are connected in parallel, so the same phasor voltage $\mathbf{V}$ appears across the impedances $Z_1, Z_2, \ldots, Z_N$. Using the phasor element constraints, the current through each impedance is $\mathbf{I}_k = \mathbf{V}/Z_k$. Next, using KCL, the total current entering the parallel connection is

$$
\begin{aligned}
\mathbf{I} &= \mathbf{I}_1 + \mathbf{I}_2 + \cdots + \mathbf{I}_N \\
&= \frac{\mathbf{V}}{Z_1} + \frac{\mathbf{V}}{Z_2} + \cdots + \frac{\mathbf{V}}{Z_N} \\
&= \left(\frac{1}{Z_1} + \frac{1}{Z_2} + \cdots + \frac{1}{Z_N}\right)\mathbf{V}
\end{aligned}
\tag{8–21}
$$

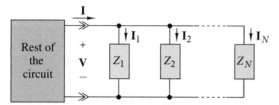

FIGURE 8–20 *Parallel connection of impedances.*

The same phasor responses $\mathbf{V}$ and $\mathbf{I}$ exist when the parallel connected elements are replaced by an equivalent impedance.

$$
\frac{1}{Z_{\mathrm{EQ}}} = \frac{\mathbf{I}}{\mathbf{V}} = \frac{1}{Z_1} + \frac{1}{Z_2} + \cdots + \frac{1}{Z_N}
\tag{8–22}
$$

These results can also be written in terms of **admittance** Y, which is defined as the reciprocal of impedance:

$$
Y = \frac{1}{Z} = G + jB
$$

The real part of Y is called **conductance** G and the imaginary part is called **susceptance** B, both of which are expressed in units of siemens (S).

Using admittances to rewrite Eq. (8–21) yields

$$
\begin{aligned}
\mathbf{I} &= \mathbf{I}_1 + \mathbf{I}_2 + \cdots + \mathbf{I}_N \\
&= Y_1\mathbf{V} + Y_2\mathbf{V} + \cdots + Y_N\mathbf{V} \\
&= (Y_1 + Y_2 + \cdots + Y_N)\mathbf{V}
\end{aligned}
\tag{8–23}
$$

Hence, the equivalent admittance of the parallel connection is

$$
Y_{\mathrm{EQ}} = \frac{\mathbf{I}}{\mathbf{V}} = Y_1 + Y_2 + \cdots + Y_N
\tag{8–24a}
$$

We can write this in terms of equivalent impedance as

$$
Z_{\mathrm{EQ}} = \frac{\mathbf{V}}{\mathbf{I}} = \frac{1}{\dfrac{1}{Z_1} + \dfrac{1}{Z_2} + \cdots + \dfrac{1}{Z_N}}
\tag{8–24b}
$$

Combining Eqs. (8–23) and (8–24a), we find that the phasor current through the kth element in the parallel connection using admittances is

$$
\mathbf{I}_k = Y_k\mathbf{V} = \frac{Y_k}{Y_{\mathrm{EQ}}}\mathbf{I}
\tag{8–25a}
$$

Similarly, using Eq. (8–24b), we can express the individual phasor currents in terms of impedances as

$$
\mathbf{I}_k = \frac{\mathbf{V}}{Z_k} = \frac{\dfrac{1}{Z_k}}{\dfrac{1}{Z_{\mathrm{EQ}}}}\mathbf{I} = \frac{\dfrac{1}{Z_k}}{\dfrac{1}{Z_1} + \dfrac{1}{Z_2} + \cdots + \dfrac{1}{Z_N}}\mathbf{I}
\tag{8–25b}
$$

Equations (8–25a) and (8–25b) are the phasor versions of the current division principle. The phasor current through any element in a parallel connection is equal to the ratio of its admittance to the equivalent admittance of the connection times the total phasor current entering the connection. Equation (8–25a) shows this in terms of admittances, while Eq. (8–25b) shows the same relationship in terms of impedances.

Note: In dealing with phasors, one finds that the concept of admittance is used more frequently than the concept of conductance is used in dc circuits. The reason is that whereas resistors in dc circuits are rarely described in terms of their conductances, ac circuits are often described in terms of their admittances. Hence, although we will emphasize impedances in this text, you should be aware that many phasor applications would use admittances as useful circuit descriptors.

EXAMPLE 8 – 9

The circuit in Figure 8–21(a) is operating in the sinusoidal steady state with $i_S(t) = 50 \cos 2000t$ mA.

(a) Transform the circuit into the phasor domain.
(b) Solve for the phasor voltage **V**.
(c) Solve for the phasor current through each element.
(d) Find the waveforms corresponding to the phasors found in (b) and (c).
(e) Sketch a phasor diagram of the resulting phasors.

FIGURE 8–21

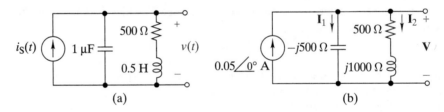

(a) (b)

SOLUTION:
(a) The phasor representing the input source current is $\mathbf{I}_S = 0.05\angle 0°$ A. The impedances of the three passive elements are

$$Z_R = R = 500\,\Omega$$

$$Z_L = j\omega L = j2000 \times 0.5 = j1000\,\Omega$$

$$Z_C = \frac{1}{j\omega C} = \frac{1}{j2000 \times 10^{-6}} = -j500\,\Omega$$

Using these results, we obtain the phasor-domain circuit in Figure 8–21(b).

(b) The admittances of the two parallel branches are

$$Y_1 = \frac{1}{-j500} = j2 \times 10^{-3}\,S$$

$$Y_2 = \frac{1}{500 + j1000} = 4 \times 10^{-4} - j8 \times 10^{-4}\,S$$

The equivalent admittance of the parallel connection is

$$Y_{EQ} = Y_1 + Y_2 = 4 \times 10^{-4} + j12 \times 10^{-4}$$

$$= 12.6 \times 10^{-4}\angle 71.6°\,S$$

and the voltage across the parallel circuit is

$$\mathbf{V} = \frac{\mathbf{I}_S}{Y_{EQ}} = \frac{0.05\angle 0°}{12.6 \times 10^{-4}\angle 71.6°}$$

$$= 39.7\angle -71.6° \, \text{V}$$

(c) The current through each parallel branch is

$$\mathbf{I}_1 = Y_1\mathbf{V} = j2 \times 10^{-3} \times 39.7\angle -71.6° = 79.4\angle 18.4° \, \text{mA}$$

$$\mathbf{I}_2 = Y_2\mathbf{V} = (4 \times 10^{-4} - j8 \times 10^{-4}) \times 39.7\angle -71.6°$$

$$= 35.5\angle -135° \, \text{mA}$$

(d) The sinusoidal steady-state waveforms corresponding to the phasors in (b) and (c) are

$$v(t) = \text{Re}\{39.7e^{-j71.6°}e^{j2000t}\} = 39.7\cos(2000t - 71.6°) \, \text{V}$$

$$i_1(t) = \text{Re}\{79.4e^{j18.4°}e^{j2000t}\} = 79.4\cos(2000t + 18.4°) \, \text{mA}$$

$$i_2(t) = \text{Re}\{35.5e^{-j135°}e^{j2000t}\} = 35.5\cos(2000t - 135°) \, \text{mA}$$

(e) The phasor diagram is shown in Figure 8–21(c) ■

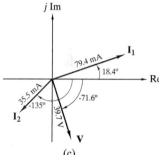

(c)

Exercise 8–14

The circuit in Figure 8–21(a) is operating in the sinusoidal steady state with $i_S(t) = 100\cos(1000t - 45°) \, \text{mA}$.

(a) Transform the circuit into the phasor domain.
(b) Solve for the phasor voltage **V**.
(c) Solve for the phasor current through each element.
(d) Find the waveforms corresponding to the phasors found in (b) and (c).

Answers:

(a) See Figure 8–21(d).
(b) $\mathbf{V} = 100\angle -45° \, \text{V}$
(c) $\mathbf{I}_R = 141.4\angle -90° \, \text{mA}$
   $\mathbf{I}_L = 141.4\angle -90° \, \text{mA}$
   $\mathbf{I}_C = 100\angle +45° \, \text{mA}$
(d) $v(t) = 100\cos(1000t - 45°) \, \text{V}$
   $i_R(t) = 141.4\cos(1000t - 90°) \, \text{mA}$
   $i_L(t) = 141.4\cos(1000t - 90°) \, \text{mA}$
   $i_C(t) = 100\cos(1000t + 45°) \, \text{mA}$

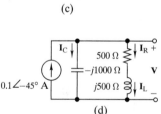

(d)

FIGURE 8–21 (Continued)

EXAMPLE 8–10

Find the steady-state currents $i(t), i_C(t)$, and $i_R(t)$ in the circuit of Figure 8–22 for $v_S = 100\cos 2000t \, \text{V}, L = 250\,\text{mH}, C = 0.5\,\mu\text{F}$, and $R = 3\,\text{k}\Omega$.

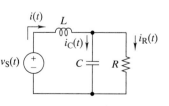

FIGURE 8–22

SOLUTION A—Classical Approach:

The phasor representation of the input voltage is $100\angle 0°$. The impedances of the passive elements are

$$Z_L = j500\,\Omega \quad Z_C = -j1000\,\Omega \quad Z_R = 3000\,\Omega$$

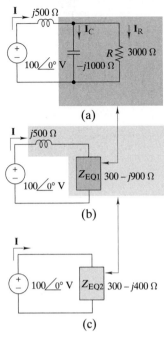

(a)

(b)

(c)

FIGURE 8–23

Figure 8–23(a) shows the phasor-domain circuit.

To solve for the required phasor responses, we reduce the circuit using a combination of series and parallel equivalence. Using parallel equivalence, we find that the capacitor and resistor can be replaced by an equivalent impedance

$$Z_{EQ1} = \frac{1}{Y_{EQ1}} = \frac{1}{\dfrac{1}{-j1000} + \dfrac{1}{3000}}$$

$$= 300 - j900 \ \Omega$$

The resulting circuit reduction is shown in Figure 8–23(b). The equivalent impedance Z_{EQ1} is connected in series with the impedance $Z_L = j500$. This series combination can be replaced by an equivalent impedance

$$Z_{EQ2} = j500 + Z_{EQ1} = 300 - j400 \ \Omega$$

This step reduces the circuit to the equivalent input impedance shown in Figure 8–23(c). The phasor input current in Figure 8–23(c) is

$$\mathbf{I} = \frac{100\angle 0°}{Z_{EQ2}} = \frac{100\angle 0°}{300 - j400} = 0.12 + j0.16 = 0.2\angle 53.1° \ \text{A}$$

Given the phasor current $\mathbf{I}$, we use current division to find $\mathbf{I_C}$:

$$\mathbf{I_C} = \frac{Y_C}{Y_C + Y_R}\mathbf{I} = \frac{\dfrac{1}{-j1000}}{\dfrac{1}{-j1000} + \dfrac{1}{3000}} 0.2\angle 53.1°$$

$$= 0.06 + j0.18 = 0.19\angle 71.6° \ \text{A}$$

By KCL, $\mathbf{I} = \mathbf{I_C} + \mathbf{I_R}$, so the remaining unknown current is

$$\mathbf{I_R} = \mathbf{I} - \mathbf{I_C} = 0.06 - j0.02 = 0.0632\angle -18.4° \ \text{A}$$

The waveforms corresponding to the phasor currents are

$$i(t) = \text{Re}\{\mathbf{I}e^{j2000t}\} = 0.2\cos(2000t + 53.1°) \ \text{A}$$
$$i_C(t) = \text{Re}\{\mathbf{I_C}e^{j2000t}\} = 0.19\cos(2000t + 71.6°) \ \text{A}$$
$$i_R(t) = \text{Re}\{\mathbf{I_R}e^{j2000t}\} = 0.0632\cos(2000t - 18.4°) \ \text{A} \qquad ■$$

SOLUTION   B—Computer-Aided Approach:

We can also solve phasor problems using OrCAD. We begin by drawing the circuit as usual in OrCAD. Since we want to do an ac phasor solution, not a transient solution, we use VAC for the source, not VSIN. With VAC, you specify the frequency component in the simulation process. For VSIN, you need to specify the frequency as a source parameter.

To have OrCAD calculate and display the phasor current (or voltage), you need to insert special devices into your schematic at the points where you want to know the current or voltage. To do this you use the IPRINT or VPRINT device found in the SPECIAL library.

In this example, since we are interested in calculating only current, we will use only the IPRINT feature. We insert the device in the schematic so that the current flows *through* the IPRINT element. You can see this in the schematic shown in

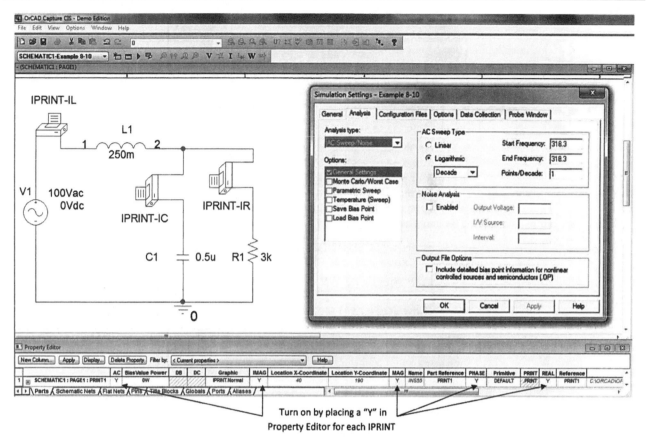

FIGURE 8-24 *Copyright © Cadence Design Systems, Inc. Used with permission.*

Figure 8–24. Note that there is a minus sign on the device. It is important to have this terminal be the exit point of the current you are trying to measure. It is necessary to tell each IPRINT device what you want it to measure. To do this, double-click on the device to open the Property Editor. In the Property Editor, place a y in those fields you want IPRINT to calculate and Probe to print. For this example, we placed y in each of AC, IMAG (imaginary), REAL (real), MAG (magnitude), and PHASE (phase). These fields will display the desired current phasor in both rectangular and polar forms. Since there are three IPRINT devices in this example, we would like to simplify their identification on the output. Hence, we label each device under the column Reference with an appropriate designator—IR, for example, for the current through the resistor.

Next, we use PSpice to set up the desired simulation. For this type of analysis, we select AC Sweep/Noise. Under General Settings, we can choose either Linear or Logarithmic for AC Sweep Type. Then we enter the frequency, in hertz, at which we want to start our sweep: 318.3 Hz (2000 rad/s). We are interested in only a single frequency since we are doing a phasor analysis, so we enter that same frequency as the end frequency. Since there is only one frequency, we enter 1 point, that is, enter a 1 in Points/Decade depending on which sweep type is chosen. We are now ready to run the simulation. See Figure 8–24 for a filled-out Simulation Setting view for this example.

Once the simulation is run, we need to select View and then Output File from the Probe window. Scroll down until you see the desired IPRINT results. The simulation for our example produced the following results, among others. We have highlighted in color the desired currents in polar form.

```
**** 02/10/11 22:00:39 ******* PSpice Lite (August 2007) ******
ID#10813 ****

**Profile: "SCHEMATIC1-Example 8-10" [C:\OrCAD\OrCAD_16.0_Demo
\tools\Example 8-10-PSpiceFiles\SCHEMATIC1\Example 8-10.sim]

****    AC ANALYSIS              TEMPERATURE = 27.000 DEG C

***************************************************************
```

FREQ	IM(V_IL)	IP(V_IL)	IR(V_IL)	II(V_IL)
3.183E+02	2.000E-01	5.313E+01	1.200E-01	1.600E-01

```
**** 02/10/11 22:00:39 ******* PSpice Lite (August 2007) ******
ID# 10813 ****

** Profile: "SCHEMATIC1-Example 8-10" [C:\OrCAD\OrCAD_16.0_Demo
\tools\Example 8-10-PSpiceFiles\SCHEMATIC1\Example 8-10.sim]

****    AC ANALYSIS              TEMPERATURE = 27.000 DEG C

***************************************************************
```

FREQ	IM(V_IC)	IP(V_IC)	IR(V_IC)	II(V_IC)
3.183E+02	1.897E-01	7.157E+01	5.999E-02	1.800E-01

```
**** 02/10/11 22:00:39 ******* PSpice Lite (August 2007) ******
ID# 10813 ****

** Profile: "SCHEMATIC1-Example 8-10" [C:\OrCAD\OrCAD_16.0_Demo
\tools\Example 8-10-PSpiceFiles\SCHEMATIC1\Example 8-10.sim]

****    AC ANALYSIS              TEMPERATURE = 27.000 DEG C

***************************************************************
```

FREQ	IM(V_IR)	IP(V_IR)	IR(V_IR)	II(V_IR)
3.183E+02	6.324E-02	-1.843E+01	6.000E-02	-2.000E-02

JOB CONCLUDED

These results compare favorably with the classical calculations. ■

Exercise 8–15

Using the values in Example 8–10, find the voltage $v_L(t)$ across the inductor in the circuit shown in Figure 8–22.

Answer: $$v_L(t) = 100 \cos(2000t + 143°) \text{ V}$$

APPLICATION EXAMPLE 8 – 11

In general, the equivalent impedance seen at any pair of terminals can be written in rectangular form as

$$Z_{EQ} = R_{EQ} + jX_{EQ} \qquad (8-26)$$

In a passive circuit the equivalent resistance R_{EQ} must always be nonnegative, that is, $R_{EQ} \geq 0$. However, the equivalent reactance X_{EQ} can be either positive (inductive) or negative (capacitive). When inductance and capacitance are both present, their reactances may exactly cancel at certain frequencies. When $X_{EQ} = 0$, the impedance is purely resistive and the circuit is said to be in **resonance**. The frequency at which this occurs is called a **resonant frequency**, denoted by ω_0.

For example, suppose we want to find the resonant frequency of the circuit in Figure 8–25. We first find the equivalent impedance of the parallel resistor and capacitor:

$$Z_{RC} = \frac{1}{Y_R + Y_C} = \frac{1}{\dfrac{1}{R} + j\omega C} = \frac{R}{1 + j\omega RC}$$

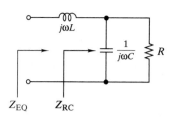

FIGURE 8–25

This expression can be put into rectangular form by multiplying and dividing by the conjugate of the denominator:

$$Z_{RC} = \frac{R}{1 + j\omega RC} \frac{1 - j\omega RC}{1 - j\omega RC} = \frac{R}{1 + (\omega RC)^2} - j\frac{\omega R^2 C}{1 + (\omega RC)^2}$$

The impedance Z_{RC} is connected in series with the inductor. Therefore, the overall equivalent impedance Z_{EQ} is

$$Z_{EQ} = Z_L + Z_{RC}$$

$$= \frac{R}{1 + (\omega RC)^2} + j\left[\omega L - \frac{\omega R^2 C}{1 + (\omega RC)^2}\right]$$

$$= R_{EQ} + jX_{EQ}$$

Note that the equivalent resistance R_{EQ} is positive for all ω. However, the equivalent reactance X_{EQ} can be positive or negative. The resonant frequency is found by setting the reactance to zero

$$X_{EQ}(\omega_0) = \omega_0 L - \frac{\omega_0 R^2 C}{1 + (\omega_0 RC)^2} = 0$$

and solving for the resonant frequency:

$$\omega_0 = \sqrt{\frac{1}{LC} - \frac{1}{(RC)^2}} \qquad (8\text{–}27)$$

Note the reactance X_{EQ} is inductive (positive) when $\omega > \omega_0$ and capacitive (negative) when $\omega < \omega_0$.

Exercise 8–16

The circuit in Figure 8–25 is operating in the sinusoidal steady state. If $R = 1\,k\Omega$, $L = 200\,mH$, and $C = 1\,\mu F$:

(a) Find the value of ω that will cause the circuit to be in resonance.
(b) What will the value of Z_{EQ} be under those conditions?

Answers:

(a) $\omega = 2\,krad/s$
(b) $Z_{EQ} = 200 + j0\,\Omega$

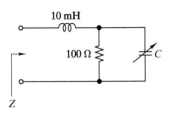

FIGURE 8–26

EXAMPLE 8–12

The circuit in Figure 8–26 is operating in the sinusoidal steady state with $\omega = 5\,\text{krad/s}$.

(a) Find the value of capacitance C that causes the input impedance Z to be purely resistive.

(b) Find the real part of the input impedance for this value of C.

SOLUTION:

(a) At the specified frequency the impedance is

$$Z = j5000 \times 0.01 + \cfrac{1}{\cfrac{1}{100} + j5000C}$$

which can be written as

$$Z = j50 + \frac{100}{1 + j5 \times 10^5 C} \times \frac{1 - j5 \times 10^5 C}{1 - j5 \times 10^5 C}$$

$$= j50 + \frac{100}{1 + (5 \times 10^5 C)^2} - j\frac{5 \times 10^7 C}{1 + (5 \times 10^5 C)^2}$$

A purely resistive input impedance means that the imaginary part of Z is zero, which requires that

$$50 - \frac{5 \times 10^7 C}{1 + (5 \times 10^5 C)^2} = 0 \quad \text{or} \quad 25 \times 10^{10} C^2 - 10^6 C + 1 = 0$$

This quadratic has a double root at $C = 2 \times 10^{-6}\,\text{F}$.

(b) For this value of capacitance the real part of Z is

$$\text{Re}\{Z\} = R_{\text{EQ}} = \frac{100}{1 + (5 \times 10^5 C)^2} = 50\,\Omega \qquad \blacksquare$$

Exercise 8–17

The circuit in Figure 8–25 is operating in the sinusoidal steady state at $\omega = 1\,\text{krad/s}$. If $R = 1\,\text{k}\Omega, L = 200\,\text{mH}$, and $C = 1\,\mu\text{F}$:

(a) Find the value of Z_{EQ} classically under those conditions.

(b) Repeat the problem using OrCAD. (*Hint*: Drive the circuit with a 1 VAC source at 1 krad/s and use IPRINT to measure the input current. The equivalent impedance then will be $Z_{\text{EQ}} = 1\angle 0°/\mathbf{I}\,\Omega$.)

Answers: For both (a) and (b) $Z_{\text{EQ}} = 500 - j300\,\Omega$.

Exercise 8–18

The circuit in Figure 8–27 is operating at $\omega = 10\,\text{krad/s}$.

(a) Find the equivalent impedance Z.

(b) What element should be connected in series with Z to make the total reactance zero?

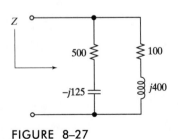

FIGURE 8–27

Answers:

(a) $Z = 256 + j195 = 322\angle37.3°\,\Omega$

(b) A capacitor with $C = 0.513\mu F$

Exercise 8–19

In Figure 8–28 $v_S(t) = 12.5\cos 1000t$ V and $i_S(t) = 0.2\cos(1000t - 36.9°)$ A. What is the impedance seen by the voltage source and what element is in the box?

Answer: $Z = 50 + j37.5\,\Omega$ and the element is a 37.5-mH inductor.

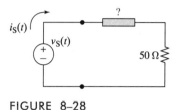

FIGURE 8–28

APPLICATION EXAMPLE 8 – 13

Consider the RC circuit in Figure 8–29(a). Find a relationship for the ratio of the output voltage phasor to the input voltage phasor. Plot the magnitude of this ratio as a function of frequency and comment on the result.

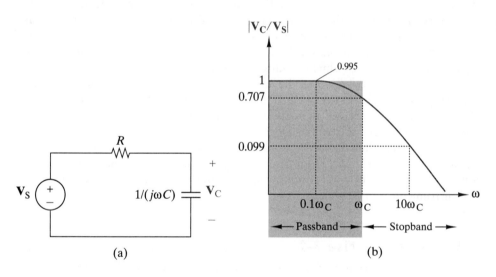

FIGURE 8–29

(a)

(b)

SOLUTION:

The output voltage can be found using a voltage divider as follows

$$\mathbf{V}_C = \frac{1/j\omega C}{R + 1/j\omega C}\mathbf{V}_S$$

$$\mathbf{V}_C = \frac{1}{j\omega CR + 1}\mathbf{V}_S$$

Studying this result one sees that this is a complex ratio and that the units of ω and $\frac{1}{RC}$ are s^{-1} or radians per second. The ratio has units of volts over volts, so it is non-dimensional. We first find the magnitude of the ratio as follows

$$\left|\frac{\mathbf{V}_C}{\mathbf{V}_S}\right| = \left|\frac{\frac{1}{RC}}{j\omega + \frac{1}{RC}}\right| = \frac{\frac{1}{RC}}{\sqrt{\omega^2 + \left(\frac{1}{RC}\right)^2}}$$

We can plot this ratio as a function of frequency by letting ω take on values from 0 to ∞. The following table shows a few salient points of the plot:

ω	$|V_C/V_S|$
0	1
$0.1/RC$	0.995
$1/RC$	$1/\sqrt{2} = 0.707$
$10/RC$	0.0995
∞	0

The graph is shown in Figure 8–29(b)

What one can see from the graph is that input signals with low frequencies appear across the capacitor relatively unscathed. High frequency signals, on the other hand, have their amplitudes significantly reduced. This makes sense because as the frequency of the circuit increases, the capacitor's impedance becomes smaller. To preserve KVL, then, the voltage across the resistor must increase as the voltage across the capacitor decreases. This behavior is significant and is a characteristic of circuits called *filters.* This simple *RC* configuration is called a *low-pass filter* because it passes low frequencies while blocking high ones. Since the curve is continuous, one might reasonably ask where does the "pass" portion of the filter stop and where does the "stop" or block portion begin? For reasons that will be explained in later chapters, the boundary between the stop and pass regions of the filter is set at the frequency where the amplitude has decreased to 0.707 of its original value. This frequency is called the *cutoff frequency* and is indicated as ω_C. For this filter, it is when $\omega = \omega_C = 1/RC$ rad/s. Filters are a major building block of electronics and are found in every type of application, from computers and communications to power and instrumentation.   ■

◑ Design Exercise 8–20

For the circuit of Figure 8–29(a), design a low-pass filter using standard parts so that the cutoff frequency is 1000 rad/s.

Answer: There are many solutions. One of them is $R = 1$ kΩ and $C = 1$ μF.

Exercise 8–21

For the circuit of Figure 8–29(a), replace the capacitor with an inductor. (a) Find the ratio $|V_L/V_S|$ and, (b) comment on its behavior as the frequency changes from 0 to ∞.

Answer:

(a) $|V_L/V_S| = |j\omega/(j\omega + R/L)| = \dfrac{\omega}{\sqrt{\omega^2 + (R/L)^2}}$

(b) The circuit blocks signals with low frequencies and passes signals with high frequencies. It is known as a high-pass filter and its cutoff frequency is $\omega_C = R/L$ rad/s.

8–4 CIRCUIT THEOREMS WITH PHASORS

In this section we treat basic properties of phasor circuits that parallel the resistance circuit theorems developed in Chapter 3. Circuit linearity is the foundation for all of

these properties. The proportionality and superposition properties are two fundamental consequences of linearity.

PROPORTIONALITY

The **proportionality** property states that phasor output responses are proportional to
the input phasor. Mathematically, proportionality means that

$$\mathbf{Y} = K\mathbf{X} \qquad (8\text{--}28)$$

where $\mathbf{X}$ is the input phasor, $\mathbf{Y}$ is the output phasor, and K is the proportionality
constant. In phasor circuit analysis, the proportionality constant is generally a
complex number.

The unit output method discussed in Chapter 3 is based on the proportionality
property and is applicable to phasors. To apply the unit output method in the phasor
domain, we assume that the output is a unit phasor $\mathbf{Y} = 1\angle 0°$. By successive
application of KCL, KVL, and the element impedances, we solve for the input
phasor required to produce the unit output. Because the circuit is linear, the
proportionality constant relating input and output is

$$K = \frac{\text{Output}}{\text{Input}} = \frac{1\angle 0°}{\text{Input phasor for unit output}}$$

Once we have the constant K, we can find the output for any input or the input
required to produce any specified output.

The next example illustrates the unit output method for phasor circuits.

EXAMPLE 8–14

Use the unit output method to find the input impedance, current $\mathbf{I}_1$, output voltage
$\mathbf{V}_C$, and current $\mathbf{I}_3$ of the circuit in Figure 8–30 for $\mathbf{V}_S = 10\angle 0° \, \text{V}$.

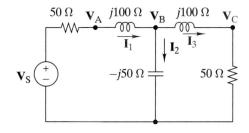

FIGURE 8–30

SOLUTION:
The following steps implement the unit output method for the circuit in Figure 8–30.

1. Assume a unit output voltage $\mathbf{V}_C = 1 + j0 \, \text{V}$.
2. By Ohm's law, $\mathbf{I}_3 = \mathbf{V}_C/50 = 0.02 + j0 \, \text{A}$.
3. By KVL, $\mathbf{V}_B = \mathbf{V}_C + (j100)\mathbf{I}_3 = 1 + j2 \, \text{V}$.
4. By Ohm's law, $\mathbf{I}_2 = \mathbf{V}_B/(-j50) = -0.04 + j0.02 \, \text{A}$.
5. By KCL, $\mathbf{I}_1 = \mathbf{I}_2 + \mathbf{I}_3 = -0.02 + j0.02 \, \text{A}$.
6. By KVL, $\mathbf{V}_S = (50 + j100)\mathbf{I}_1 + \mathbf{V}_B = -2 + j1 \, \text{V}$.

Given $\mathbf{V}_S$ and $\mathbf{I}_1$, the input impedance is

$$Z_{\text{IN}} = \frac{\mathbf{V}_S}{\mathbf{I}_1} = \frac{-2 + j1}{-0.02 + j0.02} = 75 + j25 \, \Omega$$

The proportionality factor between the input $\mathbf{V}_S$ and output voltage $\mathbf{V}_C$ is

$$K = \frac{1}{\mathbf{V}_S} = \frac{1}{-2+j} = -0.4 - j0.2$$

Given K and Z_{IN}, we can now calculate the required responses for an input $\mathbf{V}_S = 10\angle0°$:

$$\mathbf{V}_C = K\mathbf{V}_S = -4 - j2 = 4.47\angle{-153°} \text{ V}$$

$$\mathbf{I}_1 = \frac{\mathbf{V}_S}{Z_{IN}} = 0.12 - j0.04 = 0.126\angle{-18.4°} \text{ A}$$

$$\mathbf{I}_3 = \frac{\mathbf{V}_C}{50} = -0.08 - j0.04 = 0.0894\angle{-153°} \text{ A} \qquad ∎$$

Exercise 8–22

Use the unit output method to find the output current $\mathbf{I}_O$ in the circuit of Figure 8–31.

FIGURE 8–31

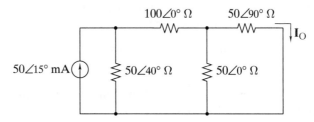

$100\angle0°$ Ω $50\angle90°$ Ω $\mathbf{I}_O$

$50\angle15°$ mA $50\angle40°$ Ω $50\angle0°$ Ω

Answer: $K = \mathbf{I}_O/\mathbf{I}_S = 0.204\angle{-24.3°}$, and therefore, $\mathbf{I}_O = 10.22\angle{-9.28°}$ mA.

SUPERPOSITION

The superposition principle applies to phasor responses only if all of the independent sources driving the circuit have the *same frequency*. That is, when the input sources have the same frequency, we can find the phasor response due to each source acting alone and obtain the total response by adding the individual phasors. If the sources have different frequencies, then superposition can still be used but its application is different. With different frequency sources, each source must be treated in a separate steady-state analysis because the element impedances change with frequency. The phasor response for each source must be changed into waveforms and then superposition applied in the time domain. In other words, the superposition principle always applies in the time domain. It also applies in the phasor domain when all independent sources have the same frequency. The following examples illustrate both cases.

EXAMPLE 8–15

Use superposition to find the steady-state voltage $v_R(t)$ in Figure 8–32 for $R = 20\,\Omega$, $L_1 = 2\,\text{mH}, L_2 = 6\,\text{mH}, C = 20\,\mu\text{F}, v_{S1} = 100\cos5000t$ V, and $v_{S2} = 120\cos(5000t+30°)$ V.

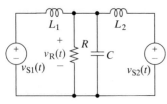

FIGURE 8–32

SOLUTION:
In this example, the two sources operate at the same frequency. Figure 8–33(a) shows the phasor-domain circuit with source 2 turned off and replaced by a short circuit. The three elements in parallel in Figure 8–33(a) produce an equivalent impedance of

$$Z_{EQ1} = \frac{1}{\dfrac{1}{20} + \dfrac{1}{-j10} + \dfrac{1}{j30}} = 7.20 - j9.60\,\Omega$$

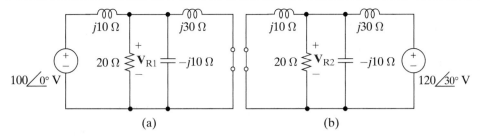

FIGURE 8–33

(a)   (b)

By voltage division, the phasor response $\mathbf{V}_{R1}$ is

$$\mathbf{V}_{R1} = \frac{Z_{EQ1}}{j10 + Z_{EQ1}} 100\angle 0°$$

$$= 92.3 - j138 = 166\angle -56.3° \text{ V}$$

Figure 8–33(b) shows the phasor-domain circuit with source 1 turned off and source 2 on. The three elements in parallel in Figure 8–33(b) produce an equivalent impedance of

$$Z_{EQ2} = \frac{1}{\dfrac{1}{20} + \dfrac{1}{-j10} + \dfrac{1}{j10}} = 20 - j0 \ \Omega$$

By voltage division, the response $\mathbf{V}_{R2}$ is

$$\mathbf{V}_{R2} = \frac{Z_{EQ2}}{j30 + Z_{EQ2}} 120\angle 30°$$

$$= 59.7 - j29.5 = 66.6\angle -26.3° \text{ V}$$

Since the sources have the same frequency, the total response can be found by adding the individual phasor responses $\mathbf{V}_{R1}$ and $\mathbf{V}_{R2}$:

$$\mathbf{V}_R = \mathbf{V}_{R1} + \mathbf{V}_{R2} = 152 - j168 = 227\angle -47.9° \text{ V}$$

The time-domain function corresponding to the phasor sum is

$$v_R(t) = \text{Re}\{\mathbf{V}_R e^{j5000t}\} = 227 \cos(5000t - 47.9°) \text{ V}$$

The overall response can also be obtained by adding the time-domain functions corresponding to the individual phasor responses $\mathbf{V}_{R1}$ and $\mathbf{V}_{R2}$:

$$v_R(t) = \text{Re}\{\mathbf{V}_{R1} e^{j5000t}\} + \text{Re}\{\mathbf{V}_{R2} e^{j5000t}\}$$

$$= 166 \cos(5000t - 56.3°) + 66.6 \cos(5000t - 26.3°) \text{ V}$$

You are encouraged to show that the two expressions for $v_R(t)$ are equivalent using the additive property of sinusoids. ∎

Exercise 8–23

The two sources in Figure 8–34 have the same frequency. Use superposition to find the phasor current $\mathbf{I}_X$.

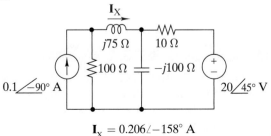

FIGURE 8–34

Answer: $\mathbf{I}_X = 0.206\angle -158° \text{ A}$

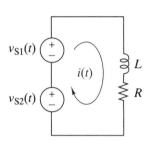

FIGURE 8–35

EXAMPLE 8–16

Use superposition to find the steady-state current $i(t)$ in Figure 8–35 for $R = 10\,\mathrm{k}\Omega$, $L = 200\,\mathrm{mH}$, $v_{S1} = 24\cos 20{,}000t$ V, and $v_{S2} = 8\cos(60{,}000t + 30°)$ V.

SOLUTION:
In this example the two sources operate at different frequencies. With source 2 off, the input phasor is $\mathbf{V}_{S1} = 24\angle0°$ V at a frequency of $\omega = 20\,\mathrm{krad/s}$. At this frequency the equivalent impedance of the inductor and resistor is

$$Z_{EQ1} = R + j\omega L = (10 + j4)\,\mathrm{k}\Omega$$

The phasor current due to source 1 is

$$\mathbf{I}_1 = \frac{\mathbf{V}_{S1}}{Z_{EQ1}} = \frac{24\angle0°}{10{,}000 + j4000} = 2.23\angle-21.8°\,\mathrm{mA}$$

With source 1 off and source 2 on, the input phasor $\mathbf{V}_{S2} = 8\angle30°$ V at a frequency of $\omega = 60\,\mathrm{krad/s}$. At this frequency the equivalent impedance of the inductor and resistor is

$$Z_{EQ2} = R + j\omega L = (10 + j12)\,\mathrm{k}\Omega$$

The phasor current due to source 2 is

$$\mathbf{I}_2 = \frac{\mathbf{V}_{S2}}{Z_{EQ2}} = \frac{8\angle30°}{10{,}000 + j12000} = 0.512\angle-20.2°\,\mathrm{mA}$$

The two input sources operate at different frequencies, so the phasors' responses $\mathbf{I}_1$ and $\mathbf{I}_2$ cannot be added to obtain the overall response. In this case the overall response is obtained by adding the corresponding time-domain functions.

$$
\begin{aligned}
i(t) &= \mathrm{Re}\{\mathbf{I}_1 e^{j20{,}000t}\} + \mathrm{Re}\{\mathbf{I}_2 e^{j60{,}000t}\} \\
&= 2.23\cos(20{,}000t - 21.8°) + 0.512\cos(60{,}000t - 20.2°)\,\mathrm{mA} \quad \blacksquare
\end{aligned}
$$

Exercise 8–24

Use superposition to find the output voltage $v_O(t)$ in the circuit of Figure 8–36 if $i_S(t) = 100\cos(10{,}000t)$ mA and $v_S(t) = 20\cos(20{,}000t - 45°)$ V.

FIGURE 8–36

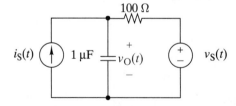

Answer: $v_O(t) = 7.07\cos(10kt - 45°) + 8.94\cos(20kt - 108.4°)\,\mathrm{V}$

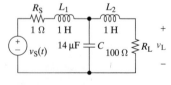

FIGURE 8–37

EVALUATION EXAMPLE 8–17

The voltage source in Figure 8–37 produces a 60-Hz sinusoid with a peak amplitude of 200 V plus a 180-Hz third harmonic with a peak amplitude of 10 V, that is, $v_S(t) = 200\cos(2\pi\,60t) + 10\cos(2\pi\,180t)$ V. The purpose of the LC T-circuit is to reduce the relative size of the third-harmonic component delivered to the 100-Ω load resistor.

Use MATLAB to calculate the voltage amplitudes across the load caused by the 60-Hz and the 180-Hz signals. For each frequency, determine the magnitude of the ratio of the output voltage to the input voltage. Is the circuit performing its task?

SOLUTION:

We will do a small amount of circuit analysis and then rely on MATLAB to perform the necessary calculations for each input frequency. Figure 8–38 shows the circuit in the phasor domain.

We can use circuit reduction and voltage division to solve for the load voltage in the phasor domain. Let Z_{EQ1} be the equivalent impedance of inductor L_2 and resistor R_L in series, and let Z_{EQ2} be the equivalent impedance of the capacitor in parallel with impedance Z_{EQ1}. We will first solve for Z_{EQ1} and Z_{EQ2} and then use voltage division to find the voltage across the capacitor. We can then apply voltage division a second time to find the load voltage for each frequency. Using the for command in MATLAB reduces the amount of code we have to write for this problem. The MATLAB code follows.

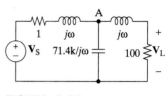

FIGURE 8–38

```
% Define the fixed parameters
Rs = 1;
L1 = 1;
L2 = 1;
C = 14e-6;
RL = 100;
% Create a list of frequencies to evaluate
wList = 2*pi*[60 180];
% List the input amplitudes and create vectors for results
AList = [200 10];
VList = zeros(size(wList));
KList = zeros(size(wList));
% Use a for loop to evaluate the response to each frequency
for n = 1:length(wList)
    w = wList(n);
    A = AList(n);
    Zeq1 = j*w*L2 + RL;
    ZC = 1/(j*w*C);
    Zeq2 = Zeq1*ZC/(Zeq1+ZC);
    VA = Zeq2*A/(Rs + j*w*L1 + Zeq2);
    VList(n) = RL*VA/Zeq1;
    KList(n) = VList(n)/A;
end
% Display the results
VList
KList
% Display the magnitudes of the results
VListMag = abs(VList)
KListMag = abs(KList)
```

The preceding code generates the following results:

```
VList =
   1.0e+002 *
   -1.9969 - 0.0880i   -0.0001 + 0.0006i
KList =
   -0.9985 - 0.0440i   -0.0005 + 0.0055i
VListMag =
   199.8841 0.0553
KListMag =
   0.9994     0.0055
```

The amplitude of the 60-Hz component is virtually unchanged as it moves through the circuit, and its amplitude at the load is 99.94% of its original value. The amplitude of the 180-Hz component is drastically reduced at the output, with a load amplitude less than 1% of its original value. This circuit is correctly filtering the higher-frequency component because the impedances of the capacitor and the inductors change with frequency ■

Exercise 8–25

Analyze the sinusoidal steady-state behavior of the circuit shown in Figure 8–37 in more detail. To do so, find the magnitude of the ratio of the output voltage to the input voltage for the range of frequencies from 1 Hz to 1 kHz. For simplicity, assume that the input signal always has a magnitude of 1 V. Examine at least 500 data points in the frequency range and space them logarithmically with the MATLAB command `logspace`. Plot the results in terms of magnitude versus frequency (in hertz) on log-log axes. Use the plot to verify the results in Example 8–17.

FIGURE 8–39

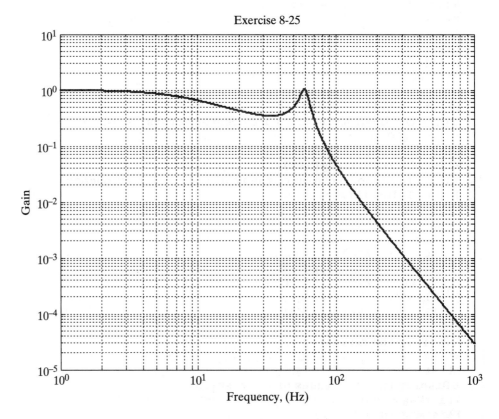

Answer: See Figure 8–39. The plot is consistent with the results presented in Example 8–17.

THÉVENIN AND NORTON EQUIVALENT CIRCUITS

In the phasor domain, a two-terminal circuit containing linear elements and sources can be replaced by the Thévenin or Norton equivalent circuits shown in Figure 8–40. The general concept of Thévenin's and Norton's theorems and their restrictions are the same as in the resistive circuit studied in Chapter 3. The important difference here is that the signals $\mathbf{V}_T, \mathbf{I}_N, \mathbf{V},$ and $\mathbf{I}$ are phasors, and Z_T and Z_L are complex numbers representing the source and load impedances.

Finding the Thévenin or Norton equivalent of a phasor circuit involves the same process as for resistance circuits, except that now we must manipulate complex numbers. The Thévenin and Norton circuits are equivalent to each other, so their circuit parameters are related as follows:

$$\mathbf{V}_{OC} = \mathbf{V}_T = \mathbf{I}_N Z_T$$

$$\mathbf{I}_{SC} = \frac{\mathbf{V}_T}{Z_T} = \mathbf{I}_N$$

$$Z_T = \frac{\mathbf{V}_{OC}}{\mathbf{I}_{SC}}$$

(8–29)

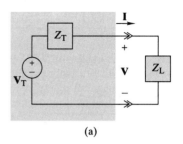

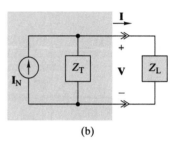

FIGURE 8–40 *Thévenin and Norton equivalent circuits in the phasor analysis.*

Algebraically, the results in Eq. (8–29) are identical to the corresponding equations for resistance circuits. The important difference is that these equations involve phasors and impedances rather than waveforms and resistances. These equations point out that we can determine a Thévenin or Norton equivalent by finding any two of the following quantities: (1) the open-circuit voltage $\mathbf{V}_{OC}$, (2) the short-circuit current $\mathbf{I}_{SC}$, and (when there are no dependent sources) (3) the impedance Z_T looking back into the source circuit with all independent sources turned off.

The relationships in Eq. (8–29) define source transformations that allow us to convert a voltage source in series with an impedance into a current source in parallel with the same impedance, or vice versa. Phasor-domain **source transformations** simplify circuits and are useful in formulating general node-voltage or mesh-current equations, discussed in the next section.

The next two examples illustrate applications of source transformation and Thévenin equivalent circuits. We will follow those examples with three additional examples exploring these concepts.

EXAMPLE 8–18

The Thévenin circuit in Figure 8–41(a) is operating in the phasor domain. If $\mathbf{V}_T = 120 \angle 30°$ V and $Z_T = 100 - j50$ Ω, perform a source transformation resulting in the Norton circuit of Figure 8–41(b).

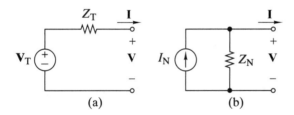

FIGURE 8–41

(a)   (b)

SOLUTION:
The source transformation is related through Ohm's law. Hence, $\mathbf{I}_N = \mathbf{V}_T/Z_T$, and the Norton impedance equals the Thévenin impedance.

$$\mathbf{I}_T = \frac{120\angle 30°}{100 - j50} = \frac{120\angle 30°}{112\angle -26.6°} = 1.07\angle 56.6° \text{ A} \qquad \blacksquare$$

$$Z_N = Z_T = 100 - j50 \ \Omega$$

Exercise 8–26

The Norton circuit in Figure 8–41(b) has a current source $\mathbf{I}_N = 300 - j400$ mA and a Norton impedance Z_N of $100 + j100$ Ω. Find the equivalent Thévenin circuit.

Answers:
$\mathbf{V}_T = 70.7 \angle -8.13°$ V, $Z_T = 141 \angle 45°$ Ω.

◗ DESIGN EXAMPLE 8–19

The circuit of Figure 8–42 is in the phasor domain. Find the Thévenin equivalent circuit that the load Z_L sees. Then design a load Z_L so that $10\angle{-90°}$ V are delivered across it.

FIGURE 8–42

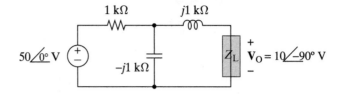

SOLUTION:
We calculate the Thévenin equivalent circuit by removing the load and looking into the open terminals, effectively looking across the capacitor. $\mathbf{V}_T$ is found using a voltage divider and Z_T using the look-back method.

$$\mathbf{V}_T = \left(\frac{-j1k}{1k - j1k}\right)(50 + j0) = \frac{1k\angle{-90°}}{1.414\angle{-45°}}50\angle{0°} = 35.35\angle{-45°}\text{V}$$

$$Z_T = \frac{(1k)(-j1k)}{1k - j1k} + j1k = \frac{1k^2\angle{-90°}}{1.414k\angle{-45°}} + j1k = 707\angle{-45°} + j1k$$

$$Z_T = 500 - j500 + j1000 = 500 + j500\,\Omega$$

We reconnect the load and expand it as $Z_L = R_L + jX_L$. Once again we use a voltage divider and solve for R_L and X_L. We set up the equation as follows

$$\mathbf{V}_L = \frac{Z_L}{Z_L + Z_T}\mathbf{V}_T$$

$$\mathbf{V}_L = 10\angle{-90°} = \frac{R_L + jX_L}{R_L + jX_L + 500 + j500}35.35\angle{-45°}$$

Cross multiplying

$$\left(\frac{10\angle{-90°}}{35.35\angle{-45°}}\right)(R_L + jX_L + 500 + j500) = R_L + jX_L$$

Expanding the left side and collecting like terms, we get

$$200 - 0.8R_L - j0.2R_L + 0.2X_L - j0.8X_L = 0$$

The real and the imaginary parts must both be zero. Hence, we get two equations in two unknowns that we can solve simultaneously.

$$200 = 0.8R_L - 0.2X_L$$

$$0 = -0.2R_L - 0.8X_L$$

Which results in $R_L = 235.3\,\Omega$ and $X_L = -58.8\Omega$. Our load, then, is $Z_L = 235.3 - j58.8\,\Omega$. ∎

Exercise 8–27 _____

Convert the Thévenin circuit found in Example 8–19 into its Norton equivalent. Then repeat the design task in that example.

Answer: $\mathbf{I_N} = 50\angle -90°$ mA, $Z_N = 500 + j500$ Ω. Since the load does not know if it is connected to a Thévenin or a Norton equivalent circuit, it performs the same. Hence, the same Z_L is required to produce the same output, $Z_L = 235.3 - j58.8$ Ω.

EXAMPLE 8–20

Both sources in Figure 8–43(a) operate at a frequency of $\omega = 5000$ rad/s. Find the steady-state voltage $v_R(t)$ using source transformations.

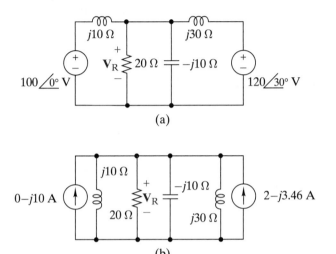

FIGURE 8–43

SOLUTION:

Example 8–15 solves this problem using superposition. In this example we use source transformations. We observe that the voltage sources in Figure 8–43(a) are connected in series with impedances and can be converted into the following equivalent current sources:

$$\mathbf{I_{EQ1}} = \frac{100\angle 0°}{j10} = 0 - j10 \text{ A}$$

$$\mathbf{I_{EQ2}} = \frac{120\angle 30°}{j30} = 2 - j3.46 \text{ A}$$

Figure 8–43(b) shows the circuit after these two source transformations. The two current sources are connected in parallel and can be replaced by a single equivalent current source:

$$\mathbf{I_{EQ}} = \mathbf{I_{EQ1}} + \mathbf{I_{EQ2}} = 2 - j13.46 = 13.6\angle -81.5° \text{ A}$$

The four passive elements are connected in parallel and can be replaced by an equivalent impedance:

$$Z_{EQ} = \frac{1}{\dfrac{1}{20} + \dfrac{1}{-j10} + \dfrac{1}{j10} + \dfrac{1}{j30}} = 16.6\angle 33.7° \text{ Ω}$$

The voltage across this equivalent impedance equals $\mathbf{V_R}$, since one of the parallel elements is the resistor R. Therefore, the unknown phasor voltage is

$$\mathbf{V_R} = \mathbf{I_{EQ}}Z_{EQ} = (13.6\angle -81.5°) \times (16.6\angle 33.7°) = 227\angle -47.9° \text{ V}$$

The value of $\mathbf{V}_R$ is the same as found in Example 8–15 using superposition. The corresponding time-domain function is

$$v_R(t) = \mathrm{Re}\{\mathbf{V}_R e^{j5000t}\} = 227\cos(5000t - 47.9°)\,\mathrm{V} \qquad \blacksquare$$

Exercise 8–28

Repeat Example 8–20 using OrCAD if both sources are operating at a frequency of $\omega = 20\,\mathrm{krad/s}$. (*Hint*: Find L_1, L_2, and C from the data in Example 8–20 first.)

Answer: $$v_R(t) = 9.196\cos(20kt - 163.8°)\,\mathrm{V}$$

EXAMPLE 8–21

Use Thévenin's theorem to find the current $\mathbf{I}_X$ in the bridge circuit shown in Figure 8–44.

FIGURE 8–44

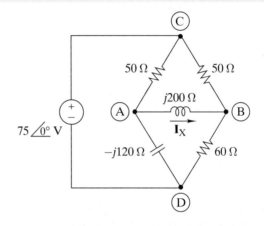

SOLUTION:
Disconnecting the impedance $j200$ from the circuit in Figure 8–44 produces the circuit shown in Figure 8–45(a). The voltage between nodes A and B is the Thévenin voltage since removing the impedance $j200$ leaves an open circuit. The voltages at nodes A and B can each be found by voltage division. Since the open-circuit voltage is the difference between these node voltages, we have

$$\mathbf{V}_T = \mathbf{V}_A - \mathbf{V}_B$$

$$= \frac{-j120}{50 - j120}75\angle0° - \frac{60}{60 + 50}75\angle0°$$

$$= 23.0 - j26.6\,\mathrm{V}$$

FIGURE 8–45

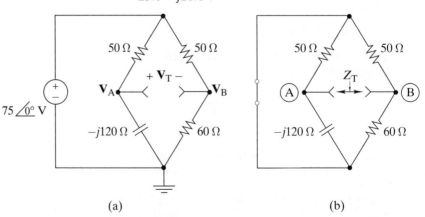

(a)   (b)

Turning off the voltage source in Figure 8–45(a) and replacing it by a short circuit produces the situation shown in Figure 8–45(b). The lookback impedance seen at the interface is a series connection of two pairs of elements connected in parallel. The equivalent impedance of the series/parallel combination is

$$Z_T = \cfrac{1}{\cfrac{1}{50} + \cfrac{1}{-j120}} + \cfrac{1}{\cfrac{1}{50} + \cfrac{1}{60}} = 69.9 - j17.8\,\Omega$$

Given the Thévenin equivalent circuit, we treat the impedance $j200$ as a load connected at the interface and calculate the resulting load current $\mathbf{I}_X$ as

$$\mathbf{I}_X = \frac{\mathbf{V}_T}{Z_T + j200} = \frac{23.0 - j26.6}{69.9 + j182} = 0.180\angle{-118°}\,\text{A} \qquad \blacksquare$$

Exercise 8–29

(a) Find the Thévenin equivalent circuit seen by the inductor in Figure 8–34.
(b) Use the Thévenin equivalent to calculate the current $\mathbf{I}_X$.

Answers:

(a) $\mathbf{V}_T = -15.4 - j22.6\,\text{V}$, $Z_T = 109.9 - j0.990\,\Omega$
(b) $\mathbf{I}_X = 0.206\angle{-158°}\,\text{A}$

EXAMPLE 8–22

In the steady state, the open-circuit voltage at an interface is observed to be

$$v_{OC}(t) = 12\cos 2000t\,\text{V}$$

When a 50-mH inductor is connected across the interface, the interface voltage is observed to be

$$v(t) = 17\cos(2000t + 45°)\,\text{V}$$

Find the Thévenin equivalent circuit at the interface.

SOLUTION:
The phasors for $v_{OC}(t)$ and $v(t)$ are $\mathbf{V}_{OC} = 12\angle 0°$ and $\mathbf{V} = 17\angle 45°$. The phasor Thévenin voltage at the interface is $\mathbf{V}_T = \mathbf{V}_{OC} = 12\angle 0°$. The impedance of the inductor is $Z_L = j\omega L = j2000 \times 0.050 = j100\,\Omega$. When the inductor load is connected across the interface, we use voltage division to express the interface voltage as

$$\mathbf{V} = \frac{Z_L}{Z_T + Z_L}\mathbf{V}_T$$

Inserting the known numerical values yields

$$17\angle 45° = \frac{j100}{Z_T + j100}12\angle 0°$$

Solving for Z_T, we have

$$Z_T = j100 \times \frac{12\angle 0°}{17\angle 45°} - j100 = 49.9 - j50.1\,\Omega$$

The Thévenin equivalent circuit at the interface is defined by $\mathbf{V}_T = 12\angle 0°$ and $Z_T = 49.9 - j50.1\,\Omega$. $\qquad \blacksquare$

Exercise 8–30

By inspection, determine the Thévenin equivalent circuit seen by the capacitor in Figure 8–30 for $\mathbf{V_S} = 10\angle 0°$ V.

Answer: $\mathbf{V_T} = 5\angle 0°$ V, $Z_T = 25 + j50\,\Omega$

Exercise 8–31

In the steady state the short-circuit current at an interface is observed to be

$$i_{SC}(t) = 0.75\sin\omega t \text{ A}$$

When a 150-Ω resistor is connected across the interface, the interface current is observed to be

$$i(t) = 0.6\cos(\omega t - 53.1°) \text{ A}$$

Find the Norton equivalent phasor circuit at the interface.

Answer: $\mathbf{I_N} = 0 - j0.75$ A, $Z_T = 0 + j200\,\Omega$

◀D DESIGN EXAMPLE 8–23

For the circuit shown in Figure 8-46(a), design an interface that converts an input voltage of $\mathbf{V_S} = 90°\angle 0°$ V into an output of $\mathbf{V_O} = 63.6\angle -45°$ V. The circuit is operating at 1000 rad/s.

FIGURE 8–46

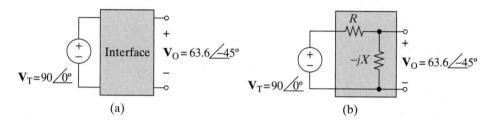

(a)   (b)

SOLUTION:
Let us see what the interface circuit is required to do. We can make a ratio of the output over the input and determine what effect the interface circuit must perform to meet the requirement.

$$\frac{\mathbf{V_O}}{\mathbf{V_S}} = \frac{63.6\angle -45°}{90\angle 0°} = 0.707\angle -45°$$

Since there is a negative phase shift, it appears that a simple voltage divider with a resistor $Z_R = R$ and a capacitor $Z_C = -jX$ could work with the output taken across the capacitor.

$$\frac{\mathbf{V_O}}{\mathbf{V_S}} = 0.707\angle -45° = \frac{-jX}{R - jX} = \frac{X\angle -90°}{\sqrt{R^2 + X^2}\,\angle\tan^{-1}\left(\frac{-X}{R}\right)}$$

Setting magnitudes and angles equal, we can solve for the necessary relationships. Thus

$$\frac{X}{\sqrt{R^2 + X^2}} = 0.707$$

and

$$\frac{\angle -90°}{\angle\tan^{-1}\left(\frac{-X}{R}\right)}\,\angle -45°$$

If we select R and X to have the same value then the magnitude of the denominator would be $\sqrt{2}X$ and the angle would be $-45°$. This would produce the desired result. The interface is shown in Figure 8–46(b).

Next, we need to select R and C. If we pick $R = 1$ kΩ we can calculate the value of the capacitor. Since the radian frequency is 1000 rad/s, C is found to be 1 μF. ■

◑ Design Exercise 8–32

Repeat the problem in Example 8–23, except you cannot use capacitors.

A n s w e r: Use an RL series circuit with the output taken across the resistor. Select $R = 1$ kΩ and $L = 1$ H. Other solutions are possible.

8–5 GENERAL CIRCUIT ANALYSIS WITH PHASORS

The previous sections discuss basic analysis methods based on equivalence, reduction, and circuit theorems. These methods are valuable because we work directly with element impedances and thereby gain insight into steady-state circuit behavior. We also need general methods, such as node and mesh analysis, to deal with more complicated circuits than the basic methods can easily handle. These general methods use node-voltage or mesh-current variables to reduce the number of equations that must be solved simultaneously.

Node-voltage equations involve selecting a reference node as ground or datum and assigning a node-to-datum voltage to each of the remaining nonreference nodes. Because of KVL, the voltage between any two nodes equals the difference of the two node voltages. This fundamental property of node voltages plus the element impedances allow us to write KCL constraints at each of the nonreference nodes.

For example, consider node A in Figure 8–47. The sum of currents leaving this node can be written as

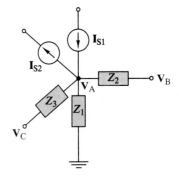

FIGURE 8–47 *An example node.*

$$\mathbf{I}_{S2} - \mathbf{I}_{S1} + \frac{\mathbf{V}_A}{Z_1} + \frac{\mathbf{V}_A - \mathbf{V}_B}{Z_2} + \frac{\mathbf{V}_A - \mathbf{V}_C}{Z_3} = 0$$

Rewriting this equation with unknowns grouped on the left and known inputs on the right yields

$$\left[\frac{1}{Z_1} + \frac{1}{Z_2} + \frac{1}{Z_3}\right]\mathbf{V}_A - \frac{1}{Z_2}\mathbf{V}_B - \frac{1}{Z_3}\mathbf{V}_C = \mathbf{I}_{S1} - \mathbf{I}_{S2}$$

Expressing this result in terms of admittances produces the following equation.

$$[Y_1 + Y_2 + Y_3]\mathbf{V}_A - [Y_2]\mathbf{V}_B - [Y_3]\mathbf{V}_C = \mathbf{I}_{S1} - \mathbf{I}_{S2}$$

This equation has a familiar pattern. The unknowns $\mathbf{V}_A$, $\mathbf{V}_B$, and $\mathbf{V}_C$ are the node-voltage phasors. The coefficient $[Y_1 + Y_2 + Y_3]$ of $\mathbf{V}_A$ is the sum of the admittances of all of the elements connected to node A. The coefficient $[Y_2]$ of $\mathbf{V}_B$ is admittance of the elements connected between nodes A and B, while $[Y_3]$ is the admittance of the elements connected between nodes A and C. Finally, $\mathbf{I}_{S1}$ and $\mathbf{I}_{S2}$ are the phasor current sources connected to node A, with $\mathbf{I}_{S1}$ directed into and $\mathbf{I}_{S2}$ directed away from the node. These observations suggest that we can write node-voltage equations for phasor circuits by inspection, just as we did with resistive circuits.

Circuits that can be drawn on a flat surface with no crossovers are called **planar** circuits. The mesh-current variables are the loop currents assigned to each mesh in a

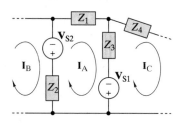

FIGURE 8-48 *An example mesh.*

planar circuit. Because of KCL, the current through any two-terminal element is equal to the difference of the two adjacent meshes. This fundamental property of mesh currents together with the element impedances allow us to write KVL constraints around each of the meshes.

For example, the sum of voltages around mesh A in Figure 8–48 is

$$Z_1 \mathbf{I}_A + Z_2 [\mathbf{I}_A - \mathbf{I}_B] + Z_3 [\mathbf{I}_A - \mathbf{I}_C] - \mathbf{V}_{S1} + \mathbf{V}_{S2} = 0$$

Arranging this equation in standard form yields

$$[Z_1 + Z_2 + Z_3]\mathbf{I}_A - [Z_2]\mathbf{I}_B - [Z_3]\mathbf{I}_C = \mathbf{V}_{S1} - \mathbf{V}_{S2}$$

This equation also displays a familiar pattern. The unknowns $\mathbf{I}_A$, $\mathbf{I}_B$, and $\mathbf{I}_C$ are mesh-current phasors. The coefficient $[Z_1 + Z_2 + Z_3]$ of $\mathbf{I}_A$ is the sum of the impedances in mesh A. The coefficient $[Z_2]$ of $\mathbf{I}_B$ is the impedance in both mesh A and mesh B, while $[Z_3]$ is the impedance common to meshes A and C. Finally, $\mathbf{V}_{S1}$ and $\mathbf{V}_{S2}$ are the phasor voltage sources in mesh A. These observations allow us to write mesh-current equations for phasor circuits by inspection.

The preceding discussion assumes that the circuit contains only current sources in the case of node analysis and voltage sources in mesh analysis. If there is a mixture of sources, we may be able to use the source transformations discussed in Sect. 8–4 to convert from voltage to current sources, or vice versa. A source transformation is possible only when there is an impedance connected in series with a voltage source or an impedance in parallel with a current source. When a source transformation is not possible, we use the phasor version of the modified node- and mesh-analysis methods described in Chapter 3.

Formulating a set of equilibrium equations in phasor form is a straightforward process involving concepts that we have used before in Chapters 3 and 4. Once formulated, we use Cramer's rule or Gaussian reduction to solve these equations for phasor responses, although this requires manipulating linear equations with complex coefficients. In principle, the solution process can be done by hand, but as a practical matter circuits with more than three nodes or meshes are best handled using computer tools. Modern hand-held scientific calculators and math analysis programs like MATLAB can deal with sets of linear equations with complex coefficients. Circuit analysis programs such as OrCAD have ac analysis options that handle steady-state circuit analysis problems.

With modern computer tools such as OrCAD and MATLAB, one may question why one should bother with hand solutions at all. Until fairly recently, the reason was that most software tools could not effectively handle symbolic operators. That is no longer an issue, with symbolic manipulators available in many software packages. MATLAB, for example, has a symbolic toolbox that permits the solution of analytical problems using symbolic variables. Nevertheless, for one to become a good engineer, it is essential to develop a sense of what looks right. This sense is not innate and must be developed with practice. The next few examples and exercises will help you to develop this sense both classically by hand and by using software tools.

EXAMPLE 8–24

The circuit in Figure 8–49(a) is operating in the sinusoidal steady state. Convert the circuit into the phasor domain and find the phasor voltages at nodes A and B.

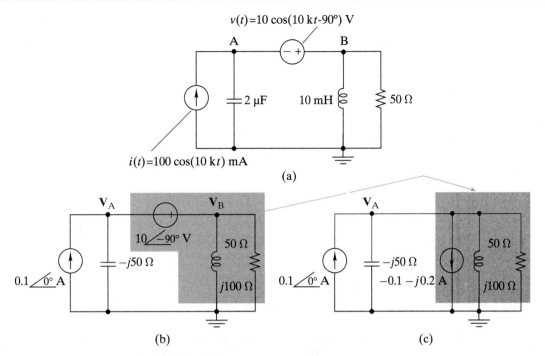

FIGURE 8–49

SOLUTION:

We start by converting the time-domain circuit into the phasor domain resulting in Figure 8–49(b). The voltage source in Figure 8–49(b) is connected in series with an impedance consisting of a resistor and inductor connected in parallel. The equivalent impedance of this parallel combination is

$$Z_{EQ} = \frac{1}{\frac{1}{50} + \frac{1}{j100}} = 40 + j20 \ \Omega$$

Applying a source transformation produces an equivalent current source of

$$I_{EQ} = \frac{10\angle -90°}{40 + j20} = -0.1 - j0.2 \ A$$

Figure 8–49(c) shows the circuit produced by the source transformation. Note that the transformation eliminates node B. The node-voltage equation at the remaining nonreference node in Figure 8–49(c) is

$$\left(\frac{1}{-j50} + \frac{1}{j100} + \frac{1}{50}\right)V_A = 0.1\angle 0° - (-0.1 - j0.2)$$

Solving for V_A yields

$$V_A = \frac{0.2 + j0.2}{0.02 + j0.01} = 12 + j4 = 12.6\angle 18.4° \ V$$

Referring to Figure 8–49(b), we see that KVL requires $V_B = V_A + 10\angle -90°$. Therefore, V_B is found to be

$$V_B = (12 + j4) + 10\angle -90° = 12 - j6 = 13.4\angle -26.6° \ V \qquad ∎$$

Exercise 8–33

Use OrCAD to solve for the node voltages $\mathbf{V}_A$ and $\mathbf{V}_B$ in the circuit of Figure 8–49(a).

FIGURE 8–50 *Copyright © Cadence Design Systems, Inc. Used with permission.*

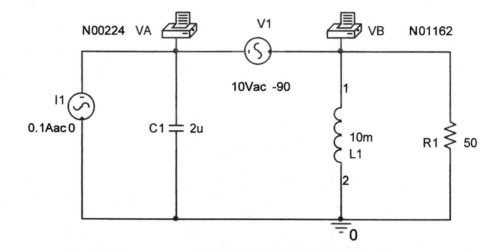

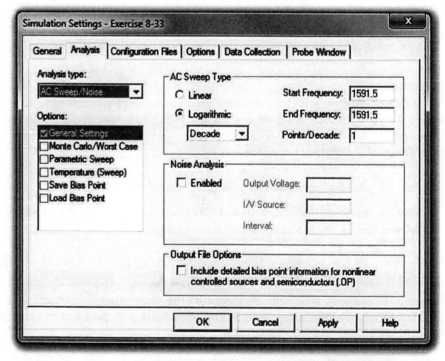

	FREQ	VM(N00224)	VP(N00224)	VR(N00224)	VI(N00224)
Node A →	1.592E+03	1.265E+01	1.844E+01	1.200E+01	4.000E+00

	FREQ	VM(N01162)	VP(N01162)	VR(N01162)	VI(N01162)
Node B →	1.592E+03	1.342E+01	-2.656E+01	1.200E+01	-6.000E+00

A n s w e r s: See Figure 8–50. $\mathbf{V}_A = 12 + j4$ V and $\mathbf{V}_B = 12 - j6$ V

EXAMPLE 8-25

Use node analysis to find the current $\mathbf{I}_X$ in Figure 8–51.

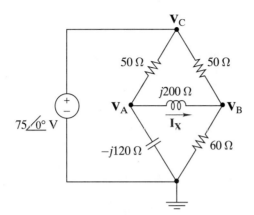

FIGURE 8–51

SOLUTION:

In this example we use node analysis on a problem solved in Example 8–21 using a Thévenin equivalent circuit. The voltage source cannot be replaced by source transformation because it is not connected in series with an impedance. By inspection, the node equations at nodes A and B are

$$\text{Node A:} \quad \frac{\mathbf{V}_A}{-j120} + \frac{\mathbf{V}_A - \mathbf{V}_B}{j200} + \frac{\mathbf{V}_A - \mathbf{V}_C}{50} = 0$$

$$\text{Node B:} \quad \frac{\mathbf{V}_B}{60} + \frac{\mathbf{V}_B - \mathbf{V}_A}{j200} + \frac{\mathbf{V}_B - \mathbf{V}_C}{50} = 0$$

A node equation at node C is not required because the voltage source forces the condition $\mathbf{V}_C = 75 \angle 0°$. Substituting this constraint into the equations of nodes A and B and arranging the equations in standard form yields two equations in two unknowns:

$$\text{Node A:} \quad \left(\frac{1}{50} + \frac{1}{-j120} + \frac{1}{j200}\right)\mathbf{V}_A - \left(\frac{1}{j200}\right)\mathbf{V}_B = \left(\frac{75\angle 0°}{50}\right)$$

$$\text{Node B:} \quad -\left(\frac{1}{j200}\right)\mathbf{V}_A + \left(\frac{1}{50} + \frac{1}{60} + \frac{1}{j200}\right)\mathbf{V}_B = \left(\frac{75\angle 0°}{50}\right)$$

Solving these equations for $\mathbf{V}_A$ and $\mathbf{V}_B$ yields

$$\mathbf{V}_A = 70.4 - j21.4 \text{ V}$$
$$\mathbf{V}_B = 38.6 - j4.33 \text{ V}$$

Using these values for $\mathbf{V}_A$ and $\mathbf{V}_B$, the unknown current is found to be

$$\mathbf{I}_X = \frac{\mathbf{V}_A - \mathbf{V}_B}{j200} = \frac{31.8 - j17.1}{j200} = 0.180\angle -118° \text{ A}$$

This value of $\mathbf{I}_X$ is the same as the answer obtained in Example 8–21. ∎

Exercise 8-34

Find the Thèvenin equivalent circuit seen by the capacitor in the circuit of Figure 8–51.

Answer: $\mathbf{V}_T = 72.52\angle 5.87° \text{ V}$, $Z_T = 47.07\angle 13.36° \text{ Ω}$.

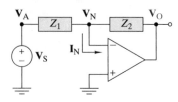

FIGURE 8–52

EXAMPLE 8 – 2 6

Use node-voltage analysis to determine the phasor input-output relationship of the OP AMP circuit in Figure 8–52.

SOLUTION:

In the sinusoidal steady state the sum of currents leaving the inverting input node is

$$\frac{\mathbf{V_N} - \mathbf{V_S}}{Z_1} + \frac{\mathbf{V_N} - \mathbf{V_O}}{Z_2} + \mathbf{I_N} = 0$$

This is the only required node equation, since the input source forces the condition $\mathbf{V_A} = \mathbf{V_S}$ and no node equation is ever required at an OP AMP output. In the time domain the i–v relationships of an ideal OP AMP are $v_P(t) = v_N(t)$ and $i_P(t) = i_N(t) = 0$. In the sinusoidal steady state these equations are written in phasor form as $\mathbf{V_P} = \mathbf{V_N}$ and $\mathbf{I_P} = \mathbf{I_N} = 0$. In the present case this means $\mathbf{V_N} = 0$ since the noninverting input is grounded. When the ideal OP AMP constraints are inserted in the node equation, we can solve for the OP AMP input-output relationship as

$$\mathbf{V_O} = -\frac{Z_2}{Z_1} \mathbf{V_S}$$

This result is the phasor-domain version of the inverting amplifier configuration. In the phasor domain, the "gain" $K = -Z_2/Z_1$ is determined by a ratio of impedances rather than resistances. Thus, the gain affects both the amplitude and the phase angle of the steady-state output. ∎

Exercise 8–35

In the circuit of Figure 8–52, Z_1 is a 1-kΩ resistor and Z_2 is the parallel combination of a 10-kΩ resistor and a 1-μF capacitor. Determine the output voltage $v_O(t)$ if the input is $v_S(t) = 1 \cos 100t$ V.

Answer: $v_O(t) = 7.07 \cos(100t + 135°)$ V.

⊟ EVALUATION EXAMPLE 8 – 27

A vendor claims that the circuit shown in Figure 8-53 is a **high-pass filter** with a cutoff frequency of $\omega_C = 1/R_1C$ and a pass-band gain of $-R_2/R_1$. Verify his claim by finding the phasor voltage ratio $\mathbf{V_O}/\mathbf{V_S}$ and determining at what radian frequency the gain falls to 0.707 of the pass-band (maximum) value.

FIGURE 8–53

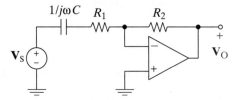

SOLUTION:

The circuit is an inverting amplifier whose input-output relationship was found in Example 8–26 as

$$\mathbf{V_O} = -\frac{Z_2}{Z_1} \mathbf{V_S}$$

$$\mathbf{V_O} = -\frac{R_2}{R_1 + \dfrac{1}{j\omega C}} \mathbf{V_S} = -\frac{j\omega C R_2}{j\omega C R_1 + 1} \mathbf{V_S} = -\left(\frac{R_2}{R_1}\right)\left(\frac{j\omega}{j\omega + \dfrac{1}{R_1 C}}\right) \mathbf{V_S}$$

The phasor voltage ratio is

$$\frac{\mathbf{V_O}}{\mathbf{V_S}} = -\left(\frac{R_2}{R_1}\right)\left(\frac{j\omega}{j\omega + \frac{1}{R_1 C}}\right)$$

At dc, ($\omega = 0,$) $\mathbf{V_O} = 0$ and as $\omega \to \infty$, $\mathbf{V_O} \to -\left(\frac{R_2}{R_1}\right)\mathbf{V_S}$. The maximum value occurs at infinite frequency with a gain of $-\left(\frac{R_2}{R_1}\right)$. This is a high-pass filter as stated, since low frequencies are blocked with a low gain and high frequencies are allowed to pass. We still need to determine the frequency where the gain falls to $0.707 \times -\left(\frac{R_2}{R_1}\right)$. The magnitude of the phasor voltage ratio is

$$\left|\frac{\mathbf{V_O}}{\mathbf{V_S}}\right| = \left|-\left(\frac{R_2}{R_1}\right)\left(\frac{j\omega}{j\omega + \frac{1}{R_1 C}}\right)\right| = \left(\frac{R_2}{R_1}\right)\frac{\omega}{\sqrt{\omega^2 + \left(\frac{1}{R_1 C}\right)^2}}$$

The value of ω that makes the output equal 0.707 of the maximum occurs at $1/R_1 C$ as stated.

$$\left.\left|\frac{\mathbf{V_O}}{\mathbf{V_S}}\right|\right|_{\omega=1/R_1 C} = \left(\frac{R_2}{R_1}\right)\frac{\frac{1}{R_1 C}}{\sqrt{\left(\frac{1}{R_1 C}\right)^2 + \left(\frac{1}{R_1 C}\right)^2}} = \left(\frac{R_2}{R_1}\right)\frac{1}{\sqrt{2}} = 0.707 \times \left(\frac{R_2}{R_1}\right)$$

The vendor accurately described his product.   ∎

◐ Design Exercise 8–36 _____

Use the circuit of Figure 8–53 to design a high-pass filter with a pass-band gain of -100 and a cutoff frequency ω_C of 10,000 rad/s. Use standard parts (See the inside rear cover).

A n s w e r :   Choose $R_1 = 1$ kΩ, then $C = 0.1$ μF and $R_2 = 100$ kΩ. Many other solutions are possible.

E X A M P L E   8 – 2 8

Find the input-output relationship $\mathbf{V_O}/\mathbf{V_S} = K$ for the circuit of Figure 8–54. Then for $R_1 = 1$ k$\Omega, C_1 = 0.1$ μF, $R_2 = 10$ kΩ, and $C_2 = 1$ μF, use MATLAB to create a plot of the log of the magnitude of the input-output relationship K versus the log of the frequency from 1 rad/s to 1 Mrad/s. Discuss the characteristics of the result.

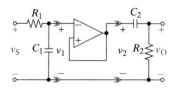

FIGURE 8–54

SOLUTION:
The OP AMP follower effectively isolates the two RC circuits. One can use the voltage division rule for each RC circuit and then simply multiply the results together to obtain the desired input-output relationship.
    The output of the first RC circuit is

$$\mathbf{V_1} = \frac{\frac{1}{j\omega C_1}}{R_1 + \frac{1}{j\omega C_1}}\mathbf{V_S} = \frac{1}{R_1 C_1 j\omega + 1}\mathbf{V_S}$$

$\mathbf{V}_1$ is the input to the follower. Hence, the output, $\mathbf{V}_2$, is equal to the input, $\mathbf{V}_1$. The output of the second RC circuit, then, is another voltage divider:

$$\mathbf{V}_O = \frac{R_2}{R_2 + \dfrac{1}{j\omega C_2}} \mathbf{V}_2 = \frac{j\omega R_2 C_2}{R_2 C_2 j\omega + 1} \mathbf{V}_2 = \frac{j\omega R_2 C_2}{R_2 C_2 j\omega + 1} \mathbf{V}_1$$

Substituting $\mathbf{V}_1$ from the first RC circuit and dividing by $\mathbf{V}_S$, we obtain our result:

$$K = \frac{\mathbf{V}_O}{\mathbf{V}_S} = \left[\frac{j\omega R_2 C_2}{R_2 C_2 j\omega + 1}\right]\left[\frac{1}{R_1 C_1 j\omega + 1}\right] = \left[\frac{j\omega}{j\omega + \dfrac{1}{R_2 C_2}}\right]\left[\frac{\dfrac{1}{R_1 C_1}}{j\omega + \dfrac{1}{R_1 C_1}}\right]$$

The magnitude of K with the values of the R's and C's substituted is given as

$$K = \left[\frac{\omega}{\sqrt{\omega^2 + \left(\dfrac{1}{R_2 C_2}\right)^2}}\right]\left[\frac{\dfrac{1}{R_1 C_1}}{\sqrt{\omega^2 + \left(\dfrac{1}{R_1 C_1}\right)^2}}\right] = \left[\frac{\omega}{\sqrt{\omega^2 + 100^2}}\right]\left[\frac{10{,}000}{\sqrt{\omega^2 + 10{,}000^2}}\right]$$

The following MATLAB code will produce the desired plot:

```
% Create the range over which we want to have MATLAB plot the
response.
% Use logspace to cover from 10^0 rad/s to 10^6 rad/s.
w=logspace(0,6);
% Define the gain of the function.
K=w./(sqrt((w.*w)+10000)).*1e4./(sqrt((w.*w)+1e8));
% Request a log-log plot of w vs K.
loglog(w,K)
grid on
xlabel('Frequency in rad/s')
ylabel('Gain (K)')
```

MATLAB returns the plot shown in Figure 8–55.

FIGURE 8–55

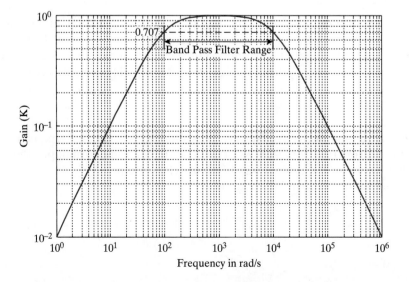

DISCUSSION: *The plot shows a prototypical characteristic of a **bandpass filter**. Signals containing frequencies less than about 100 rad/s or greater than about 10 krad/s are attenuated (decreased). Those signals that fall in the band of frequencies between these two frequencies can pass through with little or no attenuation—hence the name. The frequencies that form the boundaries between which signals pass are generally considered to be the frequency where the magnitude of the input-output relationship has dropped to 0.707 or $1/\sqrt{2}$ of the peak value.* ∎

◑ Design Exercise 8–37 $\qquad$

Design a bandpass filter with a lower frequency cutoff of 100 Hz and an upper frequency cutoff of 20 kHz.

Answer: Select $R_1 = R_2 = 1\,k\Omega$, $C_1 = 7960\,pF$, and $C_2 = 1.59\,\mu F$ in the circuit shown in Figure 8–55. Other correct solutions are possible.

EXAMPLE 8–29

The circuit in Figure 8–56 is an equivalent circuit of an ac induction motor. The current $\mathbf{I}_S$ is called the stator current, $\mathbf{I}_R$ the rotor current, and $\mathbf{I}_M$ the magnetizing current. Use the mesh-current method to solve for the branch currents $\mathbf{I}_S$, $\mathbf{I}_R$, and $\mathbf{I}_M$.

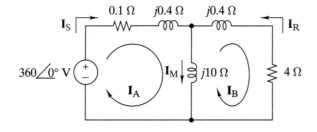

FIGURE 8–56

SOLUTION:
Applying KVL to the sum of voltages around each mesh in Figure 8–56 yields

Mesh A: $\quad -360\angle 0° + (0.1 + j0.4)\mathbf{I}_A + j10(\mathbf{I}_A - \mathbf{I}_B) = 0$

Mesh B: $\quad j10(\mathbf{I}_B - \mathbf{I}_A) + (4 + j0.4)\mathbf{I}_B = 0$

Arranging these equations in standard form yields

$$(0.1 + j10.4)\mathbf{I}_A - (j10)\mathbf{I}_B = 360\angle 0°$$
$$-(j10)\mathbf{I}_A + (4 + j10.4)\mathbf{I}_B = 0$$

Solving these equations for $\mathbf{I}_A$ and $\mathbf{I}_B$ produces

$$\mathbf{I}_A = 79.0 - j48.2\ A$$
$$\mathbf{I}_B = 81.7 - j14.9\ A$$

The required stator, rotor, and magnetizing currents are related to these mesh currents, as follows:

$$\mathbf{I}_S = \mathbf{I}_A = 92.5\angle -31.4°\ A$$
$$\mathbf{I}_R = -\mathbf{I}_B = -81.8 + j14.7 = 83.0\angle 170°\ A$$
$$\mathbf{I}_M = \mathbf{I}_A - \mathbf{I}_B = -2.68 - j33.3 = 33.4\angle -94.6°\ A$$

∎

Exercise 8-38

Use the mesh-current or node-voltage method to find the branch currents I_1, I_2, and I_3 in Figure 8–57.

FIGURE 8-57

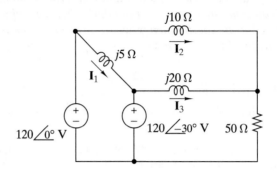

Answer: $I_1 = 12.4\angle -15°$ A, $I_2 = 3.61\angle -16.1°$ A, $I_3 = 1.31\angle 166°$ A

EXAMPLE 8-30

Use the mesh-current method to solve for output voltage V_2 and input impedance Z_{IN} of the circuit in Figure 8–58.

FIGURE 8-58

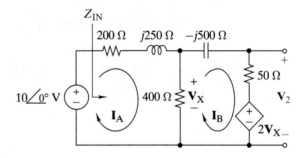

SOLUTION:

The circuit contains a voltage-controlled voltage source. We initially treat the dependent source as an independent source and use KVL to write the sum of voltages around each mesh:

Mesh A: $-10\angle 0° + (200 + j250)I_A + 400(I_A - I_B) = 0$

Mesh B: $400(I_B - I_A) + (50 - j500)I_B + 2V_X = 0$

Arranging these equations in standard form produces

Mesh A: $(600 + j250)I_A - 400I_B = 10\angle 0°$

Mesh B: $-400I_A + (450 - j500)I_B = -2V_X$

Using Ohm's law, the control voltage V_X is

$$V_X = 400(I_A - I_B)$$

Eliminating V_X from the mesh equations yields

Mesh A: $(600 + j250)I_A - 400I_B = 10\angle 0°$

Mesh B: $400I_A + (-350 - j500)I_B = 0$

Solving for the two mesh currents produces

$$\mathbf{I_A} = 10.8 - j11.1\,\text{mA}$$

$$\mathbf{I_B} = -1.93 - j9.95\,\text{mA}$$

Using these values of the mesh currents, the output voltage and input impedance are

$$\mathbf{V_2} = 2\mathbf{V_X} + 50\mathbf{I_B} = 800(\mathbf{I_A} - \mathbf{I_B}) + 50\mathbf{I_B}$$

$$= 800\mathbf{I_A} - 750\mathbf{I_B} = 10.1 - j1.42$$

$$= 10.2\angle{-8.00°}\,\text{V}$$

$$Z_{\text{IN}} = \frac{10\angle{0°}}{\mathbf{I_A}} = \frac{10\angle{0°}}{0.0108 - j0.0111} = 450 + j463\,\Omega$$

■

Exercise 8–39

Use the mesh-current or node-voltage method to find the output voltage $\mathbf{V_2}$ and input impedance Z_{IN} in Figure 8–59.

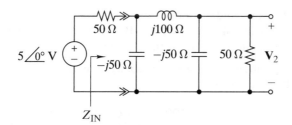

FIGURE 8–59

Answer: $\mathbf{V_2} = 1.77\angle{-135°}\,\text{V}$, $Z_{\text{IN}} = 100 - j100\,\Omega$

EXAMPLE 8–31

In the circuit in Figure 8–60, the input voltage is $v_S(t) = 10\cos 10^5 t\,\text{V}$. Use node-voltage analysis and MATLAB to find the input impedance at the input interface and the proportionality constant relating the input voltage phasor to the phasor voltage across the 50-Ω load resistor, that is, $K = \mathbf{V_O}/\mathbf{V_S}$.

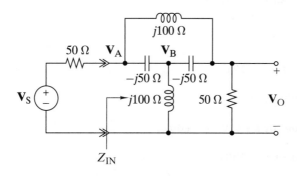

FIGURE 8–60

SOLUTION:

We will solve this example problem using two distinct approaches in MATLAB. Both approaches start by formulating the following node-voltage equations for $\mathbf{V_A}$, $\mathbf{V_B}$, and $\mathbf{V_O}$:

$$\text{Node A:}\quad \frac{\mathbf{V_A} - \mathbf{V_S}}{50} + \frac{\mathbf{V_A} - \mathbf{V_B}}{-j50} + \frac{\mathbf{V_A} - \mathbf{V_O}}{j100} = 0$$

$$\text{Node B:}\quad \frac{\mathbf{V_B} - \mathbf{V_A}}{-j50} + \frac{\mathbf{V_B}}{j100} + \frac{\mathbf{V_B} - \mathbf{V_O}}{-j50} = 0$$

$$\text{Node O:}\quad \frac{\mathbf{V_O} - \mathbf{V_A}}{j100} + \frac{\mathbf{V_O} - \mathbf{V_B}}{-j50} + \frac{\mathbf{V_O}}{50} = 0$$

In the first approach, we will convert these equations to matrix form and then use MATLAB to solve for $\mathbf{V_A}$, $\mathbf{V_B}$, and $\mathbf{V_O}$. We can then use the three voltages to solve for the input impedance and the input-output relationship for the circuit. Converting the node-voltage equations to matrix form, we get

$$\begin{bmatrix} \dfrac{1}{50} + \dfrac{1}{-j50} + \dfrac{1}{j100} & \dfrac{-1}{-j50} & \dfrac{-1}{j100} \\[2mm] \dfrac{-1}{-j50} & \dfrac{1}{-j50} + \dfrac{1}{j100} + \dfrac{1}{-j50} & \dfrac{-1}{-j50} \\[2mm] \dfrac{-1}{j100} & \dfrac{-1}{-j50} & \dfrac{1}{j100} + \dfrac{1}{-j50} + \dfrac{1}{50} \end{bmatrix} \begin{bmatrix} \mathbf{V_A} \\[2mm] \mathbf{V_B} \\[2mm] \mathbf{V_O} \end{bmatrix} = \begin{bmatrix} \dfrac{\mathbf{V_S}}{50} \\[2mm] 0 \\[2mm] 0 \end{bmatrix}$$

Since MATLAB can handle complex numbers, we can use the software to solve for $\mathbf{V_A}$, $\mathbf{V_B}$, and $\mathbf{V_O}$. The following MATLAB code will calculate the node voltages, the input impedance, and the gain of the circuit:

```
% Solve the problem using a matrix approach
% Create the matrices
Vs = 10;
A = [1/50 + 1/(-50j) + 1/(100j) -1/(-50j) -1/(100j);...
    -1/(-50j) 1/(-50j) + 1/(100j) + 1/(-50j) -1/(-50j);...
    -1/(100j) -1/(-50j) 1/(100j) + 1/(-50j) + 1/50];
B = [Vs/50; 0; 0];
% Solve for the node voltages and assign labels
V = A\B;
VA = V(1)
VB = V(2)
VO = V(3)
% Compute the input impedance and circuit gain
Iin = (Vs-VA)/50;
Zin = VA/Iin
K = VO/Vs
```

The results from the code are

```
VA =
    9.5000 + 1.5000i
VB =
    6.0000 + 2.0000i
VO =
    -0.5000 + 1.5000i
Zin =
    5.0000e+001 +3.0000e+002i
K =
    -0.0500 + 0.1500i
```

So the input impedance is $Z_{IN} = 50 + j300\,\Omega$, and the gain is $K = -0.05 + j0.15$.

We can exploit another MATLAB function to solve the same problem in a slightly different way. We will use the original node-voltage equations derived above, but this time we will allow MATLAB to perform all of the calculations by working with these equations in symbolic form. The net result is that we do not have to convert the equations into matrix form to find a solution. The required code is shown below:

```
% Create the symbolic variables and the input signal
clear all
syms Va Vb Vo
Vs = 10;
% Create the node-voltage equations
NVa = (Va-Vs)/50 + (Va-Vb)/(-50j) + (Va-Vo)/(100j);
NVb = (Vb-Va)/(-50j) + Vb/(100j) + (Vb-Vo)/(-50j);
NVo = (Vo-Va)/(100j) + (Vo-Vb)/(-50j) + Vo/50;
% Solve the node voltage equations set equal to zero for the
%symbolic variables
V = solve(NVa,NVb,NVo);
% Compute the input impedance and the proportionality constant
Vout = V.Vo;
Vin = V.Va;
Iin = (Vs-Vin)/50;
Zin = Vin/Iin;
K = Vout/Vs
```

The MATLAB results are

```
Zin =300*i + 50
K =(3*i)/20-1/20
```

which agree with our previous results using the matrix approach. Both approaches are acceptable, and either the nature of the exercise or personal preference will dictate which approach is more efficient for a particular problem. ∎

Exercise 8–40

Use MATLAB and either mesh-current or node-voltage analysis to find the current $\mathbf{I}_X$ in Figure 8–61.

Answer: $\mathbf{I}_X = 1.44\angle 171°\,A$

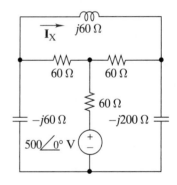

FIGURE 8–61

8–6 ENERGY AND POWER

In the sinusoidal steady state, ac power is transferred from sources to various loads. To study the transfer process, we must calculate the power delivered in the sinusoidal steady state to any specified load. It turns out that there is an upper bound on the available load power; hence, we need to understand how to adjust the load to extract the maximum power from the rest of the circuit. In this section the load is assumed to be made up of passive resistance, inductance, and capacitance. To reach our objectives, we must first study the power and energy delivered to these passive elements in the sinusoidal steady state.

In the sinusoidal steady state the current through a resistor can be expressed as $i_R(t) = I_A \cos(\omega t)$. The instantaneous power delivered to the resistor is

$$p_R(t) = Ri_R^2(t) = RI_A^2 \cos^2(\omega t)$$
$$= \frac{RI_A^2}{2}[1 + \cos(2\omega t)] \qquad (8–30)$$

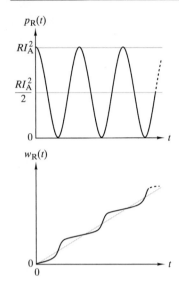

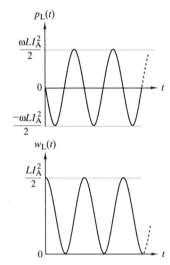

FIGURE 8–62 *Resistor power and energy in the sinusoidal steady state.*

where the identity $\cos^2(x) = \S[1 + \cos(2x)]$ is used to obtain the last line in Eq. (8–30). The energy delivered for $t \geq 0$ is found to be

$$w_R(t) = \int_0^t p_R(x)dx = \frac{RI_A^2}{2}\int_0^t dx + \frac{RI_A^2}{2}\int_0^t \cos 2\omega x \, dx$$

$$= \frac{RI_A^2}{2}t + \frac{RI_A^2}{4\omega}\sin 2\omega t$$

Figure 8–62 shows the time variation of $p_R(t)$ and $w_R(t)$. Note that the power is a periodic function with twice the frequency of the current, that both $p_R(t)$ and $w_R(t)$ are always positive, and that $w_R(t)$ increases without bound. These observations remind us that a resistor is a passive element that dissipates energy.

In the sinusoidal steady state an inductor operates with a current $i_L(t) = I_A\cos(\omega t)$. The corresponding energy stored in the element is

$$w_L(t) = \frac{1}{2}Li_L^2(t) = \frac{1}{2}LI_A^2\cos^2\omega t$$

$$= \frac{1}{4}LI_A^2(1 + \cos 2\omega t)$$

where the identity $\cos^2(x) = \frac{1}{2}[1 + \cos(2x)]$ is again used to produce the last line. The instantaneous power delivered to the inductor is

$$p_L(t) = \frac{dw_L(t)}{dt} = -\frac{\omega LI_A^2}{2}\sin(2\omega t) \qquad (8\text{–}31)$$

Figure 8–63 shows the time variation of $p_L(t)$ and $w_L(t)$. Observe that both $p_L(t)$ and $w_L(t)$ are periodic functions at twice the frequency of the ac current, that $p_L(t)$ is alternately positive and negative, and that $w_L(t)$ is never negative. Since $w_L(t) \geq 0$, the inductor does not deliver net energy to the rest of the circuit. Unlike the resistor's energy in Figure 8–62, the energy in the inductor is bounded by $\frac{1}{2}LI_A^2 \geq w_L(t)$, which means that the inductor does not dissipate energy. Finally, since $p_L(t)$ alternates signs, we see that the inductor stores energy during a positive half cycle and then returns the energy undiminished during the next negative half cycle. Thus, in the sinusoidal steady state there is a lossless interchange of energy between an inductor and the rest of the circuit.

In the sinusoidal steady state the voltage across a capacitor is $v_C(t) = V_A\cos(\omega t)$. The energy stored in the element is

$$w_C(t) = \frac{1}{2}Cv_C^2(t) = \frac{1}{2}CV_A^2\cos^2\omega t$$

$$= \frac{1}{4}CV_A^2(1 + \cos 2\omega t)$$

FIGURE 8–63 *Inductor power and energy in the sinusoidal steady state.*

The instantaneous power delivered to the capacitor is

$$p_C(t) = \frac{dw_C(t)}{dt} = -\frac{\omega CV_A^2}{2}\sin(2\omega t) \qquad (8\text{–}32)$$

Figure 8–64 shows the time variation of $p_C(t)$ and $w_C(t)$. Observe that these relationships are the duals of those found for the inductor. Thus, in the sinusoidal steady state the element power is sinusoidal and there is a lossless interchange of energy between the capacitor and the rest of the circuit.

AVERAGE POWER

We are now in a position to calculate the average power delivered to various loads. The instantaneous power delivered to any of the three passive elements is a periodic function that can be described by an average value. The **average power** in the sinusoidal steady state is defined as

$$P = \frac{1}{T_0} \int_0^{T_0} p(t)\,dt$$

The power variation of the inductor in Eq. (8–31) and capacitor in Eq. (8–32) have the same sinusoidal form. The average value of any sinusoid is zero since the areas under alternate cycles cancel. Hence, the average power delivered to an inductor or capacitor is zero:

Inductor: $P_L = 0$

Capacitor: $P_C = 0$

The resistor power in Eq. (8–30) has both a sinusoidal ac component and a constant dc component $\frac{1}{2}RI_A^2$. The average value of the ac component is zero, but the dc component yields

$$\text{Resistor: } P_R = \frac{1}{2}RI_A^2$$

To calculate the average power delivered to an arbitrary load $Z_L = R_L + jX_L$, we use phasor circuit analysis to find the phasor current $\mathbf{I}_L$ through Z_L. The average power delivered to the load is dissipated in R_L, since the reactance X_L represents the net inductance or capacitance of the load. Hence the average power to the load is

$$P = \frac{1}{2}R_L|\mathbf{I}_L|^2 \qquad (8\text{–}33)$$

Caution: When a circuit contains two or more sources, superposition applies only to the total load current and not to the total load power. You cannot find the total power to the load by summing the power delivered by each source acting alone.

The following example illustrates a power transfer calculation.

EXAMPLE 8–32

Find the average power delivered to the load to the right of the interface in Figure 8–65.

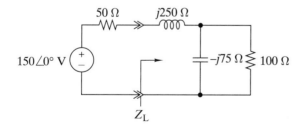

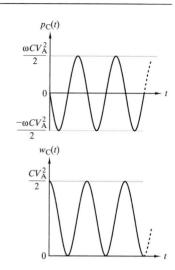

FIGURE 8–64 *Capacitor power and energy in the sinusoidal steady state.*

FIGURE 8–65

SOLUTION:
The equivalent impedance to the right of the interface is

$$Z_L = j250 + \frac{1}{\dfrac{1}{-j75} + \dfrac{1}{100}} = 36 + j202\ \Omega$$

The current delivered to the load is

$$\mathbf{I}_L = \frac{150\angle 0°}{50 + Z_L} = 0.683 \angle -66.9° \text{ A}$$

Hence, the average power delivered across the interface is

$$P = \frac{1}{2} R_L |\mathbf{I}_L|^2 = \frac{36}{2} |0.683|^2 = 8.40 \text{ W}$$

Note: All of this power goes into the 100-Ω resistor since the inductor and capacitor do not absorb average power. ∎

Exercise 8–41

Find the average power delivered to the 25-Ω load resistor in Figure 8–66.

FIGURE 8–66

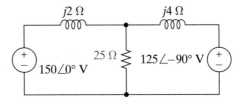

Answer: $P = 234$ W

MAXIMUM POWER

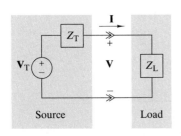

FIGURE 8–67 *A source-load interface in the sinusoidal steady state.*

To address the maximum power transfer problem, we model the source-load interface as shown in Figure 8–67. The source circuit is represented by a Thévenin equivalent circuit with source voltage $\mathbf{V}_T$ and source impedance $Z_T = R_T + jX_T$. The load circuit is represented by an equivalent impedance $Z_L = R_L + jX_L$. In the maximum power transfer problem the source parameters $\mathbf{V}_T$, R_T, and X_T are given, and the objective is to adjust the load impedance R_L and X_L so that average power to the load is a maximum.

The average power to the load is expressed in terms of the phasor current and load resistance:

$$P = \frac{1}{2} R_L |\mathbf{I}|^2$$

Then, using series equivalence, we express the magnitude of the interface current as

$$|\mathbf{I}| = \left| \frac{\mathbf{V}_T}{Z_T + Z_L} \right| = \frac{|\mathbf{V}_T|}{|(R_T + R_L) + j(X_T + X_L)|}$$

$$= \frac{|\mathbf{V}_T|}{\sqrt{(R_T + R_L)^2 + (X_T + X_L)^2}}$$

Combining the last two equations yields the average power delivered across the interface:

$$P = \frac{1}{2} \frac{R_L |\mathbf{V}_T|^2}{(R_T + R_L)^2 + (X_T + X_L)^2} \tag{8–34}$$

The quantities $|\mathbf{V}_T|$, R_T, and X_T in Eq. (8–34) are fixed. Our problem is to select R_L and X_L to maximize P.

Clearly, for every value of R_L the denominator in Eq. (8–34) is minimized and P maximized when $X_L = -X_T$. This choice of X_L is possible because a reactance can be positive or negative. When the source Thévenin equivalent has an inductive reactance ($X_T > 0$), we modify the load to have a capacitive reactance of the same magnitude, and vice versa. This step reduces the net reactance of the series connection in Figure 8–67 to zero, creating a condition in which the net impedance seen by the Thévenin voltage source is purely resistive.

When the source and load reactances cancel out, the expression for average power in Eq. (8–34) reduces to

$$P = \frac{1}{2} \frac{R_{\mathrm{L}} |\mathbf{V}_{\mathrm{T}}|^2}{(R_{\mathrm{T}} + R_{\mathrm{L}})^2} \qquad (8\text{–}35)$$

This equation has the same form encountered in Chapter 3 in dealing with maximum power transfer in resistive circuits. From the derivation in Sect. 3–5, we know P is maximized when $R_{\mathrm{L}} = R_{\mathrm{T}}$. In summary, to obtain maximum power transfer in the sinusoidal steady state, we select the load resistance and reactance so that

$$R_{\mathrm{L}} = R_{\mathrm{T}} \quad \text{and} \quad X_{\mathrm{L}} = -X_{\mathrm{T}} \qquad (8\text{–}36)$$

These conditions can be compactly expressed in the following way:

$$Z_{\mathrm{L}} = Z_{\mathrm{T}}^* \qquad (8\text{–}37)$$

The condition for maximum power transfer is called a **conjugate match**, since the load impedance is the conjugate of the source impedance. When the conjugate-match conditions are inserted into Eq. (8–34), we find that the maximum average power available from the source circuit is

$$P_{\mathrm{MAX}} = \frac{|\mathbf{V}_{\mathrm{T}}|^2}{8R_{\mathrm{T}}} \qquad (8\text{–}38)$$

where $|\mathbf{V}_{\mathrm{T}}|$ is the peak amplitude of the Thévenin equivalent voltage.

It is important to remember that conjugate matching applies when the source is fixed and the load is adjustable. These conditions arise frequently in power-limited communication systems. However, as we will see in Chapter 16, conjugate matching does not apply to electrical power systems because the power transfer constraints are different.

EXAMPLE 8–33

(a) Calculate the average power delivered to the load in the circuit shown in Figure 8–68 for $v_{\mathrm{S}}(t) = 5 \cos 10^6 t$ V, $R = 200\,\Omega$, $R_{\mathrm{L}} = 200\,\Omega$, and $C = 0.01\,\mu\text{F}$.
(b) Calculate the maximum average power available at the interface and specify the load required to draw the maximum power.

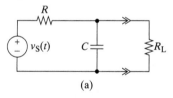

(a)

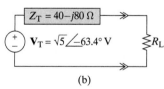

(b)

FIGURE 8–68

SOLUTION:
(a) To find the power delivered to the 200-Ω load resistor, we use a Thévenin equivalent circuit. By voltage division, the open-circuit voltage at the interface is

$$\mathbf{V}_{\mathrm{T}} = \frac{Z_C}{Z_R + Z_C} \mathbf{V}_{\mathrm{S}} = \frac{-j100}{200 - j100} 5\angle 0°$$

$$= 1 - j2 = \sqrt{5}\angle{-63.4°}\ \text{V}$$

By inspection, the short-circuit current at the interface is

$$\mathbf{I}_{\mathrm{N}} = \frac{5\angle 0°}{200} = 0.025 + j0\ \text{A}$$

Given $\mathbf{V}_{\mathrm{T}}$ and $\mathbf{I}_{\mathrm{N}}$, we calculate the Thévenin source impedance:

$$Z_{\mathrm{T}} = \frac{\mathbf{V}_{\mathrm{T}}}{\mathbf{I}_{\mathrm{N}}} = \frac{1 - j2}{0.025} = 40 - j80\ \Omega$$

Using the Thévenin equivalent shown in Figure 8–68(b), we find that the current through the 200-Ω resistor is

$$\mathbf{I} = \frac{\mathbf{V}_T}{Z_T + Z_L} = \frac{\sqrt{5}\angle -63.4°}{40 - j80 + 200} = 8.84\angle -45° \text{ mA}$$

and the average power delivered to the load resistor is

$$P = \frac{1}{2}R_L|\mathbf{I}|^2 = 100(8.84 \times 10^{-3})^2 = 7.81 \text{ mW}$$

(b) Using Eq. (8–38), the maximum average power available at the interface is

$$P_{\text{MAX}} = \frac{|\mathbf{V}_T|^2}{8R_T} = \frac{(\sqrt{5})^2}{(8)(40)} = 15.6 \text{ mW}$$

The 200-Ω load resistor in (a) draws about half of the maximum available power. To extract maximum power, the load impedance must be

$$Z_L = Z_T^* = 40 + j80 \, \Omega$$

This impedance can be obtained using a 40-Ω resistor in series with a reactance of +80 Ω. The required reactance is inductive (positive) and can be produced by an inductance of

$$L = \frac{|X_T|}{\omega} = \frac{80}{10^6} = 80 \, \mu\text{H}$$

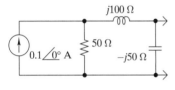

FIGURE 8–69

Exercise 8–42

Calculate the maximum average power available at the interface in Figure 8–69.

Answer: $P_{\text{MAX}} = 62.5$ mW.

SUMMARY

- A phasor is a complex number representing a sinusoidal waveform. The magnitude and angle of the phasor correspond to the amplitude and phase angle of the sinusoid. The phasor does not provide frequency information.

- The additive property states that adding phasors is equivalent to adding sinusoids of the same frequency. The derivative property states that multiplying a phasor by $j\omega$ is equivalent to differentiating the corresponding sinusoid.

- In the sinusoidal steady state, phasor currents and voltages obey Kirchhoff's laws and the element $i–v$ relationships are written in terms of impedances. Impedance can be defined as the ratio of phasor voltage over phasor current. The device and connection constraints for phasor circuit analysis have the same form as resistance circuits.

- Phasor circuit analysis techniques include series equivalence, parallel equivalence, circuit reduction,

Thévenin's and Norton's theorems, unit output method, superposition, node-voltage analysis, and mesh-current analysis.

- In the sinusoidal steady state the equivalent impedance at a pair of terminals is $Z(j\omega) = R(\omega) + jX(\omega)$, where $R(\omega)$ is called resistance and $X(\omega)$ is called reactance. A frequency at which an equivalent impedance is purely real is called a resonant frequency. Admittance is the reciprocal of impedance.

- In the sinusoidal steady state the instantaneous power to a passive element is a periodic function at twice the frequency of the driving force. The average power delivered to an inductor or capacitor is zero. The average power delivered to a resistor is $\frac{1}{2}R|\mathbf{I}_R|^2$. The maximum average power is delivered by a fixed source to an adjustable load when the source and load impedances are conjugates.

PROBLEMS

OBJECTIVE 8–1 SINUSOIDS AND PHASORS (SECT. 8–1)

Use the additive and derivative properties of phasors to convert sinusoidal waveforms into phasors and vice versa.
See Examples 8–1, 8–2, 8–3, 8–4 and Exercises 8–1, 8–2, 8–3, 8–4, 8–5, 8–6.

8–1 Transform the following sinusoids into phasor form and draw a phasor diagram. Use the additive property of phasors to find $v_1(t) + v_2(t)$.
(a) $v_1(t) = 50 \cos(\omega t - 30°)$ V
(b) $v_2(t) = 200 \cos(\omega t + 150°)$ V

8–2 Transform the following sinusoids into phasor form and draw a phasor diagram. Use the additive property of phasors to find $i_1(t) + i_2(t)$.
(a) $i_1(t) = -4 \sin(\omega t)$ A
(b) $i_2(t) = 3 \cos(\omega t - 26.6°)$ A

8–3 The sum of the two voltage phasors shown in Figure P8–3 is $\mathbf{V}_3$. If the frequency is 100 Hz, write the sum in the time domain, $v_3(t)$.

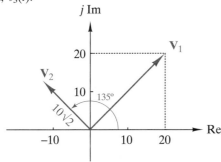

FIGURE P8–3

8–4 Convert the following phasors into sinusoidal waveforms.
(a) $\mathbf{V}_1 = 100e^{-j60°}$ V, $\omega = 10^3$ rad/s
(b) $\mathbf{V}_2 = 120e^{j45°}$ V, $\omega = 10^3$ rad/s
(c) $\mathbf{I}_1 = 240e^{j26.6°}$ mA, $\omega = 2\pi \times 10^4$ rad/s
(d) $\mathbf{I}_2 -150e^{-j45°}$ mA, $\omega = 2\pi \times 10^4$ rad/s

8–5 Use the phasors in Problem 8–4 and the additive property to find the sinusoidal waveforms
$v_3(t) = v_1(t) - v_2(t)$ and $i_3(t) = 2i_1(t) + 3i_2(t)$.

8–6 The phasor representation of a sinusoid with $\omega = 500$ rad/s is $\mathbf{V} = 15 - j10$ V. Use the phasor derivative property to find the time derivative of the sinusoid.

8–7 Convert the following phasors into sinusoids:
(a) $\mathbf{V}_1 = 4 + j5$ V, $\omega = 10$ Mrad/s
(b) $\mathbf{V}_2 = 6(8 - j8)$ V, $\omega = 200$ krad/s
(c) $\mathbf{I}_1 = 12 + j5 + \frac{5}{j}$ A, $\omega = 30$ rad/s
(d) $\mathbf{I}_2 = \frac{330 + j810}{2200 - j560}$ A, $\omega = 60$ rad/s

8–8 Use phasors to find the sinusoid $v_2(t)$, where

$$v_2(t) = \frac{1}{100}\frac{dv_1(t)}{dt} + 20\, v_1(t) \text{ and } v_1(t) = 10\cos(100t + 90°).$$

8–9 Given the sinusoids $v_1(t) = 500 \cos(\omega t - 45°)$ V and $v_2(t) = 750 \sin(\omega t)$ V use the additive property of phasors to find $v_3(t)$ such that $v_1 + v_2 + v_3 = 0$.

8–10 Graphically add the following three phasors and determine their sum:

$$\mathbf{V}_1 = 9.85 + j1.74 \text{ V}, \mathbf{V}_2 = 10\angle130°\text{V}, \mathbf{V3} = -3.42 - j9.40 \text{ V}.$$

8–11 Given a sinusoid $v_1(t)$ whose phasor is $\mathbf{V}_1 = 3 - j4$ V, use phasor methods to find a voltage $v_2(t)$ that leads $v_1(t)$ by 90° and has an amplitude of 10 V.

8–12 A new parameter Z is defined as $\mathbf{V}/\mathbf{I}$. If $\mathbf{V} = 9.85 + j1.74$ V and $i_1(t) = -4 \sin(\omega t)$ A, find Z.

8–13 Complex power S is defined as $\mathbf{VI}^*$, where $\mathbf{I}^*$ is the complex conjugate of the current phasor. If $\mathbf{V} = 1200 + j1500$ V and $\mathbf{I} = 500 - j\,250$ mA, find S.

OBJECTIVE 8–2 IMPEDANCE (SECTS. 8–2, 8–3)

Given a linear circuit in the sinusoidal steady state:
(a) Convert R, L, and C elements into impedances in the phasor domain.
(b) Use series and parallel equivalence to find the equivalent impedance at a specified pair of terminals.
See Examples 8–5, 8–6, 8–10, 8–11, 8–12 and Exercises 8–7, 8–8, 8–9, 8–10, 8–13, 8–16, 8–17, 8–18, 8–19.

8–14 A design engineer needs to know what value of R, L, or C to use in circuits to achieve a certain impedance.
(a) At what radian frequency will a 0.033 µF capacitor's impedance equal $-j100\,\Omega$?
(b) At what radian frequency will a 47 mH inductor's impedance equal $j100\,\Omega$?
(c) At what radian frequency will a 100 Ω resistor's impedance equal 100 Ω?

8–15 Find the equivalent impedance Z in Figure P8–15 when $\omega = 2000$ rad/s. Express the result in both polar and rectangular forms.

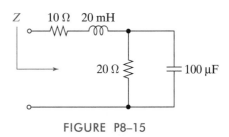

FIGURE P8–15

8–16 Find the equivalent impedance Z in Figure P8–16. If $\omega = 10$ krad/s what two elements (R, L, and/or C) could be used to replace the phasor circuit?

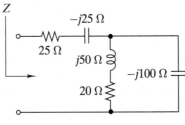

FIGURE P8–16

8–17 Find the equivalent impedance Z in Figure P8–17 when $\omega = 50$ krad/s. What two elements (R, L, and/or C) could be used to replace the phasor circuit?

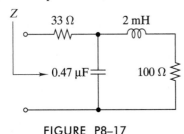

FIGURE P8–17

8–18 Find the equivalent impedance Z in Figure P8–18. If $\omega = 100$ krad/s what two elements (R, L, and/or C) could be used to replace the phasor circuit?

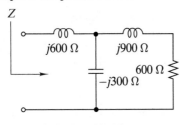

FIGURE P8–18

8–19 The circuit in Figure P8–18 is operating in the sinusoidal steady-state with $\omega = 5$ krad/s.
(a) How would the element impedances change if the steady-state frequency were reduced to 500 rad/s?
(b) What is the equivalent impedance Z at this new frequency?
(c) What two elements (R, L, and/or C) could be used to replace the phasor circuit?

8–20 The circuit in Figure P8–20 is operating in the sinusoidal steady state with $\omega = 100$ krad/s.
(a) Find the equivalent impedance Z.
(b) What circuit element can be added in series with the equivalent impedance to place the circuit in resonance?

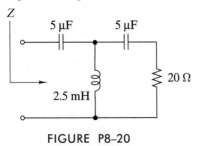

FIGURE P8–20

8–21 The circuit of Figure P8–21 is operating at 60 Hz. Find the equivalent impedance Z.

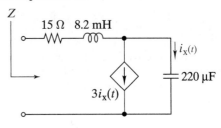

FIGURE P8–21

8–22 The equivalent impedance in Figure P8–22 is known to be $Z = 60 + j180\,\Omega$. Find the impedance of the inductor.

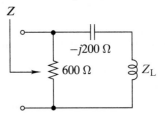

FIGURE P8–22

8–23 🔵 A capacitor C is connected in parallel with a resistor R. Select values of R and C so that the equivalent impedance of the parallel combination is $300 - j400\,\Omega$ at $\omega = 2$ Mrad/s.

8–24 🔵 The circuit in Figure P8–24 is excited by a 1 krad/s sinusoidal source. As the circuit's designer, select a capacitor C such that the impedance Z looking into the circuit is all real.

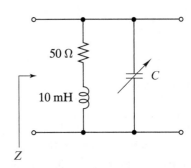

FIGURE P8–24

8–25 An 800-Ω resistor is connected in parallel with a 1000-pF capacitor. The impedance of the parallel combination is $400 - j400\,\Omega$. Find the frequency.

O BJECTIVE 8–3 B ASIC P HASOR C IRCUIT A NALYSIS AND D ESIGN (S ECTS. 8–3, 8–4)

(a) Given a linear circuit in the sinusoidal steady state, find phasor responses using equivalent circuits, circuit reduction, Thévenin or Norton equivalent circuits, proportionality, or superposition.

(b) Given a desired phasor response and a sinusoidal input, design a circuit in the sinusoidal steady state that produces the desired response.

See Examples 8–6, 8–7, 8–8, 8–9, 8–10, 8–13, 8–14, 8–15, 8–16, 8–17, 8–18, 8–19, 8–20, 8–21, 8–22, 8–23 and Exercises 8–11, 8–12, 8–13, 8–14, 8–15, 8–16, 8–20, 8–21, 8–22, 8–23, 8–24, 8–25, 8–26, 8–27, 8–28, 8–29, 8–30, 8–31, 8–32.

8–26 A voltage $v_S(t) = 30 \cos (2000t)$ V is applied to the circuit in Figure P8–26.
(a) Convert the circuit into the phasor domain.
(b) Find the phasor current flowing through the circuit and the phasor voltages across the inductor and the resistor.
(c) Plot all three phasors from (b) on a phasor diagram. Describe if the current leads or lags the inductor voltage.

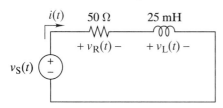

FIGURE P8–26

8–27 The circuit in Figure P8–27 is operating in the sinusoidal steady state. Find the phasor current and the two element voltages. Is the phasor voltage across the capacitor leading or lagging the current?

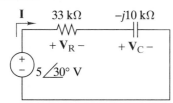

FIGURE P8–27

8–28 A voltage $v(t) = 100 \cos (3000t)$ V is applied across a series connection of a 330-Ω resistor and 110-mH inductor. Find the steady-state current $i(t)$ through the series connection.

8–29 The circuit in Figure P8–29 is operating in the sinusoidal steady state with $v_S(t) = V_A \cos (\omega t)$. Derive a general expression for the phasor response $\mathbf{I}_L$ and the voltage $\mathbf{V}_O$.

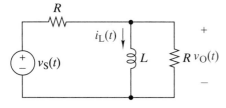

FIGURE P8–29

8–30 A current source delivering $i(t) = 120 \cos (500t)$ mA is connected across a parallel combination of a 10-kΩ resistor and a 0.2-μF capacitor. Find the steady-state current $i_R(t)$

through the resistor and the steady-state current $i_C(t)$ through the capacitor. Draw a phasor diagram showing $\mathbf{I}$, $\mathbf{I}_C$, and $\mathbf{I}_R$.

8–31 The circuit in Figure P8–31 is operating in the sinusoidal steady state with $i_S(t) = I_A \cos(\omega t)$. Derive general expressions for the steady-state responses $\mathbf{V}_R$ and $\mathbf{I}_C$.

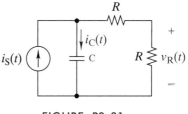

FIGURE P8–31

8–32 A practical voltage source of $v_S(t) = 120 \cos 2\pi60t$ V is in series with a 50-Ω resistor. Convert the source into the phasor domain and then do a source transformation into a current source in parallel with an impedance. Finally, convert the source back into the time domain.

8–33 A current source of $\mathbf{I}_N = 20\ \angle 33.8°$ mA is in parallel with an impedance of $Z = 100 - j50\ \Omega$. Convert the practical current source into a practical voltage source in series with an impedance.

8–34 A voltage $v(t) = 12 \cos (3030t - 45°)$ V is applied across a parallel connection of a 1.5-kΩ resistor and a 0.22-μF capacitor. Find the steady-state current $i_C(t)$ through the capacitor and the steady-state current $i_R(t)$ through the resistor. Draw a phasor diagram showing $\mathbf{V}$, $\mathbf{I}_C$, and $\mathbf{I}_R$.

8–35 The circuit in Figure P8–35 is operating in the sinusoidal steady state. Find the steady-state response $v_X(t)$.

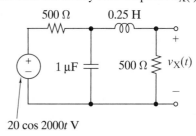

FIGURE P8–35

8–36 The circuit in Figure P8–36 is operating in the sinusoidal steady state. Find the steady-state response $v_X(t)$.

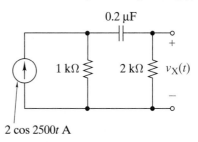

FIGURE P8–36

8–37 Use the unit-output method to find $\mathbf{V}_X$ and $\mathbf{I}_X$ in the circuit of Figure P8–37.

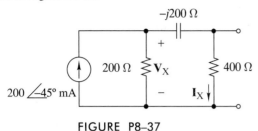

FIGURE P8–37

8–38 The circuit in Figure P8–38 is driven by a 10-krad/s source and is operating in the sinusoidal steady state. Use OrCAD to find the steady-state phasor response $\mathbf{V}_x$.

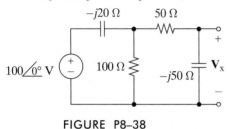

FIGURE P8–38

8–39 The circuit in Figure P8–39 is operating in the sinusoidal steady state. Use superposition to find the phasor response $\mathbf{I}_X$.

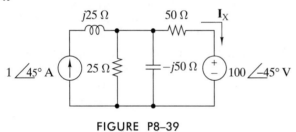

FIGURE P8–39

8–40 The circuit in Figure P8–40 is operating in the sinusoidal steady state. Use superposition to find the response $v_X(t)$. *Note:* The sources do not have the same frequency.

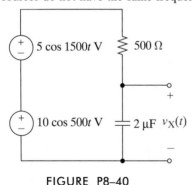

FIGURE P8–40

8–41 An *RC* series circuit is excited by a sinusoidal source $v(t) = V_A \cos{(\omega t + \phi)}$ V. Determine the effects on the magnitudes of the current, voltages, and impedances caused by changes in the source parameters. Complete the following table.

Source	$|\mathbf{V}_R|$	$|\mathbf{V}_C|$	$|\mathbf{I}|$	$|\mathbf{Z}_R|$	$|\mathbf{Z}_C|$
Increase/decrease V_A			Increase/decrease$|I|$		
Increase/decrease ω				none	
Increase/decrease φ					none

8–42 The circuit in Figure P8–42 is operating in the sinusoidal steady state. Use superposition to find the response $v_X(t)$.

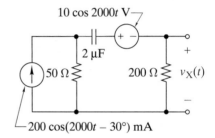

FIGURE P8–42

8–43 The circuit in Figure P8–43 is operating in the sinusoidal steady state. Use superposition to find the response $v_X(t)$. *Note:* The sources do not have the same frequency.

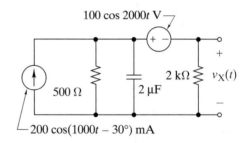

FIGURE P8–43

8–44 The circuit in Figure P8–44 is operating in the sinusoidal steady state.
(a) What impedance, if any, should be connected across $\mathbf{V}_X$ to cancel the reactance in the circuit?
(b) Is the bridge balanced, that is, $\mathbf{V}_X = 0$?

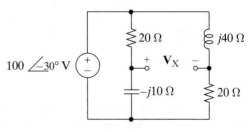

FIGURE P8–44

8–45 The circuit in Figure P8–45 is operating in the sinusoidal steady state. Use the unit output method to find the phasor responses $\mathbf{V_X}$ and $\mathbf{I_X}$.

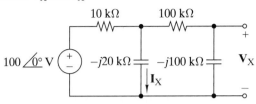

FIGURE P8–45

8–46 Find the Thévenin equivalent of the source circuit to the left of the interface in Figure P8–46. Then use the equivalent circuit to find the steady-state voltage $v(t)$ and current $i(t)$ delivered to the load. Validate you answer using OrCAD.

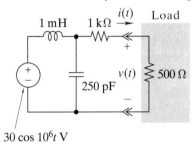

FIGURE P8–46

8–47 Find the phasor Thévenin equivalent of the source circuit to the left of the interface in Figure P8–47. Then use the equivalent circuit to find the phasor voltage $\mathbf{V}$ and current $\mathbf{I}$ delivered to the load.

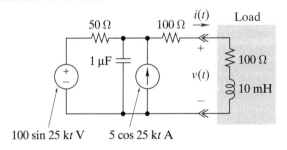

FIGURE P8–47

8–48 The circuit in Figure P8–48 is operating in the sinusoidal steady state. When $Z_L = 0$, the phasor current at the interface is $\mathbf{I} = 4.8 - j3.6$ mA. When $Z_L = -j20$ kΩ, the phasor interface current is $\mathbf{I} = 10 + j0$ mA. Find the Thévenin equivalent of the source circuit.

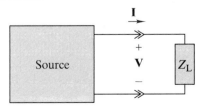

FIGURE P8–48

8–49 **D** Design a linear circuit that will deliver an output phasor $\mathbf{V_O} = 60\angle{-45°}$ V when an input phasor $\mathbf{V_S} = 240 \angle 0°$ V is applied in Figure P8–49.

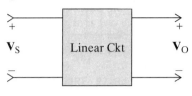

FIGURE P8–49

8–50 **D** A load of $Z_L = 100 + j100\,\Omega$ is to be driven by a phasor source $\mathbf{V_S} = 120 \angle 0°$ V. The voltage across the load needs to be $\mathbf{V_L} = 100 \angle 0°$ V. Design an interface that will meet these conditions. Validate your answer using OrCAD.

8–51 **D** Design an interface circuit so that an input voltage $v_S(t) = 100 \cos(2 \times 10^4 t)$ V delivers a steady-state output current of $i_O(t) = 10 \cos (2 \times 10^4 t - 30°)$ mA to a 1-kΩ resistive load. Validate your answer using OrCAD.

8–52 **D** Design an interface circuit so that an input voltage $v_S(t) = 15 \cos (10\,kt)$ V delivers a steady-state output voltage of $v_O(t) = 10 \cos (10\,kt - 45°)$ V.

8–53 Refer to the *RLC* series circuit shown in Figure P8–53.
  (a) What is the maximum output voltage and at what frequency does it occur? Consider using OrCAD and doing an ac sweep from 10 Hz to 1 MHz, and then narrow your sweep until you find the frequency at which the peak occurs and the output voltage at that frequency.
  (b) Bandwidth is defined as $BW = f_H - f_L$, where f_H is the higher frequency at which the magnitude of the output is exactly 0.707 of the maximum value, and f_L is the lower frequency at which the magnitude of the output is exactly 0.707 of the maximum value. What is the bandwidth of this circuit?

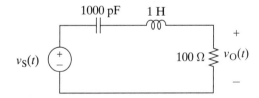

FIGURE P8–53

OBJECTIVE 8–4 GENERAL CIRCUIT ANALYSIS (SECT. 8–5)

Given a linear circuit operating in the sinusoidal steady state, find equivalent impedances and phasor responses using node-voltage or mesh-current analysis.
See Examples 8–24, 8–25, 8–26, 8–27, 8–28, 8–29, 8–30, 8–31 and Exercises 8–33, 8–34, 8–35, 8–36, 8–37, 8–38, 8–39, 8–40.

8–54 The circuit in Figure P8–54 is operating in the sinusoidal steady state with $\omega = 5$ krad/s.

Use node-voltage analysis to find the steady-state response $v_X(t)$.

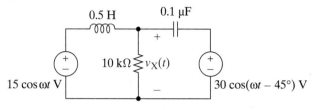

FIGURE P8–54

8–55 Use node-voltage analysis to find the steady-state phasor response $\mathbf{V}_O$ in Figure P8–55.

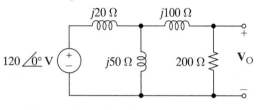

FIGURE P8–55

🖱 **8–56** The circuit in Figure P8–56 is operating in the sinusoidal steady state.
(a) Find the node voltage phasors $\mathbf{V}_A$ and $\mathbf{V}_B$.
(b) If the circuit is operating with $\omega = 10$ krad/s, use OrCAD to verify your answer in (a).

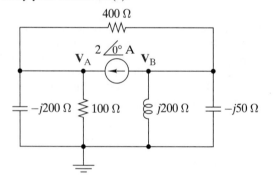

FIGURE P8–56

8–57 Use mesh-current analysis to find the phasor branch currents $\mathbf{I}_1$, $\mathbf{I}_2$, and $\mathbf{I}_3$ in the circuit shown in Figure P8–57.

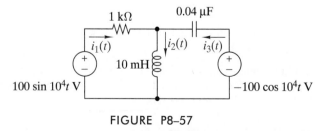

FIGURE P8–57

8–58 Use mesh-current analysis to find the phasor branch currents $\mathbf{I}_1$, $\mathbf{I}_2$, and $\mathbf{I}_3$ in the circuit shown in Figure P8–58.

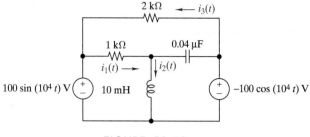

FIGURE P8–58

8–59 Use mesh-current analysis to find the phasor branch currents $\mathbf{I}_1$, $\mathbf{I}_2$, and $\mathbf{I}_3$ in the circuit shown in Figure P8–59.

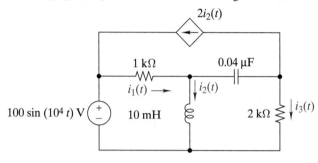

FIGURE P8–59

8–60 Use mesh-current analysis to find the phasor currents $\mathbf{I}_A$ and $\mathbf{I}_B$ in Figure P8–60.

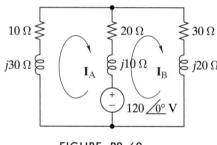

FIGURE P8–60

8–61 The OP AMP circuit in Figure P8–61 is operating in the sinusoidal steady state.
(a) Show that

$$\frac{\mathbf{V}_O}{\mathbf{V}_S} = \left(\frac{R_1 + R_2}{R_1}\right)\left(\frac{j\omega + \frac{1}{(R_1 + R_2)C}}{j\omega + \frac{1}{R_1 C}}\right)$$

(b) Find the value of the magnitude of $\mathbf{V}_O/\mathbf{V}_S$ at $\omega = 0$ and $\omega \to \infty$.

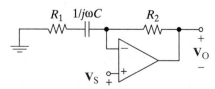

FIGURE P8–61

8–62 The circuit in Figure P8–62 is operating in the sinusoidal steady state.

(a) If $v_S(t) = 2 \cos 2128t$ V, find the output $v_O(t)$

(b) At what frequency is the magnitude of the output voltage equal to half of the magnitude of the input voltage in the circuit of Figure 8–62? Consider using OrCAD and doing an ac sweep from 1 Hz to 100 kHz, and then use OrCAD's cursor to find the desired frequency.

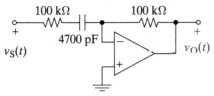

FIGURE P8–62

8–63 Use MATLAB to find the phasor current I_O in Figure P8–63.

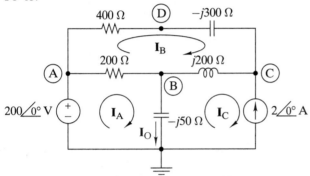

FIGURE P8–63

8–64 The circuit in Figure P8–64 is operating with $\omega = 10$ krad/s.

(a) Find the phasor outputs V_O and I_O in Figure P8–64 when $\mu = 50$ and the phasor input is $I_S = 1 + j1$ mA.

(b) Use OrCAD to verify your results above.

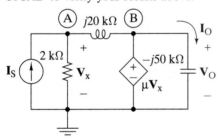

FIGURE P8–64

8–65 Find the phasor responses I_{IN} and V_O in Figure P8–65 when $V_S = 1 + j0$ V.

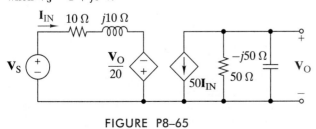

FIGURE P8–65

8–66 For the circuit of Figure P8–66 find the Thévenin equivalent circuit seen at the output.

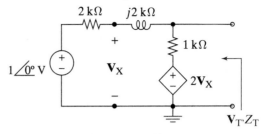

FIGURE P8–66

8–67 The OP AMP circuit in Figure P8–67 is operating in the sinusoidal steady state with $\omega = 1$ krad/s. Find the magnitude of the ratio of the output phasor V_O to the input phasor V_S. Repeat for $\omega = 10$ krad/s, $\omega = 100$ krad/s, $\omega = 1$ Mrad/s, and $\omega = 10$ Mrad/s. Use MATLAB to plot the log of $|V_O/V_S|$ versus the log of the frequency ω. Comment on the plot.

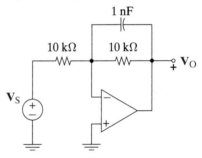

FIGURE P8–67

8–68 Find the phasor input V_S in Figure P8–68 when the phasor output is $V_O = 300 + j200$ V.

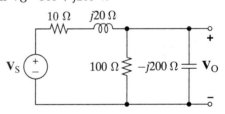

FIGURE P8–68

8–69 The dependent source circuit in Figure P8–69 is operating in the sinusoidal steady state with $\omega = 1$ krad/s and $\mu = 10^3$. Find the phasor gain $K = V_O/V_S$ and the input impedance Z_{IN} seen by V_S. Validate your answer using OrCAD.

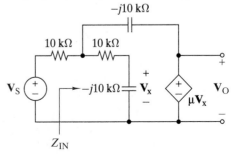

FIGURE P8–69

8–70 Find the phasor gain $K = \mathbf{V}_O/\mathbf{V}_S$ and input impedance Z_{IN} of the circuit in Figure P8–70.

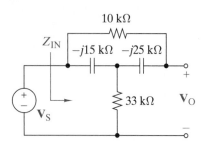

FIGURE P8–70

8–71 Find the phasor gain $K = \mathbf{V}_O/\mathbf{V}_S$, input impedance Z_{IN} of the circuit, and the capacitor current $\mathbf{I}_X$ in Figure P8–71.

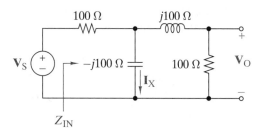

FIGURE P8–71

8–72 Given the circuit in Figure P8–72:
(a) Use node-voltage or mesh-current analysis to develop a set of matrix equations for the circuit.
(b) Use MATLAB to solve the matrix equations and then find the phasor gain $K = \mathbf{V}_O/\mathbf{V}_S$ and input impedance Z_{IN} of the circuit.
(c) Without using the matrix equations, use the MATLAB command `solve` on the original node-voltage or mesh-current equations to solve the equations and then find the phasor gain and input impedance.

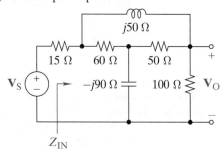

FIGURE P8–72

OBJECTIVE 8–5 AVERAGE POWER AND MAXIMUM POWER TRANSFER (SECT. 8–6)

Given a linear circuit in the sinusoidal steady state,
(a) Find the average power delivered at a specified interface.
(b) Find the maximum average power available at a specified interface.

(c) Find the load impedance required to draw the maximum available power. See Examples 8–32, 8–33 and Exercises 8–41, 8–42.

8–73 A load consisting of a 2.2-kΩ resistor in series with a 1-μF capacitor is connected across a voltage source $v_S(t) = 169.7 \cos(377t)$ V. Find the phasor voltage, current, and average power delivered to the load.

8–74 A load consisting of a 50-Ω resistor in parallel with a 0.47-μF capacitor is connected across a current source delivering $I_S(t) = 12 \cos(3000t)$ mA. Find the average power delivered to the load.

8–75 The circuit in Figure P8–75 is operating in the sinusoidal steady state at a frequency of 10 krad/s. Use OrCAD to find the average power delivered to the 100-Ω resistor.

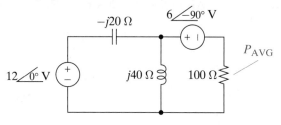

FIGURE P8–75

8–76 You have a task of designing a load that ensure maximum power is delivered to it. The load needs to be connected to a source circuit that is not readily observable, but that you can make measurements at its output terminals. You measure the open circuit voltage and read $120\angle 0°$ V. You then connect a known load of $50 - j50$ and you measure $47.1\angle 11.3°$ V across it.
(a) Design your load for maximum power transfer.
(b) Find the maximum average power delivered to your load.

8–77 **(a)** Find the average power delivered to the load in Figure P8–77.
(b) Find the maximum available average power at the interface shown in the Figure.
(c) Specify the load required to extract the maximum average power.

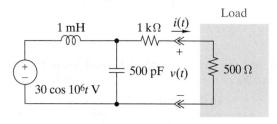

FIGURE P8–77

8–78 **(a)** Find the maximum average power available at the interface in Figure P8–60.
(b) Specify the values of R and C that will extract the maximum power from the source circuit.

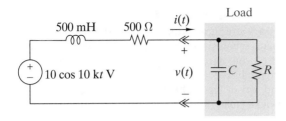

FIGURE P8–78

INTEGRATING PROBLEMS

8–79 AC **V**oltage Measurement **A**
An ac voltmeter measurement indicates the amplitude of a sinusoid and not its phase angle. The magnitude and phase can be inferred by making several measurements and using KVL. For example, Figure P8–79 shows a relay coil of unknown resistance and inductance. The following ac voltmeter readings are taken with the circuit operating in the sinusoidal steady state at $f = 60$ Hz: $|\mathbf{V_S}| = 24$ **V**, $|\mathbf{V_1}| = 10$ V, and $|\mathbf{V_2}| = 18$ V. Find R and L.

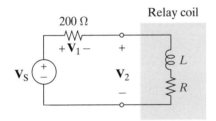

FIGURE P8–79

8–80 Home Power Distribution **A E**
The circuit of Figure P8–80 emulates a typical 60-Hz residential power system. There are three wires entering the house, two are called "hot" and the remaining one is called the return or "neutral." Each hot line is protected by a circuit breaker–but not the return. In the "house," appliances such as lights, toasters, and electronics are connected between one of the hot wires and the neutral. Large appliances like ovens and dryers are connected between the two hot wires. Appliances are designed to operate with either 120 V or 240 V within a few volts either way.
(a) Show that if R_1 and R_2 are equal, $\mathbf{I_N} = 0$ and $\mathbf{V_1} = \mathbf{V_2}$.
(b) Consider a typical power draw where a R_1 is, for example, a 130-Ω light bulb, R_2 a 40-Ω toaster, and R_3 is a 10-Ω clothes dryer. Find the phasors $\mathbf{V_1}$ and $\mathbf{V_2}$ and find $\mathbf{I_N}$.
(c) Based on your results in (a) and (b), is the neutral line even necessary? Open the neutral line, that is, force $\mathbf{I_N} = 0$, and again find the voltages $\mathbf{V_1}$ and $\mathbf{V_2}$. What is your answer? Would your home be better protected by adding a breaker to the return line?
(d) Simulate the circuit in OrCAD and validate your results. The source frequency is 60 Hz.

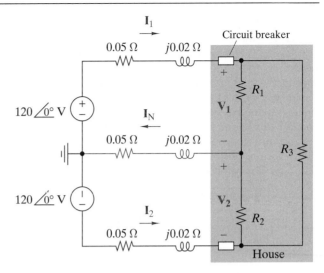

FIGURE P8–80

8–81 OP AMP Band Pass Filter **D**
Use the analysis methods discussed in Example 8–28 to find the input-output relationship **Vo/Vs** for the active band pass filter of Figure P8–81. Treat each stage separately and then multiply the input-output relationships from each stage to obtain the overall input-output relationship. Select R and C values so that the low cutoff frequency is 500 rad/s, the upper cutoff frequency is 50 krad/s, and the magnitude of the passband gain is 100.

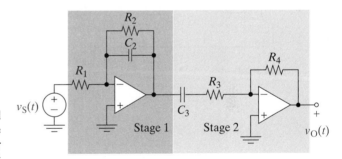

FIGURE P8–81

8–82 Power Transmission Efficiency **A**
A power transmission circuit with a source voltage of $\mathbf{V_S} = 440 + j0$ V can be modeled as shown in Figure P8–82. Find the average power produced by the source, lost in the wires, and delivered to the load. What is the transmission efficiency?

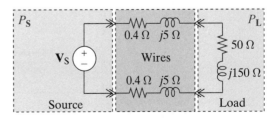

FIGURE P8–82

8–83 60 Hz Filter **E**
A 1-kΩ resistor models an important and sensitive laboratory instrument. The desired signal that the instrument measures, varies from 1 Hz to 500 Hz. However, interference from power lines in the laboratory causes the instrument to saturate. A vendor has designed a device that he claims will essentially eliminate a 60-Hz signal from the instrument with only the smallest attenuation to other frequencies. The interface circuit is shown in Figure P8–83. Connect the vendor's circuit to the model for the instrument and analyze its performance to determine if it will do the job.

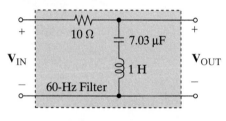

FIGURE P8–83

8–84 AC Circuit Design **D**
Select values of L and C in Figure P8–84 so that the input impedance seen by the voltage source is $50 + j0\,\Omega$ when the frequency is $\omega = 10^6$ rad/s. For these values of L and C, find the output Thévenin impedance seen by the 300-Ω load resistor.

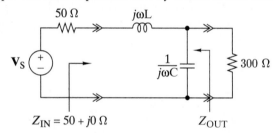

FIGURE P8–84

8–85 AC Circuit Analysis **E**
Ten years after graduating with a BSEE, you decide to go to graduate schools for a masters degree. In desperate need of income, you agree to sign on as a grader in the basic circuit analysis course. One of the problems asks the students to find $v(t)$ in Figure P8–85 when the circuit operates in the sinusoidal steady state. One of the students offers the following solution:

$$v(t) = (R + j\omega L) \times i(t)$$
$$= (20 + j20) \times 0.5 \cos 200t$$
$$= 10 \cos 200t + j10 \cos 200t$$
$$= 10\sqrt{2} \cos(200t + 45°)$$

Is the answer correct? If not what grade would you give the student? If correct, what comments would you give the student about the method of solution?

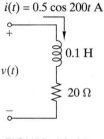

FIGURE P8–85

CHAPTER 9 LAPLACE TRANSFORMS

My method starts with a complex integral; I fear this sounds rather formidable; but it is really quite simple ... I am afraid that no physical people will ever try to make out my method: but I am hoping that it may give them confidence to try your methods.

Thomas John Bromwich, 1915,
British Mathematician

Some History Behind This Chapter

Laplace transforms have their roots in the pioneering work of the eccentric British engineer Oliver Heaviside (1850–1925). His operational calculus was essentially a collection of intuitive rules that allowed him to formulate and solve important technical problems of his day. His intuitive approach drew bitter criticism from the scientists of his day. Eventually individuals like John Bromwich recognized the importance of Heaviside's methods and began to develop the necessary mathematical foundations. A complete development was eventually discovered in the 1780 writings of the French mathematician Pierre Simon Laplace.

Why This Chapter Is Important Today

The difficulties of finding transient responses using classical differential equations are avoided when we apply the techniques of Laplace transforms. These techniques not only simplify the solution of circuit differential equations but also give us a deeper insight into circuit behavior. Transient response, frequency response, and even phasor techniques come together once Laplace transforms and techniques are understood. Laplace transforms make the analysis and design of circuits easy and fun.

Chapter Sections

Chapter Learning Objectives

9-1 Laplace Transform (Sects. 9–1, 9–2, 9–3)

Find the Laplace transform of a given signal waveform using transform properties and pairs, using the integral definition of the Laplace transformation, or using software applications. Locate the poles and zeros of the transform and construct a pole-zero diagram.

9-2 Inverse Transforms (Sects. 9–4)

(a) Find the inverse transform of a given Laplace transform using partial fraction expansion, basic transform properties and pairs, or using software tools.
(b) Given a pole-zero diagram, find the respective transform.

9-3 Circuit Response Using Laplace Transforms (Sect. 9–5)

Given a first- or second-order circuit:
(a) Determine the circuit differential equation and the initial conditions (if not given).
(b) Transform the differential equation into the *s* domain and solve for the response transform.
(c) Use the inverse transformation to find the response waveform.

9-4 Initial and Final Value Properties (Sect. 9–6)

Given the Laplace transform of a signal, find the initial and final values of the signal waveform.

9–1 SIGNAL WAVEFORMS AND TRANSFORMS

A mathematical transformation employs rules to change the form of data without altering its meaning. An example of a transformation is the conversion of numerical data from decimal to binary form. In engineering circuit analysis, transformations are used to obtain alternative representations of circuits and signals. These alternate forms provide a different perspective that can be quite useful or even essential. Examples of the transformations used in circuit analysis are the Fourier transformation, the Z-transformation, and the Laplace transformation. These methods all involve specific transformation rules, make certain analysis techniques more manageable, and provide a useful viewpoint for circuit and system design.

This chapter deals with the Laplace transformation. The discussion of the Laplace transformation follows the path shown in Figure 9–1 by the solid arrow. The process begins with a linear circuit. We derive a differential equation describing the circuit response and then transform this equation into the frequency domain, where it becomes an algebraic equation. Algebraic techniques are then used to solve the transformed equation for the circuit response. The inverse Laplace transformation then changes the frequency-domain response into the response waveform in the time domain. The dashed arrow in Figure 9–1 shows that there is another route to the time-domain response using the classical techniques discussed in Chapter 7. The classical approach appears to be more direct, but the advantage of the Laplace transformation is that solving a differential equation becomes an algebraic process and reveals properties of the circuit not easily seen in the classical approach.

Symbolically, we represent the Laplace transformation as

$$\mathscr{L}\{f(t)\} = F(s) \tag{9–1}$$

This expression states that $F(s)$ is the Laplace transform of the waveform $f(t)$. The transformation operation involves two domains: (1) the time domain, in which the signal is characterized by its **waveform** $f(t)$, and (2) the complex frequency domain, in which the signal is represented by its **transform** $F(s)$.

The symbol s stands for the complex frequency variable, a notation we first introduced in Chapter 7 in connection with the zero-state response of linear circuits.

FIGURE 9–1 *Flow diagram of dynamic circuit analysis with Laplace transforms.*

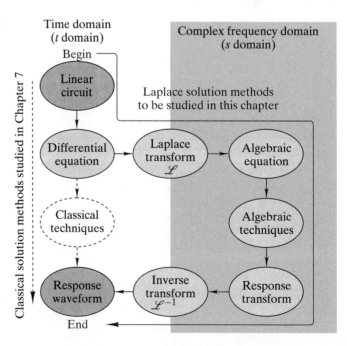

The variable s has the dimensions of reciprocal time, or frequency, and is expressed in units of radians per second. In this chapter the complex frequency variable is written as $s = \sigma + j\omega$, where $\sigma = \text{Re}\{s\}$ is the real part and $\omega = \text{Im}\{s\}$ is the imaginary part. This variable is the independent variable in the s domain, just as t is the independent variable in the time domain. Although we cannot physically measure complex frequency in the same sense that we measure time, it is an extremely useful concept that pervades the analysis and design of linear systems.

A signal can be expressed as a waveform or a transform. Collectively, $f(t)$ and $F(s)$ are called a **transform pair**, where the pair involves two representations of the signal. To distinguish between the two forms, a lowercase letter denotes a waveform and an uppercase a transform. For electrical waveforms such as current $i(t)$ or voltage $v(t)$, the corresponding transforms are denoted $I(s)$ and $V(s)$. In this chapter we will use $f(t)$ and $F(s)$ to stand for signal waveforms and transforms in general.

The **Laplace transformation** is defined by the integral

$$F(s) = \int_{0-}^{\infty} f(t)e^{-st}dt \qquad (9\text{--}2)$$

Since the definition involves an improper integral (the upper limit is infinite), we must discuss the conditions under which the integral exists (converges). The integral exists if the waveform $f(t)$ is piecewise continuous and of exponential order. **Piecewise continuous** means that $f(t)$ has a finite number of steplike discontinuities in any finite interval. **Exponential order** means that constants K and b exist such that $|f(t)| < Ke^{bt}$ for all $t > 0$. As a practical matter, the signals encountered in engineering applications meet these conditions.

From the integral definition of the Laplace transformation, we see that when a voltage waveform $v(t)$ has units of volts (V), the corresponding voltage transform $V(s)$ has units of volt-seconds (V-s). Similarly, when a current waveform $i(t)$ has units of amperes (A), the corresponding current transform $I(s)$ has units of ampere-seconds (A-s). Thus, waveforms and transforms do not have the same units. Even so, we often refer to both $V(s)$ and $v(t)$ as voltages and both $I(s)$ and $i(t)$ as currents despite the fact that they have different units. The reason is simply that it is awkward to keep adding the words *waveform* and *transform* to statements when the distinction is clear from the context.

Equation (9–2) uses a lower limit denoted $t = 0-$ to indicate a time just a whisker before $t = 0$. We use $t = 0-$ because in circuit analysis $t = 0$ is defined by a discrete event, such as closing a switch. Such an event may cause a discontinuity in $f(t)$ at $t = 0$. To capture this discontinuity, we set the lower limit at $t = 0-$, just prior to the event. Fortunately, in many situations there is no discontinuity so we will not distinguish between $t = 0-$ and $t = 0$ unless it is crucial.

Equally fortunate is the fact that the number of different waveforms encountered in linear circuits is relatively small. The list includes the three basic waveforms from Chapter 5 (the step, exponential, and sinusoid), as well as composite waveforms such as the impulse, ramp, damped ramp, and damped sinusoid. Since the number of waveforms of interest is relatively small, we do not often use the integral definition in Eq. (9–2) to find Laplace transforms. Once a transform pair has been found, it can be cataloged in a table for future reference and use. Table 9–2 in this chapter is sufficient for our purposes.

EXAMPLE 9–1

Show that the Laplace transform of the unit step function $f(t) = u(t)$ is $F(s) = 1/s$.

SOLUTION:
Applying Eq. (9–2) yields

$$F(s) = \int_0^\infty u(t)e^{-st}dt$$

Since $u(t) = 1$ throughout the range of integration this integral becomes

$$F(s) = \int_0^\infty e^{-st}dt = -\frac{e^{-st}}{s}\bigg|_0^\infty = -\frac{e^{-(\sigma+j\omega)t}}{\sigma+j\omega}\bigg|_0^\infty = \frac{1}{s} \text{ QED}$$

The last expression on the right side vanishes at the upper limit since e^{-st} goes to zero as t approaches infinity provided that $\sigma > 0$. At the lower limit the expression reduces to $1/s$. The integral used to calculate $F(s)$ is valid only in the region for which $\sigma > 0$. However, once evaluated, the result $F(s) = 1/s$ can be extended to neighboring regions provided that we avoid the point at $s = 0$ where the function blows up. ■

Exercise 9–1 _____

Find the Laplace transform of $v(t) = -7\,u(t)$ V.

Answer: $V(s) = -7/s$ V-s

EXAMPLE 9 – 2

Show that the Laplace transform $f(t) = [e^{-\alpha t}]u(t)$ is $F(s) = 1/(s+\alpha)$.

SOLUTION:
Applying Eq. (9–2) yields

$$F(s) = \int_0^\infty e^{-\alpha t}e^{-st}dt = \int_0^\infty e^{-(s+\alpha)t}dt = \frac{e^{-(s+\alpha)t}}{-(s+\alpha)}\bigg|_0^\infty = \frac{1}{s+\alpha} \text{ QED}$$

The last term on the right side vanishes at the upper limit since $e^{-(s+\alpha)t}$ vanishes as t approaches infinity provided that $\sigma > \alpha$. At the lower limit the last term reduces to $1/(s+\alpha)$. Again, the integral is valid only for a limited region, but the result $F(s) = 1/(s+\alpha)$ can be extended outside this region if we avoid the point at $s = -\alpha$. ■

Exercise 9–2 _____

Find the Laplace transform of $v(t) = 8e^{-5t}u(t)$ V.

Answer: $V(s) = \dfrac{8}{s+5}$ V-s

EXAMPLE 9 – 3

Show that the Laplace transform of the impulse function $f(t) = \delta(t)$ is $F(s) = 1$.

SOLUTION:
Applying Eq. (9–2) yields

$$F(s) = \int_{0-}^\infty \delta(t)e^{-st}dt = \int_{0-}^{0+} \delta(t)e^{-st}dt = \int_{0-}^{0+} \delta(t)dt = 1 \text{ QED}$$

The difference between $t = 0-$ and $t = 0$ is important since the impulse is zero everywhere except at $t = 0$. To capture the impulse in the integration, we take a lower limit at $t = 0-$ and an upper limit at $t = 0+$. Since $e^{-st} = 1$ and $\int \delta(t)dt = 1$ on this integration interval, we find that $F(s) = 1$. ■

Exercise 9–3 _____

Find the Laplace transform of $i(t) = 0.5\ \delta(t)$ A.

Answer: $\qquad I(s) = 0.5$ A-s

INVERSE TRANSFORMATION

So far we have used the direct transformation to convert waveforms into transforms. But Figure 9–1 points out the need to perform the inverse transformation to convert transforms into waveforms. Symbolically, we represent the inverse process as

$$\mathscr{L}^{-1}\{F(s)\} = f(t) \qquad (9\text{–}3)$$

This equation states that $f(t)$ is the inverse Laplace transform of $F(s)$. The **inverse Laplace transformation** is defined by the complex inversion integral

$$f(t) = \frac{1}{2\pi j}\int_{\alpha-j\infty}^{\alpha+j\infty} F(s)e^{st}\,ds \qquad (9\text{–}4)$$

The Laplace transformation is an integral transformation since both the direct process in Eq. (9–2) and the inverse process in Eq. (9–4) involve integrations.

Happily, formal evaluation of the complex inversion integral is not necessary because of the uniqueness property of the Laplace transformation. A symbolical statement of the **uniqueness property** is

$$\text{IF } \mathscr{L}\{f(t)\} = F(s) \text{ THEN } \mathscr{L}^{-1}\{F(s)\}(=)u(t)\,f(t)$$

The mathematical justification for this statement is beyond the scope of our treatment.[1] However, the notation $(=)$ means "equal almost everywhere." The only points where equality may not hold is at the discontinuities of $f(t)$.

If we just look at the definition of the direct transformation in Eq. (9–2), we could conclude that $F(s)$ is not affected by the values of $f(t)$ for $t < 0$. However, when we use Eq. (9–2) we are not just looking for the Laplace transform of $f(t)$, but a Laplace transform pair such that $\mathscr{L}\{f(t)\} = F(s)$ and $\mathscr{L}^{-1}\{F(s)\} = f(t)$. The inverse Laplace transformation in Eq. (9–4) always produces a causal waveform, one that is zero for $t < 0$. Hence a transform pair $[f(t) \leftrightarrow F(s)]$ is unique if and only if $f(t)$ is causal. For instance, in Example 9–1 we show that $\mathscr{L}\{u(t)\} = 1/s$; hence by the uniqueness property we know that $\mathscr{L}^{-1}\{1/s\}(=)u(t)$.

For this reason Laplace transform-related waveforms are written as $[f(t)]u(t)$ to make their causality visible. For example, in the next section we find the Laplace transform of the sinusoid waveform $\cos \beta t$. In the context of Laplace transforms this signal is not an eternal sinusoid but a causal waveform $f(t) = [\cos \beta t]u(t)$. It is important to remember that causality and Laplace transforms go hand in hand when interpreting the results of circuit analysis.

9–2 BASIC PROPERTIES AND PAIRS

The previous section gave the definition of the Laplace transformation and showed that the transforms of some basic signals can be found using the integral definition. In this section we develop the basic properties of the Laplace transformation and show how these properties can be used to obtain additional transform pairs.

---

[1]See Wilber R. LePage, *Complex Variables and Laplace Transform for Engineering*, Dover Publishing Co., New York, 1980, p. 318.

The **linearity** property of the Laplace transformation states that

$$\mathcal{L}\{A f_1(t) + B f_2(t)\} = A F_1(s) + B F_2(s) \tag{9--5}$$

where A and B are constants. This property is easily established using the integral definition in Eq. (9--2):

$$\mathcal{L}\{A f_1(t) + B f_2(t)\} = \int_0^\infty [A f_1(t) + B f_2(t)] e^{-st} dt$$

$$= A \int_0^\infty f_1(t) e^{-st} dt + B \int_0^\infty f_2(t) e^{-st} dt$$

$$= A F_1(s) + B F_2(s)$$

The integral definition of the inverse transformation in Eq. (9--4) is also a linear operation, so it follows that

$$\mathcal{L}^{-1}\{A F_1(s) + B F_2(s)\} = A f_1(t) + B f_2(t) \tag{9--6}$$

An important consequence of linearity is that for any constant K

$$\mathcal{L}\{K f(t)\} = K F(s) \text{ and } \mathcal{L}^{-1}\{K F(s)\} = K f(t) \tag{9--7}$$

The linearity property is an extremely important feature that we will use many times in this and subsequent chapters. The next two examples show how this property can be used to obtain the transform of the exponential rise waveform and a sinusoidal waveform.

E X A M P L E   9 – 4

Show that the Laplace transform of $f(t) = A(1 - e^{-\alpha t})u(t)$ is

$$F(s) = \frac{A\alpha}{s(s + \alpha)}$$

SOLUTION:
This waveform is the difference between a step function and an exponential. We can use the linearity property of Laplace transforms to write

$$\mathcal{L}\{A(1 - e^{-\alpha t})u(t)\} = A\mathcal{L}\{u(t)\} - A\mathcal{L}\{e^{-\alpha t}u(t)\}$$

The transforms of the step and exponential functions were found in Examples 9--1 and 9--2. Using linearity, we find that the transform of the exponential rise is

$$F(s) = \frac{A}{s} - \frac{A}{s + \alpha} = \frac{A\alpha}{s(s + \alpha)} \text{ QED} \qquad ■$$

Exercise 9–4 _____

Transform the response $v(t) = [10e^{-1000t} - 5]u(t)\,\text{V}$ of a particular RC circuit into the Laplace domain.

Answer:

$$V(s) = \frac{5(s - 1000)}{s(s + 1000)} \text{ V-s}$$

E X A M P L E   9 – 5

Show that the Laplace transform of the sinusoid $f(t) = A[\sin(\beta t)]u(t)$ is $F(s) = A\beta/(s^2 + \beta^2)$.

SOLUTION:
Using Euler's relationship, we can express the sinusoid as a sum of exponentials.

$$e^{+j\beta t} = \cos \beta t + j \sin \beta t$$
$$e^{-j\beta t} = \cos \beta t - j \sin \beta t$$

Subtracting the second equation from the first yields

$$f(t) = A \sin \beta t = \frac{A(e^{j\beta t} - e^{-j\beta t})}{2j} = \frac{A}{2j}e^{j\beta t} - \frac{A}{2j}e^{-j\beta t}$$

The transform pair $\mathcal{L}\{e^{-\alpha t}\} = 1/(s+\alpha)$ in Example 9–2 is valid even if the exponent α is complex. Using this fact and the linearity property, we obtain the transform of the sinusoid as

$$\mathcal{L}\{A \sin \beta t\} = \frac{A}{2j}\mathcal{L}\{e^{j\beta t}\} - \frac{A}{2j}\mathcal{L}\{e^{-j\beta t}\}$$
$$= \frac{A}{2j}\left[\frac{1}{s - j\beta} - \frac{1}{s + j\beta}\right]$$
$$= \frac{A\beta}{s^2 + \beta^2} \quad \text{QED} \qquad \blacksquare$$

Exercise 9–5

Transform the sinusoid $i(t) = 100[\sin(200t)]u(t)$ mA into the Laplace domain.

Answer:
$$I(s) = \frac{20{,}000}{s^2 + 40{,}000} \text{ mA-s}$$

INTEGRATION PROPERTY

In the time domain the $i-v$ relationships for capacitors and inductors involve integration and differentiation. Since we will be working in the s domain, it is important to establish the s-domain equivalents of these mathematical operations. Applying the integral definition of the Laplace transformation to a time-domain integration yields

$$\mathcal{L}\left[\int_0^t f(\tau)d\tau\right] = \int_0^\infty \left[\int_0^t f(\tau)d\tau\right]e^{-st}\,dt \qquad (9\text{–}8)$$

The right side of this expression can be integrated by parts using

$$y = \int_0^t f(\tau)d\tau \text{ and } dx = e^{-st}\,dt$$

These definitions result in

$$dy = f(t)dt \quad \text{and} \quad x = \frac{-e^{-st}}{s}$$

Using these factors reduces the right side of Eq. (9–8) to

$$\mathcal{L}\left[\int_0^t f(\tau)d\tau\right] = \left[\frac{-e^{-st}}{s}\int_0^t f(\tau)d\tau\right]_0^\infty + \frac{1}{s}\int_0^\infty f(t)e^{-st}\,dt \qquad (9\text{–}9)$$

The first term on the right in Eq. (9–9) vanishes at the lower limit because the integral over a zero-length interval is zero provided that $f(t)$ is finite at $t = 0$. It vanishes at the upper limit because e^{-st} approaches zero as t goes to infinity for

$\sigma > 0$. By the definition of the Laplace transformation, the second term on the right is $F(s)/s$. We conclude that

$$\mathcal{L}\left[\int_0^t f(\tau)d\tau\right] = \frac{F(s)}{s} \tag{9–10}$$

The **integration property** states that time-domain integration of a waveform $f(t)$ can be accomplished in the s domain by the algebraic process of dividing its transform $F(s)$ by s. The next example applies the integration property to obtain the transform of the ramp function.

EXAMPLE 9–6

Show that the Laplace transform of the ramp function $r(t) = tu(t)$ is $1/s^2$.

SOLUTION:
From our study of signals, we know that the ramp waveform can be obtained from $u(t)$ by integration.

$$r(t) = \int_0^t u(\tau)\, d\tau$$

In Example 9–1 we found $\mathcal{L}\{u(t)\} = 1/s$. Using these facts and the integration property of Laplace transforms, we obtain

$$\mathcal{L}\{r(t)\} = \mathcal{L}\left[\int_0^t u(\tau)\, d\tau\right] = \frac{1}{s}\mathcal{L}\{u(t)\} = \frac{1}{s^2} \quad \text{QED} \qquad\blacksquare$$

Exercise 9–6

Let $v_1(t) = V_A e^{-\alpha t} u(t)$ V. Show that the Laplace transform of $v_2(t) = \int_0^t V_A e^{-\alpha x}\, dx$ V is equal to $V_1(s)/s$.

Answer:

$$\int_0^t V_A e^{-\alpha x} dx = \frac{V_A}{-\alpha}e^{-\alpha x}\bigg|_0^t = \frac{V_A}{-\alpha}(e^{-\alpha t} - 1) = \frac{V_A}{\alpha}(1 - e^{-\alpha t})$$

$$\mathcal{L}\left\{\frac{V_A}{\alpha}(1 - e^{-\alpha t})\right\} = \frac{V_A}{\alpha}\left(\frac{1}{s} - \frac{1}{s+\alpha}\right) = \frac{V_A}{\alpha}\left[\frac{s+\alpha - s}{s(s+\alpha)}\right] = \frac{V_A}{s(s+\alpha)}$$

$$\frac{V_1(s)}{s} = \frac{V_A}{s(s+\alpha)} \quad \text{QED}$$

Exercise 9–7

If $i(t) = 6\, e^{-1000t} u(t)$ mA, find the Laplace transform of $v(t) = \dfrac{1}{10^{-6}}\displaystyle\int_0^t i(x)\, dx$ V.

Answer:

$$V(s) = \frac{I(s)}{10^{-6}s} = \frac{6000}{s(s+1000)}\text{V-s}$$

DIFFERENTIATION PROPERTY

The time-domain differentiation operation transforms into the s domain as follows:

$$\mathcal{L}\left[\frac{df(t)}{dt}\right] = \int_0^\infty \left[\frac{df(t)}{dt}\right]e^{-st}dt \tag{9–11}$$

The right side of this equation can be integrated by parts using

$$y = e^{-st} \quad \text{and} \quad dx = \frac{df(t)}{dt}\, dt$$

These definitions result in

$$dy = -se^{-st}\, dt \quad \text{and} \quad x = f(t)$$

Inserting these factors reduces the right side of Eq. (9–11) to

$$\mathscr{L}\left[\frac{df(t)}{dt}\right] = f(t)e^{-st}\Big|_{0-}^{\infty} + s\int_{0-}^{\infty} f(t)e^{-st}\, dt \qquad (9\text{–}12)$$

For $\sigma > 0$ the first term on the right side of Eq. (9–12) is zero at the upper limit because e^{-st} approaches zero as t goes to infinity. At the lower limit it reduces to $-f(0-)$. By the definition of the Laplace transform, the second term on the right side is $sF(s)$. We conclude that

$$\mathscr{L}\left[\frac{df(t)}{dt}\right] = sF(s) - f(0-) \qquad (9\text{–}13)$$

The **differentiation property** states that time-domain differentiation of a waveform $f(t)$ is accomplished in the s domain by the algebraic process of multiplying the transform $F(s)$ by s and subtracting the constant $f(0-)$. Note that the constant $f(0-)$ is the value of $f(t)$ at $t = 0-$ just prior to $t = 0$.

The s-domain equivalent of a second derivative is obtained by repeated application of Eq. (9–13). We first define a waveform $g(t)$ as

$$g(t) = \frac{df(t)}{dt} \quad \text{hence} \quad \frac{d^2 f(t)}{dt^2} = \frac{dg(t)}{dt}$$

Applying the differentiation rule to these two equations yields

$$G(s) = sF(s) - f(0-) \quad \text{and} \quad \mathscr{L}\left[\frac{d^2 f(t)}{dt^2}\right] = sG(s) - g(0-)$$

Substituting the first of these equations into the second results in

$$\mathscr{L}\left[\frac{d^2 f(t)}{dt^2}\right] = s^2 F(s) - sf(0-) - f'(0-)$$

where

$$f'(0-) = \frac{df}{dt}\bigg|_{t=0-}$$

Repeated application of this procedure produces the nth derivative:

$$\mathscr{L}\left[\frac{d^n f(t)}{dt^n}\right] = s^n F(s) - s^{n-1} f(0-) - s^{n-2} f'(0-) \cdots - f^{(n-1)}(0-) \qquad (9\text{–}14)$$

where $f^{(n-1)}(0-)$ is the $(n-1)$th derivative of $f(t)$ evaluated at $t = 0-$.

A hallmark feature of the Laplace transformation is the fact that time integration and differentiation change into algebraic operations in the s domain. This observation gives us our first hint as to why it is often easier to work with circuits and signals in the s domain. The next example shows how the differentiation rule can be used to obtain additional transform pairs.

EXAMPLE 9–7

Show that the Laplace transform of $f(t) = [\cos \beta t]u(t)$ is $F(s) = s/(s^2 + \beta^2)$.

SOLUTION:

We can express cos βt in terms of the derivative of sin βt as

$$\cos \beta t = \frac{1}{\beta} \frac{d}{dt} \sin \beta t$$

In Example 9–5 we found $\mathscr{L}\{\sin \beta t\} = \beta/(s^2 + \beta^2)$. Using these facts and the differentiation rule, we can find the Laplace transform of cos βt as follows:

$$\mathscr{L}\{\cos \beta t\} = \frac{1}{\beta} \mathscr{L}\left\{\frac{d}{dt} \sin \beta t\right\} = \frac{1}{\beta}\left[s\left(\frac{\beta}{s^2 + \beta^2}\right) - \sin(0-)\right]$$

$$= \frac{s}{s^2 + \beta^2} \text{ QED}$$

■

Exercise 9–8

Let $v_1(t)=V_A r(t)$ V. Show that the Laplace transform of $v_2(t) = dV_A r(t)/dt$ V is equal to $sV_1(s) - v_1(0-)$.

Answer:

$$\frac{dV_A r(t)}{dt} = \frac{dV_A t\, u(t)}{dt} = V_A\, u(t)$$

$$\mathscr{L}\{V_A\, u(t)\} = \frac{V_A}{s}$$

$$sV_1(s) - v_1(0-) = s\frac{V_A}{s^2} - 0 = \frac{V_A}{s} \text{ QED}$$

Exercise 9–9

If $i(t) = 30\, e^{-1200t} u(t)$ mA, find the Laplace transform of $v(t) = 0.1\dfrac{di(t)}{dt}$ V.

Answer:

$$V(s) = sI(s) - i(0-) = \frac{3s}{s + 1200} \text{ mV-s}$$

TRANSLATION PROPERTIES

The **s-domain translation property** of the Laplace transformation is

$$\text{IF } \mathscr{L}\{f(t)\} = F(s) \text{ THEN } \mathscr{L}\{e^{-\alpha t} f(t)\} = F(s + \alpha)$$

This theorem states that multiplying $f(t)$ by $e^{-\alpha t}$ is equivalent to replacing s by $s + \alpha$ (that is, translating the origin in the s plane by an amount α). In engineering applications the parameter α is always a real number, but it can be either positive or negative so the origin in the s domain can be translated to the left or right. Proof of the theorem follows almost immediately from the definition of the Laplace transformation.

$$\mathscr{L}\{e^{-\alpha t}f(t)\} = \int_0^\infty e^{-\alpha t}f(t)e^{-st}\, dt$$

$$= \int_0^\infty f(t)e^{-(s+\alpha)t}\, dt$$

$$= F(s + \alpha)$$

The s-domain translation property can be used to derive transforms of damped waveforms from undamped prototypes. For instance, the Laplace transform of the ramp, cosine, and sine functions are

$$\mathscr{L}\{tu(t)\} = \frac{1}{s^2}$$

$$\mathscr{L}\{[\cos \beta t]u(t)\} = \frac{s}{s^2 + \beta^2}$$

$$\mathscr{L}\{[\sin \beta t]u(t)\} = \frac{\beta}{s^2 + \beta^2}$$

To obtain the damped ramp, damped cosine, and damped sine functions, we multiply each undamped waveform by $e^{-\alpha t}$. Using the s-domain translation property, we replace s by $s + \alpha$ to obtain transforms of the corresponding damped waveforms.

$$\mathscr{L}\{te^{-\alpha t}u(t)\} = \frac{1}{(s+\alpha)^2}$$

$$\mathscr{L}\{[e^{-\alpha t}\cos\beta t]u(t)\} = \frac{s+\alpha}{(s+\alpha)^2 + \beta^2}$$

$$\mathscr{L}\{[e^{-\alpha t}\sin\beta t]u(t)\} = \frac{\beta}{(s+\alpha)^2 + \beta^2}$$

This completes the derivation of a basic set of transform pairs listed in Table 9–2.

The **time-domain translation property** of the Laplace transformation is

$$\text{IF } \mathscr{L}\{f(t)\} = F(s) \text{ THEN for } a > 0 \, \mathscr{L}\{f(t - a)u(t - a)\} = e^{-as}F(s)$$

The theorem states that multiplying $F(s)$ by e^{-as} is equivalent to shifting $f(t)$ to the right in the time domain by an amount $a > 0$. In other words, it is equivalent to delaying $f(t)$ in time by an amount $a > 0$. Proof of this property follows from the definition of the Laplace transformation.

$$\mathscr{L}\{f(t - a)u(t - a)\} = \int_{0-}^{\infty} f(t - a)u(t - a)e^{-st}\,dt = \int_{a}^{\infty} f(t - a)e^{-st}\,dt$$

In this equation we have used the fact that $u(t - a)$ is zero for $t < a$ and is unity for $t \geq a$. We now change the integration variable from t to $\tau = t - a$. With this change of variable the last integral in this equation takes the form

$$\mathscr{L}\{f(t - a)u(t - a)\} = \int_{0}^{\infty} f(\tau)e^{-s\tau}e^{-as}\,d\tau$$

$$= e^{-as}\int_{0}^{\infty} f(\tau)e^{-s\tau}\,d\tau$$

$$= e^{-as}F(s)$$

which confirms the statement of the time-domain translation property. A simple application of this property is finding the Laplace transform of the delayed step function.

$$\mathscr{L}\{u(t - T)\} = e^{-sT}\mathscr{L}\{u(t)\} = \frac{e^{-sT}}{s}$$

In this section we derived the basic transform properties listed in Table 9–1.

The Laplace transformation has other properties that are useful in signal-processing applications. We treat two of these properties in the last section of this chapter. However, the basic properties in Table 9–1 are used frequently in circuit analysis and are sufficient for nearly all of the applications in this book.

Similarly, Table 9–2 lists a basic set of Laplace transform pairs that is sufficient for most of the applications in this book. All of these pairs were derived in the preceding two sections. Tables 9–1 and 9–2 are repeated in the inside back cover.

All of the waveforms in Table 9–2 are causal, that is, they have a defined start or beginning. There is a $u(t)$ associated with every waveform. As a result, the Laplace transform pairs are unique and we can use the table in either direction. That is, given an $f(t)$ in the waveform column, we find its Laplace transform in the right column, or given an $F(s)$ in the right column, we find its inverse transform in the waveform column.

T A B L E **9–1** BASIC LAPLACE TRANSFORMATION PROPERTIES

PROPERTIES	TIME DOMAIN	FREQUENCY DOMAIN
Independent variable	t	s
Signal representation	$f(t)$	$F(s)$
Uniqueness	$\mathscr{L}^{-1}\{F(s)\}(=)[f(t)]u(t)$	$\mathscr{L}\{f(t)\} = F(s)$
Linearity	$Af_1(t) + Bf_2(t)$	$AF_1(s) + BF_2(s)$
Integration	$\displaystyle\int_0^t f(\tau)\,d\tau$	$\dfrac{F(s)}{s}$
Differentiation	$\dfrac{df(t)}{dt}$	$sF(s) - f(0-)$
	$\dfrac{d^2 f(t)}{dt^2}$	$s^2 F(s) - sf(0-) - f'(0-)$
	$\dfrac{d^3 f(t)}{dt^3}$	$s^3 F(s) - s^2 f(0-) - sf'(0-) - f''(0-)$
s-Domain translation	$e^{-\alpha t} f(t)$	$F(s + \alpha)$
t-Domain translation	$f(t - a)u(t - a)$	$e^{-as} F(s)$

T A B L E **9–2** BASIC LAPLACE TRANSFORM PAIRS

SIGNAL	WAVEFORM $f(t)$	TRANSFORM $F(s)$
Impulse	$\delta(t)$	1
Step function	$u(t)$	$\dfrac{1}{s}$
Ramp	$tu(t)$	$\dfrac{1}{s^2}$
Exponential	$[e^{-\alpha t}]u(t)$	$\dfrac{1}{s + \alpha}$
Damped ramp	$[te^{-\alpha t}]u(t)$	$\dfrac{1}{(s + \alpha)^2}$
Sine	$[\sin \beta t]u(t)$	$\dfrac{\beta}{s^2 + \beta^2}$
Cosine	$[\cos \beta t]u(t)$	$\dfrac{s}{s^2 + \beta^2}$
Damped sine	$[e^{-\alpha t}\sin \beta t]u(t)$	$\dfrac{\beta}{(s + \alpha)^2 + \beta^2}$
Damped cosine	$[e^{-\alpha t}\cos \beta t]u(t)$	$\dfrac{(s + \alpha)}{(s + \alpha)^2 + \beta^2}$

The last example in this section shows how to use the properties and pairs in Tables 9–1 and 9–2 to obtain the transform of a waveform not listed in the tables.

E X A M P L E 9 – 8

Find the Laplace transform of the waveform

$$f(t) = 2u(t) - 5[e^{-2t}]u(t) + 3[\cos 2t]u(t) + 3[\sin 2t]u(t)$$

SOLUTION:

Using the linearity property, we write the transform of $f(t)$ in the form

$$\mathscr{L}\{f(t)\} = 2\mathscr{L}\{u(t)\} - 5\mathscr{L}\{e^{-2t}u(t)\} + 3\mathscr{L}\{[\cos 2t]u(t)\} + 3\mathscr{L}\{[\sin 2t]u(t)\}$$

The transforms of each term in this sum are listed in Table 9–2:

$$F(s) = \frac{2}{s} - \frac{5}{s+2} + \frac{3s}{s^2+4} + \frac{6}{s^2+4}$$

Normally, a Laplace transform is written as a quotient of polynomials rather than as a sum of terms. Rationalizing the preceding sum yields

$$F(s) = \frac{16(s^2+1)}{s(s+2)(s^2+4)}$$

■

Exercise 9–10

Find the Laplace transforms of the following waveforms:

(a) $f(t) = [e^{-2t}]u(t) + 4tu(t) - u(t)$

(b) $f(t) = [2 + 2\sin 2t - 2\cos 2t]u(t)$

Answers:

(a) $F(s) = \dfrac{2(s+4)}{s^2(s+2)}$

(b) $F(s) = \dfrac{4(s+2)}{s(s^2+4)}$

Exercise 9–11

Find the Laplace transforms of the following waveforms:

(a) $f(t) = [e^{-4t}]u(t) + 5\displaystyle\int_0^t \sin 4x\,dx$

(b) $f(t) = 5[e^{-40t}]u(t) + \dfrac{d[5te^{-40t}]u(t)}{dt}$

Answers:

(a) $F(s) = \dfrac{s^3 + 36s + 80}{s(s+4)(s^2+16)}$

(b) $F(s) = \dfrac{10s + 200}{(s+40)^2}$

Exercise 9–12

Find the Laplace transforms of the following waveforms:

(a) $f(t) = A[\cos(\beta t - \phi)]u(t)$

(b) $f(t) = A[e^{-\alpha t}\cos(\beta t - \phi)]u(t)$

Answers:

(a) $F(s) = A\cos\phi \left[\dfrac{s + \beta\tan\phi}{s^2 + \beta^2}\right]$

(b) $F(s) = A\cos\phi \left[\dfrac{s + \alpha + \beta\tan\phi}{(s+\alpha)^2 + \beta^2}\right]$

Exercise 9–13

Find the Laplace transforms of the following waveforms for $T > 0$:

(a) $f(t) = Au(t) - 2Au(t - T) + Au(t - 2T)$

(b) $f(t) = Ae^{-\alpha(t-T)}u(t - T)$

Answers:

(a) $F(s) = \dfrac{A(1 - e^{-Ts})^2}{s}$

(b) $F(s) = \dfrac{Ae^{-Ts}}{s + \alpha}$

9–3 POLE-ZERO DIAGRAMS

The transforms for signals in Table 9–2 are ratios of polynomials in the complex frequency variable s. Likewise, the transform found in Example 9–8 takes the form of a ratio of two polynomials in s. These results illustrate that the signal transforms of greatest interest to us usually have the form

$$F(s) = \frac{b_m s^m + b_{m-1} s^{m-1} + \cdots + b_1 s + b_0}{a_n s^n + a_{n-1} s^{n-1} + \cdots + a_1 s + a_0} \tag{9–15}$$

If numerator and denominator polynomials are expressed in factored form, then $F(s)$ is written as

$$F(s) = K \frac{(s - z_1)(s - z_2) \cdots (s - z_m)}{(s - p_1)(s - p_2) \cdots (s - p_n)} \tag{9–16}$$

where the constant $K = b_m/a_n$ is called the **scale factor**.

The roots of the numerator and denominator polynomials, together with the scale factor K, uniquely define a transform $F(s)$. The denominator roots are called **poles** because for $s = p_i(i = 1, 2, \ldots, n)$ the denominator vanishes and $F(s)$ becomes infinite. The roots of the numerator polynomial are called **zeros** because the transform $F(s)$ vanishes for $s = z_i(i = 1, 2, \ldots, m)$. Collectively the poles and zeros are called **critical frequencies** because they are values of s at which $F(s)$ does dramatic things, like vanish or blow up.

In the s domain we can specify a signal transform by listing the location of its critical frequencies together with the scale factor K. That is, in the frequency domain we describe signals in terms of poles and zeros. The description takes the form of a **pole-zero diagram**, which shows the location of poles and zeros in the complex s plane. The pole locations in such plots are indicated by an $\times$ and the zeros by an $\bigcirc$. The independent variable in the frequency domain is the complex frequency variable s, so the poles or zeros can be complex as well. In the s plane we use a horizontal axis to plot the value of the real part of s and a vertical j-axis to plot the imaginary part. The j-axis is an important boundary in the frequency domain because it divides the s plane into two distinct half planes. The real part of s is negative in the left half plane and positive in the right half plane. As we will soon see, the sign of the real part of a pole has a profound effect on the form of the corresponding waveform.

For example, Table 9–2 shows that the transform of the exponential waveform $f(t) = e^{-\alpha t}u(t)$ is $F(s) = 1/(s + \alpha)$. The exponential signal has a single pole at $s=-\alpha$ and no finite zeros. The pole-zero diagram in Figure 9–2(a) is the s-domain portrayal of the exponential signal. In this diagram the $\times$ identifies the pole located at $s = -\alpha + j0$, a point on the negative real axis in the left half plane.

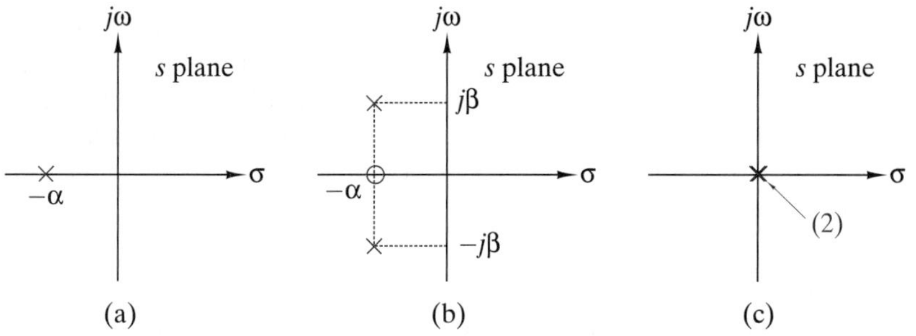

FIGURE 9–2 *Pole-zero diagrams in the s plane.*

The damped sinusoid $f(t) = [A e^{-\alpha t} \cos \beta t]u(t)$ is an example of a signal with complex poles. From Table 9–2 the corresponding transform is

$$F(s) = \frac{A(s+\alpha)}{(s+\alpha)^2 + \beta^2}$$

The transform $F(s)$ has a finite zero on the real axis at $s = -\alpha$. The roots of the denominator polynomial are $s = -\alpha \pm j\beta$. The resulting pole-zero diagram is shown in Figure 9–2(b). The poles of the damped cosine do not lie on either axis in the s plane because neither the real nor imaginary parts are zero.

Finally, the transform of a unit ramp $f(t) = tu(t)$ is $F(s) = 1/s^2$. This transform has no finite zeros and two poles at the origin ($s = 0 + j0$) in the s plane as shown in Figure 9–2(c). The poles in all of the diagrams of Figure 9–2 lie in the left half plane or on the j-axis boundary.

The diagrams in Figure 9–2 show the poles and zeros in the finite part of the s plane. Signal transforms may have poles or zeros at infinity as well. For example, the step function has a zero at infinity since $F(s) = 1/s$ approaches zero as $s \to \infty$. In general, a transform $F(s)$ given by Eq. (9–16) has a zero of order $n - m$ at infinity if $n > m$ and a pole of order $m - n$ at infinity if $n < m$. Thus, the number of zeros equals the number of poles if we include those at infinity.

The pole-zero diagram is the s-domain portrayal of the signal, just as a plot of the waveform versus time depicts the signal in the t domain. The utility of a pole-zero diagram as a description of circuits and signals will become clearer as we develop additional s-domain analysis and design concepts.

EXAMPLE 9–9

Find the poles and zeros of the waveform

$$f(t) = [e^{-2t} + \cos 2t - \sin 2t]u(t)$$

SOLUTION:
Using the linearity property and the basic pairs in Table 9–2, we write the transform in the form

$$F(s) = \frac{1}{s+2} + \frac{s}{s^2+4} - \frac{2}{s^2+4}$$

Rationalizing this expression yields $F(s)$.

$$F(s) = \frac{2s^2}{(s+2)(s^2+4)} = \frac{2s^2}{(s+2)(s+j2)(s-j2)}$$

This transform has three zeros and three poles. There are two zeros at $s = 0$ and one at $s = \infty$. There is a pole on the negative real axis at $s = -2 + j0$, and there are two

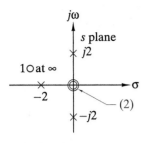

FIGURE 9–3

poles on the imaginary axis at $s = \pm j2$. The resulting pole-zero diagram is shown in Figure 9–3. Reviewing the analysis, we can trace the poles to the components of $f(t)$. The pole on the real axis at $s = -2$ came from the exponential e^{-2t}, while the complex conjugate poles on the j-axis came from the sinusoid $\cos 2t - \sin 2t$. The zeros, however, are not traceable to specific components. Their locations depend on all three components. ∎

Exercise 9–14

Find the poles and zeros of the transform of the following waveform and plot the results on a pole-zero diagram.

$$f(t) = [-2e^{-t} - t + 2]u(t)$$

Answers: Zeros: $s = 1$, two at $s = \infty$; poles: two at $s = 0$, $s = -1$.
See Figure 9–4 for the pole-zero diagram.

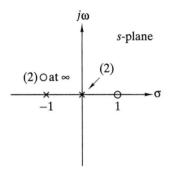

FIGURE 9–4

EXAMPLE 9–10

Find the poles and zeros of the transform of the waveform $f(t) = [4 - 3\cos 500t]u(t)$ and plot the results.

SOLUTION:
Using the linearity property and the basic pairs from Table 9–2, we write the transform in the form

$$F(s) = \frac{4}{s} - \frac{3s}{s^2 + (500)^2}$$

Rationalizing this expression results in $F(s)$.

$$F(s) = \frac{s^2 + 4 \times (500)^2}{s[s^2 + (500)^2]} = \frac{(s + j1000)(s - j1000)}{s(s + j500)(s - j500)}$$

The transform has two complex conjugate zeros at $s = \pm j1000$, and another at $s = \infty$. It has a pole at $s = 0$, and a complex conjugate pair at $s = \pm j500$. The pole at zero is caused by the step function and the two poles on the imaginary axis are caused by the cosine term. The results are plotted in Figure 9–5. ∎

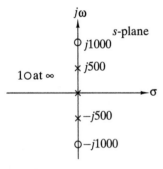

FIGURE 9–5

Exercise 9–15

Find the poles and zeros of the transform of the following waveform and plot the results on a pole-zero diagram.

$$f(t) = [e^{-10t}\cos 200t + 0.05e^{-10t}\sin 200t]u(t)$$

Answers: Zeros: $s = -20$, $s = \infty$; poles: $s = -10 \pm j\,200$.
See Figure 9–6 for the pole-zero diagram.

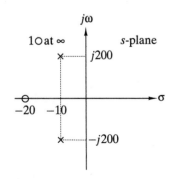

FIGURE 9–6

SOFTWARE APPLICATIONS

There are software applications that can be used to find the Laplace transform of a given waveform. In particular, MATLAB has the functions `laplace` and `ilaplace`, which operate on symbolic expressions to take the Laplace transform and inverse Laplace transform, respectively. For examples on how to use these software tools please visit web appendix C.

9–4 INVERSE LAPLACE TRANSFORMS

The inverse transformation converts a transform $F(s)$ into the corresponding waveform $f(t)$. Applying the inverse transformation in Eq. (9–4) requires knowledge of a branch of mathematics called complex analysis. Fortunately, we do not need Eq. (9–4) because the uniqueness of the Laplace transform pairs in Table 9–2 allows us to go from a transform to a waveform. This may not seem like much help since it does not take a very complicated circuit or signal before we exceed the listing in Table 9–2, or even the more extensive tables that are available. However, there is a general method of expanding $F(s)$ into a sum of terms that are listed in Table 9–2.

For linear circuits the transforms of interest are ratios of polynomials in s. In mathematics such functions are called **rational functions**. To perform the inverse transformation, we must find the waveform corresponding to rational functions of the form

$$F(s) = K \frac{(s - z_1)(s - z_2) \cdots (s - z_m)}{(s - p_1)(s - p_2) \cdots (s - p_n)} \qquad (9\text{–}17)$$

where K is the scale factor, $z_i(i = 1, 2, \ldots, m)$ are the zeros, and $p_i(i = 1, 2, \ldots, n)$ are the poles of $F(s)$.

If there are more finite poles than finite zeros ($n > m$), then $F(s)$ is called a **proper rational function**. If the denominator in Eq. (9–17) has no repeated roots ($p_i \neq p_j$ for $i \neq j$), then $F(s)$ is said to have **simple poles**. In this section we first treat the problem of finding the inverse transform of proper rational functions with simple poles. Once we have mastered this case, we will address problems with improper rational functions and multiple poles.

If a proper rational function has only simple poles, then it can be decomposed into a partial fraction expansion of the form

$$F(s) = \frac{k_1}{s - p_1} + \frac{k_2}{s - p_2} + \cdots + \frac{k_n}{s - p_n} \qquad (9\text{–}18)$$

In this case, $F(s)$ can be expressed as a linear combination of terms with one term for each of its n simple poles. The k's associated with each term are called **residues**.

Each term in the partial fraction decomposition has the form of the transform of an exponential signal. That is, we recognize that $\mathcal{L}^{-1}\{(k/s + \alpha)\} = [ke^{-\alpha t}]u(t)$. We can now write the corresponding waveform using the linearity property:

$$f(t) = [k_1 e^{p_1 t} + k_2 e^{p_2 t} + \cdots + k_n e^{p_n t}]u(t) \qquad (9\text{–}19)$$

In the time domain, the s-domain poles appear in the exponents of exponential waveforms and the residues at the poles become the amplitudes.

Given the poles of $F(s)$, finding the inverse transform $f(t)$ reduces to finding the residues. To illustrate the procedure, consider a case in which $F(s)$ has three simple poles and one finite zero.

$$F(s) = K \frac{(s - z_1)}{(s - p_1)(s - p_2)(s - p_3)} = \frac{k_1}{s - p_1} + \frac{k_2}{s - p_2} + \frac{k_3}{s - p_3}$$

We find the residue k_1 by first multiplying this equation through by the factor $(s - p_1)$:

$$(s - p_1)F(s) = K \frac{(s - z_1)}{(s - p_2)(s - p_3)} = k_1 + \frac{k_2(s - p_1)}{s - p_2} + \frac{k_3(s - p_1)}{s - p_3}$$

If we now set $s = p_1$, the last two terms on the right vanish, leaving

$$k_1 = (s - p_1)F(s)\Big|_{s = p_1} = K \frac{(s - z_1)}{(s - p_2)(s - p_3)}\Big|_{s = p_1}$$

Using the same approach for k_2 yields

$$k_2 = (s - p_2)F(s)\bigg|_{s=p_2} = K\frac{(s - z_1)}{(s - p_1)(s - p_3)}\bigg|_{s=p_2}$$

The technique generalizes so that the residue at any simple pole p_i is

$$k_i = (s - p_i)F(s)\bigg|_{s=p_i} \qquad (9\text{–}20)$$

The process of determining the residue at any simple pole is sometimes called the **cover-up algorithm** because we temporarily remove (cover up) the factor $(s = p_i)$ in $F(s)$ and then evaluate the remainder at $s = p_i$.

EXAMPLE 9–11

Find the waveform corresponding to the transform

$$F(s) = 2\frac{(s + 3)}{s(s + 1)(s + 2)}$$

SOLUTION:
$F(s)$ is a proper rational function and has simple poles at $s = 0$, $s = -1$, $s = -2$. Its partial fraction expansion is

$$F(s) = \frac{k_1}{s} + \frac{k_2}{s + 1} + \frac{k_3}{s + 2}$$

The cover-up algorithm yields the residues as

$$k_1 = sF(s)\bigg|_{s=0} = \frac{2(s + 3)}{(s + 1)(s + 2)}\bigg|_{s=0} = 3$$

$$k_2 = (s + 1)F(s)\bigg|_{s=-1} = \frac{2(s + 3)}{s(s + 2)}\bigg|_{s=-1} = -4$$

$$k_3 = (s + 2)F(s)\bigg|_{s=-2} = \frac{2(s + 3)}{s(s + 1)}\bigg|_{s=-2} = 1$$

The inverse transform $f(t)$ is

$$f(t) = [3 - 4e^{-t} + e^{-2t}]u(t) \qquad \blacksquare$$

Exercise 9–16

Find the waveforms corresponding to the following transforms:

(a) $F_1(s) = \dfrac{4}{(s + 1)(s + 3)}$

(b) $F_2(s) = e^{-5s}\left[\dfrac{2s}{(s + 1)(s + 3)}\right]$

(c) $F_3(s) = \dfrac{4(s + 2)}{(s + 1)(s + 3)}$

Answers:

(a) $f_1(t) = [2e^{-t} - 2e^{-3t}]u(t)$

(b) $f_2(t) = [-e^{-(t-5)} + 3e^{-3(t-5)}]u(t-5)$

(c) $f_3(t) = [2e^{-t} + 2e^{-3t}]u(t)$

Comment: Note that

$$F_2(s) = e^{-5s}\left[\frac{2s}{(s+1)(s+3)}\right] = e^{-5s}\left[\frac{-1}{s+1} + \frac{3}{s+3}\right]$$

so that the delay factor e^{-5s} is not involved in the partial fraction expansion but simply flags the amount by which the resulting waveform $[e^{-t} + 3e^{-3t}]u(t)$ is delayed to produce $f_2(t)$.

Exercise 9–17

Find the waveforms corresponding to the following transforms:

(a) $F(s) = \dfrac{6(s+2)}{s(s+1)(s+4)}$

(b) $F(s) = \dfrac{4(s+1)}{s(s+1)(s+4)}$

Answers:

(a) $f(t) = [3 - 2e^{-t} - e^{-4t}]u(t)$

(b) $f(t) = [1 - e^{-4t}]u(t)$

COMPLEX POLES

Special treatment is necessary when $F(s)$ has a complex pole. In physical situations the function $F(s)$ is a ratio of polynomials with real coefficients. If $F(s)$ has a complex pole $p = -\alpha + j\beta$, then it must also have a pole $p^* = -\alpha - j\beta$; otherwise the coefficients of the denominator polynomial would not be real. In other words, for physical signals the complex poles of $F(s)$ must occur in conjugate pairs. As a consequence, the partial fraction decomposition of $F(s)$ will contain two terms of the form

$$F(s) = \cdots + \frac{k}{s+\alpha - j\beta} + \frac{k^*}{s+\alpha + j\beta} + \cdots \tag{9-21}$$

The residues k and k^* at the conjugate poles are themselves conjugates because $F(s)$ is a rational function with real coefficients. These residues can be calculated using the cover-up algorithm and, in general, they turn out to be complex numbers. If the complex residues are written in polar form as

$$k = |k|e^{j\theta} \text{ and } k^* = |k|e^{-j\theta}$$

then the waveform corresponding to the two terms in Eq. (9–21) is

$$f(t) = [\cdots + |k|e^{j\theta}e^{(-\alpha + j\beta)t} + |k|e^{-j\theta}e^{(-\alpha - j\beta)t} + \cdots]u(t)$$

This equation can be rearranged in the form

$$f(t) = \left[\cdots + 2|k|e^{-\alpha t}\left\{\frac{e^{+j(\beta t + \theta)} + e^{-j(\beta t + \theta)}}{2}\right\} + \cdots\right]u(t) \tag{9-22}$$

The expression inside the brackets is of the form

$$\cos x = \left\{\frac{e^{+jx} + e^{-jx}}{2}\right\}$$

Consequently, we combine terms inside the braces as a cosine function with a phase angle:

$$f(t) = [\cdots + 2|k|e^{-\alpha t}\cos(\beta t + \theta) + \cdots]u(t) \qquad (9\text{--}23)$$

In summary, if $F(s)$ has a complex pole, then in physical applications there must be an accompanying conjugate complex pole. The inverse transformation combines the two poles to produce a damped cosine waveform. We only need to compute the residue at one of these poles because the residues at conjugate poles must be conjugates. Normally, we calculate the residue for the pole at $s = -\alpha + j\beta$ because its angle equals the phase angle of the damped cosine. Note that the imaginary part of this pole is positive, which means that the pole lies in the upper half of the s plane.

The inverse transform of a proper rational function with simple poles can be found by the partial fraction expansion method. The residues k at the simple poles can be found using the cover-up algorithm. The resulting waveform is a sum of terms of the form $[ke^{-\alpha t}]u(t)$ for real poles and $[2|k|e^{-\alpha t}\cos(\beta t + \theta)]u(t)$ for a pair of complex conjugate poles. The partial fraction expansion of the transform contains all of the data needed to construct the corresponding waveform.

EXAMPLE 9 – 1 2

Find the inverse transform of

$$F(s) = \frac{20(s+3)}{(s+1)(s^2 + 2s + 5)}$$

SOLUTION:
$F(s)$ has a simple pole at $s = -1$ and a pair of conjugate complex poles located at the roots of the quadratic factor

$$(s^2 + 2s + 5) = (s + 1 - j2)(s + 1 + j2)$$

The partial fraction expansion of $F(s)$ is

$$F(s) = \frac{k_1}{s+1} + \frac{k_2}{s+1-j2} + \frac{k_2^*}{s+1+j2}$$

The residues at the poles are found from the cover-up algorithm.

$$k_1 = \left.\frac{20(s+3)}{s^2 + 2s + 5}\right|_{s=-1} = 10$$

$$k_2 = \left.\frac{20(s+3)}{(s+1)(s+1+j2)}\right|_{s=-1+j2} = -5 - j5 = 5\sqrt{2}e^{+j5\pi/4}$$

We now have all of the data needed to construct the inverse transform.

$$f(t) = [10e^{-t} + 10\sqrt{2}e^{-t}\cos(2t + 5\pi/4)]u(t)$$

In this example we used k_2 to obtain the amplitude and phase angle of the damped cosine term. The residue k_2^* is not needed, but to illustrate a point we note that its value is

$$k_2^* = (-5 - j5)^* = -5 + j5 = 5\sqrt{2}e^{-j5\pi/4}$$

If k_2^* is used instead, we get the same amplitude for the damped sine but the wrong phase angle. *Caution:* Remember that Eq. (9–23) uses the residue at the complex pole with a positive imaginary part. In this example this is the pole at $s = -1 + j2$, not the pole at $s = -1 - j2$. ■

Exercise 9–18

Find the inverse transforms of the following rational functions:

(a) $F(s) = \dfrac{16}{(s+2)(s^2+4)}$

(b) $F(s) = \dfrac{2(s+2)}{s(s^2+4)}$

Answers:

(a) $f(t) = [2e^{-2t} + 2\sqrt{2}\cos(2t - 3\pi/4)]u(t)$

(b) $f(t) = [1 + \sqrt{2}\cos(2t \quad 3\pi/4)]u(t)$

Exercise 9–19

Find the inverse transforms of the following rational functions:

(a) $F(s) = \dfrac{8}{s(s^2+4s+8)}$

(b) $F(s) = \dfrac{4s}{s^2+4s+8}$

Answers:

(a) $f(t) = [1 + \sqrt{2}e^{-2t}\cos(2t + 3\pi/4)]u(t)$

(b) $f(t) = [4\sqrt{2}e^{-2t}\cos(2t + \pi/4)]u(t)$

SUMS OF RESIDUES

The sums of the residues of a proper rational function are subject to certain conditions that are useful for checking the calculations in a partial fraction expansion. To derive these conditions, we multiply Eqs. (9–17) and (9–18) by s and take the limit as $s \to \infty$. These operations yield

$$\lim_{s\to\infty} sF(s) = \lim_{s\to\infty} \frac{Ks^{m+1}}{s^n} = \lim_{s\to\infty}\left(\frac{k_1 s}{s+p_1} + \cdots + \frac{k_n s}{s+p_n}\right)$$

In the limit this equation reduces to

$$K\left[\lim_{s\to\infty}\frac{s^{m+1}}{s^n}\right] = k_1 + k_2 + \cdots + k_n$$

Since $F(s)$ is a proper rational function with $n > m$, the limit process in this equation yields the following conditions:

$$k_1 + k_2 + \cdots + k_n = \begin{cases} 0 \text{ if } n > m+1 \\ K \text{ if } n = m+1 \end{cases} \tag{9–24}$$

For a proper rational function with simple poles, the sum of residues is either zero or else equal to the transform scale factor K.

Exercise 9–20

Use the sum of residues to find the unknown residue in the following expansions:

(a) $\dfrac{21(s+5)}{(s+3)(s+10)} = \dfrac{6}{s+3} + \dfrac{k}{s+10}$

(b) $\dfrac{58s}{(s+2)(s^2+25)} = \dfrac{k}{s+2} + \dfrac{2+j5}{s+j5} + \dfrac{2-j5}{s-j5}$

Answers:

(a) $k = 15$

(b) $k = -4$

SOME SPECIAL CASES

Most of the transforms encountered in physical applications are proper rational functions with simple poles. The inverse transforms of such functions can be handled by the partial fraction expansion method developed in the previous section. This section covers the problem of finding the inverse transform when $F(s)$ is an improper rational function or has multiple poles. These matters are treated as special cases because they occur only for certain discrete values of circuit or signal parameters. However, some of these special cases are important, so we need to learn how to handle improper rational functions and multiple poles.

$F(s)$ is an **improper rational function** when the order of the numerator polynomial equals or exceeds the order of the denominator $(m \geq n)$. For example, the transform

$$F(s) = \frac{s^3 + 6s^2 + 12s + 11}{s^2 + 4s + 3} \tag{9–25}$$

is improper because $m = 3$ and $n = 2$. Using long division this improper rational function can be changed into the sum of a quotient plus a remainder which is a proper rational function. We proceed as follows:

$$
\begin{array}{r}
s + 2 \\
s^2 + 4s + 3 \overline{)\, s^3 + 6s^2 + 12s + 11} \\
-\left(s^3 + 4s^2 + 3s\right) \\
\hline
2s^2 + 9s + 11 \\
-\left(2s^2 + 8s + 6\right) \\
\hline
s + 5
\end{array}
$$

which yields

$$F(s) \;=\; s + 2 \;+\; \underbrace{\frac{s + 5}{s^2 + 4s + 3}}$$

$$= \text{Quotient} + \text{Remainder}$$

The remainder is a proper rational function, which can be expanded by partial fractions to produce

$$F(s) = s + 2 + \frac{2}{s + 1} - \frac{1}{s + 3}$$

All of the terms in this expansion are listed in Table 9–2 except the first term. The inverse transform of the first term is found using the transform of an impulse and the differentiation property. The Laplace transform of the derivative of an impulse is

$$\mathscr{L}\left[\frac{d\delta(t)}{dt}\right] = s\mathscr{L}[\delta(t)] - \delta(0-) = s$$

since $\mathscr{L}\{\delta(t)\} = 1$ and $\delta(0-) = 0$. By the uniqueness property of the Laplace transformation, we have $\mathscr{L}^{-1}\{s\} = d\delta(t)/dt$. The first derivative of an impulse is called a doublet. The inverse transform of the improper rational function in Eq. (9–25) is

$$f(t) = \frac{d\delta(t)}{dt} + 2\delta(t) + [2e^{-t} - e^{-3t}]u(t)$$

The method illustrated by this example generalizes in the following way. When $m = n$, long division produces a quotient K plus a proper rational function remainder. The constant K corresponds to an impulse $K\delta(t)$, and the remainder can be expanded by partial fractions to find the corresponding waveform. If $m > n$, then long division yields a quotient with terms like $s, s^2, \ldots s^{m-n}$ before a proper remainder function is obtained. These higher powers of s correspond to derivatives of the impulse. These pathological waveforms are theoretically interesting, but they do not actually occur in real circuits.

Improper rational functions can arise during mathematical manipulation of signal transforms. When $F(s)$ is improper, it is essential to reduce it by long division prior to expansion; otherwise the resulting partial fraction expansion will be incomplete.

EXAMPLE 9 – 13

Find the inverse transform of the following function

$$F(s) = \frac{s^3 + 3s^2 + 1}{s^3 + 6s^2 + 11s + 6}$$

SOLUTION:
This is an improper function since the orders of the numerator and the denominator are equal. We begin by performing a long division in order to change the function into a quotient and a remainder

$$s^3 + 6s^2 + 11s + 6 \overline{\smash{\big)}\, s^3 + 3s^2 + 0s + 1} \atop \begin{array}{r} 1 \\ \hline -(s^3 + 6s^2 + 11s + 6) \\ \hline -3s^2 - 11s - 5 \end{array}$$

which yields

$$F(s) = 1 + \frac{-3s^2 - 11s - 5}{s^3 + 6s^2 + 11s + 6} = 1 + \frac{-3s^2 - 11s - 5}{(s+1)(s+2)(s+3)}$$

We can expand the right-hand equation using partial fractions resulting in

$$F(s) = 1 + \frac{\frac{3}{2}}{s+1} + \frac{-5}{s+2} + \frac{\frac{1}{2}}{s+3}$$

The inverse transform is

$$f(t) = \delta(t) + \left(\frac{3}{2}e^{-t} - 5e^{-2t} + \frac{1}{2}e^{-3t}\right)u(t) \qquad ∎$$

Exercise 9–21

Find the inverse transforms of the following functions:

(a) $F(s) = \dfrac{s^2 + 4s + 5}{s^2 + 4s + 3}$

(b) $F(s) = \dfrac{s^2 - 4}{s^2 + 4}$

Answers:

(a) $f(t) = \delta(t) + [e^{-t} - e^{-3t}]u(t)$

(b) $f(t) = \delta(t) - [4\sin(2t)]u(t)$

Exercise 9–22

Find the inverse transforms of the following functions:

(a) $F(s) = \dfrac{2s^2 + 3s + 5}{s}$

(b) $F(s) = \dfrac{s^3 + 2s^2 + s + 3}{s + 2}$

Answers:

(a) $f(t) = 2\dfrac{d\delta(t)}{dt} + 3\delta(t) + 5u(t)$

(b) $f(t) = \dfrac{d^2\delta(t)}{dt^2} + \delta(t) + [e^{-2t}]u(t)$

MULTIPLE POLES

Under certain special conditions, transforms can have multiple poles. For example, the transform

$$F(s) = \frac{K(s - z_1)}{(s - p_1)(s - p_2)^2} \tag{9–26}$$

has a simple pole at $s = p_1$ and a pole of order 2 at $s = p_2$. Finding the inverse transform of this function requires special treatment of the multiple pole. We first factor out one of the two multiple poles.

$$F(s) = \frac{1}{s - p_2}\left[\frac{K(s - z_1)}{(s - p_1)(s - p_2)}\right] \tag{9–27}$$

The quantity inside the brackets is a proper rational function with only simple poles and can be expanded by partial fractions using the method of the previous section.

$$F(s) = \frac{1}{s - p_2}\left[\frac{c_1}{s - p_1} + \frac{k_{22}}{s - p_2}\right]$$

We now multiply through by the pole factored out in the first step to obtain

$$F(s) = \frac{c_1}{(s - p_1)(s - p_2)} + \frac{k_{22}}{(s - p_2)^2}$$

The first term on the right is a proper rational function with only simple poles, so it too can be expanded by partial fractions as

$$F(s) = \frac{k_1}{s - p_1} + \frac{k_{21}}{s - p_2} + \frac{k_{22}}{(s - p_2)^2}$$

After two partial fraction expansions, we have an expression in which every term is available in Table 9–2. The first two terms are simple poles that lead to exponential waveforms. The third term is of the form $k/(s + \alpha)^2$, which is the transform of a damped ramp waveform $[kte^{-\alpha t}]u(t)$. Therefore, the inverse transform of $F(s)$ in Eq. (9–26) is

$$f(t) = [k_1 e^{p_1 t} + k_{21} e^{p_2 t} + k_{22} t e^{p_2 t}]u(t) \tag{9–28}$$

Caution: If $F(s)$ in Eq. (9–26) had another finite zero, then the term in the brackets in Eq. (9–27) would be an improper rational function. When this occurs, long division must be used to reduce the improper rational function before proceeding to the partial-fraction expansion in the next step.

As with simple poles, the *s*-domain location of multiple poles determines the exponents of the exponential waveforms. The residues at the poles are the amplitudes of the waveforms. The only difference here is that the double pole leads to two terms rather than a single waveform. The first term is an exponential of the form e^{pt}, and the second term is a damped ramp of the form te^{pt}.

EXAMPLE 9 – 14

Find the inverse transform of

$$F(s) = \frac{4(s+3)}{s(s+2)^2}$$

SOLUTION:

The given transform has a simple pole at $s = 0$ and a double pole at $s = -2$. Factoring out one of the multiple poles and expanding the remainder by partial fractions yields

$$F(s) = \frac{1}{s+2}\left[\frac{4(s+3)}{s(s+2)}\right] = \frac{1}{s+2}\left[\frac{6}{s} - \frac{2}{s+2}\right]$$

Multiplying through by the removed factor and expanding again by partial fractions produces

$$F(s) = \frac{6}{s(s+2)} - \frac{2}{(s+2)^2} = \frac{3}{s} - \frac{3}{s+2} - \frac{2}{(s+2)^2}$$

The last expansion on the right yields the inverse transform as

$$f(t) = [3 - 3e^{-2t} - 2te^{-2t}]u(t) \qquad \blacksquare$$

Exercise 9–23

Find the inverse transform of

$$F(s) = 400\frac{(s+100)}{s(s+200)^2}$$

Answer:

$$f(t) = \left[1 - e^{-200t} + 200\,t\,e^{-200t}\right]u(t)$$

In principle, the procedure illustrated in this example can be applied to higher-order poles, although the process rapidly becomes quite tedious. For example, an *n*th-order pole would require *n* partial-fraction expansions, which is not an idea with irresistible appeal. Mathematics offers other methods of determining multiple pole residues that reduce the computational burden somewhat. However, these advanced mathematical tools are probably not worth learning because computer tools such as MATLAB readily handle multiple pole situations. Although we do encounter functions with high-order multiple poles in later chapters, we rarely need to find their inverse transforms.

Thus, for practical reasons our interest in multiple pole transforms is limited to two possibilities. First, a double pole on the negative real axis leads to the damped ramp:

$$\mathcal{L}^{-1}\left[\frac{k}{(s+\alpha)^2}\right] = [kte^{-\alpha t}]u(t) \qquad (9\text{--}29)$$

Second, a pair of double, complex poles leads to the damped cosine ramp:

$$\mathscr{L}^{-1}\left[\frac{k}{(s+\alpha-j\beta)^2}+\frac{k^*}{(s+\alpha+j\beta)^2}\right]=[2|k|te^{-\alpha t}\cos(\beta t+\angle k)]u(t) \qquad (9\text{–}30)$$

These two cases illustrate a general principle: When a simple pole leads to a waveform $f(t)$, then a double pole at the same location leads to a waveform $tf(t)$. Multiplying a waveform by t tends to cause the waveform to increase without bound unless exponential damping is present.

Exercise 9–24

Find the inverse transforms of the following functions:

(a) $F(s) = \dfrac{s}{(s+1)(s+2)^2}$

(b) $F(s) = \dfrac{16}{s^2(s+4)}$

(c) $F(s) = \dfrac{800\,s\,(s+1)}{(s+2)(s+10)^2}$

Answers:

(a) $f(t) = [-e^{-t} + e^{-2t} + 2te^{-2t}]u(t)$

(b) $f(t) = [4t - 1 + e^{-4t}]u(t)$

(c) $f(t) = [25e^{-2t} + 775e^{-10t} - 9000te^{-10t}]u(t)$

The inverse Laplace transform allows us to recover a time-domain waveform after operating on a system in the Laplace domain. Computing the inverse transformation manually is useful for understanding the relationship between the two domains, but at some point it becomes more efficient to perform the transformation calculations with software. Web Appendix C contains a discussion and numerous examples on using MATLAB to compute the Laplace transform of a waveform.

FINDING TRANSFORMS FROM POLE-ZERO DIAGRAMS

Pole-zero diagrams give great insight into a circuit's performance. In designing circuits, a designer often considers the location of poles and zeros. In some design requirements the location of poles and zeros are specified. In this subsection we discuss how one can obtain a transform from a pole-zero diagram. In subsequent chapters we will learn how to design circuits from a desired transform.

Consider the pole-zero diagram in Figure 9–7. There are three poles and three zeros. There is a pair of imaginary poles at $\pm j500$ and a single pole at -100. There are two real zeros, one at zero and one at -200. A third zero is at infinity. Starting with the poles, we can write

$$F(s) = \frac{K}{(s^2 + 500^2)(s + 100)}$$

Then including the zeros

$$F(s) = \frac{Ks(s + 200)}{(s^2 + 500^2)(s + 100)}$$

The one parameter we cannot get from the pole-zero diagram is K, the scale-factor. This needs to be specified as a separate requirement. We can then apply the rules we have learned to expand the transform. In the next chapter we will learn what circuit

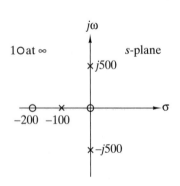

FIGURE 9–7

elements and/or signals give rise to certain poles and zeros. For now let us be satisfied with extracting a transform from a pole-zero diagram.

EXAMPLE 9–15

Find the transform $F(s)$ from the pole-zero diagram of Figure 9–8. K is 10^6.

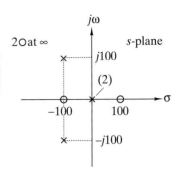

FIGURE 9–8

SOLUTION:
This diagram has four poles and four zeros, two of which are at infinity. Starting with the poles we see that two are complex conjugates at $s_1, s_2 = -100 \pm j\,100$, and there are two at $s = 0$.

$$F(s) = \frac{K\,(\text{zeros})}{s^2(s + 100 - j100)(s + 100 + j100)}$$

Including the zeros and the specified K value, we get

$$F(s) = \frac{10^6(s + 100)(s - 100)}{s^2(s + 100 - j100)(s + 100 + j100)} = \frac{10^6(s^2 - 10^4)}{s^2(s^2 + 200s + 2 \times 10^4)} \quad \blacksquare$$

Exercise 9–25

Find the transform $F(s)$ from the pole-zero diagram of Figure 9–9. K is 3×10^4.

Answer:
$$F(s) = \frac{3 \times 10^4(s^2 + 200s + 12500)}{s(s + 100)(s + 200)^2}$$

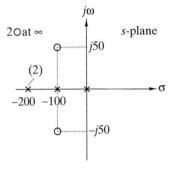

FIGURE 9–9

9–5 CIRCUIT RESPONSE USING LAPLACE TRANSFORMS

The payoff for learning about the Laplace transformation comes when we use it to find the response of dynamic circuits. The pattern for circuit analysis is shown by the solid line in Figure 9–1. The basic analysis steps are as follows:

STEP 1 Develop the circuit differential equation in the time domain.

STEP 2 Transform this equation into the s domain and algebraically solve for the response transform.

STEP 3 Apply the inverse transformation to this transform to produce the response waveform.

The first-order RC circuit in Figure 9–10 will be used to illustrate these steps.

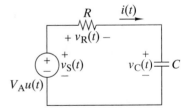

FIGURE 9–10 *First-order RC circuit.*

STEP 1

The KVL equation around the loop and the element $i-v$ relationship or element equations are

$$\text{KVL: } -v_S(t) + v_R(t) + v_C(t) = 0$$
$$\text{Source: } v_S(t) = V_A u(t)$$
$$\text{Resistor: } v_R(t) = i(t)R$$
$$\text{Capacitor: } i(t) = C\frac{dv_C(t)}{dt}$$

Substituting the $i-v$ relationships into the KVL equation and rearranging terms produces a first-order differential equation,

$$RC\frac{dv_C(t)}{dt} + v_C(t) = V_A u(t) \text{ V} \qquad (9\text{–}31)$$

with initial condition $v_C(0-) = V_0$ V.

STEP 2

The analysis objective is to use Laplace transforms to find the waveform $v_C(t)$ that satisfies this differential equation and the initial condition. We first apply the Laplace transformation to both sides of Eq. (9–31):

$$\mathscr{L}\left[RC\frac{dv_C(t)}{dt} + v_C(t)\right] = \mathscr{L}[V_A u(t)]$$

Using the linearity property leads to

$$RC\mathscr{L}\left[\frac{dv_C(t)}{dt}\right] + \mathscr{L}[v_C(t)] = V_A\mathscr{L}[u(t)]$$

Using the differentiation property and the transform of a unit step function produces

$$RC[sV_C(s) - V_0] + V_C(s) = V_A\frac{1}{s} \tag{9–32}$$

This result is an algebraic equation in $V_C(s)$, which is the transform of the response we seek. We rearrange Eq. (9–32) to the form

$$(s + 1/RC)V_C(s) = \frac{V_A/RC}{s} + V_0$$

and algebraically solve for $V_C(s)$.

$$V_C(s) = \frac{V_A/RC}{s(s + 1/RC)} + \frac{V_0}{s + 1/RC} \text{ V-s} \tag{9–33}$$

The function $V_C(s)$ is the transform of the waveform $v_C(t)$ that satisfies the differential equation and the initial condition. The initial condition appears explicitly in this equation as a result of applying the differentiation rule to obtain Eq. (9–32).

STEP 3

To obtain the waveform $v_C(t)$, we find the inverse transform of the right side of Eq. (9–33). The first term on the right is a proper rational function with two simple poles on the real axis in the s plane. The pole at the origin was introduced by the step function input. The pole at $s = -1/RC$ came from the circuit. The partial fraction expansion of the first term in Eq. (9–33) is

$$\frac{V_A/RC}{s(s + 1/RC)} = \frac{k_1}{s} + \frac{k_2}{s + 1/RC}$$

The residues k_1 and k_2 are found using the cover-up algorithm.

$$k_1 = \left.\frac{V_A/RC}{s + 1/RC}\right|_{s=0} = V_A \quad \text{and} \quad k_2 = \left.\frac{V_A/RC}{s}\right|_{s=-1/RC} = -V_A$$

Using these residues, we expand Eq. (9–33) by partial fractions as

$$V_C(s) = \frac{V_A}{s} - \frac{V_A}{s + 1/RC} + \frac{V_0}{s + 1/RC} \text{ V-s} \tag{9–34}$$

Each term in this expansion is recognizable: The first is a step function and the next two are exponentials. Taking the inverse transform of Eq. (9–34) gives

$$v_C(t) = [V_A - V_A e^{-t/RC} + V_0 e^{-t/RC}]u(t)$$
$$= [V_A + (V_0 - V_A)e^{-t/RC}]u(t)\text{V} \tag{9–35}$$

The waveform $v_C(t)$ satisfies the differential equation in Eq. (9–31) and the initial condition $v_C(0-) = V_0$. The term $V_A u(t)$ is the forced response due to the step function input, and the term $[(V_0 - V_A)e^{-t/RC}]u(t)$ is the natural response. The complete response depends on three parameters: the input amplitude V_A, the circuit time constant RC, and the initial condition V_0.

These results are identical to those found using the classical methods in Chapter 7. The outcome is the same, but the method is quite different. The Laplace transformation yields the complete response (forced and natural) by an algebraic process that inherently accounts for the initial conditions. The solid arrow in Figure 9–1 shows the overall procedure. Begin with Eq. (9–31) and relate each step leading to Eq. (9–35) to steps in Figure 9–1.

EXAMPLE 9–16

The switch in Figure 9–11 has been in position A for a long time. At $t = 0$ it is moved to position B. Find $i_L(t)$ for $t \geq 0$.

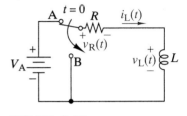

FIGURE 9–11

SOLUTION:

STEP 1 The circuit differential equation is found by combining the KVL equation and element equations with the switch in position B.

$$\text{KVL: } v_R(t) + v_L(t) = 0$$

$$\text{Resistor: } v_R(t) = i_L(t)R$$

$$\text{Inductor: } v_L(t) = L\frac{di_L(t)}{dt}$$

Substituting the element equations into the KVL equation yields

$$L\frac{di_L(t)}{dt} + Ri_L(t) = 0$$

Prior to $t = 0$, the circuit was in a dc steady-state condition with the switch in position A. Under dc conditions the inductor acts like a short circuit, and the inductor current just prior to moving the switch is $i_L(0-) = I_0 = V_A/R$.

STEP 2 Using the linearity and differentiation properties, we transform the circuit differential equation into the s domain as

$$L[sI_L(s) - I_0] + RI_L(s) = 0$$

Solving algebraically for $I_L(s)$ yields

$$I_L(s) = \frac{I_0}{s + R/L}\text{A-s}$$

STEP 3 The inverse transform of $I_L(s)$ is an exponential waveform:

$$i_L(t) = [I_0e^{-Rt/L}]u(t)\text{ A}$$

where $I_0 = V_A/R$. Substituting $i_L(t)$ back into the differential equation yields

$$L\frac{di_L(t)}{dt} + Ri_L(t) = -RI_0e^{-Rt/L} + RI_0e^{-Rt/L} = 0$$

The waveform found using Laplace transforms does indeed satisfy the circuit differential equation and the initial condition. ∎

Exercise 9–26

The inductor in Figure 9–11 is replaced by a capacitor C. The switch has been in position A for a long time. At $t = 0$ it is moved to position B. Find $V_C(s)$ and $v_C(t)$ for $t \geq 0$.

Answers:
$$V_C(s) = \frac{V_A}{s + \dfrac{1}{RC}} \text{ V-s}, \quad v_C(t) = V_A e^{-t/RC} u(t) \text{ V}$$

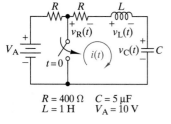

$R = 400 \,\Omega \quad C = 5 \,\mu\text{F}$
$L = 1 \,\text{H} \quad V_A = 10 \,\text{V}$

FIGURE 9–12

EXAMPLE 9–17

The switch in Figure 9–12 has been open for a long time. At $t = 0$ the switch is closed. Find $i(t)$ for $t \geq 0$.

SOLUTION:
The governing equation for the second-order circuit in Figure 9–12 is found by combining the element equations and a KVL equation around the loop with the switch closed:

$$\text{KVL: } v_R(t) + v_L(t) + v_C(t) = 0$$
$$\text{Resistor: } v_R(t) = Ri(t)$$
$$\text{Inductor: } v_L(t) = L\frac{di(t)}{dt}$$
$$\text{Capacitor: } v_C(t) = \frac{1}{C}\int_0^t i(\tau)d\tau + v_C(0)$$

Substituting the element equations into the KVL equation yields

$$L\frac{di(t)}{dt} + Ri(t) + \frac{1}{C}\int_0^t i(\tau)d\tau + v_C(0) = 0$$

Using the linearity property, the differentiation property, and the integration property, we transform this second-order integrodifferential equation into the s domain as

$$L[sI(s) - i_L(0)] + RI(s) + \frac{1}{C}\frac{I(s)}{s} + v_C(0)\frac{1}{s} = 0$$

Solving for $I(s)$ results in

$$I(s) = \frac{si_L(0) - v_C(0)/L}{s^2 + \dfrac{R}{L}s + \dfrac{1}{LC}} \text{ A-s}$$

Prior to $t = 0$, the circuit was in a dc steady-state condition with the switch open. In dc steady state, the inductor acts like a short circuit and the capacitor like an open circuit, so the initial conditions are $i_L(0-) = 0\,\text{A}$ and $v_C(0-) = V_A = 10\,\text{V}$. Inserting the initial conditions and the numerical values of the circuit parameters into the equation for $I(s)$ gives

$$I(s) = -\frac{10}{s^2 + 400s + 2 \times 10^5} \text{ A-s}$$

The denominator quadratic can be factored as $(s + 200)^2 + 400^2$ and $I(s)$ written in the following form:

$$I(s) = -\frac{10}{400}\left[\frac{400}{(s+200)^2 + (400)^2}\right] \text{ A-s}$$

Comparing the quantity inside the brackets with the entries in the $F(s)$ column of Table 9–2, we find that $I(s)$ is a damped sine with $\alpha = 200$ and $\beta = 400$. By linearity, the quantity outside the brackets is the amplitude of the damped sine. The inverse transform is

$$i(t) = [-0.025\, e^{-200t} \sin 400t] u(t) \ \text{A}$$

Substituting this result back into the circuit integrodifferential equation yields the following term-by-term tabulation:

$$L\frac{di(t)}{dt} = +5e^{-200t} \sin 400t - 10e^{-200t} \cos 400t$$

$$Ri(t) = -10e^{-200t} \sin 400t$$

$$\frac{1}{C}\int_0^t i(\tau)d\tau = +5e^{-200t} \sin 400t + 10e^{-200t} \cos 400t - 10$$

$$v_C(0) = +10$$

The sum of the right-hand sides of these equations is zero. This result shows that the waveform $i(t)$ found using Laplace transforms does indeed satisfy the circuit integrodifferential equation and the initial conditions. ■

Exercise 9–27

Find the transforms that satisfy the following equations and the given initial conditions.

(a) $\dfrac{dv(t)}{dt} + 6v(t) = 4u(t) \ \text{V} \qquad v(0-) = -3 \ \text{V}$

(b) $4\dfrac{dv(t)}{dt} + 12v(t) = 16\cos 3t \ \text{V} \qquad v(0-) = 2 \ \text{V}$

Answers:

(a) $V(s) = \dfrac{4}{s(s+6)} - \dfrac{3}{s+6} \ \text{V-s}$

(b) $V(s) = \dfrac{4s}{(s^2+9)(s+3)} + \dfrac{2}{s+3} \ \text{V-s}$

Exercise 9–28

Find the transforms that satisfy the following equations and the given initial conditions.

(a) $\displaystyle\int_0^t v(\tau)d\tau + 10v(t) = 10u(t) \ \text{V}$

(b) $\dfrac{d^2v(t)}{dt^2} + 4\dfrac{dv(t)}{dt} + 3v(t) = 5e^{-2t} \ \text{V} \qquad v'(0-) = 2 \ \text{V/s} \quad v(0-) = -2 \ \text{V}$

Answers:

(a) $V(s) = \dfrac{1}{s+0.1} \ \text{V-s}$

(b) $V(s) = \dfrac{5}{(s+1)(s+2)(s+3)} - \dfrac{2}{s+1} \ \text{V-s}$

CIRCUIT RESPONSE WITH TIME-VARYING INPUTS

It is encouraging to find that the Laplace transformation yields results that agree with those obtained by classical methods. The transform method reduces solving circuit differential equations to an algebraic process that includes the initial conditions. However, before being overcome with euphoria we must remember that the

Laplace transform method begins with the circuit differential equation and the initial conditions. It does not provide these quantities to us. The transform method simplifies the solution process, but it does not substitute for understanding how to formulate circuit equations.

The Laplace transform method is especially usefully when the circuit is driven by time-varying inputs. To illustrate, we return to the RC circuit in Figure 9–10 and replace the step function input by a general input signal denoted $v_S(t)$. The right side of the circuit differential equation in Eq. (9–31) changes to accommodate the new input by taking the form

$$RC\frac{dv_C(t)}{dt} + v_C(t) = v_S(t) \qquad (9\text{–}36)$$

with an initial condition $v_C(0) = V_0$ V.

The only change here is that the driving force on the right side of the differential equation is a general time-varying waveform $v_S(t)$. The objective is to find the capacitor voltage $v_C(t)$ that satisfies the differential equation and the initial conditions. The classical methods of solving for the forced response depend on the form of $v_S(t)$. However, with the Laplace transform method we can proceed without actually specifying the form of the input signal.

We first transform Eq. (9–36) into the s domain:

$$RC[sV_C(s) - V_0] + V_C(s) = V_S(s)$$

The only assumption here is that the input waveform is Laplace transformable, a condition met by all causal signals of engineering interest. We now algebraically solve for the response $V_C(s)$:

$$V_C(s) = \frac{V_S(s)/RC}{s + 1/RC} + \frac{V_0}{s + 1/RC} \text{ V-s} \qquad (9\text{–}37)$$

The function $V_C(s)$ is the transform of the response of the RC circuit in Figure 9–10 due to a general input signal $v_S(t)$. We have gotten this far without specifying the form of the input signal. In a sense, we have found the general solution in the s domain of the differential equation in Eq. (9–36) for any causal input signal.

All of the necessary ingredients are present in Eq. (9–37):

1. The transform $V_S(s)$ represents the applied input signal.
2. The pole at $s = -1/RC$ defines the circuit time constant.
3. The initial value $v_C(0-) = V_0$ summarizes all events prior to $t = 0$.

However, we must have a particular input in mind to solve for the waveform $v_C(t)$. The following examples illustrate the procedure for different input driving forces.

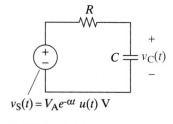

$v_S(t) = V_A e^{-\alpha t}\, u(t)$ V

FIGURE 9–13

EXAMPLE 9–18

Find $v_C(t)$ in the RC circuit in Figure 9–13 when the input is the waveform $v_S(t) = [V_A e^{-\alpha t}]u(t)$ V.

SOLUTION:
The transform of the input is $V_S(s) = V_A/(s + \alpha)$. For the exponential input the response transform in Eq. (9–37) becomes

$$V_C(s) = \frac{V_A/RC}{(s + \alpha)(s + 1/RC)} + \frac{V_0}{s + 1/RC} \text{ V-s} \qquad (9\text{–}38)$$

If $\alpha \neq 1/RC$, then the first term on the right is a proper rational function with two simple poles. The pole at $s = -\alpha$ came from the input and the pole at $s = -1/RC$ from the circuit. A partial fraction expansion of the first term has the form

$$\frac{V_A/RC}{(s + \alpha)(s + 1/RC)} = \frac{k_1}{s + \alpha} + \frac{k_2}{s + 1/RC}$$

The residues in this expansion are

$$k_1 = \left.\frac{V_A/RC}{s + 1/RC}\right|_{s = \alpha} = \frac{V_A}{1 - \alpha RC}$$

$$k_2 = \left.\frac{V_A/RC}{s + \alpha}\right|_{s = -1/RC} = \frac{V_A}{\alpha RC - 1}$$

The expansion of the response transform $V_C(s)$ is

$$V_C(s) = \frac{V_A/(1 - \alpha RC)}{s + \alpha} + \frac{V_A/(\alpha RC - 1)}{s + 1/RC} + \frac{V_0}{s + 1/RC} \text{ V-s}$$

The inverse transform of $V_C(s)$ is

$$v_C(t) = \left[\frac{V_A}{1 - \alpha RC}e^{-\alpha t} + \frac{V_A}{\alpha RC - 1}e^{-t/RC} + V_0 e^{-t/RC}\right]u(t) \text{ V}$$

The first term is the forced response, and the last two terms are the natural response. The forced response is an exponential because the input introduced a pole at $s = -\alpha$. The natural response is also an exponential, but its time constant depends on the circuit's pole at $s = -1/RC$. In this case the forced and natural responses are both exponential signals with poles on the real axis. However, the forced response comes from the pole introduced by the input, while the natural response depends on the circuit's pole.

If $\alpha = 1/RC$, then the response just given is no longer valid (k_1 and k_2 become infinite). To find the response for this condition, we return to Eq. (9–38) and replace α by $1/RC$:

$$V_C(s) = \frac{V_A/RC}{(s + 1/RC)^2} + \frac{V_0}{s + 1/RC} \text{ V-s}$$

We now have a double pole at $s = -1/RC = -\alpha$. The double pole term is the transform of a damped ramp, so the inverse transform is

$$v_C(t) = \left[V_A \frac{t}{RC}e^{-t/RC} + V_0 e^{-t/RC}\right]u(t) \text{ V}$$

When $\alpha = 1/RC$, the s-domain poles of the input and the circuit coincide and the zero-state ($V_0 = 0$) response has the form $\alpha t e^{-\alpha t}$. We cannot separate this response into forced and natural components since the input and circuit poles coincide. ∎

Exercise 9–29

The RC circuit in Figure 9-13 has $R = 10$ kΩ, $C = 0.2$ μF, and $V_0 = -5$ V. The input is $v_S(t) = 10\,e^{-1000t}u(t)$ V. Find $v_C(t)$ for $t \geq 0$.

Answer: $v_C(t) = [5e^{-500t} - 10e^{-1000t}]\,u(t)$ V.

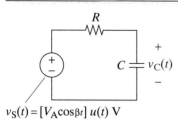

$v_S(t) = [V_A \cos \beta t]\, u(t)$ V

FIGURE 9–14

EXAMPLE 9–19

Find $v_C(t)$ when the input to the RC circuit in Figure 9–14 is $v_S(t) = [V_A \cos \beta t]u(t)$ V.

SOLUTION:
The transform of the input is $V_S(s) = V_A s/(s^2 + \beta^2)$. For a cosine input the response transform in Eq. (9–37) becomes

$$V_C(s) = \frac{sV_A/RC}{(s^2 + \beta^2)(s + 1/RC)} + \frac{V_0}{s + 1/RC} \text{ V-s}$$

The sinusoidal input introduces a pair of poles located at $s = \pm j\beta$. The first term on the right is a proper rational function with three simple poles. The partial fraction expansion of the first term is

$$\frac{sV_A/RC}{(s - j\beta)(s + j\beta)(s + 1/RC)} = \frac{k_1}{s - j\beta} + \frac{k_1^*}{s + j\beta} + \frac{k_2}{s + 1/RC}$$

To find the response, we need to find the residues k_1 and k_2:

$$k_1 = \left. \frac{sV_A/RC}{(s + j\beta)(s + 1/RC)} \right|_{s = j\beta} = \frac{V_A/2}{1 + j\beta RC} = |k_1|e^{j\theta}$$

where

$$|k_1| = \frac{V_A/2}{\sqrt{1 + (\beta RC)^2}} \quad \text{and} \quad \theta = -\tan^{-1}(\beta RC)$$

The residue k_2 at the circuit pole is

$$k_2 = \left. \frac{sV_A/RC}{s^2 + \beta^2} \right|_{s = -1/RC} = -\frac{V_A}{1 + (\beta RC)^2}$$

We now perform the inverse transform to obtain the response waveform:

$$v_C(t) = [2|k_1|\cos(\beta t + \theta) + k_2 e^{-t/RC} + V_0 e^{-t/RC}]u(t)$$

$$= \left[\frac{V_A}{\sqrt{1 + (\beta RC)^2}} \cos(\beta t + \theta) - \frac{V_A}{1 + (\beta RC)^2} e^{-t/RC} + V_0 e^{-t/RC} \right]u(t) \text{ V}$$

The first term is the forced response, and the remaining two are the natural response. The forced response is sinusoidal because the input introduces poles at $s = \pm j\beta$. The natural response is an exponential with a time constant determined by the location of the circuit's pole at $s = -1/RC$. ∎

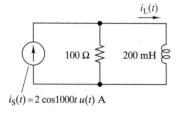

$i_S(t) = 2 \cos 1000t \, u(t)$ A

FIGURE 9–15

Exercise 9–30

The RL circuit of Figure 9–15 is in the zero state when the input $i_S(t) = [2 \cos 1000t]\, u(t)$ A is applied. Find $i_L(t)$ for $t \geq 0$.

Answer: $\qquad i_L(t) = [0.894 \cos(1000t - 63.4°) - 0.4e^{-500t}]u(t)\text{A}$

9–6 INITIAL AND FINAL VALUE PROPERTIES

The **initial value** and **final value properties** can be stated as follows:

$$\text{Initial value: } \lim_{t \to 0+} f(t) = \lim_{s \to \infty} sF(s)$$

$$\text{Final value: } \lim_{t \to \infty} f(t) = \lim_{s \to 0} sF(s)$$

(9–39)

These properties display the relationship between the origin and infinity in the time and frequency domains. The value of $f(t)$ at $t = 0+$ in the time domain (initial value) is the same as the value of $sF(s)$ at infinity in the s plane. Conversely, the value of $f(t)$ as $t \to \infty$ (final value) is the same as the value of $sF(s)$ at the origin in the s plane.

Proof of both the initial value and final value properties starts with the differentiation property:

$$sF(s) - f(0-) = \int_{0-}^{\infty} \frac{df(t)}{dt} e^{-st} \, dt \qquad (9\text{--}40)$$

To establish the initial value property, we rewrite the integral on the right side of this equation and take the limit of both sides as $s \to \infty$.

$$\lim_{s \to \infty} [sF(s) - f(0-)] = \lim_{s \to \infty} \int_{0-}^{0+} \frac{df(t)}{dt} e^{-st} \, dt + \lim_{s \to \infty} \int_{0+}^{\infty} \frac{df(t)}{dt} e^{-st} \, dt \qquad (9\text{--}41)$$

The first integral on the right side reduces to $f(0+) - f(0-)$ since e^{-st} is unity on the interval from $t = 0-$ to $t = 0+$. The second integral vanishes because e^{-st} goes to zero as $s \to \infty$. In addition, on the left side of Eq. (9–41) the $f(0-)$ is independent of s and can be taken outside the limiting process. Inserting all of these considerations reduces Eq. (9–41) to

$$\lim_{s \to \infty} sF(s) = \lim_{t \to 0+} f(t) \qquad (9\text{--}42)$$

which completes the proof of the initial value property.

Proof of the final value theorem begins by taking the limit of both sides of Eq. (9–40) as $s \to 0$:

$$\lim_{s \to 0} [sF(s) - f(0-)] = \lim_{s \to 0} \int_{0-}^{\infty} \frac{df(t)}{dt} e^{-st} dt \qquad (9\text{--}43)$$

The integral on the right side of this equation reduces to $f(\infty) - f(0-)$ because e^{-st} becomes unity as $s \to 0$. Again, the $f(0-)$ on the left side is independent of s and can be taken outside of the limiting process. Inserting all of these considerations reduces Eq. (9–43) to

$$\lim_{s \to 0} sF(s) = \lim_{t \to \infty} f(t) \qquad (9\text{--}44)$$

which completes the proof of the final value property.

A damped cosine waveform provides an illustration of the application of these properties. The transform of the damped cosine is

$$\mathscr{L}\{[Ae^{-\alpha t} \cos \beta t] u(t)\} = \frac{A(s + \alpha)}{(s + \alpha)^2 + \beta^2}$$

Applying the initial and final value limits, we obtain

$$\text{Initial value: } \lim_{t \to 0} f(t) = \lim_{t \to 0} Ae^{-\alpha t} \cos \beta t = A$$

$$\lim_{s \to \infty} sF(s) = \lim_{s \to \infty} \frac{sA(s + \alpha)}{(s + \alpha)^2 + \beta^2} = A$$

$$\text{Final value: } \lim_{t \to \infty} f(t) = \lim_{t \to \infty} Ae^{-\alpha t} \cos \beta t = 0$$

$$\lim_{s \to 0} sF(s) = \lim_{s \to 0} \frac{sA(s + \alpha)}{(s + \alpha)^2 + \beta^2} = 0$$

Note the agreement between the t-domain and s-domain limits in both cases.

There are restrictions on the initial and final value properties. The initial value property is valid when $F(s)$ is a proper rational function or, equivalently, when $f(t)$ does not have an impulse at $t = 0$. The final value property is valid when the poles of $sF(s)$ are in the left half plane or, equivalently, when $f(t)$ is a waveform that approaches a final value at $t \to \infty$. Note that the final value restriction allows $F(s)$ to have a simple pole at the origin since the limitation is on the poles of $sF(s)$.

Caution: The initial and final value properties will appear to work when the aforementioned restrictions are not met. In other words, these properties do not tell you they are giving nonsense answers when you violate their limitations. You must always check the restrictions on $F(s)$ before applying either of these properties.

For example, applying the final value property to a cosine waveform yields

$$\lim_{t \to \infty} \cos \beta t = \lim_{s \to 0} s \left[\frac{s}{s^2 + \beta^2} \right] = 0$$

The final value property appears to say that $\cos \beta t$ approaches zero as $t \to \infty$. This conclusion is incorrect since the waveform oscillates between -1 and $+1$. The problem is that the final value property does not apply to a cosine waveform because $sF(s)$ has poles on the j-axis at $s = \pm j\beta$.

EXAMPLE 9–20

Use the initial and final value properties to find the initial and final values of the waveform whose transform is

$$F(s) = 2 \frac{(s+3)}{s(s+1)(s+2)}$$

SOLUTION:
The given $F(s)$ is a proper rational function, so the initial value property can be applied as

$$f(0) = \lim_{s \to \infty} sF(s) = \lim_{s \to \infty} \left[2 \frac{(s+3)}{(s+1)(s+2)} \right] = 0$$

The poles of $sF(s)$ are located in the left half plane at $s = -1$ and $s = -2$; hence, the final value property can be applied as

$$f(\infty) = \lim_{s \to 0} sF(s) = \lim_{s \to 0} \left[2 \frac{(s+3)}{(s+1)(s+2)} \right] = 3$$

In Example 9–11 the waveform corresponding to this transform was found to be
$$f(t) = [3 - 4e^{-t} + e^{-2t}]u(t)$$
from which we find
$$f(0) = 3 - 4e^{-0} + e^{-0} = 0$$
$$f(\infty) = 3 - 4e^{-\infty} + e^{-\infty} = 3$$
which confirms the results found directly from $F(s)$. ∎

Exercise 9–31

Find the initial and final values of the waveforms corresponding to the following transforms:

(a) $F_1(s) = 100 \dfrac{s+3}{s(s+5)(s+20)}$

(b) $F_2(s) = 80 \dfrac{s(s+5)}{(s+4)(s+20)}$

Answers:

(a) Initial value = 0, final value = 3.
(b) $F_2(s)$ is not a proper rational function, final value = 0.

SUMMARY

- The Laplace transformation converts waveforms in the time domain to transforms in the s domain. The inverse transformation converts transforms into causal waveforms. A transform pair is unique if and only if $f(t)$ is causal.

- The Laplace transforms of basic signals like the step function, exponential, and sinusoid are easily derived from the integral definition. Other transform pairs can be derived using basic signal transforms and the uniqueness, linearity, time integration, time differentiation, and translation properties of the Laplace transformation.

- Proper rational functions with simple poles can be expanded by partial fractions to obtain inverse Laplace transforms. Simple real poles lead to exponential waveforms and simple complex poles to damped

sinusoids. Partial-fraction expansions of improper rational functions and functions with multiple poles require special treatment. MATLAB software can be used to find Laplace transforms from $f(t)$ and waveforms from $F(s)$. (see Web Appendix C)

- Using Laplace transforms to find the response of a linear circuit involves transforming the circuit differential equation into the s domain, algebraically solving for the response transform, and performing the inverse transformation to obtain the response waveform.

- The initial and final value properties determine the initial and final values of a waveform $f(t)$ from the value of $sF(s)$ as $s \to \infty$ and $s \to 0$, respectively. The initial value property applies if $F(s)$ is a proper rational function. The final value property applies if all of the poles of $sF(s)$ are in the left half plane.

PROBLEMS

OBJECTIVE 9–1 LAPLACE TRANSFORM (SECTS. 9–1, 9–2, 9–3)

Find the Laplace transform of a given signal waveform using transform properties and pairs, using the integral definition of the Laplace transformation, or software applications. Locate the poles and zeros of the transform and construct a pole-zero diagram.

See Examples 9–1, 9–2, 9–3, 9–4, 9–5, 9–6, 9–7, 9–8, 9–9, 9–10 and Exercises 9–1, 9–2, 9–3, 9–4, 9–5, 9–6, 9–7, 9–8, 9–9, 9–10, 9–11, 9–12, 9–13, 9–14, 9–15.

9–1 Find the Laplace transform of $f(t) = 500[1 - e^{-100t}]\,u(t)$. Locate the poles and zeros of $F(s)$.

9–2 Find the Laplace transform of $f(t) = 20\sin(377t)\,u(t)$. Locate the poles and zeros of $F(s)$.

9–3 Find the Laplace transform of $f(t) = -10\,\delta(t) + 10\,u(t)$. Locate the poles and zeros of $F(s)$.

9–4 Find the Laplace transform of $f(t) = 20[e^{-200t} - 2e^{-100t}]\,u(t)$. Locate the poles and zeros of $F(s)$.

9–5 Find the Laplace transform of $f(t) = A[(B + \alpha t)\,e^{-\alpha t}]\,u(t)$. Locate the poles and zeros of $F(s)$.

9–6 Find the Laplace transform of $f(t) = 0.005[10 - 10\cos(1000t)]\,u(t)$. Locate the poles and zeros of $F(s)$.

9–7 Find the Laplace transform of $f(t) = 5[4\cos(50t) - 5\sin(50t)]\,u(t)$. Locate the poles and zeros of $F(s)$.

9–8 Find the Laplace transform of $f(t) = \delta(t) - 200e^{-20t}\cos(200t)\,u(t)$. Locate the poles and zeros of $F(s)$.

9–9 Find the Laplace transform of $f(t) = 10[3 - 5t - 2e^{-15t}]\,u(t)$. Locate the poles and zeros of $F(s)$.

9–10 Find the Laplace transforms of the following waveforms and plot their pole-zero diagrams:
(a) $f_1(t) = [25e^{-15t} - 20e^{-10t}]\,u(t)$
(b) $f_2(t) = 10[\cos 10t + \cos 20t]\,u(t)$

9–11 Find the Laplace transforms of the following waveforms and plot their pole-zero diagrams.
(a) $f_1(t) = 2\delta(t) + [200e^{-200t} + 400e^{-400t}]\,u(t)$
(b) $f_2(t) = [15e^{-200t} + 15\cos 500t]\,u(t)$

9–12 9–12 Find the Laplace transforms of the following waveforms. Locate the poles and zeros of $F(s)$. Use MATLAB to verify your results.
(a) $f_1(t) = 5\delta(t) + (625t\,e^{-25t})\,u(t)$
(b) $f_2(t) = [10 + 5e^{-10t}(\cos 10t + \sin 10t)]\,u(t)$

9–13 9–13 Find the Laplace transforms of the following waveforms. Use MATLAB to verify your results.
(a) $f_1(t) = 5\delta(t - 2)$
(b) $f_2(t) = 10e^{-50(t-1)}u(t - 1)$
(c) $f_3(t) = 20e^{-50(t-10)}u(t - 10)$

9–14 9–14 Use MATLAB to find the Laplace transform of the following waveform
$$f(t) = [10 + 2e^{-10t}]u(t) + [5\cos 100(t - 0.05)]u(t - 0.05)$$

9–15 Find the Laplace transforms of the following waveforms.

(a) $f_1(t) = \dfrac{d}{dt}(50e^{-1000t}\cos 200kt)u(t)$

(b) $f_2(t) = \int\limits_0^t 20e^{-10x}dx + 10u(t) + 20\dfrac{de^{-10t}}{dt}u(t)$

9–16 Consider the waveform in Figure P9–16.
(a) Write an expression for the waveform $f(t)$ using step and ramp functions.

(b) Use the time-domain translation property to find the Laplace transform of the waveform $f(t)$ found in part (a).
(c) Verify the Laplace transform found in (b) by applying the definition of the Laplace transformation in Eq. (9–2) to the waveform $f(t)$ found in (a).

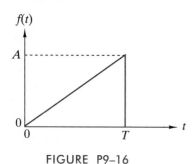

FIGURE P9–16

9–17 Consider the waveform in Figure P9–17.
(a) Write an expression for the waveform $f(t)$ in Figure P9–17 using step functions.
(b) Use the time-domain translation property to find the Laplace transform of the waveform $f(t)$ found in part (a).
(c) Verify the Laplace transform found in (b) by applying the definition of the Laplace transformation in Eq. (9–2) to the waveform $f(t)$ found in (a).

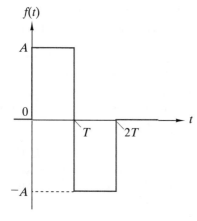

FIGURE P9–17

9–18 For the following waveform: $f(t) = [500 + 100e^{-500t}t \sin 1000t]u(t)$
(a) Find the Laplace transform of the waveform. Locate the poles and zeros of $F(s)$.
(b) Validate your result using MATLAB.

9–19 Consider the waveform in Figure P9–19.
(a) Write an expression for the waveform $f(t)$ in Figure P9–19 using a delayed exponential.
(b) Use the time-domain translation property to find the Laplace transform of the waveform $f(t)$ found in part (a).
(c) Verify the Laplace transform found in (b) by applying the definition of the Laplace transformation in Eq. (9–2) to the waveform $f(t)$ found in (a).

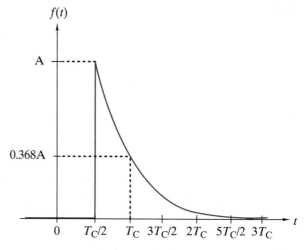

FIGURE P9–19

**O B J E C T I V E 9 – 2 I N V E R S E T R A N S F O R M S
(S E C T . 9 – 4)**

(a) Find the inverse transform of a given Laplace transform using partial fraction expansion, basic transform properties and pairs, or using software tools.
(b) Given a pole-zero diagram, find the respective transform.
See Examples 9–11, 9–12, 9–13, 9–14, 9–15 and Exercises 9–16, 9–17, 9–18, 9–19, 9–20, 9–21, 9–22, 9–23, 9–24, 9–25.

9–20 Find the inverse Laplace transforms of the following functions:

(a) $F_1(s) = \dfrac{50}{s(s+50)}$

(b) $F_2(s) = \dfrac{s+1}{(s+2)(s+3)}$

9–21 Find the inverse Laplace transforms of the following functions:

(a) $F_1(s) = \dfrac{s+30}{s(s+40)}$

(b) $F_2(s) = \dfrac{(s+10)(s+20)}{s(s+50)(s+100)}$

9–22 Find the inverse Laplace transforms of the following functions:

(a) $F_1(s) = \dfrac{5000(s+1000)}{(s+500)(s+5000)}$

(b) $F_2(s) = \dfrac{5s^2}{(s+100)(s+500)}$

9–23 Find the inverse Laplace transforms of the following functions:

(a) $F_1(s) = \dfrac{900}{(s+10)^2 + 30^2}$

(b) $F_2(s) = \dfrac{3(s+10)}{(s+10)^2 + 30^2}$

9–24 Find the inverse Laplace transfonns of the following functions and sketch their waveforms for $\beta > 0$:

(a) $F_1(s) = \dfrac{\beta(s+\beta)}{s(s^2+\beta^2)}$

(b) $F_2(s) = \dfrac{s(s+\beta)}{s^2+\beta^2}$

9–25 Find the inverse Laplace transforms of the following functions:

(a) $F_1(s) = \dfrac{\alpha^2}{s^2(s+\alpha)}$

(b) $F_2(s) = \dfrac{\alpha^2}{s(s+\alpha)^2}$

9–26 Find the inverse Laplace transforms of the following fiinctions:

(a) $F_1(s) = \dfrac{600}{(s+10)(s+20)(s+30)}$

(b) $F_2(s) = \dfrac{2(s+10)}{(s+15)(s+20)}$

9–27 Find the inverse Laplace transforms of the following functions then validate your answers using MATLAB:

(a) $F_1(s) = \dfrac{16s}{(s+3)(s^2+11s+10)}$

(b) $F_2(s) = \dfrac{5(s^2+9)}{s(s^2+25)}$

9–28 Find the inverse transforms of the following functions:

(a) $F_1(s) = \dfrac{(s+10000)(s+100000)}{s(s+1000)(s+50000)}$

(b) $F_2(s) = \dfrac{3(s^4+10s^2+4)}{s(s^2+1)(s^2+4)}$

9–29 Find the inverse transforms of the following functions:

(a) $F_1(s) = \dfrac{300(s+50)}{s(s^2+40s+300)}$

(b) $F_2(s) = \dfrac{1000s}{(s+5)(s^2+4s+8)}$

9–30 Find the inverse transforms of the following functions:

(a) $F_1(s) = \dfrac{16(s^2+256)}{s(s^2+8s+32)}$

(b) $F_2(s) = \dfrac{3(s^2+20s+400)}{s(s^2+50s+400)}$

9–31 9–31 Find the inverse Laplace transforms of the following functions then validate your answers using MATLAB:

(a) $F_1(s) = \dfrac{(s+100)^2}{(s+50)^2(s+200)}$

(b) $F_2(s) = \dfrac{(s+50)^2}{(s+100)^2(s+200)}$

9–32 A certain transform has a simple pole at $s=-20$, a simple zero at $s=-\gamma$, and a scale factor of $K=1$. Select values for γ so the inverse transform is

(a) $f(t) = \delta(t) - 5e^{-20t}$ **(b)** $f(t) = \delta(t)$ **(c)** $f(t) = \delta(t) + 5e^{-20t}$

9–33 Find the inverse transforms of the following functions:

(a) $F_1(s) = \dfrac{s(s+10)(s+20)}{(s+5)(s+30)(s+50)}$

(b) $F_2(s) = \dfrac{(s+1000)(s+5000)}{(s+2000)}$

9–34 Find the inverse transforms of the following functions:

(a) $F_1(s) = \dfrac{s^2}{(s+5)}$

(b) $F_2(s) = \dfrac{(s+1000)^2}{(s+2000)^2}$

9–35 Find the inverse transforms of the following functions:

(a) $F_1(s) = \dfrac{e^{-5s}(s+20)}{(s+10)(s+30)}$

(b) $F_2(s) = \dfrac{se^{-5s}+20}{(s+10)(s+30)}$

(c) $F_3(s) = \dfrac{s+20e^{-5s}}{(s+10)(s+30)}$

9–36 Use MATLAB to find the inverse transform and plot the poles and zeros of the following function:

$$F(s) = \dfrac{300s(s^2+30s+400)}{(s+20)(s^3+6s^2+16s+16)}$$

9–37 Use MATLAB to find the inverse transform and plot the poles and zeros of the following function:

$$F(s) = \dfrac{500(s^3+2s^2+s+2)}{s(s^3+4s^2+4s+16)}$$

9–38 Find the transform $F(s)$ from the pole-zero diagram of Figure P9–38. K is 5.

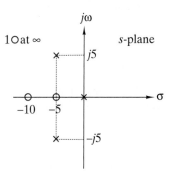

FIGURE P9–38

9–39 Find the transform $F(s)$ from the pole-zero diagram of Figure P9–39. K is 5×10^6.

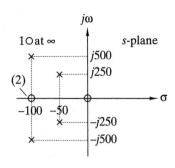

FIGURE P9–39

9–40 Find the transform $F(s)$ from the pole-zero diagram of Figure P9–40. K is 50.

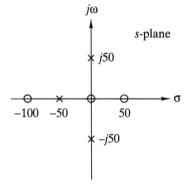

FIGURE P9–40

OBJECTIVE 9–3 CIRCUIT RESPONSE USING LAPLACE TRANSFORMS (SECT. 9–5)

Given a first- or second-order circuit:

(a) Determine the circuit differential equation and the initial conditions (if not given).

(b) Transform the differential equation into the s domain and solve for the response transform.

(c) Use the inverse transformation to find the response waveform.

See Examples 9–16, 9–17, 9–18, 9–19 and Exercises 9–26, 9–27, 9–28, 9–29, 9–30.

9–41 Use the Laplace transformation to find the $v(t)$ that satisfies the following first-order differential equations:

(a) $250\dfrac{dv(t)}{dt} + 2500v(t) = 0, \quad v(0-) = 50\,\text{V}$

(b) $\dfrac{dv(t)}{dt} + 300v(t) = 600u(t), \quad v(0-) = -150\,\text{V}$

9–42 Use the Laplace transformation to find the $i(t)$ that satisfies the following first-order differential equation:

$$\dfrac{di(t)}{dt} + 500\,i(t) = \left[0.100e^{-100t}\right]u(t), \quad i(0-) = 0\,\text{A}$$

9–43 The switch in Figure P9–43 has been open for a long time and is closed at $t = 0$. The circuit parameters are $R = 1\,\text{k}\Omega$, $L = 100\,\text{mH}$, and $V_A = 15\,\text{V}$.

(a) Find the differential equation for the inductor current $i_L(t)$ and initial condition $i_L(0)$.

(b) Solve for $i_L(t)$ using the Laplace transformation.

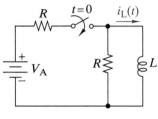

FIGURE P9–43

9–44 The switch in Figure. P9–43 has been closed for a long time and is opened at $t = 0$. The circuit parameters are $R = 50\,\Omega$, $L = 200\,\text{mH}$, and $V_A = 50\,\text{V}$.

(a) Find the differential equation for the inductor current $i_L(t)$ and initial condition $i_L(0)$.

(b) Solve for $i_L(t)$ using the Laplace transformation.

9–45 The switch in Figure P9–45 has been open for a long time. At $t = 0$ the switch is closed.

(a) Find the differential equation for the capacitor voltage and initial condition $v_C(0)$.

(b) Find $v_O(t)$ using the Laplace transformation for $v_S(t) = 25\,u(t)\,\text{V}$.

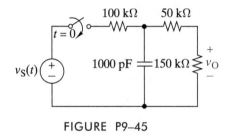

FIGURE P9–45

9–46 Repeat Problem 9–45 for the input waveform $v_S(t) = 169\,[\cos 377t]\,u(t)\,\text{V}$.

9–47 Repeat Problem 9–45 for the input waveform $v_S(t) = 24\,e^{-1000t}u(t)\,\text{V}$.

9–48 Use the Laplace transformation to find the $v(t)$ that satisfies the following second-order differential equation:

$$\dfrac{d^2v(t)}{dt^2} + 20\dfrac{dv(t)}{dt} + 1000v(t) = 0, \quad v(0-) = 20\,\text{V and } \dfrac{dv(0-)}{dt} = 0.$$

9–49 Use the Laplace transformation to find the $v(t)$ that satisfies the following second-order differential equation:

$$\dfrac{d^2v(t)}{dt^2} + 50\dfrac{dv(t)}{dt} + 400\,v(t) = 0, \quad v(0-) = 0 \text{ and } \dfrac{dv(0-)}{dt} = 1000\,\text{V/s}$$

9–50 The switch in Figure P9–50 has been open for a long time and is closed at $t = 0$. The circuit parameters are $R = 500\ \Omega$, $L = 2.5$ H, $C = 2.5\ \mu$F, and $V_A = 1000$ V.
(a) Find the circuit differential equation in $i_L(t)$ and the initial conditions $i_L(0)$ and $v_C(0)$.
(b) Use Laplace transforms to solve for $i_L(t)$ for $t \geq 0$.

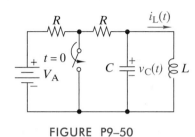

FIGURE P9–50

9–51 The switch in Figure P9–50 has been open for a long time and is closed at $t = 0$. The circuit parameters are $R = 500\ \Omega$, $L = 2.5$ H, $C = 2.5\ \mu$F, and $V_A = 50$ V.
(a) Find the circuit differential equation in $v_C(t)$ and the initial conditions $i_L(0)$ and $v_C(0)$.
(b) Use Laplace transforms to solve for $v_C(t)$ for $t \geq 0$.

9–52 The switch in Figure P9–52 has been closed for a long time and is opened at $t = 0$.
(a) Find the circuit differential equation in $v_C(t)$ and the initial conditions $i_L(0)$ and $v_C(0)$.
(b) The circuit parameters are $L = 50$ H, $C = 0.25\ \mu$F, $R_1 = 10\ k\Omega$, $R_2 = 20\ k\Omega$, and $v_S = 10u(t)$ V. Use Laplace transforms and MATLAB to solve for $v_C(t)$ for $t \geq 0$.

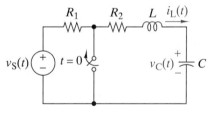

FIGURE P9–52

9–53 9–53 The switch in Figure P9–52 has been open for a long time and is closed at $t = 0$.
(a) Find the circuit differential equation in $i_L(t)$ and the initial conditions $i_L(0)$ and $v_C(0)$.
(b) The circuit parameters are $L = 50$ H, $C = 0.25\ \mu$F, $R_1 = 10\ k\Omega$, $R_2 = 10\ k\Omega$, and $v_S = 10u(t)$ V. Use Laplace transforms and MATLAB to solve for $i_L(t)$ for $t \geq 0$.

9–54 9–54 The *RLC* circuit in Figure P9–54 is in the zero state when at $t = 0$ an exponential source, $v_S(t) = V_A e^{-\alpha t}$ V, is suddenly connected to it
(a) Find the circuit integrodifferential equation that describes the behavior of the current in the circuit.
(b) If $R = 100\ \Omega$, $L = 50$ mH, $C = 10\ \mu$F, $V_A = 10$ V, and $\alpha = 500\ \text{s}^{-1}$, use Laplace transforms and MATLAB to solve for $i(t)$ for $t \geq 0$.

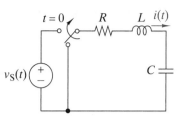

FIGURE P9–54

9–55 Repeat problem 9–54 when an exponential source, $v_S(t) = 10(1 - e^{-500t})$ V, is suddenly connected to the circuit at $t - 0$.

9–56 Find $v_C(t)$ for $t \geq 0$ when the input to the *RC* circuit shown in Figure P9–56 is $v_S(t) = V_A r(t)$ V. Assume $v_C(0-) = 0$ V.

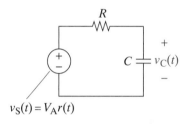

FIGURE P9–56

OBJECTIVE 9–4 INITIAL AND FINAL VALUE PROPERTIES (SECT. 9–6)

Given the Laplace transform of a signal, find the initial and final values of the signal waveform. See Example 9–20 and Exercise 9–31.

9–57 Use the initial and final value properties to find the initial and final values of the waveforms corresponding to the transforms below. If either property is not applicable, explain why.

(a) $F_1(s) = \dfrac{16(s+1)}{(s+2)(s+3)}$

(b) $F_2(s) = \dfrac{s - 100}{s(s+50)}$

9–58 Use the initial and final value properties to find the initial and final values of the waveforms corresponding to transforms below. If either property is not applicable, explain why.

(a) $F_1(s) = \dfrac{8s^2}{(s+2)(s^2 + 12s + 13)}$

(b) $F_2(s) = \dfrac{s + 1000}{s(s+50)(s+100)}$

9–59 Use the initial and final value properties to find the initial and final values of the waveforms corresponding to transforms below. If either property is not applicable, explain why.

(a) $F_1(s) = 400\dfrac{s}{(s+10)^2 + 20^2}$

(b) $F_2(s) = \dfrac{30(3s^4 + 10s^2 + 4)}{s(s^2 + 1)(s^2 + 4)}$

9–60 Use the initial and final value properties to find the initial and final values of the waveforms corresponding to the transforms below. If either property is not applicable, explain why.

(a) $F_1(s) = \dfrac{50s(s^2 + 5s + 6)}{(s+2)(s+6)(s+12)}$

(b) $F_2(s) = \dfrac{25(s^2 + 10s + 40)}{s(s^2 - 625)}$

9–61 Use the initial and final value properties to find the initial and final values of the waveforms corresponding to the following transforms. If either property is not applicable, explain why.

(a) $F_1(s) = \dfrac{s(s+6)}{s^2 + 6s + 9}$

(b) $F_2(s) = \dfrac{20(s^2 + 10s + 100)}{s(s^2 + 20s + 100)}$

9–62 Use the initial and final value properties to find the initial and final values of the waveform corresponding to the following transform. If either property is not applicable, explain why.

$$F(s) = \frac{80(s^3 + 2s^2 + s + 2)}{s(s^3 + 4s^2 + 4s + 16)}$$

9–63 The MATLAB function `limit` can be used to take the limit of a symbolic expression. Use MATLAB and the initial and final value properties to find the initial and final values of the waveforms corresponding to the following transforms. If either property is not applicable, explain why. Use MATLAB again to compute the waveforms corresponding to each transform and then take limits in the time domain to verify the answers found using the initial and final value properties.

(a) $F_1(s) = \dfrac{(s+200)^2}{(s+20)^2(s+100)}$

(b) $F_2(s) = \dfrac{(s+20)^2}{(s+100)^2(s+200)}$

INTEGRATING PROBLEMS

9–64 The Dominant Pole Approximation **A**
When a transform $F(s)$ has widely separated poles, then those closest to the j-axis tend to dominate the response because they have less damping. An approximation to the waveform can be obtained by ignoring the contributions of all except the dominant poles. We can ignore the other poles simply by discarding their terms in the partial fraction expansion of $F(s)$. The purpose of this example is to examine a dominant pole approximation of the transform

$$F(s) = 10^6 \frac{s + 4000}{(s+10000)\left[(s+25)^2 + 100^2\right]}$$

(a) Construct a partial-fraction expansion of $F(s)$ and find y $f(t)$.

(b) Construct a pole-zero diagram of $F(s)$ and identify the dominant poles.

(c) Construct a dominant pole approximation $g(t)$ by discarding the other poles in the partial fraction expansion in part (a).

(d) Plot $f(t)$ and $g(t)$ and comment on the accuracy of the approximation.

9–65 First-Order Circuit Step Response **A**
In Chapter 7 we found that the step response of a first-order circuit can be written as

$$f(t) = f(\infty) + [f(0) - f(\infty)]e^{-1/T_C}$$

where $f(0)$ is the initial value, $f(\infty)$ is the final value, and T_C is the time constant. Show that the corresponding transform has the form

$$F(s) = K\left[\frac{s+\gamma}{s(s+\alpha)}\right]$$

and relate the time-domain parameters $f(0)$, $f(\infty)$, and T_C to the s-domain parameters K, γ, and α.

9–66 Inverse Transform for Complex Poles **A**
In Section 9–4 we learned that complex poles occur in conjugate pairs and that for simple poles the partial fraction expansion of $F(s)$ will contain two terms of the form

$$F(s) = \dots \frac{k}{s+\alpha - j\beta} + \frac{k}{s+\alpha + j\beta} + \dots$$

Show that when the complex conjugate residues are written in rectangular form as

$$k = a + jb \textbf{ and } k^* = a - jb$$

the corresponding term in the waveform $f(t)$ is

$$f(t) = \dots + 2e^{-\alpha t}[a\cos(\beta t) - b\sin(\beta t)] + \dots$$

9–67 Solving State Variable Equations **A**
With zero input, a series RLC circuit can be described by the following coupled first-order equations in the inductor current $i_L(t)$ and capacitor voltage $v_C(t)$.

$$\frac{dv_C(t)}{dt} = \frac{1}{C}i_L(t)$$
$$\frac{di_L(t)}{dt} = -\frac{1}{L}v_C(t) - \frac{R}{L}i_L(t)$$

(a) Transform these equations into the s domain and solve for the transforms $I_L(s)$ and $v_C(s)$ in terms of the initial conditions $i_L(0) = I_0$ and $v_C(0) = V_0$.

(b) Find $i_L(t)$ and $v_C(t)$ for $R = 1$ kΩ, $L = 1$ H, $C = 1$ μF, $I_0 = 15$ mA and $v_0 = 10$ V.

9–68 Complex Differentiation Property **A**
The complex differentiation property of the Laplace transformation states that

$$\textbf{If } \mathscr{L}\{f(t)\} = F(s) \textbf{ then } \mathscr{L}\{tf(t)\} = -\frac{d}{ds}F(s)$$

Use this property to find the Laplace transforms *of* $f(t) = \{tg(t)\}u(t)$ when $g(t) = e^{-\alpha t}$. Repeat for $g(t) = \sin \beta t$ and $g(t) = \cos \beta t$.

9–69 Butterworth Poles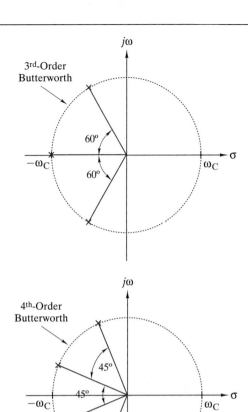

Steven Butterworth, a British engineer, 1885–1958, discovered a method of designing electric filters. He was quoted saying "An ideal electrical filter should not only completely reject the unwanted frequencies but should also have uniform sensitivity for the wanted frequencies." His algorithms are widely used in filter design, as we will see in Chapter 14. He based his design on locating the poles of his filters in a unique pattern around a circle of radius ω_C. The number of poles on the circle constitutes the order of the filter. The more poles, the better the filter. Odd -order filters include one real pole at $-\omega_C$ and pairs of complex-conjugate poles placed on a circle of radius ω_C at equal angular spacing. Even-order filters locate complex-conjugate poles placed on a circle of radius ω_C at equal angular spacing. Figure P9–69 shows the location of the poles in a third- and in a fourth-order Butterworth filter. Assuming an ω_C of 1 rad/s, what is the denominator of the Laplace transform, $F(s)$, associated with each of the two Butterworth filters shown in the figure?

FIGURE P9–69

CHAPTER 10 *s*-DOMAIN CIRCUIT ANALYSIS

The resistance operator Z is a function of the electrical constants of the circuit components and of d/dt, the operator of time-differentiation, which will in the following be denoted by p simply.

Oliver Heaviside, 1887,
British Engineer

Some History Behind This Chapter

The use of operational methods to study electric circuits was pioneered by Oliver Heaviside (1850–1925). The quotation given here was taken from his book *Electrical Papers*, originally published in 1887. His resistance operator Z, which he later called impedance, is a central theme for much of electrical engineering. Heaviside does not often receive the recognition he deserves, in part because his intuitive approach was not accepted by most Victorian scientists of his day. Mathematical justification for his methods was eventually supplied by John Bromwich and others. However, no important errors were found in Heaviside's results.

Why This Chapter Is Important Today

In this chapter we use Laplace transforms to make analyzing dynamic circuits no more difficult than dc circuits. You will see that all the circuit analysis tools learned in Chapters 1 through 4 can be extended to the study of the transient response of linear circuits. In addition, pole-zero diagrams will give you a new way to visualize and predict circuit behavior.

Chapter Sections

Chapter Learning Objectives

10-1 Equivalent Impedance (Sects. 10–1, 10–2)

Given a linear circuit, use series and parallel equivalence to find the equivalent impedance at specified terminal pairs. Select element values to obtain specified pole locations.

10-2 Basic Circuit Analysis Techniques (Sects. 10–2, 10–3)

Given a linear circuit:
(a) Determine the initial conditions (if not given) and transform the circuit into the *s* domain.
(b) Solve for zero-state and zero-input responses using circuit reduction, the unit output method, Thévenin or Norton equivalent circuits, or superposition.
(c) Identify the forced and natural poles in the responses, or select circuit parameters to place the natural poles at specified locations.

10-3 General Circuit Analysis (Sects. 10–4, 10–5, 10–6)

Given a linear circuit:
(a) Determine the initial conditions (if not given) and transform the circuit into the *s* domain.
(b) Solve for zero-state and zero-input response transforms and waveforms using node-voltage or mesh-current methods.
(c) Identify the forced and natural poles in the responses or select circuit parameters to place the natural poles at specified locations.

10–1 TRANSFORMED CIRCUITS

So far we have used the Laplace transformation to change waveforms into transforms and convert circuit differential equations into algebraic equations. These operations provide a useful introduction to the s domain. However, the real power of the Laplace transformation emerges when we transform the circuit itself and study its behavior directly in the s domain.

The solid arrow in Figure 10–1 indicates the analysis path we will be following in this chapter. The process begins with a linear circuit in the time domain. We transform the circuit into the s domain, write the circuit equations directly in that domain, and then solve these algebraic equations for the response transform. The inverse Laplace transformation then produces the response waveform. However, the s-domain approach is not just another way to derive response waveforms. This approach allows us to work directly with the circuit model using analysis tools such as voltage division and equivalence. By working directly with the circuit model, we gain insights into the interaction between circuits and signals that cannot be obtained using the classical approach indicated by the dotted path in Figure 10–1.

How are we to transform a circuit? We have seen several times that circuit analysis is based on device and connection constraints. The connection constraints are derived from Kirchhoff's laws and the device constraints from the i–v relationships used to model the physical devices in the circuit. To transform circuits, we must see how these two types of constraints are altered by the Laplace transformation.

CONNECTION CONSTRAINTS IN THE s DOMAIN

A typical KCL connection constraint could be written as

$$i_1(t) + i_2(t) - i_3(t) + i_4(t) = 0$$

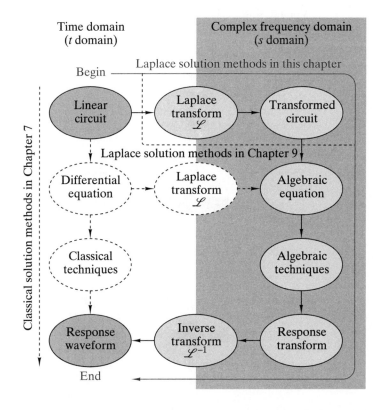

FIGURE 10–1 *Flow diagram for s-domain circuit analysis.*

This connection constraint requires that the sum of the current waveforms at a node be zero for all times *t*. Using the linearity property, the Laplace transformation of this equation is

$$I_1(s) + I_2(s) - I_3(s) + I_4(s) = 0$$

In the *s* domain the KCL connection constraint requires that the sum of the current transforms be zero for all values of *s*. This idea generalizes to any number of currents at a node and any number of nodes. In addition, this idea obviously applies to Kirchhoff's voltage law as well. The form of the connection constraints do not change because they are linear equations and the Laplace transformation is a linear operation. In summary, KCL and KVL apply to waveforms in the *t* domain and to transforms in the *s* domain.

ELEMENT CONSTRAINTS IN THE *S* DOMAIN

Turning now to the element constraints, we first deal with the independent signal sources shown in Figure 10–2. The *i*–*v* relationships for these elements are

$$\text{Voltage source: } v(t) = v_S(t)$$
$$i(t) = \text{Depends on circuit}$$
$$\text{Current source: } i(t) = i_S(t)$$
$$v(t) = \text{Depends on circuit}$$

(10–1)

Independent sources are two-terminal elements. In the *t* domain they constrain the waveform of one signal variable and adjust the unconstrained variable to meet the demands of the external circuit. We think of an independent source as a generator of a specified voltage or current waveform. The Laplace transformation of the expressions in Eq. (10–1) yields

$$\text{Voltage source: } V(s) = V_S(s)$$
$$I(s) = \text{Depends on circuit}$$
$$\text{Current source: } I(s) = I_S(s)$$
$$V(s) = \text{Depends on circuit}$$

(10–2)

FIGURE 10–2 *s-Domain models of independent sources.*

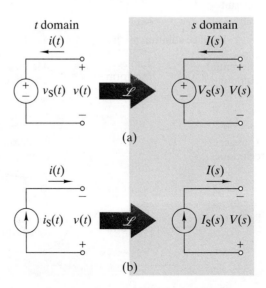

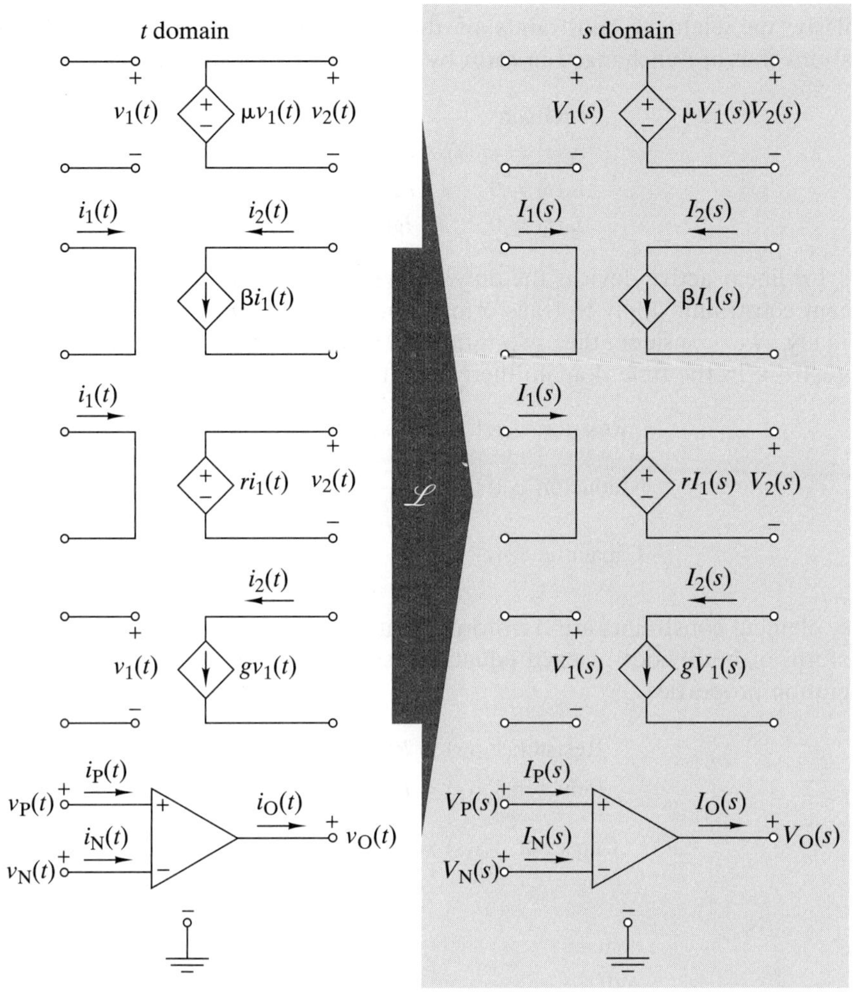

FIGURE 10–3 *s-Domain models of dependent sources and OP AMPs.*

In the s domain, independent sources function the same way as in the t domain, except that we think of them as generating voltage or current transforms rather than waveforms.

Next we consider the active elements in Figure 10–3. In the time domain, the element constraints for linear dependent sources are linear algebraic equations. Because of the linearity property of the Laplace transformation, the forms of these constraints are unchanged when they are transformed into the s domain:

	t domain	s domain	
Voltage-controlled voltage source	$v_2(t) = \mu v_1(t)$	$V_2(s) = \mu V_1(s)$	
Current-controlled current source	$i_2(t) = \beta i_1(t)$	$I_2(s) = \beta I_1(s)$	
Current-controlled voltage source	$v_2(t) = r i_1(t)$	$V_2(s) = r I_1(s)$	(10–3)
Voltage-controlled current source	$i_2(t) = g v_1(t)$	$I_2(s) = g V_1(s)$	

Similarly, the element constraints of the ideal OP AMP are linear algebraic equations that are unchanged in form by the Laplace transformation:

$$
\begin{array}{ll}
t \text{ domain} & s \text{ domain} \\
v_P(t) = v_N(t) & V_P(s) = V_N(s) \\
i_N(t) = 0 & I_N(s) = 0 \\
i_P(t) = 0 & I_P(s) = 0
\end{array}
\tag{10–4}
$$

Thus, for linear active devices the only difference is that in the s domain the ideal element constraints apply to transforms rather than waveforms.

Finally, we consider the two-terminal passive circuit elements shown in Figure 10–4. In the time domain their i–v relationships are

$$
\begin{aligned}
&\text{Resistor: } v_R(t) = R\, i_R(s) \\
&\text{Inductor: } v_L(t) = L\frac{di_L(t)}{dt} \\
&\text{Capacitor: } v_C(t) = \frac{1}{C}\int_0^t i_C(\tau)d\tau + v_C(0)
\end{aligned}
\tag{10–5}
$$

These element constraints are transformed into the s domain by taking the Laplace transform of both sides of each equation using the linearity, differentiation, and integration properties.

$$
\begin{aligned}
&\text{Resistor: } V_R(s) = RI_R(s) \\
&\text{Inductor: } V_L(s) = LsI_L(s) - Li_L(0) \\
&\text{Capacitor: } V_C(s) = \frac{1}{Cs}I_C(s) + \frac{v_C(0)}{s}
\end{aligned}
\tag{10–6}
$$

FIGURE 10–4 *s-Domain models of passive elements using voltage sources for initial conditions.*

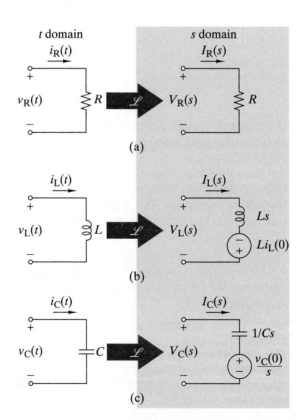

As expected, the element relationships are algebraic equations in the s domain. For the linear resistor the s domain version of Ohm's law says that the voltage transform $V_R(s)$ is proportional to the current transform $I_R(s)$. The element constraints for the inductor and capacitor also involve a proportionality between voltage and current, but include a term for the initial conditions as well.

The element constraints in Eq. (10–6) lead to the s-domain circuit models shown on the right side of Figure 10–4. The t-domain parameters L and C are replaced by proportionality factors Ls and $1/Cs$ in the s domain. The initial conditions associated with the inductor and capacitor are modeled as voltage sources in series with these elements. The polarities of these sources are determined by the sign of the corresponding initial condition terms in Eq. (10–6). These initial condition voltage sources must be included when using these models to calculate the voltage transforms $V_L(s)$ or $V_C(s)$.

IMPEDANCE AND ADMITTANCE

The concept of impedance is a basic feature of s-domain circuit analysis. For zero initial conditions the element constraints in Eq. (10–6) reduce to

$$\text{Resistor: } V_R(s) = (R)I_R(s)$$
$$\text{Inductor: } V_L(s) = (Ls)I_L(s) \tag{10–7}$$
$$\text{Capacitor: } V_C(s) = (1/Cs)I_C(s)$$

In each case the element constraints are all of the form $V(s) = Z(s)I(s)$, which means that in the s domain the voltage across the element is proportional to the current through it. The proportionality factor is called the element **impedance** $Z(s)$. Stated formally,

Impedance *is the proportionality factor relating the transform of the voltage across a two-terminal element to the transform of the current through the element with all initial conditions set to zero.*

The impedances of the three passive elements are

$$\text{Resistor: } Z_R(s) = R$$
$$\text{Inductor: } Z_L(s) = Ls \qquad \text{with } i_L(0) = 0 \tag{10–8}$$
$$\text{Capacitor: } Z_C(s) = 1/Cs \quad \text{with } v_C(0) = 0$$

It is important to remember that part of the definition of s-domain impedance is that the initial conditions are zero.

The s-domain impedance is a generalization of the t-domain concept of resistance. The impedance of a resistor is its resistance R. The impedance of the inductor and capacitor depend on the inductance L and capacitance C and the complex frequency variable s. Since a voltage transform has units of V-s and current transform has units of A-s, it follows that impedance has units of ohms since $(\text{V-s})/(\text{A-s}) = \text{V/A} = \Omega$.

Algebraically solving Eq. (10–6) for the element currents in terms of the voltages produces alternative s-domain models.

$$\text{Resistor: } I_R(s) = \frac{1}{R}V_R(s)$$
$$\text{Inductor: } I_L(s) = \frac{1}{Ls}V_L(s) + \frac{i_L(0)}{s} \tag{10–9}$$
$$\text{Capacitor: } I_C(s) = Cs\,V_C(s) - Cv_C(0)$$

FIGURE 10–5 *s-Domain models of passive elements using current sources for initial conditions.*

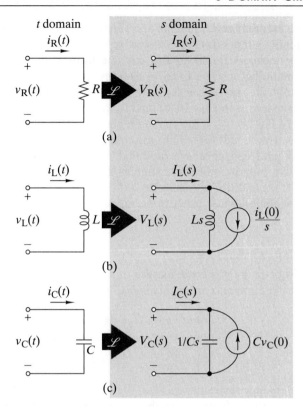

(a)

(b)

(c)

In this form, the i–v relations lead to the s-domain models shown in Figure 10–5. The reference directions for the initial condition current sources are determined by the sign of the corresponding terms in Eq. (10–9). The initial condition sources are in parallel with the element impedance. Note that the relationships shown in Eq. (10–9) are just source transformations of those shown in Eq. (10–6).

Admittance $Y(s)$ is the s-domain generalization of the t-domain concept of conductance and can be defined as the reciprocal of impedance.

$$Y(s) = \frac{1}{Z(s)} \tag{10–10}$$

Using this definition, the admittances of the three passive elements are

$$\text{Resistor: } Y_R(s) = \frac{1}{Z_R(s)} = \frac{1}{R} = G$$

$$\text{Inductor: } Y_L(s) = \frac{1}{Z_L(s)} = \frac{1}{Ls} \quad \text{with } i_L(0) = 0 \tag{10–11}$$

$$\text{Capacitor: } Y_C(s) = \frac{1}{Z_C(s)} = Cs \quad \text{with } v_C(0) = 0$$

Since $Y(s)$ is the reciprocal of impedance, its units are siemens since $\Omega^{-1} = A/V = S$.

In summary, to transform a circuit into the s domain we replace each element by an s-domain model. For independent sources, dependent sources, OP AMPs, and resistors, the only change is that these elements now constrain transforms rather than waveforms. For inductors and capacitors, we can use either the model with a series initial condition voltage source (Figure 10–4) or the model with a parallel initial condition current source (Figure 10–5). However, to avoid possible confusion we always write the inductor impedance Ls and capacitor impedance $1/Cs$ beside the transformed element regardless of which initial condition source is used.

To analyze the transformed circuit, we can use the tools developed for resistance circuits in Chapters 2 through 4. These tools are applicable because KVL and KCL apply to transforms, and the s-domain element constraints are linear equations similar to those for resistance circuits. These features make s-domain analysis of dynamic circuits an algebraic process that is akin to resistance circuit analysis or to phasor circuit analysis studied earlier.

EXAMPLE 10-1

The switch in Figure 10–6(a) has been in position 1 for a long time and is moved to position 2 at $t = 0$. For $t > 0$, transform the circuit into the s domain and use Laplace transforms to solve for the voltage $v_C(t)$.

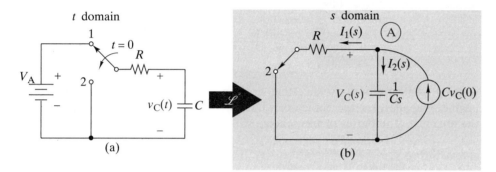

FIGURE 10-6

SOLUTION:
We have solved this type of problem before using various methods. In this example, we find the response by first transforming the circuit itself. For $t > 0$ the transformed circuit takes the form in Figure 10–6(b), where we have used a parallel current source $Cv_C(0)$ to account for the initial condition.

Applying KCL to the current transforms at node A produces

$$-I_1(s) - I_2(s) + Cv_C(0) = 0 \text{ A-s}$$

For $t < 0$ the switch in Figure 10–6(a) is in position 1 and the circuit is in a dc steady-state condition. As a result, we have $v_C(0) = V_A$. In the s-domain circuit in Figure 10–6(b), the two branch current transforms can be written in terms of the capacitor voltage and element impedances as

$$\text{Resistor: } I_1(s) = \frac{V_C(s)}{R}$$

$$\text{Capacitor: } I_2(s) = \frac{V_C(s)}{1/Cs} = CsV_C(s)$$

Substituting these observations into the KCL equation and solving for $V_C(s)$ yields

$$V_C(s) = \frac{CV_A}{Cs + \dfrac{1}{R}} = \frac{V_A}{s + \dfrac{1}{RC}} \text{ V-s}$$

Performing the inverse Laplace transformation leads to

$$v_C(t) = [V_A e^{-t/RC}]u(t) \text{ V}$$

The form of the response should be no great surprise. We could easily predict this response using the classical differential equation methods studied in Chapter 7. What is important in this example is that we obtained the response using only basic circuit concepts applied in the s domain.

One may wonder why we chose to use the parallel capacitor transform model from Figure 10–5(c) rather than the series transform model from Figure 10–4(c). In truth for this problem, it really does not matter. Let us solve the same problem using the series model.

Since we want to find the voltage across the capacitor, we must realize that the voltage includes the capacitor's initial condition as shown in Figure 10–6(c). In this case, that voltage is the same as the voltage across the resistor. We can readily find the voltage across the resistor using a voltage divider with the initial condition source providing the voltage. Thus,

$$V_C(s) = \frac{R}{R + \frac{1}{Cs}} \left(\frac{v_C(0)}{s} \right) = \frac{V_A R C s}{s(RCs + 1)} = \frac{V_A}{s + \frac{1}{RC}} \text{ V-s}$$

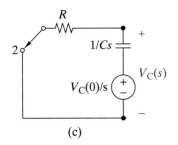

(c)

FIGURE 10–6

which transforms back to the time domain the same as above. As with most engineering analyses, the choice of which model or technique to use is left up to the engineer. The better or best choice will come with practice. ∎

Exercise 10–1

Transform the circuit of Figure 10–7(a) into the s domain and solve for the voltage $v_C(t)$ if $v_S(t) = V_A e^{-\alpha t}u(t)\text{V}$ and $v_C(0) = V_0$.

Answers:

(a) The transformed circuit is shown in Figure 10–7(b).

(b) $v_C(t) = \left[\left(V_0 - \frac{\frac{V_A}{RC}}{\frac{1}{RC} - \alpha} \right) e^{-t/RC} + \left(\frac{\frac{V_A}{RC}}{\frac{1}{RC} - \alpha} \right) e^{-\alpha t} \right] u(t) \text{ V}$

FIGURE 10–7

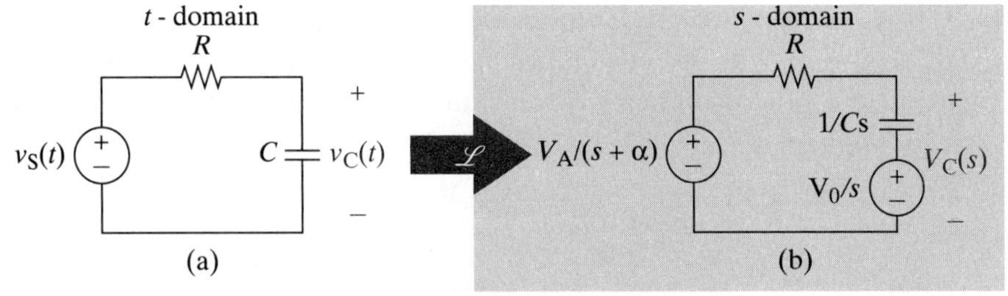

t - domain

R

$v_S(t)$ C $v_C(t)$

(a)

s - domain

R

1/Cs

$V_A/(s + \alpha)$ $V_C(s)$

V_0/s

(b)

EXAMPLE 10 – 2

(a) Transform the circuit in Figure 10–8(a) into the s domain.
(b) Solve for the current transform $I(s)$.
(c) Perform the inverse transformation to find the waveform $i(t)$.

FIGURE 10–8

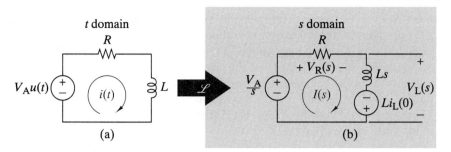

SOLUTION:

(a) Figure 10–8(b) shows the transformed circuit using a series voltage source $Li_L(0)$ to represent the inductor's initial condition. The impedances of the two passive elements are R and Ls. The independent source voltage $V_A u(t)$ transforms as V_A/s.

(b) By KVL, the sum of voltage transforms around the loop is

$$-\frac{V_A}{s} + V_R(s) + V_L(s) = 0$$

Using the impedance models, the s-domain element constraints are

$$\text{Resistor: } V_R(s) = RI(s)$$
$$\text{Inductor: } V_L(s) = LsI(s) - Li_L(0)$$

Substituting the element constraints into the KVL constraint and collecting terms yields

$$-\frac{V_A}{s} + (R + Ls)I(s) - Li_L(0) = 0$$

Solving for $I(s)$ produces

$$I(s) = \frac{V_A/L}{s(s + R/L)} + \frac{i_L(0)}{s + R/L} \text{ A-s}$$

The current $I(s)$ is the transform of the circuit response for a step function input. $I(s)$ is a rational function with simple poles at $s = 0$ and $s = -R/L$.

(c) To perform the inverse transformation, we expand $I(s)$ by partial fractions:

$$I(s) = \overbrace{\frac{V_A/R}{s}}^{\text{forced}} \overbrace{- \frac{V_A/R}{s + R/L} + \frac{i_L(0)}{s + R/L}}^{\text{natural}} \text{ A-s}$$

Taking the inverse transform of each term in this expansion gives

$$i(t) = \left[\overbrace{\frac{V_A}{R}}^{\text{forced}} \overbrace{- \frac{V_A}{R} e^{-Rt/L} + i_L(0)e^{-Rt/L}}^{\text{natural}} \right] u(t) \text{ A}$$

The forced response is caused by the step function input. The exponential terms in the natural response depend on the circuit time constant L/R. The step function and exponential components in $i(t)$ are directly related to the terms in the partial-fraction expansion of $I(s)$. The pole at the origin came from the step function input and leads to the forced response. The pole at

$s = -R/L$ came from the circuit and leads to the natural response. Thus, in the s domain the forced response is that part of the total response that has the same poles as the input excitation. The natural response is that part of the total response whose poles came from the circuit. We say that the circuit contributes the natural poles because their locations depend on circuit parameters, not on the input. In other words, poles in the response do not occur by accident. They are present because the physical response depends on two things—(1) the input and (2) the circuit.   ■

Exercise 10–2

The source of the t-domain circuit of Figure 10–8(a) is suddenly turned off. Use Laplace techniques to solve for the voltage across the resistor $v_R(t)$.

Answer:
$$v_R(t) = [V_A e^{-Rt/L}]u(t) \text{ V}$$

EXAMPLE 10–3

Find the output $v_O(t)$ for the dependent-source circuit shown in Figure 10–9(a). The capacitor has an initial voltage of V_B V.

FIGURE 10–9

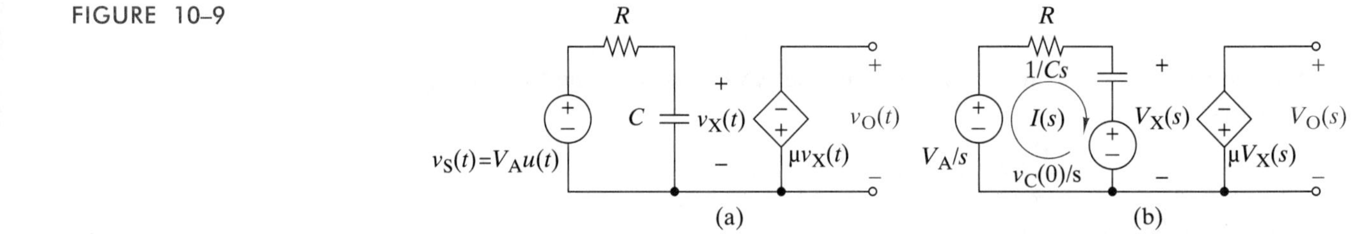

(a)   (b)

SOLUTION:
We start by transforming the circuit into the s domain as shown in Figure 10–9(b). The output transform voltage is $V_O(s) = -\mu V_X(s)$. We can solve for $V_X(s)$ by writing a loop equation around the input loop and solving for $I(s)$. Starting at the source voltage, we get

$$-\frac{V_A}{s} + RI(s) + \frac{I(s)}{Cs} + \frac{v_C(0)}{s} = 0$$

Substituting V_B for $v_C(0)$ and solving for $I(s)$

$$I(s) = \frac{(V_A - V_B)/s}{R + 1/Cs} = \frac{(V_A - V_B)/R}{s + 1/RC}$$

Recognizing that $V_X(s) = I(s)/Cs + V_B/s$, we get

$$V_X(s) = \frac{(V_A - V_B)}{RCs(s + 1/RC)} + \frac{V_B}{s}$$

We can now find $V_O(s)$

$$V_O(s) = -\mu V_X(s) = \frac{-\dfrac{\mu}{RC}(V_A - V_B)}{s(s + 1/RC)} + \frac{-\mu V_B}{s}$$

The first term can be expanded by partial fractions

$$V_O(s) = \frac{-\mu(V_A - V_B)}{s} + \frac{\mu(V_A - V_B)}{s + 1/RC} + \frac{-\mu V_B}{s} = \frac{-\mu V_A}{s} + \frac{\mu(V_A - V_B)}{s + 1/RC}$$

Taking the inverse transform of each term gives

$$v_0(t) = \left[\underbrace{-\mu V_A}_{\text{forced}} + \underbrace{\mu V_A e^{-t/RC} - \mu V_B e^{-t/RC}}_{\text{natural}} \right] u(t) \text{V}$$

The forced response is caused by the input step function. The exponential terms in the natural response are caused by the circuit's RC time constant. The dependent source, in this example, simply multiplies the input and stored voltage responses by $-\mu$. ◼

Exercise 10–3

Replace the capacitor in the circuit of Figure 10–9(a) with an inductor with an initial current of I_B A. Find $v_O(t)$ and identify the forced and natural responses.

Answer:

$$v_O(t) = \left[\underbrace{\mu R I_B e^{-Rt/L} - \mu V_A e^{-Rt/L}}_{\text{natural}} \right] u(t) \text{ V}.$$

There is no forced component–at steady state the inductor shorts out the input to the dependent source.

10–2 BASIC CIRCUIT ANALYSIS IN THE s DOMAIN

In this section we develop the s-domain versions of series and parallel equivalence, and voltage and current division. These analysis techniques are the basic tools in s-domain circuit analysis, just as they are for resistance circuit analysis. These methods apply to circuits with elements connected in series or parallel. General analysis methods using node-voltage or mesh-current equations are covered later in Sects. 10–4 and 10–5.

SERIES EQUIVALENCE AND VOLTAGE DIVISION

The concept of a series connection applies in the s domain because Kirchhoff's laws do not change under the Laplace transformation. In Figure 10–10 the two-terminal elements are connected in series; hence by KCL the same current $I(s)$ exists in impedances $Z_1(s)$, $Z_2(s)$, ..., $Z_N(s)$. Using KVL and the element constraints, the voltage across the series connection can be written as

$$\begin{aligned} V(s) &= V_1(s) + V_2(s) + \cdots + V_N(s) \\ &= Z_1(s)I(s) + Z_2(s)I(s) + \cdots + Z_N(s)I(s) \quad\quad (10\text{–}12) \\ &= [Z_1(s) + Z_2(s) + \cdots + Z_N(s)]I(s) \end{aligned}$$

The last line in this equation points out that the responses $V(s)$ and $I(s)$ do not change when the series-connected elements are replaced by an **equivalent impedance**:

$$Z_{EQ}(s) = Z_1(s) + Z_2(s) + \cdots + Z_N(s) \quad\quad (10\text{–}13)$$

FIGURE 10–10 *Series
equivalence in the* s *domain.*

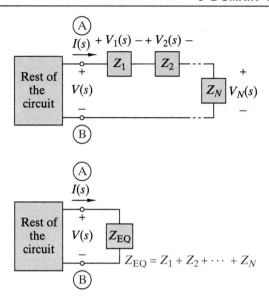

In general, the equivalent impedance $Z_{EQ}(s)$ is a quotient of polynomials in the complex frequency variable of the form

$$Z_{EQ}(s) = \frac{b_m s^m + b_{m-1} s^{m-1} + \cdots + b_1 s + b_0}{a_n s^n + a_{n-1} s^{n-1} + \cdots + a_1 s + a_0} \tag{10–14}$$

The roots of the numerator polynomial are the zeros of $Z_{EQ}(s)$, while the roots of the denominator are the poles.

Combining Eqs. (10–12) and (10–13), we can write the element voltages in the form

$$V_1(s) = \frac{Z_1(s)}{Z_{EQ}(s)} V(s) \quad V_2(s) = \frac{Z_2(s)}{Z_{EQ}(s)} V(s) \quad \cdots \quad V_N(s) = \frac{Z_N(s)}{Z_{EQ}(s)} V(s) \tag{10–15}$$

These equations are the *s*-domain **voltage division principle**:

> *Every element voltage in a series connection is equal to its impedance divided by the equivalent impedance of the connection times the voltage across the series circuit.*

This statement parallels the corresponding rule for resistance circuits given in Chapter 2.

PARALLEL EQUIVALENCE AND CURRENT DIVISION

The parallel circuit in Figure 10–11 is the dual of the series circuit discussed previously. In this circuit the two-terminal elements are connected in parallel; hence by KVL the same voltage $V(s)$ appears across admittances $Y_1(s), Y_2(s), \ldots, Y_N(s)$. Using KCL and the element constraints, the current into the parallel connection can be written as

$$\begin{aligned}
I(s) &= I_1(s) + I_2(s) + \cdots + I_N(s) \\
&= Y_1(s)V(s) + Y_2(s)V(s) + \cdots + Y_N(s)V(s) \\
&= [Y_1(s) + Y_2(s) + \cdots + Y_N(s)]V(s)
\end{aligned} \tag{10–16}$$

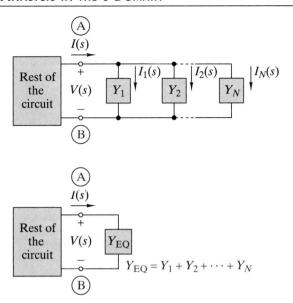

FIGURE 10–11 *Parallel equivalence in the s domain.*

The last line in this equation points out that the responses $V(s)$ and $I(s)$ do not change when the parallel connected elements are replaced by an **equivalent admittance**:

$$Y_{EQ}(s) = Y_1(s) + Y_2(s) + \cdots + Y_N(s) \qquad (10\text{–}17)$$

In general, the equivalent admittance $Y_{EQ}(s)$ is a quotient of polynomials in the complex frequency variable s. Since impedance and admittance are reciprocals, it turns out that if $Y_{EQ}(s) = p(s)/q(s)$, then the equivalent impedance at the same pair of terminals has the form $Z_{EQ}(s) = 1/Y_{EQ}(s) = q(s)/p(s)$. That is, at a given pair of terminals the poles of $Z_{EQ}(s)$ are zeros of $Y_{EQ}(s)$, and vice versa.

Combining Eqs. (10–16) and (10–17), we can write the element currents in the form

$$I_1(s) = \frac{Y_1(s)}{Y_{EQ}(s)} I(s) \quad I_2(s) = \frac{Y_2(s)}{Y_{EQ}(s)} I(s) \quad \cdots \quad I_N(s) = \frac{Y_N(s)}{Y_{EQ}(s)} I(s) \qquad (10\text{–}18)$$

These equations are the s-domain **current division principle**:

> *Every element current in a parallel connection is equal to its admittance divided by the equivalent admittance of the connection times the current into the parallel circuit.*

This statement is the dual of the results for a series circuit and parallels the current division rule for resistance circuits.

We begin to see that s-domain circuit analysis involves basic concepts that parallel the analysis of resistance circuits in the t domain. Repeated application of series/parallel equivalence and voltage/current division leads to an analysis approach called circuit reduction, discussed in Chapter 2. The major difference here is that we use impedance and admittances rather than resistance and conductance, and the analysis yields voltage and current transforms rather than waveforms.

EXAMPLE 10–4

The inductor current and capacitor voltage in Figure 10–12(a) are zero at $t = 0$.

(a) Transform the circuit into the s domain and find the equivalent impedance between terminals A and B.

(b) Use voltage division to solve for the output voltage transform $V_2(s)$.

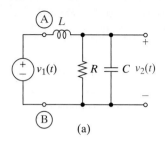

(a)

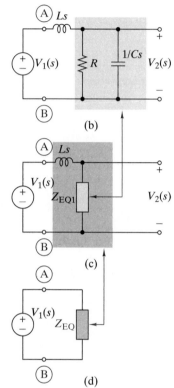

(b)

(c)

(d)

FIGURE 10–12

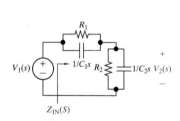

FIGURE 10–13

SOLUTION:

(a) Figure 10–12(b) shows the circuit in Figure 10–12(a) transformed into the *s* domain. As a first step we use parallel equivalence to find the equivalent impedance of the parallel resistor and capacitor.

$$Z_{EQ1}(s) = \frac{1}{Y_{EQ1}(s)} = \frac{1}{\frac{1}{R} + Cs} = \frac{R}{RCs + 1}$$

Figure 10–12(c) shows that the equivalent impedance $Z_{EQ1}(s)$ is connected in series with the inductor. This series combination can be replaced by an equivalent impedance

$$Z_{EQ}(s) = Ls + Z_{EQ1}(s) = Ls + \frac{R}{RCs + 1}$$

$$= \frac{RLCs^2 + Ls + R}{RCs + 1} \ \Omega$$

as shown in Figure 10–12(d). The rational function $Z_{EQ}(s)$ is the impedance seen between terminals A and B in Figure 10–12(b).

(b) Using voltage division in Figure 10–12(c), we find $V_2(s)$ as

$$V_2(s) = \left[\frac{Z_{EQ1}(s)}{Z_{EQ}(s)}\right] V_1(s) = \left[\frac{R}{RLCs^2 + Ls + R}\right] V_1(s)$$

Note that $Z_{EQ}(s)$ and $V_2(s)$ are rational functions of the complex frequency variable *s*.   ∎

Exercise 10–4 _____

The circuit of Figure 10–13 is in the zero state.

(a) Find the equivalent impedance $Z_{IN}(s)$ that the source sees.
(b) Find the output voltage $V_2(s)$.

Answers:

(a) $Z_{IN}(s) = \dfrac{\left[\dfrac{(C_1 + C_2)}{C_1 C_2} s + \dfrac{(R_1 + R_2)}{C_1 C_2 R_1 R_2}\right]}{\left(s + \dfrac{1}{R_1 C_1}\right)\left(s + \dfrac{1}{R_2 C_2}\right)} \ \Omega$

(b) $V_2(s) = \dfrac{\left(\dfrac{C_1}{C_1 + C_2} s + \dfrac{1}{R_1(C_1 + C_2)}\right) V_1(s)}{s + \left(\dfrac{1}{C_1 + C_2}\right)\left(\dfrac{R_1 + R_2}{R_1 R_2}\right)} \ \text{V-s}$

EXAMPLE 10 – 5

Consider the *s*-domain circuit in Figure 10–14.

(a) Find the input impedance $Z_{IN}(s)$.
(b) Find the ratio of the output $V_2(s)$ to the input $V_1(s)$.
(c) What are the poles and zeros of the input impedance and of $V_2(s)/V_1(s)$?

FIGURE 10–14

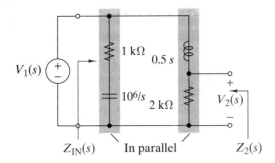

SOLUTION:

Let us start with finding $Z_{IN}(s)$. In this example there are two parallel branches: $0.5s + 2k$ and $1k + 10^6/s$. We can combine them as follows

$$Z_{IN}(s) = \frac{(0.5s + 2\,k)(1\,k + 10^6/s)}{0.5s + 2\,k + 1\,k + 10^6/s} = \frac{2s(0.5s + 2\,k)(1\,k + 10^6/s)}{2s(0.5s + 3\,k + 10^6/s}$$

$$Z_{IN}(s) = \frac{(s + 4\,k)(1000s + 10^6)}{s^2 + 6 \times 10^3 s + 2 \times 10^6} = \frac{1000(s + 4000)(s + 1000)}{s^2 + 6 \times 10^3 s + 2 \times 10^6}$$

Which can be factored as follows

$$Z_{IN}(s) = \frac{1000(s + 4000)(s + 1000)}{(s + 354)(s + 5646)} \Omega$$

The output is across a resistor that is in series with an inductor and the pair is in parallel with the source voltage. This allows us to use a simple voltage divider

$$V_2(s) = \frac{2000}{0.5s + 2000} V_1(s)$$

The ratio $V_2(s)/V_1(s)$ is

$$\frac{V_2(s)}{V_1(s)} = \frac{4000}{s + 4000}$$

The poles of the input impedance are at $s = -354$ and $s = -5646$ and zeros are at $s = -1000$ and $s = -4000$. The single pole of $V_2(s)/V_1(s)$ is at $s = -4000$, while its only zero is at infinity. ∎

Exercise 10–5

For the Figure 10–14, find the impedance $Z_2(s)$ seen looking into the $V_2(s)$ terminals with the input source removed (open circuited). What are its poles and zeros?

Answers:

$$Z_2(s) = \frac{2000(s^2 + 2000s + 2 \times 10^6)}{s^2 + 6000s + 2 \times 10^6} \Omega$$

Poles are at $s = -354$ and $s = -5646$, zeros at $s_1, s_2 = -1000 \pm j\,1000$.

D DESIGN EXAMPLE 10–6

In a circuit analysis problem we are required to find the poles and zeros of a circuit. In circuit design we are required to adjust circuit parameters to place the poles and zeros at specified s-plane locations. This example is a simple pole-placement design problem.

(a) Transform the circuit in Figure 10–15(a) into the s domain and find the equivalent impedance between terminals A and B.
(b) Select the values of R and C such that $Z_{EQ}(s)$ has a zero at $s = -5000\,\text{rad/s}$.

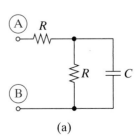

(a)

FIGURE 10–15

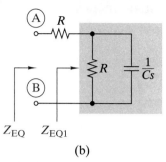

Z_{EQ} Z_{EQ1}

(b)

FIGURE 10–15 *(Continued)*

SOLUTION:

(a) Figure 10–15(b) shows the circuit transformed to the s domain. The equivalent impedance $Z_{EQ1}(s)$ is

$$Z_{EQ1}(s) = \frac{1}{Y_R + Y_C} = \frac{1}{\dfrac{1}{R} + Cs}$$

$$= \frac{R}{RCs + 1}$$

Hence the equivalent impedance between terminals A and B is

$$Z_{EQ}(s) = R + Z_{EQ1}(s) = R + \frac{R}{RCs + 1}$$

$$= R\frac{RCs + 2}{RCs + 1} \ \Omega$$

(b) For $Z_{EQ}(s)$ to have a zero at $s = -5000$ requires $2/RC = 5000$ or $RC = 4 \times 10^{-4}$. Selecting a standard value for the resistor $R = 10\,\text{k}\Omega$ in turn requires $C = 0.04\,\mu\text{F}$. ∎

◀ Design Exercise 10–6

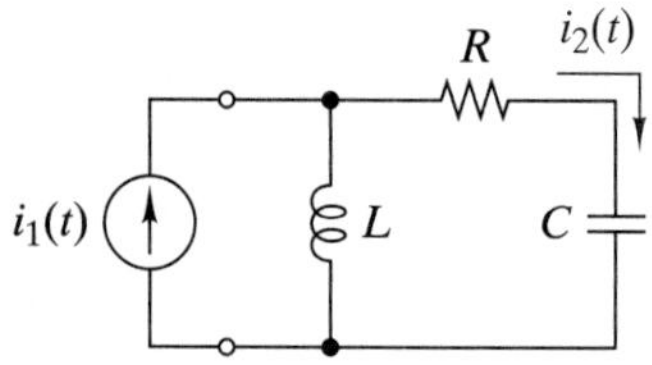

The circuit of Figure 10–16 is in the zero state.

(a) Find the output current transform $I_2(s)$.

(b) If $R = 1\,\text{k}\Omega$, select values of L and C such that $I_2(s)$ has two identical poles at -5000 rad/s.

Answers:

(a) $I_2(s) = \left[\dfrac{s^2 I_1(s)}{s^2 + \dfrac{R}{L}s + \dfrac{1}{LC}} \right]$ A-s

(b) $L = 100\,\text{mH}$, $C = 0.4\,\mu\text{F}$

FIGURE 10–16

Exercise 10–7

The inductor current and capacitor voltage in Figure 10–17 are zero at $t = 0$.

(a) Find the equivalent impedance between terminals A and B.

(b) Solve for the output voltage transform $V_2(s)$ in terms of the input voltage $V_1(s)$.

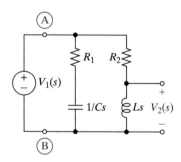

Answers:

(a) $Z_{EQ}(s) = \dfrac{(R_1 Cs + 1)(Ls + R_2)}{LCs^2 + (R_1 + R_2)Cs + 1}$

(b) $V_2(s) = \left[\dfrac{Ls}{Ls + R_2} \right] V_1(s)$

FIGURE 10–17

10–3 CIRCUIT THEOREMS IN THE s DOMAIN

In this section we study the s-domain versions of proportionality, superposition, and Thévenin/Norton equivalent circuits. These theorems define fundamental properties that provide conceptual tools for the analysis and design of linear circuits. With some modifications, all of the theorems studied in Chapter 3 apply to linear dynamic circuits in the s domain.

PROPORTIONALITY

For linear resistance circuits the **proportionality theorem** states that any output y is proportional to the input x:

$$y = Kx \qquad (10\text{–}19)$$

The same concept applies to linear dynamic circuits in the s domain except that the proportionality factor is a rational function of s rather than a constant. For instance, in Example 10–4 we found the output voltage $V_2(s)$ to be

$$V_2(s) = \left[\underbrace{\frac{R}{RLCs^2 + Ls + R}}_{K} \right] V_1(s) \qquad (10\text{–}20)$$

where $V_1(s)$ is the transform of the input voltage. The quantity inside the brackets is a rational function that serves as the proportionality factor K between the input and output transforms.

In the s-domain, rational functions that relate inputs and outputs are called **network functions**. We begin the formal study of network functions in Chapter 11. In this chapter we will simply illustrate network functions by an example.

EXAMPLE 10–7

There is no initial energy stored in the circuit in Figure 10–18. Find the network functions relating $I_R(s)$ to $V_1(s)$ and $I_C(s)$ to $V_1(s)$.

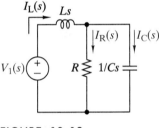

FIGURE 10–18

SOLUTION:
The equivalent impedance seen by the voltage source is

$$Z_{EQ} = Ls + \frac{1}{\dfrac{1}{R} + Cs} = \frac{RLCs^2 + Ls + R}{RCs + 1}$$

Hence we can relate the $I_L(s)$ and $V_1(s)$ as

$$I_L(s) = \frac{V_1(s)}{Z_{EQ}(s)} = \left[\frac{RCs + 1}{RLCs^2 + Ls + R} \right] V_1(s)$$

Using s-domain current division, we can relate $I_R(s)$ and $I_C(s)$ to $I_L(s)$ as

$$I_R(s) = \frac{\dfrac{1}{R}}{\dfrac{1}{R} + Cs} I_L(s) = \left[\frac{1}{RCs + 1} \right] I_L(s)$$

$$I_C(s) = \frac{Cs}{\dfrac{1}{R} + Cs} I_L(s) = \left[\frac{RCs}{RCs + 1} \right] I_L(s)$$

Finally, using these relationships plus the relationship between $I_L(s)$ and $V_1(s)$ derived previously, we obtain the required network functions.

$$I_R(s) = \left[\frac{1}{RLCs^2 + Ls + R} \right] V_1(s)$$

$$I_C(s) = \left[\frac{RCs}{RLCs^2 + Ls + R} \right] V_1(s) \qquad \blacksquare$$

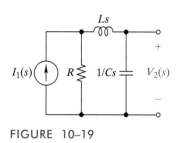

FIGURE 10-19

Exercise 10-8

In Figure 10–19 find the network function relating the output $V_2(s)$ to the input $I_1(s)$.

Answer:
$$V_2(s) = \left[\frac{R}{LCs^2 + RCs + 1}\right] I_1(s)$$

SUPERPOSITION

For linear resistance circuits, the **superposition theorem** states that any output y of a linear circuit can be written as

$$y = K_1 x_1 + K_2 x_2 + K_3 x_3 + \cdots \qquad (10\text{--}21)$$

where $x_1, x_2, x_3, \ldots$ are circuit inputs and $K_1, K_2, K_3, \ldots$ are weighting factors that depend on the circuit. The same concept applies to linear dynamic circuits in the s domain except that the weighting factors are rational functions of s rather than constants.

Superposition is usually thought of as a way to find the circuit response by adding the individual responses caused by each input acting alone. However, the principle applies to groups of sources as well. In particular, in the s domain there are two types of independent sources: (1) voltage and current sources representing the external driving forces for $t \geq 0$ and (2) initial condition voltage and current sources representing the energy stored at $t = 0$. As a result, the superposition principle states that the s-domain response can be found as the sum of two components: (1) the **zero-input response** caused by the initial condition sources with the external inputs turned off and (2) the **zero-state response** caused by the external inputs with the initial condition sources turned off. Turning a source off means replacing voltage sources by short circuits $[V_S(s) = 0]$ and current sources by open circuits $[I_S(s) = 0]$.

The zero-input response is the response of a circuit to its initial conditions when the input excitations are set to zero. The zero-state response is the response of a circuit to its input excitations when all of the initial conditions are set to zero. The term *zero input* is self-explanatory. The term *zero state* is used because there is no energy stored in the circuit at $t = 0$.

The result is that voltage and current transform in a linear circuit can be found as the sum of two components of the form

$$V(s) = V_{zs}(s) + V_{zi}(s) \qquad I(s) = I_{zs}(s) + I_{zi}(s) \qquad (10\text{--}22)$$

where the subscript zs stands for zero state and zi for zero input. An important corollary is that the time-domain response can also be partitioned into zero-state and zero-input components because the inverse Laplace transformation is a linear operation.

We analyze the circuit treated in Example 10–2 to illustrate the superposition of zero-state and zero-input responses. The transformed circuit in Figure 10–20 has two

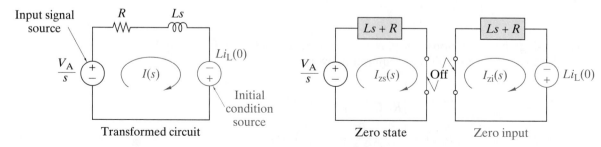

FIGURE 10-20   *Using superposition to find the zero-state and zero-input responses.*

independent voltage sources: (1) an input voltage source and (2) a voltage source representing the initial inductor current. The resistor and inductor are in series, so these two elements can be replaced by an impedance $Z_{EQ}(s) = Ls + R$.

First we turn off the initial condition source and replace its voltage source by a short circuit. Using the resulting zero-state circuit shown in Figure 10–20, we obtain the zero-state response:

$$I_{zs}(s) = \frac{V_A/s}{Z_{EQ}(s)} = \frac{V_A/L}{s(s + R/L)} \text{ A-s} \tag{10–23}$$

The pole at $s = 0$ comes from the input source and the pole at $s = -R/L$ comes from the circuit. Next, we turn off the input source and use the zero-input circuit shown in Figure 10–20 to obtain the zero-input response:

$$I_{zi}(s) = \frac{Li_L(0)}{Z_{EQ}(s)} = \frac{i_L(0)}{s + R/L} \text{ A-s} \tag{10–24}$$

The pole at $s = -R/L$ comes from the circuit. The zero-input response does not have a pole at $s = 0$ because the step function input is turned off.

Superposition states that the total response is the sum of the zero-state component in Eq. (10–23) and the zero-input component in Eq. (10–24).

$$I(s) = I_{zs}(s) + I_{zi}(s) = \frac{V_A/L}{s(s + R/L)} + \frac{i_L(0)}{s + R/L} \text{ A-s} \tag{10–25}$$

The transform $I(s)$ in this equation is the same as found in Example 10–2. To derive the time-domain response, we expand $I(s)$ by partial fractions:

$$I(s) = \underbrace{\frac{V_A/R}{s} - \frac{V_A/R}{s + R/L}}_{\text{zero state}} + \underbrace{\frac{i_L(0)}{s + R/L}}_{\text{zero input}} \text{ A-s} \tag{10–26}$$

Performing the inverse transformation on each term yields

$$i(t) = \left[\overbrace{\underbrace{\frac{V_A}{R} - \frac{V_A}{R}e^{-Rt/L}}_{\text{zero state}}}^{\text{forced}} + \overbrace{\underbrace{i_L(0)e^{-Rt/L}}_{\text{zero input}}}^{\text{natural}}\right]u(t) \text{ A} \tag{10–27}$$

Using superposition to partition the waveform into zero-state and zero-input components produces the same result as Example 10–2. The zero-state component contains the forced response. The zero-state and zero-input components both contain an exponential term due to the natural pole at $s = -R/L$ because both the external driving force and the initial condition source excite the circuit's natural response.

The superposition theorem helps us understand the response of circuits with multiple inputs, including initial conditions. It is a conceptual tool that helps us organize our thinking about s-domain circuits in general. It is not necessarily the most efficient analysis tool for finding the response of a specific multiple-input circuit.

EXAMPLE 10–8

The switch in Figure 10–21(a) has been open for a long time and is closed at $t = 0$.

(a) Transform the circuit into the s domain.
(b) Find the zero-state and zero-input components of $V(s)$.
(c) Find $v(t)$ for $I_A = 1\,\text{mA}, L = 2\,\text{H}, R = 1.5\,\text{k}\Omega,$ and $C = 1/6\,\mu\text{F}$.

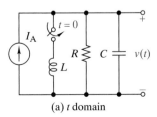

(a) t domain

FIGURE 10–21

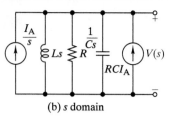

(b) *s* domain

FIGURE 10–21 *(Continued)*

SOLUTION:

(a) To transform the circuit into the *s* domain, we must find the initial inductor current and capacitor voltage. For $t < 0$ the circuit is in a dc steady-state condition with the switch open. The inductor acts like a short circuit, and the capacitor acts like an open circuit. By inspection, the initial conditions at $t = 0-$ are $i_L(0) = 0$ and $v_C(0) = I_A R$. Figure 10–21(b) shows the *s*-domain circuit for these initial conditions. The current source version for the capacitor's initial condition is used here because the circuit elements are connected in parallel. The switch and constant current source combine to produce a step function $I_A u(t)$ whose transform is I_A/s.

(b) The resistor, capacitor, and inductor can be replaced by an equivalent impedance

$$Z_{EQ} = \frac{1}{Y_{EQ}} = \frac{1}{\dfrac{1}{Ls} + \dfrac{1}{R} + Cs}$$

$$= \frac{RLs}{RLCs^2 + Ls + R}$$

The zero-state response is found with the capacitor initial condition source replaced by an open circuit and the step function input source on:

$$V_{zs}(s) = Z_{EQ}(s)\frac{I_A}{s} = \left[\frac{RLs}{RLCs^2 + Ls + R}\right]\frac{I_A}{s} = \frac{I_A/C}{s^2 + \dfrac{s}{RC} + \dfrac{1}{LC}}$$

The pole in the input at $s = 0$ is canceled by the zero at the origin in $Z_{EQ}(s)$. As a result, the zero-state response does not have a forced pole at $s = 0$. The zero-input response is found by replacing the input source by an open circuit and turning the capacitor initial condition source on:

$$V_{zi}(s) = [Z_{EQ}(s)][CRI_A] = \frac{RI_A s}{s^2 + \dfrac{s}{RC} + \dfrac{1}{LC}}$$

(c) Inserting the given numerical values of the circuit parameters and expanding the zero-state and zero-input response transforms by partial fractions yields

$$V_{zs}(s) = \frac{6000}{(s + 1000)(s + 3000)} = \frac{3}{s + 1000} + \frac{-3}{s + 3000} \text{ V-s}$$

$$V_{zi}(s) = \frac{1.5s}{(s + 1000)(s + 3000)} = \frac{-0.75}{s + 1000} + \frac{2.25}{s + 3000} \text{ V-s}$$

The inverse transforms of these expansions are

$$v_{zs}(t) = [3e^{-1000t} - 3e^{-3000t}]u(t) \text{ V}$$

$$v_{zi}(t) = [-0.75e^{-1000t} + 2.25e^{-3000t}]u(t) \text{ V}$$

Note that the circuit responses contain only transient terms that decay to zero. There is no forced response because in the dc steady state the inductor acts like a short circuit, forcing $v(t)$ to zero for $t \to \infty$. From an *s*-domain viewpoint there is no forced response because the forced pole at $s = 0$ is canceled by a zero in the network function. ∎

◑ **Design Exercise 10–9** _____

The switch in the circuit of Figure 10–21(a) has been closed for a long time. At $t = 0$ the switch is suddenly opened.

(a) Find the transform $I_R(s)$ for the current through the resistor.
(b) Select values of R, L, and C so that the current reaches at least 63% of its final value in 100 ms or less.

Answers:

(a)
$$I_R(s) = \frac{\dfrac{I_A}{RC}}{s\left(s + \dfrac{1}{RC}\right)} \text{ A-s}$$

(b) After one time constant the current reaches 63% of its final value; hence $RC < 100$ ms. $R = 56\,k\Omega$ and $C = 1\,\mu F$ are one pair of values that work; there are many others. The value of L does not matter since it is not connected in the circuit after $t = 0$.

EXAMPLE 10–9

Use superposition to find the zero-state component of $I(s)$ in the s-domain circuit shown in Figure 10–22(a).

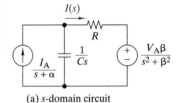

(a) s-domain circuit

SOLUTION:
Turning the voltage source off produces the circuit in Figure 10–22(b). In this circuit the resistor and capacitor are connected in parallel, so current division yields $I_1(s)$ in the form

$$I_1(s) = \frac{Y_R}{Y_C + Y_R} \frac{I_A}{(s + \alpha)} = \frac{I_A}{(RCs + 1)(s + \alpha)}$$

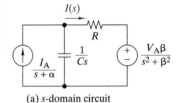

(b) Voltage source OFF

Turning the voltage source on and the current source off produces the circuit in Figure 10–22(c). In this case the resistor and capacitor are connected in series, and series equivalence gives the current $I_2(s)$ as

$$I_2(s) = \frac{1}{Z_R + Z_C} \frac{V_A\beta}{s^2 + \beta^2} = \frac{CsV_A\beta}{(RCs + 1)(s^2 + \beta^2)}$$

Using superposition, the total zero-state response is

$$I_{zs}(s) = I_1(s) - I_2(s)$$
$$= \frac{I_A}{(RCs + 1)(s + \alpha)} - \frac{CsV_A\beta}{(RCs + 1)(s^2 + \beta^2)}$$

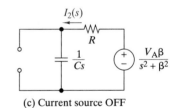

(c) Current source OFF

FIGURE 10–22

There is a minus in this equation because $I_1(s)$ and $I_2(s)$ were assigned opposite reference directions in Figures 10–22(b) and 10–22(c). The total zero-state response has four poles. The natural pole at $s = -1/RC$ came from the circuit. The forced pole at $s = -\alpha$ came from the current source, and the two forced poles at $s = \pm j\beta$ came from the voltage source.

In this example the time-domain response would have a transient component $Ke^{-t/RC}$ due to the natural pole, a forced component $Ke^{-\alpha t}$ due to the current source, and a forced component of the form $K_A \cos \beta t + K_B \sin \beta t$ due to the voltage source. We can infer these general conclusions regarding the time-domain response by simply examining the poles of the s-domain response. ∎

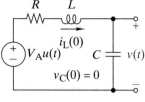

FIGURE 10–23

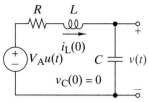

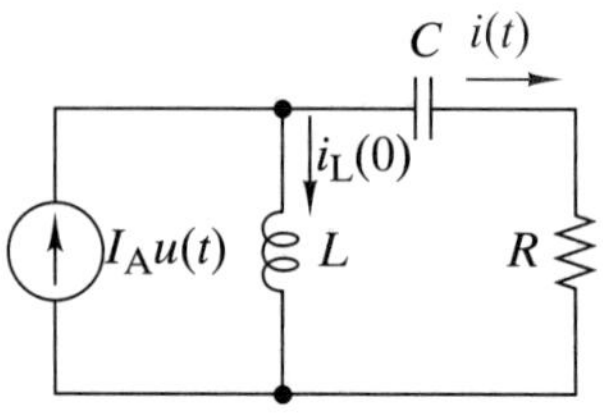

FIGURE 10–24

Exercise 10–10

The initial conditions for the circuit in Figure 10–23 are $v_C(0-) = 0$ and $i_L(0-) = I_0$. Transform the circuit into the s domain and find the zero-state and zero-input components of $V(s)$.

Answer:

$$V_{zs}(s) = \left[\frac{1}{LCs^2 + RCs + 1}\right]\frac{V_A}{s}$$

$$V_{zi}(s) = \frac{LI_0}{LCs^2 + RCs + 1}$$

Exercise 10–11

The initial conditions for the circuit in Figure 10–24 are $v_C(0-) = 0$ and $i_L(0-) = I_0$. Transform the circuit into the s domain and find the zero-state and zero-input components of $I(s)$.

Answers:

$$I_{zs}(s) = \left[\frac{LCs^2}{LCs^2 + RCs + 1}\right]\frac{I_A}{s}$$

$$I_{zi}(s) = \left[\frac{-LCs^2}{LCs^2 + RCs + 1}\right]\frac{I_0}{s}$$

EXAMPLE 10–10

A transformed dependent-source circuit is shown in Figure 10–25(a). The capacitor has an initial voltage of V_B V. Use superposition to find the output transform $V_O(s)$.

SOLUTION:

The output transform voltage is $V_O(s) = -\mu V_X(s)$. With two independent voltage sources we find $V_X(s)$ by using superposition and using voltage dividers. Starting

FIGURE 10–25

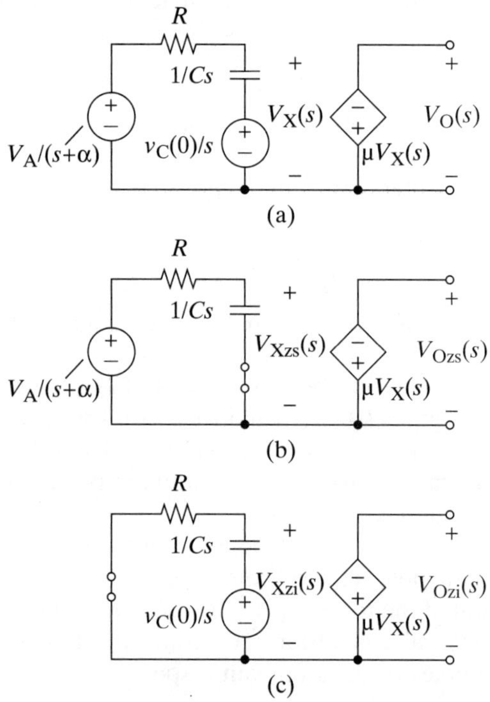

(a)

(b)

(c)

with the source voltage, we find the zero-state contribution to $V_X(s)$ as shown in Figure 10–25(b)

$$V_{Xzs}(s) = \frac{1/Cs}{R + 1/Cs}\left(\frac{V_A}{s + \alpha}\right) = \frac{1/RC}{s + 1/RC}\left(\frac{V_A}{s + \alpha}\right)$$

For the zero-input contribution to $V_X(s)$ shown in Figure 10–25(c), let $v_C(0) = V_B$, to yield

$$V_{Xzi}(s) = \frac{R}{R + 1/Cs}\left(\frac{v_C(0)}{s}\right) = \frac{1}{s + 1/RC}V_B$$

Adding these results together, we get

$$V_X(s) = \frac{1/RC}{s + 1/RC}\left(\frac{V_A}{s + \alpha}\right) + \frac{1}{s + 1/RC}V_B$$

Therefore, the output transform is

$$V_O(s) = -\mu V_X(s) = \frac{-\mu V_A/RC}{(s + 1/RC)(s + \alpha)} + \frac{-\mu V_B}{s + 1/RC}$$

We can expand the first term by partial fractions so that our total response transform is

$$V_O(s) = \left[\overbrace{\underbrace{\frac{-\mu V_A}{(1 - \alpha RC)(s + \alpha)}}_{\text{forced}} + \underbrace{\frac{\mu V_A}{(1 - \alpha RC)(s + 1/RC)}}_{\text{natural}}}^{\text{zero state}} \overbrace{- \frac{\mu V_B}{s + 1/RC}}^{\text{zero input}}\right]u(t)\text{V-s}$$

The result has two poles, a forced pole determined by the input source at $s = -\alpha$, and a second natural pole determined by the circuit at $s = -1/RC$. ∎

Exercise 10–12

For the circuit of Figure 10–25(a), let the input voltage $V_S(s) = V_A/s$, and use superposition to find the zero-state and the zero-input components of the output. Then show that $v_O(t)$ is the same as that found in Example 10–3.

Answer:
$$V_{Ozs}(s) = \frac{-\mu V_A/RC}{s(s + 1/RC)}$$

$$V_{Ozi}(s) = \frac{-\mu V_B}{(s + 1/RC)}$$

$$v_O(t) = \left[\overbrace{-\mu V_A + \mu V_A e^{-t/RC}}^{\text{zero state}} - \overbrace{\mu V_B e^{-t/RC}}^{\text{zero input}}\right]u(t)\text{V QED}$$

THÉVENIN AND NORTON EQUIVALENT CIRCUITS

In the *s* domain a two-terminal circuit containing linear elements and sources can be replaced by the Thévenin or Norton equivalent circuits in Figure 10–26. The general concept and restrictions are the same in the *s* domain as for resistive circuits. The important differences here are that the source terms $V_T(s)$ and $I_N(s)$ are transforms while Z_T and Z_N are *s*-domain impedances.

To find the Thévenin or Norton equivalent circuit, we use the same process as for resistance circuits, except that now we must manipulate rational functions of *s*.

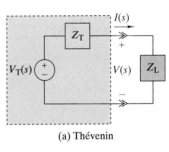

(a) Thévenin

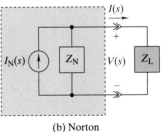

(b) Norton

FIGURE 10–26 *Thévenin and Norton equivalent circuits in the s domain.*

Since the Thévenin and Norton circuits are equivalent to each other, their circuit parameters are related to the *s*-domain open-circuit voltage $V_{OC}(s)$ and short-circuit current $I_{SC}(s)$ as

$$V_{OC}(s) = V_T(s) = I_N(s)Z_N$$

$$I_{SC}(s) = \frac{V_T(s)}{Z_T} = I_N(s)$$

$$Z_T = Z_N = \frac{V_{OC}(s)}{I_{SC}(s)}$$

(10–28)

Algebraically the results in Eq. (10–28) are identical to the corresponding equations for resistance circuits, except that these equations involve transforms and impedances rather than waveforms and resistances. Collectively these equations show that finding a Thévenin or Norton equivalent involves finding any two of the following three quantities: (1) the open-circuit voltage $V_{OC}(s)$, (2) the short-circuit current $I_{SC}(s)$, and (in the absence of dependent sources) (3) the lookback impedance with all independent sources turned off.

The relationships in Eq. (10–28) also define *source transformations* that allow us to convert a voltage source in series with an impedance into a current source in parallel with the same impedance, or vice versa. Performing *s*-domain source transformations may lead to circuit simplifications and can be useful when formulating node-voltage or mesh-current equations, as discussed in the next section.

Thévenin and Norton equivalent circuits should be regarded as important conceptual tools offering insight into how circuits operate in the *s* domain. They are not, in general, important tools for reducing the computational effort involved in *s*-domain circuit analysis.

EXAMPLE 10–11

The circuit in Figure 10–27(a) is in the zero state. Use a source transformation and voltage division to find the *s*-domain relationship between the input $I_1(s)$ and the output $V_2(s)$.

FIGURE 10–27

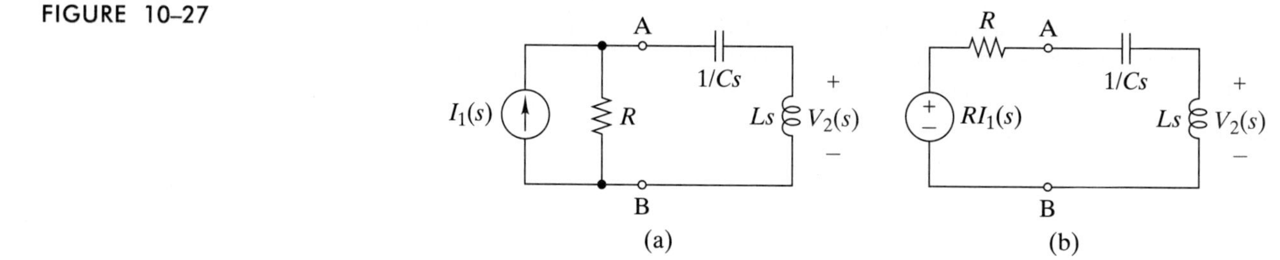

(a)   (b)

SOLUTION:
In this example we use a source transformation on the subcircuit to the left of points A and B in Figure 10–27(a). This Norton subcircuit consists of an independent current source $I_N = I_1(s)$ in parallel with an impedance $Z_N = R$. The equivalent Thévenin circuit consists of a voltage source $V_T = I_N Z_N = RI_1(s)$ in series with an impedance $Z_T = Z_N = R$. Figure 10–27(b) shows the circuit after the source transformation. Applying voltage division in the modified circuit yields the required input-output relationship.

$$V_2(s) = \left[\frac{Ls}{R + \dfrac{1}{Cs} + Ls} \right] RI_1(s) = \left[\frac{RLCs^2}{LCs^2 + RCs + 1} \right] I_1(s)$$

Exercise 10–13

The circuit of Figure 10–28 is in the zero state.

(a) Find the Thévenin equivalent circuit that the load sees.
(b) Find the Norton equivalent of the same circuit.

Answers:

(a) $V_T(s) = \dfrac{\dfrac{1}{RC_1} V_1(s)}{s + \dfrac{1}{RC_1}}$ V-s, $Z_T(s) = \dfrac{\left(\dfrac{C_1 + C_2}{C_1 C_2}\right)\left(s + \dfrac{1}{R(C_1 + C_2)}\right)}{s\left(s + \dfrac{1}{RC_1}\right)} \ \Omega$

(b) $I_N(s) = \dfrac{\left(\dfrac{C_2}{C_1 + C_2}\right)\left(\dfrac{V_1(s)}{R}\right)s}{s + \dfrac{1}{R(C_1 + C_2)}}$ A-s, $Z_N(s) = Z_T(s)$

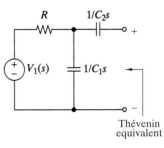

Thévenin equivalent

FIGURE 10–28

EXAMPLE 10–12

The circuit in Figure 10–29(a) is in the zero state. Use a Thévenin equivalent to find the s-domain relationship between the input $V_1(s)$ and the output $V_2(s)$.

FIGURE 10–29

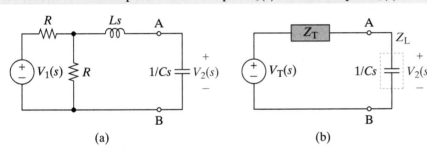

(a)　　　　　(b)

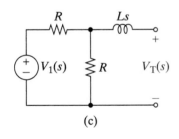

(c)

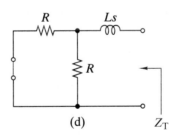

(d)　　　Z_T

SOLUTION:

In this example we treat the capacitor as a load and find the Thévenin equivalent circuit to the left of points A and B as shown in Figure 10–29(b). To obtain the required Thévenin circuit, we find the open-circuit voltage and lookback impedance. Figure 10–29(c) shows the open-circuit situation. There is no voltage across the inductor, so the open-circuit voltage $V_T(s)$ is the same as the voltage across the second resistor. Using voltage division we have

$$V_T(s) = \frac{R}{R + R} V_1(s) = \frac{V_1(s)}{2}$$

To find the lookback impedance, we turn the input voltage source off [replace $V_1(s)$ by a short circuit] to obtain the situation in Figure 10–29(d). By inspection,

$$Z_T = Ls + R\|R = Ls + \frac{R}{2}$$

Given the $V_T(s)$ and Z_T, we return to Figure 10–29(b) and use voltage division to obtain the desired input-output relationship.

$$V_2(s) = \left[\frac{Z_L}{Z_T + Z_L}\right] V_T(s) = \left[\frac{\dfrac{1}{Cs}}{Ls + \dfrac{R}{2} + \dfrac{1}{Cs}}\right] \frac{V_1(s)}{2}$$

$$= \left[\frac{1}{2LCs^2 + RCs + 2}\right] V_1(s) \qquad \blacksquare$$

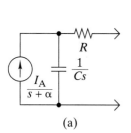

(a)

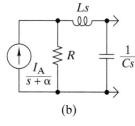

(b)

FIGURE 10–30

Exercise 10–14

Find the Norton and the Thévenin equivalents of the *s*-domain circuits in Figure 10–30.

Answers:

(a) $I_N(s) = \dfrac{I_A}{(RCs+1)(s+\alpha)}$ $\quad Z_N(s) = Z_T(s) = \dfrac{RCs+1}{Cs}$ $\quad V_T(s) = \dfrac{I_A}{Cs(s+\alpha)}$

(b) $I_N(s) = \dfrac{RI_A}{(Ls+R)(s+\alpha)}$ $\quad Z_N(s) = Z_T(s) = \dfrac{Ls+R}{LCs^2+RCs+1}$

$V_T(s) = \dfrac{RI_A}{(s+\alpha)(LCs^2+RCs+1)}$

10–4 NODE-VOLTAGE ANALYSIS IN THE *s* DOMAIN

The previous sections deal with basic analysis methods using equivalence, reduction, and circuit theorems. These methods are valuable because we work directly with the element impedances and thereby gain insight into *s*-domain circuit behavior. We also need general methods to deal with more complicated circuits that these basic methods cannot easily handle.

FORMULATING NODE-VOLTAGE EQUATIONS

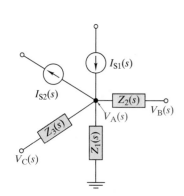

FIGURE 10–31 *An example node.*

Formulating node-voltage equations involves selecting a reference node and assigning a node-to-datum voltage to each of the remaining nonreference nodes. Because of KVL, the voltage across any two-terminal element is equal to the difference of the two node voltages. This fundamental property of node voltages, together with element impedances, allows us to write KCL constraints at each of the nonreference nodes.

For example, consider the *s*-domain circuit in Figure 10–31. The sum of currents leaving node A can be written as

$$I_{S2}(s) - I_{S1}(s) + \frac{V_A(s)}{Z_1(s)} + \frac{V_A(s) - V_B(s)}{Z_2(s)} + \frac{V_A(s) - V_C(s)}{Z_3(s)} = 0$$

Rewriting this equation with unknown node voltages grouped on the left and inputs on the right yields

$$\left[\frac{1}{Z_1(s)} + \frac{1}{Z_2(s)} + \frac{1}{Z_3(s)}\right] V_A(s) - \frac{1}{Z_2(s)} V_B(s) - \frac{1}{Z_3(s)} V_C(s) = I_{S1}(s) - I_{S2}(s)$$

Expressing this result in terms of admittances produces the following equation:

$$[Y_1(s) + Y_2(s) + Y_3(s)]V_A(s) - [Y_2(s)]V_B(s) - [Y_3(s)]V_C(s) = I_{S1}(s) - I_{S2}(s)$$

This equation has a familiar pattern. The unknowns are the node-voltage transforms $V_A(s)$, $V_B(s)$, and $V_C(s)$. The coefficient $[Y_1(s) + Y_2(s) + Y_3(s)]$ of $V_A(s)$ is the sum of the admittances of the elements connected to node A. The coefficient $[Y_2(s)]$ of $V_B(s)$ is the admittance of the elements connected between nodes A and B, while $[Y_3(s)]$ is the admittance of the elements connected between nodes A and C. Finally, $I_{S1}(s) - I_{S2}(s)$ is the sum of the source currents directed into node A. These observations suggest that we can write node-voltage equations for s-domain circuits by inspection, just as we did with resistive circuits.

The formulation method just outlined assumes that there are no voltage sources in the circuit. When transforming the circuit, we can always select the current source models to represent the initial conditions. However, the circuit may contain dependent or independent voltage sources. If so, they can be treated using the following methods:

Method 1: If there is an impedance in series with the voltage source, use a source transformation to convert it into an equivalent current source.

Method 2: Select the reference node so that one terminal of one or more of the voltage sources is connected to ground. The source voltage then determines the node voltage at the other source terminal, thereby eliminating an unknown.

Method 3: Create a supernode surrounding any voltage source that cannot be handled by method 1 or 2.

Some circuits may require more than one of these methods.

Formulating a set of equilibrium equations in the s domain is a straightforward process involving concepts developed in Chapters 3 and 4 for resistance circuits. The following example illustrates the formulation process.

EXAMPLE 10–13

Formulate s-domain node-voltage equations for the circuit in Figure 10–32(a).

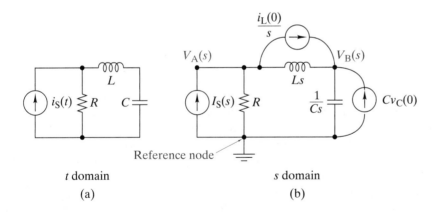

FIGURE 10–32

t domain
(a)

s domain
(b)

SOLUTION:
Figure 10–32(b) shows the circuit in the s domain. In transforming the circuit, we use current sources to represent the inductor and capacitor initial conditions. This choice facilitates writing node equations since the resulting s-domain circuit contains only current sources and fewer nodes. The sum of currents leaving nodes A and B can be written as

$$\text{Node A:} \quad \frac{V_A(s)}{R} + \frac{V_A(s) - V_B(s)}{Ls} - I_S(s) + \frac{i_L(0)}{s} = 0$$

$$\text{Node B:} \quad \frac{V_B(s)}{1/Cs} + \frac{V_B(s) - V_A(s)}{Ls} - \frac{i_L(0)}{s} - Cv_C(0) = 0$$

Rearranging these equations in the standard format with the unknowns on the left and the inputs on the right yields

$$\text{Node A:} \quad \left(\frac{1}{R} + \frac{1}{Ls}\right) V_A(s) - \left(\frac{1}{Ls}\right) V_B(s) = I_S(s) - \frac{i_L(0)}{s}$$

$$\text{Node B:} \quad -\left(\frac{1}{Ls}\right) V_A(s) + \left(\frac{1}{Ls} + Cs\right) V_B(s) = Cv_C(0) + \frac{i_L(0)}{s}$$

Note that $(1/R + 1/Ls)$ is the sum of the admittances connected to node A, $(1/Ls + Cs)$ is the sum of the admittances connected to node B, and $1/Ls$ is the admittance connected between nodes A and B. The circuit is driven by an independent current source $I_S(s)$ and two initial condition sources $Cv_C(0)$ and $i_L(0)/s$. The terms on the right side of these equations are the sum of source currents directed into each node. With practice, we learn to write these equations by inspection. ∎

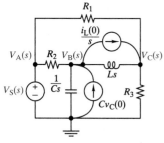

FIGURE 10–33

Exercise 10–15

Using the nodes identified, write a set of node-voltage equations for the circuit of Figure 10–33.

Answer:

$$V_B(s)\left(\frac{1}{R_2} + Cs + \frac{1}{Ls}\right) - V_C(s)\left(\frac{1}{Ls}\right) = \frac{V_S(s)}{R_2} - \frac{i_L(0)}{s} + Cv_C(0)$$

$$-V_B(s)\left(\frac{1}{Ls}\right) + V_C(s)\left(\frac{1}{Ls} + \frac{1}{R_1} + \frac{1}{R_3}\right) = \frac{V_S(s)}{R_1} + \frac{i_L(0)}{s}$$

SOLVING S-DOMAIN CIRCUIT EQUATIONS

Example 10–13 shows that node-voltage equations are linear algebraic equations in the unknown node voltages. In theory, solving these equations can be accomplished using classical techniques such as Cramer's rule or Gaussian reduction. In practice, we quickly lose interest save for a 2 × 2 or, perhaps for the stout of heart, a 3 × 3 linear set of equations, since coefficients in the equations are polynomials, making the algebra rather tedious. With today's software, it is becoming increasingly practical to use programs such as MATLAB to solve these linear equations using their symbolic analysis toolboxes. In Example 10–14, we will solve using the classical Cramer's rule a simple 2 × 2 set of equations formulated in Example 10–13. Subsequently we will solve a 3 × 3 set of equations using MATLAB in Web Appendix C. To contrast the difference and to highlight some important features of the classical solution that may be somewhat obscured in the software approach.

EXAMPLE 10–14

For the circuit of Figure 10–32, do the following:

(a) Solve the node-voltage equations from Example 10–13 and identify the zero-state and zero-input responses.
(b) Solve for the zero-state components of the waveforms $v_A(t)$ and $v_B(t)$ when $R = 1\text{k}\Omega, C = 0.2\,\mu\text{F}, L = 500\,\text{mH}$, and $i_S(t) = 10u(t)\,\text{mA}$.

SOLUTION:

(a) In Example 10–13, we formulated the following node-voltage equations for the circuit of Figure 10–32:

$$\left(\frac{1}{R} + \frac{1}{Ls}\right)V_A(s) - \frac{1}{Ls}V_B(s) = I_S(s) - \frac{i_L(0)}{s}$$

$$-\frac{1}{Ls}V_A(s) + \left(Cs + \frac{1}{Ls}\right)V_B(s) = \frac{i_L(0)}{s} + Cv_C(0)$$

Using Cramer's rule, it is convenient to find the determinant of these equations:

$$\Delta(s) = \begin{vmatrix} \dfrac{1}{R} + \dfrac{1}{Ls} & -\dfrac{1}{Ls} \\[2mm] -\dfrac{1}{Ls} & Cs + \dfrac{1}{Ls} \end{vmatrix}$$

$$= \left(\frac{1}{R} + \frac{1}{Ls}\right)\left(Cs + \frac{1}{Ls}\right) - \left(\frac{1}{Ls}\right)^2$$

$$= \frac{LCs^2 + RCs + 1}{RLs}$$

We call $\Delta(s)$ the **circuit determinant** because it depends only on the elements R, L, and C. The determinant $\Delta(s)$ characterizes the circuit and does not depend on either the input or the initial conditions. Note also that since the circuit contained no dependent sources, the determinant is symmetrical about the major diagonal. Continuing our solution, the node-voltage $V_A(s)$ is found using Cramer's rule:

$$V_A(s) = \frac{\Delta_A(s)}{\Delta(s)} = \frac{\begin{vmatrix} I_S(s) - \dfrac{i_L(0)}{s} & -\dfrac{1}{Ls} \\[2mm] \dfrac{i_L(0)}{s} + Cv_C(0) & Cs + \dfrac{1}{Ls} \end{vmatrix}}{\Delta(s)}$$

$$= \underbrace{\frac{(LCs^2 + 1)RI_S(s)}{LCs^2 + RCs + 1}}_{\text{zero state}} + \underbrace{\frac{-RLCs\,i_L(0) + RCv_C(0)}{LCs^2 + RCs + 1}}_{\text{zero input}}$$

Similarly solving for $V_B(s)$,

$$V_B(s) = \frac{\Delta_B(s)}{\Delta(s)} = \frac{\begin{vmatrix} \dfrac{1}{R} + \dfrac{1}{Ls} & I_S(s) - \dfrac{i_L(0)}{s} \\[2mm] -\dfrac{1}{Ls} & \dfrac{i_L(0)}{s} + Cv_C(0) \end{vmatrix}}{\Delta(s)}$$

$$= \underbrace{\frac{RI_S(s)}{LCs^2 + RCs + 1}}_{\text{zero state}} + \underbrace{\frac{(Ls + R)Cv_C(0) + Li_L(0)}{LCs^2 + RCs + 1}}_{\text{zero input}}$$

This solution readily provides both the zero-input and zero-state components of the response transforms $V_A(s)$ and $V_B(s)$.

DISCUSSION: *Cramer's rule results in a solution of the node voltages as a ratio of determinants of the form*

$$V_X(s) = \frac{\Delta_X(s)}{\Delta(s)}$$

*The response transform $V_X(s)$ is a rational function of s whose poles are either zeros of the circuit determinant $\Delta(s)$ or poles of the determinant $\Delta_X(s)$. This means that $V_X(s)$ has poles when $\Delta(s) = 0$ or when $\Delta_X(s) \to \infty$. The partial-fraction expansion of $V_X(s)$ will contain terms for each of these poles. We call the zeros of $\Delta(s)$ the **natural poles** because they depend only on the circuit and give rise to the natural response terms in the expansion. We call the poles of $\Delta_X(s)$ the **forced poles** because they depend on the form of the input signal and give rise to the forced response terms in the expansion. We will revisit these ideas in our discussion of circuit stability in Sect. 10–6.*

(b) Since the problem is to find the zero-state components of $v_A(t)$ and $v_B(t)$, we start by substituting the component and source values into our equations for $V_A(s)$ and $V_B(s)$ found in part (a). We use only the zero-state portions of the transforms since all initial conditions are set to zero—that is, $i_L(0) = 0$ A and $v_C(0) = 0$ V—which is by definition the circuit's zero state.

$$V_{Azs}(s) = \left(\frac{10^{-7}s^2 + 1}{10^{-10}s^2 + 0.2 \times 10^{-6}s + 10^{-3}}\right)\frac{10^{-2}}{s}$$

$$= 10\frac{s^2 + 10^7}{s[(s+1000)^2 + 3000^2]} \text{ V-s}$$

$$V_{Bzs}(s) = \left(\frac{1}{10^{-10}s^2 + 0.2 \times 10^{-6}s + 10^{-3}}\right)\frac{10^{-2}}{s}$$

$$= 10\frac{10^7}{s[(s+1000)^2 + 3000^2]} \text{ V-s}$$

Both response transforms have three poles: a forced pole at $s = 0$ and two natural poles at $s = -1000 \pm j3000$. The forced pole comes from the step function input, and the two natural poles are zeros of the circuit determinant. Expanding these rational functions as

$$V_{Azs}(s) = \frac{10}{s} - \frac{20}{3}\left(\frac{3000}{(s+1000)^2 + 3000^2}\right)$$

$$V_{Bzs}(s) = \frac{10}{s} - \frac{10}{3}\left(\frac{3000}{(s+1000)^2 + 3000^2}\right) - 10\left(\frac{s+1000}{(s+1000)^2 + 3000^2}\right)$$

and taking the inverse transforms yields the required zero-state response waveforms:

$$v_{Azs}(t) = 10u(t) - 20e^{-1000t}\left[\frac{1}{3}\sin(3000\,t)\right]u(t) \text{ V}$$

$$v_{Bzs}(t) = 10u(t) - 10e^{-1000t}\left[\frac{1}{3}\sin(3000\,t) + \cos(3000\,t)\right]u(t) \text{ V}$$

The step function in both responses is the forced response caused by the forced pole at $s = 0$. The damped sinusoids are natural responses determined by the natural poles.

Figure 10–34 shows part of an Excel spreadsheet that produces plots of $v_{Azs}(t)$ and $v_{Bzs}(t)$. Spreadsheets are useful for generating graphs, especially when we

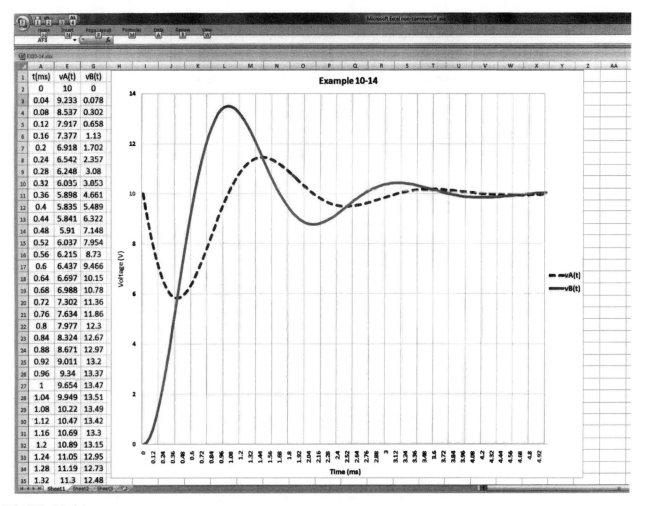

t(ms)	vA(t)	vB(t)
0	10	0
0.04	9.233	0.078
0.08	8.537	0.302
0.12	7.917	0.658
0.16	7.377	1.13
0.2	6.918	1.702
0.24	6.542	2.357
0.28	6.248	3.08
0.32	6.035	3.853
0.36	5.898	4.661
0.4	5.835	5.489
0.44	5.841	6.322
0.48	5.91	7.148
0.52	6.037	7.954
0.56	6.215	8.73
0.6	6.437	9.466
0.64	6.697	10.15
0.68	6.988	10.78
0.72	7.302	11.36
0.76	7.634	11.86
0.8	7.977	12.3
0.84	8.324	12.67
0.88	8.671	12.97
0.92	9.011	13.2
0.96	9.34	13.37
1	9.654	13.47
1.04	9.949	13.51
1.08	10.22	13.49
1.12	10.47	13.42
1.16	10.69	13.3
1.2	10.89	13.15
1.24	11.05	12.95
1.28	11.19	12.73
1.32	11.3	12.48

FIGURE 10–34

wish to compare waveforms. The two plots show that the two response waveforms are different even though they have the same poles. In other words, the basic form of a response is determined by the forced and natural poles, but the relative amplitudes are influenced by the zeros as well. ∎

Exercise 10–16

For the circuit of Figure 10–32(a), find the zero-state current transforms through each passive element.

Answer:
$$I_{R\dot{z}s}(s) = \frac{s^2 + \dfrac{1}{LC}}{s^2 + \dfrac{R}{L}s + \dfrac{1}{LC}}I_S(s) \text{ A-s} \qquad I_{Lzs}(s) = I_{Czs}(s) = \frac{\dfrac{R}{L}s}{s^2 + \dfrac{R}{L}s + \dfrac{1}{LC}}I_S(s) \text{ A-s}$$

In s-domain circuit analysis and design, the location of complex poles is often specified in terms of the undamped natural frequency (ω_0) and damping ratio (ζ) parameters introduced in our study of second-order cicuits. Using these parameters, the standard form of a second-order factor is $s^2 + 2\zeta\omega_0 s + \omega_0^2$, which locates the poles at

$$s_{1,2} = \omega_0(-\zeta \pm \sqrt{\zeta^2 - 1})$$

The quantity under the radical depends only on the damping ratio ζ. When $\zeta > 1$ the quantity is positive and the two poles are real and distinct, and the second-order factor becomes the product of two first-order terms. If $\zeta = 1$ the quantity under the radical vanishes and there is a double pole as $s = -\omega_0$. If $\zeta < 1$ the quantity under the radical is negative and the two roots are complex conjugates.

The location of complex poles is also defined in terms of the two natural frequency parameters α and β. Using these parameters, the poles are at $s_{1,2} = -\alpha \pm j\beta$, and the standard form of a second-order factor is $(s + \alpha)^2 + \beta^2$.

In *s*-domain circuit design, we often need to convert from one set of parameters to the other. First, equating their standard forms

$$s^2 + 2\alpha s + \alpha^2 + \beta^2 = s^2 + 2\zeta\omega_0 s + \omega_0^2$$

and then equating the coefficients of like powers of s yields

$$\omega_0 = \sqrt{\alpha^2 + \beta^2} \quad \text{and} \quad \zeta = \frac{\alpha}{\sqrt{\alpha^2 + \beta^2}}$$

and conversely

$$\alpha = \zeta\omega_0 \quad \text{and} \quad \beta = \omega_0\sqrt{1 - \zeta^2}$$

Figure 10–35 shows how these parameters define the locations of complex poles in the *s* plane. The natural frequency parameters α and β define the rectangular coordinates of the poles. In a sense, the parameters ω_0 and ζ define the corresponding polar coordinates. The parameter ω_0 is the radial distance from the origin to the poles. The angle θ is determined by the damping ratio ζ alone, since $\theta = \cos^{-1}\zeta$.

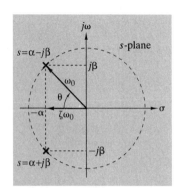

FIGURE 10–35 *s-plane geometry relating α and β to ζ and ω_0*

D DESIGN EXAMPLE 10–15

The *s*-domain circuit in Figure 10–36 is to be designed to produce a pair of complex poles defined by $\zeta = 0.5$ and $\omega_0 = 1000\,\text{rad/s}$. To simplify production the design will use equal element values $R_1 = R_2 = R$ and $C_1 = C_2 = C$. Select the values of R, C, and the gain μ so that the circuit has the desired natural poles.

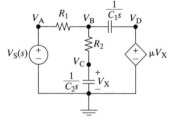

FIGURE 10–36

SOLUTION:
To locate the natural poles, we find the circuit determinant using node-voltage equations. The circuit has four nodes but only two of these involve independent variables. For the indicated reference node, the voltages at nodes A and D are $V_A(s) = V_S(s)$ and $V_D(s) = \mu V_X(s) = \mu V_C(s)$. That is, the two grounded voltage sources specify the voltages at nodes A and D. Consequently, we only need equations at nodes B and C. The sums of currents leaving these nodes are

$$\text{Node B:} \quad \frac{V_B(s) - V_S(s)}{R_1} + \frac{V_B(s) - V_C(s)}{R_2} + \frac{V_B(s) - \mu V_C(s)}{1/C_1 s} = 0$$

$$\text{Node C:} \quad \frac{V_C(s) - V_B(s)}{R_2} + \frac{V_C(s)}{1/C_2 s} = 0$$

Arranging these equations with unknown node voltages on the left and the source terms on the right yields

$$\left(C_1 s + \frac{1}{R_1} + \frac{1}{R_2}\right) V_B(s) - \left(\mu C_1 s + \frac{1}{R_2}\right) V_C(s) = \frac{V_S(s)}{R_1}$$

$$-\left(\frac{1}{R_2}\right) V_B(s) + \left(C_2 s + \frac{1}{R_2}\right) V_C(s) = 0$$

The natural poles are zeros of the circuit determinant:

$$\Delta(s) = \begin{vmatrix} \left(C_1s + \dfrac{1}{R_1} + \dfrac{1}{R_2}\right) & -\left(\mu C_1 s + \dfrac{1}{R_2}\right) \\[2ex] -\left(\dfrac{1}{R_2}\right) & \left(C_2 s + \dfrac{1}{R_2}\right) \end{vmatrix}$$

$$= C_1 C_2 s^2 + \left(\frac{C_1}{R_2} + \frac{C_2}{R_1} + \frac{C_2}{R_2} - \mu\frac{C_1}{R_2}\right)s + \frac{1}{R_1 R_2}$$

For equal resistances $R_1 = R_2 = R$ and equal capacitances $C_1 = C_2 = C$, the circuit determinant reduces to

$$\frac{\Delta(s)}{C^2} = s^2 + \left(\frac{3-\mu}{RC}\right)s + \left(\frac{1}{RC}\right)^2$$

Comparing this second-order factor to the standard form $s^2 + 2\zeta\omega_0 s + \omega_0^2$ yields the following design constraints:

$$\omega_0 = \frac{1}{RC} = 1000 \quad \text{and} \quad \zeta = \frac{3-\mu}{2} = 0.5$$

These constraints lead to the conditions $RC = 10^{-3}$ and $\mu = 2$. Selecting $R = 10\,\text{k}\Omega$ makes $C = 0.1\,\mu\text{F}$. For the specified conditions the natural poles are located at $s = -\alpha \pm j\beta$, where

$$\alpha = \zeta\omega_0 = 500\,\text{rad/s} \quad \text{and} \quad \beta = \omega_0\sqrt{1-\zeta^2} = 866\,\text{rad/s} \qquad \blacksquare$$

◗ Design Exercise 10–17 _____

Consider the circuit in Figure 10–36. Select values for the various components to produce a pair of complex poles defined by $\zeta = 0.5$ and $\omega_0 = 1\,\text{krad/s}$. To produce your design you must assume unity gain ($\mu = 1$) for the dependent source, and that $R_1 = R_2 = R$.

Answers: With the given assumptions,

$$\omega_0 = \frac{1}{R\sqrt{C_1 C_2}} \quad \text{and} \quad \zeta = \sqrt{\frac{C_2}{C_1}}$$

If we select $R = 10\,\text{k}\Omega$, then $C_1 = 0.2\,\mu\text{F}$ and $C_2 = 0.05\,\mu\text{F}$. Other correct answers are possible.

EXAMPLE 10–16

(a) For the s-domain circuit in Figure 10–37, solve for the zero-state output $V_O(s)$ in terms of a general input $V_S(s)$.
(b) Solve for the zero-state output $v_O(t)$ when the input is a unit step function $v_S(t) = u(t)\text{V}$.

FIGURE 10–37

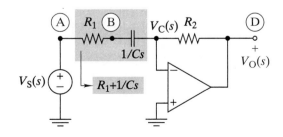

SOLUTION:

(a) We use the node-voltage method to find the OP AMP output. A node equation is not required at node A since the selected reference node makes $V_A(s) = V_S(s)$. Likewise, one is not needed at node D since it is the OP AMP output. Finally, we can avoid writing a node equation at node B by observing that the element impedances R_1 and $1/Cs$ are connected in series. We can treat this series combination as a single element with an equivalent impedance $R_1 + 1/Cs$. Using all of these observations, the sum of the currents leaving Node C is

$$\text{Node C:} \quad \frac{V_C(s) - V_S(s)}{R_1 + 1/Cs} + \frac{V_C(s) - V_O(s)}{R_2} + I_N(s) = 0$$

In the s domain the ideal OP AMP model in Eq. (10–4) requires $I_N(s) = 0$ and $V_P(s) = V_N(s)$. But $V_P(s) = 0$ since the noninverting input is grounded; hence, $V_N(s) = V_C(s) = 0$. Inserting these conditions in the node C equation and solving for the output voltage yields

$$V_{Ozs}(s) = \left[-\frac{R_2}{R_1 + 1/Cs} \right] V_S(s)$$

$$= \left[-\frac{R_2}{R_1}\left(\frac{s}{s + 1/R_1C} \right) \right] V_S(s) \text{ V-s}$$

This equation relates the zero-state output to a general input $V_S(s)$. The output transform is proportional to the input transform since the circuit is linear. The proportionality factor within the brackets is called a *network function*. In this case the network function has a natural pole at $s = -1/R_1C$ and a zero at $s = 0$.

(b) A step function input $V_S(s) = 1/s$ produces a forced pole at $s = 0$. However, the zero in the network function cancels the forced pole so that

$$v_O(t) = \mathcal{L}^{-1}\left\{ -\frac{R_2}{R_1}\left(\frac{s}{s + 1/R_1C} \right)\frac{1}{s} \right\} = \mathcal{L}^{-1}\left\{ -\frac{R_2}{R_1}\left(\frac{1}{s + 1/R_1C} \right) \right\}$$

$$= \left(-\frac{R_2}{R_1}e^{-t/R_1C} \right) u(t) \text{ V}$$

For a step function input the zero-state output has no forced pole, only a natural pole at $s = -1/R_1C$. The general principle is that the forced response can be zero even when the input is not zero. In the s domain this occurs when the network function relating output to input has zeros at the same location as forced poles. ∎

Exercise 10–18

(a) For the s-domain circuit in Figure 10–38, solve for the zero-state output $V_O(s)$ in terms of a general input $Vs(s)$.

(b) Solve for the zero-state output when the input is a step function $v_S(t) = V_Au(t)$ V.

FIGURE 10–38

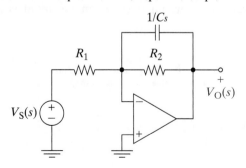

Answer:

(a) $V_O(s) = -\dfrac{1/R_1 C}{s + 1/R_2 C} V_S(s)$ V-s

(b) $v_O(t) = -\dfrac{V_A R_2}{R_1}\left[1 - e^{-t/R_2 C}\right]u(t)$ V

Exercise 10–19

Formulate node-voltage equations for the circuit in Figure 10–39 and find the circuit determinant. Assume that the initial conditions are zero.

Answer: The node equations are

Node B: $\left(\dfrac{1}{Ls} + \dfrac{1}{R_1} + \dfrac{1}{R_2}\right)V_B(s) - \left(\dfrac{1}{R_2}\right)V_C(s) = \dfrac{V_S(s)}{R_1}$

Node C: $-\left(\dfrac{1}{R_2}\right)V_B(s) + \left(Cs + \dfrac{1}{R_2}\right)V_C(s) = CsV_S(s)$

The circuit determinant is

$$\Delta(s) = \dfrac{\left(\dfrac{1}{R_1} + \dfrac{1}{R_2}\right)LCs^2 + \left(\dfrac{L}{R_1 R_2} + C\right)s + \dfrac{1}{R_2}}{Ls}$$

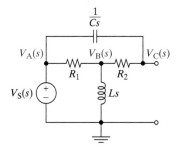

FIGURE 10–39

10–5 MESH-CURRENT ANALYSIS IN THE s DOMAIN

We can use the mesh-current method only when the circuit can be drawn on a flat surface without crossovers. Such planar circuits have special loops called meshes that are defined as closed paths that do not enclose any elements. The mesh-current variables are the loop currents assigned to each mesh in a planar circuit. Because of KCL the current through any two-terminal element can be expressed as the difference of two adjacent mesh currents. This fundamental property of mesh currents, together with the element impedances, allows us to write KVL constraints around each of the meshes.

For example, in Figure 10–40 the sum of voltages around mesh A can be written as

$$Z_1(s)I_A(s) + Z_3(s)[I_A(s) - I_C(s)] - V_{S1}(s) + Z_2(s)[I_A(s) - I_B(s)] + V_{S2}(s) = 0$$

Rewriting this equation with unknown mesh currents grouped on the left and inputs on the right yields

$$(Z_1(s) + Z_2(s) + Z_3(s))I_A(s) - Z_2(s)I_B(s) - Z_3(s)I_C(s) = V_{S1}(s) - V_{S2}(s)$$

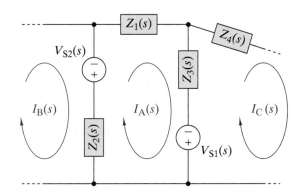

FIGURE 10–40 *An example mesh.*

This equation displays the following pattern. The unknowns are the mesh-current transforms $I_A(s)$, $I_B(s)$, and $I_C(s)$. The coefficient $[Z_1(s) + Z_2(s) + Z_3(s)]$ of $I_A(s)$ is the sum of the impedances of the elements in mesh A. The coefficients $[Z_2(s)]$ of $I_B(s)$ and $[Z_3(s)]$ of $I_C(s)$ are the impedances common to mesh A and the other meshes. Finally, $V_{S1}(s) - V_{S2}(s)$ is the sum of the source voltages around mesh A. These observations suggest that we can write mesh-current equations for s-domain circuits by inspection, just as we did with resistive circuits.

The formulation approach just outlined assumes that there are no current sources in the circuit. When writing mesh-current equations, we select the voltage source model to represent the initial conditions. If the circuit contains dependent or independent current sources, they can be treated using the following methods:

Method 1: If there is an admittance in parallel with the current source, use a source transformation to convert it into an equivalent voltage source.

Method 2: Draw the circuit diagram so that only one mesh current circulates through the current source. This mesh current is then determined by the source current.

Method 3: Create a supermesh for any current source that cannot be handled by methods 1 or 2.

Some circuits may require more than one of these methods.

The following examples illustrate the mesh-current method of s-domain circuit analysis.

EXAMPLE 10-17

(a) Formulate mesh-current equations for the circuit in Figure 10–41(a).
(b) Solve for the zero-input component of $I_A(s)$ and $I_B(s)$.
(c) Find the zero-input responses $i_A(t)$ and $i_B(t)$ for $R_1 = 100\,\Omega$, $R_2 = 200\,\Omega$, $L_1 = 50\,\text{mH}$, and $L_2 = 100\,\text{mH}$.

FIGURE 10–41

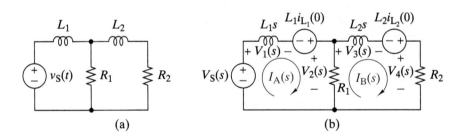

(a)   (b)

SOLUTION:
(a) Figure 10–41(b) shows the circuit transformed into the s domain. In transforming the circuit we used the voltage source model for the initial conditions. The net result is that the transformed circuit contains only voltage sources. The sum of voltages around meshes A and B can be written as

Mesh A: $-V_S(s) + L_1 s I_A(s) - L_1 i_{L_1}(0) + R_1[I_A(s) - I_B(s)] = 0$
Mesh B: $R_1[I_B(s) - I_A(s)] + L_2 s I_B(s) - L_2 i_{L_2}(0) + R_2 I_B(s) = 0$

Rearranging these equations in standard form yields

Mesh A: $(L_1 s + R_1)I_A(s) - R_1 I_B(s) = V_S(s) + L_1 i_{L_1}(0)$
Mesh B: $-R_1 I_A(s) + (L_2 s + R_1 + R_2)I_B(s) = L_2 i_{L_2}(0)$

These s-domain circuit equations are two linear algebraic equations in the two unknown mesh currents $I_A(s)$ and $I_B(s)$.

(b) To solve for the mesh equations, we first find the circuit determinant:

$$\Delta(s) = \begin{vmatrix} L_1s + R_1 & -R_1 \\ -R_1 & L_2s + R_1 + R_2 \end{vmatrix}$$
$$= L_1L_2s^2 + (R_1L_2 + R_1L_1 + R_2L_1)s + R_1R_2$$

To find the zero-input component of $I_A(s)$, we let $V_S(s) = 0$ and use Cramer's rule:

$$I_{Azi}(s) = \frac{\begin{vmatrix} L_1i_{L_1}(0) & -R_1 \\ L_2i_{L_2}(0) & L_2s + R_1 + R_2 \end{vmatrix}}{\Delta(s)}$$

$$= \frac{(L_2s + R_1 + R_2)L_1i_{L_1}(0) + R_1L_2i_{L_2}(0)}{L_1L_2s^2 + (R_1L_2 + R_1L_1 + R_2L_1)s + R_1R_2}$$

Similarly, the zero-input component in $I_B(s)$ is

$$I_{Bzi}(s) = \frac{\begin{vmatrix} L_1s + R_1 & L_1i_{L_1}(0) \\ -R_1 & L_2i_{L_2}(0) \end{vmatrix}}{\Delta(s)}$$

$$= \frac{(L_1s + R_1)L_2i_{L_2}(0) + R_1L_1i_{L_1}(0)}{L_1L_2s^2 + (R_1L_2 + R_1L_1 + R_2L_1)s + R_1R_2}$$

(c) To find the time-domain response, we insert the numerical parameters into the preceding expressions to obtain

$$I_{Azi}(s) = \frac{(0.005s + 15)i_{L_1}(0) + 10i_{L_2}(0)}{0.005s^2 + 25s + 20{,}000}$$

$$= \frac{(s + 3000)i_{L_1}(0) + 2000i_{L_2}(0)}{(s + 1000)(s + 4000)}$$

$$I_{Bzi}(s) = \frac{(0.005s + 10)i_{L_2}(0) + 5i_{L_1}(0)}{0.005s^2 + 25s + 20{,}000}$$

$$= \frac{(s + 2000)i_{L_2}(0) + 1000i_{L_1}(0)}{(s + 1000)(s + 4000)}$$

The circuit has natural poles at $s = -1000$ and -4000 rad/s. Expanding by partial fractions yields

$$I_{Azi}(s) = \frac{2}{3} \times \frac{i_{L_1}(0) + i_{L_2}(0)}{s + 1000} + \frac{1}{3} \times \frac{i_{L_1}(0) - 2i_{L_2}(0)}{s + 4000} \text{ A-s}$$

$$I_{Bzi}(s) = \frac{1}{3} \times \frac{i_{L_1}(0) + i_{L_2}(0)}{s + 1000} - \frac{1}{3} \times \frac{i_{L_1}(0) - 2i_{L_2}(0)}{s + 4000} \text{ A-s}$$

The inverse transforms of these expansions are the required zero-input response waveforms:

$$i_{Azi}(t) = \left[\frac{2}{3}[i_{L_1}(0) + i_{L_2}(0)]e^{-1000t} + \frac{1}{3}[i_{L_1}(0) - 2i_{L_2}(0)]e^{-4000t} \right] u(t) \text{ A}$$

$$i_{Bzi}(t) = \left[\frac{1}{3}[i_{L_1}(0) + i_{L_2}(0)]e^{-1000t} - \frac{1}{3}[i_{L_1}(0) - 2i_{L_2}(0)]e^{-4000t} \right] u(t) \text{ A}$$

Notice that if the initial conditions are $i_{L_1}(0) = -i_{L_2}(0)$, then both $I_A(s)$ and $I_B(s)$ have a zero at $s = -1000$. This zero effectively cancels the natural pole at $s = -1000$. As a result, this pole has zero residue in both partial fraction expansions, and the corresponding terms disappear from the time-domain responses. Likewise, if the initial conditions are $i_{L_1}(0) = 2i_{L_2}(0)$, then both $I_A(s)$ and $I_B(s)$ have a zero at $s = -4000$, and the natural pole at $s = -4000$ disappears in the s-domain responses. The general principle is that all of the circuit's natural poles may not be present in a given response. When this happens the response transform has a zero at the same location as a natural pole, and we say that the natural pole is not observable in the specified response. ■

Exercise 10–20

Solve for the zero-state components of $I_A(s)$ and $I_B(s)$ in Figure 10–41(b).

Answers:

$$I_{Azs}(s) = \frac{R_1 + R_2 + L_2 s}{L_1 L_2 s^2 + (R_1 L_2 + R_1 L_1 + R_2 L_1)s + R_1 R_2} V_S(s) \text{ A-s}$$

$$I_{Bzs}(s) = \frac{R_1}{L_1 L_2 s^2 + (R_1 L_2 + R_1 L_1 + R_2 L_1)s + R_1 R_2} V_S(s) \text{ A-s}$$

EXAMPLE 10–18

(a) Formulate mesh-current equations for the circuit in Figure 10–42(a).
(b) Solve for the zero-input component of $i_A(t)$ for $i_L(0) = 0$, $v_C(0) = 10\,\text{V}$, $L = 250\,\text{mH}, C = 1\,\mu\text{F}$, and $R = 1\,\text{k}\Omega$.

FIGURE 10–42

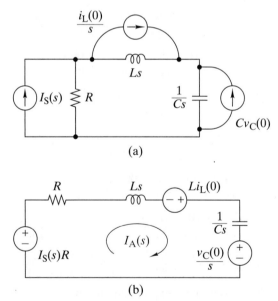

(a)

(b)

SOLUTION:
(a) Figure 10–42(a) is the s-domain circuit used in Example 10–13 to develop node equations. In this circuit each current source is connected in parallel with an impedance. Source transformations convert these current sources into the equivalent voltage sources shown in Figure 10–42(b). The circuit in Figure 10–42(b) is a

series RLC circuit of the type treated in Chapter 7. By inspection, the KVL equation for the single mesh in this circuit is

$$\left[R + Ls + \frac{1}{Cs}\right]I_A(s) = RI_S(s) + Li_L(0) - \frac{v_C(0)}{s}$$

The circuit determinant is the factor $R + Ls + 1/Cs = (LCs^2 + RCs + 1)/Cs$. The zeros of the circuit determinant are roots of the quadratic equation $LCs^2 + RCs + 1 = 0$, which we recognize as the characteristic equation of a series RLC circuit. In our study of RLC circuits we called these roots natural frequencies. Thus, the natural poles of the circuit are its natural frequencies.

(b) Solving the mesh equation for the zero-input component yields

$$I_{Azi}(s) = \frac{LCsi_L(0) - Cv_C(0)}{LCs^2 + RCs + 1} \quad \text{A-s}$$

Inserting the given numerical values produces

$$I_{Azi}(s) = -\frac{10^{-5}}{0.25 \times 10^{-6}s^2 + 10^{-3}s + 1} = -\frac{40}{s^2 + 4 \times 10^3 s + 4 \times 10^6}$$
$$= -\frac{40}{(s + 2000)^2} \quad \text{A-s}$$

The zero-state response has two natural poles, both located at $s = -2000$. The inverse transform of $I_{Azi}(s)$ is a damped ramp waveform:

$$i_{Azi}(t) = -[40te^{-2000t}]u(t) \text{ A}$$

The damped ramp response indicates a critically damped second-order circuit. The minus sign means the direction of the actual current is opposite to the reference mark assigned to $I_A(s)$ in Figure 10–42. This sign makes sense physically since the capacitor initial condition source in Figure 10–42(b) tends to drive current in a direction opposite to the assigned reference mark. ∎

Exercise 10–21

For the circuit in Figure 10–42(b), let $I_S(s) = 0.01/(s + 500)$ A-s, $L = 250$ mH, $C = 1$ μF, and $R = 1$ kΩ. Use partial fraction expansion or MATLAB to solve for the zero-state component of $i_A(t)$.

Answers:

iAZSt =
2/(225*exp(2000*t))-2/(225*exp(500*t))+(160*t)/(3*exp(2000*t))

or

$$i_{Azs}(t) = \left[\left(\frac{160t}{3} + \frac{2}{225}\right)e^{-2000t} - \frac{2}{225}e^{-500t}\right]u(t) \text{ A}$$

EXAMPLE 10–19

Formulate mesh current equations for the circuit in Figure 10–43 and solve for $I_B(s)$ in symbolic from. Locate the natural poles of the circuit for $R = 1$ kΩ, $C = 4$ μF, and $L = 1$ H.

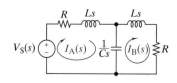

FIGURE 10–43

SOLUTION:
By inspection the two mesh-current equations are

$$\text{Mesh A:} \quad \left(R + Ls + \frac{1}{Cs}\right)I_A(s) - \frac{1}{Cs}I_B(s) = V_S(s)$$

$$\text{Mesh B:} \quad -\frac{1}{Cs}I_A(s) + \left(R + Ls + \frac{1}{Cs}\right)I_B(s) = 0$$

The circuit determinant is

$$\Delta(s) = \left(R + Ls + \frac{1}{Cs}\right)^2 - \left(\frac{1}{Cs}\right)^2$$

$$= R^2 + 2RLs + 2\frac{R}{Cs} + L^2s^2 + 2\frac{L}{C}$$

$$= \frac{(Ls + R)(LCs^2 + RCs + 2)}{Cs}$$

and the required mesh current is

$$I_B(s) = \frac{\Delta_B(s)}{\Delta(s)} = \frac{\begin{vmatrix} R + Ls + \dfrac{1}{Cs} & V_S(s) \\ -\dfrac{1}{Cs} & 0 \end{vmatrix}}{\Delta(s)}$$

$$= \frac{V_S(s)}{(Ls + R)(LCs^2 + RCs + 2)}$$

The natural poles are roots of the denominator of $I_B(s)$, namely

$$\left(s + \frac{R}{L}\right)\left(s^2 + \frac{R}{L}s + \frac{2}{LC}\right) = 0$$

For $R = 1\,\text{k}\Omega, C = 4\,\mu\text{F}$, and $L = 1\,\text{H}$, this expression factors as

$$(s + 1000)(s^2 + 1000s + 5 \times 10^5) = (s + 1000)[(s + 500)^2 + 500^2] = 0$$

so the natural poles are located at $s = -1000\,\text{rad/s}$ and $s = -500 \pm j500\,\text{rad/s}$. ∎

Exercise 10–22

(a) Formulate mesh-current equations for the circuit in Figure 10–44. Assume that the initial conditions are zero.
(b) Find the circuit determinant.
(c) Solve for the zero-state component of $I_B(s)$.

Answers:

$$\text{(a)} \quad \begin{cases} (R_1 + Ls)I_A(s) - R_1I_B(s) = V_S(s) \\ -R_1I_A(s) + (R_1 + R_2 + 1/Cs)I_B(s) = 0 \end{cases}$$

$$\text{(b)} \quad \Delta(s) = \frac{(R_1 + R_2)LCs^2 + (R_1R_2C + L)s + R_1}{Cs}$$

$$\text{(c)} \quad I_{Bzs}(s) = \frac{R_1CsV_S(s)}{(R_1 + R_2)LCs^2 + (R_1R_2C + L)s + R_1}$$

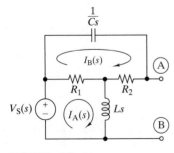

FIGURE 10–44

Exercise 10–23

Formulate mesh-current equations for the circuit in Figure 10–44 when a resistor R_3 is connected between nodes A and B. Assume that the initial conditions are zero.

Answer:

$$(R_1 + Ls)I_A(s) - R_1I_B(s) - LsI_C(s) = V_S(s)$$

$$-R_1I_A(s) + \left(R_1 + R_2 + \frac{1}{Cs}\right)I_B(s) - R_2I_C(s) = 0$$

$$-LsI_A(s) - R_2I_B(s) + (R_2 + R_3 + Ls)I_C(s) = 0$$

10–6 SUMMARY OF s-DOMAIN CIRCUIT ANALYSIS

At this point we review our progress and put s-domain circuit analysis into perspective. We have shown that linear circuits can be transformed from the time domain into the s domain. In this domain KCL and KVL apply to transforms and the passive element i–v characteristics become impedances with series or parallel initial condition sources. In relatively simple circuits we can use basic analysis methods, such as reduction, superposition, and voltage/current division. For more complicated circuits we use systematic procedures, such as the node-voltage or mesh-current methods, to solve for the circuit response.

In theory, we can perform s-domain analysis on circuits of any complexity. In practice, the algebraic burden of hand computations gets out of hand for circuits with more than three nodes or meshes. Of what practical use is an analysis method that becomes impractical at such a modest level of circuit complexity? Why not just appeal to computer-aided analysis tools in the first place?

Unquestionably, large-scale circuits are best handled by computer-aided analysis. Computer-aided analysis is probably the right approach even for small-scale circuits when numerical values for all circuit parameters are known and the desired end product is a plot or numerical listing of the response waveform. Simply put, s-domain circuit analysis is not a particularly efficient algorithm for generating numerical response data.

The purpose of s-domain circuit analysis is to gain insight into circuit behavior, not to grind out particular response waveforms. In this regard s-domain circuit analysis complements programs like MATLAB and OrCAD. It offers a way of characterizing circuits in very general terms. It provides guidelines that allow us to use computer-aided analysis tools intelligently. Some of the useful general principles derived in this chapter are summarized below.

The response transform $Y(s)$[1] is a rational function whose partial-fraction expansion leads directly to a response waveform of the form

$$y(t) = \sum_{j=1}^{\text{number of poles}} k_j e^{p_j t}$$

where k_j is the residue of the pole in $Y(s)$ located at $s = -p_j$. The location of the poles tells us a great deal about the form of the response. The pair of conjugate complex poles in Example 10–14 produced a damped sine waveform, the two distinct real poles in Example 10–17 produced exponential waveforms, and the double pole in Example 10–18 led to a damped ramp waveform. The general principle illustrated is as follows:

The poles of Y(s) are either real or complex conjugates. Simple real poles lead to exponentials, double real poles lead to a damped ramp, and complex conjugate poles lead to damped sinusoids.

---

[1] In this context $Y(s)$ is not an admittance but the Laplace transform of the circuit output $y(t)$

FIGURE 10–45 *Form of the natural response corresponding to different pole locations.*

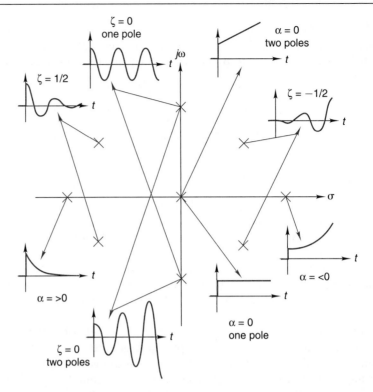

As the discussion in Example 10–14 noted, the poles in $Y(s)$ are introduced either by the circuit itself (natural poles) or by the input driving force (forced poles).

> *The natural poles are zeros of the circuit determinant and lead to the natural response. The forced poles are poles of the input $X(s)$ and lead to the forced response.*

Stability is a key concept in circuit analysis and design. For our present purposes we say that a linear circuit is stable if its natural response decays to zero as $t \to \infty$. Figure 10–45 shows the waveforms of the natural modes corresponding to different pole locations in the s plane. Poles in the left half plane give rise to waveforms that decay to zero as time increases, while those in the right half plane increase without bound. As a result, we can say that

> *A circuit is stable if all of its natural poles are located in the left half of the s plane.*

Stability requires *all* of the natural poles to be in the left half plane (LHP). The circuit is *unstable* if even one natural pole falls in the right half plane (RHP).

In Figure 10–45 the $j\omega$-axis is the boundary between the LHP (stable circuits) and RHP (unstable circuits). Poles exactly on this boundary require further discussion. As Figure 10–45 shows, *simple* j-axis poles at $s = 0$ and $s = \pm j\beta$ lead to natural modes like $u(t)$ and $\cos(\beta t)$ that neither decay to zero nor increase without bound. The figure also shows that *double* poles on the j-axis lead to natural modes like $tu(t)$ and $t\cos(\beta t)$ that increase without bound. Circuits with *simple* poles on the j-axis are sometimes said to be **marginally stable**,[2] while those with *multiple* poles on the j-axis are clearly *unstable*.

---

[2]They could just as logically be called marginally unstable. The stability status of simple j-axis poles depends on the application. For example, electronic circuits with simple poles firmly rooted on the j-axis are called stable oscillators. On the other hand, j-axis poles in audio amplifiers cause "ringing," a bad word among audiophiles.

Circuit stability is determined by natural poles, not forced poles. For example, suppose an input $x(t) = e^{10t}$ produces an output transform

$$Y(s) = \frac{12s}{\underbrace{(s+2)}_{\text{LHP}}\underbrace{(s-10)}_{\text{RHP}}}$$

This transform has a left half plane (LHP) pole and a right half plane (RHP) pole. The corresponding waveform

$$y(t) = \underbrace{2e^{-2t}}_{\substack{\text{natural} \\ \text{bounded}}} + \underbrace{10e^{10t}}_{\substack{\text{forced} \\ \text{unbounded}}} \qquad t > 0$$

has an unbounded term due to the RHP pole. Even with an unbounded response the circuit is still said to be stable because the natural pole at $s = -2$ is in the LHP and leads to a natural response that decays to zero. The unbounded part of the response waveform comes from the forced RHP pole caused by the unbounded input.

Since the natural response plays a key role, we would like to predict the number of natural poles by simply examining the circuit. Figure 10–46 summarizes examples from this chapter and leads to the following observations. Circuits with only one energy storage element (inductor or capacitor) have only one pole, circuits with two independent elements have two poles, and Example 10–19 has three poles to go with its three

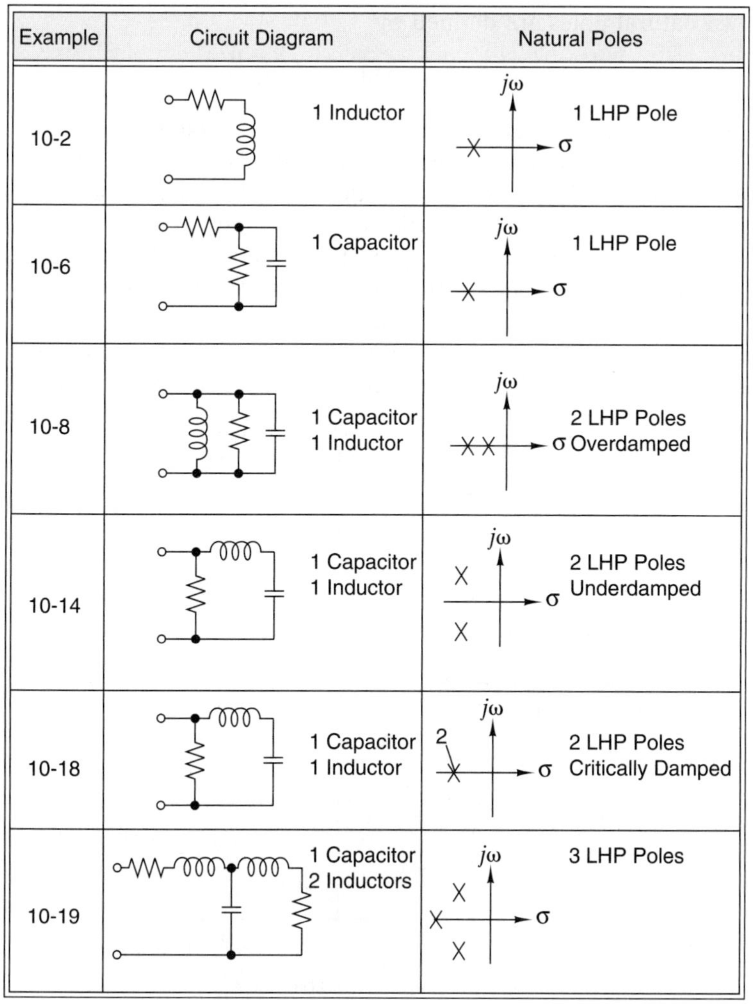

Example	Circuit Diagram	Natural Poles
10-2	1 Inductor	1 LHP Pole
10-6	1 Capacitor	1 LHP Pole
10-8	1 Capacitor 1 Inductor	2 LHP Poles Overdamped
10-14	1 Capacitor 1 Inductor	2 LHP Poles Underdamped
10-18	1 Capacitor 1 Inductor	2 LHP Poles Critically Damped
10-19	1 Capacitor 2 Inductors	3 LHP Poles

FIGURE 10–46 *Summary of Chapter 10 examples.*

energy storage elements. The conclusion appears to be that the number of natural poles is equal to the number of energy storage elements. While this rule is a useful guideline, there are exceptions (capacitors in parallel, for example). The best we can say is that

> *The number of natural poles does not exceed the number of energy storage elements.*

Another implication in Figure 10–46 comes from two additional observations. First, all of the natural poles are in the LHP; hence, all of the circuits are stable. Second, all of the circuits contain only passive resistors, capacitors, and/or inductors; that is, there are no active elements. These observations imply that

> *Circuits consisting of passive resistors, capacitors, and inductors are inherently stable.*

This conclusion makes sense physically since the passive elements can only store or dissipate energy. They cannot produce the energy needed to sustain an unbounded response.

What about circuits with active elements like dependent sources or OP AMPs? Such circuits can be unstable, as we can see by reviewing Example 10–15. In that example we analyzed the active *RC* circuit in Figure 10–36 and found the circuit determinant to be

$$\frac{\Delta(s)}{C^2} = s^2 + \left(\frac{3-\mu}{RC}\right)s + \left(\frac{1}{RC}\right)^2$$

and the two natural poles are defined by

$$\omega_0 = \frac{1}{RC} \quad \text{and} \quad \zeta = \frac{3-\mu}{2}$$

where μ is the gain of the dependent source, the active element in the circuit. For $\mu = 0$ (active element turned off), the damping ratio is $\zeta = 1.5 > 1$ and the circuit is overdamped. For $\mu = 1$, the damping ratio is $\zeta = 1$ and the circuit is critically damped. For $3 > \mu > 1$, the damping ratio is $1 > \zeta > 0$ and the circuit is underdamped. For $\mu = 3$, the damping ratio is $\zeta = 0$ and the circuit is undamped and therefore oscillates. Finally, for $\mu > 3$, the damping ratio is $\zeta < 0$ and the circuit is said to have negative damping, which is an unstable condition.

Figure 10–47 shows the loci of the natural poles as the gain increases. For $\mu > 3$, the poles move into the RHP and the circuit becomes unstable. This makes sense

FIGURE 10–47 *Pole loci on a circle of radius ω_0.*

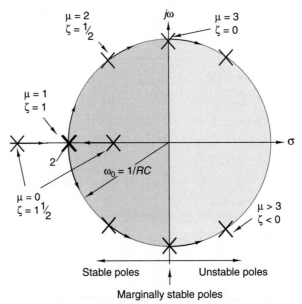

physically. When the gain is high enough, the active element can produce the energy needed to sustain an unbounded output. Since instability is almost always undesirable, we usually state the conclusion the other way around. That is, this active RC circuit is stable provided the gain $\mu < 3$. It is common for active circuits to be stable when circuit parameters are in one range and unstable when they are outside this range. For double-pole circuits the stable range can be found by relating the damping ratio to circuit parameters. For single-pole circuits the stable range ensures that the pole lies on the negative real axis.

Exercise 10–24

The circuit determinants of three circuits to be studied in subsequent chapters are given below. Determine the nature of the poles of each circuit and what conditions, if any, could cause the circuit to become unstable.

(a) $R^2 C^2 s^2 + 2RCs + 1$
(b) $R^2 C^2 s^2 + RCs + 1$
(c) $R^2 C^2 s^2 + (3 - \mu)RCs + 1$

Answers:

(a) Poles are real, negative, and equal; circuit is always stable.
(b) Poles are complex with negative real parts; circuit is always stable.
(c) Poles can vary depending on μ. If $\mu < 1$, the poles are real, distinct, and both negative, so the circuit is always stable. If $\mu = 1$, the poles are real, negative, and equal, so the circuit is always stable. If $1 < \mu < 3$, the poles are complex with negative real parts, so the circuit is always stable. If $\mu = 3$, the poles are pure imaginary and the circuit is marginally stable. If $\mu > 3$, the poles have positive real parts and the circuit is unstable.

SUMMARY

- Kirchhoff's laws apply to voltage and current waveforms in the time domain and to the corresponding transforms in the s domain.

- The s-domain models for the passive elements include initial condition sources and the element impedance or admittance. Impedance is the proportionality factor in the expression $V(s) = Z(s)I(s)$ relating the voltage and current transforms. Admittance is the reciprocal of impedance.

- The impedances of the three passive elements are $Z_R(s) = R$, $Z_L(s) = Ls$, and $Z_C(s) = 1/Cs$.

- The s-domain circuit analysis techniques closely parallel the analysis methods developed for resistance circuits. Basic analysis techniques, such as circuit reduction, Thévenin's and Norton's theorems, the unit output method, or superposition, can be used in simple circuits. More complicated networks require a general approach, such as the node-voltage or mesh-current methods.

- Response transforms are rational functions whose poles are zeros of the circuit determinant or poles of the transform of the input driving forces. Poles introduced by the circuit determinant are called natural poles and lead to the natural response. Poles introduced by the input are called forced poles and lead to the forced response.

- In linear circuits, response transforms and waveforms can be separated into zero-state and zero-input components. The zero-state component is found by setting the initial capacitor voltages and inductor currents to zero. The zero-input component is found by setting all input driving forces to zero.

- The main purpose of s-domain circuit analysis is to gain insight into circuit performance without necessarily finding the time-domain response. The natural poles reveal the form, stability, and observability of the circuit's response. The number of natural poles is never greater than the number of energy storage elements in the circuit.

- A circuit is stable if all of its natural poles are in the left half of the s plane. Passive circuits are inherently stable so their natural poles are in the left half plane. Active circuits can be stable when circuit parameters are in one range and unstable for parameters outside this range.

PROBLEMS

OBJECTIVE 10–1 EQUIVALENT IMPEDANCE (SECTS. 10–1, 10–2)

Given a linear circuit, use series and parallel equivalence to find the equivalent impedance at specified terminal pairs. Select element values to obtain specified pole locations.
See Examples 10–2, 10–4, 10–5, 10–6, and Exercises 10–4, 10–5, 10–6, 10–7.

10–1 ⓓ For a series RC circuit find $Z_{EQ}(s)$ and then select R and C so that there is a pole at zero and a zero at -1 krad/s.

10–2 ⓓ For a parallel RC circuit find $Z_{EQ}(s)$ and then select R and C so that there is a pole at -10 krad/s.

10–3 ⓓ For the circuit of Figure P10–3:
(a) Find and express $Z_{EQ}(s)$ as a rational function and locate its poles and zeros.
(b) Select values of R and C to locate a pole at -640 rad/s. Where is the resulting zero?

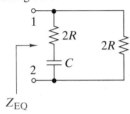

FIGURE P10–3

10–4 ⓓ For the circuit of Figure P10–4:
(a) Find and express $Z_{EQ}(s)$ as a rational function and locate its poles and zeros.
(b) Select values of R and C to locate a zero at -33 krad/s.

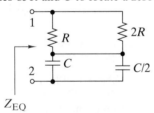

FIGURE P10–4

10–5 ⓓ For the circuit of Figure P10–5:
(a) Find and express $Z_{EQ}(s)$ as a rational function and locate its poles and zeros.
(b) If $R = 2$ kΩ and $C = 0.1$ μF, select a value of L to locate zeros at $\pm j5000$ rad/s.
(c) Where are the poles located once you have selected the inductor in part (b)?

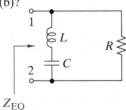

FIGURE P10–5

10–6 ⓓ For the circuit of Figure P10–6:
(a) Find and express $Z_{EQ}(s)$ as a rational function and locate its poles and zeros.
(b) Select values of R and L to locate a pole at -150 rad/s. Where is the resulting zero?

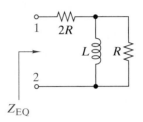

FIGURE P10–6

10–7 ⓓ For the circuit of Figure P10–7:
(a) Find and express $Z_{EQ}(s)$ as a rational function and locate its poles and zeros.
(b) Select values of R and L to locate a pole at -2 krad/s. Where are the resulting zeros?

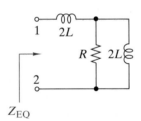

FIGURE P10–7

10–8 ⓓ For the circuit of Figure P10–8:
(a) Find and express $Z_{EQ}(s)$ as a rational function and locate its poles and zeros.
(b) If $R = 10$ kΩ, select values of L and C to locate poles at $\pm j200$ krad/s. Where are the resulting zeros?

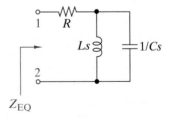

FIGURE P10–8

10–9 For the circuit of Figure P10–9:
(a) If $R = 1$ kΩ, $L = 2$ H, and $C = 0.5$ μF locate the poles and zeros of $Z_{EQ}(s)$?
(b) If we were to increase the inductance to 5 H, how would the poles and zeros change?

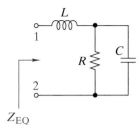

FIGURE P10–9

10–10 Find $Z_{EQ1}(s)$ and $Z_{EQ2}(s)$ for the bridge-T circuit in Figure P10–10. Express each impedance as a rational function and locate its poles and zeros.

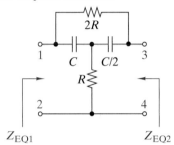

FIGURE P10–10

10–11 For the circuit of Figure P10–11:
(a) Find and express $Z_{EQ}(s)$ as a rational function and locate its poles and zeros.
(b) If its poles were located at -20 krad/s and zero, where would the poles move to if the value of R was reduced to 25% of its current value?

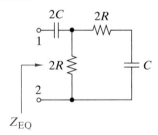

FIGURE P10–11

10–12 ⓓ For the two-port circuit of Figure P10–12:
(a) Find $Z_{EQ1}(s)$ and Z_{EQ2}, and express each impedance as a rational function and locate its poles and zeros.
(b) Select values of R and L to place a pole at -500 Hz.

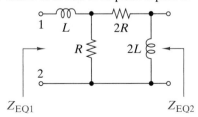

FIGURE P10–12

10–13 ⓓ Find the equivalent impedance between terminals 1 and 2 in Figure P10–13. Select values of R and L so that

$Z_{EQ}(s)$ has a pole at $s = -3000$ rad/s. Locate the zeros of $Z_{EQ}(s)$ for your choice of R and L.

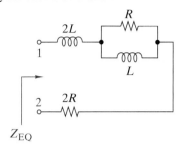

FIGURE P10–13

OBJECTIVE 10–2 BASIC CIRCUIT ANALYSIS TECHNIQUES (SECTS. 10–2, 10–3)

Given a linear circuit,
(a) Determine the initial conditions (if not given) and transform the circuit into the s domain.
(b) Solve for zero-state and zero-input responses using circuit reduction, the unit output method, Thévenin or Norton equivalent circuits, or superposition.
(c) Identify the forced and natural poles in the responses, or select circuit parameters to place the natural poles at specified locations.

See Examples 10–1, 10–2, 10–3, 10–4, 10–5, 10–7, 10–8, 10–9, 10–10, 10–11, 10–12, and Exercises 10–1, 10–2, 10–3, 10–4, 10–6, 10–7, 10–8, 10–9, 10–10, 10–11, 10–12, 10–13, 10–14.

10–14 For the circuit of Figure P10–14:
(a) Use voltage division to find $V_O(s)$.
(b) Use the look-back method to find $Z_T(s)$.

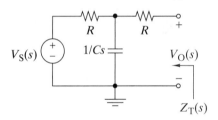

FIGURE P10–14

10–15 For the circuit of Figure P10–15:
(a) Use current division to find $I_2(s)$.
(b) Use the look-back method to find $Z_N(s)$.

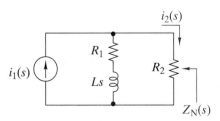

FIGURE P10–15

10–16 🅓 The circuit of Figure P10–16 needs to produce an output transform of

$$V_O(s) = \frac{V_A(RCs + 1)}{s(2RCs + 1)}$$

when the input is V_A/s. Design an appropriate circuit.

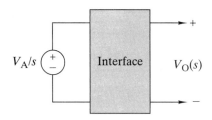

FIGURE P10–16

10–17 🅓 Find the Thévenin equivalent for the circuit in Figure P10–17. Then select values for R and L so that the Thévenin voltage has a pole at -12 krad/s.

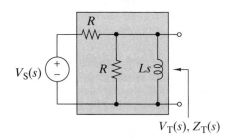

FIGURE P10–17

10–18 If the input to the RLC circuit of Figure P10–18 is $v_S(t) = u(t)$:
(a) Find the output voltage transform across each element.
(b) Compare the three outputs with regards to their respective poles and zeros.
(c) Use the initial and final value theorems to determine the value of the voltage across each element at $t = 0$ and $t = \infty$. What conclusions can one draw regarding the results?

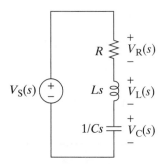

FIGURE P10–18

10–19 The switch in Figure P10–19 has been in position A for a long time and is moved to position B at $t = 0$. Transform the circuit into the s domain and solve for $I_L(s)$, $i_L(t)$, $V_O(s)$, and $v_O(t)$ in symbolic form.

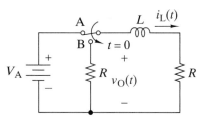

FIGURE P10–19

10–20 The switch in Figure P10–19 has been in position B for a long time and is moved to position A at $t = 0$. Transform the circuit into the s domain and solve for $I_L(s)$ and $i_L(t)$ in symbolic form.

10–21 The switch in Figure P10–21 has been in position A for a long time and is moved to position B at $t = 0$. Transform the circuit into the s domain and solve for $I_C(s)$, $i_C(t)$ $V_O(s)$, and $v_O(t)$ in symbolic form.

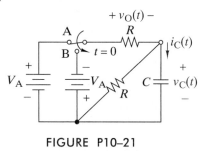

FIGURE P10–21

10–22 The switch in Figure P10–21 has been in position B for a long time and is moved to position A at $t = 0$. Transform the circuit into the s domain and solve for $V_C(s)$, $v_C(t)$, $V_L(s)$, and $v_O(t)$ in symbolic form.

10–23 Transform the circuit in Figure P10–23 into the s domain and find: $I_L(s)$, $i_L(t)$ $V_L(s)$, and $v_L(t)$ when $v_1(t) = V_A u(t)$ and $i_L(0) = I_A$.

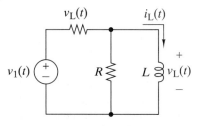

FIGURE P10–23

10–24 Transform the circuit in Figure P10–23 into the s domain and find $I_L(s)$ and $i_L(t)$ when $v_1(t) = V_A e^{-1000t} u(t)$, $R = 200\ \Omega$, $L = 100$ mH, and $i_L(0) = 0$ A.

🖱 **10–25** The switch in Figure P10–25 has been in position A for a long time and is moved to position B at $t = 0$.
(a) Transform the circuit into the s domain and solve for $I_L(s)$ in symbolic form.
(b) Repeat part (a) using MATLAB.
(c) Find $i_L(t)$ for $R_1 = R_2 = 500\ \Omega$, $R_3 = 1$ kΩ, $L = 500$ mH, $C = 0.2\ \mu$F, and $V_A = 15$ V.

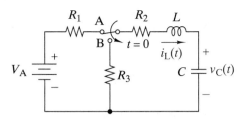

FIGURE P10–25

10–26 The switch in Figure P10–25 has been in position B for a long time and is moved to position A at $t = 0$.
(a) Transform the circuit into the s domain and solve for $V_c(s)$ in symbolic form.
(b) Find $v_C(t)$ for $R_1 = 50\,\Omega$, $R_2 = 250\,\Omega$, $R_3 = 1\,k\Omega$, $L = 500\,mH$, $C = 1\,\mu F$, and $V_A = 15\,V$. In addition, the inductor is not ideal, but has a parasitic resistance (in series) of $100\,\Omega$.
(c) Repeat part (b) using OrCAD.

10–27 The circuit in Figure P10–27 is in the zero state. The s-domain relationship between the input $I_1(s)$ and the output $I_R(s)$ is usually given as a ratio called a *network function*. Find $I_R(s)/I_1(s)$. Identify the poles and the zeros.

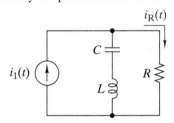

FIGURE P10–27

10–28 The circuit in Figure P10–28 is in the zero state. Find the s-domain relationship between the input $I_1(s)$ and the output $V_O(s)$. Identify the poles and the zeros.

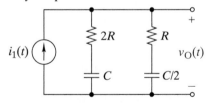

FIGURE P10–28

10–29 The initial conditions for the circuit in Figure P10–29 are $v_C(0) = 0$ and $i_L(0) = I_0$. Transform the circuit into the s domain and use superposition and voltage division to find the zero-state and zero-input components of $V_C(s)$.

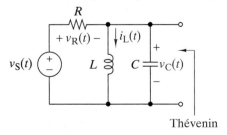

FIGURE P10–29

10–30 The initial conditions for the circuit in Figure P10–29 are $v_C(0) = 0$ and $i_L(0) = I_0$. Transform the circuit into the s domain and use superposition and voltage division to find the zero-state and zero-input components of $V_R(s)$.

10–31 The circuit in Figure P10–29 is in the zero state. Transform the circuit into the s domain and find the Thévenin equivalent circuit at the capacitor's terminals.

10–32 There is no energy stored in the capacitor in Figure P10–32 at $t = 0$. Transform the circuit into the s domain and use current division to find $v_O(t)$ when the input is $i_S(t) = 5\,e^{-500t}u(t)$ mA. Identify the forced and natural poles in $V_O(s)$.

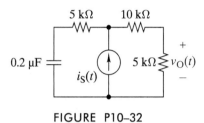

FIGURE P10–32

10–33 Repeat Problem 10–32 when $i_S(t) = 2.5\cos 2000\,t\,u(t)$ mA.

10–34 The circuit in Figure P10–34 is in the zero state. Use a Thévenin equivalent to find the s-domain relationship between the input $I_S(s)$ and the interface current $I(s)$.

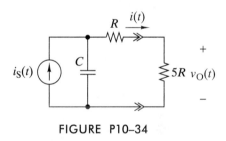

FIGURE P10–34

10–35 For the circuit of Figure 10–34:
(a) Find the Thévenin equivalent circuit that the $5R$ load resistor sees in when $v_C(0) = V_O$ V.
(b) Then find the voltage delivered to the load $v_O(t)$ if $v_C(0) = 10$ V, $i_S(t) = 10\,u(t)$ mA, $R = 1\,k\Omega$, and $C = 2\,\mu F$.
(c) Identify the forced, natural, zero state, and zero input components of $v_O(t)$.

10–36 The circuit in Figure P10–36 is in the zero state. Find the Thévenin equivalent to the left of the interface.

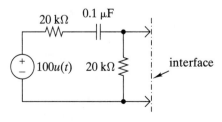

FIGURE P10–36

10–37 Ⓓ The Thévenin equivalent shown in Figure P10–37 needs to deliver

$$V_O(s) = \frac{50000}{(s+1500)(s+10)} \text{ V-s}$$

to a 1-kΩ load. Design an interface to allow that to occur.

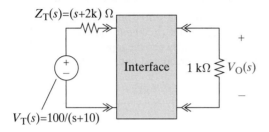

FIGURE P10–37

10–38 There is no initial energy stored in the circuit in Figure P10–38. Transform the circuit into the s domain and use superposition to find $V(s)$. Identify the forced and natural poles in $V(s)$.

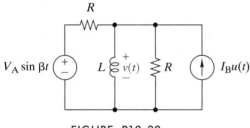

FIGURE P10–38

10–39 The equivalent impedance between a pair of terminals is

$$Z_{EQ}(s) = 2000 \left[\frac{s+3000}{s+2000} \right] \Omega$$

A voltage $v(t) = 10u(t)$ is applied across the terminals. Find the resulting current response $i(t)$.

10–40 Ⓓ There is no initial energy stored in the circuit in Figure P10–40. Use circuit reduction to find the output network function $V_2(s)/V_1(s)$. Then select values of R and C so that the poles of the network function are approximately −2618 and −382 rad/s.

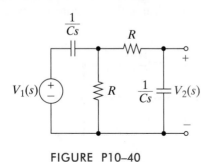

FIGURE P10–40

OBJECTIVE 10–3 GENERAL CIRCUIT ANALYSIS (SECTS. 10–4, 10–5, 10–6)

Given a linear circuit:
(a) Determine the initial conditions (if not given) and transform the circuit into the s domain.
(b) Solve for zero-state and zero-input response transforms and waveforms using node-voltage or mesh-current methods.
(c) Identify the forced and natural poles in the responses or select circuit parameters to place the natural poles at specified locations.

See Examples 10–13, 10–14, 10–15, 10–16, 10–17, 10–18, 10–19 and Exercises 10–15, 10–16, 10–17, 10–18, 10–19, 10–20, 10–21, 10–22, 10–23, 10–24.

10–41 There is no initial energy stored in the circuit in Figure P10–41.
(a) Transform the circuit into the s domain and formulate mesh-current equations.
(b) Show that the solution of these equations for $I_2(s)$ in symbolic form is

$$I_2(s) = I_B(s) = \frac{R_1 C s V_1(s)}{(R_1 + R_2)LCs^2 + (R_1 R_2 C + L)s + R_1}$$

(c) Identify the poles and zeros of $I_2(s)$.
(d) Find $i_2(t)$ for $v_1(t) = 10u(t)$ V, $R_1 = 1$ kΩ, $R_2 = 2$ kΩ, $L = 2$ H, and $C = 0.5$ μF.

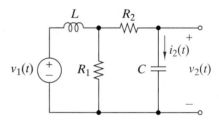

FIGURE P10–41

10–42 There is no initial energy stored in the circuit in Figure P10–41.
(a) Transform the circuit into the s domain and formulate node-voltage equations.
(b) Show that the solution of these equations for $V_2(s)$ in symbolic form is

$$V_2(s) = \frac{R_1 V_1(s)}{(R_1 + R_2)LCs^2 + (R_1 R_2 C + L)s + R_1}$$

(c) Identify the natural and forced poles of $V_2(s)$.
(d) Find $v_2(t)$ for $v_1(t) = 25u(t)$ V, $R_1 = R_2 = 500$ Ω, $L = 0.5$ H, and $C = 2$ μF.

10–43 Ⓓ There is no initial energy stored in the circuit in Figure P10–43.
(a) Transform the circuit into the s domain and formulate node-voltage equations.

(b) Solve these equations for $V_2(s)$ in symbolic form.

(c) Insert an OP AMP buffer at point A and solve for $V_2(s)$ in symbolic form. How did inserting the buffer change the denominator, and therefore, the location of the poles?

(d) Select values of R_1, R_2, C_1, and C_2 to locate a pole at 10 krad/s and a second pole at 100 krad/s. You can choose either the circuit with or the one without the OP AMP for your design.

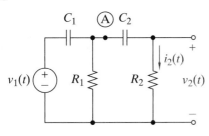

FIGURE P10–43

10–44 There is no initial energy stored in the circuit in Figure P10–43. The Thévenin equivalent circuit to the left of point A when a unit step is applied is

$$V_T(s) = \frac{1}{s + 10^3}\text{V-s, and } Z_T(s) = \frac{10^6}{s + 10^3}\,\Omega$$

Select values for R_2 and C_2 such that the output transform is

$$V_2(s) = \frac{s}{s^2 + 3000s + 10^6}\text{V-s}$$

10–45 There is no initial energy stored in the bridged-T circuit in Figure P10–45.

(a) Transform the circuit into the s domain and formulate mesh-current equations.

(b) Use the mesh-current equations to find the s-domain relationship between the input $V_1(s)$ and the output $V_2(s)$.

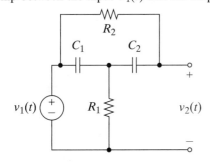

FIGURE P10–45

10–46 There is no initial energy stored in the bridged-T circuit in Figure P10–45.

(a) Transform the circuit into the s domain and formulate node-voltage equations.

(b) Use the node-voltage equations to find the s-domain relationship between the input $V_1(s)$ and the output $V_2(s)$.

10–47 Find the transform of the Thévenin equivalent circuit looking into the $V_2(t)$ terminals for the circuit of P10–45.

10–48 There is no initial energy stored in the circuit in Figure P10–48.

(a) Find the zero-state mesh currents $i_A(t)$ and $i_B(t)$ when $v_1(t) = 50\,e^{-1000t}u(t)$ V.

(b) Validate your answers using OrCAD.

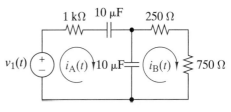

FIGURE P10–48

10–49 There is no external input in the circuit in Figure P10–49.

(a) Find the zero-input node voltages $v_A(t)$ and $V_B(t)$, and the voltage across the capacitor $v_C(t)$ when $v_C(0) = -5$ V and $i_L(0) = 0$ A.

(b) Use MATLAB to plot your results in (a).

(c) Use OrCAD to validate your results in (a).

(d) Compare the MATLAB and OrCAD plots. Are they the same?

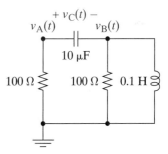

FIGURE P10–49

10–50 The two-OP AMP circuit in Figure P10–50 is a band-pass filter.

(a) Your task is to design such a filter so that the low-frequency cutoff is 1000 rad/s and the high-frequency cutoff is 100,000 rad/s. (*Hint:* See Example 10–16 and Exercise 10–18.)

(b) Show that your design is correct using OrCAD.

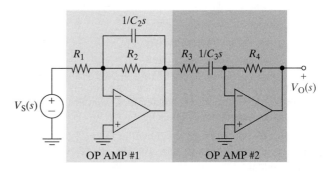

FIGURE P10–50

10–51 The circuit in Figure P10–51 is in the zero state. Use node-voltage equations to find the circuit determinant.

Select values of R, C, and μ so that the circuit has $\omega_0 = 5$ krad/s and $\zeta = 0.707$.

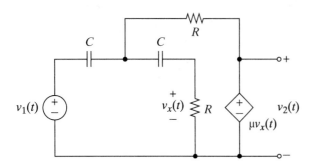

FIGURE P10–51

10–52 ◑ The circuit in Figure P10–52 is in the zero state. Use mesh-current equations to find the circuit determinant. Select values of if, R, L, and C so that the circuit has $\omega_0 = 10$ krad/s and $\zeta = 1$.

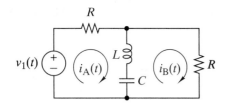

FIGURE P10–52

10–53 ◑ The OP AMP circuit in Figure P10–53 is in the zero state. Use node-voltage equations to find the circuit determinant. Select values of R, C_1, and C_2 so that the circuit has $\omega_0 = 10$ krad/s and $\zeta = 1$.

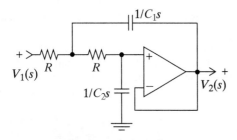

FIGURE P10–53

10–54 ◔ Compare the results of your designs of the circuits in Figures P10–52 and P10–53. Since both circuits purport to have the same response characteristics, what are the advantages and disadvantages of each?

10–55 Three node voltages are shown in Figure P10–55.
(a) Explain why only one of the node voltages is independent.
(b) Write a node voltage equation in the independent node voltage.
(c) If $V_C(s)$ is the circuit's output, find the output-input ratio or *network function*, $V_C(s)/V_S(s)$.

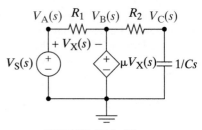

FIGURE P10–55

10–56 Three mesh currents are shown in Figure P10–56.
(a) Explain why only two of these mesh currents are independent.
(b) Write *s*-domain mesh-current equations in the two independent mesh currents.

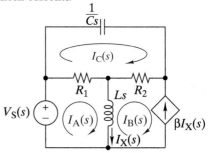

FIGURE P10–56

10–57 The switch in Figure P10–57 has been in position A for a long time and is moved to position B at $t = 0$.
(a) Write an appropriate set of node-voltage or mesh-current equations in the *s*-domain.
(b) Use MATLAB to solve for $V_C(s)$ and $v_C(t)$. Also using MATLAB, plot $v_C(t)$ and the exponential source on the same axes.
(c) Validate your results for $v_C(t)$ using OrCAD. Include the exponential source (See web Appendix C) in your Probe plot.

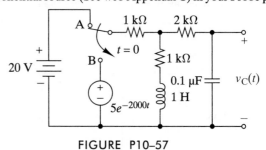

FIGURE P10–57

10–58 There is no energy stored in the circuit in Figure P10–58 at $t = 0$. Transform the circuit into the *s* domain and solve for $V_O(s)$ and $v_O(t)$

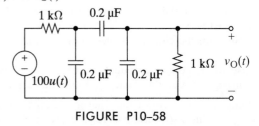

FIGURE P10–58

10–59 The switch in Figure P10–59 has been open for a long time and is closed at $t = 0$. Transform the circuit into the s domain and solve for $I_O(s)$ and $i_O(t)$

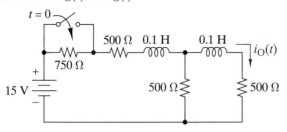

FIGURE P10–59

10–60 **D** There is no initial energy stored in the circuit in Figure P10–60.
(a) Transform the circuit into the s domain and solve for $V_O(s)$ in symbolic form.
(b) If $R_1 = R_2 = 500\ \Omega$, select values of L and C to produce $\zeta = 0.707$ and $\omega_0 = 707$ rad/s. (*Hint:* Try values of L first, i.e., 100 mH, 200 mH, etc.)

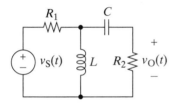

FIGURE P10–60

10–61 **D** With the circuit in the zero state, the input to the integrator shown in Figure P10–61 is $v_1(t) = \cos 1000t$ V. The desired output is $v_2(t) = -\sin 1000t$ V. Use Laplace to select values of R and C to produce the desired output. If the capacitor had 10 V across it at $t = 0$, how would that affect the output?

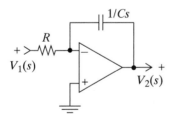

FIGURE P10–61

10–62 Show that the circuit in Figure P10–62 has natural poles at $s = -4/RC$ and $s = -2/RC \pm j2/RC$ when $L = R^2C/4$.

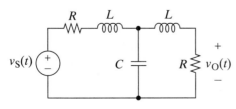

FIGURE P10–62

10–63 Find the range of the gain μ for which the circuit's output $V_X(s)$ in Figure P10–63 is stable (i.e., all poles are in the left hand side of the s plane.)

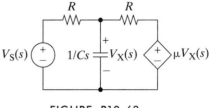

FIGURE P10–63

10–64 The circuit in Figure P10–64 is shown in the t domain with initial values for the energy storage devices.
(a) Transform the circuit into the s domain and write a set of node-voltage equations.
(b) Transform the circuit into the s domain and write a set of mesh-current equations.
(c) With the circuit in the zero state, use symbolic operations in MATLAB to solve for the node voltages.

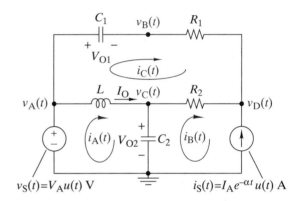

FIGURE P10–64

INTEGRATING PROBLEMS

10–65 Thévenin's Theorem from Time-domain Data **A**
A black box containing a linear circuit has an on-off switch and a pair of external terminals. When the switch is turned on, the open-circuit voltage between the external terminals is observed to be

$$v_{OC}(t) = \left(10e^{-10t} - 10e^{-50t}\right)u(t)\,\text{V}$$

The short circuit current was observed to be

$$i_{SC}(t) = \left[\left(100e^{-10t} + 200e^{-50t} + 50e^{-100t}\right)\right]u(t)\,\text{mA}$$

A 50-Ω load resistance is connected across the terminals and the switch turned on again. What is the voltage delivered to the load?

10–66 Design a Load Impedance **D**
In order to match the Thévenin impedance of a source, the load impedance in Figure P10–66 must be $Z_L(s) = \frac{s+5}{s+10}$.
(a) What impedance $Z_2(s)$ is required?

(b) How would you realize $Z_2(s)$ using only resistors, inductors, and/or capacitors? (*Hint:* Write $Z_L(s)$ as a sum of admittances, then solve for $Y_2(s)$.)

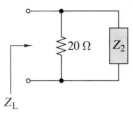

FIGURE P10–66

10–67 *RC* Circuit Analysis and Design **A D**
The *RC* circuits in Figure P10–67 represent the situation at the input to an oscilloscope. The parallel combination of R_1 and C_1 represent the probe used to connect the oscilloscope to a test point. The parallel combination of R_2 and C_2 represent the input impedance of the oscilloscope.
(a) **A** Assuming zero initial conditions, transform the circuit into the *s*-domain and find the relationship between the test point voltage $V_S(s)$ and the voltage $V_O(s)$ at the oscilloscope's input.
(b) **D** For $R_2 = 15\ \text{M}\Omega$ and $C2 = 3\ \text{pF}$, determine the values of R_1 and C_1 that make the input voltage a scaled duplicate of the test point voltage.

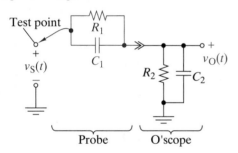

FIGURE P10–67

10–68 *s*-domain OP AMP Circuit Analysis **A**
The OP AMP circuit in Figure P10–68 is in the zero-state. Transform the circuit into the *s*-domain and use the OP AMP circuit analysis techniques developed in Section 4–4 to find the relationship between the input $V_1(s)$ and the output $V_2(s)$.

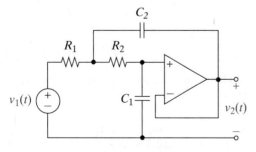

FIGURE P10–68

10–69 Pulse Conversion Circuit **E**
The purpose of the test setup in Figure P10–69 is to deliver damped sine pulses to the test load. The excitation comes from a 1-Hz square wave generator. The pulse conversion circuit must deliver damped sine waveforms with $\zeta < 0.5$ and $\omega_0 > 10\ \text{krad/s}$ to 50-Ω and 600-Ω loads. The recommended values for the pulse conversion circuit are $L = 10\ \text{mH}$ and $C = 0.1\ \mu\text{F}$. Verify that the test setup meets the specifications. (*Hint:* Compute the voltage across the load for an input signal equal to a unit step function, $u(t)$, and then again for a negative unit step function, $-u(t)$.) Note that the output of a square wave generator is the sum of a series of step functions.

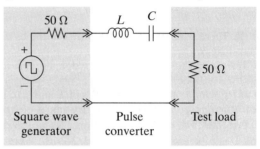

FIGURE P10–69

10–70 By-Pass Capacitor Design **D**
In transistor amplifier design, a by-pass capacitor is connected across the emitter resistor R_E to effectively short out the emitter resistor at signal frequencies. This then, improves the gain of the transistor for the desired ac signals. The circuit in Figure P10–70(a) is a common-emitter amplifier. The shaded portion is a low-frequency model of the transistor in use. In this problem the task is to design a proper by-pass capacitor so that there is a pole at $s = -300\ \text{rad/s}$. Reduce the circuit to its Thévenin equivalent as shown in Figure P10–70(b), and then select the proper capacitor. $R_S = 12\ \text{k}\Omega$, $R_\pi = 2.5\ \text{k}\Omega$, $R_E = 3.3\ \text{k}\Omega$, $R_L = 1\ \text{k}\Omega$, and $\beta = 50$.

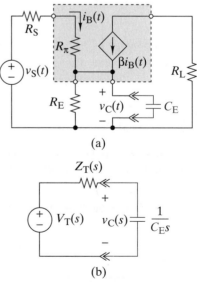

FIGURE P10–70

CHAPTER 11 NETWORK FUNCTIONS

The driving-point impedance of a network is the ratio of an impressed electromotive force at a point in a branch of the network to the resulting current at the same point.

Ronald M. Foster, 1924,
American Engineer

Some History Behind This Chapter

The network function concept emerged in the 1920s during the development of systematic methods of designing electric filters for long-distance telephone systems. The filter design effort eventually evolved into a theory known as *network synthesis*. The purpose of network synthesis is to obtain circuits that produce a desired network function. Ronald Foster along with Sidney Darlington, Hendrik Bode, Wilhelm Cauer, and Otto Brune are generally considered the founders of modern network synthesis.

Why This Chapter Is Important Today

This special chapter introduces one of the most important concepts of electrical engineering—the network function. In this chapter you will learn what network functions are and why they are important descriptors of electric circuits. Most importantly, you will learn how to design circuits that can realize a desired network function. But we don't stop there. Since design can lead to many different answers, we introduce you to the criteria used to evaluate alternative solutions.

Chapter Sections

11–1 Definition of a Network Function

11–2 Network Functions of One- and Two-Port Circuits

11–3 Network Functions and Impulse Response

11–4 Network Functions and Step Response

11–5 Network Functions and the Sinusoidal Steady-State Response

11–6 Impulse Response and Convolution

11–7 Network Function Design

Chapter Learning Objectives

11–1 Network Functions (Sects. 11–1, 11–2)

Given a linear circuit:

(a) Find specified network functions and locate their poles and zeros.

(b) Select the element values to produce specified poles and zeros.

11–2 Network Functions, Impulse Response, and Step Response (Sects. 11–3, 11–4)

(a) Given a first- or second-order linear circuit, find its impulse or step response.

(b) Given the impulse or step response of a linear circuit, find the network function.

(c) Given the impulse or step response of a linear circuit, find the response due to other inputs.

11–3 Network Functions and the Sinusoidal Steady-State Response (Sect. 11–5)

(a) Given a first- or second-order linear circuit with a specified input sinusoid, find the sinusoidal steady-state response.

(b) Given the network function, impulse response, or step response, find the sinusoidal steady-state response for a specified input sinusoid.

11–4 Network Functions and Convolution (Sect. 11–6)

(a) Given the impulse response of a linear circuit, use the convolution integral to find the response to a specified input.

(b) Use the convolution integral to derive properties of linear circuits.

11–5 Network Function Design (Sect. 11–7)

(a) Design alternative circuits that realize a given network function and meet other stated constraints.

(b) Use software to visualize and simulate alternative designs.

(c) Evaluate alternative designs using stated criteria and select the best design.

11–1   DEFINITION OF A NETWORK FUNCTION

The proportionality property of linear circuits states that the output is proportional to the input. In Chapter 10 we noted that in the s domain the proportionality factor is a rational function of s called a network function. More formally, a network function is defined as the ratio of a zero-state response transform (output) to the excitation (input) transform.

$$\text{Network function} = \frac{\text{Zero-state response transform}}{\text{Input signal transform}} \qquad (11\text{–}1)$$

Note carefully that this definition specifies zero initial conditions and implies only one input.

To study the role of network functions in determining circuit responses, we write the s-domain input-output relationship as

$$Y(s) = T(s)X(s) \qquad (11\text{–}2)$$

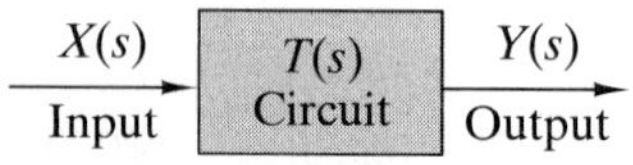

$X(s)$ Input → $T(s)$ Circuit → $Y(s)$ Output

FIGURE 11–1 *Block diagram for an s-domain input-output relationship.*

where $T(s)$ is a network function, $X(s)$ is the input signal transform, and $Y(s)$ is a zero-state response or output.[1] Figure 11–1 shows a block diagram representation of the s-domain input-output relationship in Eq. (11–2).

In an analysis problem, the circuit and input [$X(s)$ or $x(t)$] are specified. We determine $T(s)$ from the circuit, use Eq. (11–2) to find the response transform $Y(s)$, and use the inverse transformation to obtain the response waveform $y(t)$. In a design problem the circuit is unknown. The input and output are specified, or their ratio $T(s) = Y(s)/X(s)$ is given. The objective is to devise a circuit that realizes the specified input-output relationship. A linear circuit analysis problem has a unique solution, but a design problem may have one, many, or even no solutions. If more than one working solution exists, then one must ask, which design should I choose? Selecting the optimum design is one of the hallmarks of a modern engineer. Later in the chapter we will discuss how one might go about making a smart decision.

Equation (11–2) points out that the poles of the response $Y(s)$ come from either the network function $T(s)$ or the input signal $X(s)$. When there are no repeated poles, the partial-fraction expansion of the right side of Eq. (11–2) takes the form

$$Y(s) = \underbrace{\sum_{j=1}^{N} \frac{k_j}{s - p_j}}_{\substack{\text{natural} \\ \text{poles}}} + \underbrace{\sum_{\ell=1}^{M} \frac{k_\ell}{s - p_\ell}}_{\substack{\text{forced} \\ \text{poles}}} \qquad (11\text{–}3)$$

where p_j ($j = 1, 2, \ldots, N$) are the poles of $T(s)$ and $s = p_\ell$ ($\ell = 1, 2, \ldots, M$) are the poles of $X(s)$. The inverse transform of this expansion is

$$y(t) = \underbrace{\sum_{j=1}^{N} k_j e^{p_j t}}_{\substack{\text{natural} \\ \text{response}}} + \underbrace{\sum_{\ell=1}^{M} k_\ell e^{p_\ell t}}_{\substack{\text{forced} \\ \text{response}}} \qquad (11\text{–}4)$$

The poles of $T(s)$ lead to the natural response. In a stable circuit, the natural poles are all in the left half of the s plane, and all of the exponential terms in the natural response eventually decay to zero. The poles of $X(s)$ lead to the forced response. In a

---

[1]In this context $Y(s)$ is not an admittance but the transform of the output waveform $y(t)$.

stable circuit, those elements in the forced response that do not decay to zero are called the **steady-state response**.

It is important to remember that the complex frequencies in the natural response are determined by the circuit and do not depend on input. Conversely, the complex frequencies in the forced response are determined by the input and do not depend on the circuit. However, the amplitude of its part of the response depends on the residues in the partial-fraction expansion in Eq. (11–3). These residues are influenced by all of the poles and zeros, whether forced or natural. Thus, the amplitudes of the forced and natural responses depend on an interaction between the poles and zeros of $T(s)$ and $X(s)$.

The following example illustrates this discussion.

EXAMPLE 11–1

A simple series RC circuit shown in Figure 11–2 is driven by a charging exponential source. If $R = 10$ kΩ and $C = 0.01$ μF, the network function is

$$T(s) = \frac{V_2(s)}{V_1(s)} = \frac{1/RC}{s + 1/RC} = \frac{10000}{s + 10000}$$

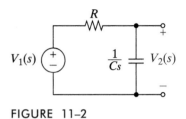

FIGURE 11–2

Find the zero-state response $v_2(t)$ when the input is $v_1(t) = 10(1 - e^{-5000t})\,u(t)$ V. Identify the natural and forced components of your answer.

SOLUTION:
The transform of the input signal is

$$V_1(s) = \frac{10}{s} - \frac{10}{s + 5000} = \frac{50000}{s(s + 5000)}$$

Using the s-domain input-output relationship in Eq. (11–2), the transform of the response is

$$V_2(s) = \frac{(50000)(10000)}{s(s + 5000)(s + 10000)}$$

Expanding by partial fractions,

$$V_2(s) = \underbrace{\frac{k_1}{s} + \frac{k_2}{s + 5000}}_{\text{forced poles}} + \underbrace{\frac{k_3}{s + 10000}}_{\text{natural pole}}$$

The two forced poles came from the input charging exponential, while the natural pole came from the RC circuit via the network function. Using the cover-up method to evaluate the residues yields

$$k_1 = \left.\frac{5 \times 10^8}{(s + 5000)(s + 10000)}\right|_{s=0} = 10$$

$$k_2 = \left.\frac{5 \times 10^8}{s(s + 10000)}\right|_{s=-5000} = -20$$

$$k_3 = \left.\frac{5 \times 10^8}{s(s + 5000)}\right|_{s=-1000} = 10$$

Collectively the residues depend on all of the poles and zeros. The inverse transform yields the zero-state response as

$$v_2(t) = \left(\underbrace{10e^{-10^4 t}}_{\text{natural response}} + \underbrace{10 - 20e^{-5 \times 10^3 t}}_{\text{forced response}} \right) u(t) \text{ V}$$

The natural pole is $s = -10{,}000$ and is located in the left-half of the s plane causing the natural response to decay to zero. The forced poles are at zero and at $s = -5000$. The pole at $s = -5000$ will decay to zero leaving a steady-state response of $10\, u(t)$. ∎

Exercise 11–1

The network function for a circuit is

$$T(s) = \frac{10s}{s + 100}$$

Find the zero-state response $v_2(t)$ when the input waveform is $v_1(t) = \cos 50t$ V.

Answer: $v_2(t) = [8e^{-100t} + 4.47 \cos(50t + 63.4°)]u(t)$ V

TEST SIGNALS

While the transfer function is a useful concept, it is clear that we cannot find the circuit response until we are given an input signal. Here, we encounter a central paradox of circuit analysis. In practice, the input signal is a carrier of information and is therefore unpredictable. We could spend a lifetime studying a circuit for various inputs and still not treat all possible signals that might be encountered in practice. What we must do is calculate the responses due to certain standard test signals. Although these test signals may never occur as real input signals, their responses tell us enough to understand the signal-processing capabilities of a circuit.

The two premier test signals used are the pulse and the sinusoid. The study of the pulse response divides into two extreme cases, short and long. When the pulse is very short compared to the circuit response time, the sudden injection of energy causes a circuit response long after the input returns to zero. The short pulse is modeled by an impulse, and the resulting *impulse response* is treated in Sect. 11–3. At the other extreme, the long pulse has a duration that greatly exceeds the circuit response time. In this case, the circuit has ample time to be driven from the zero state to a new steady-state condition. The step function is used to model the long pulse input, and the resulting *step response* is studied in Sect. 11–4.

The impulse response is of great importance because it contains all of the information needed to calculate the response due to any other input. The step response is important because it describes how a circuit response transitions from one state to another. The signal transition requirements for circuits and systems are often stated in terms of the step response using partial waveform descriptors such as rise time, fall time, propagation delay, and overshoot.

The unique properties of the sinusoid make it a useful input for characterizing the signal-processing capabilities of linear circuits and systems. When a stable linear circuit is driven by a sinusoidal input, the steady-state output is a sinusoid with the same frequency, but with a different phase angle and amplitude. The frequency-dependent relationship between the sinusoidal input and the steady-state output is called *frequency response*, a signal-processing description that is often used to specify the performance of circuits and systems. The relationship between network functions and the sinusoidal steady-state response is studied in Sect. 11–5.

11–2 NETWORK FUNCTIONS OF ONE- AND TWO-PORT CIRCUITS

The two major types of network functions are driving-point impedance and transfer functions. A **driving-point impedance** relates the voltage and current at a pair of terminals called a port. The driving-point impedance $Z(s)$ of the one-port circuit in Figure 11–3 is defined as

$$Z(s) = \frac{V(s)}{I(s)} \qquad (11\text{–}5)$$

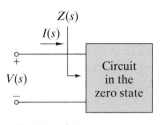

FIGURE 11–3 *A one-port circuit.*

When the one port is driven by a current source, the response is $V(s) = Z(s)I(s)$ and the natural frequencies in the response are the poles of impedance $Z(s)$. On the other hand, when the one port is driven by a voltage source, the response is $I(s) = [Z(s)]^{-1}V(s)$ and the natural frequencies in the response are the poles of $1/Z(s)$; that is, the zeros of $Z(s)$. In other words, the driving-point impedance is a network function whether upside down or right side up.

The term *driving point* means that the circuit is driven at one port and the response is observed at the same port. The element impedances defined in Sect. 10–1 are elementary examples of driving-point impedances. The equivalent impedances found by combining elements in series and parallel are also driving-point impedances. Driving-point functions are the *s*-domain generalization of the concept of the input resistance. The terms *driving-point impedance, input impedance,* and *equivalent impedance* are synonymous.

The driving-point impedance seen at a pair of terminals determines the loading effects that result when those terminals are connected to another circuit. When two circuits are connected together, these loading effects can profoundly alter the responses observed when the same two circuits operated in isolation. In an analysis situation it is important to be able to predict the response changes that occur when one circuit loads another. In design situations it is important to know when the circuits can be designed separately and then interconnected without encountering loading effects that alter their designed performance. The conditions under which loading can or cannot be ignored will be studied in this and subsequent chapters.

Transfer functions are usually of greater interest in signal-processing applications than driving-point impedances because they describe how a signal is modified by passing through a circuit. A **transfer function** relates an input and response (or output) at different ports in the circuit. Figure 11–4 shows the possible input-output configurations for a two-port circuit. Since the input and

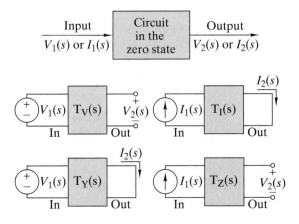

FIGURE 11–4 *Two-port circuits and transfer functions.*

output signals can be either a current or a voltage, we can define four kinds of transfer functions:

$$T_V(s) = \text{Voltage Transfer Function} = \frac{V_2(s)}{V_1(s)}$$

$$T_I(s) = \text{Current Transfer Function} = \frac{I_2(s)}{I_1(s)}$$

$$T_Y(s) = \text{Transfer Admittance} = \frac{I_2(s)}{V_1(s)}$$

$$T_Z(s) = \text{Transfer Impedance} = \frac{V_2(s)}{I_1(s)}$$

(11–6)

The functions $T_V(s)$ and $T_I(s)$ are dimensionless since the input and output signals have the same units. The function $T_Z(s)$ has units of ohms and $T_Y(s)$ has units of siemens.

The functions in Eq. (11–6) are sometimes called forward transfer functions because they relate inputs applied at port 1 to outputs occurring at port 2. There are, of course, reverse transfer functions that relate inputs at port 2 to outputs at port 1. It is important to realize that a transfer function is valid only for a specified input port and output port. For example, the voltage transfer function $T_V(s) = V_2(s)/V_1(s)$ relates the input voltage applied at port 1 in Figure 11–4 to the voltage response observed at the output port. The reverse voltage transfer function for signal transmission from output to input is *not* $1/T_V(s)$. Unlike driving-point impedance, transfer functions are not network functions when they are turned upside down.

DETERMINING NETWORK FUNCTIONS

The rest of this section illustrates analysis techniques for deriving network functions. The application of network functions in circuit analysis and design begins in the next section and continues throughout the rest of this book. But first, we illustrate ways to find the network functions of a given circuit.

The divider circuits in Figure 11–5 occur so frequently that it is worth taking time to develop their transfer functions in general terms. Using s-domain voltage division in Figure 11–5(a), we can write

$$V_2(s) = \left[\frac{Z_2(s)}{Z_1(s) + Z_2(s)} \right] V_1(s)$$

Therefore, the voltage transfer function of a voltage divider circuit is

$$T_V(s) = \frac{V_2(s)}{V_1(s)} = \frac{Z_2(s)}{Z_1(s) + Z_2(s)}$$

(11–7)

Similarly, using s-domain current division in Figure 11–5(b) yields the transfer function of a current divider circuit as

$$T_I(s) = \frac{I_2(s)}{I_1(s)} = \frac{Y_2(s)}{Y_1(s) + Y_2(s)} = \frac{Z_1(s)}{Z_1(s) + Z_2(s)}$$

(11–8)

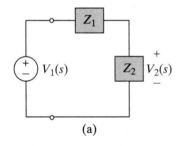

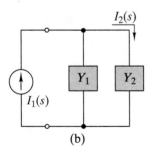

FIGURE 11–5 *Basic divider circuits. (a) Voltage divider. (b) Current divider.*

By series equivalence, the driving-point impedance at the input of the voltage divider is $Z_{EQ}(s) = Z_1(s) + Z_2(s)$. By parallel equivalence the driving-point impedance at the input of the current divider is $Z_{EQ}(s) = 1/(Y_1(s) + Y_2(s))$.

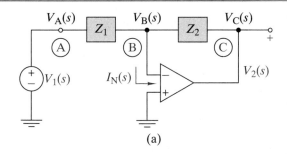

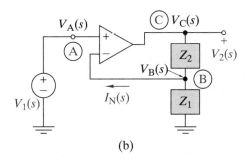

FIGURE 11–6 *Basic OP AMP circuits. (a) Inverting amplifier. (b) Noninverting amplifier.*

(a)

(b)

Two other useful circuits are the inverting and noninverting OP AMP configurations shown in Figure 11–6. To determine the voltage transfer function of the inverting circuit in Figure 11–6(a), we write the sum of currents leaving node B:

$$\frac{V_B(s) - V_A(s)}{Z_1(s)} + \frac{V_B(s) - V_C(s)}{Z_2(s)} + I_N(s) = 0$$

But the ideal OP AMP constraints require that $I_N(s) = 0$ and $V_B(s) = 0$ since the noninverting input is grounded. By definition, the output voltage $V_2(s)$ equals node voltage $V_C(s)$ and the voltage source forces $V_A(s)$ to equal the input voltage $V_1(s)$. Inserting all of these considerations into the node equations and solving for the voltage transfer function yields

$$T_V(s) = \frac{V_2(s)}{V_1(s)} = -\frac{Z_2(s)}{Z_1(s)} \tag{11–9}$$

From the study of OP AMP circuits in Chapter 4, you should recognize Eq. (11–9) as the *s*-domain generalization of the inverting OP AMP circuit gain equation, $K = -R_2/R_1$.

The driving-point impedance at the input to the inverting circuit is

$$Z_{IN}(s) = \frac{V_1(s)}{[V_A(s) - V_B(s)]/Z_1(s)}$$

But $V_A(s) = V_1(s)$ and $V_B(s) = 0$; hence the input impedance is $Z_{IN}(s) = Z_1(s)$, and should be a loading consideration.

For the noninverting circuit in Figure 11–6(b), the sum of currents leaving node B is

$$\frac{V_B(s) - V_C(s)}{Z_2(s)} + \frac{V_B(s)}{Z_1(s)} + I_N(s) = 0$$

In the noninverting configuration the ideal OP AMP constraints require that $I_N(s) = 0$ and $V_B(s) = V_1(s)$. By definition, the output voltage $V_2(s)$ equals node voltage

$V_C(s)$. Combining all of these considerations and solving for the voltage transfer function yields

$$T_V(s) = \frac{V_2(s)}{V_1(s)} = \frac{Z_1(s) + Z_2(s)}{Z_1(s)} \qquad (11\text{--}10)$$

Equation (11–10) is the s-domain version of the noninverting amplifier gain equation, $K = (R_1 + R_2)/R_1$. The transfer function of the noninverting configuration is the reciprocal of the transfer function of the voltage divider in the feedback path. The ideal OP AMP draws no current at its input terminals, so theoretically the input impedance of the noninverting circuit is infinite making it useful to solve loading issues.

The transfer functions of divider circuits and the basic OP AMP configurations are useful analysis and design tools in many practical situations. However, a general method is needed to handle circuits of greater complexity. One general approach is to formulate either node-voltage or mesh-current equations with all initial conditions set to zero. These equations are then solved for network functions using Cramer's rule for hand calculations or symbolic math analysis programs such as MATLAB. The algebra involved can be a bit tedious at times, even with MATLAB. But the tedium is reduced somewhat because we only need the zero-state response for a single input source.

The following examples illustrate methods of calculating network functions.

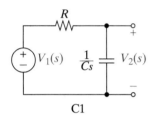

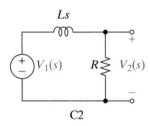

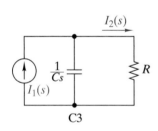

FIGURE 11–7

EXAMPLE 11–2

(a) Find the transfer functions of the circuits in Figure 11–7.
(b) Find the driving-point impedances seen by the input sources in these circuits.

SOLUTION:
(a) These are all divider circuits, so the required transfer functions can be obtained using Eq. (11–7) or (11–8).

For circuit C1: $Z_1 = R, Z_2 = 1/Cs$, and $T_V(s) = 1/(RCs + 1)$

For circuit C2: $Z_1 = Ls, Z_2 = R$, and $T_V(s) = \dfrac{1}{\dfrac{L}{R}s + 1}$

For circuit C3: $Z_1 = \dfrac{1}{Cs}$, $Z_2 = R$, and $T_I(s) = 1/(RCs + 1)$

These transfer functions are all of the form $1/(\tau s + 1)$, where τ is the circuit time constant.

(b) The driving-point impedances are found by series or parallel equivalence.

For circuit C1: $Z(s) = Z_1 + Z_2 = (RCs + 1)/Cs$

For circuit C2: $Z(s) = Z_1 + Z_2 = Ls + R$

For circuit C3: $Z(s) = 1/(Y_1 + Y_2) = 1/(Cs + G) = R/(RCs + 1)$

The three circuits have different driving-point impedances even though they have the same type of transfer functions.

The general principle illustrated here is that several different circuits can have the same transfer function. Put differently, a desired transfer function can be realized by several different circuits. This fact is important in design because circuits that produce the same transfer function offer alternatives that may differ in other features. In this example, they all have different input impedances. ∎

Exercise 11–2

For the circuit of Figure 11–8, find the voltage transfer function $T_V(s) = V_2(s)/V_1(s)$ and the driving-point impedance $Z(s)$.

Answer: $T_V(s) = \dfrac{RCs}{LCs^2 + RCs + 1}$ and $Z(s) = \dfrac{LCs^2 + RCs + 1}{Cs}$

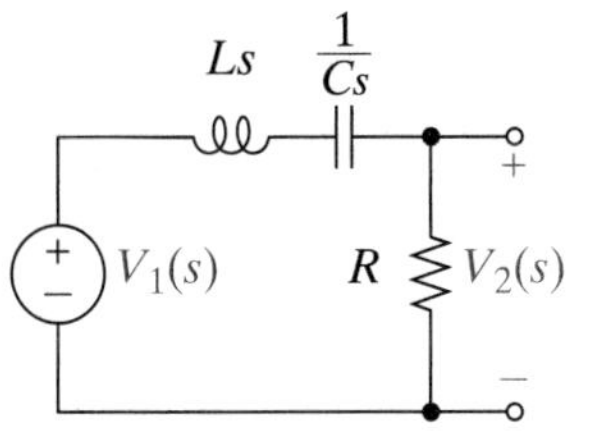

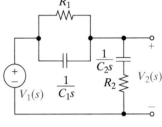

FIGURE 11–8

EXAMPLE 11–3

(a) Find the input impedance seen by the voltage source in Figure 11–9.
(b) Find the voltage transfer function $T_V(s) = V_2(s)/V_1(s)$ of the circuit.
(c) Locate the poles and zeros of $T_V(s)$ for $R_1 = 10\,\text{k}\Omega$, $R_2 = 20\,\text{k}\Omega$, $C_1 = 0.1\,\mu\text{F}$, and $C_2 = 0.05\,\mu\text{F}$.

SOLUTION:

(a) The circuit is a voltage divider. We first calculate the equivalent impedances of the two legs of the divider. The two elements in parallel combine to produce the series leg impedance $Z_1(s)$ as

$$Z_1(s) = \frac{1}{C_1 s + 1/R_1} = \frac{R_1}{R_1 C_1 s + 1}$$

FIGURE 11–9

The two elements in series combine to produce shunt leg impedance $Z_2(s)$:

$$Z_2(s) = R_2 + 1/C_2 s = \frac{R_2 C_2 s + 1}{C_2 s}$$

Using series equivalence, the driving-point impedance seen at the input is

$$Z_{\text{EQ}}(s) = Z_1(s) + Z_2(s)$$
$$= \frac{R_1 C_1 R_2 C_2 s^2 + (R_1 C_1 + R_2 C_2 + R_1 C_2)s + 1}{C_2 s (R_1 C_1 s + 1)}$$

(b) Using voltage division, the voltage transfer function is

$$T_V(s) = \frac{Z_2(s)}{Z_{\text{EQ}}(s)} = \frac{(R_1 C_1 s + 1)(R_2 C_2 s + 1)}{R_1 C_1 R_2 C_2 s^2 + (R_1 C_1 + R_2 C_2 + R_1 C_2)s + 1}$$

(c) Inserting the specified numerical values into $T_V(s)$ yields

$$T_V(s) = \frac{(10^{-3}s + 1)(10^{-3}s + 1)}{10^{-6}s^2 + 2.5 \times 10^{-3}s + 1} = \frac{(s + 1000)^2}{(s + 500)(s + 2000)}$$

which indicates a double zero at $s = -1000\,\text{rad/s}$, and simple poles at $s = -500\,\text{rad/s}$ and $s = -2000\,\text{rad/s}$. ∎

Exercise 11–3

For the circuit of Figure 11–10, (a) find the voltage transfer function $T_V(s) = V_2(s)/V_1(s)$ and the driving-point impedance $Z(s)$, and (b) locate the poles and zeros of the transfer function when $R_1 = R_2 = 1\,\text{k}\Omega$, $L = 10\,\text{mH}$, and $C = 0.1\,\mu\text{F}$.

Answers:

(a) $T_V(s) = \dfrac{R_2 Cs + 1}{LCs^2 + (R_1 + R_2)Cs + 1}$ and $Z(s) = \dfrac{LCs^2 + (R_1 + R_2)Cs + 1}{Cs}$

(b) Zeros at $s = -10^4\,\text{rad/s}$ and infinity. Poles at $s = -194{,}868\,\text{rad/s}$ and $s = -5131\,\text{rad/s}$.

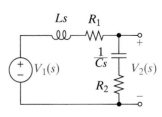

FIGURE 11–10

▶ DESIGN EXAMPLE 11–4

Find the driving-point impedance seen by the voltage source in Figure 11–11. Find the voltage transfer function $T_V(s) = V_2(s)/V_1(s)$ of the circuit. The poles of $T_V(s)$ are located at $p_1 = -1000$ rad/s and $p_2 = -5000$ rad/s. If $R_1 = R_2 = 20\,\text{k}\Omega$, what values of C_1 and C_2 are required?

FIGURE 11–11

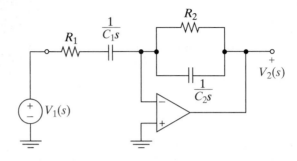

SOLUTION:
The circuit is an inverting OP AMP configuration of the form in Figure 11–6(a). The input impedance of this circuit is

$$Z_1(s) = R_1 + \frac{1}{C_1 s} = \frac{R_1 C_1 s + 1}{C_1 s}$$

The impedance Z_2 in the feedback path is

$$Z_2(s) = \frac{1}{C_2 s + 1/R_2} = \frac{R_2}{R_2 C_2 s + 1}$$

and the voltage transfer function is

$$T_V(s) = -\frac{Z_2(s)}{Z_1(s)} = -\frac{R_2 C_1 s}{(R_1 C_1 s + 1)(R_2 C_2 s + 1)}$$

The poles of $T_V(s)$ are located at $p_1 = -1/R_1 C_1 = -1000$ and $p_2 = -1/R_2 C_2 = -5000$. If $R_1 = R_2 = 20\,\text{k}\Omega$, then $C_1 = 1/1000R_1 = .05\,\mu\text{F}$ and $C_2 = 1/5000R_2 = 0.01\,\mu\text{F}$. ∎

Exercise 11–4

Suppose that capacitor C_1 in the circuit of Figure 11–11 suddenly became shorted. What effect would it have on the circuit's voltage transfer function?

Answer: The transfer function would become

$$T_V(s) = -\frac{R_2/R_1}{R_2 C_2 s + 1}$$

effectively eliminating the pole at $s = -1/R_1 C_1$.

Exercise 11–5

Suppose that capacitor C_2 in the circuit of Figure 11–11 suddenly became open-circuited. What effect would it have on the circuit's voltage transfer function?

Answer: The transfer function would become

$$T_V(s) = -\frac{R_2 C_1 s}{R_1 C_1 s + 1}$$

effectively eliminating the pole at $s = -1/R_2 C_2$.

EXAMPLE 11–5

For the circuit in Figure 11–12 find the input impedance $Z(s) = V_1(s)/I_1(s)$, the transfer impedance $T_Z(s) = V_2(s)/I_1(s)$, and the voltage transfer function $T_V(s) = V_2(s)/V_1(s)$.

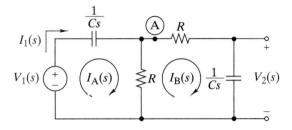

FIGURE 11–12

SOLUTION:

The circuit is not a simple voltage divider, so we use mesh-current equations to illustrate the general approach to finding network functions. By inspection, the mesh-current equations for this ladder circuit are

$$\left(R + \frac{1}{Cs}\right)I_A(s) - RI_B(s) = V_1(s)$$

$$-RI_A(s) + \left(2R + \frac{1}{Cs}\right)I_B(s) = 0$$

In terms of the mesh current, the input impedance is $Z(s) = V_1(s)/I_A(s)$. Using Cramer's rule to solve for $I_A(s)$ yields

$$I_A(s) = \frac{\Delta_A}{\Delta} = \frac{\begin{vmatrix} V_1(s) & -R \\ 0 & 2R + \dfrac{1}{Cs} \end{vmatrix}}{\begin{vmatrix} R + \dfrac{1}{Cs} & -R \\ -R & 2R + \dfrac{1}{Cs} \end{vmatrix}} = \frac{Cs(2RCs + 1)}{(RCs)^2 + 3RCs + 1}V_1(s)$$

The input impedance of the circuit is

$$Z(s) = \frac{V_1(s)}{I_A(s)} = \frac{(RCs)^2 + 3RCs + 1}{Cs(2RCs + 1)}$$

In terms of mesh current, the transfer impedance is $T_Z(s) = V_2(s)/I_A(s)$. The mesh-current equations do not yield the output voltage directly. But since $V_2(s) = I_B(s)Z_C(s)$, we can solve the second mesh equation for $I_B(s)$ in terms of as $I_A(s)$ as

$$I_B(s) = \frac{RCs}{2RCs + 1}I_A(s)$$

and obtain the specified transfer impedance as

$$T_Z(s) = \frac{I_B(s)[1/Cs]}{I_A(s)} = \frac{R}{2RCs + 1}$$

To obtain the specified voltage transfer function, we could use Cramer's rule to solve for $I_B(s)$ in terms of $V_1(s)$ and then use the fact that $V_2(s) = I_B(s)Z_C(s)$. But a moment's reflection reveals that

$$T_V(s) = \frac{V_2(s)}{V_1(s)} = \left[\frac{V_2(s)}{I_1(s)}\right]\left[\frac{I_1(s)}{V_1(s)}\right] = T_Z(s) \times \frac{1}{Z(s)}$$

Hence the specified voltage transfer function is

$$T_V(s) = \frac{R}{2RCs+1} \times \frac{Cs(2RCs+1)}{(RCs)^2 + 3RCs + 1}$$

$$= \frac{RCs}{(RCs)^2 + 3RCs + 1}$$

∎

Exercise 11–6

For the circuit shown in Figure 11–12, insert a follower at point A and find the transfer function $T_V(s) = V_2(s)/V_1(s)$. Compare your result with that found in Example 11–5 for the same transfer function.

Answer: The transfer function would become

$$T_V(s) = \frac{RCs}{(RCs)^2 + 2RCs + 1}$$

The difference appears to be slight—simply changing the middle term of the denominator from 3 to 2. However, this change affects the location of the poles and could affect the nature of the circuit's behavior. Inserting the follower eliminates the loading effect of the second RC circuit on the first.

EXAMPLE 11–6

Find the voltage transfer function $T_V(s) = V_2(s)/V_1(s)$ of the circuit in Figure 11–13.

FIGURE 11–13

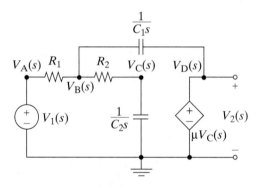

SOLUTION:
The voltage-controlled voltage source makes this an active RC circuit. We use node-voltage equations in this problem because the required output is a voltage. The circuit contains two voltage sources connected at a common node. Selecting this common node as the reference eliminates two unknowns since $V_A(s) = V_1(s)$ and $V_D(s) = \mu V_C(s) = V_2(s)$. The sums of currents leaving nodes B and C are

Node B: $$\frac{V_B(s) - V_1(s)}{R_1} + \frac{V_B(s) - V_C(s)}{R_2} + \frac{V_B(s) - \mu V_C(s)}{1/C_1 s} = 0$$

Node C: $$\frac{V_C(s) - V_B(s)}{R_2} + \frac{V_C(s)}{1/C_2 s} = 0$$

Multiplying both equations by $R_1 R_2$ and rearranging terms produces

Node B: $(R_1 + R_2 + R_1 R_2 C_1 s)V_B(s) - (R_1 + \mu R_1 R_2 C_1 s)V_C(s) = R_2 V_1(s)$

Node C: $-V_B(s) + (1 + R_2 C_2 s)V_C(s) = 0$

Using the node C equation to eliminate $V_B(s)$ from the node B equation leaves

$$(R_1 + R_2 + R_1R_2C_1s)(1 + R_2C_2s)V_C(s)$$
$$-(R_1 + \mu R_1R_2C_1s)V_C(s) = R_2V_1(s)$$

Since the output $V_2(s) = \mu V_C(s)$, the required transfer function is

$$T_V(s) = \frac{V_2(s)}{V_1(s)} = \frac{\mu}{R_1R_2C_1C_2s^2 + (R_1C_1 + R_1C_2 + R_2C_2 - \mu R_1C_1)s + 1}$$

This circuit is a member of the Sallen-Key family, often used in filter design with $R_1 = R_2 = R$ and $C_1 = C_2 = C$, in which case the transfer function reduces to

$$T_V(s) = \frac{\mu}{(RCs)^2 + (3 - \mu)RCs + 1}$$

We will encounter this result again in later chapters. ∎

Exercise 11-7 _____

(a) Find the voltage transfer function $T_V(s) = V_2(s)/V_1(s)$ of the circuit in Figure 11-14.
(b) What are the conditions on μ that will ensure the circuit is stable?

Answers:

(a) $T_V(s) = \dfrac{V_2(s)}{V_1(s)} = \dfrac{-\mu}{RCs + 1 - \mu}$

(b) For a circuit to be stable all poles must be in the left-hand portion of the s-plane, hence μ must be less than 1.

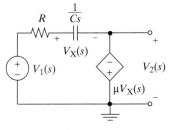

FIGURE 11-14

THE CASCADE CONNECTION AND THE CHAIN RULE

Signal-processing circuits often involve a **cascade connection** in which the output voltage of one circuit serves as the input to the next stage. In some cases, the overall voltage transfer function of the cascade can be related to the transfer functions of the individual stages by a **chain rule**

$$T_V(s) = T_{V1}(s)T_{V2}(s)\cdots T_{Vk}(s) \tag{11-11}$$

where $T_{V1}, T_{V2}, \ldots, T_{Vk}$ are the voltage transfer functions of the individual stages when operated separately. It is important to understand when the chain rule applies since it greatly simplifies the analysis and design of cascade circuits.

To illustrate the chain rule concept, consider the two-stage RC circuit in Figure 11-15. When disconnected and operated in isolation, the transfer function

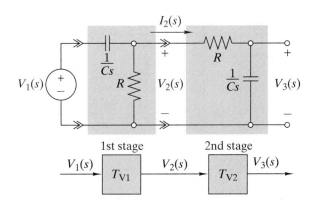

FIGURE 11-15 *Two-port circuits connected in cascade.*

of each stage can be found using voltage division as

$$T_{V1}(s) = \frac{R}{R + 1/Cs} = \frac{RCs}{RCs + 1}$$

$$T_{V2}(s) = \frac{1/Cs}{R + 1/Cs} = \frac{1}{RCs + 1}$$

When connected in cascade, the output of the first stage serves as the input to the second stage. If the chain rule applies, we would obtain the overall transfer function as

$$T_V(s) = \frac{V_3(s)}{V_1(s)} = \left(\frac{V_2(s)}{V_1(s)}\right)\left(\frac{V_3(s)}{V_2(s)}\right) = (T_{V1}(s))(T_{V2}(s))$$

$$= \underbrace{\left(\frac{RCs}{RCs + 1}\right)}_{\text{first stage}} \underbrace{\left(\frac{1}{RCs + 1}\right)}_{\text{second stage}} = \underbrace{\frac{RCs}{(RCs)^2 + 2RCs + 1}}_{\text{overall}} \qquad (11\text{--}12)$$

However, in Example 11–5, the overall transfer function of this circuit was found to be

$$T_V(s) = \frac{RCs}{(RCs)^2 + 3RCs + 1} \qquad (11\text{--}13)$$

which disagrees with the chain rule result in Eq. (11–12).

The reason for the discrepancy is that when they are connected in cascade, the second circuit "loads" the first circuit. That is, the voltage-divider rule requires that the interface current $I_2(s)$ in Figure 11–15 be zero. The no-load condition $I_2(s) = 0$ applies when the stages operate separately, but when connected in cascade, the interface current is not zero. The chain rule does not apply here because loading caused by the second stage changes the transfer function of the first stage.

However, Figure 11–16 shows how the loading problem goes away when an OP AMP voltage follower is inserted between the RC circuit stages. The follower does not draw any current from the first RC circuit [$I_2(s) = 0$] and applies $V_2(s)$ directly across the input of the second RC circuit. With this modification the chain rule in Eq. (11–11) applies because the voltage follower isolates the two circuits, thereby solving the loading problem.

In the s domain, *loading* causes the transfer function of a circuit to change when it drives the input of another circuit. In a cascade connection, loading does not occur at an interface if (1) the output (Thévenin) impedance of the driving stage is zero or (2) the input impedance of the driven stage is infinite. The voltage follower in Figure 11–16 is an example of a stage that meets both criteria (1) and (2). In general, an inverting

FIGURE 11–16 *Cascade connection with voltage follower isolation.*

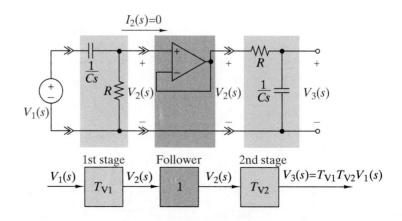

OP AMP stage meets criterion (1) but not criterion (2), while a voltage-divider stage meets neither criteria.

When analyzing or designing a cascade connection, it is important to recognize situations in which the chain rule applies. The next example illustrates this point.

E EVALUATION EXAMPLE 11–7

Figure 11–17 shows two cascade connections involving the same two stages but with their positions reversed. Do either of these connections involve loading? If not, use the chain rule to find the overall transfer function.

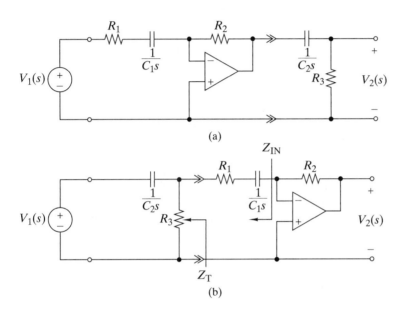

FIGURE 11–17

SOLUTION:
Both circuits involve a cascade connection of a voltage-divider stage and an inverting amplifier stage. The version in Figure 11–17(a) does not involve loading because the output impedance of the first stage is zero. Hence, connecting the second-stage voltage divider does not load the first stage and the chain rule applies. The transfer function of the inverting amplifier stage is

$$T_{V1}(s) = -\frac{Z_2(s)}{Z_1(s)} = -\frac{R_2}{R_1 + 1/C_1 s} = -\frac{R_2 C_1 s}{R_1 C_1 s + 1}$$

The second stage is a voltage divider whose transfer function is

$$T_{V2}(s) = \frac{Z_2(s)}{Z_2(s) + Z_1(s)} = \frac{R_3}{R_3 + 1/C_2 s} = \frac{R_3 C_2 s}{R_3 C_2 s + 1}$$

and the chain rule yields the overall transfer function as

$$T_V(s) = T_{V1}(s) \times T_{V2}(s) = \frac{-R_2 C_1 s}{R_1 C_1 s + 1} \times \frac{R_3 C_2 s}{R_3 C_2 s + 1}$$

$$= \frac{-R_2 C_1 R_3 C_2 s^2}{R_1 C_1 R_3 C_2 s^2 + (R_1 C_1 + R_3 C_2)s + 1}$$

The cascade connection in Figure 11–17(b) interchanges the positions of the two stages. The loading occurs in this case because the first stage is a voltage divider with a nonzero output (Thévenin) impedance of

$$Z_T = \frac{1}{\dfrac{1}{R_3} + C_2 s} = \frac{R_3}{R_3 C_2 s + 1}$$

and the inverting amplifier in the second stage has a finite input impedance of

$$Z_{IN} = Z_T + \left(R_1 + \frac{1}{C_1 s} \right) = \frac{R_3}{R_3 C_2 s + 1} + \frac{R_1 C_1 s + 1}{C_1 s} = \frac{C_1 R_3 s + (R_1 C_1 s + 1)(R_3 C_2 s + 1)}{C_1 s (R_3 C_2 s + 1)}$$

$$T_V(s) = \frac{V_2(s)}{V_T(s)} \times \frac{V_T(s)}{V_1(s)} = -\frac{Z_2}{Z_{IN}} \times \frac{R_3 C_2 s}{R_3 C_2 s + 1} \text{ since } V_1(s) = V_T(s) \left(\frac{R_3 C_2 s + 1}{R_3 C_2 s} \right)$$

This results in a voltage transfer function of

$$T_V(s) = \frac{-R_2 C_1 R_3 C_2 s^2}{R_1 C_1 R_3 C_2 s^2 + (R_1 C_1 + R_3 C_2 + R_3 C_1)s + 1}$$

which is not equal to $T_{V1}(s) \times T_{V2}(s)$. The chain rule does not apply to this connection since the second stage loads the first stage.

The effects of loading might be more visible if numbers are used. Let us look at the transfer functions if $R_1 = R_2 = R_3 = 10 \text{ k}\Omega$, $C_1 = 10 \text{ }\mu\text{F}$ and $C_2 = 1 \text{ }\mu\text{F}$. Using these element values, the transfer function for the circuit in Figure 11–17(a) is

$$T_{V1} = -\frac{s^2}{(s + 10)(s + 100)}$$

For the circuit in Figure 11–17(b), the transfer function is

$$T_{V2} = -\frac{s^2}{(s + 4.88)(s + 205)}$$

The locations of the poles are quite different for the two circuits. Avoiding unintentional loading is a sign of experienced circuit designers.   ■

▶ E Evaluation Exercise 11–8 _____

Figure 11–18 shows two cascade connections involving the same two stages but with their positions reversed. Does either of these connections involve loading? Find their voltage transfer functions and, if loading is present, determine the condition necessary to minimize the effect.

FIGURE 11–18

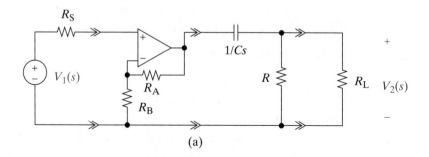

(a)

FIGURE 11–18　*(Continued)*

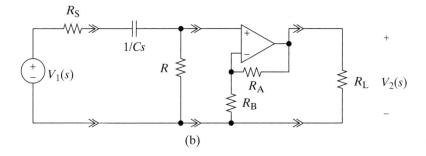

(b)

Answers:　Both circuits exhibit loading effects.

(a) $T_{Va}(s) = \dfrac{V_2(s)}{V_1(s)} = \left(\dfrac{R_A + R_B}{R_B}\right)\left(\dfrac{s}{s + \dfrac{R_L + R}{R_L R C}}\right)$

To minimize loading effects select $R \ll R_L$.

(b) $T_{Vb}(s) = \dfrac{V_2(s)}{V_1(s)} = \left(\dfrac{R_A + R_B}{R_B}\right)\left(\dfrac{\dfrac{R}{R + R_S}s}{s + \dfrac{1}{(R + R_S)C}}\right)$

To minimize loading effects select $R \gg R_S$.

11–3 NETWORK FUNCTIONS AND IMPULSE RESPONSE

The **impulse response** is the zero-state response of a circuit when the driving force is a unit impulse applied at $t = 0$. When the input signal is $x(t) = \delta(t)$, then $X(s) = \mathcal{L}\{\delta(t)\} = 1$ and the input-output relationship in Eq. (11–2) reduces to

$$Y(s) = T(s) \times 1 = T(s)$$

The impulse response transform equals the network function, and we could treat $T(s)$ as if it is a signal transform. However, to avoid possible confusion between a network function (description of a circuit) and a transform (description of a signal), we denote the impulse response transform as $H(s)$ and use $h(t)$ to denote the corresponding waveform.[2] Figure 11–19 shows this concept in a block diagram form. That is,

$$
\begin{array}{cc}
\text{Impulse Response} & \\
\text{Transform} & \text{Waveform} \\
H(s) = T(s) \times 1 & h(t) = \mathcal{L}^{-1}\{H(s)\}
\end{array}
\qquad (11\text{–}14)
$$

When there are no repeated poles, the partial-fraction expansion of $H(s)$ is

$$H(s) = \underbrace{\frac{k_1}{s - p_1} + \frac{k_2}{s - p_2} + \cdots + \frac{k_N}{s - p_N}}_{\text{natural poles}}$$

where $p_1, p_2, \ldots, p_N$ are the natural poles in the denominator of the transfer function $T(s)$. All of the poles of $H(s)$ are natural poles since the impulse excitation does not

---

[2]Not all books make this distinction. Books on signals and circuits often use $H(s)$ to represent both a transfer function and the impulse response transform.

FIGURE 11–19 *Block diagram of impulse response: (a) In the t domain. (b) In the s domain.*

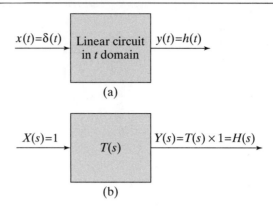

(a)

(b)

introduce any forced poles. The inverse transform gives the impulse response waveform as

$$h(t) = [\underbrace{k_1e^{p_1t} + k_2e^{p_2t} + \cdots + k_Ne^{p_Nt}}_{\text{natural response}}]u(t)$$

When the circuit is stable, all of the natural poles are in the left half plane and the impulse response waveform $h(t)$ decays to zero as $t \to \infty$. A linear circuit whose impulse response ultimately returns to zero is said to be **asymptotically stable**. Asymptotic stability means that the impulse response has a finite time duration. That is, for every $\varepsilon > 0$ there exists a finite time duration T_D such that $|h(t)| < \varepsilon$ for all $t > T_D$.

It is important to note that the impulse response $h(t)$ contains all the information needed to determine the circuit response to any other input. That is, since $\mathcal{L}\{h(t)\} = H(s) = T(s)$, we can calculate the output $y(t)$ for any Laplace transformable input $x(t)$ as

$$y(t) = \mathcal{L}^{-1}\{H(s)X(s)\}$$

This expression, known as the *convolution theorem*, states that the impulse response can be used to relate the input and output of a linear circuit. Thus, the impulse response $h(t)$ or $H(s)$ can be considered as a mathematical model of a linear circuit. Obviously it is important to be able to find the impulse response and to know how to use the impulse response to find the output for other inputs. The following examples illustrate both of these issues.

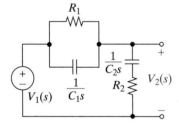

FIGURE 11–20

EXAMPLE 11–8

Find the response $v_2(t)$ in Figure 11–20 when the input is $v_1(t) = \delta(t)$. Use the element values $R_1 = 10\,\text{k}\Omega$, $R_2 = 12.5\,\text{k}\Omega$, $C_1 = 1\,\mu\text{F}$, and $C_2 = 2\,\mu\text{F}$.

SOLUTION:

In Example 11–3, the transfer function of this circuit was found to be

$$T_V(s) = \frac{V_2(s)}{V_1(s)} = \frac{(R_1C_1s + 1)(R_2C_2s + 1)}{R_1C_1R_2C_2s^2 + (R_1C_1 + R_2C_2 + R_1C_2)s + 1}$$

For the given element values, the impulse response transform is

$$H(s) = \frac{(s + 100)(s + 40)}{s^2 + 220s + 4000} = \frac{(s + 100)(s + 40)}{(s + 20)(s + 200)}$$

This $H(s)$ is not a proper rational function, so we use one step of a long division plus a partial-fraction expansion to obtain

$$H(s) = 1 + \frac{80/9}{s + 20} - \frac{800/9}{s + 200}$$

and the impulse response is

$$h(t) = \delta(t) + \frac{80}{9}\left[e^{-20t} - 10e^{-200t}\right]u(t)$$

In this case, the impulse response contains an impulse because the network function is not a proper rational function. ∎

Exercise 11–9

A certain circuit has the following voltage transfer function:

$$T_V(s) = \frac{10^5 s}{(s + 10^2)(s + 10^5)}$$

Find the circuit's impulse response $h(t)$.

Answer:
$$h(t) = [-100.1e^{-10^2 t} + 100,100e^{-10^5 t}]u(t)$$

EXAMPLE 11–9

The impulse response of a linear circuit is $h(t) = 200e^{-100t}u(t)$. Find the output when the input is a unit ramp $r(t) = tu(t)$.

SOLUTION:
The circuit impulse response transform is

$$H(s) = \mathcal{L}\{h(t)\} = \mathcal{L}\{200e^{-100t}u(t)\} = \frac{200}{s + 100}$$

The Laplace transform of the unit ramp input is $1/s^2$; hence, using the convolution theorem the response due to a ramp input is

$$\begin{aligned}
y(t) &= \mathcal{L}^{-1}\left\{H(s)\frac{1}{s^2}\right\} = \mathcal{L}^{-1}\left\{\frac{200}{(s + 100)s^2}\right\} \\
&= \mathcal{L}^{-1}\left\{\frac{1/50}{s + 100} + \frac{2}{s^2} - \frac{1/50}{s}\right\} \\
&= \frac{1}{50}(e^{-100t} + 100t - 1)u(t)
\end{aligned}$$

This example illustrates that the impulse response $h(t)$ contains all the information needed to calculate the response due to another input. ∎

Exercise 11–10

The impulse response of a circuit is $h(t) = 100e^{-20t}u(t)$. Find the output when the input is a step function $x(t) = u(t)$.

Answer:
$$y(t) = 5[1 - e^{-20t}]u(t)$$

Exercise 11–11

Find the impulse response of the circuit in Figure 11–21.

Answer:
$$h(t) = 0.1\delta(t) + [90e^{-100t}]u(t)$$

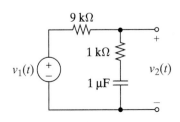

FIGURE 11–21

11–4 NETWORK FUNCTIONS AND STEP RESPONSE

The **step response** is the zero-state response of the circuit output when the driving force is a unit step function applied at $t = 0$. When the input is $x(t) = u(t)$, then $X(s) = \mathcal{L}\{u(t)\} = 1/s$ and the s-domain input-output relationship in Eq. (11–2) yields $Y(s) = T(s)/s$. The step response transform and waveform will be denoted by $G(s)$ and $g(t)$, respectively. That is,

$$\begin{array}{c} \text{Step Response} \\ \text{Transform} \qquad \text{Waveform} \\ G(s) = \dfrac{T(s)}{s} \quad g(t) = L^{-1}\{G(s)\} \end{array} \qquad (11\text{–}15)$$

The poles of $G(s)$ are the natural poles contributed by the network function $T(s)$ and a forced pole at $s = 0$ introduced by the step function input. The partial fraction expansion of $G(s)$ takes the form

$$G(s) = \underbrace{\frac{k_0}{s}}_{\substack{\text{forced} \\ \text{pole}}} + \underbrace{\frac{k_1}{s - p_1} + \frac{k_2}{s - p_2} + \cdots + \frac{k_N}{s - p_N}}_{\substack{\text{natural} \\ \text{poles}}}$$

where $p_1, p_2, \ldots, p_N$ are the natural poles in $T(s)$. The inverse transformation gives the step response waveform as

$$g(t) = \underbrace{k_0 u(t)}_{\substack{\text{forced} \\ \text{response}}} + \underbrace{[k_1 e^{p_1 t} + k_2 e^{p_2 t} + \cdots + k_N e^{p_N t}]u(t)}_{\substack{\text{natural} \\ \text{response}}}$$

When the circuit is stable, the natural response decays to zero, leaving a forced component called the **dc steady-state response**. The amplitude of the steady-state response is the residue in the partial-fraction expansion of the forced pole at $s = 0$. By the cover-up method, this residue is

$$k_0 = sG(s)|_{s=0} = T(0)$$

For a unit step input, the amplitude of the dc steady-state response equals the value of the transfer function at $s = 0$. By linearity, the general principle is that an input $Au(t)$ produces a dc steady-state output whose amplitude is $AT(0)$.

We next show the relationship between the impulse and step responses. First, combining Eqs. (11–14) and (11–15) gives

$$G(s) = \frac{H(s)}{s}$$

The step response transform is the impulse response transform divided by s. The integration property of the Laplace transform tells us that division by s in the s domain corresponds to integration in the time domain. Therefore, in the time domain we can relate the impulse and step response waveforms by integration:

$$g(t) = \int_0^t h(\tau)d\tau \qquad (11\text{–}16)$$

Using the fundamental theorem of calculus, the impulse response waveform is expressed in terms of the step response waveform

$$h(t)(=)\frac{dg(t)}{dt} \qquad (11\text{–}17)$$

where the symbol $(=)$ means equal almost everywhere, a condition that excludes those points at which $g(t)$ has a discontinuity. In the time domain, the step response

t domain

s domain

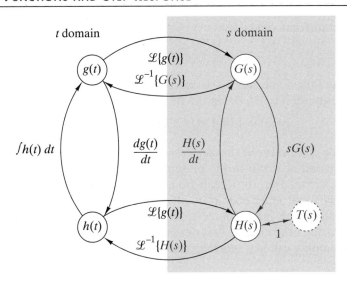

waveform is the integral of the impulse response waveform. Conversely, the impulse response waveform is (almost everywhere) the derivative of the step response waveform.

The key idea is that there are relationships between the network function $T(s)$ and the responses $H(s)$, $h(t)$, $G(s)$, and $g(t)$. If any one of these quantities is known, we can obtain any of the other four using relatively simple mathematical operations. A summary of these relationship is shown in Figure 11–22.

EXAMPLE 11–10

The element values for the circuit in Figure 11–23 are $R_1 = 10 \text{ k}\Omega$, $R_2 = 100 \text{ k}\Omega$, $C_1 = C_2 = 0.1 \, \mu\text{F}$, and $v_1(t) = u(t)$ V. Find the response $v_2(t)$.

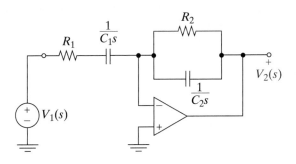

FIGURE 11–23

SOLUTION:

For a unit step input $V_1(s) = 1/s$. In Example 11–4 the transfer function of this circuit is shown to be

$$T_V(s) = \frac{V_2(s)}{V_1(s)} = -\frac{R_2 C_1 s}{(R_1 C_1 s + 1)(R_2 C_2 s + 1)}$$

Substituting the numerical element values into this expression yields the step response transform as

$$V_2(s) = \frac{T_V(s)}{s} = -\frac{1000}{(s + 100)(s + 1000)}$$

$$= \frac{-10/9}{s + 100} + \frac{10/9}{s + 1000}$$

The inverse transform yields

$$v_2(t) = \frac{10}{9}\left(-e^{-100t} + e^{-1000t}\right)u(t)$$

In this case the forced pole at $s = 0$ is canceled by a zero of $T_V(s)$ and the final value of the step response is zero. Using the final value theorem, we have

$$g(\infty) = \underset{t \to \infty}{\text{Limit }} g(t) = \underset{s \to 0}{\text{Limit }} sG(s) = T(0)$$

the final steady-state value of the step response is equal to the value of the transfer function evaluated at $s = 0$. In the present case $T_V(0) = 0$, so the dc steady-state output of the circuit is zero. Recall from Chapter 6 that in the dc steady state, capacitors can be replaced by open circuits. Replacing C_1 in Figure 11–23 by an open circuit disconnects the input source from the OP AMP, so no dc signals can be transferred through the circuit. A series capacitor that prevents the passage of dc signals is commonly called a *blocking capacitor*. ■

◗ Design Exercise 11–12

Design a circuit that will produce the following step response output:

$$v_2(t) = \left[1 - e^{-1000t}\right]u(t)\text{ V}$$

Answer: The transfer function of the desired circuit is

$$T_V(s) = \frac{10^3}{s + 10^3}$$

A simple series RC circuit with the output taken across the capacitor will produce the desired response as long as the pole at $-1/RC$ equals -1000 rad/s.

EXAMPLE 11–11

The step response of a linear circuit is $g(t) = 5\left[1 - e^{-200t}\right]u(t)$. Find the output waveform when the input is $x(t) = \left[12e^{-200t}\right]u(t)$. Use the inverse Laplace function to visualize (plot) the response $y(t)$.

SOLUTION:
Since we are given the step response $g(t)$, we start by realizing that $G(s) = H(s)/s$. We transform the step response into the s domain as follows:

$$G(s) = \frac{5}{s} - \frac{5}{s + 200} = \frac{1000}{s(s + 200)} = \frac{H(s)}{s}$$

Hence,

$$H(s) = \frac{1000}{s + 200}$$

The transform of the input signal is

$$X(s) = \frac{12}{s + 200}$$

The output is found by applying the convolution theorem

$$y(t) = \mathscr{L}^{-1}\{H(s)X(s)\}$$

$$y(t) = \mathscr{L}^{-1}\left\{\left(\frac{1000}{s + 200}\right)\left(\frac{12}{s + 200}\right)\right\}$$

$$y(t) = \mathscr{L}^{-1}\left\{\frac{12000}{(s + 200)^2}\right\} = 12000te^{-200t}\,u(t)$$

FIGURE 11–24

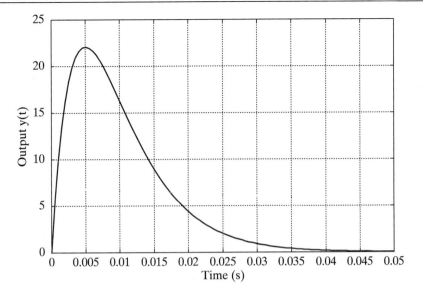

Figure 11–24 is a graph of the output. (Plotted using MATLAB.) ■

Exercise 11–13

A particular circuit has the following voltage transfer function:

$$T_V(s) = \frac{s+5}{s+10}$$

Find the circuit's step response $g(t)$, impulse response $h(t)$, step response transform $G(s)$, and impulse response transform $H(s)$.

Answer:

$$g(t) = [0.5e^{-10t} + 0.5]u(t), \quad h(t) = \delta(t) - 5e^{-10t}u(t), \quad G(s) = \frac{s+5}{s(s+10)}, \quad H(s) = \frac{s+5}{s+10}$$

APPLICATION █ EXAMPLE 11–12

Three time-domain parameters often used to describe the step response are rise time, delay time, and overshoot. **Rise time(T_R)** is the time interval required for the step response to rise from 10% to 90% of its steady-state value $g(\infty)$. **Delay time(T_D)** is the time interval required for the step response to reach 50% of its steady-state value. **Overshoot** is the difference between the peak value of the step response and its steady-state value. Overshoot is usually expressed as a percentage of the steady-state value, namely

$$\text{Overshoot} = \frac{g_{max} - g(\infty)}{g(\infty)} \times 100$$

Figure 11–25 illustrates these descriptors for a typical step response.

Step response descriptors are used to specify the performance of both analog and digital systems. Rise time governs how rapidly the system responds to an abrupt change in the input. Delay time controls the time between the application of an abrupt change and the appearance of a significant change in the output. Overshoot indicates the amount of damping present in the system. Lightly damped oscillations produce large overshoots and may cause erroneous state changes in digital systems.

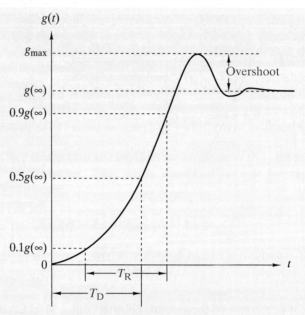

FIGURE 11–25 *Step response showing rise time (T_R), delay time (T_D), and overshoot.*

Rise time, delay time, and overshoot can be determined experimentally or calculated using modern computer tools. How to use MATLAB to find these parameteres can be found in Web appendix C.

Exercise 11–14

The impulse response of a circuit is $h(t) = 8000(600t - 1)e^{-800t}u(t)$. Find and plot the step response. Find approximate values for the rise time, delay time, and overshoot.

Answers:

$g(t) = -2.5\left[1 + (2400t - 1)\,e^{-800t}\right]u(t)$
$T_R = 329\,\mu s$
$T_D = 180\,\mu s$
Overshoot = 79%

11–5 NETWORK FUNCTIONS AND SINUSOIDAL STEADY-STATE RESPONSE

When a stable, linear circuit is driven by a sinusoidal input, the output contains a steady-state component that is a sinusoid of the same frequency as the input. This section deals with using the circuit transfer function to find the amplitude and phase angle of the sinusoidal steady-state response. To begin, we write a general sinusoidal input in the form

$$x(t) = X_A \cos(\omega t + \phi) \tag{11–18}$$

which can be expanded as

$$x(t) = X_A(\cos \omega t \cos \phi - \sin \omega t \sin \phi)$$

The waveforms $\cos \omega t$ and $\sin \omega t$ are basic signals whose transforms are given in Table 9–2 as $\mathscr{L}\{\cos \omega t\} = s/(s^2 + \omega^2)$ and $\mathscr{L}\{\sin \omega t\} = \omega/(s^2 + \omega^2)$. Therefore, the input transform is

$$X(s) = X_A \left[\frac{s}{s^2 + \omega^2} \cos \phi - \frac{\omega}{s^2 + \omega^2} \sin \phi \right]$$

$$= X_A \left[\frac{s \cos \phi - \omega \sin \phi}{s^2 + \omega^2} \right]$$

(11–19)

Equation (11–19) is the Laplace transform of the general sinusoidal waveform in Eq. (11–18).

Using Eq. (11–2), we obtain the response transform for a general sinusoidal input:

$$Y(s) = X_A \left[\frac{s \cos \phi - \omega \sin \phi}{(s - j\omega)(s + j\omega)} \right] T(s)$$

(11–20)

The response transform contains forced poles at $s = \pm j\omega$ because the input is a sinusoid. Expanding Eq. (11–20) by partial fractions,

$$Y(s) = \underbrace{\frac{k}{s - j\omega} + \frac{k^*}{s + j\omega}}_{\text{forced poles}} + \underbrace{\frac{k_1}{s - p_1} + \frac{k_2}{s - p_2} + \cdots + \frac{k_N}{s - p_N}}_{\text{natural poles}}$$

where $p_1, p_2, \ldots, p_N$ are the natural poles contributed by the transfer function $T(s)$. To obtain the response waveform, we perform the inverse transformation:

$$y(t) = \underbrace{k e^{j\omega t} + k^* e^{-j\omega t}}_{\text{forced response}} + \underbrace{k_1 e^{p_1 t} + k_2 e^{p_2 t} + \cdots + k_N e^{p_N t}}_{\text{natural response}}$$

When the circuit is stable, the natural response decays to zero, leaving a sinusoidal steady-state response due to the forced poles as $s = \pm j\omega$. The steady-state response is

$$y_{SS}(t) = k e^{j\omega t} + k^* e^{-j\omega t}$$

where the subscript SS identifies a steady-state condition.

To determine the amplitude and phase of the steady-state response, we must find the residue k. Using the cover-up method from Chapter 9, we find k to be

$$k = (s - j\omega) X_A \left[\frac{s \cos \phi - \omega \sin \phi}{(s - j\omega)(s + j\omega)} \right] T(s) \Big|_{s = j\omega}$$

$$= X_A \left[\frac{j\omega \cos \phi - \omega \sin \phi}{2 j\omega} \right] T(j\omega)$$

$$= X_A \left[\frac{\cos \phi + j \sin \phi}{2} \right] T(j\omega) = \frac{1}{2} X_A e^{j\phi} T(j\omega)$$

The complex quantity $T(j\omega)$ can be written in magnitude and angle form as $|T(j\omega)| e^{j\theta}$. Using these results, the residue becomes

$$k = \left[\frac{1}{2} X_A e^{j\phi} \right] \left| T(j\omega) \right| e^{j\theta}$$

$$= \frac{1}{2} X_A \left| T(j\omega) \right| e^{j(\phi + \theta)}$$

The inverse transform yields the steady-state response in the form

$$y_{SS}(t) = 2|k|\cos(\omega t + \angle k)$$

$$= \underbrace{X_A|T(j\omega)|}_{\text{amplitude}}\cos(\omega t + \underbrace{\phi + \theta}_{\text{phase}}) \tag{11-21}$$

In the development leading to Eq. (11–21), we treat frequency as a general variable where the symbol ω represents all possible input frequencies. In some cases the input frequency has a specific value, which we denote as ω_A. In this case the input is written as

$$x(t) = X_A\cos(\omega_A t + \phi)$$

To obtain the steady-state output, we evaluate the transfer function at the specific frequency (ω_A) of the input sinusoid, namely

$$T(j\omega_A) = |T(j\omega_A)|\angle T(j\omega_A)$$

and the steady-state output is expressed as

$$y_{SS}(t) = X_A|T(j\omega_A)|\cos[\omega_A t + \phi + \angle T(j\omega_A)]$$

This result emphasizes three things about the steady-state response:

(1) Output frequency = Input frequency = ω_A
(2) Output amplitude = Input amplitude × $|T(j\omega_A)|$ (11–22)
(3) Output phase = Input phase + $\angle T(j\omega_A)$

The block diagram in Figure 11–26 helps to clarify this result. The input $x(t)$ contributes the amplitude X_A, the frequency ω_A, and the phase ϕ to the steady-state output $y_{SS}(t)$. The linear circuit contributes a multiplier to the input amplitude and a phase shift. The multiplier is the magnitude of the transfer function $|T(s)|$ evaluated at $s = j\omega_A$ and shown as $|T(j\omega_A)|$. The phase shift is added to the input phase and is the phase of the transfer function $\angle T(s)$ evaluated at $s = j\omega_A$ and shown as $\angle T(j\omega_A)$.

FIGURE 11–26 *Block diagram of sinusoidal steady state response.*

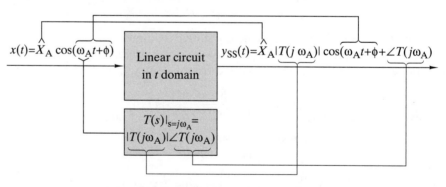

The next two examples illustrate sinusoidal steady-state response calculations. In the first example, frequency is treated as a general variable and we examine the steady-state response as frequency varies over a wide range. In the second example, we evaluate the steady-state response at two specific frequencies.

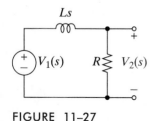

FIGURE 11–27

EXAMPLE 11–13

Find the steady-state output in Figure 11–27 for a general input $v_1(t) = V_A\cos(\omega t + \phi)$.

SOLUTION:

In Example 11–2, the circuit transfer function is shown to be

$$T(s) = \frac{R}{Ls + R}$$

The magnitude and angle of $T(j\omega)$ are

$$|T(j\omega)| = \frac{R}{\sqrt{R^2 + (\omega L)^2}}$$

$$\theta(\omega) = -\tan^{-1}\left(\frac{\omega L}{R}\right)$$

Using Eq. (11–21), the sinusoidal steady-state output is

$$v_{2SS}(t) = \frac{V_A R}{\sqrt{R^2 + (\omega L)^2}} \cos[\omega t + \phi - \tan^{-1}(\omega L/R)]$$

Note that both the amplitude and phase angle of the steady-state response depend on the frequency of the input sinusoid. In particular, at $\omega = 0$ the amplitude of the steady-state output reduces to V_A, which is the same as the amplitude of the input sinusoid. This makes sense because at dc the inductor in Figure 11–27 acts like a short circuit that directly connects the output port to the input port. At very high frequency the amplitude of the steady-state output approaches zero. This also makes sense because at very high frequency the inductor acts like an open circuit that disconnects the output port from the input port. In between these two extremes the output amplitude decreases as the frequency increases. ■

Exercise 11–15 _____

Find the steady-state output in Figure 11–28 for a general input $v_1(t) = V_A \cos(\omega t + \phi)$ V.

Answer:

$$v_{2SS}(t) = \frac{V_A}{R_1 C \sqrt{\omega^2 + \left(\frac{1}{R_2 C}\right)^2}} \cos\left[\omega t + \phi - 180° - \tan^{-1}(\omega R_2 C)\right] \text{ V}$$

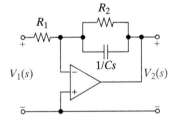

$V_1(s)$ R_1 R_2 $1/Cs$ $V_2(s)$

FIGURE 11–28

The 180° is due to the inverting OP AMP. At dc, the capacitor is an open circuit and the OP AMP is operating as a simple inverter with gain of $-R_2/R_1$. As $\omega \to \infty$, the output approaches zero.

_____

EVALUATION EXAMPLE 11–14

A sensitive piece of electronic equipment is interfered with by a 60 Hz hum. A manufacturer claims his simple circuit can effectively eliminate that noise. The manufacturer's spec sheet does not show what components make up the design, but rather, lists the transfer function as

$$T_V(s) = \frac{s^2 + 142129}{s^2 + 7.45s + 142129}$$

and claims it will reduce the interfering signal by at least 300 times, that is attenuate it by more than 50 dB without any noticeable attenuation of signals as near as 58 and 62 Hz (less than 0.5 dB).

Test the manufacturer's assertion by calculating the sinusoidal steady-state output for five frequencies: 10, 58, 60, 62, and 100 Hz.

SOLUTION:

We will proceed as follows. We will use a one volt sinusoid as the input signal, $v_1(t) = 1\cos 2\pi ft$ V as our reference. We will evaluate the magnitude and phase of the transfer function $T_V(j2\pi f)$ at the five different frequencies. Then we will find the output $v_{2SS}(t)$ for each frequency and compute the effect of the circuit on the various frequencies.

For 10 Hz

$$T_V(j2\pi 10) = \frac{|(j2\pi 10)^2 + 142129|\angle 0°}{\sqrt{[(j2\pi 10)^2 + 142129]^2 + [7.45 \times 2\pi 10]^2}\ \angle\tan^{-1}(468.1)/(138181)}$$

$$T_V(j2\pi 10) = \frac{138181\angle 0°}{138182\angle 0.194°} = 0.999\angle - 0.194°$$

$$v_{2SS}(t) = 0.999\cos(2\pi 10t - 0.194°)\ \text{V}$$

Attenuation dB $= 20\log\dfrac{0.999}{1} = -8.69 \times 10^{-3}$ dB, or less than 0.1%.

For 58 Hz

$$T_V(j2\pi 58) = \frac{|(j2\pi 58)^2 + 142129|\angle 0°}{\sqrt{[(j2\pi 58)^2 + 142129]^2 + [7.45 \times 2\pi 58]^2}\ \angle\tan^{-1}(2714.9)/(9323.6)}$$

$$T_V(j2\pi 58) = \frac{9323.6\angle 0°}{9710.8\angle 16.2°} = 0.960\angle - 16.2°$$

$$v_{2SS}(t) = 0.960\cos(2\pi 58t - 16.2°)\ \text{V}$$

Attenuation dB $= 20\log\dfrac{0.960}{1} = -0.35$ dB, or about 4%.

For 60 Hz

$$T_V(j2\pi 60) = \frac{|(j2\pi 60)^2 + 142129|\angle 0°}{\sqrt{[(j2\pi 60)^2 + 142129]^2 + [7.45 \times 2\pi 60]^2}\ \angle\tan^{-1}(2808.6)/(6.7)}$$

$$T_V(j2\pi 60) = \frac{6.7\angle 0°}{2808.6\angle 89.9°} = .00239\angle - 89.9°$$

$$v_{2SS}(t) = 2.39\cos(2\pi 60t - 89.9°)\ \text{mV}\ \text{or about a reduction of 418 times.}$$

Attenuation dB $= 20\log\dfrac{0.00239}{1} = -52.4$ dB

For 62 Hz

$$T_V(j2\pi 62) = \frac{|(j2\pi 62)^2 + 142129|\angle 180°}{\sqrt{[(j2\pi 62)^2 + 142129]^2 + [7.45 \times 2\pi 62]^2}\ \angle\tan^{-1}(2902.2)/(-9626.0)}$$

$$T_V(j2\pi 62) = \frac{9626.0\angle 180°}{10054\angle 163.3°} = 0.960\angle 16.7°$$

$$v_{2SS}(t) = 0.960\cos(2\pi 62t + 16.7°)\ \text{V}$$

Attenuation dB $= 20\log\dfrac{0.960}{1} = -0.35$ dB, or about 4%.

For 100 Hz

$$T_V(j2\pi 100) = \frac{|(j2\pi 100)^2 + 142129|\angle 180°}{\sqrt{[(j2\pi 100)^2 + 142129]^2 + [7.45 \times 2\pi 100]^2}\ \angle\tan^{-1}(4680.9)/(-252655)}$$

$$T_V(j2\pi 100) = \frac{252655\angle 180°}{252698\angle 178.94°} = 0.999\angle 1.06°$$

$v_{2SS}(t) = 0.999 \cos(2\pi 100t + 1.06°)$ V

Attenuation dB $= 20 \log \dfrac{0.999}{1} = -8.69 \times 10^{-3}$ dB, or less than 0.1%.

The manufacturer's claim was correct. ■

◐ Design Exercise 11–16

Rather than purchasing the device in Example 11–14, design a circuit to achieve the transfer function given. (*Hint:* Use a series *RLC* circuit with the output voltage taken appropriately.)

Answer. As suggested, use a series *RLC* circuit with the output taken across the *L* and *C*. Choose $R = 7.54$ Ω, $L = 1$ H, and $C = 7.04$ μF. Other answers are possible.

EXAMPLE 11–15

The impulse response of a linear circuit is $h(t) = 5000[2e^{-1000t}\cos 2000t - e^{-1000t}\sin 2000t]u(t)$.

(a) Find the sinusoidal steady-state response when $x(t) = 5\cos 1000t$.
(b) Repeat (a) when $x(t) = 5\cos 3000t$.

SOLUTION:
The transfer function corresponding to $h(t)$ is

$$T(s) = H(s) = \mathcal{L}\{h(t)\}$$

$$= 5000\left[\frac{2(s+1000)}{(s+1000)^2 + (2000)^2} - \frac{2000}{(s+1000)^2 + (2000)^2}\right]$$

$$= \frac{10^4 s}{s^2 + 2000s + 5 \times 10^6}$$

(a) At $\omega_A = 1000$ rad/s, the value of $T(j\omega_A)$ is

$$T(j1000) = \frac{10^4(j1000)}{(j1000)^2 + 2000(j1000) + 5 \times 10^6}$$

$$= \frac{j10^7}{(5 \times 10^6 - 10^6) + j2 \times 10^6} = \frac{j10}{4 + j2}$$

$$= \frac{10\,e^{j90°}}{\sqrt{20}e^{j26.6°}} = 2.24\,e^{j63.4°}$$

and the steady-state response for $x(t) = 5\cos 1000t$ is

$$y_{SS}(t) = 5 \times 2.24 \cos(1000t + 0° + 63.4°)$$

$$= 11.2 \cos(1000t + 63.4°)$$

(b) At $\omega_A = 3000$ rad/s, the value of $T(j\omega_A)$ is

$$T(j3000) = \frac{10^4(j3000)}{(j3000)^2 + 2000(j3000) + 5 \times 10^6}$$

$$= \frac{j3 \times 10^7}{5 \times 10^6 - 9 \times 10^6 + j6 \times 10^6}$$

$$= \frac{j30}{-4 + j6} = \frac{30\,e^{j90°}}{\sqrt{52}e^{j123.7°}} = 4.16\,e^{-j33.7°}$$

and the steady-state response for $x(t) = 5 \cos 3000t$ is

$$y_{SS}(t) = 5 \times 4.16 \cos(3000t + 0° - 33.7°)$$
$$= 20.8 \cos(3000t - 33.7°)$$

Again note that the amplitude and phase angle of the steady-state response depend on the input frequency. ∎

Exercise 11-17

The transfer function of a linear circuit is $T(s) = 5(s + 100)/(s + 500)$. Find the steady-state output for

(a) $x(t) = 3 \cos 100t$
(b) $x(t) = 2 \sin 500t$

Answers:
(a) $y_{SS}(t) = 4.16 \cos(100t + 33.7°)$
(b) $y_{SS}(t) = 7.21 \cos(500t - 56.3°)$

Exercise 11-18

The impulse response of a linear circuit is $h(t) = \delta(t) - 100[e^{-100t}]u(t)$. Find the steady-state output for
(a) $x(t) = 25 \cos 100t$
(b) $x(t) = 50 \sin 100t$

Answers:
(a) $y_{SS}(t) = 17.7 \cos(100t + 45°)$
(b) $y_{SS}(t) = 35.4 \cos(100t - 45°)$

NETWORK FUNCTIONS AND PHASOR CIRCUIT ANALYSIS

In this text we present two methods of finding the sinusoidal steady-state response of a linear circuit. Both methods depend on the fact that in the steady state every voltage and current in a linear circuit is a sinusoid of the same frequency as the input sinusoid. As a result, every method of sinusoidal steady-state analysis boils down to finding the amplitude and phase angle of sinusoidal waveforms, all of which have the same frequency.

The steady-state analysis method developed in this chapter involves first finding the network function $T(s)$ that relates an input and a particular output. We next form the complex quantity $T(j\omega)$ by replacing s by $j\omega$, where ω is the frequency of the input sinusoid. The magnitude $|T(j\omega)|$ and angle $\angle T(j\omega)$ then give us the amplitude and phase angle of the steady-state output via Eq. (11–21). Since $T(j\omega)$ is a function of ω, it can describe the steady-state response for a single frequency or whole range of frequencies.

The steady-state analysis method developed in Chapter 8 involves representing the amplitude and phase angles of a sinusoid by a complex number called a phasor. Although not strictly limited to a single frequency, phasor analysis works best when analyzing circuits driven by a single frequency. This method emphasizes the impedance $Z(s)$ of the individual two-terminal elements in the circuit. The element's phasor impedance $Z(j\omega)$ is found by replacing s by $j\omega$, where ω is the frequency of the input sinusoid. When numerical values of the element impedances are given, the specific value of frequency ω is not required. When the element impedances are not given, the frequency must be given so we can compute the phasor impedances of the three passive elements.

$$\begin{aligned} \text{Resistor:} \quad & Z_R = R \\ \text{Inductor:} \quad & Z_L = j\omega L \\ \text{Capacitor:} \quad & Z_C = \frac{1}{j\omega C} \end{aligned}$$

The advantage of the phasor method is that it uses familiar analysis tools like voltage or current division, series or parallel equivalence, source transformation, and mesh or node analysis to solve for the complex numbers (phasors) representing the voltages and currents in the circuit. The magnitude and angle of a phasor are the amplitude and phase angle of the corresponding sinusoidal steady-state response. Responses found using the phasor method are *identical* to those found using the network function method developed in this chapter.

Under what circumstances is one method better than the other?

Phasor circuit analysis works best when the circuit is driven at a single frequency and we need to find several voltages and currents, or the average power, such as in the electric power systems studied in Chapter 16. The network function method works best when there is a single output and the circuit is driven at many different frequencies, such as in the frequency-response and filter applications studied in Chapters 12 and 14. The network function method is the only method available when we know the impulse or step response and need to infer the frequency response of a circuit from its time-domain response. Thus, the preferred method depends on how the circuit is driven (single or multiple frequencies), how much we know about the circuit (complete circuit diagram or only the impulse response), and what we need to find out (multiple responses or a single response). Understanding both methods and how they are applied is important to engineers working with electrical circuits.

11–6 IMPULSE RESPONSE AND CONVOLUTION

In signal processing the term **convolution** refers to a process by which the impulse response of a linear system is used to determine the zero-state response due to other inputs. When the impulse response and input are given in the s domain as $H(s)$ and $X(s)$, the zero-state response is obtained from the inverse Laplace transform of their product.

$$y(t) = \mathscr{L}^{-1}\{H(s)X(s)\} \qquad (11\text{–}23)$$

The purpose of this section is to introduce the notion that the convolution process can also be viewed and carried out entirely in the time domain. Specifically, given a causal impulse response $h(t)$ and a causal input $x(t)$, the zero-state response is obtained from the **convolution integral**[3]

$$y(t) = \int_0^t h(t - \tau)x(\tau)d\tau \qquad (11\text{–}24)$$

where τ is a dummy variable of integration. The shorthand notation $y(t) = h(t) * x(t)$ is used to represent the t-domain process, where the asterisk indicates a convolution integral, not a multiplication. That is, the expression $h(t) * x(t)$ reads "$h(t)$ convolved with $x(t)$," not "$h(t)$ times $x(t)$."

Figure 11–29 indicates the parallelism between the input-output relationships in the s domain and t domain. In the s domain the impulse response $H(s)$ multiplies the input transform to produce the output transform. In the t domain the impulse response $h(t)$ is convolved with the input waveform to produce the output waveform.

The following example illustrates that the same result is achieved whether convolution is carried out in the s domain or the t domain.

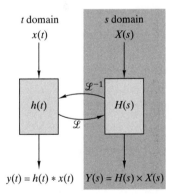

FIGURE 11–29 *Input-output relationships in the t domain and the s domain.*

---

[3]Recall from Chapter 5 that a waveform is causal if $f(t) = 0$ for all $t < 0$. If the impulse response $h(t)$ is not causal, the upper limit in the convolution integral becomes $+\infty$. If the input $x(t)$ is not causal, the lower limit becomes $-\infty$.

EXAMPLE 11-16

A linear circuit has an impulse response $h(t) = 2e^{-t}u(t)$ and an input $x(t) = e^{-2t}u(t)$. Find the zero-state response using

(a) The s-domain process in Eq. (11–23)
(b) The t-domain convolution integral in Eq. (11–24)

SOLUTION:
(a) Converting $h(t)$ and $x(t)$ to the s domain yields

$$H(s) = \mathscr{L}\{2e^{-t}\} = \frac{2}{s+1} \quad \text{and} \quad X(s) = \mathscr{L}\{e^{-2t}\} = \frac{1}{s+2}$$

Applying Eq. (11–23) produces $y(t)$ as

$$y(t) = \mathscr{L}^{-1}\{H(s)X(s)\} = \mathscr{L}^{-1}\left\{\frac{2}{(s+1)(s+2)}\right\}$$

$$= \mathscr{L}^{-1}\left\{\frac{2}{s+1} - \frac{2}{s+2}\right\} = 2e^{-t} - 2e^{-2t} \quad \text{for } t > 0$$

(b) Meanwhile, back in the t domain the convolution integral in Eq. (11–24) yields

$$y(t) = \int_0^t h(t-\tau)x(\tau)d\tau = \int_0^t 2e^{-(t-\tau)}e^{-2\tau}d\tau$$

$$= 2e^{-t}\int_0^t e^{\tau}e^{-2\tau}d\tau = 2e^{-t}\int_0^t e^{-\tau}d\tau$$

$$= 2e^{-t}(1-e^{-t}) = 2e^{-t} - 2e^{-2t} \quad \text{for } t \geq 0$$

The two methods produce the same result. The difference is that the convolution integral evaluation is carried out entirely in the time domain. ∎

Exercise 11-19

A linear circuit has an impulse response $h(t) = e^{-10t}u(t)$ and an input $x(t) = 5u(t)$. Find the zero-state response using the t-domain convolution integral.

Answer:
$$y(t) = \frac{1}{2}(1 - e^{-10t}), \quad t \geq 0$$

EQUIVALENCE OF S-DOMAIN AND T-DOMAIN CONVOLUTION

The approach used here starts out in the s domain with $Y(s) = H(s)X(s)$ and proceeds to show that $y(t) = h(t) * x(t)$. By beginning in the s domain we are assuming that the waveforms are causal. The s-domain input-output relationship can be written as

$$Y(s) = H(s)X(s) = H(s)\underbrace{\left[\int_0^{\infty} x(\tau)e^{-s\tau}d\tau\right]}_{\mathscr{L}\{x(t)\}}$$

where the bracketed term is the integral definition of $\mathscr{L}\{x(t)\}$. The impulse response can be moved inside the integration since $H(s)$ does not depend on the dummy variable τ.

$$Y(s) = \int_0^{\infty} [H(s)e^{-s\tau}]x(\tau)d\tau \tag{11–25}$$

Using the time translation property from Chapter 9 and the integral definition of the Laplace transformation, the bracketed term can be written as

$$H(s)e^{-s\tau} = \mathscr{L}\{h(t-\tau)u(t-\tau)\}$$

$$= \int_0^{\infty} h(t-\tau)u(t-\tau)e^{-st}dt$$

When the last line in this result replaces the bracketed term in Eq. (11–25), we obtain

$$Y(s) = \int_0^\infty \left[\underbrace{\int_0^\infty h(t-\tau)u(t-\tau)e^{-st}dt}_{H(s)e^{-s\tau}} \right] x(\tau)d\tau$$

Interchanging the order of integration produces

$$Y(s) = \int_0^\infty \left[\int_0^t h(t-\tau)x(\tau)\,d\tau \right] e^{-st}\,dt$$

The inner integration is now carried out with respect to the dummy variable τ. The upper limit on this integration need only extend to $\tau = t$ (rather than infinity) since $u(t-\tau) = 0$ for $\tau > t$. By definition, the outer integration (now with respect to t) yields the Laplace transform of the quantity inside the bracket. In other words, this equation is equivalent to the statement

$$Y(s) = \mathscr{L}\left[\underbrace{\int_0^t h(t-\tau)x(\tau)d\tau}_{h(t) * x(t)} \right]$$

So finally we have

$$y(t) = \mathscr{L}^{-1}\{Y(s)\} = \mathscr{L}^{-1}\{H(s)Y(s)\} = h(t) * x(t)$$

which establishes the equivalence of Eqs. (11–23) and (11–24). It is important to remember that this equivalence applies only to causal waveforms.

APPLICATIONS OF THE CONVOLUTION INTEGRAL

Given that s-domain and t-domain convolutions are equivalent, why study both methods? The best answer the authors can give at this point will have an "eat your spinach" ring to it. Suffice it to say that in subsequent courses you will encounter signals for which Laplace transforms do not exist; hence, only the t-domain method convolution is possible. Examples of such application are the noncausal waveforms used in communication systems and the discrete-time signals used in digital signal processing. We cannot treat such applications here, but rather only introduce the student to the concept of viewing convolution as a t-domain process.

EXAMPLE 11–17

A linear circuit has an impulse response $h(t) = 2e^{-t}u(t)$. Use the convolution integral to find the zero-state response for $x(t) = tu(t)$.

SOLUTION:
Direct application of Eq. (11–24) produces the following results

$$y(t) = \int_0^t 2e^{-(t-\tau)}\tau \, d\tau = 2e^{-t}\int_0^t \tau e^\tau \, d\tau$$
$$= 2e^{-t}[e^\tau(\tau-1)]_0^t$$
$$= 2(t-1+e^{-t}) \quad \text{for } t \geq 0$$

In addition to the strictly mathematical approach to evaluating the convolution integral, there is a geometric or graphical interpretation of convolution. The geometric approach offers another perspective, which may provide us with additional insight to the process or help us proceed in situations that are more complicated. ■

Exercise 11–20

A linear circuit has an impulse response $h(t) = Ae^{-at}u(t)$. Use the convolution integral to find the zero-state response for $x(t) = Be^{-bt}u(t)$. Assume $a \neq b$ and that both a and b are positive, real numbers.

Answer:

$$y(t) = \frac{AB(e^{-at} - e^{-bt})}{b - a} u(t)$$

Figure 11–30 shows a geometric interpretation of the convolution in Example 11–17. In Figures 11–30(a) and 11–30(b) the input and impulse response waveforms are plotted against the dummy variable τ. Forming $h(-\tau)$ *reflects* the impulse response across the $\tau = 0$ axis, as shown in Figure 11–30(c). Forming $h(t - \tau)$ *shifts* the reflected impulse response to the right by t seconds, as shown in Figure 11–30(d). *Multiplying* the reflected/shifted impulse response by the input produces the product $h(t - \tau) \times x(\tau)$, shown in Figure 11–30(e). The *integrating* from $\tau = 0$ to $\tau = t$ yields the area under this product, which is the value of the zero-state output at time t. At a later instant of time the reflected impulse response $h(t - \tau)$ shifts farther to the right, creating a new product $h(t - \tau) \times x(\tau)$ with a new area and a new value of $y(t)$.

Thus, the geometric interpretation of t-domain convolution involves four operations: reflecting, shifting, multiplying, and integrating. We visualize convolution as a process that reflects the impulse response across the origin and then progressively shifts it to the right as t increases. At any time t the output is the area under the product of the reflected/shifted impulse response and the input. Under this interpretation we can think of the impulse response as a weighting function. That is, when

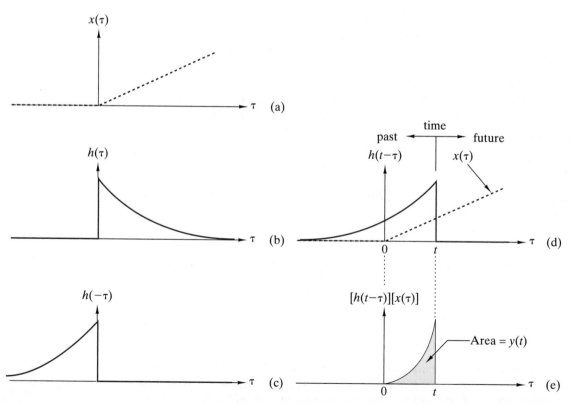

FIGURE 11–30 *Graphical interpretation of convolution.*

integrating the product $h(t - \tau) \times x(\tau)$, the impulse response tells us how much weight to assign to previous values of the input.

EXAMPLE 11–18

A certain circuit has an impulse response $h(t) = [2e^{-t}]u(t)$. Use the convolution integral to find the zero-state response for $x(t) = 5[u(t) - u(t - 2)]$.

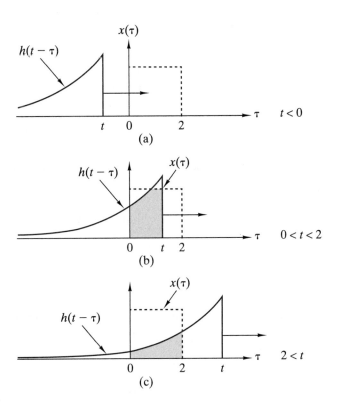

FIGURE 11–31

SOLUTION:
Evaluation of the convolution integral can be divided into the three situations shown in Figure 11–31. The situation for $t < 0$ is shown in Figure 11–31(a). For this case the reflected impulse response $h(t - \tau)$ and the input $x(\tau)$ do not overlap so the area under their product is zero. Hence $y(t) = 0$ for $t < 0$. This simply says that the zero-state response is causal when the impulse response and input are causal.

For $0 < t < 2$, the reflected impulse response and input overlap as shown in Figure 11–31(b). In this situation the area under the product $h(t - \tau) \times x(\tau)$ is found by integrating $\tau = 0$ to $\tau = t$.

$$y(t) = \int_0^t [2e^{-(t-\tau)}][5]d\tau$$

$$= 10e^{-t} \int_0^t e^\tau d\tau = 10e^{-t}(e^t - 1)$$

$$= 10(1 - e^{-t}) \quad \text{for} \quad 0 < t < 2$$

For $t > 2$, the reflected impulse response and input overlap as shown in Figure 11–31(c). In this case the product $h(t - \tau) \times x(\tau)$ is confined to the interval

$0 < \tau < 2$ since the input $x(\tau)$ is zero everywhere outside of this interval. In this situation the area under the product is found by integrating from $\tau = 0$ and $\tau = 2$.

$$y(t) = \int_0^2 [2e^{-(t-\tau)}][5]d\tau$$

$$= 10 e^{-t} \int_0^2 e^{\tau} d\tau$$

$$= 10 e^{-t}(e^2 - 1) \quad \text{for} \quad 2 \le t$$

Evaluation of the convolution integral was guided by the geometric interpretation in Figure 11–31 and leads to a zero-state response defined on three intervals.

$$y(t) = \begin{vmatrix} 0 & \text{for} & t < 0 \\ 10(1 - e^{-t}) & \text{for} & 0 \le t < 2 \\ 10e^{-t}(e^2 - 1) & \text{for} & 2 \le t \end{vmatrix}$$ ∎

Exercise 11–21

Repeat Example 11–18 by mathematically, not geometrically, computing the convolution integral. Verify that the result is equivalent to the answer found for Example 11–18. You may use MATLAB.

Answer: $y(t) = 10[1 - e^{-t}]u(t) - 10[1 - e^{-(t-2)}]u(t - 2)$. The result is equivalent to the answer in Example 11–18.

EXAMPLE 11–19

Use time-domain and s-domain convolution to find the zero-state response when

$$h(t) = x(t) = [2e^{-t}]u(t)$$

SOLUTION:
Using the convolution integral in Eq. (11–24) produces

$$y(t) = \int_0^t 2e^{-(t-\tau)}2e^{-\tau}d\tau = 4e^{-t} \int_0^t e^{\tau}e^{-\tau}d\tau$$

$$= 4e^{-t} \int_0^t d\tau = 4te^{-t} \quad \text{for} \quad t \ge 0$$

Using s-domain convolution, we have $H(s) = X(s) = 2/(s + 1)$; hence,

$$Y(s) = H(s) \times X(s) = \frac{4}{(s + 1)^2}$$

which yields $y(t) = \mathscr{L}^{-1}\{Y(s)\} = [4te^{-t}]u(t)$. ∎

Exercise 11–22

Use the convolution integral to find the zero-state response for $h(t) = 2u(t)$ and $x(t) = 5[u(t) - u(t - 2)]$.

Answer:

$$y(t) = \begin{vmatrix} 0 & \text{for} & t < 0 \\ 10t & \text{for} & 0 \le t < 2 \\ 20 & \text{for} & 2 \le t \end{vmatrix}$$

11–7 NETWORK FUNCTION DESIGN

Finding and using a network function of a given circuit is an *s*-domain **analysis** problem. An *s*-domain **synthesis** problem involves finding a circuit that realizes a given network function. For linear circuits, an analysis problem always has a unique solution. In contrast, a synthesis problem may have many solutions because different circuits can have the same network function. A transfer function design problem involves synthesizing several circuits that realize a given function and evaluating the alternative designs, using criteria such as input or output impedance, cost, and power consumption.

The design process discussed here begins with a given transfer function $T_V(s)$. We partition this transfer function into a product of simpler functions.

$$T_V(s) = T_{V1}(s)T_{V2}(s) \cdots T_{Vk}(s)$$

We then realize each of these simpler functions using basic circuit modules such as voltage dividers, inverting amplifiers, and noninverting amplifiers. The overall transfer function is then achieved by connecting the individual stages in cascade, as indicated in Figure 11–32.

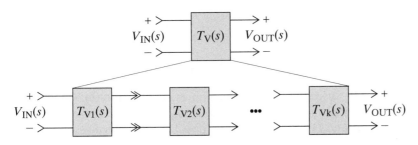

FIGURE 11–32 *Cascade connection transfer functions.*

Of course, this approach assumes that the chain rule applies. In other words, we must avoid loading when designing the stages in the cascade realization. This is accomplished by coordinating the input and output impedances of adjacent stages or using OP AMP voltage followers to isolate the individual stages.

Before turning to examples, we discuss the design of simple one-pole modules that serve as the building block stages in a cascade design.

FIRST-ORDER VOLTAGE-DIVIDER CIRCUIT DESIGN

We begin our study of transfer function design by developing a voltage-divider realization of a first-order transfer function of the form $K/(s + \alpha)$. The impedances $Z_1(s)$ and $Z_2(s)$ are related to the given transfer function using the voltage-divider relationship.

$$T_V(s) = \frac{K}{s + \alpha} = \frac{Z_2(s)}{Z_1(s) + Z_2(s)} \tag{11–26}$$

To obtain a circuit realization, we must assign part of the given $T_V(s)$ to $Z_2(s)$ and the remainder to $Z_1(s)$. There are many possible realizations of $Z_1(s)$ and $Z_2(s)$ because there is no unique way to make this assignment. For example, simply equating the numerators and denominators in Eq. (11–26) yields

$$Z_2(s) = K \quad \text{and} \quad Z_1(s) = s + \alpha - Z_2(s) = s + \alpha - K \tag{11–27}$$

Inspecting this result, we see that $Z_2(s)$ is realizable as a resistance ($R_2 = K\Omega$) and $Z_1(s)$ as an inductance ($L_1 = 1$ H) in series with a resistance [$R_1 = (\alpha - K)\Omega$]. The resulting circuit diagram is shown in Figure 11–33(a). For $K = \alpha$ the resistance R_1 can be replaced by a short circuit because its resistance is zero. A gain restriction $K \le \alpha$ is necessary because a negative R_1 is not physically realizable as a single component.

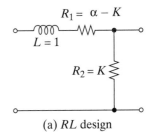

(a) *RL* design

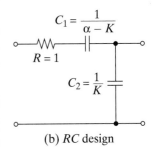

(b) *RC* design

FIGURE 11–33 *Circuit realizations of $T(s) = K/(s + \alpha)$ for $K \le \alpha$.*

An alternative synthesis approach involves factoring s out of the denominator of the given transfer function. In this case, Eq. (11–26) is rewritten in the form

$$T_V(s) = \frac{K/s}{1 + \alpha/s} = \frac{Z_2(s)}{Z_1(s) + Z_2(s)} \tag{11–28}$$

Equating numerators and denominators yields the branch impedances

$$Z_2(s) = \frac{K}{s} \quad \text{and} \quad Z_1(s) = 1 + \frac{\alpha}{s} - Z_2(s) = 1 + \frac{\alpha - K}{s} \tag{11–29}$$

In this case we see that $Z_2(s)$ is realizable as a capacitance $(C_2 = 1/K\,\text{F})$ and $Z_1(s)$ as a resistance $(R_1 = 1\,\Omega)$ in series with a capacitance $[C_1 = 1/(\alpha - K)\,\text{F}]$. The resulting circuit diagram is shown in Figure 11–33(b). For $K = \alpha$, the capacitance C_1 can be replaced by a short circuit because its capacitance is infinite. A gain restriction $K \leq \alpha$ is required to keep C_1 from being negative.

As a second design example, consider a voltage-divider realization of the transfer function $Ks/(s + \alpha)$. We can find two voltage-divider realizations by writing the specified transfer function in the following two ways:

$$T(s) = \frac{Ks}{s + \alpha} = \frac{Z_2(s)}{Z_1(s) + Z_2(s)} \tag{11–30a}$$

$$T(s) = \frac{K}{1 + \alpha/s} = \frac{Z_2(s)}{Z_1(s) + Z_2(s)} \tag{11–30b}$$

Equation (11–30a) uses the transfer function as given, while Eq. (11–30b) factors s out of the numerator and denominator. Equating the numerators and denominators in Eqs. (11–30a) and (11–30b) yields two possible impedance assignments:

Using Eq. (11–30a): $Z_2 = Ks$ and $Z_1 = s + \alpha - Z_2 = (1 - K)s + \alpha$ (11–31a)

Using Eq. (11–30b): $Z_2 = K$ and $Z_1 = 1 + \frac{\alpha}{s} - Z_2 = (1 - K) + \frac{\alpha}{s}$ (11–31b)

The assignment in Eq. (11–31a) yields $Z_2(s)$ as an inductance $L_2 = K\,\text{H}$ and $Z_1(s)$ as an inductance $[L_1 = (1 - K)\,\text{H}]$ in series with a resistance $(R_1 = \alpha\,\Omega)$. The assignment in Eq. (11–31b) yields $Z_2(s)$ as a resistance $(R_2 = K\Omega)$ and $Z_1(s)$ as a resistance $[R_1 = (1 - K)\Omega]$ in series with a capacitance $(C_1 = 1/\alpha\,\text{F})$. The two realizations are shown in Figure 11–34. Both realizations require $K \leq 1$ for the branch impedances to be realizable and both simplify when $K = 1$.

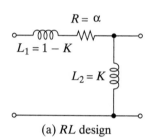

(a) *RL* design

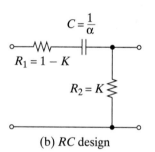

(b) *RC* design

FIGURE 11–34 *Circuit realizations of $T(s) = Ks/(s + \alpha)$ for $K \leq 1$.*

ⓘ Design Exercise 11–23

Design an *RC* circuit to realize the following transfer function

$$T(s) = \frac{200}{s + 1000}$$

Answer: Use the circuit of Figure 11–33(b) with $R = 1\,\Omega$, $C_1 = 1250\,\mu\text{F}$, and $C_2 = 5000\,\mu\text{F}$. We will learn how to scale these answers to more practical device values later in this section.

VOLTAGE-DIVIDER AND OP AMP CASCADE CIRCUIT DESIGN

The examples in Figures 11–33 and 11–34 illustrate an important feature of voltage-divider realizations. In general, we can write a transfer function as a quotient of polynomials $T(s) = r(s)/q(s)$. A voltage-divider realization requires the impedances $Z_2(s) = r(s)$ and $Z_1(s) = q(s) - r(s)$ to be physically realizable. A voltage-divider circuit usually places limitations on the gain K. This gain limitation can be overcome by using an OP AMP circuit in cascade with the divider circuit.

For example, a voltage-divider realization of the transfer function in Eq. (11–26) requires $K \leq \alpha$. When $K > \alpha$, then $T(s)$ is not realizable as a simple voltage divider, since $Z_2(s) = s + \alpha - K$ requires a negative resistance. However, the given transfer function can be written as a two-stage product:

$$T_v(s) = \frac{K}{s + \alpha} = \underbrace{\left[\frac{K}{\alpha} \right]}_{\substack{\text{first} \\ \text{stage}}} \underbrace{\left[\frac{\alpha}{s + \alpha} \right]}_{\substack{\text{second} \\ \text{stage}}}$$

When $K > \alpha$, the first stage has a positive gain greater than unity. This stage can be realized using a noninverting OP AMP circuit with a gain of $(R_1 + R_2)/R_1$. The first-stage design constraint is

$$\frac{K}{\alpha} = \frac{R_1 + R_2}{R_1}$$

Choosing $R_1 = 1\ \Omega$ requires that $R_2 = (K/\alpha) - 1$. An RC voltage-divider realization of the second stage is obtained by factoring an s out of the stage transfer function. This leads to the second-stage design constraint

$$\frac{\alpha/s}{1 + \alpha/s} = \frac{Z_2(s)}{Z_1(s) + Z_2(s)}$$

Equating numerators and denominators yields $Z_2(s) = \alpha/s$ and $Z_1(s) = 1$. Figure 11–35 shows a cascade connection of a noninverting first stage and the RC divider second stage. The chain rule applies to this circuit, since the first stage has an OP AMP output. The cascade circuit in Figure 11–35 realizes the first-order transfer function $K/(s + \alpha)$ for $K > \alpha$, a gain requirement that cannot be met by the divider circuit alone.

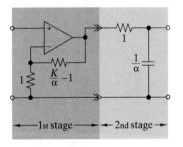

FIGURE 11–35 *Circuit realization of $T(s) = K/(s + \alpha)$ for $K > \alpha$.*

ⓓ Design Exercise 11–24 _____

Design an active RC circuit to realize the following transfer function

$$T(s) = \frac{2000}{s + 1000}$$

A n s w e r : Use the circuit shown in Figure 11–35. The OP AMP stage has a gain of 2 by making both resistors equal. Choose the components in the second stage voltage divider so that $R = 1\ \Omega$ and $C = 1000\ \mu F$. We will learn how to scale these answers to more practical device values later in this section.

ⓓ Design Exercise 11–25 _____

Design an active RL circuit to realize the following transfer function

$$T(s) = \frac{2000}{s + 1000}$$

A n s w e r : Use the circuit shown in Figure 11–35. The OP AMP stage has a gain of 2 by making both resistors equal. In the second stage, replace the resistor with an inductor and replace the capacitor with a resistor. Let the components in the second stage voltage divider be $R = 1\ k\Omega$ and $L = 1\ H$.

◗ **DESIGN EXAMPLE 11–20**

Design a circuit to realize the following transfer function using only resistors, capacitors, and OP AMPs:

$$T_V(s) = \frac{3000s}{(s + 1000)\,(s + 4000)}$$

SOLUTION:
The given transfer function can be written as a three-stage product.

$$T_V(s) = \underbrace{\left[\frac{K_1}{s + 1000}\right]}_{\substack{\text{first} \\ \text{stage}}} \underbrace{[K_2]}_{\substack{\text{second} \\ \text{stage}}} \underbrace{\left[\frac{K_3 s}{s + 4000}\right]}_{\substack{\text{third} \\ \text{stage}}}$$

where the stage gains K_1, K_2, and K_3 have yet to be selected. Factoring s out of the denominator of the first-stage transfer function leads to an RC divider realization:

$$\frac{K_1/s}{1 + 1000/s} = \frac{Z_2(s)}{Z_1(s) + Z_2(s)}$$

Equating numerators and denominators yields

$$Z_2(s) = K_1/s \quad \text{and} \quad Z_1(s) = 1 + (1000 - K_1)/s$$

The first stage $Z_1(s)$ is simpler when we select $K_1 = 1000$. Factoring s out of the denominator of the third-stage transfer function leads to an RC divider realization:

$$\frac{K_3}{1 + 4000/s} = \frac{Z_2(s)}{Z_1(s) + Z_2(s)}$$

Equating numerators and denominators yields

$$Z_2(s) = K_3 \quad \text{and} \quad Z_1(s) = 1 - K_3 + 4000/s$$

The third stage $Z_1(s)$ is simpler when we select $K_3 = 1$. The stage gains must meet the constraint $K_1 \times K_2 \times K_3 = 3000$ since the overall gain of the given transfer function is 3000. We have selected $K_1 = 1000$ and $K_3 = 1$, which requires $K_2 = 3$. The second stage must have a positive gain greater than 1 and can be realized using a noninverting amplifier with $K_2 = (R_1 + R_2)/R_1 = 3$. Selecting $R_1 = 1\Omega$ requires that $R_2 = 2\Omega$.

Figure 11–36 shows the three stages connected in cascade. The chain rule applies to this cascade connection because the OP AMP in the second stage isolates the RC voltage-divider circuits in the first and third stages. The order of the first and third stages can be swapped in this design without consequence. The circuit in Figure 11–36 realizes the given transfer function but is not a realistic design because the values of resistance and capacitance are impractical. For this reason we call this circuit a **prototype** design. We will shortly discuss how to scale a prototype to obtain practical element values.

FIGURE 11–36

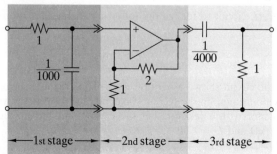

■

◑ Design Exercise 11–26

Design a circuit to realize the following transfer function using only resistors, capacitors, and no more than one OP AMP.

$$T_V(s) = \frac{10^6}{(s + 10^3)^2}$$

A n s w e r :   Figure 11–37 shows one possible prototypical solution.

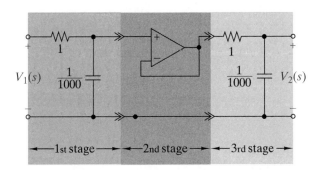

FIGURE 11–37

INVERTING OP AMP CIRCUIT DESIGN

The inverting OP AMP circuit places fewer restrictions on the form of the desired transfer function than does the basic voltage divider. To illustrate this, we will develop two inverting OP AMP designs for a general first-order transfer function of the form

$$T_V(s) = -K \frac{s + \gamma}{s + \alpha}$$

The general transfer function of the inverting OP AMP circuit is $-Z_2(s)/Z_1(s)$, which leads to the general design constraint

$$-K \frac{s + \gamma}{s + \alpha} = -\frac{Z_2(s)}{Z_1(s)} \qquad (11\text{–}32)$$

The first design is obtained by equating the numerators and denominators in Eq. (11–32) to obtain the OP AMP circuit impedances as $Z_2(s) = Ks + K\gamma$ and $Z_1(s) = s + \alpha$. Both of these impedances are of the form $Ls + R$ and can be realized by an inductance in series with a resistance, leading to the design realization in Figure 11–38(a).

A second inverting OP AMP realization is obtained by equating $Z_2(s)$ in Eq. (11–32) to the reciprocal of the denominator and equating $Z_1(s)$ to the reciprocal of the numerator. This assignment yields the impedances $Z_1(s) = 1/(Ks + K\gamma)$ and $Z_2(s) = 1/(s + \alpha)$. Both of these impedances are of the form $1/(Cs + G)$, where Cs is the admittance of a capacitor and G is the admittance of a resistor. Both impedances can be realized by a capacitance in parallel with a resistance. These impedance identifications produce the RC circuit in Figure 11–38(b).

Because it has fewer restrictions, it is often easier to realize transfer functions using the inverting OP AMP circuit. To use inverting circuits, the given transfer function must require an inversion or be realized using an even number of inverting stages. In some cases, the sign in front of the transfer function is immaterial and the required transfer function is specified as $\pm T_V(s)$. *Caution:* The input impedance of an inverting OP AMP circuit may load the source circuit.

FIGURE 11–38 *Inverting OP AMP circuit realizations of* $T(s) = -K(s + \gamma)/(s + \alpha)$.

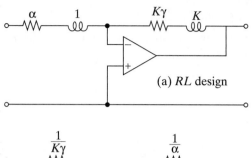

(a) *RL* design

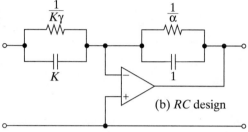

(b) *RC* design

◆ Design Exercise 11–27

Design an active *RC* prototype circuit to realize the following transfer function

$$T(s) = -100 \frac{s + 50}{s + 100}$$

Answer: See Figure 11–39.

FIGURE 11–39

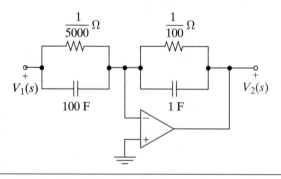

◆ **DESIGN EXAMPLE 11 – 21**

Design a circuit to realize the transfer function given in Example 11–20 using inverting OP AMP circuits.

SOLUTION:

The given transfer function can be expressed as the product of two inverting transfer functions:

$$T_V(s) = \frac{3000s}{(s + 1000)(s + 4000)} = \underbrace{\left[-\frac{K_1}{s + 1000} \right]}_{\text{first stage}} \underbrace{\left[-\frac{K_2 s}{s + 4000} \right]}_{\text{second stage}}$$

where the stage gains K_1 and K_2 have yet to be selected. The first stage can be realized in an inverting OP AMP circuit since

$$-\frac{K_1}{s + 1000} = -\frac{K_1/1000}{1 + s/1000} = -\frac{Z_2(s)}{Z_1(s)}$$

Equating the $Z_2(s)$ to the reciprocal of the denominator and $Z_1(s)$ to the reciprocal of the numerator yields

$$Z_2 = \frac{1}{1 + s/1000} \quad \text{and} \quad Z_1 = 1000/K_1$$

The impedance $Z_2(s)$ is realizable as a capacitance $(C_2 = 1/1000\,\text{F})$ in parallel with a resistance $(R_2 = 1\,\Omega)$ and $Z_1(s)$ as a resistance $(R_1 = 1000/K_1\,\Omega)$. We select $K_1 = 1000$ so that the two resistances in the first stage are equal. Since the overall gain requires $K_1 \times K_2 = 3000$, this means that $K_2 = 3$. The second-stage transfer function can also be produced using an inverting OP AMP circuit:

$$-\frac{3s}{s + 4000} = -\frac{3}{1 + 4000/s} = -\frac{Z_2(s)}{Z_1(s)}$$

Equating numerators and denominators yields $Z_2(s) = R_2 = 3$ and $Z_1(s) = R_1 + 1/C_1 s = 1 + 4000/s$.

Figure 11–40 shows the cascade connection of the RC OP AMP circuits that realize each stage. The overall transfer function is noninverting because the cascade uses an even number of inverting stages. The chain rule applies here since the first stage has an OP AMP output. The circuit in Figure 11–40 is a prototype design because the values of resistance and capacitance are impractical.

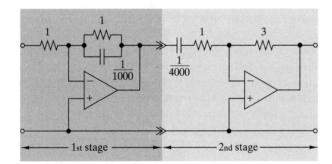

FIGURE 11–40

◑ Design Exercise 11–28

Design a circuit to realize the following transfer function using only resistors, capacitors, and no more than one OP AMP.

$$T_V(s) = \frac{-10^6}{(s + 10^3)^2}$$

Answer: Figure 11–41 shows one possible prototypical solution.

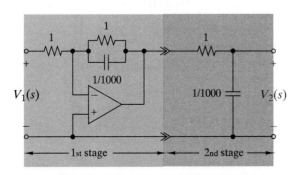

FIGURE 11–41

MAGNITUDE SCALING

The circuits obtained in Examples 11–20 and 11–21 are called prototype designs because the element values are outside of practical ranges. The allowable ranges depend on the fabrication technology used to construct the circuits. For example, monolithic integrated circuit (IC) technology limits capacitances to a few hundred picofarads, and inductors are difficult to manufacture on ICs. An OP AMP circuit should have a feedback resistance greater than 1 kΩ to keep the output current demand within the capabilities of general-purpose OP AMP devices. Other technologies and applications place different constraints on element values. For example, the power industry, where physical size is of less importance, uses devices with much larger values than the electronics industry.

There are no hard and fast rules here, but, roughly speaking, an electronic circuit is probably realizable by some means if its passive element values fall in the ranges shown in the tables on the inside rear cover, with the caveat that OP AMP circuits generally use $Rs > 1$ kΩ.

These are:

Capacitors[4]: 1 pF to 10,000 μF

Inductors[5]: 10 nH to 10 mH

Resistors[6]: 10 Ω to 10 MΩ.

The important idea here is that circuit designs like Figure 11–40 are impractical because 1-Ω resistors are too small for OP AMP circuits and 1-mF capacitors are too large physically.

It is often possible to scale the magnitude of circuit impedances so that the element values fall into practical ranges. The key is to scale the element values in a way that does not change the transfer function of the circuit. Multiplying the numerator and denominator of the transfer function of a voltage-divider circuit by a scale factor k_m yields

$$T_V(s) = \frac{k_m}{k_m} \frac{Z_2(s)}{Z_1(s) + Z_2(s)} = \frac{k_m Z_2(s)}{k_m Z_1(s) + k_m Z_2(s)} \tag{11–33}$$

Clearly, this modification does not change the transfer function but scales each impedance by a factor of k_m and changes the element values in the following way:

$$R_{\text{after}} = k_m R_{\text{before}} \quad L_{\text{after}} = k_m L_{\text{before}} \quad C_{\text{after}} = \frac{C_{\text{before}}}{k_m} \tag{11–34}$$

Equation (11–34) was derived using the transfer function of a voltage-divider circuit. It is easy to show that we would reach the same conclusion if we had used the transfer functions of inverting or noninverting OP AMP circuits.

In general, a circuit is magnitude scaled by multiplying all resistances, multiplying all inductances, and dividing all capacitances by a scale factor k_m. The scale factor must be positive, but can be greater than or less than 1. Different scale factors can be used for each stage of a cascade design, but only one scale factor can be used for each stage. These scaling operations do not change the voltage transfer function realized by the circuit.

---

[4]Recent innovations in dielectrics have enabled a large new class of electronic double-layer capacitors (EDLC) or *supercapacitors* with capacitances up to 5000 F. These devices are still relatively large for small electronic applications.
[5]Inductors up to 10 H, also called *chokes,* are possible but are quite large.
[6]Resistors are manufactured outside this range but are used only in specialty applications.

Our design strategy is first to create a prototype circuit whose element values may be unrealistically large or small. Applying magnitude scaling to the prototype produces a design with practical element values. Sometimes there may be no scale factor that brings the prototype element values into a practical range. When this happens, we must seek alternative realizations because the scaling process is telling us that the prototype is not a viable candidate.

EXAMPLE 11–22

Magnitude scale the circuit in Figure 11–40 so all resistances are at least 10 kΩ and all capacitances are less than 1 μF.

SOLUTION:
The resistance constraint requires $k_m R \geq 10^4 \, \Omega$. The smallest resistance in the prototype circuit is 1 Ω; therefore, the resistance constraint requires $k_m \geq 10^4$. The capacitance constraint requires $C/k_m \leq 10^{-6}$ F. The largest capacitance in the prototype is 10^{-3} F; therefore, the capacitance constraint requires $k_m \geq 10^3$. The resistance condition on k_m dominates the two constraints. Selecting $k_m = 10^4$ produces the scaled design in Figure 11–42. This circuit realizes the same transfer function as the prototype in Figure 11–40 but uses practical element values.

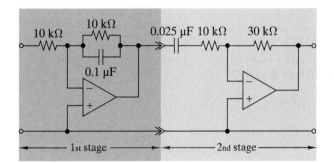

FIGURE 11–42

Exercise 11–29

Select a magnitude scale factor for each stage in Figure 11–36 so that both capacitances are 0.01 μF and all resistances are greater than 10 kΩ.

Answer: $k_m = 10^5$ for the first stage; $k_m = 10^4$ for the second stage; $k_m = 0.25 \times 10^5$ for the third stage.

Exercise 11–30

Select a magnitude scale factor for the OP AMP circuit in Figure 11–39.

Answer: $k_m = 10^8$, any larger and the feedback resistor becomes too large, any smaller and the input capacitor becomes too large.

SECOND-ORDER CIRCUIT DESIGN

An *RLC* voltage divider can also be used to realize second-order transfer functions. For example, the transfer function

$$T_V(s) = \frac{K}{s^2 + 2\zeta\omega_0 s + \omega_0^2}$$

can be realized by factoring s out of the denominator and equating the result to the voltage-divider input-output relationship:

$$T_V(s) = \frac{K/s}{s + 2\zeta\omega_0 + \omega_0^2/s} = \frac{Z_2(s)}{Z_1(s) + Z_2(s)}$$

Equating numerators and denominators yields

$$Z_2(s) = \frac{K}{s} \quad \text{and} \quad Z_1(s) = s + 2\zeta\omega_0 + \frac{\omega_0^2 - K}{s}$$

The impedance $Z_2(s)$ is realizable as a capacitance ($C_2 = 1/K$ F) and $Z_1(s)$ as a series connection of an inductance ($L_1 = 1$ H), resistance ($R_1 = 2\zeta\omega_0 \, \Omega$), and capacitance $\left[C_1 = 1/(\omega_0^2 - K) \, \text{F} \right]$. The resulting voltage-divider circuit is shown in Figure 11–43(a). The impedances in this circuit are physically realizable when $K \leq \omega_0^2$. Note that the resistance controls the damping ratio ζ because it is the element that dissipates energy in the circuit. Also note that if $K = \omega_0^2$, then the capacitor C_1 is replaced by a short circuit.

FIGURE 11–43 *Second-order circuit realizations.*

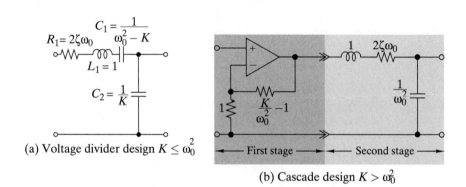

(a) Voltage divider design $K \leq \omega_0^2$

(b) Cascade design $K > \omega_0^2$

When $K > \omega_0^2$, we can partition the transfer function into a two-stage cascade of the form

$$T_V(s) = \underbrace{\left[\frac{K}{\omega_0^2} \right]}_{\substack{\text{first} \\ \text{stage}}} \underbrace{\left[\frac{\omega_0^2/s}{s + 2\zeta\omega_0 + \omega_0^2/s} \right]}_{\text{second stage}}$$

The first stage requires a positive gain greater than unity and can be realized using a noninverting OP AMP circuit. The second stage can be realized as a voltage divider with $Z_2(s) = \omega_0^2/s$ and $Z_1(s) = s + 2\zeta\omega_0$. The resulting cascade circuit is shown in Figure 11–43(b).

◑ **DESIGN EXAMPLE 11–23**

Find a second-order realization of the transfer function given in Example 11–20.

SOLUTION:
The given transfer function can be written as

$$T_V(s) = \frac{3000s}{(s + 1000)(s + 4000)} = \frac{3000s}{s^2 + 5000s + 4 \times 10^6}$$

Factoring s out of the denominator and equating the result to the transfer function of a voltage divider gives

$$\frac{3000}{s + 5000 + 4 \times 10^6/s} = \frac{Z_2(s)}{Z_1(s) + Z_2(s)}$$

Equating the numerators and denominators yields

$$Z_2(s) = 3000 \quad \text{and} \quad Z_1(s) = s + 2000 + 4 \times 10^6/s$$

Both of these impedances are realizable, so a single-stage voltage-divider design is possible. The prototype impedance $Z_1(s)$ requires a 1-H inductor, which is a bit large. A more practical value is obtained using a scale factor of $k_m = 0.1$. The resulting scaled voltage divider circuit is shown in Figure 11–44. ∎

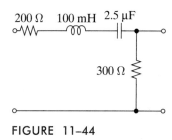

FIGURE 11–44

ⓓ Design Exercise 11–31

Design a second-order circuit to realize the following transfer function:

$$T_V(s) = \frac{10^6}{(s + 10^3)^2}$$

Answer: Figure 11–45 shows one possible solution.

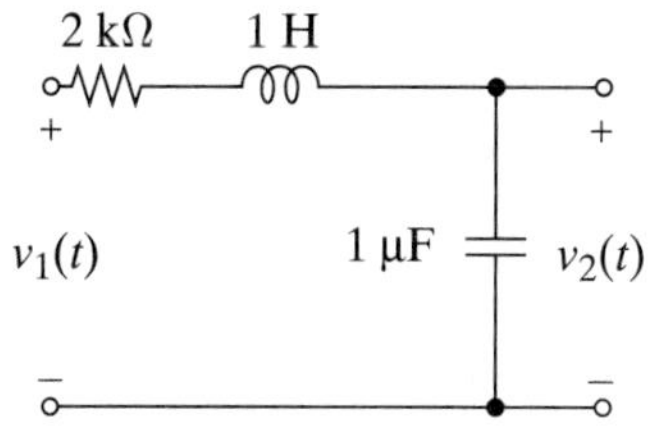

FIGURE 11–45

DESIGN EVALUATION SUMMARY

Examples 11–20, 11–21, and 11–23 show three different ways to realize the transfer function

$$T_V(s) = \frac{3000s}{(s + 1000)(s + 4000)}$$

This illustrates that a design requirement can have many solutions. Selecting the best design from among the alternatives involves additional criteria such as element count, power requirements, and output loading effects.

The element counts for each design are shown in Table 11–1. On a pure element-count basis, the RLC divider in Figure 11–44 in the best design. However, inductors have some serious drawbacks. They are heavy and lossy in low-frequency applications and are not easily fabricated in integrated circuit form. Fortunately, inductors are not essential to transfer function design, as shown by the two RC OP AMP designs.

Power requirements: The two RC OP AMP designs require external dc power supplies. The voltage divider cascade in Figure 11–36 requires less power since it uses only one OP AMP, compared with the two-OP-AMP inverting cascade. Thus, power requirements would favor the one-OP-AMP circuit over the two-OP-AMP circuit.

TABLE 11–1

EXAMPLE	FIGURE	DESCRIPTION	R	L	C	OP AMP
			|←— NUMBER OF —→|			
11–20	11–36	RC voltage-divider cascade	4	0	2	1
11–21	11–40	RC inverting cascade	4	0	2	2
11–23	11–44	RLC voltage divider	2	1	1	0

Output loading: The output impedance of the design is important if the circuit must drive a finite load of, say, 1 kΩ. The resulting loading effects could defeat the basic purpose of the circuit by changing its transfer function. Output loading considerations favor the inverting cascade in Figure 11–40 because it has an OP AMP output that has zero output impedance.

A design problem involves more than simply finding a prototype that realizes a given transfer function. In general, the first step in a design problem involves determining an acceptable transfer function, one that meets performance requirements such as the characteristics of the step or frequency response. In other words, we must first design the transfer function and then design several circuits that realize the transfer function. To deal with transfer function design we must understand how performance characteristics are related to transfer functions. The next two chapters provide some background on this issue.

◀ ▶ **D E** ⬤ **DESIGN AND EVALUATION EXAMPLE 11 – 24**

Given the step response $g(t) = \pm[1 + 4e^{-500t}]u(t)$,

(a) Find the transfer function $T(s)$.
(b) Design two *RC* OP AMP circuits that realize the $T(s)$ found in part (a).
(c) Evaluate the two designs on the basis of element count, input impedance, output impedance.

SOLUTION:
(a) The transform of the step response is

$$G(s) = \pm\mathscr{L}\{[1 + 4e^{-500t}]u(t)\} = \pm\left[\frac{1}{s} + \frac{4}{s + 500}\right] = \pm\frac{5s + 500}{s(s + 500)}$$

and the required transfer function is

$$T(s) = H(s) = sG(s) = \pm\frac{5s + 500}{s + 500}$$

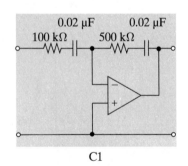

0.02 µF   0.02 µF
100 kΩ   500 kΩ

C1

(b) The first design uses an inverting OP AMP configuration. Using the minus sign on the transfer function $T(s)$ and factoring an s out of the numerator and denominator yield

$$T(s) = -\frac{5 + 500/s}{1 + 500/s} = -\frac{Z_2(s)}{Z_1(s)}$$

Equating numerators and denominators yields $Z_2(s) = 5 + 500/s$ and $Z_1(s) = 1 + 500/s$. The impedance $Z_2(s)$ is realizable as a resistance ($R_2 = 5\,\Omega$) in series with a capacitance ($C_2 = 1/500\,\mathrm{F}$) and $Z_1(s)$ as a resistance ($R_1 = 1\,\Omega$) in series with a capacitance ($C_1 = 1/500\,\mathrm{F}$). Using a magnitude scale factor $k_m = 10^5$ produces circuit C1 in Figure 11–46.

The second design uses a noninverting OP AMP configuration. Using the plus sign on the transfer function $T(s)$ and factoring an s out of the numerator and denominator yield

$$T(s) = \frac{5 + 500/s}{1 + 500/s} = \frac{Z_1(s) + Z_2(s)}{Z_1(s)}$$

Equating numerators and denominators yields

$$Z_1(s) = 1 + \frac{500}{s} \quad \text{and} \quad Z_2(s) = 5 + \frac{500}{s} - Z_1(s) = 4$$

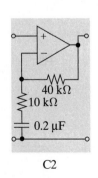

40 kΩ
10 kΩ
0.2 µF

C2

FIGURE 11–46

The impedance $Z_1(s)$ is realizable as a resistance ($R_1 = 1\,\Omega$) in series with a capacitance ($C_1 = 1/500\,\text{F}$) and $Z_2(s)$ as a resistance ($R_2 = 4\,\Omega$). Using a scale factor of $k_m = 10^4$ produces circuit C2 in Figure 11–46.

(c) Circuit C1 uses one more capacitor than circuit C2. The OP AMP output on both circuits means that they each have almost zero output impedance. The input impedance to circuit C2 is very large, because its input is the noninverting input of the OP AMP. The input impedance of circuit C1 is $Z_1(s) = k_m(1 + 500/s)$; hence, the scale factor must be selected to avoid loading the source circuit. The final design for circuit C1 in Figure 11–46 uses $k_m = 10^5$, which means that $|Z_1| > 100\,\text{k}\Omega$, which should be high enough to avoid loading the source circuit. ∎

E Evaluation Exercise 11–32

The following transfer function was realized in different ways in Figures 11–37, 11–41 and 11–45:

$$T_V(s) = \frac{\pm 10^6}{(s + 10^3)^2}$$

Compare the various designs in a table similar to Table 11–1. Which would you recommend if

(a) There was no power available?
(b) There was a desire not to invert the output and to avoid using inductors?
(c) There was a concern about loading at the output?

Answers:

(a) The RLC circuit in Figure 11–45 requires no power.
(b) The RC voltage-divider cascade in Figure 11–37 does not invert the output and does not require an inductor.
(c) None of the circuits prevents the possibility of loading at the output. One could add an OP AMP follower at the output of any of the three solutions to address loading concerns.

D DESIGN EXAMPLE 11–25

Verify that circuit C2 in Figure 11–46 meets its design requirements.

SOLUTION:
One of the important uses of computer-aided analysis is to verify that a proposed design meets the performance specifications. The circuit C2 in Figure 11–46 is designed to produce a specified step response

$$g(t) = \left[1 + 4e^{-500t}\right]u(t)\,\text{V}$$

This response jumps from zero to 5 V at $t = 0$ and then decays exponentially to 1 V at large t. The time constant of the exponential is $1/500 = 2\,\text{ms}$, which means that the final value is effectively reached after about five time constants, or 10 ms.

One can use MATLAB to better visualize the specifications of a circuit design. To have MATLAB produce the step response, we use the transfer function operator, tf, as shown in the m-file below. In this example, after we entered the circuit's transfer function, we applied the MATLAB function step to plot the desired step response of the circuit in question.

```
syms s;
s = tf('s');
H = 5*(s+100)/(s+500);
step (H)
```

FIGURE 11–47

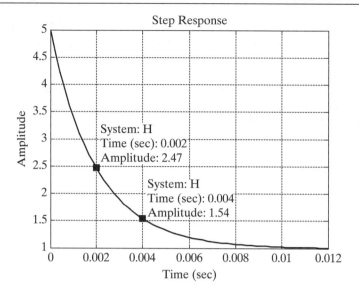

Figure 11–47 shows the step response of the circuit as plotted by MATLAB. We have selected two points for reference, namely $t = 2\,\text{ms}$ and $t = 4\,\text{ms}$.

In Figure 11–48 we have drawn the circuit in OrCAD and stimulated it using the Time Domain (Transient) analysis function. The Probe response is also shown in the figure. We have used the Probe cursor to measure the same two points so that a comparison can be made.

The theoretical values can be also calculated directly from $g(t)$ at the same two points:

$$g(0.002) = 1 + 4e^{-500 \times 0.002} = 2.4715$$

$$g(0.004) = 1 + 4e^{-500 \times 0.004} = 1.5413$$

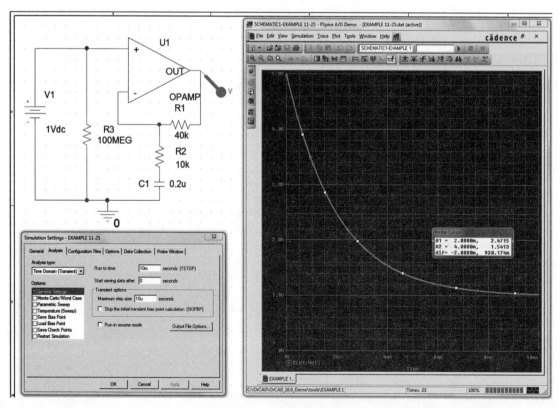

FIGURE 11–48   *Copyright © Cadence Design Systems, Inc. Used with permission.*

We summarize our results in the following table:

TIME (s) \ TECHNIQUE	HAND CALCULATION	MATLAB	ORCAD
0.002	2.4715 V	2.47 V	2.4715 V
0.004	1.5413 V	1.54 V	1.5413 V

The data show that theory and simulation agree to three significant figures.   ■

A P P L I C A T I O N   **E X A M P L E   1 1 – 2 6**

The operation of a digital system is coordinated and controlled by a periodic waveform called a clock. The *clock waveform* provides a standard timing reference to maintain synchronization between signal processing results that are generated asynchronously. Because of differences in digital circuit delays, there must be agreed-upon instants of time at which circuit outputs can be treated as valid inputs to other circuits.

Figure 11–49 shows a section of the clock distribution network in an integrated circuit. In this network the clock waveform is generated at one point and distributed to other on-chip locations by interconnections that can be modeled as lumped resistors and capacitors. Clock distribution problems arise when the *RC* circuit delays at different locations are not the same. This delay dispersion is called *clock skew*, defined as the time difference between a clock edge at one location and the corresponding edge at another location.

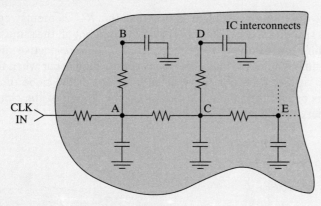

FIGURE 11–49   *Clock distribution network.*

To qualitatively calculate a clock skew, we will find the step responses in the *RC* circuit in Figure 11–50. The input $V_S(s)$ is a unit step function which simulates the leading edge of a clock pulse. The resulting step responses $V_A(s)$ and $V_B(s)$ represent the clock waveforms at points A and B in a clock distribution network. To find the step responses, we use the following s-domain node-voltage equations.

$$\text{Node A:} \quad \left(\frac{2}{R} + Cs\right) V_A(s) - \left(\frac{1}{R}\right) V_B(s) = \frac{V_S(s)}{R}$$

$$\text{Node B:} \quad -\left(\frac{1}{R}\right) V_A(s) + \left(\frac{1}{R} + Cs\right) V_B(s) = 0$$

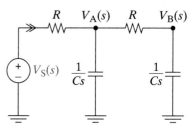

FIGURE 11–50   *Two-stage RC circuit model.*

The circuit determinant is

$$\Delta(s) = \frac{(RCs)^2 + 3(RCs) + 1}{R^2} = \frac{(RCs + 0.382)(RCs + 2.618)}{R^2}$$

which indicates that the circuit has simple poles at $s = -0.382/RC$ and $s = -2.618/RC$. Using the circuit determinant and a unit step input, we can easily solve the node equations for $V_A(s)$ and $V_B(s)$:

$$V_A(s) = \frac{RCs + 1}{s(RCs + 0.382)(RCs + 2.618)}$$

$$= \frac{1}{s} - \frac{0.7235}{s + 0.382/RC} - \frac{0.2764}{s + 2.618/RC}$$

$$V_B(s) = \frac{1}{s(RCs + 0.382)(RCs + 2.618)}$$

$$= \frac{1}{s} - \frac{1.171}{s + 0.382/RC} + \frac{0.1710}{s + 2.618/RC}$$

From these we obtain the time-domain step responses as

$$v_A(t) = 1 - 0.7235\, e^{-0.382t/RC} - 0.2764\, e^{-2.618t/RC}$$

$$v_B(t) = 1 - 1.171\, e^{-0.382t/RC} + 0.1710\, e^{-2.618t/RC} \quad \text{for} \quad t > 0$$

These two responses are plotted in Figure 11–51. For a unit step input, both responses have a final value of unity. Using the definition of step response *delay time* given in Example 11–12 (time required to reach 50% of the final value), we see that

$$T_{D_A} = 1.06/RC \quad \text{and} \quad T_{D_B} = 2.23/RC$$

The delay time skew is

$$\text{Delay Skew} = T_{D_B} - T_{D_A} = 1.17/RC$$

The clock distribution problem is not that the RC elements representing the interconnects produce time delay, but that delays are not all the same. Ideally, digital devices at different locations should operate on their respective digital inputs at exactly the same instant of time. Erroneous results may occur when the clock pulse defining that instant does not arrive at all locations at the same time. Minimizing clock skew is one of the major constraints on the design of the clock distribution network in large-scale integrated circuits.

FIGURE 11–51 *Step responses showing clock skew.*

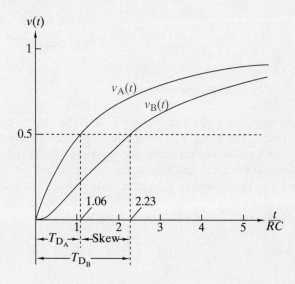

SUMMARY

- A network function is defined as the ratio of the zero-state response transform to the input transform. Network functions are either driving-point functions or transfer functions. Network functions are rational functions of s with real coefficients whose complex poles and zeros occur in conjugate pairs.

- Network functions for simple circuits like voltage and current dividers and inverting and noninverting OP AMPs are easy to derive and often useful. Node-voltage or mesh-current methods are used to find the network functions for more complicated circuits. The transfer function of a cascade connection obeys the chain rule when each stage does not load the preceding stage in the cascade.

- The impulse response is the zero-state response of a circuit for a unit impulse input. The transform of the impulse response is equal to the network function. The impulse response contains only natural poles and decays to zero in stable circuits. The impulse response of a linear, time-invariant circuit obeys the proportionality and time-shifting properties. The short pulse approximation is a useful way to simulate the impulse response in practical situations.

- The step response is the zero-state response of a circuit when the input is a unit step function. The transform of the step response is equal to the network function times $1/s$. The step response contains natural poles and a forced pole at $s = 0$ that leads to a dc steady-state response in stable circuits. The amplitude of the dc steady-state response can be found by evaluating the network function at $s = 0$. The step response waveform can also be found by integrating the impulse response waveform.

- The sinusoidal steady-state response is the forced response of a stable circuit for a sinusoidal input. With a sinusoidal input the response transform contains natural poles and forced poles at $s = \pm j\omega$ that lead to a sinusoidal steady-state response in stable circuits. The amplitude and phase angle of the sinusoidal steady-state response can be found by evaluating the network function at $s = j\omega$.

- The sinusoidal steady-state response can be found using phasor circuit analysis or directly from the transfer function. Phasor circuit analysis works best when the circuit is driven at only one frequency and several responses are needed. The transfer function method works best when the circuit is driven at several frequencies and only one response is needed.

- The convolution integral is a t-domain method relating the impulse response $h(t)$ and input waveform $x(t)$ to the zero-state response $y(t)$. Symbolically the convolution integral is represented by $y(t) = h(t) * x(t)$. Time-domain convolution and s-domain multiplication are equivalent; that is, $y(t) = h(t) * x(t) = \mathcal{L}^{-1}\{H(s) X(s)\}$. The geometric interpretation of t-domain convolution involves four operations: reflecting, shifting, multiplying, and integrating.

- First- and second-order transfer functions can be designed using voltage dividers and inverting or noninverting OP AMP circuits. Higher-order transfer functions can be realized using a cascade connection of first- and second-order circuits. Prototype designs usually require magnitude scaling to obtain practical element values.

PROBLEMS

OBJECTIVE 11–1 NETWORK FUNCTIONS
(SECTS. 11–1, 11–2)

Given a linear circuit:
(a) Find specified network functions and locate their poles and zeros.
(b) Select the element values to produce specified poles and zeros.
See Examples 11–1, 11–2, 11–3, 11–4, 11–5, 11–6, 11–7 and Exercises 11–1, 11–2, 11–3, 11–4, 11–5, 11–6, 11–7, 11–8.

11–1 Find the driving point impedance seen by the voltage source in Figure P11–1 and the voltage transfer function $T_V(s) = V_2(s)/V_1(s)$.

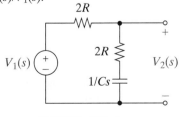

FIGURE P11–1

11–2 Find the driving point impedance seen by the voltage source in Figure P11–2 and the voltage transfer function $T_V(s) = V_2(s)/V_1(s)$.

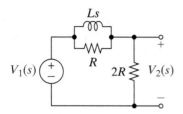

FIGURE P11–2

11–3 Find the driving point impedance seen by the voltage source in Figure P11–3 and the voltage transfer function $T_V(s) = V_2(s)/V_1(s)$.

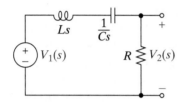

FIGURE P11–3

11–4 Find the driving point impedance seen by the voltage source in Figure P11–4 and the voltage transfer function $T_V(s) = V_2(s)/V_1(s)$.

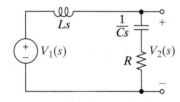

FIGURE P11–4

11–5 Find the driving point impedance seen by the voltage source in Figure P11–5 and the voltage transfer function $T_V(s) = V_2(s)/V_1(s)$.

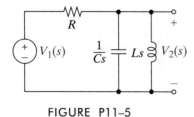

FIGURE P11–5

11–6 Find the driving point impedance seen by the voltage source in Figure P11–6 and the voltage transfer function $T_V(s) = V_2(s)/V_1(s)$.

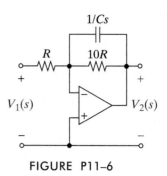

FIGURE P11–6

11–7 Find the driving point impedance seen by the voltage source in Figure P11–7 and the voltage transfer function. $T_V(s) = V_2(s)/V_1(s)$.

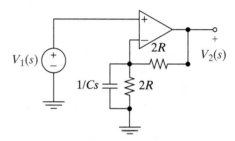

FIGURE P11–7

11–8 Find the driving point impedance seen by the voltage source in Figure P11–8 and the voltage transfer function $T_V(s) = V_2(s)/V_1(s)$.

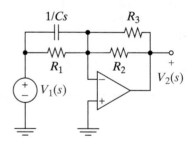

FIGURE P11–8

11–9 Find the voltage transfer function $T_V(s) = V_2(s)/V_1(s)$ in Figure P11–9.

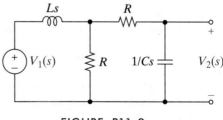

FIGURE P11–9

11–10 Find the driving point impedance seen by the voltage source in Figure P11–10 and the voltage transfer function $T_V(s) = V_2(s)/V_1(s)$. Insert a follower at A and repeat.

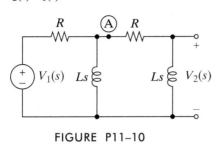

FIGURE P11–10

11–11 ◗ Find the voltage transfer function $T_V(s) = V_2(s)/V_1(s)$ in Figure P11–11. Select values of R and C so that $T_V(s)$ has a pole at $s = -100$ krad/s.

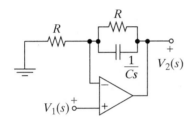

FIGURE P11–11

11–12 ◗ Find the voltage transfer function $T_V(s) = V_2(s)/V_1(s)$ in Figure P11–12. Select values of R_1, R_2, and C so that $T_V(s)$ has a pole at $s = -250$ krad/s and $R_2/R_1 = 100$.

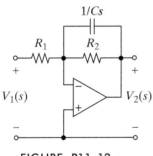

FIGURE P11–12

11–13 ◗ Find the current transfer function $T_I(s) = I_2(s)/I_1(s)$ in Figure P11–13. Select values of R and L so that $T_I(s)$ has a pole at $s = -377$ rad/s.

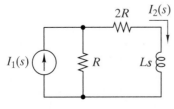

FIGURE P11–13

11–14 Find the voltage transfer function $T_V(s) = V_2(s)/V_1(s)$ of the cascade connection in Figure P11–14. Locate the poles and zeros of the transfer function.

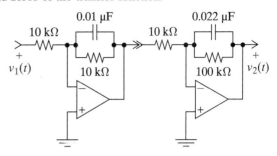

FIGURE P11–14

11–15 Find the voltage transfer function $T_V(s) = V_2(s)/V_1(s)$ of the cascade connection in Figure P11–15. Locate the poles and zeros of the transfer function.

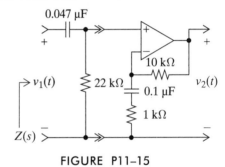

FIGURE P11–15

11–16 Find the input impedance $Z(s)$ in Figure P11–15.

Objective 11–2 Network Functions, Impulse Response, and Step Responses (Sects. 11–3, 11–4)

(a) Given a first- or second-order linear circuit, find its impulse or step response.
(b) Given the impulse or step response of a linear circuit, find the network functions.
(c) Given the impulse or step response of a linear circuit, find the response due to other inputs.

See Examples 11–8, 11–9, 11–10, 11–11, 11–12 and Exercises 11–9, 11–10, 11–11, 11–12, 11–13, 11–14.

11–17 Find the impulse response at $v_2(t)$ in Figure P11–17. Find the circuit's step response.

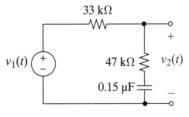

FIGURE P11–17

11–18 Find $v_2(t)$ in Figure P11–18 when $v_1(t) = \delta(t)$. Repeat for $v_1(t) = u(t)$.

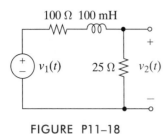

FIGURE P11–18

11–19 Find $v_2(t)$ in Figure P11–19 when $v_1(t) = \delta(t)$. Repeat for $v_1(t) = u(t)$.

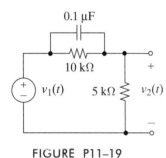

FIGURE P11–19

11–20 Find $h(t)$ and $g(t)$ for the circuit in Figure P11–20.

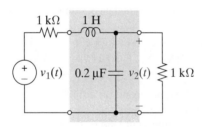

FIGURE P11–20

11–21 Find $h(t)$ and $g(t)$ for the circuit in Figure P11–21 if $R_F = 100$ kΩ.

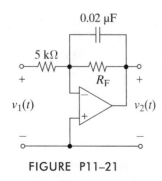

FIGURE P11–21

11–22 Ⓓ Select an appropriate R_F for the circuit of Figure P11–21 so that the step response of the circuit is $g(t) = (10e^{-1000t} - 10)\, u(t)$V.

11–23 Find $v_2(t)$ in Figure P11–23 when $v_1(t) = \delta(t)$. Repeat for $v_1(t) = u(t)$.

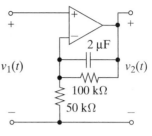

FIGURE P11–23

11–24 The impulse response of a linear circuit is $h(t) = (100e^{-200t} - 100e^{-1000t})\mu(t)$. Find the circuit's step response $g(t)$, impulse response transform $H(s)$, step response transform $G(s)$, and the circuit's transfer function $T(s)$.

11–25 The impulse response of a linear circuit is $h(t) = \delta(t) - 2000e^{-200t}\, u(t)$. Find the circuit's step response $g(t)$, impulse response transform $H(s)$, step response transform $G(s)$, and the circuit's transfer function $T(s)$.

11–26 The step response transform of a linear circuit is $G(s) = 1000/s(s + 1000)$. Find the circuit's impulse response $h(t)$, step response $g(t)$, impulse response transform $H(s)$, and the circuit's transfer function $T(s)$.

11–27 The step response of a linear circuit is $g(t) = 15(e^{-20kt} - e^{-30kt})u(t)$. Find the circuit's impulse response $h(t)$, impulse response transform $H(s)$, step response transform $G(s)$, and the circuit's transfer function $T(s)$.

11–28 Find $h(t) = \frac{dg(t)}{dt}$ when $g(t) = (3 - e^{-10t})\, u(t)$. Verify your answer by first transforming $g(t)$ into $G(s)$ and finding $H(s) = sG(s)$ and then taking the inverse transform of $H(s)$. Did you get the same answer?

11–29 The impulse response of a linear circuit is $h(t) = 1000 [e^{-1000t}]u(t)$. Find the output waveform when the input is $x(t) = 5tu(t)$.

11–30 The step response of a linear circuit is $g(t) = 0.25[1 - e^{-150t}]u(t)$. Find the output waveform when the input is $v_1(t) = [20e^{-200t}]u(t)$. Use MATLAB to find the Laplace transforms of $g(t)$ and $v_1(t)$. Then find $V_2(s)$. Finally, use the inverse Laplace function to find the waveform $v_2(t)$ and plot the results.

11–31 The step response of a linear circuit is $g(t) = 10[e^{-50t} \cos 200t]u(t)$. Find the circuit's impulse response $h(t)$, impulse response transform $H(s)$, step response transform $G(s)$, and the circuit's transfer function $T(s)$.

11–32 The impulse response transform of a linear circuit is $H(s) = (s + 2000)/(s + 1000)$. Find the output waveform when the input is $x(t) = 5e^{-1000t}u(t)$. Use MATLAB to find the Laplace transform of $x(t)$. Then find $Y(s)$. Finally, use the inverse

Laplace function to find the waveform $y(f)$ and plot the results.

11–33 The impulse response of a linear circuit is $h(t) = 20u(t) + \delta(t)$. Find the output waveform $y(t)$ when the input is $x(t) = 2[e^{-20t}]u(t)$.

OBJECTIVE 11–3 NETWORK FUNCTIONS AND THE SINUSOIDAL STEADY-STATE RESPONSE (SECT. 11–5)

(a) Given a first- or second-order linear circuit with a specified input sinusoid, find the sinusoidal steady-state response.
(b) Given the network function, impulse response, or step response, find the sinusoidal steady-state response for a specified input sinusoid.
See Examples 11–13, 11–14, 11–15 and Exercises 11–15, 11–16, 11–17, 11–18.

11–34 The circuit in Figure P11–34 is in the steady state with $v_1(t) = 5 \cos 1414.21t$ V. Find $v_{2SS}(t)$. Repeat for $v_1(t) = 5 \cos 1$ kt V. And without doing any calculations, repeat for $v_1(t) = 5$ V.

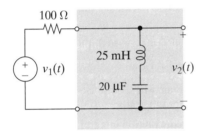

FIGURE P11–34

11–35 The circuit in Figure P11–35 is in the steady state with $v_1(t) = 10 \cos 500t$ V. Find $v_{2SS}(t)$. Repeat for $v_1(t) = 10 \cos 1$ kt V, and for $v_1(t) = 10 \cos 10$ kt V. Where is the pole located?

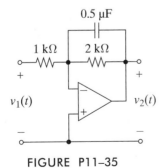

FIGURE P11–35

11–36 The circuit in Figure P11–36 is in the steady state with $v_1(t) = 25 \cos 2000t$ V. Find $v_{2SS}(t)$. Repeat for $v_1(t) = 25 \cos 10$ kt V. Where are the poles located?

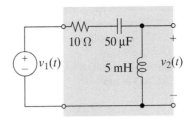

FIGURE P11–36

11–37 The circuit in Figure P11–37 is in the steady state with $i_1(t) = 10 \cos 50kt$ mA, $R_1 = 100$ Ω, $R_2 = 400$ Ω, and $L = 100$ mH. Find $i_{2SS}(t)$. Repeat for $i_1(t) = 10 \cos 5kt$ mA. Where is the pole located?

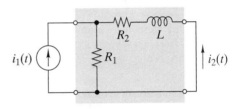

FIGURE P11–37

11–38 The circuit in Figure P11–38 is in the steady state with $i_1(t) = 5 \cos 1000t$ mA, $R = 1$ kΩ, L, = 2H, and $C = 0.5$ µF.
(a) Find $i_{2SS}(t)$.
(b) Verify your results using OrCAD. Use IAC for a source and note that the current comes out of the negative terminal (passive sign convention.). The IPRINT element on the right side of the figure is an ammeter and it comes from the Special Library. It has a polarity as indicated by the small"–" on the element. You should ensure that it will read the correct current direction. Before you can use the IPRINT element, it needs to be set up to display the current magnitude and phase. Double click on the element and in its property editor place a "y" in AC, MAG, and PHASE boxes. Click the Apply button and close the property editor. To obtain the simulation select AC Sweep/Noise. Set the start frequency and the stop frequency at 159.15 Hz (1000 rad/s). Place a 1 in Points/Decade. Run the simulation. Under View, select Output File. Scroll down until you see the IPRINT output.

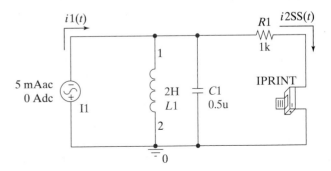

FIGURE P11–38

11–39 The circuit in Figure P11–39 is in the steady state with $i_1(t) = 10 \cos 5000t$ mA.
(a) Find the steady-state voltage $v_{2SS}(t)$. Repeat for $i_1(t) = 5 \cos 2500t$ mA.
(b) Verify your answer using OrCAD (see Problem 11–38 for help on OrCAD setup.)

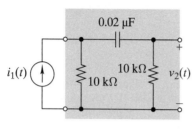

FIGURE P11–39

11–40 The impulse response transform of a circuit is $H_R(s) = V_2(s)/I_1(s) = 5000s/s + 2500$. Find $v_{2SS}(t)$ if $i_t(t) = 10 \cos 5000t$ mA. Compare your answer to that found in Problem 11–39.

11–41 The transfer function of a linear circuit is $T(s) = (s + 100)/(s + 10)$. Find the sinusoidal steady-state output for an input $x(t) = 5 \cos 100t$.

11–42 The step response of a linear circuit is $g(t) = [15e^{-500t}]u(t)$. Find the sinusoidal steady-state output for an input $x(t) = 5 \cos 1000t$.

11–43 The step response of a linear circuit is $g(t) = [2e^{+100t}]u(t)$. Find the sinusoidal steady-state output for an input $x(t) = 5 \cos 500t$.

11–44 The impulse response of a linear circuit is $h(t) = [500e^{-5000t}]u(t) - \delta(t)$. Find the sinusoidal steady-state output for an input $x(t) = 10 \cos 10 \, kt$.

11–45 The impulse response of a linear circuit is $h(t) = 800[e^{-100t} - e^{-400t}]u(t)$. Use MATLAB to find the sinusoidal steady-state output for an input $x(t) = 8 \cos 200t$. Use MATLAB to plot $y(t)$.

11–46 The step response of a linear circuit is $g(t) = [-e^{-60t} \sin 80t]u(t)$. Find the sinusoidal steady-state response for an input $x(t) = 20 \cos 100t$.

11–47 The step response of a linear circuit is $g(t) = [1 - 20te^{-10t}]u(t)$. Find the sinusoidal steady-state response for an input $x(t) = 50 \cos 10t$.

OBJECTIVE 11–4 NETWORK FUNCTIONS AND CONVOLUTION (SECT. 11–6)

(a) Given the impulse response of a linear circuit, use the convolution integral to find the response to a specified input.
(b) Use the convolution integral to derive properties of linear circuits.
See Examples 11–16, 11–17, 11–18, 11–19 and Exercises 11–19, 11–20, 11–21, 11–22.

11–48 The impulse response of a linear circuit is $h(t) = u(t)$. Use the convolution integral to find the response due to an input $x(t) = u(t)$.

11–49 The impulse response of a linear circuit is $h(t) = [u(t) - u(t - 3)]$. Use the convolution integral to find the response due to an input $x(t) = u(t - 3)$.

11–50 The impulse response of a linear circuit is $h(t) = [u(t) - u(t - 1)]$. Use the convolution integral to find the response due to an input $x(t) = u(t) - u(t - 1)$.

11–51 The impulse response of a linear circuit is $h(t) = t[u(t) - u(t - 1)]$. Use the convolution-integral to find the response due to an input $x(t) = u(t - 1)$.

11–52 The impulse response of a linear circuit is $h(t) = e^{-t}u(t)$. Use the convolution integral to find the response due to an input $x(t) = u(t)$.

11–53 The impulse response of a linear circuit is $h(t) = 10[u(t) - u(t - 1)]$. Use the convolution integral to find the response due to an input $x(t) = e^{-t}u(t)$.

11–54 The impulse response of a linear circuit is $h(t) = e^{-t}u(t)$. Use the convolution integral to find the response due to an input $x(t) = tu(t)$.

11–55 Show that $f(t) * \delta(t) = f(t)$. That is, show that convolving any waveform $f(t)$ with an impulse leaves the waveform unchanged.

11–56 Show that if $h(t) = u(t)$, then output $y(t)$ for any input $x(t)$ is $y(t) = \int_0^t x(\tau)d\tau$.

That is, a circuit whose impulse response is a step function operates as an integrator.

11–57 Use the convolution integral to show that if the input to a linear circuit is $x(t) = u(t)$ then

$$y(t) = g(t) = \int_0^t h(\tau)d\tau$$

That is, show that the step response is the integral of the impulse response.

11–58 If the input to a linear circuit is $x(t) = tu(t)$, then the output $y(t)$ is called the ramp response.
Use the convolution integral to show that

$$\frac{dy(t)}{dt} = \int_0^t h(\tau)d\tau = g(t)$$

That is, show that the derivative of the ramp response is the step response.

11–59 The impulse response of a linear circuit is $h(t) = 2u(t)$. Use MATLAB to compute the convolution integral and find the response due to an input $x(t) = t[u(t) - u(t - 1)]$.

11–60 The impulse response of a linear circuit is $h(t) = 50 e^{-5t} u(t)$ and $x(t) = tu(t)$. Use s-domain convolution to find the zero-state response $y(t)$.

11–61 The impulse responses of two linear circuits are $h_1(t) = 2e^{-2t}u(t)$ and $h_2(t) = 5e^{-5t}u(t)$. What is the impulse response of a cascade connection of these two circuits?

OBJECTIVE 11–5 NETWORK FUNCTION DESIGN (SECT. 11–7)

(a) Design alternative circuits that realize a given network function and meet other stated constraints.
(b) Use software to visualize and simulate alternative designs.
(c) Evaluate alternative designs using stated criteria and select the best design.
See Examples 11–20, 11–21, 11–22, 11–23, 11–24, 11–25, 11–26 and Exercises 11–23, 11–24, 11 -25, 11–26, 11–27, 11–28, 11–29, 11–30, 11–31, 11–32.

11–62 ◑ Design a circuit to realize the transfer function below using only resistors, capacitors, and OP AMPs.

$$T_V(s) = \frac{20000}{s + 100000}$$

Scale the circuit so that all capacitors are exactly 1000 pF.

11–63 ◑ Design a circuit to realize the transfer function below using only resistors, capacitors, and OP AMPs.

$$T_V(s) = \frac{20000}{s + 1000}$$

Scale the circuit so that all resistors are exactly 1 kΩ.

11–64 ◑ Design a circuit to realize the transfer function below using only resistors, inductors, and OP AMPs.

$$T_V(s) = \frac{s + 5000}{s}$$

Scale the circuit so that all inductors are exactly 100 mH.

11–65 ◑ Design a circuit to realize the transfer function below using only resistors, capacitors, and OP AMPs.

$$T_V(s) = \frac{-50000s}{(s + 2500)}$$

Scale the circuit so that all capacitors are exactly 0.1 μF.

11–66 ◑ Design a circuit to realize the transfer function below using only resistors, capacitors, and OP AMPs.

$$T_V(s) = \frac{50000s}{(s + 50)(s + 1000)}$$

Scale the circuit so that all capacitors are exactly 0.1 μF.

11–67 ◑ Design a circuit to realize the transfer function below using only resistors, capacitors, and OP AMPs. Scale the circuit so that all resistors are greater than 10 kΩ and all capacitors are less the 1 μF.

$$T_V(s) = \pm\frac{5 \times 10^8}{(s + 100)(s + 10,000)}$$

11–68 ◑ Design a circuit to realize the transfer function below using only resistors, capacitors and not more than one OP AMP. Scale the circuit so that all capacitors are exactly 0.01 μF.

$$T_V(s) = \pm\frac{100(s + 1000)}{(s + 100)(s + 10,000)}$$

11–69 ◑ Design a circuit to realize the transfer function below using only resistors, capacitors and not more than one OP AMP. Scale the circuit so that the final design uses only 20-kΩ resistors.

$$T_V(s) = \pm\frac{20,000s}{(s + 1000)(s + 5000)}$$

11–70 ◑ Design a *passive* circuit to realize the transfer function below using only resistors, capacitors, and inductors. Scale the circuit so that all inductors are 50 mH or less.

$$T_V(s) = \frac{s^2}{(s + 2000)^2}$$

11–71 ◑ A circuit is needed to realize the transfer function listed below.

$$T_V(s) = \pm\frac{(s + 125)(s + 500)}{(s + 250)(s + 1000)}$$

(a) Design the circuit using two OP AMPs.
(b) Design the circuit using only one OP AMP.
(c) Design the circuit using no OP AMPs.
    In all cases, scale the circuit so that all parts use practical values.

11–72 ◑ Design a circuit to realize the transfer function below using only resistors, capacitors, and OP AMPs. Use only values from the inside rear cover. Your design must be within ±10% of the desired response.

$$T_V(s) = -\frac{500(s + 100)}{s(s + 10000)}$$

11–73 ◑ A circuit is needed to realize the impulse response transform listed below. Scale the circuit so that all parts use practical values.

$$H(s) = \pm\frac{200s + 10^6}{s^2 + 200s + 10^6}$$

11–74 ◐ It is claimed that both circuits in Figure P11–74 realize the transfer function

$$T_V(s) = K\left(\frac{s + 2000}{s + 1000}\right)$$

(a) Verify that both circuits realize the specified $T_V(s)$.
(b) Which circuit would you choose if the output must drive a 1 kΩ load?
(c) Which circuit would you choose if the input comes from a 50 Ω. source?
(d) It is further claimed that connecting the two circuits in cascade produces an overall transfer function of $[T_V(s)]^2$ no matter which circuit is the first stage and which is the second stage. Do you agree or disagree? Explain.

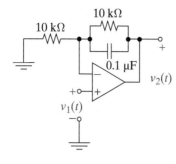

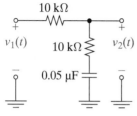

FIGURE P11–74

11–75 It is claimed that both circuits in Figure P11–75 realize the transfer function

$$T_V(s) = \frac{\pm 1000s}{(s+1000)(s+4000)}$$

(a) Verify that both circuits realize the specified $T_V(s)$.

(b) Which circuit would you choose if the output must drive a 1 kΩ load?

(c) Which circuit would you choose if the input comes from a 50 Ω source?

(d) It is further claimed that connecting the two circuits in cascade produces an overall transfer function of $[T_V(s)]^2$ no matter which circuit is the first stage and which is the second stage. Do you agree or disagree? Explain.

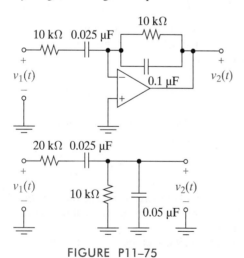

FIGURE P11–75

11–76 Design a circuit that produces the following step response.

$$g(t) = 24[1 - e^{-50t} - 50te^{-50t}]u(t)$$

11–77 A circuit is needed that will take an input of $v_1(t) = 25e^{-10t}u(t)$ mV and produce an output of $v_2(t) = 500e^{-200t}u(t)$ mV. Design such a circuit using practical parts values. Validate your design by using OrCAD.

11–78 A circuit is needed that will take an input of $v_1(t) = [1 - e^{-10,000t}]u(t)$ V and produce a constant −5 V output. Design such a circuit using practical parts values. Validate your design using OrCAD.

INTEGRATING PROBLEMS

11–79 First-Order Circuit Impulse and Step Responses **A**, **D**

Each row in the table shown in Figure P11–79 refers to a first-order circuit with an impulse response $h(t)$ and a step response $g(t)$. Fill in the missing entries in the table.

Circuit	$h(t)$	$g(t)$
$\begin{array}{c} R \\ v_1(t) \quad \dfrac{1}{\alpha R} \quad v_2(t) \end{array}$		
	$\delta(t) - [\alpha\, e^{-\alpha t}]u(t)$	
		$\left(1 + \dfrac{e^{-\alpha t}}{2}\right)u(t)$

FIGURE P11–79

11–80 OP AMP Modules and Loading **A**

Figure P11–80 shows an interconnection of three basic OP AMP modules.

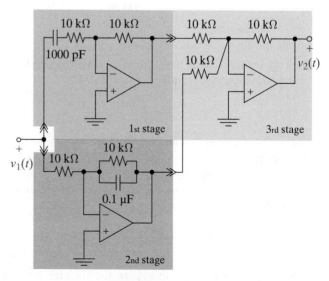

FIGURE P11–80

(a) Does this interconnection involve loading?

(b) Find the overall transfer function of the interconnection and locate its poles and zeros.

(c) Find the steady-state output $v_2(t)$ when the input is $v_1(t) = \cos 500t$ V. Repeat for $v_1(t) = \cos 10\,kt$ V and again for $v_1(t) = \cos 200\,kt$ V.

(d) Can you think of a use for this circuit?

11–81 OP AMP Modules and Stability **A**

Figure P11–81 shows an interconnection of three basic circuit modules. Does this interconnection involve loading? Find the overall transfer function of the interconnection and locate its poles and zeros. Is the circuit stable?

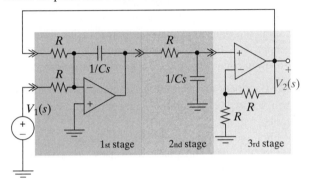

FIGURE P11–81

11–82 Step Response and Fan-Out **A**

The fan-out of a digital device is defined as the maximum number of inputs to similar devices that can be reliably driven by the device output. Figure P11–82 is a simplified

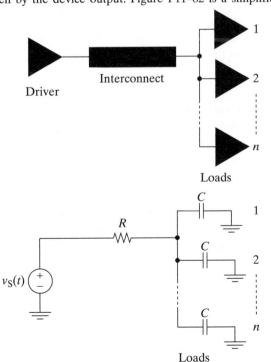

FIGURE P11–82

diagram of a device's output driving n identical capacitive inputs. To operate reliably, a 5-V step function at the device output must drive the capacitive inputs to 3.7 V in 10 ns or less. Determine the device fan-out for $R = 1$ kΩ and $C = 3$ pF.

11–83 Designing to Specifications **D**

A particular circuit needs to be designed that has the following transfer function requirements: Poles at $s = -100$ and $s = -10{,}000$; zeros at $s = 0$ and $s = -1000$; and a gain of 10 as $s \to \infty$. Find the circuit's transfer function and use MATLAB to plot its step response. Then design a circuit that will meet that requirement. Finally, use OrCAD to validate that your circuit has the same step response as found using MATLAB.

11–84 Comparison of Sinusoidal Steady-State Analysis versus Phasor Analysis **E**

A circuit designer often is faced with deciding which analysis technique to use when attempting to solve a circuit problem. In this problem we will look at the circuit in Figure P11–84 and choose which technique is the better one to use for different analysis scenarios. Explain why you selected the technique you did.

(a) You need to calculate the circuit's transfer function $T_V(s) = V_2(s)/V_1(s)$.

(b) The input is given as $v_1(t) = 5\cos 1000t$ V and you need to find $v_{2SS}(t)$.

(c) The input is given as $v_1(t) = 5\cos 1000t$ V and you need to find $i_X(t)$.

(d) The input is given as $v_1(t) = V_A \cos \omega t$ V and you need to find $v_{2SS}(t)$.

(e) The input is given as $v_1(t) = 170\cos 377t$ V and you need to find all of the voltages and currents in the circuit.

(f) You need to find the poles and zeros of the circuit.

(g) You need to find if the current leads or lags the voltage across the two resistors when the input is $5\cos 1000t$ V.

(h) You need to determine what type of filtering the circuit performs.

(i) You need to select a load for maximum power when the input is $v_1(t) = 170\cos 377t$ V.

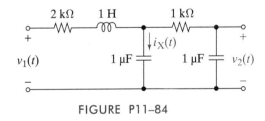

FIGURE P11–84

CHAPTER 12 FREQUENCY RESPONSE

The advantage of the straight-line approximation is, of course, that it reduces the complete characteristics to a sum of elementary characteristics.

Hendrik W. Bode, 1945,
American Engineer

Some History Behind This Chapter

Hendrik Bode spent most of his distinguished career as a member of the technical staff of the Bell Telephone Laboratories. In the 1930s Bode made major contributions to feedback amplifier theory during the development of long-distance telephone systems. His contributions included an approach to frequency response based on logarithmic plots and straight-line approximation. These so-called Bode plots are still valuable today and remain the industry-standard way to describe the frequency response of circuits and systems.

Why This Chapter Is Important Today

In Chapter 8 we used phasors to study the steady-state response at a single-frequency. In Chapter 11 we showed that network functions give the steady-state response at any frequency. In this chapter we use network functions and Bode diagrams to describe the steady-state response over a range of frequencies. Bode diagrams are used here because they allow us to quickly visualize how the poles and zeros of a network function affect the frequency response of a circuit.

Chapter Sections

Chapter Learning Objectives

12-1 First-Order Circuit Frequency Response (Sects. 12–1, 12–2, 12–3)

Given a first-order circuit or transfer function:

(a) Find and classify the frequency response.

(b) Plot the gain and phase responses using straight-line approximations and computer tools.

(c) Design circuits to produce a specified frequency response.

12-2 Bandpass and Bandstop Responses (Sect. 12–4)

Given a cascade or parallel connection of two first-order circuits:

(a) Find and classify the frequency response.

(b) Plot the gain and phase responses using straight-line approximations and computer tools.

(c) Design circuits to produce a specified frequency response.

12-3 The Frequency Response of *RLC* Circuits (Sect. 12–5)

Given an *RLC* circuit connected as a bandpass or a bandstop filter:

(a) Find the frequency-response descriptors such as Q and B.

(b) Design circuits to produce a specified frequency response.

(c) Compare the bandpass and bandstop responses obtained from *RLC* filters with similar ones obtained from first-order filters.

12-4 Bode Plots (Sects. 12–6, 12–7)

Given a linear circuit or transfer function:

(a) Plot the gain and phase responses using straight-line approximations and computer tools.

(b) Develop a transfer function from a straight-line Bode gain plot.

(c) Design a circuit that produces a given straight-line gain plot.

12-5 Frequency Response and Step Response (Sect. 12–8)

Given a circuit or a transfer function:

(a) Find the gain response corresponding to a given step response or vice versa.

(b) Use the relationship between frequency and step responses to choose the best solution for a design specification.

12–1 FREQUENCY-RESPONSE DESCRIPTORS

In Chapter 11 we learned that the sinusoidal steady-state output can be found by evaluating the transfer function $T(s)$ at $s = j\omega$, where ω is the frequency of the sinusoidal input. The function $T(j\omega)$ determines the amplitude and phase angle of the output through the **gain** function $|T(j\omega)|$ and **phase** function $\theta(\omega) = \angle T(j\omega)$. Recall Figure 11–26, which demonstrates the following

$$\text{Output amplitude} = |T(j\omega)| \times (\text{Input amplitude})$$
$$\text{Output phase} = \text{Input phase} + \theta(\omega) \tag{12-1}$$

The gain and phase functions are frequency dependent and together reveal how a circuit responds to input sinusoids of different frequencies. This frequency-dependent relationship between sinusoidal inputs and the resulting steady-state outputs is called the **frequency response** of the circuit.

Frequency-response concepts and techniques find wide applications in communication, control, and instrumentation systems. A key component in these applications is the electric **filter**, a signal processor that modifies or reshapes the frequency content of signals.

The gain and phase functions can be expressed mathematically or presented graphically, as shown in Figure 12–1. Most of the descriptive terminology of frequency response is based on the shape of the gain function. For example, the gain plot in Figure 12–1 is relatively constant at lower frequencies and decreases rapidly at higher frequencies. The range of frequencies with nearly constant gain is called a **passband**. The range of frequencies with significantly reduced gain is called a **stopband**. The frequency associated with the transition from a passband to an adjacent stopband is called the **cutoff frequency**, denoted as ω_C or f_C.

In linear circuits there is a gradual transition from a passband to a stopband, so the location of the cutoff frequency is a matter of definition. The most widely used definition assigns cutoff to the frequency at which the passband gain decreases by a factor of $1/\sqrt{2}$ from its maximum value. Under this definition ω_C is found from the condition

$$|T(j\omega_C)| = \frac{1}{\sqrt{2}} T_{\max} \tag{12-2}$$

where $T_{\max}$ is the maximum gain in the passband.

Additional terminology is based on the four basic types of gain responses in Figure 12–2. The figure also shows the sinusoidal input and outputs for each type.

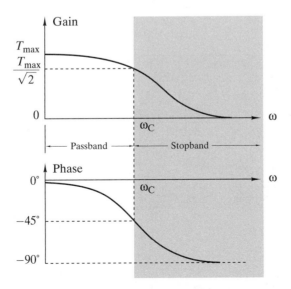

FIGURE 12–1 *Typical Frequency-response plots.*

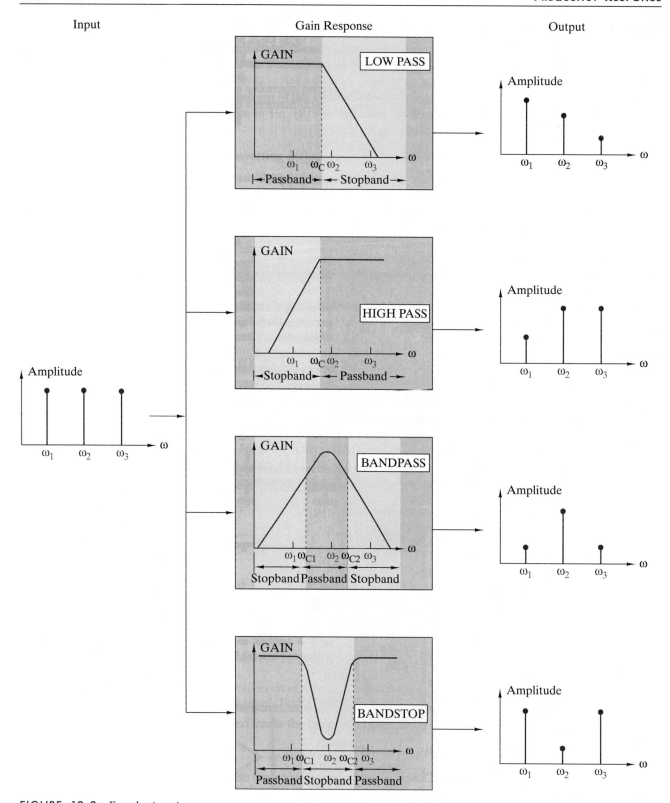

FIGURE 12–2 *Four basic gain responses.*

The input to all four is a composite signal consisting of three equal-amplitude sinusoids at distinct frequencies $\omega_1, \omega_2,$ and ω_3. The output signals all contain the same three frequencies but have amplitudes modified by the form of the gain response.

- The *low-pass* gain response has a single passband extending from zero frequency (dc) to ω_C. This type of gain passes the input at ω_1 unchanged and attenuates the inputs at ω_2 and ω_3 since they fall in the stopband above ω_C.

- The *high-pass* response has a single passband extending from ω_C to infinite frequency. This type of gain passes the inputs at ω_2 and ω_3 unchanged and attenuates the input at ω_1 since it falls in the stopband below ω_C.

- The *bandpass* response has a single passband with two adjacent stopbands—one below ω_{C1} and another above ω_{C2}. This gain response passes the input at ω_2 unchanged and attenuates the inputs at ω_1 and ω_3.

- The *bandstop* response has a single stopband with two adjacent passbands— one below ω_{C1} and another above ω_{C2}. This gain attenuates the input at ω_2 and passes the inputs at ω_1 and ω_3 unchanged.

The element impedances $Z(s)$ play a key role in circuit frequency response. Replacing s by $j\omega$ and evaluating at $Z(j\omega)$ at $\omega = 0$ (dc) and $\omega = \infty$ leads to the important conclusion shown in Table 12–1. The first row in Table 12–1 shows that the impedance of a resistor does not change with frequency. The next row shows that a capacitor has an infinite impedance (an open circuit) at dc and zero impedance (a short circuit) at infinite frequency. In the last row we see that an inductor acts like a short circuit at dc and an open circuit at infinite frequency. These conclusions are worth remembering because they often help us see how a circuit produces particular gain responses.

TABLE 12–1 PASSIVE ELEMENT IMPEDANCES AT ZERO AND INFINITE FREQUENCY

	IMPEDANCE AT	
ELEMENT	$\omega = 0$	$\omega = \infty$
Resistor $Z_R = R$	R	R
Capacitor $Z_C = 1/j\omega C$	∞ Open circuit	0 Short circuit
Inductor $Z_L = j\omega L$	0 Short circuit	∞ Open circuit

12–2 BODE DIAGRAMS

The gain $|T(j\omega)|$ is often expressed in **decibels** (dB), defined as

$$|T(j\omega)|_{dB} = 20 \log_{10}|T(j\omega)| \qquad (12–3)$$

Some understanding of the decibel scale is necessary to construct and interpret gain-response plots. A gain of $|T(j\omega)| = 10^n$ expressed in dB is $|T(j\omega)|_{dB} = 20 \log_{10}(10^n) = 20n$ dB. The gain of $T_{max}/\sqrt{2}$ expressed in dB is $20 \log_{10}(T_{max}) - 20 \log_{10}(\sqrt{2}) = |T_{max}|_{dB} - 3$ dB. In other words, the cutoff frequency occurs when the passband gain is reduced by 3 dB. For this reason, the cutoff frequency is sometimes referred to as the 3-dB down frequency.

Table 12–2 lists other values of $|T(j\omega)|$ and the corresponding values of $|T(j\omega)|_{dB}$. Note in particular that a gain of 1 is 0 dB and a gain of $2(0.5)$ is $+6(-6)$ dB. This means that the gain $2|T(j\omega)|$ expressed in dB is $|T(j\omega)|_{dB} + 6$ dB while $|T(j\omega)|/2$ is $|T(j\omega)|_{dB} - 6$ dB.

The frequency range of interest is often so wide that a linear frequency scale tends to mask important features for the response. For this reason frequency-response plots almost always use a logarithmic scale for the frequency variable. The use of log-frequency scales is a standard practice, and the resulting response plots are called **Bode diagrams**.

Bode diagrams are plots of the gain $|T(j\omega)|_{dB}$ and phase $\theta(\omega)$ versus log-frequency.

TABLE 12–2 VALUES OF GAIN AND GAIN IN DB

$|T(j\omega)|$	$|T(j\omega)|_{dB}$
10^3	60
10^2	40
10	20
2	6
$\sqrt{2}$	3
1	0
$1/\sqrt{2}$	-3
0.5	-6
10^{-1}	-20
10^{-2}	-40
10^{-3}	-60

The use of log-frequency scales involves some special terminology. An **octave** is any frequency range whose end points have a 2:1 ratio, and a **decade** is any range with a 10:1 ratio. For example, the frequency range from 10 Hz to 20 Hz is one octave, as is the range from 20 MHz to 40 MHz. The standard UHF (ultrahigh frequency) band spans a one-decade range from 0.3 GHz to 3 GHz. The audio range from 20 Hz to 20 kHz spans three decades.

In summary, Bode diagrams, or Bode plots, as they are sometimes called, are used to describe the frequency response of circuits and systems. Although it is convenient to use software to create Bode plots—such as MATLAB, to analyze transfer functions, or OrCAD to simulate a circuit's behavior—it is useful to understand how poles, zeros, and different circuit elements give rise to the various features in a Bode diagram. In the sections that follow, we will spend some time relating the circuit's behavior to the resulting Bode diagram. We will also see how a circuit's damping coefficient ζ can significantly affect its frequency response. Understanding these behaviors will help us in learning to analyze, design, and evaluate filters.

APPLICATION   **E X A M P L E   1 2 – 1**

The use of the decibel as a measure of performance pervades the literature and folklore of electrical engineering. The decibel originally came from the definition of power ratios in **bels**.[1]

$$\text{Number of bels} = \log_{10}\frac{P_{\text{OUT}}}{P_{\text{IN}}}$$

The decibel (dB) is more commonly used in practice. The number of decibels is 10 times the number of bels:

$$\text{Number of dB} = 10 \times (\text{Number of bels}) = 10\log_{10}\frac{P_{\text{OUT}}}{P_{\text{IN}}}$$

When the input and output powers are delivered to equal input and output resistances R, then the power ratio can be expressed in terms of voltages across the resistances.

$$\text{Number of dB} = 10\log_{10}\frac{v_{\text{OUT}}^2/R}{v_{\text{IN}}^2/R} = 20\log_{10}\frac{v_{\text{OUT}}}{v_{\text{IN}}}$$

or in terms of currents through the resistances:

$$\text{Number of dB} = 10\log_{10}\frac{i_{\text{OUT}}^2 \times R}{i_{\text{IN}}^2 \times R} = 20\log_{10}\frac{i_{\text{OUT}}}{i_{\text{IN}}}$$

The definition of gain in dB in Eq. (12–3) is consistent with these results, since in the sinusoidal steady state the transfer function equals the ratio of output amplitude to input amplitude. The preceding discussion is not a derivation of Eq. (12–3) but simply a summary of its historical origin. In practice Eq. (12–3) is applied when the input and output are not measured across resistances of equal value.

When the chain rule applies to a cascade connection, the overall transfer function is a product

$$T(j\omega) = T_1 \times T_2 \times \cdots \times T_N$$

---

[1]The name of the unit honors Alexander Graham Bell (1847–1922), the inventor of the telephone.

where $T_1, T_2, \ldots, T_N$ are the transfer functions of the individual stages in the cascade. Expressed in dB, the overall gain is

$$
\begin{aligned}
|T(j\omega)|_{\mathrm{dB}} &= 20\log_{10}(|T_1| \times |T_2| \times \cdots |T_N|) \\
&= 20\log_{10}|T_1| + 20\log_{10}|T_2| + \cdots + 20\log_{10}|T_N| \\
&= |T_1|_{\mathrm{dB}} + |T_2|_{\mathrm{dB}} + \cdots + |T_N|_{\mathrm{dB}}
\end{aligned}
$$

Because of the logarithmic definition, the overall gain (in dB) is the sum of the gains (in dB) of the individual stages in a cascade connection. The effect of altering a stage or adding an additional stage can be calculated by simply adding or subtracting the change in dB. Since summation is simpler than multiplication, the enduring popularity of the dB comes from its logarithmic definition, not its somewhat tenuous relationship to power ratios.

Exercise 12–1 ─────────────────────────────

A transfer function has a passband gain of 25. At a particular frequency in its stopband, the gain of the transfer function is only 0.0006. By how many decibels does the gain of the passband exceed that of that frequency in the stopband?

Answer: 92.4 dB

Exercise 12–2 ─────────────────────────────

A particular filter is said to be 83 dB down at a desired stop frequency. How many times reduced is a signal at that frequency compared to a signal in the filter's passband?

Answer: By 14,125 times.

12–3   FIRST-ORDER LOW-PASS AND HIGH-PASS RESPONSES

FIRST-ORDER LOW-PASS RESPONSE

We begin the study of frequency response with the first-order low-pass transfer function:

$$
T(s) = \frac{K}{s + \alpha} \tag{12–4}
$$

The constants K and α are real. The constant K can be positive or negative, but α must be positive so that the natural pole at $s = -\alpha$ is in the left half of the s plane to ensure that the circuit is stable. Remember, the concepts of sinusoidal steady state and frequency response do not apply to unstable circuits that have poles in the right half of the s plane or on the j-axis.

To describe the frequency response of the low-pass transfer function, we replace s by $j\omega$ in Eq. (12–4)

$$
T(j\omega) = \frac{K}{\alpha + j\omega} \tag{12–5}
$$

and express the gain and phase functions as

$$
|T(j\omega)| = \frac{|K|}{\sqrt{\omega^2 + \alpha^2}} \tag{12–6}
$$

$$
\theta(\omega) = \angle K - \tan^{-1}(\omega/\alpha)
$$

FIGURE 12–3 *First-order low-pass Bode plots.*

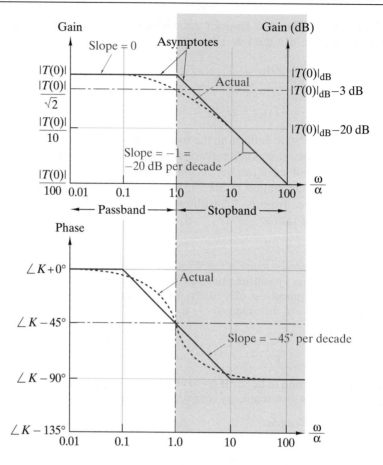

The gain function is a positive number. Since K is real, the angle of $K(\angle K)$ is either $0°$ when $K > 0$ or $\pm 180°$ when $K < 0$. An example of a negative K occurs in an inverting OP AMP configuration where $T(s) = -Z_2(s)/Z_1(s)$.

Figure 12–3 shows Bode plots of the gain and phase of the first-order low-pass function using a log scale for the normalized frequency ω/α. The gain plot displays a low-pass characteristic with a passband at low frequency and a stopband at high frequency. The maximum passband gain occurs at $\omega = 0$, where $T_{max} = |K|/\alpha$. The gain gradually decreases as frequency increases until at $\omega = \alpha$ we have

$$|T(j\alpha)| = \frac{|K|}{\sqrt{\alpha^2 + \alpha^2}} = \frac{|K|/\alpha}{\sqrt{2}} = \frac{1}{\sqrt{2}} T_{max} \qquad (12\text{–}7)$$

Referring to Eq. (12–2), we conclude that $\omega_C = \alpha$ is the cutoff frequency marking the boundary between a low-frequency passband and a high-frequency stopband. This boundary occurs when $\omega/\alpha = 1$ as shown in Figure 12–3.

The low- and high-frequency gain asymptotes shown in Figure 12–3 are especially important. The low-frequency asymptote is the horizontal line and the high-frequency asymptote is the sloped line. At low frequencies ($\omega \ll \alpha$) the gain approaches $|T(j\omega)| \rightarrow |K|/\alpha$. At high frequencies ($\omega \gg \alpha$) the gain approaches $|T(j\omega)| \rightarrow |K|/\omega$. The intersection of the two asymptotes occurs when $|K|/\alpha = |K|/\omega$. The intersection forms a "corner" at $\omega = \alpha$, so the cutoff frequency is also called the **corner frequency**.

The high-frequency gain asymptote decreases by a factor of 10 (-20 dB) whenever the frequency increases by a factor of 10 (one decade). As a result, the high-frequency asymptote has a slope of -1 or -20 dB per decade and the low-frequency asymptote has a slope of 0 or 0 dB/decade. These two asymptotes provide a straight-line approximation to the gain response that differs from the true response by a maximum of 3 dB or a factor of $\sqrt{2}$ at the corner frequency.

The semilog plot of the phase shift of the first-order low-pass transfer function is shown in Figure 12–3. At $\omega = \alpha$ the phase angle in Eq. (12–6) is

$$\theta(\omega_C) = \angle K - \tan^{-1}\left(\frac{\alpha}{\alpha}\right)$$

$$= \angle K - 45°$$

At low frequency ($\omega \ll \alpha$) the phase angle approaches $\angle K$ and at high frequencies ($\omega \gg \alpha$) the phase approaches $\angle K - 90°$. Almost all of the $-90°$ phase change occurs in the two-decade range from $\omega/\alpha = 0.1$ to $\omega/\alpha = 10$. The straight-line segments in Figure 12–3 provide an approximation of the phase response. The phase approximation below $\omega/\alpha = 0.1$ is $\theta = \angle K$ and above $\omega/\alpha = 10$ is $\theta = \angle K - 90°$. Between these values the phase approximation is a straight line that begins at $\theta = \angle K$, passes through $\theta = \angle K - 45°$ at the cutoff frequency, and reaches $\theta = \angle K - 90°$ at $\omega/\alpha = 10$. The slope of this line segment is $-45°$/decade since the total phase change is $-90°$ over a two-decade range.

To construct the straight-line approximations for a first-order low-pass transfer function, we need two parameters, the value of $T(0)$ and α. The parameter α defines the cutoff frequency and the value of $T(0)$ defines the passband gain $|T(0)|$ and the low-frequency phase $\angle T(0)$. The required quantities $T(0)$ and α can be determined directly from the transfer function $T(s)$ and can often be estimated by inspecting the circuit itself.

Using logarithmic scales in Bode plots allows us to make straight-line approximations to both the gain and phase responses. These approximations provide a useful way of visualizing a circuit's frequency response. Often such graphical estimates are adequate for developing analysis and design approaches. For example, the frequency response of the first-order low-pass function can be characterized by calculating the gain and phase over a two-decade band from one decade below to one decade above the cutoff frequency.

EXAMPLE 12–2

Consider the circuit in Figure 12–4. Find the transfer function $T(s) = V_2(s)/V_1(s)$ and construct the straight-line approximations to the gain and phase responses.

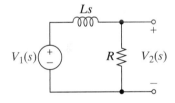

FIGURE 12–4

SOLUTION:
Applying voltage division, the voltage transfer function for the circuit is

$$T(s) = \frac{R}{Ls + R} = \frac{R/L}{s + R/L}$$

Comparing this with Eq. (12–4), we see that the circuit has a low-pass gain response with $\alpha = R/L$ and $T(0) = 1$. Therefore, $|T(0)|_{dB} = 0\ dB$, $\omega_C = R/L$, and $\angle K = 0°$. Given these quantities, we construct the straight-line approximations shown in Figure 12–5. Note that the frequency scale in Figure 12–5 is normalized by multiplying ω by $L/R = 1/\alpha$.

Circuit interpretation: The low-pass response in Figure 12–5 can be explained in terms of circuit behavior. At zero frequency the inductor acts like a short circuit that directly connects the input port to the output port to produce a passband gain of 1 (or 0 dB). At infinite frequency the inductor acts like an open circuit, effectively disconnecting the input and output ports and leading to a gain of zero. Between these two extremes the impedance of the inductor gradually increases, causing the

FIGURE 12–5

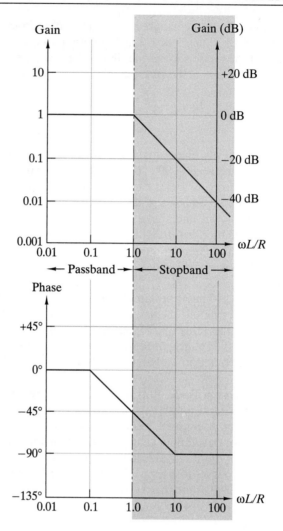

circuit gain to decrease. In particular, at the cutoff frequency we have $\omega L = R$, the impedance of the inductor is $j\omega L = jR$, and the transfer function reduces to

$$T(j\omega_C) = \frac{R}{R + jR} = \frac{1}{\sqrt{2}} \angle -45°$$

In other words, at the cutoff frequency the gain is -3 dB and the phase shift $-45°$. Obviously, the changing impedance of the inductor gives the circuit its low-pass gain features.   ∎

▶ Design Exercise 12–3 _____

Select values of R and L for the circuit of Figure 12–4 so that the cutoff frequency occurs at 10 kHz.

Answer: $R = 100\,k\Omega, L = 1.591$ H. Many other design options are possible.

▶ ⬛ DESIGN EXAMPLE 1 2 – 3 ⬛

Design an RC low-pass filter with a cutoff frequency of 1 krad/s and a passband gain of 1.

SOLUTION:

The transfer function for a low-pass filter was shown in Eq. (12–4) to be

$$T(s) = \frac{K}{s + \alpha}$$

where K is a constant to be determined and α is the cutoff frequency. We can design this circuit using a simple RC voltage divider with the output taken across the capacitor.

$$T(s) = \frac{1/Cs}{R + 1/Cs} = \frac{1/RC}{s + 1/RC}$$

We can see that $\alpha = 1/RC = 1000\,\text{rad/s}$, and for the passband gain to be 1, K must equal α. We need to select R and C so the reciprocal of their product equals $1000\,\text{rad/s}$. Using standard values we select $R = 1\,\text{k}\Omega$ and $C = 1\,\mu\text{F}$. Many other solutions are possible. ■

◗ Design Exercise 12–4

Design an RC low-pass filter with a cutoff of $100\,\text{rad/s}$ and a passband gain of $+4$.

Answer: See Figure 12–6. A different approach is shown in Design Example 12–4.

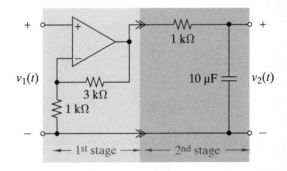

$v_1(t)$ $3\,\text{k}\Omega$ $1\,\text{k}\Omega$ $1\,\text{k}\Omega$ $10\,\mu\text{F}$ $v_2(t)$

← 1st stage → ← 2nd stage →

FIGURE 12–6

◗ DESIGN EXAMPLE 12–4

(a) Show that the transfer function $T(s) = V_2(s)/V_1(s)$ in Figure 12–7 has a low-pass gain characteristic.
(b) Select element values so the passband gain is -4 and the cutoff frequency is 100 rad/s.
(c) Use OrCAD to simulate the frequency response of the results in part (b).

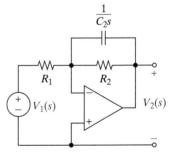

FIGURE 12–7

SOLUTION:

(a) The circuit is an inverting amplifier configuration with

$$Z_1(s) = R_1 \quad \text{and} \quad Z_2(s) = \frac{1}{C_2 s + \dfrac{1}{R_2}} = \frac{R_2}{R_2 C_2 s + 1}$$

The circuit transfer function is found as

$$T(s) = -\frac{Z_2(s)}{Z_1(s)} = -\frac{R_2}{R_1} \times \frac{1}{R_2 C_2 s + 1}$$

Rearranging the standard low-pass form in Eq. (12–4) as

$$T(s) = \frac{K/\alpha}{s/\alpha + 1}$$

shows that the circuit transfer function has a low-pass form with

$$\omega_C = \alpha = \frac{1}{R_2 C_2} \quad \text{and} \quad T(0) = -\frac{R_2}{R_1}$$

This is an inverting circuit, so the $-90°$ phase swing of the low-pass form runs from $\angle T(0) = -180°$ to $\angle T(\infty) = -270°$, passing through $\angle T(j\omega_C) = -225°$ along the way.

Circuit interpretation: The low-pass response is easily deduced from known circuit performance. At dc the capacitor acts like an open circuit and the circuit in Figure 12–7 reduces to a resistance inverting amplifier with $K = T(0) = -R_2/R_1$. At infinite frequency the capacitor acts like a short circuit that connects the OP AMP output directly to the inverting input. This connection results in zero output since the node voltage at the inverting input is necessarily zero. In between these two

FIGURE 12–8 *Copyright ©
Cadence Design Systems, Inc.
Used with permission.*

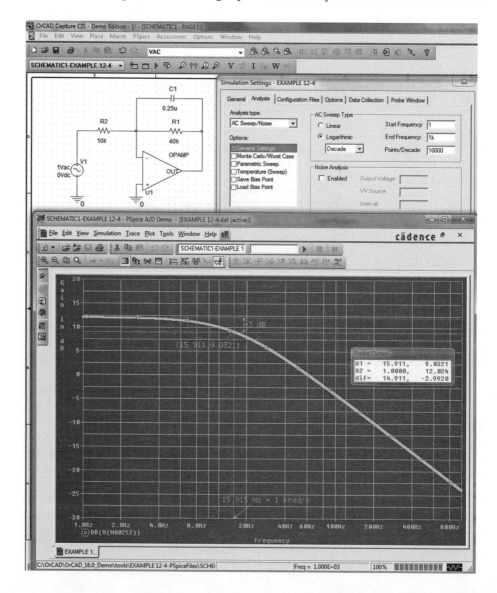

extremes the gain gradually decreases as the decreasing capacitor impedance gradually pulls the OP AMP output down to zero at infinite frequency.

(b) The design constraints require that $\omega_C = 1/R_2C_2 = 100$ and $|T(0)| = R_2/R_1 = 4$. Selecting $R_1 = 10\,k\Omega$ implies that $R_2 = 40\,k\Omega$ and $C_2 = 0.25\,\mu F$.

(c) To create the circuit in OrCAD, use the VAC source with a value of 1 V and no offset. This source allows one to run an AC frequency sweep. It does not require you to select a frequency; you will do that when you set up the simulation profile. After the circuit is drawn in OrCAD, call up the simulation profile and select AC Sweep/Noise under "Analysis type." Since we want a log-log plot of the gain response, we select "Logarithmic" and "Decade" for the sweep type. In deciding where to start and stop our sweep, we generally want to capture the critical points of the circuit's frequency performance. Since OrCAD uses only cyclic frequencies, we divide the cutoff frequency 100 rad/s by 2π to convert our values to hertz, namely 15.915 Hz. Since log graphs look better in multiples of 10, we select our sweep to start at 1 Hz and end at 1 kHz. To make the sweep look smooth and the data more accurate, we ask Probe to calculate 10,000 points per decade. We also ask Probe to display the y-axis as a decibel of the output by selecting "Trace", then under "Add Trace" select the "DB()" function and insert the output node's name between the parenthesis, DB(N00252) in this case. The circuit, simulation profile, and Probe output are shown in Figure 12–8. The cursor in Probe is set as close to the cutoff frequency as possible—15.911 Hz versus 15.915 Hz—and the gain is displayed as 9.0321 dB, which is very close to 3 dB down from the passband gain of $20\log_{10}(4) = 12.041$ dB. The circuit performed as desired. ∎

D Design Exercise 12–5

Design a low-pass active filter that has a gain of -10 and a cutoff frequency of 10 krad/s.

Answer: Use the circuit of Figure 12–7 with $R_1 = 10\,k\Omega$, $R_2 = 100\,k\Omega$, and $C = 1000\,pF$. Other designs are possible.

APPLICATION   EXAMPLE 12 – 5

In terms of frequency response, the ideal OP AMP model introduced in Chapter 4 assumes that the device has an infinite gain and an infinite bandwidth. A more realistic model of the device is shown in Figure 12–9(a). The controlled source gain in Figure 12–9(a) is a low-pass transfer function with a dc gain of A and a cutoff frequency ω_C. The straight-line asymptotes of controlled source gain are shown in Figure 12–9(b). The **gain-bandwidth product** $(G = A\omega_C)$ is the basic performance parameter of this model.

With no feedback the OP AMP transfer function is the same as the controlled-source transfer function. The gain-bandwidth product of the open-loop transfer function is

$$G = A\omega_C \quad \text{(open loop)}$$

The closed-loop transfer function of the circuit in Figure 12–9(a) is found by writing the following device and connection equations:

$$\text{Device equation: } V_O(s) = \frac{A}{s/\omega_C + 1}(V_P(s) - V_N(s))$$

$$\text{Input connection: } V_P(s) = V_S(s)$$

$$\text{Feedback connection: } V_N(s) = V_O(s)$$

Substituting the connection equations into the OP AMP device equation yields

$$V_O(s) = \frac{A}{s/\omega_C + 1}(V_S(s) - V_O(s))$$

FIGURE 12–9

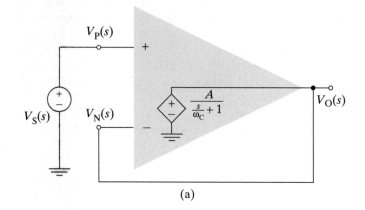

(a)

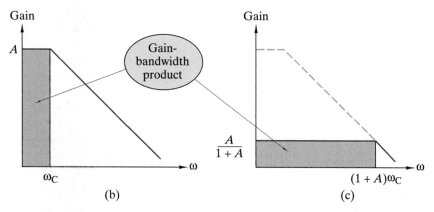

(b)   (c)

Solving for the closed-loop transfer function produces

$$T(s) = \frac{V_O(s)}{V_S(s)} = \frac{A}{A+1}\left[\frac{1}{\dfrac{s}{(A+1)\omega_C}+1}\right]$$

The straight-line asymptotes of the closed-loop transfer function are shown in Figure 12–9(c).

The closed-loop circuit has a low-pass transfer function with a dc gain of $A/(A+1)$ and a cutoff frequency of $(A+1)\omega_C$. The gain-bandwidth product of the closed-loop circuit is

$$G = \left[\frac{A}{A+1}\right][(A+1)\omega_C] = A\omega_C \quad \text{(closed loop)}$$

which is the same as the open-loop case. In other words, the gain-bandwidth product is invariant and is not changed by feedback. It can be shown that this result is a general one that applies to all linear OP AMP circuits, regardless of the circuit configuration.

The gain-bandwidth product is a fundamental parameter that limits the frequency response of OP AMP circuits. For example, an OP AMP with a gain-bandwidth product of $G = 10^6$ Hz is connected as a noninverting amplifier with a closed-loop gain of 20. The frequency response of the resulting closed-loop circuit has a low-pass characteristic with a passband gain of 20 and a cutoff frequency of

$$f_C = \frac{10^6}{20} = 50\,\text{kHz}$$

A particular uA741 OP AMP, a common all-purpose amplifier, has a guaranteed minimum gain-bandwidth product of 0.7 MHz. Using a straight-line approximation, what is the highest frequency that can be amplified by 40 dB without any loss of gain?

Answer: $f_{MAX} = 7\,kHz.$

FIRST-ORDER HIGH-PASS RESPONSE

We next treat the first-order high-pass transfer function

$$T(s) = \frac{Ks}{s + \alpha} \tag{12–8}$$

The high-pass function differs from the low-pass case by the introduction of a zero at $s = 0$. Replacing s by $j\omega$ in $T(s)$ and solving for the gain and phase functions yields

$$|T(j\omega)| = \frac{|K|\omega}{\sqrt{\omega^2 + \alpha^2}} \tag{12–9}$$

$$\theta(\omega) = \angle K + 90° - \tan^{-1}(\omega/\alpha)$$

Figure 12–10 shows Bode plots for the first-order high-pass function plotted using a log scale for the normalized frequency ω/α. The gain diagram displays a high-pass characteristic with a passband at high frequency and a stopband at low frequency. The maximum passband gain occurs at high frequency $(\omega \gg \alpha)$ where the gain

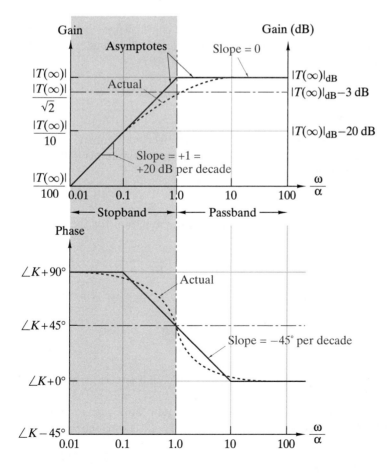

FIGURE 12–10 *First-order high-pass Bode plots.*

$|T(j\omega)| \to T_{\max} = |K|$. In the passband the gain gradually decreases as frequency decreases until at $\omega = \alpha$ we have

$$|T(j\alpha)| = \frac{|K|\alpha}{\sqrt{\alpha^2 + \alpha^2}} = \frac{|K|}{\sqrt{2}} = \frac{1}{\sqrt{2}} T_{\max} \qquad (12-10)$$

Again invoking Eq. (12–2), we find that $\omega_C = \alpha$ is the cutoff frequency marking the boundary between a high-frequency passband and a low-frequency stopband. This boundary is shown as $\omega/\alpha = 1$ in Figure 12–10.

The low- and high-frequency gain asymptotes approximate the gain response in Figure 12–10. The high-frequency asymptote ($\omega \gg \alpha$) is the horizontal line whose ordinate is $|K|$ (slope $= 0$ or $0\,\mathrm{dB/decade}$). The low-frequency asymptote ($\omega \ll \alpha$) is a line of the form $|K|\omega/\alpha$ (slope $= +1$ or $+20\,\mathrm{dB/decade}$). The intersection of these two asymptotes occurs when $|K| = |K|\omega/\alpha$, which defines a corner frequency at $\omega = \alpha$.

The semilog plot of the phase shift of the first-order high-pass function is shown in Figure 12–10. The phase shift approaches $\angle K$ at high frequency, passes through $\angle K + 45°$ at the cutoff frequency, and approaches $\angle K + 90°$ at low frequency. Most of the 90° phase change occurs over the two-decade range centered on the cutoff frequency. The phase shift can be approximated by the straight-line segments shown in Figure 12–10. As in the low-pass case, $\angle K$ is 0° when K is positive and $\pm 180°$ when K is negative.

Like the low-pass function, the first-order high-pass Bode plots can be approximated by straight-line segments. To construct these lines we need two parameters, $T(\infty)$ and α. The parameter α defines the cutoff frequency, and the quantity $T(\infty)$ gives the passband gain $|T(\infty)|$ and the high-frequency phase angle $\angle T(\infty)$. The quantities $T(\infty)$ and α can be determined directly from the transfer function or estimated directly from the circuit in some cases. The straight line shows that the first-order high-pass response can be characterized by calculating the gain and phase over a two-decade band from one decade below to one decade above the cutoff frequency.

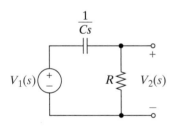

FIGURE 12–11

EXAMPLE 12–6

Show that the transfer function $T(s) = V_2(s)/V_1(s)$ in Figure 12–11 has a high-pass gain characteristic. Construct the straight-line approximations to the gain and phase responses of the circuit.

SOLUTION:
Applying voltage division, the voltage transfer function for the circuit is

$$T(s) = \frac{R}{R + 1/Cs} = \frac{RCs}{RCs + 1}$$

Rearranging Eq. (12–8) as

$$T(s) = \frac{K(s/\alpha)}{s/\alpha + 1}$$

shows that the circuit has a high-pass gain characteristic with $\alpha = 1/RC$ and $T(\infty) = 1$. Therefore, $|T(\infty)|_{\mathrm{dB}} = 0\,\mathrm{dB}$, $\omega_C = 1/RC$, and $\angle T(\infty) = 0°$. Given these quantities, we construct the straight-line gain and phase approximations in Figure 12–12. The frequency scale in Figure 12–12 is normalized by multiplying ω by $RC = 1/\alpha$.

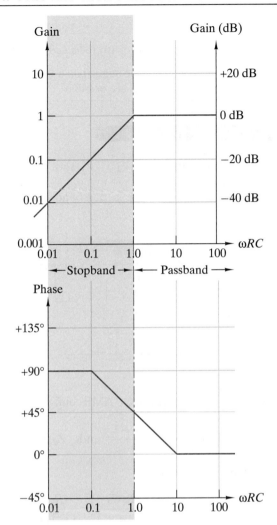

Circuit interpretation: The high-pass response in Figure 12–12 can be understood in terms of known circuit behavior. At zero frequency the capacitor acts like an open circuit that effectively disconnects the input signal source, leading to zero gain. At infinite frequency the capacitor acts like a short circuit that directly connects the input to the output, leading to a passband gain of 1 (or 0 dB). Between these two extremes, the impedance of the capacitor gradually decreases, causing the gain to increase. In particular, at the cutoff frequency we have $1/\omega C = R$, the impedance of the capacitor is $1/j\omega C = -jR$, and the transfer function is

$$T(j\omega_C) = \frac{R}{R - jR} = \frac{1}{\sqrt{2}}\angle{+45°}$$

In other words, at the cutoff frequency the gain is −3 dB and the phase shift +45°. Obviously, the decreasing impedance of the capacitor gives the circuit its high-pass gain characteristics. ■

Exercise 12–7 _____

The circuit shown in Figure 12–11 has $R = 2.2$ kΩ and $C = 0.33$ μF. What is the gain of the circuit at $\omega = 1$ krad/s in dB?

Answer: $|T(j1000)|_{dB} = -4.62$ dB.

Design Exercise 12–8 _____

Design an RL high-pass filter with a cutoff of 10 krad/s and a passband gain of 1.

Answer: Use an RL voltage divider with the output taken across the inductor. Select $R = 1\ k\Omega$ and $L = 100$ mH. Many other designs are possible.

---

① ⬛ DESIGN EXAMPLE 12–7

(a) Show that the transfer function $T(s) = V_2(s)/V_1(s)$ of the circuit in Figure 12–13 has a high-pass gain characteristic.
(b) Select the element values to produce a passband gain of -4 and a cutoff frequency of 40 krad/s.

FIGURE 12–13

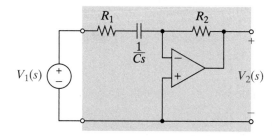

SOLUTION:

(a) The branch impedances of the inverting OP AMP configuration in Figure 12–13 are

$$Z_1(s) = R_1 + \frac{1}{Cs} = \frac{R_1 Cs + 1}{Cs} \quad \text{and} \quad Z_2(s) = R_2$$

and the voltage transfer function is

$$T(s) = -\frac{Z_2(s)}{Z_1(s)} = -\frac{R_2 Cs}{R_1 Cs + 1} = \frac{(-R_2/R_1)s}{s + 1/R_1 C}$$

This results in a high-pass transfer function of the form $Ks/(s + \alpha)$ with $K = -R_2/R_1$ and $\alpha = \omega_C = 1/R_1 C$.

Circuit interpretation: The high-pass response of this circuit is easily understood in terms of element impedances. At dc the capacitor in Figure 12–13 acts like an open circuit that effectively disconnects the input source, resulting in zero gain. At infinite frequency the capacitor acts like a short circuit that reduces the circuit to an inverting amplifier with $K = T(\infty) = -R_2/R_1$. As the frequency varies from zero to infinity, the gain gradually increases as the capacitor impedance decreases.

(b) The design requirements specify that $1/R_1 C = 4 \times 10^4$ and $R_2/R_1 = 4$. Selecting $R_1 = 10\ k\Omega$ requires $R_2 = 40\ k\Omega$ and $C = 2500$ p. ∎

---

① Design Exercise 12–9 _____

Design a high-pass filter that has the following transfer function:

$$T(s) = -\frac{200s}{s + 5000}$$

Answer: Use the circuit of Figure 12–13 with $R_1 = 1\ k\Omega$, $R_2 = 200\ k\Omega$, and $C = 0.2\ \mu F$. Other solutions are possible.

Your company issued a request for proposals listing the following design requirements and evaluation criteria.

Design requirements call for a high-pass filter with a passband gain of unity and a cutoff frequency of 150 Hz ± 10%. The filter input is driven by a sensor with a 50-Ω source resistance.

Evaluation criteria are filter performance, parts count, power consumption, and cost. The three vendors have responded with the designs shown in Figure 12–14. As a junior engineer, you have been asked to evaluate the designs and identify the best design. Which vendor would you recommend and why?

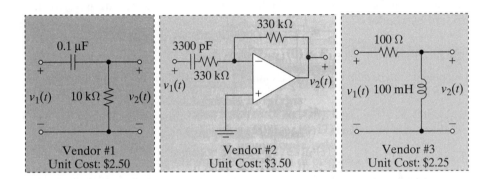

FIGURE 12–14

SOLUTION:
The design requirements describe a high-pass transfer function of the form

$$T(s) = \frac{Ks}{s+\alpha} = \frac{\pm 1s}{s+2\pi 150} = \frac{\pm s}{s+942.5}$$

The first task is to analyze the performance of the three proposed filters.
   Vendor #1: Using voltage division to find $T(s)$ yields

$$T(s) = \frac{10^4}{10^4 + \dfrac{1}{10^{-7}s}} = \frac{10^{-3}s}{10^{-3}s+1} = \frac{s}{s+1000}$$

This is a high-pass response with $|T(\infty)| = 1$ and $f_C = 1000/(2\pi) = 159.2$ Hz.
   Vendor #2: The circuit is an inverting amplifier whose transfer function is

$$T(s) = \frac{-330 \times 10^3}{330 \times 10^3 + \dfrac{1}{3.3 \times 10^{-9}s}} = \frac{-1.089 \times 10^{-3}s}{1.089 \times 10^{-3}s+1} = \frac{-s}{s+918.3}$$

This is a high-pass response with $|T(\infty)| = 1$ and $f_C = 918.3/(2\pi) = 146.2$ Hz.
   Vendor #3: Using voltage division to find $T(s)$ yields

$$T(s) = \frac{0.1s}{0.1s+100} = \frac{s}{s+1000}$$

This is a high-pass response with $|T(\infty)| = 1$ and $f_C = 1000/(2\pi) = 159.2$ Hz.
   Evaluation discussion: The analysis summary in Table 12–3 shows that all three filters meet the basic passband gain (±1) and cutoff frequency (150 Hz ± 10%) requirements. The vendor #2 design can be eliminated because it has the largest part count, requires a dc power supply for its OP AMP, and has the highest cost. The vendor #3 filter has a serious loading problem that offsets its lower cost. This filter has a 100-Ω input resistor that loads the specified 50-Ω input source. The actual

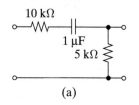

(a)

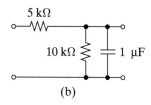

(b)

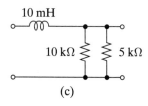

(c)

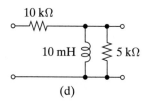

(d)

FIGURE 12–15

FIGURE 12–16

T A B L E **12–3** ANALYSIS SUMMARY

	PASSBAND GAIN	CUTOFF FREQ. (Hz)	PARTS COUNT	POWER CONSUMPTION	COST	SOURCE LOADING
Vendor #1	1	159.2	2	Low	$2.50	No
Vendor #2	−1	146.2	4	Medium	$3.50	No
Vendor #3	1	159.2	2	Low	$2.25	Yes

cutoff frequency of the source/filter combination is $(R_{source} + R_{filter})/L = (50 + 100)/0.1 = 1500$ rad/s or 238.7 Hz, not the 159.2 Hz found by analyzing the filter in isolation. The input impedance of the vendor #1 design is greater than 10 kΩ, so it does not load the 50-Ω source. All factors considered, the best design is the filter proposed by vendor #1. ∎

◀ **E** Evaluation Exercise 12–10 _____

Suppose that the evaluation task of Example 12–8 included the requirement that the filter feed a 500-Ω recorder. Would the choice be different?

A n s w e r : Yes. The 500-Ω recorder would load both of the passive filters. Only the OP AMP solution from vendor #2 would work.

Exercise 12–11 _____

For each circuit in Figure 12–15, identify whether the gain response has low-pass or high-pass characteristics and find the passband gain and cutoff frequency.

A n s w e r s :

(a) High pass, $|T(\infty)| = 1/3$, $\omega_C = 66.7$ rad/s
(b) Low pass, $|T(0)| = 2/3$, $\omega_C = 300$ rad/s
(c) Low pass, $|T(0)| = 1$, $\omega_C = 333$ krad/s
(d) High pass, $|T(\infty)| = 1/3$, $\omega_C = 333$ krad/s

◀ **D** Design Exercise 12–12 _____

Consider the two circuits shown in Figure 12–16. Determine if each has a low-pass or high-pass characteristic. Then design each to have a cutoff frequency of 10 krad/s and a gain of 200. The input resistance must be ≥ 1 kΩ.

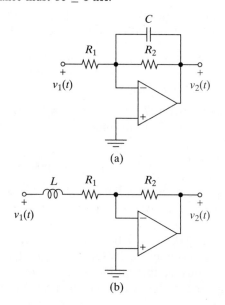

(a)

(b)

Answers:

(a) This is a low-pass filter as shown in Figure 12–7. Select $R_1 = 2\,\text{k}\Omega$, $R_2 = 400\,\text{k}\Omega$, and $C = 250$ pF.

(b) This is also a low-pass filter with $\omega_C = R_1/L$ and gain $= -R_2/R_1$. Select $R_1 = 2\,\text{k}\Omega$, $R_2 = 400\,\text{k}\Omega$, and $L = 200$ mH. In both cases, other designs are possible.

Exercise 12–13

State whether the following transfer functions have low-pass or high-pass gain characteristics and find the passband gain and cutoff frequency.

(a) $T_1(s) = \dfrac{1}{10s^{-1} + 10^{-3}}$

(b) $T_2(s) = \dfrac{10^2}{25s + 10^3}$

(c) $T_3(s) = \dfrac{20/s}{50 + 20/s}$

Answers:

(a) High pass, passband gain $= 1000$, $\omega_C = 10\,\text{krad/s}$
(b) Low pass, passband gain $= 0.1$, $\omega_C = 40\,\text{rad/s}$
(c) Low pass, passband gain $= 1$, $\omega_C = 0.4\,\text{rad/s}$

12–4 BANDPASS AND BANDSTOP RESPONSES

Bandpass and bandstop responses can be obtained using first-order high-pass and low-pass circuits as building blocks. Figure 12–17 shows a cascade connection of first-order high-pass and low-pass circuits. When the second stage does not load the first, the overall transfer function can be found by the chain rule as

$$T(s) = T_1(s) \times T_2(s) = \underbrace{\left(\frac{K_1 s}{s + \alpha_1}\right)}_{\text{high pass}} \underbrace{\left(\frac{K_2}{s + \alpha_2}\right)}_{\text{low pass}} \tag{12–11}$$

Replacing s by $j\omega$ in Eq. (12–11) and solving for the gain response yields

$$|T(j\omega)| = \underbrace{\left(\frac{|K_1|\omega}{\sqrt{\omega^2 + \alpha_1^2}}\right)}_{\text{high pass}} \underbrace{\left(\frac{|K_2|}{\sqrt{\omega^2 + \alpha_2^2}}\right)}_{\text{low pass}} \tag{12–12}$$

Note that the overall gain is zero at $\omega = 0$ and again at infinite frequency. This pattern suggests a bandpass response.

The overall gain in Eq. (12–12) is bandpass when the high-pass cutoff frequency is much lower than the low-pass cutoff frequency ($\alpha_1 \ll \alpha_2$). To see why, we develop the gain asymptotes for Eq. (12–12) in three frequency ranges.

- **Low frequency** ($\omega \ll \alpha_1 \ll \alpha_2$): This range falls in the stopband of the high-pass gain and the passband of the low-pass gain. As a result, the overall gain approaches

$$|T(j\omega)| \rightarrow \underbrace{\left(\frac{|K_1|\omega}{\alpha_1}\right)}_{\text{high pass}} \underbrace{\left(\frac{|K_2|}{\alpha_2}\right)}_{\text{low pass}} = \frac{|K_1||K_2|\omega}{\alpha_1\alpha_2}$$

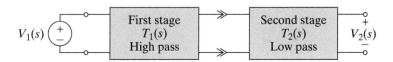

$V_1(s)$ → First stage $T_1(s)$ High pass → Second stage $T_2(s)$ Low pass → $V_2(s)$

FIGURE 12–17 *Cascade connection of high-pass and low-pass circuits.*

FIGURE 12–18 *Bandpass gain characteristic.*

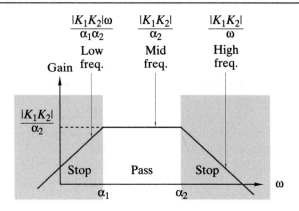

- **High frequency** ($\alpha_1 \ll \alpha_2 \ll \omega$): This range falls in the passband of the high-pass gain and the stopband of the low-pass gain. In this range the overall gain approaches

$$|T(j\omega)| \rightarrow \underbrace{\left(\frac{|K_1|\omega}{\omega}\right)}_{\text{high pass}} \underbrace{\left(\frac{|K_2|}{\omega}\right)}_{\text{low pass}} = \frac{|K_1||K_2|}{\omega}$$

- **Mid-frequency** ($\alpha_1 \ll \omega \ll \alpha_2$): This range falls in the passband of both first-order gains, so the overall gain approaches

$$|T(j\omega)| \rightarrow \underbrace{\left(\frac{|K_1|\omega}{\omega}\right)}_{\text{high pass}} \underbrace{\left(\frac{|K_2|}{\alpha_2}\right)}_{\text{low pass}} = \frac{|K_1||K_2|}{\alpha_2}$$

Figure 12–18 shows a plot of the low-, mid-, and high-frequency gain asymptotes. The low-frequency and mid-frequency asymptotes intersect when $|K_1K_2|\omega/\alpha_1\alpha_2 = |K_1K_2|/\alpha_2$. This occurs at $\omega = \alpha_1$, which is the cutoff frequency of the high-pass stage. The high-frequency and mid-frequency asymptotes intersect when $|K_1K_2|/\omega = |K_1K_2|/\alpha_2$. This occurs at $\omega = \alpha_2$ which is the cutoff frequency of the low-pass stage. The straight-line gain plot based on these asymptotes indicates a passband between cutoff frequencies $\omega_{C1} = \alpha_1$ and $\omega_{C2} = \alpha_2$. The mid-frequency gain applies in this passband. Finally, there are two stopbands; one below ω_{C1} and the other above ω_{C2}.

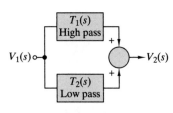

FIGURE 12–19 *Parallel connection of high-pass and low-pass circuits.*

The input signal to the bandpass cascade must pass through both a high-pass and a low-pass stage to reach the output. In the parallel connection in Figure 12–19, an input can reach the output via either a high-pass or a low-pass path. As a result, the overall transfer function is the sum of the high-pass and low-pass transfer functions.

$$T(s) = T_1(s) + T_2(s) = \underbrace{\left(\frac{K_1 s}{s + \alpha_1}\right)}_{\text{high pass}} + \underbrace{\left(\frac{K_2}{s + \alpha_2}\right)}_{\text{low pass}}$$

Since the overall gain is a sum, we must consider each of these paths separately. Replacing *s* by *jω* and solving for the individual path gains gives

$$|T_1(j\omega)| = \underbrace{\left(\frac{|K_1|\omega}{\sqrt{\omega^2 + \alpha_1^2}}\right)}_{\text{high pass}} \quad \text{and} \quad |T_2(j\omega)| = \underbrace{\left(\frac{|K_2|}{\sqrt{\omega^2 + \alpha_2^2}}\right)}_{\text{low pass}} \tag{12–13}$$

The overall gain has a bandstop characteristic when the high-pass cutoff frequency is much higher than the low-pass cutoff frequency ($\alpha_2 \ll \alpha_1$). To see why, we develop the gain asymptotes of Eq. (12–13) in three frequency ranges.

- **Low frequency** ($\omega \ll \alpha_2 \ll \alpha_1$): This range falls in the stopband of the high-pass gain and the passband of the low-pass gain. In this range the low-pass path dominates, so the overall gain approaches $|T(j\omega)| \rightarrow |K_2|/\alpha_2$.

- **High frequency** ($\alpha_2 \ll \alpha_1 \ll \omega$): This range falls in the passband of the high-pass gain and the stopband of the low-pass gain. In this range the high-pass path dominates, so the overall gain approaches $|T(j\omega)| \rightarrow |K_1|$.

Thus, there is a low-frequency passband and a high-frequency passband. In a bandstop response these passbands normally have the same gain, so that $|K_1| = |K_2|/\alpha_2$. A mid-frequency stopband lies in between these two passbands.

- **Mid-frequency** ($\alpha_2 \ll \omega \ll \alpha_1$): This range falls in the stopband of both first-order gains. The stopband asymptotes of the low-pass and high-pass gains are $|K_2|/\omega$ and $|K_1|\omega/\alpha_1$, respectively. These asymptotes intersect when $|K_2|/\omega = |K_1|\omega/\alpha_1$. Since $|K_1| = |K_2|/\alpha_2$ the intersection frequency turns out to be $\omega = \sqrt{\alpha_1\alpha_2}$.

Figure 12–20 is a plot of the low-, mid-, and high-frequency gain asymptotes. The low-pass gain dominates at frequencies below the intersection frequency $\sqrt{\alpha_1\alpha_2}$ and the high-pass gain dominates above the intersection. The straight-line gain plot indicates a stopband between cutoff frequencies at $\omega_{C1} = \alpha_2$ and $\omega_{C2} = \alpha_1$ and two passbands: one below ω_{C1} and the other above ω_{C2}.

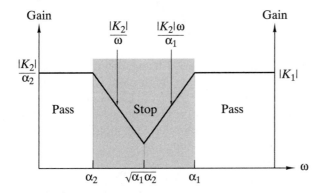

FIGURE 12–20 *Bandstop gain characteristic.*

This analysis shows that bandpass and bandstop responses can be obtained using first-order circuit building blocks. More importantly, the straight-line gain approximations to the first-order gains help us *understand* how the building blocks interact to produce other types of responses. This understanding explains the enduring utility of Bode plots. The Bode plots in Figures 12–18 and 12–20 are a reasonably good approximation of the actual gain as long as the two first-order cutoff frequencies are widely separated. Although *widely separated* has no precise definition, the straight-line gain works fairly well when $\alpha_2 > 10\alpha_1$ for a bandpass response and $\alpha_1 > 10\alpha_2$ for a bandstop response.

◑ DESIGN EXAMPLE 12–9

Design a bandpass circuit with a passband gain of 10 and cutoff frequencies at 20 Hz and 20 kHz.

FIGURE 12–21

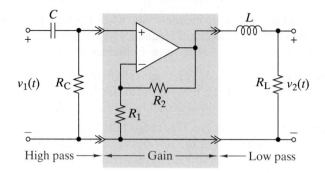

SOLUTION:

Our design uses a cascade connection of first-order low- and high-pass building blocks. The required transfer function has the form in Eq. (12–11) with the following constraints

$$\omega_{C1} = \alpha_1 = 2\pi(20) = 40\pi \text{ rad/s}$$

$$\omega_{C2} = \alpha_2 = 2\pi(20 \times 10^3) = 4\pi \times 10^4 \text{ rad/s}$$

$$\frac{|K_1 K_2|}{\alpha_2} = 10$$

Inserting these numerical values in Eq. (12–11) allows us to write the required transfer function as

$$T(s) = \underbrace{\left(\frac{s}{s + 40\pi}\right)}_{\text{high pass}} \underbrace{(10)}_{\text{gain}} \underbrace{\left(\frac{4\pi \times 10^4}{s + 4\pi \times 10^4}\right)}_{\text{low pass}}$$

This transfer function can be realized using the three-stage cascade circuit in Figure 12–21. The first stage is the RC high-pass circuit from Example 12–6 and the third stage is the RL low-pass circuit from Example 12–2. The noninverting OP AMP circuit in the second stage serves two purposes: (1) It isolates the first and third stages so the chain rule applies, and (2) it supplies the mid-band gain. Using the chain rule, the circuit transfer function is

$$T(s) = \underbrace{\left(\frac{s}{s + 1/R_C C}\right)}_{\text{high pass}} \underbrace{\left(\frac{R_1 + R_2}{R_1}\right)}_{\text{gain}} \underbrace{\left(\frac{R_L/L}{s + R_L/L}\right)}_{\text{low pass}}$$

Comparing the circuit transfer function to the required transfer function leads to the constraints and stage designs listed in Table 12–4. As always, many other solutions are possible.

T A B L E **12–4**

STAGE	CONSTRAINT	STAGE DESIGN
High pass	$R_C C = 1/40\pi$	Let $R_C = 100\,\text{k}\Omega$; then $C = 0.0796\,\mu\text{F}$
Gain	$1 + R_2/R_1 = 10$	Let $R_1 = 10\,\text{k}\Omega$; then $R_2 = 90\,\text{k}\Omega$
Low pass	$L/R_L = 1/40{,}000\pi$	Let $R_L = 200\,\text{k}\Omega$; then $L = 1.592\,\text{H}$

Exercise 12–14

Select the element values in Figure 12–21 so that the passband gain is 6 dB and the cutoff frequencies are 1 krad/s and 50 krad/s.

A n s w e r s : $R_C = 5\,k\Omega$; $C = 0.2\,\mu F$; $R_1 = R_2$; $R_L = 20\,k\Omega$; $L = 400\,mH$

Exercise 12–15

Two first-order circuits in a cascade connection have the following transfer functions.

$$T_1(s) = \frac{20}{\dfrac{s}{2000}+1} \quad \text{and} \quad T_2(s) = \frac{s}{s+40}$$

What are the cutoff frequencies and the passband gain? Assume the chain rule applies.

A n s w e r : $\omega_{C1} = 40\,rad/s$; $\omega_{C2} = 2000\,rad/s$; passband gain $= 20$

E EVALUATION EXAMPLE 12–10

There is a need for a design of a bandstop filter centered at 100 Hz that has the following transfer function ±2%:

$$T(s) = \frac{4(s^2 + 200s + 4 \times 10^5)}{(s+100)(s+4000)}$$

A vendor has submitted the circuit of Figure 12–22, claiming that the design meets the specifications within ±1%. Is the vendor's claim accurate?

SOLUTION:
Let us start by calculating the transfer function from the circuit. If the transfer function is as desired then we can determine if the transfer function meets the specifications. The circuit consists of a high-pass filter, a low-pass filter and an inverting summer. The transfer function of the filter circuits after substituting the

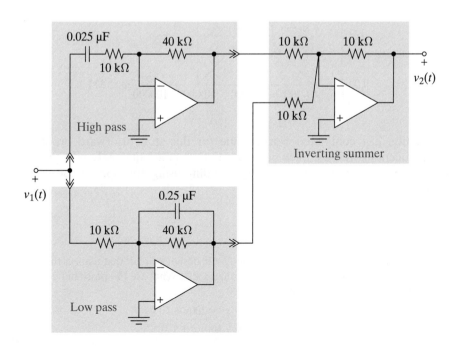

FIGURE 12–22

values on the circuit are

$$T_{\text{HP}}(s) = -\frac{(R_2/R_1)s}{s + 1/R_1C} = -\frac{4s}{s + 4000}$$

$$T_{\text{LP}}(s) = -\frac{1/R_1C}{s + 1/R_2C} = -\frac{400}{s + 100}$$

The summer adds the two and inverts the signs

$$T(s) = -(T_{\text{HP}}(s) + T_{\text{LP}}(s))$$

$$T(s) = -\left[\left(-\frac{4s}{s + 4000}\right) + \left(-\frac{400}{s + 100}\right)\right]$$

$$T(s) = \frac{4(s^2 + 200s + 4 \times 10^5)}{(s + 100)(s + 4000)}$$

The circuit realizes the desired transfer function. Now we need to determine if the transfer function actually meets the desired specifications to ±1%.

We can determine if the specifications are met graphically by plotting the low-pass and high-pass frequency characteristics and determining where the minimum occurs. Figure 12–23 displays the two transfer functions plotted on a log-log plot. We know that first-order filters, as these are, will have corner frequencies at their poles and slopes of positive or negative one, depending on whether they are high-pass or low-pass designs. The low-pass filter has a corner frequency at 100 rad/s, a passband magnitude of 4, a slope of −1, and crosses 10 krad/s at a magnitude of 0.04. The high-pass filter has a corner frequency of 4000 rad/s, also a passband magnitude of 4, a slope of +1, and crosses 40 rad/s at a magnitude of 0.04. The high- and low-pass curves cross at approximately 630 rad/s, which equals 100.3 Hz and is well within the ±1% required by the specification and the vendor's claim.

FIGURE 12–23

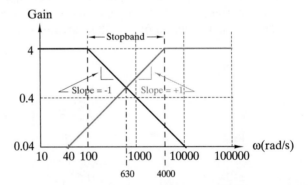

The solution just completed worked fine for this straightforward problem. For transfer functions with complex poles or multiple poles, the analysis requires more effort. We will study alternate means, including using appropriate software tools later in this chapter. ■

◗ Design Exercise 12–16

Following the analysis pattern in Example 12–10, design a circuit that realizes the following transfer function. Use no resistor smaller than 1 kΩ. What are the passband gain and the cutoff frequencies of the filter?

$$T(s) = \frac{200(s^2 + 200s + 10^6)}{(s + 100)(s + 10^4)}$$

A n s w e r : See Figure 12–24 for one design solution. The passband gains are 200 and the low-pass filter's cutoff frequency is 100 rad/s, while the high-pass filter's cutoff frequency is 10 krad/s.

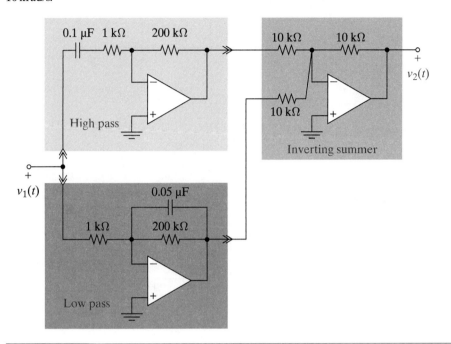

FIGURE 12–24

12–5 THE FREQUENCY RESPONSE OF *RLC* CIRCUITS

A second-order circuit can take many forms as long as it has two energy storing elements. However, the properties of second-order circuits are traditionally introduced using simple series and parallel *RLC* circuits. In Chapter 7 these circuits were used to highlight the role of ζ and ω_0 in the transient response of second-order circuits. In this section they are used to introduce the role of impedance in second-order circuit frequency response. These canonic circuits are useful introductory vehicles because they give us physical insight into the relationship between circuit parameters and circuit response.

SERIES *RLC* BANDPASS CIRCUIT

The transfer function of the *RLC* voltage divider in Figure 12–25 is

$$T(s) = \frac{V_2(s)}{V_1(s)} = \frac{R}{R + Ls + 1/Cs} = \frac{R}{R + Z_{LC}(s)} \qquad (12\text{–}14)$$

where $Z_{LC}(s) = Ls + 1/Cs$ is the impedance of the series leg of the voltage divider. To describe the frequency response of the circuit, we replace s by $j\omega$ to obtain

$$T(j\omega) = \frac{R}{R + Z_{LC}(j\omega)}$$

The variation of the impedance $Z_{LC}(j\omega)$ is the key to understanding *RLC* circuit frequency response.

Figure 12–26 shows how the impedance $Z_{LC}(j\omega) = j(\omega L - 1/\omega C)$ produces a bandpass gain response. At $\omega = 0$ (dc) the capacitor acts like an open circuit and $Z_{LC}(0) = \infty$. At $\omega = \infty$ the inductor acts like an open circuit and $Z_{LC}(\infty) = \infty$. In

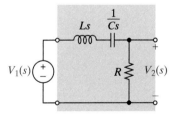

FIGURE 12–25 *Series RLC bandpass circuit.*

FIGURE 12–26 *Effect of Z_{LC} on the series RLC bandpass response.*

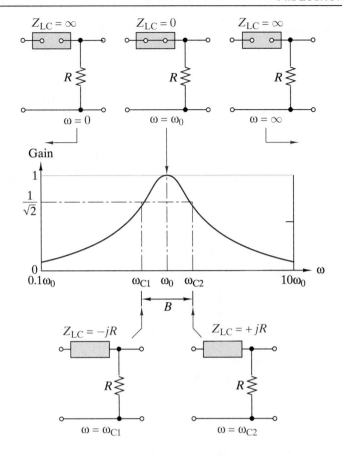

either case, the open circuit effectively disconnects the input source, making $V_2 = 0$ and producing zero-gain stopbands at low and high frequencies.

At $\omega = \omega_0 = 1/\sqrt{LC}$, the impedance $Z_{LC}(j\omega_0) = 0$, since

$$Z_{LC}(j\omega_0) = j\left(\frac{L}{\sqrt{LC}} - \frac{\sqrt{LC}}{C}\right) = j\left(\sqrt{\frac{L}{C}} - \sqrt{\frac{L}{C}}\right) = 0$$

That is, the inductor and capacitor combine to produce a short circuit that makes a direct connection between the input and the output ports. This connection makes $V_2 = V_1$, producing a voltage gain of $|T(j\omega_0)| = T_{max} = 1$. Clearly T_{max} is the maximum voltage gain available from this voltage-divider circuit. The cutoff frequencies occur when $Z_{LC}(j\omega) = \pm jR$, since

$$|T(j\omega)| = \left|\frac{R}{R \pm jR}\right| = \frac{1}{\sqrt{2}} = \frac{1}{\sqrt{2}}T_{max}$$

so that the voltage gain is reduced by a factor of $1/\sqrt{2}$ from its maximum value at $\omega = \omega_0$.

Several parameters are used to describe the bandpass gain response. First, the maximum gain occurs at the **center frequency** ω_0 located at

$$\omega_0 = \frac{1}{\sqrt{LC}} \qquad (12\text{–}15)$$

The cutoff frequencies occur when $Z_{LC}(j\omega) = \pm jR$, which requires that

$$\left(\omega L - \frac{1}{\omega C}\right) = \pm R$$

and leads to the quadratic equation

$$LC\omega^2 - (\pm R)C\omega - 1 = 0$$

Because of the $\pm$ sign, this quadratic has four roots, only two of which have physical meaning, namely

$$\omega_{C1} = -\frac{R}{2L} + \sqrt{\left(\frac{R}{2L}\right)^2 + \frac{1}{LC}}$$

$$\omega_{C2} = +\frac{R}{2L} + \sqrt{\left(\frac{R}{2L}\right)^2 + \frac{1}{LC}}$$

$$(12\text{--}16)$$

where ω_{C1} and ω_{C2} are the **cutoff frequencies** shown in Figure 12–26. The product of the two cutoff frequencies in Eq. (12–16) is

$$\omega_{C1}\,\omega_{C2} = -\left(\frac{R}{2L}\right)^2 + \left(\frac{R}{2L}\right)^2 + \frac{1}{LC} = \frac{1}{LC} = \omega_0^2$$

In other words, the center frequency $\omega_0 = \sqrt{\omega_{C1}\,\omega_{C2}}$ is the geometric mean of the two cutoff frequencies. Finally, using the results in Eq. (12–16), the **bandwidth** B in Figure 12–26 is found to be

$$B = \omega_{C2} - \omega_{C1} = \frac{R}{L} \qquad (12\text{--}17)$$

In summary, the descriptive parameters $\omega_0, \omega_{C1}, \omega_{C2}$, and B are related to the series *RLC* circuit parameters by Eqs. (12–15), (12–16), and (12–17).

For historical reasons, it is traditional to add a fifth descriptive parameter called the **quality factor** Q, defined as the ratio of center frequency over bandwidth, namely

$$Q = \frac{\omega_0}{B} = \frac{\sqrt{L/C}}{R} \qquad (12\text{--}18)$$

Figure 12–27 shows the effect of Q on the bandpass response characteristics. When $Q \gg 1$, the response is said to be *narrow band*, since $B \ll \omega_0$. Conversely, when $Q \ll 1$, the response is said to be *wide band*, since $B \gg \omega_0$. Thus, $Q = 1$ is the dividing point between narrow-band or high-Q responses and wide-band or low-Q responses. But regardless of the value of Q, the maximum gain is always $T_{\max} = 1$ (or 0 dB) at $\omega = \omega_0$.

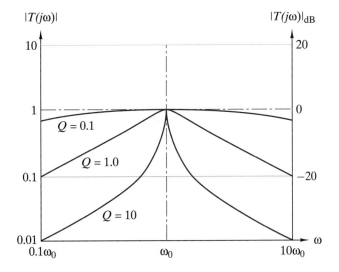

FIGURE 12–27 *Effect of Q on the series RLC bandpass response.*

A high-Q bandpass circuit is sometimes called a *tuned filter*. This terminology applies when the center frequency is carefully adjusted (tuned) to select a narrow band of signal frequencies while rejecting a much broader range of signal frequencies outside of the passband. At the center frequency $Z_{LC}(j\omega_0) = 0$, so the filter input impedance is $R + j0$, a condition known as *resonance*. For this reason the center frequency is also called the *resonant frequency*.

EXAMPLE 12–11

A series RLC circuit has a center frequency of $\omega_0 = 20\,\text{krad/s}$, a quality factor of $Q = 5$, and a resistance $R = 50\,\Omega$. Find the values of L, C, B, ω_{C1}, and ω_{C2}.

SOLUTION:
Using the definition of Q, the bandwidth is $B = \omega_0/Q = 4\,\text{krad/s}$. For a series RLC circuit, $B = R/L$ and the inductance is $L = R/B = 50/4000 = 12.5\,\text{mH}$. Using this inductance and the center frequency, the capacitance is found as $C = 1/\omega_0^2 L = 0.2\,\mu\text{F}$. Inserting these results into Eq. (12–16) yields the lower cutoff frequency as

$$\omega_{C1} = -2000 + \sqrt{2000^2 + 20{,}000^2} = 18.1\,\text{krad/s}$$

Using the definition of bandwidth, we get $\omega_{C2} = \omega_{C1} + B = 22.1\,\text{krad/s}$. ■

Exercise 12–17

A series RLC circuit has a center frequency of $\omega_0 = 500\,\text{krad/s}$, a resistance of $40\,\Omega$, and a bandwidth of $50\,\text{krad/s}$. Find Q, L, C, ω_{C1}, and ω_{C2}.

A n s w e r s : $Q = 10$, $L = 0.8\,\text{mH}$, $C = 5000\,\text{pF}$, $\omega_{C1} = 475{,}625\,\text{rad/s}$ and $\omega_{C2} = 525{,}625\,\text{rad/s}$.

⊕ DESIGN EXAMPLE 12–12

Design a bandpass circuit using a series RLC circuit that meets the filter requirements in Example 12–9. Compare this RLC design with the circuit developed in Example 12–9.

SOLUTION:
The bandpass filter requirements given in Example 12–9 call for passband gain of 10 and cutoff frequencies at 20 Hz and 20 kHz. The series RLC circuit can produce the required cutoff frequencies. However, we also need a gain stage since the RLC circuit has a maximum gain of 1. The required bandpass response and gain can be obtained using the cascade connection in Figure 12–28. The chain rule applies to this

FIGURE 12–28

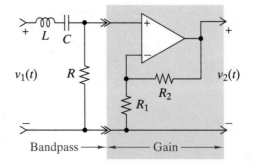

Bandpass ⟶   ⟵ Gain ⟶

FIGURE	DESCRIPTION	R	L	C	OP AMP
12–21	First-order cascade	4	1	1	1
12–28	Series *RLC* cascade	3	1	1	1

circuit since the gain stage has an infinite input impedance that does not load the output of the *RLC* stage.

Given that $f_{C1} = 20\,\text{Hz}$ and $f_{C2} = 20\,\text{kHz}$, the descriptive parameters of the bandpass stage are

$$\omega_0 = 2\pi\sqrt{f_{C1}\,f_{C2}} = 3.974\,\text{krad/s}$$

$$B = 2\pi(f_{C2} - f_{C1}) = 125.5\,\text{krad/s}$$

$$Q = \frac{\omega_0}{B} = 0.0317$$

Since $Q \ll 1$, the design requirements here and in Example 12–9 describe a wideband filter. Selecting $R = 10\,\text{k}\Omega$, the inductance in the series *RLC* circuit is $L = R/B = 79.66\,\text{mH}$. Using this inductance and the center frequency gives the capacitance as $C = 1/\omega_0^2 L = 0.795\,\mu\text{F}$. The gain stage is a noninverting amplifier for which a gain of 10 requires $1 + R_2/R_1 = 10$. Selecting $R_1 = 10\,\text{k}\Omega$ makes $R_2 = 90\,\text{k}\Omega$.

Evaluation discussion: The design developed here uses the series *RLC* cascade in Figure 12–28. The design developed in Example 12–9 uses the cascade of first-order circuits shown in Figure 12–21. The element count for the two designs is summarized in Table 12–5.

The minor difference in element count is not significant. What is significant is the difference in possible output stage loading. The circuit in Figure 12–28 is not subject to loading since the output is an OP AMP whose output impedance is zero. The first-order cascade in Figure 12–21 is susceptible to loading since its output stage is a passive *RL* circuit with a finite output impedance. ∎

▶ Design Exercise 12–18 _____

Design a series *RLC* bandpass circuit that has a center frequency of $\omega_0 = 10\,\text{krad/s}$, a maximum gain of 20 dB, and a bandwidth of 5 krad/s.

A n s w e r : Same design as in Figure 12–28, with $L = 100\,\text{mH}$, $C = 0.1\,\mu\text{F}$, $R = 500\,\Omega$, $R_1 = 10\,\text{k}\Omega$, and $R_2 = 90\,\text{k}\Omega$. Other answers are possible.

THE SERIES *RLC* BANDSTOP CIRCUIT

The bandpass circuit in Figure 12–25 and the bandstop circuit in Figure 12–29 are both series *RLC* circuits. The difference is that the bandpass circuit takes its output across the resistor while the bandstop circuit takes its output across the inductor and capacitor in series. The voltage transfer function of the bandstop voltage-divider circuit in Figure 12–29 is

$$T(s) = \frac{V_2(s)}{V_1(s)} = \frac{Ls + 1/Cs}{R + Ls + 1/Cs} = \frac{Z_{LC}(s)}{R + Z_{LC}(s)} \qquad (12\text{–}19)$$

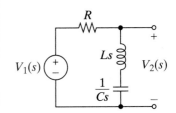

FIGURE 12–29 *Series RLC bandstop circuit.*

where $Z_{LC}(s) = Ls + 1/Cs$ is the impedance of the shunt leg of the voltage divider. This impedance is the key to understanding the shape of *RLC* circuit frequency response.

Figure 12–30 shows how the impedance $Z_{LC}(j\omega) = j(\omega L - 1/\omega C)$ produces a bandstop gain response. At $\omega = 0$, the capacitor acts like an open circuit so that the

FIGURE 12–30 *Effect of Z_{LC} on the series RLC bandstop response.*

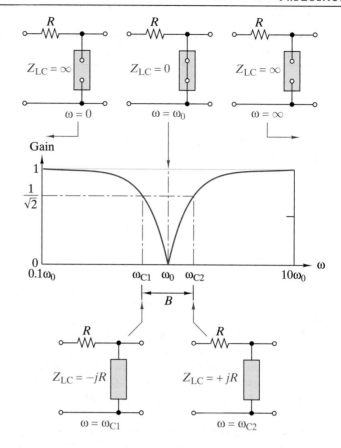

output voltage is $V_2 = V_{OC} = V_1$. At $\omega = \infty$, the inductor acts like an open circuit and again we have $V_2 = V_{OC} = V_1$. In either case, the condition $V_2 = V_1$ produces maximum gains of $T_{max} = 1$ in the low- and high-frequency passbands.

At $\omega = \omega_0 = 1/\sqrt{LC}$, the impedance $Z_{LC}(j\omega) = 0$. The resulting short circuit makes $V_2 = 0$, reducing the gain to zero and producing the null or notch in the gain response. The cutoff frequencies occur when $Z_{LC}(j\omega) = \pm jR$, since

$$|T(j\omega)| = \left| \frac{\pm jR}{R \pm jR} \right| = \frac{1}{\sqrt{2}} = \frac{1}{\sqrt{2}} T_{max}$$

so that the voltage gain is reduced by a factor of $1/\sqrt{2}$ from its maximum value in the two passbands. In sum, the values of $Z_{LC}(j\omega)$ that produce the bandpass response in Figure 12–26 produce the bandstop response shown in Figure 12–30.

Since Z_{LC} controls bandstop gain response, the equations relating the descriptive parameters to circuit parameters are the same as those governing the bandpass case. Specifically, the notch occurs when $Z_{LC} = 0$ at a frequency of

$$\omega_0 = \frac{1}{\sqrt{LC}}$$

The passband cutoff frequencies occur when $Z_{LC} = \pm jR$ at frequencies of

$$\omega_{C1} = -\frac{R}{2L} + \sqrt{\left(\frac{R}{2L}\right)^2 + \frac{1}{LC}}$$

$$\omega_{C2} = +\frac{R}{2L} + \sqrt{\left(\frac{R}{2L}\right)^2 + \frac{1}{LC}}$$

Finally, the width of the stopband is

$$B = \omega_{C2} - \omega_{C1} = \frac{R}{L}$$

These equations are the same as Eqs. (12–15), (12–16), and (12–17) for the bandpass circuit, except that here they describe the location and width of the stopband rather than the passband.

The *RLC* bandstop circuit is best suited to narrow-band applications aimed at eliminating a single frequency. Narrow-band notch circuits are often used to eliminate power line noise at 60 Hz (50 Hz in many other countries), especially in biomedical instrumentation, where signal levels are usually very low.

▶ DESIGN EXAMPLE 12–13

A voltage source with a Thévenin resistance of 50 Ω has a spurious (undesirable) conducted emission at 25 krad/s. Connecting an inductor and capacitor in series across the 50-Ω source produces a bandstop filter that can eliminate the troublesome signal. To avoid reducing nearby useful signals, the stopband bandwidth must be less than 1 krad/s. Select values of L and C.

SOLUTION:
The stopband bandwidth limitation requires $B = R/L < 1\,\mathrm{krad/s}$. Since $R = 50\,\Omega$, this constraint means that $L > R/1000 = 50\,\mathrm{mH}$. To eliminate the spurious emission, the bandstop notch must be located at

$$\omega_0 = \frac{1}{\sqrt{LC}} = 25 \times 10^3 \text{ rad/s}$$

Selecting $L = 100\,\mathrm{mH}$ makes the notch bandwidth $B = R/L = 0.5\,\mathrm{krad/s} < 1\,\mathrm{krad/s}$, which meets the bandwidth limitation. Given this inductance, the required notch frequency calls for a capacitance of $C = 1/\omega_0^2 L = 0.016\,\mu\mathrm{F}$. ∎

▶ Design Exercise 12–19

Design a bandstop filter to eliminate a 13.5 kHz signal. The bandwidth of the notch should not exceed 10 kHz.

Answer: Use a series *RLC* bandstop design with $L = 0.1\,\mathrm{H}, C = 1390\,\mathrm{pF}, R \leq 6.28\,\mathrm{k}\Omega$. Other designs are possible.

PARALLEL *RLC* BANDPASS CIRCUIT

Using current division, the current transfer function of the parallel *RLC* circuit in Figure 12–31 is written as

$$T(s) = \frac{I_2(s)}{I_1(s)} = \frac{1/R}{1/R + Cs + 1/Ls} = \frac{1/R}{1/R + Y_{LC}(s)} \qquad (12\text{–}20)$$

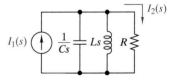

FIGURE 12–31 *Parallel RLC bandpass circuit.*

where $Y_{LC}(s) = Cs + 1/Ls$ is the admittance of the inductor and capacitor in parallel. The bandpass features of the parallel *RLC* circuit are controlled by this admittance.

The variation of the admittance $Y_{LC}(j\omega) = j(\omega C - 1/\omega L)$ produces a bandpass response for the following reasons. At $\omega = 0$ and $\omega = \infty$, the admittance $Y_{LC}(j\omega)$ is infinite, which is equivalent to a zero-impedance short circuit. This short circuit shunts all of the input current around the resistor in Figure 12–31, making $I_2 = 0$ and producing zero-gain stopbands at low and high frequencies.

At $\omega = \omega_0 = 1/\sqrt{LC}$, the admittance $Y_{LC}(j\omega) = 0$, since

$$Y_{LC}(j\omega_0) = j\left(\frac{C}{\sqrt{LC}} - \frac{\sqrt{LC}}{L}\right) = j\left(\sqrt{\frac{C}{L}} - \sqrt{\frac{C}{L}}\right) = 0$$

A zero admittance is the same as an infinite-impedance open circuit. Because of this open circuit, all of the input current passes through the resistor, making $I_2 = I_1$. This creates a current gain of $|T(j\omega_0)| = T_{max} = 1$. Clearly T_{max} is the maximum current gain available from this current-divider circuit.

The cutoff frequencies occur when $Y_{LC}(j\omega) = \pm j/R$, since

$$|T(j\omega)| = \left|\frac{1/R}{1/R \pm j/R}\right| = \frac{1}{\sqrt{2}} = \frac{1}{\sqrt{2}} T_{max}$$

and the current gain is reduced by a factor of $1/\sqrt{2}$ from its maximum value at $\omega = \omega_0$. Thus, the current gain of the parallel circuit in Figure 12–31 displays a passband response centered at ω_0 with stopbands at both high and low frequencies.

The admittance $Y_{LC}(j\omega)$ plays the same role in the parallel RLC circuit that the impedance $Z_{LC}(j\omega)$ plays in the series RLC circuit. In fact, comparing the transfer functions in Eqs. (12–14) and (12–20) reveals that the two circuits are duals. That is, one transfer function can be converted into the other using the following duality interchanges.

Series RLC		Parallel RLC
R	$\longleftrightarrow$	$1/R$
L	$\longleftrightarrow$	C
C	$\longleftrightarrow$	L

Duality also means that these interchanges convert Eqs. (12–15), (12–16), (12–17), and (12–18) for the series RLC circuit into the corresponding relationships for the parallel RLC circuit.

Starting with Eq. (12–15), the interchange leaves the center frequency unchanged at

$$\omega_0 = \frac{1}{\sqrt{LC}} \tag{12–21}$$

The two cutoff frequencies in Eq. (12–16) become

$$\omega_{C1} = -\frac{1}{2RC} + \sqrt{\left(\frac{1}{2RC}\right)^2 + \frac{1}{LC}}$$

$$\omega_{C2} = +\frac{1}{2RC} + \sqrt{\left(\frac{1}{2RC}\right)^2 + \frac{1}{LC}} \tag{12–22}$$

It is easy to show that the center frequency $\omega_0 = \sqrt{\omega_{C1}\omega_{C2}}$ is the geometric mean of these two cutoff frequencies. Given these cutoff frequencies, the bandwidth and quality factor of the parallel RLC circuit are

$$B = \omega_{C2} - \omega_{C1} = \frac{1}{RC}$$

$$Q = \frac{\omega_0}{B} = R\sqrt{C/L} \tag{12–23}$$

Equations (12–21), (12–22), and (12–23) relate the descriptive parameters of a bandpass response to the circuit parameters of the parallel RLC circuit.

Finally, the gain plots in Figure 12–27 also apply to the current gain of the parallel RLC circuit. As we noted with the series circuit, the response is narrow band when

$Q \gg 1$ and is wide band when $Q \ll 1$. Thus, high-Q and narrow-band are synonymous terms in both the series and parallel *RLC* circuits. But regardless of the value of Q, the maximum current gain is always $T_{max} = 1$ (or 0 dB) at $\omega = \omega_0$.

A high-Q parallel *RLC* circuit is sometimes called a *tank circuit*. This terminology apparently comes from the vacuum tube era, when the tunable *LC* components in frequency-selective amplifiers were packaged in a metal "tanklike" container. The center frequency is also the resonant frequency of the parallel *RLC* circuit, since

$$Z_{IN}(j\omega_0) = \cfrac{1}{\cfrac{1}{R} + j\underbrace{\left(\omega_0 C - \cfrac{1}{\omega_0 L}\right)}_{\substack{\text{at resonance} \\ \omega_0 = \frac{1}{\sqrt{LC}}}}} = R + j0$$

So, again, the passband center frequency is also called the *resonant frequency*.

EXAMPLE 12–14

A parallel *RLC* circuit has a bandwidth of $B = 12\,\text{krad/s}$, a quality factor of $Q = 3$, and a 1-mH inductance. Find the values of R, C, ω_0, ω_{C1}, and ω_{C2}.

SOLUTION:
Using the definition of Q, we find the center frequency as $\omega_0 = BQ = 36\,\text{krad/s}$. Using the given inductance and the center frequency, we get the capacitance as $C = 1/\omega_0^2 L = 0.772\,\mu\text{F}$. For a parallel *RLC* circuit, $B = 1/RC$ and the resistance is $R = 1/BC = 108\,\Omega$. Inserting these results into Eq. (12–22) yields the lower cutoff frequency as

$$\omega_{C1} = -6000 + \sqrt{6000^2 + 36,000^2} = 30.5\,\text{krad/s}$$

Then, using the definition of bandwidth, we get $\omega_{C2} = \omega_{C1} + B = 42.5\,\text{krad/s}$. ■

Exercise 12–20

Verify with OrCAD that the circuit designed in Example 12–14 indeed meets the specifications.

Answer: Simulating the parallel *RLC* circuit described, the Probe output shown in Figure 12–32 yields $\omega_0 = 2\pi f_0 = 35.99\,\text{krad/s}$, $\omega_{C1} = 30.49\,\text{krad/s}$, $\omega_{C2} = 42.58\,\text{krad/s}$, and $Q = 2.98$.

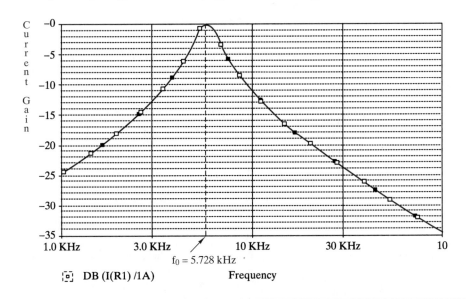

FIGURE 12–32

Ⓓ DESIGN EXAMPLE 12−15

Design a parallel RLC circuit with cutoff frequencies at 12 kHz and 16 kHz.

SOLUTION:
The descriptive parameters of the required circuit are

$$\omega_0 = 2\pi\sqrt{f_{C1}\,f_{C2}} = 87.1\,\text{krad/s}$$

$$B = 2\pi(f_{C2} - f_{C1}) = 25.1\,\text{krad/s}$$

Selecting $R = 1\,\text{k}\Omega$ gives $C = 1/RB = 0.0398\,\mu\text{F}$ and $L = 1/\omega_0^2 C = 3.32\,\text{mH}$.   ∎

Exercise 12–21

A series RLC circuit has cutoff frequencies at 100 rad/s and 10 krad/s. Find the values of B, ω_0, and Q. Does the circuit have a wide-band or narrow-band response?

Answers: $B = 9.9\,\text{krad/s}$; $\omega_0 = 1\,\text{krad/s}$; $Q = 0.101$; wide-band

---

12–6 BODE DIAGRAMS WITH REAL POLES AND ZEROS

The purpose of this section is to offer a method for understanding the effects of poles and zeros on the frequency response of circuits described by their transfer functions. In general, few people still draw Bode plots by hand. Software tools such as MATLAB and OrCAD easily permit one to generate Bode plots from either transfer functions or circuits, respectively. So, one may ask, why spend any time at all learning of the effects of poles and zeros? Software tools are extremely useful, but they cannot tell you if you made an error or, if the results are not what you expected, how to go about improving them. With some practice, one can tell by looking at a Bode plot where poles and zeros are located, and this can help in making changes to improve the performance of a circuit. Finally, understanding the effects of poles and zeros on a Bode plot can help one focus on critical features and aid in further analysis.

We will begin by studying how a straight-line approximation of a Bode plot is constructed when the poles and zeros are located on the real axis in the s plane. As a place to start, consider the following transfer function:

$$T(s) = \frac{Ks(s + \alpha_1)}{(s + \alpha_2)(s + \alpha_3)} \tag{12–24}$$

where $K, \alpha_1, \alpha_2,$ and α_3 are real. This function has zeros at $s = 0$ and $s = -\alpha_1$, and poles at $s = -\alpha_2$ and $s = -\alpha_3$. All of these critical frequencies lie on the real axis in the s plane. When making Bode plots, we put $T(j\omega)$ in a standard format obtained by factoring out $\alpha_1, \alpha_2,$ and α_3:

$$T(j\omega) = \left(\frac{K\alpha_1}{\alpha_2\alpha_3}\right)\frac{j\omega(1 + j\omega/\alpha_1)}{(1 + j\omega/\alpha_2)(1 + j\omega/\alpha_3)} \tag{12–25}$$

Using the following example notation:

$$\text{Magnitude} = M = |1 + j\omega/\alpha| = \sqrt{1 + (\omega/\alpha)^2}$$

$$\text{Angle} = \theta = \angle(1 + j\omega/\alpha) = \tan^{-1}(\omega/\alpha) \tag{12–26}$$

$$\text{Scale factor} = K_0 = \frac{K\alpha_1}{\alpha_2\alpha_3}$$

we can write the transfer function in Eq. (12–25) in the form

$$T(j\omega) = K_0\,\frac{(\omega\,e^{j90°})(M_1\,e^{j\theta_1})}{(M_2\,e^{j\theta_2})(M_3\,e^{j\theta_3})} = \frac{|K_0|\omega M_1}{M_2 M_3}\,e^{j(\angle K_0 + 90° + \theta_1 - \theta_2 - \theta_3)} \tag{12–27}$$

The gain (in dB) and phase responses are

$$|T(j\omega)|_{dB} = \underbrace{20\log_{10}|K_0|}_{\text{scale factor}} + \underbrace{20\log_{10}\omega}_{\text{zero}} + \underbrace{20\log_{10}M_1}_{\text{zero}} - \underbrace{20\log_{10}M_2}_{\text{pole}} - \underbrace{20\log_{10}M_3}_{\text{pole}}$$

$$\theta(\omega) = \overbrace{\angle K_0}^{} + \overbrace{90°}^{} + \overbrace{\theta_1}^{} - \overbrace{\theta_2}^{} - \overbrace{\theta_3}^{}$$

$$(12\text{–}28)$$

The terms in Eq. (12–28) caused by zeros have positive signs and increase the gain and phase angle, while the pole terms have negative signs and decrease the gain and phase.

The summations in Eq. (12–28) illustrate a general principle. In a Bode plot, the gain and phase responses are determined by the following types of factors:

1. The scale factor K_0
2. A factor of the form $j\omega$ due to a zero or a pole at the origin
3. Factors of the form $(1 + j\omega/\alpha)$ caused by a zero or pole at $s = -\alpha$

We can construct Bode plots by considering the contributions of these three factors.

The scale factor K_0. The gain and phase contributions of the scale factor are constants that are independent of frequency. In dB, the gain contribution $20\log_{10}|K_0|$ is positive when $|K_0| > 1$ and negative when $|K_0| < 1$. The phase contribution $\angle K_0$ is $0°$ when K_0 is positive and $\pm 180°$ when K_0 is negative.

The factor $(j\omega)$. If there were no critical points at the origin, the slope of the gain plot would be 0 — that is, a horizontal line — and the phase would be $0°$. A simple zero or pole at the origin contributes $\pm 20\log_{10}\omega$ to the gain or a slope of ± 1 and $\pm 90°$ to the phase, where the plus sign applies to a zero and the minus to a pole. When $T(s)$ has a factor s^n in the numerator (denominator), it has a zero (pole) of order n at the origin. Multiple zeros or poles at $s = 0$ contribute $\pm 20\, n\log_{10}\omega$ to the gain or a slope of $\pm n$ and $\pm n90°$ to the phase. Figure 12–33 shows that the gain factors contributed by zeros and poles at the origin are straight lines that pass through a gain of 1 (0 dB) at $\omega = 1.$[2]

The factor $(1 + j\omega/\alpha)$. The gain contributions of first-order zeros and poles are shown in Figure 12–34. Like the first-order transfer functions studied earlier in this

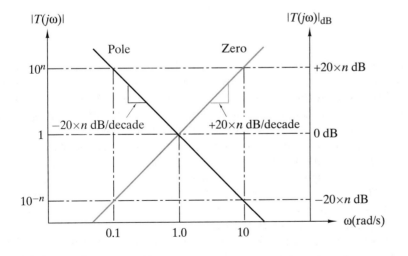

FIGURE 12–33 *Gain responses of poles and zeros at $s = 0$.*

---

[2]Strictly speaking, a circuit with a natural pole at the origin is unstable and does not have a sinusoidal steady-state response. Nevertheless, it is traditional to treat poles at the origin in Bode diagrams because there are practical applications in which such poles are important considerations.

FIGURE 12–34 *Gain responses of poles and zeros at s = −α.*

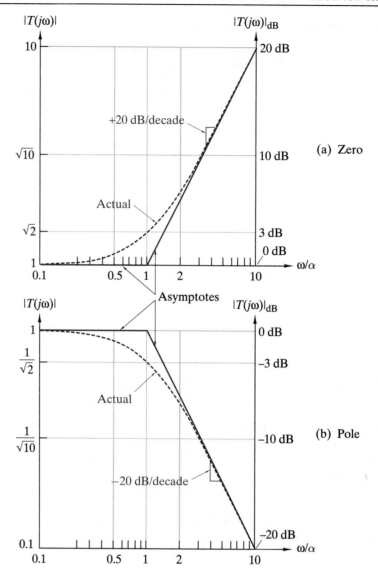

chapter, these factors produce straight-line gain asymptotes at low and high frequency. In a Bode plot of a low-pass filter, the low-frequency ($\omega \ll \alpha$) asymptotes are horizontal lines at a gain of 1 (0 dB). The high-frequency ($\omega \gg \alpha$) asymptotes are straight lines of the form $\pm\omega/\alpha$ or $\pm 20\log(\omega/\alpha)$ dB, where the plus sign applies to a zero and the minus to a pole. The high-frequency gain asymptote is proportional to the frequency ω (slope $= +1$ absolute or $+20$ dB/decade) for a zero and proportional to $1/\omega$ (slope $= -1$ absolute or -20 dB/decade) for a pole. In either case, the low- and high-frequency asymptotes intersect at the corner frequency $\omega = \alpha$.

To construct straight-line gain approximations we develop a piecewise linear function $|T(j\omega)|_{SL}$ where the subscript SL refers to the straight-line approximation, defined by the asymptote of each factor in $T(j\omega)$. The function $|T(j\omega)|_{SL}$ has a corner frequency at each of the critical frequencies of the transfer function. At frequencies below a corner, ($\omega < \alpha$) a first-order factor is represented by a gain of 1. Above the corner frequency ($\omega > \alpha$) the factor is represented by its high-frequency asymptote ω/α. To generate $|T(j\omega)|_{SL}$, we start out below the lowest critical frequency, usually at least one decade, with a low-frequency baseline that accounts for the scale factor K_0 and any poles or zeros at the origin. We increase frequency and change the slope of $|T(j\omega)|_{SL}$ whenever we pass a corner frequency. We proceed upward in frequency until we have gone beyond the highest critical frequency, at which point we have a complete expression for $|T(j\omega)|_{SL}$.

The following example illustrates the process.

EXAMPLE 12–16

(a) Construct the Bode plot of the straight-line approximation of the gain of the transfer function

$$T(s) = \frac{12{,}500\,(s+10)}{(s+50)\,(s+500)}$$

(b) Find the point at which the high-frequency gain falls below the dc gain.

SOLUTION:
(a) Written in the standard form for a Bode plot, the transfer function is

$$T(j\omega) = \frac{5(1+j\omega/10)}{(1+j\omega/50)\,(1+j\omega/500)}$$

The scale factor is $K_0 = 5$ and the corner frequencies are at $\omega_C = 10$ (a zero), 50 (a pole), and 500 (a pole) rad/s. At low frequency ($\omega < 10$ rad/s), all of the first-order factors are represented by their low-frequency asymptotes. As a result $|T(j\omega)| \approx 5(1)/[(1)(1)] = 5$, so the low-frequency baseline is $|T(j\omega)|_{SL} = 5$ for $\omega \leq 10$.

What this means is that since there are no critical points at $\omega = 0$, the slope of the gain asymptote is zero entering our Bode plot from the left at a gain of $K_0 = 5$. At $\omega = 10$ rad/s we encounter the first critical frequency: a zero. Beginning at this point, the slope changes by $+1$ absolute or $+20$ dB/decade. The straight-line gain becomes $|T(j\omega)|_{SL} = 5(\omega/10) = \omega/2$. After this point the line continues upward with the $+1$ slope until it reaches the next critical point: a pole at $\omega = 50$ rad/s. The pole causes the slope of the Bode plot to change by -1 at the pole, or -20 dB/decade. The plot that was growing with a slope of $+1$ becomes horizontal again with a slope of zero. After this point the gain contribution due to the pole factor $1/(1+j\omega/50)$ is represented by its high-frequency asymptote $\omega/50$, and $|T(j\omega)|_{SL} = 5(\omega/10)/(\omega/50) = 25$. The horizontal line continues until we reach the last critical point: another pole at $\omega = 500$ rad/s. The slope changes one last time by -1 or -20 dB/decade, exiting our graph down to the right. The gain contribution of the last pole is approximated by $\omega/500$, and the gain rolls off with a slope of -1 as $|T(j\omega)|_{SL} = 5(\omega/10)/(\omega/50)(\omega/500) = 12{,}500/\omega$. In summary, the straight-line approximation to the gain is

$$|T(j\omega)|_{SL} = \begin{vmatrix} 5 & \text{if} & 0 < \omega \leq 10 \\ \omega/2 & \text{if} & 10 < \omega \leq 50 \\ 25 & \text{if} & 50 < \omega \leq 500 \\ 12{,}500/\omega & \text{if} & 500 < \omega \end{vmatrix}$$

Given this function, we can easily plot the straight-line gain response in Figure 12–35. At low frequency ($\omega < 10$) the gain is flat at a value of 5 (14 dB). At $\omega = 10$ the zero causes the gain to increase as ω (slope $= +1$ or $+20$ dB/decade). This increasing gain continues until $\omega = 50$, where the first pole cancels the effect of the zero and the gain is flat at a value of 25 (28 dB). The gain remains flat until the final pole causes a corner at $\omega = 500$. Thereafter the gain falls off as $1/\omega$ (slope $= -1$ or -20 dB/decade).

FIGURE 12–35

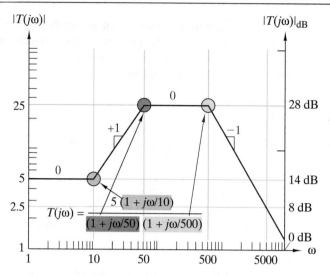

(b) The dc gain is 5. A quick look at the sketch in Figure 12–35 shows that the high-frequency gain falls below the dc gain in the region above $\omega = 500$, where the straight-line gain is $12{,}500/\omega$. Hence we estimate the required frequency to be $\omega = 12{,}500/5 = 2500 \, \text{rad/s}$. ∎

Exercise 12–22

Use MATLAB to graph the Bode magnitude plot of the transfer function in Example 12–16.

Answer: The required MATLAB code is:

```
T=tf([12500 125000], [1 550 25000]);
w = logspace(0,4,1000);
bodemag(T,w);grid
```

The Bode plot is shown in Figure 12–36.

FIGURE 12–36

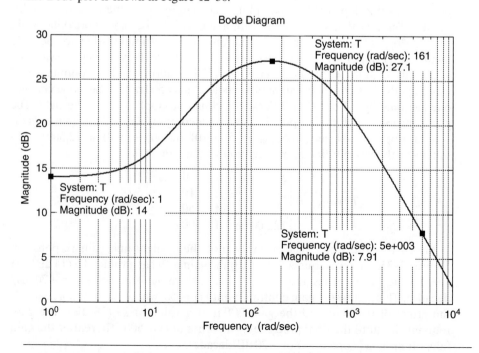

You need to design a circuit that has the transfer function indicated by the straight-line Bode plot shown in Figure 12–37.

(a) Develop a transfer function $T(s)$ from the straight-line graph.
(b) Validate your results using MATLAB.
(c) Design a cascade circuit that realizes the $T(s)$ found in part (a).
(d) Validate your design using OrCAD.

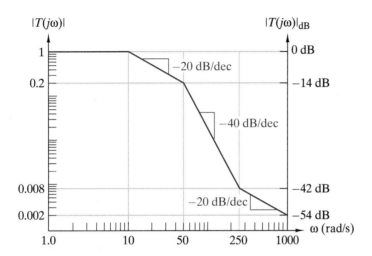

FIGURE 12–37

SOLUTION:

(a) In previous examples we were given a transfer function and required to construct a straight-line gain plot. Here we are given a straight-line gain plot and asked to find the transfer function. The gain plot shows a finite gain at dc and corner frequencies at $\omega = 10, 50,$ and 250 rad/s. The slope of the straight-line gain response between the first corner at $\omega = 10$ and 50 rad/s is

$$m = (-14 - 0)/(\log 50 - \log 10) = -20 \text{ dB/decade}$$

The -20 dB/decade slope means that $T(s)$ has a pole at $s = -10$. The gain slope between the second corner at $\omega = 50$ and 250 rad/s is

$$m = [-42 - (-14)]/(\log 250 - \log 50) = -40 \text{ dB/decade}$$

The slope decreases by another -20 dB/decade, so $T(s)$ also has a pole at $s = -50$. The gain slope between the last corner at $\omega = 250$ and 1000 rad/s is

$$m = [-54 - (-42)]/(\log 1000 - \log 250) = -20 \text{ dB/decade}$$

The gain slope increases by $+20$ dB/decade, so $T(s)$ must have zero at $s = -250$. These critical frequencies account for all of the corner frequencies shown in the figure, and $T(s)$ has the form

$$T(s) = \frac{K(s + 250)}{(s + 10)(s + 50)}$$

The dc gain of this transfer function is $T(0) = K/2$. From Figure 12–37, the dc gain $T(0) = 1$. Hence, we find $K = 2$, so that the desired transfer function becomes

$$T(s) = \frac{2(s + 250)}{(s + 10)(s + 50)}$$

FIGURE 12–38

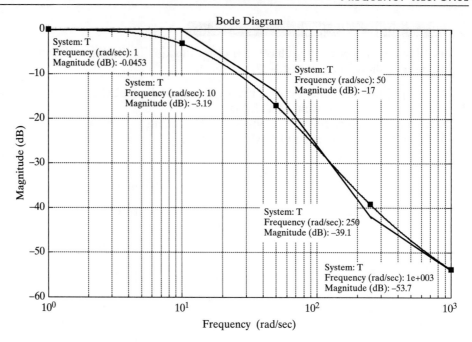

(b) We use MATLAB to produce the Bode magnitude plot of the transfer function we just found and compare it with the desired straight-line graph. The following MATLAB code produces the Bode plot shown in Figure 12–38:

```
T = tf([2 500],[1 60 500])
w = logspace(0,3,1000);
bodemag(T,w); grid
```

We superimposed the straight-line graph on the MATLAB plot. It appears to fit well, so we can now design our circuit.

(c) The transfer function found in part (a) can be expressed as the product of two transfer functions:

$$T(s) = \underbrace{\left(\dfrac{1}{\dfrac{s}{10}+1} \right)}_{T_1(s)} \underbrace{\left(\dfrac{\dfrac{s}{250}+1}{\dfrac{s}{50}+1} \right)}_{T_2(s)}$$

Both of these transfer functions can be realized by voltage dividers. Factoring s out of the denominator of $T_1(s)$ and equating the result to a voltage divider yields

$$\underbrace{\dfrac{1/s}{1/10 + 1/s}}_{T_1(s)} = \underbrace{\dfrac{Z_2(s)}{Z_1(s) + Z_2(s)}}_{\text{voltage divider}}$$

Equating the numerators and denominators yields a voltage divider with $Z_2 = 1/C_1 s$ and $Z_1 = R_1$, where $C_1 = 1\,\text{F}$ and $R_1 = 1/10\,\Omega$.

Factoring s out of the numerator and denominator of $T_2(s)$ and equating the result to a voltage divider yields

$$\underbrace{\dfrac{1/250 + 1/s}{1/50 + 1/s}}_{T_2(s)} = \underbrace{\dfrac{Z_2(s)}{Z_1(s) + Z_2(s)}}_{\text{voltage divider}}$$

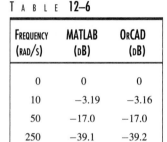

FIGURE 12–39

Equating the numerators and denominators leads to a voltage divider with $Z_2 = R_2 + 1/C_2 s$ and $Z_1 = R_3$, where $R_2 = 1/250\,\Omega$, $C_2 = 1\,$F, and $R_3 = 4/250\,\Omega$.

Figure 12–39 shows a cascade circuit using the two voltage dividers separated by a voltage follower to ensure that the chain rule applies. The element values in the prototype version of this circuit are $R_1 = 1/10\,\Omega$, $C_1 = C_2 = 1\,$F, $R_2 = 1/250\,\Omega$, and $R_3 = 4/250\,\Omega$. Using a magnitude scale factor of $k_m = 10^6$ to get practical values produces a final design with $R_1 = 100\,$kΩ, $C_1 = C_2 = 1\,\mu$F, $R_2 = 4\,$kΩ, and $R_3 = 16\,$kΩ. Many other designs are possible.

(d) We use OrCAD to simulate the circuit in Figure 12–39 using an ac sweep. Figure 12–40 shows the circuit as drawn in OrCAD and the response characteristic. To ensure compliance with the specifications, we compared key points with the MATLAB simulation in Figure 12–38. These key points are tabulated and compared in Table 12–6. The circuit meets all of the desired specifications.

Frequency (RAD/S)	MATLAB (DB)	OrCAD (DB)
0	0	0
10	−3.19	−3.16
50	−17.0	−17.0
250	−39.1	−39.2
1000	−53.7	−53.8

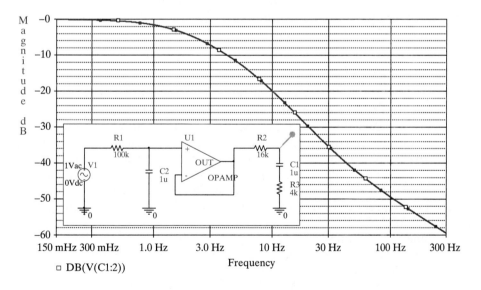

FIGURE 12–40

□ DB(V(C1:2))

Frequency

Exercise 12–23

(a) Derive an expression for the straight-line approximation to the gain response of the following transfer function:

$$T(s) = \frac{500(s + 50)}{(s + 20)(s + 500)}$$

(b) Find the straight-line gains at $\omega = 10, 30$, and $100\,\text{rad/s}$.

(c) Find the frequency at which the high-frequency gain asymptote falls below $-20\,\text{dB}$.

(d) Compare your answers in parts (b) and (c) with MATLAB.

Answers:

(a) $$|T(j\omega)|_{\text{SL}} = \begin{vmatrix} 2.5 & \text{if} & 0 < \omega \le 20 \\ 50/\omega & \text{if} & 20 < \omega \le 50 \\ 1 & \text{if} & 50 < \omega \le 500 \\ 500/\omega & \text{if} & 500 < \omega \end{vmatrix}$$

(b) 8 dB, 4.4 dB, 0 dB

(c) 5 krad/s

(d) 7.16 dB, 4.16 dB, 0.628 dB, and 4.975 krad/s.

In many situations the straight-line Bode plot tells us all we need to know. When greater accuracy is needed, the straight-line gain plot can be refined by adding gain corrections in the neighborhood of the corner frequency. Figure 12–34 shows that the actual gains and the straight-line approximations differ by ± 3 dB at the corner frequency. They differ by roughly ± 1 dB an octave above or below the corner frequency. When these "corrections" are included, we can sketch the actual response and achieve somewhat greater accuracy. However, making the graphical gain corrections is usually not worth the trouble. First, the gain corrections overlap unless the corner frequencies are separated by more than two octaves. More important, the purpose of a straight-line gain analysis is to provide insight, not to generate accurate frequency-response data. The straight-line plots are useful in preliminary analysis and in the early stages of design. At some point accurate response data will be needed, in which case it is better to use computer-aided analysis rather than trying to "correct the errors" graphically in a straight-line plot.

STRAIGHT-LINE PHASE ANGLE PLOTS

Figure 12–41 shows the phase contributions of first-order zeros and poles. The straight-line approximations are similar to the gain asymptotes except that there are two slope changes. The first occurs a decade below the gain corner frequency, and the second occurs a decade above. The total phase changes by $90°$ over this two-decade range, so the straight-line approximations have slopes of $\pm 45°$ per decade, where the plus sign applies to a zero and the minus to a pole. Poles and zeros at the origin contribute a constant phase angle of $\pm n90°$, where n is the order of the critical frequency and the plus (minus) sign applies to zeros (poles).

To generate a straight-line phase plot, we begin with the low-frequency phase asymptote. This low-frequency baseline accounts for the effect of the scale factor K_0 and any poles or zeros at the origin. We account for the effect of other critical frequencies by introducing a slope change of $\pm 45°$/decade one decade below and one decade above each gain corner frequency. These slope changes generate a straight-line phase plot as we proceed from the low-frequency baseline to a high frequency that is at least a decade above the highest gain corner frequency.

It is important to remember that a decade above the highest corner frequency the phase asymptote is a constant value with zero slope. That is, at high frequency the straight-line phase plot is a horizontal line at $\theta(j\omega) = (m - n)90°$, where m is the number of finite zeros and n is the number of poles.

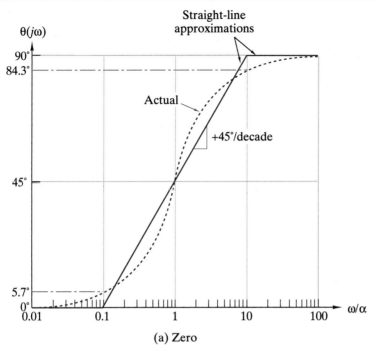

(a) Zero

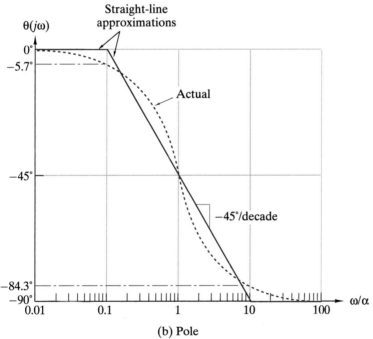

(b) Pole

FIGURE 12–41 *Phase responses of poles and zeros at s = −α.*

EXAMPLE 12–18

Find the straight-line approximation to the phase response of the transfer function in Example 12–16. Verify your solution using MATLAB.

SOLUTION:

In Example 12–16 the standard form of $T(j\omega)$ is shown to be

$$T(j\omega) = \frac{5(1 + j\omega/10)}{(1 + j\omega/50)(1 + j\omega/500)}$$

The scale factor is $K_0 = 5$ and the corner frequencies are $\omega_C = 10$ (zero), 50 (pole), and 500 (pole) rad/s. At low frequency $T(j\omega) \to K_0 = 5$, so the low-frequency phase asymptote is $\theta(\omega) \to \angle K_0 = 0°$. Proceeding from one decade below the lowest corner frequency (1 rad/s) to one decade above the highest corner frequency (5000 rad/s), we encounter the slope changes listed in Table 12–7.

T A B L E **12–7**

FREQUENCY	CAUSED BY	SLOPE CHANGE	NET SLOPE
1	zero at $s = -10$	$+45°$/decade	$+45°$/decade
5	pole at $s = -50$	$-45°$/decade	$0°$/decade
50	pole at $s = -500$	$-45°$/decade	$-45°$/decade
100	zero at $s = -10$	$-45°$/decade	$-90°$/decade
500	pole at $s = -50$	$+45°$/decade	$-45°$/decade
5000	pole at $s = -500$	$+45°$/decade	$0°$/decade

FIGURE 12–42

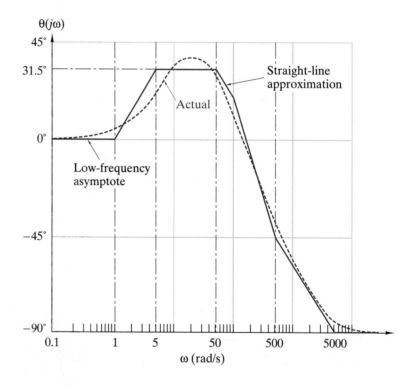

Figure 12–42 shows the straight-line approximation and the actual phase response from MATLAB. ■

Exercise 12–24

Construct a Bode plot of the straight-line approximation to the phase response of the transfer function in Exercise 12–23. Use the plot to estimate the phase angles at $\omega = 1, 15, 300,$ and 10^4 rad/s. Compare your results with those obtained using MATLAB.

A n s w e r s : $0°, -18°, -45°, -90°$; MATLAB: $-1.8°, -21.9°, -36.6°, -87.3°$

12–7 BODE DIAGRAMS WITH COMPLEX POLES AND ZEROS

The frequency response of transfer functions with complex poles and zeros can be analyzed using straight-line gain plots. However, complex critical frequencies may produce resonant peaks (or valleys) where the actual gain response departs significantly from the straight-line approximation. Straight-line plots can be used to define a starting place for describing the frequency response of these highly resonant circuits.

Complex poles and zeros occur in conjugate pairs that appear as quadratic factors of the form

$$s^2 + 2\zeta \omega_0 s + \omega_0^2 \qquad (12\text{–}29)$$

where ζ and ω_0 are the damping ratio and undamped natural frequency. In a Bode diagram the appropriate standard form of the quadratic factor is obtained by factoring out ω_0^2 and replacing s by $j\omega$ to obtain

$$1 - (\omega/\omega_0)^2 + j2\zeta(\omega/\omega_0) \qquad (12\text{–}30)$$

In a Bode diagram this quadratic factor introduces gain and phase terms of the following form:

$$|T(j\omega)|_{dB} = \pm 20 \log_{10} \sqrt{[1 - (\omega/\omega_0)^2]^2 + (2\zeta \omega/\omega_0)^2} \qquad (12\text{–}31a)$$

$$\theta(\omega) = \pm \tan^{-1} \frac{2\zeta \omega/\omega_0}{1 - (\omega/\omega_0)^2} \qquad (12\text{–}31b)$$

where the plus sign applies to complex zeros of $T(s)$ and the minus sign to complex poles.

Figure 12–43 shows the gain contribution of complex poles and zeros for several values of the damping ratio ζ. The low-frequency ($\omega \ll \omega_0$) gain asymptotes for these plots are unity (0 dB). The high-frequency ($\omega \gg \omega_0$) gain asymptotes are of the form $(\omega/\omega_0)^{\pm 2}$. Expressed in dB, the high-frequency asymptote is $\pm 40 \log_{10}(\omega/\omega_0)$, which in a Bode diagram is a straight line with a slope of ± 2 or ± 40 dB/decade, where again the plus sign applies to zeros and the minus to poles.

These asymptotes intersect at a corner frequency of $\omega = \omega_0$. The gains in the neighborhood of the corner frequency are strong functions of the damping ratio. For $\zeta > 1/\sqrt{2}$ the actual gain lies entirely above the asymptotes for complex zeros and entirely below the asymptotes for complex poles. For $\zeta < 1/\sqrt{2}$ the gain is a minimum at $\omega = \omega_0 \sqrt{1 - 2\zeta^2}$ for complex zeros and a maximum for complex poles. These valleys (for zeros) and peaks (for poles) are not particularly conspicuous until $\zeta < 0.5$.

To develop a straight-line gain plot for complex critical frequencies we insert a corner frequency at $\omega = \omega_0$. Below this corner frequency we use the low-frequency asymptote to approximate the gain, and above the corner we use the high-frequency asymptote. The actual gain around the corner frequency depends on ζ. But generally speaking, the straight-line gain is within ± 3 dB of the actual gain for ζ in the range from about 0.3 to about 0.7. When ζ falls outside this range, we can calculate the actual gain at the corner frequency and perhaps a few points on either side of the corner frequency. These gains may give us a better picture of the gain plot in the vicinity of the corner frequency.

However, we should keep in mind that the purpose of straight-line gain analysis is insight into the major features of a circuit's frequency response. If greater

FIGURE 12–43 *Gain responses of complex poles and zeros.*

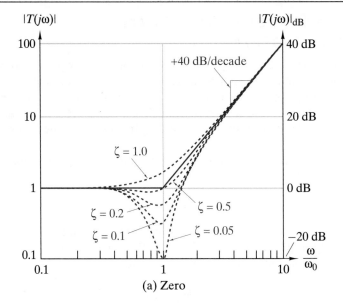

(a) Zero

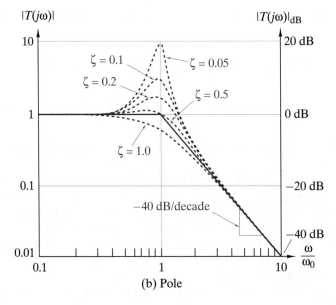

(b) Pole

accuracy is required, then computer-aided analysis is the best approach. The straight-line gain gives useful results when the resonant peaks and valleys are not too abrupt. When a circuit has lightly damped critical frequencies, the straight-line approach may not be particularly helpful.

EXAMPLE 12–19

(a) Construct the straight-line gain plot for the transfer function

$$T(s) = \frac{5000(s + 100)}{s^2 + 400s + (500)^2}$$

(b) Verify the solution using MATLAB.

SOLUTION:

(a) The transfer function has a real zero at $s = -100$ rad/s and a pair of complex poles with $\zeta = 0.4$ and $\omega_0 = 500$ rad/s. This damping ratio falls in the range (0.3 to 0.7) in which the resonant peak due to the complex poles is not too pronounced. Hence we expect the straight-line gain to give a useful approximation. Written in standard form, $T(j\omega)$ is

$$T(j\omega) = 2\left(\frac{1 + j\omega/100}{1 - (\omega/500)^2 + j0.8\omega/500}\right)$$

The scale factor is $K_0 = 2$, and there are corner frequencies at $\omega = 100$ rad/s due to the zero and $\omega = 500$ rad/s due to the pair of complex poles. At low frequency $T(j\omega) \to 2$, so the low-frequency ($\omega < 100$) baseline is $|T(j\omega)|_{SL} = 2$. This gain applies until we pass the first critical frequency at $\omega = 100$. Beginning at that point, the zero is represented by its high-frequency asymptote ($\omega/100$) and the straight-line gain becomes

$$|T(j\omega)|_{SL} = 2(\omega/100) = \omega/50$$

This linearly increasing gain applies until we pass the critical frequency at $\omega_0 = 500$. Thereafter the complex poles are represented by their high-frequency asymptote ($500^2/\omega^2$). After this point the gain rolls off as

$$|T(j\omega)|_{SL} = 2(\omega/100)(500^2/\omega^2) = 5000/\omega$$

In summary, the straight-line gain function is

$$|T(j\omega)|_{SL} = \begin{vmatrix} 2 & \text{if} & 0 < \omega \leq 100 \\ \omega/50 & \text{if} & 100 < \omega \leq 500 \\ 5000/\omega & \text{if} & 500 < \omega \end{vmatrix}$$

Figure 12–44 shows a plot of the straight-line gain. We expect to see a gain peak around $\omega = 500$ rad/s due to the complex poles. The plot in Figure 12–44 shows that the zero at $s = -100$ rad/s causes the gain to bend upward prior to the corner frequency at $\omega = 500$ rad/s. This upward bend enhances the height of the resonant peak caused by the complex poles.

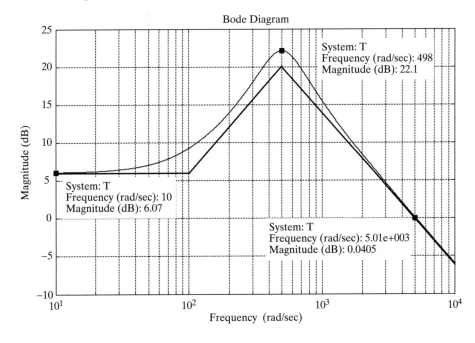

Bode Diagram

System: T
Frequency (rad/sec): 498
Magnitude (dB): 22.1

System: T
Frequency (rad/sec): 10
Magnitude (dB): 6.07

System: T
Frequency (rad/sec): 5.01e+003
Magnitude (dB): 0.0405

FIGURE 12–44

(b) In addition to the straight-line approximation, Figure 12–44 shows the magnitude response created with the following MATLAB code:

```
T=tf([5000 500000],[1 400 250000]);
w = logspace(1,4,1000);
bodemag(T,w);grid
```

The straight-line results are reasonable approximations of the true response, but there is a larger error observed than in transfer functions with only real poles and zeros. In this case, the errors are especially pronounced between 100 rad/s and the peak at 500 rad/s. Practically, it is usually better to use MATLAB to plot the Bode plots when there are complex poles. ∎

Exercise 12–25

Construct a straight-line graph of the gain function of the following transfer function. Then use MATLAB to plot the actual Bode magnitude plot.

$$T(s) = \frac{20s}{s^2 + 2s + 2500}$$

Answer:

$$|T(j\omega)|_{\text{SL}} = \begin{vmatrix} \omega/125 & \text{for } 0 < \omega \le 50 \\ 20/\omega & \text{for } \omega > 50 \end{vmatrix}$$

The required MATLAB code is

```
T=tf([20 0],[1 2 2500]);
w = logspace(0,3,1000);
bodemag(T,w);grid
```

Both the straight-line graph and the MATLAB Bode magnitude plot are shown in Figure 12–45.

FIGURE 12–45

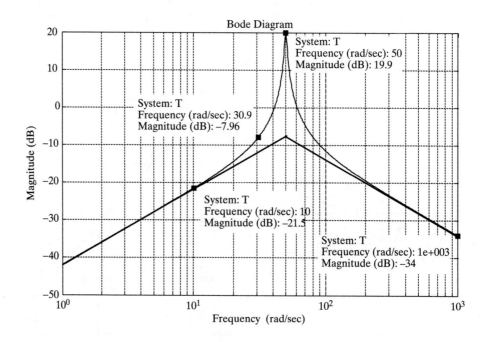

PHASE PLOTS FOR COMPLEX CRITICAL FREQUENCIES

Figure 12–46 shows the phase contribution from complex poles or zeros for several values of ζ. The low-frequency phase asymptotes are $0°$ and the high-frequency limits are $\pm 180°$. The phase is always $\pm 90°$ at $\omega = \omega_0$ regardless of the value of the damping ratio. The total phase change is $\pm 180°$, and most of this change occurs in a two-decade range from $\omega_0/10$ to $10\omega_0$. As a result, the straight lines in Figure 12–46 offer crude approximations to the phase shift. The shapes of the phase curves change radically with the damping ratio, so these straight-line approximations are of little use except at ω_0 and around the end points.

The net result is that phase angle plots for complex critical frequencies are best generated by computer-aided analysis. In practical applications we can often derive useful information from the gain plot alone without generating the phase response. However, the converse is not true. When we need phase response, we usually need gain as well.

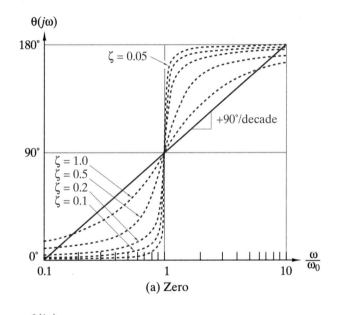

(a) Zero

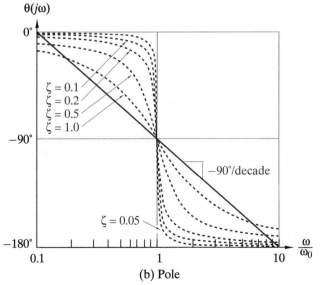

(b) Pole

FIGURE 12–46 *Phase responses of complex poles and zeros.*

Exercise 12–26

Use MATLAB to plot the actual Bode phase plot for the transfer function in Exercise 12–25.

Answer: The required MATLAB code is

```
T=tf([20 0],[1 2 2500]);
w = logspace(0,3,1000); bode(T,w);grid
```

Figure 12–47 shows the resulting Bode magnitude and phase plots.

FIGURE 12–47

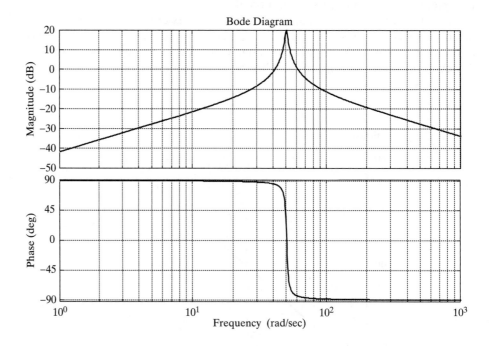

12–8 FREQUENCY RESPONSE AND STEP RESPONSE

The step response $g(t)$ and frequency response $|T(j\omega)|$ are alternative ways to describe the performance of a filter. The transient and frequency responses are not independent. In fact, one can be derived from the other. Filter performance is usually specified in terms of frequency response. Hence, it is important to understand the time-domain consequences of frequency-domain specifications, and vice versa.

As a starting place, consider the first-order low-pass filter defined by

$$T(s) = \frac{K}{s + \alpha}$$

Section 12–3 shows that this filter has a passband (dc) gain of K/α and a cutoff frequency of $\omega_C = \alpha$. Recall from Chapter 11 that the step response of a circuit is $G(s) = T(s)/s$. Hence, the step response of this low-pass filter is

$$G(s) = \frac{K}{s(s + \alpha)} = \frac{K/\alpha}{s} - \frac{K/\alpha}{s + \alpha}$$

The time-domain step response $g(t)$ is

$$g(t) = \mathcal{L}^{-1}\{G(s)\} = \frac{K}{\alpha}\left(1 - e^{-\alpha t}\right) \quad \text{for } t \geq 0$$

The time-domain response has a final value $(t \to \infty)$ of K/α and a time constant of $T_C = 1/\alpha$.

Specifying the gain and cutoff frequency of the frequency response also determines the amplitude and duration of the transient response, and vice versa. For instance, if the duration of the transient response is specified to be less than then 20 μs then $5T_C = 5/\alpha = 5/\omega_C < 20 \times 10^{-6}$ and hence $\omega_C > 250\,\text{krad/s}$ or 39.8 kHz. This cutoff frequency might not satisfy the frequency-response requirements. In any case, we need to look at the response in both domains.

Today's users of communication technology generally recognize that wide band and high speed are synonymous. The first-order low-pass filter illustrates the connection, since $\omega_C T_C = 1$. If we want a short response time (small T_C), then a wide bandwidth (large ω_C) is required. On the other hand, a fixed bandwidth (given ω_C) limits the speed of response, since $T_C = 1/\omega_C$. This illustrates a signal-processing principle called *reciprocal spreading*. Shrinking the response in one domain causes the response in the other domain to spread out.

The important point is that evaluating filter performance involves both the frequency domain and the time domain. We will encounter this idea again in Chapter 14, where we design much more complex filters. In any case is it clear that it is important to be able to go from one domain to the other. Changing domains involves concepts learned in several previous chapters, as illustrated in the following examples.

E X A M P L E   1 2 – 2 0

You are given a second-order low-pass filter with the following transfer function:

$$T(s) = \frac{10{,}000}{s^2 + 200\zeta s + 10{,}000}$$

Characterize the effects of changing ζ on the step response and the magnitude of the frequency response.

SOLUTION:

We will consider three values of ζ, namely 0.25, 1, and 5. These represent the three different damping cases considered in Chapter 7. Figure 12–48 shows both the step responses and the Bode magnitude plots for the three cases. The required MATLAB code to generate the plots is

```
T1 = tf([10000],[1  1000  10000]);
T2 = tf([10000],[1  200  10000]);
T3 = tf([10000],[1  50  10000]);
step(T1,T2,T3); grid
figure
w = logspace(0,3,1000);
bodemag(T1,T2,T3,w); grid
```

The frequency response is clearly that of a second-order low-pass filter. The undamped natural frequency is 100 rad/s. As ζ decreases, the filter's roll-off becomes steeper near the critical frequency. In many filter applications, a steeper roll-off indicates a "better" filter. The resonant peak that occurs with a small ζ is not always desirable, since signals passing through the filter are amplified unequally in the passband. However, in general, one would like the steeper roll-off that a smaller ζ offers.

Now consider the step response of the same circuit. The effect of ζ is quite different. As the damping coefficient decreases, the response becomes more underdamped and oscillates more as it approaches its final value. This additional oscillation is not

FIGURE 12–48

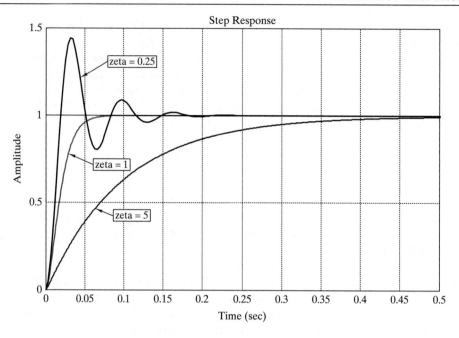

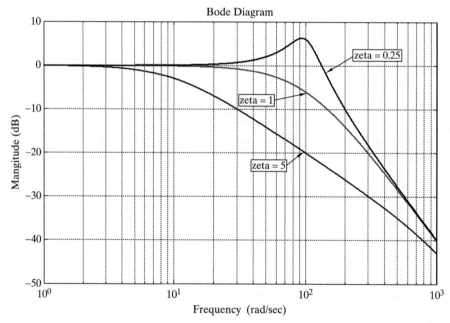

desirable if one is processing an audio signal, since underdamped circuits introduce "ringing" and distortion. When the circuit is critically damped ($\zeta = 1$), it is said to offer the fastest rise time without an overshoot. It also offers a good compromise regarding filtering, as seen in the frequency response. We will see other characteristics in Chapter 14, but for now, we can see the effects of ζ on two important circuit behavior properties. A circuit designer must evaluate these competing properties and decide on which is best for the application under consideration. ∎

Exercise 12–27

There is a need for a passive notch filter at 100 rad/s. The narrower the notch, the better, but there should be minimal ringing of signals passing through. Shown below are the transforms of three filters submitted for consideration are shown below. Which would you recommend and why?

$$T_1(s) = \frac{s^2 + 10s + 10{,}000}{s^2 + 150s + 10{,}000}$$

$$T_2(s) = \frac{s^2 + 20s + 10{,}000}{s^2 + 250s + 10{,}000}$$

$$T_3(s) = \frac{s^2 + 10{,}000}{s^2 + 50s + 10{,}000}$$

A n s w e r : $T_1(s)$ offers the best compromise—a relatively narrow notch with only a slight overshoot. $T_2(s)$ would be the best choice if no ringing at all was desired, but the notch would be wider.

EXAMPLE 12–21

For $t \geq 0$, the step response of a filter is

$$g(t) = -5e^{-40t} + 25e^{-200t}$$

Is this a low-pass, high-pass, bandpass, or bandstop filter? What is the passband gain and the approximate cutoff frequency?

SOLUTION:
Recall from Chapter 11 that $T(s) = sG(s)$ where $G(s) = \mathcal{L}\{g(t)\}$. The filter transfer function is found as follows:

$$T(s) = s\mathcal{L}\{g(t)\} = -\frac{5s}{s+40} + \frac{25s}{s+200}$$

$$= \frac{20s^2}{(s+40)(s+200)}$$

This is a high-pass filter with $T(0) = 0$ and $T(\infty) = 20$. The high-frequency passband gain is 20 or, equivalently, 26 dB. The filter has a corner frequency (pole) at $\omega = 40$ rad/s and another at $\omega = 200$ rad/s The low-frequency gain asymptote is $20\omega^2/(40 \times 200) = \omega^2/400$. This asymptote has a slope of 40 dB/decade and a value of 4 (12 dB) at the first corner. Above the first corner, the gain asymptote is $20\omega/(200) = \omega/10$, which has a slope of 20 dB/decade and a value of 20 (26 dB) at the second corner. Given these results, we can easily sketch the straight-line gain shown in Figure 12–49. Clearly the cutoff frequency is in the vicinity of the second corner frequency, hence $\omega_C \approx 200$ rad/s.

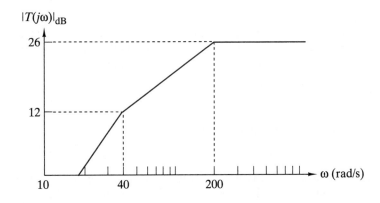

FIGURE 12–49

Exercise 12–28

The impulse response of a particular filter is $h(t) = 10^4 t e^{-100t} u(t)$. What type of filter is this? What is its passband bandwidth?

Answer: It is a second-order low-pass filter. Its cutoff frequency is about 64 rad/s; hence, that is also its bandwidth.

EXAMPLE 12–22

The straight-line gain response of a filter is shown in Figure 12–50. What are the initial and final values of the step response? What is the approximate duration of the transient response?

FIGURE 12–50

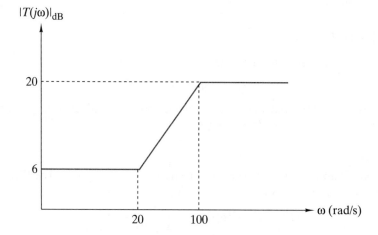

SOLUTION:

In this case we must derive the transfer function from the gain plot. The straight-line gain has corner frequencies at $\omega = 20$ rad/s and another at $\omega = 100$ rad/s. The first corner increases the gain slope by $(20 - 6)/(\log 100 - \log 20) = +20$ dB/ decade, so there is a simple zero at $s = -20$. The second corner decreases the slope by -20 dB/ decade, so there is a simple pole at $s = -100$. The two critical frequencies account for all of the corner frequencies shown in the figure, so the necessary transfer function has the form

$$T(s) = K \frac{s + 20}{s + 100}$$

The infinite frequency gain of this transfer function is $T(\infty) = K$. The high-frequency gain in Figure 12–50 is +20 dB; hence, $K = 10$. In general the step response is $G(s) = T(s)/s$; hence, for this filter we have

$$G(s) = \frac{10}{s} \frac{s + 20}{s + 100} = \frac{2}{s} + \frac{8}{s + 100}$$

The time-domain step response $g(t)$ is

$$g(t) = \mathcal{L}^{-1}\{G(s)\} = 2 + 8e^{-100t} \quad t > 0$$

The initial value is $g(0) = 10$, the final value is $g(\infty) = 2$, and the time constant of the transient term is $T_C = 1/100$ s. The duration of the transient response is about $5T_C = 50$ ms. ∎

Exercise 12–29

A certain RLC series bandpass circuit has the following transfer function:

$$T(s) = \frac{200s}{s^2 + 200s + 640{,}000}$$

Is it possible to alter a circuit element to keep the same center frequency while minimizing the ringing?

Answer: Yes. Increasing the resistance will make the middle term in the denominator, $R/L = 2\zeta\omega_0 = 200$, larger, which will increase ζ but leave the center frequency at 800 rad/s. Multiplying the value of the resistance by 8 will increase the damping ratio to make it equal to 1 and all ringing will stop. Note that the bandwidth will increase with increasing ζ.

E EVALUATION EXAMPLE 12 – 23

Modern piezoelectric pressure transducers are fabricated as an integrated circuit that includes both the sensing crystal and the signal-conditioning electronics. The step response of such a transducer takes the form

$$v_{TR}(t) = K_V e^{-t/T_C} p(t)$$

where $v_{TR}(t)$ is the transducer output voltage caused by an abrupt change in pressure $p(t) = P_A u(t)$. The transducer transient response is defined by the sensitivity K_V (usually in mV/psi), and the discharge time constant T_C (usually in seconds). A transducer is to be installed in an environment with sinusoidal pressure variations in the range from 2 mHz to 2 kHz. A vendor offers transducers with the characteristics presented in Table 12–10. Select the model that meets the frequency-response requirements and produces the largest output signal for any given pressure change.

TABLE 12–10

MODEL	261	265	266	268	269
K_V (mV/psi)	50	10	5	1	0.5
T_C (s)	10	50	100	500	1000

SOLUTION:
The specified response $v_{TR}(t)$ is for an input $P_A u(t)$. The unit step response $g(t)$ is derived from this response by dividing out the amplitude of the input step P_A.

$$g(t) = \frac{1}{P_A} v_{TR}(t) = K_V e^{-t/T_C} u(t)$$

In the s domain the step response is

$$G(s) = \frac{K_V}{s + 1/T_C}$$

and the pressure-to-voltage transfer function is found to be

$$T(s) = sG(s) = \frac{K_V s}{s + 1/T_C}$$

This transfer function has a first-order high-pass characteristic with a passband gain of K_V and a cutoff frequency of $\omega_C = 1/T_C$. The specified frequency range from

2 mHz to 2 kHz must fall in the transducer passband, which means that the cutoff frequency must be less than 2 mHz. In equation form this requires

$$\omega_C = \frac{1}{T_C} < 2\pi \times 0.002 \quad \text{or} \quad T_C > \frac{250}{\pi} = 79.6\,\text{s}$$

Models 266, 268, and 269 meet the frequency-response requirement, since they have discharge time constants T_C greater than 79.6 s. Model 266 is the best choice because it has the most sensitivity (passband gain) at $K_V = 5\,\text{mV/psi}$. ■

SUMMARY

- The frequency response of a circuit is defined by the variation of the gain $|T(j\omega)|$ and phase $\angle T(j\omega)$ with frequency. The gain function is usually expressed in dB in frequency-response plots. Logarithmic frequency scales are used on frequency-response plots of the gain and phase functions.

- A passband is a range of frequencies over which the steady-state output is essentially constant with very little attenuation. A stopband is a range of frequencies over which the steady-state output is significantly attenuated. The cutoff frequency is the boundary between a passband and the adjacent stopband.

- Circuit gain responses are classified as low pass, high pass, bandpass, and bandstop depending on the number and location of the stopbands and passbands. The performance of devices and circuits is often specified in terms of frequency-response descriptors such as bandwidth, passband gain, and cutoff frequency.

- The low- and high-frequency gain asymptotes of a first-order circuit intersect at a corner frequency determined by the location of its pole. The total phase change from low to high frequency is $\pm 90°$. First-order

circuits can be connected to produce bandpass and bandstop responses.

- Series and parallel *RLC* circuits provide bandpass and bandstop gain characteristics that are easily related to the circuit parameters.

- Bode plots are graphs of the gain (in dB) and phase angle (in degrees) versus log-frequency scales. Straight-line approximations to the gain and phase can be constructed using the corner frequencies defined by the poles and zeros of $T(s)$. The purpose of the straight-line approximations is to develop a conceptual understanding of frequency response. The straight-line plots do not necessarily provide accurate data at all frequencies, especially for circuits with complex poles.

- Computer-aided circuit analysis programs can accurately generate and plot frequency-response data. The user must have a rough idea of the gain and frequency ranges of interest to use these tools intelligently.

- It is often necessary to relate the time-domain characteristics of a circuit to its frequency response, and vice versa. The transfer function provides a link between the frequency-domain and time-domain responses.

PROBLEMS

OBJECTIVE 12–1 FIRST-ORDER CIRCUIT FREQUENCY RESPONSE (SECTS. 12–1, 12–2, 12–3)

Given a first-order circuit or transfer function:
(a) Find and classify the frequency response.
(b) Plot the gain and phase responses using straight-line approximations and computer tools.
(c) Design circuits to produce a specified frequency response.
See Examples 12–2, 12–3, 12–4,12–5,12–6,12–7, 12–8 and Exercises 12–3, 12–4, 12–5, 12–6, 12–7, 12–8, 12–9, 12–10, 12–11, 12–12, 12–13.

12–1 A transfer function has a passband gain of 500. At a particular frequency in its stopband, the gain of the transfer

function is only 0.00025. By how many decibels does the gain of the passband exceed that of that frequency in the stopband?

12–2 A particular filter is said to be 56 dB down at a desired stop frequency. How many times reduced is a signal at that frequency compared to a signal in the filter's pass-band?

12–3 A certain low-pass filter has the Bode diagram shown in Figure P12–3.
(a) How many dB down is the filter at 5000 rad/s?
(b) Estimate where the cutoff frequency occurs, then determine how many dB down is the filter at one decade after the cutoff frequency?

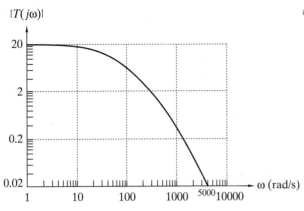

FIGURE P12–3

12–4 Find the transfer function $T_V(s) = V_2(s)/V_1(s)$ of the circuit in Figure P12–4.

(a) Find the dc gain, infinite frequency gain, and cutoff frequency. Identify the type of gain response.

(b) Sketch the straight-line approximations of the gain and phase responses.

(c) Calculate the gain at $\omega = 0.5\omega_C$, ω_C, and $2\omega_C$.

(d) Use OrCAD to plot the Bode magnitude gain response of the circuit. Validate your answers for part (c).

(e) How many dB down from the passband is the filter at one octave past the cutoff?

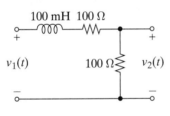

FIGURE P12–4

12–5 Find the transfer function $T_V(s) = V_2(s)/V_1(s)$ of the circuit in Figure P12–5.

(a) Find the dc gain, infinite frequency gain, and cutoff frequency. Identify the type of gain response.

(b) Sketch the straight-line approximation of the gain response.

(c) Calculate the gain at $\omega = 0.5\omega_C$, ω_C, and $2\omega_C$.

(d) Use OrCAD to plot the Bode magnitude gain response of the circuit. Validate your answers for part (c).

(e) How many dB down from the passband is the filter at one octave before the cutoff?

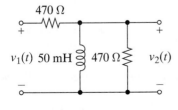

FIGURE P12–5

12–6 Find the transfer function $T_V(s) = V_2(s)/V_1(s)$ of the circuit in Figure P12–6.

(a) Find the dc gain, infinite frequency gain, and cutoff frequency. Identify the type of gain response.

(b) Sketch the straight-line approximation of the gain response.

(c) Calculate the gain at $\omega = 0.1\,\omega_C$, ω_C, and $10\omega_C$.

(d) Use OrCAD to plot the Bode magnitude gain response of the circuit. Validate your answers for part (c).

(e) What element values would you change to increase the passband gain to 10?

(f) How many dB down from the passband is the filter at one decade past the cutoff?

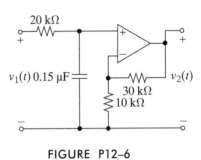

FIGURE P12–6

12–7 Find the transfer function $T_V(s) = V_2(s)/V_1(s)$ of the circuit in Figure P12–7.

(a) Find the dc gain, infinite frequency gain, and cutoff frequency. Identify the type of gain response.

(b) Sketch the straight-line approximation of the gain response.

(c) Calculate the gain at $\omega = 0.1\omega_C$, ω_C, and $10\omega_C$.

(d) Use OrCAD to plot the Bode magnitude gain response of the circuit. Validate your answers for part (c).

(e) What element values would you change to double the passband gain without changing the cutoff frequency?

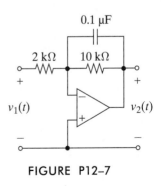

FIGURE P12–7

12–8 **E** You task is to connect the modules in Figure P12–8 so that the gain of the transfer function is 4 and the cutoff frequency of the filter is 500 rad/s when connected between the source and the load. Repeat if the cutoff frequency is 625 rad/s and the gain did not matter.

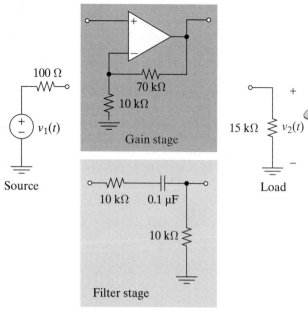

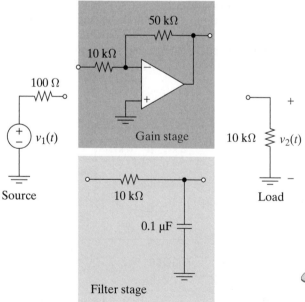

FIGURE P12–8

12–9 **E** A young designer needed to design a low-pass filter with a cutoff of 1 krad/s and a gain of −5. The filter is to fit as an interface between the source and the load. The designer was perplexed when no matter how the stages shown in Figure P12–9 are connected the results are not what were expected. Explain the problem and suggest a way to achieve the desired results.

FIGURE P12–9

12–10 **D** Design a low-pass first-order filter with a cutoff frequency of 2000 rad/s and a passband gain of 1.

12–11 **D** Design an RC low-pass first-order filter with a cutoff frequency of 10,000 rad/s and a passband gain of $+10$.

12–12 **D** Design an RL high-pass first-order filter with a cutoff frequency of 120 Hz and a passband gain of -15.

12–13 **D** Design an RC high-pass first-order filter with a cutoff frequency of 2000 rad/s and a passband gain of -100.

12–14 Find the transfer function $T_V(s) = V_2(s)/V_1(s)$ of the circuit in Figure P12–14.

(a) Find the dc gain, infinite frequency gain, and cutoff frequency. Identify the type of gain response.

(b) Use OrCAD to plot the Bode magnitude gain response of the circuit.

(c) What element value would you change to increase the cutoff frequency by one decade?

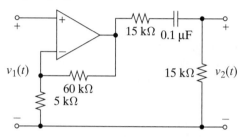

FIGURE P12–14

12–15 **D** Find the transfer function $T_V(s) = V_2(s)/V_1(s)$ of the circuit in Figure P12–15. What type of gain response does the circuit have? What is the passband gain? Select practical, standard values of R and C from the inside rear cover so that the cutoff frequency is 300 kHz $\pm 5\%$.

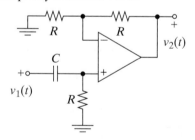

FIGURE P12–15

12–16 A first-order high-pass circuit has a passband gain of 20 dB and a cutoff frequency of 1000 rad/s. Find the gain (in dB) at $\omega = 0$, 500, 1000, and 5000 rad/s.

12–17 A first-order low-pass circuit has a passband gain of 0 dB and a cutoff frequency of 2 krad/s. Find the gain (in dB) at $\omega = 0$, 200 rad/s, 20 krad/s, and 200 krad/s.

12–18 **D** The transfer function of a first-order circuit is

$$T(s) = \frac{10,000}{s + 2000}$$

(a) Identify the type of gain response. Find the cutoff frequency and the passband gain.

(b) Use MATLAB to plot the magnitude of the Bode gain response.

(c) Design a circuit to realize the transfer function.
(d) Use OrCAD to validate your circuit design by comparing its frequency response to the MATLAB output.

12–19 (D) The transfer function of a first-order circuit is

$$T(s) = \frac{5s}{s + 15000}$$

(a) Identify the type of gain response. Find the cutoff frequency and the passband gain.
(b) Use MATLAB to plot the magnitude of the Bode gain response.
(c) Design a circuit to realize the transfer function.
(d) Use OrCAD to validate your circuit design by comparing its frequency response to the MATLAB output

12–20 (D) A student decided that she needed a low-pass filter that had a roll-off of -2 or -40dB/decade, with a cutoff frequency of 1000 rad/s. She correctly designed two identical passive RC filters, each with a cutoff of 1000 rad/s, and connected them in cascade or one after the other. She cleverly separated them with a buffer to avoid loading. When she measured the cutoff frequency she was dismayed to discover that it was not 1000 rad/s although the roll-off was correct.
(a) Simulate the cascaded pair and determine where the actual cutoff frequency occurs.
(b) Help the student by altering her design so that the two identical RC filters actually produce the desired outcome. Explain why the discrepancy occurred.

OBJECTIVE 12–2 BANDPASS AND BANDSTOP RESPONSES (SECT. 12–4)

Given a cascade or parallel connection of two first-order circuits:
(a) Find and classify the frequency response.
(b) Plot the gain and phase responses using straight-line approximations and computer tools.
(c) Design circuits to produce a specified frequency response.
See Examples 12–9, 12–10 and Exercises 12–14, 12–15, 12–16.

12–21 (D) The circuit in Figure P12–21 produces a bandpass response for a suitable choice of element values. Identify the elements that control the two cutoff frequencies. Select the element values so that the passband gain is 100 and the cutoff frequencies are 1000 rad/s and 40 krad/s. Use practical element values with $R \geq 10$ kΩ and $C \leq 1$ μF. Use OrCAD to validate your design.

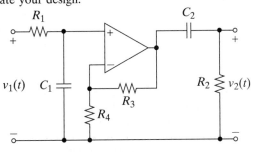

FIGURE P12–21

12–22 (D) Repeat Problem 12–21 for the circuit in Figure P12–22. Use OrCAD to validate your design.

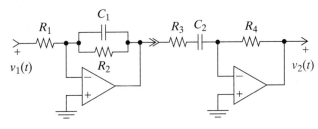

FIGURE P12–22

12–23 (E) Suppose the circuits you designed in problems 12–21 and 12–22 had to feed a 500 Ω instrument. Which circuit would you select and why?

12–24 (D) The circuit in Figure P12–24 produces a bandstop response for a suitable choice of element values.
(a) Find the circuit's transfer function.
(b) Identify the elements that control the two cutoff frequencies. Select the element values so that the cutoff frequencies are 200 krad/s and 4000 krad/s. Use practical element values with $Rs \geq 10$ kΩ and $Cs \leq 1$ μF. Design your passband gain to be $+20$ dB.
(c) Use MATLAB to plot the Bode magnitude plot for the values you selected.
(d) Simulate your circuit using OrCAD and compare the results to the MATLAB output.

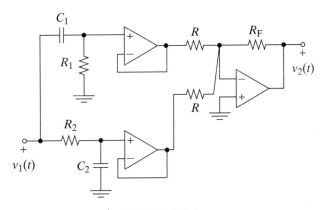

FIGURE P12–24

12–25 (D) Design an audio amplifier that amplifies signals from 20 Hz to 20 kHz. Your approach should be to use a cascade connection of two first-order passive circuits separated by a non-inverting OP AMP. The source has a 50-Ω series resistor and the output of the filter feeds a 10-kΩ audio transducer. Design a bandpass circuit with the following specifications all $\pm 5\%$: $f_{CL} = 20$ Hz, $f_{CH} = 20$ kHz, $B = 19980$ Hz. Passband gain $+40$ dB. Use OrCAD to validate your results.

12–26 (D)(E) Design an audio amplifier that amplifies signals from 2 kHz to 10 kHz. Your approach should be to use a cascade connection of two first-order passive circuits separated by a non-inverting OP AMP. The source has a 50-Ω series resistor and the output of the filter feeds a 10-kΩ audio transducer. Design a bandpass circuit with the following specifications all ± 5%: $f_{CL} = 2\,kHz$, $f_{CH} = 10\,kHz$, $B = 8\,kHz$. Passband gain +40 dB. Use OrCAD to validate your results. Even though problems 12–25 and 12–26 share similar constructs one cannot achieve the desired specifications in 12–26. What is the reason for the problem?

12–27 **D** Design an audio amplifier that amplifies signals from 20 Hz to 20 kHz. Your approach should be to use a cascade connection of two first-order active OP AMP circuits. The source has a 50-Ω series resistor and the output of the filter feeds a 10-kΩ audio transducer. Design a bandpass circuit with the following specifications all ± 5%: $f_{CL} = 20\,Hz$, $f_{CH} = 20\,kHz$, $B = 19980\,Hz$. Passband gain should be adjustable (use a potentiometer) from 0dB to +40 dB. Use OrCAD to validate your results.

12–28 **E** A student needed to design a bandstop filter that was to block frequencies between 1000 rad/s and 10000 rad/s with unity gain in the passbands. His design is shown in Figure P12–28. As a teaching assistant, you are required to grade his design. What grade would you assign and what critique would you give him?

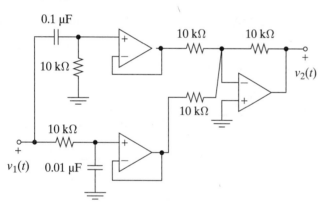

FIGURE P12–28

OBJECTIVE 12–3 THE FREQUENCY RESPONSE OF RLC CIRCUITS (SECT. 12–5)

Given an *RLC* circuit connected as a bandpass or a bandstop filter:
(a) Find the frequency-response descriptors such as Q and B.
(b) Design circuits to produce a specified frequency response.
(c) Compare the bandpass and bandstop responses obtained from *RLC* filters with similar ones obtained from first-order filters.
See Examples 12–11, 12–12, 12–13, 12–14, 12–15 and Exercises 12–17, 12–18, 12–19, 12–20, 12–21

12–29 Determine the filter type for the circuit in Figure P12–29. Then find Q, $B, \omega_{C1}, \omega_{C2}$, and ω_0. Is the circuit a narrow-band or a wide-band filter?

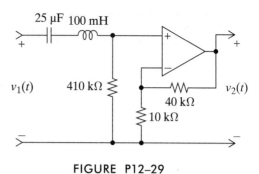

FIGURE P12–29

12–30 **D** Design an *RLC* bandstop filter with a center frequency of 33 krad/s and a bandwidth of 3.3 krad/s. The passband gain is 0 dB. Use practical values for *R*, *L*, and *C*.

12–31 **D** Design an *RLC* bandpass filter with a center frequency of 1000 rad/s and a *Q* of 20. The passband gain is +20 dB. Use practical values for *R*, *L*, and *C*.

12–32 A series *RLC* bandpass circuit with $R = 30\,\Omega$ is designed to have a bandwidth of 5 Mrad/s and a center frequency of 50 Mrad/s. Find *L*, *C*, *Q*, and the two cutoff frequencies. Could you design this circuit using a cascade connection of two first-order filters separated by a follower? Why or why not?

12–33 A parallel *RLC* bandpass circuit with $C = 0.01\,\mu F$ and $Q = 10$ has a center frequency of 500 krad/s. Find *R*, *L*, and the two cutoff frequencies. Could you design this circuit using a cascade connection of two first-order filters separated by a follower? Why or why not?

12–34 A 20-mH inductor with an internal series resistance of 25 Ω is connected in series with a capacitor and a voltage source with a Thévenin resistance of 50 Ω.
(a) What value of *C* is needed to produce $\omega_0 = 5\,krad/s$?
(b) Find the bandwidth and quality factor of the circuit.

12–35 In a series *RLC* circuit, which element would you adjust (and by how much) to
(a) Double the bandwidth without changing the center frequency?
(b) Double the center frequency without changing the bandwidth?
(c) Repeat parts (a) and (b) for a parallel *RLC* circuit.

12–36 A parallel *RLC* circuit with $R = 1\,k\Omega$ has a center frequency of 50 kHz and a bandwidth of 50 kHz. Find the values of *L* and *C*. Could you design this circuit using a cascade connection of two first-order filters separated by a follower? Why or why not?

12–37 **D** A series *RLC* bandpass filter is required to have resonance at $f_0 = 50\,kHz$. The circuit is driven by a sinusoidal source with a Thévenin resistance of 60 Ω. The following standard capacitors are available in the stock room: 1 μF, 0.68 μF, 0.47 μF, 0.33 μF, 0.2 μF, and 0.12 μF. The inductor will be custom-designed to match the capacitor used. Select the available capacitor that minimizes the filter bandwidth.

12–38 A series RLC bandstop circuit is to be used as a notch filter to eliminate a bothersome 120-Hz hum in an audio channel. The signal source has a Thévenin resistance of 300 Ω. Select values of L and C so the upper cutoff frequency of the stopband is below 180 Hz. Use OrCAD to verify your design.

12–39 Find the transfer function $T_V(s) = V_2(s)/V_1(s)$ for the bandpass circuit in Figure P12–39. Use MATLAB to visualize the Bode characteristics if $R = 50\,Ω$, $L = 500\,μH$, and $C = 0.002\,μF$. Design an active circuit to meet those characteristics. Verify your design using OrCAD.

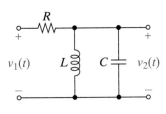

FIGURE P12–39

12–40 Show that the transfer function $T_V(s) = V_2(s)/V_1(s)$ of the circuit in Figure P12–40 has a bandstop filter characteristic. Derive expressions relating the notch frequency and the cutoff frequencies to R, L, and C. Then select values of R, L, and C so that the bandwidth and the center frequency are both 10 krad/s. Validate you design using OrCAD.

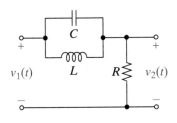

FIGURE P12–40

12–41 Figure P12–41 shows an RLC filter with an input current and an output voltage. The purpose of this problem is to determine the filter type using informal circuit analysis. Use the element impedances and basic analysis tools to find the magnitude of the output voltage $|V_2(j\omega)|$ at $\omega = 0$, $\omega = \infty$, and $\omega = \omega_0 = 1/\sqrt{LC}$. What is the filter type?

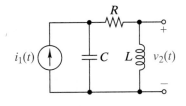

FIGURE P12–41

12–42 A student was designing a passive tuned filter using a series RLC circuit with $R = 10\,Ω$, $L = 10\,μH$, and $C =$

0.025 μF. When he connected the output of the filter across the capacitor he was surprised by an output near the cutoff point that was greater than the input signal. Since the filter was passive he was quite perplexed. Explain how this can be.

OBJECTIVE 12–4 BODE PLOTS (SECTS. 12–6, 12–7)

Given a linear circuit or transfer function:
(a) Plot the gain and phase responses using straight-line approximations and computer tools.
(b) Develop a transfer function from a straight-line Bode gain plot.
(c) Design a circuit that produces a given straight-line gain plot. See Examples 12–16, 12–17, 12–18, 12–19 and Exercises 12–22, 12–23, 12–24, 12–25, 12–26.

12–43 Find the transfer function $T_V(s) = V_2(s)/V_1(s)$ for the circuit in Figure P12–43.
(a) Construct the straight-line Bode plot of the gain and phase of the transfer function. Use the straight-line plots to estimate the amplitude and phase of the steady-state output for $v_1(t) = 10 \sin 100t$ V.
(b) Calculate the actual output amplitude and phase for this input and compare the two results.
(c) Use MATLAB to generate a Bode plot of your transfer function. Compare your answers from the straight-line plot and your hand calculations, with that produced by MATLAB.

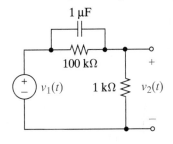

FIGURE P12–43

12–44 Repeat Prob. 12–43 using the circuit in Figure P12–44.

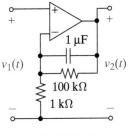

FIGURE P12–44

12–45 For the following transfer function

$$T(s) = \frac{10(s + 100)}{(s + 1)}$$

(a) Construct the straight-line Bode plot of the gain. Is this a low-pass, high-pass, bandpass, or bandstop function? Estimate the cutoff frequency and passband gain.

(b) Use MATLAB to plot the Bode magnitude of the transfer function.

(c) Design a circuit using practical values for the components that has the same transfer function.

(d) Use OrCAD to compare the frequency response of your designed circuit with the MATLAB Bode plot.

12–46 **D** For the following transfer function

$$T(s) = \frac{100(s + 10)}{s(s + 100)}$$

(a) Construct the straight-line Bode plot of the gain. Is this a low-pass, high-pass, bandpass, or bandstop function? Estimate the cutoff frequency and passband gain.

(b) Use MATLAB to plot the Bode magnitude of the transfer function.

(c) Design a circuit using practical values for the components that has the same transfer function.

(d) Use OrCAD to compare the frequency response of your designed circuit with the MATLAB Bode plot.

12–47 **D** For the following transfer function

$$T(s) = \frac{10^6 s^2}{(s + 1000)^2}$$

(a) Construct the straight-line Bode plot of the gain. Is this a low-pass, high-pass, bandpass, or bandstop function? Estimate the cutoff frequency and passband gain.

(b) Use MATLAB to plot the Bode magnitude of the transfer function.

(c) Design a circuit using practical values for the components that has the same transfer function.

(d) Use OrCAD to compare the frequency response of your designed circuit with the MATLAB Bode plot.

12–48 **D** For the following transfer function

$$T(s) = \frac{100s(s + 300)}{(s + 30)(s + 3000)}$$

(a) Construct the straight-line Bode plot of the gain. Is this a low-pass, high-pass, bandpass, or bandstop function? Estimate the cutoff frequency and passband gain.

(b) Use MATLAB to plot the Bode magnitude of the transfer function.

(c) Design a circuit using practical values for the components that has the same transfer function.

(d) Use OrCAD to compare the frequency response of your designed circuit with the MATLAB Bode plot.

12–49 **D** For the following transfer function $T_V(s) = V_2(s)/V_1(s)$

$$T_V(s) = \frac{10(s + 10)(s + 100)}{(s + 1)(s + 1000)}$$

(a) Construct the straight-line Bode plot of the gain. Is this a low-pass, high-pass, bandpass, or bandstop function? Estimate the cutoff frequency and passband gain.

(b) Use MATLAB to plot the Bode magnitude and phase of the transfer function.

(c) What is the output $v_2(t)$, if $v_1(t) = 10 \cos (200t + 45°)$ V?

12–50 **D** For the following transfer function $T_V(s) = V_2(s)/V_1(s)$

$$T(s) = \frac{10^8(s + 100)^2}{(s + 1000)^4}$$

(a) Construct the straight-line Bode plot of the gain. Is this a low-pass, high-pass, bandpass, or bandstop function? Estimate the cutoff frequency(ies) and passband gain.

(b) Use MATLAB to plot the Bode magnitude and phase of the transfer function.

(c) What is the output $v_2(t)$, if $v_1(t) = 10 \cos (1000t - 60°)$ V?

12–51 **D** For the following transfer function $TV(s) = V_2(s)/V_1(s)$

$$T_V(s) = \frac{4s}{0.04s^2 + 0.2s + 1}$$

(a) Construct the straight-line Bode plot of the gain. Is this a low-pass, high-pass, bandpass, or bandstop function?

(b) Use the straight-line plot to estimate the maximum gain and the frequency at which it occurs.

(c) Use MATLAB to plot the Bode magnitude and phase of the transfer function.

12–52 **D** Consider the gain plot in Figure P12–52.

(a) Find the transfer function corresponding to the straight-line gain plot.

(b) Use MATLAB to plot the Bode magnitude of the transfer function.

(c) Compare the straight-line gain and the actual gain at $\omega = 20$ and 200 rad/s.

(d) Design a circuit to realize the Bode plot.

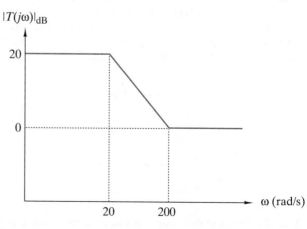

FIGURE P12–52

12–53 ⓓ Consider the gain plot in Figure P12–53.
(a) Find the transfer function corresponding to the straight-line gain plot.
(b) Use MATLAB to plot the Bode magnitude of the transfer function.
(c) Compare the straight-line gain and the actual gain at $\omega = 100$ and $500\,\text{rad/s}$.
(d) Design a circuit that will realize the transfer function found in part (a).
(e) Use OrCAD to verify your circuit design.

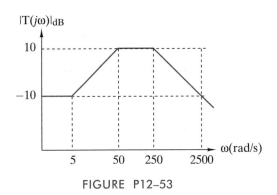

FIGURE P12–53

12–54 Consider the gain plot in Figure P12–54.
(a) Find a transfer function corresponding to the straight-line gain plot. Note that the magnitude of the actual frequency response must be exactly 5 at the geometric mean of the two cutoff frequencies (245 rad/s).
(b) Use MATLAB to plot the Bode magnitude of the transfer function.

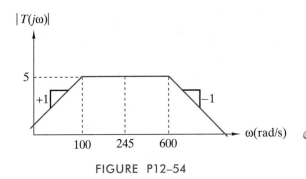

FIGURE P12–54

12–55 ⓓ Consider the gain plot in Figure P12–55.
(a) Find the transfer function corresponding to the straight-line gain plot.
(b) Use MATLAB to plot the Bode magnitude of the transfer function.
(c) Design a circuit that will realize the transfer function found in part (a).
(d) Use OrCAD to verify your circuit design.

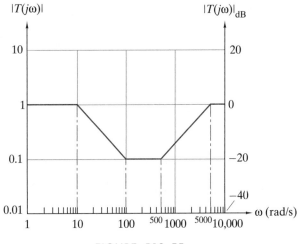

FIGURE P12–55

12–56 ⓓ Consider the gain plot in Figure P12–56.
(a) Find the transfer function corresponding to the straight-line gain plot.
(b) Use MATLAB to plot the Bode magnitude of the transfer function.
(c) Design a circuit that will realize the transfer function found in part (a).
(d) Use OrCAD to verify your circuit design.

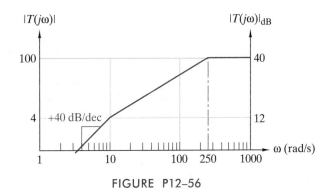

FIGURE P12–56

12–57 ⓓ Consider the gain plot in Figure P12–57. The goal is to design a circuit that will result in the dashed curve shown on the plot.
(a) Find the transfer function corresponding to the straight-line gain plot.
(b) Use MATLAB to plot the Bode magnitude of the transfer function.
(c) Adjust the poles so that the transfer function results in the dashed line. (*Hint:* multiply the two poles into a quadratic expression. Then adjust the Q of the circuit to attain the desired result.)
(d) Design a circuit that will realize the transfer function found in part (c).
(e) Use OrCAD to verify your circuit design.

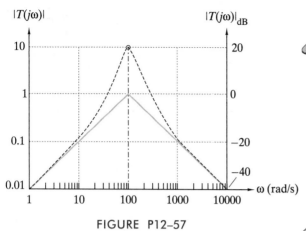

FIGURE P12–57

12–58 For the following transfer function $T_V(s) = V_2(s)/V_1(s)$

$$T_V(s) = \frac{10(s + 100)}{s + 100}$$

(a) Construct the straight-line Bode plot of the phase.
(b) Use the straight-line phase diagram to estimate the phase at $\omega = 1, 10, 100,$ and 1000 rad/s.
(c) Use MATLAB to plot the Bode gain and phase and compare the phase plot to the straight-line estimate.

12–59 For the following transfer function $T_V(s) = V_2(s)/V_1(s)$

$$T_V(s) = \frac{10(s + 50)}{(s + 10)(s + 100)}$$

(a) Construct the straight-line Bode plot of the phase.
(b) Use the straight-line phase diagram to estimate the phase at $\omega = 1, 10, 100,$ and 1000 rad/s.
(c) Use MATLAB to plot the Bode gain and phase and compare the phase plot to the straight-line estimate.

OBJECTIVE **12–5 F**REQUENCY **R**ESPONSE AND **S**TEP **R**ESPONSE **(S**ECT. **12–8)**

Given a circuit or a transfer function:
(a) Find the gain response corresponding to a given step response or vice-versa.
(b) Use the relationship between frequency and step responses to choose the best solution for a design specification.
See Examples 12–20, 12–21, 12–22, 12–23 and Exercises 12–27, 12–28, 12–29.

12–60 The step response of a linear circuit is

$$g(t) = \left[6 - 4e^{-100t} - 2e^{-1000t}\right]u(t)$$

(a) Is the circuit a low-pass, high-pass, bandpass, or bandstop filter?
(b) Construct the straight-line Bode gain plot and estimate the cutoff frequency and passband gain.
(c) Use MATLAB to plot the Bode magnitude and step responses.

(d) Design a circuit to achieve the transfer function.
(e) Use OrCAD to verify the step and frequency responses.

12–61 The step response of a linear circuit is

$$g(t) = \left[2 - 2e^{-20t} + 2e^{-500t}\right]u(t)$$

(a) Is the circuit a low-pass, high-pass, bandpass, or bandstop filter?
(b) Construct the straight-line Bode gain plot and estimate the cutoff frequency and passband gain.
(c) Use MATLAB to plot the Bode magnitude and step responses.
(d) Design a circuit to achieve the transfer function.
(e) Use OrCAD to verify the step and frequency responses.

12–62 The impulse response transform of a linear circuit is

$$H(s) = \frac{2500s^2}{(s + 1000)^2}$$

(a) Is the circuit a low-pass, high-pass, bandpass, or bandstop filter?
(b) Construct the straight-line Bode gain plot and estimate the cutoff frequency and passband gain.
(c) Use MATLAB to plot the Bode magnitude and step responses.
(d) Design a circuit to achieve the transfer function.
(e) Use OrCAD to verify the step and frequency responses.

12–63 The straight-line gain response of a linear circuit is shown in Figure P12–52. What are the initial and final values of the circuit step response? What is the approximate duration of the transient response?

12–64 The straight-line gain response of a linear circuit is shown in Figure P12–54. What are the initial and final values of the circuit step response? What is the approximate duration of the transient response?

12–65 There is a need for a passive notch filter at 10 krad/s. The narrower the notch the better, but there should be minimal ringing of the signals passing through. The transforms of three filters were submitted for consideration. Which would you recommend and why?

$$T_1(s) = \frac{s^2 + 100s + 10^8}{s^2 + 10^4 s + 10^8}$$

$$T_2(s) = \frac{s^2 + 500s + 10^8}{s^2 + 2500s + 10^8}$$

$$T_3(s) = \frac{s^2 + 10^8}{s^2 + 5000s + 10^8}$$

12–66 There is a need for a passive tuned filter at 10 krad/s. The higher the Q the better, but there should be no ringing of the signals passing through. The transform of a prototype filter is shown. Design the filter by selecting the middle term of the denominator to maximize the Q while assuring there is no ringing.

$$T_{tuned}(s) = \frac{10^8 s}{s^2 + xs + 10^8}$$

INTEGRATING PROBLEMS

12–67 Step Response of an *RLC* bandpass circuit **A D**
The step response of a series *RLC* bandpass circuit is

$$g(t) = \left[\frac{4}{5}e^{-200t}\sin(500t)\right]u(t)$$

(a) Find the passband center frequency and the two cutoff frequencies.

(b) Design a circuit that would possess the above step response.

12–68 A Tunable Tank Circuit **A**
The *RLC* circuit in Figure P12–68 (often called a tank circuit) has $R = 4.7$ kΩ, $C = 680$ pF, and an adjustable (tunable) L ranging from 64 to 640 μH.

(a) Show that the circuit is a bandpass filter.

(b) Find the frequency range (in Hz) over which the center frequency can be tuned.

(c) Find the bandwidth (in Hz) at the end points of this range.

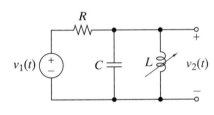

FIGURE P12–68

12–69 Filter Design Specification **D**
Construct a transfer function whose gain response lies entirely within the non-shaded region in Figure P12–69.

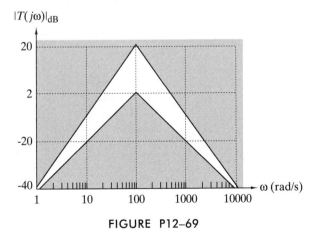

FIGURE P12–69

12–70 Chip *RC* Networks **E**
Integrated circuit (chip) *RC* networks are used at parallel data ports to suppress radio frequency noise. In a certain application, RF noise at 3.2 MHz is interfering with a 4-bit parallel data signal operating at 1.1 MHz. A chip *RC* network

is to be used to reduce the RF noise on the parallel bus by at least 7 dB without reducing the data signals by more than 2 dB. A vendor offers a family of chip *RC* networks connected as shown in Figure P12–70. The available circuit parameters are shown in Table P12–70. Select the part number that best meets the noise suppression requirements.

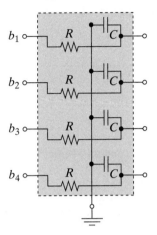

FIGURE P12–70

$R(\Omega)$	C(pF)	Part No.	$R(\Omega)$	C(pF)	Part No.
47	220	ZAEN14	220	220	ZAEN34
	470	ZAEN15		470	ZAEN35
	1000	ZAEN16		1000	ZAEN36
100	220	ZAEN24	470	220	ZAEN44
	470	ZAEN25		470	ZAEN45
	1000	ZAEN26		1000	ZAEN46

12–71 Design Evaluation **E**
Your company issued a request for proposals listing the following design requirements and evaluation criteria.
Design Requirements: Design a low-pass filter with a passband gain of $9 \pm 10\%$ and a cutoff frequency of $90 \pm 10\%$ krad/s. A sensor drives the filter input with a 1-kΩ

First-Order Filter Company

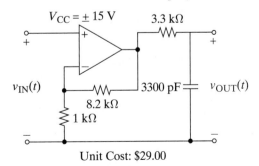

Unit Cost: $29.00

FIGURE P12–71

Simply Filters, Ltd

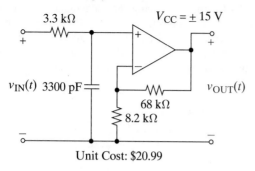

Unit Cost: $20.99

FIGURE P12–71 *(Continued)*

source resistance and an open-circuit voltage range of $\pm$ 1.6 V.

Evaluation Criteria: Filter performance, parts count, use of standard parts, and cost. The two vendors have responded with the designs shown in Figure P12–71. As a junior engineer, the project manager asks you evaluate the designs and recommend a vendor. Which vendor would you recommend and why?

12–72 Design Evaluation

In a research laboratory, you need a bandpass filter to meet the following requirements:

Design Requirements: Passband gain: $10 \pm 5\%$, $B = 10$ krad/s $\pm 5\%$, $\omega_0 = 5$ krad/s $\pm 2\%$, $\omega_{CL} = 2$ krad/s $\pm 10\%$.

Evaluation Criteria: Filter performance, parts count, use of standard parts, and cost.

The two vendors have responded with the designs shown in Figure P12–72.

As a research assistant, your supervisor asks you to evaluate the two designs and recommend a vendor. Which vendor would you recommend and why?

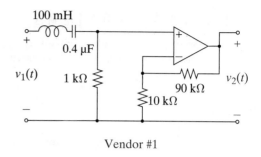

Vendor #1

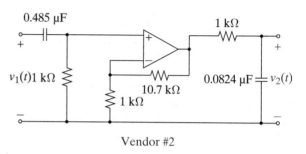

Vendor #2

FIGURE P12–72

12–73 Design Evaluation

In a cable service distribution station, you need a bandstop filter to meet the following requirements:

Design Requirements: Passband gain: $10 \pm 5\%$, $B = 3.3$ kHz $\pm 5\%$, $f_0 = 500$ Hz $\pm 2\%$, $f_{CL} = 75$ Hz $\pm 10\%$. Filter must interface with a 50-Ω source and a 500-Ω load.

Evaluation Criteria: Filter performance including depth of notch, parts count, power usage, ease of maintenance, use of standard parts, and cost.

The two vendors have responded with the designs shown in Figure P12–73. As an engineering cable guy, your supervisor asks you to evaluate the two designs and recommend a vendor. Which vendor would you recommend and why?

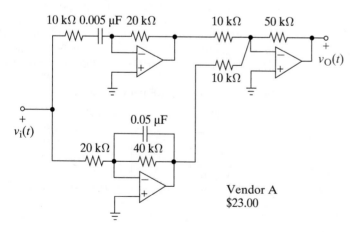

Vendor A
$23.00

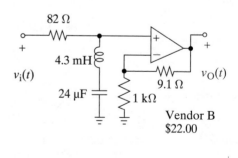

Vendor B
$22.00

FIGURE P12–73

CHAPTER 13 FOURIER ANALYSIS

The series formed of sines or cosines of multiple arcs are therefore adapted to represent between definite limits all possible functions, and the ordinates of lines or surfaces whose form is discontinuous.

Jean Baptiste Joseph Fourier, 1822,
French Physicist

Some History Behind This Chapter

The analysis techniques in this chapter have their roots in the works of the French physicist J. B. J. Fourier (1768–1830). Fourier found that discontinuous functions could be represented by an infinite series of harmonic sinusoids. Fourier was guided by his physical intuition and presented empirical evidence to support his claim but left the question of series convergence unanswered. This question was later settled by the German mathematician P. G. L. Dirichlet (1805–1859). Fourier's methods were eventually accepted and are now widely used.

Why This Chapter Is Important Today

The word *spectrum* denotes both a natural resource and a mathematical concept. As a natural resource, it is critical to modern communications and its use is regulated by national and international agencies. As a mathematical concept, it describes the range of frequencies needed to reproduce a signal. Fourier theory suggests that perfect reproduction requires an infinite spectrum, a practical impossibility. The theory also shows that a finite range of frequencies carry most of the energy in periodic signals. Intentionally limiting a signal to this range makes the spectrum available to other users and allows today's profusion of communication systems.

Fourier transforms do for aperiodic signals what Fourier series do for periodic ones. Fourier transforms let us understand the frequency content of aperiodic signals and apply that understanding to communication systems and other applications.

Chapter Sections

Chapter Learning Objectives

13-1 The Fourier Series (Sects. 13–1, 13–2, 13–3)

(a) Given an equation or graph of a periodic waveform, derive expressions for the Fourier coefficients.

(b) Given a_0, a_n, and b_n calculate the Fourier coefficients of a given periodic waveform.

(c) Given a Fourier series of a periodic waveform, determine properties of the waveform and plot its amplitude and phase spectra.

13-2 Fourier Series and Circuit Analysis (Sect. 13–4)

(a) Given a linear circuit with a periodic input waveform, find the Fourier series of a steady-state response.

(b) Given a network function with a periodic input, find the amplitude and phase spectra of the steady-state output.

13-3 RMS Value and Average Power (Sect. 13–5)

(a) Given a periodic waveform, find the rms value of the waveform and the average power delivered to a specified load.

(b) Given the Fourier series of a periodic waveform, find the fraction of the average power carried by specified components and estimate the average power delivered to a specified load.

13-4 The Fourier Transform (Sect. 13–6)

(a) Given an equation or graph of an aperiodic waveform, derive an expression for the Fourier transform.

(b) Given an equation or graph of a Fourier transform, derive an expression for the corresponding waveform.

(c) Understand the relationship between Fourier transforms and Laplace transforms and know the conditions under which Laplace transform techniques can be used to compute Fourier transforms.

13-5 Fourier Transforms and Circuit Analysis (Sect. 13–7, 13–8, 13–9)

(a) Given a linear circuit, use Fourier transform techniques to determine the output response for aperiodic input signals.

(b) Given a linear circuit, use Parseval's theorem to determine the energy content of selected signals.

13–1 OVERVIEW OF FOURIER ANALYSIS

In this chapter, we develop a method of finding the steady-state response of circuits to periodic and aperiodic signals. Periodic waveforms can be written as a Fourier series consisting of an infinite sum of harmonically related sinusoids. Figure 13–1 provides examples of common periodic signals. If the signal is aperiodic, then a Fourier transform is required to characterize the frequency content of the signal. Sections 13–6 through 13–9 will explore Fourier transforms in more detail.

If a signal $f(t)$ is periodic with period T_0 and is reasonably well behaved, then $f(t)$ can be expressed as a **Fourier series** of the form

$$f(t) = a_0 + a_1 \cos(2\pi f_0 t) + a_2 \cos(2\pi 2 f_0 t) + \cdots + a_n \cos(2\pi n f_0 t) + \cdots \\ + b_1 \sin(2\pi f_0 t) + b_2 \sin(2\pi 2 f_0 t) + \cdots + b_n \sin(2\pi n f_0 t) + \cdots \tag{13–1}$$

or, more compactly,

$$f(t) = \underbrace{a_0}_{\text{dc}} + \underbrace{\sum_{n=1}^{\infty} [a_n \cos(2\pi n f_0 t) + b_n \sin(2\pi n f_0 t)]}_{\text{ac}} \tag{13–2}$$

FIGURE 13–1 *Examples of periodic waveforms.*

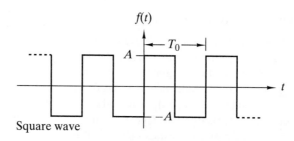

Square wave

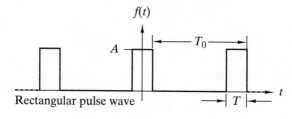

Rectangular pulse wave

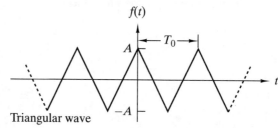

Triangular wave

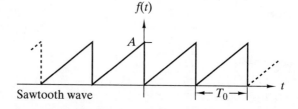

Sawtooth wave

The coefficient a_0 is the dc component or average value of $f(t)$. The constants a_n and b_n ($n = 1, 2, 3, \ldots$) are the **Fourier coefficients** of the sinusoids in the ac component. The lowest frequency in the ac component occurs for $n = 1$ and is called the **fundamental frequency**, defined as $f_0 = 1/T_0$. The other frequencies are integer multiples of the fundamental called the second harmonic ($2f_0$), third harmonic ($3f_0$), and, in general, the nth harmonic (nf_0).

Since Eq. (13–2) is an infinite series, there is always a question of convergence. We have said that the series converges as long as $f(t)$ is reasonably well behaved. Basically, this means that $f(t)$ is single valued, the integral of $|f(t)|$ over a period is finite, and $f(t)$ has a finite number of discontinuities in any one period. These requirements, called the **Dirichlet conditions**, are sufficient to assure convergence. Every periodic waveform that meets the Dirichlet conditions has a convergent Fourier series. However, there are waveforms that do not meet the Dirichlet conditions that also have convergent Fourier series. That is, while the Dirichlet conditions are sufficient, they are not necessary and sufficient. This limitation does not present a serious problem because the Dirichlet conditions are satisfied by the waveforms generated in physical systems. All of the periodic waveforms in Figure 13–1 meet the Dirichlet requirements.

It is important to have an overview of how a Fourier series is used in circuit analysis. In our previous study we learned how to find the steady-state responses due to a dc or an ac input. The Fourier series resolves a periodic input into a dc component and an infinite sum of ac components. We treat each of the terms in the Fourier series as a separate input source and use our circuit analysis tools to find the steady-state responses due to each term acting alone. The complete steady-state response is found by superposition, that is, adding up the responses due to each term acting alone. The net result is that the output is a modified version of the Fourier series of the periodic input.

At first glance it may seem complicated to find the complete response by finding the individual responses due to infinitely many inputs. However, each step simply involves circuit analysis tools we have already mastered. The end result tells us how the circuit transforms the input series into the output series. The distribution of amplitudes and phase angles in a Fourier series is called the **spectrum** of a periodic waveform. Frequency-domain signal processing involves modifying a given input spectrum to produce a desired output spectrum. Thus, finding the Fourier series of periodic waveforms is not an end in itself, but the first step in the study of frequency-domain signal processing.

13–2 FOURIER COEFFICIENTS

The Fourier coefficients for any periodic waveform $f(t)$ satisfying the Dirichlet conditions can be obtained from the equations

$$a_0 = \frac{1}{T_0} \int_{-T_0/2}^{+T_0/2} f(t)dt$$

$$a_n = \frac{2}{T_0} \int_{-T_0/2}^{+T_0/2} f(t)\cos(2\pi nt/T_0)dt \qquad (13\text{–}3)$$

$$b_n = \frac{2}{T_0} \int_{-T_0/2}^{+T_0/2} f(t)\sin(2\pi nt/T_0)dt$$

The integration limits in these equations extend from $-T_0/2$ to $+T_0/2$. However, the limits can span any convenient interval as long as it is exactly one period. For example, the limits could be from 0 to T_0 or $-T_0/4$ to $3T_0/4$. We will show where Eq. (13–3) comes from in a moment, but first we use these equations to obtain the Fourier coefficients of the sawtooth wave.

EXAMPLE 13–1

Find the Fourier coefficients for the sawtooth wave in Figure 13–1.

SOLUTION:
An expression for a sawtooth wave on the interval $0 \leq t \leq T_0$ is

$$f(t) = \frac{At}{T_0} \qquad 0 \leq t < T_0$$

For this definition of $f(t)$ we use 0 and T_0 as the limits in Eq. (13–3). The first expression in Eq. (13–3) yields a_0 as

$$a_0 = \frac{1}{T_0} \int_0^{T_0} \frac{At}{T_0} dt = \frac{At^2}{2T_0^2} \Big|_0^{T_0} = \frac{A}{2}$$

This result states that the average or dc value is $A/2$, which is easy to see because the area under one cycle of the sawtooth wave, a triangle, is $AT_0/2$. The second expression in Eq. (13–3) yields a_n as

$$
\begin{aligned}
a_n &= \frac{2}{T_0} \int_0^{T_0} \frac{At}{T_0} \cos(2\pi nt/T_0) dt \\
&= \frac{2A}{T_0^2} \left[\frac{\cos(2\pi nt/T_0)}{(2\pi n/T_0)^2} + \frac{t \times \sin(2\pi nt/T_0)}{(2\pi n/T_0)} \right]_0^{T_0} \\
&= \frac{2A}{T_0^2} \left[\frac{\cos(2\pi n) - \cos(0)}{(2\pi n/T_0)^2} \right] = 0 \qquad \text{for all } n
\end{aligned}
$$

Since $a_n = 0$ for all n, there are no cosine terms in the series. The b_n coefficients are found using the third expression in Eq. (13–3):

$$
\begin{aligned}
b_n &= \frac{2}{T_0} \int_0^{T_0} \frac{At}{T_0} \sin(2\pi nt/T_0) dt \\
&= \frac{2A}{T_0^2} \left[\frac{\sin(2\pi nt/T_0)}{(2\pi nt/T_0)^2} - \frac{t \times \cos(2\pi nt/T_0)}{(2\pi n/T_0)} \right]_0^{T_0} \\
&= \frac{2A}{T_0^2} \left[\frac{-T_0 \cos(2\pi n)}{(2\pi n/T_0)} \right] = -\frac{A}{n\pi} \qquad \text{for all } n
\end{aligned}
$$

Given the coefficients a_n and b_n found above, the Fourier series for the sawtooth wave is

$$f(t) = \frac{A}{2} + \sum_{n=1}^{\infty} \left[-\frac{A}{n\pi} \right] \sin(2\pi nf_0 t)$$

■

Exercise 13–1

Find the Fourier coefficients for the rectangular pulse wave in Figure 13–1.

Answer:

$$a_0 = \frac{AT}{T_0}, a_n = \frac{2A}{n\pi} \sin\left(\frac{n\pi T}{T_0}\right), b_n = 0$$

EXAMPLE 13 – 2

In this example we use MATLAB to show that a truncated Fourier series approximates a periodic waveform. The waveform is a sawtooth with $A = 10$ and $T_0 = 2\,\text{ms}$. Calculate the Fourier coefficients of the first 20 harmonics and plot the truncated series representation of the waveform using the first 5 harmonics and repeat for the first 10 harmonics.

SOLUTION:

From Example 13–1 the Fourier coefficients for the sawtooth wave are

$$a_0 = A/2 \quad a_n = 0 \quad \text{and} \quad b_n = -\frac{A}{n\pi} \quad \text{for all } n$$

The following MATLAB code will create the requested harmonics and generate the required plots:

```
% Create the symbolic variables
syms t
% Define the waveform parameters
A = 10;
T0 = 0.002;
n = 1:20;
% Calculate the coefficients
a0 = A/2;
bn = -A./n/pi;
an = zeros(size(bn));
% Create vectors for the cosine and sine terms
cnt = cos(2*pi*n*t/T0);
snt = sin(2*pi*n*t/T0);
% Create the partial sums
N = 5;
ft5 = a0 + an(1:N)*cnt(1:N)' + bn(1:N)*snt(1:N)';
N = 10;
ft10 = a0 + an(1:N)*cnt(1:N)' + bn(1:N)*snt(1:N)';
% Substitute in a time vector
tt = 0:T0/500:2*T0;
ftt5 = subs(ft5,t,tt);
ftt10 = subs(ft10,t,tt);
% Create the plots
figure
plot(tt,ftt5,'b','LineWidth',2)
xlabel('Time (sec)')
ylabel('f(t)')
title('dc + first 5 harmonics')
grid on
figure
plot(tt,ftt10, 'b', 'LineWidth', 2)
xlabel('Time (sec)')
ylabel('f(t)')
title('dc + first 10 harmonics')
grid on
```

In the MATLAB code, note that the partial sums were created using vector multiplication in the second and third terms. For the vector multiplication, the first vector multiplies the transpose of the second vector to complete the multiplication and summation in a single MATLAB operation.

Figure 13–2 displays the resulting waveforms. The waveform generated with the first 5 harmonics in Figure 13–2 (a) has the correct general shape, amplitude, and period.

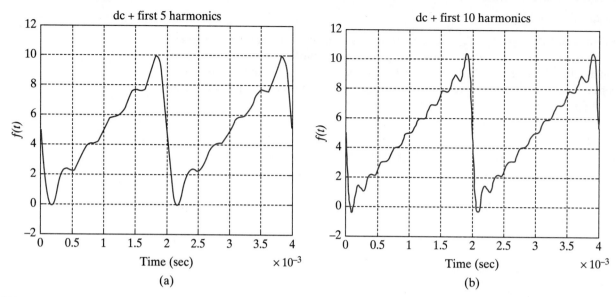

FIGURE 13–2

When we increase the number of harmonics to 10 in Figure 13–2 (b), the shape of the waveform improves and we note that the transition from 10 to 0 in the waveform at $t = 2$ ms is steeper and a better approximation of the actual vertical transition.   ∎

Exercise 13–2

Given a rectangular pulse waveform as shown in Figure 13–1, let $A = 10, T_0 = 5$ ms, and $T = 2$ ms. (a) Use MATLAB to calculate the Fourier coefficients for the first 10 harmonics. (b) Use the 10 harmonics to plot a truncated series representation of the waveform.

Answers:
(a) $a_0 = 4$, $a_1 = 6.054$, $a_2 = 1.871$, $a_3 = -1.247$, $a_4 = -1.514$, $a_5 = 0$, $a_6 = 1.009$, $a_7 = 0.535$, $a_8 = -0.468$, $a_9 = -0.673$, $a_{10} = 0$, $b_n = 0$ for all n.
(b) See Figure 13–3.

FIGURE 13–3

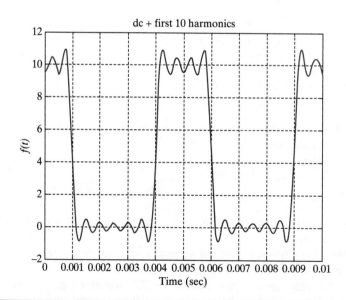

DERIVING EQUATIONS FOR a_n AND b_n

The sawtooth wave example shows how to calculate the Fourier coefficients using Eq. (13–3). We now turn to the derivation of these equations. An equation for a Fourier coefficient is derived by multiplying both sides of Eq. (13–2) by the sinusoid associated with the coefficient and then integrating the result over one period. This multiply and integrate process isolates one coefficient because it turns out that all of the integrations produce zero except one.

The following derivation makes use of the fact that the area under a sine or cosine wave over an integer number of cycles is zero. That is,

$$\int_{-T_0/2}^{+T_0/2} \sin(2\pi k\, f_0 t)dt = 0 \qquad \text{for all } k$$

$$\int_{-T_0/2}^{+T_0/2} \cos(2\pi k\, f_0 t)dt = 0 \qquad \text{for } k \neq 0 \qquad \text{(13–4)}$$

$$= T_0 \qquad \text{for } k = 0$$

where k is an integer. These equations state that integrating a sinusoid over $k \neq 0$ cycles produces zero, since the areas under successive half-cycles cancel. The single exception occurs when $k = 0$, in which case the cosine function reduces to one and the net area for one period is T_0.

We derive the equation for the amplitude of the dc component a_0 by integrating both sides of Eq. (13–2):

$$\int_{-T_0/2}^{+T_0/2} f(t)dt$$

$$= a_0 \int_{-T_0/2}^{+T_0/2} dt + \sum_{n=1}^{\infty} \left[a_n \int_{-T_0/2}^{+T_0/2} \cos(2\pi n f_0 t)dt + b_n \int_{-T_0/2}^{+T_0/2} \sin(2\pi n f_0 t)dt \right] \qquad \text{(13–5)}$$

$$= a_0 T_0 \qquad + \qquad\qquad\qquad 0 \qquad + \qquad\qquad\qquad 0$$

The integrals of the ac components vanish because of the properties in Eq. (13–4), and the right side of this expression reduces to $a_0 T_0$. Solving for a_0 yields the first expression in Eq. (13–3).

To derive the expression for a_n we multiply Eq. (13–2) by $\cos(2\pi m f_0 t)$ and integrate over the interval from $-T_0/2$ to $+T_0/2$:

$$\int_{-T_0/2}^{+T_0/2} f(t) \cos(2\pi m f_0 t)dt = a_0 \int_{-T_0/2}^{+T_0/2} \cos(2\pi m f_0 t)dt$$

$$+ \sum_{n=1}^{\infty} \left[a_n \int_{-T_0/2}^{+T_0/2} \cos(2\pi m f_0 t) \cos(2\pi n f_0 t)dt \right. \qquad \text{(13–6)}$$

$$\left. + b_n \int_{-T_0/2}^{+T_0/2} \cos(2\pi m f_0 t) \sin(2\pi n f_0 t)dt \right]$$

All of the integrals on the right side of this equation are zero except one. To show this we use identities

$$\cos(x)\cos(y) = \frac{1}{2}\cos(x - y) + \frac{1}{2}\cos(x + y)$$

$$\cos(x)\sin(y) = \frac{1}{2}\sin(x - y) + \frac{1}{2}\sin(x + y)$$

to change Eq. (13–6) into the following form:

$$
\int_{-T_0/2}^{+T_0/2} f(t) \cos(2\pi m f_0 t)\,dt = a_0 \int_{-T_0/2}^{+T_0/2} \cos(2\pi m f_0 t)\,dt
$$

$$
+ \sum_{n=1}^{\infty} \left\{ \frac{a_n}{2} \left[\int_{-T_0/2}^{+T_0/2} \cos[2\pi(m-n)f_0 t]\,dt + \int_{-T_0/2}^{+T_0/2} \cos[2\pi(m+n)f_0 t]\,dt \right] \right\} \quad (13\text{–}7)
$$

$$
+ \sum_{n=1}^{\infty} \left\{ \frac{b_n}{2} \left[\int_{-T_0/2}^{+T_0/2} \sin[2\pi(m-n)f_0 t]\,dt + \int_{-T_0/2}^{+T_0/2} \sin[2\pi(m+n)f_0 t]\,dt \right] \right\}
$$

All of the integrals are now in the form of expressions in Eq. (13–4). Consequently, we see all of the integrals on the right side of Eq. (13–7) vanish, except for one cosine integral when $m = n$. This one survivor corresponds to the $k = 0$ case in Eq. (13–4), and the right side of Eq. (13–7) reduces to

$$
\int_{-T_0/2}^{T_0/2} f(t) \cos(2\pi n f_0 t)\,dt = \frac{a_n}{2} \int_{-T_0/2}^{T_0/2} \cos[2\pi(n-n)f_0 t]\,dt
$$

$$
= \frac{a_n}{2} T_0
$$

Solving Eq. (13–7) for a_n yields the second expression in Eq. (13–3).

To obtain the expression for b_n we multiply Eq. (13–2) by $\sin(2\pi m f_0 t)$ and integrate over the interval $t = -T_0/2$ to $+T_0/2$. The derivation steps then parallel the approach used to find a_n. The end result is that the dc component integral vanishes and the ac component integrals reduce to $b_n T_0/2$, which yields the expression for b_n in Eq. (13–3).

The derivation of Eq. (13–3) focuses on the problem of finding the Fourier coefficients of a given periodic waveform. Some experience and practice are necessary to understand the implications of this procedure. On the other hand, it is not necessary to go through these mechanics for every newly encountered periodic waveform because tables of Fourier series expansions are available. For our purposes the listing in Figure 13–4 on the next page will suffice. For each waveform defined graphically, the figure lists the expressions for a_0, a_n, and b_n as well as restrictions on the integer n.

EXAMPLE 13–3

Verify the Fourier coefficients given for the square wave in Figure 13–4 and write the first three nonzero terms in its Fourier series.

SOLUTION:

An expression for a square wave on the interval $0 < t < T_0$ is

$$
f(t) = \begin{cases} A & 0 < t < T_0/2 \\ -A & T_0/2 < t < T_0 \end{cases}
$$

Using the first expression in Eq. (13–3) to find a_0 yields

$$
a_0 = \frac{1}{T_0} \int_0^{T_0/2} A\,dt + \frac{1}{T_0} \int_{T_0/2}^{T_0} (-A)\,dt
$$

$$
= \frac{A}{T_0} \left[\frac{T_0}{2} - 0 - T_0 + \frac{T_0}{2} \right] = 0
$$

Waveform	Fourier Coefficients	Waveform	Fourier Coefficients
Constant (dc) 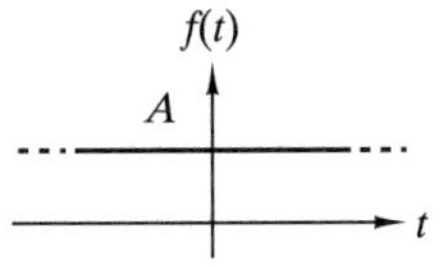f(t) A t	$a_0 = A$ $a_n = 0$ all n $b_n = 0$ all n	Sawtooth wave 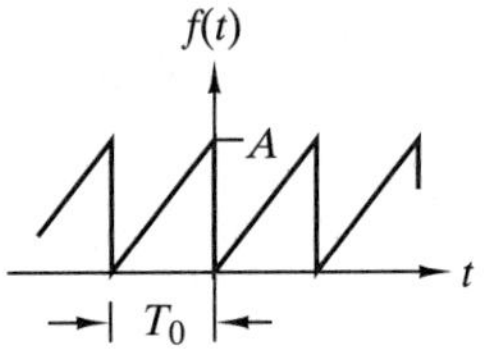f(t) A t T_0	$a_0 = \frac{A}{2}$ $a_n = 0$ all n $b_n = -\frac{A}{n\pi}$ all n
Cosine wave 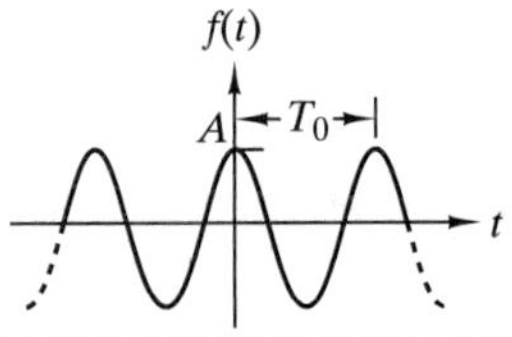f(t) A T_0 t	$a_0 = 0$ $a_1 = A$ $a_n = 0$ $n \neq 1$ $b_n = 0$ all n	Triangular wave 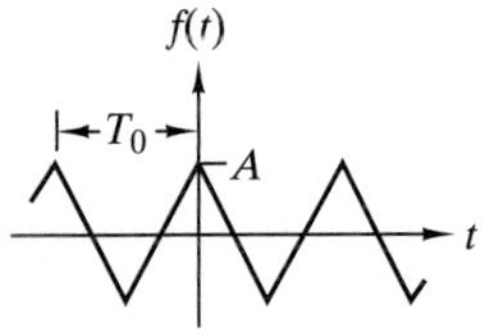f(t) T_0 A t	$a_0 = 0$ $a_n = \frac{8A}{(n\pi)^2}$ n odd $a_n = 0$ n even $b_n = 0$ all n
Sine wave 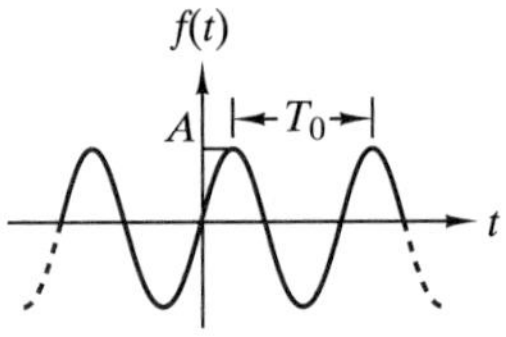f(t) A T_0 t	$a_0 = 0$ $a_n = 0$ all n $b_1 = A$ $b_n = 0$ $n \neq 1$	Half-wave rectified sine wave 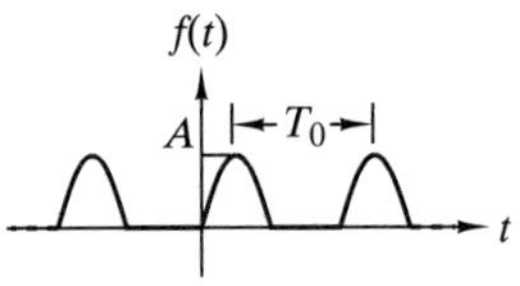f(t) A T_0 t	$a_0 = \frac{A}{\pi}$ $a_n = \frac{2A/\pi}{1-n^2}$ n even $a_n = 0$ n odd $b_1 = \frac{A}{2}$ $n = 1$ $b_n = 0$ $n \neq 1$
Square wave 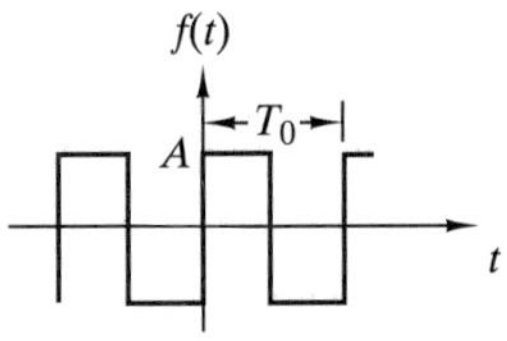f(t) T_0 A t	$a_0 = 0$ $a_n = 0$ all n $b_n = \frac{4A}{n\pi}$ n odd $b_n = 0$ n even	Full-wave rectified sine wave 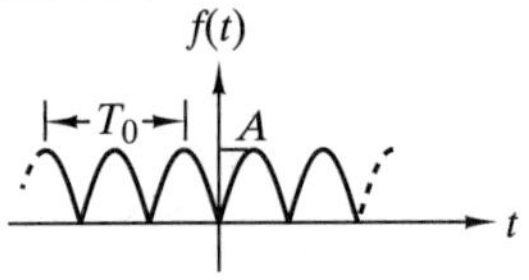f(t) T_0 A t	$a_0 = 2A/\pi$ $a_n = \frac{4A/\pi}{1-n^2}$ n even $a_n = 0$ n odd $b_n = 0$ all n
Rectangular pulse 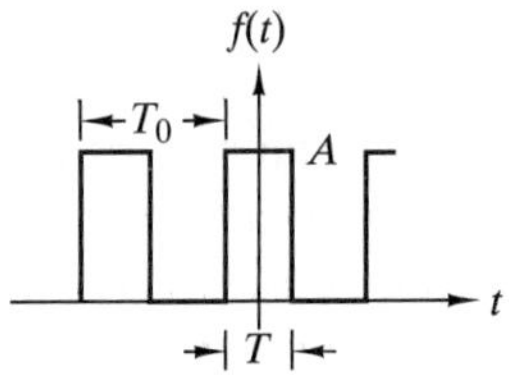f(t) T_0 A t T	$a_0 = \frac{AT}{T_0}$ $a_n = \frac{2A}{n\pi}\sin\left(\frac{n\pi T}{T_0}\right)$ $b_n = 0$ all n	Parabolic wave 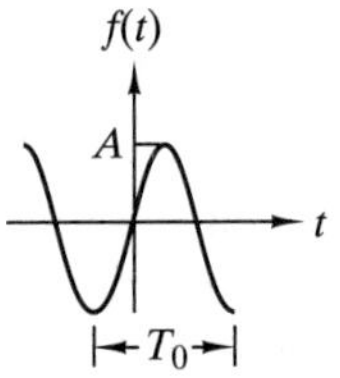f(t) A t T_0	$a_0 = 0$ $a_n = 0$ all n $b_n = \frac{32A}{(n\pi)^3}$ n odd $b_n = 0$ n even

FIGURE 13–4 *Fourier coefficients for some periodic waveforms.*

The result $a_0 = 0$ means that the dc value of the square wave is zero, which is easy to see because the area under a positive half-cycle cancels the area under a negative half-cycle. Using the second expression in Eq. (13–3) to find a_n produces

$$\begin{aligned} a_n &= \frac{2}{T_0}\int_0^{T_0/2} A\cos(2\pi nt/T_0)dt + \frac{2}{T_0}\int_{T_0/2}^{T_0}(-A)\cos(2\pi nt/T_0)dt \\ &= \frac{2A}{T_0}\left[\frac{\sin(2\pi nt/T_0)}{2\pi n/T_0}\right]_0^{T_0/2} - \frac{2A}{T_0}\left[\frac{\sin(2\pi nt/T_0)}{2\pi n/T_0}\right]_{T_0/2}^{T_0} \\ &= \frac{A}{n\pi}[\sin(n\pi) - \sin(0) - \sin(2n\pi) + \sin(n\pi)] = 0 \end{aligned}$$

Since $a_n = 0$ for all n, there are no cosine terms in the series. This makes some intuitive sense because a sinewave with the same fundamental frequency as the square wave fits nicely inside the square wave with zeros crossing at the same points, whereas a cosine with the same frequency does not fit at all. The b_n coefficients for the sine terms are found using the third expression in Eq. (13–3):

$$
\begin{aligned}
b_n &= \frac{2}{T_0}\int_0^{T_0/2} A\sin(2\pi nt/T_0)dt + \frac{2}{T_0}\int_{T_0/2}^{T_0}(-A)\sin(2\pi nt/T_0)dt \\
&= \frac{2A}{T_0}\left[-\frac{\cos(2\pi nt/T_0)}{2\pi n/T_0}\right]_0^{T_0/2} - \frac{2A}{T_0}\left[-\frac{\cos(2\pi nt/T_0)}{2\pi n/T_0}\right]_{T_0/2}^{T_0} \\
&= \frac{A}{n\pi}[-\cos(n\pi)+\cos(0)+\cos(2n\pi)-\cos(n\pi)] \\
&= \frac{2A}{n\pi}[1-\cos(n\pi)]
\end{aligned}
$$

The term $[1-\cos(n\pi)]=2$ if n is odd and zero if n is even. Hence b_n can be written as

$$
b_n = \begin{cases} \dfrac{4A}{n\pi} & n\ \text{odd} \\[2mm] 0 & n\ \text{even} \end{cases}
$$

The first three nonzero terms in the Fourier series of the square wave are

$$
f(t) = \frac{4A}{\pi}\left[\sin 2\pi f_0 t + \frac{1}{3}\sin 2\pi 3 f_0 t + \frac{1}{5}\sin 2\pi 5 f_0 t + \cdots\right]
$$

Note that this series contains only odd harmonic terms. ∎

Exercise 13–3

The triangular wave in Figure 13–4 has a peak amplitude of $A=10$ and $T_0=2\,\text{ms}$. Calculate the Fourier coefficients of the first nine harmonics.

Answers: $a_0=0$, $a_1=8.11$, $a_2=0$, $a_3=0.901$, $a_4=0$, $a_5=0.324$, $a_6=0$, $a_7=0.165$, $a_8=0$, $a_9=0.100$, $b_n=0$ for all n.

ALTERNATIVE FORM OF THE FOURIER SERIES

The series in Eq. (13–1) can be written in several alternative yet equivalent forms. From our study of sinusoids in Chapter 5, we recall that the Fourier coefficients determine the amplitude and phase angle of the general sinusoid. Thus, we can write a general Fourier series in the form

$$
\begin{aligned}
f(t) = A_0 &+ A_1\cos(2\pi f_0 t + \phi_1) + A_2\cos(2\pi 2 f_0 t + \phi_2) + \cdots \\
&+ A_n\cos(2\pi n f_0 t + \phi_n) + \cdots
\end{aligned} \tag{13–8}
$$

where

$$
A_n = \sqrt{a_n^2 + b_n^2} \quad \text{and} \quad \phi_n = \tan^{-1}\frac{-b_n}{a_n} \tag{13–9}
$$

The coefficient A_n is the amplitude of the nth harmonic and ϕ_n is its phase angle.[1]

Note that the amplitude A_n and phase angle ϕ_n contain all of the information needed to construct the Fourier series in the form of Eq. (13–8). Figure 13–5 shows

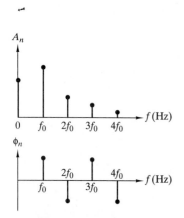

FIGURE 13–5 *Amplitude and phase spectra.*

---

[1] There is a 180° ambiguity in the value returned by the inverse tangent function in most computational tools. The ambiguity is resolved by the following rule: $b_n>0$ implies that the angle is in the range 0 to -180°, while $b_n<0$ implies the 0 to $+180^\circ$ range.

how plots of this information are used to display the spectral content of a periodic waveform $f(t)$. The plot of A_n versus nf_0 (or $n\omega_0$) is called the **amplitude spectrum**, while the plot of ϕ_n versus nf_0 (or $n\omega_0$) is called the **phase spectrum**. Both plots are **line spectra** because spectral content can be represented as a line at discrete frequencies.

In theory, a Fourier series includes infinitely many harmonics, although the harmonics tend to decrease in amplitude at high frequency. For example, the summary in Figure 13–4 shows that the amplitudes of the square wave decrease as $1/n$, the triangular wave as $1/n^2$, and the parabolic wave as $1/n^3$. The $1/n^3$ dependence means that the amplitude of the fifth harmonic in a parabolic wave is less than 1% of the amplitude of the fundamental (actually 1/125th of the fundamental). In practical signals the harmonic amplitudes decrease at high frequency so that at some point the higher-order components become negligibly small. This means that we can truncate the series at some finite frequency and still retain the important features of the signal. This is an important consideration in systems with finite bandwidth.

EXAMPLE 13–4

Derive expressions for the amplitude A_n and phase angle ϕ_n of the Fourier series of the sawtooth wave in Figure 13–4. Sketch the amplitude and phase spectra of a sawtooth wave with $A = 5$ and $T_0 = 4\,\text{ms}$.

SOLUTION:

Figure 13–4 gives the Fourier coefficients of the sawtooth wave as

$$a_0 = \frac{A}{2} \quad a_n = 0 \quad b_n = -\frac{A}{n\pi} \quad \text{for all } n$$

Using Eq. (13–9) yields

$$A_n = \sqrt{a_n^2 + b_n^2} = \begin{cases} \dfrac{A}{2} & n = 0 \\ \dfrac{A}{n\pi} & n > 0 \end{cases}$$

and

$$\phi_n = \tan^{-1}\frac{-b_n}{a_n} = \begin{cases} \text{undefined} & n = 0 \\ 90° & n > 0 \end{cases}$$

For $A = 5$ and $f_0 = 1/T_0 = 250$ Hz the first four nonzero terms in the series are

$$f(t) = 2.5 + 1.59\ \cos(2\pi250t + 90°)$$
$$+ 0.796\ \cos(2\pi500t + 90°) + 0.531\ \cos(2\pi750t + 90°)$$

Figure 13–6 shows the amplitude and phase spectra for this signal. Note that the lines in the amplitude spectrum are inversely proportional to frequency. As

FIGURE 13–6

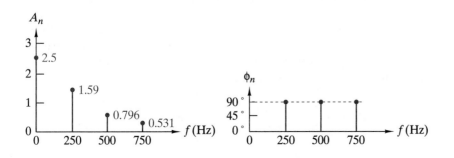

frequency increases, the amplitudes decrease so that the high-frequency components become negligible.

■

Exercise 13–4

Derive expressions for the amplitude A_n and phase angle ϕ_n for the triangular wave in Figure 13–4 and write an expression for the first three nonzero terms in the Fourier series with $A = \pi^2/8$ and $T_0 = 2\pi/5000\,\text{s}$.

Answers:

$$A_n = \frac{8A}{(n\pi)^2} \quad \phi_n = 0° \quad n\,\text{odd}$$

$$A_n = 0 \quad \phi_n\,\text{undefined} \quad n\,\text{even}$$

$$f(t) = \cos(5000t) + \frac{1}{9}\cos(15{,}000t) + \frac{1}{25}\cos(25{,}000t) + \cdots$$

13–3 WAVEFORM SYMMETRIES

Many of the Fourier coefficients are zero when a periodic waveform has certain types of symmetries. It is helpful to recognize these symmetries, since they may simplify the calculation of the Fourier coefficients.

The first expression in Eq. (13–3) shows that the amplitude of the dc component a_0 is the average value of the periodic waveform $f(t)$. If the waveform has equal area above and below the time axis, then the integral over one cycle vanishes, the average value is zero, and $a_0 = 0$. The square wave, triangular wave, and parabolic wave in Figure 13–4 are examples of periodic waveforms with zero average value.

A waveform is said to have **even symmetry** if $f(-t) = f(t)$. The cosine wave, rectangular pulse, and triangular wave in Figure 13–4 are examples of waveforms with even symmetry. The Fourier series of an even waveform is made up entirely of cosine terms: that is, all of the b_n coefficients are zero. To show this we write the Fourier series for $f(t)$ in the form

$$f(t) = a_0 + \sum_{n=1}^{\infty}[a_n \cos(2\pi nf_0 t) + b_n \sin(2\pi nf_0 t)] \tag{13–10}$$

Given the Fourier series for $f(t)$, we use the identities $\cos(-x) = \cos(x)$ and $\sin(-x) = -\sin(x)$ to write the Fourier series for $f(-t)$ as follows:

$$f(-t) = a_0 + \sum_{n=1}^{\infty}[a_n \cos(2\pi nf_0 t) - b_n \sin(2\pi nf_0 t)] \tag{13–11}$$

For even symmetry $f(t) = f(-t)$ and the right sides of Eqs. (13–10) and (13–11) must be equal. Comparing the Fourier coefficients term by term, we find that $f(t) = f(-t)$ requires $b_n = -b_n$. The only way this can happen is for $b_n = 0$ for all n.

A waveform is said to have **odd symmetry** if $-f(-t) = f(t)$. The sine wave, square wave, and parabolic wave in Figure 13–4 are examples of waveforms with this type of symmetry. The Fourier series of odd waveforms are made up entirely of sine terms: that is, all of the a_n coefficients are zero. Given the Fourier series for $f(t)$ in Eq. (13–10), we use the identities $\cos(-x) = \cos(x)$ and $\sin(-x) = -\sin(x)$ to write the Fourier series for $-f(-t)$ in the form

$$-f(-t) = -a_0 + \sum_{n=1}^{\infty}[-a_n \cos(2\pi nf_0 t) + b_n \sin(2\pi nf_0 t)] \tag{13–12}$$

With odd symmetry $f(t) = -f(-t)$ and the right sides of Eq. (13–10) and (13–12) must be equal. Comparing the Fourier coefficients term by term, we find that odd symmetry requires $a_0 = -a_0$ and $a_n = -a_n$. The only way this can happen is for $a_n = 0$ for all n, including $n = 0$.

A waveform is said to have **half-wave symmetry** if $-f(t - T_0/2) = f(t)$. This requirement states that inverting the waveform $[-f(t)]$ and then time shifting by half a cycle $(T_0/2)$ must produce the same waveform. Basically, this means that successive half-cycles have the same waveshape but opposite polarities. In Figure 13–4 the sine wave, cosine wave, square wave, triangular wave, and parabolic wave have half-wave symmetry. The sawtooth wave, half-wave sine, rectangular pulse train, and full-wave sine do not have this symmetry.

With half-wave symmetry the amplitudes of all even harmonics are zero. To show this we use the identities $\cos(x - n\pi) = (-1)^n \cos(x)$ and $\sin(x - n\pi) = (-1)^n \sin(x)$ to write the Fourier series of $-f(t - T_0/2)$ in the form

$$-f(t - T_0/2) = -a_0 + \sum_{n=1}^{\infty}[-(-1)^n a_n \cos(2\pi n f_0 t) \\ - (-1)^n b_n \sin(2\pi n f_0 t)] \qquad (13\text{–}13)$$

For half-wave symmetry the right sides of Eqs. (13–10) and (13–13) must be equal. Comparing the coefficients term by term, we find that equality requires $a_0 = -a_0$, $a_n = -(-1)^n a_n$, and $b_n = -(-1)^n b_n$. The only way this can happen is for $a_0 = 0$ and for $a_n = b_n = 0$ when n is even. In other words, the only nonzero Fourier coefficients occur when n is odd.

A waveform may have more than one symmetry. For example, the triangular wave in Figure 13–4 has even symmetry and half-wave symmetry, while the square wave has both odd and half-wave symmetries. The sawtooth wave in Figure 13–4 is an example where an underlying odd symmetry is masked by a dc component. A symmetry that is not apparent until the dc component is removed is sometimes called a **hidden symmetry**.

Finally, whether a waveform has even or odd symmetry (or neither) depends on where we choose to define $t = 0$. For example, the triangular wave in Figure 13–4 has even symmetry because the $t = 0$ vertical axis is located at a local maximum. If the axis is shifted to a zero crossing, the waveform has odd symmetry and the cosine terms in the series are replaced by sine terms. If the vertical axis is shifted to a point between a zero cross and a maximum, then the resulting waveform is neither even nor odd and its Fourier series contains both sine and cosine terms.

EXAMPLE 13–5

Given that $f(t)$ is a square wave of amplitude A and period T_0, use the Fourier coefficients in Figure 13–4 to find the Fourier coefficients of $g(t) = f(t + T_0/4)$.

SOLUTION:
Figure 13–7 compares the square waves $f(t)$ and $g(t) = f(t + T_0/4)$. The square wave $f(t)$ has odd symmetry (sine terms only) and half-wave symmetry (odd harmonics only). Using the coefficients in Figure 13–4, the Fourier series for $f(t)$ is

$$f(t) = \sum \frac{4A}{n\pi} \sin(2\pi n t/T_0) \quad n \text{ odd}$$

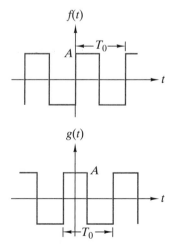

FIGURE 13–7

The Fourier series for $g(t) = f(t + T_0/4)$ can be written in the form

$$g(t) = f(t + T_0/4) = \sum \frac{4A}{n\pi} \sin[2\pi n(t + T_0/4)/T_0] \quad n \text{ odd}$$

$$= \sum \frac{4A}{n\pi} \sin(2\pi nt/T_0 + n\pi/2)$$

$$= \sum \frac{4A}{n\pi} \cos(2\pi nt/T_0) \sin(n\pi/2)$$

$$= \sum \frac{4A}{n\pi} \cos(2\pi nt/T_0)(-1)^{(n-1)/2} \quad n \text{ odd}$$

Figure 13–7 shows that $g(t)$ has even and half-wave symmetry, so its Fourier series has only cosine terms and odd harmonics. The Fourier coefficients for $g(t)$ are

$$a_0 = 0 \quad a_n = \begin{cases} 0 & n \text{ even} \\ \left[\dfrac{4A}{n\pi}\right](-1)^{(n-1)/2} & n \text{ odd} \end{cases}$$

$$b_n = 0 \quad \text{all } n$$

Shifting the time origin alters the even or odd symmetry properties of a periodic waveform because these symmetries depend on values of $f(t)$ on opposite sides of the vertical axis at $t = 0$. The half-wave symmetry of a waveform is not changed by time shifting because this symmetry only requires successive half-cycles to have the same form but opposite polarities. ∎

Exercise 13–5

(a) Identify the symmetries in the waveform $f(t)$ whose Fourier series is

$$f(t) = \frac{2\sqrt{3}A}{\pi}\left[\cos(\omega_0 t) - \frac{1}{5}\cos(5\,\omega_0 t) + \frac{1}{7}\cos(7\,\omega_0 t)\right.$$

$$\left. - \frac{1}{11}\cos(11\,\omega_0 t) + \frac{1}{13}\cos(13\,\omega_0 t) + \cdots\right]$$

(b) Write the corresponding terms of the function $g(t) = f(t - T_0/4)$.

Answers:

(a) Even symmetry; half–wave symmetry; zero average value.

(b)
$$g(t) = \frac{2\sqrt{3}A}{\pi}\left[\sin(\omega_0 t) - \frac{1}{5}\sin(5\,\omega_0 t) - \frac{1}{7}\sin(7\,\omega_0 t)\right.$$

$$\left. + \frac{1}{11}\sin(11\,\omega_0 t) + \frac{1}{13}\sin(13\,\omega_0 t) + \cdots\right]$$

13–4 CIRCUIT ANALYSIS USING THE FOURIER SERIES

Up to this point we have concentrated on finding the Fourier series description of periodic waveforms. We are now in a position to address circuit analysis problems of the type illustrated in Figure 13–8. This first-order *RL* circuit is driven by a periodic sawtooth voltage, and the objective is to find the steady-state current $i(t)$.

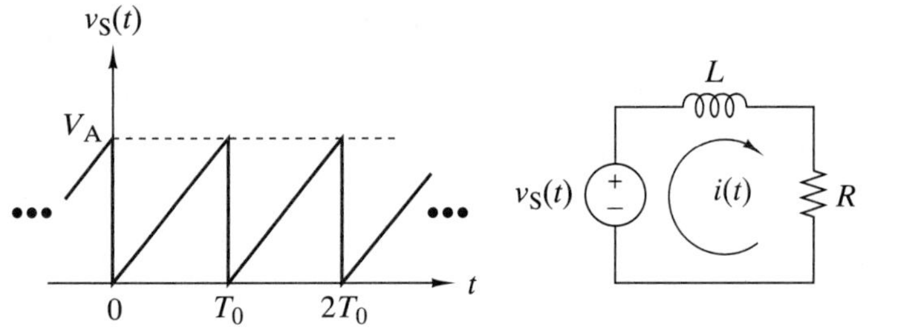

FIGURE 13–8 *Linear circuit with a periodic input.*

We begin by using the results in Example 13–4 to express the input voltage as a Fourier series in the form

$$v_s(t) = \underbrace{\frac{V_A}{2}}_{\text{dc}} + \underbrace{\sum_{n=1}^{\infty} \frac{V_A}{n\pi} \cos(n\omega_0 t + 90°)}_{\text{ac}} \qquad (13\text{–}14)$$

This result expresses the input driving force as the sum of a dc component plus ac components at harmonic frequencies $n\omega_0 = 2\pi n f_0$, $n = 1, 2, 3, \ldots$. Since the circuit is linear, we find the steady-state response caused by each component acting alone and then obtain the total response by superposition.

In the dc steady state the inductor acts like a short circuit, so the steady-state current due to the dc input $V_A/2$ is simply $I_0 = V_A/(2R)$. The amplitude and phase angle of the nth harmonic of the sawtooth input are

$$V_n = \frac{V_A}{n\pi} \quad \text{and} \quad \phi_n = 90°$$

In Chapter 11 we found that the sinusoidal steady-state response at frequency ω_A can be found directly from a network function $T(s)$ as

$$\text{Output amplitude} = \text{Input amplitude} \times |T(j\omega_A)|$$
$$\text{Output phase} = \text{Input phase} + T(j\omega_A)$$

In the present case the input is the nth harmonic at $\omega_A = n\omega_0$, the output is the current $I(s) = V(s)/Z(s)$, and hence the network function of interest is $T(s) = 1/(Ls + R)$. Applying the Chapter 11 method here yields the amplitude and phase of the nth harmonic current as

$$\text{Amplitude} = V_n \times \left| \frac{1}{jn\omega_0 L + R} \right| = \frac{V_A}{n\pi R} \frac{1}{\sqrt{1 + (n\omega_0 L/R)^2}}$$

$$\text{Phase} = \phi_n + \text{angle}\left(\frac{1}{jn\omega_0 L + R} \right) = 90° - \tan^{-1}(n\omega_0 L/R)$$

Defining $\theta_n = \tan^{-1}(n\omega_0 L/R)$, we get the waveform of the steady-state response to the nth harmonic input.

$$i_n(t) = \underbrace{\frac{V_A}{n\pi R} \frac{1}{\sqrt{1 + (n\omega_0 L/R)^2}}}_{\text{amplitude}} \cos(\underbrace{n\omega_0 t + 90° - \theta_n}_{\text{phase}})$$

We have now found the steady-state response of the circuit due to the dc component acting alone and the nth harmonic ac component acting alone. Since the circuit is linear, superposition applies and we find that the steady-state response caused the sawtooth input by summing the contributions of each of these sources:

$$i(t) = I_0 + \sum_{n=1}^{\infty} i_n(t)$$

$$= \frac{V_A}{2R} + \frac{V_A}{R} \sum_{n=1}^{\infty} \frac{1}{n\pi\sqrt{1 + (n\omega_0 L/R)^2}} \cos(n\omega_0 t + 90° - \theta_n) \qquad (13-15)$$

The Fourier series in Eq. (13–15) represents the steady-state current due to a sawtooth driving force whose Fourier series is in Eq. (13–14).

EXAMPLE 13 – 6

Find the first four nonzero terms of the Fourier series in Eq. (13–15) for $V_A = 25\,\text{V}, R = 50\,\Omega, L = 40\,\mu\text{H}, \omega_0 = 1\,\text{Mrad/s}$.

SOLUTION:
Equation (13–15) gives a Fourier series of the form

$$i(t) = I_0 + \sum_{n=1}^{\infty} I_n \cos(n\omega_0 t + \Psi_n)$$

where

$$I_0 = \frac{V_A}{2R} \quad I_n = \frac{V_A}{R} \frac{1}{n\pi\sqrt{1 + (n\omega_0 L/R)^2}} \quad \Psi_n = \frac{\pi}{2} - \tan^{-1}(n\omega_0 L/R)$$

Inserting the numerical values leads to

$$I_0 = \frac{1}{4} \quad I_n = \frac{1}{2n\pi\sqrt{1 + (n \times 0.8)^2}} \quad \Psi_n = \frac{\pi}{2} - \tan^{-1}(n \times 0.8)$$

The first four nonzero terms in the Fourier series include the dc component plus the first three harmonics. For $n = 0, 1, 2, 3$, we have

$$I_0 = 0.25$$
$$I_1 = 0.124 \quad \Psi_1 = 0.896 \quad \text{rad} = 51.3°$$
$$I_2 = 0.0422 \quad \Psi_2 = 0.559 \quad \text{rad} = 32.0°$$
$$I_3 = 0.0204 \quad \Psi_3 = 0.395 \quad \text{rad} = 22.6°$$

Hence, the desired expression is

$$i(t) = 0.25 + 0.124\cos(10^6 t + 51.3°) + 0.0422\cos(2 \times 10^6 t + 32°)$$
$$+ 0.0204\cos(3 \times 10^6 t + 22.6°) \cdots \qquad \blacksquare$$

Exercise 13–6

(a) Given the circuit in Figure 13–8, use Laplace-domain techniques to find the current $i(t)$ when the voltage source is a scaled ramp function $v_s(t) = (V_A/T_0)tu(t)$.
    (*Hint:* Recall the relationship between the impulse response, step response, and ramp response.)

(b) Using the same values as in Example 13–6, on the same axes plot the current $i(t)$ found above and the approximation for $i(t)$ found in Example 13–6 over one period.

Answers:

(a) $i(t) = \dfrac{V_A}{RT_0}\left[t - \dfrac{L}{R}\left(1 - e^{-Rt/L}\right)\right]u(t)$

(b) See Figure 13–9 for the plot.

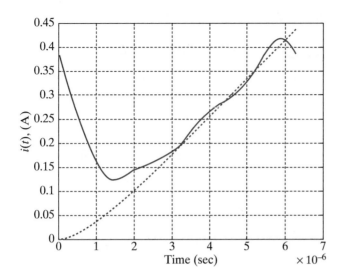

FIGURE 13–9

The dashed line is the ramp response and the solid line is the plot of the approximate response using the truncated series from Example 13–6.

The results in Example 13–6 and Exercise 13–6 require some additional scrutiny to appreciate fully the Fourier analysis.

First note that if $L = 0$, Eq. (13–15) reduces to

$$i(t) = \frac{V_A/R}{2} + \sum_{n=1}^{\infty} \frac{V_A/R}{n\pi} \cos(n\omega_0 t + 90°)$$

which is the Fourier series of a sawtooth wave of amplitude V_A/R. This makes sense because without the inductor the circuit in Figure 13–8 is a simple resistive circuit in which $i(t) = v_s(t)/R$, so the input and response must have the same waveform.

When $L \neq 0$ the response is not a sawtooth, but we can infer some features of its waveform if we examine the amplitude spectrum. At high frequency $[(n\omega_0 L/R) \gg 1]$ Eq. (13–15) points out that the amplitudes of the ac components are approximately

$$I_n \approx \frac{V_A}{R}\frac{1}{n^2\pi\omega_0 L/R} \tag{13–16}$$

In the steady-state response the amplitudes of the high-frequency ac components decrease as $1/n^2$, whereas the ac components in the input sawtooth decrease as $1/n$. In other words, the relative amplitudes of the high-frequency components are much smaller in the response than in the input. This makes sense because the inductor's impedance increases with frequency and thereby reduces the amplitudes of the high-frequency ac currents. We would expect the circuit to filter out the high-frequency components in the input and produce a response without the sharp corners and discontinuities in the input sawtooth.

This is exactly the response we see in Figure 13–9 with the plot of the truncated series from Example 13–6 (solid line). The plot approximates a rounded version of the sawtooth waveform, which indicates the high-frequency components have been removed. The high-frequency components are responsible for the sharp transitions, which appear in the original sawtooth waveform, but not in the waveform representing the steady-state current $i(t)$ in the circuit.

The next example demonstrates the fact that a square wave, for example, really consists of individual sinusoids that add together to form the periodic waveform. The subsequent example illustrates how the Fourier components of a waveform allow us to better understand the performance of a realistic circuit application.

◑ DESIGN EXAMPLE 13–7

Design a series RLC tuned filter to pass only the third harmonic of a 5-V 200 kHz square wave. Show that only the third harmonic is the dominant frequency at the output of the filter.

SOLUTION:
A 5-V 200 kHz square wave can be represented by the following Fourier series

$$v_1(t) = a_0 + \sum_{1}^{\infty} (a_n \cos 2\pi n f_0 t + b_n \sin 2\pi n f_0 t) \text{ V}$$

For a 5-V square wave with odd symmetry, Figure 13–4 indicates that

$$a_0 = 0, a_n = 0, b_n = \frac{4V_A}{n\pi} \text{ for } n = \text{odd and } b_n = 0 \text{ for } n = \text{even}$$

Substituting the values for V_A, f_0, and n, we can write the series of the first three terms as $v_1(t) = 6.366 \sin 1.256\text{M}t + 2.122 \sin 3.770\text{M}t + 1.273 \sin 6.283\text{M}t + \ldots$ V

A properly designed tuned filter would allow the third harmonic to exit the filter with no attenuation while severely attenuating the other frequencies. Higher harmonics will be increasingly attenuated since they are naturally weaker (magnitude divided by n) and the filter attenuates these frequencies more as they are further removed from the tuned frequency. The transfer function of a series RLC tuned filter is

$$T(s) = \frac{\frac{R}{L}s}{s^2 + \frac{R}{L}s + \frac{1}{LC}}$$

We know we want the center frequency (ω_0) to be equal to the frequency of the third harmonic, namely 3.770 Mrad/s. For a series RLC circuit, the center frequency is given as $\omega_0 = 1/\sqrt{LC}$. In order to select out just the third harmonic we will need a modestly high Q. Again for a series RLC tuned filter, $Q = \frac{\sqrt{L/C}}{R}$. Since L and C will be selected to meet the center or resonant frequency, R will be chosen to create the Q. Selecting $L = 1$ mH, a standard value, results in a $C = 70.362$ pF, a value obtained using a standard 50-100 pF trimmer capacitor. A standard value R of 100 Ω yields a Q of 37.7, a reasonably high Q. This circuit is shown in Figure 13–10(a).

The resulting transfer function is

$$T(s) = \frac{10^5 s}{s^2 + 10^5 s + 1.421 \times 10^{13}}$$

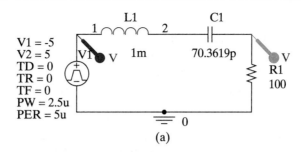

(a)

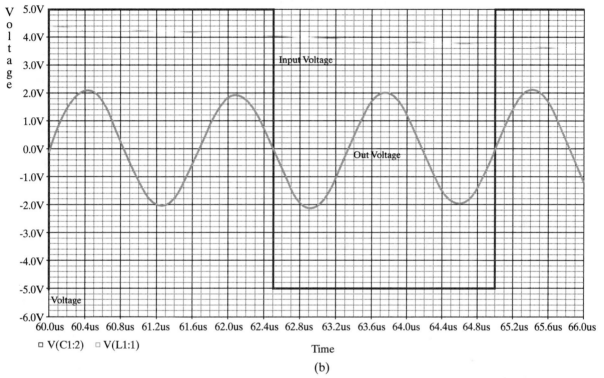

□ V(C1:2)   □ V(L1:1)

Time

(b)

FIGURE 13–10

We will now calculate the resulting magnitudes of the sinusoidal steady state output of the fundamental, and the third and fifth harmonics.

$$|T(j1.256M)| = \frac{10^5 \times 1.256 \times 10^6}{\sqrt{(1.421 \times 10^{13} - (1.256 \times 10^6)^2)^2 + (10^5 \times 1.256 \times 10^6)^2}}$$

$$|T(j1.256M)| = 9.942 \times 10^{-3}$$

$$|T(j3.770M)| = \frac{10^5 \times 3.770 \times 10^6}{\sqrt{(1.421 \times 10^{13} - (3.770 \times 10^6)^2)^2 + (10^5 \times 3.770 \times 10^6)^2}}$$

$$|T(j3.770M)| = 0.999$$

$$|T(j6.283M)| = \frac{10^5 \times 6.283 \times 10^6}{\sqrt{(1.421 \times 10^{13} - (6.283 \times 10^6)^2)^2 + (10^5 \times 6.283 \times 10^6)^2}}$$

$$|T(j6.283M)| = 24.86 \times 10^{-3}$$

Our output becomes

$$v_2(t) = 0.0632 \sin 1.256 \, \mathrm{M}t + 2.119 \sin 3.770 \mathrm{M}t + 0.0316 \sin 6.283 \, \mathrm{M}t + \cdots \mathrm{V}$$

This output and the input are shown as an OrCAD simulation in Figure 13–10(b). The fundamental has been attenuated by 40 dB, while the fifth harmonic has been attenuated by 32 dB. If we were to increase the Q, we could be more selective; however, a higher Q value requires more accurate values for L and C. While this is not a problem in a simulation, this becomes a challenge in actually picking values of real components and would entail a lot of "tweaking" to get the resonant frequency to fall on the desired harmonic in the laboratory. ∎

◑ Design Exercise 13–7

Design a first-order low pass filter to allow only the fundamental of the square wave of Design Example 13–7 to pass with an attenuation of ≤ 3 dB. Verify your results by calculating the magnitude of the fundamental and of the third harmonic.

Answers: Select $\omega_C = 1.256$ Mrad/s, the frequency of the fundamental. Use a series RC circuit with the output taken across C. Select $R = 1$ kΩ, $C = 796$ pF. Magnitude of fundamental = 4.50 V (-3 dB). Magnitude of third harmonic = 0.671 V (-10 dB). Note: A first-order filter has a roll-off of -20 dB per decade. A third harmonic is only half a decade on a log scale. If better isolation is desired, then a higher-order filter is needed, which is the topic of Chapter 14.

APPLICATION EXAMPLE 13–8

Figure 13–11 shows a block diagram of a dc power supply. The ac input is a sinusoid that is converted to a full-wave sine by the rectifier. The filter passes the dc component in the rectified sine and suppresses the ac components. The result is an output consisting of a small residual ac ripple riding on top of a much larger dc signal.

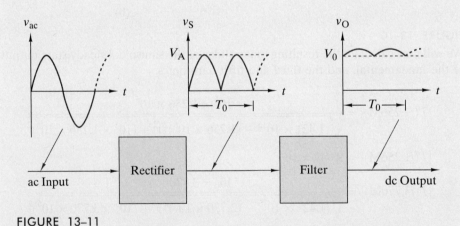

FIGURE 13–11

Calculate and plot the first 10 harmonics in amplitude spectra of the filter input and output for $V_A = 23.6$ V, $T_0 = 1/60$ s, and a low-pass filter transfer function of

$$T(s) = \frac{(200)^2}{s^2 + 70s + (200)^2}$$

SOLUTION:
The amplitude spectrum of the filter input is obtained using the Fourier coefficients for the full-wave rectified sine in Figure 13–4:

$$a_0 = 2V_A/\pi = 15.02\ \text{V} \quad a_n = \begin{cases} 0 & n\ \text{odd} \\ \left| \dfrac{4V_A/\pi}{1 - n^2} \right| = \dfrac{30.04}{n^2 - 1} & n\ \text{even} \end{cases} \tag{13–17}$$

The magnitude of the transfer function at each of these discrete frequencies is

$$|T(jn\omega_0)| = \frac{(200)^2}{\sqrt{[(200)^2 - (n\omega_0)^2]^2 + (70\,n\omega_0)^2}} \tag{13–18}$$

To obtain the specified output spectrum, we must generate the product of the input amplitude times the transfer function magnitude for $n = 0, 1, 2, 3, \ldots, 10$.

Spreadsheets are ideally suited for making repetitive calculations of this type. Figure 13–12 shows an Excel spreadsheet that implements the required calculations. Column A gives the index n and Column B gives the corresponding frequencies in Hz. Column C calculates the frequencies in rad/s. Column D calculates the input amplitudes using Eq. (13–17). Column E calculates the magnitude of the transfer function at the frequency in question—the Excel equation for the calculation in Eq. (13–18) is shown in the f_x window above the column designators. Finally, Column F computes the output amplitudes by taking the product of the input amplitude and the magnitude of the transfer function for each frequency. Excel then is used to produce the bar graph shown. Since the low-pass filter has unity gain at zero frequency, the dc components in the input and output are equal. The second harmonic at 120 Hz is the first non-zero component at the output. It has a 10-V input amplitude but only a 754-mV output amplitude. By the fourth harmonic at 240 Hz, the 2-V input amplitude has been reduced to less than 36 mV. This and subsequent harmonics are entirely negligible. ■

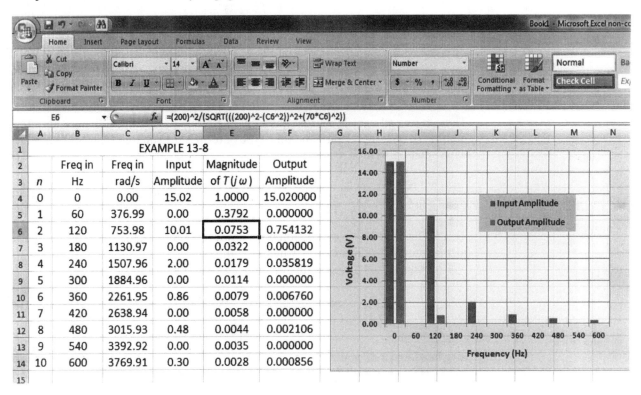

FIGURE 13–12

Exercise 13–8

Derive an expression for the first three nonzero terms in the Fourier series of the steady–state output voltage in Example 13–8.

Answer: $v_O(t) = 15.02 + 0.758 \cos(2\pi\,120t + 5.7°) + 0.036 \cos(2\pi\,240t + 2.7°)$ V

13–5 RMS VALUE AND AVERAGE POWER

In Chapter 5 we introduced the rms value of a periodic waveform as a descriptor of the average power carried by a signal. In this section we relate the rms value of the waveform to the amplitudes of the dc and ac components in its Fourier series. The rms value of a periodic waveform is defined as

$$F_{\text{rms}} = \sqrt{\frac{1}{T_0}\int_0^{T_0} [f(t)]^2 dt}$$ (13–19)

The waveform $f(t)$ can be expressed as a Fourier series in the amplitude and phase form as

$$f(t) = A_0 + \sum_{n=1}^{\infty} A_n \cos(n\omega_0 t + \phi_n)$$

Substituting this expression into Eq. (13–19), we can write F_{rms}^2 as

$$F_{\text{rms}}^2 = \frac{1}{T_0}\int_0^{T_0}\left[A_0 + \sum_{n=1}^{\infty} A_n \cos(n\omega_0 t + \phi_n)\right]^2 dt$$ (13–20)

Squaring and expanding the integrand on the right side of this equation produces three types of terms. The first is the square of the dc component:

$$\frac{1}{T_0}\int_0^{T_0} [A_0]^2 dt = A_0^2$$ (13–21)

The second is the cross product of the dc and ac components, which takes the form

$$\frac{1}{T_0}\sum_{n=1}^{\infty} 2A_0 \int_0^{T_0} A_n \cos(n\omega_0 t + \phi_n)dt = 0$$ (13–22)

These terms all vanish because they involve integrals of sinusoids over an integer number of cycles. The third and final type of term is the square of the ac components, which can be written as

$$\frac{1}{T_0}\sum_{n=1}^{\infty}\sum_{m=1}^{\infty} \int_0^{T_0} A_n \cos(n\omega_0 t + \phi_n)A_m \cos(m\omega_0 t + \phi_m)dt = \frac{1}{2}\sum_{n=1}^{\infty} A_n^2$$ (13–23)

This rather formidable expression boils down to a simple sum of squares because all of the integrals vanish except when $m = n$.

Combining Eqs. (13–19) through (13–23), we obtain the rms value as

$$F_{\text{rms}} = \sqrt{A_0^2 + \sum_{n=1}^{\infty} \frac{A_n^2}{2}}$$

$$= \sqrt{A_0^2 + \sum_{n=1}^{\infty} \left(\frac{A_n}{\sqrt{2}}\right)^2}$$ (13–24)

Since the rms value of a sinusoid of amplitude A is $A/\sqrt{2}$, we conclude that

The rms value of a periodic waveform is equal to the square root of the sum of the square of the dc value and the square of the rms value of each of the ac components.

In Chapter 5 we found that the average power delivered to a resistor is related to its rms voltage or current as

$$P = \frac{V_{rms}^2}{R} = I_{rms}^2 R$$

Combining these expressions with the result in Eq. (13–24), we can write the average power delivered by a periodic waveform in terms of the average power delivered by each of its Fourier components:

$$P = \frac{V_0^2}{R} + \sum_{n=1}^{\infty} \frac{V_n^2}{2R} = I_0^2 R + \sum_{n=1}^{\infty} \frac{I_n^2}{2} R$$

$$= P_0 + \sum_{n=1}^{\infty} P_n \tag{13–25}$$

where P_0 is the average power delivered by the dc component and P_n is the average power delivered by the nth ac component. This additive feature is important because it means we can find the total average power by adding the average power carried by the dc component to that carried by each of the ac components.

Caution: In general, we cannot find the total power by adding the power delivered by each component acting alone because the superposition principle does not apply to power. However, the average power carried by harmonic sinusoids is additive because they belong to a special class call orthogonal signals.

EXAMPLE 13–9

Derive an expression for the average power delivered to a resistor by a sawtooth voltage of amplitude V_A and period T_0. Then calculate the fraction of the average power carried by the dc component plus the first three ac components.

SOLUTION:
An equation for the sawtooth voltage is $v(t) = V_A t/T_0$ for the range $0 < t < T_0$. The square of the rms value of a sawtooth is

$$V_{rms}^2 = \frac{1}{T_0} \int_0^{T_0} \left(\frac{V_A t}{T_0}\right)^2 dt = V_A^2 \left[\frac{t^3}{3T_0^3}\right]_0^{T_0} = \frac{V_A^2}{3}$$

The average power delivered to a resistor is

$$P = \frac{V_{rms}^2}{R} = \frac{V_A^2}{3R} = 0.333 \frac{V_A^2}{R}$$

This result is obtained directly from the sawtooth waveform without having to sum an infinite series. The same answer could be obtained by summing an infinite series. The question this example asks is, how much of the average power is carried by the first four components in the Fourier series of the sawtooth wave? The amplitude spectrum of the sawtooth wave (see Example 13–4) is

$$A_n = \begin{cases} \dfrac{V_A}{2} & n = 0 \\[2mm] \dfrac{V_A}{n\pi} & n > 0 \end{cases}$$

From Eq. (13–25), the average power in terms of amplitude spectrum is

$$P = \frac{(V_A/2)^2}{R} + \sum_{n=1}^{\infty} \frac{(V_2/n\pi)^2}{2R}$$

which can be arranged in the form

$$P = \frac{V_A^2}{R} \left\{ \underbrace{\underbrace{\frac{1}{(2)^2} + \frac{1}{2(\pi)^2} + \frac{1}{2(2\pi)^2} + \frac{1}{2(3\pi)^2}}_{0.319(96\%)} + \frac{1}{2(4\pi)^2} + \cdots}_{0.333} \right\}$$

The infinite series within the braces sums to 0.333 and matches the average power we calculated directly from the waveform itself. The dc component plus the first three ac terms contribute 0.319 to the infinite sum. In other words, these four components alone deliver 96% of the average power carried by the sawtooth wave.   ■

Exercise 13–9

The full–wave rectified sine wave shown in Figure 13–4 has an rms value of $A/\sqrt{2}$. What fraction of the average power that the waveform delivers to a resistor is carried by the first two nonzero terms in its Fourier series?

Answer: Fraction $= 88/9\pi^2$ or 99.07%

EXAMPLE 13 – 10

Figure 13–8 shows a series RL circuit driven by a sawtooth voltage source. Estimate the average power delivered to the resistor for $V_A = 25\,\text{V}, R = 50, \Omega, L = 40\,\mu\text{H}$, $T_0 = 5\,\mu\text{s}$, and $\omega_0 = 2\pi/T_0 = 1.26\,\text{Mrad/s}$.

SOLUTION:
The Fourier series of the current in a series RL circuit is given in Eq. (13–15) as

$$i(t) = I_0 + \sum_{n=1}^{\infty} I_n \cos(n\omega_0 t + \Psi_n)$$

where

$$I_0 = \frac{V_A}{2R} \quad I_n = \frac{V_A}{R} \frac{1}{n\pi\sqrt{1 + (n\omega_0 L/R)^2}}$$

We cannot directly calculate the rms value of this current without a closed form expression for its waveform. However, we can get an estimate of the average power from the Fourier series. In terms of its Fourier series, the average power delivered by the current is

$$P = I_0^2 R + \sum_{n=1}^{\infty} \frac{I_n^2}{2} R$$

When $n\omega_0 L/R \gg 1$, the amplitude I_n decreases as $1/n^2$ and its contribution to the average power falls off as $1/n^4$. In other words, the infinite series for the average power converges very rapidly. To show how rapidly the series converges, we calculate the first few terms using the specified numerical values

$$P = I_0^2 R \quad + \frac{I_1^2}{2} R \quad + \frac{I_2^2}{2} R \quad + \frac{I_3^2}{2} R \quad + \cdots$$

$$= 3.125 + 0.315 + 0.0314 + 0.00697 + \cdots$$

$$\longleftarrow 3.44 \longrightarrow |$$
$$\longleftarrow\!\!\!\!\!\!\!\!\!\!\!\!\!\!\!\! 3.471 \longrightarrow$$
$$\longleftarrow\! 3.478 \longrightarrow |$$

These results indicate that $P = 3.48$ W is a reasonable estimate of the average power, an estimate obtained using only four terms in the infinite series. The important point is that the high-frequency ac components are not important contributors to the average power carried by a signal. ∎

Exercise 13–10

In the rectangular-pulse waveform shown in Figure 13–4, the width of the pulse is one-third the period, $T = T_0/3$. The waveform is to pass through a low-pass filter and then through a resistive load. The load must receive at least 97% of the average power in the original waveform. Determine the minimum value of N such that if the filter passes components $V_0, V_1, V_2, \ldots, V_N$, the load will receive the required amount of power.
(*Hint:* You may want to use MATLAB to perform the calculations.)

Answer: $N = 10$.

13–6 FOURIER TRANSFORMS

We have examined Fourier series as a tool to resolve a periodic waveform into an infinite series of sinusoids at discrete harmonic frequencies nf_0, $n = 1, 2, \ldots, \infty$. The Fourier series allows us to construct the line spectrum of a periodic waveform from the distribution of amplitudes and phase angles across the harmonic frequencies. Since not all signals are periodic, we now extend the idea of a frequency spectrum to accommodate aperiodic waveforms that occur only once. The Fourier transform of an aperiodic signal is analogous to the Fourier series representation of a periodic waveform. Both representations involve infinitely many frequencies. The difference is that the frequencies in a Fourier series are discrete multiples of f_0, whereas the frequencies in a Fourier transform span a continuum of values.

DEFINITION OF FOURIER TRANSFORMS

Fourier transforms are integral transforms like Laplace transforms. Like Laplace transforms, the Fourier transforms involve two domains: (1) the time domain in which the signal is represented by its *waveform f(t)*, and (2) the frequency domain in which the signal is characterized by its Fourier *transform F(ω)*. In general $F(\omega)$ is a complex-valued function of the *real frequency variable* ω. The magnitude and phase angle of $F(\omega)$ are called the *amplitude spectrum* and *phase spectrum*, respectively.

The Fourier transform of a waveform $f(t)$ is defined by the integral

$$F(\omega) = \int_{-\infty}^{\infty} f(t)e^{-j\omega t}dt \qquad (13\text{--}26)$$

As you would expect from our study of Laplace transforms, there is an inverse procedure which converts a Fourier transform into a time-domain waveform. Inverse Fourier transforms are defined by the integral

$$f(t) = \frac{1}{2\pi} \int_{-\infty}^{\infty} F(\omega)e^{j\omega t} d\omega \qquad (13\text{--}27)$$

which yields $f(t)$ when given $F(\omega)$. The waveform $f(t)$ and transform $F(\omega)$ comprise a **Fourier transform pair**. We say that $F(\omega)$ is the Fourier transform of $f(t)$, and conversely, $f(t)$ is the inverse transform of $F(\omega)$. The short-hand notation $F(\omega) = \mathscr{F}\{f(t)\}$ and $f(t) = \mathscr{F}^{-1}\{F(\omega)\}$ is used to denote these operations.

The limits on these integrals indicate important features of Fourier transforms. The direct transform in Eq. (13–26) is a time integration across the interval $-\infty < t < \infty$. This points out that Fourier transforms exist for either causal or noncausal waveforms.[2] The inverse transform in Eq. (13–27) is an integration on the real frequency variable ω across the range $-\infty < \omega < \infty$. This points out that ω is a continuous variable that takes on positive and negative values. Thus, Fourier transforms have spectral content across a continuous range of frequencies, whereas the spectral content of a Fourier series exists only at isolated, harmonic frequencies.

The infinite limits in Eq. (13–26) should also alert the reader to the question of convergence and existence of $F(\omega)$. Existence is assured by the **Dirichlet conditions**, which require that the waveform $f(t)$ have a finite number of discontinuities and be absolutely integrable, that is,

$$\int_{-\infty}^{\infty} |f(t)| dt < \infty \qquad (13\text{--}28)$$

A number of useful waveforms meet the Dirichlet conditions, for example, the rectangular pulse $[u(t) - u(t - T)](|T| < \infty)$ and the causal exponential $u(t)e^{-\alpha t}(\alpha > 0)$. While the Dirichlet conditions are mathematically sufficient, they are not necessary *and* sufficient—there being many useful signals with Fourier transforms that fail to meet them.

Finally, whenever $F(\omega)$ exists there is a unique, one-to-one correspondence between a waveform $f(t)$ and its Fourier transform $F(\omega)$. Uniqueness means that a transform pair can be used to go from waveform to transform or transform to waveform. Once we have developed a basic set of pairs, we will summarize our results in a table of Fourier transform pairs similar to our table of Laplace transform pairs.

EXAMPLE 13–11

Find the Fourier transform of the rectangular pulse in Figure 13–13 and plot its amplitude spectrum.

SOLUTION:
The specified waveform meets the Dirichlet conditions since it is clearly absolutely integrable and has only two isolated discontinuities. For this waveform the

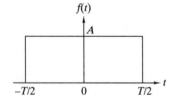

FIGURE 13–13

integration limits in Eq. (13–26) are $-T/2$ and $+T/2$ since $f(t)$ is zero everywhere outside this range. As a result, $F(\omega)$ is found as

$$
\begin{aligned}
F(\omega) &= \int_{-T/2}^{T/2} Ae^{-j\omega t}\,dt = -\frac{A}{j\omega}e^{-j\omega t}\bigg|_{-T/2}^{T/2} \\
&= \frac{A}{\omega/2}\left[\frac{e^{j\omega T/2}-e^{-j\omega T/2}}{j2}\right] \\
&= AT\frac{\sin(\omega T/2)}{\omega T/2}
\end{aligned}
$$

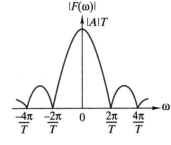

FIGURE 13–14

Figure 13–14 shows a plot of the amplitude spectrum $|F(\omega)|$ versus ω.[3] The spectrum of a rectangular pulse is a continuous function of a continuous frequency variable. There is some spectral content in every nonzero frequency range no matter how small. The isolated nulls in the spectral plot occur at the discrete frequencies $\omega = 2k\pi/T\,(k = \pm1, \pm2, \ldots)$ where $\sin(\omega T/2) = 0$. ∎

Exercise 13–11

Use MATLAB to compute the Fourier transform of the rectangular pulse in Figure 13–13. Confirm the results of Example 13–11.

Answer:

The required MATLAB code is

```
% Create the symbolic variables
syms t A T w
% Compute the Fourier transform with two techniques
Fw1 = int(A*exp(-j*w*t),t,-T/2,T/2)
Fw2 = fourier(A*(heaviside(t+T/2)-heaviside(t-T/2)),t,w)
```

Both options for calculating the Fourier transform yield answers that agree with the results in Example 13–11.

EXAMPLE 13–12

Find the Fourier transform of the causal exponential $f(t) = Ae^{-\alpha t}u(t)$ and plot its amplitude and phase spectra. Assume $\alpha > 0$.

SOLUTION:
This waveform also meets the Dirichlet conditions since it has only one isolated discontinuity and is absolutely integrable in the sense of Eq. (13–28). The waveform is causal, so the integration extends from $t = 0$ to $t = +\infty$.

$$
\begin{aligned}
F(\omega) &= \int_0^\infty Ae^{-\alpha t}e^{-j\omega t}\,dt = A\int_0^\infty e^{-(\alpha+j\omega)t}\,dt \\
&= -A\frac{e^{-(\alpha+j\omega)t}}{\alpha+j\omega}\bigg|_0^\infty
\end{aligned}
$$

---

[3]The transform in this example involves a function usually denoted as $\mathrm{sinc}(x) = \sin x/x$. At first glance it may appear that this function blows up at $x = 0$. However, a single application of l'Hôpital's rule shows that $\mathrm{sinc}(0) = 1$.

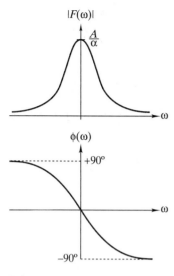

FIGURE 13–15

For $\alpha > 0$ the integral vanishes at the upper limit and $F(\omega)$ becomes

$$F(\omega) = \frac{A}{\alpha + j\omega} \quad (\alpha > 0)$$

The amplitude and phase spectra of $F(\omega)$ are

$$|F(\omega)| = \frac{|A|}{\sqrt{\alpha^2 + \omega^2}} \quad \text{(Amplitude)}$$

and

$$\phi(\omega) = F(\omega) = -\tan^{-1}(\omega/\alpha) \quad \text{(Phase)}$$

Figure 13–15 shows plots of the amplitude and phase spectra. The amplitude spectrum of this aperiodic waveform is a continuous function of a continuous frequency variable. There is some spectral content at every frequency, including negative frequencies. ■

Exercise 13–12

Find the Fourier transform of the waveform in Figure 13–16.

Answer:

$$F(\omega) = -jAT\frac{1 - \cos(\omega T/2)}{\omega T/2}$$

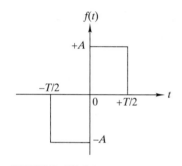

FIGURE 13–16

INVERSE FOURIER TRANSFORMS

The formal inversion integral for Laplace transforms is difficult to use. In contrast, the inversion integral for Fourier transforms is comparatively easy to use. As an example, consider the transform

$$F(\omega) = 2\pi\delta(\omega - \beta) \tag{13–29}$$

The transform is a frequency-domain impulse located at $\omega = \beta$ with an area of 2π. The inverse transform is found by applying the inversion integral in Eq. (13–27). Because $F(\omega)$ is zero everywhere except at $\omega = \beta$, the integration limits need only span the range from $\omega = \beta-$ to $\omega = \beta+$, just slightly above and below the frequency $\omega = \beta$.

$$f(t) = \frac{1}{2\pi}\int_{\beta-}^{\beta+} 2\pi\delta(\omega - \beta)e^{j\omega t}d\omega$$

$$= e^{j\beta t}\int_{\beta-}^{\beta+} \delta(\omega - \beta)d\omega = e^{j\beta t} \tag{13–30}$$

Straightforward use of the inversion integral yields $\mathscr{F}^{-1}\{2\pi\delta(\omega - \beta)\} = e^{j\beta t}$. Because the Fourier transformation is unique, we also know that $\mathscr{F}\{e^{j\beta t}\} = 2\pi\delta(\omega - \beta)$.

In particular, if $\beta = 0$, then $f(t) = e^0 = 1$ for all time t and $F(\omega) = 2\pi\delta(\omega)$. In other words, the Fourier transform of a dc waveform with amplitude $A = 1$ is an impulse of area 2π at $\omega = 0$. Figure 13–17 shows the waveform and Fourier transform of a dc signal. Note that the dc waveform is noncausal so it cannot be represented by a Laplace transform. Likewise, the dc waveform does not meet the Dirichlet conditions because it is not absolutely integrable.

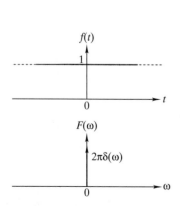

FIGURE 13–17

EXAMPLE 13–13

Use the inversion integral to find the waveform corresponding to the transform

$$F(\omega) = \frac{\pi A}{\beta}[u(\omega + \beta) - u(\omega - \beta)]$$

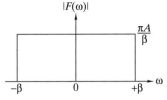

SOLUTION:
For this transform the integration limits on the inversion integral run from $\omega = -\beta$ to $\omega = +\beta$ because $F(\omega)$ is zero everywhere except in this range. Hence the inverse transform is

$$f(t) = \frac{1}{2\pi}\int_{-\beta}^{\beta}\pi\frac{A}{\beta}e^{j\omega t}d\omega = \frac{A}{2\beta}\frac{e^{j\omega t}}{jt}\Big|_{-\beta}^{\beta}$$

$$= \frac{A}{\beta t}\frac{e^{j\beta t} - e^{-j\beta t}}{j2} = A\frac{\sin(\beta t)}{\beta t}$$

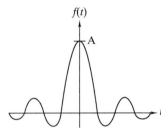

FIGURE 13–18

Figure 13–18 shows the given amplitude spectrum $|F(\omega)|$ and the resulting inverse transform $f(t)$. The waveform $f(t) = A\sin(\beta t)/(\beta t)$ is another example of the function $\mathrm{sinc}(x) = \sin(x)/x$, which plays an important role in Fourier analysis. Note that $f(t)$ is a noncausal waveform for which there is no Laplace transform. ■

Exercise 13–13

Use the inversion integral to find the inverse transform of $F(\omega) = j\omega\pi[u(\omega + 1) - u(\omega - 1)]$.

Answer:
$$f(t) = \frac{\cos t}{t} - \frac{\sin t}{t^2}$$

LAPLACE TRANSFORMS AND FOURIER TRANSFORMS

For certain signals there is a simple relationship between the Laplace and Fourier transforms. In Chapter 9 the integral definition of Laplace transforms is given as

$$\mathscr{L}\{f(t)\} = \int_0^\infty f(t)e^{-st}dt = F(s) \tag{13–31}$$

where $s = \sigma + j\omega$ is the complex frequency variable. The lower limit of this integration reminds us that $f(t)$ must be causal for a unique transform pair to exist. If $f(t)$ is absolutely integrable in the sense defined in Eq. (13–28), then the Laplace transform integration converges when $\sigma = 0$, in which case Eq. (13–31) becomes

$$\mathscr{L}\{f(t)\} = \int_0^\infty f(t)e^{-j\omega t}dt = F(s)|_{\sigma = 0} \tag{13–32}$$

On the other hand, when $f(t)$ is causal and absolutely integrable, its Fourier transform exists and is found from Eq. (13–26) to be

$$\mathscr{F}\{f(t)\} = \int_0^\infty f(t)e^{-j\omega t}dt = F(\omega) \tag{13–33}$$

Comparing Eqs. (13–32) and (13–33), we conclude that

$$F(\omega) = F(s)|_{\sigma=0} \tag{13–34}$$

provided that $f(t)$ is causal and absolutely integrable.

For $f(t)$ to be absolutely integrable it must have finite duration or decay to zero as $t \to \infty$. A sufficient condition for $f(t)$ to decay to zero is that all of the poles of $F(s)$ lie in the left half plane. For example, the Laplace transform of the causal exponential is $\mathscr{L}\{Ae^{-\alpha t}u(t)\} = A/(s+\alpha)$. The transform has a pole at $s = -\alpha$, which lies in the left half of the s plane for $\alpha > 0$. When $\alpha > 0$, the Fourier transform of the causal exponential is found from its Laplace transform to be

$$F(\omega) = F(s)|_{\sigma=0} = \frac{A}{j\omega + \alpha}$$

The $F(\omega)$ obtained here using Laplace transforms agrees with Example 13–12, where the integral definition of Fourier transforms gave the same result.

We can use Eq. (13–34) to find $F(\omega)$ provided the poles of $F(s)$ avoid both the right half plane and the j-axis boundary. For example, the Fourier transform of a causal sinusoid cannot be found from its Laplace transform. The reason is that the transform

$$F(s) = \mathscr{L}\{u(t)\sin(\beta t)\} = \beta/(s^2 + \beta^2)$$

has poles on the j axis at $s = \pm j\beta$.

The Laplace transform method can be used to find the Fourier transform of the unit impulse because $\delta(t)$ is causal and absolutely integrable. Applying Eq. (13–34), we obtain

$$\mathscr{F}\{\delta(t)\} = \mathscr{L}\{\delta(t)\}|_{\sigma=0} = 1 \tag{13–35}$$

Figure 13–19 shows the impulse waveform $\delta(t)$ and its constant-amplitude spectrum $|\Delta(\omega)|$. Note the duality between Figures 13–19 and 13–17. Figure 13–17 shows that a constant time-domain waveform leads to an impulse in the frequency domain. The dual result in Figure 13–19 shows that an impulse in the time domain leads to a constant in the frequency domain.

Laplace transform concepts can be used to find inverse Fourier transforms as well. Given a Fourier transform $F(\omega)$, we form a Laplace transform $F(s)$ by replacing $j\omega$ by s, or equivalently, replacing ω by $-js$. If the poles of $F(s)$ all lie in the left half of the s plane, then by the uniqueness property of Fourier and Laplace transforms, we have

$$f(t) = \mathscr{F}^{-1}\{F(\omega)\} = \mathscr{L}^{-1}\{F(s)\} \quad (t > 0) \tag{13–36}$$

Partial fraction expansion can be used to find the inverse transform of $F(s)$ in this equation. However, keep in mind that the Laplace transform method requires either (1) $f(t)$ to be causal and absolutely integrable, or (2) all of the poles of $F(s)$ to be in the left half of the s plane.

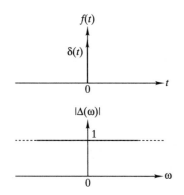

FIGURE 13–19 *Waveform and Fourier transform of the unit impulse signal.*

Exercise 13–14

Use Laplace transforms to find the Fourier transforms of the following causal waveforms. Assume that $\alpha > 0$.

(a) $f(t) = A[e^{-\alpha t}\sin\beta t]u(t)$

(b) $f(t) = A[\alpha t e^{-\alpha t}]u(t)$

Answers:

$$\text{(a)} \quad F(\omega) = \frac{A\beta}{(j\omega + \alpha)^2 + \beta^2}$$

$$\text{(b)} \quad F(\omega) = \frac{A\alpha}{(j\omega + \alpha)^2}$$

EXAMPLE 13–14

Use Laplace transforms to find the inverse Fourier transform of

$$F(\omega) = \frac{10}{(j\omega + 2)(j\omega + 4)}$$

SOLUTION:
Replacing $j\omega$ by s yields

$$F(s) = \frac{10}{(s + 2)(s + 4)}$$

$F(s)$ is a rational function of s with poles at $s = -2$ and $s = -4$. Both poles are in the left half plane, so the inverse transform of $F(s)$ is the inverse transform of $F(\omega)$. Expanding $F(s)$ by partial fractions yields

$$F(s) = \frac{k_1}{s + 2} + \frac{k_2}{s + 4}$$

The poles are simple so the residues k_1 and k_2 are

$$k_1 = (s + 2)F(s)|_{s=-2} = 5$$
$$k_2 = (s + 4)F(s)|_{s=-4} = -5$$

and the inverse transform of $F(\omega)$ is

$$\mathcal{F}^{-1}\{F(\omega)\} = \mathcal{L}^{-1}\{F(s)\} = \mathcal{L}^{-1}\left\{\frac{5}{s + 2}\right\} + \mathcal{L}^{-1}\left\{\frac{-5}{s + 4}\right\}$$
$$= [5e^{-2t} - 5e^{-4t}]u(t) \quad\blacksquare$$

Exercise 13–15

Use the Laplace transform method to find the causal waveform corresponding to the following Fourier transform:

$$F(\omega) = \frac{5(j\omega - 3)}{(j\omega + 1)(j\omega + 5)}$$

Answer:

$$f(t) = [-5e^{-t} + 10e^{-5t}]u(t)$$

Exercise 13–16

Explain why Laplace transforms cannot be used to find the Fourier transforms of the following waveforms. Assume $\alpha > 0$.

(a) $f(t) = A\alpha t u(t)$
(b) $f(t) = Ae^{-\alpha|t|}$
(c) $f(t) = A \sin \alpha t$

Answers:

(a) $f(t)$ is not absolutely integrable or, equivalently, $F(s)$ has a double pole at $s = 0$.
(b), (c) $f(t)$ is not causal, or equivalently (b) has a pole in the RHP, and (c) has poles on the $j\omega$ axis.

Exercise 13–17

Explain why Laplace transforms cannot be used to find the inverse transforms of the following functions. Assume $\alpha > 0$.

$$\text{(a)} \quad F(\omega) = \frac{2\alpha}{\alpha^2 + \omega^2}$$

$$\text{(b)} \quad F(\omega) = e^{-\alpha|\omega|}$$

Answers:

(a) $F(s)$ has a right half plane pole at $s = +\alpha$.
(b) $F(s)$ is not a rational function.

BASIC FOURIER TRANSFORM PROPERTIES AND PAIRS

We will introduce eight properties of the Fourier transformation, and use these properties to derive additional transform pairs.

Linearity

Both Laplace and Fourier transforms are linear transformations, which means that the transform of a sum of waveforms is the sum of their transforms. Stated formally, the linearity property is

$$\mathcal{F}\{Af_1(t) + Bf_2(t)\} = AF_1(\omega) + BF_2(\omega) \tag{13–37}$$

where A and B are constants. Proof of this property follows directly from the integral definition of Fourier transforms.

$$\mathcal{F}\{Af_1(t) + Bf_2(t)\} = \int_{-\infty}^{\infty} [Af_1(t) + Bf_2(t)]e^{-j\omega t}dt$$

$$= A\int_{-\infty}^{\infty} f_1(t)e^{-j\omega t}dt + B\int_{-\infty}^{\infty} f_2(t)e^{-j\omega t}dt$$

$$= A\,F_1(\omega) + B\,F_1(\omega)$$

The following example is an application of linearity.

EXAMPLE 13–15

Find the Fourier transforms of the eternal sinusoids $\cos \beta t$ and $\sin \beta t$.

SOLUTION:
The Laplace transforms method doesn't work here because these waveforms are not causal. We previously found the Fourier transform of complex exponential $e^{j\beta t}$ to be

$$\mathcal{F}\{e^{j\beta t}\} = 2\pi\delta(\omega - \beta)$$

Using Euler's relationship for $\cos \beta t$ and the linearity property, we can write

$$\mathscr{F}\{\cos \beta t\} = \mathscr{F}\left\{\frac{e^{j\beta t} + e^{-j\beta t}}{2}\right\}$$

$$= \frac{1}{2}\mathscr{F}\{e^{j\beta t}\} + \frac{1}{2}\mathscr{F}\{e^{-j\beta t}\}$$

$$= \pi[\delta(\omega - \beta) + \delta(\omega + \beta)]$$

Similarly, for $\sin \beta t$ we have

$$\mathscr{F}\{\sin \beta t\} = \mathscr{F}\left\{\frac{e^{j\beta t} - e^{-j\beta t}}{j2}\right\}$$

$$= \frac{1}{j2}\mathscr{F}\{e^{j\beta t}\} - \frac{1}{j2}\mathscr{F}\{e^{-j\beta t}\}$$

$$= -j\pi[\delta(\omega - \beta) - \delta(\omega + \beta)]$$

Here we see that the transforms of $\cos \beta t$ and $\sin \beta t$ involve impulses located at $\omega = \pm\beta$. In general, the spectral content of a Fourier transform is distributed across a continuous range of frequencies. The spectral content of a sinusoid is concentrated in frequency-domain impulses. ∎

Exercise 13–18

Use MATLAB to find the Fourier transform of $f(t) = Ae^{-|\alpha t|}\cos(\beta t)$ for $\alpha > 0$.

Answer:

The required MATLAB code is

```
syms A B t w real
syms a positive
Fw = simplify(int(A*exp(a*t)*cos(B*t)*exp(-j*w*t),t,-inf,0}...
+int(A*exp(-a*t)*cos(B*t)*exp(-j*w*t),t,0,inf))
```

The result is

$$F(\omega) = \frac{2A\alpha(\omega^2 + \beta^2 + \alpha^2)}{(\omega^2 + 2\beta\omega + \beta^2 + \alpha^2)(\omega^2 - 2\beta\omega + \beta^2 + \alpha^2)}$$

Time Differentiation and Integration

The **time differentiation** property of Fourier transforms states that differentiating $f(t)$ in the time domain corresponds to multiplying $F(\omega)$ by $j\omega$ in the frequency domain. That is,

$$\mathscr{F}\left\{\frac{df(t)}{dt}\right\} = j\omega F(\omega) \tag{13–38}$$

Derivation of this property of Fourier transforms begins with the inversion integral:

$$f(t) = \frac{1}{2\pi}\int_{-\infty}^{\infty} F(\omega)e^{j\omega t}d\omega$$

We first differentiate both sides of this equation with respect to time:

$$\frac{df(t)}{dt} = g(t) = \frac{d}{dt}\left[\frac{1}{2\pi}\int_{-\infty}^{\infty} F(\omega)e^{j\omega t}d\omega\right]$$

Assuming the order of integration and differentiation on the right side of this equation can be interchanged, we obtain

$$\frac{df(t)}{dt} = g(t) = \frac{1}{2\pi} \int_{-\infty}^{\infty} \left[\frac{d}{dt} F(\omega) e^{j\omega t} \right] d\omega$$

$$= \frac{1}{2\pi} \int_{-\infty}^{\infty} j\omega F(\omega) e^{j\omega t} d\omega$$

The right side of the last line in this equation implies $\mathscr{F}\{g(t)\} = j\omega F(\omega)$. Since $g(t) = df/dt$, we conclude that

$$\mathscr{F}\left\{ \frac{df(t)}{dt} \right\} = j\omega F(\omega)$$

as expected.

Given the form of the differentiation property, it is reasonable to expect that integrating $f(t)$ in the time domain corresponds to dividing $F(\omega)$ by $j\omega$ in the frequency domain. This proves to be correct, except that integrating a waveform may produce a constant offset or dc component. As we have seen, the Fourier transform of a dc waveform is an impulse at $\omega = 0$

As a result, the statement of the **integration property** is

$$\mathscr{F}\left\{ \int_{-\infty}^{t} f(x)dx \right\} = \frac{F(\omega)}{j\omega} + \pi F(0)\delta(\omega) \tag{13–39}$$

Integrating $f(t)$ in the time domain leads to division of $F(\omega)$ by $j\omega$ plus an additive term $\pi F(0)\delta(\omega)$ to account for the possibility of a dc component that may appear in the integral of $f(t)$.

EXAMPLE 13–16

Use the integration property to find the Fourier transform of the step function $u(t)$.

SOLUTION:

The Fourier transform of an impulse was previously found to be $\mathscr{F}\{\delta(t)\} = \Delta(\omega) = 1$. Since the step function is the integral of an impulse, the integration property yields the Fourier transform of the unit step function.

$$\mathscr{F}\{u(t)\} = \mathscr{F}\left\{ \int_{-\infty}^{t} \delta(x)dx \right\} = \frac{\Delta(\omega)}{j\omega} + \pi\Delta(0)\delta(\omega)$$

$$= \frac{1}{j\omega} + \pi\delta(\omega) \qquad\blacksquare$$

Exercise 13–19

Use symbolic operations in MATLAB to confirm the time differentiation property of Fourier transforms.

Answer:

The required MATLAB code is

```
syms t w real
Fw = fourier(diff(sym('f(t)')),t,w)
```

The result is

```
Fw = w*transform::fourier (f (t), t, -w) *i
```

which is consistent with the property. The MATLAB expression "`transform:: fourier`" is part of the symbolic calculation engine and its frequency variable "w" is defined with the opposite sign of the frequency variable for the "`fourier`" command.

Reversal

The **reversal** property states that

$$\text{If } \mathscr{F}\{f(t)\} = F(\omega), \text{ then } \mathscr{F}\{f(-t)\} = F(-\omega) \qquad \text{(13–40)}$$

Simply stated, reversing $f(t)$ in the time domain reverses $F(\omega)$ in the frequency domain.

Deriving this property begins with the integral definition of Fourier transforms.

$$F(\omega) = \int_{-\infty}^{\infty} f(t)e^{-j\omega t}dt \qquad \text{(13–41)}$$

Defining $\mathscr{F}\{f(-t)\} = G(\omega)$, we can write

$$G(\omega) = \int_{-\infty}^{\infty} f(-t)e^{-j\omega t}dt$$

Changing the dummy variable of integration to $\tau = -t$ produces

$$G(\omega) = -\int_{\infty}^{-\infty} f(\tau)e^{j\omega \tau}d\tau$$
$$= \int_{-\infty}^{\infty} f(\tau)e^{-j(-\omega)\tau}d\tau = F(-\omega)$$

We conclude that $\mathscr{F}\{f(-t)\} = G(\omega) = F(-\omega)$, as stated in the reversal property.

The reversal property is used to derive the Fourier transforms of waveforms of the type in Figure 13–20. The mathematical expressions for these two waveforms are

$$\text{sgn}(t) = u(t) - u(-t)$$
$$e^{-\alpha|t|} = u(t)e^{-\alpha t} + u(-t)e^{-\alpha(-t)} \qquad \text{(13–42)}$$

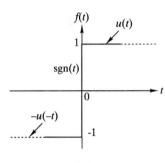

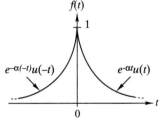

FIGURE 13–20 *Waveforms of the signum and double-sided exponential signals.*

The first waveform is called the **signum function**, which is defined as positive 1 for $t > 0$, negative 1 for $t < 0$, and zero at $t = 0$. As a result, $\text{sgn}(t)$ can be expressed as the difference between a step $u(t)$ to a reversed step $u(-t)$, as indicated in Eq. (13–42).

The second waveform in Figure 13–20 is called a **double-sided** exponential, which is defined as the sum of a causal exponential and a reversed causal exponential. This waveform is 1 at $t = 0$ and decays exponentially to zero in both directions along the time axis.

The signum and double-sided exponential are examples of noncausal waveforms that can be written as $f(t) = g(t) \pm h(-t)$, where $g(t)$ and $h(t)$ are causal. If the Fourier transforms of $g(t)$ and $h(t)$ are known, then the reversal property yields $F(\omega) = G(\omega) \pm H(-\omega)$.

EXAMPLE 13–17

Use the reversal property to find the Fourier transform of $f(t) = \text{sgn}(t)$.

SOLUTION:

The signum function is expressed as $\mathrm{sgn}(t) = u(t) - u(-t)$. In Example 13–16 we found that $\mathscr{F}\{u(t)\} = 1/j\omega + \pi\delta(\omega)$. Using the reversal property yields

$$\mathscr{F}\{\mathrm{sgn}(t)\} = \mathscr{F}\{u(t)\} - \mathscr{F}\{u(-t)\}$$

$$= \left[\frac{1}{j\omega} + \pi\delta(\omega)\right] - \left[\frac{1}{j(-\omega)} + \pi\delta(-\omega)\right]$$

$$= \frac{2}{j\omega}$$

In the last line of this equation we have used the fact that $\delta(\omega) = \delta(-\omega)$ since both impulses occur at $\omega = 0$. ■

Exercise 13–20 _____

Use the reversal property to find the Fourier transform of $f(t) = e^{-\alpha|t|}$. Assume $\alpha > 0$.

Answer:
$$F(\omega) = \frac{2\alpha}{\omega^2 + \alpha^2}$$

Duality

The **duality property** states that

$$\text{If } \mathscr{F}\{f(t)\} = F(\omega), \text{ then } \mathscr{F}\{F(t)\} = 2\pi f(-\omega) \tag{13–43}$$

Proof of this property begins with the inversion integral written in the form

$$2\pi f(t) = \int_{-\infty}^{\infty} F(\omega)e^{j\omega t}d\omega$$

Replacing t by $-t$ yields

$$2\pi f(-t) = \int_{-\infty}^{\infty} F(\omega)e^{-j\omega t}d\omega$$

Now interchanging t and ω produces

$$2\pi f(-\omega) = \int_{-\infty}^{\infty} F(t)e^{-j\omega t}dt = \mathscr{F}\{F(t)\}$$

Thus, $\mathscr{F}\{F(t)\} = 2\pi f(-\omega)$, as stated in the duality property.

Figure 13–21 illustrates the waveforms/transforms duality for rectangular functions. The upper pair comes from Example 13–11, where we found that a rectangular waveform has a wiggly Fourier transform. The lower pair comes from Example 13–13, where we found that a rectangular transform produces a wiggly waveform. In sum, a rectangular function in one domain leads to an oscillatory $\mathrm{sinc}(x)$ function in the other domain.

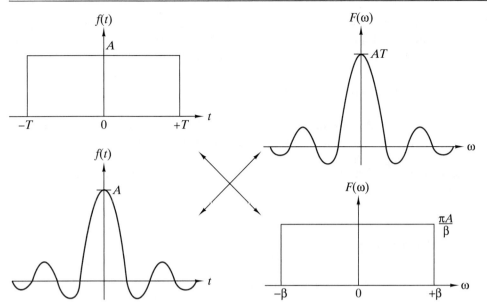

FIGURE 13–21 *Example of symmetry property of Fourier transforms.*

EXAMPLE 13–18

Given that $\mathcal{F}\{e^{-\alpha|t|}\} = 2\alpha/(\alpha^2 + \omega^2)$, use the duality property to find the Fourier transform of $g(t) = \alpha^2/(\alpha^2 + t^2)$.

SOLUTION:
If we define $f(t) = e^{-\alpha|t|}$, then $F(\omega) = 2\alpha/(\alpha^2 + \omega^2)$. According to duality,

$$\mathcal{F}\{F(t)\} = 2\pi f(-\omega) = 2\pi e^{-\alpha|-\omega|} = 2\pi e^{-\alpha|\omega|}$$

But since

$$F(t) = \frac{2\alpha}{\alpha^2 + t^2} = \frac{2}{\alpha}\left[\frac{\alpha^2}{\alpha^2 + t^2}\right] = \frac{2}{\alpha}g(t)$$

it follows from linearity that

$$\mathcal{F}\{g(t)\} = \frac{\alpha}{2}\mathcal{F}\{F(t)\} = \pi\alpha e^{-\alpha|\omega|} \qquad\blacksquare$$

Exercise 13–21

Use the duality property to find the Fourier transform of $g(t) = 1/t$.

Answer:
$$G(\omega) = j\pi\,\text{sgn}(-\omega)$$

Time and Frequency Shifting
There are two shifting properties of the Fourier transforms:

Time domain   If $\mathcal{F}\{f(t)\} = F(\omega)$, then $\mathcal{F}\{f(t-T)\} = e^{-j\omega T}F(\omega)$

Frequency domain   If $\mathcal{F}^{-1}\{F(\omega)\} = f(t)$, then $\mathcal{F}^{-1}\{F(\omega - \beta)\} = e^{j\beta T}f(t)$ (13–44)

Proofs of these properties follow directly from the integral definitions in Eqs. (13–26) and (13–27).

The time-domain statement indicates that shifting a waveform by an amount T is equivalent to multiplying its transform by $e^{-j\omega T}$. One of the results of this is that delaying a waveform $f(t)$ does not change its amplitude spectrum since

$$|\mathscr{F}\{f(t-T)\}| = |e^{-j\omega T}F(\omega)| = |e^{-j\omega T}||F(\omega)| = |F(\omega)|$$

The frequency-domain statement says that shifting the location of a transform by an amount β is equivalent to multiplying the waveform by $e^{j\beta t}$. The frequency-domain shifting property is the basis for a signal processing operation called *modulation*, in which the spectrum of a signal is shifted from one frequency range to another.

Scaling

The **scaling property** of the Fourier transformation is

$$\text{If } \mathscr{F}\{f(t)\} = F(\omega), \text{ then } \mathscr{F}\{f(at)\} = \frac{1}{|a|}F(\omega/a) \qquad (13\text{–}45)$$

where a is a real constant.

Figure 13–22 shows the Fourier transforms of two rectangular pulses with different pulse durations. The figure points out that shortening the pulse duration $(a > 1)$ causes the transform to spread out in the frequency domain, and vice versa. For this reason the scaling property is sometimes called the *reciprocal spreading* property. That is, compressing a signal in one domain causes it to spread out in the other domain. An important implication is that reducing pulse duration to increase data rate will increase the bandwidth requirement since the signal spectrum spreads out in the frequency domain.

FIGURE 13–22 *Example of scaling property of Fourier transforms.*

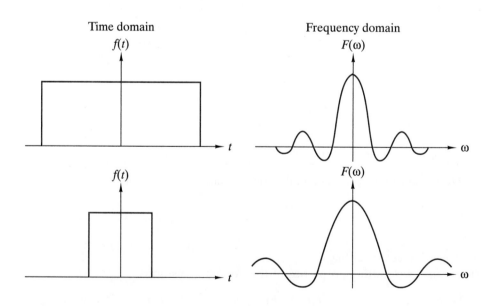

Summary Tables

Table 13–1 lists a basic set of Fourier transform pairs. Because of the uniqueness property, this table can be used in either direction. Additional pairs are easy to

T A B L E 13–1 BASIC FOURIER TRANSFORM PAIRS

SIGNAL	WAVEFORM $f(t)$	TRANSFORM $F(\omega)$
Impulse	$\delta(t)$	1
Constant (dc)	1	$2\pi\delta(\omega)$
Step function	$u(t)$	$\dfrac{1}{j\omega}+\pi\delta(\omega)$
Signum	$\mathrm{sgn}(t)$	$\dfrac{2}{j\omega}$
Causal exponential	$[e^{-\alpha t}]u(t)$	$\dfrac{1}{\alpha + j\omega}$
Two-sided exponential	$e^{-\alpha|t|}$	$\dfrac{2\alpha}{\alpha^2 + \omega^2}$
Complex exponential	$e^{j\beta t}$	$2\pi\delta(\omega - \beta)$
Cosine (ac)	$\cos\beta t$	$\pi[\delta(\omega - \beta) + \delta(\omega + \beta)]$
Sine (ac)	$\sin\beta t$	$-j\pi[\delta(\omega - \beta) - \delta(\omega + \beta)]$
Sinc	$\dfrac{\sin\beta t}{\pi t}$	$u(\omega + \beta) - u(\omega - \beta)$

T A B L E 13–2 BASIC FOURIER TRANSFORM PROPERTIES

PROPERTY	TIME DOMAIN	FREQUENCY DOMAIN
Linearity	$Af_1(t) + Bf_2(t)$	$AF_1(\omega) + BF_2(\omega)$
Differentiation	$\dfrac{df(t)}{dt}$	$j\omega F(\omega)$
Integration	$\int_{-\infty}^{t} f(x)dx$	$\dfrac{F(\omega)}{j\omega}+\pi F(0)\delta(\omega)$
Reversal	$f(-t)$	$F(-\omega)$
Duality	$F(t)$	$2\pi f(-\omega)$
Time shift	$f(t - T)$	$e^{-j\omega T}F(\omega)$
Frequency shift	$e^{j\beta t}f(t)$	$F(\omega - \beta)$
Scaling	$|a| f(at)$	$F(\omega/a)$

derive using the pairs in Table 13–1 together with the transform properties summarized in Table 13–2.

Exercise 13–22

Use Tables 13–1 and 13–2 to find the Fourier transforms of

(a) $f_1(t) = \mathrm{sgn}(t) + 2$
(b) $f_2(t) = \cos(2t) + 2$

Answers:

(a) $F_1(\omega) = \dfrac{2}{j\omega} + 4\pi\delta(\omega)$

(b) $F_2(\omega) = \pi[\delta(\omega - 2) + 4\delta(\omega) + \delta(\omega + 2)]$

Exercise 13–23

Use Tables 13–1 and 13–2 to find the inverse Fourier transforms of

$$\text{(a)} \quad F_1(\omega) = \frac{1}{j\omega} + 3\pi\delta(\omega)$$

$$\text{(b)} \quad F_2(\omega) = \frac{2}{4 + j\omega} - 2 + 4\pi\delta(\omega + 2)$$

Answers:

(a) $f_1(t) = u(t) + 1$
(b) $f_2(t) = 2e^{-4t}u(t) - 2\delta(t) + 2e^{-j2t}$

13–7 CIRCUIT ANALYSIS USING FOURIER TRANSFORMS

Because of our previous study of Laplace transforms it should come as no surprise that Fourier transforms can be used in linear circuit analysis. To see how the Fourier transforms apply to circuit analysis, we must examine how they affect connection and device constraints.

In the time domain a typical KVL constraint might be

$$v_1(t) + v_2(t) - v_3(t) = 0$$

Because of the linearity property, the Fourier transform of this equation is

$$V_1(\omega) + V_2(\omega) - V_3(\omega) = 0$$

This example obviously generalizes to any KVL or KCL constraint. The Fourier transformation changes waveforms into transforms but leaves the form of the connection constraints unchanged.

The frequency-domain element constraints are found by transforming the time-domain i–v characteristics of the passive elements. Using the differentiation and linearity properties of Fourier transforms we write the time-domain and frequency-domain element constraints as

Time Domain	Frequency Domain
Resistor: $v(t) = Ri(t)$	$V(\omega) = RI(\omega)$
Inductor: $v(t) = L\dfrac{di}{dt}$	$V(\omega) = j\omega LI(\omega)$
Capacitor: $i(t) = C\dfrac{dv}{dt}$	$I(\omega) = j\omega CV(\omega)$

As might be expected, in the frequency domain the element constraints are algebraic equations similar in form to Ohm's law. The proportionality factors relating the voltage and current transforms are the ac impedances of the passive elements, namely

$$Z_R = R \qquad Z_L = j\omega L \qquad Z_C = \frac{1}{j\omega C}$$

The forms of the frequency-domain connection and element constraints have a familiar ring. The key points are: (1) the Fourier transformation does not change the form of the connection constraints; and (2) the element constraints are similar to Ohm's law. We have seen analogous conclusions before in our study of sinusoidal steady-state (phasor) circuit analysis and again in s-domain circuit analysis. As a result, we know that our repertoire of algebraic circuit analysis technique can be applied in the frequency domain.

To use Fourier methods in circuit analysis, we represent currents and voltages as transforms and passive elements as ac impedances. We then use analysis techniques such as voltage division, current division, equivalence, or even node/mesh analysis to solve for the unknown current or voltage transforms. Inverse Fourier transforms are then used to obtain the response waveform in the time domain.

This discussion closely parallels the steps in s-domain circuit analysis, except for one thing—there is no mention of initial conditions. The reason for this omission is that the lower limit on the defining integral for Fourier transforms is $t = -\infty$. In other words, Fourier transforms account for the entire history of a circuit beginning at $t = -\infty$. With Laplace transforms the lower limit on the defining integral is $t = 0$, and the circuit's history for $t < 0$ is accounted for by the initial conditions at $t = 0$.

In summary, the Fourier transform method of circuit analysis explicitly takes into account all inputs including those that happened prior to $t = 0$. Fourier transforms exist for noncausal waveforms, and the method yields responses that are valid for $-\infty < t < \infty$. As a result, Fourier transform methods cannot handle circuit analysis problems in which initial conditions are used to account for inputs prior to $t = 0$.

EXAMPLE 13–19

Use Fourier transforms to find $v_2(t)$ in the circuit in Figure 13–23 for $v_1(t) = u(t)$.

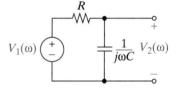

FIGURE 13–23

SOLUTION:
This example asks us to find the step response of a first-order RC circuit. In Chapter 7 we solved this problem using classical differential equation methods. The problem was revisited in Chapters 9 and 11 using Laplace transforms. From this previous experience we already know that the answer is $v_2(t) = [1 - e^{-t/RC}]u(t)$. Our purpose here is simply to show how Fourier transforms produce the same result.

The circuit in Figure 13–23 is shown in the frequency domain, so we use voltage division to relate the input and output transforms.

$$V_2(\omega) = \left[\frac{\frac{1}{j\omega C}}{R + \frac{1}{j\omega C}}\right] V_1(\omega) = \left[\frac{1/RC}{j\omega + 1/RC}\right] V_1(\omega)$$

For a unit step function input $V_1(\omega) = \mathscr{F}\{u(t)\} = 1/j\omega + \pi\delta(\omega)$ and the output transform is

$$V_2(\omega) = \left[\frac{1/RC}{j\omega + 1/RC}\right]\frac{1}{j\omega} + \left[\frac{1/RC}{j\omega + 1/RC}\right]\pi\delta(\omega)$$

The first term on the right side can be expanded by partial fractions as

$$\frac{1/RC}{(j\omega + 1/RC)j\omega} = \frac{1}{j\omega} - \frac{1}{j\omega + 1/RC}$$

The second term on the right side reduces to

$$\left[\frac{1/RC}{j\omega + 1/RC}\right]\pi\delta(\omega) = \left[\frac{1/RC}{1/RC}\right]\pi\delta(\omega) = \pi\delta(\omega)$$

That is, in general $F(\omega)\delta(\omega) = F(0)\delta(\omega)$ because the impulse $\delta(\omega)$ is zero everywhere except at $\omega = 0$. Combining these results the output transform can be arranged as

$$V_2(\omega) = \underbrace{\frac{1}{j\omega} + \pi\delta(\omega)}_{\substack{\text{step} \\ \text{function}}} - \underbrace{\frac{1}{j\omega + 1/RC}}_{\substack{\text{causal} \\ \text{exponential}}}$$

Each term in this expansion is listed in Table 13–1, leading to

$$v_2(t) = \mathcal{F}^{-1}\left\{\frac{1}{j\omega} + \pi\delta(\omega)\right\} - \mathcal{F}^{-1}\left\{\frac{1}{j\omega + 1/RC}\right\}$$

$$= \qquad\qquad u(t) \qquad - \qquad e^{-t/RC}u(t)\text{V}$$

As expected, the step response waveform is $v_2(t) = [1 - e^{-t/RC}]u(t)\text{V}$. This example illustrates a rather subtle point. For causal inputs Fourier transforms inherently yield only the zero-state response, whereas Laplace transforms yield both the zero-state and zero-input responses. ∎

Exercise 13–24

Use Fourier transforms to find $v_2(t)$ in the circuit in Figure 13–23 for $v_1(t) = Ae^{-\alpha t}u(t)$ V.

Answer: $$v_2(t) = \frac{A}{\alpha RC - 1}\left[e^{-t/RC} - e^{-\alpha t}\right]u(t) \text{ V}$$

EXAMPLE 13–20

Use Fourier transforms to find $v_2(t)$ in the circuit in Figure 13–23 for $v_1(t) = \text{sgn}(t)$.

SOLUTION:
In this example we find the output of a first-order RC circuit for a noncausal input. From Table 13–1 the input transform is $V_1(\omega) = 2/j\omega$. Using the voltage division result from Example 13–19 we write

$$V_2(\omega) = \left[\frac{1/RC}{j\omega + 1/RC}\right]\frac{2}{j\omega}$$

Expanding by partial fractions,

$$V_2(\omega) = \frac{2/RC}{j\omega(j\omega + 1/RC)} = \underbrace{\frac{2}{j\omega}}_{\text{signum}} - \underbrace{\frac{2}{j\omega + 1/RC}}_{\text{exponential}}$$

Each term in this expansion is listed in Table 13–1. The inverse transform produces

$$v_2(t) = \mathcal{F}^{-1}\left\{\frac{2}{j\omega}\right\} - \mathcal{F}^{-1}\left\{\frac{2}{j\omega + 1/RC}\right\}$$

$$= \qquad \text{sgn}(t) - \qquad\qquad 2e^{-t/RC}u(t)\text{V}$$

Both a noncausal signum waveform and a causal exponential are required to describe $v_2(t)$ on the interval $-\infty < t + \infty$. Figure 13–24 shows how these waveforms combine to produce $v_2(t)$. A physical interpretation of this response is that signum waveform applies an input $v_1(t) = -1$ V at $t = -\infty$ (a long time ago). After about five time constants ($t = -\infty + 5RC$, still a long time ago), the circuit output reaches a steady-state condition of $v_2(t) = -1$ V. Thus, for all practical purposes $v_2(t) = -1$ V for all $t < 0$. At $t = 0$ the signum input jumps from $v_1(t) = -1$ to $+1$ V. This drives the output from $v_2(t) = -1$ V at $t = 0$ to $v_2(t) = +1$ V for $t > 5RC$. ■

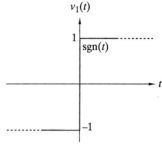

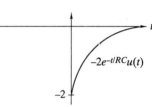

Exercise 13–25

Use Fourier transforms to find $v_2(t)$ in the circuit in Figure 13–23 for $v_1(t) = u(t) - 1$ V.

Answer:
$$v_2(t) = -u(-t) - e^{-t/RC}u(t)\,\text{V}$$

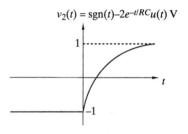

FIGURE 13–24

13–8 IMPULSE RESPONSE AND CONVOLUTION

The previous section shows that Fourier transforms can be used to analyze linear circuits. The process is straightforward once we accept the idea of noncausal waveforms. However, the important applications of Fourier transforms are found in system analysis and signal processing rather than traditional circuit analysis.

Fourier transforms find applications in these areas because they can represent idealized models of signal processors. These ideal models often involve noncausal responses that cannot be represented by Laplace transforms. Of course real physical systems must have causal responses so these models can only be approximated by engineering hardware. Nevertheless, ideal models are useful in feasibility studies and as "gold standards" for evaluating real systems.

Signal processors are often described in the frequency domain by the Fourier transform equivalent of the s-domain transfer function. To develop the Fourier equivalent, we begin with the convolution integral from Chapter 11:

$$y(t) = \int_0^t h(t - \tau)x(\tau)\,d\tau$$

In this expression $y(t)$ is the output of a system whose impulse response is $h(t)$ when the input is $x(t)$. The limits of integration are $\tau = 0$ to $\tau = t$ because in Chapter 11 $h(t)$ and $x(t)$ are causal waveforms. To remove this limitation the integration limits are extended backward to $\tau = -\infty$ to accommodate a noncausal input $x(\tau)$ and forward to $t = +\infty$ to accommodate a noncausal $h(t)$. The result is

$$y(t) = \int_{-\infty}^{\infty} h(t - \tau)x(\tau)\,d\tau \qquad (13\text{–}46)$$

This equation is a general form of the convolution integral that applies to causal and noncausal waveforms alike.

We now take the Fourier transformation on both sides of Eq. (13–46).

$$\mathscr{F}\{y(t)\} = Y(\omega) = \mathscr{F}\left\{\int_{\tau=-\infty}^{\infty} h(t - \tau)x(\tau)\,d\tau\right\}$$
$$= \int_{t=-\infty}^{\infty}\left[\int_{\tau=-\infty}^{\infty} h(t - \tau)x(\tau)\,d\tau\right]e^{-j\omega t}dt$$

Next we interchange the order of integration and factor out $x(\tau)$, since it does not depend on t.

$$Y(\omega) = \int_{\tau=-\infty}^{\infty} x(\tau) \left[\underbrace{\int_{t=-\infty}^{\infty} h(t-\tau)e^{-j\omega t}dt}_{H(\omega)e^{-j\omega\tau}} \right] d\tau$$

Using the time-shifting property the quantity within the bracket is $\mathscr{F}\{h(t-\tau)\} = H(\omega)e^{-j\omega\tau}$, and this equation reduces to

$$Y(\omega) = \int_{\tau=-\infty}^{\infty} x(\tau)H(\omega)e^{-j\omega\tau}d\tau$$

$$= H(\omega) \left[\underbrace{\int_{\tau=-\infty}^{\infty} x(\tau)e^{-j\omega\tau}d\tau}_{X(\omega)} \right]$$

By definition the remaining integral inside the bracket is $\mathscr{F}\{x(\tau)\} = X(\omega)$. We conclude that

$$Y(\omega) = H(\omega)X(\omega) \qquad (13\text{--}47)$$

Thus, in the frequency domain convolution is accomplished by multiplying the Fourier transform of the system impulse response by the Fourier transform of the input.

Equation (13–47) looks suspiciously like the s-domain relationship $Y(s) = T(s)X(s)$, where $T(s)$ is the s-domain transfer function. Since the two input–output relationships have the same form, we call $H(\omega)$ the **Fourier-domain transfer function**. Key differences between $T(s) = \mathscr{L}\{h(t)\}$ and $H(\omega) = \mathscr{F}\{h(t)\}$ are:

1. $H(\omega)$ exists for noncausal impulse responses for which $T(s)$ does not exist.

2. $T(s)$ exists for unstable impulse responses for which $H(\omega)$ does not exist.

EXAMPLE 13–21

The impulse response of a system is $h(t) = e^{-|t|}$. Use Fourier transform convolution to find the output $y(t)$ when $x(t) = \text{sgn}(t)$.

SOLUTION:
Using transform pairs in Table 13–1, we have

$$H(\omega) = \frac{2}{1+\omega^2} \quad \text{and} \quad X(\omega) = \frac{2}{j\omega}$$

Using convolution in the frequency domain, the output transform is

$$Y(\omega) = H(\omega)X(\omega) = \frac{4}{(1+\omega^2)j\omega}$$

Expanding $Y(\omega)$ by partial fractions,

$$Y(\omega) = \frac{4}{(1+j\omega)(1-j\omega)j\omega} = \frac{-2}{1+j\omega} + \frac{2}{1-j\omega} + \frac{4}{j\omega}$$

Using Table 13–1, the inverse transform of the first term in the expansion is

$$\mathcal{F}^{-1}\left\{\frac{-2}{1+j\omega}\right\} = -2\,e^{-t}u(t)$$

According to the reversal property $\mathcal{F}^{-1}\{F(-\omega)\} = f(-t)$; hence, the inverse transform fo the second term is

$$\mathcal{F}^{-1}\left\{\frac{2}{1+j(-\omega)}\right\} = 2\,e^{t}u(-t)$$

The inverse transform of the third term is $\mathcal{F}^{-1}\{4/j\omega\} = 2\,\text{sgn}(t)$. So finally, the output waveform is

$$y(t) = -2\,e^{-t}u(t) + 2\,e^{t}u(-t) + 2\,\text{sgn}(t)$$

Notice that we cannot use Laplace transforms here because the input, output, and impulse response are all noncausal waveforms. ■

Exercise 13–26

The impulse response of a system is $h(t) = 2e^{-4t}u(t)$. Find the output $y(t)$ when $x(t) = -2 + 4u(t)$.

Answer:
$$y(t) = \text{sgn}(t) - 2e^{-4t}u(t)$$

The Fourier domain transfer function $H(\omega)$ represents the system frequency response, including noncausal systems. For example, consider the $H(\omega)$ shown in Figure 13–25. In the filter terminology, the gain $|H(\omega)|$ is an *ideal low-pass filter* with a passband gain of 1, a stopband gain of zero, and a bandwidth of β. In Example 13–13 we found that the waveform corresponding to a rectangular transform is the noncausal $h(t)$ shown in Figure 13–25. The noncausal $h(t)$ means that an ideal low-pass filter responds *before* the impulse is applied at $t = 0$. Such anticipatory behavior is physically impossible, which means that we can't actually build an ideal filter using physical hardware. Nevertheless, the ideal low-pass filter model is used in conceptual design work because it is easily implemented in the system simulation software and provides a "gold standard" for evaluating the performance of real systems.

Fourier transforms highlight the relationship between impulse response and frequency response. Digital filter design involves defining an impulse response $h(t)$ whose Fourier transform $H(\omega)$ has a desired frequency response. The signal processing action is actually carried out using software that convolves $h(t)$ and $x(t)$ in the time domain. Fast algorithms for performing time-domain convolution are readily available in commercial software. Thus, Fourier methods play a key role in digital signal processing because they provide a computationally efficient way to implement signal filtering.

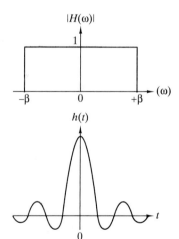

FIGURE 13–25 *Impulse response of an ideal low-pass filter.*

EXAMPLE 13–22

Describe the frequency response of a system whose impulse response is

$$h(t) = \frac{\alpha}{2}e^{-\alpha|t|}$$

SOLUTION:

The given $h(t)$ is a two-sided exponential whose transform from Table 13–1 is

$$H(\omega) = \frac{\alpha}{2}\mathcal{F}\{e^{-\alpha|t|}\} = \frac{\alpha^2}{\alpha^2 + \omega^2}$$

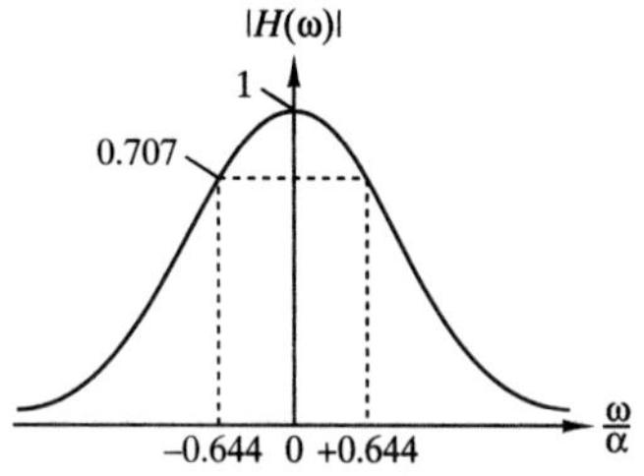

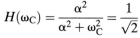

FIGURE 13–26

Figure 13–26 shows the plot of gain $|H(\omega)|$ versus normalized frequency. The system acts like a low-pass filter with a dc gain of $H(0) = 1$ and the high-frequency gain rolls off α^2/ω^2, or $-40\,\text{dB/decade}$, so to speak. The passband cutoff frequency occurs when $|H(\omega_C)| = H(0)/\sqrt{2}$, when requires

$$H(\omega_C) = \frac{\alpha^2}{\alpha^2 + \omega_C^2} = \frac{1}{\sqrt{2}}$$

yielding

$$\omega_C = \pm\alpha\sqrt{\sqrt{2}-1} = \pm 0.644\,\alpha \qquad \blacksquare$$

Exercise 13–27

(a) Find the transfer function of a system whose impulse response is $h(t) = \delta(t) - \alpha e^{-\alpha|t|}$.

(b) Describe the frequency response of the system.

Answers:

(a) $H(\omega) = (\omega^2 - \alpha^2)/(\omega^2 + \alpha^2)$.

(b) System has a bandstop characteristic with a notch at $\omega = \pm\alpha$.

13–9 PARSEVAL'S THEOREM

Parseval's theorem relates the energy carried by a waveform to the amplitude spectrum of its Fourier transform. The total energy carried by any waveform is defined to be

$$W_T = \int_{-\infty}^{+\infty} p(t)\,dt$$

where $p(t)$ is the power the waveform delivers to a specified load. For a resistive load $p(t) = v^2(t)/R = i^2(t)R$, so the energy delivered to a 1-Ω resistive load can be written in the form

$$W_{1\Omega} = \int_{-\infty}^{+\infty} f^2(t)\,dt \qquad (13\text{–}48)$$

where $f(t)$ can be either a voltage or a current waveform. Equation (13–48) is often used as the definition of the energy carried by a waveform, although an implied 1-Ω resistance is required for the result to have the dimensions of energy.

Parseval's theorem states that the total energy carried by a signal can be calculated in either the time domain or the frequency domain.

$$W_{1\Omega} = \int_{-\infty}^{+\infty} f^2(t)\,dt = \frac{1}{2\pi}\int_{-\infty}^{+\infty} |F(\omega)|^2\,d\omega \qquad (13\text{–}49)$$

That is, the total 1-Ω energy can be found from either the waveform $f(t)$ or its transform $F(\omega)$.

Parseval's theorem assumes that the integrals in Eq. (13–49) converge to a finite value. Signals for which $W_{1\Omega}$ is finite are called **energy signals**. Examples of energy signals are the causal exponential, the two-sided exponential, the rectangular pulse, and the damped sine. Finite energy is a stronger requirement than being absolutely integrable. As a result, not all signals that have Fourier transforms are energy signals. For example, the impulse, step, signum, and eternal sinusoid all have Fourier transforms but are not energy signals.

The derivation of Parseval's theorem begins with energy in the time domain. Assuming the Fourier transform of $f(t)$ exists, this energy can be written in the form

$$W_{1\Omega} = \int_{-\infty}^{+\infty} [f(t)][f(t)]dt = \int_{-\infty}^{+\infty} [f(t)] \underbrace{\left[\frac{1}{2\pi} \int_{-\infty}^{+\infty} F(\omega)e^{j\omega t}d\omega \right]}_{f(t) = \mathscr{F}^{-1}\{F(\omega)\}} dt$$

The integration within the brackets does not involve time, so $f(t)$ can be moved inside the second integral and the $1/2\pi$ moved outside the first integral to produce

$$W_{1\Omega} = \int_{-\infty}^{+\infty} [f(t)][f(t)]dt = \frac{1}{2\pi} \int_{-\infty}^{+\infty} \int_{-\infty}^{+\infty} f(t)F(\omega)e^{j\omega t}d\omega\, dt$$

Reversing the order of integration and factoring out $F(\omega)$ produces

$$W_{1\Omega} = \int_{-\infty}^{+\infty} [f(t)][f(t)]dt = \frac{1}{2\pi} \int_{-\infty}^{+\infty} F(\omega) \underbrace{\left[\int_{-\infty}^{+\infty} f(t)e^{-j(-\omega)t}dt \right]}_{F(-\omega)} d\omega$$

By definition the function inside the bracket is $F(-\omega)$. But $F(-\omega)$ is formed by replacing $j\omega$ by $j(-\omega) = -j\omega$, which means that $F(-\omega) = F(\omega)^*$, the conjugate of $F(\omega)$. Thus, the product $F(\omega)F(-\omega)$ is the square of the magnitude of $F(\omega)$ and this equation reduces to

$$W_{1\Omega} = \int_{-\infty}^{+\infty} f(t)^2 dt = \frac{1}{2\pi} \int_{-\infty}^{+\infty} |F(\omega)|^2 d\omega$$

as stated in Parseval's theorem.

Since $|F(\omega)|^2$ is an even function, the integral from $-\infty$ to 0 is the same as the integral from 0 to $+\infty$. Hence,

$$\frac{1}{2\pi} \int_{-\infty}^{+\infty} |F(\omega)|^2 d\omega = \frac{1}{\pi} \int_{0}^{+\infty} |F(\omega)|^2 d\omega$$

and Parseval's theorem can also be written as

$$W_{1\Omega} = \int_{-\infty}^{+\infty} f(t)^2 dt = \frac{1}{\pi} \int_{0}^{+\infty} |F(\omega)|^2 d\omega \qquad (13\text{--}50)$$

EXAMPLE 13–23

Derive an expression for the 1-Ω energy in the causal exponential $v(t) = V_A e^{-\alpha t}u(t)$ V (a) in the time domain and (b) in the frequency domain.

SOLUTION:

(a) The given waveform is causal, so the time integration extends from $t = 0$ to $t = +\infty$.

$$W_{1\Omega} = \int_{0}^{+\infty} (V_A e^{-\alpha t})^2 dt = \frac{V_A^2}{-2\alpha} e^{-2\alpha t} \Big|_{0}^{+\infty} = \frac{V_A^2}{2\alpha}$$

This result applies if $\alpha > 0$ since for $\alpha < 0$ the waveform is unbounded and the integral does not converge. In other words, an unbounded exponential is not an energy signal.

(b) The Fourier transform of the signal is $V(\omega) = V_A/(j\omega + \alpha)$ provided $\alpha > 0$. In the frequency domain the 1-Ω energy is found to be

$$W_{1\Omega} = \frac{1}{\pi} \int_0^{+\infty} \frac{V_A^2}{\alpha^2 + \omega^2} d\omega = \frac{V_A^2}{\pi\alpha} \tan^{-1}(\omega/\alpha) \Big|_0^{+\infty}$$

$$= \frac{V_A^2}{\pi\alpha} \left[\frac{\pi}{2} - 0 \right] = \frac{V_A^2}{2\alpha}$$

∎

which is the same as the result in (a).

Exercise 13–28

Derive an expression for the 1-Ω energy in the waveform $v(t) = A[u(t + T) - u(t - T)]$ V (a) in the time domain and (b) in the frequency domain.

Answers:

(a) $W_{1\Omega} = 2A^2 T$
(b) $W_{1\Omega} = 2A^2 T$

APPLICATIONS OF PARSEVAL'S THEOREM

The square amplitude spectrum $|F(\omega)|^2$ is called the **energy spectral density** because it describes how the total energy carried by $f(t)$ is distributed among the frequencies in its spectrum. In this sense we think of the integral

$$W_{1\Omega} = \frac{1}{\pi} \int_{\omega_1}^{\omega_2} |F(\omega)|^2 d\omega$$

as the amount of energy carried by frequencies in the band $\omega_1 < \omega < \omega_2$. Notice that the energy distribution for a Fourier transform is fundamentally different than for a Fourier series. With a Fourier transform the energy carried by $f(t)$ is distributed across all of the frequencies between ω_1 and ω_2. With a Fourier series the energy is located only at the discrete harmonic frequencies.

EXAMPLE 13–24

(a) Find the percentage of the 1-Ω energy in $v(t) = V_A e^{-\alpha t} u(t)$ V that is carried in the frequency band $|\omega| < \alpha$.
(b) What frequency band carries 90% of the 1-Ω energy?

SOLUTION:

(a) The energy in the band $|\omega| < \alpha$ is

$$W_\alpha = \frac{1}{\pi} \int_0^\alpha \frac{V_A^2}{\alpha^2 + \omega^2} d\omega = \frac{V_A^2}{\pi\alpha} \tan^{-1}(\omega/\alpha) \Big|_0^\alpha$$

$$= \frac{V_A^2}{\pi\alpha} \left(\frac{\pi}{4} - 0 \right) = \frac{V_A^2}{4\alpha}$$

In Example 13–23 we found that the total energy is $W_{1\Omega} = (V_A)^2/2\alpha$; hence, the energy ratio is

$$\frac{W_\alpha}{W_{1\Omega}} = \frac{V_A^2/4\alpha}{V_A^2/2\alpha} = \frac{1}{2}$$

That is, 50% of the total energy is carried in the frequency band $|\omega| < \alpha$.

(b) Let the frequency band $|\omega| < \beta$ carry 90% of the total energy. The energy in this band is

$$W_\beta = \frac{1}{\pi} \int_0^\beta \frac{V_A^2}{\alpha^2 + \omega^2} d\omega = \frac{V_A^2}{\pi\alpha} \tan^{-1}(\omega/\alpha) \Big|_0^\beta$$

$$= \frac{V_A^2}{\pi\alpha} \tan^{-1}(\beta/\alpha) = 0.9 \frac{V_A^2}{2\alpha}$$

The equality sign in the last line of this equation requires that

$$\tan^{-1}(\beta/\alpha) = 0.9 \frac{\pi}{2} \quad \text{or} \quad \beta = \alpha \tan(9\pi/20) = 6.31\,\alpha$$

Hence, the band $|\omega| < 6.31\,\alpha$ carries 90% of the total energy.   ∎

Exercise 13–29

Find the percentage of the 1-Ω energy in $v(t) = A[u(t + T) - u(t - T)]$ V that is carried in the frequency band $|\omega| < \pi/T$, which represents the main lobe of the spectrum.

Answer: 90.28%

The energy viewpoint offers a way to quantify the selectivity of filters. Since the output of a filter is $Y(\omega) = H(\omega)X(\omega)$, the energy spectral density of the output signal is

$$|Y(\omega)|^2 = |H(\omega)|^2 |X(\omega)|^2 \qquad (13\text{--}51)$$

Combining Eqs. (13–49) and (13–51) yields the energy in the output signal as

$$W_{1\Omega} = \frac{1}{\pi} \int_0^{+\infty} |Y(\omega)|^2 d\omega = \frac{1}{\pi} \int_0^{+\infty} |H(\omega)|^2 |X(\omega)|^2 d\omega \qquad (13\text{--}52)$$

We can view filtering as a process that shapes the output energy spectral density by accepting input energy at frequencies in the passband and rejecting input energy at frequencies in the stopband. Note that we calculate the energy processing from the Fourier transforms rather than the signal waveforms.

EXAMPLE 13–25

The signal $x(t) = 10e^{-20t}u(t)$ is input to an ideal low-pass filter with $H(\omega) = u(\omega + 100) - u(\omega - 100)$. Find the percentage of the input energy in the output signal.

SOLUTION:
The 1-Ω energy in the input signal is

$$W_{1\Omega,\text{IN}} = \int_0^{+\infty} 100\, e^{-40t} dt = -\frac{100\, e^{-40t}}{40} \Big|_0^\infty = 2.5\,\text{J}$$

The Fourier transform of the input signal is $10/(20 + j\omega)$. The frequency response of the ideal low-pass filter is $|H(\omega)| = 1$ for $|\omega| < 100$ and $|H(\omega)| = 0$ elsewhere. The 1-Ω energy in the output signal is

$$W_{1\Omega,\text{out}} = \frac{1}{\pi} \int_0^{100} 1 \times |X(\omega)|^2 d\omega$$

$$= \frac{1}{\pi} \int_0^{100} \frac{10^2}{20^2 + \omega^2} d\omega = \frac{100}{20\,\pi} \tan^{-1}(\omega/20) \Big|_0^{100}$$

$$= \frac{5}{\pi} \tan^{-1}(5) = 2.19 \, \text{J}$$

The fraction of the input energy in the output signal is 2.19/2.5, or about 87.6%. ∎

Exercise 13–30

(a) Find the total 1-Ω energy carried by a double-sided exponential $f(t) = A\,e^{-\alpha|t|}$.
(b) Find the fraction of the total energy carried by the frequencies band $|\omega| < \alpha$.

Answers:

(a) A^2/α
(b) $0.5 + 1/\pi = 0.818$

EXAMPLE 13–26

The transfer function of a first-order low-pass filter is $H(\omega) = 100/(100 + j\omega)$. The input signal to the filter is $x(t) = 10\,e^{-20t}u(t)$.

(a) Find the percentage of the input energy in the filter output signal.
(b) Find the percentage of the output energy that lies within the filter passband.

SOLUTION:
(a) The input signal is the same as in Example 13–25, where we found $W_{1\Omega,\text{IN}} = 2.5\,\text{J}$.

The following commented MATLAB code performs all of the computations associated with finding the percentage of the input energy in the filter output signal:

```
% Create symbolic variables
syms t w real
% Define the input signal and find its Fourier transform
xt = 10*exp(-20*t)*heaviside(t);
Xw = fourier(xt,t,w);
% Define the transfer function
Hw = 100(100+j*w);
% Find the input energy
Win = double(int(xt^2,t,-inf,inf))
% Compute the output signal
Yw = Hw*Xw;
% Compute the output energy
Wout = double(int(Yw*conj(Yw),w,0,inf)pi)
% Find the percentage of input energy in the output
Wpercent = double(WoutWin)
```

The results are

```
Win =         2.5000
Wout =        2.0833
Wpercent =    0.8333
```

Therefore, 83.3% of the input energy is in the filter output signal.

(b) The cutoff frequency of the low-pass filter is 100 rad/s. The following MATLAB code computes the percentage of the output energy in the passband, given that we have already executed the MATLAB code presented in part (a).

```
% Find the output energy in the passband
Woutpb = double(int(Yw*conj(Yw),w,0,100)pi)
% Find the percentage of output energy in the passband
Wpbpercent = double(WoutpbWout)
```

The results are:

```
Woutpb =      2.0165
Wpbpercent =  0.9679
```

Therefore, 96.8% of the output energy lies within the filter passband. ∎

Exercise 13–31

What percentage of the total energy in $f(t) = 5e^{-10t}u(t)$ is carried by the frequency band $|\omega| > 5$ rad/s?

Answer: 70.5%

APPLICATION   **E X A M P L E   1 3 – 2 7**

The Fourier series serves as an introduction to the concept that a signal can be described by a spectrum that gives the distribution of amplitudes (and sometimes phases) of the sinusoidal components in a waveform. Radio, television, cell phones, satellite communication, and radar systems must confine their signal spectra to an allocated portion of the available electromagnetic spectrum. Spectral allocations are regulated by various governmental agencies since there are many potential users and a limited spectral resource. As a result, users must design their systems to operate within specified spectral limitations.

These limitations are often specified using the concept of **signal bandwidth**, defined as the frequency interval outside of which the amplitude spectrum is less than some specified value. Accordingly, signal bandwidth (B) can be quantified by the expression

$$B = f_U - f_L$$

where f_U and f_L are the upper and lower limits on the bandwidth interval. Bandwidth is a partial signal descriptor that places an upper bound on the spectral content outside of the interval $f_L < f < f_U$. *Caution*: The spectral content inside the interval can be less than this upper bound at some frequencies but must be less at every frequency outside the interval.

For a periodic waveform the lower frequency bound is $f_L = f_0$ when there is no dc component ($a_0 = 0$) and is $f_L = 0$ when a_0 is not zero. The upper bound is another matter. In general, the Fourier series is an infinite series, so that in theory its spectral content extends to infinite frequency. However, we have seen the amplitudes of the higher *harmonics* become negligibly small and make little contribution to the waveform's energy content. In other words, an acceptable approximation to a periodic waveform is obtained when its Fourier series is terminated at some designated $f_U = nf_0$.

One simple way to select nf_0 is to compare the amplitudes of the high-frequency harmonics with the amplitude of the largest ac component, usually the fundamental. Harmonics whose amplitudes are less than a specified fraction (say 5%) of the fundamental are ignored. Applying a 5% criterion to waveforms whose Fourier coefficients decrease as $1/n$ (square wave, rectangular pulse, sawtooth wave; see Figure 13–4) yields $f_U = 20f_0$. Waveforms whose harmonics fall off as $1/n^2$ (triangular wave, rectified sine waves) meet a 5% bound at about $f_U = 5f_0$. Regardless of the criteria used, the basic idea is that signal bandwidth is an important design consideration that in turn influences the bandwidth required in system components.

For example, accurate representation of arterial blood pressure waveforms requires the first eight harmonics of the pressure signal. To get a useful record of the arterial pressure waveform, the bandwidth (in Hz) of a measurement system must be at least eight times the maximum heart rate (in beats/second). Thus, designing a recording system for a maximum heart rate of 180 beats/minute requires a minimum bandwidth of 24 Hz.

APPLICATION   E X A M P L E   1 3 – 2 7

Digital signal processing uses samples of an analog waveform, as contrasted with analog processing, which operates on the entire waveform. Sampling refers to the process of selecting discrete values of a time-varying analog waveform for further processing. By far the most common method of sampling is to record the waveform amplitudes at equally spaced time intervals. The set of samples $v_k\{k = 1, 2, 3, \ldots\}$ of an analog waveform $v(t)$ is defined as

$$v_k = v(kT_S)$$

where k is an integer and T_S is the time interval between successive samples. Figure 13–27 shows an example of sampling an analog waveform.

The digitized samples can be stored in a computer or some other medium such as a compact disc. These samples become the only means of describing the original analog signal. How good of a description can they be? From our experience we know that a reasonable facsimile can be obtained if the samples are closely spaced. Intuitively it might seem that exact reconstruction of the analog waveform would require the time between samples to approach zero. Surprisingly, exact reproduction is possible from samples taken at finite intervals.

> *An analog waveform $v(t)$ whose spectral content falls below f_{max} can be reproduced exactly from samples $v_k = v(kT_s)$ if the sampling rate $f_S = 1/T_S$ is greater than $2f_{max}$.*

This statement, known as the **sampling theorem**, is one of the key principles of signal processing. The theorem states that exact recovery requires a minimum sampling rate of $2f_{max}$.[4] Analog waveforms are usually sampled at a rate higher than the minimum. For example, the industry standard sampling rate for recording digital

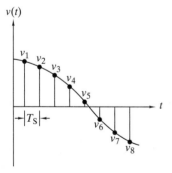

FIGURE 13–27 *Sampled signal.*

---

[4]This minimum sampling rate is called the *Nyquist rate*. The name honors Harry Nyquist, who along with Claude Shannon and others made several key breakthroughs in signal processing in the era from 1920 to 1950 while working at the Bell Telephone Laboratories.

audio signals is $f_S = 44.1$ kHz, which is slightly more than twice the generally accepted upper limit on human hearing at $f_{max} = 20$ kHz.

The important point is that the minimum sampling rate is determined by the spectral content of the original analog waveform. Waveforms whose sinusoidal components all fall in the band $0 \leq f \leq f_{max}$ are said to be **strictly band-limited**. For example, the waveform

$$v(t) = 8 + 5\cos(2\pi 200t) + 3\cos(2\pi 400t) + 1.5\cos(2\pi 1200t)$$

is strictly band-limited with $f_{max} = 1.2$ kHz and can be reproduced exactly from samples taken at any $f_S > 2.4$ kHz.

Most analog signals are not strictly band-limited but usually have an upper frequency limit beyond which the spectral content is of no interest. For example, a periodic waveform is not strictly band-limited because its Fourier series is an infinite sum of harmonics at nf_0. However, the amplitudes of the higher harmonic become negligibly small as $n \to \infty$. A band-limited approximation to a periodic waveform is obtained by defining $f_{max} = nf_0$, where nf_0 is the harmonic beyond which the spectral amplitudes are less than some specified value.

Examples of the maximum frequency of interest for some biomedical signals are:

Type	f_{max}
Electrocardiogram (ECG)	250 Hz
Blood flow	25 Hz
Respiratory rate	10 Hz
Electromyogram (EMG)	10 kHz

For these signals, the spectral content above the listed f_{max} does not contain useful diagnostic information and can be ignored. As a result, biomedical signals have minimum sampling rates ranging from $f_S = 20$ Hz to $f_S = 20$ kHz.

But we cannot simply ignore the spectral content above f_{max}. It turns out that sampling signals that are not strictly band-limited causes *aliasing*, a process by which seemingly negligible out-of-band spectral content reappears as in-band distortion. The answer to the aliasing problem is to filter the analog signal prior to sampling. These *anti-aliasing filters* must pass spectral content up to the *highest frequency of interest*, f_{max} and suppress the spectral content above the *lowest aliasing frequency* at $f_S - f_{max}$ where f_S is the sampling frequency.

Figure 13–28 shows the gain response of an anti-aliasing filter used in telecommunication. This low-pass filter has 0dB gain at dc, −3dB gain at $f_{max} = 3.3$ kHz and less than −60 dB gain at $2f_{max}$. This type of gain response cannot be achieved by the first- and second-order filters studied in Chapter 12. This type of performance calls for higher-order filters of the type discussed in Chapter 14.

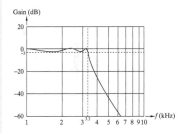

FIGURE 13–28 *Antialiasing filter*

SUMMARY

- The Fourier series resolves a periodic waveform into a dc component plus an ac component containing an infinite sum of harmonic sinusoids. The dc component is equal to the average value of the waveform. The amplitudes of the sine and cosine terms in the ac component are called Fourier coefficients.

- Waveform symmetries cause the amplitudes of some terms in a Fourier series to be zero. Even symmetry causes all of the sine terms in the ac component to be zero. Odd symmetry causes all of the cosine terms to be zero. Half-wave symmetry causes all of the even harmonics to be zero.

- An alternative form of the Fourier series represents each harmonic in the ac component by its amplitude and phase angle. A plot of amplitudes versus frequency is called the amplitude spectrum. A plot of phase angles versus frequency is called the phase spectrum.

• The steady-state response of a linear circuit for a periodic driving force can be found by first finding the steady-state response due to each term in the Fourier series of the input. The Fourier series of the steady-state response is then found by adding (superposing) responses due to each term acting alone. The individual responses can be found using either phasor or s-domain analysis.

• The rms value of a periodic waveform is equal to the square root of the sum of the square of the dc value and the square of the rms value of each of the ac components. The average power delivered by a periodic waveform is equal to the average power delivered by the dc component plus the sum of average power delivered by each of the ac components.

• The Fourier transformation applies to aperiodic waveforms that may be causal or noncausal.

• Sufficient conditions for the existence of $F(\omega)$ are that $f(t)$ be absolutely integrable and have a finite number of discontinuities. Signals not meeting these conditions may still have a Fourier transform. When it exists, a Fourier transform is unique.

• If $f(t)$ is causal and absolutely integrable, then its Fourier transform can be obtained from its Laplace transform $F(s)$ by replacing s by $j\omega$. A causal waveform $f(t)$ is absolutely integrable if all of the poles of $F(s)$ lie in the left half of the s plane.

• Circuit analysis using Fourier transforms involves finding the input transform, transforming the circuit into the frequency domain, solving for the unknown response transform, and performing the inverse transform to obtain the response waveform.

• The frequency-domain transfer function $H(\omega)$ is the Fourier transform of the system impulse response. The system output transform is $Y(\omega) = H(\omega)X(\omega)$, where $X(\omega)$ is the Fourier transform of the input.

• Parseval's theorem relates the energy carried by a waveform $f(t)$ to its Fourier transform $F(\omega)$. The theorem allows us to find the percentage of the total input energy in the output of a filter and to find the percentage of the output energy carried in specified frequency bands.

PROBLEMS

13–1 Derive expressions for the Fourier coefficients of the periodic waveform in Figure P13–1.

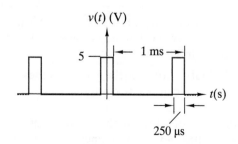

FIGURE P13–1

13–2 The equation for the first cycle ($0 \le t \le T_0$) of a periodic pulse train is

$$v(t) = V_A[2u(t) - 2u(t - T_0/2)]V$$

(a) Sketch the first two cycles of the waveform.
(b) Derive expressions for the Fourier coefficients a_n and b_n.

13–3 Derive expressions for the Fourier coefficients of the periodic waveform in Figure P13–3.

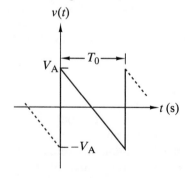

FIGURE P13–3

13–4 The equation for the first cycle ($0 \le t \le T_0$) of a periodic waveform is

$$v(t) = V_A e^{-t/2T_0} V$$

(a) Sketch the first two cycles of the waveform.

(b) Derive expressions for the Fourier coefficients a_n and b_n.

13–5 Derive expressions for the Fourier coefficients of the periodic waveform in Figure P13–5.

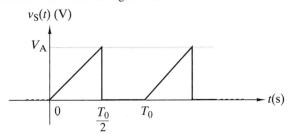

FIGURE P13–5

13–6 Use the results in Figure 13–4 to calculate the Fourier coefficients of the square wave in Figure P13–6. Write an expression for the first four nonzero terms in the Fourier series.

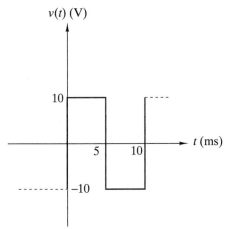

FIGURE P13–6

13–7 Use the results in Figure 13–4 to calculate the Fourier coefficients of the shifted triangular wave in Figure P13–7. Write an expression for the first four nonzero terms in the Fourier series.

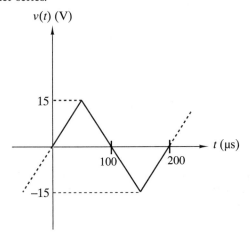

FIGURE P13–7

13–8 Derive expressions for the Fourier coefficients of the periodic waveform in Figure P13–8.

(a) Write an expression for the first four nonzero terms in the Fourier series.

(b) Plot the spectrum of the Fourier coefficients a_n and b_n.

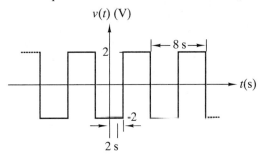

FIGURE P13–8

13–9 A particular periodic waveform with a period of 10 ms has the following Fourier coefficients

$$a_0 = 4, \quad a_n = \frac{8}{n\pi} \sin\frac{n\pi}{2} \cos\frac{n\pi}{4}, \quad b_n = \frac{-8}{n\pi} \sin\frac{n\pi}{2} \sin\frac{n\pi}{4}$$

(a) Write an expression for the terms in the Fourier series a_0 and $n = 1$ through 8.

(b) Convert your expression to amplitude and phase form and plot its spectrum.

13–10 Use the results in Figure 13–4 to calculate the Fourier coefficients of the full-wave rectified sine wave in Figure P13–10. Use MATLAB to verify your results. Write an expression for the first four nonzero terms in the Fourier series.

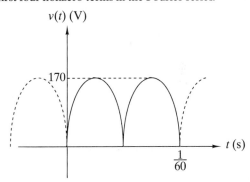

FIGURE P13–10

13–11 A half-wave rectified sine wave has an amplitude of 339 V and a fundamental frequency of 50 Hz. Use the results in Figure 13–4 to write an expression for the first four nonzero terms in the Fourier series. Use MATLAB to plot the amplitude spectrum of the signal.

13–12 The waveform $f(t)$ is a 2-kHz triangular wave with a *peak-to-peak* amplitude of 8 V. Use the results in Figure 13–4 to write an expression for the first four nonzero terms in the Fourier series of $g(t) = 4 + f(t)$ and plot its amplitude spectrum. Use MATLAB to plot two periods of $g(t)$ and an estimate for $g(t)$ using the first four nonzero terms in the

Fourier series. Comment on the differences between the two waveforms.

13–13 A sawtooth wave has *peak-to-peak* amplitude of 24 V and a fundamental frequency of 2 kHz. Use the results in Figure 13–4 to write an expression for the first four nonzero terms in the Fourier series and plot the amplitude spectrum of the signal. Use MATLAB to plot two periods of the original signal and an estimate for the signal using the first four nonzero terms in the Fourier series. Comment on the differences between the two waveforms.

13–14 The equation for the first cycle $(0 \leq t \leq T_0)$ of a periodic pulse train is

$$v(t) = V_A[u(3t - T_0) - u(3t - 2T_0)] \text{ V}$$

(a) Sketch the first two cycles of the waveform and identify a related signal in Figure 13–4.
(b) Use the Fourier series of the related signal to find the Fourier coefficients of $v(t)$.

13–15 The equation for a periodic waveform is

$$v(t) = V_A[\sin(4\pi/T_0) + |\sin(4\pi t/T_0)|]$$

(a) Sketch the first two cycles of the waveform and identify a related signal in Figure 13–4.
(b) Use the Fourier series of the related signal to find the Fourier coefficients of $v(t)$.
(c) Use MATLAB to sketch an estimate for the signal using the Fourier coefficients for $n \leq 6$.

13–16 The first four terms in the Fourier series of a periodic waveform are

$$v(t) = 15\left[\sin(150\pi t) - \frac{1}{9}\sin(450\pi t) + \frac{1}{25}\sin(750\pi t) - \frac{1}{49}\sin(1050\pi t)\right] \text{V}$$

(a) Find the period and fundamental frequency in rad/s and Hz. Identify the harmonics present in the first four terms.
(b) Identify the symmetry features of the waveform.
(c) Write the first four terms in the Fourier series for the waveform $v(t - T_0/4)$.

13–17 The first five terms in the Fourier series of a periodic waveform are

$$v(t) = 5 + 25\left[\frac{\pi}{4}\cos(500t) - \frac{1}{3}\cos(1000t) - \frac{1}{15}\cos(1500t) - \frac{1}{35}\cos(2500t)\right] \text{V}$$

(a) Find the period and fundamental frequency in rad/s and Hz. Identify the harmonics present in the first five terms.
(b) Use MATLAB to plot two periods of $v(t)$.
(c) Identify the symmetry features of the waveform.
(d) Write the first five terms of the Fourier series for the waveform $v(-t)$.

13–18 The equation for a full-wave rectified cosine is $v(t) = V_A$ $|\cos(2\pi t/T_0)|$ V.
(a) Sketch $v(t)$ for $-T_0 \leq t \leq T_0$.
(b) Compute the Fourier coefficients for $v(t)$.
(c) Use the Fourier coefficients to plot an estimate for $v(t)$.

13–19 Find the Fourier series for the waveform in Figure 13–19.

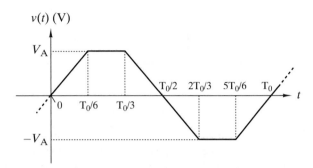

FIGURE P13–19

OBJECTIVE 13–2 FOURIER SERIES AND CIRCUIT ANALYSIS (SECT. 13–4)

(a) Given a linear circuit with a periodic input waveform, find the Fourier series of a steady-state response.
(b) Given a network function with a periodic input, find the amplitude and phase spectra of the steady-state output.

See Examples 13–6, 13–7, 13–8 and Exercises 13–6, 13–7, 13–8.

13–20 The periodic pulse train in Figure P13–20 is applied to the *RL* circuit shown in the figure.
(a) Use the results in text Figure 13–4 to find the Fourier coefficients of the input for $V_A = 10$ V, $T_0 = \pi$ ms, and $T = T_0/4$.
(b) Find the first four nonzero terms in the Fourier series of $v_0(t)$ for $R = 400 \, \Omega$, and $L = 100$ mH.

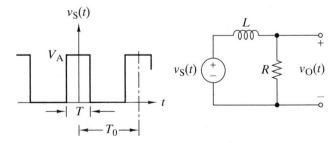

FIGURE P13–20

13–21 The periodic triangular wave in Figure P13–21 is applied to the *RC* circuit shown in the figure. The Fourier coefficients of the input are:

$$a_0 = 0 \quad a_n = 0 \quad b_n = \frac{8V_A}{(n\pi)^2}\sin\left(\frac{n\pi}{2}\right)$$

If $V_A = 20$ V and $T_0 = 2\pi$ ms, find the first four nonzero terms in the Fourier series of $v_O(t)$ for $R = 10$ kΩ, and $C = 0.05$ μF.

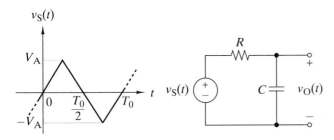

FIGURE P13–21

13–22 The periodic sawtooth wave in Figure P13–22 drives the OP AMP circuit shown in the figure.
(a) Use the results in text Figure 13–4 to find the Fourier coefficients of the input for $V_A = 5$V and $T_0 = 4\pi$ ms.
(b) Find the first four nonzero terms in the Fourier series of $v_O(t)$ for $R_1 = 20$ kΩ, $R_2 = 200$ kΩ, and $C = 0.1$ μF.

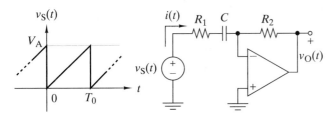

FIGURE P13–22

13–23 The periodic sawtooth wave in Figure P13–22 above drives the OP AMP circuit shown in the figure.
(a) Use the results in text Figure 13–4 to find the Fourier coefficients of the input for $V_A = 8$ V and $T_0 = 20\pi$ ms
(b) Find the first four nonzero terms in the Fourier series of $i(t)$ for $R_1 = 100$ kΩ, $R_2 = 50$ kΩ, and $C = 0.4$ μF.
(c) Find the first four nonzero terms in the Fourier series of $v_O(t)$ with the same element values as in part (b).

13–24 **D** **(a)** Design a low-pass OP AMP circuit to pass only the fundamental and the next nonzero harmonic of a 20π ms square wave. The gain of the OP AMP should be +10.
(b) Find the first four nonzero terms in the Fourier series of the output of your filter.

13–25 **D** **(a)** Design a passive low-pass RC filter to block the fundamental and all harmonics from a full-wave rectified sinusoidal waveform. Use the results of text Figure 13–4 to find the Fourier coefficients of the input for $V_A = 170$ V, $T_0 = 16.6$ ms.
(b) Find the first four nonzero terms in the Fourier series of the output of your filter.

13–26 The periodic triangular wave in Figure P13–26 is applied to the RLC circuit shown in the figure.
(a) Use the results in text Figure 13–4 to find the Fourier coefficients of the input for $V_A = 10$ V and $T_0 = 400\pi$ μs.

(b) Find the amplitude of the first five nonzero terms in the Fourier series for $i(t)$ when $R = 1$ Ω, $L = 8$ mH, and $C = 0.2$ μF. What term in the Fourier series tends to dominate the response? Explain.

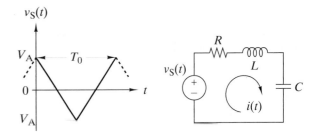

FIGURE P13–26

13–27 A sawtooth wave with $V_A = 12$ V and $T_0 = 10\pi$ ms drives a circuit with a transfer function $T(s) = s/(s + 400)$. Find the amplitude of the first five nonzero terms in the Fourier series of the steady-state output. Construct plots of the amplitude spectra for the input and output waveforms and comment on any differences.

13–28 Repeat Problem 13–27 for $T(s) = 400/(s + 400)$.

13–29 **D** Design a tuned RLC filter to pass the third harmonic of a triangular wave.
(a) Use the results in Figure 13–4 to find the Fourier coefficients of the input for $V_A = 8$ V and $T_0 = 20\pi$ ms. Design your filter with a Q of 10.
(b) Compare the magnitudes of the fundamental and of the fifth harmonic with that of the third harmonic at the input and output of the tuned filter.

13–30 **D** Design a notch RLC filter to block the third harmonic of a triangular wave.
(a) Use the results in Figure 13–4 to find the Fourier coefficients of the input for $V_A = 8$ V and $T_0 = 20\pi$ ms. Design your filter with a Q of 20.
(b) Compare the magnitudes of the fundamental and of the fifth harmonic with that of the third harmonic at the input and output of the tuned filter.

13–31 The voltage across a 1000-pF capacitor is a triangular wave with $V_A = 150$ V and $f_0 = 1$ kHz. Construct plots of the amplitude spectra of the capacitor voltage and current. Discuss any differences in spectral content.

13–32 An ideal time delay is a signal processor whose output is $v_O(t) = v_{IN}(t - T_D)$. Write an expression for $v_O(t)$ for $T_D = 0.5$ ms and

$$v_{IN}(t) = 10 + 10\cos(2\pi 500t) + 2.5\cos(2\pi 1000t)$$
$$+ 0.625\cos(2\pi 4000t) \text{ V}$$

Discuss the spectral changes caused by the time delay.

13–33 A sawtooth wave with $V_A = 10$ V and $T_0 = 20\pi$ ms drives a circuit whose transfer function is

$$T(s) = \frac{100s}{(s + 50)^2 + 400^2}$$

(a) Find the amplitude of the first four nonzero terms in the Fourier series of the steady-state output. What term in the Fourier series tends to dominate the response? Explain.
(b) Repeat Part (a) for a new transfer function $X(s) = T(s)/s$. What is the key difference between $T(s)$ and $X(s)$?

OBJECTIVE 13–3 RMS VALUE AND AVERAGE POWER (SECT. 13–5)

(a) Given a periodic waveform, find the rms value of the waveform and the average power delivered to a specified load.
(b) Given the Fourier series of a periodic waveform, find the fraction of the average power carried by specified components and estimate the average power delivered to a specified load.
See Examples 13–9, 13–10 and Exercises 13–9, 13–10.

13–34 The current through a 500-Ω resistor is

$$i(t) = 50 + 36\cos(120\pi t - 30°) - 12\cos(360\pi t + 45°)\,\text{mA}$$

Find the rms value of the current and the average power delivered to the resistor.

13–35 The voltage across a 50-Ω resistor is

$$v(t) = 60 + 24\sin(200\pi t) - 8\sin(600\pi t) + 4.8\sin(1000\pi t)\,\text{V}$$

(a) Find expressions for the current through the resistor and the power dissipated by the resistor.
(b) Find the average of the power expression by integrating over one period of the waveform and dividing by the period.
(c) Find the rms value of the voltage by applying Eq (13–19) to the expression for $v(t)$.
(d) Find the rms value of the voltage by applying Eq (13–24) to the Fourier coefficient of $v(t)$ and compare the result to the answer in Part (c).
(e) Compute the average power dissipated by the resistor using $P = V_{\text{rms}}^2/R$ and compare the result to the answer in Part (b).

13–36 Find the rms value of a square wave. Find the fraction of the total average power carried by the first three nonzero ac components in the Fourier series.

13–37 Find the rms value of a triangular wave. Find the fraction of the total average power carried by the first three nonzero ac components in the Fourier series. Compare with the results found in Problem 13–36.

13–38 Find the rms value of a parabolic wave. Find the fraction of the total average power carried by the first three nonzero ac components in the Fourier series. Compare with the results found in Problem 13–36.

13–39 Use MATLAB to find the rms value of a half-wave rectified sine wave. Find the fraction of the total average power carried by the dc component plus the first three nonzero ac components in the Fourier series.

13–40 Find the rms value of the periodic waveform in Figure P13–40 and the average power the waveform delivers to a resistor. Find the dc component of the waveform and the average power carried by the dc component. What

fraction of the total average power is carried by the dc component? What fraction is carried by the ac components?

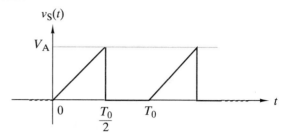

FIGURE P13–40

13–41 Repeat Problem 13–40 for the periodic waveform in Figure P13–41.

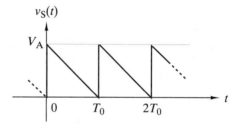

FIGURE P13–41

13–42 A first-order low-pass filter has a cutoff frequency of 600 rad/s and a passband gain of 20 dB. The input to the filter is $v(t) = 10\cos 300t + 6\cos 1200t$ V. Find the rms value of the steady-state output.

13–43 Repeat Problem 13–42 for a first-order high-pass filter with the same cutoff frequency and passband gain.

13–44 Estimate the rms value of the periodic voltage

$$v(t) = V_A\left[2 - \cos(\omega_0 t) + \frac{1}{3}\cos(3\omega_0 t) - \frac{1}{5}\cos(5\omega_0 t)\right.$$
$$\left. + \frac{1}{7}\cos(7\omega_0 t) - \ldots\right]\text{V}$$

13–45 The input to the circuit in Figure P13–45 is the voltage

$$v_s(t) = 15\cos(2\pi 2000t) + 5\cos(2\pi 6000t)\,\text{V}$$

Calculate the average power delivered to the 50-Ω load resistor.

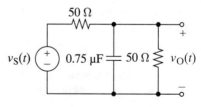

FIGURE P13–45

13–46 Find an expression for the average power delivered to a resistor R by a triangular wave voltage with amplitude V_A and period T_0. How many components of the Fourier series are required to account for 98% of the average power carried by the waveform?

OBJECTIVE 13–4 THE FOURIER TRANSFORM (SECT. 13–6)

(a) Given an equation or graph of an aperiodic waveform, derive an expression for the Fourier transform.
(b) Given an equation or graph of a Fourier transform, derive an expression for the corresponding waveform.
(c) Understand the relationship between Fourier transforms and Laplace transforms and know the conditions under which Laplace transform techniques can be used to compute Fourier transforms. See Examples 13–11, 13–12, 13–13, 13–14, 13–15, 13–16, 13–17, 13–18 and Exercises 13–11, 13–12, 13–13, 13–14, 13–15, 13–16, 13–17, 13–18, 13–19, 13–20, 13–21, 13–22, 13–23.

13–47 Use the defining integral to find the Fourier transform of $f(t) = A[u(t) - u(t - 3)]$.

13–48 Use the defining integral to find the Fourier transform of $f(t) = A[u(t + T_0) - u(t - T_0)]$.

13–49 Use the defining integral to find the Fourier transform of $f(t) = At[u(t) - u(t - 1)]$.

13–50 Use MATLAB and the defining integral to find the Fourier transform of the following waveform:

$$f(t) = A\cos(\pi t/4)[u(t) - u(t - 4)]$$

13–51 Use the inversion integral to find the inverse transform of the following function:

$$F(\omega) = 10\pi[u(\omega + 1) - u(\omega - 1)]$$

13–52 Use MATLAB and the inversion integral to find the inverse transform of the following function:

$$F(\omega) = j\pi 8[2u(\omega + 2) - 4u(\omega) + 2u(\omega - 2)]$$

13–53 Use MATLAB and the inversion integral to find the inverse transform of the following function:

$$F(\omega) = \sin(\pi\omega/4)[u(\omega + 4) - u(\omega - 4)]$$

13–54 Find the inverse transforms of the following functions:

(a) $F_1(\omega) = \dfrac{5000}{(j\omega + 50)(j\omega + 100)}$

(b) $F_2(\omega) = \dfrac{40j\omega}{(j\omega + 20)(j\omega + 40)}$

13–55 Find the inverse transforms of the following functions:

(a) $F_1(\omega) = \dfrac{1600}{j\omega(j\omega + 20)(j\omega + 40)}$

(b) $F_2(\omega) = \dfrac{-\omega^2}{j\omega(j\omega + 20)(j\omega + 40)}$

13–56 Find the Fourier transforms of the following waveforms:
(a) $f_1(t) = 2u(-t) - 2$
(b) $f_2(t) = -2\,\text{sgn}(t) - 2u(-t)$
(c) $f_3(t) = -\text{sgn}(t) - 1$

13–57 Find the Fourier transforms of the following waveforms:
(a) $f_1(t) = 2e^{-2t}\,u(t) - 2\,\text{sgn}(t)$
(b) $f_2(t) = 2e^{-2t}\,u(t) - 2u(t)$

13–58 Find the Fourier transforms of the following waveforms:
(a) $f_1(t) = 2\sin(t) - 2\cos(t)$
(b) $f_2(t) = (2/t)\sin(t) - 2\cos(t)$

13–59 Find the Fourier transforms of the following waveforms:
(a) $f_1(t) = 5\cos[2\pi(t - 3)]$
(b) $f_2(t) = 2e^{j4t}\text{sgn}(t)$

13–60 Find the inverse transforms of the following functions:
(a) $F_1(\omega) = 4\pi\delta(\omega) + 4\pi\delta(\omega - 2) + 4\pi\delta(\omega - 4)$
(b) $F_2(\omega) = 4\pi\delta(\omega) - j2/\omega + 4\pi\delta(\omega - 2)$
(c) $F_3(\omega) = 2\pi\delta(\omega) - j2/\omega$

13–61 Use the duality property to find the inverse transforms of the following functions:
(a) $F_1(\omega) = 3\cos(10\omega)$
(b) $F_2(\omega) = 6u(\omega) - 3$
(c) $F_3(\omega) = 6e^{-|3\omega|}$

13–62 Use the time-shifting property to find the inverse transforms of the following functions:
(a) $F_1(\omega) = [6\pi\delta(\omega) - j3/\omega]e^{-j3\omega}$
(b) $F_2(\omega) = 9e^{-j4\omega}/(j\omega + 5)$
(c) $F_3(\omega) = 2\cos(5\omega)/j\omega$

13–63 Given that the Fourier transform of $f(t)$ is

$$F(\omega) = \frac{1600}{(j\omega + 20)(j\omega + 40)}$$

use the integration property to find the waveform

$$g(t) = \int_{-\infty}^{t} f(x)dx$$

13–64 Use the reversal property to show that

$$\mathscr{F}\left\{Ae^{-\alpha|t|}\,\text{sgn}(t)\right\} = \frac{-2Aj\omega}{\omega^2 + \alpha^2}$$

13–65 Use the frequency shifting property to prove the modulation property

$$\mathscr{F}\{f(t)\sin(\beta t)\} = \frac{F(\omega - \beta)}{2j} - \frac{F(\omega + \beta)}{2j}$$

13–66 Use the frequency shifting property to show that

$$\mathscr{F}\{\cos(\beta t)u(t)\} = \frac{j\omega}{\beta^2 - \omega^2} + \frac{\pi}{2}[\delta(\omega - \beta) + \delta(\omega + \beta)]$$

OBJECTIVE 13–5 FOURIER TRANSFORMS AND CIRCUIT ANALYSIS (SECT. 13–7, 13–8, 13–9)

(a) Given a linear circuit, use Fourier transform techniques to determine the output response for aperiodic input signals.
(b) Given a linear circuit, use Parseval's theorem to determine the energy content of selected signals.
See Examples 13–19, 13–20, 13–21, 13–22, 13–23, 13–24, 13–25, 13–26, 13–27, 13–28 and Exercises 13–24, 13–25, 13–26, 13–27, 13–28, 13–29, 13–30, 13–31.

13–67 The input in Figure P13–67 is $v_1(t) = 5e^{-|t|}$ V. Use Fourier transforms to find $v_2(t)$.

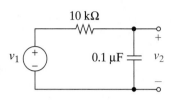

FIGURE P13–67

13–68 The input in Figure P13–67 is $v_1(t) = 10 \, \text{sgn}(t)$ V. Use Fourier transforms to find $v_2(t)$.

13–69 The input in Figure P13–69 is $v_1(t) = 2 \, \text{sgn}(t)$ V. Use Fourier transforms to find $v_2(t)$.

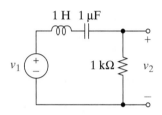

FIGURE P13–69

13–70 The input in Figure P13–70 is $v_1(t) = 10e^{4t}u(-t)$ V. Use Fourier transforms to find $v_2(t)$.

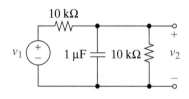

FIGURE P13–70

13–71 The input in Figure P13–70 is $v_1(t) = 10 \, \text{sgn}(t)$ V. Use Fourier transforms to find $v_2(t)$.

13–72 The input in Figure P13–72 is $v_1(t) = 5u(-t)$ V. Use Fourier transforms to find $v_2(t)$.

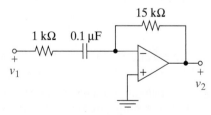

FIGURE P13–72

13–73 The input in Figure P13–72 is $v_1(t) = 10e^{-4|t|}$ V. Use Fourier transforms to find $v_2(t)$.

13–74 The impulse response of a linear system is $h(t) = e^{-2t}u(t)$. Find the output for an input $x(t) = u(-t)$.

13–75 The impulse response of a linear system is $h(t) = e^{-2|t|}$. Find the output for an input $x(t) = u(-t)$.

13–76 The impulse response of a linear system is $h(t) = 2\delta(t) - 4e^{-t}u(t)$. Use MATLAB and Fourier transform techniques to find the output for an input $x(t) = \text{sgn}(-t)$.

13–77 The impulse response of a linear system is $h(t) = A[\delta(t) - \alpha e^{-\alpha t}u(t)]$, with $\alpha > 0$. Let $A = 5$ and $\alpha = 2$ and use MATLAB to plot $|H(\omega)|$. On the same axes, plot $|H(\omega)|$ for $A = 5$ and $\alpha = 4$. Describe the system frequency response and the influence of the parameter α.

13–78 The impulse response of a linear system is $h(t) = A[\delta(t) - \sin(\beta t)/\pi t]$. Let $A = 5$ and $\beta = 2$ and use MATLAB to plot $|H(\omega)|$. On the same axes, plot $|H(\omega)|$ for $A = 5$ and $\beta = 4$. Describe the system frequency response and the influence of the parameter β. You may need to use the following equality in your solution: $\frac{\sin(\beta t)}{\pi t} = \frac{1}{2j}\left(e^{j\beta t} - e^{-j\beta t}\right)$.

13–79 The impulse response of a linear system is $h(t) = -Ae^{-\alpha t}u(t) + Ae^{-\alpha(-t)}u(-t)$, with $\alpha > 0$. Let $A = 10$ and $\alpha = 2.5$ and use MATLAB to plot $|H(\omega)|$. On the same axes, plot $|H(\omega)|$ for $A = 10$ and $\alpha = 5$. Describe the system frequency response and the influence of the parameter α.

13–80 The frequency response of a linear system is shown in Figure P13–80. Find the system impulse response $h(t)$.

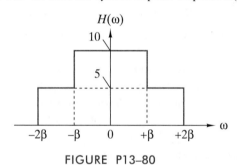

FIGURE P13–80

13–81 The frequency response of a linear system is shown in Figure P13–81. Find the system impulse response $h(t)$.

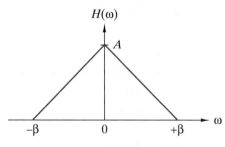

FIGURE P13–81

13–82 Find the 1-Ω energy carried by the signal $F(\omega) = 8/(\omega^2 + 16)$.

13–83 Compute the 1-Ω energy carried by the signal $f(t) = 5e^{2.5t}u(-t)$.

13–84 Find the 1-Ω energy carried by the signal

$$F(\omega) = \frac{j\omega A}{\omega^2 + \alpha^2}$$

Then find the percentage of the 1-Ω energy carried in the frequency band $|\omega| \leq \alpha$.

13–85 The transfer function of an ideal high-pass filter is $H(\omega) = 1$ for $|\omega| \geq 2$ krad/s. The filter input signal is $x(t) = 10\,e^{-2000t}\,u(t)$. Find the 1-Ω energy carried by the input signal and the percentage of the input energy that appears in the output.

13–86 The impulse response of a filter is $h(t) = 100\,e^{-400t}u(t)$. Find the 1-Ω energy in the output signal when the input is $x(t) = 5\,e^{-100t}u(t)$.

13–87 The impulse response of a filter is $h(t) = 100\,e^{-10t}u(t)$. Find the 1-Ω energy in the output signal when the input is $x(t) = \delta(t)$.

13–88 The current in a 100-Ω resistor is $i(t) = -4\,u(t+2) + 8\,u(t) - 4\,u(t-2)$. Find the total energy delivered to the resistor.

13–89 The transfer function of an ideal bandpass filter is $H(\omega) = 1$ for $500 \leq \omega \leq 1000$ rad/s. Use MATLAB to find the 1-Ω energy carried by the output signal when the input is $x(t) = 15\,e^{-1000t}u(t)$. What percentage of the input signal energy.

INTEGRATING PROBLEMS

13–90 Fourier Series from a Bode Plot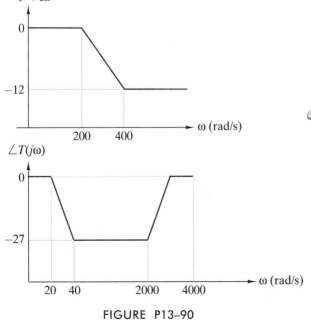
The transfer function of a linear circuit has the straight-line gain and phase Bode plots in Figure P13–90. The first four terms in the Fourier series of a periodic input $v_1(t)$ to the circuit are

$$v_1(t) = 42\cos(100t) + 14\cos(300t) + 8.4\cos(500t) + 6\cos(700t)\ \text{V}$$

Estimate the amplitudes and phase angles of the first four terms in the Fourier series of the steady-state output $v_2(t)$.

FIGURE P13–90

13–91 Spectrum of a Periodic Impulse Train
A periodic impulse train can be written as

$$x(t) = T_0 \sum_{n=-\infty}^{\infty} \delta(t - nT_0)$$

Find the Fourier coefficients of $x(t)$. Plot the amplitude spectrum and comment on the frequencies contained in the impulse train.

13–92 Power Supply Filter Design
The input to a power supply filter is a full-wave rectified sine wave with $f_0 = 60$ Hz. The filter is a first-order low pass with unity dc gain. Select the cutoff frequency of the filter so that all of the ac components in the filter output are all less than 1% of the dc component.

13–93 Spectrum Analyzer Calibration
A certain spectrum analyzer measures the average power delivered to a calibrated resistor by the individual harmonics of periodic waveforms. The calibration of the analyzer has been checked by applying a 1-MHz square wave and the following results reported

f(MHz)	1	3	5	7	9	11
P(dBm)	12.1	2.56	−1.88	−4.80	−6.98	−8.73

The reported power in dBm is $P = 10\,\text{Log}(P_n)$, where P_n is the average power delivered by the nth harmonic in mW. Is the spectrum analyzer correctly calibrated? (note: dbm is the power ratio of the output relative to a 1 mW input.)

13–94 Virtual Keyboard Design
Electronic keyboards are designed using the following equation that assigns particular frequencies to each of the 88 keys in a standard piano keyboard,

$$f(n) = 440\left(\sqrt[12]{2}\right)^{n-49}\ \text{Hz}$$

where n is the key number. There is a need for an amplifier that can pass middle C, key 40, but block keys 39 and 41. Design such a filter using an RLC tuned circuit with a gain K of 100.

13–95 Fourier Series and Fourier Transforms.
Given a rectangular pulse as shown in Figure 13–4, with amplitude A, width T, and period T_0, we can compute and plot the coefficients in the corresponding Fourier series. If we allow T_0 to increase to infinity, the waveform is a single pulse and the Fourier series approaches a scaled version of the Fourier transform. To see this graphically, use MATLAB to create the following series of plots. Let $A = 5$, $T = 1$, and $T_0 = 2$. Compute the Fourier series coefficients for $n = 0, 1, 2, \ldots\ 10\ T_0$. Create a stem plot of $(a_n \times T_0)$ on the vertical axis versus the (n/T_0) on the horizontal axis. Increment T_0 by one and repeat the stem plot. Create plots up until $T_0 = 20$ and comment on the behavior of the results. Now compute the Fourier transform of $f(t) = A[u(t+T/2) - u(t-T/2)]$. Evaluate $F(\omega)$ for $\omega = 0$ to 20π rad/s. On the same

axes as your final stem plot, plot $2F(\omega)$ versus $\omega/2\pi$. Comment on the results.

13–96 Impulse Generator ▲

Theoretically, an impulse has an amplitude spectrum that is constant at all frequencies. In practice, a constant spectrum across an infinite bandwidth cannot be achieved, nor is it really necessary. What is required is an amplitude spectrum that is "essentially" constant over the frequency range of interest. Under this concept, an impulse generator is a signal source that produces a pulse waveform whose amplitude spectrum does not vary more than a prescribed amount over a specified frequency range.

Consider a rectangular pulse:

$$f(t) = A[u(t + T/2) - u(t - T/2)]$$

Find $F(\omega)$ and sketch its amplitude spectrum $|F(\omega)|$. Select the pulse duration T such that the amplitude spectrum does not change by more than 10% over a frequency range from 1 MHz to 10 MHz.

CHAPTER 14 ACTIVE FILTER DESIGN

In its usual form the electric wave-filter transmits currents of all frequencies lying within one or more specified ranges, and excludes currents of all other frequencies.

George A. Campbell, 1922,
American Engineer

Some History Behind This Chapter

The electric filter was independently invented during World War I by George Campbell in the United States and by K. W. Wagner in Germany. Electric filters and vacuum tube amplifiers were key technologies that triggered the growth of telephone and radio communication systems in the 1920s and 1930s. The emergence of semiconductor electronics in the 1960s, especially the integrated circuit OP AMP, allowed the functions of filtering and amplification to be combined into what are now called active filters.

Why This Chapter Is Important Today

This is a fun chapter dedicated to the design of practical filters. After studying this chapter, you will be able to design a wide range of analog filters that find applications in instrumentation systems, audio systems, communication systems, and even digital systems. You will also learn how to evaluate different solutions to a filter design problem.

Chapter Sections

14–1 Active Filters
14–2 Second-Order Low-Pass and High-Pass Filters
14–3 Second-Order Bandpass and Bandstop Filters
14–4 Low-Pass Filter Design
14–5 Low-Pass Filter Evaluation
14–6 High-Pass Filter Design
14–7 Bandpass and Bandstop Filter Design

Chapter Learning Objectives

14-1 Second-Order Filter Analysis (Sects. 14–1, 14–2, 14–3)
(a) Given a second-order filter circuit, find a specified transfer function.
(b) Given the transfer function of a second-order circuit, develop a method of selecting the element values to achieve specified filter characteristics.

14-2 Second-Order Filter Design (Sects. 14–2, 14–3)
(a) Construct a second-order transfer function with specified filter characteristics.
(b) Design a second-order circuit with specified filter characteristics.

14-3 Low-Pass Filter Design (Sects 14–4, 14–5)
Given a low-pass filter specification:
(a) Construct a transfer function that meets the specification.
(b) Design a cascade of first- and second-order circuits that implements a given transfer function.
(c) Select the best design from competitive filter approaches based on specified frequency and step response criteria.

14-4 High-Pass, Bandpass, and Bandstop Filter Design (Sects. 14–6, 14–7)
Given a high-pass, bandpass, or bandstop filter specification:
(a) Construct a transfer function that meets the specification.
(b) Design a cascade or parallel connection of first- and second-order circuits that implements a given transfer function.
(c) Select the best design from competitive filter approaches based on specified frequency and step response criteria.

14–1 ACTIVE FILTERS

An electric filter is a signal processor that amplifies, attenuates, or otherwise reshapes the frequency content of input signals. There is a multitude of applications for filters. Engineers use filters in communication systems for noise suppression, to isolate a single communication band from many, to keep signals from one communication band from spilling over onto an adjacent band, and to recover modulated signals. In instrumentation systems, engineers use filters to select desired frequency components or eliminate undesired ones. In addition, instrumentation systems require filters to bandwidth-limit analog inputs prior to converting them to digital signals. They also need filters to convert digital signals back to analog representations. In audio systems, audio engineers use filters in crossover networks to send frequencies to different speakers, or in recording and playback applications where fine control of frequency components is crucial to the music industry. Biomedical systems use filters to interface physiological sensors with data-logging and diagnostic equipment.

Filters are classified as passive or active depending on the components used in their physical realization. **Passive filter** circuits contain only resistors, capacitors, and inductors. The *RLC* bandpass and bandstop circuits studied in Chapter 12 are passive filters. These circuits can be highly selective when losses are low, and the response is highly resonant. However, they cannot supply passband gains greater than one and suffer from loading effects that can nullify the chain rule in a cascade design.

In this chapter we emphasize an **active filter** as a circuit that contains only resistors, capacitors, and OP AMPs. These filters offer several advantages:

- They provide frequency selectivity comparable to passive *RLC* circuits plus passband gains greater than one.
- They have OP AMP outputs, which means that the chain rule applies in a cascade design.
- They do not require inductors, which can be large, lossy, and expensive in low-frequency applications.

Our study of active filter design will begin by building on our introductory knowledge from Chapter 12. We will introduce filter prototypes that easily can be designed and cascaded. We will look at the low-pass and high-pass Sallen-Key[1] configurations and learn how to design each using two different approaches. We will also look at some *RC* stand-alone, active tuned and notch filter configurations that can rival the *RLC* filters studied in Chapter 12. After gaining confidence on designing first- and second-order filters, we will learn how to cascade them to achieve steep roll-offs using three common cascade algorithms for active, multi-pole filter design.

Active filter design involves devising circuits that realize a given transfer function $T(s)$. Our design strategy is based on the familiar cascade connection in Figure 14–1. Under the chain rule, the overall transfer function is

$$T(s) = T_1(s) \times T_2(s) \times T_3(s) \times \cdots \times T_n(s)$$

The stages in the cascade are either first-order or second-order active filters. The real poles in $T(s)$ are produced using the first-order building block developed in Chapter 12. The complex poles are produced by second-order building blocks developed in the next two sections. The complex pole locations are specified by the customary second-order parameters—our old friends the damping ratio ζ and

---

[1]Both circuits belong to a family of circuits originally proposed by R. P. Sallen and E. L. Key in "A Practical Method of Designing *RC* Active Filters," *IRE Transactions on Circuit Theory*, Vol. CT-2, pp. 74–85, 1955. In 1955 the controlled sources were designed using vacuum tubes.

First stage   Second stage   Third stage

undamped natural frequency ω_0. Thus, we design an active filter by controlling the poles introduced by each stage in a cascade connection.

In a second-order transfer function, the denominator controls the location of the poles and thus the critical frequencies. The numerator controls the zeros and thus determines if a transfer function is a low-pass, high-pass, band-pass, or band-reject filter. Hence, the following transfer functions summarize the different filter types. Low-pass filters have the following form

$$T(s) = \frac{K}{(s/\omega_0)^2 + 2\zeta(s/\omega_0) + 1} \tag{14–1}$$

High-pass filters have the following form

$$T(s) = \frac{K(s/\omega_0)^2}{(s/\omega_0)^2 + 2\zeta(s/\omega_0) + 1} \tag{14–2}$$

Band-pass filters have the following form

$$T(s) = \frac{K(s/\omega_0)}{(s/\omega_0)^2 + 2\zeta(s/\omega_0) + 1} \tag{14–3}$$

Finally, band-stop filters have the following form

$$T(s) = \frac{K[(s/\omega_0)^2 + 1]}{(s/\omega_0)^2 + 2\zeta(s/\omega_0) + 1} \tag{14–4}$$

It is also useful to recall from Chapter 12 that the denominator of all of the different second-order filters can be written using different parameters as

$$T(s) = \frac{\text{numerator}}{(s/\omega_0)^2 + 2\zeta(s/\omega_0) + 1} = \frac{\text{numerator}}{(s/\omega_0)^2 + \frac{B}{\omega_0}(s/\omega_0) + 1} = \frac{\text{numerator}}{(s/\omega_0)^2 + \frac{1}{Q}(s/\omega_0) + 1}$$

where ζ is the damping ratio, B is the bandwidth and Q is the quality factor of the circuit. Different types of filter circuits are specified using different parameters. All filters specify critical frequencies and passband gains. In addition, high-pass and low-pass filters often specify damping ratios ζ, notch and tuned filters require certain B and Q, and wide bandpass filters often insist on a particular B. We will discuss these parameters in more detail as we cover each filter type.

For future reference, note that replacing s/ω_0 by ω_0/s converts the low-pass transfer function $T(s)$ in Eq. (14–1) into the high-pass transfer function $T(s)$ in Eq. (14–2). We will make good use of this low-pass to high-pass transformation in a later section.

14–2 SECOND-ORDER LOW-PASS AND HIGH-PASS FILTERS

The second-order building blocks developed in this section are the counterparts of the first-order low-pass and high-pass filters in Chapter 12. These second-order circuits all have the following features: (1) an OP AMP output (to avoid loading),

(2) two capacitors (to get two poles), and (3) at least one feedback path (to make the poles complex). We begin our development with the low-pass case.

SECOND-ORDER LOW-PASS FILTERS

From Eq. (14–1) the transfer function of a second-order low-pass filter has the form

$$T(s) = \frac{K}{(s/\omega_0)^2 + 2\zeta(s/\omega_0) + 1}$$

The gain response is found by substituting $j\omega$ for s to obtain

$$|T(j\omega)| = \frac{|K|}{\sqrt{[1 - (\omega/\omega_0)^2]^2 + 4\zeta^2(\omega/\omega_0)^2}} \tag{14–5}$$

At low frequency ($\omega \ll \omega_0$) the gain approaches $|T(j\omega)| \rightarrow |K| = |T(0)|$. At high frequency ($\omega \gg \omega_0$) the gain approaches $|T(j\omega)| \rightarrow |K|(\omega_0/\omega)^2$. Figure 14–2 presents Bode plots of these gain asymptotes (solid lines) and the actual gain (dashed curves) for several values of ζ. The key points to remember are the following:

- The low- and high-frequency gain asymptotes intersect at $\omega = \omega_0$.
- For $\omega < \omega_0$ the asymptotic gain equals the dc gain $|T(0)| = |K|$.
- For $\omega > \omega_0$ the slope of the asymptotic gain is $-40\,\text{dB/decade}$.
- The actual gain at $\omega = \omega_0$ is $|T(j\omega_0)| = |K|/2\zeta$.

From a design perspective, we say that ω_0 locates the corner frequency and ζ controls the gain in the vicinity of the corner.

The circuit in Figure 14–3(a) is analyzed in Chapter 11 (Example 11–6), where its voltage transfer function is shown to be

$$T(s) = \frac{V_2(s)}{V_1(s)} = \frac{\mu}{R_1 R_2 C_1 C_2 s^2 + (R_1 C_1 + R_1 C_2 + R_2 C_2 - \mu R_1 C_1)s + 1} \tag{14–6}$$

This is a second-order low-pass function with a dc gain of $|T(0)| = \mu$.

FIGURE 14–2 *Second-order low-pass gain responses.*

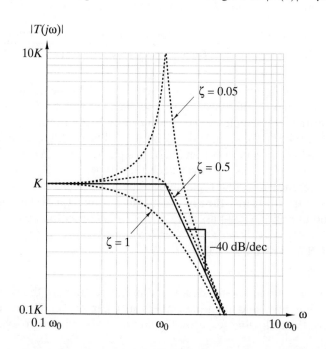

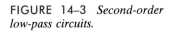

FIGURE 14–3 *Second-order low-pass circuits.*

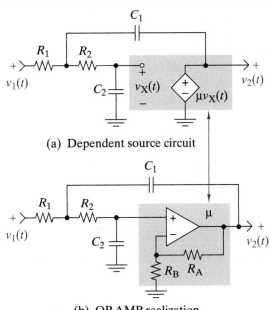

(a) Dependent source circuit

(b) OP AMP realization

An OP AMP RC filter is obtained by replacing the dependent source in Figure 14–3(a) by the OP AMP circuit in Figure 14–3(b). The gain of this noninverting amplifier gain is $1 + R_A/R_B$. Thus, modification simulates the dependent source when $1 + R_A/R_B = \mu$ or, equivalently, $R_A = (\mu - 1)R_B$. The modified circuit is our first second-order building block. Note that it has an OP AMP output, two capacitors, and a feedback path.

Our objective is to find methods of determining the gain μ and the four element values that produce specified values of ω_0 and ζ. Comparing the denominators in Eqs. (14–1) and (14–6) gives

$$\sqrt{R_1 R_2 C_1 C_2} = 1/\omega_0 \quad \text{and} \quad R_1 C_2 + R_2 C_2 + (1 - \mu)R_1 C_1 = 2\zeta/\omega_0$$

Using the first equation to eliminate ω_0 from the second equation leads to

$$\sqrt{R_1 R_2 C_1 C_2} = \frac{1}{\omega_0} \quad \text{and} \quad \sqrt{\frac{R_1 C_2}{R_2 C_1}} + \sqrt{\frac{R_2 C_2}{R_1 C_1}} + (1 - \mu)\sqrt{\frac{R_1 C_1}{R_2 C_2}} = 2\zeta \qquad (14–7)$$

Two methods of selecting element values to meet these design goals are discussed below.

The *equal element* method requires that $R_1 = R_2 = R$ and $C_1 = C_2 = C$. Inserting these conditions into Eq. (14–7) leads to

$$RC = \frac{1}{\omega_0} \quad \text{and} \quad \mu = 3 - 2\zeta \qquad (14–8)$$

Using this method, we select values of R (or C) and R_B, then solve for C (or R) and $R_A = (\mu - 1)R_B$. The dc gain achieved by this method is $|T(0)| = \mu = 3 - 2\zeta$, and the method is valid for $\zeta < 1.5$.

The *unity gain* method requires that $R_1 = R_2 = R$ and $\mu = 1$. Inserting these conditions in Eq. (14–7) leads to

$$R\sqrt{C_1 C_2} = \frac{1}{\omega_0} \quad \text{and} \quad \frac{C_2}{C_1} = \zeta^2 \qquad (14–9)$$

Using this method, we select a value of C_1 and calculate $C_2 = \zeta^2 C_1$ and $R = (\omega_0 \sqrt{C_1 C_2})^{-1}$. To get a gain $\mu = 1$, we make the noninverting OP AMP circuit a voltage follower. That is, we replace R_A by a short circuit and R_B by an open circuit. This eliminates the need for R_A and R_B but requires two different capacitors. Obviously the dc gain achieved by this design method is $|T(0)| = \mu = 1$. The unity gain method does not place any restrictions on the value of the damping coefficient, ζ.

The equal element and unity gain methods provide alternative ways to design an active low-pass filter with prescribed values of ω_0 and ζ. However, the dc gains produced by these methods are predetermined and are not adjustable design parameters. An additional gain correction stage may be needed when ω_0, ζ, and the dc gain are all three prescribed.

▶ DESIGN EXAMPLE 14–1

Develop a second-order low-pass transfer function with a corner frequency at $\omega_0 = 1$ krad/s and with corner frequency gain equal to the dc gain. Use MATLAB to help visualize the Bode plots of the desired transfer function. Then design two competing circuits using the equal element and unity gain design techniques. Use OrCAD to simulate the frequency responses of the designed circuits. Compare the results and comment on any differences.

SOLUTION:
The required transfer function has the form

$$T(s) = \frac{K}{(s/1000)^2 + 2\zeta(s/1000) + 1}$$

The requirement that the corner frequency gain equal the dc gain results in $|T(j1000)| = K/2\zeta$. These two gains are equal when $\zeta = 0.5$.

Equal element design: Inserting $\omega_0 = 1$ krad/s and $\zeta = 0.5$ into Eq. (14–8) yields $RC = 10^{-3}$ s and $\mu = 2$. The resulting transfer function then is

$$T_1(s) = \frac{2 \times 10^6}{s^2 + 1000s + 10^6}$$

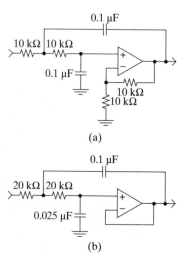

(a)

(b)

FIGURE 14–4

Selecting $R = R_B = 10\,\text{k}\Omega$ requires $C = 0.1\,\mu\text{F}$ and $R_A = (\mu - 1)R_B = 10\,\text{k}\Omega$. The resulting circuit is shown in Figure 14–4(a) and has a dc gain of $|T(0)| = \mu = 2$.

Unity gain design: Inserting $\omega_0 = 1$ krad/s and $\zeta = 0.5$ into Eq. (14–9) yields the following transfer function:

$$T_1(s) = \frac{10^6}{s^2 + 1000s + 10^6}$$

With the recognition that $R\sqrt{C_1 C_2} = 10^{-3}$ s and $C_2 = \zeta^2 C_1 = 0.25\,C_1$, selecting $C_1 = 0.1\,\mu\text{F}$ dictates that $C_2 = 0.025\,\mu\text{F}$ and $R = 20\,\text{k}\Omega$. The dc gain in this circuit is, by design, 1 and Figure 14–4(b) shows the resulting circuit.

We can visualize our transfer function designs by using the bodemag function of MATLAB. The following code will generate the Bode magnitude plots:

```
T1=tf([2000000],[1 1000 1000000]);
T2=tf([1000000],[1 1000 1000000]);
w = logspace(1,4,1000);
bodemag(T1,T2,w);grid
```

FIGURE 14–5

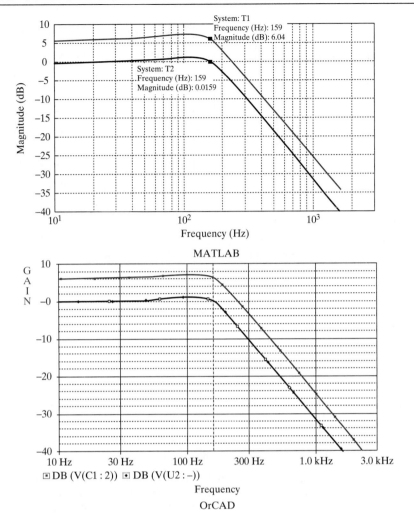

MATLAB returns the graphs shown in Figure 14–5(a), plotted in dB versus Hz.

The circuits are built in OrCAD and simulated using AC Sweep. The results are shown in Figure 14–5(b), plotted in dB versus Hz. Clearly, both circuit designs implement their transfer functions extremely well.

Comparing the two circuit designs shows that the equal element design has a greater gain and uses equal-value components, but requires two extra resistors. The unity gain design is simpler, but requires two different capacitors. The preferred design depends on the application and other nonstated requirements. ∎

D Design Exercise 14–1 _____

Develop a second-order low-pass transfer function with a corner frequency of 50 rad/s, a dc gain of 2, and a gain of 4 at the corner frequency. Validate your result by using MATLAB to plot the transfer function's absolute gain versus frequency.

Answer: The desired transfer function is

$$T(s) = \frac{5000}{s^2 + 25s + 2500}$$

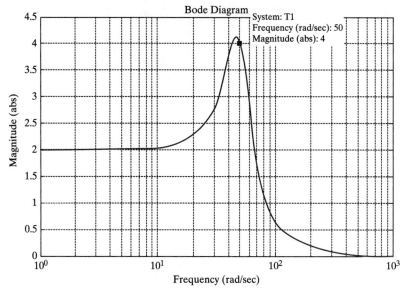

FIGURE 14–6

Figure 14–6 shows the MATLAB results.

◖ Design Exercise 14–2 _____

Design circuits using both the equal element design and unity gain design techniques to realize the transfer function in Exercise 14–1. Use OrCAD to simulate your designs and compare them to the MATLAB results shown in Figure 14–6.

Answer: The two designs are shown in Figure 14–7 along with the Probe output. Note that both designs required adjustments to their gains to get an overall gain of 2. The filter from the equal element design required a dc gain of 2.5 to achieve the necessary damping ratio. To get the overall circuit gain of 2, a voltage divider was connected to the output of the filter. In the unity gain design, a second amplifier stage was required to boost the overall gain to 2.

FIGURE 14–7

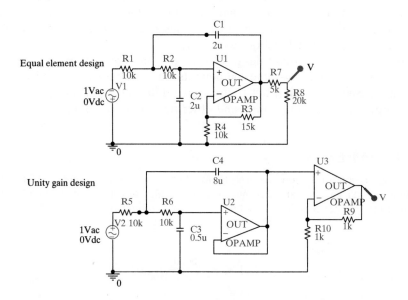

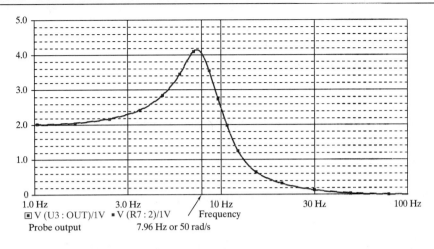

FIGURE 14–7 *(Continued)*

V (U3 : OUT)/1V ▪ V (R7 : 2)/1V
Probe output   7.96 Hz or 50 rad/s

SECOND-ORDER HIGH-PASS FILTERS

The transfer function of a second-order high-pass filter from Eq. (14-2) has the form

$$T(s) = \frac{K(s/\omega_0)^2}{(s/\omega_0)^2 + 2\zeta(s/\omega_0) + 1}$$

This transfer function has two poles and a double zero at $s = 0$. **Substituting** $j\omega$ for s yields the gain response as

$$|T(j\omega)| = \frac{|K|(\omega/\omega_0)^2}{\sqrt{[1 - (\omega/\omega_0)^2]^2 + 4\zeta^2(\omega/\omega_0)^2}} \tag{14–10}$$

At high frequency ($\omega \gg \omega_0$) the gain approaches $|T(j\omega)| \rightarrow |K| = |T(\infty)|$. At low frequency ($\omega \ll \omega_0$) the gain approaches $|T(j\omega)| \rightarrow |K|(\omega/\omega_0)^2$. Figure 14–8 presents

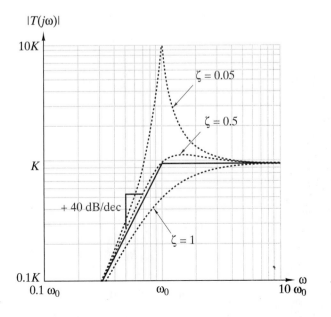

FIGURE 14–8 *Second-order high-pass gain responses.*

Bode plots of these gain asymptotes (solid lines) and the actual gain (dashed curves) for several values of ζ. The key points to remember are the following:

- The high- and low-frequency gain asymptotes intersect at $\omega = \omega_0$.
- For $\omega > \omega_0$ the asymptotic gain equals the infinite frequency gain $|T(\infty)| = |K|$.
- For $\omega < \omega_0$ the slope of the asymptotic gain is $+40\,\text{dB/decade}$.
- The actual gain at $\omega = \omega_0$ is $|T(j\omega_0)| = |K|/2\zeta$.

As in the low-pass case, the design parameter ω_0 locates the corner frequency and ζ controls the actual gain around the corner frequency.

The circuit in Figure 14–9(a) has a transfer function of the form

$$T(s) = \frac{V_2(s)}{V_1(s)} = \frac{\mu R_1 R_2 C_1 C_2 s^2}{R_1 R_2 C_1 C_2 s^2 + (R_2 C_2 + R_1 C_1 + R_1 C_2 - \mu R_2 C_2)s + 1} \qquad (14\text{--}11)$$

This is a second-order high-pass function with an infinite-frequency gain of $|T(\infty)| = \mu$. The high-pass circuit in Figure 14–9(a) is obtained from the low-pass circuit in Figure 14–3(a) by interchanging the locations of the resistors and capacitors. In Problem 14–1 the student will discover how this interchange converts the low-pass $T(s)$ in Eq. (14–6) into the high-pass $T(s)$ in Eq. (14–11).

As in the low-pass case, we replace the dependent source in Figure 14–9(a) by the noninverting OP AMP circuit in Figure 14–9(b). The OP AMP circuit simulates the dependent source when $1 + R_A/R_B = \mu$ or, equivalently, $R_A = (\mu - 1)R_B$. The modified high-pass circuit is a second-order building block with an OP AMP output, two capacitors, and a feedback path.

Again our goal is to find methods of determining the gain μ and the four element values that produce specified values of ω_0 and ζ. Comparing the denominators in Eqs. (14–2) and (14–11) gives

$$\sqrt{R_1 R_2 C_1 C_2} = 1/\omega_0 \quad \text{and} \quad R_1 C_1 + R_1 C_2 + (1 - \mu)R_2 C_2 = 2\zeta/\omega_0$$

FIGURE 14–9 *Second-order high-pass circuits.*

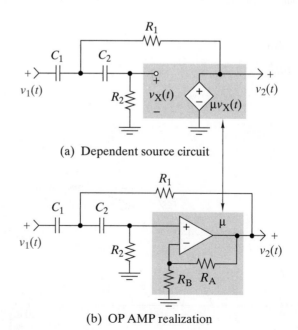

(a) Dependent source circuit

(b) OP AMP realization

Using the first equation to eliminate ω_0 from the second equation leads to

$$\sqrt{R_1 R_2 C_1 C_2} = \frac{1}{\omega_0} \quad \text{and} \quad \sqrt{\frac{R_1 C_1}{R_2 C_2}} + \sqrt{\frac{R_1 C_2}{R_2 C_1}} + (1 - \mu)\sqrt{\frac{R_2 C_2}{R_1 C_1}} = 2\zeta \qquad (14\text{--}12)$$

To select the element values in Eq. (14–12), we use methods similar to those used to design our low-pass building block.

The *equal element method* requires that $R_1 = R_2 = R$ and $C_1 = C_2 = C$. Inserting these conditions in Eq. (14–12) leads to

$$RC - \frac{1}{\omega_0} \quad \text{and} \quad \mu = 3 - 2\zeta \qquad (14\text{--}13)$$

Under this method we select values of R (or C) and R_B, and solve for C (or R) and $R_A = (\mu - 1)R_B$. The infinite-frequency gain achieved by this method is $|T(\infty)| = \mu = 3 - 2\zeta$, and the method is valid for $\zeta < 1.5$.

The *unity gain method* requires that $C_1 = C_2 = C$ and $\mu = 1$. Inserting these conditions in Eq. (14–12) gives

$$C\sqrt{R_1 R_2} = \frac{1}{\omega_0} \quad \text{and} \quad \frac{R_1}{R_2} = \zeta^2 \qquad (14\text{--}14)$$

Using this method, we select a value of R_2 and calculate $R_1 = \zeta^2 R_2$ and $C = (\omega_0 \sqrt{R_1 R_2})^{-1}$. To get a gain $\mu = 1$, we make the noninverting OP AMP circuit into a voltage follower. That is, we replace R_A by a short circuit and R_B by an open circuit, thereby eliminating the need for these two resistors. Obviously the infinite-frequency gain achieved by this method is $|T(\infty)| = \mu = 1$. As with the low-pass filter, the unity gain method does not place any restrictions on the value of the damping coefficient, ζ.

The equal element and unity gain methods are alternative ways to design an active high-pass filter with prescribed values of ω_0 and ζ. As we found in the low-pass case, the passband gains produced by these methods are predetermined and are not adjustable design parameters. An additional gain correction stage may be needed when ω_0, ζ, and the infinite-frequency gain are all three prescribed.

D DESIGN EXAMPLE 14–2

Develop a second-order high-pass transfer function with a corner frequency at $\omega_0 = 20\,\text{krad/s}$, an infinite-frequency gain of 0 dB, and a corner frequency gain of -3 dB. Use MATLAB to help visualize the Bode magnitude plot of the desired transfer function. Then design two competing circuits using the equal element and unity gain design techniques. Use OrCAD to simulate the frequency responses of the designed circuits. Compare the results and comment on the differences.

SOLUTION:
The required transfer function has the form

$$T(s) = \frac{K(s/20{,}000)}{(s/20{,}000)^2 + 2\zeta(s/20{,}000) + 1}$$

The infinite-frequency gain is $T(\infty) = K$ and the corner frequency gain is $|T(j20{,}000)| = K/2\zeta$. A gain of 0 dB at infinite frequency requires $K = 1$. A

FIGURE 14–10

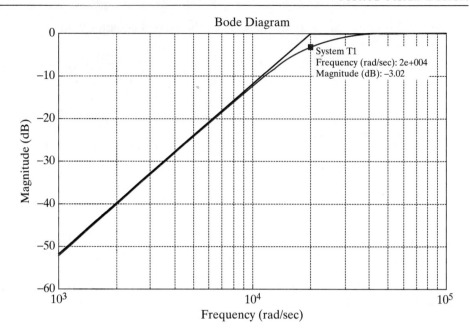

gain of −3 dB at the corner frequency requires $|T(j20{,}000)| = 1/\sqrt{2}$, which in turn requires that $\zeta = 1/\sqrt{2} = 0.707$. Therefore, the desired transfer function is

$$T(s) = \frac{s^2}{s^2 + 28{,}280s + 400 \times 10^6}$$

We can visualize our transfer function design by using the bodemag function of MATLAB as follows:

```
T=tf([1 0 0],[1 28280 400e6]);
w = logspace(3,5,10000);
bodemag(T,w);grid
```

Figure 14–10 shows the resulting plot.

Equal element design: Inserting $\omega_0 = 20$ krad/s and $\zeta = 0.707$ into Eq. (14–13) yields $RC = 5 \times 10^{-5}$ s and $\mu = 3 - \sqrt{2} = 1.586$. Selecting $C = 0.005\,\mu$F and $R_B = 50\,k\Omega$ requires $R = 10\,k\Omega$ and $R_A = (\mu - 1)R_B = 29.3\,k\Omega$. The high-frequency gain of this design is $|T(\infty)| = \mu = 1.586$, which is more than the specified value of 1 (0 dB). We add a gain correction stage with a gain of $1/1.586 = 0.6305$ to bring the overall gain down to 0 dB. Figure 14–11(a) shows the resulting two-stage design.

Unity gain design: Inserting $\omega_0 = 20$ krad/s and $\zeta = 0.707$ into Eq. (14–14) yields $C\sqrt{R_1 R_2} = 5 \times 10^{-5}$ s and $R_1 = \zeta^2 R_2 = 0.5R_2$. Selecting $R_2 = 10\,k\Omega$ requires that $R_1 = 5\,k\Omega$ and $C = 7071$ pF. The $|T(\infty)|$ gain of this circuit is 1, which matches the desired 0 dB. The resulting single-stage design is shown in Figure 14–11(b).

The circuits were created in OrCAD and simulated using AC Sweep. Their Probe results are shown in Figure 14–12. Clearly, both circuit designs implement the transfer function extremely well and meet all three design specifications.

Comparing the two circuit designs shows that the *equal element design* requires an extra OP AMP used as a buffer and four more resistors than the *unity gain design* because its gain is greater than 1. The voltage divider could be placed after the filter thereby eliminating the need for the buffer, but raising a concern about

FIGURE 14–11

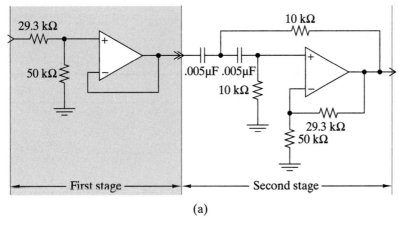

(a)

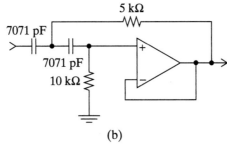

(b)

FIGURE 14–12

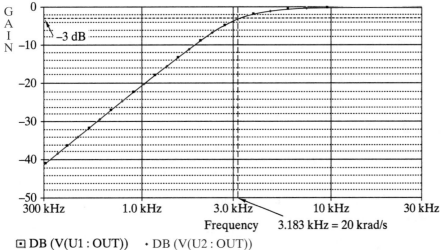

loading at the output. The unity gain design appears to be the better choice in this case. ∎

◗ Design Exercise 14–3

Construct a second-order high-pass transfer function with a corner frequency of 20 rad/s, an infinite-frequency gain of 4, and a gain of 2 at the corner frequency.

Answer:
$$T(s) = \frac{4s^2}{s^2 + 40s + 400}$$

The design requirements for the Sallen-Key second-order filters studied are summarized in Table 14–1:

TABLE 14–1

	EQUAL ELEMENT	UNITY GAIN
Low Pass	$R_1 = R_2 = R$ $C_1 = C_2 = C$ $\omega_0 = {}^1/_{RC}$ $\mu = 3 - 2\zeta$	$R_1 = R_2 = R$ $\mu = 1$ $\omega_0 = 1/R\sqrt{C_1 C_2}$ $\zeta = \sqrt{C_2/C_1}$
High Pass	$R_1 = R_2 = R$ $C_1 = C_2 = C$ $\omega_0 = {}^1/_{RC}$ $\mu = 3 - 2\zeta$	$C_1 = C_2 = C$ $\mu = 1$ $\omega_0 = 1/C\sqrt{R_1 R_2}$ $\zeta = \sqrt{R_1/R_2}$

14–3 SECOND-ORDER BANDPASS AND BANDSTOP FILTERS

The active RC filters in this section are the counterparts of the passive RLC circuits studied in Chapter 12. These active filters achieve the frequency selectivity of the passive RLC circuits without the need for inductors. We continue our development of active filter building blocks with the bandpass case.

SECOND-ORDER BANDPASS FILTERS

The transfer function of a second-order bandpass filter from Eq. (14–3) has the form

$$T(s) = \frac{K(s/\omega_0)}{(s/\omega_0)^2 + 2\zeta(s/\omega_0) + 1}$$

This transfer function has two poles and a zero at $s = 0$. The gain response is found in the usual way as

$$|T(j\omega)| = \frac{|K||(\omega/\omega_0)|}{\sqrt{[1 - (\omega/\omega_0)^2]^2 + 4\zeta^2(\omega/\omega_0)^2}} \qquad (14\text{–}15)$$

At low frequency $(\omega \ll \omega_0)$ the gain approaches $|T(j\omega)| \to |K|(\omega/\omega_0)$. At high frequency $(\omega \gg \omega_0)$ the gain approaches $|T(j\omega)| \to |K|(\omega/\omega_0)$. Figure 14–13 presents Bode plots of these gain asymptotes (solid lines) and the actual gain (dashed curves) for several values of ζ. The key points to remember are the following:

- The low- and high-frequency gain asymptotes intersect at $\omega = \omega_0$.
- For $\omega < \omega_0$ the slope of the asymptotic gain is $+20$ dB/decade.
- For $\omega > \omega_0$ the slope of the asymptotic gain is -20 dB/decade.
- The actual gain at $\omega = \omega_0$ is $|T(j\omega_0)| = |K|/2\zeta$.

Note again that the design parameter ω_0 locates the center frequency and ζ controls the actual gain near the center frequency.

The circuit in Figure 14–14 has a transfer function of the form

$$T(s) = \frac{V_2(s)}{V_1(s)} = \frac{-R_2 C_2 s}{R_1 R_2 C_1 C_2 s^2 + (R_1 C_1 + R_1 C_2)s + 1} \qquad (14\text{–}16)$$

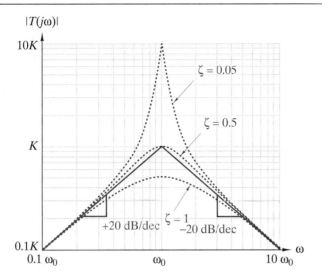

FIGURE 14–13 *Second-order bandpass gain responses.*

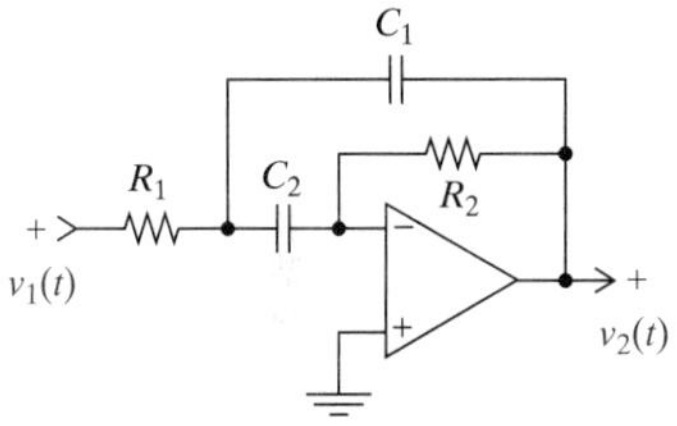

FIGURE 14–14 *Second-order bandpass circuit.*

This circuit produces a second-order bandpass function. It has an OP AMP output, two capacitors, and two feedback paths—one via resistor R_2 and the other through the capacitor C_2. The dual feedback identifies this circuit as a member of the multiple-feedback family of active filters.[2]

Derivation of Eq. (14–16) using node analysis is easy, as the student can readily verify by working Problem 14–2.

Our design goal is to select element values to achieve specified values of ω_0 and ζ. Comparing the denominators in Eqs. (14–3) and (14–16) gives

$$\sqrt{R_1 R_2 C_1 C_2} = 1/\omega_0 \quad \text{and} \quad R_1 C_1 + R_1 C_2 = 2\zeta/\omega_0$$

Using the first equation to eliminate ω_0 from the second equation produces

$$\sqrt{R_1 R_2 C_1 C_2} = \frac{1}{\omega_0} \quad \text{and} \quad \sqrt{\frac{R_1 C_1}{R_2 C_2}} + \sqrt{\frac{R_1 C_2}{R_2 C_1}} = 2\zeta \qquad (14\text{–}17)$$

In this case there are two design constraints and four unknown element values.

An *equal-capacitor method* $(C_1 = C_2 = C)$ can be used to reduce the number of unknowns. Inserting $C_1 = C_2 = C$ into Eq. (14–17) yields

$$\sqrt{R_1 R_2}\,C = \frac{1}{\omega_0} \quad \text{and} \quad \frac{R_1}{R_2} = \zeta^2 \qquad (14\text{–}18)$$

Under this method we select a value of R_2 and solve for $R_1 = \zeta^2 R_2$ and $C = (\omega_0\sqrt{R_1 R_2})^{-1}$. Since this method uses $C_1 = C_2$, the center frequency gain found from Eq. (14–16) is $|T(j\omega_0)| = R_2/2R_1 = 1/2\zeta^2$. Note that the center frequency gain is greater than 1 when $\zeta < 1/\sqrt{2}$. For example, when $\zeta = 0.1$, the gain is 50. This contrasts with the passive RLC bandpass circuit, whose center frequency gain is always 1.

The key descriptive parameters of a second-order bandpass filter are its **center frequency** ω_0 and **bandwidth** $B = 2\zeta\omega_0$. It is customary to add a third parameter called the **quality factor**, defined as $Q = \omega_0/B$. From this definition it is clear that Q

[2]For an extensive discussion of the multiple-feedback family, see Wai-Kai Chen, Ed., *The Circuits and Filters Handbook*, CRC Press, 1995, Chapter 76, pp. 2372 ff. The bandpass circuit in Figure 14–14 is sometimes called the Delyannis-Friend circuit. See M. E. Van Valkenburg, *Analog Filter Design*, Holt, Rinehart & Winston, Inc, 1982, p. 203.

and ζ are both dimensionless parameters related as $Q = 1/2\zeta$. Either parameter can be used to characterize filter bandwidth. When $Q > 1$ ($\zeta < 0.5$), the filter is said to be narrow-band because the bandwidth is less than the center frequency. When $Q < 1$ ($\zeta > 0.5$), the filter is said to be wide-band. The active bandpass building block developed here is best suited to narrow-band applications. Filters with a high Q are also referred to as **tuned** filters. The design of wide-band active bandpass filters is discussed in Sect. 14–7.

◑ DESIGN EXAMPLE 14 – 3

Use the active RC circuit in Figure 14–14 to design a bandpass filter with a center frequency at 10 kHz and a bandwidth of 4 kHz. Find the center frequency gain for the design. Use MATLAB to show the filter's gain characteristics.

SOLUTION:
The specified filter parameters define the center frequency and bandwidth as

$$\omega_0 = 2\pi \times 10^4 = 62.8 \text{ krad/s} \quad \text{and} \quad B = 2\pi \times 4000 = 25.1 \text{ krad/s}$$

The required damping ratio is $\zeta = B/2\omega_0 = 0.2$. Inserting the values of ω_0 and ζ into Eq. (14–18) yields $R_1 = 0.04R_2$ and $C = 1.592 \times 10^{-5}/\sqrt{R_1 R_2}$. Selecting $R_2 = 100\,\text{k}\Omega$ requires $R_1 = 4\,\text{k}\Omega$ and $C = 796\,\text{pF}$. The center frequency gain for this design is $|T(j\omega_0)| = 1/2\zeta^2 = 12.5$. The transfer function for the design is

$$T(s) = \frac{-314 \times 10^3 s}{s^2 + 25.1 \times 10^3 s + 3.9456 \times 10^9}$$

The required MATLAB code is

```
T=tf([-314e3 0],[1 25.1e3 3.9456e9]);
w = logspace(4,6,10000);
bodemag(T,w);grid
```

FIGURE 14–15

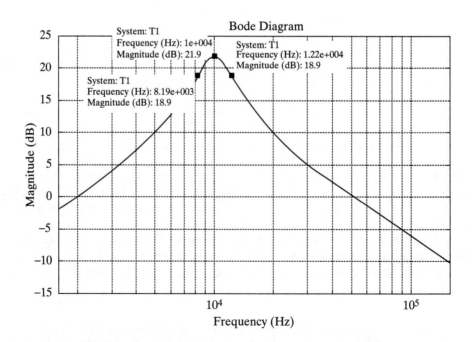

The Bode magnitude plot is shown in Figure 14–15.

⏻ Design Exercise 14–4 _____

Rework the design in Example 14–3, starting with $C = 2000\,pF$.

Answers: $R_1 = 1.59\,k\Omega$; $R_2 = 39.8\,k\Omega$

⏻ Design Exercise 14–5 _____

Construct a second-order bandpass transfer function with a corner frequency of 50 rad/s, a bandwidth of 10 rad/s and a center frequency gain of 4.

Answer:
$$T(s) = \frac{-40s}{s^2 + 10s + 2500}$$

SECOND-ORDER BANDSTOP FILTERS

The transfer function of a second-order bandstop filter from Eq. (14–4) has the form

$$T(s) = \frac{K[(s/\omega_0)^2 + 1]}{(s/\omega_0)^2 + 2\zeta(s/\omega_0) + 1}$$

This transfer function has two zeros at $s = \pm j\omega_0$ along with the two poles defined by ζ and ω_0. The resulting gain response is

$$|T(j\omega)| = \frac{|K||1 - (\omega/\omega_0)^2|}{\sqrt{[1 - (\omega/\omega_0)^2]^2 + 4\zeta^2(\omega/\omega_0)^2}} \qquad (14\text{–}19)$$

At low frequency ($\omega \ll \omega_0$) and high frequency ($\omega \gg \omega_0$) the gain approaches $|T(0)| = |T(\infty)| \to |K|$. At $\omega = \omega_0$ the gain is $|T(jw_0)| = 0$. Figure 14–16 shows the gain predicted by Eq. (14–19) for some representative values of ζ. Key points to remember are the following:

- The low- and high-frequency gains are $|T(0)| = |T(\infty)| = |K|$.

- At $\omega = \omega_0$ there is a zero-gain notch in the gain response.

The notch is caused by the j-axis zeros in $T(s)$. From a design perspective, we say that the zeros at $s = \pm j\omega_0$ locate the notch and ζ controls the width of the notch.

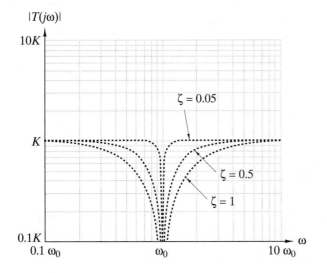

FIGURE 14–16   *Second-order bandstop gain responses.*

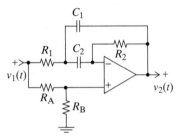

FIGURE 14–17 *Second-order bandstop circuit.*

The circuit in Figure 14–17 is a modification of the bandpass circuit in Figure 14–14. The modified circuit retains the two feedback paths and adds a second input path via the voltage divider R_A and R_B. The two input and feedback paths combine to produce an overall circuit transfer function of

$$T(s) = \frac{V_2(s)}{V_1(s)} = \frac{R_B}{R_A + R_B} \left[\frac{R_1 R_2 C_1 C_2 s^2 + (R_1 C_1 + R_1 C_2 - R_2 C_2 R_A/R_B)s + 1}{R_1 R_2 C_1 C_2 s^2 + (R_1 C_1 + R_1 C_2)s + 1} \right] \quad (14\text{–}20)$$

Derivation of this transfer function involves straightforward (if laborious) node analysis and is left for the enjoyment of the student in Problem 14–3.

The denominator in Eq. (14–20) is the same as the denominator of the bandpass filter in Eq. (14–15). Hence, the design constraints in Eq. (14–16) apply here as well, namely

$$\sqrt{R_1 R_2 C_1 C_2} = \frac{1}{\omega_0} \quad \text{and} \quad \sqrt{\frac{R_1 C_1}{R_2 C_2}} + \sqrt{\frac{R_1 C_2}{R_2 C_1}} = 2\zeta \quad (14\text{–}21)$$

These constraints control the locations of the bandstop filter poles. Comparing the numerators in Eqs. (14–4) and (14–20) gives

$$R_1(C_1 + C_2) - R_2 C_2 R_A/R_B = 0 \quad (14\text{–}22)$$

This constraint ensures that the circuit has the requisite *j*-axis zeros to produce the bandstop notch. Taken together, Eqs. (14–21) and (14–22) give us three design constraints and five unknown element values.

The *equal-capacitor method* $(C_1 = C_2 = C)$ reduces the number of unknowns in the design constraints. When $C_1 = C_2 = C$, the pole location constraints in Eq. (14–21) reduce to

$$\sqrt{R_1 R_2} C = \frac{1}{\omega_0} \quad \text{and} \quad \frac{R_1}{R_2} = \zeta^2 \quad (14\text{–}23)$$

Using this design method, we select a value of R_2 and solve for $R_1 = \zeta^2 R_2$ and $C = (\omega_0 \sqrt{R_1 R_2})^{-1}$. When $C_1 = C_2 = C$, the bandstop notch requirement in Eq. (14–22) reduces to

$$\frac{R_A}{R_B} = \frac{2R_1}{R_2} \quad (14\text{–}24)$$

An obvious way to meet this requirement is to let $R_A = 2R_1$ and $R_B = R_2$.

Second-order bandstop filters are customarily described in terms of the **notch frequency** ω_0 and the **notch bandwidth** $B = 2\zeta\omega_0$. The bandstop circuit in Figure 14–17 is best suited for narrow-band applications aimed at eliminating a narrow range of frequencies or even a single frequency. Filters with a narrow bandwidth of this type are often referred to as **notch** filters. The design of wide-band bandstop filters is discussed in Sect. 14–7.

Narrow-band notch circuits are often used to eliminate power line noise at 60 Hz (50 Hz in many other countries). The next example illustrates such an application.

▶ **DESIGN EXAMPLE 14–4**

Use the active *RC* circuit in Figure 14–17 to design a bandstop filter with a notch frequency at 60 Hz and a notch bandwidth of 12 Hz. Find the circuit's transfer function and use MATLAB to plot the filter's gain characteristic and to estimate the attenuation of the notch.

SOLUTION:

The specified filter parameters define the notch frequency and bandwidth as

$$\omega_0 = 2\pi \times 60 = 377\,\text{rad/s} \quad \text{and} \quad B = 2\pi \times 12 = 75.4\,\text{rad/s}$$

The required damping ratio is $\zeta = B/2\omega_0 = 0.1$. Using equal capacitors $(C_1 = C_2 = C)$ and selecting $R_2 = 100\,\text{k}\Omega$, we use Eqs. (14–23) and (14–24) to calculate the remaining values as follows:

$$R_1 = \zeta^2 R_2 = 1\,\text{k}\Omega$$

$$C = \frac{1}{\omega_0 \sqrt{R_1 R_2}} = 0.265\,\mu\text{F}$$

$$R_A = 2R_1 = 2\,\text{k}\Omega$$

$$R_B = R_2 = 100\,\text{k}\Omega$$

The transfer function for the design is

$$T(s) = 0.98\,\frac{s^2 + 142.4 \times 10^3}{s^2 + 75.47s + 142.4 \times 10^3}$$

The required MATLAB code is

```
T=tf([0.98 0 0.98*142.4e3],[1 75.47 142.4e3]);
w = logspace(2,3,10000);
bodemag(T,w);grid
```

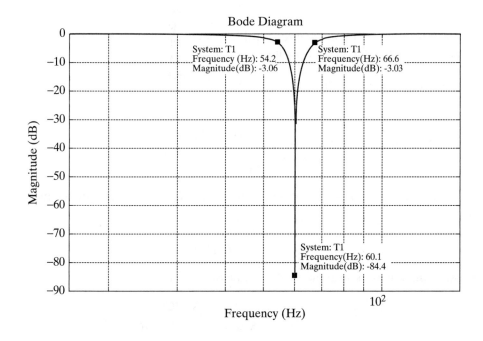

Bode Diagram

FIGURE 14–18

Figure 14–18 shows the Bode magnitude plot. The center frequency is at 60.1 Hz and the bandwidth is approximately 12.4 Hz. An estimate of the notch attenuation is −84.4 dB. Note that the simulated attenuation at the center frequency will depend on the number of values selected for the frequency variable. That is, the more calculations you ask MATLAB to make the more accurate the depth of the notch will be found. We used 10,000 points over 1 decade. ∎

⭕ Design Exercise 14–6 _____

Rework the circuit design in Example 14–4 starting with $C_1 = C_2 = C = 0.2\,\mu F$.

Answers: $R_1 = 1.33\,k\Omega; R_2 = 133\,k\Omega; R_A = 2.66\,k\Omega; R_B = 133\,k\Omega$

⭕ Design Exercise 14–7 _____

Construct a second-order bandstop transfer function with a notch frequency of 50 rad/s, a notch bandwidth of 10 rad/s, and passband gains of 5.

Answer:
$$T(s) = \frac{5(s^2 + 50^2)}{s^2 + 10s + 50^2}$$

The tuned and notch filters in Figures 14–14 and 14–17 are efficient circuits that are easy to design and simulate. Let us analyze another pair of realizations for tuned and notch filters. Consider Figure 14–19.

FIGURE 14–19

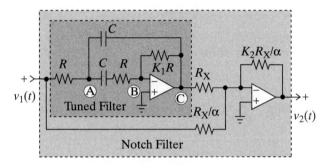

This circuit consists of a tuned circuit inside a notch circuit. We have labeled three nodes to help us write node equations.

at Node A:
$$0 = \frac{V_A(s) - V_1(s)}{R} + \frac{V_A(s) - V_C(s)}{\frac{1}{Cs}} + \frac{V_A(s) - V_B(s)}{R + \frac{1}{Cs}}$$

at Node B:
$$0 = \frac{V_B(s) - V_A(s)}{R + \frac{1}{Cs}} + \frac{V_B(s) - V_C(s)}{K_1 R}$$

We note that $V_B(s) = 0$ by the OP AMP i–v relationship. Using the Node B equation we find

$$V_A(s) = -\frac{V_C(s)(1 + RCs)}{K_1 RCs}$$

Substituting this back into the Node A equation yields

$$\frac{V_C(s)}{V_1(s)} = T_{\text{tuned}}(s) = -\frac{K_1 RCs}{R^2 C^2 s^2(1 + K_1) + 3RCs + 1}$$

$$= -\frac{\left(\dfrac{K_1}{K_1 + 1}\right)\left(\dfrac{1}{RC}\right)s}{s^2 + \underbrace{\dfrac{3}{(K_1 + 1)RC}}_{2\zeta\omega_0}s + \underbrace{\dfrac{1}{(K_1 + 1)R^2 C^2}}_{\omega_0^2}} \qquad (14\text{--}25)$$

This, then, is the basic equation for designing a tuned filter. Note that the filter's gain is not a controllable parameter; rather, it is a result of selecting a particular damping coefficient and resonant frequency. If one needs a particular gain it needs to be adjusted with either a cascading gain stage or a voltage divider.

The notch filter takes the output of the tuned filter and adds the input in such a way as to create the band-stop characteristic in Eq. (14–4). The output of the overall circuit $V_2(s)$ is found as

$$V_2(s) = -\frac{\dfrac{K_2 R_X}{\alpha}}{R_X} V_C(s) - \frac{\dfrac{K_2 R_X}{\alpha}}{R_X/\alpha} V_1(s)$$

$$V_2(s) = -\frac{K_2}{\alpha}\left(-\frac{\left(\dfrac{K_1}{K_1+1}\right)\left(\dfrac{1}{RC}\right)sV_1(s)}{s^2 + \dfrac{3}{(K_1+1)RC}s + \dfrac{1}{(K_1+1)R^2C^2}} \right) - K_2 V_1(s)$$

$$\frac{V_2(s)}{V_1(s)} = \frac{\left(\dfrac{K_2}{\alpha}\right)\left(\dfrac{K_1}{K_1+1}\right)\left(\dfrac{1}{RC}\right)s - K_2\left(s^2 + \dfrac{3}{(K_1+1)RC}s + \dfrac{1}{(K_1+1)R^2C^2}\right)}{s^2 + \dfrac{3}{(K_1+1)RC}s + \dfrac{1}{(K_1+1)R^2C^2}}$$

With some effort we find that this reduces to

$$\frac{V_2(s)}{V_1(s)} = \frac{-K_2 s^2 + \overbrace{\left[\dfrac{-3K_2}{(K_1+1)RC} + \dfrac{K_1 K_2}{\alpha(K_1+1)RC}\right]}^{=0}s - \dfrac{K_2}{(K_1+1)R^2C^2}}{s^2 + \dfrac{3}{(K_1+1)RC}s + \dfrac{1}{(K_1+1)R^2C^2}}$$

The coefficient of s in the numerator must equal zero for this to be a notch filter. Setting those terms to zero and solving for α yields

$$\frac{-3K_2}{(K_1+1)RC} + \frac{K_1 K_2}{\alpha(K_1+1)RC} = 0$$

$$\alpha = K_1/3$$

This relationship is a gain condition applied to resistors that enables the inverting summer to eliminate the s-term in the numerator and turn the tuned filter into a notch filter. Note that the overall gain K_2 is controllable for the notch filter case. The transfer function for the notch filter is

$$\frac{V_2(s)}{V_1(s)} = T_{\text{notch}}(s) = -\frac{K_2\left(s^2 + \dfrac{1}{(K_1+1)R^2C^2}\right)}{s^2 + \dfrac{3}{(K_1+1)RC}s + \dfrac{1}{(K_1+1)R^2C^2}} \qquad (14\text{–}26)$$

◀▶ **D E DESIGN AND EVALUATION EXAMPLE 14 – 5**

Use the active RC circuit in Figure 14–19 to design a bandstop filter with a notch frequency at 60 Hz and a notch bandwidth of 12 Hz. Find the circuit's transfer function and use OrCAD to plot the filter's gain characteristic and to estimate the attenuation of the notch. Compare your results with that of Example 14–4.

SOLUTION:
The specified filter parameters define the notch frequency and bandwidth as

$$\omega_0 = 2\pi \times 60 = 377 \text{ rad/s and } B = 2\pi \times 12 = 75.4 \text{ rad/s}$$

Equating the s-term and the last term in the denominator to these parameters yields

$$B = \frac{3}{(K_1 + 1)RC} = 24\pi$$

$$\omega_0^2 = \frac{1}{(K_1 + 1)R^2C^2} = (120\pi)^2$$

Solving for RC in the first equation and substituting it into the second equation allows for K_1 to be found.

$$120\pi = \frac{24\pi(K_1 + 1)}{3\sqrt{K_1 + 1}}$$

$$K_1 = (15)^2 - 1 = 224$$

We can now find R and C by substituting back into one of the equations

$$\frac{3}{75.4(K_1 + 1)} = RC = 1.77 \times 10^{-4}$$

Selecting $C = 1\ \mu\text{F}$, results in $R = 177\ \Omega$.

We have just designed the tuned filter. However, our task is to design a notch filter and compare it with the notch filter designed in Example 14–4. We, therefore, need to find the parameter α that will eliminate the s-term in the numerator.

$$\alpha = \frac{K_1}{3} = \frac{224}{3} = 74.67$$

One last thing before we design the circuit. Unlike the circuit designed in Example 14–4, we can control the gain of the notch filter. In order to compare it we need to make the gain of this filter K_2 equal to the gain of the filter in Example 14–4. The pass-band gain of that circuit simulates to 0.98. The OrCAD circuits used in this comparison are shown in Figure 14–20.

An ac logarithmic sweep was conducted from 30 to 100 Hz with 10000 points per decade. The two results fell atop each other, but slight differences were noted. Circuit (a), the two OP AMP notch filter, had an f_0 of 60.009 Hz and a maximum notch of −52.15 dB. Circuit (b) had a notch frequency of 59.930 Hz and a maximum attenuation of 68.29 dB. From simulation data it would appear that the simpler circuit (b) was a better choice since it had a deeper notch and fewer parts. The notch error was just a little worse than that of circuit (a). The major advantage of circuit (a) is the ability to provide gain. A second stage could be added to circuit (b) to provide gain; however, this would then mitigate the parts-count advantage. One should also do a laboratory comparison to see which can be easier and more stable to build and test.

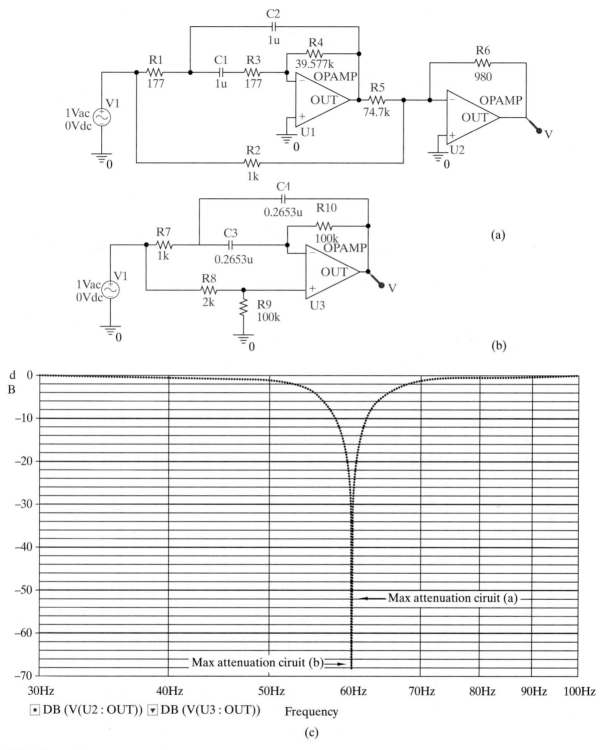

FIGURE 14–20

◑ Design Exercise 14–8 _____

Design a notch filter using the realization in Figure 14–19 to achieve a notch at 200 krad/s, a B of 20 krad/s, and a passband gain of 10.

Answer: $K_1 = 899, K_2 = 10, C = 1000 \, pF, R = 166 \, \Omega, K_1R = 149.2 \, k\Omega, \alpha = 299.7, R_X = 299.7$ $k\Omega, R_X/\alpha = 1 \, k\Omega, K_2R_X/\alpha = 10 \, k\Omega$. These values simulate to ω_0 of 200.7 krad/s, a notch of $-34.1 \, dB$, and a passband gain of $+20 \, dB$ or 10. Other design options are possible.

14–4 LOW-PASS FILTER DESIGN

We are about to start our study of cascaded filter design. There are many different types of active filters. Among the most useful are the First-Order Cascade, Butterworth, Chebyshev types 1 and 2, and Elliptic or Cauer. These filters are usually chosen on the basis of their magnitude selectivity. However, there are applications when one must also be concerned about the phase response of these filters. Figure 14–21 shows the basic magnitude characteristics of these filter types.

The First-Order Cascade filter is very simple in its design but requires a very high order filter to achieve significant selectivity. It does have excellent transient response

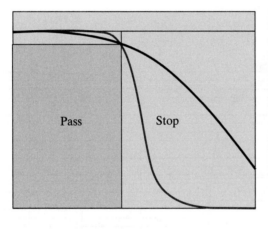

Butterworth ——— and
First-Order Cascade ———

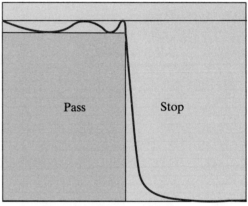

Chebyshev Type 1

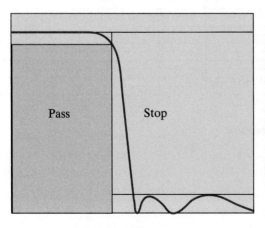

Chebyshev Type 2

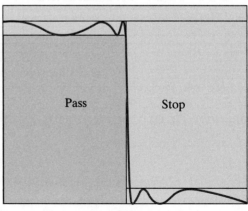

Elliptic (Cauer)

FIGURE 14–21

characteristics making it useful in applications that do not require high selectivity. Butterworth filters, named after British engineer Stephen Butterworth, are also called "maximally flat" filters because they have an excellent passband response. They have much better selectivity than First-Order Cascade filters and a moderate transient response. Chebyshev filters, named after Russian mathematician Pafnuty Chebyshev, are also called "equiripple" filters because they trade off a steep selectivity with adding a ripple in either the pass-band, type 1, or in the stop-band, type 2. They have a very complex transient response and phase response. Elliptic or Cauer filters, named after German network theorist Wilhelm Cauer, offer the steepest roll-off but at the expense of equiripples in both the pass- and stop-bands. In general, if a steep cutoff response is required, then a high-order filter is needed. A higher order filter requires a more complex and costly circuit. High-order filters also have the most complex transient and phase responses. As a relative example, a particular design specification that might require a fifth-order Elliptic filter would require a tenth-order Chebyshev or a 35th-order Butterworth. It would not even be possible with a First-Order Cascade regardless of the order. In this chapter, we will focus on the design of the first three types, namely First-Order Cascade, Butterworth, and Chebyshev type 1. You should gain sufficient understanding of cascaded active filter design to apply it to any filter type.

In this section we learn how to design multi-pole filters with prescribed low-pass characteristics. The Bode plot in Figure 14–22 shows how low-pass filter characteristics are specified. To satisfy the specification, the filter must be designed so that its gain response lies within the blue region. There are many different gain responses that meet this requirement, as illustrated by the two shown in the figure. All such gain responses are constrained by the conditions imposed in three adjacent frequency bands.

In the **passband** $(0 \leq \omega \leq \omega_C)$ the gain response must be in the range

$$\frac{T_{\text{MAX}}}{\sqrt{2}} \leq |T(j\omega)| \leq T_{\text{MAX}}$$

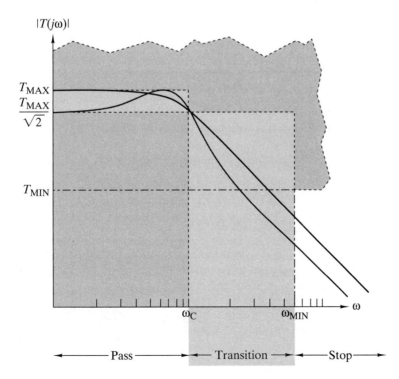

FIGURE 14–22 *Low-pass filter specification and responses.*

In the **stopband**$(\omega_{MIN} \leq \omega)$ the gain response must be in the range $|T(j\omega)| \leq T_{MIN}$.

In this chapter on filters, we want to define a new band between the passband and the stopband. The **transition band**$(\omega_C \leq \omega \leq \omega_{MIN})$ is the most significant part of filter design. The main filtering action occurs here. It is in this region that a filter designer wants the gain response to change dramatically with frequency. Two ratios describe this change: T_{MAX}/T_{MIN} defines how much the gain must change, and ω_{MIN}/ω_C specifies how rapidly the change must occur. As we study varying approaches to filter design, we will see that there are design approaches that can increase the filter's attenuation rate but possibly at a price, both in the "flatness" of the passband and in the filter's transient performance—its step response. For some applications these consequences are a small price to pay for better filter performance, while in other cases these compromises are not acceptable.

In summary, a low-pass filter design problem is defined by specifying the four parameters: $T_{MAX}, \omega_C, T_{MIN},$ and ω_{MIN}. Transfer functions (there may be several) that meet these requirements are of the form

$$T(s) = \frac{K}{q_n(s)}$$

where $q_n(s)$ is an nth-order polynomial whose roots define the poles of $T(s)$. From our previous design experience, we can see that filter design involves two related tasks: (1) selecting $q_n(s)$ and K so that the $|T(j\omega)|$ meets the filter specification and (2) devising a circuit that realizes the transfer function $T(s)$. We will discuss three methods of dealing with the first task. The second task is accomplished using our active filter building blocks to design a cascade circuit.

FIRST-ORDER CASCADE FILTERS

A simple way to produce a multi-pole low-pass filter is to connect n identical first-order low-pass filters in cascade. The transfer function of such a cascade connection is written as

$$T(s) = \underbrace{\left[\frac{K}{s/\alpha + 1}\right] \times \left[\frac{K}{s/\alpha + 1}\right] \times \cdots \times \left[\frac{K}{s/\alpha + 1}\right]}_{n \text{ stages}} = \frac{K^n}{(s/\alpha + 1)^n} \qquad (14\text{--}27)$$

To design such a filter, we must select $K, \alpha,$ and n such that the gain response of this transfer function meets a filter specification defined by $T_{MAX}, T_{MIN}, \omega_C,$ and ω_{MIN}.

The gain response of the transfer function in Eq. (14–27) is

$$|T(j\omega)| = \frac{|K|^n}{\left[\sqrt{1 + (\omega/\alpha)^2}\right]^n} \qquad (14\text{--}28)$$

The maximum gain in Eq. (14–28) occurs at $\omega = 0$, where the gain is $|T(0)| = |K|^n$. To meet the passband requirement of the specification, we set $|K|^n = T_{MAX}$ and evaluate the gain in Eq. (14–28) at the prescribed cutoff frequency ω_C.

$$|T(j\omega_C)| = \frac{T_{MAX}}{\left[\sqrt{1 + (\omega_C/\alpha)^2}\right]^n} = \frac{T_{MAX}}{\sqrt{2}}$$

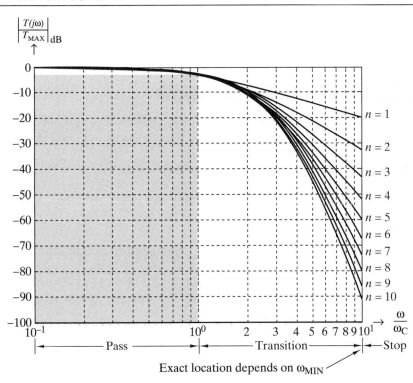

FIGURE 14-23 *First-order low-pass cascade filter responses.*

Equating the denominators in this equation and then solving for α yields

$$\alpha = \frac{\omega_C}{\sqrt{2^{1/n} - 1}} \tag{14-29}$$

This equation relates the cutoff frequency of each first-order stage (α) to the cutoff frequency of a cascade connection (ω_C) of n first-order stages. Each of these stages has a gain of $K = (T_{MAX})^{1/n}$ and a cutoff frequency of α. To complete a design, we need to know the number of stages required.

Figure 14–23 shows normalized gain responses defined by Eq. (14–28) for $n = 1$ to $n = 10$ where $T_{MAX} = |K|^n$ and α is given by Eq. (14–29). All of these responses meet the passband requirements, and, more importantly, the transition band gain decreases as we increase n. Obviously, we can improve the transition band performance of $T(s)$ by increasing n, but this increase adds more stages to the cascade circuit that realizes $T(s)$. There is a trade-off between transition band performance and circuit complexity. What we need to know is the smallest value of n that meets the filter specification.

We can estimate the smallest value of n from the gain plots in Figure 14–23. For example, suppose the transition band gain must decrease by 40 dB ($T_{MAX}/_{MIN} = 100$) in the decade above cutoff ($\omega_{MIN}/\omega_C = 10$). In Figure 14–23 we see that at $\omega/\omega_C = 10$, the normalized gain $|T(j\omega)|/T_{MAX}$ is -32 dB for $n = 2$ and -42 dB for $n = 3$. The $n = 3$ curve identifies the smallest value of n that reduces the gain by at least 40 dB in a one-decade transition band.

In summary, given values of T_{MAX}, T_{MIN}, ω_C, and ω_{MIN}, we construct a first-order cascade transfer function as follows. We use Figure 14–23 (or trial and error) to find the smallest integer n that meets the transition band requirements defined by T_{MAX}/T_{MIN} and ω_{MIN}/ω_C. Given n and ω_C, we calculate α using Eq. (14–29), $K = (T_{MAX})^{1/n}$ and use Eq. (14–27) to get the required transfer function. The transfer function is partitioned into a product of n identical first-order functions, each of which is realized using a first-order low-pass circuit. The next example illustrates this design procedure.

D DESIGN EXAMPLE 14–6

(a) Construct a first-order cascade transfer function that meets the following requirements: $T_{MAX} = 10\,dB$, $\omega_C = 200\,rad/s$, $T_{MIN} = -10\,dB$, and $\omega_{MIN} = 800\,rad/s$. Use MATLAB to visualize the gain plot.

(b) Design a cascade of active RC circuits that realizes the transfer function developed in (a). Use OrCAD to simulate the expected frequency response. Compare your result with the MATLAB plot.

SOLUTION:

(a) The specification requires the gain to decrease by 20 dB in a transition band with $\omega_{MIN}/\omega_C = 4$. Figure 14–23 shows that at $\omega/\omega_C = 4$, the normalized gain is about $-17\,dB$ for $n = 2$ and about $-22\,dB$ for $n = 3$. Thus, $n = 3$ is the smallest integer that meets the transition band requirement. Given n and ω_C, we calculate α using Eq. (14–29):

$$\alpha = \frac{\omega_C}{\sqrt{2^{1/n} - 1}} = \frac{200}{\sqrt{2^{1/3} - 1}} = 392\,rad/s$$

Since $T_{MAX} = 10\,dB$ (factor of $\sqrt{10}$), we write $K = (\sqrt{10})^{1/3} = 1.468$. So, finally, the required first-order cascade transfer function is

$$T(s) = \left(\frac{1.468}{s/392 + 1}\right)^3$$

Note that the cutoff frequency of each stage ($\alpha = 392\,rad/s$) is greater than the cutoff frequency of the n-stage transfer function ($\omega_C = 200\,rad/s$). A quick look at Eq. (14–29) reveals that $\alpha > \omega_C$ for all $n > 1$.

FIGURE 14–24

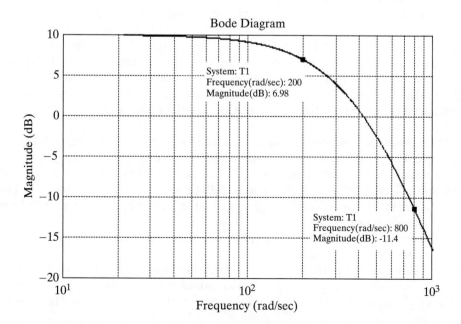

We expand the transfer function to place it in MATLAB as follows:

$$T(s) = \frac{190.56 \times 10^6}{s^3 + 1176s^2 + 461.0 \times 10^3 s + 60.24 \times 10^6}$$

MATLAB's bodemag returns the plot shown in Figure 14–24. The filter has the desired cutoff at 200 rad/s, and at 800 rad/s it is down 21.4 dB from the passband gain, which exceeds the requirement. We can now proceed to design our filter.

(b) The transfer function developed in part (a) can be partitioned into a product of three identical first-order functions of the form $1.468/(s/392 + 1)$. Each of these can be realized using the first-order low-pass circuit in Figure 14–25. The transfer function of this circuit is

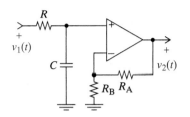

FIGURE 14–25 *First-order low-pass circuit.*

$$T(s) = \frac{1 + R_A/R_B}{RC_S + 1}$$

The design constraints for each stage are $1 + R_A/R_B = 1.468$ and $RC = 1/392$. Selecting $R_B = R = 100\,\text{k}\Omega$ leads to $R_A = 46.8\,\text{k}\Omega$ and $C = 0.0255\,\mu\text{F}$.

The OrCAD simulation of three cascaded stages produces the low-pass frequency response characteristic shown in Figure 14–26. The Probe results show that the design meets the target specifications.

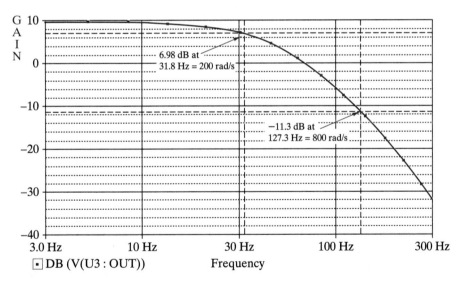

FIGURE 14–26

Design Exercise 14–9

Construct a first-order cascade transfer function that meets the following requirements: $T_{\text{MAX}} = 0\,\text{dB}$, $T_{\text{MIN}} = -30\,\text{dB}$, $\omega_C = 200\,\text{rad/s}$, and $\omega_{\text{MIN}} = 1\,\text{krad/s}$.

Answer:
$$T(s) = \left(\frac{460}{s + 460}\right)^4$$

BUTTERWORTH LOW-PASS FILTERS

All Butterworth low-pass filters have transfer functions of the form $T(s) = K/q_n(s)$, whose gain response is

$$|T(j\omega)| = \frac{|K|}{\sqrt{1 + (\omega/\omega_C)^{2n}}} \tag{14-30}$$

where ω_C is the cutoff frequency and n is the order of the denominator polynomial, that is, the number of poles. By inspection, the maximum gain occurs at dc, where $|T(0)| = |K| = T_{MAX}$. At the cutoff frequency the gain is $|T(j\omega_C)| = |K|/\sqrt{2} = T_{MAX}/\sqrt{2}$ for all values of n. When we make $K = T_{MAX}$, the gain response in Eq. (14-30) meets the passband requirements of a filter specification for all values of n. At high frequency ($\omega \gg \omega_C$) the gain approaches an asymptote of $|K|(\omega_C/\omega)^n$, which has a slope of $-20n$ dB/dec. Thus, we can decrease the gain in the transition band by increasing the number of poles.

Figure 14-27 compares Butterworth and first-order cascade gain responses for $n = 4$. Both responses have high-frequency asymptotes whose slopes are -80 dB/dec. However, the Butterworth gain approaches its asymptote at a lower frequency, so it has less gain in the transition band. The reduced gain means that the Butterworth response has better transition band performance than the first-order cascade.

FIGURE 14–27 *First-order cascade and Butterworth low-pass filter responses for $n = 4$.*

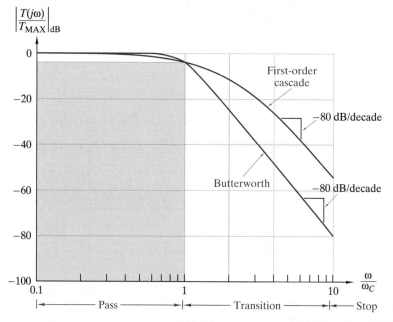

Figure 14–28 shows normalized plots of the Butterworth gain responses defined by Eq. (14–30), where $T_{MAX} = |K|$. Clearly the transition band gain decreases rapidly as we increase n, the filter order. Increasing n means that $T(s) = K/q_n(s)$ has more poles, which in turn means more stages in a cascade circuit realizing $T(s)$. As we discovered with the first-order cascade filter, there is a trade-off between reducing the transition band gain and circuit complexity. What we need to know is the smallest value of n that meets a given filter specification.

Evaluating the gain in Eq. (14–30) at $\omega = \omega_{MIN}$ with $T_{MAX} = |K|$ produces the inequality

$$|T(j\omega_{MIN})| = \frac{T_{MAX}}{\sqrt{1 + (\omega_{MIN}/\omega_C)^{2n}}} \leq T_{MIN}$$

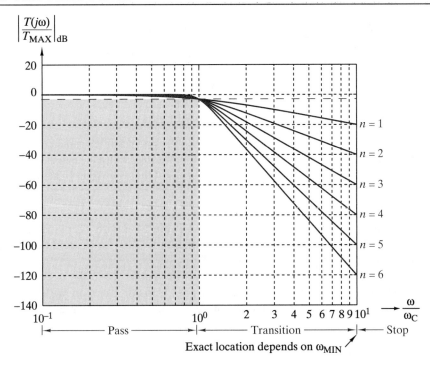

FIGURE 14–28 *Butterworth low-pass filter responses.*

Solving the inequality for n yields the constraint

$$n \geq \frac{1}{2} \frac{\ln\left[(T_{MAX}/T_{MIN})^2 - 1\right]}{\ln\left[\omega_{MIN}/\omega_C\right]}$$ (14–31)

The right side of this equation is a lower bound on the filter order n. Note that the lower bound is determined by the transition band ratios T_{MAX}/T_{MIN} and ω_{MIN}/ω_C. In other words, the smallest n depends on how much and how rapidly the gain must change in the transition band.

For example, if the transition band gain must decrease by 30 dB ($T_{MAX}/T_{MIN} = 10^{3/2}$) in the two octaves above cutoff ($\omega_{MIN}/\omega_C = 4$), then Eq. (14–31) yields

$$n \geq \frac{1}{2} \frac{\ln\left[(10^{3/2})^2 - 1\right]}{\ln[4]} = 2.49$$

Since filter order must be an integer, the smallest value is $n = 3$. The gain plots in Figure 14–28 confirm this result, since a normalized gain below -30 dB at $\omega/\omega_C = 4$ cannot be achieved by any order less than $n = 3$. For Butterworth low-pass filters, the smallest value of n can be determined analytically using Eq. (14–31) or graphically using Figure 14–28.

Once the required value of n is known, we need an nth-order polynomial $q_n(s)$ to construct a transfer function with Butterworth gain characteristics. A method of obtaining such polynomials is given in web appendix B and using that method we generated the normalized ($\omega_C = 1$) polynomials $q_n(s)$ in Table 14–2. All of these polynomials have the property that $q_n(0) = 1$.

Using these polynomials, we obtain an nth-order Butterworth transfer function with a dc gain of K and cutoff frequency of ω_C as

$$T(s) = \frac{K}{q_n(s/\omega_C)}$$ (14–32)

T A B L E **14–2**   **NORMALIZED POLYNOMIALS THAT PRODUCE BUTTERWORTH RESPONSES**

ORDER	NORMALIZED DENOMINATOR POLYNOMIALS
1	$(s+1)$
2	$(s^2 + 1.414s + 1)$
3	$(s+1)(s^2 + s + 1)$
4	$(s^2 + 0.7654s + 1)(s^2 + 1.848s + 1)$
5	$(s+1)(s^2 + 0.6180s + 1)(s^2 + 1.618s + 1)$
6	$(s^2 + 0.5176s + 1)(s^2 + 1.414s + 1)(s^2 + 1.932s + 1)$

where $q_n(s/\omega_C)$ is the nth-order polynomial in Table 14–2 with s replaced by s/ω_C. Since $q_n(0) = 1$, the dc gain of this transfer function is $|T(0)| = |K|$ and the poles of $T(s)$ are roots of the equation $q_n(s/\omega_C) = 0$.

For example, for $n = 3$ the polynomial in Table 14–2 is $q_3(s) = (s + 1)$ $(s^2 + s + 1)$. A third-order Butterworth low-pass transfer function with a dc gain of $0\,\mathrm{dB}$ ($K = 1$) and a cutoff frequency of $\omega_C = 1$ krad/s is written as

$$T(s) = \frac{1}{q_3(s/1000)} = \frac{1}{(s/1000 + 1)\left[(s/1000)^2 + (s/1000) + 1\right]}$$

$$= \underbrace{\left[\frac{10^3}{s + 10^3}\right]}_{\text{first order}} \underbrace{\left[\frac{10^6}{s^2 + 1000s + 10^6}\right]}_{\text{second order}} = \frac{10^9}{s^3 + 2000s^2 + 2 \times 10^6 s + 10^9}$$

Figure 14–29 shows how the Butterworth realization works. The first-order low-pass filter's transfer function is multiplied by the second-order filter's transfer function to produce the desired response.

This transfer function has a real pole at $s = -1000$ rad/s and a pair of complex poles with $\omega_0 = 1000$ rad/s and $\zeta = 0.5$. The three poles are all located a distance of $\omega_C = 1000$ from the origin in the s plane. This illustrates a general principle of Butterworth poles—*Butterworth poles are all located on a circle of radius ω_C in the left-half of the s plane.*

FIGURE 14–29 *How a third-order Butterworth response is constructed.*

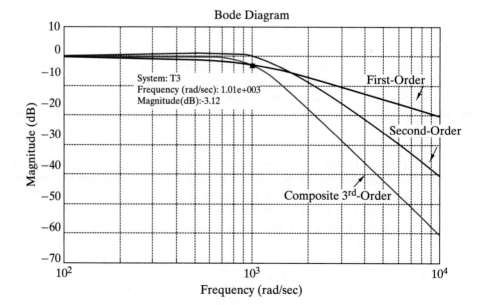

Once we have a transfer function $T(s)$ whose gain meets the filter specification, we partition it into a product of first- and second-order functions. Each of these functions is then realized using one of our active filter building blocks. A cascade connection of these building blocks then produces the required transfer function. The next example illustrates the procedure.

◗ **DESIGN EXAMPLE 14-7**

(a) Construct a Butterworth low-pass transfer function that meets the following requirements:

$T_{MAX} = 20\,dB, \omega_C = 1\,krad/s, T_{MIN} = -20\,dB$, and $\omega_{MIN} = 4\,krad/s$.

(b) Design a cascade of active RC circuits that realizes the transfer function found in part (a). Validate your design using OrCAD.

SOLUTION:

(a) The specification requires the gain to decrease by 40 dB in the transition band between $\omega_C = 1\,krad/s$ and $\omega_{MIN} = 4\,krad/s$. The transition band ratios for this specification are $T_{MAX}/T_{MIN} = 100$ and $\omega_{MIN}/\omega_C = 4$. Inserting these ratios in Eq. (14–31) yields

$$n \geq \frac{1}{2}\frac{\ln[100^2 - 1]}{\ln[4]} = 3.32$$

Thus, $n = 4$ is the lowest-order Butterworth filter that meets the filter specification. Using $K = T_{MAX} = 10$ and $q_4(s)$ from Table 14–2, the required fourth-order Butterworth low-pass transfer function is

$$T(s) = \frac{K}{q_4(s/1000)}$$

$$= \frac{10}{\left[\left(\dfrac{s}{1000}\right)^2 + 0.7654\left(\dfrac{s}{1000}\right) + 1\right]\left[\left(\dfrac{s}{1000}\right)^2 + 1.848\left(\dfrac{s}{1000}\right) + 1\right]}$$

(b) The fourth-order Butterworth transfer function in part (a) can be partitioned as

$$T(s) = T_1(s)T_2(s)$$

$$= \left[\frac{K_1}{\left(\dfrac{s}{1000}\right)^2 + 0.7654\left(\dfrac{s}{1000}\right) + 1}\right]\left[\frac{K_2}{\left(\dfrac{s}{1000}\right)^2 + 1.848\left(\dfrac{s}{1000}\right) + 1}\right]$$

where K_1 and K_2 are to be determined. This partitioning calls for a cascade of two second-order low-pass building blocks.

Figure 14–30 on the next page shows a design sequence for circuits that realize these transfer functions. The first two rows in the figure show the required transfer functions and the stage parameters ω_0 and ζ. The stage protoype in the third row is the low-pass building block circuit in Figure 14–3. Using the equal element design method for this circuit [see Eq. (14–8)] leads to the design constraints in the fourth row. The element values selections in the fifth row produce the two-stage design in the last row of Figure 14–30.

As discussed in Sect. 14–2, the equal element design method does not allow us to control K_1 and K_2. The values achieved in our two-stage design are $K_1 = 2.235$ and $K_2 = 1.152$. Together the two stages produce an overall dc

Item	First Stage	Second Stage
Prototype transfer function	$\dfrac{K_1}{(s/1000)^2 + 0.7654(s/1000) + 1}$	$\dfrac{K_2}{(s/1000)^2 + 1.848(s/1000) + 1}$
Stage parameters	$\omega_0 = 1000 \quad \zeta = 0.7654/2 = 0.3827$	$\omega_0 = 1000 \quad \zeta = 1.848/2 = 0.924$
Stage prototype	(circuit diagram with R, R, C, C, R_B, R_A)	(circuit diagram with R, R, C, C, R_B, R_A)
Design constraints	$RC = \dfrac{1}{\omega_0} = 0.001$ $K_1 = 3 - 2\zeta = 2.2346$ $R_A = (K_1 - 1)R_B$	$RC = \dfrac{1}{\omega_0} = 0.001$ $K_2 = 3 - 2\zeta = 1.152$ $R_A = (K_2 - 1)R_B$
Element values	Let $R = 100 \text{ k}\Omega$, then $C = 0.01 \text{ μF}$ Let $R_B = 100 \text{ k}\Omega$, then $R_A = 123 \text{ k}\Omega$	Let $R = 100 \text{ k}\Omega$, then $C = 0.01 \text{ μF}$ Let $R_B = 100 \text{ k}\Omega$, then $R_A = 15.2 \text{ k}\Omega$
Final designs	(circuit diagram: 0.01 μF, 100 kΩ 100 kΩ, 0.01 μF, 123 kΩ, 100 kΩ, $K_1 = 2.2346$)	(circuit diagram: 0.01 μF, 100 kΩ 100 kΩ, 0.01 μF, 15.2 kΩ, 100 kΩ, $K_2 = 1.152$)

FIGURE 14–30 *Design sequence for Example 14–7.*

gain of $K = K_1 K_2 = 2.235 \times 1.152 = 2.575$, which is less than the $K = 10$ required by the specification. We deal with this shortfall by adding a gain correction stage with a gain of $10/2.575 = 3.883$. Figure 14–31 shows the final three-stage design with the added gain correction stage.

FIGURE 14–31

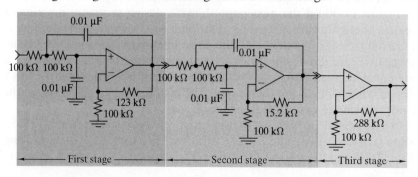

FIGURE 14–32

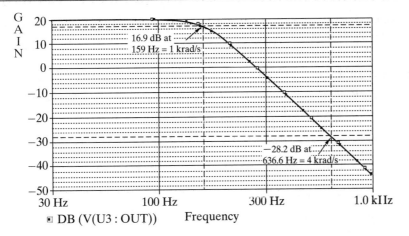

▣ DB (V(U3 : OUT))   Frequency

The OrCAD simulation of the circuit in Figure 14–31 is shown in Figure 14–32. It meets or exceeds all of the design requirements. ■

❶ Design Exercise 14–10

The circuit design in Example 14–7 used the *equal element* method. Rework the problem using the *unity gain* technique. Use OrCAD to validate your design. Comment on the two approaches.

Answer: Figure 14–33(a) shows the unity gain redesign of the fourth-order low-pass Butterworth filter. The Probe results in Figure 14–33(b) show that the design meets all of the specifications.

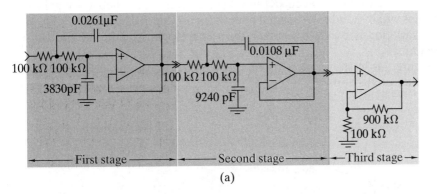

First stage ← → Second stage ← → Third stage →

(a)

FIGURE 14–33

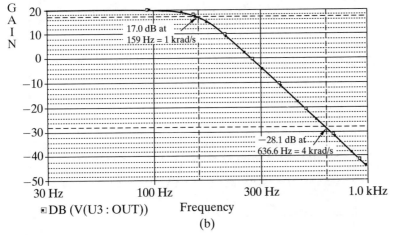

▣ DB (V(U3 : OUT))   Frequency

(b)

The unity gain technique requires fewer resistors but requires precise capacitors that may be difficult to find.

▶D Design Exercise 14–11

Construct a Butterworth low-pass transfer function that meets the following requirements: $T_{MAX} = 0\,dB$, $T_{MIN} = -40\,dB$, $\omega_C = 250\,rad/s$, and $\omega_{MIN} = 1.5\,krad/s$.

Answer:
$$T(s) = \frac{250^3}{(s + 250)(s^2 + 250s + 250^2)}$$

CHEBYCHEV LOW-PASS FILTERS

All Chebychev type 1 low-pass filters have a gain response of the form

$$|T_n(j\omega)| = \frac{|K|}{\sqrt{1 + C_n^2(\omega/\omega_C)}} \tag{14–33}$$

where $C_n(x)$ is an nth-order Chebychev polynomial defined by

$$C_n(x) = \cos\left[n \times \cos^{-1}(x)\right] \quad x \leq 1 \tag{14–34a}$$

and

$$C_n(x) = \cosh\left[n \times \cosh^{-1}(x)\right] \quad x > 1 \tag{14–34b}$$

In the passband ($x = \omega/\omega_C \leq 1$) $C_n(x)$ is the cosine function in Eq. (14–34a), and the term $1 + C_n^2(\omega/\omega_C)$ in the denominator of Eq. (14–33) varies between 1 (when $C_n = 0$) and 2 (when $C_n = \pm 1$). This means that in the passband the gain $|T(j\omega)|$ in Eq. (14–33) varies between $|K|$ and $|K|/\sqrt{2}$. Thus, the Chebychev gain variation remains within standard passband bounds for all values of n. However, the passband gain variation is oscillatory rather than smooth and steady like the Butterworth response.

Figure 14–34 shows plots of Eq. (14–33) for $n = 4$ and $n = 5$. The oscillatory variation of the term $1 + C_n^2(\omega/\omega_C)$ produces a sequence of resonant peaks and

FIGURE 14–34 *Chebychev passband gain responses.*

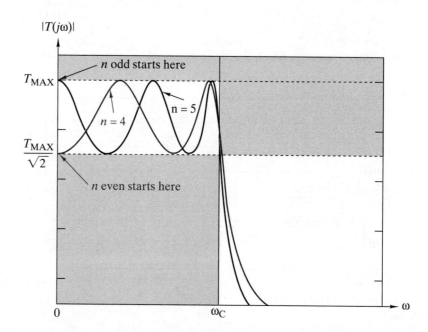

valleys given the descriptive name *ripple*. The gain at the top of every peak is $|K|$ and the gain at the bottom of every valley is $|K|/\sqrt{2}$. The Chebychev passband gain is called an *equal-ripple response* because the upper bound is always $|K|$ and the lower bound $|K|/\sqrt{2}$. Because of the equal-ripple response, the Chebychev gain has $T_{MAX} = |K|$ and $|T(j\omega_C)| = |K|/\sqrt{2}$ for all values of n.

However, the Chebychev dc gain is $|K|$ when n is odd and $|K|/\sqrt{2}$ when n is even. In other words, T_{MAX} does not occur at dc when n is even. We must account for this difference in dc gain when constructing low-pass transfer functions with Chebychev poles.

Figure 14–35 compares the Butterworth and Chebychev gain responses for $n = 4$. Both responses have high-frequency asymptotes with slopes of -80 dB/dec. The Butterworth response is relatively flat in the passband and sedately approaches its high-frequency asymptote in the transition band. In contrast, the last resonant peak in the Chebychev passband produces an initial slope in the transition band that is steeper than -80 dB/dec.[3] As a result, a Chebychev response has less gain in the transition band than a Butterworth response of the same order.

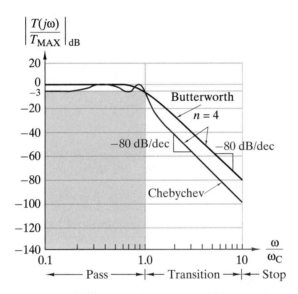

FIGURE 14–35 *Butterworth and Chebychev low-pass filter responses for $n = 4$.*

Figure 14–36 shows normalized plots of the Chebychev gain responses with $T_{MAX} = |K|$. As with Butterworth responses, the Chebychev transition band gain decreases rapidly as we increase n, the filter order. Again to minimize circuit complexity, we need to know the smallest value of n that meets a given filter specification. Evaluating the Chebychev gain in Eq. (14–33) at $\omega = \omega_{MIN}$ with $T_{MAX} = |K|$ produces the inequality

$$|T(j\omega_{MIN})| = \frac{T_{MAX}}{\sqrt{1 + C_n^2(\omega_{MIN}/\omega_C)}} \leq T_{MIN}$$

Solving this constraint for $C_n(\omega_{MIN}/\omega_C)$ yields

$$C_n(\omega_{MIN}/\omega_C) \geq \sqrt{\left(\frac{T_{MAX}}{T_{MIN}}\right)^2 - 1}$$

[3]It can be shown that at $\omega = \omega_C$ the slope of a Chebychev gain is n times as steep as a Butterworth gain. See Aram Budak, *Passive and Active Network Analysis and Synthesis*, Houghton Mifflin, 1974, p. 516.

FIGURE 14–36 *Chebychev low-pass filter responses.*

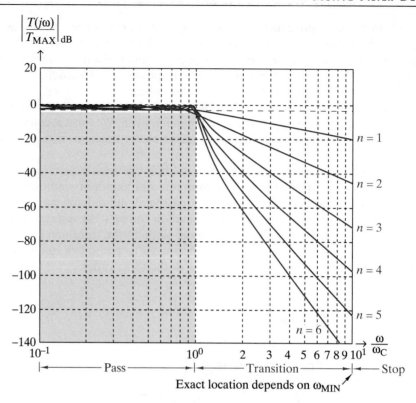

In the stopband $(x = \omega/\omega_C \geq 1)$ the function $C_n(x)$ is defined by the hyperbolic cosine function in Eq. (14–34b). Inserting this definition into the preceding inequality and then solving for n yields

$$n \geq \frac{\cosh^{-1}\left(\sqrt{(T_{MAX}/T_{MIN})^2 - 1}\right)}{\cosh^{-1}(\omega_{MIN}/\omega_C)} \qquad (14\text{–}35)$$

The right side of this equation is a lower bound on the filter order n. As might be expected, the lower bound is determined by the transition band ratios T_{MAX}/T_{MIN} and ω_{MIN}/ω_C. Based on our experience with first-order cascade, Butterworth, and now Chebychev filters, it seems reasonable to infer the following general principle of filter design.

> *Filter order is determined by how much and how rapidly the gain must decrease in the transition band.*

The smallest order for a Chebychev filter can be determined analytically using Eq. (14–35) or graphically using the gain plots in Figure 14–36. Once the required order is known, we need an nth-order polynomial $q_n(s)$ to construct a transfer function with a Chebychev gain response. A method of obtaining such a polynomial is given in web Appendix B. Using the method given there, we can generate the normalized $(\omega_C = 1)$ polynomials $q_n(s)$ in Table 14–3. Note that these polynomials all have the property that $q_n(0) = 1$.

Using these polynomials, an nth-order Chebychev transfer function with a dc gain of K and cutoff frequency of ω_C is written as

$$T(s) = \frac{K}{q_n(s/\omega_C)} \qquad n \text{ odd} \qquad (14\text{–}36a)$$

TABLE 14–3   NORMALIZED POLYNOMIALS THAT PRODUCE CHEBYCHEV RESPONSES

ORDER	NORMALIZED DENOMINATOR POLYNOMIALS
1	$(s+1)$
2	$[(s/0.8409)^2 + 0.7654(s/0.8409) + 1]$
3	$[(s/0.2980) + 1][(s/0.9159)^2 + 0.3254(s/0.9159) + 1]$
4	$[(s/0.9502)^2 + 0.1789(s/0.9502) + 1)][(s/0.4425)^2 + 0.9276(s/0.4425) + 1]$
5	$[(s/0.1772) + 1][(s/0.9674)^2 + 0.1132(s/0.9674) + 1][(s/0.6139)^2 + 0.4670(s/0.6139) + 1]$
6	$[(s/0.9771)^2 + 0.0781(s/0.9771) + 1][(s/0.7223)^2 + 0.2886(s/0.7223) + 1][(s/0.2978)^2 + 0.9562(s/0.2978) + 1]$

or

$$T(s) = \frac{K/\sqrt{2}}{q_n(s/\omega_C)} \quad n \text{ even} \qquad\qquad (14\text{–}36b)$$

where $q_n(s/\omega_C)$ is the nth-order polynomial in Table 14–3 with s replaced by s/ω_C. Since $q_n(0) = 1$, the scale factors in Eqs. (14–36a) and (14–36b) ensure that the dc gains are $T(0) = |K|$ when n is odd and $T(0) = K/\sqrt{2}$ when n is even. These scale factor adjustments are required to ensure that $T_{MAX} = |K|$ whether n is even or odd. Without these adjustments, we cannot use Eq. (14–35) or Figure 14–36 when n is even. In either case, however, the poles of $T(s)$ are roots of the equation $q_n(s/\omega_C) = 0$.

As an example of using Table 14–3, we write a second-order Chebychev low-pass transfer function with a dc gain of 20 dB ($K = 10$) and a cutoff frequency of $\omega_C = 1\,\text{krad/s}$ as

$$T(s) = \frac{10/\sqrt{2}}{q_2(s/1000)} = \frac{7.07}{(s/840.9)^2 + 0.7654(s/840.9) + 1}$$

This transfer function has $T_{MAX} = 10$, a cutoff frequency at $\omega_C = 1000\,\text{rad/s}$, and a pair of complex poles with $\omega_0 = 840.9\,\text{rad/s}$ and $\zeta = 0.3827$. This illustrates two important properties of Chebychev poles: (1) Their natural frequencies are less than the filter cutoff frequency ($\omega_0 < \omega_C$), and (2) their damping ratios are less than those of Butterworth poles of the same order. As a result, the Chebychev poles produce resonant peaks in the gain response at frequencies below ω_C. Put differently, Chebychev poles are specifically located to produce an equal-ripple response in the passband.

Once we have a $T(s)$ that meets the filter specification, we realize it using a cascade of first- and second-order building blocks. The next example illustrates the design of a Chebychev low-pass filter.

D DESIGN EXAMPLE 14–8

(a) Construct a Chebychev low-pass transfer function that meets the following requirements: $T_{MAX} = 20\,\text{dB}$, $\omega_C = 10\,\text{rad/s}$, $T_{MIN} = -30\,\text{dB}$,   and $\omega_{MIN} = 50\,\text{rad/s}$.

(b) Design a cascade of active RC circuits that produces the transfer function found in part (a).

SOLUTION:

(a) This specification requires the gain to decrease by 50 dB in the transition band between $\omega_C = 10$ rad/s and $\omega_{MIN} = 50$ rad/s. The corresponding transition band ratios are $T_{MAX}/T_{MIN} = 10^{5/2}$ and $\omega_{MIN}/\omega_C = 5$. Using these ratios in Eq. (14–35) yields

$$n \geq \frac{\cosh^{-1}\left(\sqrt{(10^{5/2})^2 - 1}\right)}{\cosh^{-1}(5)} = 2.81$$

The smallest integer meeting the specification is $n = 3$. Using the polynomial $q_3(s)$ from Table 14–3, we construct the required Chebychev low-pass transfer function using Eq. (14–36a).

$$T(s) = \frac{K}{q_3(s/10)}$$

$$= \frac{10}{\left[\left(\dfrac{s}{2.98}\right) + 1\right]\left[\left(\dfrac{s}{9.159}\right)^2 + 0.3254\left(\dfrac{s}{9.159}\right) + 1\right]}$$

No scale factor adjustment is needed here because $n = 3$ is odd.

(b) The Chebychev transfer function in (a) can be partitioned as

$$T(s) = T_1(s)\,T_2(s)$$

$$= \left[\frac{K_1}{\dfrac{s}{2.98} + 1}\right]\left[\frac{K_2}{\left(\dfrac{s}{9.159}\right)^2 + 0.3254\left(\dfrac{s}{9.159}\right) + 1}\right]$$

where K_1 and K_2 are to be determined. This partition calls for a cascade of a first-order and a second-order low-pass filter.

The first row in the design sequence in Figure 14–37 shows the required transfer functions. The second, third, and fourth rows show the stage parameters, stage prototypes, and design constraints. The second-order prototype is the low-pass circuit in Figure 14–3. The design constraints for this circuit use the equal element design method [see Eq. (14–8)]. The equal element method produces a dc gain of $K_2 = 3 - 2\zeta = 2.675$. The first-order prototype is the low-pass circuit in Figure 14–25. Its pole location is determined by the RC product and its gain K_1 can be adjusted without changing the pole location, so we are free to set $K_1 = 10/K_2 = 3.738$, which makes $K = K_1 K_2 = 10$ as required. The assigned element values in the fifth row produce the final design in the last row. A cascade connection of these two stages meets all design requirements without an added gain correction stage.

Figure 14–38 demonstrates the effects of the Chebyshev polynomial in building a filter with a steeper roll-off than the order of the filter in the transition band. We have used MATLAB to plot the effects of combining a first-order and a second-order stage to produce the composite third-order result. Note that the first-order filter crosses the -3 dB (0.707) point at 3 rad/s while the second-order filter has a cutoff at 14 rad/s. The composite, however, reaches the -3 dB point exactly at the desired 10 rad/s. The absolute and normalized scale highlights the ripple nature of the Chebyshev realization.

Item	First Stage	Second Stage
Prototype transfer function	$\dfrac{K_1}{(s/2.98)+1}$	$\dfrac{K_2}{(s/9.159)^2 + 0.3254(s/9.159)+1}$
Stage parameters	$\omega_0 = 2.980$	$\omega_0 = 9.159 \quad \zeta = 0.3254/2 = 0.1627$
Stage prototype	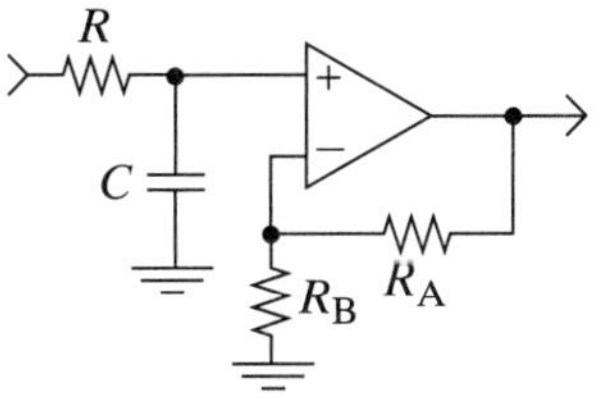 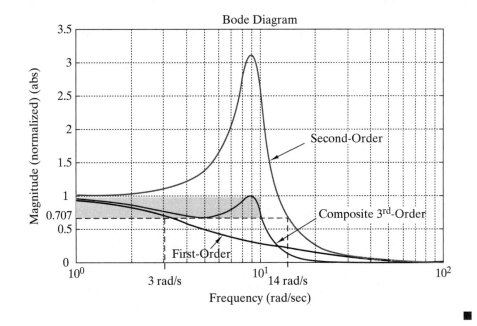	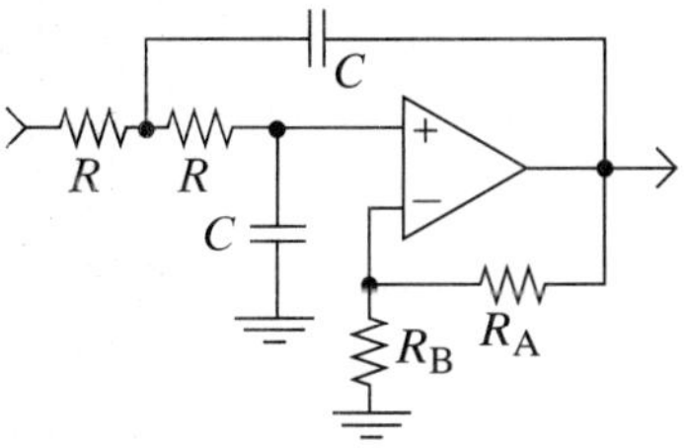
Design constraints	$RC = \dfrac{1}{\omega_0} = 0.3356$ $K_1 = 10/K_2 = 3.74$ $R_A = (K_1 - 1)R_B$	$RC = \dfrac{1}{\omega_0} = 0.1092$ $K_2 = 3 - 2\zeta = 2.67$ $R_A = (K_2 - 1)R_B$
Element values	Let $R = 100$ kΩ, then $C = 3.36$ µF Let $R_B = 100$ kΩ, then $R_A = 274$ kΩ	Let $R = 100$ kΩ, then $C = 1.092$ µF Let $R_B = 100$ kΩ, then $R_A = 167$ kΩ
Final designs	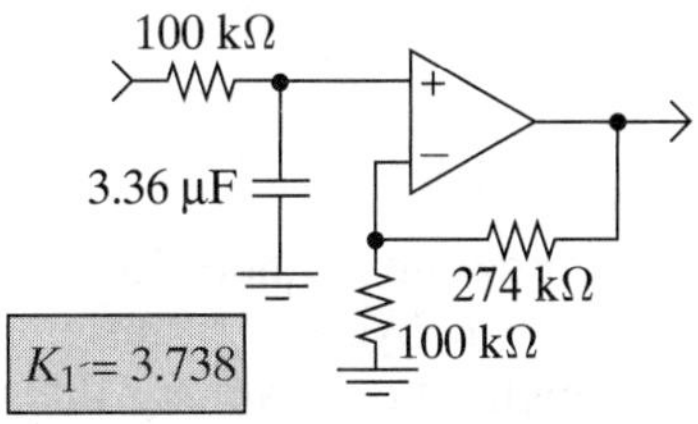 $K_1 = 3.738$	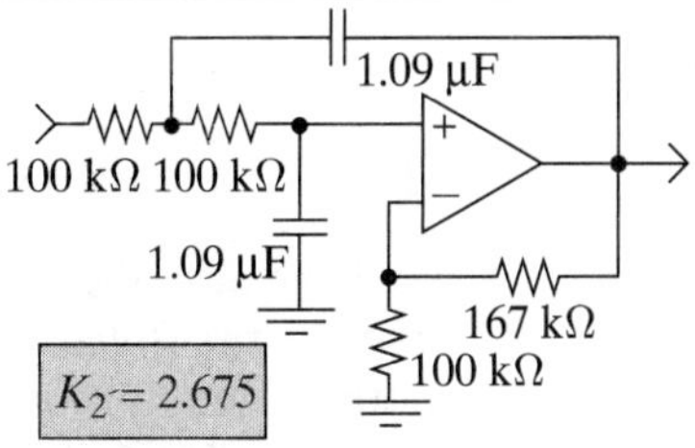 $K_2 = 2.675$

FIGURE 14–37 *Design sequence for Example 14–8.*

FIGURE 14–38 *How a third-order Chebychev response is constructed.*

Bode Diagram

Second-Order

Composite 3rd-Order

First-Order

3 rad/s

14 rad/s

0.707

Magnitude (normalized) (abs)

Frequency (rad/sec)

◐ Design Exercise 14–12

Rework the design in Example 14–8 using the unity gain method in Sect. 14–2 to design the required third-order low-pass circuit.

FIGURE 14–39

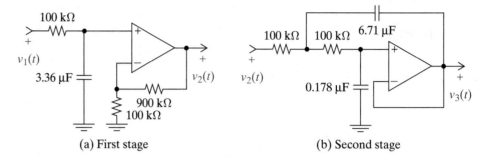

(a) First stage   (b) Second stage

Answer: Figure 14–39 shows the unity gain method redesigns of first- and second-order stages in Figure 14–37.

◐ Design Exercise 14–13

Construct a Chebychev low-pass transfer function that meets the following requirements: $T_{MAX} = 0\,dB$, $T_{MIN} = -30\,dB$, $\omega_C = 250\,rad/s$, and $\omega_{MIN} = 1.5\,krad/s$.

Answer:

$$T(s) = \frac{210^2/\sqrt{2}}{s^2 + 161s + 210^2}$$

14–5 LOW-PASS FILTER EVALUATION

We have described low-pass filter design methods for three responses, the first-order cascade, the Butterworth, and the Chebychev. At this point we want to compare the methods and discuss how we might choose among them. In filter applications the gain response is obviously important. Figure 14–40 shows three straight-line asymptotes (solid lines) for components of the three filter types and the total gain responses (dashed curves) for $n = 4$. All three of the total responses meet the same passband requirements, have the same cutoff frequency, and have high-frequency asymptotes with slopes of $-20n = -80\,dB/decade$. However, the three response components use different corner frequencies to achieve the same total response. At one extreme the corner frequency of the first-order cascade response is above the cutoff frequency, so

FIGURE 14–40 *First-order cascade, Butterworth, and Chebychev step responses for n = 4.*

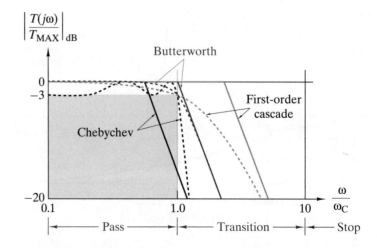

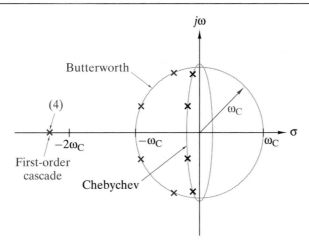

FIGURE 14–41 *First-order cascade, Butterworth, and Chebychev pole locations for* $n = 4$.

its actual gain response approaches its asymptote very gradually. At the other extreme the Chebychev corner frequency lies below the cutoff frequency and the actual response has a resonant peak that fills the gap between the corner frequency and the cutoff frequency. This resonance causes the Chebychev response to decrease rapidly in the neighborhood of the cutoff frequency. The Butterworth response has its corner frequency at the cutoff frequency, so its gain response falls between these two extremes.

The differences in gain response can be understood by examining the pole-zero diagram in Figure 14–41. The Butterworth poles are evenly distributed on a circle of radius ω_C. The Chebychev poles lie on an ellipse whose minor axis is much smaller than ω_C. As a result, the Chebychev poles are closer to the j-axis, have lower damping ratios, and produce a gain response with pronounced resonant peaks. These resonant peaks lead to the equal-ripple response in the passband and the steep gain slope in the neighborhood of the cutoff frequency. At the other extreme the first-order cascade response has a fourth-order pole (quadruple pole) located on the negative real axis. The distance from the j-axis to the first-order cascade poles is much larger than ω_C, which explains the rather leisurely way its gain response transitions from the passband to stopband asymptote. As might be expected, the Butterworth poles fall between these two extremes.

This discussion illustrates the following principle. For any given value of n, the Chebychev response produces more transition band attenuation than the Butterworth response, which, in turn, produces more than the first-order cascade response. If transition band performance is the only consideration, then we should choose the Chebychev response. However, the Chebychev response comes at a price.

Figure 14–42 shows the step response of these three low-pass filters for $n = 4$. The step response of the Chebychev filter has lightly damped oscillations that produce a large overshoot and a long settling time. These undesirable features of the Chebychev step response are a direct result of the low-damping-ratio complex poles that produce the desirable features of its gain response. At the other extreme the step response of the first-order cascade filter rises rapidly to its final value without overshooting. This result should not be surprising since the remote poles of the first-order cascade produce exponential waveforms that have relatively short durations. In other words, the desirable features of the first-order cascade step response are a direct result of the remote real poles that produce the undesirable features of its gain response. Not surprisingly, the step response of the Butterworth filter lies between these two extremes.

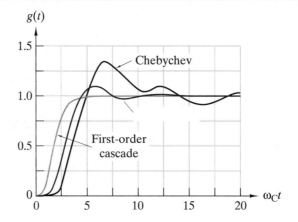

Finally, consider the element values in the circuit realizations of these filters. Examination of Table 14–3 reveals that each pair of complex poles in a Chebychev filter has a different ω_0 and a different ζ. These parameters define the constraints on the element values for each stage in the filter. As a result, each stage in a cascade realization of a Chebychev filter has a different set of element values. In contrast, the stages in a first-order cascade filter can be exactly the same. From a manufacturing point of view, it may be better to produce and stock identical circuits rather than uniquely different circuits.

The essential point is that transition band attenuation is important, but it does not tell the whole story. Filter design, and indeed all design, involves trade-offs between conflicting requirements. The choice of a design approach is driven by the weight assigned to conflicting requirements.

D E **DESIGN AND EVALUATION EXAMPLE 14 – 9**

A low-pass filter is required that will process high-fidelity audio signals meeting the following criteria:

$f_0 = 20\,\text{kHz}$

$T_{\text{MAX}} = 20\,\text{dB}$

$T_{\text{MIN}} = -50\,\text{dB}$

$f_{\text{MIN}} = 200\,\text{kHz}$

The filter should have less than ± 1.5 dB gain variation over most of the passband. Design an appropriate circuit and justify your choice.

SOLUTION:

We have three possible design approaches: first-order cascade, Butterworth, and Chebyshev. We need to determine the filter order of each design that will accomplish the task. The filter will need to provide 70 dB of attenuation from the passband to f_{MIN}. It will have one decade from the cutoff frequency to accomplish it. From Figure 14–23 we determine that the filter order for a first-order cascade to achieve the desired response is $n = 7$. From Figure 14–28, we can determine that a fourth-order Butterworth filter will accomplish the same task. Finally, from Figure 14–36, a third-order Chebyshev filter will meet the roll-off requirement.

Although it appears that the Chebyshev solution is the most efficient, we need to look at other factors as well. The application is for an audio system. Ringing caused by a poor step response is a major consideration for an audio system. Chebyshev

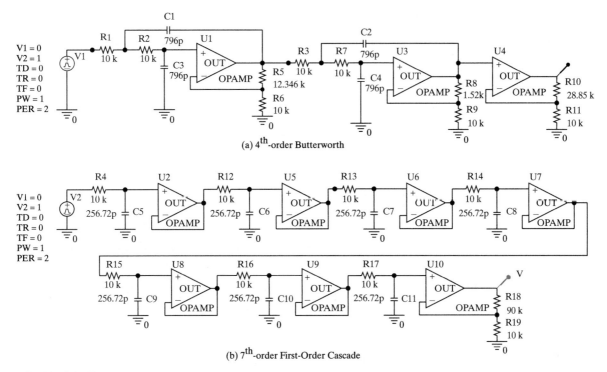

(a) 4th-order Butterworth

(b) 7th-order First-Order Cascade

FIGURE 14–43

filters have different small damping ratios, which cause a tortured step response with multiple ringing frequencies. Furthermore, for an audio application the 3-dB ripple in the passband would result in varying gains for different audio frequencies—a real concern for serious audiophiles, who can detect signal differences of ±1.5 dB. A Butterworth filter has a maximally flat response in the passband—a desirable quality for audio applications. However, it has a step response that is less than ideal. Since a first-order cascade has only real poles, it will have the best step response characteristic. The decision, then, focuses on the trade-off between simplicity in design and a very flat passband, favoring the Butterworth solution or the better step response of the first-order cascade. To see how each performs, both circuits are designed and simulated in OrCAD. Figure 14–43 shows the circuit designs used.

The gain characteristics shown in Figure 14–44(a) favor the Butterworth solution. The first-order cascade begins to roll off well before reaching the cutoff frequency; for example, it is 1.5 dB down at 14 kHz while the Butterworth solution does not reach 1.5 dB down until about 18 kHz. The roll-off is more rapid for the Butterworth filter than the first-order cascade in the transition band—a better performance for a low-pass filter. They both reach –60 dB at about 200 kHz, as required. Figure 14–44(b) shows the step response of both circuits. The first-order cascade has a very fast rise time without any overshoot—good for reproducing sounds from percussion instruments. The Butterworth filter has an 11% overshoot, a somewhat slower rise time, and a damped natural frequency of about 19 kHz—within the filter's passband. These qualities tend to diminish the overall audio performance.

The Butterworth filter is cheaper, uses less power, and has better frequency response characteristics. The first-order cascade has a better step response. Before one decides, it would be better to build breadboards of both circuits and expose the results to an audiophile to see which performs better. The final decision would include that result along with costs and manufacturing issues. Technology alone is

FIGURE 14–44

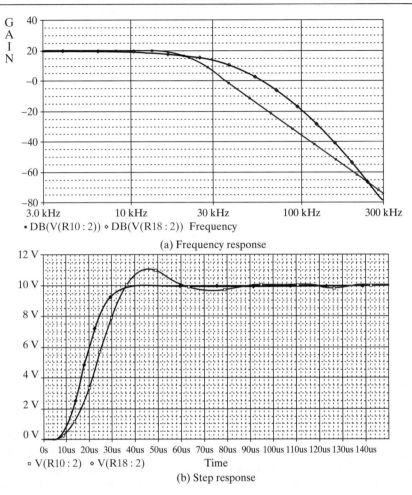

(a) Frequency response

(b) Step response

often not an obvious or sufficient basis for determining which of two or more competing solutions is best. ∎

14–6 HIGH-PASS FILTER DESIGN

An ideal high-pass filter has a constant gain above a cutoff frequency ω_C and zero gain below ω_C. Filter design specifications define the degree to which real filters are required to approach the ideal. Figure 14–45 is a Bode plot specifying the allowable region for high-pass filter gain responses. The allowable region is defined by four familiar parameters: T_{MAX}, T_{MIN}, ω_C, and ω_{MIN}. We are acquainted with the definitions and use of these parameters from our study of low-pass filter design.

The difference here is that $\omega_{MIN} < \omega_C$, that is, the transition band and stopband are below the passband. In fact, if we look carefully we see that Figure 14–45 is the mirror image of Figure 14–22 for low-pass filters. This suggests that low-pass filter design methods apply to high-pass filter design when we interchange the positions of the passband and the stopband. Mathematically the interchange is called a low-pass to high-pass transformation. The transformation is carried out by inverting the frequency variables in the low-pass filter relationships. We demonstrate this first for first-order cascade responses.

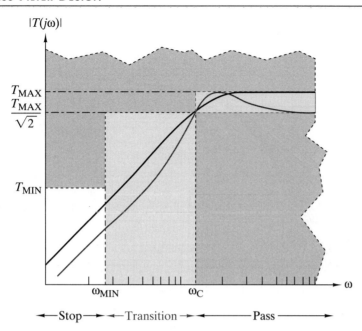

FIGURE 14–45 *High-pass filter specification and responses.*

FIRST-ORDER CASCADE

First-order cascade low-pass gain responses are defined in Eq. (14–28) using the normalized frequency variable ω/α. Using the inverted frequency variable α/ω in Eq. (14–28) produces the definition of first-order cascade high-pass responses:

$$|T(j\omega)| = \frac{|K|^n}{\left[\sqrt{1 + (\alpha/\omega)^2}\right]^n} \qquad (14\text{–}37)$$

Following the same process as for the low-pass case, the cutoff frequency α for the cascaded filter is found as

$$\alpha = \omega_C \sqrt{2^{1/n} - 1} \qquad (14\text{–}38)$$

This equation relates the cutoff frequency of each first-order stage (α) to the desired cutoff frequency of the cascade connection (ω_C) of n first-order stages. Each of these stages has a gain of $K = (T_{MAX})^{1/n}$ and a cutoff frequency of α. To determine the number of stages required to obtain a particular roll-off, we can use the normalized gain responses for $n = 1$ to $n = 11$ in Figure 14–46.

◁ DESIGN EXAMPLE 14–10

Design a cascade of active RC high-pass filter circuits to create a filter with a K of 0 dB, a cutoff frequency of 500 rad/s, a T_{MIN} of $-40\,dB$, and a ω_{MIN} of 100 rad/s. Simulate the output using OrCAD to verify your design.

SOLUTION:
From Figure 14–46 we find that the filter order to meet our requirement is 8. Using Eq. (14–38), we find that the cutoff frequency of each filter is 150.422 rad/s. If we select $R = 10\,k\Omega$, C is 0.6648 μF, which we will write equivalently as 664.8 nF, in this case, to

FIGURE 14–46 *First-order high-pass cascade filter responses.*

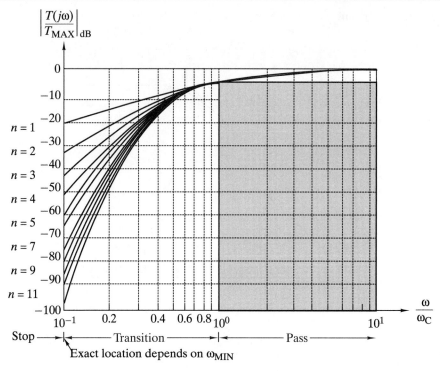

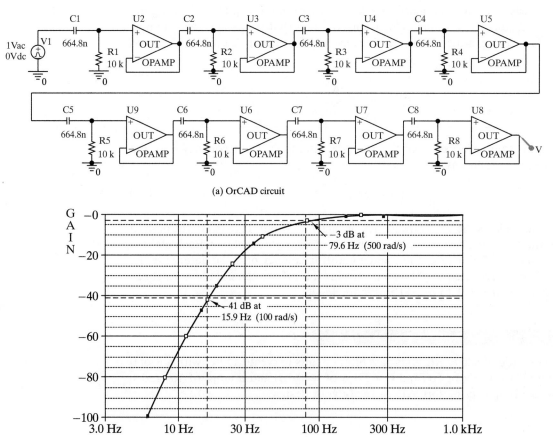

(a) OrCAD circuit

(b) Frequency response

FIGURE 14–47

save space in the circuit diagram. Figure 14–47(a) shows the circuit constructed in OrCAD and simulated using an AC Sweep. Figure 14–47(b) shows the resulting frequency response. From the Probe output, the cutoff frequency is 79.6 Hz (500 rad/s), as required, and at 15.9 Hz (100 rad/s) the attenuation is 41 dB, exceeding the −40 dB requirement. ∎

◐ Design Exercise 14–14

Design a high-pass, first-order cascade filter with a cutoff frequency of 100 krad/s, a T_{MIN} of −65 dB, a ω_{MIN} of 10 krad/s, and a passband gain of 100.

Answer: Use 13 stages of a first-order RC high-pass filter consisting of a 10-kΩ resistor and a 4273-pF capacitor. Separate the stages using noninverting OP AMPs, each with a gain of 1.4251.

BUTTERWORTH HIGH-PASS FILTERS

Butterworth low-pass gain responses are defined in Eq. (14–30) using the normalized frequency variable ω/ω_C. Using an inverted frequency variable ω_C/ω in Eq. (14–30) produces the definition of Butterworth high-pass gain responses:

$$|T(j\omega)| = \frac{|K|}{\sqrt{1 + (\omega_C/\omega)^{2n}}} \quad \text{Butterworth high-pass} \tag{14–39}$$

where n is the filter order. Figure 14–48 shows normalized plots of Eq. (14–39) for $n = 1$ to 6, where $T_{MAX} = |K|$. These plots confirm that the inverted variable ω_C/ω does in fact interchange the passband and stopband while leaving the cutoff frequency unchanged at $\omega/\omega_C = 1$. The inversion causes low-pass gain responses

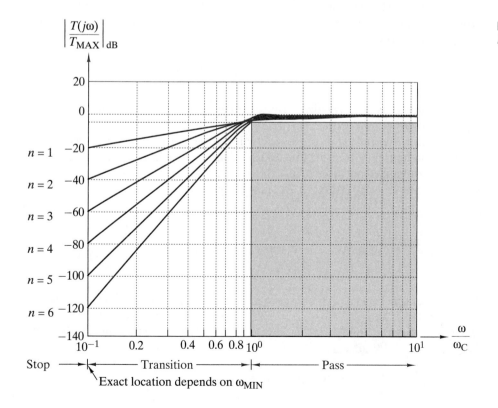

FIGURE 14–48 *Butterworth high-pass filter responses.*

that occur below ω_C to occur as high-pass responses above ω_C, and vice versa. In particular, Figure 14–48 shows that gains in the high-pass transition band decrease rapidly as n increases.

To limit circuit complexity, we need to know the smallest n that meets a given high-pass filter specification. For Butterworth low-pass filters, Eq. (14–31) provides a lower bound on n in terms of the transition band ratios. For Butterworth high-pass filters, the lower bound is found by substituting the inverted variable ω_C/ω in the derivation of Eq. (14–31). The modified derivation is straightforward and leads to the following equation:

$$n \geq \frac{1}{2} \frac{\ln[(T_{\text{MAX}}/T_{\text{MIN}})^2 - 1]}{\ln[\omega_C/\omega_{\text{MIN}}]} \quad \text{High-pass order} \tag{14–40}$$

The lower bound on the right side of this equation depends on two transition band ratios, namely $T_{\text{MAX}}/T_{\text{MIN}}$ and $\omega_C/\omega_{\text{MIN}}$. There is no real surprise here. We have seen several times that filter order depends on how much and how abruptly the transition band gain must decrease to meet a given specification.

To give a concrete example, suppose the transition band gain of a Butterworth high-pass filter must decrease by 40 dB ($T_{\text{MAX}}/T_{\text{MIN}} = 10^2$) when $\omega_C/\omega_{\text{MIN}} = 5$. Equation (14–40) yields a lower bound of

$$n \geq \frac{1}{2} \frac{\ln[(10^2)^2 - 1]}{\ln[5]} = 2.86$$

The smallest integer meeting this bound is $n = 3$. This result is confirmed by the plots in Figure 14–48, which show that a normalized gain below -40 dB at $\omega/\omega_C = 0.2$ cannot be produced by any order less than $n = 3$.

Given the value of n, we need to construct a high-pass transfer function with a Butterworth gain response. A Butterworth low-pass transfer function is constructed using a polynomial $q_n(s)$ from Table 14–2 in Eq. (14–32). An nth-order high-pass transfer function is found by inverting the frequency variable in Eq. (14–32) to get

$$T(s) = \frac{K}{q_n(\omega_C/s)} \quad \text{High-pass transfer function} \tag{14–41}$$

where $q_n(\omega_C/s)$ is a normalized Butterworth polynomial in Table 14–2 with s replaced by ω_C/s.

To see how the inversion process actually works, let us develop a second-order Butterworth high-pass transfer function with $\omega_C = 1$ krad/s and $T_{\text{MAX}} = 20$ dB. For $n = 2$ the normalized polynomial in Table 14–2 is $q_2(s) = s^2 + 1.414s + 1$. Using $\omega_C = 1000$ rad/s and $K = 10$ (20 dB) in Eq. (14–41) produces

$$T(s) = \frac{10}{\left(\dfrac{1000}{s}\right)^2 + 1.414\left(\dfrac{1000}{s}\right) + 1} = \frac{10\left(\dfrac{s}{1000}\right)^2}{\left(\dfrac{s}{1000}\right)^2 + 1.414\left(\dfrac{s}{1000}\right) + 1}$$

This transfer function has a double zero at $s = 0$ and a pair of Butterworth poles with $\omega_0 = 1000$ rad/s and $\zeta = 1.414/2 = 0.707$. The double zero makes the dc gain $|T(0)| = 0$ and the infinite-frequency gain $|T(\infty)| = 10$. The gain at the cutoff frequency $\omega_C = 1$ krad/s is

$$|T(j1000)| = 10/1.414 = |T(\infty)|/\sqrt{2}$$

as required. In sum, the transfer function $T(s)$ has the properties of a second-order Butterworth high-pass filter.

Designing an nth-order Butterworth high-pass filter to meet prescribed values of T_{MAX}, T_{MIN}, ω_C, and ω_{MIN} involves the following steps. We first determine the smallest n that meets the specification using Eq. (14–40) or the graphs in Figure 14–48. Taking the polynomial $q_n(s)$ from Table 14–2 and using $K = T_{MAX}$, we obtain $T(s)$ using Eq. (14–41). The polynomial $q_n(\omega_C/s)$ in the denominator of $T(s)$ supplies n Butterworth poles and n zeros at the origin. We then partition $T(s)$ into a product of first- and second-order high-pass functions and use a cascade of high-pass building blocks to get the overall filter circuit.

The next example illustrates the Butterworth high-pass design process.

D DESIGN EXAMPLE 14–11

(a) Construct a Butterworth high-pass transfer function that meets the following requirements: $T_{MAX} = 20\,dB$, $\omega_C = 10\,rad/s$, $T_{MIN} = -10\,dB$, and $\omega_{MIN} = 3\,rad/s$. Use MATLAB to plot the gain response.
(b) Design a cascade of active RC circuits that realizes the transfer function found in (a). Validate your design using OrCAD.

SOLUTION:
(a) The specification requires the gain to decrease by 30 dB in the transition band between $\omega_{MIN} = 3\,rad/s$ and $\omega_C = 10\,rad/s$. The transition band ratios for this specification are $T_{MAX}/T_{MIN} = 10^{3/2}$ and $\omega_C/\omega_{MIN} = 3.33$. Inserting these ratios in Eq. (14–40) yields

$$n \geq \frac{1}{2}\frac{\ln[(10^{3/2})^2 - 1]}{\ln(3.33)} = 2.87$$

Hence, $n = 3$ is the lowest-order Butterworth response that meets the transition band requirement. For $n = 3$, Table 14–2 lists $q_3(s) = (s+1)(s^2+s+1)$. Using $\omega_C = 10$ and $K = 10$ (20 dB) in Eq. (14–41) gives the required high-pass transfer function:

$$T(s) = \frac{10}{q_3(10/s)} = \frac{10}{\left[\left(\dfrac{10}{s}\right)+1\right]\left[\left(\dfrac{10}{s}\right)^2+\left(\dfrac{10}{s}\right)+1\right]}$$

$$= \frac{10\left(\dfrac{s}{10}\right)^3}{\left[\left(\dfrac{s}{10}\right)+1\right]\left[\left(\dfrac{s}{10}\right)^2+\left(\dfrac{s}{10}\right)+1\right]}$$

This third-order high-pass function has three zeros at $s = 0$, a real pole at $s = -10\,rad/s$, and a pair of complex poles with $\omega_0 = 10\,rad/s$ and $\zeta = 0.5$. Note that these high-pass poles all lie on a circle of radius ω_C, as did the poles in Butterworth low-pass filters.

The transfer function was expanded for use in MATLAB, and the code below was used to produce the gain curve shown in Figure 14–49. The result shows that the transfer function meets the required specifications.

FIGURE 14–49

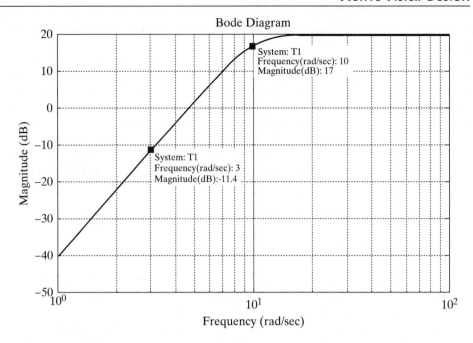

```
T=tf([10 0 0 0],[1 20 200 1000]);
w = logspace(0,2,10000);
bodemag(T,w);grid
```

(b) The Butterworth high-pass function developed in part (a) can be partitioned as follows:

$$T(s) = T_1(s)T_2(s)$$

$$= \left[\frac{(s/10)^2}{(s/10)^2 + s/10 + 1}\right]\left[\frac{10(s/10)}{(s/10) + 1}\right]$$

In this partition all of the passband gain $T_{MAX} = 10$ (20 dB) has been assigned to the first-order transfer function. This makes the passband gain of the second-order transfer function 1 (0 dB) so that it can be realized using the unity gain design method [see Eq. (14–14)].

   Figure 14–50 shows a design sequence based on this partitioning. The stage transfer functions in the first row lead to the stage parameters in the second row. The two stage prototypes are a second-order high-pass circuit with unity gain and a first-order high-pass circuit with an adjustable gain. The design constraints for the second-order stage use the unity gain design method. The design constraints for the first-order stage locate the real pole using the RC product and adjust the OP AMP's feedback to get the required gain. Using these constraints together with the element value assignments in the fifth row produces the final designs in the last row of Figure 14–50. The required passband gain is provided by the first-order stage, so no additional gain correction stage is needed in this example.

   The final design in Figure 14–50 was built in OrCAD and simulated. OrCAD produced the frequency response shown in Figure 14–51. The design meets or exceeds the specifications for the filter.

Item	First Stage	Second Stage
Prototype transfer function	$\dfrac{(s/10)^2}{(s/10)^2 + (s/10) + 1}$	$\dfrac{10\,(s/10)}{(s/10) + 1}$
Stage parameters	$\omega_0 = 10 \quad \zeta = 0.5 \quad K_1 = 1$	$\omega_0 = 10 \quad K_2 = 10$
Stage prototype	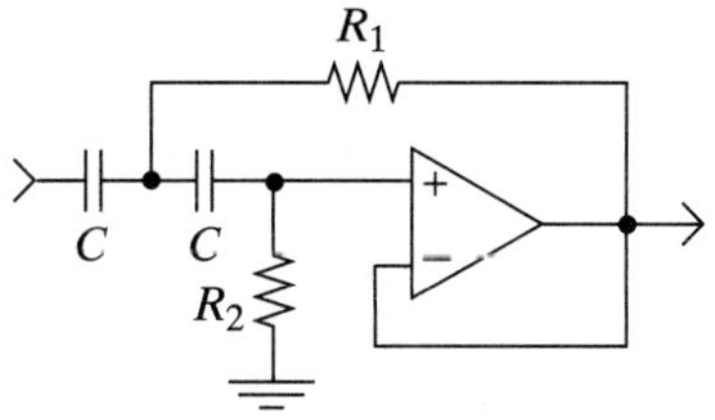 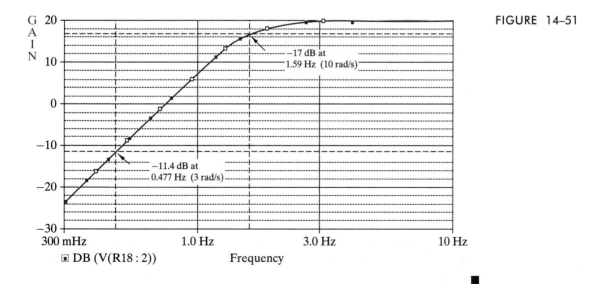	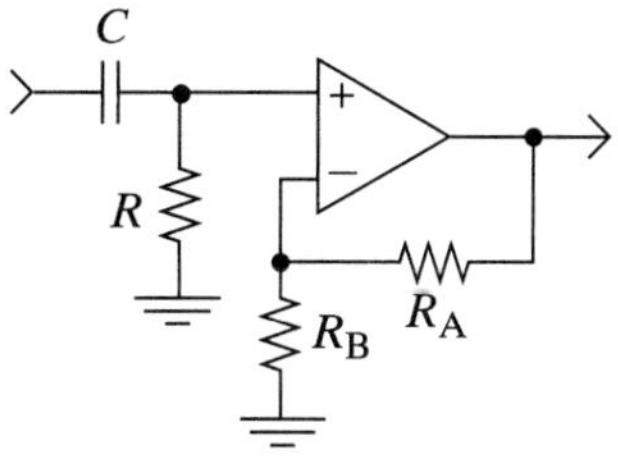
Design constraints	$\sqrt{R_1 R_2}\,C = 1/\omega_0 = 0.1$ $R_1/R_2 = \zeta^2 = 0.25$	$RC = 1/\omega_0 = 0.1$ $R_A/R_B = K_2 - 1 = 9$
Element values	Let $R_2 = 100\ \text{k}\Omega$, then $R_1 = 25\ \text{k}\Omega$ and $C = 2\ \mu\text{F}$	Let $R = 50\ \text{k}\Omega$, then $C = 2\ \mu\text{F}$ Let $R_B = 10\ \text{k}\Omega$, then $R_A = 90\ \text{k}\Omega$
Final designs	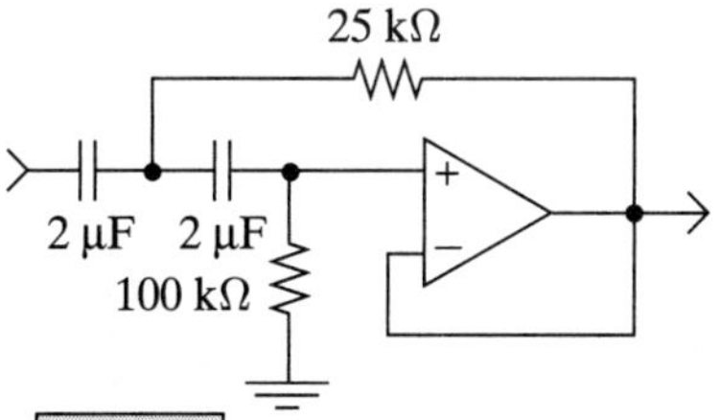 $K_1 = 1$	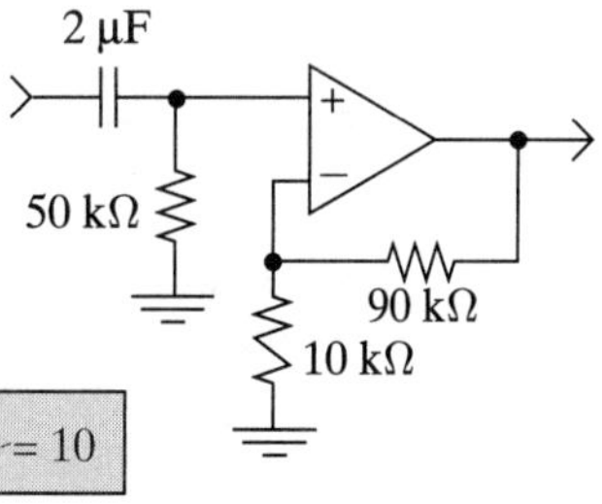 $K_2 = 10$

FIGURE 14–50 *Design sequence for Example 14–11.*

FIGURE 14–51

−17 dB at 1.59 Hz (10 rad/s)

−11.4 dB at 0.477 Hz (3 rad/s)

DB (V(R18 : 2))   Frequency

◗ Design Exercise 14–15 _____

Design a Butterworth high-pass filter using the equal element configuration that meets the following conditions: passband gain 100, $\omega_0 = 5$ krad/s, $\omega_{MIN} = 500$ rad/s, $T\text{min} = -40$ dB ± 1 dB. You must use 10-kΩ resistors as much as possible.

Answer: The solution requires a fourth-order filter. Use two-stage equal element, high-pass filters followed by a non-inverting gain stage. See Figure 14–52(a) for a working design.

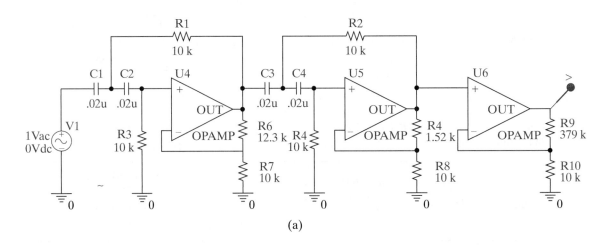

(a)

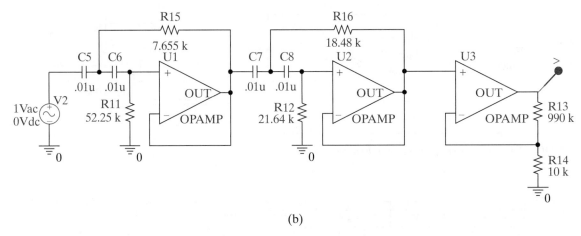

(b)

FIGURE 14–52

◗ Design Exercise 14–16 _____

Repeat Exercise 14–15, but design a Butterworth high-pass filter using the unity gain configuration. You must use 0.01 μF capacitors.

Answer: The solution requires a fourth-order filter. Use two-stage unity-gain, high-pass filters followed by a non-inverting gain stage. See Figure 14–52(b) for a working design.

CHEBYCHEV HIGH-PASS FILTERS

Our treatment of Chebychev high-pass filters can be brief because the low-pass to high-pass transformation used to develop Butterworth high-pass filters works equally well here. The Chebychev low-pass gains are defined in Eq. (14–33) in

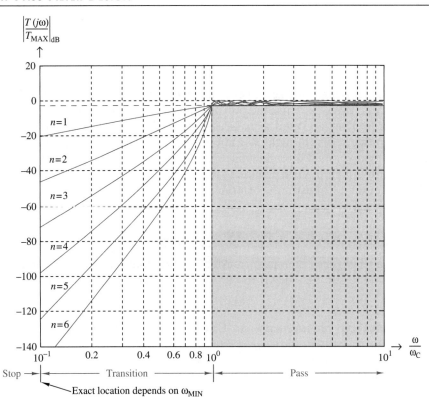

FIGURE 14–53 *Chebychev high-pass filter responses.*

terms of a frequency variable ω/ω_C. Using the inverted variable ω_C/ω in this definition produces the definition of Chebychev high-pass gain responses.

$$|T_n(j\omega)| = \frac{|K|}{\sqrt{1 + C_n^2(\omega_C/\omega)}} \quad \text{Chebyschev high-pass} \qquad (14\text{–}42)$$

where $C_n(x)$ is an nth-order Chebychev polynomial. The normalized plots of the Chebychev high-pass gains in Figure 14–53 show that the passband and stopband have been interchanged. These Chebychev high-pass responses have equal-ripple responses in the passband above ω_C and transition band gains below ω_C that decrease rapidly as n increases.

Equation (14–35) provides a lower bound on n for Chebychev low-pass filters. Using the inverted variable ω_C/ω in the derivation of this equation yields a lower bound on the Chebychev high-pass order as

$$n \geq \frac{\cosh^{-1}\left(\sqrt{(T_{MAX}/T_{MIN})^2 - 1}\right)}{\cosh^{-1}(\omega_C/\omega_{MIN})} \quad \text{High-pass order} \qquad (14\text{–}43)$$

Once again we see that the lower bound on filter order depends on how much (T_{MAX}/T_{MIN}) and how abruptly (ω_C/ω_{MIN}) the transition band gain must decrease to meet the specification.

For any given n, a Chebychev low-pass transfer function is obtained using a polynomial $q_n(s/\omega_C)$ in Eq. (14–36a) or (14–36b). To obtain a Chebychev high-pass transfer function, we invert the frequency variable and write

$$T(s) = \frac{K}{q_n(\omega_C/s)} \quad n \text{ odd} \qquad (14\text{–}44a)$$

or

$$T(s) = \frac{K/\sqrt{2}}{q_n(\omega_C/s)} \quad n \text{ even} \tag{14-44b}$$

where $q_n(\omega_C/s)$ is a normalized Chebychev denominator polynomial in Table 14–3 with s replaced by ω_C/s. Thus, the polynomials in Table 14–2 and Table 14–3 serve two purposes. They are used as $q_n(s/\omega_C)$ in the design of low-pass filters and as $q_n(\omega_C/s)$ in the design of high-pass filters. For the Chebychev high-pass case, the order of the polynomial is found analytically using Eq. (14–43) or graphically using the normalized gain plots in Figure 14–53.

The next example illustrates constructing Chebychev high-pass transfer functions to meet a given specification.

▶ DESIGN EXAMPLE 14–12

Construct a Chebychev high-pass transfer function that meets the requirements of Example 14–11. Use MATLAB to plot the gain of the transfer function.

SOLUTION:
The specification requirements in Example 14–11 are $T_{MAX} = 20\,\text{dB}$, $\omega_C = 10\,\text{rad/s}$, $T_{MIN} = -10\,\text{dB}$, and $\omega_{MIN} = 3\,\text{rad/s}$. The transition band gain must decrease by 30 dB ($T_{MAX}/T_{MIN} = 10^{3/2}$) on the range $\omega_C/\omega_{MIN} = 3.33$. Inserting these ratios in Eq. (14–43) yields

$$n \geq \frac{\cosh^{-1}\left[\sqrt{(10^{3/2})^2 - 1}\right]}{\cosh^{-1}(3.33)} = 2.21$$

Hence, $n = 3$ is the lowest-order Chebychev response that meets the design requirement. The third-order polynomial in Table 14–3 is

$$q_3(s) = (s/0.2980 + 1)[(s/0.9159)^2 + 0.3254(s/0.9159) + 1]$$

Using $\omega_C = 10$ and $K = 10$ ($T_{MAX} = 20\,\text{dB}$) in Eq. (14–44a) gives the required high-pass transfer function:

$$T_{HP}(s) = \frac{10}{q_3(10/s)}$$

$$= \frac{10}{\left[\left(\dfrac{10}{0.2980s}\right) + 1\right]\left[\left(\dfrac{10}{0.9159s}\right)^2 + 0.3254\left(\dfrac{10}{0.9159s}\right) + 1\right]}$$

$$= \frac{10s^3}{(s + 33.55)(s^2 + 3.553s + 119.2)}$$

In Example 14–11 we meet this specification using a third-order Butterworth high-pass with poles located on a circle of radius $\omega_C = 10\,\text{rad/s}$. The third-order Chebychev high-pass function constructed here has three zeros at $s = 0$, a real pole at $s = -33.55\,\text{rad/s}$, and a pair of complex poles with $\omega_0 = \sqrt{119.2} = 10.92\,\text{rad/s}$ and $\zeta = 0.1627$. These poles all lie outside of a circle of radius $\omega_C = 10\,\text{rad/s}$. The reason for this is easy to understand. A Chebychev high-pass function has an equal-ripple response in the passband *abo e* ω_C. Hence, the Chebychev poles that produce the ripple must have natural frequencies greater than ω_C.

Figure 14–54 displays the Bode magnitude plot derived using MATLAB. The design meets or exceeds all of the specifications. Note that at ω_{MIN} the Chebyshev

FIGURE 14–54

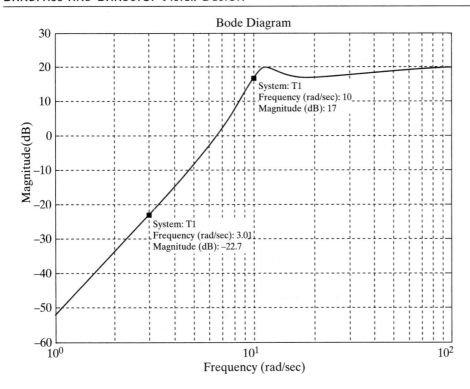

Bode Diagram

System: T1
Frequency (rad/sec): 10
Magnitude (dB): 17

System: T1
Frequency (rad/sec): 3.01
Magnitude (dB): –22.7

approach has a far greater attenuation ($-22.7\,\mathrm{dB}$) than the Butterworth approach ($-11.4\,\mathrm{dB}$). The price for this better filtering is the ripple in the passband. Furthermore, exceeding a specification generally does not offer a significant advantage over simply meeting it. ∎

◗ Design Exercise 14–17

Construct Butterworth and Chebychev high-pass transfer functions that meet the following requirements: $T_{\mathrm{MAX}} = 10\,\mathrm{dB}$, $\omega_C = 50\,\mathrm{rad/s}$, $T_{\mathrm{MIN}} = -40\,\mathrm{dB}$, and $\omega_{\mathrm{MIN}} = 10\,\mathrm{rad/s}$.

Answer:

$$T_{\mathrm{BU}}(s) = \frac{\sqrt{10}s^4}{(s^2 + 38.3s + 50^2)(s^2 + 92.4s + 50^2)}$$

$$T_{\mathrm{CH}}(s) = \frac{\sqrt{10}s^3}{(s + 168)(s^2 + 17.8s + 54.6^2)}$$

14–7 BANDPASS AND BANDSTOP FILTER DESIGN

In Chapter 12 we found that the cascade connection in Figure 14–55 can produce a bandpass filter. When the cutoff frequency of the low-pass filter (ω_{CLP}), is higher than the cutoff frequency of the high-pass filter (ω_{CHP}), the interval between the two frequencies is a passband separating two stopbands. The low-pass filter provides the high-frequency stopband and the high-pass filter the low-frequency stopband. Frequencies between the two cutoffs fall in the passband of both filters and are

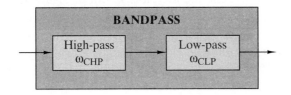

BANDPASS

| High-pass ω_{CHP} | Low-pass ω_{CLP} |

FIGURE 14–55 *Cascade connection of high-pass and low-pass filters.*

transmitted through the cascade connection, producing the passband of the resulting bandpass filter.

When $\omega_{CLP} \gg \omega_{CHP}$, two bandpass cutoff frequencies are approximately $\omega_{C1} \approx \omega_{CHP}$ and $\omega_{C2} \approx \omega_{CLP}$. Under these conditions the center frequency and bandwidth of the bandpass filter are

$$\omega_0 = \sqrt{\omega_{CHP}\omega_{CLP}} \quad \text{and} \quad B = \omega_{CLP} - \omega_{CHP}$$

and the ratio of the center frequency over the bandwidth is approximately

$$\frac{\omega_0}{B} = Q \approx \sqrt{\frac{\omega_{CHP}}{\omega_{CLP}}} \ll 1$$

Since the quality factor is less than 1, this method of bandpass filter design produces a wide-band filter, as contrasted with the narrow-band response $(Q > 1)$ produced by the active RC bandpass circuit studied earlier in Sect. 14–3.

D DESIGN EXAMPLE 14–13

Use second-order Butterworth low-pass and high-pass functions to obtain a fourth-order bandpass function with a passband gain of 0 dB and cutoff frequencies at $\omega_{C1} = 10 \, \text{rad/s}$ and $\omega_{C2} = 50 \, \text{rad/s}$. Use MATLAB to show a complete Bode plot of the transfer function.

SOLUTION:
The upper cutoff frequency at 50 rad/s is produced by a low-pass function. Using second-order Butterworth poles, the required low-pass function is

$$T_{LP}(s) = \frac{1}{(s/50)^2 + \sqrt{2}(s/50) + 1}$$

The lower cutoff frequency at 10 rad/s is to be produced by a second-order high-pass function. Using second-order Butterworth poles, the required high-pass function is

$$T_{HP}(s) = \frac{(s/10)^2}{(s/10)^2 + \sqrt{2}(s/10) + 1}$$

When circuits realizing these two transfer functions are connected in cascade, the overall transfer function is

$$T_{HP}(s) \times T_{LP}(s) = \left[\frac{(s/10)^2}{(s/10)^2 + \sqrt{2}(s/10) + 1}\right] \times \left[\frac{1}{(s/50)^2 + \sqrt{2}(s/50) + 1}\right]$$

$$= \frac{2500s^2}{s^4 + 60\sqrt{2}s^3 + 3600s^2 + 30{,}000\sqrt{2}s + 250{,}000}$$

We can use MATLAB to plot a complete Bode diagram of our transfer function. We use the following code to generate the plots shown in Figure 14–56:

```
T=tf([2500 0 0], . . .
     [1 60*sqrt(2)  3600  30000*sqrt(2) 250000]);
w = logspace(0,3,10000);
bode(T,w);
grid
```

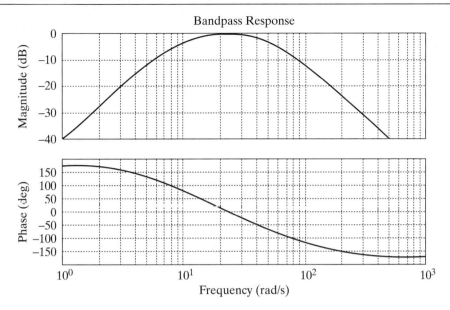

FIGURE 14–56

The gain response shows that the transfer function indeed has a bandpass characteristic. The phase shifts over $360°$, from $180°$ at low frequencies to $-180°$ at high frequencies, passing through $0°$ at the center frequency of $\sqrt{10 \times 50} = 22.36$ rad/s. At 10 rad/s and 50 rad/s, the cutoffs are as desired, but the gain is shy of 0 dB, reaching a maximum of -0.341 dB. This small error is probably insignificant. As the cutoff frequencies get closer together and the bandwidth decreases, the two filter components interact to decrease the gain at the center frequency. This interaction will cause the gain to fail to attain the desired value. The Q of this filter is $\omega_0/B = 22.36/40 = 0.56$. As the Q approaches 1, a different design approach is needed, such as the tuned filter design shown in Figure 14–14 or Figure 14–19. ∎

Figure 14–57 shows the dual situation in which a high-pass and a low-pass filter are connected in parallel to produce a bandstop filter. When $\omega_{CLP} \ll \omega_{CHP}$, the region between the two cutoff frequencies is a stopband separating two passbands. The low-pass filter provides the low-frequency passband via the lower path and the high-pass filter provides the high-frequency passband via the upper path. Frequencies between the two cutoffs fall in the stopband of both filters and are not transmitted through either path in the parallel connection. As a result, the two filters produce the stopband of the resulting bandstop filter. When $\omega_{CHP} \gg \omega_{CLP}$, two cutoff frequencies are approximately $\omega_{C1} \approx \omega_{CLP}$ and $\omega_{C2} \approx \omega_{CHP}$.

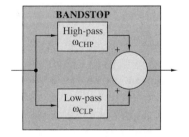

FIGURE 14–57 *Parallel connection of high-pass and low-pass filters.*

D DESIGN EXAMPLE 14–14

Use second-order Butterworth low-pass and high-pass functions to obtain a fourth-order bandstop function with passband gains of 0 dB and cutoff frequencies at $\omega_{C1} = 10$ rad/s and $\omega_{C2} = 50$ rad/s. Use MATLAB to show a complete Bode plot of the transfer function.

SOLUTION:
In Example 14–13 we obtained a Butterworth bandpass response using a cascade connection of a second-order high-pass function and a low-pass function with

$\omega_{\text{CLP}} = 50 \gg \omega_{\text{CHP}} = 10$. To obtain a bandstop response, we interchange the cutoff frequencies and write the two functions as

$$T_{\text{LP}}(s) = \frac{1}{(s/10)^2 + \sqrt{2}(s/10) + 1}$$

$$T_{\text{HP}}(s) = \frac{(s/50)^2}{(s/50)^2 + \sqrt{2}(s/50) + 1}$$

We now have $\omega_{\text{CLP}} = 10 \ll \omega_{\text{CHP}} = 50$, which leads to a bandstop response. For the parallel connection the overall transfer function is the sum of the two transfer functions.

$$T_{\text{LP}}(s) + T_{\text{HP}}(s) = \frac{1}{(s/10)^2 + \sqrt{2}(s/10) + 1} + \frac{(s/50)^2}{(s/50)^2 + \sqrt{2}(s/50) + 1}$$

$$= \frac{s^4 + 10\sqrt{2}s^3 + 200s^2 + 5000\sqrt{2}s + 250{,}000}{s^4 + 60\sqrt{2}s^3 + 3600s^2 + 30{,}000\sqrt{2}s + 250{,}000}$$

As with the bandpass example, we can use MATLAB to plot a complete Bode diagram of our transfer function. We use the following code to generate the plots shown in Figure 14–58:

```
T=tf([1 10*sqrt(2)  200  5000*sqrt(2)  250000],. . .
      [1 60*sqrt(2) 3600  30000*sqrt(2)  250000]);
w = logspace(0,3,10000);
bode(T,w);
grid
```

The gain response shows that the transfer function indeed has a bandstop characteristic. The phase shifts from $0°$ at low frequencies to $-360°$ at high frequencies, passing through $180°$ at the center frequency of $\sqrt{10 \times 50} = 22.36$ rad/s. At 10 rad/s and 50 rad/s, the cutoffs are as desired, with the filter reaching a maximum attenuation of 10.2 dB. As the cutoff frequencies get closer together and the bandwidth of the stopband decreases, the two components of the filter will interact to decrease the amount of attenuation at the center frequency. In other words, as the

FIGURE 14–58

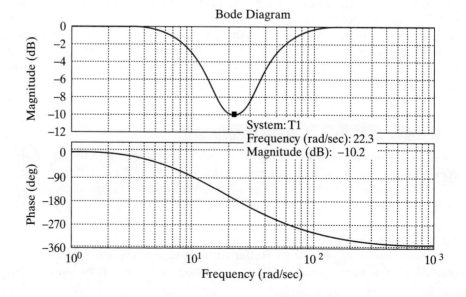

bandwidth decreases, the filter's notch will not be as deep. The Q of this filter is $\omega_0/B = 22.36/40 = 0.56$. As the Q approaches 1, a different design approach is needed, such as the notch filter design shown in Figure 14–17 or Figure 14–19. ∎

When the two cutoff frequencies are widely separated, we can realize wide-band bandpass and bandstop filters using a cascade or parallel connection of low-pass and high-pass filters. The design problem reduces to designing separate low-pass and high-pass filters and then connecting them in cascade or parallel to obtain the required overall response.

◗ Design Exercise 14–18

Construct Butterworth low-pass and high-pass transfer functions whose cascade connection produces a bandpass function with cutoff frequencies at 20 rad/s and 500 rad/s, a passband gain of 0 dB, and a stopband gain less than −20 dB at 5 rad/s and 2000 rad/s.

Answer:
$$T(s) = \left[\frac{500^2}{s^2 + 707s + 500^2}\right]\left[\frac{s^2}{s^2 + 28.3s + 400}\right]$$

◗ Design Exercise 14–19

Develop Butterworth low-pass and high-pass transfer functions whose parallel connection produces a bandstop filter with cutoff frequencies at 2 rad/s and 800 rad/s, passband gains of 20 dB, and stopband gains less than −30 dB at 20 rad/s and 80 rad/s.

Answer:
$$T_{LP}(s) = \frac{80}{(s^2 + 2s + 4)(s + 2)}$$
$$T_{HP}(s) = \frac{10s^3}{(s^2 + 800s + 800^2)(s + 800)}$$

SUMMARY

- A filter design problem is defined by specifying attributes of the gain response such as a straight-line gain plot, cutoff frequency, passband gain, and stopband attenuation. The first step in the design process is to construct a transfer function $T(s)$ whose gain response meets the specification requirements.

- In the cascade design approach, the required transfer function is partitioned into a product of first- and second-order transfer functions, which can be independently realized using basic active RC building blocks.

- Transfer functions with real poles and zeros can be realized using the voltage divider, noninverting amplifier, or inverting amplifier building blocks. Transfer functions with complex poles can be realized using second-order active RC circuits.

- Transfer functions meeting low-pass filter specifications can be constructed using first-order cascade, Butterworth, or Chebychev poles. First-order cascade

filters are easy to design but have poor stopband performance. Butterworth responses produce maximally flat passband responses and more stopband attenuation than a first-order cascade with the same number of poles. The Chebychev responses produce equal-ripple passband responses and more stopband attenuation than the Butterworth response with the same number of poles.

- Filter designers should also consider the transient or step response behavior of the filters they design. First-order cascade filters generally have better step response performance than either Butterworth or Chebyshev filters because the latter's complex poles give rise to ringing and an overshoot.

- A high-pass transfer function can be constructed from a low-pass prototype by replacing s with ω_C/s. Band-pass (bandstop) filters can be constructed using a cascade (parallel) connection of a low-pass and a high-pass filter.

PROBLEMS

OBJECTIVE 14–1 SECOND-ORDER FILTER ANALYSIS (SECTS. 14–1, 14–2, 14–3)

(a) Given a second-order filter circuit, find a specified transfer function.

(b) Given the transfer function of a second-order circuit, develop a method of selecting the element values to achieve specified filter characteristics.

See Examples 14–1, 14–2, 14–3, 14–4, 14–5 and Exercises 14–1, 14–2, 14–3, 14–4, 14–5, 14–6, 14–7, 14–8.

14–1 Interchanging the positions of the resistors and capacitors converts the low-pass filter in Figure 14–3(a) into the high-pass filter in Figure 14–9(a). This *CR-RC* interchange involves replacing R_k by $1/C_k s$ and $C_k s$ by $1/R_k$. Show that this interchange converts the low-pass transfer function in Eq. (14–6) into the high-pass function in Eq. (14–11).

14–2 Show that the circuit in Figure 14–14 has the bandpass transfer function in Eq. (14–16).

14–3 Show that the circuit in Figure 14–17 has the bandstop transfer function in Eq. (14–20).

14–4 **D** Show that the active filter in Figure P14–4 has a transfer function of the form

$$T(s) = \frac{V_2(s)}{V_1(s)} = \frac{1}{R_1 R_2 C_1 C_2 s^2 + R_2 C_2 s + 1}$$

Using $C_1 = C_2 = C$, develop a method of selecting values for C, R_1, and R_2. Then select values so that the filter has a cutoff frequency of 100 krad/s and a ζ of 0.6. Use MATLAB to plot the filter's Bode diagram. Determine the type of filter it is and its roll-off.

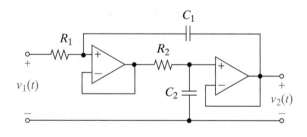

FIGURE P14–4

14–5 **D** The circuit in Figure 14–3(b) has a low-pass transfer function given in Eq. (14–6) and repeated below

$$T(s) = \frac{V_2(s)}{V_1(s)}$$
$$= \frac{\mu}{R_1 R_2 C_1 C_2 s^2 + (R_1 C_1 + R_1 C_2 + R_2 C_2 - \mu R_1 C_1)s + 1}$$

In Section 14–2, we developed *equal element* and *unity gain* design methods for this circuit. This problem explores an

equal time constant design method. Using $R_1 C_1 = R_2 C_2$ and $\mu = 2$, develop a method of selecting values for C_1, C_2, R_1, and R_2. Then select values so that the filter has a cutoff frequency of 2 krad/s and a ζ of 0.05. Use MATLAB to plot the filter's Bode diagram. Determine the location in rad/s and magnitude in dB of the peak in the frequency response.

14–6 **D** The active filter in Figure P14–6 has a transfer function of the form

$$T(s) = \frac{V_2(s)}{V_1(s)}$$
$$= \frac{-R_2/R_1}{R_2 R_3 C_1 C_2 s^2 + [R_3 C_2 + R_2 C_2(1 + R_3/R_1)]s + 1}$$

Using $R_1 = R_2 = R_3 = R$, develop a method of selecting values for C_1, C_2 and R. Then select values so that the filter has a cutoff frequency of 150 krad/s and a ζ of 0.01. Use MATLAB to plot the filter's Bode diagram. Build and simulate your circuit in OrCAD and compare the results with MATLAB's. Determine the location in rad/s and magnitude in dB of the peak in the frequency response.

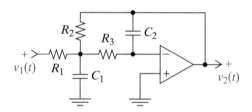

FIGURE P14–6

14–7 **D** Show that the active filter in Figure P14–7 has a transfer function of the form

$$T(s) = \frac{V_2(s)}{V_1(s)} = \frac{R_1 R_2 C_1 C_2 s^2}{R_1 R_2 C_1 C_2 s^2 + R_1 C_1 s + 1}$$

Using $C_1 = C_2 = C$, develop a method of selecting values for C, R_1, and R_2. Then select values so that the filter has a cutoff frequency of 1 krad/s and a ζ of 0.1. Use MATLAB to plot the filter's Bode diagram. Determine the type of filter it is and its roll-off How does the circuit and this transfer function compare with that of Problem 14–4?

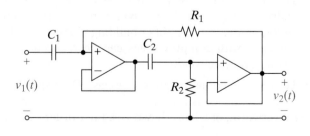

FIGURE P14–7

14–8 Ⓓ The circuit in Figure 14–9(b) has a high-pass transfer function given in Eq. (14–11) and repeated below

$$T(s) = \frac{V_2(s)}{V_1(s)}$$

$$= \frac{\mu R_1 R_2 C_1 C_2\, s^2}{R_1 R_2 C_1 C_2\, s^2 + (R_2 C_2 + R_1 C_1 + R_1 C_2 - \mu R_2 C_2)s + 1}$$

In Section 14–2, we developed *equal element* and *unity gain* design methods for this circuit. This problem explores an *equal time constant* design method. Using $R_1 C_1 = R_2 C_2$ and $\mu = 2$, develop a method of selecting values for C_1, C_2, R_1, and R_2. Then select values so that the filter has a cutoff frequency of 500 krad/s and a ζ of 0.25. Use MATLAB to plot the filter's Bode diagram.

14–9 Ⓓ Show that the active filter in Figure P14–9 has a transfer function of the form

$$T(s) = \frac{V_2(s)}{V_1(s)} = \frac{-R_1 C_1 s}{R_1 R_2 C_1 C_2 s^2 + (R_1 C_2 + R_2 C_2)s + 1}$$

Using $R_1 = R_2 = R$, develop a method of selecting values for C_1, C_2, and R. Then select values so that the filter has a ω_0 of 50 krad/s and a ζ of 0.1. Use MATLAB to plot the filter's Bode diagram. What type of filter is this?

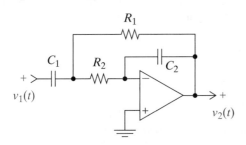

FIGURE P14–9

14–10 Ⓓ The active filter in Figure P14–10 has a transfer function of the form

$$T(s) = \frac{V_2(s)}{V_1(s)}$$

$$= \frac{-R_3 C_2 s}{R_1 R_3 C_1 C_2 s^2 + (R_1 C_1 + R_1 C_2)s + 1 + R_1/R_2}$$

Using $C_1 = C_2 = C$ and $R_1 = R_2 = R$, develop a method of selecting values for C and R. Then select values so that the filter has a ω_0 of 10 krad/s and a ζ of 0.5. Use MATLAB to plot the filter's Bode diagram. What type of filter is this?

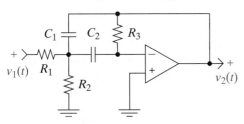

FIGURE P14–10

14–11 Ⓓ The active filter in Figure P14–11 has a transfer function of the form

$$T(s) = \frac{V_2(s)}{V_1(s)} = \frac{(RCs)^2 + 1}{(RCs)^2 + 2RCs + 1}$$

Select values of R and C so that the filter has a ω_0 of 377 krad/s. Use OrCAD to plot the filter's Bode magnitude diagram. What type of filter is this? With the gain equal to 1, is ζ selectable? What is ζ in this filter?

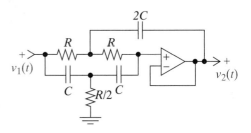

FIGURE P14–11

OBJECTIVE 14–2 SECOND-ORDER FILTER DESIGN (SECTS. 14–2, 14–3)

(a) Construct a second-order transfer function with specified filter characteristics.
(b) Design a second-order circuit with specified filter characteristics.
See Examples 14–1, 14–2, 14–3, 14–4, 14–5 and Exercises 14–1, 14–2, 14–3, 14–4, 14–5, 14–6, 14–7, 14–8.

14–12 Ⓓ Ⓔ The task is to design a second-order low-pass Filter using the three approaches shown in Problems 14–4, 14–5, and 14–6. The filter specs are a cutoff frequency of 100 krad/s and a ζ of 0.25. Using OrCAD, build your three designs and select the best one based on how well each meets the specs, the number of parts, and the gain.

For Problems 14–13 to 14–17, construct second-order transfer functions that meet the following requirements. Use MATLAB to plot the transfer function's Bode diagram and validate the requirements.

PROBLEM	TYPE	ω_0 (rad/s)	ζ	$|T(j\omega_0)|$	CONSTRAINTS
14–13	Low pass	100,000	*	50	dc gain of 10
14–14	High pass	100	0.5	*	infinite frequency gain of 20
14–15	Bandpass	15,000	*	5	$B = 250$ rad/s
14–16	Bandpass	632	15.8	10	dc gain of zero
14–17	Bandstop	377	0.01	0	pass-band gain of 10

*not specified.

For Problems 14–18 to 14–28, design second-order active filters that meet the following requirements. Simulate your designs in OrCAD to validate the requirements.

PROBLEM	TYPE	ω_0 (rad/s)	ζ	CONSTRAINTS
14–18 🅓	Low pass	20 k	0.5	Use 10-kΩ resistors
14–19 🅓	Low pass	2500	0.8	dc gain of 20 dB
14–20 🅓	Low pass	100 k	0.25	Use 0.2-μF capacitors
14–21 🅓	High pass	2500	0.75	Use 0.2-μF capacitors
14–22 🅓	High pass	25 k	0.25	High-frequency gain of 40 dB
14–23 🅓	Bandpass	5000	*	Center-frequency gain of 20 dB
14–24 🅓	Bandpass	3974	*	Bandwidth of 125.5 krad/s.
14–25 🅓	Tuned	3.45 M	0.005	Use 500-pF capacitors
14–26 🅓	Notch	377	0.01	Use 0.022-μF capacitors
14–27 🅓	Bandstop	5000	2	Use 0.2-μF capacitors
14–26 🅓	Notch	754	0.05	pass-band gain of 20 dB

*not specified.

For Problems 14–29 to 14–33, construct the lowest-order transfer functions that meet the following low-pass filter specifications. Calculate the gain (in dB) of the transfer function at $\omega = \omega_C$ and ω_{MIN}. Use MATLAB to validate that your transfer function meets the design specifications.

PROBLEM	POLE TYPE	ω_C (rad/s)	T_{MAX}	ω_{MIN} (rad/s)	T_{MIN}
14–29	First-order Cascade	5000	60 dB	50 k	−20 dB
14–30	Butterworth	10 k	0 dB	40 k	−50 dB
14–31	Chebychev	25 k	20 dB	250 k	−80 dB
14–32	Butterworth	300 k	10 dB	600 k	−10 dB
14–33	Chebychev	2 M	20 dB	4 M	−10 dB

14–34 🅓 Design an active low-pass filter to meet the specification in Problem 14–29. Use OrCAD to verify that your design meets the specifications.

14–35 🅓 Design an active low-pass filter to meet the specification in Problem 14–30. Use OrCAD to verify that your design meets the specifications.

14–36 🅓 Design an active low-pass filter to meet the specification in Problem 14–31. Use OrCAD to verify that your design meets the specifications.

14–37 🅓 Design an active low-pass filter to meet the specification in Problem 14–32. Use OrCAD to verify that your design meets the specifications.

14–38 🅓 Design an active low-pass filter to meet the specification in Problem 14–33. Use OrCAD to verify that your design meets the specifications.

14–39 🅓 A low-pass filter is needed to suppress the harmonics in a periodic waveform with $f_0 = 1$ kHz. The filter must have unity passband gain, less than −50 dB gain at the 3rd harmonic, and less than −80 dB gain at the 5th harmonic. Since power is at a premium, choose a filter approach that minimizes the number of OP AMPs. Design a filter that meets these requirements. Verify your design using OrCAD.

14–40 🅓 Design a low-pass filter with 6 dB passband gain, a cutoff frequency of 2 kHz, and a stopband gain of less than −14 dB at 6 kHz. The filter must not have an overshoot greater than 13%. Verify your design using OrCAD.

14–41 🅓 Design a low-pass filter with 0 dB passband gain, a cutoff frequency of 1 kHz, and a stopband gain of less than −50 dB at 4 kHz. The filter must not have an overshoot greater than 12%. Verify your design using OrCAD.

14–42 🅓 Design a low-pass filter with 10 dB passband gain, a cutoff frequency of 1 kHz, and a stopband gain of less than −20 dB at 2 kHz. Overshoot is not a problem, but a low filter order, least number of parts, and a maximum of two OP AMPs is desired. Verify your design using OrCAD.

14–43 🅓 A pesky signal at 100 kHz is interfering with a desired signal at 20 kHz. A careful analysis suggests that reducing the interfering signal by 72 dB will eliminate the problem, provided the desired signal is not reduced by more than 3 dB. Design an active *RC* filter that meets these requirements. Verify your design using OrCAD.

14–44 **D** A strong signal at 2.45 MHz is interfering with an AM signal at 980 kHz. Design a filter that will attenuate the undesired signal by at least 60 dB. Verify your design using OrCAD.

14–45 **D** A 10 kHz square wave must be bandwidth-limited by attenuating all harmonics after the third. Design a low-pass filter that attenuates the fifth harmonic and greater by at least 20 dB. The fundamental and third harmonic should not be reduced by more than 3 dB and the overshoot cannot exceed 13%.

OBJECTIVE 14–4 HIGH PASS, BANDPASS, AND BANDSTOP FILTER DESIGN (SECTS. 14–6, 14–7)

Given a high-pass, bandpass, or bandstop filter specification:
(a) Construct a transfer function that meets the specification.
(b) Design a cascade or parallel connection of first- and second-order circuits that implements a given transfer function.
(c) Select the best design from competitive filter approaches based on specified frequency and step response criteria.
See Examples 14–10, 14–11, 14–12, 14–13, 14–14 and Exercises 14–14, 14–15, 14–16, 14–17, 14–18, 14–19.

Construct the lowest order, high-pass transfer functions that meet the following filter specifications. Calculate the gain (in dB) of the transfer function at $\omega = \omega_C$ and ω_{MIN}. Use MATLAB to validate that your transfer function meets the design specifications.

PROBLEM	POLE TYPE	ω_C (rad/s)	T_{MAX}	ω_{MIN} (rad/s)	T_{MIN}
14–46	First-order cascade	10 k	40 dB	1000	0 dB
14–47	Butterworth	10 k	20 dB	1000	−30 dB
14–48	Chebychev	10 k	0 dB	5000	−40 dB
14–49	You Decide	10 k	10 dB	2000	−50 dB

14–50 **D** Design an active high-pass filter to meet the specification in Problem 14–46. Use OrCAD to verify that your design meets the specifications.

14–51 **D** Design an active high-pass filter to meet the specification in Problem 14–47. Use OrCAD to verify that your design meets the specifications.

14–52 **D** Design an active high-pass filter to meet the specification in Problem 14–48. Use OrCAD to verify that your design meets the specifications.

14–53 **D** Design an active high-pass filter to meet the specification in Problem 14–49. Use OrCAD to verify that your design meets the specifications.

14–54 **E** A certain instrumentation system for a new hybrid car needs a bandpass filter to limit its output bandwidth prior to digitization. The filter must meet the following specifications:

$$T_{MAX} = +20 \pm 1 \text{ dB} \quad \omega_{CH} = 5.5 \text{ krad/s} \ (875.4 \text{ Hz}) \pm 10\%$$

$$\omega_{CL} = 5 \text{ krad/s} \ (795.8 \text{ Hz}) \pm 10\%$$

$$T_{MIN} \leq -20 \text{ dB} \quad \omega_{CHMIN} = 55 \text{ krad/s} \ (8.754 \text{ kH}z)$$

$$\omega_{CLMIN} = 500 \text{ rad/s} \ (79.6 \text{ Hz})$$

Two vendors have submitted proposed solutions in the form of OrCAD graphics, shown in Figure P14–54. As the requirements engineer on the project, you need to select the best one. You must check whether each circuit meets or fails every specification. You should also consider (1) parts count, (2) ease of maintainability and implementation (fewer parts, number of similar parts, adjustments of potentiometers, standard values), (3) frequency-domain response, and (4) cost. You should address each item for each design at least qualitatively. Attach whatever software output you think will support your decision.

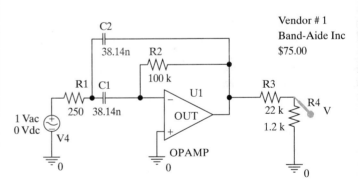

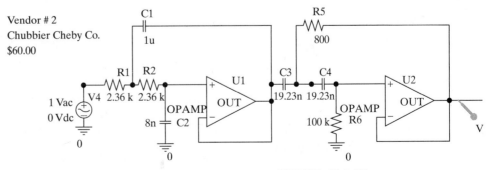

FIGURE P14–54

14–55 ▣ You are working at an aircraft manufacturing plant on an altitude sensor that eventually will be used to retrofit dozens of similar sensors on an upgrade to a current airframe. You are required to find a quality notch filter to eliminate an undesirable interference at 400 Hz. You find a vendor and ask for a suitable product specification and a bid.

The specifications you require are:

1. *Must block signals at* $400 \pm 2\%$ *Hz*
2. *Pass-band gain should be* 0 *dB* ± 1 *dB.*
3. *Bandwidth should be* 20 *Hz* $\pm 5\%$ *Hz*

The vendor's proposal, shown as an OrCAD drawing in Figure P14–55, claims to meet all of the design requirements. Determine if it passes all of the stated requirements. Then explain if you would buy the filter and why or why not?

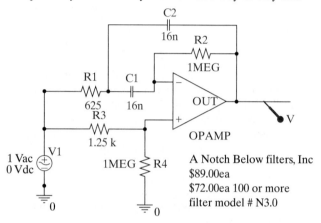

FIGURE P14–55

14–56 ▣ An amplified portion of the radio spectrum is shown in Figure P14–56. You need to hear all of the signals from 1.0 MHz to 2.0 MHz, but there is an interfering signal at 1.8 MHz. Design a notch filter to reduce that signal by at least 50 dB and not reduce the desired signal at 1.7 MHz by more than 6dB. Use OrCAD to validate your design.

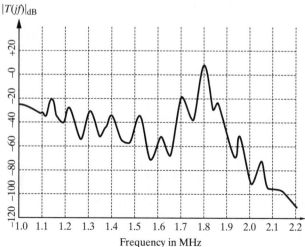

FIGURE P14–56

14–57 ▣ A amplified portion of the radio spectrum is shown in Figure P14–56. You want to select the signal at 1.22 MHz but it is barely above the background noise. Design a tuned filter that has a Q of at least 50 and amplifies the signal by 20 dB. Use OrCAD to validate your design.

14–58 ▣ The portion of the radio spectrum shown in Figure P14–56 is the result you want after you design a suitable filter and amplifier. The passband gain desired is 100 dB and the "shoulders" of your filter should have a roll-off of -120 dB per decade. Design a wide-band filter and amplifier that is flat in the pass band and has a bandwidth of 1 MHz with a lower cutoff frequency of 1 MHz that meets all of these criteria. Use OrCAD to validate your design.

INTEGRATING PROBLEMS

14–59 Design Evaluation ▣

A need exists for a third-order Butterworth low-pass filter with a cutoff frequency of 2 krad and a dc gain of 0 dB. The design department has proposed the circuit in Figure P14–59. As a junior engineer in the manufacturing department, you have been asked to verify the design and suggest modifications that would simplify production.

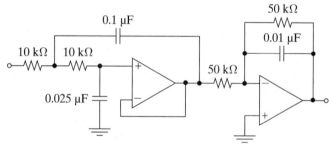

FIGURE P14–59

14–60 Modifying an Existing Circuit ▣

One of your company's products includes the passive *RLC* filter and OP AMP buffer circuit in Figure P14–60. The supplier of the inductor is no longer in business and a suitable replacement is not available, even on *eBay* or *Craig's List*. You have been asked to design a suitable inductorless replacement. To minimize production changes, your design must use the existing OP AMP as is and either the 1-kΩ resistor or the 0.1 μF capacitor or both, if possible.

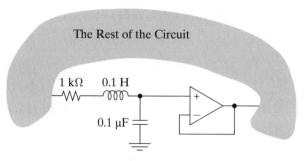

FIGURE P14–60

14–61 What's a High-Pass Filter ◀**A**▶

Ten years after earning a BSEE, you return for a master's degree and sign on as the laboratory instructor for the basic circuit analysis course. One experiment asks the students to build the active filter in Figure P14–61 and measure its gain response over the range from 150 Hz to 15 kHz. The lab instructions say the circuit is a high-pass filter with a $\omega_0 = 10$ krad/s and an infinite frequency gain of 0 dB. Everything goes well until a student, intrigued by the concept of infinite frequency, inputs 1 MHz and measures a gain of only 0.7. The student then inputs 2 MHz and measures a gain of 0.45. The high-pass filter appears to be a bandpass filter! Motivated by an insatiable thirst for understanding, the student asks you for an explanation. You first check the student's circuit and find it to be correct. You next replace the OP AMP and get almost the same results. Desperate for an explanation (your credibility is on the line here), you read the course textbook (it is Thomas, Rosa, and Toussaint), and find the answer in Chapter 12 (Example 12–5). What do you tell the student?

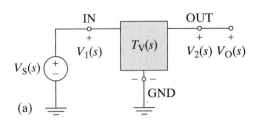

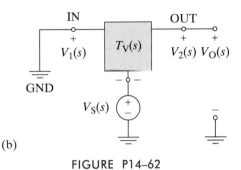

FIGURE P14–62

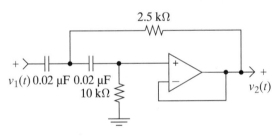

FIGURE P14–61

14–62 Bandpass to Bandstop Transformation ◀**A**▶

The three-terminal circuit in Figure P14–62(a) has a bandpass transfer function of the form

$$T(s) = \frac{V_O(s)}{V_S(s)} = \frac{2\zeta(s/\omega_0)}{(s/\omega_0)^2 + 2\zeta(s/\omega_0) + 1}$$

Show that the circuit in Figure P14–62(b) has a bandstop transfer function of the form

$$T(s) = \frac{V_O(s)}{V_S(s)} = \frac{(s/\omega_0)^2 + 1}{(s/\omega_0)^2 + 2\zeta(s/\omega_0) + 1}$$

That is, show that interchanging the input and ground terminals changes a unity-gain bandpass circuit into a unity-gain bandstop circuit.

14–63 Third-Order Butterworth Circuit ◀**A**▶

Show that the circuit in Figure P14–63 produces a third-order Butterworth low-pass filter with a cutoff frequency of $\omega_C = 1/RC$ and a passband gain of $K = 4$.

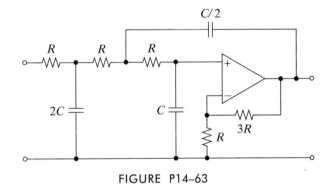

FIGURE P14–63

14–64 Notch Filter Comparison ◀**E**▶

To eliminate an interfering signal at 10 krad/s on a new product design, your consulting firm needs to purchase a notch filter with the following specifications:

  Center frequency = 10 krad/s ±0.5%
  Bandwidth = 200 rad/s ± 2%
  Depth of notch (attenuation at ω_0) = 50 dB min.

Two vendors have submitted proposed solutions as OrCAD graphics shown in Figure P14–64. As the owner, you want to select the best one. You must check that each circuit meets or fails every specification. You should also consider: (1) parts count, (2) ease of maintainability and implementation (fewer parts, number of similar parts, adjustments of potentiometers, standard values, and power usage), (3) frequency domain response, and (4) cost. You should address each item for each design at least qualitatively. Attach whatever software outputs you think will support your decision.

Simplicity Plus

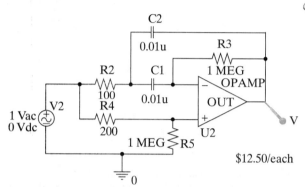

$15.00/each

Filters-R-Us

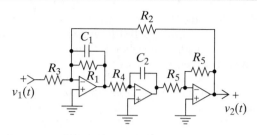

FIGURE P14–65

14–66 Crystal Filters **A**

Although not an active filter, crystal (Quartz) filters are very high-Q filters. Some can have Q's approaching 100,000. High Q means high selectivity; hence, crystal filters are used extensively in communications where fine tuning is essential. In its simplest form, a quartz crystal is a rectangular cut crystal with metal plates attached to its flattest sides – imagine a parallel-plate capacitor with the crystal as the dielectric. An electronic equivalent of a quartz filter is shown in Figure P14–66. C_O is called the shunt capacitance and $C_1, L_1,$ and R_1 the mechanical capacitance, inductance, and resistance of the crystal, respectively. A particular crystal has $C_O = 1.0\,\text{pF}$, $C_1 = 2.6\,\text{fF}$, $L_1 = 6.0$ mH, and $R_1 = 35\,\Omega$. Assume $R_L = 10\,\Omega$. Analyze this filter and determine its Q and ω_0. Use MATLAB or OrCAD to examine this circuit and help calculate the requested parameters.

FIGURE P14–64

14–65 Biquad Filter **A**

A biquad filter has the unique properties of having the ability to alter the filter's parameters, namely, gain K, quality factor Q, and resonant frequency ω_0. This is done in each case by simply adjusting one of the circuit's resistors. Furthermore, adjusting any of the three parameters leaves the others unchanged. Analyze the biquad circuit in Figure P14–65 and determine which resistors influence each parameter. Explain how you can adjust three resistors in a specific order to set all three parameters.

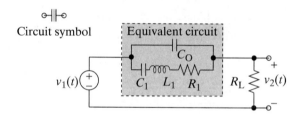

Circuit symbol   Equivalent circuit

FIGURE P14–66

CHAPTER 15 MUTUAL INDUCTANCE AND TRANSFORMERS

From the foregoing facts, it appears that a current of electricity is produced, for an instant, in a helix of copper wire surrounding a piece of soft iron whenever magnetism is induced in the iron; also that an instantaneous current in one or the other direction accompanies every change in the magnetic intensity of the iron.

Joseph Henry, 1831,
American Physicist

Some History Behind This Chapter

The discovery of electromagnetic induction led Michael Faraday to wind two separate insulated wires around an iron ring. One winding was connected to a battery via a switch and the other to a galvanometer. He observed that the current in the first winding did indeed induce a current in the second, but only when the switch was opened or closed. From this he correctly deduced that it was the change in current that produced the inductive effect. Faraday's coils and iron ring were actually a crude transformer. Practical ac power transformers were perfected by the British engineers Lucien Gauland and Josiah Gibbs in the early 1880s.

Why This Chapter Is Important Today

Coupled coils are found in applications such as power systems, communications, and high-quality audio. In this chapter you will learn how to model the magnetic coupling between coils using the concept of mutual inductance. This concept leads to a new circuit element called a transformer, which provides impedance matching, electrical isolation, and changes in the voltage level in power systems. The large power transformers used in such systems are very efficient, with internal losses often less than 1%.

Chapter Sections

15–1 Coupled Inductors
15–2 The Dot Convention
15–3 Energy Analysis
15–4 The Ideal Transformer
15–5 Linear Transformers

Chapter Learning Objectives

15-1 Mutual Inductance (Sects. 15–1, 15–2, 15–3)

Given the current through or voltage across two coupled inductors, find other currents or voltages.

15-2 The Ideal Transformer (Sect. 15–4)

Given a circuit containing ideal transformers:
(a) Find specified voltages, currents, powers, and equivalent circuits.
(b) Select the turns ratio to meet prescribed conditions.

15-3 Linear Transformers (Sect. 15–5)

Given a linear circuit with a transformer operating in the sinusoidal steady-state, find phasor voltages and currents, average powers, and equivalent impedances.

15–1   COUPLED INDUCTORS

The i–v characteristic of the inductor results from the magnetic field produced by current through a coil of wire. A constant current produces a constant magnetic field that forms closed loops of magnetic flux lines in the vicinity of the inductor. A changing current causes these closed loops to expand or contract, thereby "cutting" the turns in the winding that makes up the inductor. Faraday's law states that voltage across the inductor is equal to the time rate of change of the total flux linkage. In earlier chapters we expressed this relationship between a time-varying current and an induced voltage in terms of a circuit parameter called inductance L.

Now suppose that a second inductor is brought close to the first so that the flux from the first inductor links with the turns of the second inductor. If the current in the first inductor is changing, then this flux linkage will generate a voltage in the second inductor. The magnetic coupling between the changing current in one inductor and the voltage generated in a second inductor produces **mutual inductance**.

i–v CHARACTERISTICS

The i–v characteristics of coupled inductors unavoidably involve describing the effects observed in one inductor due to causes occurring in the other. We will use a double-subscript notation because it clearly identifies the various cause-and-effect relationships. The first subscript indicates the inductor in which the effect takes place, and the second identifies the inductor in which the cause occurs. For example, $v_{12}(t)$ is the voltage across inductor 1 due to causes occurring in inductor 2, whereas $v_{11}(t)$ is the voltage across inductor 1 due to causes occurring in inductor 1 itself.

We begin by assuming that the inductors are far apart, as shown in Figure 15–1(a). Under these circumstances there is no magnetic coupling between the two. A current $i_1(t)$ passes through the N_1 turns of the first inductor and $i_2(t)$ through N_2 turns in the second. Each inductor produces a flux

$$\begin{array}{cc} \text{Inductor 1} & \text{Inductor 2} \\ \phi_1(t) = k_1 N_1 i_1(t) & \phi_2(t) = k_2 N_2 i_2(t) \end{array} \qquad (15\text{–}1)$$

where k_1 and k_2 are proportionality constants. The flux linkage in each inductor is proportional to the number of turns:

$$\begin{array}{cc} \text{Inductor 1} & \text{Inductor 2} \\ \lambda_{11}(t) = N_1 \phi_1(t) & \lambda_{22}(t) = N_2 \phi_2(t) \end{array} \qquad (15\text{–}2)$$

By Faraday's law the voltage across an inductor is equal to the time rate of change of the flux linkage. Using Eqs. (15–1) and (15–2) together with the relationship between voltage and time rate of change of flux linkage gives

$$\text{Inductor 1: } v_{11}(t) = \frac{d\lambda_{11}(t)}{dt} = N_1 \frac{d\phi_1(t)}{dt} = \left[k_1 N_1^2\right] \frac{di_1(t)}{dt}$$

$$\text{Inductor 2: } v_{22}(t) = \frac{d\lambda_{22}(t)}{dt} = N_2 \frac{d\phi_2(t)}{dt} = \left[k_2 N_2^2\right] \frac{di_2(t)}{dt} \qquad (15\text{–}3)$$

Equations (15–3) provide the i–v relationships for the inductors when there is no mutual coupling. These results are the same as previously found in Chapter 6.

Now suppose the inductors are brought close together so that part of the flux produced by each inductor intercepts the other, as indicated in Figure 15–1(b). That is, part (but not necessarily all) of the fluxes $\phi_1(t)$ and $\phi_2(t)$ in Eq. (15–1) intercept the opposite inductor. We describe the cross coupling using the double-subscript notation:

$$\begin{array}{cc} \text{Inductor 1} & \text{Inductor 2} \\ \phi_{12}(t) = k_{12} N_2 i_2(t) & \phi_{21}(t) = k_{21} N_1 i_1(t) \end{array} \qquad (15\text{–}4)$$

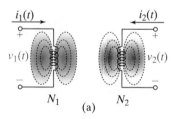

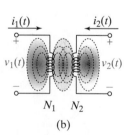

FIGURE 15–1   *(a) Inductors separated, only self-inductance present. (b) Inductors coupled, both self- and mutual inductance present.*

The quantity $\phi_{12}(t)$ is the flux intercepting inductor 1 due to the current in inductor 2, and $\phi_{21}(t)$ is the flux intercepting inductor 2 due to the current in inductor 1. The total flux linkage in each inductor is proportional to the number of turns:

$$\text{Inductor 1} \qquad \text{Inductor 2}$$
$$\lambda_{12}(t) = N_1\phi_{12}(t) \qquad \lambda_{21}(t) = N_2\phi_{21}(t) \tag{15–5}$$

By Faraday's law, the voltage across a winding is equal to the time rate of change of the flux linkage. Using Eqs. (15–4) and (15–5) together with derivative relationship, the time rate of change of flux linkages and voltages gives

$$\text{Inductor 1: } v_{12}(t) = \frac{d\lambda_{12}(t)}{dt} = N_1\frac{d\phi_{12}(t)}{dt} = [k_{12}N_1N_2]\frac{di_2(t)}{dt}$$

$$\text{Inductor 2: } v_{21}(t) = \frac{d\lambda_{21}(t)}{dt} = N_2\frac{d\phi_{21}(t)}{dt} = [k_{21}N_1N_2]\frac{di_1(t)}{dt} \tag{15–6}$$

The expressions in Eq. (15–6) are the i–v relationships describing the cross coupling between inductors when there is mutual coupling.

When the magnetic medium supporting the fluxes is linear, the superposition principle applies, and the total voltage across the inductors is the sum of the results in Eqs. (15–3) and (15–6):

$$\text{Inductor 1: } v_1(t) = v_{11}(t) + v_{12}(t) = \left[k_1N_1^2\right]\frac{di_1(t)}{dt} + [k_{12}N_1N_2]\frac{di_2(t)}{dt}$$

$$\text{Inductor 2: } v_2(t) = v_{21}(t) + v_{22}(t) = [k_{21}N_1N_2]\frac{di_1(t)}{dt} + \left[k_2N_2^2\right]\frac{di_2(t)}{dt} \tag{15–7}$$

We can identify four inductance parameters in these equations:

$$L_1 = k_1N_1^2 \qquad L_2 = k_2N_2^2 \tag{15–8}$$

and

$$M_{12} = k_{12}N_1N_2 \qquad M_{21} = k_{21}N_1N_2 \tag{15–9}$$

The two inductance parameters in Eq. (15–8) are the **self-inductance** of the inductors. The two parameters in Eq. (15–9) are the **mutual inductances** between the two inductors. In a linear magnetic medium, $k_{12} = k_{21} = k_M$. As a result, we can define a single mutual inductance parameter M as

$$M = M_{12} = M_{21} = k_MN_1N_2 \tag{15–10}$$

Using the definitions in Eqs. (15–8) and (15–10), the i–v characteristics of two coupled inductors are

$$\text{Inductor 1: } v_1(t) = L_1\frac{di_1(t)}{dt} + M\frac{di_2(t)}{dt}$$

$$\text{Inductor 2: } v_2(t) = M\frac{di_1(t)}{dt} + L_2\frac{di_2(t)}{dt} \tag{15–11}$$

Coupled inductors involve three inductance parameters, the two self-inductances L_1 and L_2 and the mutual inductance M.

The preceding development assumes that the cross coupling is additive. Additive coupling means that a positive rate of change of current in inductor 2 induces a positive voltage in inductor 1, and vice versa. The additive assumption produces the positive sign on the mutual inductance terms in Eq. (15–11). Unhappily, it is possible for a positive rate of change of current in one inductor to induce a negative voltage in the other. To account for additive and subtractive coupling, the general form of the

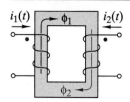

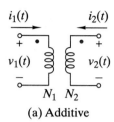

(a) Additive

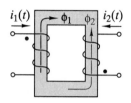

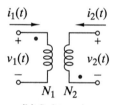

(b) Subtractive

FIGURE 15–2 *Winding orientations and corresponding reference dots.*

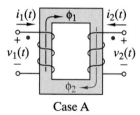

Case A

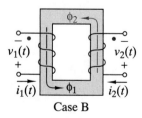

Case B

FIGURE 15–3 *All possible current and voltage reference marks for a fixed winding orientation.*

coupled inductor i–v characteristics includes a $\pm$ sign on the mutual inductance terms:

$$\text{Inductor 1: } v_1(t) = L_1\frac{di_1(t)}{dt} \pm M\frac{di_2(t)}{dt}$$

$$\text{Inductor 2: } v_2(t) = \pm M\frac{di_1(t)}{dt} + L_2\frac{di_2(t)}{dt}$$

(15–12)

When applying these element equations, it is necessary to know when to use a plus sign and when to use a minus sign.

15–2 THE DOT CONVENTION

The parameter M is positive, so the question is. What sign should be placed in front of this positive parameter in the i–v relationships in Eq. (15–12)? The correct sign depends on two things: (1) the spatial orientation of the two windings and (2) the reference marks given to the currents and voltages.

Figure 15–2 shows the additive and subtractive spatial orientation of two coupled inductors. In either case, the direction of the flux produced by a current is found using the right-hand rule treated in physics courses. In the additive case, currents i_1 and i_2 both produce clockwise fluxes ϕ_1 and ϕ_2. In the subtractive case, the currents produce opposing fluxes because ϕ_1 is clockwise and ϕ_2 is counterclockwise. The sign for the mutual inductance term is positive for the additive orientation and negative for the subtractive case.

In general, it is awkward to show the spatial features of the windings in circuit diagrams. The dots shown near one terminal of each winding in Figure 15–2 are special reference marks indicating the relative orientation of the windings. The reference directions for the currents and voltages are arbitrary. They can be changed as long as we follow the passive sign convention. However, the dots indicate physical attributes of the windings that make up the coupled inductors. They are not arbitrary. They cannot be changed.

The correct sign for the mutual inductance term hinges on how the reference marks for currents and voltages are assigned relative to the dots. For a given winding orientation, Figure 15–3 shows all four possible current and voltage reference assignments under the passive sign convention. In cases A and B the fluxes are additive, so the mutual inductance term is positive. In cases C and D the fluxes are subtractive and the mutual inductance term is negative. From these results we derive the following rule:

> *Mutual inductance is additive when both current reference directions point toward or both point away from dotted terminals; otherwise, it is subtractive.*

Because the current reference directions can be changed, a corollary of this rule is that we can always assign reference directions so that the positive sign applies to the mutual inductance. This corollary is important because a positive sign is built into the mutual inductance models in circuit analysis programs like SPICE.

The following examples and exercises illustrate selecting the correct sign and applying the i–v characteristics in Eq. (15–12).

EXAMPLE 15–1

For the coupled inductors in Figure 15–4:

(a) Write the i–v characteristics using the reference marks shown in the figure.
(b) For $v_S(t) = 200\sin 400t$ V, find $v_2(t)$ when the output terminals are open-circuited ($i_2 = 0$).

SOLUTION:

(a) The two current references directions both point toward the dotted terminals. Therefore, the mutual inductance is additive and the i–v equations are

$$v_1(t) = L_1 \frac{di_1(t)}{dt} + M \frac{di_2(t)}{dt} = 10^{-2} \frac{di_1(t)}{dt} + 2 \times 10^{-3} \frac{di_2(t)}{dt}$$

$$v_2(t) = M \frac{di_1(t)}{dt} + L_2 \frac{di_2(t)}{dt} = 2 \times 10^{-3} \frac{di_1(t)}{dt} + 10^{-2} \frac{di_2(t)}{dt}$$

(b) For $v_1(t) = v_S(t) = 200 \sin 400t$ and $i_2(t) = 0$, these equations reduce to

$$200 \sin 400t = 10^{-2} \frac{di_1(t)}{dt}$$

$$v_2(t) = 2 \times 10^{-3} \frac{di_1(t)}{dt}$$

Solving the first equation for $di_1(t)/dt$ yields

$$\frac{di_1(t)}{dt} = 20{,}000 \sin 400t$$

Substituting this result in the second equation produces

$$v_2(t) = 2 \times 10^{-3} (20{,}000 \sin 400t) = 40 \sin 400t \text{ V} \quad \blacksquare$$

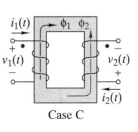

Case C

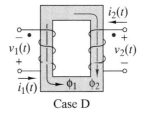

Case D

FIGURE 15–3 *(Continued)*

Exercise 15–1

In Figure 15–4 $i_1(t) = -20 \cos 8000t \text{ mA}$ and $i_2(t) = 0$. Find $v_1(t)$ and $v_2(t)$.

Answers: $v_1(t) = 1.6 \sin 8000t \text{ V}$; $v_2(t) = 0.32 \sin 8000t \text{ V}$

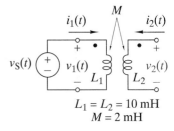

$L_1 = L_2 = 10 \text{ mH}$
$M = 2 \text{ mH}$

FIGURE 15–4

EXAMPLE 15–2

For the coupled inductors in Figure 15–5:

(a) Write the i–v characteristics using the reference marks shown in the figure.
(b) For $i_S(t) = 2 \cos 500t \text{ A}$, find $v_1(t)$ and $i_2(t)$ when the output terminals are short-circuited ($v_2 = 0$).

$L_1 = 50 \text{ mH}, L_2 = 75 \text{ mH}, M = 60 \text{ mH}$

FIGURE 15–5

SOLUTION:

(a) The $i_1(t)$ reference points toward and the $i_2(t)$ reference points away from the dotted terminals, so the mutual inductance is subtractive and the i–v equations are

$$v_1(t) = L_1 \frac{di_1(t)}{dt} - M \frac{di_2(t)}{dt} = 5 \times 10^{-2} \frac{di_1(t)}{dt} - 6 \times 10^{-2} \frac{di_2(t)}{dt}$$

$$v_2(t) = -M \frac{di_1(t)}{dt} + L_2 \frac{di_2(t)}{dt} = -6 \times 10^{-2} \frac{di_1(t)}{dt} + 7.5 \times 10^{-2} \frac{di_2(t)}{dt}$$

(b) For $i_1(t) = i_S(t) = 2 \cos 500t$ and $v_2(t) = 0$, these equations reduce to

$$v_1(t) = 5 \times 10^{-2} \frac{d(2 \cos 500t)}{dt} - 6 \times 10^{-2} \frac{di_2(t)}{dt}$$

$$0 = -6 \times 10^{-2} \frac{d(2 \cos 500t)}{dt} + 7.5 \times 10^{-2} \frac{di_2(t)}{dt}$$

Solving the second equation for di_2/dt yields

$$\frac{di_2(t)}{dt} = 0.8 \frac{d(2 \cos 500t)}{dt}$$

or

$$i_2(t) = 1.6 \cos 500t \text{ A}$$

Substituting this result in the first equation produces

$$v_1(t) = 5 \times 10^{-2} \frac{d(2 \cos 500t)}{dt} - 6 \times 10^{-2} \frac{d(1.6 \cos 500t)}{dt}$$

$$= -2 \sin 500t \text{ V}$$

■

Exercise 15–2

Find $v_1(t)$ and $v_2(t)$ for the circuit in Figure 15–6.

Answers:
$$v_1(t) = -10 \sin 10^4 t - 3 \cos 5 \times 10^3 t \text{ V}$$
$$v_2(t) = -15 \sin 10^4 t - 5 \cos 5 \times 10^3 t \text{ V}$$

FIGURE 15–6

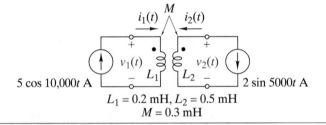

$$L_1 = 0.2 \text{ mH}, L_2 = 0.5 \text{ mH}$$
$$M = 0.3 \text{ mH}$$

15–3 ENERGY ANALYSIS

Calculating the total energy stored in a pair of coupled inductors reveals a fundamental limitation on allowable values of the self- and mutual inductances. To uncover this limitation, we first calculate the total power absorbed. Multiplying the first equation in Eq. (15–12) by $i_1(t)$ and the second equation by $i_2(t)$ produces

$$p_1(t) = v_1(t)i_1(t) = L_1 i_1(t) \frac{di_1(t)}{dt} \pm M i_1(t) \frac{di_2(t)}{dt}$$

$$p_2(t) = v_2(t)i_2(t) = \pm M i_2(t) \frac{di_1(t)}{dt} + L_2 i_2(t) \frac{di_2(t)}{dt}$$

(15–13)

The quantities $p_1(t)$ and $p_2(t)$ are the powers absorbed with inductors 1 and 2. The total power is the sum of the individual inductor powers:

$$p(t) = p_1(t) + p_2(t)$$

$$= L_1 \left[i_1(t) \frac{di_1(t)}{dt} \right] \pm M \left[i_1(t) \frac{di_2(t)}{dt} + i_2(t) \frac{di_1(t)}{dt} \right] + L_2 \left[i_2(t) \frac{di_2(t)}{dt} \right]$$

(15–14)

Each of the bracketed terms in Eq. (15–14) is a perfect derivative. Specifically,

$$i_1(t) \frac{di_1(t)}{dt} = \frac{1}{2} \frac{di_1^2(t)}{dt}$$

$$i_2(t) \frac{di_2(t)}{dt} = \frac{1}{2} \frac{di_2^2(t)}{dt}$$

(15–15)

$$i_1(t) \frac{di_2(t)}{dt} + i_2(t) \frac{di_1(t)}{dt} = \frac{di_1(t)i_2(t)}{dt}$$

Therefore, the total power in Eq. (15–14) is

$$p(t) = \frac{d}{dt}\left[\frac{1}{2}L_1 i_1^2(t) \pm M i_1(t)i_2(t) + \frac{1}{2}L_2 i_2^2(t)\right] \tag{15–16}$$

Because power is the time rate of change of energy, the quantity inside the brackets in Eq. (15–16) is the total energy stored in the two inductors. That is,

$$w(t) = \frac{1}{2}L_1 i_1^2(t) \pm M i_1(t)i_2(t) + \frac{1}{2}L_2 i_2^2(t) \tag{15–17}$$

In Eq. (15–17) the self-inductance terms are always positive. However, the mutual inductance term can be either positive or negative. At first glance it appears that the total energy could be negative. But the total energy must be positive; otherwise, the coupled inductors could deliver net energy to the rest of the circuit.

The condition $w(t) \geq 0$ places a constraint on the values of the self- and mutual inductances. First, if $i_2(t) = 0$, then $w(t) \geq 0$ in Eq. (15–17) requires $L_1 > 0$. Next, if $i_1(t) = 0$, then $w(t) \geq 0$ in Eq. (15–17) requires $L_2 > 0$. Finally, if $i_1(t) \neq 0$ and $i_2(t) \neq 0$, then we divide Eq. (15–17) by $[i_2(t)]^2$ and define a variable $x = i_1/i_2$. With these changes, the energy constraint $w(t) > 0$ becomes

$$\frac{w(t)}{i_2^2(t)} = f(x) = \frac{1}{2}L_1 x^2 \pm Mx + \frac{1}{2}L_2 \geq 0 \tag{15–18}$$

The minimum value of $f(x)$ occurs when

$$\frac{df(x)}{dx} = L_1 x \pm M = 0 \quad \text{hence} \quad x_{\min} = \mp \frac{M}{L_1} \tag{15–19}$$

The value $x_{\min}$ yields the minimum of $f(x)$ because the second derivative of $f(x)$ is positive. Substituting $x_{\min}$ back into Eq. (15–18) yields the condition

$$f(x_{\min}) = \frac{1}{2}L_1\frac{M^2}{L_1^2} - \frac{M^2}{L_1} + \frac{1}{2}L_2 = \frac{1}{2}\left[-\frac{M^2}{L_1} + L_2\right] \geq 0 \tag{15–20}$$

The constraint in Eq. (15–20) means that the stored energy in a pair of coupled inductors is positive if

$$L_1 L_2 \geq M^2 \tag{15–21}$$

Energy considerations dictate that in any pair of coupled inductors, the product of the self-inductances must exceed the square of the mutual inductance.

The constraint in Eq. (15–21) is usually written in terms of a new parameter called the **coupling coefficient** k:

$$k = \frac{M}{\sqrt{L_1 L_2}} \leq 1 \tag{15–22}$$

The parameter k ranges from 0 to 1. If $M = 0$, then $k = 0$ and the coupling between the inductors is zero. The condition $k = 1$ requires **perfect coupling** in which all of the flux produced by one inductor links the other. Perfect coupling is physically impossible, although careful design can produce coupling coefficients of 0.99 and higher.

EXAMPLE 15–3

A pair of coupled inductors have self-inductances $L_1 = 2.5$ H, $L_2 = 1.6$ H, and a coupling coefficient of $k = 0.8$. When the terminals of L_2 are short-circuited

$(v_2(t) = 0)$ the short-circuit current is observed to be $i_2(t) = -50\sin(2000t)$ mA. Find the input voltage $v_1(t)$ for additive coupling.

SOLUTION:

The given self-inductances and coupling coefficient yield the mutual inductance as

$$M = k\sqrt{L_1 L_2} = 0.8\sqrt{2.5 \times 1.6} = 1.6\,\text{H}$$

For additive **coupling** the i–v equations are

$$v_1(t) = L_1\frac{di_1(t)}{dt} + M\frac{di_2(t)}{dt} = 2.5\frac{di_1(t)}{dt} + 1.6\frac{di_2(t)}{dt}$$

$$v_2(t) = M\frac{di_1(t)}{dt} + L_2\frac{di_2(t)}{dt} = 1.6\frac{di_1(t)}{dt} + 1.6\frac{di_2(t)}{dt}$$

For $v_2(t) = 0$ and $i_2(t) = -50 \times 10^{-3}\sin 2000t$ these equations reduce to

$$v_1(t) = 2.5\frac{di_1(t)}{dt} + 1.6\frac{d\left[-50 \times 10^{-3}\sin 2000t\right]}{dt}$$

$$0 = 1.6\frac{di_1(t)}{dt} + 1.6\frac{d\left[-50 \times 10^{-3}\sin 2000t\right]}{dt}$$

Solving the second equation for $di_1(t)/dt$ yields

$$\frac{di_1(t)}{dt} = -\frac{d\left[-50 \times 10^{-3}\sin 2000t\right]}{dt} = 100\cos 2000t$$

Substituting this result into the first equation produces

$$v_1(t) = 2.5[100\cos 2000t] + 1.6[-100\cos 2000t] = 90\cos 2000t\ \text{V} \qquad \blacksquare$$

Exercise 15–3

A pair of coupled inductors have self-inductances $L_1 = 4.5$ mH and $L_2 = 8$ mH. What is the maximum possible mutual inductance?

Answers: $M = 6$ mH

15–4 THE IDEAL TRANSFORMER

A **transformer is an** electrical device that utilizes magnetic coupling between two inductors. Transformers find application in virtually every type of electrical system, but especially in power supplies and commercial power grids. Some example devices from these applications are shown in Figure 15–7.

In Figure 15–8 the transformer is shown as an interface device between a source and a load. The winding connected to the source is called the **primary winding**, and the one connected to the load is called the **secondary winding**. In most applications the transformer is a coupling device that transfers signals (especially power) from the source to the load. The basic purpose of the device is to change voltage and current levels so that the conditions at the source and load are compatible.

Transformer design involves two primary goals: (1) to maximize the magnetic coupling between the two windings and (2) to minimize the power loss in the

www.acutran.com

(a)

(b)

(c)

www.butlerwinding.com

(d)

FIGURE 15–7 *Examples of transformer devices.*
(a) Powerline step-down.
(b) Iron core. (c) Toroidal.
(d) Surface mount.

windings. The first goal produces nearly perfect coupling ($k \approx 1$) so that almost all of the flux in one winding links the other. The second goal produces nearly zero power loss so that almost all of the power delivered to the primary winding transfers to the load. The **ideal transformer** is a circuit element in which coupled inductors are assumed to have perfect coupling and zero power loss. Using these two idealizations, we can derive the i–v characteristics of an ideal transformer.

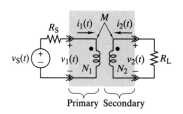

FIGURE 15–8 *Transformer connected at a source-load interface.*

PERFECT COUPLING

Perfect coupling means that all of the flux in the first winding links the second, and vice versa. Equation (15–1) defines the total flux in each winding as

$$\begin{array}{cc} \text{Winding 1} & \text{Winding 2} \\ \phi_1(t) = k_1 N_1 i_1(t) & \phi_2(t) = k_2 N_2 i_2(t) \end{array} \qquad (15\text{–}23)$$

where k_1 and k_2 are proportionality constants. Equation (15–4) defines the cross coupling using the double subscript notation:

$$\begin{array}{cc} \text{Winding 1} & \text{Winding 2} \\ \phi_{12}(t) = k_{12} N_2 i_2(t) & \phi_{21}(t) = k_{21} N_1 i_1(t) \end{array} \qquad (15\text{–}24)$$

In this equation $\phi_{12}(t)$ is the flux intercepting winding 1 due to the current in winding 2, and $\phi_{21}(t)$ is the flux intercepting winding 2 due to the current in winding 1. Perfect coupling means that

$$\phi_{21}(t) = \phi_1(t) \quad \text{and} \quad \phi_{12}(t) = \phi_2(t) \qquad (15\text{–}25)$$

Comparing Eqs. (15–23) and (15–24) shows that perfect coupling requires $k_1 = k_{21}$ and $k_2 = k_{12}$. But in a linear magnetic medium $k_{12} = k_{21} = k_M$, so perfect coupling implies

$$k_1 = k_2 = k_{12} = k_{21} = k_M \qquad (15\text{–}26)$$

Substituting the perfect coupling conditions in Eq. (15–26) into the i–v characteristics in Eq. (15–7) gives

$$v_1(t) = \left[k_M N_1^2 \right] \frac{di_1(t)}{dt} \pm \left[k_M N_1 N_2 \right] \frac{di_2(t)}{dt}$$
$$v_2(t) = \pm \left[k_M N_1 N_2 \right] \frac{di_1(t)}{dt} + \left[k_M N_2^2 \right] \frac{di_2(t)}{dt} \qquad (15\text{–}27)$$

Factoring N_1 out of the first equation and $\pm N_2$ out of the second produces

$$v_1(t) = N_1 \left(\left[k_M N_1 \right] \frac{di_1(t)}{dt} \pm \left[k_M N_2 \right] \frac{di_2(t)}{dt} \right)$$
$$v_2(t) = \pm N_2 \left(\left[k_M N_1 \right] \frac{di_1(t)}{dt} \pm \left[k_M N_2 \right] \frac{di_2(t)}{dt} \right) \qquad (15\text{–}28)$$

Dividing the second equation by the first shows that perfect coupling implies

$$\frac{v_2(t)}{v_1(t)} = \pm \frac{N_2}{N_1} = \pm n \qquad (15\text{–}29)$$

where the plus sign applies for additive mutual inductance and the minus sign for subtractive.

The parameter n in Eq. (15–29) is called the **turns ratio**. When $n > 1$ the secondary voltage is larger than the primary voltage and the device is called a **step-up transformer**. Conversely, when $n < 1$ the primary voltage is larger than the secondary voltage and the device is called a **step-down transformer**. The ability to increase or decrease ac voltages is a dominant feature and application of transformers. In commercial power systems ac voltages are stepped up to several hundred kilovolts for transmission and then stepped down to safer levels for use by customers.

EXAMPLE 15–4

A transformer with perfect coupling has $N_1 = 250$ turns and $N_2 = 25$ turns. The voltage across the primary winding is $v_1(t) = 120 \sin 377t$ V, and a resistive load $R_L = 50 \, \Omega$ is connected across the secondary winding. Find the current, voltage, and average power delivered to the load. Assume additive coupling.

SOLUTION:
For additive coupling Eq. (15–29) gives

$$v_2(t) = \frac{N_2}{N_1} v_1(t) = \frac{120}{10} \sin 377t$$
$$= 12 \sin 377t \text{ V}$$

The current delivered to the load across the secondary is

$$i_L(t) = \frac{v_2(t)}{R_L} = \frac{12}{50} \sin 377t$$
$$= 0.24 \sin 377t \text{ A}$$

and the average power delivered to the load is

$$P = \frac{1}{2}(0.24)^2 \times 50 = 1.44 \text{ W}$$ ∎

Exercise 15–4

For a transformer with perfect coupling, find the secondary voltage $v_2(t)$ when the primary voltage is $v_1(t) = 50 \cos 3000t$ V, $n = 50$, and the mutual inductance is subtractive.

Answer: $v_2(t) = -2500 \cos 3000t$ V

ZERO POWER LOSS

The ideal transformer model also assumes that there is no power loss in the transformer. With the passive sign convention, the quantity $v_1(t)i_1(t)$ is the power in the primary winding and $v_2(t)i_2(t)$ is the power in the secondary winding. Zero power loss requires

$$v_1(t)i_1(t) + v_2(t)i_2(t) = 0 \tag{15–30}$$

This equation states that whatever power enters the transformer on one winding immediately leaves via the other winding. This not only means that there is no energy lost in the ideal transformer, but also that there is no energy stored within the element. The zero-power-loss constraint can be rearranged as

$$\frac{i_2(t)}{i_1(t)} = -\frac{v_1(t)}{v_2(t)} \tag{15–31}$$

But under the perfect coupling assumption, $v_2(t)/v_1(t) = \pm n$. With zero power loss and perfect coupling the primary and secondary currents are related as

$$\frac{i_2(t)}{i_1(t)} = \mp \frac{1}{n} \tag{15–32}$$

The correct sign in this equation depends on the orientation of the current reference directions relative to the dots describing the transformer structure.

With both perfect coupling and zero power loss, the secondary current is inversely proportional to the turns ratio. A step-up transformer ($n > 1$) increases the voltage and decreases the current, which improves transmission line efficiency because the i^2R losses in the conductors are smaller.

i–v CHARACTERISTICS

Equations (15–29) and (15–32) define the i–v characteristics of the ideal transformer circuit element.

$$v_2(t) = \pm nv_1(t)$$
$$i_2(t) = \mp \frac{1}{n}i_1(t) \tag{15-33}$$

Rules for selecting the correct sign in these equations are discussed below.

Equation (15–33) applies to time-varying signals since only the changes in one coupled inductor have any effect on the other. Transformers do not pass constant voltages or currents since coupled inductors act like short circuits to dc signals. The sinusoid is the most common time-varying signal of interest. For the purpose of ac circuit analysis, it is convenient to convert Eq. (15–33) into the phasor domain as

$$\mathbf{V}_2 = \pm n\mathbf{V}_1$$
$$\mathbf{I}_2 = \mp \frac{1}{n}\mathbf{I}_1 \tag{15-34}$$

These equations are based on the passive sign convention. Under this convention the rules for selecting the signs in these equations are

When the reference directions for the currents are both toward or both away from the dotted terminals, then $\mathbf{V}_2 = +n\mathbf{V}_1$ and $\mathbf{I}_2 = -\mathbf{I}_1/n$; otherwise $\mathbf{V}_2 = -n\mathbf{V}_1$ and $\mathbf{I}_2 = +\mathbf{I}_1/n$.

Figure 15–9 shows the circuit symbol for the ideal transformer and several examples of the application of these rules. When using these rules remember that the dots represent physical features of a transformer; they are part of the circuit symbol and cannot be changed. The reference directions for the currents and voltages can be changed as long as the assignment follows the passive convention. This means that the current reference directions can always be assigned so that both are toward a dotted terminal.

In the typical application shown in Figure 15–10 an ac source is connected across the primary (input) side of the transformer and a load Z_L across the secondary (output) side. An important feature of this arrangement is that the transformer input

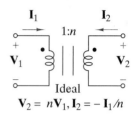
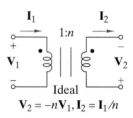
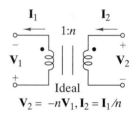
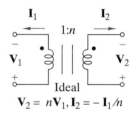

FIGURE 15–9

FIGURE 15–10

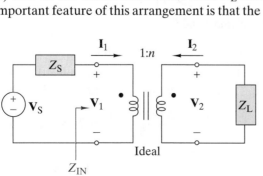

impedance Z_{IN} is related to the load impedance Z_L on the output. On the primary side we have

$$Z_{IN} = \frac{\mathbf{V}_1}{\mathbf{I}_1}$$

On the secondary side, the reference marks for Z_L do not follow the passive sign convention, so

$$Z_L = \frac{\mathbf{V}_2}{-\mathbf{I}_2}$$

The current reference directions in Figure 15–10 are both toward dotted terminals, so according to our rule for selecting signs $\mathbf{V}_2 = +n\mathbf{V}_1$ and $\mathbf{I}_2 = -\mathbf{I}_1/n$. Hence

$$Z_L = \frac{\mathbf{V}_2}{-\mathbf{I}_2} = \frac{n\mathbf{V}_1}{-(-\mathbf{I}_1/n)} = n^2 \frac{\mathbf{V}_1}{\mathbf{I}_1} = n^2 Z_{IN}$$

Therefore the transformer input impedance is

$$Z_{IN} = \frac{Z_L}{n^2} \tag{15–35}$$

Thus, Eq. (15–33) or (15–34) shows how an ideal transformer changes the voltage and current levels, while Eq. (15–35) shows how the impedance level is changed. Taken together, these equations allow us to analyze ac circuits containing ideal transformers.

EXAMPLE 15–5

The turns ratio of the ideal transformer in Figure 15–10 is $n = 5$. The source and load impedance are $Z_S = 2.5 + j1.5\,\Omega$ and $Z_L = 75 + j10\,\Omega$. Find $\mathbf{I}_1, \mathbf{V}_1, \mathbf{I}_2$, and $\mathbf{V}_2$ when the input is $\mathbf{V}_S = 220\,\angle 0°\,$V.

SOLUTION:
Using Eq. (15–35), the transformer input impedance is

$$Z_{IN} = \frac{Z_L}{n^2} = \frac{75 + j10}{25} = 3 + j0.4\,\Omega$$

The impedance seen by the voltage source is $Z_S + Z_{IN}$; hence

$$\mathbf{I}_1 = \frac{\mathbf{V}_S}{Z_S + Z_{IN}} = \frac{220\angle 0°}{5.5 + j1.9}$$

$$= \frac{220\,\angle 0°}{5.82\,\angle 19.1°} = 37.8\,\angle{-19.1°}\,\text{A}$$

The voltage on the primary side is

$$\mathbf{V}_1 = \mathbf{I}_1\,Z_{IN} = (37.8\,\angle{-19.1°})(3.03\,\angle 7.6°) = 114\,\angle{-11.5°}\,\text{V}$$

The reference directions for both $\mathbf{I}_1$ and $\mathbf{I}_2$ are toward dotted terminals, so the correct signs in Eq. (15–34) are $\mathbf{V}_2 = +n\mathbf{V}_1$ and $\mathbf{I}_2 = -\mathbf{I}_1/n$. Hence, we get

$$\mathbf{I}_2 = \frac{-\mathbf{I}_1}{n} = \frac{-37.8\,\angle{-19.1°}}{5} = 7.56\,\angle 160.9°\,\text{A}$$

and

$$\mathbf{V}_2 = n\mathbf{V}_1 = 570\,\angle{-11.5°}\,\mathbf{V}$$

Exercise 15–5

The turns ratio of the ideal transformer in Figure 15–10 is $n = 1/4$ and the source and load impedances are $Z_S = 50 + j0\ \Omega$ and $Z_L = 5 - j2\ \Omega$. Find the impedance seen by the voltage source.

Answer: $Z_S + Z_{IN} = 130 - j32\ \Omega$

EXAMPLE 15–6

Since a transformer changes impedance levels, it can be used to match a source and load to achieve maximum power transfer. Figure 15–11 shows a circuit model of an audio amplifier with an output impedance of 600 Ω feeding an 8-Ω speaker. Find the transformer turns ratio needed to achieve maximum power transfer.

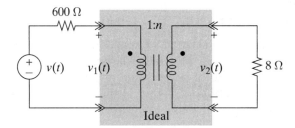

FIGURE 15–11

SOLUTION:

The maximum power transfer theorem states that power transfer is maximized when the source and load resistance are equal (matched). The purpose of the transformer is to create matched conditions at the output of the amplifier. Using Eq. (15–35), the input resistance on the primary side of the transformer is

$$R_{IN} = \frac{R_L}{n^2} = \frac{8}{n^2}$$

To achieve matched conditions we need $R_{IN} = 600 = 8/n^2$, which means that the required turns ratio is $n = \sqrt{8/600} = 1/8.66$. ∎

Exercise 15–6

The input impedance on the primary side of an ideal transformer is 1250 Ω when the load connected to the secondary is 50 Ω. What is the transformer turns ratio?

Answer: $n = 1/5$

APPLICATION EXAMPLE 15–7

In a transformer the primary and secondary windings are magnetically coupled but are usually electrically isolated. Transformer performance in some applications can be improved by electrically connecting the two magnetically coupled windings in a configuration called an **autotransformer**. Figure 15–12 shows a two-winding ideal transformer connected in an autotransformer step-up configuration.

The rating of the ideal transformer when connected in the usual electrically isolated two-winding configuration is $P_{rated} = v_1(t)i_1(t) = -v_2(t)i_2(t)$. Connected in the autotransformer configuration in Figure 15–12, the power delivered to the load is

$$P_{load} = (v_1(t) + v_2(t))(-i_2(t))$$

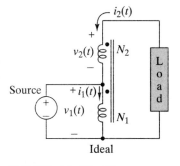

FIGURE 15–12 *The autotransformer connection.*

For an ideal transformer $v_2(t) = nv_1(t)$ and $i_2(t) = -i_1(t)/n$, where $n = N_2/N_1$. Hence the power delivered to the load is

$$P_{\text{load}} = (v_1(t) + nv_1(t))(i_1(t)/n) = \left(1 + \frac{1}{n}\right)v_1(t)i_1(t)$$

$$= \left(1 + \frac{1}{n}\right)P_{\text{rated}}$$

The autotransformer configuration delivers more power to the load than the power rating of the two-winding transformer. Put differently, the autotransformer can supply a specified load power using a transformer with a lower power rating. Autotransformers are normally used when the turns ratio is less than 3:1, so this advantage can be significant. A disadvantage is that the electrical isolation provided by the usual transformer configuration is lost.

15–5 LINEAR TRANSFORMERS

By far the most common application of transformers occurs in electric power systems, where they operate in the sinusoidal steady state. In this context we describe transformers in terms of phasors and impedances and deal with average power transfer. The ac analysis of a transformer begins with the time-domain element equations for a pair of coupled inductors

$$v_1(t) = L_1\frac{di_1(t)}{dt} \pm M\frac{di_2(t)}{dt}$$

$$v_2(t) = \pm M\frac{di_1(t)}{dt} + L_2\frac{di_2(t)}{dt}$$

Transforming these equations into the phasor domain involves replacing waveforms by phasors using the derivative property to obtain the phasors for di_1/dt and di_2/dt:

$$\mathbf{V}_1 = j\omega L_1\mathbf{I}_1 \pm j\omega M\mathbf{I}_2$$
$$\mathbf{V}_2 = \pm j\omega M\mathbf{I}_1 + j\omega L_2\mathbf{I}_2 \qquad (15\text{–}36)$$

where

1. $\mathbf{V}_1$ and $\mathbf{I}_1$ are the phasors representing ac voltage and current of the first winding, and $j\omega L_1$ is the impedance of the self-inductance of first winding.

2. $\mathbf{V}_2$ and $\mathbf{I}_2$ are the phasors representing ac voltage and current of the second winding, and $j\omega L_2$ is the impedance of the self-inductance of second winding.

3. $j\omega M$ is the impedance of the mutual inductance between the two windings.

The self-inductance impedances relate phasor voltage and current at the same pair of terminals, while the mutual inductance impedance relates the phasor voltage at one pair of terminals to the phasor current at the other pair. The phasor-domain equations can also be written in terms of reactances as

$$\mathbf{V}_1 = jX_1\mathbf{I}_1 \pm jX_M\mathbf{I}_2$$
$$\mathbf{V}_2 = \pm jX_M\mathbf{I}_1 + jX_2\mathbf{I}_2$$

where the three reactances are $X_1 = \omega L_1, X_2 = \omega L_2$, and $X_M = \omega M$. The degree of coupling between windings is indicated by the coupling coefficient, which can be written in terms of reactances as

$$k = \frac{M}{\sqrt{L_1L_2}} = \frac{X_M}{\sqrt{X_1X_2}}$$

In either form, energy considerations dictate that $0 \le k \le 1$.

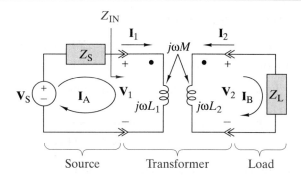

FIGURE 15–13 *Phasor circuit model of the two-winding transformer.*

Figure 15–13 shows the phasor-domain version of transformer coupling between a source and a load. Following our previous notation, the winding connected to the source is called the **primary**, and the winding connected to the load is called the **secondary**. Although the transformer is a bilateral device, we normally think of signal and power transfer as passing from the primary to the secondary winding.

Our immediate objective is to write circuit equations for the transformer using the mesh currents $\mathbf{I}_A$ and $\mathbf{I}_B$ in Figure 15–13. Applying KVL around the primary circuit (mesh A) and secondary circuit (mesh B), we obtain the following equations:

$$\text{Mesh A:} \quad Z_S\mathbf{I}_A + \mathbf{V}_1 = \mathbf{V}_S$$
$$\text{Mesh B:} \quad -\mathbf{V}_2 + Z_L\mathbf{I}_B = 0 \tag{15–37}$$

The reference directions for the inductor currents in Figure 15–13 are both into the dotted terminals, so the mutual inductance coupling is additive and the plus signs in Eq. (15–36) apply. Using KCL we see that the reference directions for the currents lead to the relations $\mathbf{I}_1 = \mathbf{I}_A$ and $\mathbf{I}_2 = -\mathbf{I}_B$. The I–V relationships of the coupled inductors in terms of the mesh currents are

$$\mathbf{V}_1 = +j\omega L_1\mathbf{I}_A + j\omega M(-\mathbf{I}_B)$$
$$\mathbf{V}_2 = +j\omega M\mathbf{I}_A + j\omega L_2(-\mathbf{I}_B) \tag{15–38}$$

Substituting the inductor voltages from Eq. (15–38) into the KVL equations in Eq. (15–37) yields

$$\text{Mesh A:} \quad (Z_S + j\omega L_1)\mathbf{I}_A - j\omega M\mathbf{I}_B = \mathbf{V}_S$$
$$\text{Mesh B:} \quad -j\omega M\mathbf{I}_A + (Z_L + j\omega L_2)\mathbf{I}_B = 0 \tag{15–39}$$

This set of mesh equations provides a complete description of the circuit ac response. Once we solve for the mesh currents, we can calculate every phasor voltage and current using Kirchhoff's laws and element equations.

EXAMPLE 15–8

The source circuit in Figure 15–13 has $Z_S = 0 + j20\ \Omega$ and $\mathbf{V}_S = 2500\angle 0°$ V at $\omega = 377$ rad/s. The transformer has $L_1 = 2$ H, $L_2 = 0.2$ H, and $M = 0.6$ H. The load impedance is $Z_L = 25 + j15\ \Omega$. Find $\mathbf{I}_A, \mathbf{I}_B, \mathbf{V}_1, \mathbf{V}_2, Z_{IN}$, and the average power delivered by the source at input interface.

SOLUTION:
The transformer impedances are

$$j\omega L_1 = j377 \times 2 = j754\ \Omega$$
$$j\omega L_2 = j377 \times 0.2 = j75.4\ \Omega$$
$$j\omega M = j377 \times 0.6 = j226\ \Omega$$

Using these impedances in Eq. (15–39) yields the following mesh equations:

Mesh A: $(j20 + j754)\mathbf{I_A} - j226\mathbf{I_B} = 2500 \angle 0°$

Mesh B: $-j226\mathbf{I_A} + (25 + j15 + j75.4)\mathbf{I_B} = 0$

The result is two linear equations in the two unknown mesh currents. Routine analysis yields

$$\mathbf{I_A} = 4.393 - j7.499 = 8.691 \angle -59.64° \text{ A}$$
$$\mathbf{I_B} = 15.0 - j14.606 = 20.959 \angle -44.18° \text{ A}$$

Given the mesh currents, the winding voltages are found to be

$$\mathbf{V_1} = \mathbf{V_S} - Z_S\mathbf{I_A} = 2350 - j87.9 = 2352 \angle -2.14° \text{ V}$$
$$\mathbf{V_2} = Z_L\mathbf{I_B} = 594.9 - j139.7 = 611 \angle -13.22° \text{ V}$$

The input impedance seen by the source circuit is

$$Z_{IN} = \frac{\mathbf{V_1}}{\mathbf{I_A}} = R_{IN} + jX_{IN} = 145.4 + j228.2 \ \Omega$$

The average power delivered by the source at the input interface is

$$P_{IN} = \frac{1}{2}|\mathbf{I_A}|^2 R_{IN} = \frac{(8.691)^2}{2} 145.4 = 5.491 \text{ kW} \qquad ∎$$

Exercise 15–7

Using the values of the mesh currents found in Example 15–8, find the average power delivered to the load Z_L.

Answers: 5.491 kW

The method used in the preceding example illustrates a general approach to the analysis of transformer circuits. The steps in the method are as follows:

STEP 1   Write KVL equations around the primary and secondary circuits using assigned mesh currents, source voltages, and inductor voltages.

STEP 2   Write the **I–V** characteristics of the coupled inductors in terms of the mesh currents using the dot convention to determine whether the coupling is additive or subtractive.

STEP 3   Use the **I–V** relationships from step 2 to eliminate the inductor voltages from the KVL equations obtained in step 1 to obtain mesh-current equations.

The next two examples illustrate this method of formulating mesh equations.

EXAMPLE 15–9

Find $\mathbf{I_A}, \mathbf{I_B}, \mathbf{V_1}, \mathbf{V_2}$, and the impedance seen by the voltage source in Figure 15–14.

FIGURE 15–14

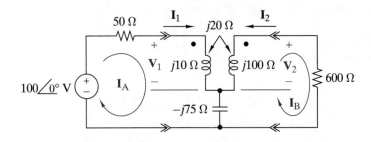

SOLUTION:

STEP 1 The KVL equations around meshes A and B are

$$\text{Mesh A:}\quad (50 - j75)\mathbf{I}_A - (-j75)\mathbf{I}_B + \mathbf{V}_1 = 100\angle 0°$$
$$\text{Mesh B:}\quad -(-j75)\mathbf{I}_A + (600 - j75)\mathbf{I}_B - \mathbf{V}_2 = 0$$

STEP 2 For the assigned reference directions, the coupling is additive. By KCL we have $\mathbf{I}_1 = \mathbf{I}_A$ and $\mathbf{I}_2 = -\mathbf{I}_B$. Hence, the element equations for the coupled inductors in terms of the mesh currents are

$$\mathbf{V}_1 = j10\mathbf{I}_A + j20(-\mathbf{I}_B)$$
$$\mathbf{V}_2 = j20\mathbf{I}_A + j100(-\mathbf{I}_B)$$

STEP 3 Using these equations to eliminate the inductor voltages from the KVL equations found in step 1 yields the following mesh equations:

$$\text{Mesh A:}\quad (50 - j75 + j10)\mathbf{I}_A + (j75 - j20)\mathbf{I}_B \quad = 100\angle 0°$$
$$\text{Mesh B:}\quad (j75 - j20)\mathbf{I}_A + (600 - j75 + j100)\mathbf{I}_B = 0$$

Solving these equations for the mesh currents produces

$$\mathbf{I}_A = 0.756 + j0.896 = 1.17\angle 49.8° \text{ A}$$
$$\mathbf{I}_B = 0.0791 - j0.0726 = 0.107\angle -42.5° \text{ A}$$

Given the mesh currents, we find the inductor voltages from the **I–V** relations:

$$\mathbf{V}_1 = j10\mathbf{I}_A - j20\mathbf{I}_B = -10.4 + j5.98 = 12.0\angle 150° \text{ V}$$
$$\mathbf{V}_2 = j20\mathbf{I}_A - j100\mathbf{I}_B = -25.2 + j7.21 = 26.2\angle 164° \text{ V}$$

Finally, the impedance seen by the input voltage source is

$$Z_{IN} = \frac{\mathbf{V}_S}{\mathbf{I}_A} = 55.0 - j65.2 \ \Omega$$ ∎

EXAMPLE 15–10

Figure 15–15 shows an ideal transformer connected as an autotransformer. Find the voltage and average power delivered to the load Z_L for $\mathbf{V}_S = 500\angle 0°$ V, $Z_S = j10\ \Omega, Z_L = 50 + j0\ \Omega, N_1 = 200$, and $N_2 = 280$.

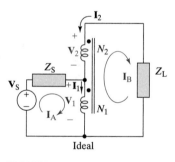

FIGURE 15–15

SOLUTION:

The three-step method of writing mesh equations can be used here with a modification to the second step:

STEP 1 The KVL equations around meshes A and B are

$$\text{Mesh A:}\quad Z_S\mathbf{I}_A + \mathbf{V}_1 = \mathbf{V}_S$$
$$\text{Mesh B:}\quad -\mathbf{V}_1 - \mathbf{V}_2 + Z_L\mathbf{I}_B = 0$$

STEP 2 For the assigned reference directions, the coupling is additive so the ideal transformer voltages and currents are related as

$$\mathbf{V}_2 = n\mathbf{V}_1 \quad\text{and}\quad \mathbf{I}_1 = -n\mathbf{I}_2$$

By KCL we have $\mathbf{I}_1 = \mathbf{I}_A - \mathbf{I}_B$ and $\mathbf{I}_2 = -\mathbf{I}_B$. Hence, the ideal transformer constrains the two mesh currents as

$$\mathbf{I}_A = (n + 1)\mathbf{I}_B$$

STEP 3 Using these results to eliminate $\mathbf{I}_A$ and $\mathbf{V}_2$ from the KVL equation in step 1 yields

$$\text{Mesh A:}\quad (n + 1)Z_S\mathbf{I}_B + \mathbf{V}_1 = \mathbf{V}_S$$
$$\text{Mesh B:}\quad -(n + 1)\mathbf{V}_1 + Z_L\mathbf{I}_B = 0$$

Notice that we cannot eliminate both $\mathbf{V}_1$ and $\mathbf{V}_2$ from the KVL equations since voltage and current are independent in an ideal transformer. Substituting the numerical values produces

$$\text{Mesh A:} \qquad j24\mathbf{I}_B + \mathbf{V}_1 = 500 \angle 0°$$
$$\text{Mesh B:} \quad -2.4\mathbf{V}_1 + 50\mathbf{I}_B = 0$$

Solving these equations for $\mathbf{I}_B$ and $\mathbf{V}_1$ yields

$$\mathbf{I}_B = 10.3 - j11.9 = 15.7 \angle{-49.0°} \text{ A}$$
$$\mathbf{V}_1 = 215 - j247 = 328 \angle{-49.0°} \text{ A}$$

Given the $\mathbf{I}_B$, we find the output quantities as

$$\mathbf{V}_L = Z_L\mathbf{I}_B = 516 - j594 = 787 \angle{-49.0°} \text{ A}$$

$$P_L = \frac{|\mathbf{I}_B|^2}{2}R_L = \frac{15.7^2}{2}50 = 6.16 \text{ kW} \qquad ■$$

Exercise 15–8

Using the results in Example 15–10, find the input impedance seen by the source in Figure 15–15.

Answer: $8.66 + j10.0 \, \Omega$

EXAMPLE 15–11

The linear transformer in Figure 15–16 is in the sinusoidal steady-state with reactances of $X_1 = 25 \, \Omega$, $X_2 = 16 \, \Omega$, $X_M = 18 \, \Omega$, and a load impedance of $Z_L = 25 - j10 \, \Omega$. Find $\mathbf{V}_2$ and $\mathbf{I}_2$ when $\mathbf{V}_S = 100\angle 0° \text{ V}$.

FIGURE 15–16

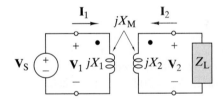

SOLUTION:
This problem can be solved without the formal use of standard mesh equations.

STEP 1   The coupling is additive so the i–v equations for the transformer are

$$\mathbf{V}_1 = jX_1\mathbf{I}_1 + jX_M\mathbf{I}_2 = j25\mathbf{I}_1 + j18\mathbf{I}_2$$
$$\mathbf{V}_2 = jX_M\mathbf{I}_1 + jX_2\mathbf{I}_2 = j18\mathbf{I}_1 + j16\mathbf{I}_2$$

STEP 2   By KVL the two coil voltages are

$$\mathbf{V}_1 = \mathbf{V}_S = 100$$
$$\mathbf{V}_2 = \mathbf{V}_L = -Z_L\mathbf{I}_2 = (-25 + j10)\mathbf{I}_2$$

STEP 3   Substituting these expressions for $\mathbf{V}_1$ and $\mathbf{V}_2$ into the i–v equations in STEP 1 gives

$$100 = j25\mathbf{I}_1 + j18\mathbf{I}_2$$
$$(-25 + j10)\mathbf{I}_2 = j18\mathbf{I}_1 + j16\mathbf{I}_2$$

which can be rearranged as

$$100 = j25\mathbf{I}_1 + j18\mathbf{I}_2$$
$$0 = j18\mathbf{I}_1 + (25 + j6)\mathbf{I}_2$$

Solving these two equations in two unknowns yields

$$\mathbf{I}_1 = 1.92 - j3.46 \text{ A}$$
$$\mathbf{I}_2 = -2.67 - j0.744 \text{ A}$$

and finally

$$\mathbf{V}_2 = -Z_L\mathbf{I}_2 = 74.3 - j8.13 \text{ V}$$

Exercise 15-9

Using the results found in Example 15–11, find the input impedance seen by the source in Figure 15–16.

Answer: $12.26 + j22.10 \ \Omega$

APPLICATION   EXAMPLE 15–12

A transformer transfers signals or power from one circuit to another strictly by magnetic induction without conductive paths between the circuits. Such circuits are said to be *electrically isolated* since there is no electric current flowing directly from one to the other. Any transformer provides isolation, as well as the common functions of changing voltage levels or matching impedances. Transformers that have isolation as their primary purpose are called **isolation transformers.**

One such purpose is the elimination of what are called *ground loops*. In a circuit diagram all ground symbols are assumed to be at the same potential because of an implied zero-resistance connection between them. In reality this may not be the case and Figure 15–17 illustrates how this results in a ground loop. A ground loop is a closed conductive path formed in part by an intentional connection between the two circuits and in part by an unintentional (and uncontrolled) connection formed when two "grounds" are not at the same potential.

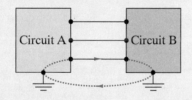

FIGURE 15–17   *A Ground Loop*

For example, if circuit B in the figure is a metal ship moored in port and supplied with ac power from an onshore source in Circuit A. The ship's ground (its metal structure) may not be at the same potential as the "earth" ground of the onshore source. This forms a ground loop and allows an errant return current to flow through an uncontrolled path. This current causes undesirable galvanic corrosion when it passes through the interface between the water and the ship's metal hull. One solution is to introduce isolation transformers in the intentional ship-to-shore connections thus electrically isolating the ship and eliminating the conductive paths needed to form ground loops.

Isolation transformers come in many sizes for different applications. They are often one-to-one ($n = 1$) transformers since their purpose is electrical isolation rather than changing voltage levels. For example, Figure 15–18(a) shows a schematic of a 16-pin dual inline package containing three tiny one-to-one isolation transformers and Figure 15–18(b) shows a photo of an actual device. Such a package finds application in high-speed digital circuits in part because it is compatible with PC board manufacturing methods.

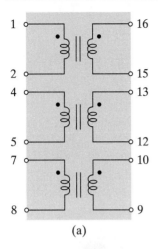

(a)

www.ctparts.com

(b)

FIGURE 15–18   *16-Pin Isolation Transformer Package (a) Pin layout, (b) Photo of actual device.*

The typical mast antenna in Figure 15–19 offers an interesting application of an isolation transformer. The mast is made of conductive metal and is mounted on a non-conductive base that isolates it from earth ground. The mast is driven by a radio frequency (RF) source that may raise its metal structure to several hundred kV above ground potential. Because of its height, a mast antenna is equipped with attached air-traffic obstacle warning lamps that are powered by ordinary 60-Hz ac power. The wires carrying ac power to the lamps are of necessity exposed to the antenna's radiated fields. The RF currents induced in these wires naturally seek a conductive path to earth ground in the lamp power source. To avoid this diversion of RF radiated power, an isolation transformer is inserted between the lamp power source and the circuit bringing power to the lamps. This type of isolation transformer requires special design features and is usually called an Austin transformer, so named for an early patent holder, Arthur O. Austin.

FIGURE 15–19   *A mast antenna.*

SUMMARY

- Mutual inductance describes magnetic coupling between two inductors. The mutual inductance parameter relates the voltage induced in one inductor to the rate of change of current in the other inductor. The induced voltage can be either positive (additive coupling) or negative (subtractive coupling).

- The dot convention describes the physical orientation of the two magnetically coupled inductors. Mutual inductance coupling is additive when both current reference arrows point toward or away from dotted terminals; otherwise, it is subtractive. The reference directions for the currents can always be selected so that the coupling is additive.

- The degree of coupling is indicated by the coupling coefficient. Energy analysis shows that the coupling coefficient must lie between zero and one. A coupling coefficient of unity is called perfect coupling and means that all of the flux produced by first winding links with a second winding, and vice versa. A coupling coefficient of zero indicates no magnetic linkage between the windings.

- A transformer is an electrical device based on the mutual inductance coupling between two windings. Transformers find applications in almost all electrical systems, especially in power supplies and the electrical power grid. The transformer winding connected to the power source is called the primary winding, and the winding connected to the load is called the secondary winding.

- The ideal transformer is a circuit element in which the primary and secondary windings are assumed to be perfectly coupled and to have no power loss. In an ideal transformer the voltages and currents in the primary and secondary windings are related by the turns ratio, which is the ratio of the number of turns in the secondary winding to that in the primary winding.

- In the sinusoidal steady state, transformers can be analyzed using phasors to determine steady-state currents, voltages, impedances, and average power transfer. The phasor analysis of transformers is carried out using a modified mesh-current approach.

PROBLEMS

OBJECTIVE 15–1 MUTUAL INDUCTANCE (SECT. 15–1, 15–2, 15–3)

Given the current through or voltage across two coupled inductors, find other currents or voltages.
See Examples 15–1, 15–2, 15–3 and Exercises 15–1, 15–2, 15–3.

15–1 In Figure P15–1 $L_1 = 120\,\text{mH}$, $L_2 = 400\,\text{mH}$, $M = 180\,\text{mH}$, and $v_2(t) = 60 \sin 500t$ V.
(a) Write the i–v relationships for the coupled inductors using the reference marks in the figure.
(b) Find the source voltage $v_S(t)$ when the output terminals are open-circuited ($i_2(t) = 0$).

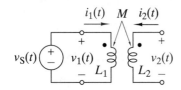

FIGURE P15–1

15–2 In Figure P15–1 $L_1 = 10\,\text{mH}$, $L_2 = 5\,\text{mH}$, $M = 7\,\text{mH}$, and $v_S(t) = 100 \sin 1000t$ V.
(a) Write the i–v relationships for the coupled inductors using the reference marks in the figure.

(b) Solve for $i_2(t)$ when the output terminals are short-circuited ($v_2 = 0$). Assume that $i_2(t)$ has no dc component.

15–3 In Figure P15–1 $L_1 = 10\,\text{mH}$, $L_2 = 10\,\text{mH}$, $M = 7\,\text{mH}$, $i_1(t) = 2 \sin 1000t$ A, and $i_2(t) = -2 \sin 1000t$ A.
(a) Write the i–v relationships for the coupled inductors using the reference marks given.
(b) Find the input voltage $v_1(t)$ and the output voltage $v_2(t)$.

15–4 In Figure P15–4 $L_1 = 30\,\text{mH}$, $L_2 = 25\,\text{mH}$, and $M = 25\,\text{mH}$. The output current is observed to be $i_2(t) = 5 \sin 100t$ A when the output terminals are short-circuited ($v_2(t) = 0$). Find $v_1(t)$ and $i_S(t)$, assuming that $i_S(t)$ has no dc component.

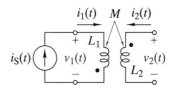

FIGURE P15–4

15–5 In Figure P15–4 $L_1 = 25\,\text{mH}$, $L_2 = 30\,\text{mH}$, and $M = 25\,\text{mH}$. The output voltage is observed to be $v_2(t) = 50 \cos 1000t$ V when the output terminals are open-circuited ($i_2(t) = 0$). Find $v_1(t)$.

15–6 A pair of coupled inductors have $L_1 = 3.6$ H, $L_2 = 2.5$ H, and $k = 1.0$. When the output terminals are open-circuited ($i_2(t) = 0$), the output voltage is observed to be $v_2(t) = 30 \sin 1000t$ V. Find the input voltage $v_1(t)$ for additive coupling.

15–7 In Figure P15–7 $L_1 = 6$ H, $L_2 = 3$ H, $M = 4$ H, and $i_1(t) = 5 \sin (1000t)$ mA. Find the input voltage $v_X(t)$.

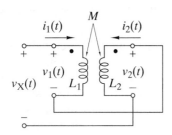

FIGURE P15–7

15–8 In Figure P15–8 show that $L_{EQ} = L_1(1 - k^2)$, where k is the coupling coefficient.

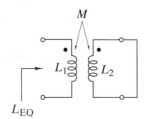

FIGURE P15–8

15–9 In Figure P15–9 show that the indicated open-circuit voltage is $v_{OC}(t) = \left(k\sqrt{L_2/L_1} \right) v_1(t)$, where k is the coupling coefficient.

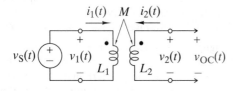

FIGURE P15–9

15–10 In Figure P15–10 show that the indicated short-circuit current is $i_{SC}(t) = \left(k\sqrt{L_1/L_2} \right) i_1(t)$, where k is the coupling coefficient. Assume that $i_1(t)$ has no dc component.

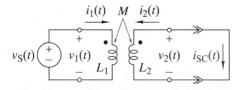

FIGURE P15–10

OBJECTIVE 15–2 THE IDEAL TRANSFORMER (SECT. 15–4)

Given a circuit containing ideal transformers:
(a) Find specified voltages, currents, powers, and equivalent circuits.
(b) Select the turns ratio to meet prescribed conditions.
See Examples 15–4, 15–5, 15–6, 15–7 and Exercises 15–4, 15–5, 15–6.

15–11 The primary voltage of an ideal transformer is a 120-V, 60-Hz sinusoid. The secondary voltage is a 24-V, 60-Hz sinusoid. The secondary winding is connected to an 800-Ω resistive load.
(a) Find the transformer turns ratio.
(b) Write expressions for the primary current and voltage.

15–12 The number of turns in the primary and secondary of an ideal transformer are $N_1 = 50$ and $N_2 = 400$. The primary winding is connected to a 120-V, 60 Hz source with a source resistance of 50 Ω. The secondary winding is connected to a 1600-Ω load. Find the primary and secondary currents.

15–13 The turns ratio of the second ideal transformer in Figure P15–13 is $n = 5$. Find the equivalent resistance indicated in the figure.

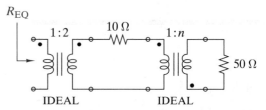

FIGURE P15–13

15–14 The equivalent resistance in Figure P15–13 is $R_{EQ} = 15\Omega$. Find the turns ratio of the second ideal transformer.

15–15 Figure P15–15 shows an ideal transformer connected as an autotransformer. Find $i_L(t)$ and $i_S(t)$ when $v_S(t) = 120 \sin 400t$ and $R_L = 60$ Ω.

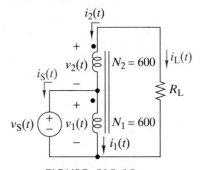

FIGURE P15–15

15–16 The primary winding of an ideal transformer with $n = 1/5$ is connected to a voltage source with a source resistance of 2 kΩ. Find the load resistance connected across the secondary winding that will draw maximum power from the source.

15–17 🄳 Select the turns ratio n of an ideal transformer in the interface circuit shown in Figure P15–17 so that the input resistance R_{IN} seen by the voltage source is 150 Ω.

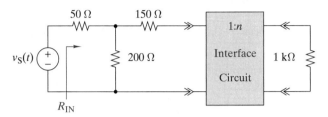

FIGURE P15–17

15–18 A voltage source with Thévenin parameters $v_T(t) = 5 \sin 1000t$ V and $R_T = 25$ Ω drives the primary winding of an ideal transformer with $n = 2$. Write an expression for the instantaneous power delivered to a 300-Ω load connected across the secondary winding.

15–19 Find $i_S(t)$ in Figure P15–19 when $v_S(t) = 15 \sin 377t$ V.

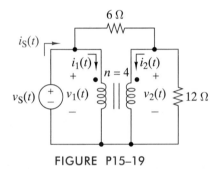

FIGURE P15–19

15–20 Show that the equivalent resistance in Figure P15–20 is

$$R_{EQ} = \left(\frac{N_1 + N_2}{N_2}\right)^2 R_L$$

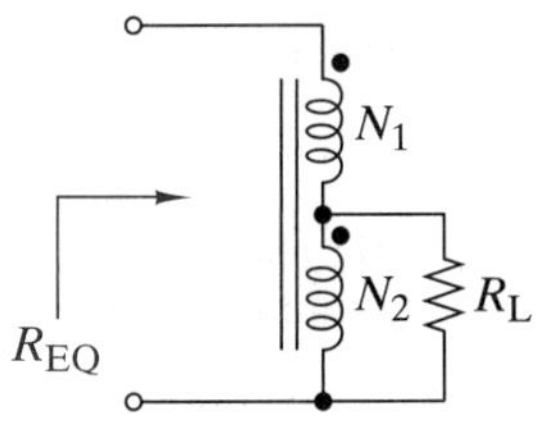

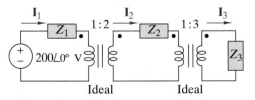

FIGURE P15–20

15–21 In Figure P15–21 the impedances are $Z_1 = 25 -j45$ Ω, $Z_2 = 45 + j30$ Ω, and $Z_3 = 360 + j270$ Ω. Find $\mathbf{I}_1$, $\mathbf{I}_2$, and $\mathbf{I}_3$.

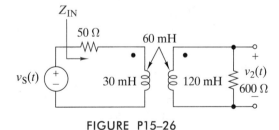

FIGURE P15–21

15–22 In Figure P15–21 the impedances are $Z_1 = 35 + j20$ Ω, $Z_2 = 70 + j20$ Ω, and $Z_3 = 270 - j90$ Ω. Find the average power delivered to Z_3.

15–23 An ideal transformer has a turns ratio of $n = 10$. The secondary winding is connected to a load $Z_L = 600 + j100$ Ω. The primary is connected to a voltage source with a peak amplitude of 400 V and an internal impedance of $Z_S = j7$ Ω. Find the average power delivered to the load.

15–24 The primary winding of an ideal transformer with $N_1 = 100$ and $N_2 = 250$ is connected to a 480-V source. A load impedance of $Z_L = 150 + j200$ Ω is connected across the secondary windings. Find the amplitudes of the primary and secondary currents.

15–25 A transformer that can be treated as ideal has 480 turns in the primary winding and 240 turns in the secondary winding. The primary is connected to a 60-Hz source with peak amplitude of 440 V. The secondary winding delivers an average power of 5 kW to a resistive load. Find the amplitudes of the primary and secondary currents and the transformer input resistance.

OBJECTIVE 15–3 LINEAR TRANSFORMERS (SECT. 15–5)

Given a linear transformer with a transformer operating in the sinusoidal steady-state, find phasor voltages and currents, average powers, and equivalent impedances.
See Examples 15–8, 15–9, 15–10, 15–11, 15–12 and Exercises 15–7, 15–8, 15–9

15–26 The input voltage to the transformer in Figure P15–26 is a sinusoid $v_S(t) = 60 \cos 2500t$ V. With the circuit operating in the sinusoidal steady-state, use mesh-current analysis to find the phasor output voltage $\mathbf{V}_2$ and the input impedance Z_{IN}.

FIGURE P15–26

15–27 Repeat Problem 15–26 with $v_S(t) = 60 \cos 500t$ V.

15–28 A transformer has self-inductances $L_1 = 200$ mH, $L_2 = 200$ mH, and a coupling coefficient of $k = 0.99$. The transformer is operating in the sinusoidal steady state with $\omega = 500$ rad/s with a short-circuit connected across the secondary winding. Find the transformer input impedance. Assume additive coupling.

15–29 A transformer operating in the sinusoidal steady state with $\omega = 377$ rad/s has self inductances $L_1 = 200$ mH, $L_2 = 400$ mH, and $k = 0.95$. The load connected across the secondary winding is $Z_L = 75 + j150$ Ω. Find the transformer input impedance. Assume additive coupling.

15–30 The circuit in Figure P15–30 is in the sinusoidal steady-state with $\mathbf{V}_S = 500\,\text{V}$ and $Z_L = 16 + j12\,\Omega$. Use mesh-current analysis to find the amplitude of the output voltage, the input impedance seen by the source, and the average power delivered to Z_L.

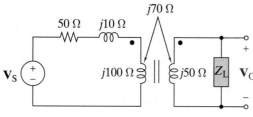

FIGURE P15–30

15–31 Repeat Problem 15–30 with $Z_L = 16 - j12\,\Omega$.

15–32 Find the phasor current $\mathbf{I}$ and impedance Z_{IN} in the circuit in Figure P15–32.

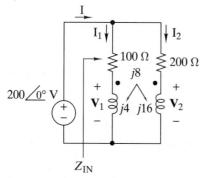

FIGURE P15–32

15–33 The circuit in Figure P15–33 is in the sinusoidal steady state with $\mathbf{V}_S = 100\angle 0°\,\text{V}$ and $R_L = 45\,\Omega$. Use mesh-current analysis to find $\mathbf{V}_O$ and the average power delivered to R_L.

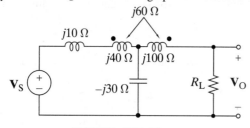

FIGURE P15–33

15–34 In Figure P15–34 find $\mathbf{I}_A$, $\mathbf{I}_B$, and the input impedance seen by the voltage source.

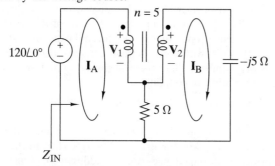

FIGURE P15–34

15–35 The ideal transformer in Figure P15–35 is connected as an auto transformer. Find $\mathbf{I}_S$ and the average power delivered to R_L when $N_1 = N_2 = 500$ turns, $R_L = 60\,\Omega$, and $\mathbf{V}_S = 120\angle 0°\,\text{V}$

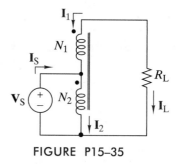

FIGURE P15–35

15–36 The transformer in Figure P15–36 is operating in the ac steady-state with a voltage source connected at the input and the output shorted. Show that the input impedance is $Z_{IN} = jX_1(1 - k^2)$, where k is the coupling coefficient. Hint: $k = X_M/\sqrt{X_1 X_2}$.

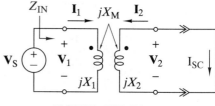

FIGURE P15–36

15–37 The transformer in Figure P15–36 is operating in the ac steady-state with a voltage source connected at the input and the output shorted. Show that the short-circuit current is

$$\mathbf{I}_{SC} = \left(\frac{k^2}{1-k^2}\right)\frac{\mathbf{V}_S}{jX_M}$$

Hint: use the result derived in Problem 15–36 to find $\mathbf{I}_1$.

15–38 The linear transformer in Figure P15–38 is in the sinusoidal steady-state with reactances of $X_1 = 32\,\Omega$, $X_2 = 48\,\Omega$, $X_M = 36\,\Omega$, and a load impedance of $Z_L = 150 - j75\,\Omega$. Find the input impedance seen by the input voltage source.

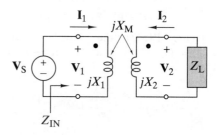

FIGURE P15–38

15–39 The linear transformer in Figure P15–38 is in the sinusoidal steady-state with reactances of $X_1 = 15\,\Omega$, $X_2 = 60\,\Omega$, $X_M = 27\,\Omega$. Find the transformer secondary response $\mathbf{V}_2$ and $\mathbf{I}_2$ when $Z_L = 200 - j100\,\Omega$ and $\mathbf{V}_S = 200\angle 0°\,\text{V}$.

15–40 The self and mutual inductances of a transformer can be calculated from measurements of the steady-state ac voltages and currents with the secondary winding open-circuited and short-circuited. Suppose the measurements are $|\mathbf{V}_1| = 120\,\text{V}$, $|\mathbf{I}_1| = 120\,\text{mA}$, and $|\mathbf{V}_2| = 240\,\text{V}$ when the secondary is open. When the secondary is shorted, the measurements are $|\mathbf{I}_1| = 10\,\text{A}$ and $|\mathbf{I}_2| = 2.2\,\text{A}$. All measurements were made at $f = 400\,\text{Hz}$. Find L_1, L_2, and M.

INTEGRATING PROBLEMS

15–41 Transformer Sawtooth Response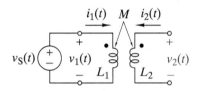
The linear transformer in Figure P15–41 has inductances of $L_1 = 25\,\text{mH}$, $L_2 = 100\,\text{mH}$, and $M = 50\,\text{mH}$. The input voltage $v_S(t)$ has a sawtooth waveform with a peak amplitude of 5 V and a period of 50 μs. Derive an expression for the first three terms of the Fourier series of the open-circuit ($i_2(t) = 0$) output voltage $v_2(t)$.

FIGURE P15–41

15–42 Transformer Thévenin Equivalent
In the time-domain, the i–v relationships for a linear transformer are

$$v_1(t) = L_1 \frac{di_1(t)}{dt} + M \frac{di_2(t)}{dt}$$

$$v_2(t) = M \frac{di_1(t)}{dt} + L_2 \frac{di_2(t)}{dt}$$

Assuming zero initial conditions, transform these equations into the s-domain and show that the s-domain parameters of the Thévenin equivalent at the output are

$$Z_T(s) = \left(1 - k^2\right) L_2 s \quad \text{and} \quad V_T(s) = k\sqrt{\frac{L_2}{L_1}} V_1(s)$$

where k is the coupling coefficient.

15–43 Perfectly Coupled Transformer
Figure P15–43 is an equivalent circuit of a perfectly coupled transformer. This model is the basis for the transformer equivalent circuits used in the analysis of power systems. The inductance L_m is called the magnetizing inductance. The current through this inductance represents the current needed to establish the magnetic field in the transformer at no load ($i_2(t) = 0$). Show that the i–v equations for this circuit are:

$$v_1(t) = L_m \frac{di_1(t)}{dt} + nL_m \frac{di_2(t)}{dt}$$

$$v_2(t) = nL_m \frac{di_1(t)}{dt} + n^2 L_m \frac{di_2(t)}{dt}$$

Use these equations to show that $k = 1$ and $v_2(t)/v_1(t) = n$.

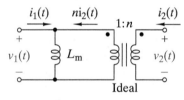

FIGURE P15–43

CHAPTER 16 AC POWER SYSTEMS

George Westinghouse was in my opinion, the only man on the globe who could take my alternating current (power) system under the circumstances then existing and win the battle against prejudice and money power.

Nikola Tesla, 1932
American Engineer

Some History Behind This Chapter

The 1890s saw a competition between the dc power system developed by Thomas Edison and the newly emerging ac system. Initially the main drawback of the ac approach was a lack of practical motors. The Serbian immigrant Nikola Tesla (1856–1943) invented the three-phase ac induction motor that met this need. In 1887, Tesla founded a company to develop his inventions, eventually producing some 40 patents on three-phase equipment. George Westinghouse recognized the importance of this work and purchased the rights to Tesla's patents. The ac-versus dc-competition was settled when an ac system was chosen for a large hydroelectric power station at Niagara Falls, New York.

Why This Chapter Is Important Today

Rolling blackouts and collapsing power grids remind us that the reliable flow of electrical power is essential in a modern society. Although you may never work directly in the electrical power field, some understanding of its concepts and limitations is important in many adjacent areas of technology. Among these concepts are complex power, single-phase and three-phase circuits, and power flow. Have you ever wondered why the electric transmission lines marching across the countryside have three wires each suspended from a large insulator? You will find the answer here.

Chapter Sections

Chapter Learning Objectives

16-1 Complex Power (Sects. 16–1, 16–2)

Given a linear circuit in the sinusoidal steady state:
(a) Find the average, reactive, and instantaneous power for a specified voltage and current.
(b) Find the load impedance for a specified load power flow.

16-2 Single-Phase Circuit Analysis (Sect. 16–3)

Given a single-phase circuit operating in the ac steady state, find the power produced by sources or delivered to specified loads.

16-3 Single-Phase Power Flow (Sect. 16–4)

Given a specified load power in a single-phase circuit:
(a) Find the required source outputs.
(b) Find the parallel capacitance needed to produce a specified power factor.

16-4 Balanced Three-Phase Circuits (Sect. 16–5)

For a balanced three-phase circuit:
(a) Find all of the phase and line voltage phasors for a given phase reference.
(b) Find equivalent Y- or -connected sources and loads.

16-5 Three-Phase Circuit Analysis (Sect. 16–6)

For a given balanced three-phase circuit:
(a) Find the line and phase current phasors for a specified phase reference.
(b) Find the source or load power using the scalars V_L and I_L.

16-6 Three-Phase Power Flow (Sect. 16–7)

Given a single-line diagram of a balanced three-phase system, find the source outputs and bus voltages that produce a prescribed load power flow.

16–1 AVERAGE AND REACTIVE POWER

We begin our study of electric power circuits with the two-terminal interface in Figure 16–1. In power applications we normally think of one circuit as the source and the other as the load. Our objective is to describe the flow of power across the interface when the circuit is operating in the sinusoidal steady state. To this end, we write the interface voltage and current in the time domain as sinusoids of the form

$$v(t) = V_A \cos(\omega t + \theta) \text{ V}$$
$$i(t) = I_A \cos \omega t \text{ A}$$

(16–1)

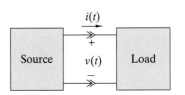

FIGURE 16–1 *A two-terminal interface.*

In Eq. (16–1) V_A and I_A are real, positive numbers representing the peak amplitudes of the voltage and current, respectively.

In Eq. (16–1) we have selected the $t = 0$ reference at the positive maximum of the current $i(t)$ and assigned a phase angle to $v(t)$ to account for the fact that the voltage maximum may not occur at the same time. In the phasor domain the angle $\theta = \phi_V - \phi_I$ is the angle between the phasors $\mathbf{V} = V_A \angle \phi_V$ and $\mathbf{I} = I_A \angle \phi_I$. In effect, choosing $t = 0$ at the current maximum shifts the phase reference by an amount $-\phi_I$ so that the voltage and current phasors become $\mathbf{V} = V_A \angle \theta$ and $\mathbf{I} = I_A \angle 0$.

A method of relating power to phasor voltage and current will be presented in the next section, but at the moment we write the instantaneous power in the time domain.

$$p(t) = v(t) \times i(t)$$
$$= V_A I_A \cos(\omega t + \theta) \cos \omega t$$

(16–2)

This expression for instantaneous power contains dc and ac components. To separate the components, we first use the identity $\cos(x + y) = \cos x \cos y - \sin x \sin y$ to write $p(t)$ in the form

$$p(t) = V_A I_A [\cos \omega t \cos \theta - \sin \omega t \sin \theta] \cos \omega t$$
$$= [V_A I_A \cos \theta]\cos^2 \omega t - [V_A I_A \sin \theta] \cos \omega t \sin \omega t$$

(16–3)

Using the identities $\cos^2 x = \frac{1}{2}(1 + \cos 2x)$ and $\cos x \sin x = \frac{1}{2} \sin 2x$, we write $p(t)$ in the form

$$p(t) = \underbrace{\left[\frac{V_A I_A}{2} \cos \theta\right]}_{\text{dc component}} + \underbrace{\left[\frac{V_A I_A}{2} \cos \theta\right] \cos 2\omega t - \left[\frac{V_A I_A}{2} \sin \theta\right] \sin 2\omega t}_{\text{ac component}}$$

(16–4)

Written this way, we see that the instantaneous power is the sum of a dc component and a double-frequency ac component.

Note that instantaneous power in Eq. (16–4) is periodic. In Chapter 5 we defined the average value of a periodic waveform as

$$P = \frac{1}{T} \int_0^T p(t)dt$$

where $T = 2\pi/2\omega$ is the period of $p(t)$. In Chapter 5 we also showed that the average value of a sinusoid is zero. Therefore, the **average value** of $p(t)$, denoted as P, is equal to the constant or dc term in Eq. (16–4):

$$P = \frac{V_A I_A}{2} \cos \theta \text{ W}$$

(16–5)

The amplitude of the $\sin 2\omega t$ term in Eq. (16–4) has a form much like the average power in Eq. (16–5), except it involves $\sin\theta$ rather than $\cos\theta$. This amplitude factor is called the **reactive power** of $p(t)$, where reactive power Q is defined as

$$Q = \frac{V_A I_A}{2}\sin\theta \text{ VAR} \tag{16–6}$$

and has the units of Volt-Amperes Reactive or VAR.

Substituting Eqs. (16–5) and (16–6) into Eq. (16–4) yields the instantaneous power in terms of the average power and reactive power:

$$p(t) = \underbrace{P(1 + \cos 2\omega t)}_{\text{unipolar}} - \underbrace{Q\sin 2\omega t}_{\text{bipolar}} \tag{16–7}$$

The energy transferred across the interface during one cycle $T = 2\pi/2\omega$ of $p(t)$ is

$$\begin{aligned}
W &= \int_0^T p(t)dt \\
&= P\underbrace{\int_0^T (1+\cos 2\omega t)dt}_{\text{net energy}} - Q\underbrace{\int_0^T \sin 2\omega t\, dt}_{\text{no net energy}} \tag{16–8} \\
&= \qquad P \times T \qquad - \qquad 0
\end{aligned}$$

The unipolar term in Eq. (16–7) provides a net energy transfer of $P \times T$ per cycle. The bipolar term is a sinusoidal power oscillation of amplitude Q that provides no net energy transfer. The average power P indicates a unidirectional flow of energy while the reactive power Q indicates bidirectional interchange of energy.

The flow of average and reactive power is the central issue in ac power systems. Although these two kinds of power have the same dimensions (volts × amperes), they represent quite different effects. For this reason, they are given different units: average power is measured in watts (W) and reactive power in Volt-Amperes Reactive (VAR).

EXAMPLE 16–1

The ac steady-state inputs to the load in Figure 16–1 are

$$v(t) = 166\cos(500t + 55°)\text{ V}$$
$$i(t) = 3.5\cos(500t - 25°)\text{ A}$$

Find the average power, reactive power, and instantaneous power carried by these waveforms.

SOLUTION:
The power factor angle for this case is

$$\theta = \phi_V - \phi_I$$
$$= 55° - (-25°) = 80°$$

The average and reactive powers are

$$P = 0.5 \times 166 \times 3.5 \times \cos(80°) = 50.4\text{ W}$$
$$Q = 0.5 \times 166 \times 3.5 \times \sin(80°) = 286\text{ VAR}$$

Using Eq. (16–7), the instantaneous power is

$$p(t) = P[1 + \cos(2\omega t)] - Q\sin(2\omega t)$$
$$= 50.4[1 + \cos(1000t)] - 286\sin(1000t)\ \text{W}$$

Since $P > 0$, a net energy of $P \times T$ per cycle is transferred into the load. ■

Exercise 16–1

Using the reference marks in Figure 16–1, calculate the average and reactive power for the following voltages and currents.

(a) $v(t) = 168\cos(377t + 45°)\ \text{V}, i(t) = 0.88\cos 377t\ \text{A}$
(b) $v(t) = 285\cos(2500t + 68°)\ \text{V}, i(t) = 0.66\cos 2500t\ \text{A}$

Answers:

(a) $P = +52.3\ \text{W}, \quad Q = +52.3\ \text{VAR}$
(b) $P = +35.2\ \text{W}, \quad Q = +87.2\ \text{VAR}$

16–2 COMPLEX POWER

It is important to relate average and reactive power to phasor quantities because steady-state analysis is conveniently carried out using phasors. In our previous work the magnitude of a phasor represented the peak amplitude of a sinusoid. However, in power circuit analysis it is convenient to express phasor magnitudes in rms (root-mean-square) values. In this chapter phasor voltages and currents are expressed as

$$\mathbf{V} = V_{\text{rms}}e^{j\phi_V} \quad \text{and} \quad \mathbf{I} = I_{\text{rms}}e^{j\phi_I} \tag{16–9}$$

Notice that the phasor magnitudes are the rms amplitude of the corresponding sinusoid.

Equations (16–5) and (16–6) express average and reactive power in terms of peak amplitudes V_A and I_A. In Chapter 5 we showed that the peak and rms values of a sinusoid are related by $V_{\text{rms}} = V_A/\sqrt{2}$. The expression for average power can be easily converted to rms amplitudes since we can write Eq. (16–5) as

$$P = \frac{V_A I_A}{2}\cos\theta = \frac{V_A}{\sqrt{2}}\frac{I_A}{\sqrt{2}}\cos\theta$$
$$= V_{\text{rms}}I_{\text{rms}}\cos\theta \tag{16–10}$$

where $\theta = \phi_V - \phi_I$ is the angle between the voltage and current phasors. By similar reasoning, Eq. (16–6) becomes

$$Q = V_{\text{rms}}I_{\text{rms}}\sin\theta \tag{16–11}$$

Using rms phasors, the **complex power** (S) at a two-terminal interface is defined as follows:

$$S = \mathbf{VI}^* \tag{16–12}$$

That is, the complex power at an interface is the product of the voltage phasor times the conjugate of the current phasor. Substituting Eq. (16–9) into this definition yields

$$S = \mathbf{VI}^* = V_{\text{rms}}e^{j\phi_V}I_{\text{rms}}e^{-j\phi_I}$$
$$= [V_{\text{rms}}I_{\text{rms}}]e^{j(\phi_V - \phi_I)} \tag{16–13}$$

Using Euler's relationship and the fact that the angle is $\theta = \phi_V - \phi_I$, we can write complex power as

$$
\begin{aligned}
S &= [V_{rms} I_{rms}] e^{j\theta} \\
&= [V_{rms} I_{rms}] \cos\theta + j[V_{rms} I_{rms}] \sin\theta \\
&= P + jQ \text{ VA}
\end{aligned}
\tag{16-14}
$$

The real part of the complex power S is the average power, and the imaginary part is the reactive power. Although S is a complex number, it is not a phasor. However, it is a convenient variable for keeping track of the two components of power when the voltage and currents are expressed as phasors. Its units are volt-amperes.

The power triangles in Figure 16-2 provide a convenient way to remember complex power relationships and terminology. We confine our study to cases in which net energy is transferred from source to load. In such cases $P > 0$ and the power triangles fall in the first or fourth quadrant, as indicated in Figure 16-2.

FIGURE 16-2 *Power triangles.*

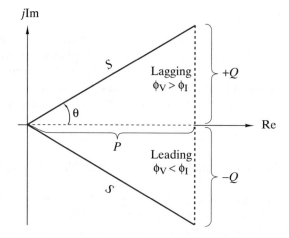

The magnitude $|S| = V_{rms} I_{rms}$ is called **apparent power** and is expressed using the unit volt-ampere (VA). The ratio of the average power to the apparent power is called the **power factor** (pf). Using Eq. (16-10), we see that the power factor is

$$
\text{pf} = \frac{P}{|S|} = \frac{V_{rms} I_{rms} \cos\theta}{V_{rms} I_{rms}} = \cos\theta
\tag{16-15}
$$

Since pf $= \cos\theta$, the angle θ is called the **power factor angle**.

When the power factor is unity, the phasors **V** and **I** are in phase ($\theta = 0°$) and the reactive power is zero since $\sin\theta = 0$. When the power factor is less than unity, the reactive power is not zero and its sign is indicated by the modifiers *lagging* or *leading*. The term *lagging power factor* means the current phasor lags the voltage phasor so that $\theta = \phi_V - \phi_I > 0$. For a lagging power factor, S falls in the first quadrant in Figure 16-2 and the reactive power is positive since $\sin\theta > 0$. The term *leading power factor* means the current phasor leads the voltage phasor so that $\theta = \phi_V - \phi_I < 0$. In this case S falls in the fourth quadrant in Figure 16-2 and the reactive power is negative since $\sin\theta < 0$. Most industrial and residential loads have lagging power factors.

EXAMPLE 16-2

Find the average power, reactive power, and apparent power for the following voltage and current phasors. Find the power factor and state whether it is lagging or leading.

$$\mathbf{V} = 350\angle 45° \text{ V(rms)}, \quad \mathbf{I} = 6\angle 65° \text{ A(rms)}$$

SOLUTION:

$$
\begin{aligned}
S = \mathbf{VI}^* &= (350\angle 45°)(6\angle -65°) \\
&= 2100\angle -20° = 2100[\cos(-20°) + j\sin(-20°)] \\
&= 1973 - j718 \text{ VA}
\end{aligned}
$$

The average, reactive, and apparent powers are $P = 1973$ W, $Q = -718$ VAR, and $|S| = 2100$ VA. The power factor is

$$\text{pf} = \frac{P}{|S|} = \frac{1973}{2100} = 0.939$$

The power factor is leading since Q is negative. ∎

Exercise 16-2

Determine the average power, reactive power, and apparent power for the following voltage and current phasors. State whether the power factor is lagging or leading.

(a) $\mathbf{V} = 208\angle -90°$ V(rms), $\mathbf{I} = 1.75\angle -75°$ A(rms)
(b) $\mathbf{V} = 277\angle +90°$ V(rms), $\mathbf{I} = 11.3\angle 0°$ A(rms)

Answers:

(a) $P = 352$ W; $Q = -94.2$ VAR; $|S| = 364$ VA; leading
(b) $P = 0$ W; $Q = +3.13$ kVAR; $|S| = 3.13$ kVA; lagging

COMPLEX POWER AND LOAD IMPEDANCE

Figure 16–3 shows the general case for a two-terminal load. For the assigned reference directions the load produces the element constraint $\mathbf{V} = Z\mathbf{I}$. Using this constraint in Eq. (16–12), we write the complex power of the load as

$$
\begin{aligned}
S = \mathbf{V} \times \mathbf{I}^* = Z\mathbf{I} \times \mathbf{I}^* &= Z|\mathbf{I}|^2 \\
&= (R + jX)I_{\text{rms}}^2
\end{aligned}
$$

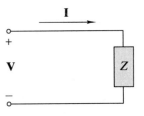

FIGURE 16–3 *A two-terminal impedance.*

where R and X are the resistance and reactance of the load, respectively. Since $S = P + jQ$, we conclude that

$$R = \frac{P}{I_{\text{rms}}^2} \quad \text{and} \quad X = \frac{Q}{I_{\text{rms}}^2} \qquad (16\text{–}16)$$

The load resistance and reactance are proportional to the average and reactive power of the load, respectively.

Equation (16–16) reveals several important properties of loads made up of passive resistors, capacitors, and inductors:

a. The load resistance requires $R \geq 0$, which means that P is nonnegative.

b. For an inductive load $X = \omega L > 0$, which means that Q is positive.

c. For a capacitive load $X = -1/\omega C < 0$, which means that Q is negative.

The terms *inductive load, positive reactive power,* and *lagging power factor* are equivalent statements, as are the terms *capacitive load, negative reactive power,* and *leading power factor*.

EXAMPLE 16–3

At 440 V (rms) a two-terminal load draws 3 kVA of apparent power at a lagging power factor of 0.9. Find

(a) I_{rms}
(b) P
(c) Q
(d) the load impedance

Draw the power triangle for the load.

SOLUTION:
(a) $I_{rms} = |S|/V_{rms} = 3000/440 = 6.82$ A (rms)
(b) $P = V_{rms}I_{rms}\cos\theta = 3000 \times 0.9 = 2.7$ kW
(c) For $\cos\theta = 0.9$ lagging, $\sin\theta = 0.436$ and $Q = V_{rms}I_{rms}\sin\theta = 1.31$ kVAR
(d) $Z = (P + jQ)/(I_{rms})^2 = (2700 + j1310)/46.5 = 58.0 + j28.2\,\Omega$.

Figure 16–4 shows the power triangle for this load.

FIGURE 16–4

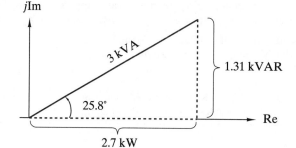

Exercise 16–3

Find the impedance of a two-terminal load under the following conditions.

(a) $\mathbf{V} = 120\angle30°$ V (rms) and $\mathbf{I} = 20\angle75°$ A (rms)
(b) $|S| = 3.3$ kVA, $Q = -1.8$ kVAR, and $I_{rms} = 7.5$ A

Answers:

(a) $Z = 4.24 - j4.24\,\Omega$
(b) $Z = 49.2 - j32\,\Omega$

16–3 SINGLE-PHASE CIRCUIT ANALYSIS

Single-phase circuit analysis deals with linear circuits in the sinusoidal steady state. The term *single-phase* means that all of the ac sources have the same phase angle. The analysis objective is to find the complex power produced by the sources or delivered to specified loads. Our analysis strategy is straightforward. We use the methods of linear ac circuit analysis from Chapter 8 to find the voltage across and current through an element. We then calculate complex power as $S = \mathbf{VI}^*$ or, in the case of a load impedance, $S = |\mathbf{I}|^2 Z$. In terms of circuit analysis concepts, the only thing new here is that the final result is complex power rather than current and voltage phasors.

We can sometimes make good use of the principle of the **conservation of complex power**, which states that

In a linear circuit the sum of the complex powers produced by all the sources is equal to the sum of the complex powers delivered to all of the passive loads.

To apply this principle we must be clear on the meaning of "produced by" and "delivered to.'' In ac power systems, we modify our practice of always using the passive sign convention. We continue to use the passive convention (current directed in at the + voltage reference) for all passive loads, but switch to the active convention (current directed out at the + voltage reference) for all sources. In either case we calculate complex power as $S = \mathbf{V}\mathbf{I}^* = P + jQ$. The net result is that the average power "produced by" a source or "delivered to" a load has the same sign.

EXAMPLE 16-4

A load consisting of a 2.5-kΩ resistor in parallel with a 2-μF capacitor is connected across a 440-V (rms), 60-Hz voltage source. Find the complex power delivered to the load and the load power factor. State whether the power factor is leading or lagging.

SOLUTION:
The source angular frequency is $\omega = 2\pi60 = 377$ rad/s. The impedance of the load is

$$Z_L = \frac{R \times (1/j\omega C)}{R + (1/j\omega C)} = \frac{2500}{2500 \times j377 \times 2 \times 10^{-6} + 1}$$
$$= 549 - j1035 \,\Omega$$

The magnitude (rms value) of the current through the load is found as

$$|\mathbf{I}| = \frac{|\mathbf{V}|}{|Z_L|} = \frac{440}{|549 - j1035|}$$
$$= 0.375 \text{ A (rms)}$$

The complex power delivered to the load and the load power factor are

$$S_L = |\mathbf{I}|^2 Z_L = P_L + jQ_L$$
$$= (0.375)^2 \times (549 - j1035)$$
$$= 77.4 - j146 \text{ VA}$$

$$\text{pf} = \frac{P_L}{|S_L|} = \frac{77.4}{|77.4 - j146|} = 0.469$$

The power factor is leading since Q_L is negative. ∎

EXAMPLE 16-5

In Figure 16–5 the two parallel loads are connected across a 15-V (rms) source.

(a) Find the complex power delivered to each load.
(b) Find the complex power produced by the source.

FIGURE 16–5

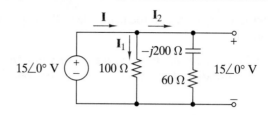

SOLUTION:

(a) The voltage across each load is $15\angle0°$ V and the load impedances are $Z_1 = 100\,\Omega$ and $Z_2 = 60 - j200\,\Omega$. The load currents shown in the figure are

$$\mathbf{I}_1 = \frac{15\angle0°}{100} = 0.15\angle0°\ \text{A (rms)}$$

$$\mathbf{I}_2 = \frac{15\angle0°}{60 - j200} = 0.0718\angle73.3°\ \text{A (rms)}$$

The complex power delivered to each load is

$$S_1 = |\mathbf{I}_1|^2 Z_1 = (0.15)^2(100 + j0) = 2.25 + j0\ \text{VA}$$

$$S_2 = |\mathbf{I}_2|^2 Z_2 = (0.0718)^2(60 - j200) = 0.309 - j1.03\ \text{VA}$$

(b) We could calculate the source power by first finding $\mathbf{I} = \mathbf{I}_1 + \mathbf{I}_2$ and then calculating $S = \mathbf{V}\mathbf{I}^*$. However, the conservation of complex power allows us to calculate the power "produced by" the source as the sum of the powers "delivered to" the loads.

$$S = S_1 + S_2 = P + jQ$$
$$= 2.56 - j1.03\ \text{VA}$$

The power factor of the source is

$$\text{pf} = \frac{P}{|S|} = \frac{2.56}{|2.56 - j1.03|} = 0.928$$

The power factor is leading since Q is negative. ◾

Exercise 16–4

A load consisting of a 50-Ω resistor in parallel with an inductor whose reactance is 75 Ω is connected across a 500-V (rms) source. Find the complex power delivered to the load and the load power factor. State whether the power factor is leading or lagging.

Answer: $S_L = 5 + j3.33\ \text{kVA}$, pf = 0.832, lagging

EXAMPLE 16–6

In Figure 16–6 the load Z_L is a 100-Ω resistor in series with a capacitor whose reactance is $-60\,\Omega$. The source voltage is 880 V (rms). Find the complex power delivered to the load and the load power factor.

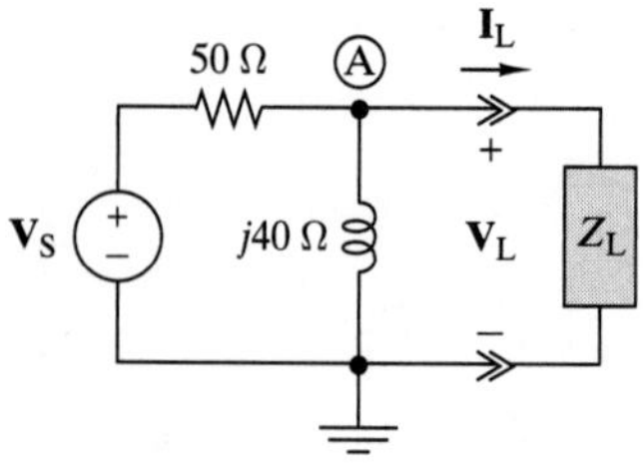

SOLUTION:

By inspection the node-voltage equation at node A is

$$\frac{\mathbf{V}_L - \mathbf{V}_S}{50} + \frac{\mathbf{V}_L}{j40} + \frac{\mathbf{V}_L}{Z_L} = 0$$

FIGURE 16–6

Solving for $\mathbf{V}_L$ with $\mathbf{V}_S = 880\angle 0°$ V and $Z_L = 100 - j60\,\Omega$ yields $\mathbf{V}_L = 411 + j309$ V (rms). The magnitude of the load current is

$$|\mathbf{I}_L| = \frac{|\mathbf{V}_L|}{|Z_L|} = \frac{|411 + j309|}{|100 - j60|} = 4.41 \text{ A (rms)}$$

The complex power delivered to the load and the load power factor are

$$S_L = |\mathbf{I}_L|^2 Z_L = 4.41^2(100 - j60)$$
$$= (1.95 - j1.17) \times 10^3 \text{ VA}$$
$$\text{pf} = \frac{P_L}{|S_L|} = \frac{1.95 \times 10^3}{|1.95 \times 10^3 - j1.17 \times 10^3|} = 0.857$$

The power factor is leading since Q_L is negative. ∎

Exercise 16–5

In Figure 16–6 the load Z_L is an 80-Ω resistor and the source voltage is 220 V (rms). Find the complex power produced by the *source*.

Answer: $S_S = 594 + j288$ VA

EVALUATION EXAMPLE 16–7

Figure 16–7 shows the residential power distribution circuit used in the United States. The circuit is called three-wire, single-phase service. The term *three-wire* refers to the three lines (A, B, and neutral) connecting the sources to the loads Z_1, Z_2, and Z_3. The term *single-phase* means that the two voltage sources $\mathbf{V}_{S1} = 110\angle 0°$ and $\mathbf{V}_{S2} = 110\angle 0°$ are in phase. The loads Z_1 and Z_2 are connected between line A or B and the neutral line. These impedances represent small-appliance and lighting loads that require 110 V (rms) service. The load Z_3 is connected between line A and line B and represents heavier loads that require 220 V (rms) service. The impedances Z_W and Z_N are line impedances that are normally much smaller than the load impedances.

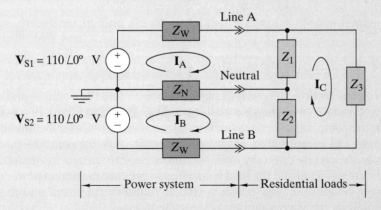

FIGURE 16–7 *Residential power distribution circuit.*

Suppose each 110-V source has a rated output capacity of 5 kVA and the three loads are $Z_1 = 30 + j12\,\Omega$, $Z_2 = 24 + j8\,\Omega$, and $Z_3 = 10 + j2\,\Omega$. In addition, the two line impedances are $Z_W = 0.05 + j0\,\Omega$ and $Z_N = 0.14 + j0\,\Omega$. The evaluation problem is to determine whether the two sources can supply these loads without exceeding their rated output. The mesh current $\mathbf{I}_A$ in Figure 16–7 is the current out of the + terminal of the upper source $\mathbf{V}_{S1}$; hence the complex power produced by this source is $S_{upper} = \mathbf{V}_{S1}(\mathbf{I_A})^*$. Similarly, mesh current $\mathbf{I}_B$ is the current out of the + terminal of the lower source $\mathbf{V}_{S2}$; hence it produces $S_{lower} = \mathbf{V}_{S2}(\mathbf{I_B})^*$. To complete the evaluation we need to find $\mathbf{I}_A$ and $\mathbf{I}_B$.

By inspection, mesh-current equations for the circuit in Figure 16–7 are

$$\text{Mesh A:} \quad (Z_W + Z_1 + Z_N)\mathbf{I}_A - Z_N\mathbf{I}_B - Z_1\mathbf{I}_C = \mathbf{V}_{S1}$$
$$\text{Mesh B:} \quad -Z_N\mathbf{I}_A + (Z_W + Z_2 + Z_N)\mathbf{I}_B - Z_2\mathbf{I}_C = \mathbf{V}_{S2}$$
$$\text{Mesh C:} \quad -Z_1\mathbf{I}_A - Z_2\mathbf{I}_B + (Z_1 + Z_2 + Z_3)\mathbf{I}_C = 0$$

Using MATLAB to solve for the mesh currents, we first assign values to the known circuit parameters.

```
Z1 = 30 +12j; Z2 = 24 +8j; Z3 = 10 +2j;
ZW = 0.05 +0j; ZN = 0.14 +0j;
VS1 = 110 +0j; VS2 = 110 +0j;
```

We formulate the problem in matrix form using the $\mathbf{Ax} = \mathbf{B}$ structure as follows:

```
A=[(ZW +Z1 +ZN) -ZN -Z1;-ZN (ZW +Z2 +ZN)-Z2;-Z1 -Z2 (Z1 +Z2 +Z3)];
B=[VS1;VS2;0];
x = A\B;
IA = x(1); IB = x(2);
```

We now have the mesh currents $\mathbf{I}_A$ and $\mathbf{I}_B$ and can calculate the required complex powers as

```
Supper = VS1*conj(IA)
Supper =
    2.6468e+003 +5.9138e+002j
Slower = VS2*conj(IB)
Slower =
    2.7506e+003 +6.0261e+002j
```

In round numbers the apparent powers produced by the two sources are

$$|S_{upper}| = |2647 + j591| = 2.71 \text{ kVA} \quad \text{and} \quad |S_{lower}| = |2751 + j603| = 2.82 \text{ kVA}$$

Both apparent power outputs are within the 5-kVA rating.

16–4 SINGLE-PHASE POWER FLOW

In the previous section, the analysis objective was to find the voltages and currents needed to calculate an *unknown* complex power delivered to a load. In a **power flow problem**, the complex power delivered to a load is specified, and the unknowns are the voltages and currents that will make this power flow happen. The power flow problem is completely different from maximum power transfer. In the latter situation, the source is fixed and the load is adjusted to produce maximum power transfer. In a power system, the load power is fixed by customer demand and the system operator adjusts the source output to meet the demand.

We illustrate power flow analysis using the **two-wire, single-phase** power system in Figure 16–8. This system consists of ac source $\mathbf{V}_S$ supplying power to a load Z_L through a two-wire transmission line with an impedance of Z_W in each wire. The

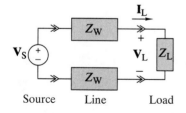

FIGURE 16–8 *A simple electrical power system.*

analysis objective is to find the source outputs required to deliver a prescribed complex power to the load.

There are several ways to specify the complex power at the load. The most obvious is simply to specify the average power P_L and reactive power Q_L. We can also specify the apparent power $|S_L|$ and the power factor pf $= \cos\theta$, in which case the complex load power is

$$
\begin{aligned}
S_L &= |S_L| \times (\cos\theta \pm j\sin\theta) \\
&= |S_L| \times (\text{pf} \pm j\sqrt{1 - \text{pf}^2})
\end{aligned}
$$

where the plus sign applies to lagging power factors and the minus sign to leading power factors. Given the average power P_L and the power factor, we calculate the apparent power as $|S_L| = P_L/\text{pf}$ and use the equation above to obtain S_L. Finally, the complex power delivered by a given current $\mathbf{I}_L$ through a load Z_L is

$$
S_L = |\mathbf{I}_L|^2 Z_L = (I_{\text{rms}})^2 Z_L
$$

EXAMPLE 16-8

The average power delivered to the load in Figure 16–9 is 20 kW at a lagging power factor of 0.8. The load voltage magnitude is 480 V (rms) and the two-wire line has an impedance of $Z_W = 0.1 + j0.6\ \Omega$ per wire. Find the required apparent power output and rms voltage of the source.

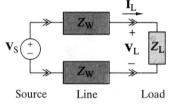

Source   Line   Load

FIGURE 16–9

SOLUTION:
For $P_L = 20 \times 10^3$ W at a power factor of pf = 0.8, the complex power delivered to the load is

$$
\begin{aligned}
S_L &= \frac{P_L}{\text{pf}} \times (\text{pf} + j\sqrt{1 - \text{pf}^2}) \\
&= 25 \times 10^3 (0.8 + j0.6) \\
&= (20 + j15) \times 10^3\ \text{VA}
\end{aligned}
$$

Given $|\mathbf{V}_L| = 480$ V, the magnitude (rms value) of the load current is

$$
\begin{aligned}
|\mathbf{I}_L| &= \frac{|S_L|}{|\mathbf{V}_L|} = \frac{25 \times 10^3}{480} \\
&= 52.1\ \text{A (rms)}
\end{aligned}
$$

The single-phase system in Figure 16–9 has a *two-wire* transmission line, so the total complex power absorbed by the line is

$$
\begin{aligned}
S_W &= 2 \times \left[|\mathbf{I}_L|^2 Z_W\right] \\
&= 2 \times (52.1)^2 \times (0.1 + j0.6) \\
&= 542 + j3260\ \text{VA}
\end{aligned}
$$

The source in Figure 16–9 supplies both S_W and S_L; hence by complex power conservation,

$$
\begin{aligned}
S_S &= S_W + S_L \\
&= (20.5 + j18.3) \times 10^3 \\
&= 27.5 \times 10^3\ \underline{/41.6^\circ}\ \text{VA}
\end{aligned}
$$

The required source apparent power is $|S_S| = 27.5\,\text{kVA}$. The rms value of the source voltage is

$$\begin{aligned}
|\mathbf{V}_S| &= \frac{|S_S|}{|\mathbf{I}_L|} = \frac{27.5 \times 10^3}{52.1} \\
&= 528\,\text{V (rms)}
\end{aligned}$$

∎

EXAMPLE 16–9

Repeat Example 16–8 when the load power factor is increased to 0.95 and all other conditions remain the same.

SOLUTION:
For $P_L = 20 \times 10^3$ W at a power factor of pf $= 0.95$, the complex power delivered to the load is

$$\begin{aligned}
S_L &= \frac{P_L}{\text{pf}} \times \left(\text{pf} + j\sqrt{1 - \text{pf}^2}\right) \\
&= 21.05 \times 10^3 (0.95 + j0.312) \\
&= (20 + j6.57) \times 10^3\,\text{VA}
\end{aligned}$$

For $|\mathbf{V}_L| = 480$ V the magnitude (rms value) of the load current is

$$\begin{aligned}
|\mathbf{I}_L| &= \frac{|S_L|}{|\mathbf{V}_L|} = \frac{21.05 \times 10^3}{480} \\
&= 43.86\,\text{A}
\end{aligned}$$

The total complex power absorbed by the two-wire line is

$$\begin{aligned}
S_W &= 2 \times \left(|\mathbf{I}_L|^2 Z_W\right) \\
&= 2 \times (43.86)^2 \times (0.1 + j0.6) \\
&= 385 + j2308\,\text{VA}
\end{aligned}$$

The power produced by the source is

$$\begin{aligned}
S_S &= S_W + S_L \\
&= (20.4 + j8.88) \times 10^3 \\
&= 22.2 \times 10^3 \, \underline{/23.54^\circ}\,\text{VA}
\end{aligned}$$

and the required source power is $|S_S| = 22.24$ kVA. The required rms source voltage is

$$\begin{aligned}
|\mathbf{V}_S| &= \frac{|S_S|}{|\mathbf{I}_L|} = \frac{22.2 \times 10^3}{43.86} \\
&= 507\,\text{V (rms)}
\end{aligned}$$

In Example 16–8 the load power factor and reactive power are pf $= 0.8$ and $Q_L = 15\,\text{kVAR}$. In this example the load power factor and reactive power are pf $=0.95$ and $Q_L = 6.57$ kVAR. The decrease in the load reactive power reduces the required source output from 27.5 kVA to 22.2 kVA. The apparent power capacity of the source is a limiting factor in a power system. The reactive power of the load affects the system-generating capacity even though it transfers no net energy to the load. ∎

Exercise 16–6

A single-phase source supplies a load through a two-wire line with an impedance of $Z_W = 2 + j10\,\Omega$ per wire. The rms load voltage is 2.4 kV and the load receives an apparent power of 25 kVA at a lagging power factor of 0.85. Find the required source power and rms voltage.

Answers: 26.6 kVA and 2.55 kV (rms)

POWER FACTOR CORRECTION

An ac power system generally operates at a lagging power factor since almost all loads are inductive. The system sources must have the kVA capacity to furnish both the average power and the positive reactive power the inductive loads demand. As we saw in Example 16–9, this reactive power flow increases the kVA demand on the source. For this reason, large industrial customers may pay a premium for the average power consumed by inductive loads with low power factors.

Power factor correction is a process that increases the power factor without changing the power flow to an inductive load. The correction is achieved by adding capacitance in parallel with the inductive load Z_L, as shown in Figure 16–10. To see how this improves the power factor, we first find the reactive power of the capacitor. The voltage across the capacitor is $\mathbf{V}_L$ since it is in parallel with the load Z_L. The capacitor current is $\mathbf{I}_C = j\omega C\mathbf{V}_L$, so the reactive power delivered to the capacitor is

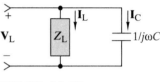

FIGURE 16–10

$$Q_C = |\mathbf{I}_C|^2 X_C = |j\omega C\mathbf{V}_L|^2\left(\frac{-1}{\omega C}\right)$$
$$= -\omega C|\mathbf{V}_L|^2$$

The reactive power Q_C is always negative while Q_L for an inductive load is always positive. The reactive power supplied to the parallel combination is $Q_L + Q_C$, which means that Q_C can cancel part (or all) of Q_L. The net decrease in reactive power means that the power factor of the parallel combination is higher than the power factor of the inductive load acting alone.

The parallel capacitor does not change the power flow to the inductive load since voltage $\mathbf{V}_L$ and current $\mathbf{I}_L$ in Figure 16–10 are unchanged. In terms of power flow, the local capacitor "supplies" part (or all) of the reactive power required by the inductive load. Physically, reactive power represents a periodic interchange of energy. With power factor correction, this interchange occurs between an inductive load and a local capacitor rather than between the inductive load and a distant power source.

EXAMPLE 16–10

The inductive load Z_L in Figure 16–10 draws an apparent power of 5 kVA at a lagging power factor of 0.75 when the load voltage is 1.2 kV (rms) at 60 Hz. Find the power factor of the parallel combination when $C = 5\,\mu F$.

SOLUTION:
For $|S_L| = 5$ kVA at a lagging power factor of 0.75, the complex power delivered to Z_L is

$$S_L = 5000 \times (0.75 + j\sqrt{1 - 0.75^2})$$
$$= 3750 + j3307 \text{ VA}$$

When $|\mathbf{V}_L| = 1200$ V (rms) at 60 Hz, the reactive power of a 5-μF capacitor is

$$Q_C = -2\pi f C|\mathbf{V}_L|^2 = -2\pi 60 \times 5 \times 10^{-6}(1200)^2$$
$$= -2714\,\text{VAR}$$

The total complex power delivered to the parallel combination is

$$S_P = S_L + jQ_C = 3750 + j3307 - j2714$$
$$= 3750 + j593 \ \text{VA}$$

and the "corrected" power factor of the parallel combination is

$$\text{pf}_{\text{cor}} = \frac{3750}{|3750 + j593|} = 0.988$$

■

▶ ◖ D E S I G N   E X A M P L E   1 6 – 1 1 ◗

The inductive load Z_L in Figure 16–10 draws an apparent power of 2 kVA at a lagging power factor of 0.8 when the rms load voltage is 880 V (rms) at 60 Hz. Find the value of the capacitance C needed to raise the power factor of the combination to 0.95 lagging.

SOLUTION:
For $|S_L| = 2$ kVA at a lagging power factor of 0.8, the complex power delivered to Z_L is

$$S_L = 2000 \times \left(0.8 + j\sqrt{1 - 0.8^2}\right)$$
$$= 1600 + j1200\,\text{VA}$$

Since $P_L = 1600$ W, the parallel combination must draw an average power of 1600 W because the capacitor does not draw any average power. To deliver 1600 W at a power factor of 0.95, the complex power delivered to the parallel combination must be

$$S_P = \frac{1600}{0.95} \times \left(0.95 + j\sqrt{1 - 0.95^2}\right)$$
$$= 1600 + j526 \ \text{VA}$$

The capacitive reactive power needed to produce this result is the difference between the reactive power in S_P and S_L. That is,

$$Q_C = \text{Im}(S_P) - \text{Im}(S_L)$$
$$= 526 - 1200 = -674\,\text{VAR}$$

Since $Q_C = -2\pi f C|\mathbf{V}_L|^2$, the required capacitance is found to be

$$C = \frac{-Q_C}{2\pi f|\mathbf{V}_L|^2}$$
$$= \frac{-(-674)}{2\pi 60(880)^2} = 2.31 \times 10^{-6}\,\text{F}$$

■

◐ Design Exercise 16–7 _____

For the load conditions in Example 16–11, find the capacitance needed to raise the power factor to unity.

A n s w e r : $C = 4.11\,\mu\text{F}$

_____

FIGURE 16–11 *Three-phase power transmission tower.*

16–5 BALANCED THREE-PHASE CIRCUITS

Electricity is transferred from generating power plants be they coal, gas, nuclear, hydro, wind, photovoltaic, etc., using high-voltage transmission lines to substations located near users of the electricity, such as, major factories, population centers, and the like. Subsequently, the electricity is transferred from the substations to the end users via a distribution network. Transmission lines typically use three-phase alternating current (ac) at high-voltages, and are carried on towers similar to that pictured in Figure 16–11. Smaller towers at lower voltages are used for the distribution network that delivers electricity to industrial users using three-phase power. Residential and small commercial consumers use split-phase 120 V/240 V power. In this section we will concentrate on three-phase power.

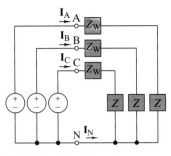

FIGURE 16–12

In a single-phase circuit, all of the voltage sources have the same phase angle. A three-phase circuit contains three single-phase circuits with each source at a different phase angle. Figure 16–12 shows a three-phase power circuit made up of three single-phase circuits. The power flows from the sources to the loads though three lines labeled A, B, and C. The phasor currents in these lines, labeled I_A, I_B, and I_C, are called line currents. There is also a neutral line labeled N shown carrying a current I_N. As we shall see, $I_N = 0$ in balanced three-phase circuits, so this line plays no role in the power flow from source to load.

BALANCED THREE-PHASE VOLTAGES

There are six voltages associated with the three-phase sources in Figure 16–12, namely three line-to-line voltages and three line-to neutral voltages. These six voltages are defined using two subscripts to identify the two points across which the voltage is measured. In this double subscript notation there is an implied + reference mark at the first subscript and an implied − at the second.[1] For example, V_{AB} is the voltage between line A and line B with an implied + at A and a − at B. Using this convention, the other two line-to-line voltages are denoted V_{BC} and V_{CA}. Collectively, these three line-to-line voltages are called simply the **line voltages**. Using the same convention, the three line-to-neutral voltages are denoted V_{AN}, V_{BN}, and V_{CN} and are collectively called the **phase voltages**.

These two sets of voltages are not independent since they are constrained by KVL. Applying KVL in Figure 16–12 around the closed path from point N to A to B and back to N yields $-V_{AN} + V_{AB} + V_{BN} = 0$ or

$$V_{AB} = V_{AN} - V_{BN} \tag{16-17a}$$

Applying the same logic to the paths N to B to C to N and N to C to A to N leads to

$$V_{BC} = V_{BN} - V_{CN}$$
$$V_{CA} = V_{CN} - V_{AN} \tag{16-17b}$$

These equations show that the line and phase voltages are not independent.

The balanced three-phase source in Figure 16–12 produces phase voltages of the form

$$V_{AN} = V_P \angle 0° \text{ V (rms)}$$
$$V_{BN} = V_P \angle -120° \text{ V (rms)} \tag{16-18}$$
$$V_{CN} = V_P \angle -240° \text{ V (rms)}$$

That is, a balanced source produces phase voltages that are (1) separated in phase by $120°$ and (2) have the same magnitude V_P, where the scalar V_P is the rms value of each phase voltage. Figure 16–13(a) is the phasor diagram of the phase voltages in Eq. (16–18). This arrangement is called the **positive phase sequence** or ABC sequence since V_{AN} leads V_{BN} by $120°$ and V_{BN} in turn leads V_{CN} by $120°$.

Figure 16–13(b) shows the ACB or **negative phase sequence** obtained by interchanging the positions of V_{BN} and V_{CN} in the ABC sequence. Physically, the negative phase (ACB) sequence merely switches the labels on lines B and C in Figure 16–12.[2] In what follows we will always use the positive (ABC) sequence,

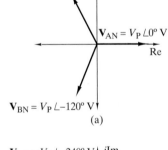

(a)

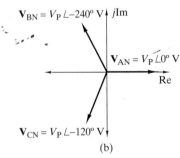

(b)

FIGURE 16–13 *Two possible phase sequences: (a) Positive. (b) Negative.*

---

[1] If we defined all six voltages using the usual + and − reference marks, a three-phase circuit diagram would be hopelessly cluttered and confusing.

[2] When connecting real three-phase equipment it is essential to maintain the identity of lines A, B, and C. For example, reversing any two leads to a three-phase motor changes the phase sequence and causes it to rotate in the opposite direction.

except in a few homework problems where the negative sequence is specifically called for.

The sum of the phase voltages in Eq. (16–18) is

$$\mathbf{V}_{AN} + \mathbf{V}_{BN} + \mathbf{V}_{CN} = V_P\angle0° + V_P\angle{-120°} + V_P\angle{-240°}$$

$$= V_P\left(1 - 0.5 - j\frac{\sqrt{3}}{2} - 0.5 + j\frac{\sqrt{3}}{2}\right)$$

$$= 0$$

This zero-sum property is also evident graphically in Figure 16–13, where we see that the sum of any two of the phasors is the exact opposite of the third. Any set of three phasors is said to be **balanced** when the phasors have equal amplitudes and 120° phase separation. The sum of any balanced set of phasors is always zero.

Inserting the phase voltages $\mathbf{V}_{AN}$ and $\mathbf{V}_{BN}$ from Eq. (16–18) into Eq. (16–17a) yields the line voltage V_{AB}.

$$\mathbf{V}_{AB} = \mathbf{V}_{AN} - \mathbf{V}_{BN} = V_P\angle0° - V_P\angle{-120°}$$

$$= V_P - V_P\left(-\frac{1}{2} - j\frac{\sqrt{3}}{2}\right)$$

$$= V_P\left(\frac{3}{2} + j\frac{\sqrt{3}}{2}\right) \tag{16–19a}$$

$$= \sqrt{3}V_P\angle30° \text{ V}$$

Similarly, inserting Eq. (16–18) into Eq. (16–17b) gives the other two line voltages:

$$\mathbf{V}_{BC} = \sqrt{3}V_P\angle{-90°} \text{ V (rms)}$$

$$\mathbf{V}_{CA} = \sqrt{3}V_P\angle{-210°} \text{ V (rms)} \tag{16–19b}$$

The three line voltages in Eq. (16–19) are balanced since they are separated in phase by 120° and all have a magnitude of $\sqrt{3}V_P$. Figure 16–14 shows the phasor diagram of the line voltages in Eq. (16–19) and the phase voltages in Eq. (16–18). Note that each line voltage leads a corresponding phase voltage by 30°. We will often make use of this 30° phase lead.

In a balanced system the scalar V_L denotes the magnitude (rms value) of the line voltages. Obviously, V_L is related to the phase voltage magnitude V_P as

$$V_L = \sqrt{3}V_P \tag{16–20}$$

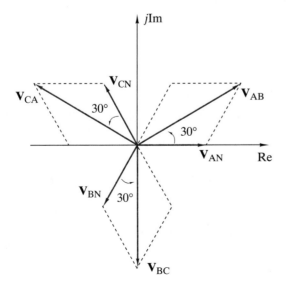

FIGURE 16–14 *Phasor diagram showing phase and line voltages for the positive phase sequence.*

In a three-phase circuit, the line-to-line voltages are $\sqrt{3}$ times as large as the line-to-neutral voltages. This ratio appears in three-phase equipment ratings, such as 120/208 V, where 120 is the phase voltage and 208 the line voltage.

It may seem that keeping track of six voltages all at different phase angles makes three-phase circuits much more complicated than single-phase circuits. Actually, *balanced* three-phase circuits are not that complicated because the six voltages are not independent and their phasors have a great deal of symmetry. In fact, given the phase sequence and any one of the six voltages, we can get the other five.

EXAMPLE 16–12

In a balanced three-phase circuit the line voltages have an rms value of $V_{\mathrm{L}} = 480$ V (rms). Using $\angle\mathbf{V}_{\mathrm{AB}}$ as the phase reference, find all of the line and phase voltages for a positive phase sequence.

SOLUTION:
The specified phase reference means that we arbitrarily assign $\angle\mathbf{V}_{\mathrm{AB}} = 0°$. This assignment together with $V_{\mathrm{L}} = 480$ allows us to write $\mathbf{V}_{\mathrm{AB}} = 480\angle 0°$ V (rms). The other two line voltages have the same rms value, $V_{\mathrm{L}} = 480$, and phase angles that lag $\angle\mathbf{V}_{\mathrm{AB}}$ at $120°$ intervals. For a positive (ABC) phase sequence these voltages are $\mathbf{V}_{\mathrm{BC}} = 480\angle{-}120°$ V (rms) and $\mathbf{V}_{\mathrm{CA}} = 480\angle{-}240°$ V (rms). The rms value of the phase voltages is $V_{\mathrm{P}} = 480/\sqrt{3} = 277$ V (rms). In Figure 16–14, we see that the line voltage $\mathbf{V}_{\mathrm{AB}}$ leads the phase voltage $\mathbf{V}_{\mathrm{AN}}$ by $30°$. Since $\angle\mathbf{V}_{\mathrm{AB}} = 0°$ is the phase reference, we can write $\mathbf{V}_{\mathrm{AN}} = 277\angle{-}30°$ V (rms). The other two phase voltages have the same rms value and lag $\angle\mathbf{V}_{\mathrm{AN}} = {-}30°$ at $120°$ intervals. For a positive phase sequence, these voltages are $\mathbf{V}_{\mathrm{BN}} = 277\angle{-}150°$ V (rms) and $\mathbf{V}_{\mathrm{CN}} = 277\angle{-}270°$ V (rms).

There is nothing sacred about assigning $\angle\mathbf{V}_{\mathrm{AB}} = 0°$ as the phase reference. A three-phase circuit has an abundance of voltages and currents all with different phase angles. We have to start somewhere by choosing one of the phasors as the phase reference. The choice is arbitrary, but we often use $\angle\mathbf{V}_{\mathrm{AN}} = 0°$ as a phase reference. Had we done so in this example, the phasor magnitudes V_{P} and V_{L} would be unchanged but all phase angles would increase by $30°$. ∎

Exercise 16–8

In a balanced three-phase circuit the rms line voltage is $V_{\mathrm{L}} = 7.2$ kV (rms). Find all of the phase and line voltages for a positive phase sequence using $\angle\mathbf{V}_{\mathrm{AN}} = 0°$ as the phase reference.

Answers:
Phase voltages: $\mathbf{V}_{\mathrm{AN}} = 4160\angle 0°$ V (rms), $\mathbf{V}_{\mathrm{BN}} = 4160\angle{-}120°$ V (rms), $\mathbf{V}_{\mathrm{CN}} = 4160\angle{-}240°$ V (rms).
Line voltages: $\mathbf{V}_{\mathrm{AB}} = 7200\angle{+}30°$ V (rms), $\mathbf{V}_{\mathrm{BC}} = 7200\angle{-}90°$ V (rms), $\mathbf{V}_{\mathrm{CA}} = 7200\angle{-}210°$ V (rms).

Exercise 16–9

In a balanced three-phase circuit $\mathbf{V}_{\mathrm{BC}} = 480\angle{-}120°$ V (rms). Find the phase voltages for a positive phase sequence.

Answers:
Phase voltages: $\mathbf{V}_{\mathrm{AN}} = 277\angle{-}30°$ V (rms), $\mathbf{V}_{\mathrm{BN}} = 277\angle{-}150°$ V (rms), $\mathbf{V}_{\mathrm{CN}} = 227\angle{-}270°$ V (rms).

FIGURE 16–15

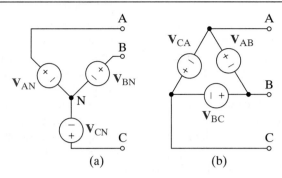

(a)   (b)

BALANCED THREE-PHASE CONNECTIONS

A balanced three-phase source can be represented by three voltage sources. The sources can be either Y[3]-connected, as in Figure 16–15(a), or Δ-connected, as in Figure 16–15(b). Similarly, a balanced three-phase load can be represented by three equal impedances that can be either Y-connected as in Figure 16–16(a) or Δ-connected as in Figure 16–16(b). These connections may be arranged various ways in a circuit diagram, but a Y-connection always involves three (and only three) elements tied together at a single node, while a Δ-connection always involves a loop containing three (and only three) elements.

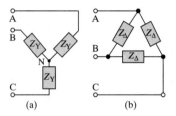

FIGURE 16–16

Since the source and load can be either Y- or Δ-connected, it may seem that we must deal with four possible circuit configurations.

Configuration	Source	Load
Y-Y	Y-connected	Y-connected
Y-Δ	Y-connected	Δ-connected
Δ-Y	Δ-connected	Y-connected
Δ–Δ	Δ-connected	Δ-connected

Keeping track of the various three-phase voltages and currents in four different configurations may look like a daunting task. Fortunately, there is a way to deal with all of these cases using an equivalent Y-Y circuit.

To develop this equivalence, we start with balanced three-phase sources. The Y-connected sources in Figure 16–15(a) produce the phase voltages $\mathbf{V}_{AN}$, $\mathbf{V}_{BN}$, and $\mathbf{V}_{CN}$. The Δ-connected sources in Figure 16–15(b) produce the line voltages $\mathbf{V}_{AB}$, $\mathbf{V}_{BC}$, and $\mathbf{V}_{CA}$. In a balanced system, these voltages are not independent. Given any one of them, we can easily calculate the other five. Specifically, if a balanced Δ-connected source produces $\mathbf{V}_{AB} = V_L\angle\phi$, then a Y-connected source with

$$\mathbf{V}_{AN} = \frac{V_L}{\sqrt{3}}\angle\phi - 30°\ \text{V (rms)}$$

$$\mathbf{V}_{BN} = \frac{V_L}{\sqrt{3}}\angle\phi - 150°\ \text{V (rms)} \qquad (16\text{–}21)$$

$$\mathbf{V}_{CN} = \frac{V_L}{\sqrt{3}}\angle\phi - 270°\ \text{V (rms)}$$

---

[3]In power jargen "Y" is at times referred to as "Wye"

would produce the same line-to-line voltages as the Δ-connected source. In other words, a balanced Δ-connected source can be replaced by an equivalent Y-connected source without changing the response of the rest of the circuit.

Actually, Δ-connected generators are rare because of practical limitations. If the three sources in Figure 16–15(b) are perfectly balanced, then $\mathbf{V}_{AB} + \mathbf{V}_{BC} + \mathbf{V}_{CA} = 0$ and KVL is satisfied around the loop of voltage sources. However, even a slight imbalance can produce large losses due to a current circulating in this loop whose magnitude is limited only by the internal impedances of the sources. In addition, in circuit theory a loop of ideal voltage sources is called a pathological circuit because there is no unique solution for the current produced by each source. For both practical and theoretical reasons we will not treat Δ-connected sources further in our study.

An equivalent Y-connected load can replace a Δ-connected load. First, observe that the impedance seen between any two terminals of the Y-connected load in Figure 16–16(a) is $Z_Y + Z_Y = 2Z_Y$. Second, note that the impedance seen between any two terminals of the Δ-connected load in Figure 16–16(b) is $Z_\Delta \| (Z_\Delta + Z_\Delta) = Z_\Delta \| (2Z_\Delta)$. The two loads are equivalent when these impedances are equal, that is, when

$$2Z_Y = Z_\Delta \| (2Z_\Delta) = \frac{Z_\Delta \times 2Z_\Delta}{Z_\Delta + 2Z_\Delta} = \frac{2Z_\Delta}{3}$$

Solving for Z_Y yields

$$Z_Y = \frac{Z_\Delta}{3} \tag{16–22}$$

Any balanced Δ-connected load with phase impedance Z_Δ can be replaced by an equivalent Y-connected load whose phase impedance is $Z_Y = Z_\Delta/3$ without changing the response of the rest of the circuit.

The change from a Δ-connected load to an equivalent Y-connected load does not change the angle of the phase impedance. That is, Eq. (16–22) is a scalar change only so that

$$\angle Z_Y = \angle Z_\Delta = \theta$$

This means that equivalent Y- and Δ-connected loads have the same power factor, $pf = \cos \theta$.

EXAMPLE 16–13

Figure 16–17 shows a balanced Δ-connected load in parallel with a balanced Y-connected load. The two-phase impedances are $Z_\Delta = 120 + j40 \ \Omega$ and $Z_Y = 50 + j30 \ \Omega$. Find the phase impedance of an equivalent Y-connected load.

SOLUTION:

STEP 1 We first convert the Y-connected load in Figure 16–17 into an equivalent Δ-connected load using Eq. (16–22). The conversion yields phase impedances of $3Z_Y$ and produces the circuit configuration in Figure 16–18(a).

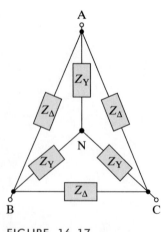

FIGURE 16–17

STEP 2 Each phase impedance $3Z_Y$ is in parallel with phase impedance Z_Δ. Combining these parallel impedances as $Z_\Delta \| 3Z_Y$ produces the equivalent Δ-connected load in Figure 16–18(b), where

$$Z_{\Delta EQ} = Z_\Delta \| 3Z_Y = \frac{Z_\Delta \times 3Z_Y}{Z_\Delta + 3Z_Y}$$

$$= \frac{(120 + j40)(150 + j90)}{120 + j40 + 150 + j90}$$

$$= 67.6 + j29.7 \, \Omega$$

STEP 3 We use Eq. (16–22) again, this time to convert the load in Figure 16–18(b) into the equivalent Y connected load in Figure 16–18(c), where

$$Z_{YEQ} = \frac{Z_{\Delta EQ}}{3} = \frac{67.6 + j29.7}{3}$$

$$= 22.5 + j9.9 \, \Omega$$

In sum, Z_{YEQ} is the phase impedance of a balanced Y-connected load that is equivalent to the two parallel loads in Figure 16–17. ∎

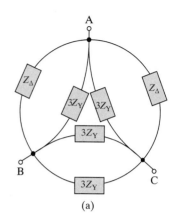

(a)

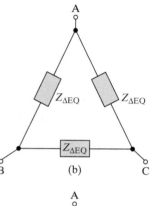

(b)

Exercise 16–10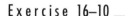

Two balanced Δ-connected loads are connected in parallel. Their phase impedances are $Z_{\Delta 1} = 50 + j24 \, \Omega$ and $Z_{\Delta 2} = 60 + j25 \, \Omega$. Find the equivalent Y-connected load for the two parallel loads.

Answer: $Z_{YEQ} = 9.10 + j4.10 \, \Omega$

16–6 THREE-PHASE CIRCUIT ANALYSIS

The analysis of the Y-Y circuit in Figure 16–19 is the key to understanding balanced three-phase circuits. The source and load in this circuit are Y-connected since both involve three elements tied together at a single node represented by the neutral line. This configuration involves three separate single-phase circuits with the source phase angles separated by 120°. The usual analysis objectives are to determine the line currents $\mathbf{I}_A$, $\mathbf{I}_B$, and $\mathbf{I}_C$ and the power delivered to the load.

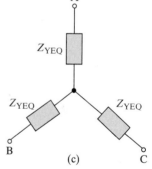

(c)

FIGURE 16–18

Each phase of Y-Y circuit has a load connected across a voltage source, so the three line currents are easily seen to be

$$\mathbf{I}_A = \frac{\mathbf{V}_{AN}}{Z_Y} \qquad \mathbf{I}_B = \frac{\mathbf{V}_{BN}}{Z_Y} \qquad \mathbf{I}_C = \frac{\mathbf{V}_{CN}}{Z_Y}$$

The line currents $\mathbf{I}_A$, $\mathbf{I}_B$, and $\mathbf{I}_C$ form a balanced set of phasors because the source voltages $\mathbf{V}_{AN}$, $\mathbf{V}_{BN}$, and $\mathbf{V}_{CN}$ are balanced. As a result, the line currents all have the same amplitude and are separated in phase by 120°. By calculating one line current—for example, $\mathbf{I}_A$—we can easily construct $\mathbf{I}_B$ and $\mathbf{I}_C$.

Using $\angle\mathbf{V}_{AN} = 0°$ as the phase reference, the source voltage is written as $\mathbf{V}_{AN} = V_P\angle0°$. Writing the phase impedance as $Z_Y = |Z_Y|\angle\theta$ leads to the line current $\mathbf{I}_A$ as

$$\mathbf{I}_A = \frac{V_P\angle0°}{|Z_Y|\angle\theta} = \frac{V_P}{|Z_Y|}\angle{-\theta} = I_L\angle{-\theta} \text{ A (rms)} \qquad (16\text{–}23)$$

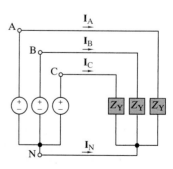

FIGURE 16–19

FIGURE 16–20 *Line currents and phase voltages in a Y-Y circuit.*

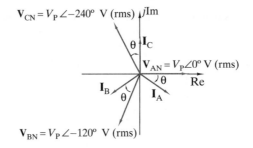

The other two line currents lag $\mathbf{I}_A$ at $120°$ intervals; hence for a positive phase sequence $\mathbf{I}_B = I_L\angle-120°$ A (rms) and $\mathbf{I}_C = I_L\angle-240°$ A (rms). In general usage, the scalar I_L denotes the magnitude (rms value) of the line currents in any balanced three-phase circuit. For a balanced Y-connected load I_L is defined by

$$I_L = \frac{V_P}{|Z_Y|} \qquad (16\text{--}24)$$

Figure 16–20 shows a phasor diagram of the line currents and phase voltages in a Y-Y circuit.

Applying KCL at node N in Figure 16–19 yields $\mathbf{I}_A + \mathbf{I}_B + \mathbf{I}_C + \mathbf{I}_N = 0$. Since the line currents are balanced their sum is $\mathbf{I}_A + \mathbf{I}_B + \mathbf{I}_C = 0$, hence KCL requires $\mathbf{I}_N = 0$. There is no current in the neutral connection of a balanced Y-Y circuit. The response of the circuit is not changed when this line is replaced by any impedance whatsoever, including infinite. Put differently, the power flow does not change when the neutral line in Figure 16–19 is removed. As a visual aid, however, we may continue to show the neutral line in Y-Y circuits because it serves as a reference node for the phase voltages.

EXAMPLE 16–14

In a balanced Y-Y circuit the line voltage is $V_L = 480$ V(rms) and the phase impedance is $Z_Y = 24 + j9\,\Omega$ per phase. Using $\angle\mathbf{V}_{AN} = 0°$ as the phase reference, find the line current and line voltage phasors for a positive phase sequence.

SOLUTION:
The magnitude of the phase voltage is $V_P = V_L/\sqrt{3} = 277$ V (rms). For the given phase reference we have $\mathbf{V}_{AN} = 277\angle0°$ and the phase A line current is found to be

$$\mathbf{I}_A = \frac{\mathbf{V}_{AN}}{Z_Y} = \frac{277\angle0°}{|24 + j9|\angle20.6°} = 10.8\angle-20.6°\ A\ (\text{rms})$$

The other two line currents have the same magnitude and are separated in phase by $120°$. For a positive phase sequence these currents are $\mathbf{I}_B = 10.8\angle-140.6°$ A (rms) and $\mathbf{I}_C = 10.8\angle-260.6°$ A (rms). In a positive phase sequence $\mathbf{V}_{AB}$ leads $\mathbf{V}_{AN}$ by $30°$; hence $\mathbf{V}_{AB} = V_L\angle30° = 480\angle30°$ V (rms). The other two line voltages are $\mathbf{V}_{BC} = 480\angle-90°$ V (rms) and $\mathbf{V}_{CA} = 480\angle-210°$ V (rms). ∎

Exercise 16–11

A balanced Y-Y circuit operates with $V_L = 4160$ V (rms) and phase impedances of $Z_Y = 100 + j40\,\Omega$ per phase. Using $\angle\mathbf{V}_{AB} = 0°$ as the phase reference, find $\mathbf{I}_A$ and $\mathbf{V}_{AN}$ for a positive phase sequence.

Answers: $\mathbf{I}_A = 22.3\angle-51.8°$ A (rms), $\mathbf{V}_{AN} = 2400\angle-30°$ V (rms)

Figure 16–21 shows a Y-Y circuit with three equal impedances Z_W added to represent a three-wire transmission line connecting the source to the load. Adding line impedances does not unduly complicate the analysis since the overall Y-Y circuit is still balanced. The line impedances can be treated as part of an augmented Y-connected load whose phase impedance is $Z_Y + Z_W$.

The main difference is that the line voltages at the source and at the load are not equal because of the voltage drop in the line. For this reason, the lines are labeled with uppercase letters (ABCN) at the source and with lowercase letters (abcn) at the load. For example, the line voltages are denoted $\mathbf{V}_{AB}$, $\mathbf{V}_{BC}$, and $\mathbf{V}_{CA}$ at the source and $\mathbf{V}_{ab}$, $\mathbf{V}_{bc}$, and $\mathbf{V}_{ca}$ at the load.

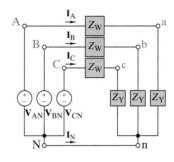

FIGURE 16–21

EXAMPLE 16–15

In Figure 16–21, the load impedance and line impedances are $Z_Y = 10 + j5\,\Omega$ per phase and $Z_W = 0.15 + j0.85\,\Omega$ per phase, respectively. The magnitude of the line voltage at the source is $V_L = 208\,\text{V}$ (rms). Using $\angle\mathbf{I}_A = 0°$ as the phase reference, find the line current phasors and the line voltage phasors at the load for a positive phase sequence.

SOLUTION:
In this example the line voltage at the source is $V_L = 208\,\text{V}$ (rms). The phase voltage magnitude at the source is $V_P = V_L/\sqrt{3} = 120\,\text{V}$ (rms). This phase voltage appears across the combined impedance $Z_W + Z_Y$, so the line current magnitude is found to be

$$I_L = \frac{V_P}{|Z_Y + Z_W|} = \frac{120}{|10.15 + j5.85|} = 10.24\,\text{A (rms)}$$

The specified phase reference means that $\mathbf{I}_A = 10.24\angle 0°\,\text{A (rms)}$. For a positive phase sequence the other two line currents are $\mathbf{I}_B = 10.24\angle -120°\,\text{A (rms)}$ and $\mathbf{I}_C = 10.24\angle -240°\,\text{A (rms)}$. The phase voltages at the load are balanced and denoted $\mathbf{V}_{an}$, $\mathbf{V}_{bn}$, and $\mathbf{V}_{cn}$. The first of these is found as

$$\mathbf{V}_{an} = \mathbf{I}_A Z_Y = 10.24\angle 0° \times (10 + j5) = 115\angle 26.6°\,\text{V (rms)}$$

Line voltages are $\sqrt{3}$ times as large as phase voltages and lead the phase voltages by $30°$. Hence, we have

$$\mathbf{V}_{ab} = \mathbf{V}_{an} \times \sqrt{3}\angle 30° = 199\angle 56.6°\,\text{V (rms)}$$

The other two line voltages are $\mathbf{V}_{bc} = 199\angle -63.4°$ and $\mathbf{V}_{ca} = 199\angle -183.4°\,\text{V (rms)}$. The line voltage magnitude at the load (199 V) is less than the line voltage at the source (208 V) due to the voltage drop in the line impedances. ∎

Exercise 16–12

In a balanced Y-Y circuit the load and line impedances are $Z_Y = 16 + j12\,\Omega$ per phase and $Z_W = 0.25 + j1.5\,\Omega$ per phase. The line current is $I_L = 14.2\,\text{A}$ (rms). Find the line voltage phasors at the source using $\angle\mathbf{I}_A = 0°$ as the phase reference.

Answers: $\mathbf{V}_{AB} = 520\angle 69.7°\,\text{V (rms)}$, $\mathbf{V}_{BC} = 520\angle -50.3°\,\text{V (rms)}$, $\mathbf{V}_{CA} = 520\angle -170.3°\,\text{V (rms)}$

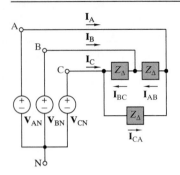

FIGURE 16–22

ANALYSIS OF THE Y-Δ CIRCUIT

We now extend our analysis of three-phase circuits to the Y-Δ circuit in Figure 16–22. In this circuit, the three sources are Y-connected because they are tied together at node N. The three load impedances are Δ-connected since they form a three-element loop. A neutral line is not shown in this case since the delta load has no place to connect a neutral line.

When the analysis objective is to determine the line currents and the power flow, we replace the Δ-connected load by an equivalent Y-connected load and analyze the resulting Y-Y circuit. However, in some cases we may also need to know the currents labeled $\mathbf{I}_{AB}$, $\mathbf{I}_{BC}$, and $\mathbf{I}_{CA}$ shown in Figure 16–22. These are the **phase currents** that exist in each load impedance of the Δ-connected load.

The impedances Z_Δ are connected from line to line so the phase currents are easily seen to be

$$\mathbf{I}_{AB} = \frac{\mathbf{V}_{AB}}{Z_\Delta} \qquad \mathbf{I}_{BC} = \frac{\mathbf{V}_{BC}}{Z_\Delta} \qquad \mathbf{I}_{CA} = \frac{\mathbf{V}_{CA}}{Z_\Delta}$$

The phase currents $\mathbf{I}_{AB}$, $\mathbf{I}_{BC}$, and $\mathbf{I}_{CA}$ are balanced since the line voltages $\mathbf{V}_{AB}$, $\mathbf{V}_{BC}$, and $\mathbf{V}_{CA}$ are balanced. Since they are balanced, the phase currents all have the same amplitude and are separated by 120°. Hence, we can calculate one of them, say $\mathbf{I}_{AB}$, and use it to get the other two.

For a positive phase sequence using $\angle \mathbf{V}_{AN} = 0^\circ$ as the phase reference, we have $\mathbf{V}_{AB} = V_L \angle 30^\circ$. Writing the phase impedance as $Z_\Delta = |Z_\Delta| \angle \theta$, the phase current $\mathbf{I}_{AB}$ is

$$\mathbf{I}_{AB} = \frac{V_L \angle 30^\circ}{|Z_\Delta| \angle \theta} = \frac{V_L}{|Z_\Delta|} \angle -\theta + 30^\circ \text{ A (rms)} \qquad (16\text{--}25)$$

The other two phase currents lag $\mathbf{I}_{AB}$ at 120° intervals; hence for a positive phase sequence $\mathbf{I}_{BC} = I_P \angle -\theta - 90^\circ$ A (rms) and $\mathbf{I}_{CA} = I_P \angle -\theta - 210^\circ$ A (rms). In a balanced Δ-connected load, the scalar I_P denotes the magnitude (rms value) of the phase currents and is defined as

$$I_P = \frac{V_L}{|Z_\Delta|} \qquad (16\text{--}26)$$

It turns out that there is a simple relationship between line and phase currents. To show this we calculate the line current $\mathbf{I}_A$ in the equivalent Y-Y circuit. Using $\angle \mathbf{V}_{AN} = 0^\circ$ as the phase reference, the appropriate phase voltage is $\mathbf{V}_{AN} = (V_L/\sqrt{3}) \angle 0^\circ$ V (rms). Writing the phase impedance in the equivalent Y-connected load as $Z_Y = |Z_\Delta/3| \angle \theta$, we express $\mathbf{I}_A$ as

$$\mathbf{I}_A = \frac{(V_L/\sqrt{3}) \angle 0^\circ}{|Z_\Delta/3| \angle \theta} = \sqrt{3} \frac{V_L}{|Z_\Delta|} \angle -\theta = \sqrt{3} I_P \angle -\theta \text{ A (rms)} \qquad (16\text{--}27)$$

Comparing Eqs. (16–25) and (16–27) reveals two important results:

1. $\sqrt{3} I_P = I_L$, that is, the line currents are $\sqrt{3}$ times as large as the phase currents.

2. $\angle \mathbf{I}_{AB} = \angle \mathbf{I}_A + 30^\circ$, that is, phase currents lead line currents by 30°.

The $\sqrt{3}$ magnitude factor and 30° phase shift provide a simple way to get the phase currents from the line currents, or vice versa. Since both sets are balanced, we really only need to calculate one of them to get the other five.

The next two examples illustrate that phasor responses in a balanced Y-Δ circuit can be found using $\mathbf{I}_A$ in the equivalent Y-Y circuit.

EXAMPLE 16–16

The line voltage at a Δ-connected load with $Z_\Delta = 40 + j30\,\Omega$ per phase is $V_L = 2.4\,\text{kV (rms)}$. Find the line and phase current phasors using $\angle \mathbf{V}_{AN} = 0°$ as the phase reference.

SOLUTION:
We first calculate the line current $\mathbf{I}_A$ in the equivalent Y-connected load. The phase voltage at the load is $V_P = 2400/\sqrt{3} = 1386\,\text{V (rms)}$ and the load impedance is $Z_Y = Z_\Delta/3 = 13.33 + j10\,\Omega$. For the specified phase reference we have $\mathbf{V}_{AN} = 1386\angle 0°\,\text{V (rms)}$, and the phase A line current is calculated as

$$\mathbf{I}_A = \frac{\mathbf{V}_{AN}}{Z_Y} = \frac{1386\angle 0°}{|13.33 + j10|\angle 36.9°} = 83.1\angle -36.9°\,\text{A (rms)}$$

The other two line currents lag $\mathbf{I}_A$ at $-120°$ intervals. For a positive phase sequence these currents are $\mathbf{I}_B = 83.1\angle -156.9°\,\text{A (rms)}$ and $\mathbf{I}_C = 83.1\angle -276.9°\,\text{A (rms)}$. The magnitude of the phase current is $I_P = 83.1/\sqrt{3} = 48\,\text{A (rms)}$. For a positive phase sequence the phase current $\mathbf{I}_{AB}$ leads $\mathbf{I}_A$ by $30°$; hence $\mathbf{I}_{AB} = 48\angle -6.9°\,\text{A (rms)}$ and the other two phase currents are $\mathbf{I}_{BC} = 48\angle -126.9°\,\text{A (rms)}$ and $\mathbf{I}_{CA} = 48\angle -246.9°\,\text{A (rms)}$.

Another approach is to calculate $\mathbf{I}_{AB}$ directly in the Δ-connected load. Using $\angle \mathbf{V}_{AN} = 0°$ as the phase reference, the appropriate line voltage is $\mathbf{V}_{AB} = V_L\angle 30° = 2400\angle 30°\,\text{V (rms)}$. This voltage appears across the impedance Z_Δ; hence $\mathbf{I}_{AB}$ is found to be

$$\mathbf{I}_{AB} = \frac{\mathbf{V}_{AB}}{Z_\Delta} = \frac{2400\angle 30°}{|40 + j30|\angle 36.9°} = 48\angle -6.9°\,\text{A (rms)}$$

which is the same as the result derived using the line current $\mathbf{I}_A$. ∎

Exercise 16–13

The line voltage at a Δ-connected load with $Z_\Delta = 520 + j400\,\Omega$ per phase is $V_L = 1300\,\text{V (rms)}$. Find $\mathbf{I}_A$ and $\mathbf{I}_{AB}$ using $\angle \mathbf{V}_{AB} = 0°$ as the phase reference.

Answers: $\mathbf{I}_A = 3.43\angle -67.6°\,\text{A (rms)}$, $\mathbf{I}_{AB} = 1.98\angle -37.6°\,\text{A (rms)}$

EXAMPLE 16–17

A balanced three-phase source with $V_L = 200\,\text{V (rms)}$ feeds a Δ-connected load with $Z_\Delta = 12 + j6\,\Omega$ per phase through a three-wire line with $Z_W = 0.1 + j0.55\,\Omega$ per phase. Find the line current and phase current phasors using $\angle \mathbf{I}_A = 0°$ as the phase reference.

SOLUTION:
The phase impedance of the equivalent Y-connected load $Z_Y = Z_\Delta/3 = 4 + j2\,\Omega$. The phase voltage magnitude at the source is $V_P = 200/\sqrt{3} = 115.5\,\text{V (rms)}$. In each phase the voltage V_P appears across the series combination of $Z_W + Z_Y$; hence the line current magnitude is

$$I_L = \frac{V_P}{|Z_W + Z_Y|} = \frac{115.5}{|4.1 + j2.55|} = 23.9\,\text{A (rms)}$$

Using $\angle \mathbf{I}_A$ as the phase reference means $\mathbf{I}_A = 23.9 \angle 0° \text{ A (rms)}$ and the other two line currents lag $\mathbf{I}_A$ at $-120°$ intervals. For a positive phase sequence these currents are $\mathbf{I}_B = 23.9 \angle -120° \text{ A (rms)}$ and $\mathbf{I}_C = 23.9 \angle -240° \text{ A (rms)}$. The phase current magnitude is $I_P = 23.9/\sqrt{3} = 13.8 \text{ A (rms)}$. The phase current $\mathbf{I}_{AB}$ leads $\mathbf{I}_A$ by $30°$; hence $\mathbf{I}_{AB} = 13.8 \angle +30° \text{ A (rms)}$ and the other two phase currents are $\mathbf{I}_{BC} = 13.8 \angle -90° \text{ A (rms)}$ and $\mathbf{I}_{CA} = 13.7 \angle -210° \text{ A (rms)}$.

Again, we can easily get the phase currents in the Δ-connected load by first finding the line current $\mathbf{I}_A$ in the equivalent Y-Y circuit. ∎

Exercise 16–14

The phase B line current in a Δ-connected load with $Z_\Delta = 14 + j9\,\Omega$ per phase is $\mathbf{I}_B = 26 \angle -165° \text{ A (rms)}$. Find $\mathbf{I}_{AB}$ and $\mathbf{V}_{AB}$ for a positive phase sequence.

Answers: $\mathbf{I}_{AB} = 15 \angle -15° \text{ A (rms)}$, $\mathbf{V}_{AB} = 250 \angle 17.7° \text{ V (rms)}$

The phasor responses in a balanced three-phase circuit can be found in several ways using various methods of ac circuit analysis. However, the previous examples suggest a shortcut method based on the phasor symmetries in balanced circuits. Because of these symmetries phasor responses can be derived from $\mathbf{I}_A$ (or $\mathbf{I}_{AB}$) and $\mathbf{V}_{AN}$ (or $\mathbf{V}_{AB}$) using a $\sqrt{3}$ magnitude change and a $30°$ phase shift. As a matter of analysis strategy, a shortcut method of finding phasor responses is to find first $\mathbf{I}_A$ and $\mathbf{V}_{AN}$ in the equivalent Y-Y circuit. Then use the inherent phasor symmetries in balanced circuits to find other phasor voltages and currents as needed.

These symmetries, however, are lost when the circuit is unbalanced. For example, the line current symmetries are lost when the load is made up of three unequal impedances. Such simple unbalanced circuits can be handled by ac circuit analysis methods, including PSpice. However, circuit models for a power system involve levels of complexity that are extremely difficult to manage using ordinary ac circuit analysis. For this reason, unbalanced three-phase circuits are analyzed using symmetrical components.

The method of symmetrical components is treated in detail in courses on power system analysis. Briefly, this method divides an unbalanced circuit into three parts:

1. A *balanced* three-phase circuit with a positive phase sequence
2. A *balanced* three-phase circuit with a negative phase sequence
3. Three identical *single-phase* circuits

Responses in the unbalanced circuit are then calculated as a superposition of the responses found in each of these parts. Thus, the method of symmetrical components treats unbalanced circuits using methods developed in this chapter, namely the analysis of *single-phase* and *balanced* three-phase circuits.

ANALYSIS OF POWER IN THREE-PHASE CIRCUITS

The analysis of power focuses on the flow of complex power rather than the phasor voltages and currents in the circuit. The circuit in Figure 16–23 is used to illustrate the analysis of three-phase power flow. The circuit consists of a balanced three-phase source supplying power to a balanced three-phase load through a three-wire power line with a wire impedance Z_W in each line. Our immediate objective is to find the complex power delivered to the load for specified conditions at the source or load terminals. The load can be either Y- or Δ-connected but we do not need to know how the source is connected. We only need to know that it is balanced.

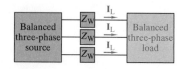

FIGURE 16–23

The power delivered to the load in Figure 16–23 can be found using the scalars I_L and V_L rather than phasors. The scalar I_L is the magnitude (rms value) of the current in the lines connecting the source and load. The scalar V_L is the magnitude (rms value) of the line-to-line voltage at the load. However, remember that the voltages at the source and at the load are not the same due to the voltage drops in the line impedances.

When the load in Figure 16–23 is Y-connected, an rms current I_L passes through each phase impedance Z_Y. The complex power delivered to each phase is $I_L^2 Z_Y$, so the total load power is

$$S_L = 3I_L^2 Z_Y$$

Writing the definition of I_L in Eq. (16–24) in terms of V_L as

$$I_L = \frac{V_P}{|Z_Y|} = \frac{V_L/\sqrt{3}}{|Z_Y|}$$

allows us to express the total complex power as

$$S_L = 3(I_L)(I_L)Z_Y$$
$$= 3\frac{V_L/\sqrt{3}}{|Z_Y|}I_L Z_Y$$
$$= \sqrt{3}V_L I_L \frac{Z_Y}{|Z_Y|}\sqrt{3}V_L I_L e^{j\theta}$$

where $\theta = \angle Z_Y$. Thus, the complex power delivered to the Y-connected load depends on the line voltage, line current, and the angle of the phase impedance.

$$SL = 3VLILe j\theta = PL + jQL \tag{16–28}$$

where $P_L = \sqrt{3}\, V_L I_L \cos\theta$ and $Q_L = \sqrt{3}\, V_L I_L \sin\theta$ are the average and reactive powers delivered to the load.

When the load in Figure 16–23 is Δ-connected, an rms current I_P passes through each phase impedance Z_Δ. The total complex power delivered to the load is

$$S_L = 3I_P^2 Z_\Delta$$

Using the definition of I_P in Eq. (16–26) and the fact the $I_P = I_L/\sqrt{3}$, the total power can be written as

$$S_L = 3(I_P)(I_P)Z_\Delta = 3\left(\frac{V_L}{|Z_\Delta|}\right)\left(\frac{I_L}{\sqrt{3}}Z_\Delta\right) = \sqrt{3}V_L I_L \frac{Z_\Delta}{|Z_\Delta|} = \sqrt{3}V_L I_L e^{j\theta}$$

where $\theta = \angle Z_\Delta$. The final expression for S_L found here is the same as Eq. (16–28) for a Y-connected load. Thus, Eq. (16–28) applies to the balanced load in Figure 16–23 whether it is Y-connected or Δ-connected.

The generality of Eq. (16–28) is a key result. We will make frequent use of this result in the analysis of three-phase power flow, especially the fact that the apparent power delivered to any balanced load is $|S_L| = \sqrt{3}\, V_L I_L$ since $|e^{j\theta}| = 1$.

EXAMPLE 16–18

The load in Figure 16–23 is Y-connected with a phase impedance of $Z_Y = 12 + j5\ \Omega$/phase and the line voltage at the load is $V_L = 440$ V(rms). Find the line current I_L and the complex power delivered to the load.

SOLUTION:

Since $V_L = 440$ V(rms), the phase voltage at the load is $V_P = 440/\sqrt{3} = 254$ V(rms). In a Y-connected load, this voltage appears across the phase impedance Z_Y and the line current is found to be

$$I_L = \frac{V_P}{|Z_Y|} = \frac{254}{|12 + j5|} = 19.5 \text{ A(rms)}$$

The scalar I_L is the rms current in all three phase impedances Z_Y. Hence, the total complex power delivered to the load is

$$\begin{aligned} S_L &= 3I_L^2 Z_Y = 3(19.5)^2(12 + j5) \\ &= 13.7 \times 10^3 + j5.73 \times 10^3 \text{ VA} \end{aligned}$$

No phasors were needed to solve this problem. The power flow in a balanced three-phase circuit can usually be found using only the scalars V_L and I_L together with the line and load impedances. ∎

Exercise 16–15

In Figure 16–23 the load is Δ-connected with a phase impedance of $Z_\Delta = 26 + j8 \ \Omega$/phase and the line voltage at the load is $V_L = 1.0$ kV(rms). Find the line current I_L and the complex power delivered to the load.

Answers: $I_L = 63.7$ A(rms), $S_L = 105 \times 10^3 + j32.4 + 10^3$ VA

EXAMPLE 16–19

In Figure 16–23 the load is Y-connected with a phase impedance of $Z_Y = 15 + j6 \ \Omega$/phase and the line current is $I_L = 10$ A(rms). Find the line voltage V_L and the complex power delivered to the load.

SOLUTION:

In this example the line current I_L and phase impedance Z_Y are given, hence the complex power delivered to the load is

$$\begin{aligned} S_L &= 3I_L^2 Z_Y = 3(10)^2(15 + j6) \\ &= 4.5 + j1.8 \text{ kVA} \end{aligned}$$

The apparent power delivered to any balanced load can be written as $|S_L| = \sqrt{3} \, V_L I_L$. Solving for the line voltage gives

$$V_L = \frac{|S_L|}{\sqrt{3} \, I_L} = \frac{|4.5 \times 10^3 + j1.8 \times 10^3|}{\sqrt{3} \times 10} = 280 \text{ V(rms)}$$

Again, no phasors are needed to solve this problem. ∎

Exercise 16–16

In a balanced three-phase circuit, the line voltage at the load is 4160 V(rms) and the apparent power delivered to the load is 60 kVA. Find the line current.

Answer: $I_L = 8.33$ A(rms)

EXAMPLE 16–20

An average power of 20 kW is delivered to a balanced Δ-connected load with $Z_\Delta = 30 + j45\ \Omega/\text{phase}$. Find the line voltage V_L at the load and the complex power delivered to the load.

SOLUTION:
In a balanced Δ-connected load the phase current I_P passes through all three of the phase impedances. The total average power delivered to the load is $P_L = 3I_P^2 R_\Delta$, where R_Δ is the resistive part of the phase impedance Z_Δ. Solving for I_P yields

$$I_P = \sqrt{\frac{P_L}{3R_\Delta}} = \sqrt{\frac{20 \times 10^3}{3 \times 30}} = 14.9\ \text{A(rms)}$$

The total complex power delivered to a delta load is then found as

$$S_L = 3I_P^2 Z_\Delta = 3(14.9)^2(30 + j45)$$
$$= 20 \times 10^3 + j30 \times 10^3\ \text{VA}$$

Given the phase current, the line current is $I_L = \sqrt{3} \times 14.9 = 25.8\ \text{A(rms)}$. The apparent power delivered to any balanced load is $|S_L| = \sqrt{3}\,V_L I_L$. Solving for the line voltage gives

$$V_L = \frac{|S_L|}{\sqrt{3}\,I_L} = \frac{|20 \times 10^3 + j30 \times 10^3|}{\sqrt{3} \times 25.8} = 807\ \text{V(rms)}\qquad\blacksquare$$

EXAMPLE 16–21

In Figure 16–23 the three-phase source produces an apparent power of 3.5 kVA at a power factor of 0.8 lagging and a line current of $I_L = 4.6\ \text{A(rms)}$. The three lines connecting the source to the load have impedances of $Z_W = 1 + j6\ \Omega/\text{phase}$. Find the complex power delivered to the load and the line voltage at the load.

SOLUTION:
The complex power produced by the source is

$$S_S = |S_S|\left(\text{pf} + j\sqrt{1 - \text{pf}^2}\right) = 3.5 \times 10^3\left(0.8 + j\sqrt{1 - 0.8^2}\right)$$
$$= (2.8 + j2.1)\ \text{kVA}$$

The complex power lost in any one wire is $I_L^2 Z_W$, so the total line loss is

$$S_W = 3I_L^2 Z_W = 3 \times (4.6)^2(1 + j6)$$
$$= 63.5 + j381\ \text{VA}$$

The complex power delivered to the load equals the power produced by the source minus the line losses. Hence

$$S_L = S_S - S_W = 2.74 + j1.72\ \text{kVA}$$

Accordingly, we get the line voltage at the load as

$$V_L = \frac{|S_L|}{\sqrt{3}\,I_L} = \frac{|2.74 \times 10^3 + j1.72 \times 10^3|}{\sqrt{3} \times 4.6} = 406\ \text{V(rms)}\qquad\blacksquare$$

Exercise 16–17

In Figure 16–23 the line current is $I_L = 10\,\text{A(rms)}$, the line impedance is $Z_W = 0.6 + j3.7\,\Omega/\text{phase}$, and the phase impedance of the load is $Z_Y = 15 + j28\,\Omega/\text{phase}$. Find the complex power produced by the source.

Answer: $S_S = 4.68 + j9.51\,\text{kVA}$

Exercise 16–18

In Figure 16–23 the line current is $I_L = 5\,\text{A(rms)}$, the line impedance is $Z_W = 2 + j6\,\Omega/\text{phase}$ and the load absorbs $S_L = 3 + j2\,\text{kVA}$. Find the complex power produced by the source.

Answer: $S_S = 3.15 + j2.45\,\text{kVA}$

INSTANTANEOUS POWER IN THREE-PHASE CIRCUITS

Earlier in this chapter, we found that the instantaneous power in a single-phase circuit has a dc component and an ac component. The ac components cause the instantaneous power flow to oscillate about an average value. One of the advantages of balanced three-phase circuits is that the total instantaneous power does not have an ac component.

We previously developed Eq. (16–7) to relate the instantaneous power $p(t)$ in a single-phase circuit to the two components of complex power P and Q. Applying this equation to a balanced three-phase circuit yields the instantaneous power in any phase of a load:

$$p_{\text{phase}}(t) = P[1 + \cos(2\omega t + 2\phi)] - Q[\sin(2\omega t + 2\phi)]$$

where P and Q are the average and reactive powers carried by the phase, and the angle ϕ is $0°$ for phase A, $-120°$ for phase B, and $-240°$ for phase C.

The total instantaneous power is the sum of the power in each phase.

$$p_{\text{total}}(t) = p_A(t) + p_B(t) + p_C(t)$$
$$= 3P + P[\cos(2\omega t) + \cos(2\omega t - 240°) + \cos(2\omega t - 480°)]$$
$$- Q[\sin(2\omega t) + \sin(2\omega t - 240°) + \sin(2\omega t - 480°)]$$

Both bracketed terms in this equation contain three sinusoids with equal amplitudes and separated in phase by $120°$ ($-480°$ is the same as $-120°$). Like their phasor counterparts, the sum of a balanced set of three sinusoidal waveforms is zero. Hence, the bracketed terms both vanish, leaving

$$p_{\text{total}}(t) = 3P = P_L = \sqrt{3}V_L I_L \cos\theta$$

The fact that the total instantaneous power flow is constant rather than oscillating means that three-phase systems operate smoothly with less vibration at their mechanical inputs and outputs.

16–7 THREE-PHASE POWER FLOW

In ordinary circuit analysis, the goal is to find the power delivered to the load for specified source conditions. In power flow analysis, the power delivered to the load is specified and the source conditions are the unknowns. We illustrate power flow analysis using the three-phase power system in Figure 16–24. The system consists of two balanced three-phase loads each of which draws a specified complex power from the system. The power is supplied by a balanced source through a pair of three-wire transmission lines with impedances of Z_{W1} and Z_{W2}. In a power system, a **bus** is defined as a set of three nodes at which three-phase elements are tied together. The analysis problem is to determine the system voltages and currents that produce a specified power flow to the loads.

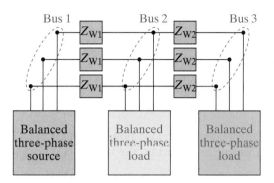

FIGURE 16–24

Even in a simple system like Figure 16–24, it seems clear that showing all three of the phases contributes little to our understanding of the power flow. Actual power systems may have hundreds of sources and loads all interconnected by a web of transmission lines. Including all three of the phases would make the system diagram hopelessly cluttered. In a balanced system, each phase carries the same complex power, so we do not need to show all of them.

A simpler representation is to use a single line to represent all three of the phases. Figure 16–25 is a single-line representation of the power system in Figure 16–24 with the three buses shown as short horizontal lines. In a single-line diagram, a bus can be thought of as a single node at which sources, loads, and lines are tied together. The source at bus 1 is shown as a circle with an arrow indicating power flow into the system. The loads at bus 2 and bus 3 are shown as arrows indicating power flow out of the system. The two transmission lines are represented by their impedances Z_{W1} and Z_{W2}.

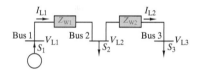

FIGURE 16–25

By power flow we mean complex power, $S = P + jQ$. We use a subscript to identify the bus at which power enters or exits the system. In Figure 16–25, the symbol S_1 is the power supplied to the system by the source at bus 1, while S_2 and S_3 are the powers extracted from the system by the loads at bus 2 and bus 3. In single-line diagrams the line currents and line voltages at represented by the scalars I_L and V_L, with subscripts added for clarity as needed. In general, we do not use load impedances since this information is implicit in the specified complex power flow. We do not need to know whether sources and loads are Y- or Δ-connected. We need only know that they are balanced.

The line voltage at each of its buses and the impedances of the lines connecting the buses control the power flow in a system. Solving a power flow problem involves finding a set of bus voltages and line currents that produce a prescribed system load flow. Specifically for Figure 16–25, this would mean finding a set of bus voltages V_{L1}, V_{L2}, V_{L3} and line currents I_{L1}, I_{L2} that produce the specified load powers S_2 and S_3.

In power systems with several hundred buses, the power flow problem is a very significant computational challenge. Although our examples are very elementary, they do give some insight into the computational methods involved.

EXAMPLE 16–22

In Figure 16–26 the source at bus 1 and the load at bus 2 are interconnected by a transmission line with $Z_W = 1.5 + j8.5 \, \Omega/\text{phase}$. The load at bus 2 draws a complex power of $S_2 = 70 + j35 \, \text{kVA}$. Assuming that $V_{L2} = 2400 \, \text{V(rms)}$, find the complex power produced by the source and the line voltage at bus 1.

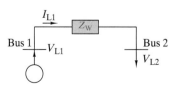

FIGURE 16–26

SOLUTION:

For the given values of V_{L2}, and S_2 we find the line current as

$$I_{L1} = \frac{|S_2|}{\sqrt{3}V_{L2}} = \frac{|(70 + j35) \times 10^3|}{\sqrt{3} \times 2400} = 18.83 \text{ A(rms)}$$

The power lost in the line is

$$S_W = 3I_L^2 Z_W = 3 \times (18.83)^2 (1.5 + j8.5)$$
$$= 1.6 + j9.0 \text{ kVA}$$

The source at bus 1 must supply the load power at bus 1 plus the losses in the line. Hence, the complex power produced by the source is

$$S_1 = S_2 + S_W = 71.6 + j44 \text{ kVA}$$

Now that we have I_{L1}, and S_1, we find the line voltage at bus 1 as

$$V_{L1} = \frac{|S_2|}{\sqrt{3}\,I_{L1}} = \frac{|(71.6 + j44) \times 10^3|}{\sqrt{3} \times 18.83} = 2.577 \text{ kV(rms)}$$

In round numbers, the conditions $V_{L1} = 2.58 \text{ kV}$, $I_{L1} = 18.8 \text{ A}$, and $V_{1.2} = 2.4 \text{ kV}$ will produce the required load power flow. This set of conditions is not unique and many other solutions exist.   ■

Exercise 16–19

In Figure 16–26 the load at bus 2 draws a complex power of $S_2 = 125 + j60 \text{ kVA}$. The line current is $I_L = 40 \text{ A(rms)}$, and the line impedance is $Z_W = 2 + j10 \text{ }\Omega/\text{phase}$. Find the line voltages at bus 1 and bus 2.

Answers: $V_{L1} = 2.49 \text{ kV(rms)}$, $V_{L2} = 2.00 \text{ kV(rms)}$

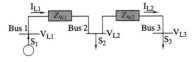

FIGURE 16–27

EXAMPLE 16–23

In Figure 16–27 the source at bus 1 supplies two load buses through transmission lines with $Z_{W1} = 45 + j250 \text{ }\Omega/\text{phase}$ and $Z_{W2} = 50 + j330 \text{ }\Omega/\text{phase}$. Line 1 connects the source bus to a load at bus 2 that draws a complex power of $S_2 = 1.5 + j0.5 \text{ MVA}$. Line 2 connects bus 2 to a load at bus 3 that draws a complex power of $S_3 = 2 + j1.5 \text{ MVA}$. Assuming that the line voltage at bus 3 is $V_{L3} = 115 \text{ kV(rms)}$, find the bus voltages V_{L1} and V_{L2}, the line currents I_{L1} and I_{L2}, and the source power output needed to produce the specified load powers S_2 and S_3.

SOLUTION:

For the given values of V_{L3} and S_3, we find the line current in line 2 as

$$I_{L2} = \frac{|S_3|}{\sqrt{3}V_{L3}} = \frac{|(2 + j1.5) \times 10^6|}{\sqrt{3} \times 115 \times 10^3} = 12.55 \text{ A(rms)}$$

The power lost in the line 2 is

$$S_{W2} = 3I_{L2}^2 Z_{W2} = 3 \times (12.55)^2 (50 + j330)$$
$$= 23.6 + j156 \text{ kVA}$$

At bus 2 the input power to line 2 supplies its losses S_{W2} plus the bus 3 load S_3, namely

$$S_{W2} + S_3 = 2.024 + j1.656 \, \text{MVA}$$

At bus 2 the line current into line 2 is I_{L2}. Since the input power to line 2 is $S_{W2} + S_3$, the line voltage at bus 2 is found to be

$$V_{L2} = \frac{|S_{W2} + S_3|}{\sqrt{3} \, I_{L2}} = \frac{|(2.024 + j1.656) \times 10^6|}{\sqrt{3} \times 12.55} = 120.3 \, \text{kV(rms)}$$

At bus 2 the output power of line 1 supplies the bus 2 load plus the input power into line 2, namely

$$S_2 + (S_{W2} + S_3) = 3.524 + j2.156 \, \text{MVA}$$

The line voltage at the output of line 1 is V_{L2}, hence the line current in line 1 is

$$I_{L1} = \frac{|S_2 + S_{W2} + S_3|}{\sqrt{3} V_{L2}} = \frac{|(3.524 + j2.156) \times 10^6|}{\sqrt{3} \times 120.3 \times 10^3} = 19.83 \, \text{A(rms)}$$

Given the line current I_{L1} the total complex power lost in line 1 is

$$S_{W1} = 3I_{L1}^2 Z_{W1} = 3 \times (19.83)^2 (45 + j250)$$
$$= 53.1 + j295 \, \text{kVA}$$

Finally we arrive at bus 1, where the input power to line 1 must equal its losses S_{W1}, plus its output at the bus 2. This power is supplied by the source at bus 1, hence

$$S_1 = S_{W1} + [S_2 + (S_{W2} + S_3)] = 3.577 + j2.451 \, \text{MVA}$$

At bus 1 the line current is I_{L1}, hence the line voltage is found to be

$$V_{L1} = \frac{|S_1|}{\sqrt{3} \, I_{L1}} = \frac{|(3.577 + j2.451) \times 10^6|}{\sqrt{3} \times 19.83} = 126.2 \, \text{kV(rms)}$$

In round numbers, the conditions $V_{L1} = 126.2 \, \text{kV(rms)}$, $I_{L1} = 19.8 \, \text{A(rms)}$, $V_{L2} = 120.3 \, \text{kV(rms)}$, $I_{L2} = 12.5 \, \text{A(rms)}$, and $V_{L3} = 115 \, \text{kV(rms)}$ will produce the required load power flow. This set of conditions is not unique. For example, the set $V_{L1} = 153 \, \text{kV(rms)}$, $I_{L1} = 15.9 \, \text{A(rms)}$, $V_{L2} = 148 \, \text{kV(rms)}$, $I_{L2} = 10 \, \text{A(rms)}$, and $V_{L3} = 144 \, \text{kV(rms)}$ will produce the same load power flow. ∎

S U M M A R Y

- For sinusoidal voltages and currents the instantaneous power is the sum of a dc component and an ac component. The dc component is the average power that produces a net transfer of energy. The ac component produces no net energy transfer.

- In ac power analysis the amplitudes of voltage and current phasors are expressed in rms values. Complex power is defined as $S = \mathbf{V}\mathbf{I}^* = P + jQ$. The real part P is the average power in watts (W), the imaginary part Q is the reactive power in volt-amperes reactive

(VAR), and the magnitude $|S|$ is the apparent power in volt-amperes (VA).

- The power factor is the ratio of the average power to the apparent power or the cosine of the phase angle between the voltage and current phasors. The power factor is said to be lagging when the current lags the voltage $(Q > 0)$ and leading when the current leads the voltage $(Q < 0)$.

- In a single-phase circuit all sources have the same phase angle. The objective of single-phase circuit analysis is to find the power delivered to a specified load for given source conditions. The objective of power flow analysis is to find the source conditions needed to deliver a specified power to a load.

- Power factor correction increases the power factor without changing the power flow to an inductive load. The correction is achieved by adding capacitance in parallel with the inductive load.

- A balanced three-phase source produces line-to-neutral voltages $\mathbf{V}_{AN}$, $\mathbf{V}_{BN}$, and $\mathbf{V}_{CN}$ with $120°$ phase separation and the same amplitude V_P. The line-to-line voltages $\mathbf{V}_{AB}$, $\mathbf{V}_{BC}$, and $\mathbf{V}_{CA}$ have $120°$ phase separation and the same amplitude $V_L = \sqrt{3} V_P$. In a positive phase sequence $\mathbf{V}_{AB}$ leads $\mathbf{V}_{AN}$ by $30°$.

- A balanced Y-connected load has three equal impedances Z_Y, and a balanced Δ-connected load has three equal impedances Z_Δ. A Δ-connected load can be replaced by an equivalent Y-connected load whose impedances are $Z_Y = Z_\Delta/3$.

- A shortcut way to analyze a balanced three-phase circuit is convert it into an equivalent Y-Y configuration. Analyzing phase A of the Y-Y equivalent yields the phasor responses needed to get the phasor responses in any other phase.

- In a balanced three-phase circuit (1) the neutral wire carries no current, (2) the line currents $\mathbf{I}_A$, $\mathbf{I}_B$, and $\mathbf{I}_C$ have $120°$ phase separation and the same amplitude I_L, (3) in a Δ-connected load the phase currents $\mathbf{I}_{AB}$, $\mathbf{I}_{BC}$, and $\mathbf{I}_{CA}$ have $120°$ phase separation and the same amplitude $I_P = I_L/\sqrt{3}$, (4) in a positive phase sequence $\mathbf{I}_{AB}$ leads $\mathbf{I}_A$ by $30°$, and (5) the complex power delivered to a Y- or a Δ-connected load is $S_L = P_L + jQ_L = \sqrt{3} V_L \times I_L \angle\theta$, where θ is the angle of the phase impedance.

- The power flow in a balanced three-phase system is determined by the line voltage at each of its buses and the impedances of the lines connecting the buses. The net power flow in all of the lines connected to a bus determines the power flow into or out of a bus. A power flow problem involves finding a set of bus voltages and line currents that produce a prescribed load power flow.

PROBLEMS

OBJECTIVE 16–1 COMPLEX POWER (SECTS. 16–1, 16–2)

Given a linear circuit in the sinusoidal steady state:
(a) Find the average, reactive, and instantaneous powers for a specified voltage and current.
(b) Find the load impedance for a specified load power flow.
See Examples 16–1, 16–2, 16–3 and Exercises 16–1, 16–2, 16–3

16–1 The following sets of $v(t)$ and $i(t)$ apply to the load circuit in Figure P16–1. Find the average power, reactive power, and instantaneous power delivered to the load.
(a) $v(t) = 1500 \cos(\omega t - 45°)$V, $i(t) = 2\cos(\omega t + 50°)$A
(b) $v(t) = 90 \cos(\omega t + 60°)$V, $i(t) = 10.5\cos(\omega t - 20°)$A

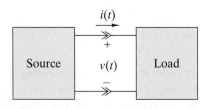

FIGURE P16–1

16–2 The following sets of $v(t)$ and $i(t)$ apply to the load circuit in Figure P16–1. Calculate the average power and the reactive power.
(a) $v(t) = 135 \cos(\omega t)$V, $i(t) = 2 \cos(\omega t + 30°)$ A
(b) $v(t) = 370 \sin(\omega t)$V, $i(t) = 10 \cos(\omega t + 20°)$ A

16–3 The following voltage and current phasors apply to the circuit in Figure P16–3. Calculate the average power and reactive power delivered to the impedance Z. Find the power factor and state whether the power factor is lagging or leading.
(a) $\mathbf{V} = 250\angle 0°$ V(rms), $\mathbf{I} = 0.25\angle -15°$ A(rms)
(b) $\mathbf{V} = 120\angle 135°$ V(rms), $\mathbf{I} = 12.5\angle 165°$ A(rms)

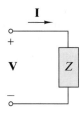

FIGURE P16–3

16–4 The following sets of $\mathbf{V}$ and $\mathbf{I}$ apply to the circuit in Figure P16–3. Calculate the complex power and the power factor. State whether the power factor is lagging or leading.
(a) $\mathbf{V} = 120\angle 30°$ V(rms), $\mathbf{I} = 3.3\angle -15°$A(rms)
(b) $\mathbf{V} = 480\angle 45°$ V(rms), $\mathbf{I} = 8.5\angle 90°$ A(rms)

16–5 The conditions in this problem apply to the circuit in Figure P16–3. Calculate the complex power for each condition given below.
(a) $\mathbf{V} = 15\angle 45°$ kV(rms), $Z = 500\angle -15°\Omega$
(b) $Z = 40 - j30\Omega$, $\mathbf{I} = 10\angle 25°$ A(rms)

16–6 Find the power factor under the following conditions. State whether the power factor is lagging or leading.
(a) $S = 1000 + j250$ kVA
(b) $|S| = 15$ kVA, $P = 12$ kW, $Q < 0$.

16–7 A load draws an apparent power of 30 kVA at a power factor of 0.8 lagging from a 2400-V (rms) source. Find P, Q, and the load impedance.

16–8 A load draws 8 kW at a power factor of 0.8 leading from a 880-V(rms) source. Find Q and the load impedance.

16–9 A load draws 15 A (rms), 5 kW, and 2.5 kVARS (lagging) from a 60-Hz source. Find the load power factor and impedance.

16–10 Find the impedance of a load that is rated at 440 V (rms), 5 A (rms), and 2.2 kW.

OBJECTIVE 16–2 SINGLE-PHASE CIRCUIT ANALYSIS (SECT. 16–3)

Given a single-phase circuit operating in the ac steady state, find the power produced by sources or delivered to specified loads.
See Examples 16–4, 16–5, 16–6, 16–7 and Exercises 16–4, 16–5

16–11 A load made up of a 100-Ω resistor in series with a 150-mH inductor is connected across a 240-V(rms), 60-Hz voltage source. Find the complex power delivered to the load and the load power factor. State whether the power factor is lagging or leading.

16–12 A load made up of a 50-Ω resistor in parallel with a 10-μF capacitor is connected across a 400-Hz source that delivers 110 V(rms). Find the complex power delivered to the load and the load power factor. State whether the power factor is lagging or leading.

16–13 A load made up of a resistor R in series with an inductor L draws a complex power of $1200 + j800$VA when connected across a 440-V (rms), 60-Hz voltage source. Find the value of the value of R and L.

16–14 In Figure P16–14 the load Z_L, is a 60-Ω resistor in series with a capacitor whose reactance is -30Ω. The source voltage is 440 V (rms). Find the complex power produced by the source and the complex power delivered to the load.

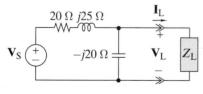

FIGURE P16–14

16–15 Repeat Problem 16–14 when Z_L is a 50-Ω resistor in parallel with an impedance of $40 - j60\,\Omega$.

16–16 In Figure P16–16 the load Z_L is a 100-Ω resistor and the source voltage is 240 V(rms). Find the complex power produced by the source.

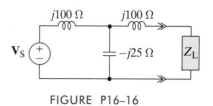

FIGURE P16–16

16–17 Repeat Problem 16–16 when the load Z_L, is a 50-Ω resistor in series with a capacitor whose reactance is $-100\,\Omega$.

16–18 In Figure P16–18 the three load impedances are: $Z_1 = 20 + j15\,\Omega$, $Z_2 = 25 + j10\,\Omega$, and $Z_3 = 75 + j50\,\Omega$. Find the total complex power produced by the two sources and the overall circuit power factor.

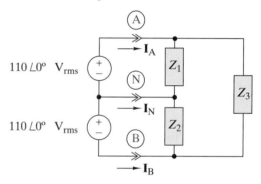

FIGURE P16–18

16–19 In Figure P16–18 the complex powers delivered to each load are: $S_1 = 400 + j270$ VA, $S_2 = 550 + j150$ VA, and $S_3 = 1000 + j0$ VA. Find the line currents $\mathbf{I}_A$, $\mathbf{I}_N$, and $\mathbf{I}_B$.

16–20 Two loads are connected in parallel across an 880 V(rms) line. The first load draws an average power of 20 kW at a lagging power factor of 0.8. The second load draws 16 kW at a lagging power factor of 0.9. Find the overall power factor of the circuit and the rms current drawn from the line.

OBJECTIVE 16–3 SINGLE-PHASE POWER FLOW (SECT. 16–4)

Given a specified load power in a single phase circuit:
(a) Find the required source outputs.
(b) Find the parallel capacitance needed to produce a specified power factor.
See Examples 16–8, 16–9, 16–10, 16–11 and Exercise 16–6, 16–7

16–21 The apparent power delivered to the load Z_L in Figure P16–21 is 46 kV A at a lagging power factor of 0.84. The load voltage is 2.4 kV(rms) and the line has an impedance of $Z_W = 1 + j8\,\Omega$/wire. Find the required apparent source power, the source power factor, and the magnitude of the source voltage.

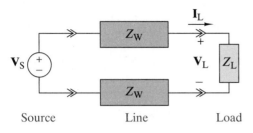

FIGURE P16–21

16–22 Repeat Problem 16–21 with the load power factor increased to 0.95.

16–23 The average power delivered to the load Z_L in Figure P16–21 is 250 kW at a lagging power factor of 0.85. The load voltage is 7.2 kV(rms) and the line has an impedance of $Z_W = 2 + j12 \, \Omega/\text{wire}$. Find the apparent power supplied by the source and magnitude of the source voltage.

16–24 The complex power delivered to the load Z_L in Figure P16–21 is $20 + j15$ kVA. The source produces an average power of 21 kW and the line has an impedance of $Z_W = 2.1 + j12 \, \Omega/\text{wire}$. Find the magnitude of the source and load voltages.

16–25 In Figure P16–25 the voltage across the two loads is $|\mathbf{V}_L| = 4.8 \, \text{kV(rms)}$. The load Z_1 draws an average power of 12 kW and a power factor of 0.85 lagging. The load Z_2 draws an apparent power of 15 kVA and at a lagging power factor of 0.8. The line has an impedance of $Z_W = 9 + j50 \, \Omega/\text{wire}$. Find the apparent power produced by the source and the rms value of the source voltage.

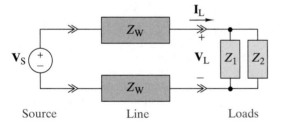

FIGURE P16–25

16–26 The two loads in Figure P16–25 draw apparent powers of $|\mathbf{S}_1| = 16 \, \text{kVA}$ at a lagging power factor of 0.8 and $|\mathbf{S}_2| = 25 \, \text{kVA}$ at unity power factor. The voltage across the loads is 3.8 kV and the line has an impedance of $Z_W = 5 + j26 \, \Omega/\text{wire}$. Find the apparent power produced by the source and the rms value of the source voltage.

16–27 A 60-Hz voltage source feeds a two-wire line with $Z_W = 0.6 + j3.4 \, \Omega/\text{wire}$. The load at the receiving end of the line draws an apparent power of 5 kVA at a leading power factor 0.8. The voltage across the load is 500 V(rms). Find the apparent power produced by the source and the rms value of the source voltage.

16–28 In Figure P16–28 the load voltage is $|\mathbf{V}_L| = 4160 \, \text{V(rms)}$ at 60 Hz and the load Z_L draws an average power of 10 kW at a lagging power factor of 0.725. Find the overall power factor of the combination if the parallel capacitance is 1 μF. Find the value of the capacitance needed to raise the overall power factor to unity.

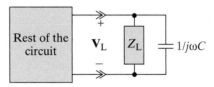

FIGURE P16–28

16–29 In Figure P16–28 the load voltage is $|\mathbf{V}_L| = 2400 \, \text{V(rms)}$ at 60 Hz. The load Z_L draws an apparent power of 25 kVA at a lagging power factor of 0.7. Find the value of the capacitance required to raise the overall power factor of the parallel combination to 0.95. Repeat this problem for an overall power factor of unity.

16–30 A load draws 5 A (rms) and 4 kW at a power factor 0.8 (lagging) from a 60-Hz source. Find the capacitance needed in parallel with the load to raise the overall power factor to unity.

OBJECTIVE **16–4 B**ALANCED **T**HREE**-P**HASE **C**IRCUITS **(S**ECT**. 16–5)**

For a balanced three-phase circuit:
(a) Find all of the phase and line voltage phasors for a given phase reference.
(b) Find equivalent Y- or Δ-connected sources and loads.
See Examples 6-12, 16–13 and Exercises 16–8, 16–9, 16–10

16–31 In a balanced three-phase circuit the phase voltage magnitude is $V_P = 277 \, \text{V(rms)}$. For a positive phase sequence:
(a) Find all of the line and phase voltage phasors using $\mathbf{V}_{AN}$ as the phase reference.
(b) Sketch a phasor diagram of the line and phase voltages.

16–32 In a balanced three-phase circuit the line voltage magnitude is $V_L = 2.4 \, \text{kV(rms)}$. For a positive phase sequence:
(a) Find all of the line and phase voltage phasors using $\mathbf{V}_{AB}$ as the phase reference.
(b) Sketch a phasor diagram of the line and phase voltages.

16–33 In a balanced three-phase circuit $\mathbf{V}_{BC} = j208 \, \text{V(rms)}$. Find all the line and phase voltage phasors in polar form for a positive phase sequence.

16–34 In a balanced three-phase circuit $\mathbf{V}_{BN} = 200 + j150 \, \text{V(rms)}$. Find all of the line and phase voltage phasors in polar form for a positive phase sequence.

16–35 A balanced Δ-connected three phase source has $\mathbf{V}_{AB} = 208\angle 30° \, \text{V(rms)}$ and a positive phase sequence. Find the three source voltages of the equivalent Y-connected source.

16–36 In a balanced three-phase circuit $\mathbf{V}_{BN} = V_P\angle 90°$ V(rms). Show that $\mathbf{V}_{BC} = \sqrt{3}V_P\angle 120°$ V(rms) for a positive phase sequence.

16–37 A balanced Y-connected load with $Z_Y = 10 - j10\ \Omega$/phase is connected in parallel with a balanced Δ-connected load with $Z_\Delta = 60 + j30\ \Omega$/phase. Find the phase impedance of an equivalent Δ-connected load.

16–38 A balanced Y-connected load with $Z_Y = 30 + j30\ \Omega$/phase is connected in parallel with a balanced Δ-connected load with $Z_\Delta = 60 - j900\ \Omega$/phase. Find the phase impedance of an equivalent Y-connected load.

16–39 A balanced Y-connected load with $Z_{Y1} = 12 + j6\ \Omega$/phase is connected in parallel with a second Y-connected load with $Z_{Y2} = 24 + j6\ \Omega$/phase. Find the phase impedance of the equivalent Y-connected load.

16–40 In a balanced Δ-Δ circuit the Δ-connected source produces $\mathbf{V}_{AB} = 2400\angle 45°$ V(rms) and a positive phase sequence. The phase impedance of the load is $Z_\Delta = 200\angle 45°\ \Omega$/phase. Find the three source voltages and the phase impedance in the equivalent Y-Y circuit.

OBJECTIVE 16–5 THREE-PHASE CIRCUIT ANALYSIS (SECT. 16–6)

For a given balanced three-phase circuit:
(a) Find the line and phase current phasor for a specified phase reference.
(b) Find the source or load power, using the scalars V_L and I_L.
See Examples 16–14, 16–15, 16–16, 16–17, 16–18, 16–19, 16–20, 16–21 and Exercises 16–11, 16–12, 16–13, 16–14, 16–15, 16–16, 16–17, 16–18

16–41 In a balanced Δ-Y circuit the line voltage and phase impedance are $V_L = 560$ V(rms) and $Z_Y = 25 + j10\ \Omega$/phase. Using $\angle\mathbf{V}_{AB} = 0°$ as the phase reference, find the line current and phase voltage phasors in polar form for a positive phase sequence.

16–42 In a balanced Y-Y circuit the line voltage is $V_L = 480$ V(rms). The phase impedance is $Z_Y = 40 + j25\ \Omega$/phase. Using $\angle\mathbf{I}_A = 0°$ as the phase reference, find $\mathbf{I}_A$ and $\mathbf{V}_{AB}$ in polar form for a positive phase sequence.

16–43 In a balanced Y-Δ circuit the line voltage and phase impedance are $V_L = 440$ V(rms) and $Z_\Delta = 16 + j12\ \Omega$/phase. Using $\angle\mathbf{V}_{AN} = 0°$ as the phase reference, find the line current and phase current phasors in polar form for a positive phase sequence.

16–44 In a balanced Y-Δ circuit the line impedances connecting the source and load are $Z_W = 1.5 + j6.5\ \Omega$/phase. The phase impedances of the Δ-connected load are $Z_\Delta = 15 + j8\ \Omega$/phase and the line voltage at the source is $V_L = 250$ V(rms). Using $\angle\mathbf{V}_{AB} = 0°$ as the phase reference, find the line current $\mathbf{I}_A$ and phase current $\mathbf{I}_{AB}$ for a positive phase sequence.

16–45 In a balanced Y-Δ circuit the line voltage and phase impedance are $V_L = 4.16$ kV(rms) and $Z_\Delta = 250\angle 30°\ \Omega$/phase. Using $\angle\mathbf{V}_{AB} = 0°$ as the phase reference, find the phase current $\mathbf{I}_{AB}$ and line current $\mathbf{I}_A$ for a positive phase sequence.

16–46 A balanced three-phase source with $\mathbf{V}_{AB} = 440\angle 30°$ V(rms) supplies a balanced Δ-connected load with a phase impedance of $Z_\Delta = 20\angle -60°\ \Omega$/phase. Find the phase current $\mathbf{I}_{AB}$ and line current $\mathbf{I}_A$ for a positive phase sequence.

16–47 In a balanced Y-connected load the line current and phase impedance are $I_L = 6$ A(rms) and $Z_Y = 20 + j15\ \Omega$/phase. Using $\angle\mathbf{V}_{AB} = 0°$ as the phase reference, find the line current $\mathbf{I}_A$ and phase voltage $\mathbf{V}_{AN}$ for a positive phase sequence.

16–48 In a balanced Δ-connected load the phase current and phase impedance are $I_P = 12$ A(rms) and $Z_\Delta = 200\angle 60°\ \Omega$/phase. Using $\angle\mathbf{V}_{AN} = 0°$ as the phase reference, find the line current $\mathbf{I}_A$ and line voltage $\mathbf{V}_{AB}$ for a positive phase sequence.

16–49 An average power of 6 kW is delivered to a balanced three-phase load with a phase impedance of $Z_Y = 40 + j30\ \Omega$/phase. Find V_L and the complex power delivered to the load.

16–50 An apparent power of 12 kVA is delivered to balanced three-phase load with a phase impedance of $Z_Y = 120 + j90\ \Omega$/phase. Find I_L, V_L, and the complex power delivered to the load.

16–51 A balanced three-phase load has a phase impedance of $Z_Y = 60 - j20\ \Omega$/phase. The line voltage at the load is $V_L = 440$ V(rms). Find I_L and the complex power delivered to the load.

16–52 A balanced three-phase load has a phase impedance of $Z_\Delta = 200 + j100\ \Omega$/phase. The line voltage at the load is $V_L = 2.4$ kV(rms). Find I_L and the complex power delivered to the load.

16–53 A balanced three-phase load has a phase impedance is $Z_Y = 200 + j100\ \Omega$/phase. The line current at the load is $I_L = 16$ A(rms). Find V_L and the complex power delivered to the load.

16–54 The balanced three-phase source in Figure P16–54 produces an average power of 50 kW. The line impedance is $Z_W = 1 + j6\ \Omega$/phase and the balanced load has a phase impedance $Z_Y = 200 + j100\ \Omega$/phase. Find the line voltage V_L at the load and complex power delivered to the load.

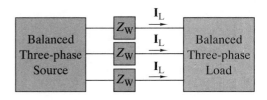

FIGURE P16–54

16–55 The balance three-phase source in Figure P16–54 produces an average power of 15 kW at a power factor of 0.85 lagging. The line impedance is $Z_W = 3 + j15 \, \Omega$/phase and the balanced load has a phase impedance of $Z_\Delta = 250 + j55 \, \Omega$/phase. Find the line voltage V_L at the load and complex power delivered to the load.

16–56 The balanced three-phase source in Figure P16–54 produces an apparent power of 25 kVA at a power factor 0.9 lagging. The line impedance is $Z_W = 2 + j10 \, \Omega$/phase and line current is 10 A(rms). Find the line voltage V_L at the load and complex power delivered to the load.

16–57 The balanced three-phase source in Figure P16–54 produces a average power of 50 kW at a power factor of 0.75 lagging. The line impedance is $Z_W = 2 + j12 \, \Omega$/phase and line voltage at the source is 4160 V(rms). Find the line voltage V_L at the load and complex power delivered to the load.

16–58 Two balanced three-phase loads are connected in parallel. The first load absorbs 25 kW at a lagging power factor of 0.9. The second load absorbs an apparent power of 30 kVA at a leading power factor of 0.1. The line voltage at the parallel loads is $V_L = 880$ V(rms). Find the line current into the combined load.

16–59 The average power delivered to a balanced Y-connected load is 20 kW at a lagging power factor of 0.8. The line voltage at the load is $V_L = 480$ V(rms). Find the phase impedance Z_Y of the load.

16–60 The apparent power delivered to a balanced Δ-connected load is 30 kVA at a lagging power factor of 0.72. The line voltage at the load is $V_L = 2.4$ kV(rms). Find the phase impedance Z_Δ of the load.

OBJECTIVE 16–6 THREE-PHASE POWER FLOW (SECT. 16–7)

Given a single-line diagram of a balanced three-phase system, find the source outputs and bus voltages that produce a prescribed load power flow.
See Examples 16–22, 16–23, and Exercise 16–19

16–61 In Figure P16–61 the source and load busses are interconnected by a transmission line with $Z_W = 70 + j400 \, \Omega$/phase. The load at bus 2 draws an apparent power of $|S_2| = 3$ MVA at a lagging power factor of 0.85 and the line voltage at bus 2 is $V_{L2} = 230$ kV(rms). Find the apparent power produced by the source at bus 1, the source power factor, and the line voltage at bus 1.

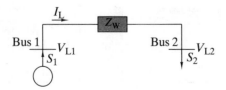

FIGURE P16–61

16–62 In Figure P16–61 the source and load busses are interconnected by a transmission line with $Z_W = 2 + j10 \, \Omega$/phase. The load at bus 2 draws an average power of $P_2 = 50$ kW at a lagging power factor of 0.8 and the line voltage at bus 2 is $V_{L2} = 4.16$ kV(rms). Find the apparent power produced by the source at bus 1, the source power factor, and the line voltage at bus 1.

16–63 In Figure P16–61 the source and load busses are interconnected by a transmission line with $Z_W = 10 + j75 \, \Omega$/phase. The load at bus 2 draws an average power of $P_2 = 600$ kW at a lagging power factor of 0.8 and the line current is $I_{L1} = 14$ A(rms). Find the source power factor and the line voltages at bus 1 and bus 2.

16–64 In Figure P16–64 the three buses are interconnected by transmission lines with wire impedances of $Z_{W1} = 100 + j600 \, \Omega$/phase and $Z_{W2} = 120 + j800 \, \Omega$/phase. The source at bus 2 produces an apparent power of $|S_2| = 300$ kVA at a lagging power factor of 0.85. The load at bus 3 draws an apparent power of $|S_3| = 600$ kVA at a lagging power factor of 0.8. The line voltage at bus 3 is $V_{L3} = 161$ kV(rms). Find the apparent power produced by the source at bus 1, the source power factor, and the line voltages at bus 1 and bus 2.

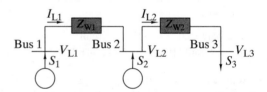

FIGURE P16–64

16–65 In Figure P16–64 the three buses are interconnected by transmission lines with wire impedances of $Z_{W1} = 120 + j800 \, \Omega$/phase and $Z_{W2} = 200 + j1200 \, \Omega$/phase. The source at bus 2 produces an apparent power of $|S_2| = 400$ kVA at a leading power factor of 0.9. The load at bus 3 draws an apparent power of $|S_3| = 650$ kVA at a lagging power factor of 0.95. The line current in line 2 is $I_{L2} = 3.25$ A(rms). Find the apparent power produced by the source at bus 1, the source power factor, and the line voltages at bus 1, bus 2, and bus 3.

16–66 In Figure P16–66 the three buses are interconnected by transmission lines with wire impedances of $Z_{W1} = 100 + j850 \, \Omega$/phase and $Z_{W2} = 50 + j250 \, \Omega$/phase. The load at bus 1 draws an apparent power of $|S_1| = 400$ kVA at a lagging power factor of 0.85. The line voltage at bus 1 is $V_{L1} = 138$ kV(rms). The load at bus 3 draws an apparent power of $|S_3| = 475$ kVA at a lagging power factor of 0.95. The line current in line 2 is $I_{L2} = 2.0$ A(rms). Find the apparent power produced by the source at bus 2, the source power factor, and the line voltages at bus 2 and bus 3.

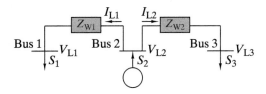

FIGURE P16–66

16–67 In Figure P16–67 the source at bus 1 supplies two load buses through transmission lines with wire impedances of $Z_{W1} = 6 + j33\ \Omega/\text{phase}$ and $Z_{W2} = 3 + j15\ \Omega/\text{phase}$. The load at bus 2 draws an apparent power 4 MVA at a lagging power factor of 0.95. The load at bus 3 draws an apparent power of 3 MVA at a lagging power factor of 0.9. The line current in line 2 is $I_{L2} = 12.5\ \text{A(rms)}$. Find the apparent power produced by the source at bus 1, the source power factor, and the line voltages at bus 1, bus 2, and bus 3.

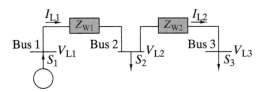

FIGURE P16–67

INTEGRATING PROBLEMS

16–68 Three-phase Voltages in the Time Domain A
A balanced 60-Hz three-phase system has a line voltage of $V_L = 208\ \text{V(rms)}$. Using $\angle \mathbf{V}_{AB} = 0°$ as the phase reference, derive time-domain expressions for the phase voltages $v_{AN}(t)$, $v_{BN}(t)$, and $v_{CN}(t)$ for a positive phase sequence.

16–69 Single-Phase Motor Power Factor A
When a $\frac{1}{2}$-hp (1 hp = 746 W) single-phase induction motor delivers its rated mechanical output it is 65% efficient and draws a current of 4.8 A(rms) from a 200 V(rms) line. Find the power factor of the motor.

16–70 Three-Phase Transformer A
Figure P16–70 shows three identical ideal transformers connected to form a three-phase transformer. The windings on the primary (source) side are Δ-connected and the windings on the secondary (load) side are Y-connected. On the secondary side of the transformer $V_{L2} = 830\ \text{V(rms)}$ and $I_{L2} = 40\ \text{A(rms)}$. The turns ratio of each transformer

is $n = 0.2$. Find the line voltage and line current on the primary side.

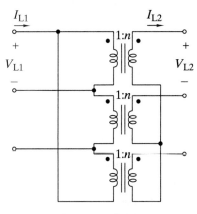

FIGURE P16–70

16–71 Phase Converter Efficiency A
Three-phase motors are often used in equipment because they are more efficient and reliable than single-phase motors. Such equipment may be installed in locations where only single-phase power is available and the cost of installing three-phase service is prohibitive. The rotary phase converter in Figure P16–71 is one way of providing three-phase power from a single-phase source. Simply stated, the converter is a rotating transformer that shifts the phase of a portion of the single-phase input to produce a balanced set of three-phase voltages. In a certain application a converter supplies three-phase power to a 30-hp motor (1 hp = 746 W) that is 85% efficiency at full load. At full load the single-phase inputs are 220 V (rms) and 130 A (rms) at a power factor of 0.95 lagging. Find the efficiency of the converter.

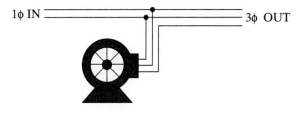

FIGURE P16–71

CHAPTER 17 TWO-PORT NETWORKS

"Among the tools of thought and artifices by which man forces his mind to give him more service, perhaps the most intensely useful are the simple mathematical rules of inversion known as reciprocity theorems"

Stuart Ballantine, 1928
American Radio Pioneer

Some History Behind This Chapter

The purpose of this chapter is to introduce a special class of network functions that characterize linear two-port circuits. These network functions involve both the driving point and transfer functions as defined in Chapter 11. The added constraint here is that they are defined under open-circuit and short-circuit conditions. These special functions known collectively as *reciprocity theorems* have their roots from the earliest days of communications and power transmission and find continued use today.

Why This Chapter Is Important Today

Power and communication networks as well as the design of electronic amplifiers are often described as two-port circuits, devices, or networks. Each application uses a set of parameters that help define how each operates and are interconnected. This chapter discusses four basic sets of parameters that are most commonly used in these ubiquitous applications.

Chapter Sections

Chapter Learning Objectives

17-1 Impedance and Admittance Parameters (Sects. 17–2, 17–3)

(a) Given a linear two-port network, find the impedance or admittance parameters.

(b) Given the impedance or admittance parameters of a linear two-port network, find responses at the input and output ports for specified external connections.

17-2 Hybrid and Transmission Parameters (Sects. 17–4, 17–5)

(a) Given a linear two-port network, find the hybrid or transmission parameters.

(b) Given the hybrid or transmission parameters of a linear two-port network, find responses at the input and output ports for specified external connections.

17-3 Two-port Conversions and Connections (Sect. 17–6)

(a) Given a set of two-port parameters, find other two-port parameters.

(b) Given the parameters of several two port networks, find the parameters of an interconnection of the two ports.

17–1 INTRODUCTION

The purpose of this chapter is to study methods of characterizing and analyzing two-port networks. A **port** is a terminal pair where energy can be supplied or extracted. A **two-port network** is a four-terminal circuit in which the terminals are paired to form an input port and an output port. Figure 17–1 shows the customary way of defining the port voltages and currents. Note that the reference marks for the port variables comply with the passive sign convention. The linear circuit connecting the two ports is assumed to be in the zero state and to be free of any independent sources. In other words, there is no initial energy stored in the circuit and the box in Figure 17–1 contains only resistors, capacitors, inductors, mutual inductance, and dependent sources.

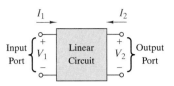

FIGURE 17–1

A four-terminal network qualifies as a two-port network if *the net current entering each terminal pair is zero.* This means that the current exiting the lower port terminals in Figure 17–1 must be equal to the currents entering the upper terminals. One way to meet this condition is to always connect external sources and loads between the input terminal pair or between the output terminal pair.

The first task is to identify circuit parameters that characterize a two-port. In the two-port approach the only available variables are the port voltages V_1 and V_2, and the port currents I_1 and I_2. A set of two-port parameters is defined by expressing two of these four port variables in terms of the other two variables. In this chapter we study the four ways in Table 17–1.

Note that each set of parameters is defined by two equations, one for each of the two dependent port variables. Each equation involves a sum of two terms, one for each of the two independent port variables. Each term involves a proportionality because the two port is a linear circuit and superposition applies. The names given the parameters indicate their dimensions (impedance and admittance), a mixture of dimensions (hybrid), or their original application (transmission lines). With double subscripted parameters, the first subscript indicates the port at which the dependent variable appears and the second subscript the port at which the independent variable appears.

Regardless of their dimensions, all two-port parameters are network functions. In general the parameters are functions of the complex frequency variable and *s*-domain circuit analysis applies. For sinusoidal steady-state problems, we replace s by $j\omega$ and use phasor circuit analysis. For purely resistive circuits, the two-port parameters are real constants and we use resistive circuit analysis.

Before turning to specific parameters, it is important to specify the objectives of two-port network analysis. Briefly, these objectives are:

1. Determine two-port parameters of a given circuit.

2. Use two-port parameters to find port variable responses for specified input sources and output loads.

In principle, the port variable responses can be found by applying node or mesh analysis to the internal circuitry connecting the input and output ports. So

TABLE 17–1 **TWO-PORT PARAMETERS**

NAME	EXPRESS	IN TERMS OF	DEFINING EQUATIONS
Impedance	V_1, V_2	I_1, I_2	$V_1 = z_{11}I_1 + z_{12}I_2$ and $V_2 = z_{21}I_1 + z_{22}I_2$
Admittance	I_1, I_2	V_1, V_2	$I_1 = y_{11}V_1 + y_{12}V_2$ and $I_2 = y_{21}V_1 + y_{22}V_2$
Hybrid	V_1, I_2	I_1, V_2	$V_1 = h_{11}I_1 + h_{12}V_2$ and $I_2 = h_{21}I_1 + h_{22}V_2$
Transmission	V_1, I_1	$V_2, -I_2$	$V_1 = AV_2 - BI_2$ and $I_1 = CV_2 - DI_2$

why adopt the two-port point of view? Why not use straight-forward circuit analysis?

There are several reasons. First, two-port parameters can be determined experimentally without resorting to circuit analysis. Second, there are applications in power systems and microwave circuits in which input and output ports are the only places that signals can be measured or observed. Finally, once two-port parameters of a circuit are known, it is a relatively simple matter to find port variable responses for different input sources and/or different output loads.

17–2 IMPEDANCE PARAMETERS

The impedance parameters are obtained by expressing the port voltages V_1 and V_2 in terms of the port currents I_1 and I_2.

$$V_1 = z_{11}I_1 + z_{12}I_2$$
$$V_2 = z_{21}I_1 + z_{22}I_2$$

(17–1)

The network functions z_{11}, z_{12}, z_{21}, and z_{22} are called the **impedance parameters** or simply the **z-parameters**. These equations have the matrix form

$$\begin{bmatrix} V_1 \\ V_2 \end{bmatrix} = \begin{bmatrix} z_{11} & z_{12} \\ z_{21} & z_{22} \end{bmatrix} \begin{bmatrix} I_1 \\ I_2 \end{bmatrix} = [\mathbf{z}] \begin{bmatrix} I_1 \\ I_2 \end{bmatrix}$$

(17–2)

where the matrix $[\mathbf{z}]$ is called the **impedance matrix** of a two-port network.

To measure or compute the impedance parameters we apply excitation at one port and leave the other port open circuited. When we drive at port 1 with port 2 open ($I_2 = 0$), the expressions in Eq. (17–1) reduce to one term each, and yield the definitions of z_{11} and z_{21}.

$$z_{11} = \left. \frac{V_1}{I_1} \right|_{I_2=0} = \text{input impedance with the output port open.}$$

(17–3a)

$$z_{21} = \left. \frac{V_2}{I_1} \right|_{I_2=0} = \text{forward transfer impedance with the output port open.}$$

(17–3b)

Conversely, when we drive at port 2 with port 1 open ($I_1 = 0$), the expressions in Eq. (17–1) reduce to one term each that define z_{12} and z_{22} as

$$z_{12} = \left. \frac{V_1}{I_2} \right|_{I_1=0} = \text{reverse transfer impedance with the input port open.}$$

(17–4a)

$$z_{22} = \left. \frac{V_2}{I_2} \right|_{I_1=0} = \text{output impedance with the input port open.}$$

(17–4b)

All of these parameters are impedances with dimensions of ohms.

A two-port network is said to be **reciprocal** when the open-circuit voltage measured at one port due to a current excitation at the other port is unchanged when the measurement and excitation ports are interchanged. A two-port network that fails this test is said to be **nonreciprocal**. Circuits containing resistors, capacitors, and inductors (including mutual inductance) are always reciprocal. Adding dependent sources to the mix usually makes the two-port network nonreciprocal.

If a two-port network is reciprocal, then $z_{12} = z_{21}$. To prove this we apply an excitation $I_1 = I_x$ at the input port and observe that Eq. (17–1) gives the open-circuit

$(I_2 = 0)$ voltage at the output port as $V_{2OC} = z_{21}I_x$. Reversing the excitation and observation ports, we find that an excitation $I_2 = I_x$ produces an open-circuit $(I_1 = 0)$ voltage at the input port of $V_{1OC} = z_{12}I_x$. Reciprocity requires that $V_{1OC} = V_{2OC}$, which can only happen if $z_{12} = z_{21}$.

EXAMPLE 17−1

Find the impedance parameters of the resistive circuit in Figure 17–2.

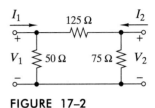

FIGURE 17–2

SOLUTION:

We start with an open circuit at port 2 $(I_2 = 0)$. The resistance looking in at port 1 is

$$z_{11} - 50 \, \|(125 + 75) = 40 \, \Omega$$

To find the forward transfer impedance, we use current division to express the current through the 75-Ω resistor in terms of I_1.

$$I_{75} = \frac{50}{50 + 125 + 75} I_1 = 0.2 I_1$$

By Ohm's law the open-circuit voltage at port 2 is $V_2 = I_{75} \times 75$. Therefore, the forward transfer impedance is

$$z_{21} = \left. \frac{V_2}{I_1} \right|_{I_2=0} = \frac{(0.2 I_1) \times 75}{I_1} = 15 \, \Omega$$

Next we assume that port 1 is open $(I_1 = 0)$. The resistance looking in at port 2 is

$$z_{22} = 75 \| (125 + 50) = 52.5 \, \Omega$$

To find the reverse transfer impedance, we first express the current through the 50-Ω resistor in terms of I_2. Using current division again

$$I_{50} = \frac{75}{50 + 125 + 75} I_2 = 0.3 I_2$$

By Ohm's law the open-circuit voltage at port 1 is $V_1 = I_{50} \times 50$. Therefore, the reverse transfer impedance is

$$z_{12} = \left. \frac{V_1}{I_2} \right|_{I_1=0} = \frac{(0.3 I_2) \times 50}{I_2} = 15 \, \Omega$$

Note that since $z_{12} = z_{21} = 15 \, \Omega$, the two-port network is reciprocal.   ∎

EXAMPLE 17−2

A 5-V voltage source is connected at port 1 in Figure 17–2 and a 50-Ω resistor is connected at port 2. Find the port currents I_1 and I_2.

SOLUTION:

Using the impedance parameters found in Example 17–1, the i-v relationships of the two-port network in Figure 17–2 are

$$V_1 = 40 I_1 + 15 I_2$$
$$V_2 = 15 I_1 + 52.5 I_2$$

The external connections at the input and output port mean that $V_1 = 5$ and $V_2 = -50I_2$. Substituting these constraints into the two-port i-v relationships yields

$$5 = 40I_1 + 15I_2$$
$$-50I_2 = 15I_1 + 52.5I_2$$

The result is two linear equations in the two unknown port currents. Solving these equations simultaneously yields $I_1 = 132\,\text{mA}$ and $I_2 = -19.4\,\text{mA}$. The minus sign here means that the output current I_2 actually flows out of port 2 as you would expect. ■

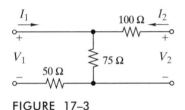

FIGURE 17-3

Exercise 17-1

Find the impedance parameters of the circuit in Figure 17–3.

Answers: $z_{11} = 125\,\Omega$, $z_{12} = 75\,\Omega$, $z_{21} = 75\,\Omega$, $z_{22} = 175\,\Omega$.

Exercise 17-2

The impedance parameters of a two-port network are $z_{11} = 25\,\Omega$, $z_{12} = 50\,\Omega$, $z_{21} = 75\,\Omega$, and $z_{22} = 75\,\Omega$. Find the port currents I_1 and I_2 when a 15-V voltage source is connected at port 1 and port 2 is short circuited.

Answers: $I_1 = -0.6$ A, $I_2 = 0.6$ A

17-3 ADMITTANCE PARAMETERS

The admittance parameters are obtained by expressing the port currents I_1 and I_2 in terms of the port voltages V_1 and V_2. The resulting two-port i-v relationships are

$$I_1 = y_{11}V_1 + y_{12}V_2$$
$$I_2 = y_{21}V_1 + y_{22}V_2 \tag{17-5}$$

The network functions y_{11}, y_{12}, y_{21}, and y_{22} are called the **admittance parameters,** or simply the **y-parameters**. In matrix form these equations are

$$\begin{bmatrix} I_1 \\ I_2 \end{bmatrix} = \begin{bmatrix} y_{11} & y_{12} \\ y_{21} & y_{22} \end{bmatrix} \begin{bmatrix} V_1 \\ V_2 \end{bmatrix} = [\mathbf{y}] \begin{bmatrix} V_1 \\ V_2 \end{bmatrix} \tag{17-6}$$

where the matrix $[\mathbf{y}]$ is called the **admittance matrix** of a two-port network.

To measure or compute the admittance parameters, we apply excitation at one port and short circuit the other port. When we drive at port 1 with port 2 shorted ($V_2 = 0$), the expressions in Eq. (17–5) reduce to one term each that define y_{11} and y_{21} as

$$y_{11} = \left.\frac{I_1}{V_1}\right|_{V_2=0} = \text{input admittance with the output port shorted.} \tag{17-7a}$$

$$y_{21} = \left.\frac{I_2}{V_1}\right|_{V_2=0} = \text{foward transfer admittance with the output port shorted.} \tag{17-7b}$$

Conversely, when we drive at port 2 with port 1 shorted ($V_1 = 0$), the expressions in Eq. (17–5) reduce to one term each that define y_{22} and y_{12} as

$$y_{12} = \frac{I_1}{V_2}\bigg|_{V_1=0} = \text{reverse transfer admittance with the input port shorted.} \qquad (17\text{–}8a)$$

$$y_{22} = \frac{I_2}{V_2}\bigg|_{V_1=0} = \text{output admittance with the input port shorted.} \qquad (17\text{–}8b)$$

All of these network functions are admittances with dimensions of siemens. If a two-port network is reciprocal, then $y_{12} = y_{21}$. This can be proved using the same thought process applied to the z-parameters.

The admittance parameters express port currents in terms of port voltages, whereas the impedance parameters express the port voltages in terms of the port currents. In effect these parameter are inverses. To see this mathematically, we multiply Eq. (17–2) by $[\mathbf{z}]^{-1}$, the inverse of the impedance matrix.

$$[\mathbf{z}]^{-1}\begin{bmatrix} V_1 \\ V_2 \end{bmatrix} = [\mathbf{z}]^{-1}[\mathbf{z}]\begin{bmatrix} I_1 \\ I_2 \end{bmatrix} = \begin{bmatrix} I_1 \\ I_2 \end{bmatrix}$$

In other words

$$\begin{bmatrix} I_1 \\ I_2 \end{bmatrix} = [\mathbf{z}]^{-1}\begin{bmatrix} V_1 \\ V_2 \end{bmatrix}$$

Comparing this result with Eq. (17–6) we conclude that $[\mathbf{y}] = [\mathbf{z}]^{-1}$. That is, the admittance matrix of a two-port network is the inverse of its impedance matrix. This means that the admittance parameters can be derived from the impedance parameters, provided $[\mathbf{z}]^{-1}$ exists. We will return to this idea in a later section. For the moment remember that admittance and impedance parameters are not independent descriptions of a two-port network.

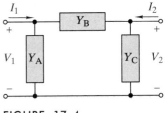

FIGURE 17–4

EXAMPLE 17–3

Find the admittance parameters of the two-port circuit in Figure 17–4.

SOLUTION:
A short circuit at port 2 connects admittances Y_A and Y_B in parallel. Hence, the admittance looking in at port 1 is $y_{11} = Y_A + Y_B$. The short-circuit current at port 2 is $I_2 = -Y_B V_1$, hence $y_{21} = -Y_B$. A short circuit at port 1 connects Y_B and Y_C in parallel so the admittance looking in at port 2 is $y_{22} = Y_B + Y_C$. The short-circuit current at port 1 is $I_1 = -Y_B V_2$, which means $y_{12} = -Y_B$. Thus, the admittance parameters of a general pi-circuit are

$$y_{11} = Y_A + Y_B$$
$$y_{21} = y_{12} = -Y_B$$
$$y_{22} = Y_B + Y_C \qquad \blacksquare$$

EXAMPLE 17–4

The impedance parameters of a two-port network are $z_{11} = 30\,\Omega$, $z_{12} = z_{21} = 10\,\Omega$, and $z_{22} = 20\,\Omega$. Find the admittance parameters of the network.

SOLUTION:
The impedance matrix for the two-port network is

$$[\mathbf{z}] = \begin{bmatrix} 30 & 10 \\ 10 & 20 \end{bmatrix}$$

The admittance matrix is the inverse of the impedance matrix and is found as

$$[\mathbf{y}] = [\mathbf{z}]^{-1} = \frac{\text{adj}[\mathbf{z}]}{\det[\mathbf{z}]} = \frac{\begin{bmatrix} 20 & -10 \\ -10 & 30 \end{bmatrix}}{500}$$

$$= \begin{bmatrix} 0.04 & -0.02 \\ -0.02 & 0.06 \end{bmatrix}$$

Hence, the admittance parameters are $y_{11} = 40\,\text{mS}$, $y_{12} = y_{21} = -20\,\text{mS}$, and $y_{22} = 60\,\text{mS}$. ∎

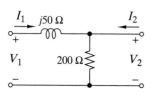

FIGURE 17–5

Exercise 17–3

Find the admittance parameters of the circuit in Figure 17–5.

Answers: $y_{11} = -j20\,\text{mS}$, $y_{21} = y_{12} = j20\,\text{mS}$, and $y_{22} = 5 - j20\,\text{mS}$.

Exercise 17–4

The admittance parameters of a two-port network are $y_{11} = 20\,\text{mS}$, $y_{12} = 0$, $y_{21} = 100\,\text{mS}$, and $y_{22} = 40\,\text{mS}$. Find the output voltage V_2 when a 5-V voltage source is connected at port 1 and port 2 is connected to a 100-Ω load resistor.

Answer: $V_2 = -10$ V

17–4 HYBRID PARAMETERS

The hybrid parameters are defined in terms of a mixture of port variables. Specifically, these parameters express V_1 and I_2 in terms of I_1 and V_2. The resulting two-port i-v relationships are

$$V_1 = h_{11}I_1 + h_{12}V_2$$
$$I_2 = h_{21}I_1 + h_{22}V_2 \tag{17–9}$$

Where h_{11}, h_{12}, h_{21}, and h_{22} are called the **hybrid parameters** or simply the **h-parameters**. In matrix form these equations are

$$\begin{bmatrix} V_1 \\ I_2 \end{bmatrix} = \begin{bmatrix} h_{11} & h_{12} \\ h_{21} & h_{22} \end{bmatrix} \begin{bmatrix} I_1 \\ V_2 \end{bmatrix} = [\mathbf{h}] \begin{bmatrix} I_1 \\ V_2 \end{bmatrix} \tag{17–10}$$

where the matrix [**h**] is called the **h-matrix** of a two-port network.

The h-parameters can be measured or calculated as follows. When we drive at port 1 with port 2 shorted ($V_2 = 0$), the expressions in Eq. (17–9) reduce to one term each, and yield the definitions of h_{11} and h_{21}.

$$h_{11} = \left. \frac{V_1}{I_1} \right|_{V_2=0} = \text{input impedance with the output port shorted.} \tag{17–11a}$$

$$h_{21} = \left. \frac{I_2}{I_1} \right|_{V_2=0} = \text{forward current transfer function with the output port shorted.} \tag{17–11b}$$

When we drive at port 2 with port 1 open ($I_1 = 0$), the expressions in Eq. (17–9) reduce to one term each, and yield the definitions of h_{12} and h_{22}.

$$h_{12} = \frac{V_1}{V_2}\bigg|_{I_1=0} = \text{reverse voltage transfer function with the input port open.} \qquad (17\text{–}12a)$$

$$h_{22} = \frac{I_2}{V_2}\bigg|_{I_1=0} = \text{output admittance with the input port open.} \qquad (17\text{–}12b)$$

These network functions have a mixture of dimensions: h_{11} is an impedance in ohms, h_{22} is an admittance in siemens, and h_{21} and h_{12} are dimensionless transfer functions. If a two-port network is reciprocal, then $h_{12} = -h_{21}$. This can be proved by the same thought process applied to the z-parameters.

EXAMPLE 17–5

The circuit in Figure 17–6 is a small-signal model of a standard CMOS amplifier cell. Find the h-parameters of the circuit.

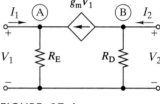

FIGURE 17–6

SOLUTION:
The sum of currents at nodes A and B can be written as

$$\text{Node A:} \quad I_1 + g_m V_1 - \frac{V_1}{R_E} = 0$$

$$\text{Node B:} \quad I_2 - g_m V_1 - \frac{V_2}{R_D} = 0$$

Solving the node A equation for V_1 yields

$$V_1 = \left[\frac{R_E}{1 - g_m R_E}\right] I_1 + [0] V_2$$
$$= \quad h_{11} I_1 \quad + \quad h_{12} V_2$$

Comparing this result and h-parameter definitions we have $h_{11} = R_E/(1 - g_m R_E)$ and $h_{12} = 0$. Inserting the expression for V_1 from node A into the node B equation and solving for I_2 yields

$$I_2 = \left[\frac{g_m R_E}{1 - g_m R_E}\right] I_1 \quad + \quad \left[\frac{1}{R_D}\right] V_2$$
$$= \quad h_{21} I_1 \quad + \quad h_{22} V_2$$

which means that $h_{21} = g_m R_E/(1 - g_m R_E)$ and $h_{22} = 1/R_D$. ∎

EXAMPLE 17–6

The h-parameters of a two-port network are $h_{11} = 2k\Omega$, $h_{12} = -2$, $h_{21} = 10$, and $h_{22} = 500\,\mu S$. A 10-V voltage source is connected at the input port. Find the Norton equivalent circuit at the output port.

SOLUTION:
The given h-parameters give the i-v relationships of the two port as

$$V_1 = 2000 I_1 - 2 V_2$$
$$I_2 = 10 I_1 + 5 \times 10^{-4} V_2$$

The voltage source connected at the input port makes $V_1 = 10$ V. Inserting this value into the first h-parameter equation and solving for I_1 yields

$$I_1 = \frac{1}{200} + \frac{1}{1000} V_2$$

Substituting this into the second h-parameter equation produces

$$I_2 = \frac{1}{20} + \left(\frac{1}{100} + 5 \times 10^{-4} \right) V_2$$

This equation gives the actual i-v relationship at the output port with the 10-V source connected at the input. Figure 17–7 shows the desired Norton equivalent circuit. Summing the currents at node A yields the i-v relationship of the desired Norton circuit as

$$I_2 = -I_N + G_N V_2$$

Comparing the Norton and the actual characteristics we conclude that

$$I_N = -\frac{1}{20} = -50 \, \text{mA}$$

$$G_N = \frac{1}{100} + 5 \times 10^{-4} = 10.5 \, \text{mS}$$

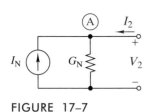

FIGURE 17–7

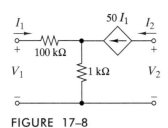

FIGURE 17–8

Exercise 17–5

Find the h-parameters of the circuit in Figure 17–8.

Answers: $h_{11} = 151 \, \text{k}\Omega$, $h_{12} = 0$, $h_{21} = 50$, and $h_{22} = 0$.

Exercise 17–6

The h-parameters of a two-port network are $h_{11} = j200 \, \Omega$, $h_{12} = 1$, $h_{21} = -1$ and $h_{22} = 2 \, \text{mS}$. Find the input impedance when the output port is open and the output admittance when the input port is short circuited.

Answer: $Z_{IN} = 500 + j200\Omega$, $Y_{OUT} = 2 - j5 \, \text{mS}$.

17–5   TRANSMISSION PARAMETERS

The transmission parameters express the input-port variables V_1, and I_1, in terms of the output-port variables V_2 and $-I_2$. The resulting two-port i-v relationships are

$$V_1 = AV_2 - BI_2$$
$$I_1 = CV_2 - DI_2$$

(17–13)

Where A, B, C, and D are called the **transmission parameters** or simply the **t-parameters**. In matrix form these equations are

$$\begin{bmatrix} V_1 \\ I_1 \end{bmatrix} = \begin{bmatrix} A & B \\ C & D \end{bmatrix} \begin{bmatrix} V_2 \\ -I_2 \end{bmatrix} = [\textbf{t}] \begin{bmatrix} V_2 \\ -I_2 \end{bmatrix}$$

(17–14)

where the matrix [**t**] is called the **transmission-matrix** of a two-port network.

The matrix equation explicitly shows that the independent variables are V_2 and $-I_2$. In other words, the minus signs in Eq. (17–13) are associated with I_2 and not with the parameters B and D. In effect, the minus sign reverses the reference direction of the output current in Figure 17–1. The reason for this convention is partly historical. The t-parameters originated in the analysis of power transmission lines, where the traditional positive reference for the receiving end current is defined in the direction of the power flow.

The transmission parameters are measured or calculated with a short circuit or an open circuit at the output port. Applying the conditions for a short-circuit $(V_2 = 0)$ or open-circuit $(-I_2 = 0)$ to Eqs. (17–13) leads to the following parameter identifications.

$$\frac{1}{A} = \left.\frac{V_2}{V_1}\right|_{-I_2=0} = \text{voltage transfer function with the output port open.} \qquad (17\text{–}15\mathrm{a})$$

$$\frac{1}{B} = \left.\frac{-I_2}{V_1}\right|_{V_2=0} = \text{negative transfer admittance with the output shorted.} \qquad (17\text{–}15\mathrm{b})$$

$$\frac{1}{C} = \left.\frac{V_2}{I_1}\right|_{-I_2=0} = \text{transfer impedance with the output port open.} \qquad (17\text{–}15\mathrm{c})$$

$$\frac{1}{D} = \left.\frac{-I_2}{I_1}\right|_{V_2=0} = \text{negative current transfer function with the output shorted.} \qquad (17\text{–}15\mathrm{d})$$

These results are expressed as reciprocals to conform with the transfer function definitions we have previously used. Arranged in this way, we see that the reciprocals of the transmission parameters are all forward (input-to-output) transfer functions. If a two-port network is reciprocal, then $AD - BC = 1$; a result that can be proved using the same method applied to z-parameters.

The transmission parameters are particularly useful when two-port networks are connected in cascade, as shown in Figure 17–9. The matrix equations for individual two-port networks N_a and N_b are

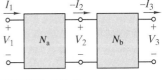

FIGURE 17–9

$$\begin{bmatrix} V_1 \\ I_1 \end{bmatrix} = \begin{bmatrix} A_a & B_a \\ C_a & D_a \end{bmatrix} \begin{bmatrix} V_2 \\ -I_2 \end{bmatrix} \quad \text{and} \quad \begin{bmatrix} V_2 \\ -I_2 \end{bmatrix} = \begin{bmatrix} A_b & B_b \\ C_b & D_b \end{bmatrix} \begin{bmatrix} V_3 \\ -I_3 \end{bmatrix}$$

Note that the output-port variables for N_a are the input-port variables for N_b. This happens because the two networks are connected in an output-to-input cascade, and because the minus signs in Eq. (17–13) effectively reverse the reference directions of the output currents. Substituting the N_b equations into the N_a equations yields

$$\begin{bmatrix} V_1 \\ I_1 \end{bmatrix} = \overbrace{\begin{bmatrix} A_a & B_a \\ C_a & D_a \end{bmatrix}}^{N_a} \overbrace{\begin{bmatrix} A_b & B_b \\ C_b & D_b \end{bmatrix}}^{N_b} \begin{bmatrix} V_3 \\ -I_3 \end{bmatrix}$$

$$\begin{bmatrix} V_1 \\ I_1 \end{bmatrix} = \begin{bmatrix} A & B \\ C & D \end{bmatrix} \begin{bmatrix} V_3 \\ -I_3 \end{bmatrix}$$

Thus, the transmission matrix of the overall network is the matrix product of the transmission matrices of the individual two-port networks in the cascade connection.

$$\begin{bmatrix} A & B \\ C & D \end{bmatrix} = \begin{bmatrix} A_a & B_a \\ C_a & D_a \end{bmatrix} \begin{bmatrix} A_b & B_b \\ C_b & D_b \end{bmatrix} \qquad (17\text{–}16)$$

This result can be generalized for any number of two-port networks in cascade. This generalization is quite useful since electrical power systems and communication system are made up of many two-port networks connected in cascade. In such applications, the individual matrices must be placed in the matrix product in the same order that the two-port networks are connected in the cascade. The reason being that matrix multiplication is not necessarily commutative.

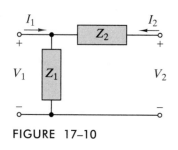

FIGURE 17–10

EXAMPLE 17–7

Find the t-parameters of the two-port network in Figure 17–10.

SOLUTION:

With the output open $(-I_2 = 0)$, we use KVL to write $V_1 = V_2$ and $V_2 = Z_1 I_1$. Hence

$$A = \left.\frac{V_1}{V_2}\right|_{-I_2=0} = 1 \quad \text{and} \quad C = \left.\frac{I_1}{V_2}\right|_{-I_2=0} = \frac{1}{Z_1}$$

With the output shorted $(V_2 = 0)$, we use current division and KVL to write

$$-I_2 = \frac{Z_1}{Z_1 + Z_2} I_1 \quad \text{and} \quad V_1 = Z_2(-I_2)$$

hence

$$D = \left.\frac{I_1}{-I_2}\right|_{V_2=0} = \frac{Z_1 + Z_2}{Z_1} \quad \text{and} \quad B = \left.\frac{V_1}{-I_2}\right|_{V_2=0} = Z_2$$

Note that this two-port network is reciprocal since

$$AD - BC = \frac{Z_1 + Z_2}{Z_1} - \frac{Z_2}{Z_1} = 1$$

∎

EXAMPLE 17–8

The t-parameters of a two-port network are $A = -1$, $B = j50\,\Omega$, $C = j20\,\text{mS}$, and $D = 0$. Find the input impedance when a 50-Ω resistor is connected across the output port.

SOLUTION:

The i-v relationships of the two-port network are

$$V_1 = -V_2 + j50(-I_2)$$
$$I_1 = j20 \times 10^{-3} V_2$$

The 50-Ω resistor connected at the output means that $V_2 = 50(-I_2)$. Substituting this constraint into the two-port i-v characteristics yields the input impedance as

$$Z_{\text{IN}} = \frac{V_1}{I_1} = \frac{-50(-I_2) + j50(-I_2)}{(j20 \times 10^{-3})50(-I_2)} = 50 + j50\,\Omega$$

∎

Exercise 17–7

The t-parameters of a two-port network are $A = 2$, $B = 200\,\Omega$, $C = 10\,\text{mS}$, and $D = 1.5$. Find the Thévenin equivalent circuit at the output port when a 10-V voltage source is connected at the input port.

Answers: $V_T = 5\,\text{V}$, $R_T = 100\,\Omega$.

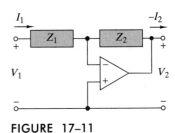

FIGURE 17-11

Exercise 17-8

Find the *t*-parameters of the OPAMP circuit in Figure 17–11.

Answers: $A = -Z_1/Z_2$, $B = 0$, $C = -1/Z_2$, $D = 0$.

17-6 TWO-PORT CONVERSIONS AND CONNECTIONS

Two-port conversion refers to the process of relating one set of parameters to a different set. We have already seen an example. In Section 17–3 we found that the admittance matrix is the inverse of the impedance matrix. Thus, the *y*-parameters are related to the *z*-parameters by matrix inversion.

$$\begin{bmatrix} y_{11} & y_{12} \\ y_{21} & y_{22} \end{bmatrix} = [\mathbf{z}]^{-1} = \frac{\text{adj}[\mathbf{z}]}{\det[\mathbf{z}]} = \begin{bmatrix} \dfrac{z_{22}}{\Delta_z} & \dfrac{-z_{12}}{\Delta_z} \\ \dfrac{-z_{21}}{\Delta_z} & \dfrac{z_{11}}{\Delta_z} \end{bmatrix} \qquad (17\text{--}17)$$

where $\Delta_z = \det[\mathbf{z}] = z_{11}z_{22} - z_{12}z_{21}$.

For a second example, we begin with the *z*-parameter *i-v* relationships

$$V_1 = z_{11}I_1 + z_{12}I_2$$
$$V_2 = z_{21}I_1 + z_{22}I_2$$

We now rearrange these equations to conform to *h*-parameter equations. First, solving the second *z*-parameter equation for I_2 yields

$$I_2 = \frac{-z_{21}}{z_{22}}I_1 + \frac{1}{z_{22}}V_2$$
$$= h_{21}I_1 \quad + h_{22}V_2$$

Using this result to eliminate I_2 for the first *z*-parameter equation produces

$$V_1 = z_{11}I_1 + z_{12}\left[\frac{-z_{21}}{z_{22}}I_1 + \frac{1}{z_{22}}V_2\right]$$
$$= \frac{\Delta_z}{z_{22}}I_1 + \frac{z_{12}}{z_{22}}V_2 = h_{11}I_1 + h_{12}V_2$$

Taken together, the two rearranged equations provide relationships between *h*-parameters and *z*-parameters that are summarized by the following matrix equation.

$$\begin{bmatrix} h_{11} & h_{12} \\ h_{21} & h_{22} \end{bmatrix} = \begin{bmatrix} \dfrac{\Delta_z}{z_{22}} & \dfrac{z_{12}}{z_{22}} \\ \dfrac{-z_{21}}{z_{22}} & \dfrac{1}{z_{22}} \end{bmatrix} \qquad (17\text{--}18)$$

The steps used to produce Eqs. (17–17) and (17–18) are typical of the derivations used to produce the parameter conversion formulas in Table 17–2. To convert from one set to another, we enter the table in the column for the given parameters and find the appropriate conversion formulas in the row for the desired parameters.

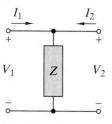

FIGURE 17–12

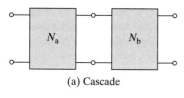

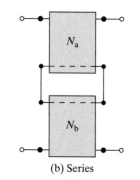

(a) Cascade

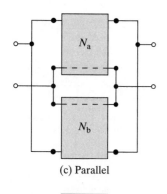

(b) Series

(c) Parallel

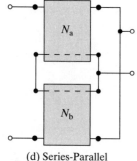

(d) Series-Parallel

FIGURE 17–13

T A B L E 17–2 TWO-PORT PARAMETER CONVERSION TABLE

DESIRED PARAMETERS	GIVEN PARAMETERS			
	[z]	**[y]**	**[h]**	**[t]**
[z]	$\begin{bmatrix} z_{11} & z_{12} \\ z_{21} & z_{22} \end{bmatrix}$	$\begin{bmatrix} \dfrac{y_{22}}{\Delta_y} & \dfrac{-y_{12}}{\Delta_y} \\ \dfrac{-y_{21}}{\Delta_y} & \dfrac{y_{11}}{\Delta_y} \end{bmatrix}$	$\begin{bmatrix} \dfrac{\Delta_h}{h_{22}} & \dfrac{h_{12}}{h_{22}} \\ \dfrac{-h_{21}}{h_{22}} & \dfrac{1}{h_{22}} \end{bmatrix}$	$\begin{bmatrix} \dfrac{A}{C} & \dfrac{\Delta_t}{C} \\ \dfrac{1}{C} & \dfrac{D}{C} \end{bmatrix}$
[y]	$\begin{bmatrix} \dfrac{z_{22}}{\Delta_z} & \dfrac{-z_{12}}{\Delta_z} \\ \dfrac{-z_{21}}{\Delta_z} & \dfrac{z_{11}}{\Delta_z} \end{bmatrix}$	$\begin{bmatrix} y_{11} & y_{12} \\ y_{21} & y_{22} \end{bmatrix}$	$\begin{bmatrix} \dfrac{1}{h_{11}} & \dfrac{-h_{12}}{h_{11}} \\ \dfrac{h_{21}}{h_{11}} & \dfrac{\Delta_h}{h_{11}} \end{bmatrix}$	$\begin{bmatrix} \dfrac{D}{B} & \dfrac{-\Delta_t}{B} \\ \dfrac{-1}{B} & \dfrac{A}{B} \end{bmatrix}$
[h]	$\begin{bmatrix} \dfrac{\Delta_z}{z_{22}} & \dfrac{z_{12}}{z_{22}} \\ \dfrac{-z_{21}}{z_{22}} & \dfrac{1}{z_{22}} \end{bmatrix}$	$\begin{bmatrix} \dfrac{1}{y_{11}} & \dfrac{-y_{12}}{y_{11}} \\ \dfrac{y_{21}}{y_{11}} & \dfrac{\Delta_y}{y_{11}} \end{bmatrix}$	$\begin{bmatrix} h_{11} & h_{12} \\ h_{21} & h_{22} \end{bmatrix}$	$\begin{bmatrix} \dfrac{B}{D} & \dfrac{\Delta_t}{D} \\ \dfrac{-1}{D} & \dfrac{C}{D} \end{bmatrix}$
[t]	$\begin{bmatrix} \dfrac{z_{11}}{z_{21}} & \dfrac{\Delta_z}{z_{21}} \\ \dfrac{1}{z_{21}} & \dfrac{z_{22}}{z_{21}} \end{bmatrix}$	$\begin{bmatrix} \dfrac{-y_{22}}{y_{21}} & \dfrac{-1}{y_{21}} \\ \dfrac{-\Delta_y}{y_{21}} & \dfrac{-y_{11}}{y_{21}} \end{bmatrix}$	$\begin{bmatrix} \dfrac{-\Delta_h}{h_{21}} & \dfrac{-h_{11}}{h_{21}} \\ \dfrac{-h_{22}}{h_{21}} & \dfrac{-1}{h_{21}} \end{bmatrix}$	$\begin{bmatrix} A & B \\ C & D \end{bmatrix}$

$\Delta_z = z_{11}z_{22} - z_{12}z_{21}$ $\Delta_y = y_{11}y_{22} - y_{12}y_{21}$ $\Delta_h = h_{11}h_{22} - h_{12}h_{21}$ $\Delta_t = AD - BC$

For example, converting from y-parameters to t-parameters is accomplished by the matrix equation

$$\begin{bmatrix} A & B \\ C & D \end{bmatrix} = \begin{bmatrix} \dfrac{-y_{22}}{y_{21}} & \dfrac{-1}{y_{21}} \\ \dfrac{-\Delta_y}{y_{21}} & \dfrac{-y_{11}}{y_{21}} \end{bmatrix}$$

The derivations of the formulas in Table 17–2 assume that all four sets of parameters exist. Certain two-port parameters do not exist for some circuits. For example, the y-parameters do not exist for the circuit in Figure 17–12. By inspection, the z-parameters exist for this circuit and are $z_{11} = z_{12} = z_{21} = z_{22} = Z$. As a result $\Delta_z = 0$, and mathematically the y-parameters do not exist because $[\mathbf{z}]^{-1}$ does not exist. Physically the y-parameters don't exist because a short circuit applied at either port in Figure 17–12 produces an infinite driving-point admittance at the other port.

In some applications it is useful to view a circuit as an interconnection of subcircuits, each of which is treated as a two-port network. Figure 17–13 shows four possible interconnections of two-port subcircuits N_a and N_b. We regard the subcircuits as building blocks that are interconnected to form a composite circuit. We need to relate the two-port parameters of the composite interconnection to the two-port parameters of the subcircuit building blocks.

The *cascade connection* in Figure 17–13(a) was addressed in Section 17–5, where we showed that the transmission matrix of the interconnection is

$$[\mathbf{t}] = [\mathbf{t_a}][\mathbf{t_b}] \tag{17–19}$$

where $[\mathbf{t_a}]$ and $[\mathbf{t_b}]$ are the transmission matrices of the subcircuits N_a and N_b, respectively. The circuit in Figure 17–13(b) is called a *series connection* of two-ports networks. It is easy to show that the impedance matrix of the inter-connection is

$$[\mathbf{z}] = [\mathbf{z}_a] + [\mathbf{z}_b] \tag{17-20}$$

where $[\mathbf{z}_a]$ and $[\mathbf{z}_b]$ are the impedance matrices of subcircuits N_a and N_b, respectively. The circuit in Figure 17–13(c) is called *a parallel connection* of two-port networks. As you might expect from network duality, the admittance matrix of this inter-connection is

$$[\mathbf{y}] = [\mathbf{y}_a] + [\mathbf{y}_b] \tag{17-21}$$

where $[\mathbf{y}_a]$ and $[\mathbf{y}_b]$ are the admittance matrices of N_a and N_b, respectively. Finally, Figure 17–13(d) is called a *series-parallel connection* whose hybrid-parameter matrix is

$$[\mathbf{h}] = [\mathbf{h}_a] + [\mathbf{h}_b] \tag{17-22}$$

where $[\mathbf{h}_a]$ and $[\mathbf{h}_b]$ are the hybrid matrices of N_a and N_b respectively.

Equations (17–19) through (17–22) show how to obtain a two-port matrix of an interconnection using the two-port matrices of subcircuits N_a and N_b. These equations assume that the port i-v characteristics of N_a and N_b are unchanged by their interconnection. One way for this to hold is for N_a and N_b to have a "common ground" as indicated by the dashed lines in Figure 17–13. In what follows we assume these equations always apply.

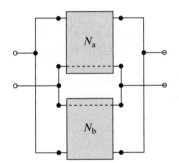

FIGURE 17–14

EXAMPLE 17–9

The networks N_a and N_b in Figure 17–14 are connected in parallel and have two-port matrices

$$[\mathbf{z}_a] = \begin{bmatrix} 15 & 5 \\ 5 & 10 \end{bmatrix} \Omega \text{ and } [\mathbf{y}_b] = \begin{bmatrix} 0.2 & -0.2 \\ -0.2 & 0.7 \end{bmatrix} \text{S}$$

Find the y-parameters of the parallel connection.

SOLUTION:

To get the desired y-parameters we use the fact that $[y] = [y_a] + [y_b]$ in a parallel connection. We are given $[z_a]$ and $[y_b]$, so we need to convert the z-matrix of N_a into a y-matrix. Equation (17–17) indicates that $[y] = [z]^{-1}$, so the desired y-matrix is found to be

$$[\mathbf{y}] = [\mathbf{z}_a]^{-1} + [\mathbf{y}_b] = \begin{bmatrix} 15 & 5 \\ 5 & 10 \end{bmatrix}^{-1} + \begin{bmatrix} 0.2 & -0.2 \\ -0.2 & 0.7 \end{bmatrix}$$

$$= \begin{bmatrix} 0.08 & -0.04 \\ -0.04 & 0.12 \end{bmatrix} + \begin{bmatrix} 0.2 & -0.2 \\ -0.2 & 0.7 \end{bmatrix} = \begin{bmatrix} 0.28 & -0.24 \\ -0.24 & 0.82 \end{bmatrix} \text{S}$$

Given the y-matrix of the parallel connection, the y-parameters are seen to be $y_{11} = 0.28$ S, $y_{12} = y_{21} = -0.24$ S, and $y_{22} = 0.82$ S. ∎

Exercise 17–9 _____

The h-parameter of a two-port network are $h_{11} = 1\,\text{k}\Omega$, $h_{12} = 0.02$, $h_{21} = -50$, and $h_{22} = 10^{-4}$ S. Find the t-parameters of the two-port network.

Answers: $A = 0.022$, $B = 20\,\Omega$, $C = 2\,\mu\text{S}$, and $D = 0.02$.

VOLTAGE AND CURRENT GAIN

Our study of two-port networks to this point can be summarized as follows. First, we used the port i-v characteristics to define four sets of two-port parameters. Next, we used basic circuit analysis to find the two-port parameters of given circuits. We showed that when two-port parameters are known, the defining i-v characteristics can be used to find voltage and current responses at the input and output ports. Finally, Table 17–2 allows us to convert from one set of parameters to a different set, while Eqs. (17–19) through (17–21) show us how to find a two-port matrix of an interconnection of two-port subcircuits.

These results are the basic tools of two-port circuit analysis. They are used at the next level of analysis to study global properties of two-port circuits such as voltage gain, current gain, Thévenin equivalents, and the effects of feedback and loading. We limit our study to voltage and current gain.

Voltage and current gain are defined in terms of the input port variables (V_1, I_1) and output port variables (V_2, I_2) as follows.

$$\text{Voltage Gain} = T_{\text{V}} = \left.\frac{V_2}{V_1}\right|_{I_2=0}$$

$$\text{Current Gain} = T_{\text{I}} = \left.\frac{I_2}{I_1}\right|_{V_2=0}$$

These gains are both forward (input to output) transfer functions for prescribed conditions at the output port. For T_{V} the output is open-circuited ($I_2 = 0$) and for T_{I} the output is short-circuited ($V_2 = 0$).

T_{V} and T_{I} can be related to two-port parameters by imposing the output conditions $I_2 = 0$ or $V_2 = 0$ on the defining i-v equations. For example, the z-parameter i-v equations are

$$V_1 = z_{11}I_1 + z_{12}I_2$$

$$V_2 = z_{21}I_1 + z_{22}I_2$$

For the $I_2 = 0$ condition these equations reduce to $V_1 = z_{11}I_1$ and $V_2 = z_{21}I_1$. Dividing the second reduced equation by the first gives $T_{\text{V}} = V_2/V_1 = z_{21}/z_{11}$. The $V_2 = 0$ condition reduces the second i-v equation to

$$0 = z_{21}I_1 + z_{22}I_2$$

Solving for I_2 leads to $T_{\text{I}} = I_2/I_1 = -z_{21}/z_{22}$

As a second example, consider the i-v equations of the t-parameters.

$$V_1 = AV_2 - BI_2$$

$$I_1 = CV_2 - DI_2$$

The $I_2 = 0$ condition reduces the first equation to $V_1 = AV_2$, which leads to $T_{\text{V}} = V_2/V_1 = 1/A$. Imposing $V_2 = 0$ reduces the second equation to $I_1 = -DI_2$, which lead to $T_{\text{I}} = I_2/I_1 = -1/D$.

These examples illustrate that voltage and current gain can be expressed in terms of two-port parameters by imposing the condition $V_2 = 0$ or $I_2 = 0$ on the defining i-v equations. Table 17–3 summarizes these relationships for the four parameter sets defined in this chapter. These relationships apply to individual two-port networks or to interconnections of two-port networks when we know the parameters of the interconnection. The next example illustrates this point.

TABLE 17–3   VOLTAGE GAIN AND CURRENT GAIN IN TERMS OF TWO-PORT PARAMETERS

DEFINITION	GIVEN PARAMETERS			
	[z]	[y]	[H]	[T]
$T_V = \dfrac{V_2}{V_1}\Big|_{I_2=0}$	$T_V = \dfrac{z_{21}}{z_{11}}$	$T_V = \dfrac{-y_{21}}{y_{22}}$	$T_V = \dfrac{-h_{21}}{\Delta_h}$	$T_V = \dfrac{1}{A}$
$T_I = \dfrac{I_2}{I_1}\Big|_{V_2=0}$	$T_I = \dfrac{-z_{21}}{z_{22}}$	$T_I = \dfrac{y_{21}}{y_{11}}$	$T_I = h_{21}$	$T_I = \dfrac{-1}{D}$

EXAMPLE 17–10

The networks N_a and N_b in Figure 17–15 are connected in cascade and have two-port matrices

$$[t_a] = \begin{bmatrix} 3 & 25 \\ 0.2 & 2 \end{bmatrix} \text{ and } [z_b] = \begin{bmatrix} 200 & 0 \\ -2000 & 20 \end{bmatrix} \Omega$$

Find the current gain of this cascade connection.

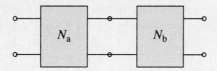

FIGURE 17–15

SOLUTION:

To get a two-port description of this connection we use the t-parameters since $[t] = [t_a][t_b]$ in a cascade connection. We are given the t-matrix of N_a and the z-matrix of N_b. Given the matrix $[z_b]$, the z-parameters of N_b are seen to be $z_{11b} = 200\,\Omega$, $z_{12b} = 0$, $z_{21b} = -2000\,\Omega$, $z_{2b} = 20\,\Omega$, and $\Delta_{zb} = z_{11b}z_{22b} - z_{12b}z_{21b} = 4000$. To get the matrix $[t_b]$, we use the $[z]$ to $[t]$ conversion shown in Table 17–2.

$$[t_b] = \begin{bmatrix} \dfrac{z_{11b}}{z_{21b}} & \dfrac{\Delta_{zb}}{z_{21b}} \\ \dfrac{1}{z_{21b}} & \dfrac{z_{22b}}{z_{21b}} \end{bmatrix} = \begin{bmatrix} -0.1 & -2 \\ -5 \times 10^{-4} & -0.01 \end{bmatrix}$$

Now using $[t] = [t_a][t_b]$ yields the transmission matrix of the cascade connection as

$$[t] = [t_a][t_b] = \begin{bmatrix} 3 & 25 \\ 0.2 & 2 \end{bmatrix} \begin{bmatrix} -0.1 & -2 \\ -5 \times 10^{-4} & -0.01 \end{bmatrix} = \begin{bmatrix} -0.313 & -6.25 \\ -0.021 & -0.420 \end{bmatrix}$$

Given the matrix $[t]$, the t-parameters of the connection are seen to be $A = -0.313$, $B = -6.25\,\Omega$, $C = -0.021\,\text{S}$, and $D = -0.420$. Using the appropriate relationship in Table 17–3 gives us the current gain of the connection.

$$T_I = \dfrac{-1}{D} = \dfrac{-1}{-0.420} = 2.381$$

The t-matrices of networks N_a and N_b in Figure 17–15 are

$$[\mathbf{t_a}] = \begin{bmatrix} 3 & 200 \\ 0.01 & 0 \end{bmatrix} \text{ and } [\mathbf{t_b}] = \begin{bmatrix} 2 & 0 \\ 0.04 & 1 \end{bmatrix}$$

Find the t-parameters of the cascade connection.

Answers: $A = 14$, $B = 200\,\Omega$, $C = 0.1\,S$, and $D = 2$.

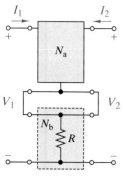

FIGURE 17–16

EXAMPLE 17–11

The two-port network N_a in Figure 17–16 is an amplifier with h-parameters $h_{11a} = 500\,\Omega$, $h_{12a} = 0$, $h_{21a} = -500$, and $h_{22a} = 0.1\,S$. The two-port N_b is a passive resistance circuit that provides feedback around the amplifier. The purpose of this example is to find the voltage gain of the overall network first without feedback ($R = 0$) and then with feedback ($R = 1\,\text{k}\Omega$).

SOLUTION:
To get a two-port description of this connection we use z-parameters since $[\mathbf{z}] = [\mathbf{z_a}] + [\mathbf{z_b}]$ in a series connection. We are given the h-parameters of N_a, namely $h_{11a} = 500\,\Omega$, $h_{12a} = 0$, $h_{21a} = -500$, $h_{22b} = 0.1\,S$, and $\Delta_{ha} = h_{11a}h_{22a} - h_{12a}h_{21a} = 50$. To get the impedance matrix $[\mathbf{z_a}]$ we use the $[\mathbf{h}]$ to $[\mathbf{z}]$ conversion in Table 17–2.

$$[\mathbf{z_a}] = \begin{bmatrix} \dfrac{\Delta_{ha}}{h_{22a}} & \dfrac{h_{12a}}{h_{22a}} \\[2ex] \dfrac{-h_{12a}}{h_{22a}} & \dfrac{1}{h_{22a}} \end{bmatrix} = \begin{bmatrix} 500 & 0 \\ 5000 & 10 \end{bmatrix} \Omega$$

Given the matrix $[\mathbf{z_a}]$, the z-parameters of N_a are seen to be $z_{11a} = 500\,\Omega$, $z_{12a} = 0$, $z_{21a} = 5000\,\Omega$, and $z_{22a} = 10\,\Omega$. Network N_b is a simple shunt resistor R so it is easy to see that its z-parameters are. $z_{11b} = z_{12b} = z_{21b} = z_{22b} = R$. Table 17–3 expresses the voltage gain T_V in terms of z-parameters as $T_V = z_{21}/z_{11}$. Again, since $[\mathbf{z}] = [\mathbf{z_a}] + [\mathbf{z_b}]$ in a series connection, the overall voltage gain is

$$T_V = \frac{z_{21}}{z_{11}} = \frac{z_{21a} + z_{21b}}{z_{11a} + z_{11b}} = \frac{5000 + R}{500 + R}$$

Without feedback ($R = 0$) the voltage gain is $T_V = 5000/500 = 10$. With feedback ($R = 1000\,\Omega$) the gain is $T_V = 6000/1500 = 4$.   ■

What is the current gain of the series connection in Figure 17–16 when $R = 0$ and $R = 1\,\text{k}\Omega$?

Answers: $T_I = -500$ and $T_I = -5.94$.

SUMMARY

- A **port** is a terminal pair where energy can be supplied or extracted. A **two-port network** is a four-terminal circuit with the terminals paired to form an input port and an output port.

- The two-port method applies to linear circuits with no independent sources, no initial energy storage, and a net current of zero at both ports.

- The only accessible variables are the port voltages V_1 and V_2, and the port currents I_1 and I_2. Two-port parameters are defined by expressing two of these four port variables in terms of the other two variables.

- The most often used two-port parameters are the impedance, admittance, hybrid, and transmission parameters. Each set of two-port parameters defines two simultaneous linear equations in the port variables.

- In general, two-port parameters are network functions of the complex frequency variable s. Setting $s = j\omega$

yields the sinusoidal steady state values of two-port parameters. Setting $s = 0$ yields the dc values of two-port parameters.

- A two-port network is reciprocal if the voltage response observed at one port due to a current excitation at the other port is unchanged when the response and excitation ports are interchanged.

- Every two-port parameter can be calculated or measured by applying excitation at one of the ports and connecting a short circuit or an open circuit at the other port. The relationships between different sets of parameters are given in Table 17–2.

- There are simple relationships between the two-port parameters of an interconnection of two-port networks and the two-port parameters of its subcircuits. These relationships assume that the interconnections do not modify the two-port parameters of the subcircuits.

PROBLEMS

OBJECTIVE 17–1 IMPEDANCE AND ADMITTANCE PARAMETERS (SECTS. 17–2, 17–3)

(a) Given a linear two-port network, find the impedance or admittance parameters.

(b) Given the impedance or admittance parameters of a linear two-port network, find responses at the input and output ports for specified external connections.

See Examples 17–1, 17–2, 17–3, 17–4, and Exercises 17–1, 17–2, 17–3, 17–4.

17–1 Find the z-parameters of the two-port network in Figure P17–1.

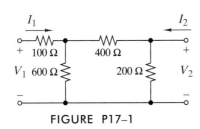

FIGURE P17–1

17–2 Find the y-parameters of the two-port network in Figure P17–1.

17–3 Find the z-parameters of the two-port network in Figure P17–3.

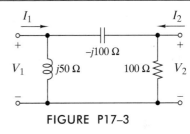

FIGURE P17–3

17–4 Find the y-parameters of the two-port network in Figure P17–3.

17–5 Find the z-parameters of the two-port network in Figure P17–5.

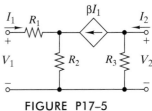

FIGURE P17–5

17–6 Find the y-parameters of the two-port network in Figure P17–5.

17–7 The z-parameters of a two-port circuit are $z_{11} = 1\text{ k}\Omega$, $z_{12} = z_{21} = 500\ \Omega$, and $z_{22} = 1.5\text{ k}\Omega$. Find the port currents I_1 and I_2 when a 12-V voltage source is connected across the input port and a load resistor $R_L = 250\ \Omega$ is connected across the output port.

17–8 The z-parameters of a two-port circuit are $z_{11} = 80\,\text{k}\Omega$, $z_{12} = 3\,\text{M}\Omega$, $z_{21} = -400\,\text{k}\Omega$, and $z_{22} = 5\,\text{k}\Omega$. Find the open-circuit ($I_2 = 0$) voltage gain $T_V = V_2/V_1$.

17–9 The y-parameters of a two-port circuit are $y_{11} = 4\,\text{mS}$, $y_{12} = y_{21} = -2\,\text{mS}$, and $y_{22} = 2\,\text{mS}$. A 15-V voltage source is connected at the input port and a load resistor $R_L = 2500\,\Omega$ is connected across the output port. Find the port variable responses V_2, I_1, and I_2.

17–10 The y-parameters of a two-port circuit are $y_{11} = 15 + j20\,\text{mS}$, $y_{12} = y_{21} = -j20\,\text{mS}$, and $y_{22} = 40 + j20\,\text{mS}$. Find the short-circuit ($V_2 = 0$) current gain $T_I = I_2/I_1$.

17–11 In Figure P17–11 a load impedance Z_L is connected across the output port. Show that the input impedance $Z_{IN} = V_1/I_1$ is

$$Z_{IN} = z_{11} - \frac{z_{12}z_{21}}{Z_L + z_{22}}$$

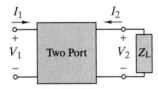

FIGURE P17–11

17–12 In Figure P17–11 a load impedance Z_L is connected across the output port. Show that the voltage gain $T_V = V_2/V_1$ is

$$T_V = \frac{-y_{21}}{Y_L + y_{22}}$$

OBJECTIVE 17–2 HYBRID AND TRANSMISSION PARAMETERS (SECTS. 17–4, 17–5)

(a) Given a linear two-port network, find the hybrid or transmission parameters.
(b) Given the hybrid or transmission parameters of a linear two-port network, find responses at the input and output ports for specified external connections.
See Examples 17–5, 17–6, 17–7, 17–8, and Exercises 17–5, 17–6, 17–7, 17–8.

17–13 Find the h-parameters of the two-port networks in Figure P17–13.

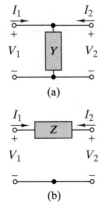

(a)

(b)

FIGURE P17–13

17–14 Find the t-parameters of the two-port networks in Figure P17–13.

17–15 Find the h-parameters of the two-port network in Figure P17–15.

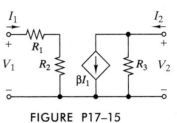

FIGURE P17–15

17–16 Find the t-parameters of the two-port network in Figure P17–15.

17–17 Find the h-parameters of the two-port network in Figure P17–17.

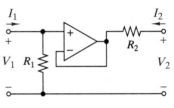

FIGURE P17–17

17–18 Find the t-parameters of the two-port network in Figure P17–17.

17–19 The h-parameters of a two-port network are $h_{11} = 500\,\Omega$, $h_{12} = 1$, $h_{21} = -1$ and $h_{22} = 2\,\text{mS}$. Find the Thévenin equivalent circuit at the output port when a 12-V voltage source is connected at the input port.

17–20 The h-parameters of a two-port network are $h_{11} = 6\,\text{k}\Omega$, $h_{12} = 0$, $h_{21} = 50$, and $h_{22} = 0.2\,\text{mS}$. Find the voltage gain V_2/V_1 when a 20-kΩ load resistor is connected across the output port.

17–21 The t-parameters of a two-port network are $A = 2$, $B = 400\,\Omega$, $C = 2.5\,\text{mS}$, and $D = 1$.
(a) Find the output resistance V_2/I_2 when the input port is short-circuited.
(b) Find the output resistance V_2/I_2 when the input port is open-circuited.

17–22 The t-parameters of a two-port network are $A = 0$, $B = -j100\,\Omega$, $C = -j20\,\text{mS}$, and $D = 1 - j0.25$. Find the voltage gain V_2/V_1 when a 50-Ω load resistor is connected across the output port.

17–23 When a voltage V_X is applied across the input port, the short-circuit current at the output port is I_{2SC}. When the same voltage is applied across the output port, the short-circuit current at the input port is I_{1SC}. Show that reciprocity ($I_{1SC} = I_{2SC}$) requires that $h_{12} = -h_{21}$.

17–24 When a load impedance Z_L is connected across the output port, show that the input impedance is

$$Z_{IN} = \frac{AZ_L + B}{CZ_L + D}$$

OBJECTIVE 17–3 TWO-PORT CONVERSIONS AND CONNECTIONS (SECT. 17–6)

(a) Given a set of two-port parameters, find other two-port parameters.

(b) Given the parameters of several two-port networks, find the parameters of an interconnection of the two-ports.

See Examples 17–9, 17–10, 17–11 and Exercises 17–9, 17–10, 17–11.

17–25 Starting with the h-parameter i-v relationships in Eq. (17–9), show that:

$$A = \frac{-\Delta_h}{h_{21}}, \quad B = \frac{-h_{11}}{h_{21}}, \quad C = \frac{-h_{22}}{h_{21}}, \quad D = \frac{-1}{h_{21}}$$

where $\Delta_h = h_{11}h_{22} - h_{12}h_{21}$.

17–26 Starting with the t-parameter i-v relationships in Eq. (17–13), show that

$$y_{11} = \frac{D}{B}, \quad y_{12} = \frac{-\Delta_t}{B}, \quad y_{21} = \frac{-1}{B}, \quad y_{22} = \frac{A}{B}D$$

where $\Delta_t = AD - BC$.

17–27 Starting with the h-parameter i-v relationships in Eq. (17–9), show that for $I_2 = 0$ the voltage gain is

$$T_V = \frac{V_2}{V_1} = \frac{-h_{21}}{\Delta_h}$$

17–28 Starting with the t-parameter i-v relationships in Eq. (17–13), show that for $V_2 = 0$ the current gain is $T_I = -1/D$.

$$T_I = \frac{I_2}{I_1} = \frac{-1}{D}$$

17–29 The t-parameters of the two-port networks N_a and N_b in Figure P17–29 are

$$[\mathbf{t_a}] = \begin{bmatrix} 60 & 10 \\ 5 & 40 \end{bmatrix} \text{ and } [\mathbf{t_b}] = \begin{bmatrix} 10 & 10 \\ 2.5 & 5 \end{bmatrix}$$

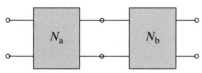

FIGURE P17–29

Find the t-parameters of the cascade connection.

17–30 The t-parameters of the two-port networks N_a and N_b in Figure P17–29 are

$$[\mathbf{t_a}] = \begin{bmatrix} 4 & 25 \\ 0.01 & 4 \end{bmatrix} \text{ and } [\mathbf{t_b}] = \begin{bmatrix} 10 & 500 \\ 0.2 & 10 \end{bmatrix}$$

Find the impedance looking into the input port of N_a when a 50-Ω resistive load is connected across the output port of N_b. Note: the result in Problem 17–24 will prove useful.

17–31 The cascade connection in Figure P17–29 is a two-stage amplifier with identical two-port stages each having the h-parameters $h_{11} = 1\,\text{k}\Omega, h_{12} = 0, h_{21} = -10,$ and $h_{22} = 1\,\text{mS}.$

Find the t-parameters of the cascade connection and then calculate the current gain of the two-stage amplifier.

17–32 Find the voltage gain of the two-stage amplifier defined in Problem 17–31.

17–33 The two-port parameters of the series connection in Figure P17–33 are

$$[\mathbf{z_a}] = \begin{bmatrix} 60 & 10 \\ 10 & 10 \end{bmatrix} \Omega \text{ and } [\mathbf{y_b}] = \begin{bmatrix} 60 & -40 \\ -20 & 80 \end{bmatrix} \text{mS}$$

Find the z-parameters of the series connection. Is the network reciprocal?

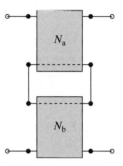

FIGURE P17–33

17–34 ▶ In Figure 17–33 the network N_a is an active device with h-parameters $h_{11} = 20\,\text{k}\Omega, h_{12} = 0, h_{21} = -5 \times 10^3,$ and $h_{22} = 50\,\text{mS}.$ The network N_b is a resistive feedback circuit with z-parameters $z_{11} = 2R$ and $z_{12} = z_{21} = z_{22} = R.$ Find a value of R such that the voltage gain of the series connection is $T_V = 2.$

17–35 The two-port parameters of the parallel connection in Figure P17–35 are

$$[\mathbf{z_a}] = \begin{bmatrix} 30 & 10 \\ 10 & 20 \end{bmatrix} \Omega \text{ and } [\mathbf{y_b}] = \begin{bmatrix} 60 & -40 \\ -40 & 80 \end{bmatrix} \text{mS}$$

Find the y-parameters of the parallel connection. Is the network reciprocal?

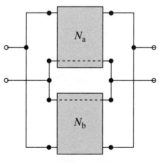

FIGURE P17–35

17–36 The h-parameters of a two-port amplifier are $h_{11} = 10\,\text{k}\Omega,$ $h_{12} = 0, h_{21} = -40,$ and $h_{22} = 1\,\text{mS}.$ Find the forward current gain of a parallel connection of two such amplifiers. Assume the connection does not change the parameters of either amplifier.

INTEGRATING PROBLEMS

17–37 Unilateral Two-port Network ◄A

A two-port network is said to be **unilateral** if excitation applied at the input port produces a response at the output port, but the same excitation applied at the output port produces no response at the input port. Show that a two-port is unilateral if $AD - BC = 0$.

17–38 Input Impedance of a Two-port Network ◄A

A load impedance Z_L is connected at the output of a two-port network. Show that the input impedance $Z_{IN} = V_1/I_1 = Z_L$ when $A = D$ and $C = B/(Z_L)^2$. *Hint*: the result shown in Problem 17–24 is the place to start.

17–39 Maximum Power ◄D

◄D The y-parameters of a two-port network operating in the sinusoidal steady state are $y_{11} = 20 - j15$ mS, $y_{12} = y_{21} = -j12$ ms, and $y_{22} = 10 - j20$ mS. The input port is driven by an ac voltage source $\mathbf{V}_S = 120\angle 25°$ V(rms). Find the output port load impedance Z_L that will draw the maximum average power from the network.

17–40 Two-port Parameters of a Transformer ◄A

In the phasor-domain the **I-V** relationships of a linear transformer with positive coupling are

$$\mathbf{V}_1 = j\omega L_1 \mathbf{I}_1 + j\omega M \mathbf{I}_2$$
$$\mathbf{V}_2 = j\omega M \mathbf{I}_1 + j\omega L_2 \mathbf{I}_2$$

Find the phasor-domain transmission parameters of the linear transformer.

Appendix A COMPLEX NUMBERS

Using complex numbers to represent signals and circuits is a fundamental tool in electrical engineering. This appendix reviews complex-number representations and arithmetic operations. These procedures, though rudimentary, must be second nature to all who aspire to be electrical engineers. Exercises are provided to confirm your mastery of these essential skills.

COMPLEX-NUMBER REPRESENTATIONS

A complex number z can be written in rectangular form as

$$z = x + jy \qquad (A–1)$$

where j represents $\sqrt{-1}$. Mathematicians customarily use i to represent $\sqrt{-1}$, but i represents current in electrical engineering, so we use the symbol j instead.

The quantity z is a two-dimensional number represented as a point in the complex plane, as shown in Figure A–1. The x component is called the **real part** and y (not jy) the **imaginary part** of z. A special notation is sometimes used to indicate these two components:

$$x = \mathrm{Re}\{z\} \quad \text{and} \quad y = \mathrm{Im}\{z\} \qquad (A–2)$$

where $\mathrm{Re}\{z\}$ means the real part and $\mathrm{Im}\{z\}$ the imaginary part of z.

Figure A–1 also shows the polar representation of the complex number z. In polar form a complex number is written

$$z = M\angle\theta \qquad (A–3)$$

where M is called the **magnitude** and θ the **angle** of z. A special notation is also used to indicate these two components.

$$|z| = M \quad \text{and} \quad \angle z = \theta \qquad (A–4)$$

where $|z|$ means the magnitude and $\angle z$ the angle of z.

The real and imaginary parts and magnitude and angle of z are all shown geometrically in Figure A–1. The relationships between the rectangular and polar forms are easily derived from the geometry in Figure A–1:

$$\text{Rectangular to polar} \quad M = \sqrt{x^2 + y^2} \quad \theta = \tan^{-1}\frac{y}{x}$$
$$\text{Polar to rectangular} \quad x = M\cos\theta \quad y = M\sin\theta \qquad (A–5)$$

The inverse tangent relation for θ involves an ambiguity that can be resolved by identifying the correct quadrant in the z-plane using the signs of the two rectangular components. *Caution:* In calculating $\tan^{-1}\left(\dfrac{y}{x}\right)$ hand-held calculators do not recognize if a minus sign belongs to the real or to the imaginary part of the ratio or if the signs cancel. The calculated angle is either in the first or fourth quadrant and may be off by $180°$. [See Exercise A–1(b) and (c).]

Another version of the polar form is obtained using Euler's relationship:

$$e^{j\theta} = \cos\theta + j\sin\theta \qquad (A–6)$$

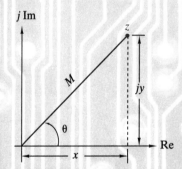

FIGURE A–1 *Graphical representation of complex numbers.*

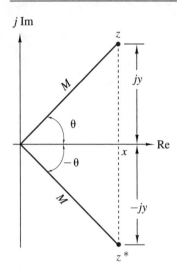

$j\,\mathrm{Im}$

Re

FIGURE A–2 *Graphical representation of conjugate complex numbers.*

We can write the polar form as

$$z = Me^{j\theta} = M\cos\theta + jM\sin\theta \qquad (A–7)$$

This polar form is equivalent to Eq. (A–3), since the right side yields the same polar-to-rectangular relationships as Eq. (A–5). Thus, a complex number can be represented in three ways:

$$z = x + jy \quad z = M\angle\theta \quad z = Me^{j\theta} \qquad (A–8)$$

The relationships between these forms are given in Eq. (A–5).

The quantity z^* is called the conjugate of the complex number z. The asterisk indicates the **conjugate** of a complex number formed by reversing the sign of the imaginary component. In rectangular form the conjugate of $z = x + jy$ is written as $z^* = x - jy$. In polar form the conjugate is obtained by reversing the sign of the angle of z, $z^* = Me^{-j\theta}$. The geometric interpretation in Figure A–2 shows that conjugation simply reflects a complex number across the real axis in the complex plane.

Exercise A–1

Convert the following complex numbers to polar form:

(a) $1 + j\sqrt{3}$ (b) $-10 + j20$ (c) $-2000 - j8000$ (d) $60 - j80$

Answers: (a) $2e^{j60°}$ (b) $22.4e^{j117°}$ (c) $8246e^{j256°}$ (d) $100e^{j307°}$

Exercise A–2

Convert the following complex numbers to rectangular form:

(a) $12e^{j90°}$ (b) $3e^{j45°}$ (c) $400\angle\pi$ (d) $8e^{-j60°}$ (e) $15e^{j\pi/6}$ (f) $25\angle-120°$

Answers: (a) $0 + j12$ (b) $2.12 + j2.12$ (c) $-400 + j0$ (d) $4 - j6.93$
(e) $13 + j7.5$ (f) $-12.5 - j21.65$

Exercise A–3

Evaluate the following expressions:

(a) $\mathrm{Re}(12e^{j\pi})$ (b) $\mathrm{Im}(100\angle 60°)$ (c) $\angle(-2 + j6)$ (d) $\mathrm{Im}[(4e^{j\frac{\pi}{4}})^*]$

Answers: (a) -12 (b) 86.6 (c) $108.4°$ (d) -2.83

ARITHMETIC OPERATIONS: ADDITION AND SUBTRACTION

Addition and subtraction are defined in terms of complex numbers in rectangular form. Two complex numbers

$$z_1 = x_1 + jy_1 \quad \text{and} \quad z_2 = x_2 + jy_2 \qquad (A–9)$$

are added by separately adding the real parts and imaginary parts. The sum $z_1 + z_2$ is defined as

$$z_1 + z_2 = (x_1 + x_2) + j(y_1 + y_2) \qquad (A–10)$$

Subtraction follows the same pattern except that the components are subtracted:

$$z_1 - z_2 = (x_1 - x_2) + j(y_1 - y_2) \qquad (A–11)$$

Figure A–3 shows a geometric interpretation of addition and subtraction. In particular, note that $z + z^* = 2x$ and $z - z^* = j2y$.

$j\,\mathrm{Im}$

Re

FIGURE A–3 *Graphical representation of addition and subtraction of complex numbers.*

MULTIPLICATION AND DIVISION

Multiplication and division of complex numbers can be accomplished with the numbers in either rectangular or polar form. For complex numbers in rectangular form, the multiplication operation yields

$$
\begin{aligned}
z_1 z_2 &= (x_1 + jy_1)(x_2 + jy_2) \\
&= (x_1 x_2 + j^2 y_1 y_2) + j(x_1 y_2 + x_2 y_1) \\
&= (x_1 x_2 - y_1 y_2) + j(x_1 y_2 + x_2 y_1)
\end{aligned}
\tag{A-12}
$$

For numbers in polar form, the product is

$$
\begin{aligned}
z_1 z_2 &= (M_1 e^{j\theta_1})(M_2 e^{j\theta_2}) \\
&= (M_1 M_2) e^{j(\theta_1 + \theta_2)}
\end{aligned}
\tag{A-13}
$$

Multiplication is somewhat easier to carry out with the numbers in polar form, although both methods should be understood. In particular, the product of a complex number z and its conjugate z^* is the square of its magnitude, which is always positive.

$$
zz^* = (M e^{j\theta})(M e^{-j\theta}) = M^2
\tag{A-14}
$$

For complex numbers in polar form the division operation yields

$$
\begin{aligned}
\frac{z_1}{z_2} &= \frac{M_1 e^{j\theta_1}}{M_2 e^{j\theta_2}} \\
&= \left(\frac{M_1}{M_2}\right) e^{j(\theta_1 - \theta_2)}
\end{aligned}
\tag{A-15}
$$

When the numbers are in rectangular form, the numerator and denominator of the quotient are multiplied by the conjugate of the denominator.

$$
\frac{z_1}{z_2} \frac{z_2^*}{z_2^*} = \frac{(x_1 + jy_1)(x_2 - jy_2)}{(x_2 + jy_2)(x_2 - jy_2)}
$$

Applying the multiplication rule from Eq. (A–12) to the numerator and denominator yields

$$
\frac{z_1}{z_2} = \frac{(x_1 x_2 + y_1 y_2) + j(x_2 y_1 - x_1 y_2)}{x_2^2 + y_2^2}
\tag{A-16}
$$

Complex division is easier to carry out with the numbers in polar form, although both methods should be understood.

Exercise A-4

Evaluate the following expressions using $z_1 = 3 + j4, z_2 = 5 - j7, z_3 = -2 + j3$, and $z_4 = 5\angle-30°$:

(a) $z_1 z_2$ (b) $z_3 + z_4$ (c) $z_2 z_3 / z_4$ (d) $z_1^* + z_3 z_1$ (e) $z_2 + (z_1 z_4)^*$

Answers:

(a) $43 - j$ (b) $2.33 + j0.5$ (c) $-0.995 + j6.12$ (d) $-15 - j3$ (e) $28 - j16.8$

Exercise A-5

Given $z = x + jy = M e^{j\theta}$, evaluate the following statements:

(a) $z + z^*$ (b) $z - z*$ (c) z/z^* (d) z^2 (e) $(z^*)^2$ (f) zz^*

Answers:

(a) $2x$ (b) $j2y$ (c) $e^{j2\theta}$ (d) $x^2 - y^2 + j2xy$ (e) $x^2 - y^2 - j2xy$ (f) $x^2 + y^2$

Exercise A–6

Given $z_1 = 1$, $z_2 = -1$, $z_3 = j$, and $z_4 = -j$, evaluate (a) z_1/z_3 (b) z_1/z_4 (c) $z_3 z_4$
(d) $z_3 z_3$ (e) $z_4 z_4$ (f) $z_2 z_3^*$

Answers:

(a) $-j$ (b) $+j$ (c) 1   (d) -1 (e) -1 (f) j

Exercise A–7

Evaluate the expression $T(\omega) = j\omega/(j\omega + 10)$ at $\omega = 5, 10, 20, 50, 100$.

Answers: $0.447\angle63.4°, 0.707\angle45°, 0.894\angle26.6°, 0.981\angle11.3°, 0.995\angle5.71°$

Exercise A–8

If $Z_1 = -j20$, $Z_2 = j10$, and $Z_3 = 10 + j5$, find $Z_{EQ} = Z_2 + \dfrac{1}{1/Z_1 + 1/Z_3}$

Answer: $Z_{EQ} = 14.94\angle34.5° = 12.31 + j8.46$

ANSWERS TO SELECTED PROBLEMS

CHAPTER 01

1–1 (a) $1,000,000 \, \text{Hz} = 1 \times 10^6 \, \text{Hz} = 1 \, \text{MHz}$
    (b) $102.5 \times 10^9 \, \text{W} = 103 \, \text{GW}$
    (c) $0.333 \times 10^{-7} \, \text{s} = 33.3 \times 10^{-9} \, \text{s} = 33.3 \, \text{ns}$
    (d) $10 \times 10^{-12} \, \text{F} = 10 \, \text{pF}$

1–3 $11.88 \, \text{MC}$

1–5 (a) 10^{-6}
    (b) 10^3
    (c) 10^{-3}
    (d) 10^6

1–7 $4 \, \text{mA}$

1–9 $i(0) = 0 \, \text{A}, \ i(0.5) = 3 \, \text{A}, \ i(1) = 6 \, \text{A}$

1–11 $q = 100 \, \text{C}$

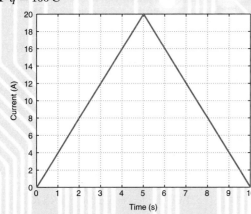

1–13 (a) $i = 16.667 \, \text{A}, t = 6 \, \text{hours}$
     (b) $i = 25 \, \text{A}, t = 4 \, \text{hours}$

1–15 $q(t) = \frac{3}{2}\left(1 - e^{-2t}\right)\text{C}, \ t \geq 0$

1–17

v (V)	i (A)	p (W)	Absorbing/Delivering
-2	-9.8647	19.7293	Absorbing
2	-2.6109	-5.2219	Delivering
3	10.0855	30.2566	Absorbing

1–19 $i_{\text{Max}} = 27.778 \, \text{mA}$

1–21

Case	v	i	p	Power Transfer
(a)	$+11 \, \text{V}$	$-1.1 \, \text{A}$	$-12.1 \, \text{W}$	B to A
(b)	$+80 \, \text{V}$	$+20 \, \text{mA}$	$+1.6 \, \text{W}$	A to B
(c)	$-120 \, \text{V}$	$-12 \, \text{mA}$	$+1.44 \, \text{W}$	A to B
(d)	$-1.5 \, \text{V}$	$-600 \, \mu\text{A}$	$900 \, \mu\text{W}$	A to B

1–23 $v_4 = 5 \, \text{V}$; absorbing

1–25 $q_{\text{Total}} = 20 \, \text{mC}$

1–27 MATLAB script:

```
t = 5e-3
vt = 170*cos(377*t)
it = 2*sin(377*t)
pt = vt*it
Absorbing = pt>0
```

```
t = 10e-3
vt = 170*cos(377*t)
it = 2*sin(377*t)
pt = vt*it
Absorbing = pt>0
```

t (ms)	$v(t)$ (V)	$i(t)$ (mA)	$p(t)$ (W)	Absorbing/Delivering
5	−52.54	1.90	−99.94	Delivering
10	−137.52	−1.18	161.69	Absorbing

1–29 66.67%

1–31 Sum the magnitudes of the powers, $p = 122.25$ W; the result is reasonable if maximum values are not required continuously

CHAPTER 02

2–1 $v = 123.2$ V

2–3 $i = 1$ mA

2–5 $R_x = 9$ kΩ

2–7 No, the resistor is 33 kΩ ± 10% and should have a silver tolerance stripe instead of a gold stripe

2–9 $v = 111.803$ V

2–11 (a)

Voltage	Current	Power
−800.0000e−003	−200.0000e−018	160.0000e−018
−400.0000e−003	−200.0000e−018	80.0000e−018
−200.0000e−003	−199.9329e−018	39.9866e−018
−100.0000e−003	−196.3369e−018	19.6337e−018
0.0000e+000	0.0000e+000	0.0000e+000
100.0000e−003	10.7196e−015	1.0720e−015
200.0000e−003	595.9916e−015	119.1983e−015
400.0000e−003	1.7772e−009	710.8888e−012
800.0000e−003	15.7926e−003	12.6341e−003

(b) Nonlinear, nonbilateral, and passive

(c) For $v = 5$ V, $i = 1.45 \times 10^{71}$ A and $p = 7.23 \times 10^{71}$ W. The model is not valid.

(d) For $v = -5$ V, $i = -2.00 \times 10^{-16}$ A and $p = 1.00 \times 10^{-15}$ W. The model is valid.

2–13 $i_1 = 5$ A, $i_4 = -3$ A

2–15 (a) Nodes: A, B, and C; loops: 1–2; 2–3–4; 1–3–4

(b) Series: 3 and 4; parallel: 1 and 2

(c) KCL   Node A $-i_1 - i_2 - i_3 = 0$
            Node B $i_3 - i_4 = 0$
            Node C $i_1 + i_2 + i_4 = 0$
    KVL   Loop 1 2 $-v_1 + v_2 = 0$
            Loop 2 3 4 $-v_2 + v_3 + v_4 = 0$
            Loop 1 3 4 $-v_1 + v_3 + v_4 = 0$

2–17 (a) Nodes: A, B, C, D; loops: 1–3–2; 2–4–5; 3-6-4; 1-6-5; 2-3-6-5

(b) Series: none; parallel: none

(c) KCL   Node A $-i_2 - i_3 - i_4 = 0$
            Node B $-i_1 + i_3 - i_6 = 0$
            Node C $i_1 + i_2 + i_5 = 0$
            Node D $i_4 - i_5 + i_6 = 0$
    KVL   Loop 1 3 2 $-v_1 - v_3 + v_2 = 0$
            Loop 2 4 5 $-v_2 + v_4 + v_5 = 0$
            Loop 3 6 4 $v_3 + v_6 - v_4 = 0$

2–19 Problem 2–18 is no longer valid because there is a conflict with the voltages

2–21 $v_1 = 30$ V, $v_3 = 20$ V, $v_6 = 15$ V

2–23 (a) KVL   Loop 1 2 $v_1 + v_2 = 0$
                    Loop 2 4 5 $-v_2 + v_4 - v_5 = 0$
                    Loop 3 4 $v_3 + v_4 = 0$

(b) If $v_4 = 0$ V, then $v_3 = 0$ V, $v_1 = -v_2$, $v_2 = -v_5$

2–25 $i_x = -500$ μA, $v_x = -34$ V

2–27 $i_x = 200$ μA, $v_x = 13.6$ V, i_y is not zero because there is a voltage across the 33-kΩ resistor

2–29 $i_x = 1$ A, $v_x = 11$ V

2–31 $p_S = -16.667$ mW

2–33 $R = 90$ kΩ

2–35 $R_{EQ} = 15$ Ω

2–37 $R_{EQ} = 22.99$ kΩ

2–39 Switch open: $R_{EQ} = 150$ Ω; switch closed: $R_{EQ} = 100$ Ω

2–41

Resistance (Ω)	Terminal 1	Terminal 2
100	A	D
200	A	B
150	A	C
50	C	D
25	A+B+C	D
33.3	B+C	D
133.3	A	B+C

2–43 $R_{AB} = 33$ kΩ; $R_{AC} = 33$ kΩ; $R_{AD} = 83$ kΩ; $R_{BC} = 0$ Ω; $R_{BD} = 50$ kΩ; $R_{CD} = 50$ kΩ

2–45

R_{EQ} (Ω)	Combination of 1-kΩ Resistors
2000	Two resistors in series: $R + R$
500	Two resistors in parallel: $R \parallel R$
1500	One resistor in series with a parallel combination of two resistors: $R + (R \parallel R)$
333	Three resistors in parallel: $R \parallel R \parallel R$
250	Four resistors in parallel: $R \parallel R \parallel R \parallel R$
400	Two resistors in series in parallel with two resistors in parallel: $(R + R) \parallel R \parallel R$

2–47 $V_S = 50$ V, $R_S = 10\,\Omega$

2–49 $R_x = 21.78$ kΩ

2–51 $R = R_L/3$

2–53 $R_{EQ} = R/3$

2–55 $v_L = \dfrac{R_L v_S}{R + 2R_L}$

2–57 $v_L = R_L i_L = \dfrac{R R_L i_S}{2R + R_L}$

2–59 $v_O = 1.875$ V

2–61 $v_{O,\text{MIN}} = 0$ V, $v_{O,\text{MAX}} = 21.4286$ V

2–63 $R_x = \dfrac{R_B R_C}{R_A}$

2–65 $R_1 = 1.7$ k$\Omega, R_2 = 2.3$ k$\Omega, R_3 = 1$ kΩ

2–67 For $v_L = 2$ V, $R_x = 33.3\,\Omega$; for $v_L = 4$V, $R_x = \infty\,\Omega$; for $v_L = 6$ V, no solution

2–69 $v_x = 558.46$ V, $i_x = 126.92$ mA, $p_x = 406.15$ W

2–71 $v_x = 9.375$ V, $i_x = -12.5$ mA, $p_x = 3.125$ W

2–73 $i_x = 100$ mA

2–75 $v_O = (v_1 + v_2 + v_3)/3$

2–77 $R_x = 500\,\Omega$

2–79 $v_1 = 24$ V, $v_2 = 16$ V, $v_3 = 8$ V, $v_4 = 1.6$ V, $v_5 = 6.4$ V$i_1 = -2$ mA, $i_2 = 2$ mA, $i_3 = 1.6$ mA, $i_4 = 0.4$ mA, $i_5 = 0.4$ mA

2–81 $v_1 = 120$ V, $v_2 = 82.12$ V, $v_3 = 37.88$ V, $v_4 = 25.52$ V, $v_5 = 12.36$ V, $v_6 = 7.26$ V, $v_7 = 5.10$ V, $i_1 = -547.46\,\mu$A, $i_2 = 547.46\,\mu$A, $i_3 = 172.19\,\mu$A, $i_4 = 375.28\,\mu$A, $i_5 = 220.75\,\mu$A, $i_6 = 154.53\,\mu$A, $i_7 = 154.53\,\mu$A

2–83 (b) Nonlinear, passive, bilateral
(c) -8.7706 V $< v < 8.7706$ V
(d) $|v| > 87.7058$ V

CHAPTER 03

3–1

$$\begin{bmatrix} \left(\dfrac{1}{R_1} + \dfrac{1}{R_5}\right) & -\dfrac{1}{R_1} & -\dfrac{1}{R_5} \\ -\dfrac{1}{R_1} & \left(\dfrac{1}{R_1} + \dfrac{1}{R_2} + \dfrac{1}{R_3}\right) & -\dfrac{1}{R_3} \\ -\dfrac{1}{R_5} & -\dfrac{1}{R_3} & \left(\dfrac{1}{R_3} + \dfrac{1}{R_4} + \dfrac{1}{R_5}\right) \end{bmatrix} \begin{bmatrix} v_A \\ v_B \\ v_C \end{bmatrix} = \begin{bmatrix} i_S \\ 0 \\ 0 \end{bmatrix}$$

3–3 (a)

$$\begin{bmatrix} \left(\dfrac{1}{5} + \dfrac{1}{10} + \dfrac{1}{8}\right) & -\dfrac{1}{8} \\ -\dfrac{1}{8} & \left(\dfrac{1}{8} + \dfrac{1}{4}\right) \end{bmatrix} \begin{bmatrix} v_A \\ v_B \end{bmatrix} = \begin{bmatrix} 0 \\ 2 \end{bmatrix}$$

(b) $v_A = 1.7391$ V, $v_B = 5.913$ V
(c) $v_x = 5.913$ V, $i_x = 347.826$ mA

3–5 (a)

$$\begin{bmatrix} \left(\dfrac{1}{4000} + \dfrac{1}{4000}\right) & 0 \\ 0 & \left(\dfrac{1}{1000} + \dfrac{1}{4000}\right) \end{bmatrix} \begin{bmatrix} v_A \\ v_B \end{bmatrix} = \begin{bmatrix} 11.25 \times 10^{-3} \\ -5 \times 10^{-3} \end{bmatrix}$$

(b) $v_A = 22.5$ V, $v_B = -4$ V
(c) $v_x = 26.5$ V, $i_x = -4.625$ mA

3–7

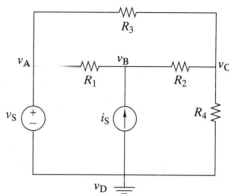

3–9 (a)

$$\left(\dfrac{1}{R_1} + \dfrac{1}{R_4}\right) v_A - \dfrac{1}{R_1} v_B - \dfrac{1}{R_4} v_C = i_{S1}$$

$$-\dfrac{1}{R_1} v_A + \left(\dfrac{1}{R_1} + \dfrac{1}{R_2}\right) v_B = i_{S2}$$

$$-\dfrac{1}{R_4} v_A + \left(\dfrac{1}{R_3} + \dfrac{1}{R_4}\right) v_C = -i_{S2}$$

(b)

$$v_A = \dfrac{(R_1 R_3 + R_1 R_4 + R_2 R_3 + R_2 R_4)i_{S1} + (R_2 R_4 - R_1 R_3)i_{S2}}{R_1 + R_2 + R_3 + R_4}$$

$$v_B = \dfrac{R_2[(R_3 + R_4)i_{S1} + (R_1 + R_4)i_{S2}]}{R_1 + R_2 + R_3 + R_4}$$

$$v_C = \dfrac{R_3[(R_1 + R_2)i_{S1} - (R_1 + R_4)i_{S2}]}{R_1 + R_2 + R_3 + R_4}$$

(c) $v_A = 5.9812$ V, $v_B = 5.3402$ V, $v_C = -1.4103$ V

3–11 (a)

$$\left(\dfrac{1}{R_1} + \dfrac{1}{R_2} + \dfrac{1}{R_x}\right) v_A - \dfrac{1}{R_x} v_B = \dfrac{v_S}{R_1}$$

$$-\dfrac{1}{R_x} v_A + \left(\dfrac{1}{R_3} + \dfrac{1}{R_4} + \dfrac{1}{R_x}\right) v_B = \dfrac{v_S}{R_3}$$

(b) $v_x = -4.3548$ V, $i_x = 10.6452$ mA

3–13 (a) $(10000 + 5000)i_A - 5000 i_B = 5$
$-5000 i_A + (5000 + 10000)i_B = -10$
(b) $i_A = 125\,\mu$A, $i_B = -625\,\mu$A
(c) $v_x = 1.25$ V, $i_x = 750\,\mu$A

3–15 (a)

$$\begin{bmatrix} (4000 + 1000 + 4000) & -4000 \\ -4000 & (4000 + 1000 + 4000) \end{bmatrix} \begin{bmatrix} i_A \\ i_B \end{bmatrix} = \begin{bmatrix} 0 \\ 12 \end{bmatrix}$$

(b) $i_A = 738.4615\,\mu A$, $i_B = 1.6615\,mA$

(c) $v_x = 6.6462\,V$, $i_x = 923.0769\,\mu A$

3–17 (a) $(R_A + R_B)i_A - R_B i_B - R_A i_C = v_S$
$-R_B i_A + (R_B + R_C + R_D)i_B - R_C i_C = 0$
$-R_A i_A - R_C i_B + (R_A + R_C + R_E)i_C = 0$

(b)
$$\left(\frac{1}{R_A} + \frac{1}{R_B} + \frac{1}{R_C}\right)v_A - \frac{1}{R_C}v_B = \frac{v_S}{R_A}$$
$$-\frac{1}{R_C}v_A + \left(\frac{1}{R_C} + \frac{1}{R_D} + \frac{1}{R_E}\right)v_B = \frac{v_S}{R_E}$$

(c) Node-voltage equations, because there are fewer unknowns

(d)
$$v_x = \frac{R_B(R_A R_D + R_C R_D + R_C R_E + R_D R_E)v_S}{D_x}$$
$$i_x = -\frac{(R_A R_B + R_A R_C + R_B R_C + R_B R_E)v_S}{D_x}$$
$$D_x = R_A R_B R_D + R_A R_B R_E + R_A R_C R_D + R_A R_C R_E$$
$$+ R_B R_C R_D + R_A R_D R_E + R_B R_C R_E + R_B R_D R_E$$

3–19 (a)
$$i_A = -\frac{R_4 i_{S1} + R_2 i_{S2}}{R_1 + R_2 + R_4}$$
$$i_B = \frac{(R_1 + R_2)i_{S1} - R_2 i_{S2}}{R_1 + R_2 + R_4}$$
$$i_C = \frac{-R_4 i_{S1} + (R_1 + R_4)i_{S2}}{R_1 + R_2 + R_4}$$

(b) $v_x = \dfrac{-R_4[(R_1 + R_2)i_{S1} - R_2 i_{S2}]}{R_1 + R_2 + R_4}$

3–21 (a)

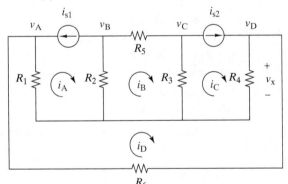

(b) $i_A = -i_{S1}$
$-R_2 i_A + (R_2 + R_3 + R_5)i_B - R_3 i_C = 0$
$i_C = i_{S2}$
$-R_1 i_A - R_4 i_C + (R_1 + R_4 + R_6)i_D = 0$

(c) $v_x = 224\,V$

3–23 $R_{IN} = 2R$

3–25 (a) $i_A = \dfrac{v_S(R_2 + R_3 + R_4)}{R_1 R_2 + R_1 R_3 + R_1 R_4 + R_2 R_3 + R_2 R_4}$

(b) $R_{IN} = \dfrac{v_S}{i_A} = \dfrac{R_1 R_2 + R_1 R_3 + R_1 R_4 + R_2 R_3 + R_2 R_4}{R_2 + R_3 + R_4}$

(c) $R_{EQ} = \dfrac{R_1 R_2 + R_1 R_3 + R_1 R_4 + R_2 R_3 + R_2 R_4}{R_2 + R_3 + R_4}$

3–27 (a)
$$\begin{bmatrix} 139000 & -100000 & 0 & 0 \\ -100000 & 132000 & -22000 & 0 \\ 0 & -22000 & 171000 & -68000 \\ 0 & 0 & -68000 & 68000 \end{bmatrix}\begin{bmatrix} i_A \\ i_B \\ i_C \\ i_D \end{bmatrix} = \begin{bmatrix} -100 \\ 0 \\ 0 \\ 150 \end{bmatrix}$$

(b) $i_A = -1.2380\,mA$, $i_B = -720.8214\,\mu A$,
$i_C = 1.3023\,mA$, $i_D = 3.5082\,mA$

(c)
$$\begin{bmatrix} \left(\dfrac{1}{39000} + \dfrac{1}{100000} + \dfrac{1}{10000}\right) & -\dfrac{1}{10000} \\[2mm] -\dfrac{1}{10000} & \left(\dfrac{1}{10000} + \dfrac{1}{22000} + \dfrac{1}{81000}\right) \end{bmatrix}\begin{bmatrix} v_A \\ v_B \end{bmatrix}$$
$$= \begin{bmatrix} \dfrac{50}{39000} + \dfrac{150}{100000} \\[2mm] \dfrac{150}{22000} \end{bmatrix}$$

3–29 $K = 1/5000\,S$

3–31
$$K = \frac{i_O}{i_S} = \frac{R_1 R_2}{R_1 R_2 + R_1 R_3 + R_1 R_4 + R_2 R_3 + R_2 R_4}$$

3–33 $K = 12.3095\,k\Omega$, $v_O = 123.095\,V$

3–35 $K = 5.5325\,mS$, $i_O = 553.25\,mA$

3–37 $i_O = 6\,mA$

3–39 $v_O = 20\,V$

3–41 $v_O = \dfrac{v_1 + v_2}{6}$

3–43 $p_L = 9\,W$

3–45 $v_O = v_{S1} + 2v_{S2} - 3v_{S3} - 2v_{S4}$

3–47 Thévenin: $v_T = 15\,V$, $R_T = 11\,k\Omega$; Norton: $i_N = 1.3636\,mA$, $R_N = 11\,k\Omega$

3–49 (a) $v_T = 30\,V$, $R_T = 11.6\,k\Omega$
(b) $i_L = 892.857\,\mu A$

3–51 $v_T = 10\,V$, $R_T = 25\,\Omega$; for $R_L = 5\,\Omega$, $v_L = 1.6667\,V$; for $R_L = 10\,\Omega$, $v_L = 2.8571\,V$; for $R_L = 20\,\Omega$, $v_L = 4.4444\,V$

3–53 $v_T = 3.0775\,V$, $R_T = 23.8182\,k\Omega$

3–55 (a) $i_N = 221.6\,\mu A$, $R_N = 36.93\,k\Omega$
(b) $p_L = 453.34\,\mu W$

3–57 $v_L = 20\,V$

3–59 $v_L = 16.667\,V$

3–61 $R_L = 151\,\Omega$

3–63 The circuit can support any number of loads connected in parallel

3–65 $v_L = 2.582\,V$, $i_L = 4.836\,mA$, $p_L = 12.486\,mW$

3–67 (a) $R_L = 13.33\,\Omega$

(b) Not possible

(c) $R_L = 208.0776\,\Omega$ or $R_L = 1.9224\,\Omega$

3–69 $v_T = 10\,\text{V}$, $R_T = 30\,\Omega$

3–71 (a) $R_L = \infty\,\Omega$, $v_L = 33.6\,\text{V}$

(b) $R_L = 0\,\Omega$, $i_L = 14.483\,\text{mA}$

(c) $R_L = 2.32\,\text{k}\Omega$, $p_L = 121.655\,\text{mW}$

3–73 $p_{MAX} = 1.125\,\text{W}$

3–75 (a) $R_L = \infty\,\Omega$, $v_L = 7.5\,\text{V}$

(b) $R_L = 0\,\Omega$, $i_L = 2.1429\,\text{mA}$

(c) $R_L = 3.5\,\text{k}\Omega$, $p_L = 4.0179\,\text{mW}$

3–77 $p_{MAX} = 1.35\,\text{W}$

3–79 Use series resistance with $R_S = 1.5\,\text{k}\Omega$

3–81 Use series resistance with $R_S = 1.3 + 2 + 6.2 = 9.5\,\text{k}\Omega$

3–83 Team A

3–85 Use series resistance with $R_S = 11.8034\,\Omega$

3–87 $R_1 = 66.67\,\Omega$ in series with the source resistance and $R_2 = 100\,\Omega$ in parallel with the load

3–89 $R_1 = 82\,\Omega$ in parallel with the series combination of $R_2 = 270\,\Omega$ and the load resistor

3–91 The claim is true; for $R_L = [16, 8, 4]\,\Omega$, $R_{IN} = [600, 604, 606]\,\Omega$, and $R_{OUT} = [15.79, 7.95, 3.99]\,\Omega$

3–93 Use a T-interface with one 50-Ω resistor in series with the load, one in parallel with those two resistors, and one in series with the source resistor

3–95 Circuits 2 meets the voltage requirement and the source power rating

3–97 Both designs can meet the specifications; for Circuit 1, $R_1 = 66.67\,\Omega$ and $R_2 = 100\,\Omega$; for Circuit 2, $R_1 = 300\,\Omega$ and $R_2 = 100\,\Omega$

3–99 $v_L = 2\left[\dfrac{R_2 v_{S1} + R_1 v_{S2}}{R_1 + R_2}\right]$

CHAPTER 04

4–1 $K_V = -8$, $K_I = -2$

4–3 $K_V = 2.4$, $K_I = 4.8$, $p_S = 6.25\,\text{mW}$, $p_L = 72\,\text{mW}$

4–5 $K_V = \dfrac{50}{49}$

4–7 $v_O = 4.087\,\text{V}$

4–9 $K_V = \dfrac{R_O(1 - gR_S)}{R_O + R_S - gR_OR_S}$

4–11 $g = 4.545\,\text{S}$

4–13 With R_F, $v_T = \dfrac{-\mu R_F v_S}{R_F + R_S(1 + \mu)}$ and $R_T = R_P$; for $R_F = \infty$, $v_T = -\mu v_S$ and $R_T = R_P$

4–15 $R_{IN} = R + r$

4–17 $v_T = \dfrac{-\beta R_O v_S}{R_S + R_x}$, $R_T = R_O$

4–19 For $v_S = 1\,\text{V}$, $i_C = 300\,\mu\text{A}$, $v_{CE} = 14.01\,\text{V}$; for $v_S = 5\,\text{V}$, $i_C = 4.3\,\text{mA}$, $v_{CE} = 0.810\,\text{V}$

4–21 For $v_S = 0.8\,\text{V}$, $i_C = 600\,\mu\text{A}$, $v_{CE} = 6\,\text{V}$; for $v_S = 2\,\text{V}$, $i_C = 1.5\,\text{mA}$, $v_{CE} = 0\,\text{V}$

4–23 (a) $K = -10$

(b) $K = 11$

4–25 (a) The claim is true

(b) The claim is not true for Circuit 2

4–27 $K = -5.025$

4–29 Use a noninverting amplifier with a 1-kΩ resistor connected to ground and a potientiometer for the feedback resistor with a range from 9 to 99 kΩ

4–31 (a) $v_O = 10v_S$

(b) For $v_S = 0.5\,\text{V}$, $i_O = 45.45\,\mu\text{A}$; for $v_S = 2\,\text{V}$, $i_O = 161.818\,\mu\text{A}$

4–33 (a) Use a subtractor OP AMP circuit with v_1 connected to the positive input side of the OP AMP and v_2 connected to the negative input side of the OP AMP; select $R_1 = 1\,\text{k}\Omega$, $R_2 = 3\,\text{k}\Omega$, and $R_3 = R_4 = 1\,\text{k}\Omega$

4–35 $R_1 = 100\,\text{k}\Omega$, $R_2 = 10\,\text{k}\Omega$, $R_3 = 1\,\text{k}\Omega$

4–37 Switch open, $v_O = -2v_{S1}$; switch closed $v_O = -2v_{S1} - 2v_{S2}$

4–39 The claim is false; switch closed: $v_O = -v_S$; switch open: $v_O = v_S$

4–41 (a) $K = -\dfrac{R_1R_2 + R_1R_3 + R_2R_3}{RR_2}$

(b) $R = R_1 = R_2 = 10\,\text{k}\Omega$, $R_3 = 25\,\text{k}\Omega$

4–43 $v_O = 22v_1 - 3v_2$

4–45 $v_O = -\frac{1}{3}(10v_1 - 2v_2)$

4–47 $v_O = 5v_S - 10\,\text{V}$

4–49 $v_2 = -6v_1 - 2\,\text{V}$

4–51 The student's circuit has the same gain, but a different input resistance

4–53 Use a noninverting amplifier with a 10–kΩ resistor connected to ground and four 10–kΩ resistors in series in the feedback path

4–55 Use a subtractor design with $R_1 = 5\,\text{k}\Omega$, $R_2 = 150\,\text{k}\Omega$, $R_3 = 5\,\text{k}\Omega$, $R_4 = 150\,\text{k}\Omega$

4–57 Use the design in Figure 4–45 with equal input resistors and an OP AMP gain of 1000

4–59 Use an inverting amplifier with a gain of -5000 connected to an inverting summer with a gain of -4000 connected to the first amplifier and a gain of -1 connected to a -3.5-V source

4–61 Using the circuit in Figure 4–53, swap the input sources and use $R_1 = R_3 = 499\,\text{k}\Omega$, $R_2 = 2\,\text{k}\Omega$

4–63 Use a cascade connection of an inverting summer and an inverting amplifier; the summer has gains of -150 and -0.1 and the amplifier has a gain of -1

4–65 Circuit 1 meets the specifications

4–67 The design meets the specifications

4–69 $v_O = 1.5873\,\text{V}$; resolution is $79.3651\,\text{mV}$

4–71 $|v_{MAX}| = 9.6875\,\text{V}$; resolution is $0.3125\,\text{V}$

4–73 $K = +119$, $V_b = -952\,\text{mV}$

4–75 $K = -16667$, $V_b = 11.667$ V

4–77 $K = 4.1667$, $V_b = 0$ V

4–79 Purchase the circuit if the vendor will add a buffer at the circuit input without increasing the cost.

4–81 For $v_S = 2$ V, $v_O = 0$ V; for $v_S = -2$ V, $v_O = 0$ V; for $v_S = 6$ V, $v_O = 5$ V

4–83 Use a comparator with $v_P = v_S$, $v_N = 1.8127$ V, and $+V_{CC} = 5$ V

4–89 $v_O = v_1 - v_2$

CHAPTER 05

5–1

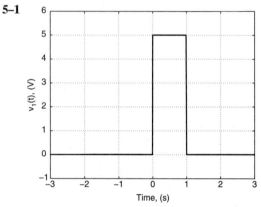

(a)

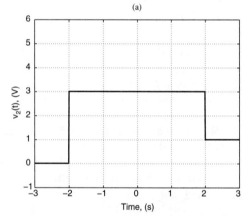

(b)

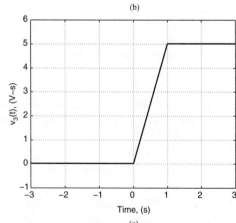

(c)

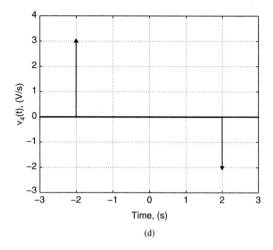

(d)

5–3

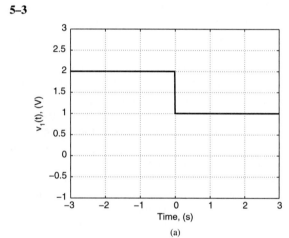

(a)

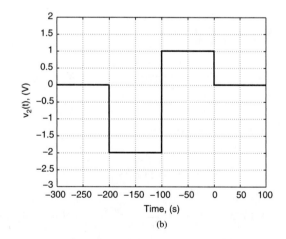

(b)

5–5 (a) $v_1(t) = 5u(1 - t) - 2u(2 - t)$
      (b) $v_2(t) = -6r(t) + 9r(t - 2) - 3r(t - 6)$

5–7 (a) $v_1(t) = 5r(t + 1) - 10r(t) + 10r(t - 2) - 5r(t - 3)$
      (b) $v_2(t) = 5u(t) - 2.5r(t) + 5r(t - 2) - 2.5r(t - 4) - 5(t - 4)$

5–9

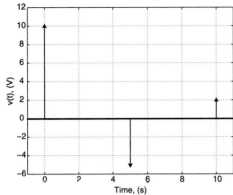

5–11

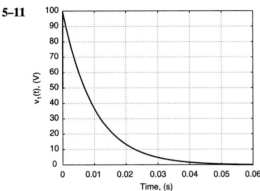

(a) $T_C = 10$ ms, $A = 100$ V

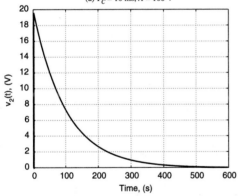

(b) $T_C = 10$ ms, $A = 19.6$ V

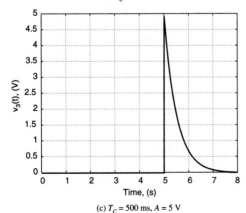

(c) $T_C = 500$ ms, $A = 5$ V

(d) $T_C = 100$ μs, $A = 50$ V

5–13 $T_C = 36.0674$ ms

5–15 $T_C = 5.566$ ms

5–17 $v(t) = 7\left(1 - e^{-250t}\right)u(t)$

5–19 (a) $T_0 = 16.67$ ms, $f_0 = 60$ Hz, $A = 339.41$ V, $T_S = 2.0833$ ms, $\phi = -45°$

(b) $T_0 = 20$ μs, $f_0 = 50$ kHz, $A = 50$ V, $T_S = 7.9517$ μs, $\phi = -143.13°$

5–21 $v(t) = 5\cos(200\pi t - \pi/2)$ V $= 5\cos(200\pi t - 90°)$ V, $\phi = -\pi/2$ radians $= -90°$, $T_S = 2.5$ ms

5–23 (a) $a = 95.96$ V, $b = -72.05$ V, $f_0 = 60$ Hz, $\omega_0 = 376.99 = 120\pi$ rad/s

(b) $a = 0$ A, $b = -12.00$ A, $f_0 = 60$ Hz, $\omega_0 = 376.99 = 120\pi$ rad/s

5–25 (a) $a = 86.60$ V, $b = -50.00$ V, $f_0 = 400$ Hz, $\omega_0 = 800\pi = 2513.3$ rad/s

(b)

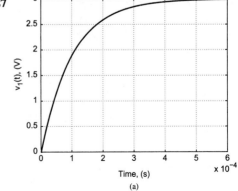

5–27

(a)

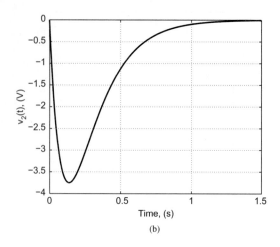

(b)

5–29 The maximum is 12.263 mV and occurs at $t = T_C = 1/30$ s.

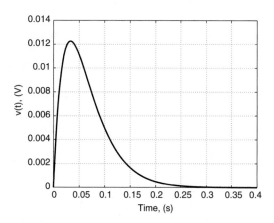

5–31 $v(t) = -6 + 14\cos(200\pi t - 180°)$ V

5–33 $V_{MAX} = 15$ V, $V_{MIN} = -5$ V, $\beta = 1000\pi$ rad/s

5–35 $v(t) = 2\{1 + e^{-t/T_C}[u(t) - u(t - 0.005)]$
$-e^{-(t-0.005)/T_C}u(t - 0.005)$ V

5–37 $v(t) = 5e^{-100t}\cos(200\pi t - \pi/2)$V $= 5e^{-100t}\sin(200\pi t)$ V

5–39(a) $V_p = 169.71$ V, $V_{PP} = 339.41$ V,
$V_{avg} = 0$ V, $V_{rms} = 120$ V
(b) $V_p = 50$ V, $V_{pp} = 100$ V,
$V_{avg} = 0$ V, $V_{rms} = 35.36$ V
(c) $V_p = 34$ V, $V_{pp} = 48$ V, $V_{avg} = 10$ V,
$V_{rms} = 19.70$ V

5–41 $V_p = V_{pp} = 15$ V, $V_{avg} = 7.5$ V,
$V_{rms} = 9.3541$ V, $T_0 = 20$ ms

5–43 $V_p = 120$ V, $V_{pp} = 235$ V, $V_{avg} = 0$ V,
$V_{rms} = 72.11$ V, $T_0 = 2$ ms

5–45 $V_{avg} = V_0$, $V_{rms} = \sqrt{V_0^2 + \dfrac{V_A^2}{2}}$

5–47 $V_{max} = 3$ V, $V_{min} = -3$ V, $V_p = 3$ V, $V_{pp} = 6$ V,
$V_{avg} = 0$ V

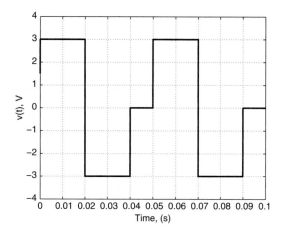

5–49

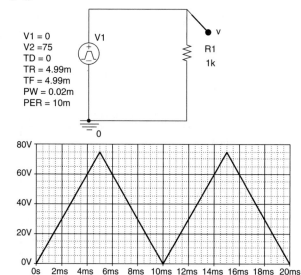

5–51 $v(t) = 30\cos(4.8\pi t)[u(t + 2) - u(t - 3)]$

5–53 $i(t) = 35e^{-211.82t}[u(t) - 2u(t - 0.004) + u(t - 0.008)]$ A

5–55 The compensation will occur in time if the system can detect the condition within 2 ms.

CHAPTER 06

6–1 For $t \geq 0$, $i_C(t) = 0$ A and $p_C(t) = 0$ W, so it is neither absorbing nor delivering power

6–3 $i_C(t) = -\pi\sin(2\pi 10^4 t)$ mA, $p_C(t) = -25\pi\sin(4\pi 10^4 t)$ mW, it is both absorbing and delivering power

6–5 $C = 1000$ pF

6–7 The capacitor both absorbs and delivers power

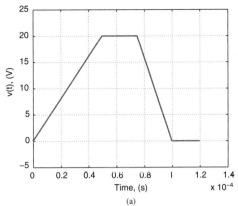

(a)

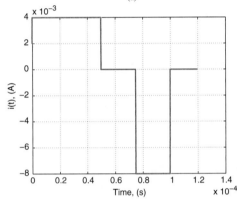

(b)

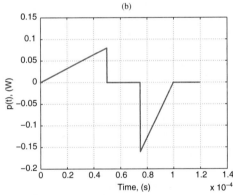

(c)

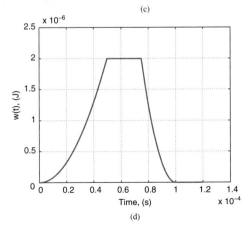

(d)

6–9

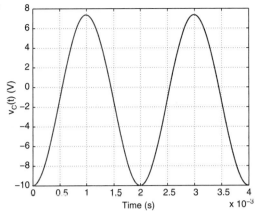

6–11 $v_L(t) = -5.6e^{-10000t}$ V, $p_L(t) = -5.6e^{-20000t}$ W, $w_L(t) = 280e^{-20000t}$ μJ, delivering power

6–13 $i_L(t) = \dfrac{1}{10n}\sin(1000nt)$; as $n \to \infty$, the inductor acts like an open circuit

6–15 $v_L(t) = (40 - 80000t)e^{-2000t}$ V for $t \ge 0$; the inductor is both absorbing and delivering power

6–17 (a) $v_C(0) = 0$ V

(b) $v_L(t) = -LC\dfrac{d^2 v_L(t)}{dt^2}$, $v_C(0) = v_L(0) = 0$ V, $i_L(0) = 25$ mA

(d) Sinusoidal

6–19 $i_L(t) = 80e^{-1000t} - 50$ mA, $p_L(t) = 400e^{-1000t} - 640e^{-2000t}$ μW; both absorbing and delivering power

6–21 $i_R(t) = -30.30e^{-1000t}$ μA

6–23 Inductor, $L = 2.5$ H, $i_L(0) = 16$ mA

6–25

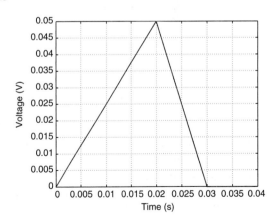

6–27 (a) Differentiator with $RC = 10$

(b) Integrator with $RC = 0.1$

(c) Amplifier with $K = -2000$

6–29 $v_O(t) = \dfrac{\cos(\omega t) - 1}{(11 \times 10^{-6})(\omega)}$; $\omega \ge 12.1212$ krad/s

6–31

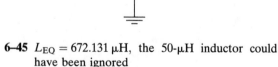

A plot of Voltage (V) versus Time (s) (× 10⁻³). The curve starts near −3.2 V, rises to about 6.2 V at t = 2, stays at 6.2 V until t = 3, then drops to 0 V.

6–33 $R = 15\,\text{k}\Omega$, $C = 0.015\,\mu\text{F}$

6–35 $\dfrac{d(v_O + v_S)}{dt} + \dfrac{v_O}{RC} = 0$

6–37 Use node-voltage equations to show
$$\dfrac{dv_O(t)}{dt} = \dfrac{v_S(t)}{RC}$$

6–39 The top branch is an inverting differentiator with $R = 1\,\Omega$ for the input resistor and $L = 100\,\text{H}$ for the feedback path; the bottom branch is an inverting integrator with $L = 100\,\text{H}$ for the input path and $R = 1\,\Omega$ for the feedback path; the two branches are connected by an inverting summer with gains of -1 on each path; note the resistor and inductor values are not practical

6–41 Use an inverting summer with gains of -1 and -15 for the top and bottom, respectively; the top input is an RC integrator with $R = 10\,\text{k}\Omega$ and $C = 1\,\mu\text{F}$; the bottom input is $v_S(t)$

6–43 $R_1 = 66.7\,\text{k}\Omega$, $R_2 = 50\,\text{k}\Omega$, $C = 1\,\mu\text{F}$

Circuit diagram with inputs v_{S1} through R_1, v_{S2} through R_2, capacitor C in the feedback, and output v_O from an op-amp.

6–45 $L_{EQ} = 672.131\,\mu\text{H}$, the 50-$\mu$H inductor could have been ignored

6–49 $C_{EQ} = 0.0086924\,\mu\text{F}$, $v_C(0) = 35\,\text{V}$

6–51 One capacitor in parallel with one inductor with $C = 12.4\,\mu\text{F}$ and $L = 3\,\text{H}$

6–53 Connect three 22–μF capacitors in series and repeat the series connection three times in parallel, using a total of nine capacitors

6–55 For Circuit 1, $i = 27.27\,\text{mA}$; for Circuit 2, $i = 0\,\text{mA}$

6–57 For
$$\omega = 0\,\text{rad/s}, \; K = -R_2/R_1;\; \text{as } \omega \to \infty,\; K \to 0$$

6–59 $C = 166.67\,\text{pF}$

6–61 $C = 2\,\text{F}$

6–63 Design #2 because it is simpler and meets the requirements

6–65 $C_{EQ} = C$

CHAPTER 07

7–1 $i(t) = 15e^{-50t}\,\text{mA}$

7–3 For Circuit C1, $T_C = 1\,\text{ms}$; for Circuit C2 $T_C = 200\,\mu\text{s}$

7–5 For Circuit C1, position A, $T_C = \dfrac{(C_1 + C_2)R_2R_3}{R_2 + R_3}$;

for Circuit C1, position B, $T_C = \dfrac{(C_1 + C_2)R_1R_3}{R_1 + R_3}$;

for Circuit C2, closed, $T_C = \dfrac{CR_1R_3R_4}{R_1R_3 + R_1R_4 + R_3R_4}$;

for Circuit C2, open, $T_C = \dfrac{CR_1R_3(R_2 + R_4)}{R_1R_2 + R_1R_3 + R_1R_4 + R_2R_3 + R_3R_4}$

7–7 $i_L(t) = 3e^{-38775t}\,\text{mA}$, $v_O(t) = -58.162e^{-38775t}\,\text{V}$

7–9 $v_C(t) = 12.5(1 - e^{-10^5 t})\,\text{V}$, $t \geq 0$

7–11 $i_L(t) = 2.4194e^{-4700t}\,\text{mA}$, $t \geq 0$

7–13 $v_O(t) = V_A(1 - e^{-Rt/L})\,\text{V}$, $t \geq 0$; the forced response is the constant term and the natural response is the exponential term

7–15 $v_O(t) = 10e^{-5000t}\,\text{V}$, $t \geq 0$; the natural response is the exponential term and there is no forced response

7–17 $v_C(t) = \left[12 - 4e^{-10000t}\right]\left[u(t) - u(t - 0.0001)\right]$
$$+ \left[8 + 2.5285e^{-7500(t-0.0001)}\right]u(t - 0.0001)\,\text{V}, \; t \geq 0$$

7–19 $v(t) = \dfrac{5000}{2499}\left(e^{-0.02t} - e^{-50t}\right)\text{V}, \; t \geq 0$

7–21 (a) $v_L(t) = 5.879e^{-194t}\,\text{V}$, $t \geq 0$
(b) $v_L(t) = 1.2341\cos(100t) - 2.3942\sin(100t)$
$\quad\quad + 4.6447e^{-194t}\,\text{V}$, $t \geq 0$
(c) $v_L(t) = 12.133e^{-194t} - 6.254e^{-100t}\,\text{V}$, $t \geq 0$

7–23 $i_L(t) = 1.35e^{-10^6 t} + 0.15\,\text{mA}, \; t \geq 0$

7–25 $v_C(t) = \left[50 + 30e^{-100t}\right]\left[u(t) - u(t - 0.01)\right]$
$$+ \left[80 - 18.964e^{-62.5(t-0.01)}\right]u(t - 0.01)\,\text{V}$$

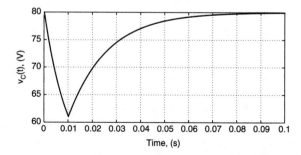

A plot of $v_C(t)$, (V) versus Time, (s). The curve starts at 80 V, decreases to about 61 V at t = 0.01, then rises asymptotically back toward 80 V.

7–27 $v_C(t) = 100e^{-50t}[u(t) - u(t - 0.02)] + 36.788e^{-200(t-0.02)}$
$[u(t - 0.02)]$ V

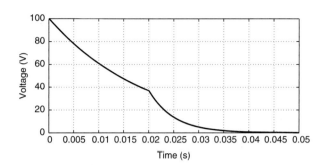

7–29 For $v_S(t) = 10\cos(\omega t)u(t)$ V, with $\omega = 100$ rad/s, $v_C(t) = 9.9504\cos(100t - 5.71°)$ V; with $\omega = 1000$ rad/s, $v_C(t) = 7.0711\cos(1000t - 45°)$ V; with $\omega = 10000$ rad/s, $v_C(t) = 0.995\cos(10000t - 84.29°)$ V; as the frequency increases, the amplitude decreases and the phase shift approaches $-90°$

7–31 (a) $C = 0.01\,\mu F$; $i_C(t) = -2e^{-10000t}$ mA
(b) $w_C(0.002) = 8.4967 \times 10^{-24}$ J

7–33 $R = 500\,\Omega$, $C = 1\,\mu F$

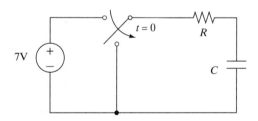

7–35 $R = 4\,k\Omega$, $C = 1\,\mu F$

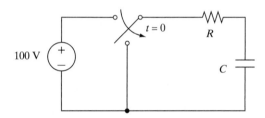

7–37 $v_C(0) = -5$ V, $v_C(\infty) = 5$ V, $T_C = 1$ ms, $R = 667\,\Omega$, $i_C(t) = 15e^{-1000t}$ mA

7–39 $i_L(0) = -5$ mA, $i_L(\infty) = 5$ mA, $T_C = 500\,\mu s$ $R = 100\,\Omega$, $L = 50$ mH

7–41 After the switch moves, the interface is a 100-Ω resistor in series with the source and a 2.67-μF capacitor in parallel with the 300-Ω resistor

7–43 Vendor B is less expensive and meets the specification

7–45 $v(t) = 3.667e^{-2.5t}\sin(5.4544t)$ V

7–47 $v(t) = 2 + e^{-5t}[8\cos(10t) + 6.5\sin(10t)]$ V

7–49 (a) $v_C(t) = e^{-500t}[10\cos(1323t) + 3.78\sin(1323t)]$ V, $i_L(t) = -7.56e^{-500t}\sin(1323t)$ mA
(b) Underdamped

7–51 As R increases, the response becomes less damped

7–53 (a) $v_C(t) = 5.164e^{-1000t}\sin(3873t)$ V, $i_L(t) = 1 - e^{-1000t}[0.258\sin(3873t) + \cos(3873t)]$ mA
(b) Underdamped

7–55 (a) $v_C(t) = -8e^{-100t}\sin(700t)$ V, $i_L(t) = -6 + e^{-100t}[\sin(700t) + 7\cos(700t)]$ mA
(b) Underdamped

7–57 (a) $v_C(t) = e^{-440t}[15\cos(1009.2t) + 6.5397\sin(1009.2t)]$ V, $i_L(t) = -59.45e^{-440t}\sin(1009.2t)$ mA
(b) Underdamped

7–59 $v_O(t) = 10.247e^{-12890t} - 0.2467e^{-310.33t}$ V

7–61 $v_O(t) = 24 - 16.971e^{-1333t}\sin(3771t)$ V

7–63 $\omega_0 = \dfrac{1}{\sqrt{LC}}$, $\zeta = \dfrac{1}{6R}\sqrt{\dfrac{L}{C}}$

7–65 (a) $R = 440\,\Omega$, $C = 0.2525\,\mu F$
(b) $v_C(t) = 26.4e^{-3000t} - 66e^{-6000t}$ V, $t \geq 0$

7–67 $v_C(t) = e^{-1000t}[2\cos(2000t) - 3\sin(2000t)]$ V, $t \geq 0$

7–69 $R = 500\,\Omega$, $L = 100$ mH, $C = 0.1\,\mu F$

7–71 $R = 20\,k\Omega$, $L = 100$ mH, $C = 0.001\,\mu F$, critically damped

7–73 Change the resistor to be $R = 6.25\,k\Omega$

7–75 $790.6 \times 10^{-6} \leq \zeta \leq 25 \times 10^{-3}$

7–77 $v_O(t) = -\dfrac{R_2 V_A}{R_1}e^{-t/R_1 C_1}$, $t \geq 0$

7–79

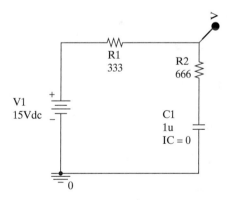

7–81 $R = 1.74477\,M\Omega$

7–83 (a) Amplitude increases and frequency remains the same
(b) Amplitude remains the same and frequency increases

7–85 Engineer C is correct; use the 10-mA fuse

7–87 The 6.8-mH inductor provides the lowest damping ratio

CHAPTER 08

8–1 $\mathbf{V}_1 = 50e^{-j30°} = 43.3 - j25$ V,
$\mathbf{V}_2 = 200e^{j150°} = -173.2 + j100$ V,
$\mathbf{V}_1 + \mathbf{V}_2 = 150e^{j150°} = -129.9 + j75$ V

8–3 $v_3(t) = 31.62 \cos(200\pi t + 71.57°)$ V

8–5 $v_3(t) = 174.96 \cos(1000t - 101.49°)$
V, $i_3(t) = 918 \cos(20000\pi t - 35.5°)$ mA

8–7 $v_1(t) = 6.4 \cos(10^7 t + 51.34°)$ V, $v_2(t) = 67.88 \cos(200000t - 45°)$ V, $i_1(t) = 12 \cos(30t)$ A,
$i_2(t) = 1.4537 \cos(60t + 136.39°)$ A

8–9 $v_3(t) = 1159 \cos(\omega t + 107.8°)$ V

8–11 $\mathbf{V}_2 = 8 + j6 = 10\underline{/36.87°}$ V

8–13 $S = 225 + j1050 = 1074\underline{/77.9°}$ VA

8–15 $Z = 11.177 + j35.29 = 37.02\underline{/72.43°}$ Ω

8–17 $Z = 46.6 - j50.4 = 68.6\underline{/-47.2°}$ Ω,
$R = 46.6$ Ω, $C = 0.397$ μF

8–19 (a) inductor impedances would decrease by a factor of 10 and the capacitor impedance would be multiplied by a factor of 10
(b) $Z = 611.7 + j26.66 = 612.2\underline{/2.5°}$ Ω
(c) $R = 611.68$ Ω, $L = 53.33$ mH

8–21 $Z = 15 + j0.077 = 15\angle 0.294°$ Ω

8–23 $R = 833$ Ω, $C = 800$ pF

8–25 $\omega = 1.25$ Mrad/s

8–27 $\mathbf{I} = 145\angle 46.9°$ μA, $\mathbf{V}_R = 4.785\angle 46.9°$ V,
$\mathbf{V}_C = 1.45\angle -43.1°$ V; the capacitor voltage lags the current

8–29 $\mathbf{I}_L = \dfrac{V_A}{R + j2\omega L}$, $\mathbf{V}_O = \dfrac{j\omega L V_A}{R + j2\omega L}$

8–31 $\mathbf{V}_R = \dfrac{RI_A}{1 + j2\omega RC}$, $\mathbf{I}_C = \dfrac{j2\omega RC I_A}{1 + j2\omega RC}$

8–33 $\mathbf{V}_S = 2.236\angle 7.23°$ V, $R = 100 - j50$ Ω

8–35 $v_X(t) = 8.944 \cos(2000t - 63.43°)$ V

8–37 $\mathbf{V}_X = 28.28\angle -53.13°$ V, $\mathbf{I}_X = 63.25\angle -26.6°$ mA

8–39 $\mathbf{I}_X = 1.304\angle 130.6°$ A

8–41

Source	$|\mathbf{V}_R|$	$|\mathbf{V}_C|$	$|\mathbf{I}|$	$|\mathbf{Z}_R|$	$|\mathbf{Z}_C|$
Increase/decrease V_A	Inc/dec $|\mathbf{V}_R|$	Inc/dec $|\mathbf{V}_C|$	Inc/dec $|\mathbf{I}|$	none	none
Increase/decrease ω	Inc/dec $|\mathbf{V}_R|$	Dec/inc $|\mathbf{V}_C|$	Inc/dec $|\mathbf{I}|$	none	Dec/inc $|\mathbf{Z}_C|$
Increase/decrease ϕ	none	none	none	none	none

8–43 $v_X(t) = 2.27 \cos(1000t - 68.7°) + 94.8 \cos(2000t - 174.6°)$ V

8–45 $\mathbf{V}_X = 59.66\angle -72.6°$ V, $\mathbf{I}_X = 4.22\angle 62.4°$ mA

8–47 $\mathbf{V} = 143.9\angle -50.7°$ V, $\mathbf{I} = 534.4\angle -118.9°$ mA

8–49 Use an L-interface with $Z_1 = 6.4654 + j10$ kΩ in series with the source and $Z_2 = 3.5355$ kΩ in parallel with the output

8–51 Use $R = 7.66$ kΩ and $L = 250$ mH in series with the source and load

8–53 (a) $|v_{O, MAX}| = |v_S|$, $f = 5.0329$ kHz
(b) $B = 15.8$ Hz

8–55 $\mathbf{V}_O = 64.62\angle -36.9°$ V

8–57 $\mathbf{I}_1 = 99.55\angle -98.33°$ mA, $\mathbf{I}_2 = 145.05\angle -95.95°$ mA,
$\mathbf{I}_3 = 45.77\angle -90.75°$ mA

8–59 $\mathbf{I}_1 = 97.53\angle -92.41°$ mA, $\mathbf{I}_2 = 48.32\angle -121.95°$ mA,
$\mathbf{I}_3 = 76.97\angle 19.39°$ mA

8–61 (b) For $\omega = 0$, $K = 1$; as $\omega \to \infty$, $K = \dfrac{R_1 + R_2}{R_1}$

8–63 $\mathbf{I}_O = 3.473\angle 7.125°$ A

8–65 $\mathbf{I}_{IN} = 11.17\angle 35.91°$ mA, $\mathbf{V}_O = 19.75\angle 170.91°$ V

8–67 For $\omega = [1, 10, 100, 1000, 10000]$ krad/s,
$\mathbf{V}_O/\mathbf{V}_S = [0.999, 0.995, 0.707, 0.0995, 0.00999]$

8–69 $K = j$, $Z_{IN} = 9990 + j10$ Ω

8–71 $K = 0.2 - j0.4$, $Z_{IN} = 200 - j100$ Ω,
$\mathbf{I}_X = \dfrac{\mathbf{V}_S(1 + j3)}{500}$

8–73 $\mathbf{V}_L = 169.7\angle 0°$ V, $\mathbf{I}_L = 49.2\angle 50.3°$ mA,
$P = 2.6674$ W,

8–75 $P = 2.6381$ W

8–77 (a) $P_L = 144$ mW
(b) $P_{MAX} = 450$ mW
(c) $Z_L = 1 - j2$ kΩ

8–79 $R = 152$ Ω, $L = 865.6$ mH

8–81 $R_1 = R_3 = 20$ kΩ, $R_2 = R_4 = 200$ kΩ,
$C_2 = 100$ pF, $C_3 = 0.1$ μF

8–85 The answer is correct, but the approach is not valid. Student mixed time and phasor domains.

CHAPTER 09

9–1 $F(s) = \dfrac{50000}{s(s + 100)}$, zeros at $s = [\infty, \infty]$,
poles at $s = [0, -100]$

9–3 $F(s) = \dfrac{10(1 - s)}{s}$, zeros at $s = 1$, poles at $s = 0$

9–5 $F(s) = \dfrac{ABs + A\alpha(B + 1)}{(s + \alpha)^2}$, zeros at $s = \left[-\dfrac{\alpha(B + 1)}{B}, \infty \right]$,
poles at $s = [-\alpha, -\alpha]$

9–7 $F(s) = \dfrac{20s - 1250}{s^2 + 2500}$, zeros at $s = [62.5, \infty]$,
poles at $s = \pm j50$

9–9 $F(s) = \dfrac{10(s^2 + 40s - 75)}{s^2(s + 15)}$, zeros at $s = [-41.79, 1.79, \infty]$,
poles at $s = [0, 0, -15]$

9–11 (a) $F_1(s) = \dfrac{2(s^2 + 900s + 160000)}{(s + 200)(s + 400)}$

(b) $F_2(s) = \dfrac{30(s^2 + 100s + 125000)}{(s + 200)(s^2 + 500^2)}$

9–13 (a) $F_1(s) = 5e^{-2s}$

(b) $F_2(s) = \dfrac{10e^{-s}}{s + 50}$

(c) $F_3(s) = \dfrac{20e^{-10s}}{s + 50}$

9–15 (a) $F_1(s) = \dfrac{-50000(s + 10001000)}{s^2 + 2000s + 40001000000}$

(b) $F_2(s) = \dfrac{-10(19s - 12)}{s(s + 10)}$

9–17 (a) $f(t) = Au(t) - 2Au(t - T) + Au(t - 2T)$

(b) $F(s) = \dfrac{A}{s}\left[1 - 2e^{-Ts} + e^{-2Ts}\right]$

9–19 (a) $f(t) = Ae^{-(t - T_C/2)/T_C}u(t - T_C/2)$

(b) $F(s) = \dfrac{Ae^{-T_Cs/2}}{s + 1/T_C}$

9–21 (a) $f_1(t) = \dfrac{1}{4}\left[3 + e^{-40t}\right]u(t)$

(b) $f_2(t) = \dfrac{1}{25}\left[1 - 12e^{-50t} + 36e^{-100t}\right]u(t)$

9–23 (a) $f_1(t) = 30e^{-10t}\sin(30t)u(t)$

(b) $f_2(t) = 3e^{-10t}\cos(30t)u(t)$

9–25 (a) $f_1(t) = [-1 + \alpha t + e^{-\alpha t}]u(t)$

(b) $f_2(t) = [1 - \alpha te^{-\alpha t} - e^{-\alpha t}]u(t)$

9–27 (a) $f_1(t) = \left[\dfrac{24}{7}e^{-3t} - \dfrac{8}{9}e^{-t} - \dfrac{160}{63}e^{-10t}\right]u(t)$

(b) $f_2(t) = \dfrac{1}{5}[9 + 16\cos(5t)]u(t)$

9–29 (a) $f_1(t) = \left[10e^{-30t} - 60e^{-10t} + 50\right]u(t)$

(b) $f_2(t) = \left\{\dfrac{5000}{13}e^{-2t}[\cos(2t) - \sin(2t)] - \dfrac{5000}{13}e^{-5t}\right\}u(t)$

9–31 (a) $f_1(t) = \left[\dfrac{5}{9}e^{-50t} + \dfrac{4}{9}e^{-200t} + \dfrac{50}{3}te^{-50t}\right]u(t)$

(b) $f_2(t) = \left[\dfrac{9}{4}e^{-200t} - \dfrac{5}{4}e^{-200t} + 25te^{-100t}\right]u(t)$

9–33 (a) $f_1(t) = \delta(t) + \left[12e^{-30t} - \dfrac{1}{3}e^{-5t} - \dfrac{200}{3}e^{-50t}\right]u(t)$

(b) $f_2(t) = \dfrac{d\delta(t)}{dt} + 4000\delta(t) - 3000000e^{-2000t}u(t)$

9–35 (a) $f_1(t) = \dfrac{1}{2}\left[e^{-10(t-5)} + e^{-30(t-5)}\right]u(t - 5)$

(b) $f_2(t) = -\dfrac{1}{2}\left[e^{-10(t-5)} - 3e^{-30(t-5)}\right]u(t - 5)$
$\qquad + \left[e^{-10t} - e^{-30t}\right]u(t)$

(c) $f_3(t) = -\dfrac{1}{2}\left[e^{-10t} - 3e^{-30t}\right]u(t) +$
$\qquad \left[e^{-10(t-5)} - e^{-30(t-5)}\right]u(t - 5)$

9–37 $f(t) = 225\cos(2t) + \dfrac{425}{2}e^{-4t} - 75\sin(2t) + \dfrac{125}{2}$

9–39 $F(s) = \dfrac{5 \times 10^6(s)(s + 100)^2}{\left[(s + 50)^2 + (250)^2\right]\left[(s + 100)^2 + (500)^2\right]}$

9–41 (a) $v(t) = 50e^{-10t}u(t)$
(b) $v(t) = \left[2 - 152e^{-300t}\right]u(t)$

9–43 (a) $\dfrac{di_L(t)}{dt} + 5000i_L(t) = 75u(t),\ i_L(t) = 0\,\text{mA}$

(b) $i_L(t) = 15\left[1 - e^{-5000t}\right]u(t)\,\text{mA}$

9–45 (a) $\dfrac{dv_C(t)}{dt} + 15000v_C(t) = 250000u(t),\ v_C(0) = 0\,\text{V}$

(b) $v_O(t) = 12.5\left[1 - e^{-15000t}\right]u(t)\,\text{V}$

9–47 (a) $\dfrac{dv_C(t)}{dt} + 15000v_C(t) = 240000e^{-1000t}u(t),\ v_C(0) = 0\,\text{V}$

(b) $v_O(t) = \dfrac{90}{7}\left[e^{-1000t} - e^{-15000t}\right]u(t)\,\text{V}$

9–49 $v(t) = \dfrac{100}{3}\left[e^{-10t} - e^{-40t}\right]u(t)\,\text{V}$

9–51 (a) $\dfrac{d^2v_C(t)}{dt^2} + 800\dfrac{dv_C(t)}{dt} + 160000v_C(t) = 0,$
$\qquad v_C(0) = 0, i_L(0) = 50\,\text{mA},$

(b) $v_C(t) = -20000te^{-400t}u(t)\,\text{V}$

9–53 (a) $\dfrac{d^2i_L(t)}{dt^2} + 200\dfrac{i_L(t)}{dt} + 80000i_L(t) = 0,$
$\qquad i_L(0) = 0, v_C(0) = v_S$

(b) $i_L(t) = -\dfrac{\sqrt{7}}{3500}e^{-100t}\sin(100\sqrt{7}t)u(t)\,\text{A}$

9–55 $i(t) = 80e^{-500t} - e^{-1000t}[80\cos(1000t) - 40\sin(1000t)]\,\text{mA},$
$t \geq 0$

9–57 (a) $f_1(0) = 16, f_1(\infty) = 0$

(b) $f_2(0) = 1, f_2(\infty) = -2$

9–59 (a) $f_1(0) = 400, f_1(\infty) = 0$

(b) $f_2(0) = 90, f_2(\infty)$ is undefined with poles on the j-axis

9–61 (a) $f_1(0)$ is not defined, $f_1(\infty) = 0$

(b) $f_2(0) = 20, f_2(\infty) = 20$

9–63 (a) $f_1(0) = 1, f_1(\infty) = 0$

(b) $f_2(0) = 1, f_2(\infty) = 0$

9–65 $K = f(0), \gamma = \dfrac{f(\infty)}{f(0)T_C}, \alpha = \dfrac{1}{T_C}$

9–67 $v_C(t) = 10e^{-500t}\left[\cos(500\sqrt{3}t) + \dfrac{4\sqrt{3}}{3}\sin(500\sqrt{3}t)\right]$ V, $t \geq 0$

$i_L(t) = \dfrac{3}{200}e^{-500t}\left[\cos(500\sqrt{3}t) - \dfrac{7\sqrt{3}}{9}\sin(500\sqrt{3}t)\right]$ A, $t \geq 0$

9–69 $D_3 = (s+1)(s^2+s+1), D_4 = (s^2+1.848s+1)(s^2+0.7654s+1)$

CHAPTER 10

10–1 $Z_{EQ}(s) = \dfrac{RCs+1}{Cs}$, $R = 1$, $C = 1\,\mu F$

10–3 (a) $Z_{EQ}(s) = \dfrac{Rs + \dfrac{1}{2C}}{s + \dfrac{1}{4RC}}$, zero at $s = -\dfrac{1}{2RC}$,

pole at $s = -\dfrac{1}{4RC}$

(b) $R = 1\,k\Omega, C = 0.391\,\mu F$, zero at $s = -1280$ rad/s

10–5 (a) $Z_{EQ}(s) = \dfrac{RLCs^2+R}{LCs^2+RCs+1}$, zero at $s = \pm j\dfrac{1}{\sqrt{LC}}$,

poles at

$s = \dfrac{-RC \pm \sqrt{R^2C^2-4LC}}{2LC}$

(b) $L = 400$ mH
(c) poles at $s = -2500 \pm j4330$ rad/s

10–7 (a) $Z_{EQ}(s) = \dfrac{4Ls(Ls+R)}{2Ls+R}$, zero at $s = -\dfrac{R}{L}$,

pole at $s = -\dfrac{R}{2L}$

(b) $R = 1\,k\Omega$, $L = 250$ mH, zero at $s = -4000$ rad/s

10–9 (a) zeros at $s = [-1000, -1000]$ rad/s,
pole at $s = -2000$ rad/s

(b) zeros at $s = [-1775, -225.4]$ rad/s,
pole at $s = -2000$ rad/s

10–11 (a) $Z_{EQ}(s) = \dfrac{8R^2C^2s^2+8RCs+1}{2Cs(4RCs+1)}$,

zeros at $s = \dfrac{-2 \pm \sqrt{2}}{4RC}$,

poles at $s = \left[0, -\dfrac{1}{4RC}\right]$

(b) poles at $s = [0, -80]$ krad/s

10–13 $Z_{EQ}(s) = \dfrac{2L^2s^2+5RLs+2R^2}{Ls+R}, R = 3\,k\Omega$,

$L = 1$ H, zeros at $s = [-1500, -6000]$ rad/s

10–15 (a) $I_2(s) = \dfrac{R_1+Ls}{R_1+R_2+Ls}I_1(s)$

(b) $Z_N(s) = \dfrac{R_2(R_1+Ls)}{R_1+R_2+Ls}$

10–17 $V_T(s) = \dfrac{Ls}{R+2Ls}V_S(s)$, $Z_T(s) = \dfrac{RLs}{R+2Ls}$,

$R = 3\,k\Omega$, $L = 125$ mH

10–19 $I_L(s) = \dfrac{\dfrac{V_A}{R}}{s+\dfrac{2R}{L}}, i_L(t) = \dfrac{V_A}{R}e^{-2Rt/L}u(t)$,

$V_O(s) = -\dfrac{V_A}{s+\dfrac{2R}{L}}, v_O(t) = -V_Ae^{-2Rt/L}u(t)$

10–21 $I_C(s) = -\dfrac{2\dfrac{V_A}{R}}{s+\dfrac{2}{RC}}, i_C(t) = -2\dfrac{V_A}{R}e^{-2t/RC}u(t)$,

$V_O(s) = -\dfrac{\dfrac{V_A}{2}}{s} - \dfrac{V_A}{s+\dfrac{2}{RC}}, v_O(t) = -\left[\dfrac{V_A}{2}+V_Ae^{-2t/RC}\right]u(t)$

10–23 $I_L(s) = \dfrac{\dfrac{V_A}{R}}{s} + \dfrac{I_A-\dfrac{V_A}{R}}{s+\dfrac{R}{2L}}, i_L(t) = \left[\dfrac{V_A}{R}+\left(I_A-\dfrac{V_A}{R}\right)e^{-Rt/2L}\right]u(t)$,

$V_L(s) = \dfrac{\dfrac{1}{2}(V_A-RI_A)}{s+\dfrac{R}{2L}}, v_L(t) = \dfrac{1}{2}(V_A-RI_A)e^{-Rt/2L}u(t)$

10–25 (a) $I_L(s) = -\dfrac{\dfrac{V_A}{L}}{s^2+\dfrac{R_2+R_3}{L}s+\dfrac{1}{LC}}$

(b) $i_L(t) = -10.78e^{-1500t}\sin(2784t)u(t)$ mA

10–27 $\dfrac{I_R(s)}{I_1(s)} = \dfrac{LCs^2+1}{LCs^2+RCs+1}$, zeros at $s = \pm\dfrac{j}{\sqrt{LC}}$,

poles at $s = \dfrac{-RC \pm \sqrt{R^2C^2-4LC}}{2LC}$

10–29 $V_{Czs}(s) = \dfrac{\dfrac{1}{RC}sV_s(s)}{s^2+\dfrac{1}{RC}s+\dfrac{1}{LC}}, V_{Czi}(s) = \dfrac{-\dfrac{1}{C}I_O}{s^2+\dfrac{1}{RC}s+\dfrac{1}{LC}}$

10–31 $V_T(s) = \dfrac{\dfrac{1}{RC}sV_s(s)}{s^2+\dfrac{1}{RC}s+\dfrac{1}{LC}}, Z_T(s) = \dfrac{\dfrac{1}{C}s}{s^2+\dfrac{1}{RC}s+\dfrac{1}{LC}}$

10–33 $v_O(t) = -0.14423e^{-250t} + 3.2692\cos(2000t)$
$+1.1538\sin(2000t)$ V, $t \geq 0$, natural pole $s = -250$
rad/s, forced poles $s = \pm j2000$ rad/s

10–35 (a) $V_T(s) = \dfrac{I_s(s)+CV_0}{Cs}, Z_T(s) = \dfrac{RCs+1}{Cs}$

(b) $v_O(t) = \left[50 + (8.333-50)e^{-83.3t}\right]u(t) = \left[50-41.667e^{-83.3t}\right]u(t)$

(c) $v_{O,zs}(t) = \left[50-50e^{-83.3t}\right]u(t)$ V, $v_{O,zi}(t) = \left[8.333e^{-83.3t}\right]u(t)$ V,
$v_{O,forced}(t) = 50u(t)$ V, $v_{O,natural}(t) = -41.667e^{-83.3t}u(t)$ V

10–37 Use a 1-H inductor in series with $Z_T(s)$ the 1-kΩ resistor

10–39 $i(t) = \dfrac{1}{600}\left[2 + e^{-3000t}\right]u(t)$ A

10–41 (a) $(Ls + R_1)I_A - R_1I_B = V_1(s)$

(b) $-R_1I_A + \left(R_1 + R_2 + \frac{1}{Cs}\right)I_B = 0$

(c) zero at $s = 0$, poles at

$$s = \dfrac{-(L + R_1R_2C) \pm \sqrt{(L + R_1R_2C)^2 - 4(R_1 + R_2)LCR_1}}{2(R_1 + R_2)LC}$$

(d) $i_2(t) = \dfrac{\sqrt{3}}{300}e^{-500t}\sin\left(\dfrac{500\sqrt{3}}{3}t\right)u(t)$ A

10–43 (a) $\left(C_1s + \frac{1}{R_1} + C_2s\right)V_A - C_2sV_B = C_1sV_1(s)$

$-C_2sV_A + \left(C_2s + \frac{1}{R_2}\right)V_B = 0$

(b) $V_2(s) = \dfrac{R_1R_2C_1C_2s^2V_1(s)}{R_1R_2C_1C_2s^2 + (R_1C_1 + R_1C_2 + R_2C_2)s + 1}$

(c) $V_2(s) = \dfrac{R_1R_2C_1C_2s^2V_1(s)}{R_1R_2C_1C_2s^2 + (R_1C_1 + R_2C_2)s + 1}$

(d) Use the OP AMP design with $R_1 = R_2 = 1$ kΩ, $C_1 = 0.1$ μF, $C_2 = 0.01$ μF

10–45 (a) $\left(R_1 + \frac{1}{C_1s}\right)I_A - \left(\frac{1}{C_1s}\right)I_B = V_1(s)$

$-\left(\frac{1}{C_1s}\right)I_A + \left(R_2 + \frac{1}{C_1s} + \frac{1}{C_2s}\right)I_B = 0$

(b) $\dfrac{V_2(s)}{V_1(s)} = \dfrac{R_1R_2C_1C_2s^2 + (R_1C_1 + R_1C_2)s + 1}{R_1R_2C_1C_2s^2 + (R_1C_1 + R_1C_2 + R_2C_2)s + 1}$

10–47 $V_T(s) = \dfrac{[R_1R_2C_1C_2s^2 + (R_1C_1 + R_1C_2)s + 1]V_1(s)}{R_1R_2C_1C_2s^2 + (R_1C_1 + R_1C_2 + R_2C_2)s + 1}$,

$Z_T(s) = \dfrac{R_2[(R_1C_1 + R_1C_2)s + 1]}{R_1R_2C_1C_2s^2 + (R_1C_1 + R_1C_2 + R_2C_2)s + 1}$

10–49 $v_A(t) = -2.5e^{-500t}[\cos(500t) + \sin(500t)]u(t)$ V,

$v_B(t) = 2.5e^{-500t}[\cos(500t) - \sin(500t)]u(t)$ V,

$v_C(t) = -5e^{-500t}\cos(500t)u(t)$ V

10–51 $R = 2$ kΩ, $C = 0.1$ μF, $\mu = 3 - \sqrt{2} = 1.5858$

10–53 $R = 1$ kΩ, $C_1 = C_2 = 0.1$ μF

10–55 (a) The two voltage sources control two of the node voltages, so only one voltage, $V_C(s)$, can be set independently

(b) $\left(\dfrac{1}{R_2} + Cs\right)V_C(s) - \dfrac{1}{R_2}\mu V_X(s) = 0$, which can be rewritten as

$$V_C(s) = \dfrac{\left(\dfrac{\mu}{1 + \mu}\right)V_S(s)}{R_2Cs + 1}$$

(c) $\dfrac{V_C(s)}{V_S(s)} = \dfrac{\left(\dfrac{\mu}{1 + \mu}\right)}{R_2Cs + 1}$

10–57 (a) Mesh-current equations

$-\dfrac{5}{s + 2000} + (1000 + 1000 + s)I_A(s) - (1000 + s)I_B(s) - 0.01 = 0$

$0.01 - (1000 + s)I_A(s) + \left(1000 + 2000 + s + \dfrac{10^7}{s}\right)I_B(s) + \dfrac{10}{s} = 0$

(b)

$V_C(s) = \dfrac{10(3s^2 + 6000s - 5000000)}{(s + 2000)(3s^2 + 15000s + 20000000)}$,

$v_C(t) = \left[35e^{-2500t}(\cos(645.5t) - 0.5533\sin(645.5t)) - 25e^{-2000t}\right]u(t)$

10–59 $I_O(s) = \dfrac{s^2 + 20000s + 1.5 \times 10^8}{200s(s^2 + 20000s + 7.5 \times 10^7)}$,

$i_O(t) = (2.5e^{-15000t} - 7.5e^{-5000t} + 10)u(t)$ mA

10–61 $R = 1$ kΩ, $C = 1$ μF, with the initial capacitor voltage, the output voltage shifts by 10 V

10–63 $\mu < 2$

10–65 $v_L(t) = \left[3.333e^{-10t} + (307.69t - 4.4379)e^{-50t}\right.$

$\left. + 1.1045e^{-87.143t}u(t)\right]$ V

10–67 (a) $\dfrac{V_O(s)}{V_S(s)} = \dfrac{R_1R_2C_1s + R_2}{(R_1R_2C_2 + R_1R_2C_1)s + R_1 + R_2}$

(b) $R_1 = 15$ MΩ, $C_1 = 3$ pF

CHAPTER 11

11–1 $Z(s) = \dfrac{4RCs + 1}{Cs}$, $T_V(s) = \dfrac{2RCs + 1}{4RCs + 1}$

11–3 $Z(s) = \dfrac{LCs^2 + RCs + 1}{Cs}$, $T_V(s) = \dfrac{RCs}{LCs^2 + RCs + 1}$

11–5 $Z(s) = \dfrac{RLCs^2 + Ls + R}{LCs^2 + 1}$, $T_V(s) = \dfrac{Ls}{RLCs^2 + Ls + R}$

11–7 $Z(s) = \infty$, $T_V(s) = 2RCs + 2$

11–9 $T_V(s) = \dfrac{R}{2RLCs^2 + (R^2C + L)s + R}$

11–11 $T_V(s) = \dfrac{RCs + 2}{RCs + 1}$, $R = 10$ kΩ, $C = 0.001$ μF

11–13 $T_I(s) = \dfrac{R}{Ls + 3R}$, $R = 100$ Ω, $L = 796$ mH

11–15 $T_V(s) = \dfrac{s(11s + 10000)}{(s + 10000)(s + 967.1)}$,

zeros at $s = 0$, -909.1 rad/s, poles at $s = -10000, -967.1$ rad/s

11–17 $h(t) = \dfrac{47}{80}\delta(t) + \dfrac{275}{8}e^{-250t/3}u(t)$,

$g(t) = \left(1 - \frac{33}{80}e^{-250t/3}\right)u(t)$

11–19 $h(t) = \delta(t) - 2000e^{-3000t}u(t),$

$$g(t) = \frac{1}{3}\left(1 + 2e^{-3000t}\right)u(t)$$

11–21 $h(t) = -10000e^{-500t}u(t),\ g(t) = 20\left(e^{-500t} - 1\right)u(t)$

11–23 $h(t) = \delta(t) + 10e^{-5t}u(t),\ g(t) = \left(3 - 2e^{-5t}\right)u(t)$

11–25 $g(t) = \left(-9 + 10e^{-200t}\right)u(t),$

$$H(s) = T(s) = \frac{s - 1800}{s + 200},\ G(s) = \frac{s - 1800}{s(s + 200)}$$

11–27 $h(t) = 150000\left(3e^{-30000t} - 2e^{-20000t}\right)u(t),$

$$H(s) = T(s) = \frac{150000s}{(s + 20000)(s + 30000)},$$
$$G(s) = \frac{150000}{(s + 20000)(s + 30000)}$$

11–29 $y(t) = \dfrac{1}{200}\left(1000t - 1 + e^{-1000t}\right)u(t)$

11–31 $h(t) = 10\delta(t) - 500e^{-50t}[\cos(200t) + 4\sin(200t)]u(t),$

$$H(s) = T(s) = \frac{10s(s + 50)}{(s + 50)^2 + 40000},\ G(s) = \frac{10(s + 50)}{(s + 50)^2 + 40000}$$

11–33 $y(t) = 2u(t)$

11–35 $v_1(t) = 17.89\cos(500t + 153°)\ \text{V},$
$v_1(t) = 14.14\cos(1000t + 135°)\ \text{V},$
$v_1(t) = 1.99\cos(10000t + 95.7°)\ \text{V},$
pole at $s = -1000\ \text{rad/s}$

11–37 $i_{2SS}(t) = 199\cos(50000t + 95.7°)\ \mu\text{A},$
$i_{2SS}(t) = 1.414\cos(5000t + 135°)\ \text{mA},$
pole at $s = -5000\ \text{rad/s}$

11–39 $v_{2SS}(t) = 44.72\cos(5000t + 26.57°)\ \text{V},$
$v_{2SS}(t) = 17.68\cos(2500t + 45°)\ \text{V}$

11–41 $y_{SS}(t) = 7.036\cos(100t - 39.3°)$

11–43 The system is not stable, so there is no steady-state response

11–45 $y_{SS}(t) = 19.2\cos(200t - 90°) = 19.2\sin(200t)$

11–47 $y_{SS}(t) = 0$

11–49 $y(t) = r(t - 3) - r(t - 6)$

11–51 $y(t) = \frac{1}{2}\left(t^2 - 2t + 1\right)u(t - 1) - \frac{1}{2}\left(t^2 - 2t\right)u(t - 2)$

11–53 $y(t) = 10(1 - e^{-t})u(t) + 10\left(e^{-(t-1)} - 1\right)u(t - 1)$

11–55 $y(t) = \displaystyle\int_0^t f(t - \tau)\delta(\tau)d\tau = \int_0^t f(t)\delta(\tau)d\tau = f(t)\int_0^t \delta(\tau)d\tau =$
$f(t)(1) = f(t)$

11–57 $y(t) = x(t) * h(t) = \displaystyle\int_0^t x(t - \tau)h(\tau)d\tau = \int_0^t u(t - \tau)h(\tau)d\tau$

$$= \left[\int_0^t (1)h(\tau)d\tau\right]u(t) = g(t)u(t)$$

11–59 $y(t) = t^2 u(t) - \left(t^2 - 1\right)u(t - 1)$

11–61 $h(t) = \dfrac{10}{3}\left(e^{-2t} - e^{-5t}\right)u(t)$

11–63 $R = 1\,\text{k}\Omega,\ C_1 = 1\,\mu\text{F},\ C_2 = 0.05\,\mu\text{F}$

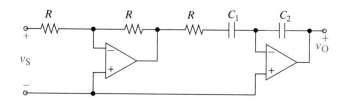

11–65 Use three stages in cascade; the first stage is a noninverting amplifier with a gain of 100; the second stage is a passive RC high-pass filter with $C = 0.1\,\mu\text{F}$ and $R = 4\,\text{k}\Omega$; the third stage is an inverting amplifier with a gain of -500

11–67 Use two active low-pass RC OP AMP filters in cascade; the first filter has an input resistor $R = 10\,\text{k}\Omega$, a feedback resistor $R = 5\,\text{M}\Omega$, and a feedback capacitor $C = 0.002\,\mu\text{F}$; the second filter has an input resistor $R = 10\,\text{k}\Omega$, a feedback resistor $R = 10\,\text{k}\Omega$, and a feedback capacitor $C = 0.01\,\mu\text{F}$

11–69 Use three stages in cascade; the first stage is a passive RC low-pass filter with $R = 20\,\text{k}\Omega$ and $C = 0.01\,\mu\text{F}$; the second stage is a noninverting amplifier with a gain of 4 by using three 20-kΩ resistors in the feedback path and one 20-kΩ resistor connected to ground; the third stage is a passive RC high-pass filter with $R = 20\,\text{k}\Omega$ and $C = 0.05\,\mu\text{F}$

11–71 (a) $R_1 = 8\,\text{k}\Omega,\ R_2 = 4\,\text{k}\Omega,\ R_3 = 2\,\text{k}\Omega,\ R_4 = 1\,\text{k}\Omega,$
$\qquad C_1 = C_2 = C_3 = C_4 = 1\,\mu\text{F}$

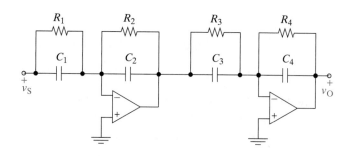

(b) $R_2 = 8\,\text{k}\Omega,\ R_4 = 2\,\text{k}\Omega,\ C_1 = C_2 = C_3 = C_4 = 1\,\mu\text{F}$

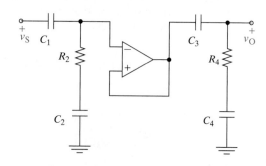

(c) $R_1 = R_2 = 6.25\,\text{k}\Omega$, $C_1 = C_2 = 1.6\,\mu\text{F}$, $L = 10$

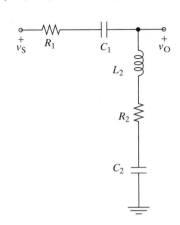

11–73 Use a passive RLC series circuit with the output taken across the resistor and capacitor, $R = 200\,\Omega$, $L = 1\,\text{H}$, $C = 1\,\mu\text{F}$

11–75 (b) Choose the first circuit to avoid loading at the output

(c) Either circuit will work without significant loading issues

(d) The claim is false; if the passive circuit is first, there will be loading issues

11–77 $R_1 = 1\,\text{k}\Omega$, $R_2 = 19\,\text{k}\Omega$, $R_4 = 100\,\text{k}\Omega$, $C_3 = 0.05263\,\mu\text{F}$, $C_4 = 1\,\mu\text{F}$

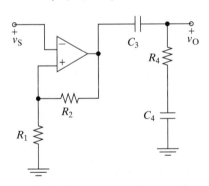

11–79 First row: $h(t) = \alpha e^{-\alpha t} u(t)$, $g(t) = (1 - e^{-\alpha t})u(t)$

Second row: passive RC high-pass filter with $C = \dfrac{1}{\alpha R}$ and $R = R$; $g(t) = e^{-\alpha t}u(t)$

Third row: passive RC circuit with $C_1 = \dfrac{2}{\alpha R}$ in series with the input and a series combination of $R = R$ and $C_2 = \dfrac{2}{\alpha R}$ in parallel with the output;

$h(t) = \delta(t) - \dfrac{\alpha}{2} e^{-\alpha t} u(t)$

11–81 There is no loading; $T(s) = \dfrac{-2}{R^2 C^2 s^2 + RCs + 2}$,

zeros at $s = -\infty$, $-\infty$, poles at $s = \dfrac{-1 \pm j\sqrt{7}}{2RC}$; the circuit is stable

11–83 $T(s) = \dfrac{10s(s + 1000)}{(s + 100)(s + 10000)}$

$R_1 = 10\,\text{k}\Omega$, $R_3 = 1\,\text{k}\Omega$, $R_4 = 9\,\text{k}\Omega$, $R_6 = 10\,\text{k}\Omega$,

$C_1 = 1\,\mu\text{F}$, $C_5 = 0.01111\,\mu\text{F}$, $C_6 = 0.1\,\mu\text{F}$

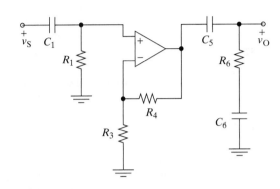

CHAPTER 12

12–1 $K = 126.02\,\text{dB}$

12–3 (a) 60 dB

(b) $\omega_C = 50\,\text{rad/s}$, $-23\,\text{dB}$

12–5 $T_V(s) = \dfrac{\frac{1}{2}s}{s + 4700}$

(a) $|T(j0)| = 0$, $|T(j\infty)| = \frac{1}{2}$, $\omega_C = 4700\,\text{rad/s}$, high-pass filter

(c) $|T(j0.5\omega_C)| = -13.01\,\text{dB}$, $|T(j\omega_C)| = -9.03\,\text{dB}$, $|T(j2\omega_C)| = -6.99\,\text{dB}$

(e) $-6.99\,\text{dB}$

12–7 $T_V(s) = \dfrac{-5000}{s + 1000}$

(a) $|T(j0)| = 5$, $|T(j\infty)| = 0$, $\omega_C = 1000\,\text{rad/s}$, low-pass filter

(c) $|T(j0.1\omega_C)| = 13.94\,\text{dB}$, $|T(j\omega_C)| = 10.97\,\text{dB}$, $|T(j10\omega_C)| = -6.06\,\text{dB}$

(e) replace the 2-kΩ resistor with a 1-kΩ resistor

12–9 The problem is loading with the filter stage and either the source or the load; connect the source to the gain stage, connect the gain stage to the filter stage, connect the filter stage to a buffer and then the load

12–11 Use two stages connected in cascade; the first stage is a passive RC low-pass filter with $R = 10\,\text{k}\Omega$ and $C = 0.01\,\mu\text{F}$; the second stages is a noninverting amplifier with a gain of 10

12–13 Use a single stage, active RC high-pass filter; the input resistor is $R = 1\,\text{k}\Omega$ and the input capacitor is $C = 0.5\,\mu\text{F}$; the feedback resistor is $R = 100\,\text{k}\Omega$

12–15 $T_V(s) = \dfrac{2s}{s + \frac{1}{RC}}$; high-pass filter with a passband gain of 2; $R = 15 + 1 = 16\,\text{k}\Omega$, $C = 33\,\text{pF}$

12–17 $|T(j0)| = 0$ dB, $|T(j200)| = -0.043$ dB, $|T(j20000)| = -20.04$ dB, $|T(j200000)| = -40.00$ dB

12–19 (a) High-pass filter, $\omega_C = 15000$ rad/s, $K = 5$
(c) Use two stages connected in cascade; the first stage is a passive RC high-pass filter with $R = 6.67$ kΩ and $C = 0.01$ μF; the second stage is a noninverting amplifier with a gain of 5

12–21 R_1 and C_1 are the low-pass filter and control the larger cutoff frequency; R_2 and C_2 are the high-pass filter and control the smaller cutoff frequency; $R_1 = 25$ kΩ, $C_1 = 0.001$ μF, $R_2 = 10$ kΩ, $C_2 = 0.1$ μF, $R_3 = 99$ kΩ, $R_4 = 1$ kΩ

12–23 Use the circuit in Problem 12–22, since it will not cause any loading at the output

12–25 The input stage is a passive RC low-pass filter with $R = 800$ kΩ and $C = 10$ pF; the second stage is a noninverting amplifier with a gain of 10; the output stage is a passive RC high-pass filter with $C = 0.8$ μF connected to the $R = 10$ kΩ audio transducer

12–27 The first stage is an active RC low-pass filter, the input resistor is $R = 800$ kΩ, the feedback resistor is $R = 800$ kΩ, and the feedback capacitor is $C = 10$ pF; the second stage is an active RC high-pass filter, the input resistor is $R = 10$ kΩ, the input capacitor is $C = 0.8$ μF, and the feedback resistor is a potentiometer with a range from 10 kΩ to 1 MΩ

12–29 The circuit is a passive, series RLC bandpass filter in cascade with noninverting amplifier with a gain of 5; $Q = 0.1543$, $B = 4100$ rad/s, $\omega_{C1} = 95.34$ rad/s, $\omega_{C2} = 4195.3$ rad/s, $\omega_0 = 632.5$ rad/s, the filter is wide band

12–31 Use two stages connected in cascade; the first stage is a passive, series RLC circuit with the output taken across the resistor and component values $R = 50$ Ω, $L = 1$ H, and $C = 1$ μF; the second stage is a noninverting amplifier with a gain of 10

12–33 $R = 2$ kΩ, $L = 400$ μH, $\omega_{C1} = 475.6$ krad/s, $\omega_{C2} = 525.6$ krad/s; there is no way to design this filter with a cascade connection of two first-order filters because $Q = 10$ is too large. You cannot get a $Q > 1$ using first-order circuits

12–35 (a) Replace R with $2R$
(b) Replace C with $C/4$
(c) Replace R with $R/2$; replace L with $L/4$

12–37 $C = 0.12$ μF

12–39 $T_V(s) = \dfrac{Ls}{RLCs^2 + Ls + R}$;

use two stages connected in cascade; the first stage is a first-order, active, low-pass RC filter with an input resistor $R = 99$ kΩ, a feedback resistor $R = 99$ kΩ, and a feedback capacitor $C =$

1 pF; the second stage is a first-order, active, high-pass RC filter with an input resistor $R = 1.01$ kΩ, an input capacitor $C = 0.01$ μF, and a feedback resistor $R = 1.01$ kΩ

12–41 $|V_2(j0)| = 0$, $|V_2(j\infty)| = 0$, $|V_2(j\omega_0)| = \frac{L}{RC}|I_1(j\omega_0)|$; bandpass filter

12–43 $T_V(s) = \dfrac{R_1R_2C_1s + R_2}{R_1R_2C_1s + R_1 + R_2}$

(a) $v_{2ss}(t) = \cos(100t)$ V
(b) $v_{2ss}(t) = 0.99\cos(100t - 11.37°)$ V; the amplitude matches the estimate within 1%, but the phase differs by more than 11°

12–45 (a) The function is a modified low-pass filter with a cutoff frequency of 1 rad/s and passband gain of 60 dB
(c) $R_1 = 1$ kΩ, $C_1 = 10$ μF, $R_2 = 1$ MΩ, $C_2 = 1$ μF

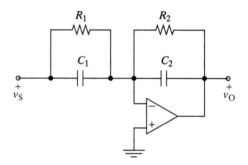

12–47 (a) The function is a high-pass filter with a cutoff frequency of approximately 1560 rad/s and a passband gain of 120 dB; note that having both poles at $\omega = 1000$ rad/s shifts the cutoff frequency to be higher than 1000 rad/s
(c) Use two identical stages in cascade; each stage is an active, first-order, high-pass RC filter with a 1-kΩ input resistor, a 1-μF input capacitor, and a 1-MΩ feedback resistor

12–49 (a) The function is a bandstop filter with cutoff frequencies of approximately 1 rad/s and 1000 rad/s and a passband gain of 20 dB
(c) $v_{2ss}(t) = 21.95\cos(200t + 94.55°)$ V

12–51 (a) Bandstop function
(b) The maximum gain of 26 dB occurs at $\omega = 5$ rad/s

12–53 (a) $T(s) = \dfrac{790.57(s + 5)}{(s + 50)(s + 250)}$
(c) For $\omega = 100$ rad/s, the straight-line gain is 10 dB and the actual gain is 8.3972 dB; for $\omega = 500$ rad/s, the straight-line gain is 3.9794 dB and the actual gain is 2.9675 dB; the straight-line gain is a reasonably good approximation at these two frequencies

(d) $R_1 = 1\,\text{k}\Omega$, $R_2 = 3.16\,\text{k}\Omega$, $R_3 = 200\,\text{k}\Omega$, $C_3 = 1\,\mu\text{F}$, $R_4 = 20\,\text{k}\Omega$, $C_4 = 1\,\mu\text{F}$, $R_5 = 10\,\text{k}\Omega$, $C_5 = 0.4\,\mu\text{F}$

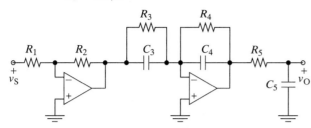

12–55 (a) $T(s) = \dfrac{(s+100)(s+500)}{(s+10)(s+5000)}$

(c) $R_1 = 10\,\text{k}\Omega$, $R_2 = 100\,\text{k}\Omega$, $R_3 = 10\,\text{k}\Omega$, $R_4 = 1\,\text{k}\Omega$, $C_1 = C_2 = 1\,\mu\text{F}$, $C_3 = C_4 = 0.2\,\mu\text{F}$

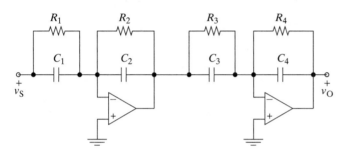

12–57 (a) $T(s) = \dfrac{200s}{s^2 + 200s + 10000}$

(c) $T(s) = \dfrac{100s}{s^2 + 10s + 10000}$

(d) Use two stages connected in cascade; the first stage is a passive RLC series circuit with the output taken across the resistor and $R = 10\,\Omega$, $L = 1\,\text{H}$, and $C = 100\,\mu\text{F}$; the second stage is a noninverting amplifier with a gain of 10

12–59 (a)

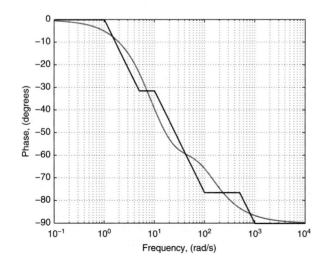

(b) For $\omega = [1,\ 10,\ 100,\ 1000]$ rad/s, the estimate phase angles are $[0°,\ -31.5°,\ -76.5°,\ -90°]$ and the actual angles are $[-5.1°,\ -39.4°,\ -65.9°,\ -86.6°]$

12–61 (a) Bandstop

(b) $\omega_{C1} = 20$ rad/s, $\omega_{C2} = 500$ rad/s, $K = 6$ dB

(d) $R_1 = 480\,\Omega$, $R_2 = 40\,\Omega$, $L_2 = 1\,\text{H}$, $C_2 = 100\,\mu\text{F}$, $R_3 = R_4 = 1\,\text{k}\Omega$

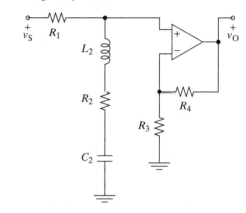

12–63 $g(0) = 1$, $g(\infty) = 10$, the duration is approximately 250 ms

12–65 $T_2(s)$ has the narrowest notch and the lowest amplitude ringing

12–67 (a) $\omega_0 = 538.52$ rad/s, $\omega_{C1} = 374.46$ rad/s, $\omega_{C2} = 774.46$ rad/s

(b) Use a series RLC filter with the output taken across the resistor and $R = 400\,\Omega$, $L = 1\,\text{H}$, and $C = 3.448\,\mu\text{F}$

12–69 $T_V(s) = \dfrac{100s}{s^2 + 50s + 10000}$

12–71 Select the circuit from the First-Order Filter Company because it is the only one that meets specifications

12–73 Select the circuit from Vendor A because it is the only one that meets specifications

CHAPTER 13

13–1 $a_0 = \dfrac{5}{4}$, $a_n = \dfrac{10}{n\pi}\sin\left(\dfrac{n\pi}{4}\right)$, $b_n = 0$ for all n

13–3 $a_0 = 0$, $a_n = 0$, $b_n = \dfrac{2V_A}{n\pi}$ for all n

13–5 $a_0 = \dfrac{V_A}{4}$, $a_n = \dfrac{V_A[\cos(n\pi) - 1]}{n^2\pi^2}$, $b_n = \dfrac{-V_A\cos(n\pi)}{n\pi}$

13–7 $a_0 = 0$, $a_n = 0$, $b_n = \dfrac{120\sin(\pi n/2)}{(\pi n)^2}$

$v(t) \approx 12.1585\sin(10000\pi t) - 1.3509\sin(30000\pi t) + 0.4863\sin(50000\pi t) - 0.2481\sin(70000\pi t)$ V

13–9 (a) $v(t) \approx 4 + 1.8\cos(200\pi t) - 1.8\sin(200\pi t)$
$+ 0.6\cos(600\pi t) + 0.6\sin(600\pi t) - 0.36\cos(1000\pi t)$
$+ 0.36\sin(1000\pi t) - 0.257\cos(1400\pi t)$
$-0.257\sin(1400\pi t)$ V

(b)

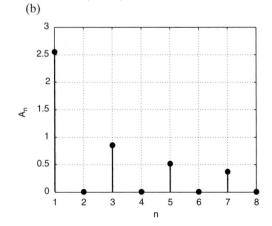

13–11 $v(t) \approx 107.91 + 169.5\sin(100\pi t) - 71.94\cos(200\pi t)$
$-14.39\cos(400\pi t) - 6.17\cos(600\pi t)$ V

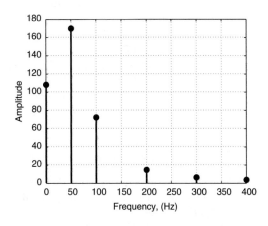

13–13 $v(t) \approx 12 - 7.6394\sin(4000\pi t) - 3.8197\sin(8000\pi t)$
$-2.5465\sin(12000\pi t) - 1.9099\sin(16000\pi t)$ V

13–15 (a) The waveform is a half-wave rectified sine wave with an amplitude of $2V_A$ and a period of $T_0/2$

(b) $a_0 = \dfrac{2V_A}{\pi}$, $a_n = \dfrac{4V_A/\pi}{1-n^2}$ for n even, $a_n = 0$ for n odd,

$b_1 = V_A$, and $b_n = 0$ for $n \ne 1$

13–17 (a) $T_0 = 12.57$ ms, $f_0 = \dfrac{250}{\pi} = 79.578$ Hz,

$\omega_0 = 500$ rad/s, the first, second, third, and fifth harmonics are present

(b) The waveform has even symmetry

(c) Since the waveform has even symmetry, $v(t) = v(-t)$

13–19 $a_0 = 0$, $a_n = 0$,

$$b_n = \frac{V_A[6\sin(\pi n/3) + 3\sin(2\pi n/3) - 3\sin(4\pi n/3)}{-\pi n\cos(\pi n/3) + \pi n\cos(5\pi n/3)]}{(n\pi)^2}$$

13–21 $v_O(t) = 14.5\cos(1000t - 116.6°) + 0.999\cos(3000t + 33.7°)$
$+ 0.241\cos(5000t - 158.2°) + 0.091\cos(7000t$
$+15.95°)$ V

13–23 (a) $a_0 = 4$, $a_n = 0$, $b_n = \dfrac{-8}{n\pi}$

(b) $i(t) = 24.7\cos(100t + 104.0°) + 12.63\cos(200t + 97.1°)$
$+ 8.46\cos(300t + 94.76°) + 6.35\cos(400t + 93.58°)$ μA

(c) $v_O(t) = 1.235\cos(100t - 75.96°)$
$+0.632\cos(200t - 82.88°) + 0.423\cos(300t - 85.24°)$
$+0.318\cos(400t - 86.42°)$ V

13–25 (a) Use a passive low-pass RC filter with $R = 100\,\text{k}\Omega$ and $C = 1\,\mu\text{F}$;

$$a_0 = \frac{340}{\pi} = 108.225, \quad a_n = \frac{680/\pi}{1-n^2} = \frac{216.45}{1-n^2}$$

(b) $v_O(t) = 108.225 + 0.957\cos(240\pi t + 90.76°)$
$+ 0.0957\cos(480\pi t + 90.38°)$
$+0.0273\cos(720\pi t + 90.25°)$ V

13–27

ω (rad/s):	200	400	600	800	1000
$\lvert I_{SS}(\omega)\rvert$ (A):	1.7082	1.3505	1.0594	0.8541	0.7093

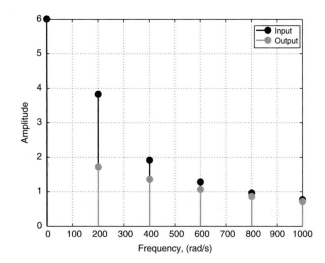

13–29 Use a tuned RLC filter with $R = 30\,\Omega$, $L = 1$ H, and $C = 11.11\,\mu\text{F}$

(a) $a_0 = 0$, $a_n = \dfrac{64}{(n\pi)^2}$ for n odd, $a_n = 0$,

for n even, $b_n = 0$

(b) The first and fifth harmonics are severely attenuated and the third harmonic is not changed between the input and output; specifically input and output values for the first harmonic are 6.48 and 0.243, for the third harmonic are 0.721 and 0.721, and for the fifth harmonic are 0.259 and 0.024

13–31 The harmonics of the voltage waveform decay at a rate of $\frac{1}{n^2}$, but the harmonics in the current waveform decay at a rate of $\frac{1}{n}$; the higher frequencies components of the current are more significant than those in the voltage

13–33 (a) The amplitudes of the first four components are 0.208, 0.256, 0.406, and 0.794 V; the amplitudes decrease after the fourth component; the fourth component dominates the response because the transfer function is a bandpass filter

(b) The amplitudus of the first four components are 3.08, 2.08, 1.28, 1.35 mV; in this case, the dc component dominates the response because the new transfer function represents a low-pass filter

13–35 (a) $i(t) = 1.2 + 0.48\sin(200\pi t) - 0.16\sin(600\pi t) + 0.096\sin(1000\pi t)$ A, $p(t) = v(t)i(t)$

(b) $P = 78.63$ W

(c) $V_{rms} = 62.70$ V

(d) $V_{rms} = 62.70$ V, the result is the same

(e) $P = 78.63$ W, the result is the same

13–37 $F_{rms} = \sqrt{\dfrac{A^2}{3}}$, 99.93%

13–39 $F_{rms} = \dfrac{A}{2}$, 99.895%

13–41 $V_{rms} = \dfrac{V_A}{\sqrt{3}}$, $P = \dfrac{V_A^2}{3R}$, $V_{dc} = \dfrac{V_A}{2}$, $P_{dc} = \dfrac{V_A^2}{4R}$; dc component carries 75% of the total average power and the ac components carry 25% of the total average power

13–43 $Y_{rms} = 49.40$ V

13–45 $P = 574.6$ mW

13–47 $F(\omega) = \dfrac{A}{j\omega}\left(1 - e^{-j3\omega}\right)$

13–49 $F(\omega) = \dfrac{A}{\omega^2}\left[(1 + j\omega)e^{-j\omega} - 1\right]$

13–51 $f(t) = \dfrac{10\sin(t)}{t}$

13–53 $f(t) = \dfrac{j4\sin(4t)}{\pi^2 - 16t^2}$

13–55 (a) $f_1(t) = \text{sgn}(t) + \left[-4e^{-20t} + 2e^{-40t}\right]u(t)$

(b) $f_2(t) = \left[-e^{-20t} + 2e^{-40t}\right]u(t)$

13–57 (a) $F_1(\omega) = \dfrac{-j2\omega - 8}{j\omega(j\omega + 2)}$

(b) $F_2(\omega) = -2\pi\delta(\omega) - \dfrac{4}{j\omega(j\omega + 2)}$

13–59 (a) $F_1(\omega) = 5\pi e^{-j3\omega}[\delta(\omega - 2\pi) + \delta(\omega + 2\pi)]$

(b) $F_2(\omega) = \dfrac{4}{j(\omega - 4)}$

13–61 (a) $f_1(t) = \dfrac{3}{2}[\delta(t - 10) + \delta(t + 10)]$

(b) $f_2(t) = \dfrac{3j}{t\pi}$

(c) $f_3(t) = \dfrac{18}{\pi(9 + t^2)}$

13–63 $g(t) = \left[2 - 4e^{-20t} + 2e^{-40t}\right]u(t)$

13–65 $\mathscr{F}\{f(t)\sin(\beta t)\} = \mathscr{F}\left\{\dfrac{1}{2j}f(t)\left[e^{j\beta t} - e^{-j\beta t}\right]\right\}$
$= \dfrac{1}{2j}[F(\omega - \beta) - F(\omega + \beta)]$

13–67 $v_2(t) = 4.995e^t u(-t) + 5.005e^{-t}u(t) - 0.01e^{-1000t}u(t)$ V

13–69 $v_2(t) = \dfrac{40\sqrt{3}}{3}e^{-500t}\sin(500\sqrt{3}t)u(t)$ V

13–71 $v_2(t) = 5\,\text{sgn}(t) - 10e^{-200t}u(t)$ V

13–73 $v_2(t) = 0.06e^{-4t}u(t) - 0.06e^{4t}u(-t) - 0.12e^{-10000t}u(t)$ V

13 75 $y(t) = u(-t) + \dfrac{1}{2}e^{-2t}u(t) - \dfrac{1}{2}e^{2t}u(-t)$

13–77 $H_1(\omega) = \dfrac{j5\omega}{j\omega + 2}$, $H_2(\omega) = \dfrac{j5\omega}{j\omega + 4}$; the frequency response is a high-pass filter with a cutoff frequency of α

13–79 $H_1(\omega) = \dfrac{20j\omega}{6.25 + \omega^2}$, $H_2(\omega) = \dfrac{20j\omega}{25 + \omega^2}$; the frequency response is a bandpass filter, as α increases, the center frequency increases and the passband gain decreases

13–81 $h(t) = \dfrac{2A\sin^2(\beta t/2)}{\pi\beta t^2}$

13–83 $W_{1\Omega} = 5$ J

13–85 $W_{1\Omega} = \dfrac{1}{40}$ J; 50%

13–87 $W_{1\Omega} = 500$ J

13–89 $W_{1\Omega} = 23.04$ mJ; 20.48%

13–91 $a_0 = 1$, $a_n = 2$, $b_n = 0$ for all n; the impulse train contains equal amounts of energy at every nonzero frequency

13–93 The expected output power levels agree closely with the measured values; the spectrum analyzer is calibrated correctly

CHAPTER 14

14–1 $T_{LP}(s) = \dfrac{\mu}{R_1 R_2 C_1 C_2 s^2 + (R_1 C_1 + R_1 C_2 + R_2 C_2 - \mu R_1 C_1)s + 1}$

substitute:

$T_{HP}(s) = \dfrac{\mu}{\dfrac{1}{C_1 s}\dfrac{1}{C_2 s}\dfrac{1}{R_1}\dfrac{1}{R_2} + \dfrac{1}{C_1 s}\dfrac{1}{R_1} + \dfrac{1}{C_1 s}\dfrac{1}{R_2} + \dfrac{1}{C_2 s}\dfrac{1}{R_2} - \mu\dfrac{1}{C_1 s}\dfrac{1}{R_1} + 1}$

simplify:

$T_{HP}(s) = \dfrac{\mu R_1 R_2 C_1 C_2 s^2}{1 + (R_2 C_2 + R_1 C_2 + R_1 C_1 - \mu R_2 C_2)s + R_1 R_2 C_1 C_2 s^2}$

14–3 Solve node-voltage equations for V_2 in terms of V_1:

Node 1 : $\dfrac{V_A - V_1}{R_1} + \dfrac{V_A - V_2}{1/C_1 s} + \dfrac{V_A - V_N}{1/C_2 s} = 0$

Node 2 : $\dfrac{V_N - V_A}{1/C_2 s} + \dfrac{V_N - V_2}{R_2} = 0$

Node 3 : $V_N = V_P = \dfrac{R_B V_1}{R_A + R_B}$

14–5 Select C_1 then calculate $R_1 = 1/(\omega_0 C_1)$, $C_2 = 2\zeta/(R_1 \omega_0) = 2\zeta C_1$, and $R_2 = R_1 C_1 / C_2$; $R_1 = 20\,\text{k}\Omega$, $R_2 = 200\,\text{k}\Omega$, $C_1 = 0.025\,\mu\text{F}$, $C_2 = 0.0025\,\mu\text{F}$; frequency response peak occurs at 2000 rad/s and has a magnitude of 26 dB

14–7 Solve node-voltage equations for V_2 in terms of V_1:

Node 1 : $\dfrac{V_A - V_1}{1/C_1 s} + \dfrac{V_A - V_2}{R_1} = 0$

Node 2 : $\dfrac{V_2 - V_A}{1/C_2 s} + \dfrac{V_2}{R_2} = 0$

Select C, then calculate $R_2 = \dfrac{1}{2\zeta\omega_0 C}$ and $R_1 = \dfrac{1}{\omega_0^2 R_2 C^2}$

$C = 0.1\,\mu\text{F}$, $R_1 = 2\,\text{k}\Omega$, $R_2 = 50\,\text{k}\Omega$; high-pass filter with a -40 dB/decade roll-off

14–9 Solve node-voltage equations for V_2 in terms of V_1:

Node 1 : $\dfrac{V_A - V_1}{1/C_1 s} + \dfrac{V_A - V_2}{R_1} + \dfrac{V_A}{R_2} = 0$

Node 2 : $\dfrac{-V_A}{R_2} + \dfrac{-V_2}{1/C_2 s} = 0$

Select R, then calculate $C_1 = \dfrac{1}{\zeta\omega_0 R}$ and $C_2 = \dfrac{1}{\omega_0^2 C_1 R^2}$

$R = 2\,\text{k}\Omega$, $C_1 = 0.1\,\mu\text{F}$, $C_2 = 1000\,\text{pF}$; bandpass filter

14–11 $R = 2.6525\,\text{k}\Omega$, $C = 1000\,\text{pF}$; bandstop filter; with a gain of 1, ζ is not selectable and $\zeta = 1$

14–13 $T(s) = \dfrac{10^{12}}{s^2 + 2 \times 10^5 s + 10^{10}}$

14–15 $T(s) = \dfrac{1250 s}{s^2 + 250 s + 225 \times 10^6}$

14–17 $T(s) = \dfrac{10(s^2 + 377^2)}{s^2 + 7.54 s + 377^2}$

14–19 Use the circuit in Figure 14–3(b) with $R_1 = R_2 = 50\,\text{k}\Omega$, $R_A = 0\,\Omega$, $R_B = \infty\,\Omega$, $C_1 = 0.01\,\mu\text{F}$, $C_2 = 0.0064\,\mu\text{F}$ followed in cascade by a noninverting amplifier with a gain of 10

14–21 Use the circuit in Figure 14–9(b) with $R_1 = R_2 = 2\,\text{k}\Omega$, $R_A = 5\,\text{k}\Omega$, $R_B = 10\,\text{k}\Omega$, $C_1 = C_2 = 0.2\,\mu\text{F}$

14–23 Use the circuit in Figure 14–14 with $R_1 = 5\,\text{k}\Omega$, $R_2 = 100\,\text{k}\Omega$, $C_1 = C_2 = 8944.3\,\text{pF}$

14–25 Use the circuit in Figure 14–14 with $R_1 = 2.8986\,\Omega$, $R_2 = 115.942\,\text{k}\Omega$, $C_1 = C_2 = 500\,\text{pF}$

14–27 Use the circuit in Figure 14–17 with $R_1 = 2\,\text{k}\Omega$, $R_2 = 500\,\Omega$, $R_A = 4\,\text{k}\Omega$, $R_B = 500\,\Omega$, $C_1 = C_2 = 0.2\,\mu\text{F}$

14–29 $T(s) = \left[\dfrac{(2.3714)(16620)}{s + 16620}\right]^8$

14–31 $T(s) = \dfrac{10(4430)(584.9 \times 10^6)(235.5 \times 10^6)}{(s + 4430)(s^2 + 2738 s + 584.9 \times 10^6)} \\ \times (s^2 + 7167 s + 235.5 \times 10^6)$

14–33 $T(s) = \dfrac{(7.07)(3.6115 \times 10^{12})(783.23 \times 10^9)}{(s^2 + 339982 s + 3.6115 \times 10^{12})} \\ \times (s^2 + 820926 s + 783.23 \times 10^9)$

14–35 Use three active, low-pass OP AMP filters in cascade; the first stage is a first-order filter with $R_1 = R_2 = 10\,\text{k}\Omega$ and $C = 0.01\,\mu\text{F}$; the second stage is a second-order, unity-gain filter with $R_1 = R_2 = 32.363\,\text{k}\Omega$, $C_1 = 0.01\,\mu\text{F}$, and $C_2 = 954.8\,\text{pF}$; the third stage is a second-order, unity gain filter with $R_1 = R_2 = 12.361\,\text{k}\Omega$, $C_1 = 0.01\,\mu\text{F}$, and $C_2 = 6545\,\text{pF}$

14–37 Use three active, low-pass OP AMP filters in cascade; the first stage is a noninverting amplifier with a gain of $\sqrt{10} = 3.1623$; the second stage is a second-order, unity-gain filter with $R_1 = R_2 = 8.71\,\text{k}\Omega$, $C_1 = 0.001\,\mu\text{F}$, and $C_2 = 146.5\,\text{pF}$; the third stage is a second-order, unity gain filter with $R_1 = R_2 = 3.608\,\text{k}\Omega$, $C_1 = 0.001\,\mu\text{F}$, and $C_2 = 853.8\,\text{pF}$

14–39 Use a fifth-order Chebychev design with two active, second-order low-pass OP AMP filters followed by a first-order passive RC low-pass filter; the first stage is a second-order, unity-gain filter with $R_1 = R_2 = 29.067\,\text{k}\Omega$, $C_1 = 0.1\,\mu\text{F}$, and $C_2 = 320.4\,\text{pF}$; the second stage is a second-order, unity gain filter with $R_1 = R_2 = 11.103\,\text{k}\Omega$, $C_1 = 0.1\,\mu\text{F}$, and $C_2 = 5452\,\text{pF}$; the third stage is a passive RC low-pass filter with $R = 8.982\,\text{k}\Omega$ and $C = 0.1\,\mu\text{F}$

14–41 Use a fifth-order Butterworth design with three active, low-pass OP AMP filters in cascade; the first stage is a first-order filter with $R_1 = R_2 = 1.592\,\text{k}\Omega$ and $C = 0.1\,\mu\text{F}$; the second stage is a second-order, unity-gain filter with $R_1 = R_2 = 5.151\,\text{k}\Omega$, $C_1 = 0.1\,\mu\text{F}$, and $C_2 = 9548\,\text{pF}$; the third stage is a second-order, unity gain filter with $R_1 = R_2 = 1.967\,\text{k}\Omega$, $C_1 = 0.1\,\mu\text{F}$, and $C_2 = 0.06545\,\mu\text{F}$

14–43 Use a fourth-order, low-pass Chebychev design with three OP AMP stages; the first stage is a noninverting amplifier with a gain of 7.07; the second stage is a second-order, unity-gain filter with $R_1 = R_2 = 9.363\,\text{k}\Omega$, $C_1 = 0.01\,\mu\text{F}$, and

$C_2 = 80\,\text{pF}$; the third stage is a second-order, unity gain filter with $R_1 = R_2 = 3.877\,\text{k}\Omega$, $C_1 = 0.01\,\mu\text{F}$, and $C_2 = 2151\,\text{pF}$

14-45 Use a fifth-order Butterworth design with three active, low-pass OP AMP filters in cascade; the first stage is a first-order filter with $R_1 = R_2 = 5.305\,\text{k}\Omega$ and $C = 0.001\,\mu\text{F}$; the second stage is a second-order, unity-gain filter with $R_1 = R_2 = 17.169\,\text{k}\Omega$, $C_1 = 0.001\,\mu\text{F}$, and $C_2 = 95.48\,\text{pF}$; the third stage is a second-order, unity gain filter with $R_1 = R_2 = 6.558\,\text{k}\Omega$, $C_1 = 0.001\,\mu\text{F}$, and $C_2 = 654.5\,\text{pF}$

14-47 $T(s) = \dfrac{10s^3}{(s+10^4)(s^2+10^4s+10^8)}$

14-49 Use a fifth-order Butterworth design;

$$T(s) = \frac{\sqrt{10}s^5}{(s+10^4)(s^2+6180s+10^8)(s^2+16180s+10^8)}$$

14-51 Use two active, high-pass OP AMP filter stages connected in cascade; the first stage is a first-order filter with $R_1 = 1\,\text{k}\Omega$, $R_2 = 10\,\text{k}\Omega$, and $C_1 = 0.1\,\mu\text{F}$; the second stage is a second-order, unity gain filter with $C_1 = C_2 = 0.02\,\mu\text{F}$, $R_1 = 2.5\,\text{k}\Omega$, and $R_2 = 10\,\text{k}\Omega$

14-53 Use a fifth-order Butterworth design with three active, high-pass OP AMP filters in cascade; the first stage is a first-order filter with $R_1 = 1\,\text{k}\Omega$, $R_2 = 3.1623\,\text{k}\Omega$, and $C = 0.1\,\mu\text{F}$; the second stage is a second-order, unity-gain filter with $C_1 = C_2 = 0.03236\,\mu\text{F}$, $R_1 = 954.81\,\Omega$, and $R_2 = 10\,\text{k}\Omega$; the third stage is a second-order, unity gain filter with $C_1 = C_2 = 0.01236\,\mu\text{F}$, $R_1 = 6.5448\,\text{k}\Omega$, and $R_2 = 10\,\text{k}\Omega$

14-55 The filter has a notch located at 397.9 Hz with a gain of -48.7 dB; the passband gain is 0 dB; the cutoff frequencies occur at 388.0 Hz and 408.0 Hz, so the bandwidth is 20 Hz; the filter meets all of the specifications

14-57 Use an active, second-order bandpass OP AMP filter followed by a voltage divider in cascade; the bandpass filter has $C_1 = C_2 = 1\,\text{pF}$, $R_1 = 1.3045\,\text{k}\Omega$, and $R_2 = 13.045\,\text{M}\Omega$; the voltage divider reduces the signal by a factor of 1/500 with a 499-kΩ resistor followed by a 1-kΩ resistor

14-59 The design meets the specification; the design could be simplified by replacing the second OP AMP stage with a passive RC low-pass filter

14-61 Real OP AMPs have a finite gain-bandwidth product, typically on the order of 1 MHz, which limits the gain at higher frequencies

CHAPTER 15

15-1 (a) $v_1(t) = 0.12\dfrac{di_1(t)}{dt} + 0.18\dfrac{di_2(t)}{dt}$

$v_2(t) = 0.18\dfrac{di_1(t)}{dt} + 0.4\dfrac{di_2(t)}{dt}$

(b) $v_S(t) = 40\sin(500t)$ V

15-3 (a) $v_1(t) = 0.01\dfrac{di_1(t)}{dt} + 0.007\dfrac{di_2(t)}{dt}$

$v_2(t) = 0.007\dfrac{di_1(t)}{dt} + 0.01\dfrac{di_2(t)}{dt}$

(b) $v_1(t) = 6\cos(1000t)$ V; $v_2(t) = -6\cos(1000t)$ V

15-5 $v_1(t) = 50\cos(1000t)$ V

15-7 $v_X(t) = 85\cos(1000t)$ V

15-9 $i_2(t) = 0$; $v_1(t) = L_1\dfrac{di_1(t)}{dt}$; $v_2 = M\dfrac{di_1(t)}{dt}$

$= M\dfrac{v_1(t)}{L_1} = \dfrac{k\sqrt{L_1L_2}}{L_1}v_1(t) = k\sqrt{\dfrac{L_2}{L_1}}v_1(t)$

15-11 (a) $n = \dfrac{1}{5}$

(b) $v_1(t) = 120\cos(120\pi t)$ V; $i_1(t) = 6\cos(120\pi t)$ mA

15-13 $R_{EQ} = 3\,\Omega$

15-15 $i_L(t) = 4\sin(400t)$ A; $i_S(t) = -8\sin(400t)$ A

15-17 $n = 4.4721$

15-19 $i_S(t) = 42.5\sin(377t)$ A

15-21 $I_1 = 3.044 + j1.947$ A; $I_2 = 1.522 + j0.987$ A; $I_3 = 0.507 + j0.329$ A

15-23 $P_L = 4.8\,\text{kW}$

15-25 $|I_1| = 22.727$ A; $|I_2| = 45.455$; $R_{IN} = 19.36\,\Omega$

15-27 $V_2 = 12.414 + j31.035$ V; $Z_{IN} = 51.485 + j14.852\,\Omega$

15-29 $Z_{IN} = 8.0082 + j43.282\,\Omega$

15-31 $|V_O| = 176.63$ V, $Z_{IN} = 96.12 + j0.471\,\Omega$, $P_L = 623.97$ W

15-33 $V_O = 114.623 - j63.679$ V, $P_L = 191.038$ W

15-35 $I_S = 8$ A, $P_L = 480$ W

15-37 From Problem 15-36, $Z_{IN} = jX_1(1-k^2)$, so
$V_1 = V_S = jX_1(1-k^2)I_1$; $I_1 = \dfrac{V_S}{jX_1(1-k^2)}$;
with the output short circuited, $V_2 = 0$, so we have $jX_2I_2 = -jX_MI_1$; substitute for I_1 and apply $X_1X_2 = X_M^2/k^2$ and $I_{SC} = -I_2$ to prove:
$I_{SC} = -I_2 = \dfrac{X_MV_S}{jX_1X_2(1-k^2)} = \dfrac{X_MV_S}{j\left(\dfrac{X_M^2}{k^2}\right)(1-k^2)} = \dfrac{k^2}{1-k^2}\dfrac{V_S}{jX_M}$

15-39 $V_2 = 367.599 - j17.154$ V, $I_2 = -1.505 - j0.6666$ A

15-41 $v_2(t) = \dfrac{-10}{\pi}\left[\sin(40000\pi t) + \dfrac{1}{2}\sin(80000\pi t) + \dfrac{1}{3}\sin(120000\pi t)\right]$ V

CHAPTER 16

16-1 (a) $P = -130.73$ W, $Q = -1494.3$ VAR,
$p(t) = -130.73[1 + \cos(2\omega t)] + 1494.3\sin(2\omega t)$ W

(b) $P = 82.05$ W, $Q = 465.32$ VAR, $p(t) = 82.05[1 + \cos(2\omega t)] - 465.32\sin(2\omega t)$ W

16-3 (a) $P = 60.37$ W, $Q = 16.18$ VAR, pf $= 0.9659$, lagging

(b) $P = 1299$ W, $Q = -750$ VAR, pf $= 0.866$, leading

16–5 (a) $S = 434.67 - j116.47\,\text{kVA}$
(b) $S = 4 - j3\,\text{kVA}$

16–7 $P = 24\,\text{kW}, Q = 18\,\text{kVAR}, Z_\text{L} = 153.6 + j115.2\,\Omega$

16–9 pf $= 0.894, Z = 22.22 + j11.11\,\Omega$

16–11 $S = 436.44 + j246.80\,\text{VA}$, pf $= 0.8704$ lagging

16–13 $R = 111.69\,\Omega, L = 197.52\,\text{mH}$

16–15 $S_\text{S} = 6.217 + j2.738\,\text{kVA}, S_\text{L} = 1.450 - j0.604\,\text{kVA}$

16–17 $S = 88.62 + j708.92\,\text{VA}$

16–19 $\mathbf{I}_\text{A} = 8.18 - j2.45\,\text{A}, \mathbf{I}_\text{N} = 1.36 + j1.09\,\text{A}, \mathbf{I}_\text{B} = -9.55 + j1.36\,\text{A}$

16–21 $|S_\text{S}| = 50.01\,\text{kVA}$, pf $= 0.787, |\mathbf{V}_\text{S}| = 2.609\,\text{kV}$

16–23 $|S_\text{S}| = 322.3\,\text{kVA}, |\mathbf{V}_\text{S}| = 7.891\,\text{kV}$

16–25 $|S_\text{S}| = 31.82\,\text{kVA}, |\mathbf{V}_\text{S}| = 5.251\,\text{kV}$

16–27 $|S_\text{S}| = 4.728\,\text{kVA}, |\mathbf{V}_\text{S}| = 472.83\,\text{V}$

16–29 For pf $= 0.95, C = 5.573\,\mu\text{F}$; for pf $= 1, C = 8.222\,\mu\text{F}$

16–31 $\mathbf{V}_\text{AB} = 480\angle 30°\,\text{V}, \mathbf{V}_\text{BC} = 480\angle -90°\,\text{V}, \mathbf{V}_\text{CA} = 480\angle -210°\,\text{V}$
$\mathbf{V}_\text{AN} = 277\angle 0°\,\text{V}, \mathbf{V}_\text{BN} = 277\angle -120°\,\text{V}, \mathbf{V}_\text{CN} = 277\angle -240°\,\text{V}$

16–33 $\mathbf{V}_\text{AB} = 208\angle 0°\,\text{V}, \mathbf{V}_\text{BC} = 208\angle -120°\,\text{V}, \mathbf{V}_\text{CA} = 208\angle -240°\,\text{V}$
$\mathbf{V}_\text{AN} = 120\angle -30°\,\text{V}, \mathbf{V}_\text{BN} = 120\angle -150°\,\text{V}, \mathbf{V}_\text{CN} = 120\angle -270°\,\text{V}$

16–35 $\mathbf{V}_\text{AN} = 120\angle -0°\,\text{V}, \mathbf{V}_\text{BN} = 120\angle -120°\,\text{V}, \mathbf{V}_\text{CN} = 120\angle -240°\,\text{V}$

16–37 $Z_{\Delta\text{EQ}} = 30 - j10\,\Omega$

16–39 $Z_\text{YEQ} = 8.1 + j3.3\,\Omega$

16–41 $\mathbf{I}_\text{A} = 12\angle -51.8°\,\text{A(rms)}, \mathbf{I}_\text{B} = 12\angle -171.8°\,\text{A(rms)}$,
$\mathbf{I}_\text{C} = 12\angle -291.8°\,\text{A(rms)}, \mathbf{V}_\text{AN} = 323.3\angle -30°\,\text{V}$,
$\mathbf{V}_\text{BN} = 323.3\angle -150°\,\text{V}, \mathbf{V}_\text{CN} = 323.3\angle -270°\,\text{V}$

16–43 $\mathbf{I}_\text{A} = 38.1\angle -36.9°\,\text{A (rms)}, \mathbf{I}_\text{B} = 38.1 \angle -156.9°\,\text{A (rms)}$,
$\mathbf{I}_\text{C} = 38.1\angle -276.9°\,\text{A(rms)}, \mathbf{I}_\text{AB} = 22\angle -6.9°\,\text{A (rms)}$,
$\mathbf{I}_\text{BC} = 22\angle -126.9°\,\text{A (rms)}, \mathbf{I}_\text{CA} = 22\angle -246.9°\,\text{A (rms)}$

16–45 $\mathbf{I}_\text{AB} = 16.64\angle -30°\,\text{A (rms)}, \mathbf{I}_\text{A} = 28.82\angle -60°\,\text{A (rms)}$

16–47 $\mathbf{I}_\text{A} = 6\angle -66.9°\,\text{A (rms)}, \mathbf{V}_\text{AN} = 150\angle -30°\,\text{V}$

16–49 $V_\text{L} = 612.4\,\text{V (rms)}, S_\text{L} = 6 + j4.5\,\text{kVA}$

16–51 $I_\text{L} = 4.017\,\text{A (rms)}, S_\text{L} = 2904 - j968\,\text{VA}$

16–53 $V_\text{L} = 6197\,\text{V (rms)}, S_\text{L} = 153.6 + j76.8\,\text{kVA}$

16–55 $V_\text{L} = 1125\,\text{V (rms)}, S_\text{L} = 14.48 + j3.185\,\text{kVA}$

16–57 $V_\text{L} = 4011\,\text{V (rms)}, S_\text{L} = 49.49 + j41.01\,\text{kVA}$

16–59 $Z_\text{Y} = 7.37 + j5.53\,\Omega$

16–61 $|S_1| = 3.046\,\text{MVA}, \text{pf}_1 = 0.841, V_\text{L1} = 233.6\,\text{kV}$

16–63 $\text{pf}_1 = 0.775, V_\text{L1} = 32.24\,\text{kV}, V_\text{L2} = 30.93\,\text{kV}$

16–65 $|S_1| = 504.8\,\text{kVA}, \text{pf}_1 = 0.527, V_\text{L1} = 121.9\,\text{kV}$,
$V_\text{L2} = 118.8\,\text{kV}, V_\text{L3} = 115.5\,\text{kV}$

16–67 $|S_1| = 7.034\,\text{MVA}, \text{pf}_1 = 0.926, V_\text{L1} = 139.7\,\text{kV}$,
$V_\text{L2} = 138.8\,\text{kV}, V_\text{L3} = 138.6\,\text{kV}$

16–69 pf $= 0.5978$

16–71 Efficiency $\eta = 96.9\%$

CHAPTER 17

17–1 $z_{11} = 400\,\Omega, z_{12} = z_{21} = 100\,\Omega, z_{22} = 166.7\,\Omega$

17–3 $z_{11} = 20 + j60\,\Omega, z_{12} = z_{21} = -20 + j40\,\Omega, z_{22} = 20 - j40\,\Omega$

17–5 $z_{11} = R_1 + (\beta + 1)R_2, z_{12} = 0, z_{21} = -\beta R_3, z_{22} = R_3$

17–7 $I_1 = 14\,\text{mA}, I_2 = -4\,\text{mA}$

17–9 $V_2 = 12.5\,\text{V}, I_1 = 35\,\text{mA}, I_2 = -5\,\text{mA}$

17–11 The load impedance implies $V_2 = -Z_\text{L}I_2$;
$V_2 = -Z_\text{L}I_2 = z_{21}I_1 + z_{22}I_2$; solve for $I_2 = -z_{21}I_1/(Z_\text{L} + z_{22})$;

$$V_1 = z_{11}I_1 + z_{12}I_2 = z_{11}I_1 + \frac{-z_{12}z_{21}}{Z_\text{L} + z_{22}}I_1;$$

$$Z_\text{IN} = \frac{V_1}{I_1} = z_{11} - \frac{z_{12}z_{21}}{Z_\text{L} + z_{22}}$$

17–13 (a) $h_{11} = 0, h_{12} = 1, h_{21} = -1, h_{22} = Y$
(b) $h_{11} = Z, h_{12} = 1, h_{21} = -1, h_{22} = 0$

17–15 $h_{11} = R_1 + R_2, h_{12} = 0, h_{21} = \beta, h_{22} = \dfrac{1}{R_3}$

17–17 $h_{11} = R_1, h_{12} = 0, h_{21} = -\dfrac{R_1}{R_2}, h_{22} = \dfrac{1}{R_2}$

17–19 $V_\text{T} = 6\,\text{V}, R_\text{T} = 250\,\Omega$

17–21 (a) $Z_\text{OUT} = 200\,\Omega$
(b) $Z_\text{OUT} = 400\,\Omega$

17–23 If $V_1 = V_\text{x}$ and $V_2 = 0$, then $I_2 = I_\text{2SC}$; applying the hybrid parameter equations, we get $V_1 = V_\text{x} = h_{11}I_1 + h_{12}V_2 = h_{11}I_1$ and $I_2 = I_\text{2SC} = h_{21}I_1 + h_{22}V_2 = h_{21}I_1$; substitute $I_1 = I_\text{2SC}/h_{21}$ into $V_\text{x} = h_{11}I_1 = h_{11}I_\text{2SC}/h_{21}$ to get $I_\text{2SC} = h_{21}V_\text{x}/h_{11}$; if $V_2 = V_\text{x}$ and $V_1 = 0$, then $I_1 = I_\text{1SC}$; applying the first hybrid parameter equation, we get $V_1 = 0 = h_{11}I_\text{1SC} + h_{12}V_\text{x}$, which implies $I_\text{1SC} = -h_{12}V_\text{x}/h_{11}$; comparing the results for the two cases, for reciprocity $(I_\text{1SC} = I_\text{2SC})$, we must have $h_{12} = -h_{21}$

17–25 Solve $I_2 = h_{21}I_1 + h_{22}V_2$ for I_1 to get $I_1 = -h_{22}V_2/h_{21} + I_2/h_{21}$, which provides the relationships for the C and D parameters; substitute I_1 into $V_1 = h_{11}I_1 + h_{12}V_2$ to get $V_1 = -h_{11}h_{22}V_2/h_{21} + h_{11}I_2/h_{21} + h_{12}V_2$

$= \left[\dfrac{h_{12}h_{21} - h_{11}h_{22}}{h_{21}}\right]V_2 + \dfrac{h_{11}}{h_{21}}I_2$, which provides the relationships for the A and B paramters

17–27 Start with $V_1 = h_{11}I_1 + h_{21}V_2$ and $I_2 = h_{21}I_1 + h_{22}V_2$, set $I_2 = 0$ and solve for $I_1 = -h_{22}V_2/h_{21}$; substitute into the first equation

$V_1 = \dfrac{-h_{11}h_{22}}{h_{21}}V_2 + h_{12}V_2 = \dfrac{h_{12}h_{21} - h_{11}h_{22}}{h_{21}}V_2$, so we have

$\dfrac{V_2}{V_1} = \dfrac{h_{21}}{h_{12}h_{21} - h_{11}h_{22}} = \dfrac{-h_{21}}{h_{11}h_{22} - h_{12}h_{21}} = \dfrac{-h_{21}}{\Delta_\text{h}}$

17–29 $A = 625, B = 650\,\Omega, C = 150\,\text{S}, D = 250$

17–31 $A = 0.02, B = 20\,\Omega, C = 20\,\mu\text{S}, D = 0.02; I = -50$

17–33 $z_{11} = 80\,\Omega, z_{12} = 20\,\Omega, z_{21} = 15\,\Omega, z_{22} = 25\,\Omega$; the network is not reciprocal

17–35 $y_{11} = 100\,\text{mS}, y_{12} = -60\,\text{mS}, y_{21} = -60\,\text{mS}, y_{22} = 140\,\text{mS}$ the network is reciprocal

17–39 $Z_\text{L} = 20 - j40\,\Omega$

INDEX

STANDARD VALUES

THE RESISTOR COLOR CODE

Color	First Significant Digit	Second Significant Digit	Multiplier	Tolerance
None				±20%
Silver			0.01	±10%
Gold			0.1	±5%
Black	0	0	1	
Brown	1	1	10	±1%
Red	2	2	100	±2%
Orange	3	3	1k	
Yellow	4	4	10k	
Green	5	5	100k	
Blue	6	6	1000k	
Violet	7	7		
Gray	8	8		
White	9	9		

STANDARD VALUES FOR RESISTORS

Value	Tolerances	Value	Tolerances	Value	Tolerances
10	±5%, ±10%, ±20%	22	±5%, ±10%, ±20%	47	±5%, ±10%, ±20%
11	±5%	24	±5%	51	±5%
12	±5%, ±10%	27	±5%, ±10%	56	±5%, ±10%
13	±5%	30	±5%	62	±5%
15	±5%, ±10%, ±20%	33	±5%, ±10%, ±20%	68	±5%, ±10%, ±20%
16	±5%	36	±5%	75	±5%
18	±5%, ±10%	39	±5%, ±10%	82	±5%, ±10%
20	±5%	43	±5%	91	±5%

STANDARD VALUES FOR CAPACITORS

pF	pF	pF	pF	µF	µF	µF	µF	µF	µF	µF
1.0	10	100	1000	0.01	0.1	1.0	10	100	1000	10,000
1.1	11	110	1100							
1.2	12	120	1200							
1.3	13	130	1300							
1.5	15	150	1500	0.015	0.15	1.5	15	150	1500	
1.6	16	160	1600							
1.8	18	180	1800							
2.0	20	200	2000							
2.2	22	220	2200	0.022	0.22	2.2	22	220	2200	
2.4	24	240	2400							
2.7	27	270	2700							
3.0	30	300	3000							
3.3	33	330	3300	0.033	0.33	3.3	33	330	3300	
3.6	36	360	3600							
3.9	39	390	3900							
4.3	43	430	4300							
4.7	47	470	4700	0.047	0.47	4.7	47	470	4700	
5.1	51	510	5100							
5.6	56	560	5600							
6.2	62	620	6200							
6.8	68	680	6800	0.068	0.68	6.8	68	680	6800	
7.5	75	750	7500							
8.2	82	820	8200							
9.1	91	910	9100							

STANDARD VALUES FOR INDUCTORS

nH, µH			
1.0	10	100	1000
1.1	11	110	1100
1.2	12	120	1200
1.3	13	130	1300
1.5	15	150	1500
1.6	16	160	1600
1.8	18	180	1800
2.0	20	200	2000
2.2	22	220	2200
2.4	24	240	2400
2.7	27	270	2700
3.0	30	300	3000
3.3	33	330	3300
3.6	36	360	3600
3.9	39	390	3900
4.3	43	430	4300
4.7	47	470	4700
5.1	51	510	5100
5.6	56	560	5600
6.2	62	620	6200
6.8	68	680	6800
7.5	75	750	7500
8.2	82	820	8200
8.7	87	870	8700
9.1	91	910	9100